Ionic and Organometallic-Catalyzed Organosilane Reductions

Ionic and Organometallic-Catalyzed Organosilane Reductions

GERALD L. LARSON and JAMES L. FRY

WILEY

A JOHN WILEY & SONS, INC., PUBLICATION

Published by John Wiley & Sons, Inc., Hoboken, New Jersey

Published simultaneously in Canada.

Library of Congress Catalog Card Number: 42-20265
ISBN 978-0-470-54787-8

Printed in the United States of America

10 9 8 7 6 5 4 3 2 1

CONTENTS

FOREWORD

Chemical synthesis is an intellectually and technically challenging enterprise. Over the many decades of progress in this discipline, spectacular advances in methods have made once intimidating transformations now routine. However, as the frontier advances and the demands for ready access to greater molecular complexity increases, so does the sophistication of the chemical reactions needed to achieve these goals. With this greater sophistication (and the attendant expectation of enhanced generality, efficiency, and selectivity) comes the challenge of adapting these technologies to the specific applications needed by the practitioner. In its 67-year history, *Organic Reactions* has endeavored to meet this challenge by providing focused, scholarly, and comprehensive overviews of a given transformation.

By any yardstick, the reduction of organic compounds is one of the most important methods for the manipulation of functional groups. No fewer than 27 of the 245 chapters in the *Organic Reactions* series are dedicated to reduction in one if its manifold forms. Among the mildest of these transformations is the use of ubiquitous organosilanes. Because of the wide range of organosilanes available and the different mechanisms by which they can be activated to transfer the hydride, these reagents have found extensive application in organic synthesis. A startling array of different functional groups can be reduced by organosilanes with high selectivity. The oxidized organosilane byproducts are easily removed and non-toxic. Moreover, some of the most exciting recent developments involve the enantioselective reduction of double-bonded functional groups.

The *Organic Reactions* series is fortunate to have published a timely chapter on this important process that single-handedly constituted Volume 71. This comprehensive chapter was authored by two of the internationally recognized leaders in this field, Dr. Gerald L. Larson and Prof. James L. Fry. Although many reviews and book chapters have been written on reductions with organosilanes, this massive work constitutes the definitive treatise in the field. Thus, in keeping with our educational mission, the Board of Editors of *Organic Reactions* has decided to publish this chapter as a separate, soft cover book to make the work available to a wider audience of chemists. In addition, to keep pace with the rapid development of this field, Drs. Larson and Fry have provided updated references that bring the literature coverage up to March 2009. These references are appended at the end of the original reference section and organized by the Tabular presentation of the different substrates for reduction.

The publication of this book represents the third soft cover reproduction of single-volume *Organic Reactions* chapters. The success of the first two soft cover books ("The Stille Reaction" taken from Volume 50 and "Handbook of Nucleoside Synthesis" taken from Volume 55) has convinced us that the availability of low-cost, high-quality publications that cover broadly useful transformations is addressing an unmet need in the organic synthesis community. Thus we will continue to identify candidates for the compilation of such individual volumes as opportunities present themselves.

Scott E. Denmark
Urbana, Illinois

PREFACE

The reduction of organic functional groups ranks as one of the most important transformations in synthetic organic chemistry. The range of reagents for organic reductions spans from the highly-reactive, highly-nucleophilic metal hydrides at one extreme, to the inert dihydrogen molecule at the other. In view of the variety of organic functional groups that are subject to reduction, highly selective reducing agents are needed. Organosilanes are particularly well suited to meet this challenge because they can be electronically and sterically modified to finely tune their reactivity and thus achieve a number of chemoselective reductions. The substrates in ionic organosilane reductions require activation by an acid or organometallic catalyst. Thus, the reaction selectivity can be further optimized by the choice of activator. Organosilane reduction of essentially all reducible organic functional groups including aromatic rings, aryl halides, carboxylic acids, and amides have been achieved. Moreover, the use of chiral ligands and chiral organometallic catalysts in conjunction with organosilanes has allowed the enantioselective reduction of ketones, imines, enamines, enones, enals, and α,β-unsaturated esters with extremely high selectivities.

Experimental reaction conditions are discussed and numerous specific experimental procedures are provided from the literature for a wide spectrum of representative functional group transformations using this method. A tabular survey of all examples of organosilicon hydrides used in organic synthesis is presented in 34 tables organized by type of substrate reduced with listings within each table according to increasing carbon number of the substrates.

The literature is covered through March 2009, supplementing the coverage used in the earlier hard cover chapter. These latest literature references have been collected in separate sections according to the sequence of the tables in the tabular survey section. In each of the sections, the individual citations have been arranged in alphabetic order of the author names.

Ionic and Organometallic-Catalyzed Organosilane Reductions is intended to be a useful, easily read tool for all practitioners of organic synthesis.

IONIC AND ORGANOMETALLIC-CATALYZED ORGANOSILANE REDUCTIONS

GERALD L. LARSON

Gelest, Inc., Morrisville, Pennsylvania 19067

JAMES L. FRY

6016 Trotwood Court, Fort Collins, Colorado 80524

CONTENTS

jlarson@gelest.com
jameslfry@msn.com
Ionic and Organometallic-Catalyzed Organosilane Reductions,
by Gerald L. Larson and James L. Fry

ACKNOWLEDGMENTS

The authors thank Mr. Thomas C. Johns of E. I. du Pont de Nemours and Company, Mr. Christopher W. Fry, Ms. Diane Micham, and the Department of Chemistry, University of Pennsylvania, for valuable assistance in the literature survey.

INTRODUCTION

The purpose of this chapter is to present a critical review of synthetically useful variations of ionic methods for hydrogenation of organic compounds. In practice, ionic hydrogenation involves the formal introduction of hydride from a donor source to an electron-deficient carbon center. The electrophilic centers can be formed by the departure of a leaving group (nucleofuge) from a saturated center or by the addition of an electrophile to a multiple bond. In the former mechanism, substitution of hydrogen for the leaving group is the net chemical consequence. In the latter, addition across the multiple bond is the result.

In this chapter, we cover the use of organosilicon hydrides as the source of ionic hydride with the goal of completing and updating earlier review works on the subject.[1–4] Similar chemistry is observed when molecular hydrogen[5,6] or various hydrocarbons[7–9] are used as hydride sources; these methods have been reviewed previously and are not covered herein. The use of organosilicon hydride-metal catalyst mixtures[10–12] for effecting reductions is included in this review, but use of trichlorosilane-tertiary amine combinations[13] is not.

Organosilicon compounds with at least one Si-H bond (called hydrosilanes, organosilicon hydrides, or simply silanes) have the ability to serve as mild air- and water-stable sources of hydride and thus have reducing properties. For example, triethylsilane is reported to reduce a variety of inorganic metal salts directly to the free metals.[14,15] Even the hexachloroantimonate anion can be reduced to Sb(0) upon contact with this silane.[16] Organotellurium chlorides are reduced to tellurium metal by a number of organosilicon hydrides.[17] Reaction of organosilicon hydrides with strong Brønsted acids leads to decomposition of the silane and the production of hydrogen gas.[14] In general, organosilicon hydrides do not undergo spontaneous reactions with organic compounds unless the organic substrate is a reasonably strong electrophile or the silane has been first activated by the interaction of a nucleophilic species with the silicon center. The organosilicon hydrides are covalent compounds that have little or no nucleophilic properties of their own. Aside from the parent silane, SiH_4, which is pyrophoric, the organosilicon hydrides are fairly innocuous compounds whose physical properties bear resemblance to their hydrocarbon analogs. Thus, their physical and chemical reducing properties differ from those of many familiar metal hydride reducing agents.[18,19] The use of organosilicon hydrides often provides a means of effecting reductions of organic substrates under very mild conditions and with excellent functional group selectivity.

Consideration of the nature of the Si–H bond provides insight into the chemical behavior of organosilicon hydrides. Comparison of the bond strengths as represented by bond dissociation energy (BDE) of hydrosilanes with those of hydrocarbon analogs shows that, in general, the Si–H bond is not much weaker than the C–H bond. Thus, the BDE values for the respective Si–H bonds in TMS–H and $(C_2H_5)_3Si–H$ are 90.3[20] and 90.1 kcal/mol[21] compared with a value of approximately 92 kcal/mol[20] for the tertiary C–H bond in $(CH_3)_3C–H$. On the other hand, there is a significant difference between the polarization characteristics of the Si–H and C–H bonds.[22] Compared to the Pauling electronegativity of hydrogen (2.20), the electronegativity of carbon (2.50) is greater and that of silicon (1.90) is less.[23] Carbon-hydrogen bonds are thus polarized in the direction $C^{\delta-}–H^{\delta+}$, whereas Si–H bonds are $Si^{\delta+}–H^{\delta-}$. As will be seen, this enhanced hydridic nature manifests itself in the chemical behavior of essentially all hydrosilanes.

Limited studies of the germanium and tin hydride analogs of the silicon hydrides show that they share this ability to function as hydride sources in ionic hydrogenations; however, their relatively greater reactivity toward acids appears to restrict their practical applications in organic synthesis.[24,25]

MECHANISM

General Considerations

The mechanistic discussion of silane reductions will be limited to those of cationic reductions, thus excluding the many silane reductions that involve metal catalysis.

Since tetravalent organosilicon hydrides intrinsically lack nucleophilicity, they react only with atomic centers that are substantially electron deficient, for example, carbocations. Because of this, organosilicon hydride reductions are potentially very selective. The "ionic" reductions of organic compounds by organosilicon hydrides are understood on the basis of two mechanisms. In the first, substitution by hydrogen of a leaving group bonded to a saturated carbon occurs. This path may be called a σ-route as it involves the stepwise cleavage of a σ-bond to a saturated center and the intervention of a carbocation intermediate that is captured by donation of hydride from the organosilicon hydride (Eq. 1). Alternatively, addition of an electrophile/hydride pair takes place across a multiple bond. This path may be termed a π-route (Eq. 2). Complexation of an electrophile to one end of a π-bond is followed by capture of the intermediate cation or complex by organosilicon hydride. The electrophile may be as simple as a proton or be one of a variety of Lewis acids or alkylating agents. The group Y can be C, O, N, or S. Sometimes the product of Eq. 2 can continue reacting by way of Eq. 1, with the moiety Y–E acting as a leaving group. When this occurs, the net effect is to replace the C=Y functionality with CH_2. The normal caveats regarding carbocation behavior such as the possible occurrences of eliminations, skeletal isomerizations, and bimolecular reactions prior to capture by hydride must be expected in all of these scenarios.

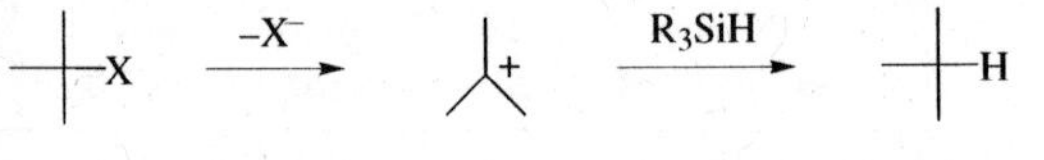

(Eq. 1)

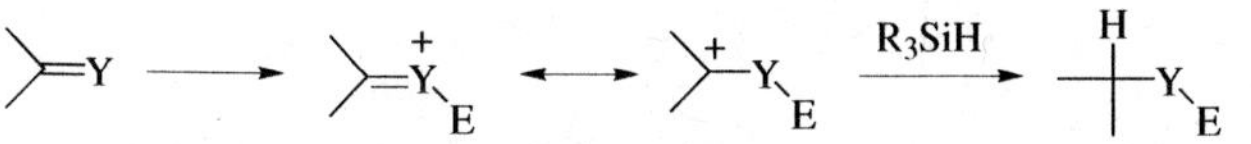

(Eq. 2)

It is necessary for the intermediate cation or complex to bear considerable carbocationic character at the carbon center in order for effective hydride transfer to be possible. By carbocationic character it is meant that there must be a substantial deficiency of electron density at carbon or reduction will not occur. For example, the sesquixanthydryl cation **1**,[26] dioxolenium ion **2**,[27] boron-complexed imines **3**, and O-alkylated amide **4**,[28] are apparently all too stable to receive hydride from organosilicon hydrides and are reportedly not reduced (although the behavior of **1** is in dispute[29]). This lack of reactivity by very stable cations toward organosilicon hydrides can enhance selectivity in ionic reductions.

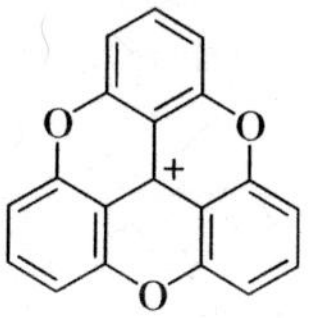

1

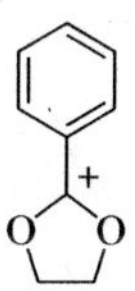

2

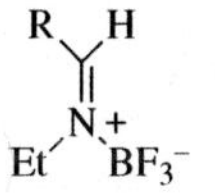

3

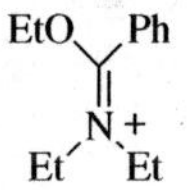

4

Role of Trivalent Silicon Species

The overall stoichiometry of hydride transfer from a silicon center to an electron-deficient carbon center is quite straightforward. Almost without exception, it appears that there is simple interchange of hydride to the carbocation while the silicon center receives the elements of the carbocation's counterion (Eq. 3).

$$R_3C^+ X^- + R'_3SiH \longrightarrow R_3CH + R'_3SiX \qquad \text{(Eq. 3)}$$

When the counterion is complex, for example metal-halogen anions such as BF_4^-, the most electronegative portion of the counterion becomes attached to the silicon center. Because of this attachment, it is natural to consider the intermediacy of a silicenium cation (silylium or silylenium ion) intermediate in such reactions (Eq. 4). Bond energies derived from electron impact studies indicate that Eq. 4 is exothermic in the gas phase by about 8 kcal/mol.[26,29] There seems little doubt that trivalent silicon-centered cationic species do exist in the gas phase[30,31] or that processes similar to that shown in Eq. 4 do occur there.[32,33]

$$Me_3C^+ + (CH_3)_3SiH \longrightarrow (CH_3)_3CH + (CH_3)_3Si^+ \qquad \text{(Eq. 4)}$$

The existence of trivalent silicenium cations as reactive species in solution is more controversial. Many early attempts to demonstrate the solution-phase existence of stable silicenium ions by using techniques analogous to those successfully applied to carbocation formation failed.[34-36] Other reports of attempts to generate silicenium ions in solution under stable ion conditions[37–46] and in solvolyses[47] are more convincing, but not without controversy.[48–51] The single-crystal X-ray determination of a non-planar triethylsilylium moiety paired in the crystalline state with the tetrakis(pentafluorophenyl)borate gegenion and toluene solvate stirred much debate about its interpretation and extension to reaction systems.[52,53] Recent crystallographic evidence supports the notion of a three-coordinate structure of a trimesitylsilylium cation paired with a carborane anion in the solid phase.[54,55] The balance of experimental evidence seems to indicate that, whereas trivalent silicenium cations may have fleeting existence as reaction intermediates, it is unlikely that they exist as stable, long-lived species in solution.[56] The failure to observe such trivalent species in solution is related to the very strong ability of electron-deficient silicon centers to coordinate with the media in which they are formed.[53,57,58] This is true even in solvents that exhibit little or no nucleophilicity toward carbocations and is further enhanced by the relatively long bond lengths to silicon centers that allow a close approach by coordinating species.[53]

The available experimental information is suggestive, but not unambiguously conclusive, of the intervention of electron-deficient silicon-centered species that may resemble silicenium cations in simple hydride exchanges occurring in solvents with low coordinating abilities. For example, substituted triarylmethane derivatives such as chlorotriphenylmethane (trityl chloride) undergo reduction through halogen-hydride exchange with organosilicon hydrides. The reactions proceed more rapidly in solvents with high ionizing power, but are kinetically first order with respect to both organosilicon hydride and triarylmethane derivative in benzene solvent.[59] In benzene, the exchange of halogen with hydride occurs with retention of configuration at the silicon center.[60] These results have led to the suggestion that the exchanges proceed by way of a four-center transition state **5**, in which there is simultaneous attack by the halide of the carbocation-halide ion pair on the silicon center as hydride undergoes transfer to the carbocation center.[60]

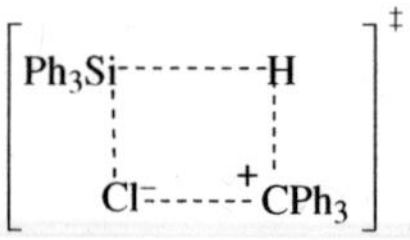

5

Simple variation of the solvent has a very significant effect on the stereochemistry at the silicon center for these exchange reactions. The stereochemistry changes from essentially complete retention to inversion and even to racemization. For example, in dichloromethane the halogen is delivered to the silicon center with complete racemization.[61] This implies that the degree of "tightness" of the carbocation-counterion pair must change depending on the solvent.

Organosilicon hydride reductions of preformed stable carbocations such as triphenylmethyl (trityl) tetrafluoroborate and hexafluoroantimonate salts are rapid and essentially quantitative.[62,63] Reductions of these and similar stable ions in dichloromethane/trifluoroacetic acid (TFA) show primary deuterium kinetic isotope effects in the range $k_H/k_D = 1.27$[64] to 1.89[65] at room temperature, whereas effects equal to $k_H/k_D = 1.50–2.33$ are seen for the reduction of diarylcarbenium ions with deuteriosilanes at $-70°$.[66] The kinetic rate dependence for similar reactions in acetic acid is first order in both the cation and the silane. The rates of a series of substituted arylsilanes correlate with σ constants, but not with σ^+ constants, to produce Hammett plots with $\rho = -1.84$ for triarylsilanes and $\rho = -1.01$[65] to -2.46[66] for aryldimethylsilanes. These results are interpreted to mean that the reactions occur through a four-center transition state in which the silicon center assumes a trigonal-bipyramidal shape with hydride exiting from an equatorial position while the carbocation's counterion approaches axially.[65]

Trityl and tropylium (cycloheptatrienyl) cation salts having complex metal-halide anions such as SbX_6^-, AsF_6^-, PF_6^-, $FeCl_4^-$, and BF_4^- undergo reduction with trialkylsilanes and aryldialkylsilanes at rates that are independent of the nature of the anion or of ring strain in the organosilicon hydride, are kinetically first order with respect to both cation and organosilicon hydride, and that display primary deuterium kinetic isotope effects of $k_H/k_D = 1.41–1.49$ in dichloromethane.[67,68] It is argued that these reactions proceed by way of a three-step mechanism involving a rate-determining single-electron transfer step[69] to create a charge-transfer complex between the carbocation and the organosilicon hydride followed by a faster transfer of hydride to the carbon center and the creation of a silicenium ion intermediate that is then rapidly captured by the counterion present (Eq. 5).[68] Others regard this argument as doubtful compared to the polar mechanism in which Si–H bond cleavage is rate determining.[66]

$$R'_3SiH + R_3\overset{+}{C} \rightleftharpoons [R'_3SiH\text{---}\overset{+}{C}R_3] \xrightarrow{\text{slow}} R'_3\overset{+}{Si}\text{----}H\text{–}CR_3 \longrightarrow R'_3\overset{+}{Si} + R_3CH \quad \text{(Eq. 5)}$$

Uncertainties in understanding the exact mechanistic details of these reactions are sure to stimulate continued work to define the nature of trivalent silicon cations in ionic reductions by organosilicon hydrides.

Role of Hypervalent Silicon Species

It is well known that strong electrophiles such as carbocations are reduced by organosilicon hydrides (Eq. 1).[3,70,71] On the other hand, simple mixtures of organosilicon hydrides and compounds with weakly electrophilic carbon centers such as ketones and aldehydes are normally unreactive unless the electrophilicity of the carbon center is enhanced by complexation of the carbonyl oxygen with Brønsted acids[3,70–73] or certain Lewis acids (Eq. 2).[1,70,71,74,75] Using these acids, hydride transfer from the silicon center to carbon may then occur to give either alcohol-related or hydrocarbon products.

Alternatively, unreactive mixtures of organosilicon hydrides and carbonyl compounds react by hydride transfer from the silicon center to the carbon center when certain nucleophilic species with a high affinity for silicon are added to the mixture.[76–94] This outcome likely results from the formation of valence-expanded, pentacoordinate hydrosilanide anion reaction intermediates that have stronger hydride-donating capabilities than their tetravalent precursors (Eq. 6).[22,95–101]

$$R_3SiH + Nu^- \rightleftharpoons R_3SiHNu^- \xrightarrow[2.\ H_2O]{1.\ R'_2C=O} R_3SiNu + R'_2CHOH \qquad \text{(Eq. 6)}$$

The bicyclic silatrane molecule **6**, which has a strong degree of coordination between the silicon center and the nitrogen bridgehead, has been shown to have unusually strong reducing properties compared to normal tetravalent organosilicon hydrides.[102] The hypothesis that valence-expanded pentacoordinate silicon species are the actual reducing species[76,77,83] is plausible, for such species are well known.[96,99,101,103–107] Other examples are known of the enhanced reducing powers of organosilicon hydrides that undergo intramolecular coordination and expansion to pentavalent[82] and even hexavalent states.[84,101,108–111]

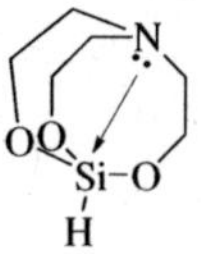

6

A variety of nucleophilic species cause valence expansion of organosilanes and the enhancement of reducing reactivity. These include formate, thiocyanate, tartrate, and phthalate salts,[78] as well as alkoxides[91,92,96,99,107] and catecholates.[93] The strong propensity of fluoride ion to cause silicon centers to undergo valence expansion[95,112] makes it especially effective in activating organosilicon hydrides as reducing agents. Aldehydes, ketones, and esters may all be reduced by such a technique, frequently with excellent functional group and stereochemical selectivity.[76–89]

The valence-expanded silicon intermediate retains some measure of stereochemical integrity up to the point of hydride transfer as evidenced by the small degrees of asymmetric induction that are observed in the reduction of prochiral ketones coupled with similar degrees of chirality found at the silicon center.[85] The transfer of hydride from the silicon center to the carbonyl carbon takes place in the rate-determining step as judged by the primary deuterium kinetic isotope effect of 1.50 observed in the fluoride-induced reduction of acetophenone with dimethylphenylsilane-d_1.[88] There is also evidence that the pentacoordinate silicon hydride can serve as a single-electron-transfer donor since radical-coupling products are sometimes obtained, although the general importance of this process is open to question.[89,99]

A diaryldihydrosilane with a hexacoordinated silicon center, produced through intramolecular coordination, is reported not to react with benzaldehyde, although the silane is capable of reducing silver ion to silver metal.[113] There is also a report of a heptacoordinate silicon hydride species with the ability to transfer hydride to trityl cation while remaining inactive toward methanol.[108,114]

Role of O/N-Silylated Cationic Intermediates

An interesting variation of the reaction mechanisms discussed above has been offered following studies of the hydrosilation reductions of aldehydes, ketones, and esters to their corresponding silyl ethers and acetals, respectively, when catalyzed by tris(pentafluorophenyl)borane, $(C_6F_5)_3B$,[115,116] and related boranes.[117] This mechanism proposes a pathway in which the first step is the reversible formation of a linear silane-borane adduct that undergoes subsequent nucleophilic attack by the carbonyl compound of the substrate to yield an O-silylated cationic intermediate along with a boron hydride anion.[118] The boron hydride ion then transfers a hydride to the carbon center of the O-silylated cation to yield the reduction product and regenerated free borane. A simplified view of the suggested mechanism is shown below (Scheme 1). Similar reaction paths have been proposed for the hydrosilation of enones and silyl ethers[119] as well as imines.[120]

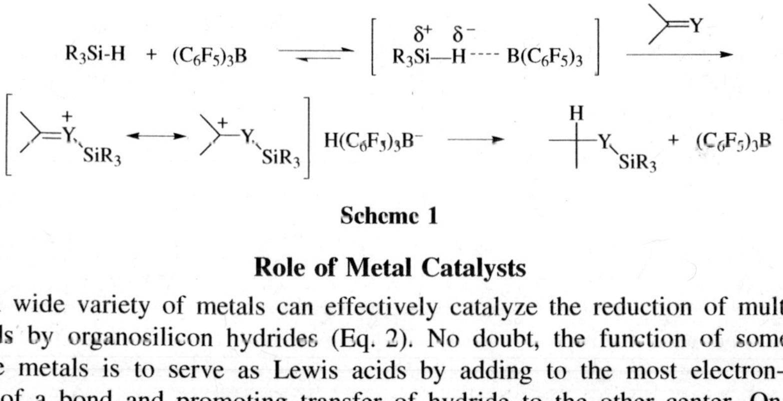

Scheme 1

Role of Metal Catalysts

A wide variety of metals can effectively catalyze the reduction of multiple bonds by organosilicon hydrides (Eq. 2). No doubt, the function of some of these metals is to serve as Lewis acids by adding to the most electron-rich end of a bond and promoting transfer of hydride to the other center. On the other hand, it is clear that many transition metal complexes function through significantly different and more complex catalytic pathways to promote silane reductions. A common reaction stage suggested for many of the catalytic cycles is the creation of a reactive intermediate having a metal-hydrogen bond that is formed by hydrogen transfer from the silane to the catalytic metal center.[116] This reducing center, often with appropriate coordinating ligands, subsequently delivers hydrogen to the substrate and the metal center is freed for additional catalytic cycles. When the catalytic metal ligands are chiral, this process can lead to very high degrees of enantiomeric selectivity in the reduction of prochiral substrates.[121–125]

SCOPE AND LIMITATIONS

Reduction of Substituted Alkanes

Alcohols to Alkanes. Many alcohols are converted directly into hydrocarbons when treated with acid in the presence of organosilicon hydrides (Eq. 7). The mechanism normally follows the pathway shown in Eq. 1.

$$\mathrm{ROH} \xrightarrow[\mathrm{R'_3SiH}]{\text{"acid"}} \mathrm{RH} \qquad \text{(Eq. 7)}$$

The reaction generally proceeds cleanly and in high yields (70–100%) when the starting alcohol permits the formation of reasonably stable carbocation intermediates. Alcohols capable of producing carbenium ions spanning a range of stabilities of more than 24 pK_{R+} units undergo this reduction.[26] Depending on the reaction conditions, secondary[126] and tertiary[127] aliphatic alcohols, secondary and tertiary benzylic alcohols,[26,126] some ring-substituted primary benzylic alcohols,[26,128,129] and cyclopropylcarbinols[130] are reduced to the corresponding alkanes. However, olefinic and rearrangement products can occur from side reactions under these acid conditions.[126,131,132] Phenols are not reduced under the same conditions.

Almost any organosilicon hydride causes reduction of the cations produced, although the order of reactivity of simple alkyl and aryl-substituted silanes is observed to be triethyl > trioctyl ~ diethyl > diphenyl ~ triphenyl.[26] A detailed quantitative study of the reactions of organosilicon hydrides with diarylcarbenium ions in dichloromethane at $-70°$ indicates a relative reactivity order of $R_3SiH > R_2SiH_2 > RSiH_3$, with alkyl substituent groups generally producing greater reactivity than aryl substituents.[61] Use of a deuterated silane yields the corresponding deuterated hydrocarbon (Eq. 8).[127,133,66]

$$\mathrm{ROH} \xrightarrow[\mathrm{R'_3SiD}]{\text{"acid"}} \mathrm{RD} \qquad \text{(Eq. 8)}$$

Normally, only a small stoichiometric excess (2–30 mol%) of silane is necessary to obtain good preparative yields of hydrocarbon products. However, because the capture of carbocation intermediates by silanes is a bimolecular occurrence, in cases where the intermediate may rearrange or undergo other unwanted side reactions such as cationic polymerization, it is sometimes necessary to use a large excess of silane in order to force the reduction to be competitive with alternative reaction pathways. An extreme case that illustrates this is the need for eight equivalents of triethylsilane in the reduction of benzyl alcohol to produce only a 40% yield of toluene; the mass of the remainder of the starting alcohol is found to be consumed in the formation of oligomers by bimolecular Friedel-Crafts-type side reactions that compete with the capture of the carbocations by the silane.[129]

Both Brønsted and Lewis acids are effective in coordinating with the hydroxyl oxygen to induce heterolysis of the C–O bond and cause formation of the necessary carbocation intermediate. The reactions are frequently conducted under homogeneous conditions in inert solvents such as dichloromethane or chloroform. Conditions used include the treatment of alcohols with organosilicon hydrides in neat acetic acid,[26,29] neat trifluoroacetic acid[134] or trifluoroacetic acid/ammonium fluoride[135] as well as mixtures of trifluoroacetic acid,[26] methanesulfonic acid,[126] or triflic (trifluoromethanesulfonic) acid with triflic anhydride[126] in dichloromethane or chloroform, mixtures of acetic acid and sulfuric or *p*-toluenesulfonic acid,[134] acetic acid and hydrogen chloride/aluminum chloride,[136] and boron trifluoride[126] or boron trifluoride etherate in dichloromethane[137,133] or chloroform.[138] The use of very strong Brønsted acids such as methanesulfonic and triflic acids may cause decomposition of the organosilane through hydrogen production[14] and/or cleavage of Si–C bonds[139] which compete with the desired reduction of the alcohol.[126] These undesirable side reactions may be avoided or reduced by running the reaction at $-78°$.[140] Sulfuric acid may cause undesirable oxidations to occur.[134] On balance, the most commonly chosen set of conditions for the reduction of alcohols is triethylsilane and trifluoroacetic acid (Et_3SiH/TFA) in dichloromethane solution.

The experimental evidence is convincing, at least with benzyl alcohols, that a "free" carbenium ion intermediate devoid of influence from its progenitor is the species that is captured by the non-nucleophilic organosilicon hydride. When optically active 2-phenyl-2-butanol is treated with Et_3SiH/TFA in chloroform, the 2-phenylbutane product is formed with complete racemization.[26] When a dichloromethane solution of the same alcohol is treated with trifluoroacetic acid in the presence of enantiomerically enriched 1-naphthylphenylmethylsilane, the 2-phenylbutane product obtained shows a small, but reproducible enantiomeric excess of 2–3%.[141] The predominant enantiomer formed in the product is dependent only on the predominant enantiomer of silane used as the reducing agent and is independent of whether one of the pure enantiomers or the racemic alcohol is used as substrate.[142] The same stereochemical results are obtained in the hydrocarbon product when the alkene 2-phenyl-1-butene is the precursor to the carbenium ion intermediate (π-route, Eq. 2) instead of the tetrahedral alcohol (σ-route, Eq. 1).[142] A similar conclusion is reached from a study of the reduction of optically active 1-phenylethanol to phenylethane-d_1 with boron trifluoride etherate and triethylsilane-d_1.[133] These experiments illustrate the lack of nucleophilicity or S_N2-like behavior of the organosilicon hydrides in these reactions and presage the stereochemistry expected from such transformations.

Primary Alkyl Alcohols. Primary alkyl alcohols do not undergo reduction when treated with Brønsted acids and organosilicon hydrides under usual laboratory conditions.[143] This reflects the relative instability of primary alkyl carbenium ions in the condensed phase and the weak intrinsic nucleophilicity of organosilicon hydrides. On the other hand, the combination of excess Et_3SiH and catalytic amounts (5–10 mol%) of $(C_6F_5)_3B$ reduces primary aliphatic alcohols to the alkanes in high yields (Eq. 9), but the reaction stops at the non-reductive silylation of the alcohol with only a single equivalent of the silane.[144,145] This type of reaction is thought to proceed via a direct nucleophilic displacement rather than by way of a carbenium ion mechanism.[145]

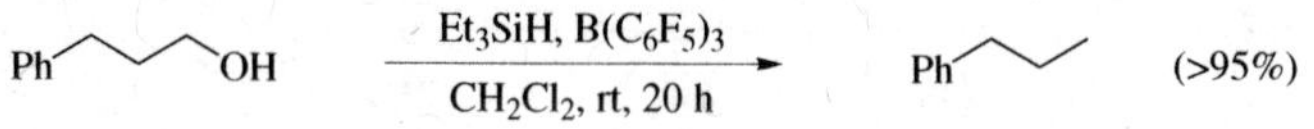

(Eq. 9)

Secondary Alkyl Alcohols. Treatment of secondary alkyl alcohols with trifluoroacetic acid and organosilicon hydrides results only in the formation of the trifluoroacetate esters; no reduction is reported to occur.[1,2] Reduction of secondary alkyl alcohols does take place when very strong Lewis acids such as boron trifluoride[126,129] or aluminum chloride[136,146] are used. For example, treatment of a dichloromethane solution of 2-adamantanol and triethylsilane (1.3 equivalents) with boron trifluoride gas at room temperature for 15 minutes gives upon workup a 98% yield of the hydrocarbon adamantane along with fluorotriethylsilane (Eq. 10).[129]

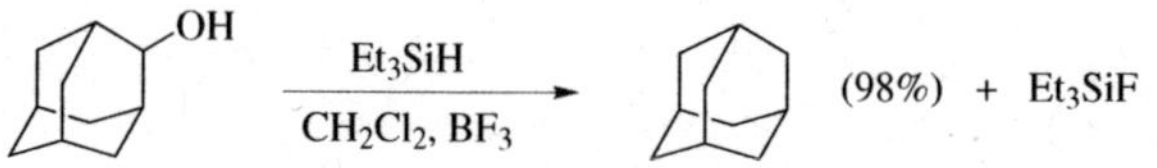

(Eq. 10)

In contrast, when boron trifluoride etherate is substituted for the free boron trifluoride, only a trace of the hydrocarbon is formed, even after weeks of reaction.[143] The unique effectiveness of boron trifluoride gas in promoting these reductions is believed to be due to several factors, including the ability of the coordinatively unsaturated boron center to rapidly and tightly coordinate with oxygen centers and to the thermodynamically favorable creation of a Si–F bond.[1] A slight pressure of boron trifluoride gas must be maintained over the surface of the solution throughout the reaction because boron trifluoride has only limited solubility in the weakly coordinating dichloromethane solvent.

The formation of alkenes and alkene-related polymerization products can seriously reduce the yields of desired alkane products from secondary alcohols, which can undergo elimination reactions. For example, reduction of 2-octanol at 0° with boron trifluoride gas in dichloromethane containing 1.2 equivalents of triethylsilane gives only a 58% yield of *n*-octane after 75 minutes (Eq. 11).[129] The remainder of the hydrocarbon mass comprises nonvolatile polymeric material.[126]

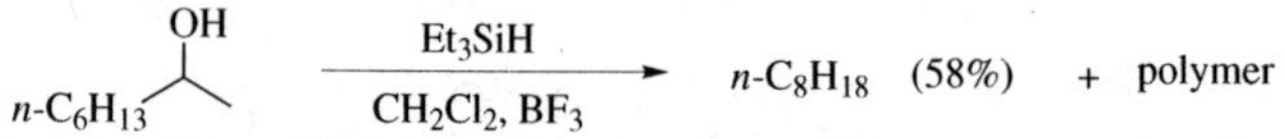

(Eq. 11)

Aluminum chloride, used either as a stoichiometric reagent or as a catalyst with gaseous hydrogen chloride, may be used to promote silane reductions of secondary alkyl alcohols that otherwise resist reduction by the action of weaker acids.[136] For example, cyclohexanol is not reduced by organosilicon hydrides in the presence of trifluoroacetic acid in dichloromethane, presumably because of the relative instability and difficult formation of the secondary cyclohexyl carbocation. By contrast, treatment of cyclohexanol with an excess of hydrogen chloride gas in the presence of a three-to-four-fold excess of triethylsilane and 1.5 equivalents of aluminum chloride in anhydrous dichloromethane produces 70% of cyclohexane and 7% of methylcyclopentane after a reaction time of 3.5 hours at room temperature (Eq. 12).[136] The cyclohexane is presumably formed by capture of the secondary cyclohexyl cation, whereas the methylcyclopentane must arise from hydride capture of the more stable tertiary methylcyclopentyl cation formed by rearrangement of the cyclohexyl cation.[147,148] Diminishing the amount of aluminum chloride to only 0.5 equivalents results in no reaction after one-half hour and the formation of only 8% of cyclohexane after four hours reaction time. The reaction proceeds slowly in the absence of hydrogen chloride, producing 53% of cyclohexane and 6% of methylcyclopentane after 16.5 hours using two equivalents of aluminum chloride.

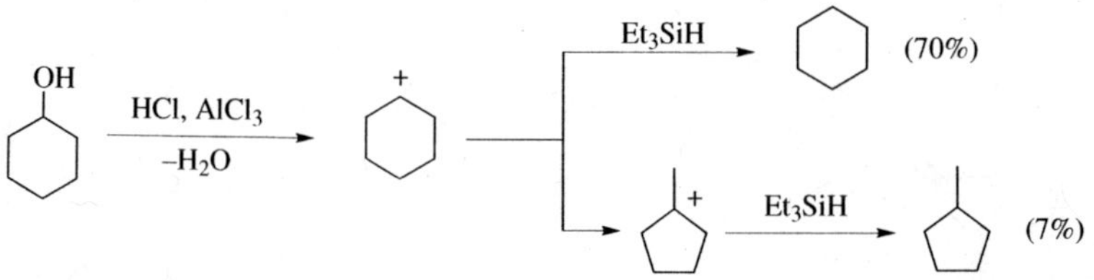

(Eq. 12)

Tertiary Alkyl Alcohols. Tertiary alkyl alcohols generally undergo facile reduction when treated with acids in the presence of organosilicon hydrides.[127,136] This comparative ease of reduction reflects the enhanced stability and ease of formation of tertiary alkyl carbenium ions compared with primary and secondary carbenium ions. Thus, treatment of 1-methylcyclohexanol with mixtures of triethylsilane and aluminum chloride in dichloromethane produces near quantitative yields of methylcyclohexane with or without added hydrogen chloride in as little as 30 minutes at room temperature, in contrast to the more vigorous conditions needed for the reduction of the secondary alcohol cyclohexanol.[136]

Similarly, and in contrast to the behavior of its secondary isomer, 2-adamantanol, 1-adamantanol undergoes smooth, quantitative reduction to adamantane in less than an hour at room temperature in dichloromethane solution containing triethylsilane under the catalysis of either free boron trifluoride[129] or boron trifluoride etherate (Eq. 13).[143]

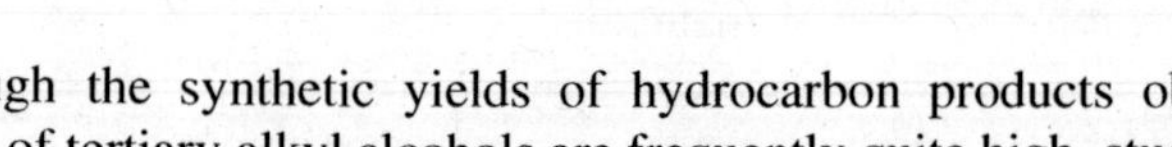

(Eq. 13)

Although the synthetic yields of hydrocarbon products obtained from the reduction of tertiary alkyl alcohols are frequently quite high, studies show that the reaction pathways taken by the reactants are not always as direct or straightforward as might be suggested by the structural relationships between reactants and products. For example, preparative-scale treatment of a dichloromethane solution of 3-ethylpentan-3-ol and triphenylsilane (1.2 equivalents) with excess trifluoroacetic acid (1.5 M) at room temperature for 24 hours gives 3-ethylpentane in 78% yield (Eq. 14).[127] Under these reaction conditions, the alcohol rapidly undergoes elimination to 3-ethyl-2-pentene, which is the actual species undergoing reduction.

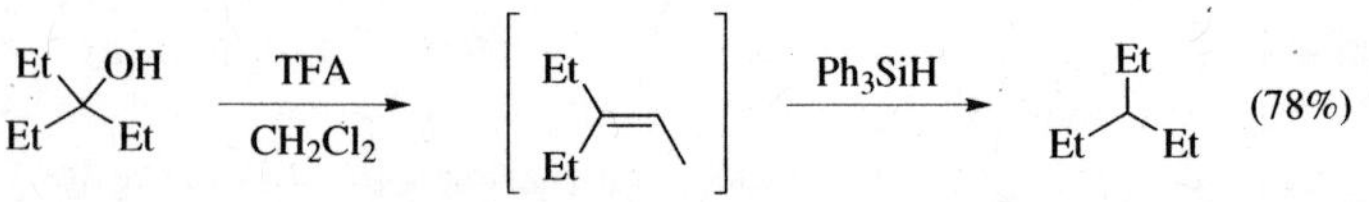

(Eq. 14)

The tertiary alcohol *cis,cis,trans*-perhydro-9*b*-phenalenol (**7**) is converted stereospecifically and in high yield (92%) to *trans,trans,trans*-perhydrophenalene (**10**) when treated with either triethylsilane or triphenylsilane and trifluoroacetic acid in dichloromethane (Eq. 15). Studies indicate that the reaction path follows the cation rearrangement **8** → **9** and that the trans trifluoroacetate ester related to cation **9** is an intermediate, which accumulates during the reaction.[127]

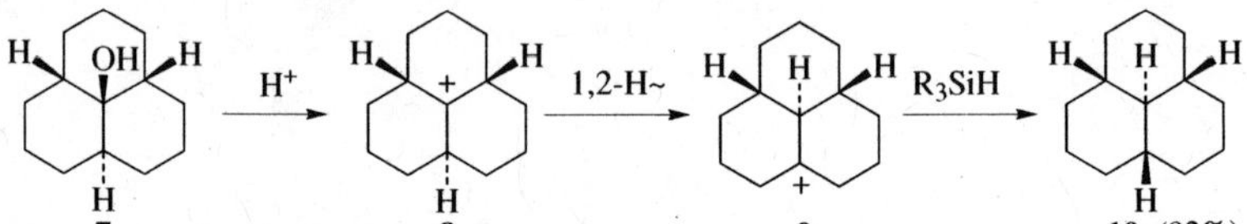

7 **8** **9** **10** (92%)

(Eq. 15)

The conversion of alcohols directly into the structurally related hydrocarbons by ionic hydrogenation can provide a means of synthesis for compounds that would be extremely difficult or impossible to obtain by other methods. A good example is the synthesis of 2-*tert*-butyladamantane (**12**, R = Me). This interesting, highly strained compound may be synthesized in moderate overall yield by a conventional multiple-step route.[149] Alternatively, it is obtained in 90% isolated yield upon treatment of a dichloromethane solution of the readily available 2-*tert*-butyl-2-adamantanol (**11**, R = Me)[150] and one equivalent of either tri-*n*-hexylsilane[151,152] or triethylsilane[153] with trifluoroacetic acid at room temperature (Eq. 16).

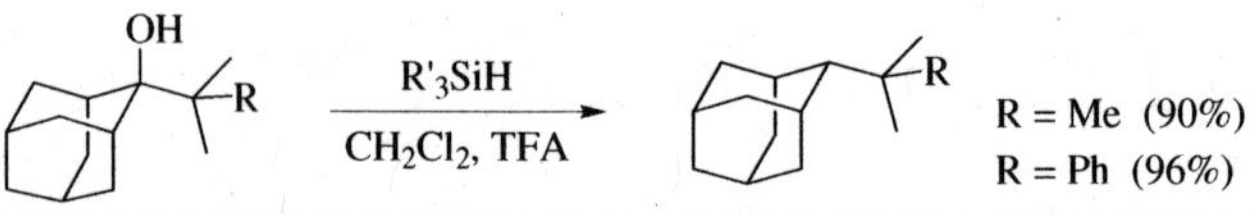

11 **12**

(Eq. 16)

In a similar fashion, 2-cumyladamantane (**12**, R = Ph) is formed in nearly quantitative yield upon treatment of the easily synthesized 2-cumyl-2-adamantanol (**11**, R = Ph)[154] with triethylsilane and methanesulfonic acid in dichloromethane at $-78°$.[155] The high yield of a single very strained hydrocarbon product in each reaction is quite surprising in view of the very complex interconversions of carbocations known to take place from the alcohol precursors.[140,151,152,156]

The remarkable chemoselectivity of this reductive technique is demonstrated by the conversion of the functionally rich compound **13** into **14** in 86% yield upon treatment with Et_3SiH/TFA at room temperature for two hours (Eq. 17).[157]

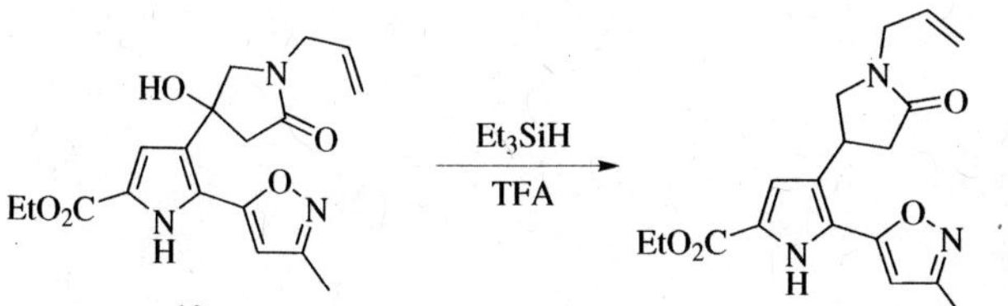

13 14

(86%) (Eq. 17)

Several sterically congested aryldiadamantylmethanols are reduced to atropisomeric diastereomeric mixtures of the corresponding aryldiadamantylmethanes with Et_3SiH/TFA (Eq. 18).[158–161]

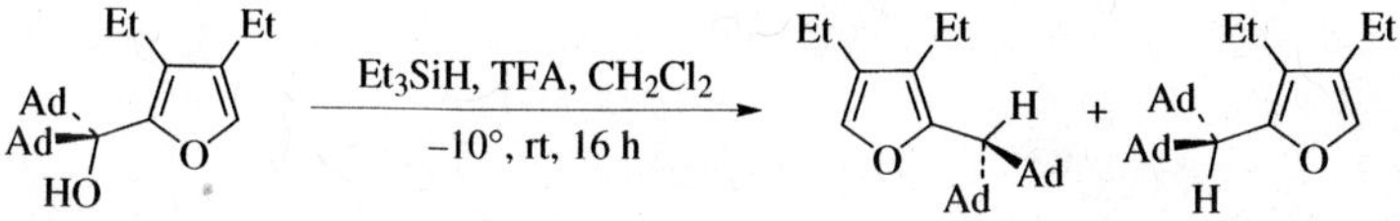

(—) 1:1 (Eq. 18)

This reagent combination reduces a tertiary alcohol in the presence of a quinone moiety (Eq. 19).[162] Tertiary alcohols are also reduced with the reagent combinations $Et_3SiH/MeSO_3H$[140] and $Et_3SiH/AlCl_3/HCl$.[136]

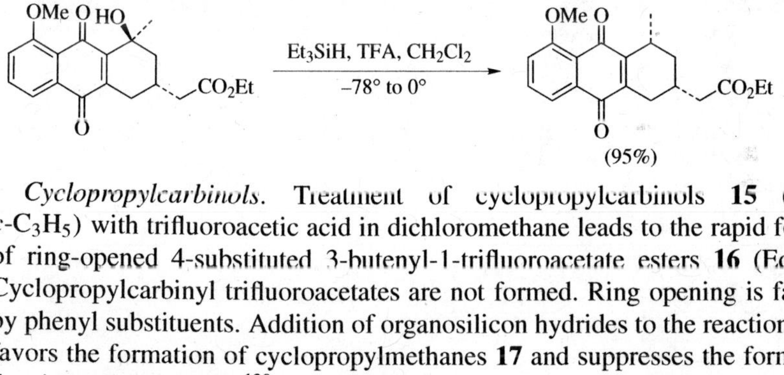

(95%) (Eq. 19)

Cyclopropylcarbinols. Treatment of cyclopropylcarbinols **15** (R = Ph, c-C_3H_5) with trifluoroacetic acid in dichloromethane leads to the rapid formation of ring-opened 4-substituted 3-butenyl-1-trifluoroacetate esters **16** (Eq. 20).[130] Cyclopropylcarbinyl trifluoroacetates are not formed. Ring opening is facilitated by phenyl substituents. Addition of organosilicon hydrides to the reaction mixture favors the formation of cyclopropylmethanes **17** and suppresses the formation of the ring-opened esters.[130]

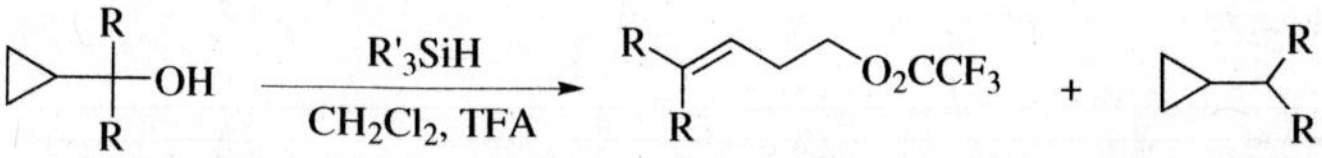

15 **16** **17** (Eq. 20)

R = c-C_3H_5	R' = Et	**16 + 17** (—), **16**:**17** = 1:99
R = Ph	R' = Et	**16 + 17** (—), **16**:**17** = 19:81

Triethylsilane and diethylsilane are somewhat more effective than triphenylsilane at increasing the amount of reduced product **17**.[130] Yields of **17** in excess of 90% may be obtained. The remainder of the product is butenyl ester **16**. Hydrogenolysis of the cyclopropyl rings does not occur under these conditions. A better yield of **17** is obtained when the reaction is carried out at −15° than at room temperature. Under the same set of reaction conditions (dichloromethane, 0.5 M trifluoroacetic acid, 0.5 hour, room temperature), the amount of hydrocarbon product **17** (R = Ph) obtained from diphenylcyclopropylcarbinol changes from 16% with triphenylsilane as the hydride-donating reagent to 45% with triphenylgermane, 85% with triphenylstannane, and 78% with tri-*n*-butylstannane.[24]

Benzyl Alcohols. Benzyl alcohols of nearly all kinds undergo reduction when treated with acid in the presence of organosilicon hydrides. The most obvious exception to this is the behavior of benzyl alcohol itself. It resists reduction by the action of trifluoroacetic acid and triethylsilane, even after extended reaction times.[26] Reducing systems consisting of triethylsilane and sulfuric acid/acetic acid or *p*-toluenesulfonic acid/acetic acid mixtures also fail to reduce benzyl alcohol to toluene.[134] As previously mentioned, substitution of boron trifluoride for trifluoroacetic acid results in the formation of modest yields of toluene, but only when a very large excess of the silane is used in order to capture the benzyl cation intermediate and suppress Friedel-Crafts oligomerization processes.[129,143]

Ring-substituted benzyl alcohols sometimes undergo such reduction more effectively than unsubstituted alcohols. For example, treatment of a dichloromethane solution of 2,4,6-trimethylbenzyl alcohol with trifluoroacetic acid and triphenylsilane produces a 41% isolated (89% by GLC) yield of isodurene.[26] Treatment of 2-methyl-4,6-di-*tert*-butylbenzyl alcohol with a three-fold excess of triethylsilane and trifluoroacetic acid in dichloromethane at room temperature gives an 85% yield of 2-methyl-4,6-di-*tert*-butyltoluene together with 15% of 3,5-di-*tert*-butyltoluene. The latter is presumably formed by loss of protonated formaldehyde from the C1 ring-protonated substrate.[128] Similar treatment of 2,4,6-tri-*tert*-butylbenzyl alcohol produces a 90% yield of 2,4,6-tri-*tert*-butyltoluene within one hour (Eq. 21).[128]

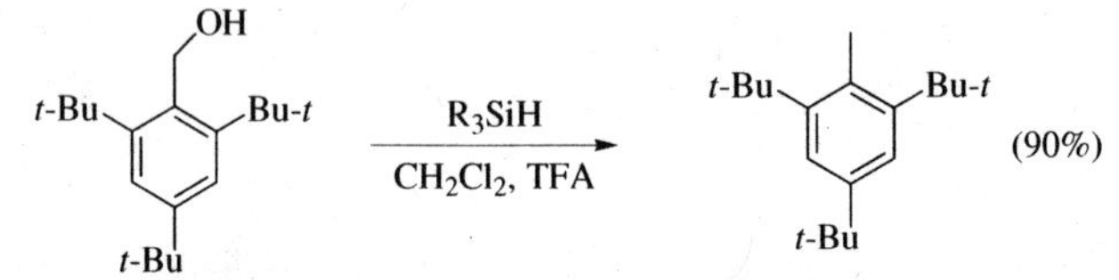

(Eq. 21)

The reduction of 2-(hydroxymethyl)-1,4,6,8-tetramethylazulene to 1,2,4,6,8-pentamethylazulene occurs quantitatively upon treatment with triethylsilane and trifluoroacetic acid at 60° for 19 hours (Eq. 22).[163]

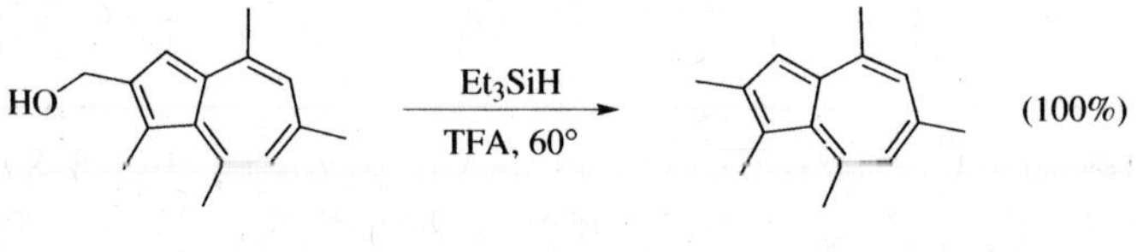

(Eq. 22)

Treatment of either the cis or trans isomer of 4-*tert*-butyl-1-phenylcyclohexanol with trifluoroacetic acid and one of a variety of organosilicon hydrides in dichloromethane yields a mixture of *cis*- and *trans*-4-*tert*-butyl-1-phenylcyclohexane and the elimination product, 4-*tert*-butyl-1-phenylcyclohexene

(22–72%) (Eq. 23).[26] More elimination product is obtained from the cis than from the trans alcohol. The trans/cis ratio of reduced products is independent of the isomer of starting alcohol used and depends only on the nature of the silane used. This ratio is ca. 1.8 for triorganosilanes (e.g., triethyl, tri-*n*-octyl, triphenyl) and ca. 4.0 for diorganosilanes (e.g., diethyl, diphenyl) and phenylsilane. The most important factor for the stereoselectivity of product formation seems to be the degree of steric bulk provided by the organic groups bonded to silicon, rather than the electronic nature of the substituents. The smaller the effective steric bulk of the reducing agent, the greater is the trans/cis product ratio.[26] Replacement of the silanes with germanes or stannanes as the hydride donors causes a decrease in the amount of elimination product formed so that it becomes a minor product (10–38%).[24] The trans/cis ratio of reduced products is ~2 when triphenylgermane is used, ~1.4 with triphenylstannane, and ~0.85 with tri-*n*-butylstannane.[24]

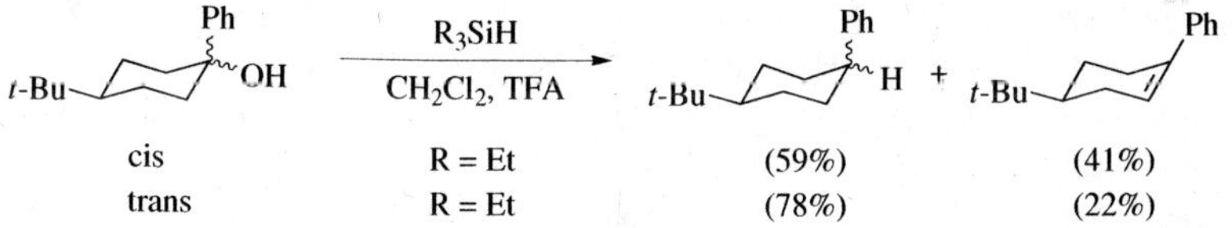

(Eq. 23)

Reduction of either the exo or endo isomer of 2-phenyl-2-norbornanol with trifluoroacetic acid and triethylsilane, triphenylsilane, or phenylsilane in dichloromethane gives endo-2-phenylnorbornane quantitatively (Eq. 24).[164] The stereospecific formation of only the endo-hydrocarbon can be understood on the basis that only exo approach by organosilicon hydride toward the 2-phenylnorbornyl cation intermediate is kinetically competitive for product formation.[164]

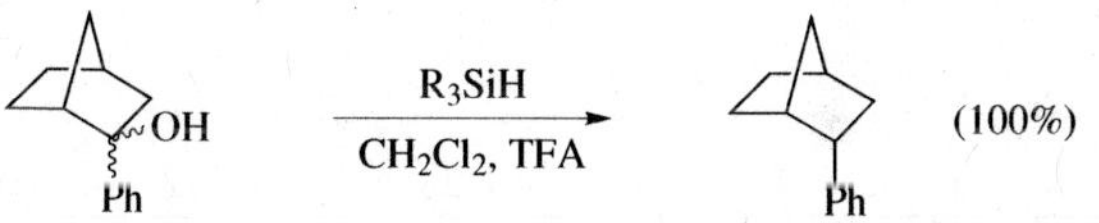

(Eq. 24)

The bornyl system is less subject than the norbornyl system to exclusive exo approach by organosilicon hydrides and related reducing agents because of the steric restraints imposed by the additional methyl groups. Thus, treatment of a dichloromethane solution of *p*-anisylisoborneol (**18**) with trifluoroacetic acid and triphenylsilane quantitatively provides the isomeric reduced products *p*-bornylanisole (**19**) and *p*-isobornylanisole (**20**) in a ratio of 68 : 32 (Eq. 25).[24] Replacement of the triphenylsilane with triphenylstannane produces 98% of **19** and 2% of **20**. Use of the sterically less demanding phenylsilane gives **19** as the exclusive product. By comparison, trapping of the cation derived from **18** with borohydride gives 87% of **19** and 13% of **20**.[165]

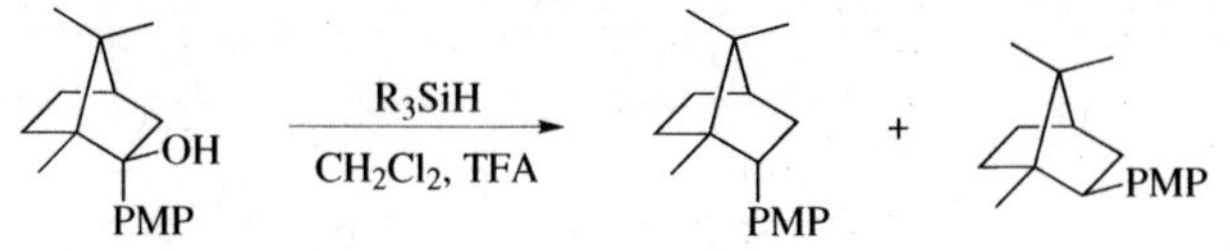

(Eq. 25)

18 **19** **20**

PMP = 4-$MeOC_6H_4$ R = Ph **19** + **20** (100%), **19**:**20** = 68:32

Reductions of tertiary or benzylic alcohols do not always take place as quickly and simply as might be expected. A study of the reduction of 2-methyl-1-phenylpropan-1-ol (**21**, R = OH) and its isomer, 2-methyl-1-phenylpropan-2-ol (**22**, R = OH), illustrates this observation.[166] Both of these alcohols are reduced in high yield (98%) by the action of acid and triphenylsilane to the same hydrocarbon, 2-methyl-1-phenylpropane (**21** or **22**, R = H). The immediate products formed from either alcohol in a 55% solution of trifluoroacetic acid in nitrobenzene are the trifluoroacetate esters (R = CF_3CO_2). Surprisingly, at 25° in the absence of organosilicon hydride, ester **21** (R = CF_3CO_2) undergoes complete isomerization to ester **22** (R = CF_3CO_2) within two hours. Use of triphenylsilane-d_1 as the reducing agent indicates that alcohol **21** (R = OH) actually produces a mixture of the two isotope-position isomers of the reduction product (**21** and **22**, R = D), with isomer **22** favored by a factor of 2 to 3 over isomer **21**. Similar results are found when the starting alcohol is **22**. The conclusion is reached that the species actually captured by the organosilicon hydride consists of a dynamic mixture of the two cations derived from **21** and **22**.[166]

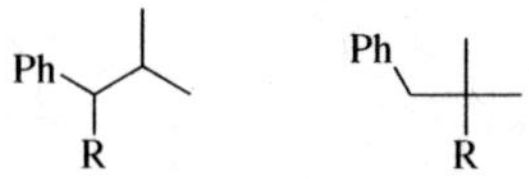

21 **22**

In some instances, treatment of polyfunctional benzylic alcohols with acid in the presence of organosilicon hydrides causes multiple functional group transformations to occur simultaneously. This phenomenon is illustrated by the reduction of the secondary benzylic alcohol function and concomitant loss of the methoxymethyl protecting group of 2-(1-hydroxydecyl)-5-methoxy-1-(methoxymethyleneoxy)naphthalene upon treatment with Et_3SiH/TFA in dichloromethane (Eq. 26).[167]

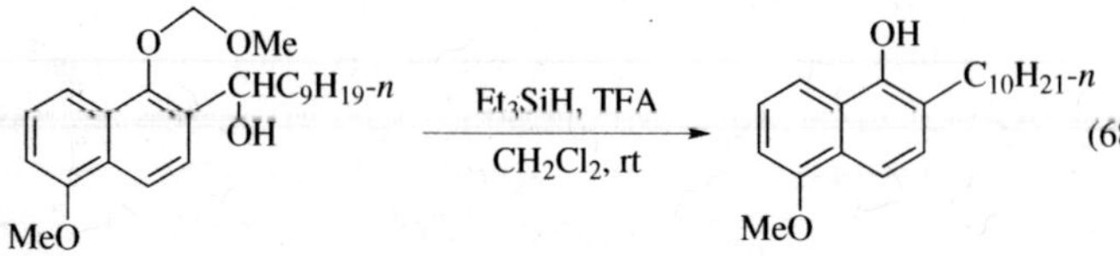

(68%) (Eq. 26)

An example of an exclusive chemoselective reduction of a benzylic hydroxy function in a polyfunctional compound is seen in the conversion of **23** into **24** in 76% yield using Et_3SiH/TFA (Eq. 27).[168] The benzylic hydroxy group in a complex polypeptide derived from lysobactin is selectively reduced with the same reagents (Eq. 28).[169]

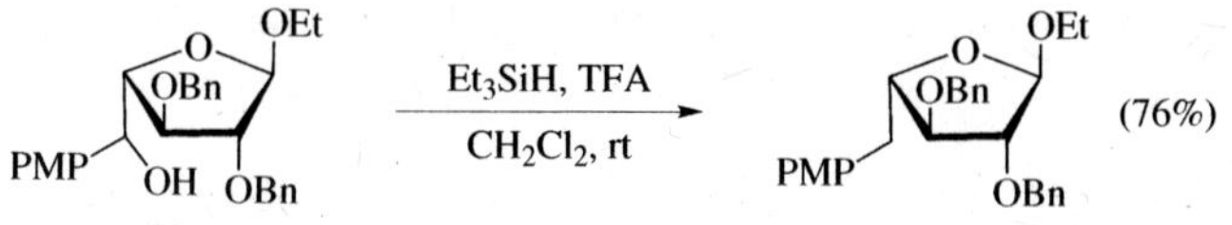

(Eq. 27)

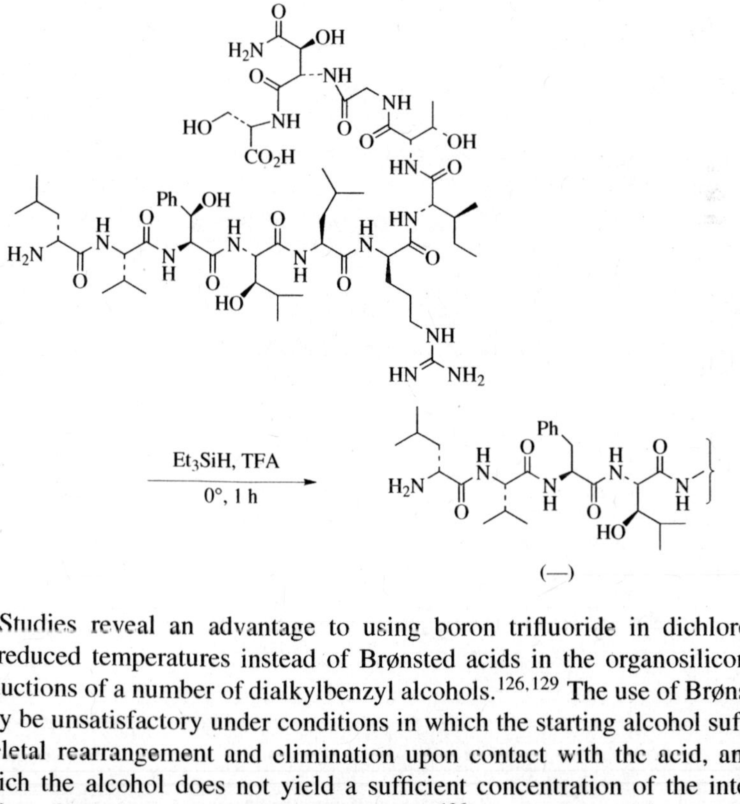

(Eq. 28)

Studies reveal an advantage to using boron trifluoride in dichloromethane at reduced temperatures instead of Brønsted acids in the organosilicon hydride reductions of a number of dialkylbenzyl alcohols.[126,129] The use of Brønsted acids may be unsatisfactory under conditions in which the starting alcohol suffers rapid skeletal rearrangement and elimination upon contact with the acid, and also in which the alcohol does not yield a sufficient concentration of the intermediate carbocation when treated with protic acids.[126]

An example of an alcohol that can undergo rapid skeletal rearrangement is 3,3-dimethyl-2-phenyl-2-butanol (Eq. 29). Attempts to reduce this alcohol in dichloromethane solution with 1-naphthyl(phenyl)methylsilane yield only a mixture of the rearranged elimination products 3,3-dimethyl-2-phenyl-1-butene and 2,3-dimethyl-3-phenyl-1-butene when trifluoroacetic acid or methanesulfonic acid is used. Use of a 1 : 1 triflic acid/triflic anhydride mixture with a 50 mol% excess of the silane gives good yields of the unrearranged reduction product 3,3-dimethyl-2-phenylbutane, but also causes extensive decomposition of the silane.[126] In contrast, introduction of boron trifluoride gas into a dichloromethane solution of the alcohol and a 10 mol% excess of the silane at −60° produces 86.5% of the desired, structurally intact hydrocarbon, 3,3-dimethyl-2-phenylbutane, along with 13.5% of the methyl-shifted product, 2,3-dimethyl-2-phenylbutane, within only six minutes. Clean formation of the fluorosilane related to the organosilicon hydride accompanies the reduction. The workup consists of quenching with solid potassium carbonate followed by addition of water, drying of the dichloromethane solution, and normal product isolation.[126]

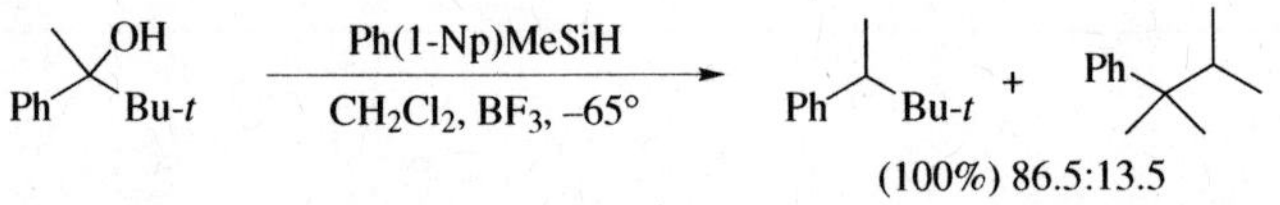

(Eq. 29)

A variety of para-substituted 2-phenyl-2-butanols undergo quick and efficient reductions to the corresponding 2-phenylbutanes when they are dissolved in dichloromethane and a 2–10% excess of phenylmethylneopentylsilane and boron trifluoride is introduced at 0° (Eq. 30).[126] Several reactions deserve mention. For example, when R = CF_3, use of trifluoroacetic acid produces no hydrocarbon product, even after two hours of reaction time. In contrast, addition of boron trifluoride catalyst provides an 80% yield of product after only two minutes. When R = MeO, both trifluoroacetic acid and boron trifluoride produce a quantitative yield of the hydrocarbon within two minutes. However, when R = NO_2, attempts to promote the reduction with either trifluoroacetic acid or even methanesulfonic acid fail; even after reaction periods of up to eight hours, only recovered starting alcohol is obtained. Use of boron trifluoride provides a quantitative conversion into 2-(*p*-nitrophenyl)butane after only ten minutes. It is significant that the normally easily reducible nitro group survives these conditions entirely intact.[126,129] Triethylsilane may be used as the silane.[143]

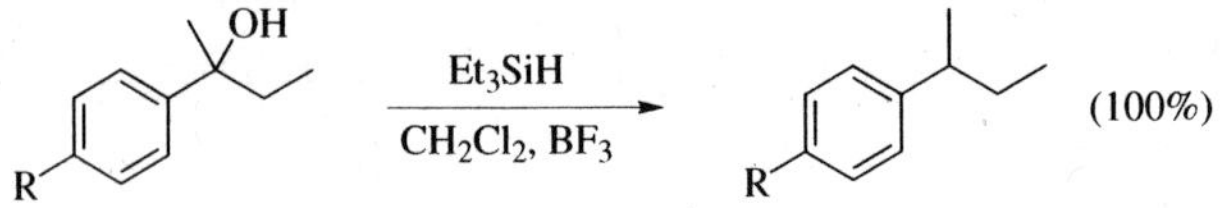

(Eq. 30)

Treatment with triethylsilane and boron trifluoride etherate allows a variety of methyl β-hydroxy-β-arylpropionates to be reduced to methyl β-arylpropionates in yields of 85–100% as part of a synthetic sequence leading to the preparation of indanones (Eq. 31).[170] Small amounts of dehydration products formed simultaneously are reduced to the methyl β-arylpropionates by mild catalytic hydrogenation.[170]

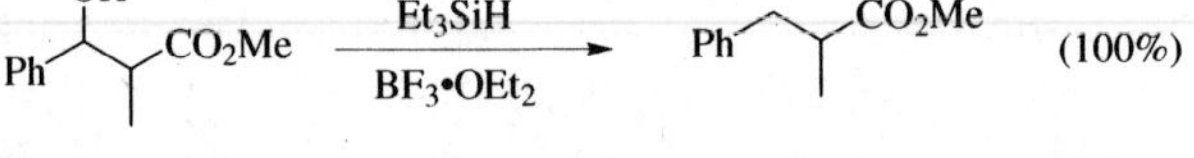

(Eq. 31)

Diaryl and triaryl benzylic alcohols generally undergo smooth reduction to the corresponding hydrocarbons. Thus, both diphenyl- and triphenylcarbinol quickly give good to excellent yields of the corresponding substituted methanes when treated with triphenylsilane or Et_3SiH/trifluoroacetic acid in dichloromethane[26] or with triethylsilane and mixtures of either sulfuric or *p*-toluenesulfonic acids in acetic acid.[136] The reductions do not occur with the parent, unsubstituted carbinols using only acetic acid; however, tri-*p*-anisylcarbinol, 2,2′,2″,6,6′,6″-hexamethoxytriphenylcarbinol, 9-phenyl-9*H*-xanthen-9-ol, 9-*p*-anisyl-9*H*-xanthen-9-ol,[26] and 9-(2,6-dimethoxyphenyl)-1,8-dimethoxy-9*H*-xanthen-9-ol[29] all undergo smooth conversion to the respective hydrocarbons when treated with acetic acid containing triethylsilane.

In fact, the use of acetic acid as both solvent and catalyst may be the method of choice in effecting the reductions of very electron-rich benzylic alcohols and those that form acid-labile reduction products. When 2,2′,2″,6,6′,6″-hexamethoxytriphenylcarbinol is treated with Et_3SiH/trifluoroacetic acid in dichloromethane, the reduction goes beyond the triarylmethane stage to produce one equivalent of 2,2′,6,6′-tetramethoxydiphenylmethane and one equivalent of *m*-dimethoxybenzene.[26] These additional products are thought to arise from protonation of one of the electron-rich rings of the initially formed 2,2′,2″,6,6′,6″-hexamethoxytriphenylmethane at C1 by the trifluoroacetic acid. Expulsion of a molecule of *m*-dimethoxybenzene followed by hydride capture of the 2,6,2′,6′-tetramethoxydiphenylmethyl "daughter" cation formed accounts for the final mixture of products.[26] Sesquixanthydrol **25** undergoes reduction to the hydrocarbon only reluctantly, presumably because of the great stability of the sesquixanthydryl cation (**1**). An early report indicates that the alcohol is able to resist reduction upon treatment with trifluoroacetic acid and excess triethylsilane in dichloromethane for 24 hours and to remain unreduced when dissolved in acetic acid containing triethylsilane.[26] A later report indicates formation of the hydrocarbon in 89% yield (Eq. 32).[29]

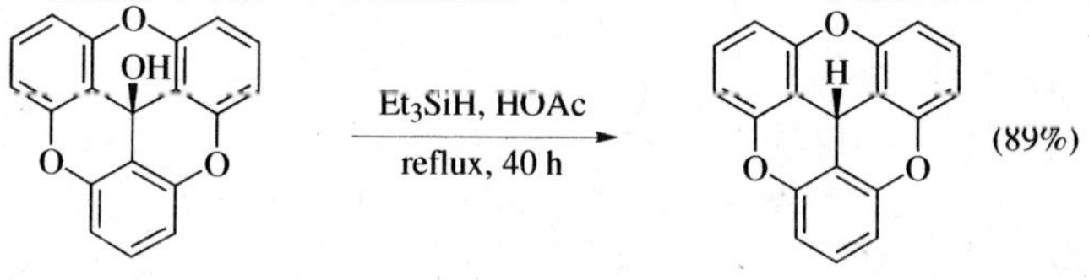

25

(Eq. 32)

Intramolecular Friedel-Crafts reactions can sometimes compete with organosilicon hydride reductions of benzylic-type alcohols to cause formation of undesired products. An example is the attempted reduction of alcohol **26** to the corresponding hydrocarbon. When **26** is treated with triethylsilane in trifluoroacetic acid at room temperature for 15 hours, a mixture of the two fluorene isomers **27** and **28** is obtained in a combined yield of 45%. None of the hydrocarbon structurally related to the substrate alcohol **26** is obtained.[171] Whether this problem could be circumvented by running the reduction at a lower temperature or with a different acid remains subject to experimentation.

Both benzylic and secondary aliphatic alcohols are reduced with the combination of Ph_2ClSiH and a catalytic amount of indium trichloride. This combination

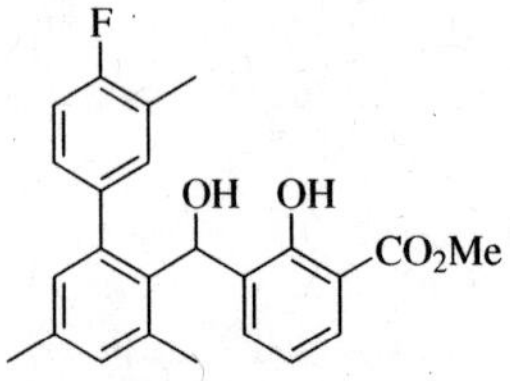

26

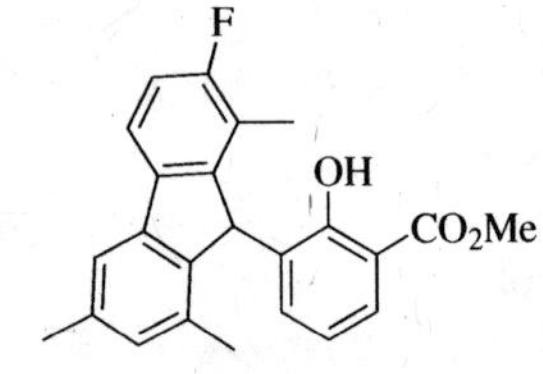

27

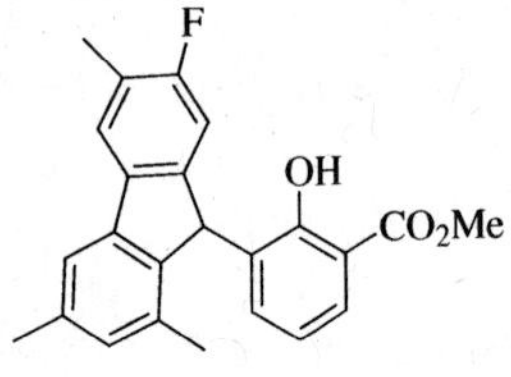

28

chemoselectively reduces benzyl alcohols in the presence of both ester and halide functions (Eq. 33).[172]

(Eq. 33)

R	Conditions	
Et	$ClCH_2CH_2Cl$, 50°, 0.5 h	(68%)
CH_2Cl	CH_2Cl_2, rt, 3 h	(95%)

The combination of excess Et_3SiH and catalytic amounts (5–10 mol%) of $(C_6F_5)_3B$ reduces benzylic alcohols to the hydrocarbons (Eq. 34), although the reaction stops at the non-reductive simple silylation of the alcohol with only a single equivalent of the silane.[144,145]

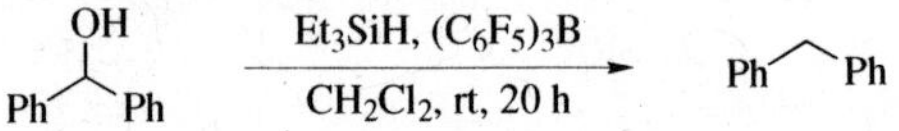

(98%) (Eq. 34)

Allyl Alcohols. Secondary cyclic allylic alcohols are reduced with the combination of Et_3SiH and ethereal $LiClO_4$, even in the presence of a tertiary alcohol (Eq. 35) or ketal function.[173] Primary allylic alcohols do not undergo deoxygenation under similar conditions.[173]

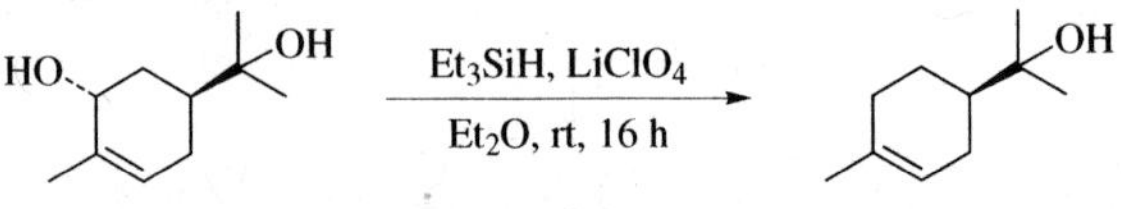

(52%) (Eq. 35)

Treatment of 1-[2-(2-methoxy-5-isopropylphenyl)-1-hydroxyethyl]-2,6,6-trimethylcyclohexene with triethylsilane and boron trifluoride etherate in dichloromethane at −10° leads to its reduction to 2-(2,6,6-trimethyl-1-cyclohexenyl)-1-(2-methoxy-5-isopropylphenyl)ethane in 69% yield (Eq. 36).[174]

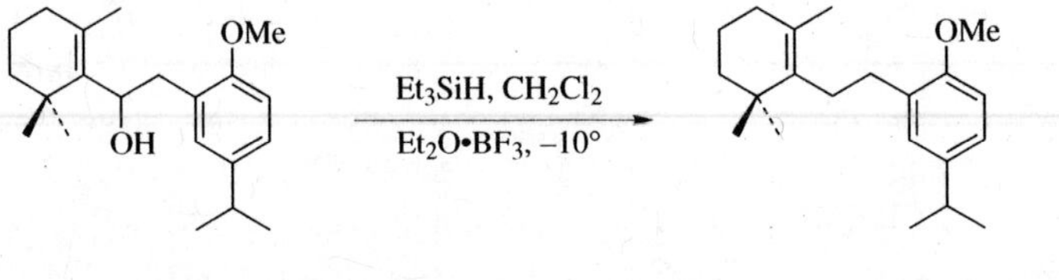

(69%) (Eq. 36)

Metal-Complexed Alcohols. It is well known that carbocations are frequently stabilized when organotransition metal centers are present in adjacent portions of the molecule.[175–177] It is thus not surprising that alcohols possessing such centers are prone to undergo facile reduction upon treatment with acids and organosilicon hydrides. Perhaps it is more surprising that the coordinated metal centers survive the reduction conditions so well.

Methylferrocenylcarbinols bearing several functional groups (R = H, Cl, CO_2Me, CN) on the distal C_5 ring undergo reduction to the corresponding ethylferrocenes when treated with an excess of Et_3SiH/TFA in acetic acid solution at 20° (Eq. 37).[178] The yields of reduced product are no less than 85% within three hours, except when R = CN. Then the conversion into the ethylferrocene is only 55% complete after 20 hours of reaction time, reflecting the destabilizing effect of the cyano group on the intermediate carbocation. In a similar fashion, symmetrically disubstituted ferrocenylcarbinols undergo facile double deoxidative reduction in yields of more than 80% within three hours at 20° when dissolved in trifluoroacetic acid and treated with two equivalents of triethylsilane (Eq. 38).[179] The "half-sandwich" cyclopentadienylmanganese tricarbonyl (cymantrene) carbinol undergoes reduction in a similar way to that of its ferrocenylcarbinol analogs (Eq. 39).[180]

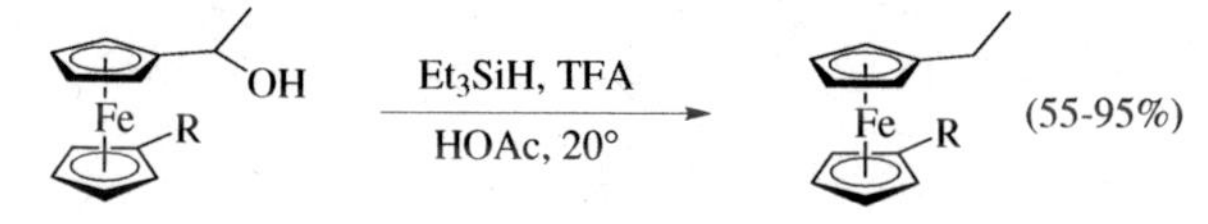

(Eq. 37)

R = H, Cl, CO_2Me, CN

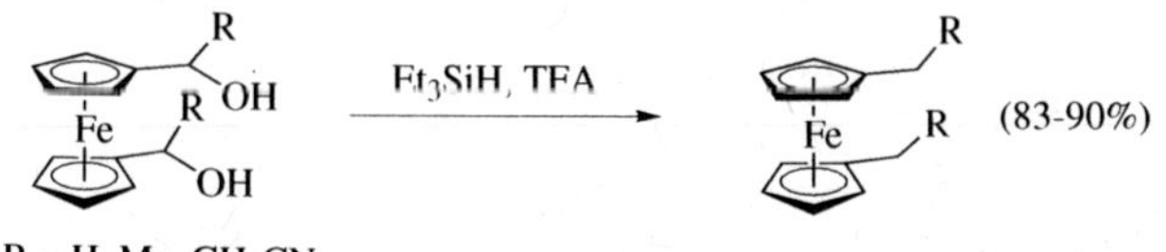

(Eq. 38)

R = H, Me, CH_2CN

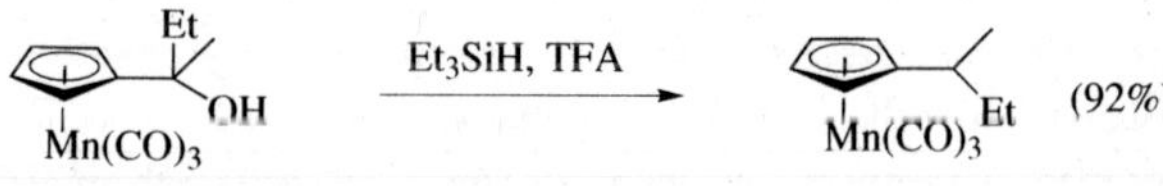

(Eq. 39)

Highly diasteroselective and chemoselective reductions may be performed on the hydroxy functions of (η^6-arene)-tricarbonylchromium complexes. Treatment of the chromium-complexed benzylic alcohol **29** with triethylsilane and boron trifluoride etherate in dichloromethane at −78° to 0° gives only diastereomer **30** in 75% yield (Eq. 40).[181] In a similar fashion, treatment of the complexed exo-allyl-endo-benzylic alcohol **31** with an excess of Et_3SiH/TFA in dichloromethane at room temperature under nitrogen produces only the endo-allyl product **32** in 92% yield after 1.5 hours (Eq. 41). It is noteworthy that no reduction of the isolated double bond occurs.[182]

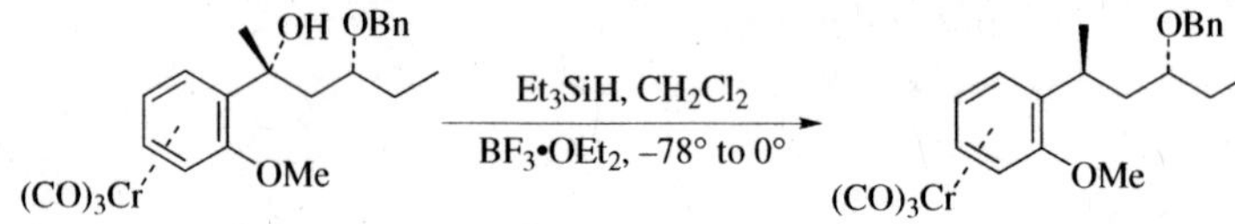

29 30

(75%) (Eq. 40)

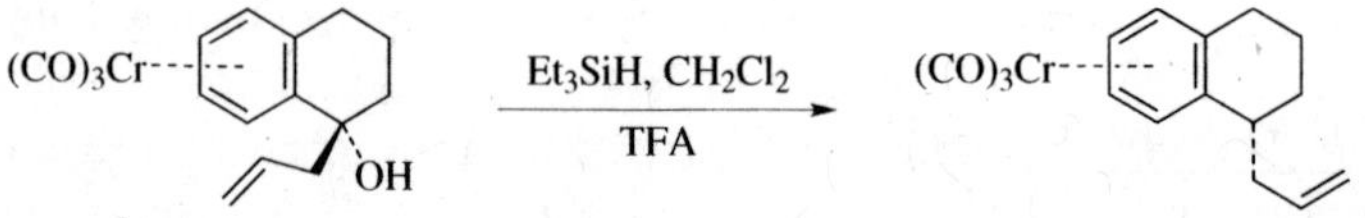

31 32

(92%) (Eq. 41)

Treatment of α-hydroxyalkylidynetricobalt nonacarbonyl complexes of type **33** with strong acids produces the related highly stabilized carbocations **34**.[183] As expected, heating a tetrahydrofuran solution of the methyl compound (**33**, R = Me) with trifluoroacetic acid and triethylsilane at reflux for four hours produces the related hydrocarbon complex (**35**, R = Me) in 72% yield. Somewhat surprising, however, is the report that the hexafluorophosphate salt of the phenyl-substituted carbocation (**34**, R = Ph, $X^- = PF_6^-$), preformed by treatment of the corresponding alcohol with hexafluorophosphoric acid, produces only 7% of the related hydrocarbon complex when exposed to triethylsilane in tetrahydrofuran for 1.5 hours at room temperature. This lower yield of hydrocarbon complex from the phenyl system compared with the methyl analog is probably a reflection of the greater stability and lower reactivity of the intermediate cation.

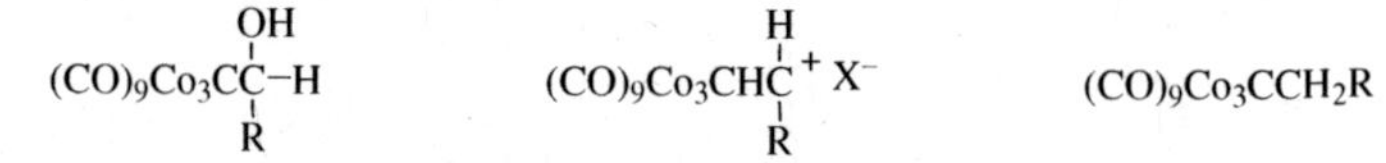

33 **34** **35**

33–35 R = Me, Ph

Polyfunctional Hydroxy Compounds. Different classes of alcohols can serve as the precursors to carbocations that have different stabilities and different degrees of ease of formation. It is thus no surprise that selective acid-catalyzed organosilicon hydride reductions of alcohols of one type may be effected in the presence of others if the proper reaction conditions are employed. For example, either tertiary or secondary benzylic hydroxy groups may be replaced by hydrogen without affecting primary aliphatic-type hydroxy groups in the same molecule when boron trifluoride etherate and triethylsilane are used.[137] This $Et_3SiH/BF_3 \cdot OEt_2$ reagent combination is also selective for a benzylic alcohol over an aliphatic alcohol function.[137]

Treatment of 1,1-diphenylpropane-1,3-diol with two equivalents each of boron trifluoride etherate and triethylsilane in dichloromethane at 0° gives a 90% yield of 3,3-diphenylpropan-1-ol after 30 minutes (Eq. 42). Replacement of the terminal CH_2OH group by a CO_2Et group and similar treatment produces a product mixture containing 50% of the reduced product and 18% of the corresponding alkene elimination product. The carboethoxy group is unaffected by these reduction conditions.[137]

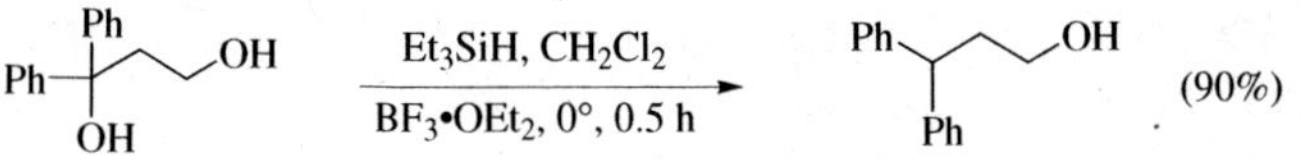

(Eq. 42)

The secondary benzylic alcohol 1-phenylethan-1,2-diol requires 20 hours of treatment at room temperature to produce a 64% yield of 2-phenylethanol (Eq. 43).[137] Under the same conditions, methyl mandelate fails to undergo reduction, presumably because of the greater carbocation-destabilizing effect of a neighboring carboalkoxy compared to a hydroxymethyl group (Eq. 43).[137]

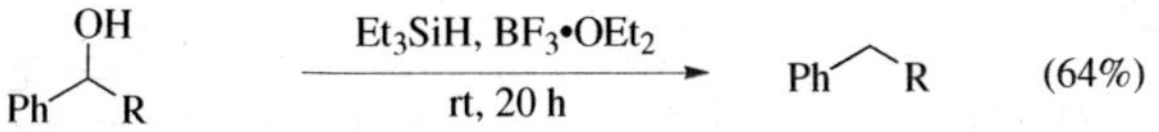

R	
CH_2OH	(64%)
CO_2Me	(0%)

(Eq. 43)

Triethylsilane/boron trifluoride etherate in chloroform at room temperature reduces only the benzylic 12-hydroxy group of the polyfunctional compound **36** to form (±)-homochelidonine **37** in 92% yield (Eq. 44).[138]

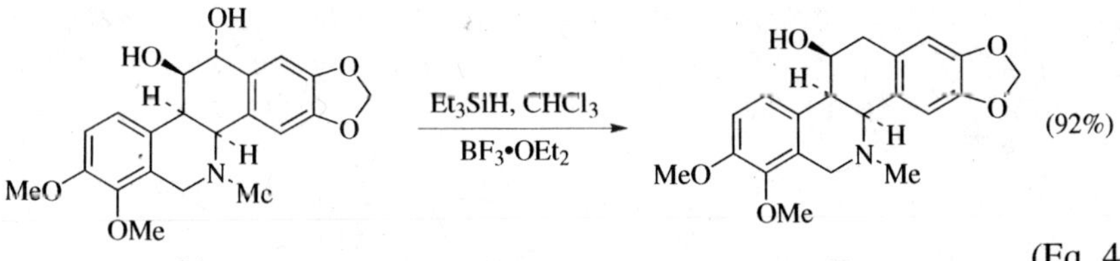

36 **37**

(Eq. 44)

It is possible to effect reduction of tertiary benzylic hydroxy functions in the presence of primary halogens. Treatment of 1,1-diphenyl-1-hydroxy-2-haloethanes in chloroform with a slight excess of triethylsilane and a 9- to 10-fold excess of trifluoroacetic acid yields the corresponding 2,2-diphenyl-1-haloethanes (Eq. 45). The yield of the chloride is 77% after one hour at −15°, whereas that of the bromide is 66% following one hour at 0°.[184]

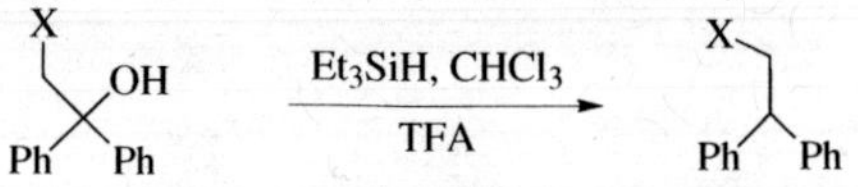

X	
Cl	(77%)
Br	(66%)

(Eq. 45)

Alkyl Halides and Triflates to Alkanes. The normal requirements for conversion of alkyl halides (and triflates) to alkanes using organosilicon hydrides are essentially the same as those needed for the reduction of the corresponding alcohols, namely, the substrates must generally be able to serve as precursors to carbocations that may be captured by the hydride. The reduction of alkyl halides has been accomplished with triethylsilane/aluminum chloride. Substrates that undergo reduction under these conditions include primary alkyl halides,[146,185,186] secondary alkyl halides,[146,187,188] gem-dihalides,[189] vicinal dihalides,[189] and tertiary alkyl halides.[187,188] As expected, haloarenes generally do not undergo such reductions, even under vigorous conditions.[146,190]

An exception to the need for carbocation formation is found when the silyl hydride functional group is part of a valence-expanded hydrosiliconate species. For example, potassium tetraethoxyhydridosilicate[104] is capable of reducing primary alkyl and benzylic bromides and chlorides directly to the corresponding hydrocarbons without the need for additional catalysis (Eq. 46).[95] The reaction is not a simple nucleophilic displacement of hydride for halide, however, since dimers can be formed as part of the reaction product mixture. In addition, when 6-bromo-1-hexene is used as the substrate, 4.4% of methylcyclopentane is obtained along with 63% of 1-hexene product.[95] The presence of the ring-closed product is suggestive of the operation of single-electron transfer (SET) processes.[191]

$$[(EtO)_4SiH]^- K^+ + RX \xrightarrow{\text{THF, 48 h}} RH + RR + Si(OEt)_4 \qquad \text{(Eq. 46)}$$

R = primary alkyl, benzyl

Alkyl Halides. Commonly, reductions with liquid silanes and liquid alkyl halides do not require the use of a solvent.[186] When the alkyl halide is a solid, either pentane[186] or dichloromethane may be used as solvent.[192] No significant difference in reactivities is observed between alkyl chloride and bromide substrates,[186] but allyl halides are more reactive than 2-halopropanes, which, in turn, are more reactive than 1-halopropanes.[190,146]

With halides having a strong propensity to undergo ionization, such as trityl halides, reductions may occur in the absence of added Lewis acids.[29,54] Otherwise, the presence of Lewis acids is required. Catalytic amounts of aluminum bromide and aluminum chloride seem to be equally effective unless there are other Lewis base centers such as oxygen in the molecule to compete with the halogen for complexation with the Lewis acid.[192] Then it is necessary to add more than one equivalent of the Lewis acid to effect reduction of the carbon-halogen function.[136,146] Skeletal rearrangements may occur during these reductions, as in the reduction of bromocycloheptane (Eq. 47).[146,185]

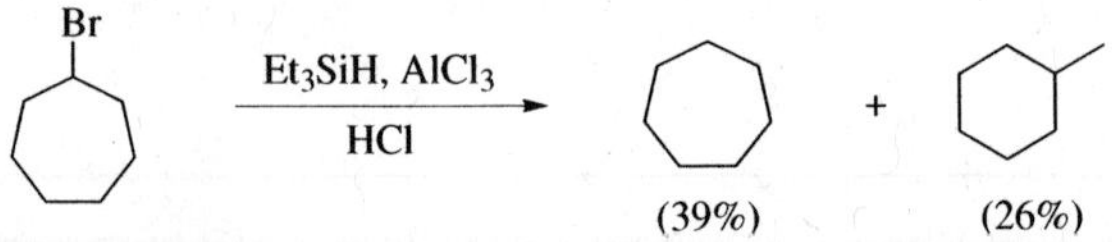

(39%) (26%)

(Eq. 47)

In contrast to the behavior of primary alcohols, which resist reduction by organosilicon hydrides even in the presence of very strong acids, primary halo alkanes, including methyl iodide and ethyl bromide,[186] undergo reduction when treated with aluminum chloride and organosilicon hydrides.[146,185,186] Slow addition of a catalytic amount of aluminum chloride to a nearly equimolar mixture of 1-chlorohexane and triethylsilane produces a vigorous reaction that, after 28 hours and simple distillation, gives a 57% yield of *n*-hexane and an 88% yield of chlorotriethylsilane (Eq. 48).[185] Similar treatment of 2,2-dimethyl-1-chloropropane gives 2-methylbutane in 37% yield (Eq. 49), whereas treatment of 3,3-dimethyl-1-chlorobutane gives a 50% yield of 2-methylbutane (Eq. 50).[185] Polymeric hydrocarbon by-products accompany the products of the latter two reactions. The structures of the products are clear evidence of the occurrence of 1,2-alkyl shifts leading to more stable carbocationic intermediates.

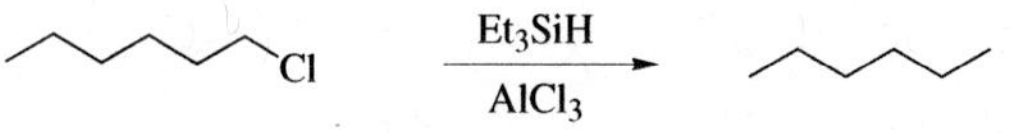
(57%) (Eq. 48)

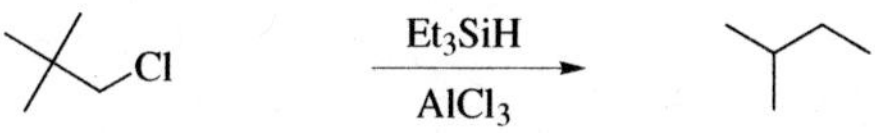
(37%) (Eq. 49)

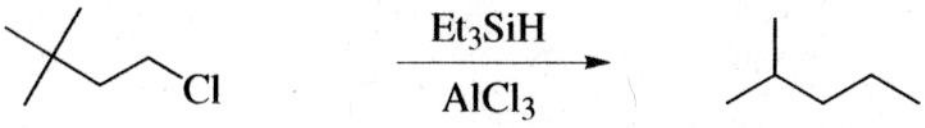
(50%) (Eq. 50)

The use of a deuterium-labeled organosilicon hydride and location of the deuterium isotope in the reduced product shows that 1,2-hydride shifts also occur. Thus, reduction of 1-bromohexane with triethylsilane-d_1 yields hexane with all of the deuterium at C2 (Eq. 51); similar treatment of cyclohexylmethyl bromide produces methylcyclohexane-1-d_1 (Eq. 52).[186]

(—) (Eq. 51)

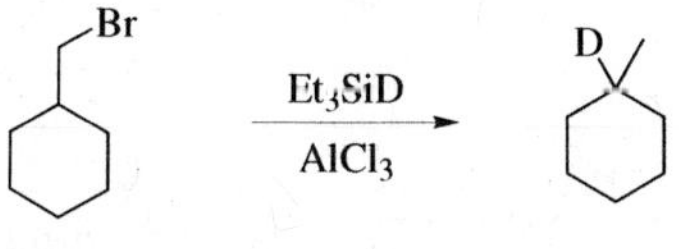
(—) (Eq. 52)

Trialkylsilanes are generally more effective than dialkyl- or monoalkylsilanes in minimizing isomerizations. The reduction of 2-bromododecane to dodecane proceeds under aluminum chloride catalysis in 82% yield using *n*-butylsilane and in 87% yield with tri-*n*-butylsilane.[186] However, similar treatment of bromocycloheptane with triethylsilane yields a mixture of 39% cycloheptane and 26% methylcyclohexane. The same substrate yields 65% methylcyclohexane and less than 1% cycloheptane when *n*-butylsilane is the reducing agent.[186]

Total reduction of unbranched open-chain and cyclic derivatives of dichloro and dibromo alkanes occurs at room temperature within 30 minutes in dichloromethane solutions containing ca. 2.5 equivalents of triethylsilane and ca. 0.25 equivalents of aluminum chloride.[189] The reaction occurs equally well with geminal, vicinal, and ω-dihalo alkanes. For example, 1,5-dibromopentane gives *n*-pentane in 85% yield when treated in this way (Eq. 53).[189]

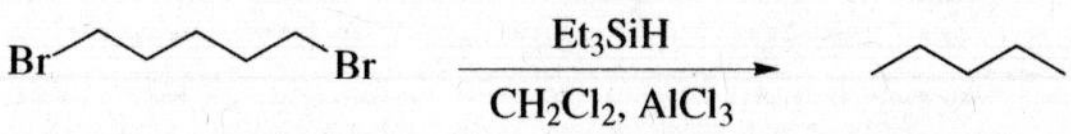
(85%) (Eq. 53)

Chlorocyclohexane is converted into cyclohexane in dichloromethane using ethyldichlorosilane as reducing agent.[192] The product yield is 40% with 25 mol% aluminum chloride and 45% with aluminum bromide. 1-Chloro-1-methylcyclohexane gives a 94% yield of methylcyclohexane using aluminum chloride and a 92% yield with aluminum bromide. Ethyldichlorosilane is superior as a hydride donor to either cumene or dicumylmethane.[192]

2-Bromoadamantane and 1-bromoadamantane are reduced to adamantane in yields of 84% and 79%, respectively, when treated with triethylsilane and catalytic amounts of aluminum chloride.[186] Similar treatment of benzhydryl chloride and *exo*-2-bromonorbornane gives the related hydrocarbons in yields of 100% and 96%, respectively.[186] In contrast, 2-bromo-1-phenylpropane gives only a 43% yield of 1-phenylpropane; the remainder consists of Friedel-Crafts alkylation products.[186] Some alkyl halides resist reduction by this method, even when forcing conditions are employed. These include *p*-nitrobenzyl bromide, 3-bromopropanenitrile, and 5-bromopentanenitrile.[186]

The reduction of 4-chloro-4-methyltetrahydropyran with triethylsilane requires more than a catalytic amount of aluminum chloride. No 4-methyltetrahydropyran is obtained after 20 hours at room temperature even when 0.75 equivalents of the catalyst is used, but a 92% yield is obtained after only 30 minutes when two equivalents of catalyst and three equivalents of triethylsilane are used.[136,146] This is presumably a result of the ability of the Lewis acid to coordinate at the ring oxygen as well as at the chlorine. The introduction of alkyl groups at C2 appears to introduce enough steric hindrance near the ring oxygen to enable less than one equivalent of aluminum chloride to effect reduction, but also makes the products unstable to the reaction conditions so that the synthetic yields decline compared with the unsubstituted substrate.[136]

Dichloromethane solutions of some sterically congested benzyl chlorides and triethylsilane need only the addition of excess trifluoroacetic acid to promote rapid conversion of the chlorides to the related hydrocarbons.[128] Thus 2,4,6-trimethylbenzyl chloride produces a 79% yield of isodurene at room temperature after 2.5 hours, 2-methyl-4,6-di-*tert*-butylbenzyl chloride gives 50% 1, 2-dimethyl-4,6-di-*tert*-butylbenzene after 40 minutes at reflux, and 2,4,6-*tert*-butylbenzyl chloride gives a 100% yield of 2,4,6-tri-*tert*-butyltoluene within 17 minutes at reflux (Eq. 54). The unsubstituted parent benzyl chloride remains unreacted under these conditions even after 30 days.[128]

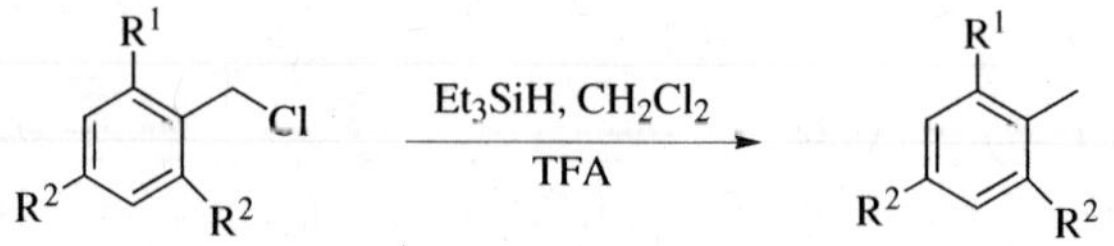

(Eq. 54)

R^1	R^2	
Me	Me	(79%)
Me	*t*-Bu	(50%)
t-Bu	*t*-Bu	(100%)

It is clear that the ionizing power of the solvent used is important in many of these reductions. When 2,4,6-trimethylbenzyl chloride is heated with diphenylsilane in nitrobenzene at temperatures as high as 130°, no isodurene is formed.[193] Not unexpectedly, the same lack of reactivity is reported for a series of benzyl fluorides, chlorides, and bromides substituted in the para position with nitro or methyl groups or hydrogen when they are heated in nitrobenzene solutions with triethylsilane, triethoxysilane, or diphenylsilane.[193]

The combination of boron trifluoride etherate and triethylsilane can cause the reduction of tertiary fluoride centers even in polyfunctional compounds (Eq. 55).[194]

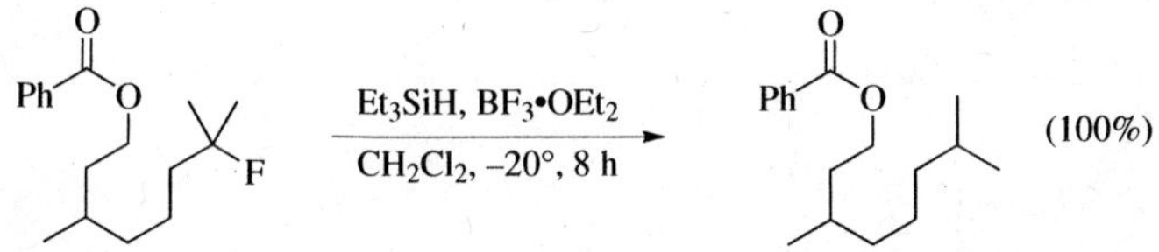

(Eq. 55)

Alkyl iodides, benzyl chlorides, benzyl bromides, and adamantyl bromides and iodides undergo reduction with triethylsilane/palladium chloride.[195] The reduction of a β-chloro ether occurs in excellent yield with this system (Eq. 56).[195]

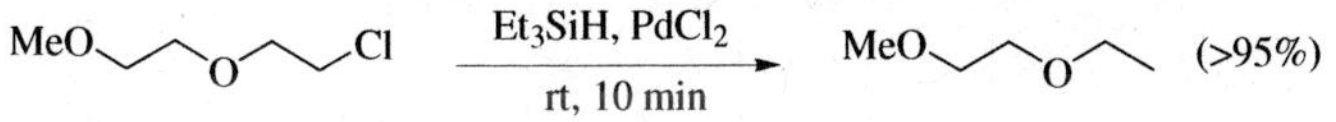

(Eq. 56)

Allyl Halides. Reduction of a polyfunctional allyl chloride occurs without rearrangement and without reduction of the tosylate using $Ph_2SiH_2/ZnCl_2/Pd(PPh_3)_4$ (Eq. 57).[196]

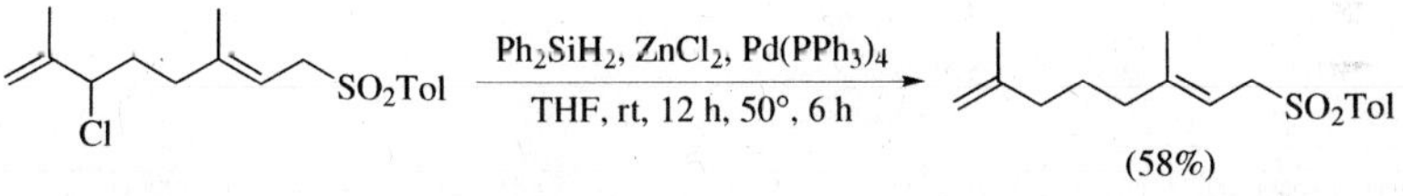

(Eq. 57)

α-Halocarbonyl Compounds. The reduction of α-chloro and α-bromo ketones and esters has been accomplished with combinations of $PhSiH_3/Mo(CO)_6$,[197] $Ph_2SiH_2/ZnCl_2/Pd(PPh_3)_4$,[197] $Ph_2SiH_2/Pd(OAc)_2$,[197] and $Et_3SiH/PdCl_2$,[195] with the first reagent combination giving the best results.[197] One example of an α-chloro amide reduction is reported.[198] 2-Bromopropiophenone is reduced to propionic acid with polymethylhydrosiloxane (PMHS, **38**), an inexpensive industrial commodity, and $Pd(PPh_3)_4$ in 35% yield.[199] This reagent combination also reduces α-halo ketones in high yields (Eq. 58).[199]

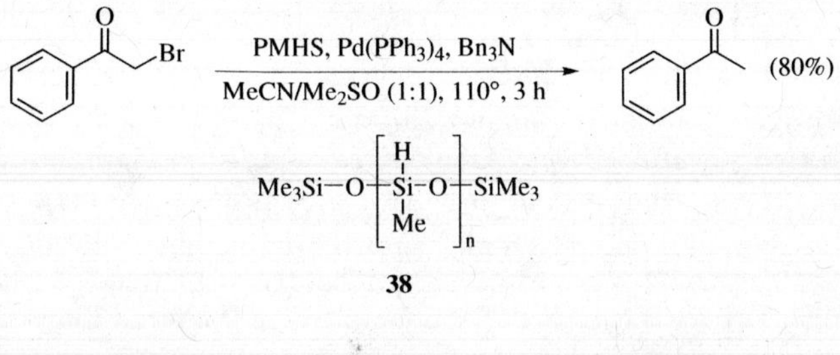

(Eq. 58)

38

Vinyl and Aryl Halides and Triflates. The organosilane reduction of aryl halides is possible in high yields with triethylsilane and palladium chloride.[195] The reaction is equally successful with aryl chlorides, bromides, and iodides. Aryl bromides and iodides, but not chlorides, are reduced with PMHS/$Pd(PPh_3)_4$ in moderate to excellent yields.[199] This system also reduces vinyl bromides.[199] *p*-Chlorobenzophenone is reduced to benzophenone with *sym*-tetramethyldisiloxane and Ni/C in excellent yield (Eq. 59).[200] There is a report of the organosilane reduction of aryl and vinyl triflates in very high yields with the combination of $Et_3SiH/Pd(OAc)_2$/dppp (1,3-bis(diphenylphosphino)propane) (Eq. 60).[201]

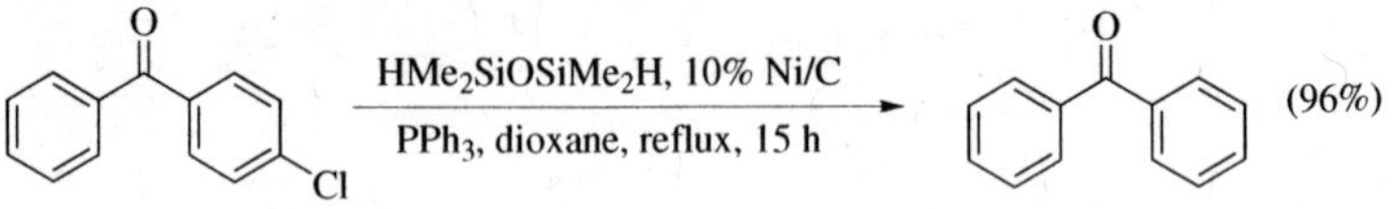

(Eq. 59)

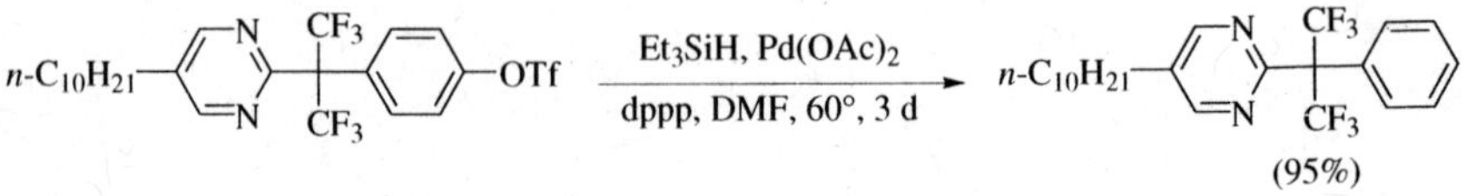

(Eq. 60)

Reduction of Unsaturated Hydrocarbons

Alkenes to Alkanes. The "ionic hydrogenation" of many compounds containing carbon-carbon double bonds is effected with the aid of strong acids and organosilicon hydrides following the π-route outlined in Eq. 2. A number of factors are important to the successful application of this method. These include the degree and type of substituents located around the double bond as well as the nature and concentrations of the acid and the organosilicon hydride and the reaction conditions that are employed.

The most common reaction conditions for alkene reductions use excess trifluoroacetic acid and triethylsilane either neat[202–204] or in an inert solvent such as nitrobenzene,[134] 2-nitropropane,[205] carbon tetrachloride,[206] chloroform,[207] or dichloromethane.[127,164] Reaction temperatures from $-78°$ to well over $100°$ are reported. Ambient or ice-bath temperatures are most commonly used, but variations of these conditions abound.

Among other silicon hydrides reported are *n*-butylsilane, diethylsilane, triisopentylsilane, tricyclopentylsilane, triphenylsilane, tri-*sec*-butylsilane, di-*tert*-butylsilane, di-*tert*-butylmethylsilane, tri-*tert*-butylsilane,[204] phenylsilane, diethylmethylsilane,[202] diphenylsilane,[134,208,209] dichloroethylsilane,[192] PMHS,[77] and polyethylhydrosiloxane.[207]

Acids that are used in addition to trifluoroacetic acid include trifluoroacetic acid with added sulfuric acid[203] or boron trifluoride etherate,[210,211] perfluorobutyric acid,[212] hydrogen chloride/aluminum chloride,[136,146,213] perchloric acid in chloroform,[214] *p*-toluenesulfonic acid alone[134] or with aluminum bromide or aluminum chloride,[192] concentrated sulfuric acid in two-phase systems with dichloromethane, alcohol, or ether solvents,[209,215] trifluoromethanesulfonic acid,[216] chlorodifluoroacetic acid,[134] and the monohydrate of boron trifluoride

($BF_3{\cdot}OH_2$).[217] The use of a sulfonated phenol-formaldehyde polymer in conjunction with formic acid is also reported.[208] Acids that are ineffective include phosphoric,[208] trichloroacetic, dichloroacetic, and acetic acids.[134] It is reported that addition of lithium perchlorate to the reaction mixture improves product yields.[193,205]

Other organosilane/acid reagent combinations that are used in the reduction of olefins to alkanes include Et_3SiH/NH_4F/TFA,[135] $Et_3SiH/HClO_4$,[214] $Et_3SiH/TiCl_4$,[218] PMHS/Pd-nanocomposite,[219] Et_3SiH/TFA/$HClO_4$,[205] $Et_3SiH/PdCl_2$,[220] polyethylhydrogensiloxane (PEHS)/TFA,[207] Et_3SiH/TMSOTf,[216] and Et_3SiH/HCO_2H.[208]

The triethylsilane/trifluoroacetic acid reagent system reduces alkenes to alkanes in poor to excellent yields depending largely on the ability of the alkene to form carbocations upon protonation. Under these conditions the more substituted olefins are reduced in better yields and styrene double bonds are reduced in high yields.[127,202,207,221–228] The reduction of 1,2-dimethylcyclohexene with this reagent gives a mixture of *cis*- and *trans*-1,2-dimethylcyclohexane.[229] The formation of the trifluoroacetate esters is a side reaction.[205,230]

Potential problems associated with double bond reduction by this method may be understood in terms of Eq. 61. Protonation of the double bond leads to the formation of the more stable carbocation. This carbocation may rearrange by a first-order process or react competitively with either indigenous nucleophiles or added silicon hydride by second-order processes. If strong nucleophiles such as those associated with weak Brønsted acids are present, then the limited degree of reversibility of carbocation regeneration following nucleophilic capture may lead to diversion of the desired reduction products to unwanted nucleophilic substitution products.[209] Another problem exists if bimolecular polymerization reactions compete with carbocation capture by organosilicon hydrides because of the proximity of carbocations and unprotonated alkene substrate. When this occurs, yields of reduced product suffer. The yields of hydrocarbons from alkenes are, in fact, frequently lower than those of the same products derived from the corresponding alcohols because of this problem.[134,142]

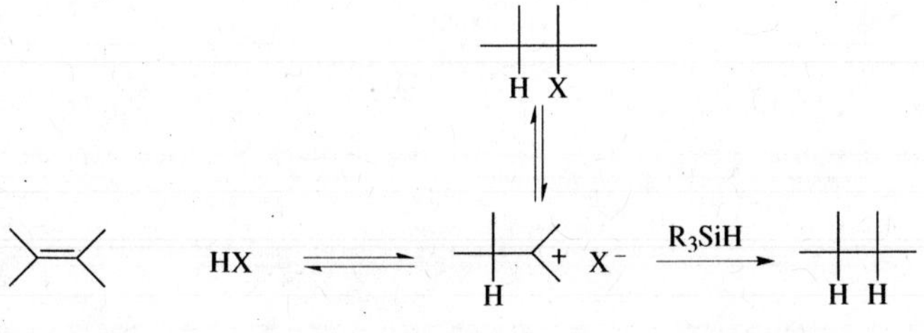

(Eq. 61)

When trifluoroacetic acid is used as the source of protons, it is known that rapid formation of trifluoroacetate esters precedes reduction to hydrocarbons.[134,204,206] Use of acetic acid in place of trifluoroacetic acid, for example, would be expected to fail to produce good conversion to reduced product because of the combination of decreased acidity and increased nucleophilicity of acetic acid relative to trifluoroacetic acid as well as its weaker ionizing power as a solvent. This is consistent with experimental observations.[134,209]

The relative stability of the carbenium ion resulting from double bond protonation is a controlling factor in the limitation of this method of hydrogenation. On a practical level, only alkenes that can produce carbenium ions at least as stable as tertiary aliphatic ones undergo reduction to alkanes in useful yields. This distinction serves as a basis for selectivity of reduction. Under essentially every set of conditions reported, 1-methylcyclohexene, which forms a tertiary aliphatic carbenium ion upon protonation, undergoes reduction to methylcyclohexane in good to excellent yields, whereas cyclohexene, which can only form a secondary aliphatic carbenium ion intermediate upon protonation, does not normally undergo reduction. Indeed, treatment of an equimolar mixture of cyclohexene and 1-methylcyclohexene with two equivalents of triethylsilane and four equivalents of trifluoroacetic acid at 50° gives, after 10 hours, a 70% yield of methylcyclohexane together with completely recovered, unreacted cyclohexene.[231]

An exception is reported when the reactions are conducted using a twofold excess of dichloroethylsilane with equal equivalents of either aluminum chloride or aluminum bromide and *p*-toluenesulfonic acid at 40° for two hours in dichloromethane. Under these conditions, 1-methylcyclohexene affords methylcyclohexane in 65–75% yield, whereas cyclohexene gives cyclohexane in 17–23% yield.[192]

The use of deuterated organosilicon hydrides in conjunction with proton acids permits the synthesis of site-specific deuterium-labeled compounds.[59,126,221] Under such conditions, the deuterium atom in the final product is located at the charge center of the ultimate carbocation intermediate (Eq. 62). With the proper choice of a deuterated acid and organosilicon hydride, it may be possible to use ionic hydrogenation in a versatile manner to give products with a single deuterium at either carbon of the original double bond, or with deuterium atoms at both carbon centers.[127]

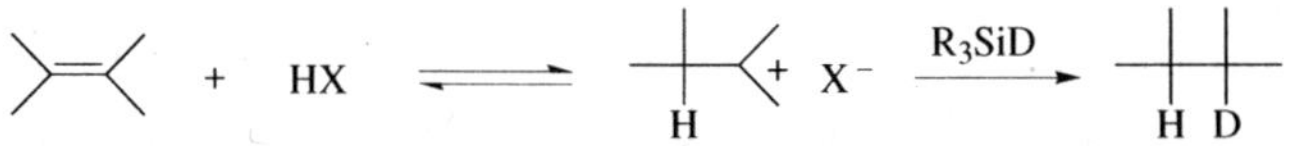

(Eq. 62)

Monosubstituted Alkenes. Simple unbranched terminal alkenes that have only alkyl substituents, such as 1-hexene,[203] 1-octene,[209] or allylcyclohexane[230] do not undergo reduction in the presence of organosilicon hydrides and strong acids, even under extreme conditions.[1,2] For example, when 1-hexene is heated in a sealed ampoule at 140° for 10 hours with triethylsilane and excess trifluoroacetic acid, only a trace of hexane is detected.[203] A somewhat surprising exception to this pattern is the formation of ethylcyclohexane in 20% yield upon treatment of vinylcyclohexane with trifluoroacetic acid and triethylsilane.[230] Protonation of the terminal carbon is thought to initiate a 1,2-hydride shift that leads to the formation of the tertiary 1-ethyl-1-cyclohexyl cation.[230]

On the other hand, if the single substituent can stabilize an adjacent carbocation center following protonation of the alkene, then reduction may occur.

Styrene is reported to undergo reduction upon treatment with trifluoroacetic acid and triethylsilane,[203] although competing polymerization reactions limit the yield of ethylbenzene to only 30% (Eq. 63).[70] Vinylcyclopropane is reduced to ethylcyclopropane within 30 minutes under similar conditions (Eq. 64).[232] It is important to note that the cyclopropane ring of ethylcyclopropane can be opened under these reaction conditions, albeit with longer reaction times, to give some *trans*-2-pentene in the final reaction mixture.[233]

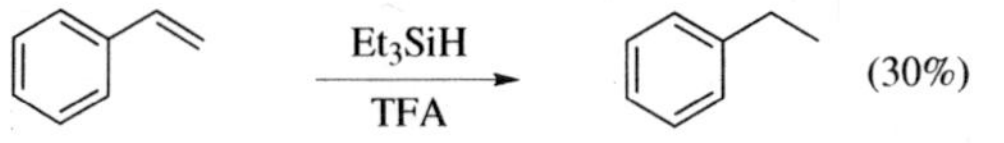

(Eq. 63)

(Eq. 64)

Examples of the behavior of other substituted vinyl substrates upon exposure to the action of trifluoroacetic acid and triethylsilane are known. For example, *n*-butyl vinyl ether, when reacted at 50° for 10 hours, gives *n*-butyl ethyl ether in 80% yield (Eq. 65).[234] In contrast, *n*-butyl vinyl thioether gives only a 5% yield of *n*-butyl ethyl sulfide product after 2 hours and 15% after 20 hours of reaction.[234] It is suggested that this low reactvity is the result of the formation of a very stable sulfur-bridged carbocation intermediate that resists attack by the organosilicon hydride (Eq. 66).

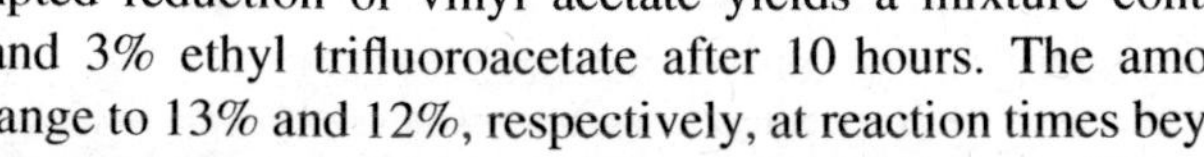

(Eq. 65)

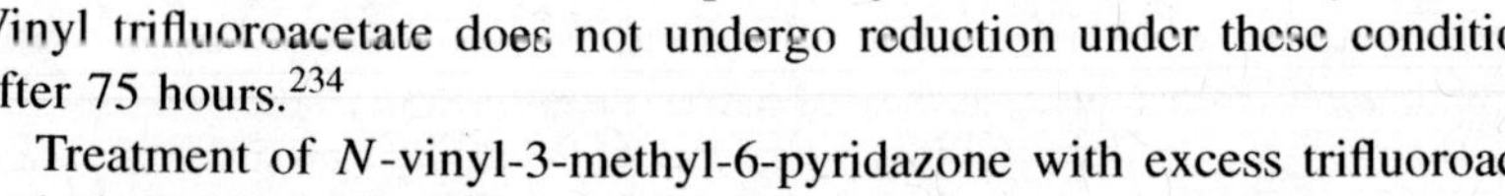

(Eq. 66)

Attempted reduction of vinyl acetate yields a mixture containing 8% ethyl acetate and 3% ethyl trifluoroacetate after 10 hours. The amounts of the two esters change to 13% and 12%, respectively, at reaction times beyond 60 hours.[234] Vinyl trifluoroacetate does not undergo reduction under these conditions, even after 75 hours.[234]

Treatment of *N*-vinyl-3-methyl-6-pyridazone with excess trifluoroacetic acid and triethylsilane at 65° for 25 hours yields 67% of the reduced product *N*-ethyl-3-methyl-6-pyridazone (Eq. 67).[235] It is noteworthy that only the vinyl group in this compound undergoes reduction under these conditions, and not the ring or carbonyl sites. Examination of a solution of the starting *N*-vinyl-3-methyl-6-pyridazone in neat trifluoroacetic acid by ^{1}H NMR spectroscopy shows the existence of the trifluoroacetate ester expected from the carbocation formed by protonation of the vinyl group at the terminal carbon. It is of interest that a similar compound, *N*-allyl-3-methyl-6-pyridazone, is inert under these conditions (Eq. 68). This reflects the differences of the relative stabilities of the carbocations formed upon protonation of the C=C groups in each reaction.

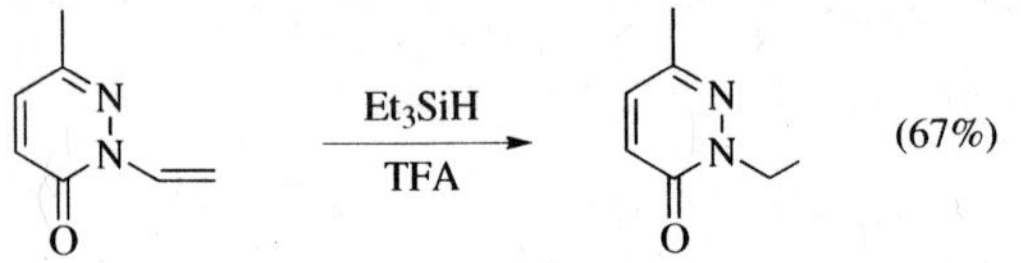

(Eq. 67)

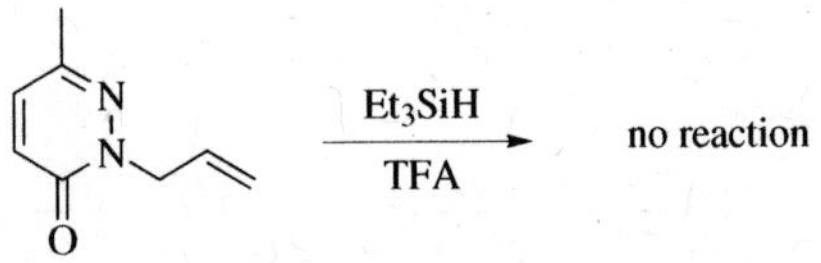

(Eq. 68)

Disubstituted Alkenes. Simple 1,2-disubstituted alkenes such as 2-octene or cyclohexene, which produce only secondary aliphatic carbocation reaction intermediates, do not undergo reduction upon treatment with a Brønsted acid and an organosilicon hydride. Even when extreme conditions are employed, only traces of reduction products are detected.[192,203,207–210,214] An exception is the report that 4-methyl-2-pentene forms 2-methylpentane in 70% yield when heated to 50° for 20 hours with a mixture of Et_3SiH/TFA containing a catalytic amount of sulfuric acid. It is believed that 4-methyl-2-pentene is isomerized to 2-methyl-2-pentene prior to reduction.[203]

Unlike cyclohexene, its oxa analog, 3,4-dihydro-2*H*-pyran, undergoes facile reduction to tetrahydropyran in yields ranging from 70 to 92% when treated with a slight excess of triethylsilane and an excess of either trifluoroacetic acid or a combination of hydrogen chloride and aluminum chloride (Eq. 69).[146] This difference in behavior can be understood in terms of the accessibility of the resonance-stabilized oxonium ion intermediate formed upon protonation.

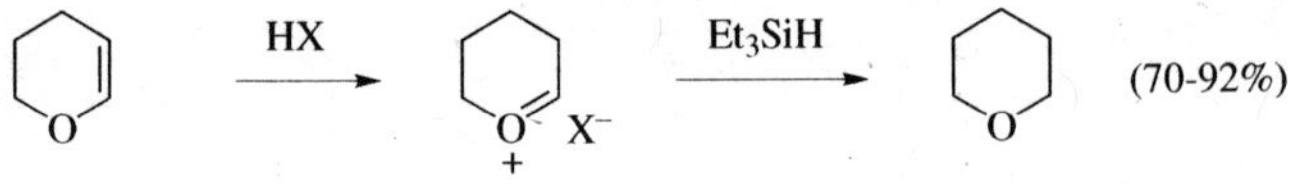

(Eq. 69)

The behavior of the isomeric dihydronaphthalenes emphasizes the importance of the relative stabilities of carbocation intermediates in ionic hydrogenations. Treatment of 1,2-dihydronaphthalene with Et_3SiH/TFA at 50–60° gives a 90% yield of tetralin after one hour. Under the same conditions, the 1,4-dihydronaphthalene isomer gives less than 5% of tetralin after 70 hours.[224] This difference in reactivity is clearly related to the relatively accessible benzylic cation formed upon protonation of the 1,2-isomer compared to the less stable secondary cation formed from the 1,4-isomer.[224]

The behavior of members of the bicyclo[2.2.1]heptene family is also different from that of other common 1,2-disubstituted alkenes.[230] The parent bicyclo[2.2.1]heptene gives bicyclo[2.2.1]heptane in only 3.5% yield when it is treated with Et_3SiH/TFA. The major product is reported to be a 2-bicyclo[2.2.1]heptyl trifluoroacetate of unspecified configuration (Eq. 70).[230] The carbocation intermediate is presumably the 2-norbornyl cation. Addition of small amounts of boron trifluoride etherate to the reaction mixture causes the yield of hydrocarbon product to rise to 22% after a reaction time of 24 hours at room temperature. Further exposure of the reaction mixture to the reaction conditions does not result in additional hydrocarbon formation from the ester.

(Eq. 70)

A mixture of exo- and endo-isomers of 5-methylbicyclo[2.2.1]hept-2-ene is hydrogenated with the aid of five equivalents of triethylsilane and 13.1 equivalents of trifluoroacetic acid to produce a 45% yield of *endo*-2-methylbicyclo[2.2.1] heptane (Eq. 71). The same product is formed in 37% yield after only five minutes. The remainder of the reaction products is a mixture of three isomeric secondary exo-methylbicyclo[2.2.1]heptyl trifluoroacetates that remains inert to the reaction conditions. Use of triethylsilane-1-d_1 gives the *endo*-2-methylbicyclo-[2.2.1]heptane product with an *exo*-deuterium at the tertiary carbon position shared with the methyl group. This result reflects the nature of the internal carbocation rearrangements that precede capture by the silane.[230]

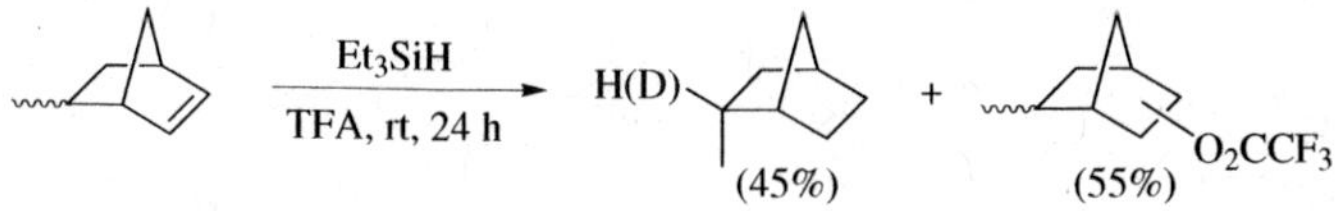

(Eq. 71)

Alkenes with a 1,1-disubstitution pattern form tertiary carbocations upon treatment with a Brønsted acid. Consequently, such compounds are often easily reduced (Eq. 72). An example of this is the formation of 2-methylpentane in 93% yield after only 5 minutes when a dichloromethane solution of 2-methyl-1-pentene and 1.4 equivalents of triethylsilane is treated with 1.4 equivalents of trifluoromethanesulfonic acid at −75°.[216] Similar treatment of 2,3-dimethyl-1-butene gives a 96% yield of 2,3-dimethylbutane.[216]

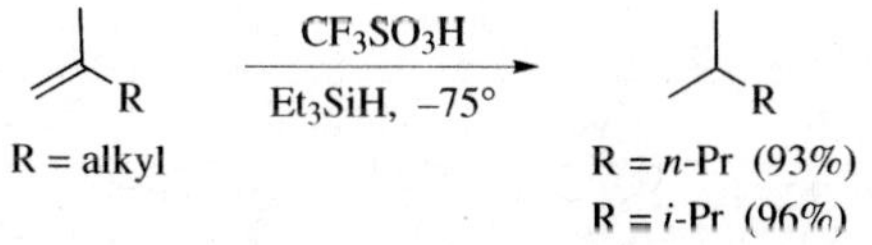

(Eq. 72)

Use of deuterated silane and/or acid with this method leads to site-specific deuterium incorporation in the reduced products. Thus, treatment of 2-methyl-1-pentene with one equivalent of deuterated triethylsilane and two equivalents of trifluoroacetic acid at 50° for 24 hours gives 2-methylpentane-2-d_1 in 90% yield (Eq. 73).[221] In the same way, isopropenylcyclopropane gives an 80% yield of deuterated isopropylcyclopropane after 30 minutes at −10° (Eq. 74).[221]

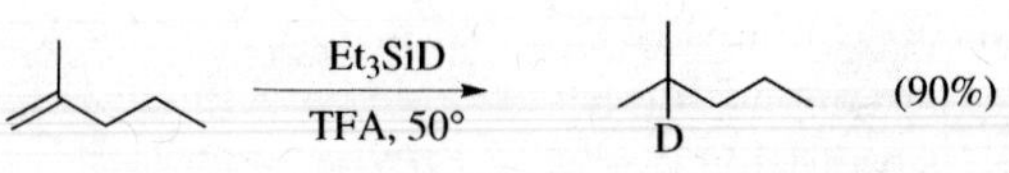

(Eq. 73)

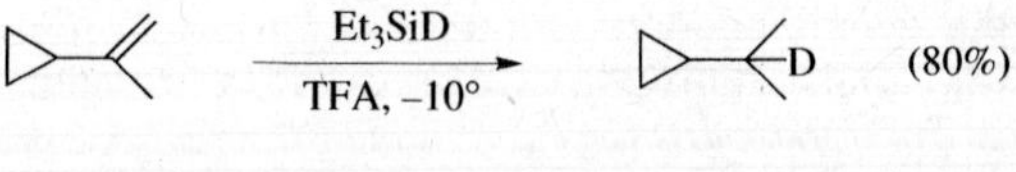

(Eq. 74)

Preferential protonation of oxygen in comparison to carbon prevents 4-methyl-enetetrahydropyran from undergoing reduction to 4-methyltetrahydropyran even when held at 70° for 10 hours in the presence of triethylsilane and a 20-fold excess of trifluoroacetic acid.[146] However, when the reaction conditions are changed so that a dichloromethane solution of the same substrate is treated with a mixture of four equivalents of triethylsilane and three equivalents of aluminum chloride in the presence of excess hydrogen chloride, a 40% yield of 4-methyltetrahydropyran product is obtained at room temperature after one hour (Eq. 75).[136]

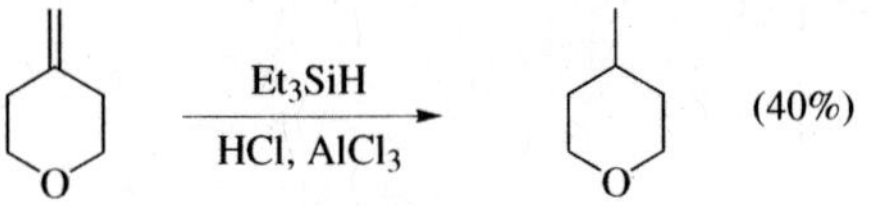

(Eq. 75)

The cis-to-trans ratios of the isomeric 4-*tert*-butyl-1-methylcyclohexanes derived from treatment of 4-*tert*-butyl-1-methylenecyclohexane with trifluoroacetic acid vary with the steric features of the organosilicon hydrides that are used (Eq. 76).[204] The ratio is 0.04 with *n*-butylsilane, 0.09 with diethylsilane, 0.11 with triethylsilane, 0.10 with triisopentylsilane, and 0.19 with either tri-*sec*-butylsilane or di-*tert*-butylsilane.

(Eq. 76)

Trisubstituted Alkenes. With very few exceptions, trisubstituted alkenes that are exposed to Brønsted acids and organosilicon hydrides rapidly undergo ionic hydrogenations to give reduced products in high yields. This is best illustrated by the broad variety of reaction conditions under which the benchmark compound 1-methylcyclohexene is reduced to methylcyclohexane.[134,146,192,202,203,207–210,214,234]

When 1-methylcyclohexene is reduced with one equivalent of deuterated triethylsilane and two equivalents of trifluoroacetic acid at 50°, methylcyclohexane-1-d_1 is obtained in 80% yield after 24 hours (Eq. 77).[221] Under similar conditions, 2-methyl-2-butene gives 2-methylbutane-2-d_1 (90%) and 1-methylcyclopentene gives methylcyclopentane-1-d_1 (60%).[221]

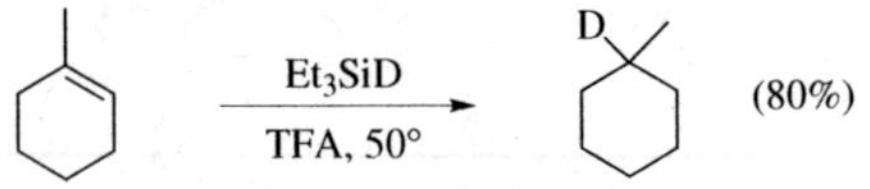

(Eq. 77)

Surprisingly, α-cyanoacrylic acid is reported to react spontaneously with triethylsilane in the absence of any additional acid to give a quantitative yield of the triethylsilyl ester of α-cyanopropionic acid.[236] Ethyl α-cyanoacrylate requires the presence of trifluoroacetic acid to undergo reduction to ethyl 2-cyanopropionate.[236] Many of these reductions are highly stereoselective. For example, treatment of

2-phenylnorbornene with a solution of trifluoroacetic acid and triethylsilane in dichloromethane is reported to yield only *endo*-2-phenylnorbornane (Eq. 78).[164]

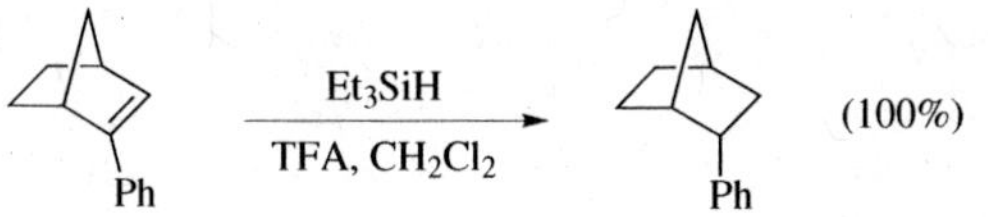

(Eq. 78)

A mixture of Et_3SiH/TFA in dichloromethane reduces 3-methyl-5-α-cholest-2-ene to give the pure equatorial methyl isomeric product, 3β-methyl-5α-cholestane, in 66% yield (Eq. 79).[126] On the other hand, attempts to reduce cholest-5-ene using the same technique yield neither 5α-cholestane nor 5β-cholestane, but instead an isomeric mixture of rearranged olefins. This result is presumably because of the inability of hydride attack to compete with carbocation skeletal isomerization and elimination.[126]

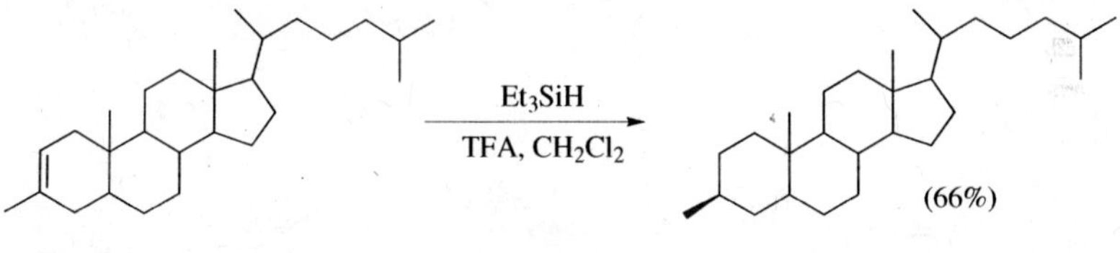

(Eq. 79)

Treatment with trifluoroacetic acid and triethylsilane causes octahydro-6,7,8, 12,13,14,16,17-15*H*-cyclopenta[*a*]phenanthrene to form decahydro-6,7,8,9,11,12, 13,14,16,17-15*H*-cyclopenta[*a*]phenanthrene by reduction of the conjugated double bond.[237] Similar treatment of the 3-methyl ether of $\Delta^{9(11)}$-dehydro-D-homoestrol gives the 3-methyl ether of estradiol in better than 50% yield.[226]

A similar transformation occurs as a critical step in the total synthesis of (+)-estrone by a Diels-Alder cycloaddition-cycloreversion pathway (Eq. 80).[227] It is worth noting that in this reaction the conjugated double bond is stereoselectively reduced while both an isolated double bond and a ketone carbonyl are preserved.

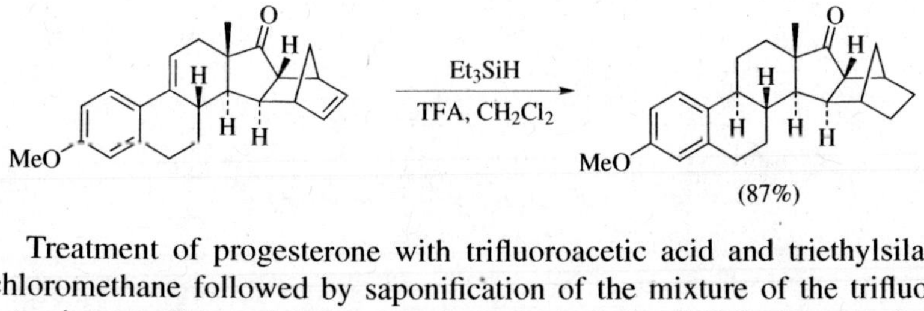

(Eq. 80)

Treatment of progesterone with trifluoroacetic acid and triethylsilane in dichloromethane followed by saponification of the mixture of the trifluoroacetate ester intermediates of 5-β-pregnane-3α,20β-diol and 5-β-pregnane-3α,20α-diol and Jones oxidation yields 5-β-pregnanedione in 65% yield (Eq. 81).[238]

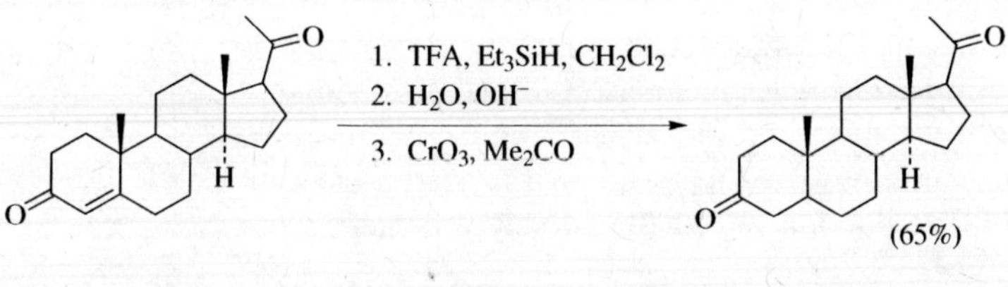

(Eq. 81)

Hydrogenation of the carbon-carbon double bond occurs without alteration of the ester function when citronellyl acetate is treated with 2.5 equivalents of trifluoroacetic acid and two equivalents of triethylsilane in 2-nitropropane.[205] The reduced product is obtained in 90% yield after 22 hours at room temperature in the presence of one equivalent of added lithium perchlorate (Eq. 82). The yields are lower in the absence of this added salt. Similar reduction of an unsaturated phenolic chroman derivative occurs to give an 85% yield of product with only the carbon-carbon double bond reduced (Eq. 83).[205]

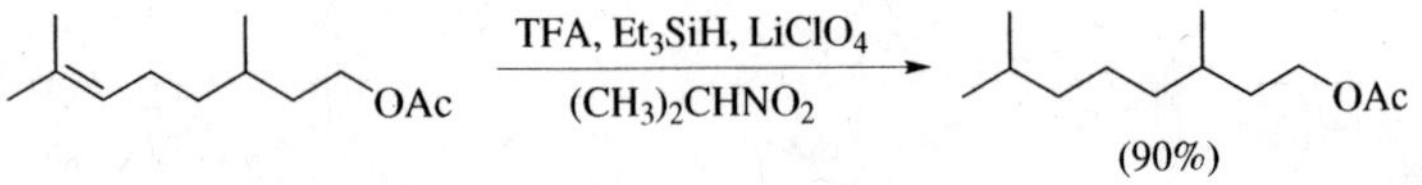

(Eq. 82)

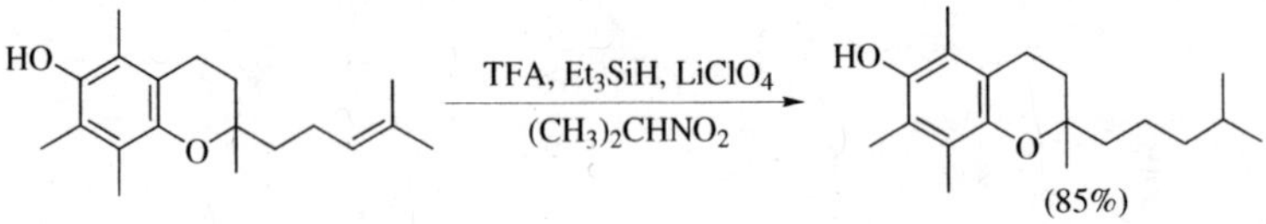

(Eq. 83)

A dichloromethane solution of 4-methyl-5,6-dihydro-2*H*-pyran gives 4-methyltetrahydropyran in 35% yield when treated with a mixture of five equivalents of triethylsilane and 2.5 equivalents of aluminum chloride in the presence of excess hydrogen chloride at room temperature for one hour (Eq. 84).[136] This behavior is essentially the same as that exhibited by the disubstituted 4-methylenetetrahydropyran isomer under similar conditions.[136]

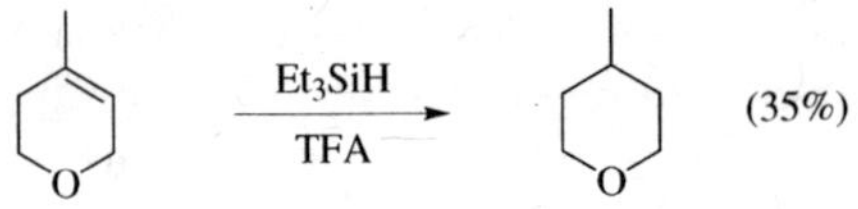

(Eq. 84)

Exceptions to the generally facile ionic hydrogenation of trisubstituted alkenes include the resistance of both 2-methyl-1-nitropropene ($R = NO_2$) and 3,3-dimethylacrylic acid ($R = CO_2H$) to the action of a mixture of triethylsilane and excess trifluoroacetic acid at 50° (Eq. 85).[234] The failure to undergo reduction is clearly related to the unfavorable effects caused by the electron-withdrawing substituents on the energies of the required carbocation intermediates.

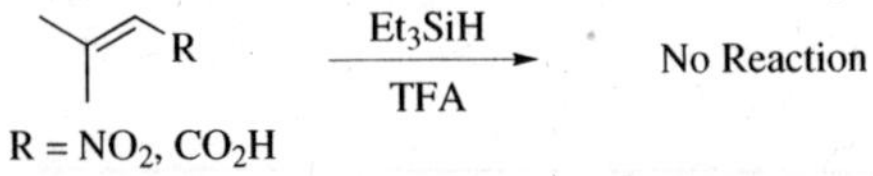

(Eq. 85)

Tetrasubstituted Alkenes. Tetrasubstituted alkenes lacking electron-withdrawing substituents undergo facile ionic hydrogenation to alkanes in very good yields. Simple examples include 2,3-dimethyl-2-butene,[208,214] 1,2-dimethylcyclopentene, 1,2-dimethylcyclohexene,[229] and $\Delta^{9(10)}$-octalin.[126,204,212]

Interesting variations are observed in the stereoselectivities of these ionic hydrogenations. Reduction of 1,2-dimethylcyclopentene with Et_3SiH/TFA near room temperature gives 1,2-dimethylcyclopentane with a cis to trans ratio of 0.083, compared to a ratio of 0.63 for 1,2-dimethylcyclohexene.[229]

The reduction of $\Delta^{9(10)}$-octalin to *cis*- and *trans*-decalins occurs with cis to trans stereoselectivities that vary with the nature of the organosilicon hydride employed. The ratios are 0.28–0.59 with *n*-butylsilane, 0.67 with diethylsilane,[204] 0.34[212] or 0.72[204] with triethylsilane, 0.67 with diphenylsilane, 0.77 with diphenylmethylsilane,[212] 1.38[204]–1.80[127,212] with triphenylsilane, 0.54 with triisopentylsilane, 1.17 with tricyclopentylsilane, 2.57 with tri-*sec*-butylsilane, 3.35 with di-*tert*-butylsilane, 4.88 with di-*tert*-butylmethylsilane, and 13.3 with tri-*tert*-butylsilane.[204] Opinions differ about the mechanistic significance of these changes in isomer ratios.[204,212]

Treatment of $\Delta^{8(9)}$-dehydroestradiol with trifluoroacetic acid and triethylsilane gives estradiol in 96% yield (Eq. 86).[239] The 3-methyl ether is similarly reduced to the 3-methyl ether of estradiol in >50% yield.[239] The structurally related 18-ethyl and 18-propyl 17-keto compounds experience reduction of the $\Delta^{8(9)}$ function in excess of 70% yield without concomitant reduction of the 17-keto group.[239]

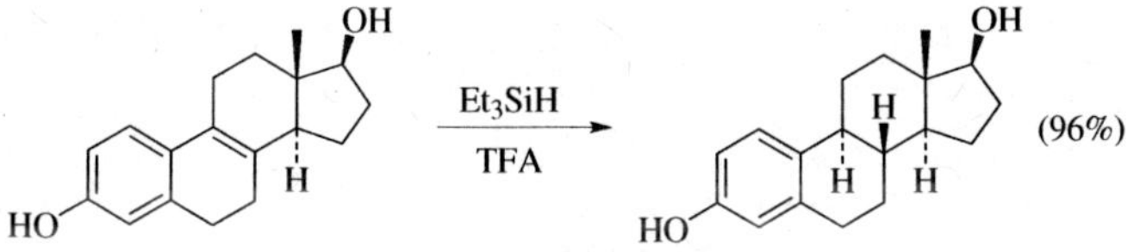

(Eq. 86)

Treatment of $\Delta^{8(9)}$-dehydro-D-homoestradiol (**39**, R = H) (or its 3-methyl ether, R = Me) with Et_3SiH/TFA followed by saponification of the trifluoroacetate ester intermediate leads to D-homoestradiol (**40**) (or its 3-methyl ether) containing 2–15% D-homoequilenol (**41**) (or its 3-methyl ether).[240] By contrast, reduction and saponification of 3,17-diacetyl-$\Delta^{8(9)}$-dehydro-D-homoestradiol (**39**, R = AcO) gives a 60% yield of D-homoestradiol without the presence of any D-homoequilenol (Eq. 87).[240]

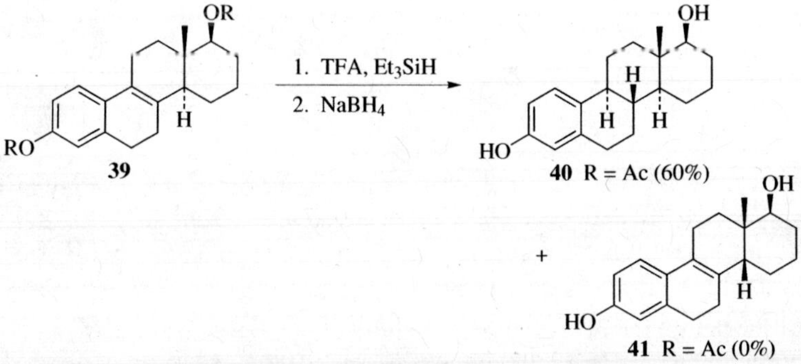

(Eq. 87)

Polyenes. The behavior of substrates with multiple carbon-carbon double bonds toward the conditions employed for ionic hydrogenations with organosilicon hydrides depends heavily on the number and kinds of substituents and whether or not the multiple double bonds are conjugated. In the absence of conjugation, the individual double bonds react independently.

The full reduction of 1,3-dienes with Et_3SiH/TFA occurs in certain systems although the yields are only modest.[231] For example, 1,3-cyclohexadiene gives a 65% yield of cyclohexyl trifluoroacetate, presumably by way of cyclohexene (Eq. 88).[211] On the contrary, 1,4-cyclohexadiene fails to undergo reaction with 10 equivalents of triethylsilane and 20 equivalents of trifluoroacetic acid even after 24 hours at room temperature (Eq. 89).

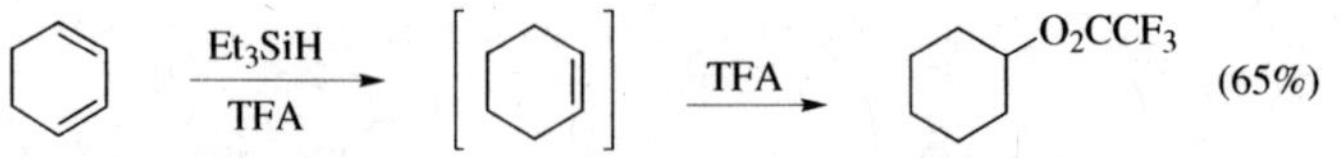

(Eq. 88)

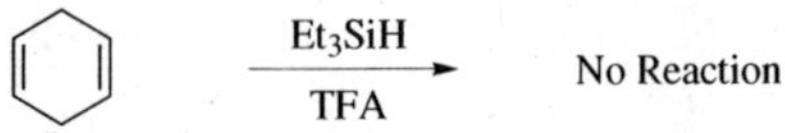

(Eq. 89)

Additional evidence of this pattern of behavior is shown upon treatment of the conjugated diene 1-propenylcyclohexene with two equivalents of triethylsilane and three equivalents of trifluoroacetic acid at 50°. This diene gives a 70% yield of completely reduced propylcyclohexane after 10 hours (Eq. 90).[231] No partially reduced intermediates are found.

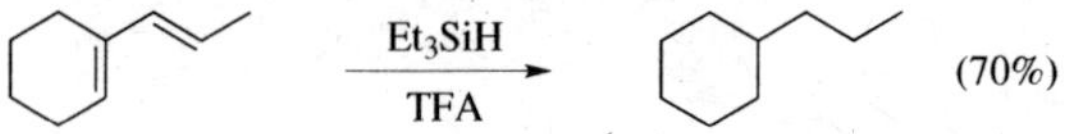

(Eq. 90)

Similar treatment of the isomeric, nonconjugated 1-(3-propenyl)cyclohexene gives a mixture of products containing 55% of the partially reduced 3-propenylcyclohexane and 15% of the completely reduced propylcyclohexane (Eq. 91).[231] The yield of the latter product increases to 25% when the amounts of Et_3SiH/TFA used are raised to 6.5 and 12.1 equivalents, respectively, and the reaction time is increased to 24 hours.[230] The nonconjugated 1-(3-butenyl)cyclohexene gives a 65% yield of partially hydrogenated 3-butenylcyclohexane under identical conditions.[231]

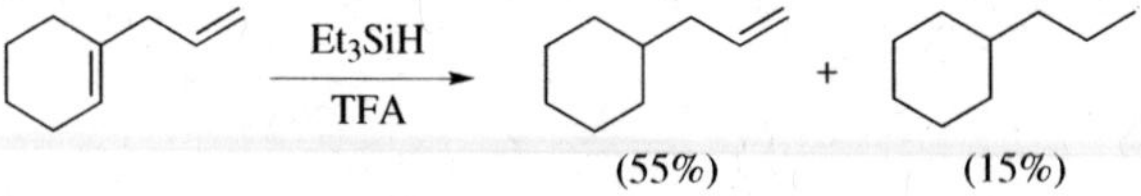

(Eq. 91)

Reduction of dienes incorporated into steroid structures may lead to different configurations in the products. For example, treatment of 8(9),14(15)-bisdehydroestrone **42** (R = H) for four hours at room temperature with twenty equivalents of trifluoroacetic acid and two equivalents of triethylsilane leads to an ionic hydrogenation product mixture containing the natural $8\beta,9\alpha,14\alpha$-estrone **43** as a minor component (11%) and the $8\alpha,9\beta,14\beta$-isomer **44** as the major component (83%) (Eq. 92).[241] The related methyl ether (**42**, R = Me) behaves in a similar fashion.[241] The yield of natural isomer **46** formed from the methyl ether of $\Delta^{8(9),14(15)}$-bisdehydroestradiol analog **45** increases from 22 to 34%, and that of isomer **47** decreases from 78 to 66%, when the solvent is changed from benzene to dichloromethane (Eq. 93).[242]

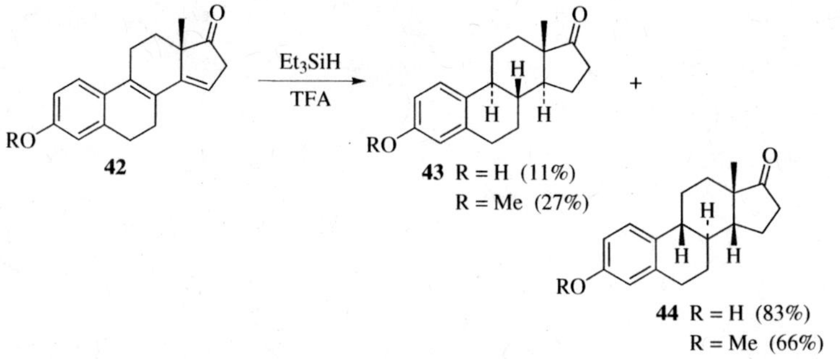

(Eq. 92)

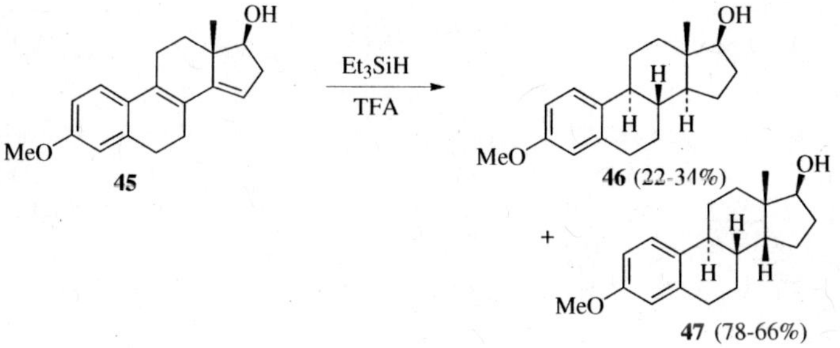

(Eq. 93)

Treatment of linalyl *p*-tolyl sulfone (R = $SO_2C_6H_4Me$-*p*) with 2.5 equivalents of trifluoroacetic acid and two equivalents of triethylsilane in 2-nitropropane containing one equivalent of lithium perchlorate gives, after 20 hours at room temperature, an 87% yield of the product in which only the double bond distal to the sulfone function is reduced (Eq. 94).[205]

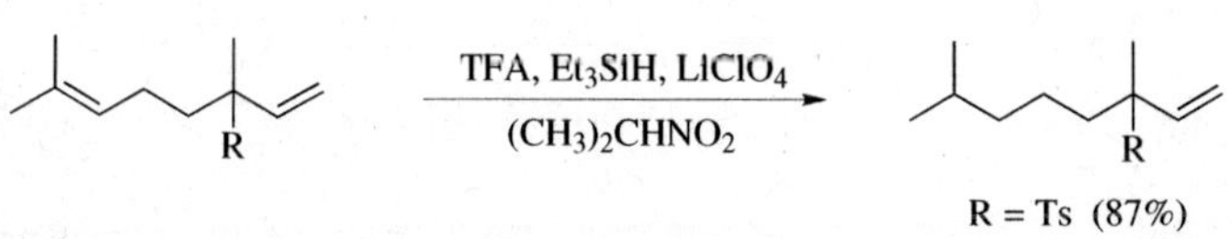

(Eq. 94)

Surprisingly, linalyl acetate (R = OAc) fails to undergo reduction under these conditions; instead, it rapidly decomposes through cyclization and polymerization pathways.[205] The same reaction conditions transform geranyl *p*-tolyl sulfone (R = $SO_2C_6H_4Me$-*p*) into a mixture of 7% reduced and 93% cyclized products within 20 hours, whereas geranyl acetate (R = OAc) gives only a 20% yield of cyclized and no reduced product (Eq. 95).[205]

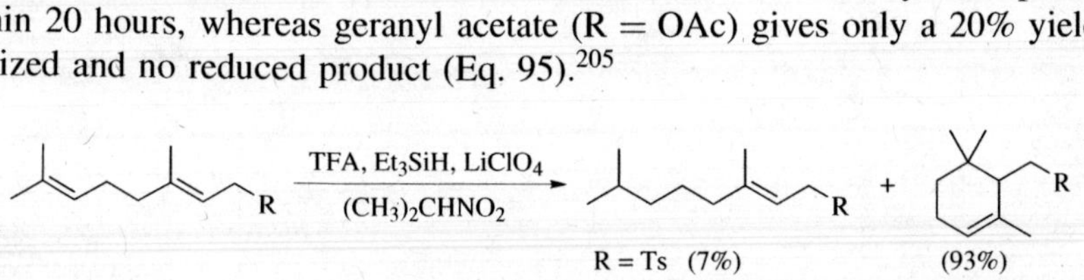

(Eq. 95)

Homoconjugation results in enhanced reactivity of substrates toward ionic hydrogenation. Bicyclo[2.2.1]hepta-2,5-diene forms a mixture of the trifluoroacetate esters of bicyclo[2.2.1]hepten-2-ol, tricyclo[2.2.1.0^{2,6}]heptan-3-ol, and bicyclo[2.2.1]heptan-2-ol in a 62 : 20 : 17 ratio on treatment with 10 equivalents of triethylsilane and 20 equivalents of trifluoroacetic acid for 24 hours at room temperature (Eq. 96).[230]

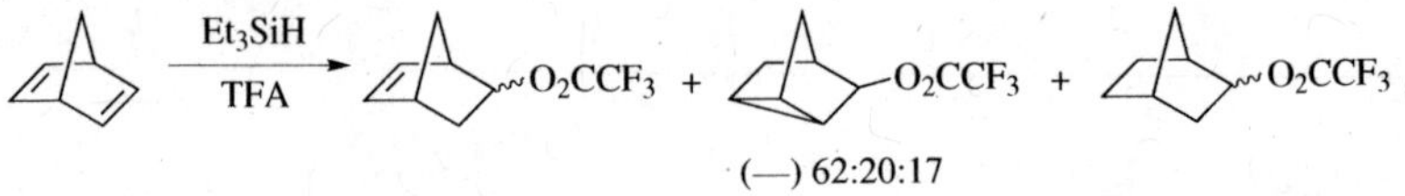

(Eq. 96)

(—) 62:20:17

Treatment of 5-methylenebicyclo[2.2.1]hept-2-ene with 10 equivalents of triethylsilane and 20 equivalents of trifluoroacetic acid either for 24 hours at room temperature or 3 hours at 50° gives an 85% yield of completely hydrogenated *endo*-2-methylbicyclo[2.2.1]heptane (Eq. 97). The combination of Et_3SiH/TFA/$BF_3{\cdot}OEt_2$ gives this product in 80% yield.[230] The reaction presumably proceeds by way of 2-methyltricyclo[2.2.1.0^{2,6}]heptane as a reaction intermediate, since this compound is expected to rapidly give the same final product when it is subjected to these reaction conditions.[230] The analogous stereospecific behavior is exhibited by 5-ethylidenebicyclo[2.2.1]hept-2-ene.[230]

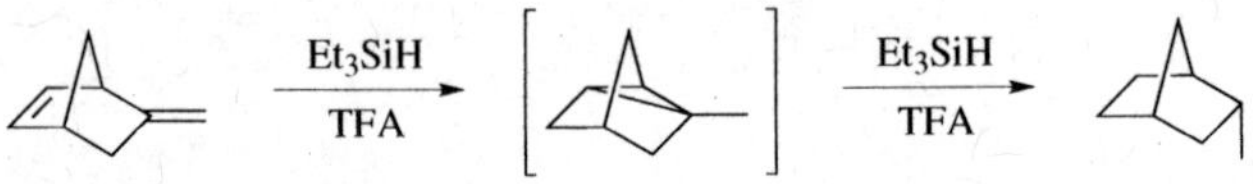

(85%) (Eq. 97)

Transannular interactions lead to ring closures and reductions to adamantane compounds when dienes of the bicyclo[3.3.1]nonane family are treated with Brønsted acids and triethylsilane. Compounds **48–51** form reaction mixtures containing various amounts of products **52–54** (R = OH, O_2CCF_3, Cl) under such conditions.[243] The best yields of hydrocarbon **52** occur when the dienes are treated with a 25% excess of sulfuric acid and a 50% excess of triethylsilane in dichloromethane at 20°.[243] The stereospecific nature of these transannular reductions is demonstrated by the observation that the enantiomeric purity of the chiral diene **55** is retained in the chiral hydrocarbon product **56** (Eq. 98).[243] Dienes of

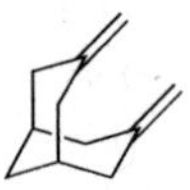
48

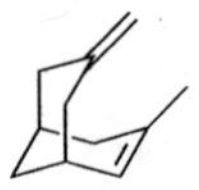
49

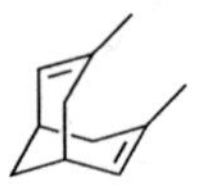
50

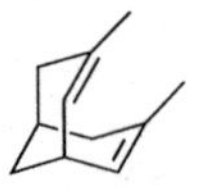
51

52

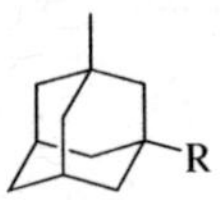

53

R = OH, OTFA, Cl

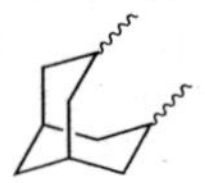
54

the type shown can be reduced to the chlorides (Eq. 99).[243] When HCl is replaced with TFA, methyladamantane and methyladamantanol are formed (Eq. 100).

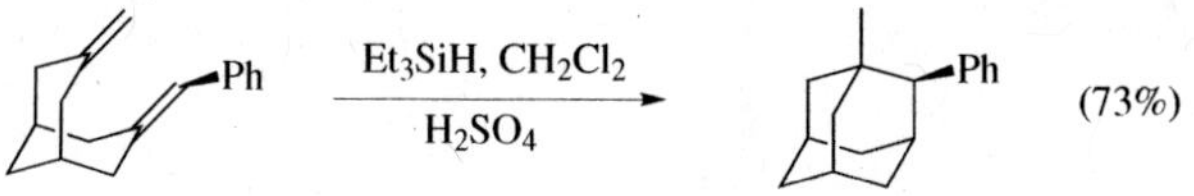

55 **56**

(Eq. 98)

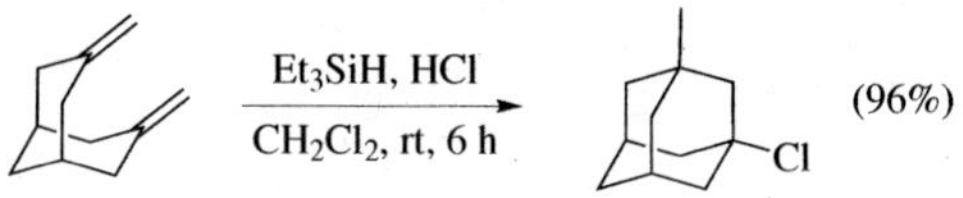

(Eq. 99)

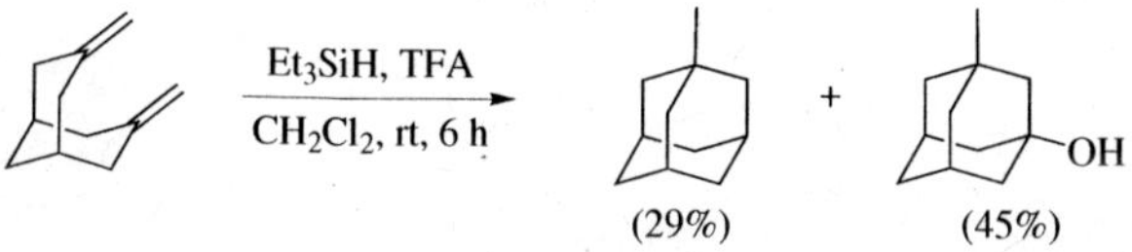

(Eq. 100)

Alkynes to Alkanes. In contrast to the facile ionic hydrogenations that many alkenes undergo, alkynes as a group are very resistant to reduction with the organosilicon hydride/acid combinations. Only those alkynes having an electron-rich aryl group in conjugation with the carbon-carbon triple bond give even modest amounts of reduced products as seen in the example of *p*-tolylacetylene (Eq. 101).[244] Alkenes are not observed as products.[244]

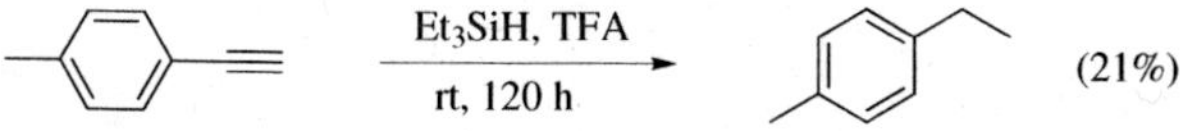

(Eq. 101)

The use of stronger acid conditions provides somewhat better synthetic yields of alkanes from alkynes. A useful method consists of treatment of the substrate with a combination of triethylsilane, aluminum chloride, and excess hydrogen chloride in dichloromethane.[146] Thus, treatment of phenylacetylene with 5 equivalents of triethylsilane and 0.2 equivalents of aluminum chloride in this way at room temperature yields 50% of ethylbenzene after 1.5 hours. Diphenylacetylene gives a 50% yield of bibenzyl when treated with 97 equivalents of triethylsilane and 2.7 equivalents of aluminum chloride after 2.8 hours. Even 1-hexyne gives a mixture of 44% *n*-hexane and 7% methylpentane of undisclosed structure when treated with 10 equivalents of triethylsilane and 0.5 equivalent of aluminum chloride for 0.5 hour.[146]

The reductive cyclization of enynes has been used to prepare exo-methylenecycloalkanes. Two systems have proven successful in this transformation, namely PMHS/$Pd_2(dba)_3$•$CHCl_3$[245] (Eq. 102) and Et_3SiH/Pd(dppe)Cl_2/HCO_2H (Eq. 103).[246]

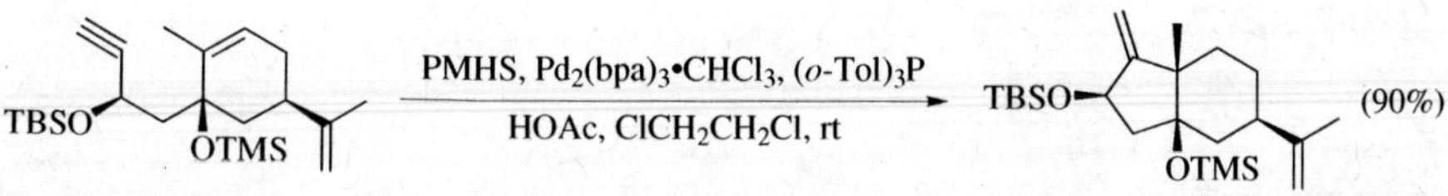

(Eq. 102)

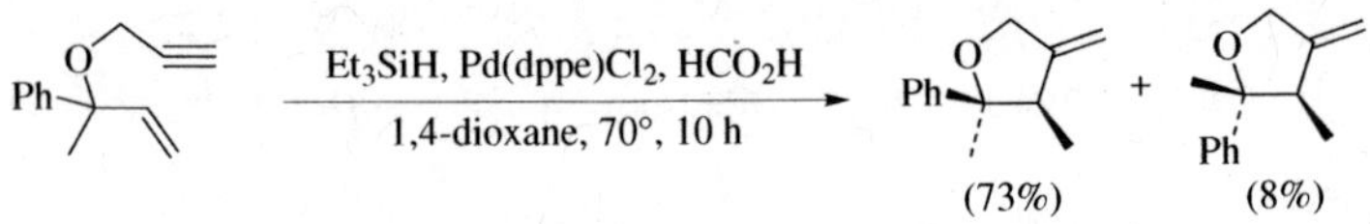

(Eq. 103)

The triethylsilane/$Pd_2(dba)_3$ combination is also used for these reductive cyclizations, although lower yields are reported.[247] 1,6-Diynes are reductively cyclized to 1,2-dialkylidenecyclopentanes in good yields with $Et_3SiH/Pd_2(dba)_3{\cdot}CHCl_3$ (Eq. 104).[248]

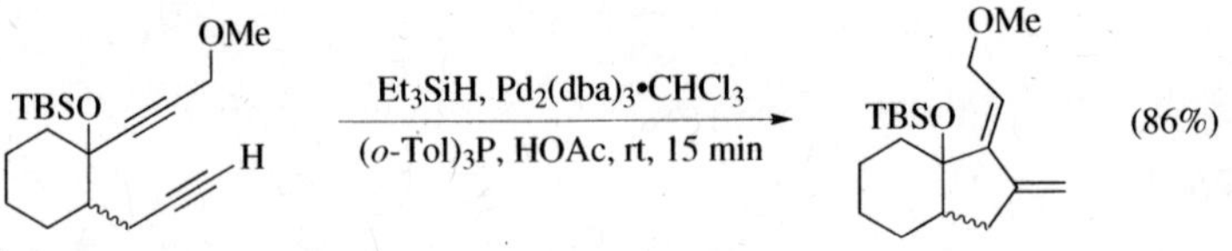

(Eq. 104)

o-Bromobenzyl alkynylalkyl ethers can be reductively cyclized in modest yields with $Et_3SiH/Pd(PPh_3)_4/Cs_2CO_3$ as shown in Eq. 105.[249] In a like manner, enynes with a vinyl bromide as the olefin function undergo reductive cyclization (Eq. 106).[249]

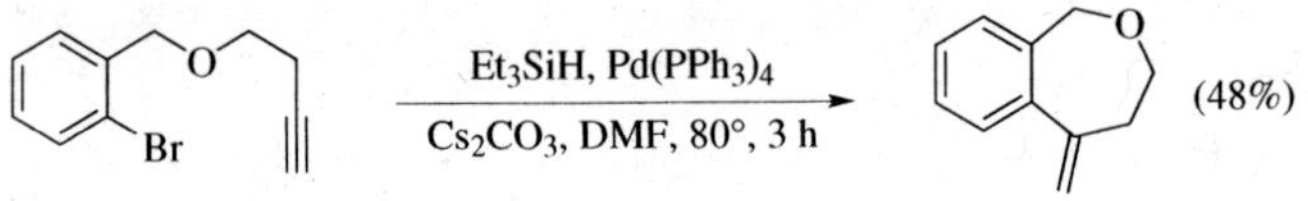

(Eq. 105)

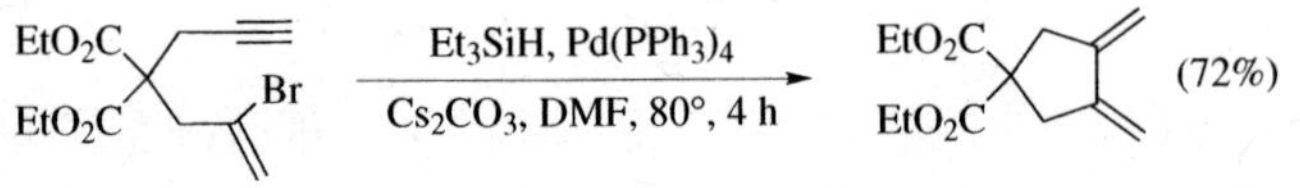

(Eq. 106)

Cyclopropanes to Alkanes. Cyclopropanes that can form ring-opened tertiary aliphatic or benzylic carbenium ion intermediates undergo ionic hydrogenation with reasonable ease when treated with Brønsted acids and organosilicon hydrides. Ring opening occurs preferentially between the most and least highly substituted ring carbons. For example, treatment of 1,1,2-trimethylcyclopropane with one equivalent of triethylsilane and two equivalents of trifluoroacetic acid gives a mixture of 2,3-dimethylbutane (75%) and 2-methylpentane (25%) (Eq. 107).[233] The conversion into the hydrocarbon mixture is 15% after 15 minutes, 65% after 12 hours, and complete after 16 hours at room temperature.[222,232] Essentially the same results are obtained when 2,3-dimethyl-2-butene is used as the substrate.[222] The disubstituted isomer 1-methyl-2-ethylcyclopropane gives an alkane reaction mixture consisting primarily of 3-methylpentane along with 2-methylpentane (Eq. 108).[222]

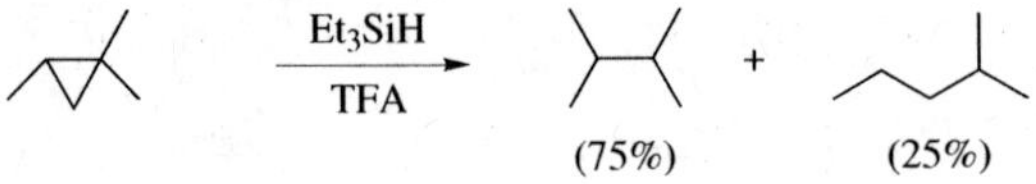

(Eq. 107)

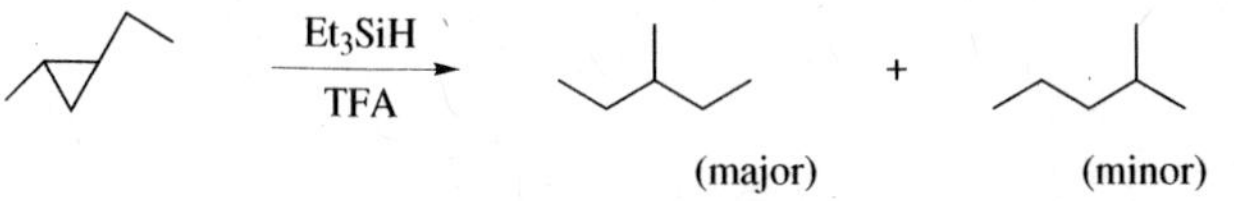

(Eq. 108)

Unlike its disubstituted isomer, the monosubstituted isopropylcyclopropane undergoes reduction to 2-methylpentane to the extent of only 50% after 24 hours (Eq. 109),[232] a result similar to that observed when 2-methyl-1-pentene is the substrate.[222] It is interesting that deuterated triethylsilane produces 2-methylpentane that contains the deuterium label only at the C2 position.[250] This label position suggests that in this reaction ring protonation and opening are followed by a 1,2-hydride shift that precedes capture by the silyl hydride of any initially formed carbocation intermediates.[250] Ethylcyclopropane, with an unbranched side chain, shows no sign of reduction under these conditions even after 200 hours.[232] Phenylcyclopropane is reduced to 1-phenylpropane.[222]

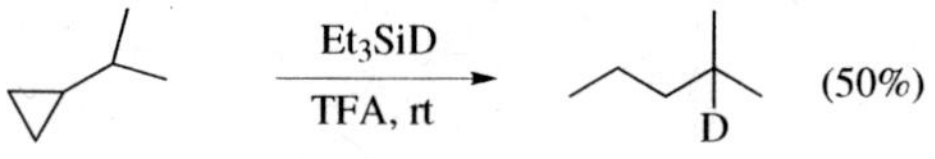

(Eq. 109)

Bicyclic hydrocarbons that contain a three-membered ring slowly undergo ionic hydrogenation when treated with at least one equivalent of triethylsilane and an excess of trifluoroacetic acid at room temperature.[229] Thus, bicyclo[3.1.0]-hexane gives a product mixture containing methylcyclopentane (28%) and cyclohexane (3%) when reacted with one equivalent of triethylsilane and four equivalents of trifluoroacetic acid for 140 hours (Eq. 110). The main products are the trifluoroacetates of cyclohexanol and *cis*-and *trans*-2-methylcyclopentanol in a ratio of 10 : 35.[229] Under the same conditions, bicyclo[4.1.0]heptane yields a mixture containing mainly methylcyclohexane (79%) with some cycloheptane (5%) and the corresponding trifluoroacetates (16%) (Eq. 111).[229]

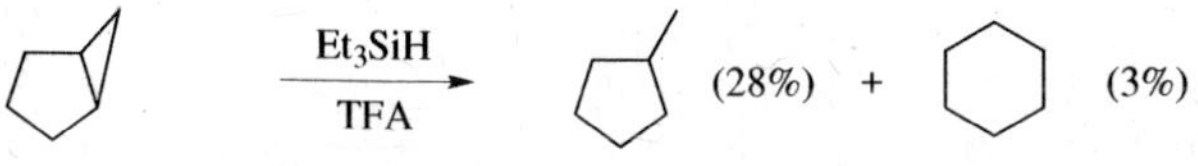

(Eq. 110)

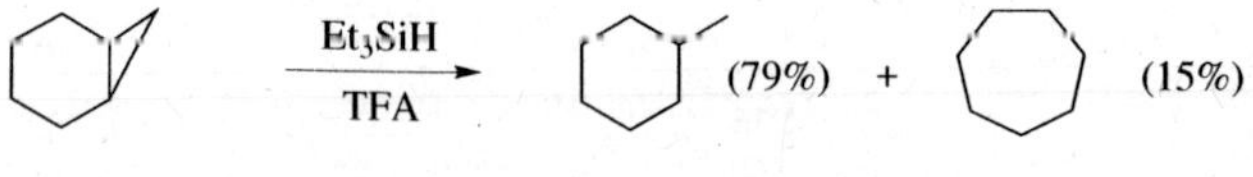

(Eq. 111)

After ten days at room temperature in the presence of one equivalent of triethylsilane and two equivalents of trifluoroacetic acid, both 1-methylbicyclo-[3.1.0]hexane and 1-methylbicyclo[4.1.0]heptane form mixtures of the two isomers of their respective 1,2-dimethylcycloalkanes (Eqs. 112 and 113).[229]

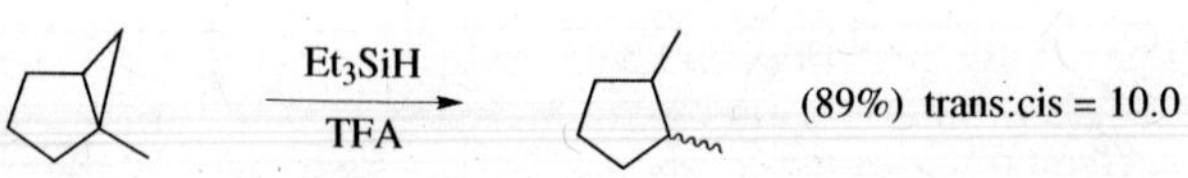

(Eq. 112)

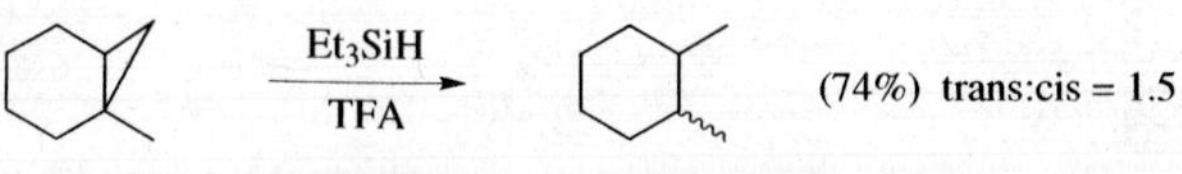

(Eq. 113)

Based on the few reported examples, the pattern of ring cleavage that accompanies the ionic hydrogenation of alkylidenencyclopropanes seems to be related to the pattern and degree of substitution on both the ring and the double bond.[233] Thus, treatment of 1,1-dimethyl-2-methylenecyclopropane with two equivalents of triethylsilane and four equivalents of trifluoroacetic acid for 90 hours at room temperature yields 65% of 2,3-dimethylbutane (Eq. 114).[229] Exposure of 1,1-dimethyl-2-isopropylidenecyclopropane to the same ratio of reactants at 50° for 16 hours produces a complex mixture containing 63% of 2,5-dimethylhexane, 18.5% of 2,5-dimethyl-3-hexene, 1.6% of 2,5-dimethyl-2-hexene, and 7% of 2,5-dimethyl-2-hexyl trifluoroacetate (Eq. 115).[229]

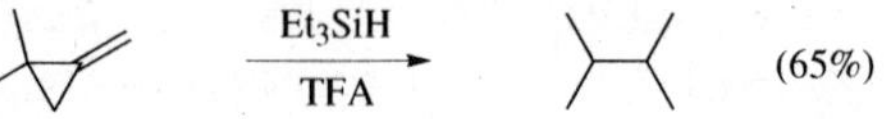

(Eq. 114)

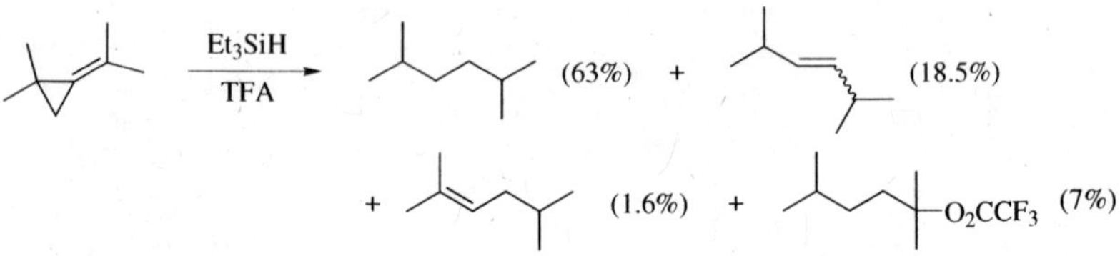

(Eq. 115)

Aromatic Substrates. Aromatic hydrocarbons can be reduced with organosilanes to dienes, alkenes, or alkanes. The combination of $Et_3SiH/TFA/BF_3{\cdot}OEt_2$ reduces furans to tetrahydrofurans in good yields (Eq. 116).[211] In general, poor yields are obtained with the Et_3SiH/TFA reduction of benzofurans,[251] but the C3-substituted benzofuran shown undergoes reduction of the furan ring in excellent yield with this reagent (Eq. 117).[252] Similarly, benzothiophenes are reduced in 60 to 90% yields under the same conditions.[253] The Et_3SiH/TFA system reduces thiophenes to tetrahydrothiophenes in good yields.[254–257] In α-hydroxy thiophenes, both double bonds of the thiophene unit and the hydroxy group are reduced (Eq. 118).[258]

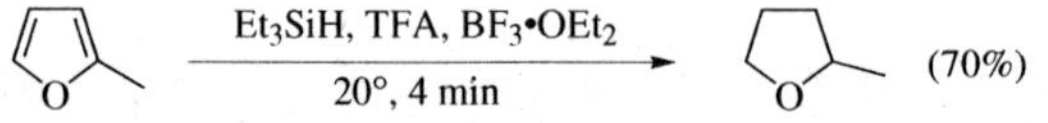

(Eq. 116)

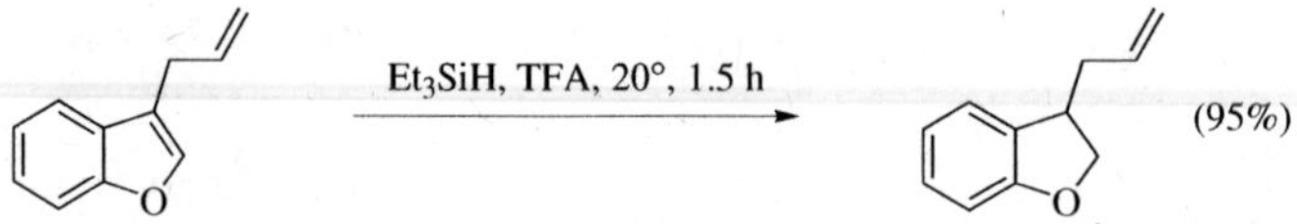

(Eq. 117)

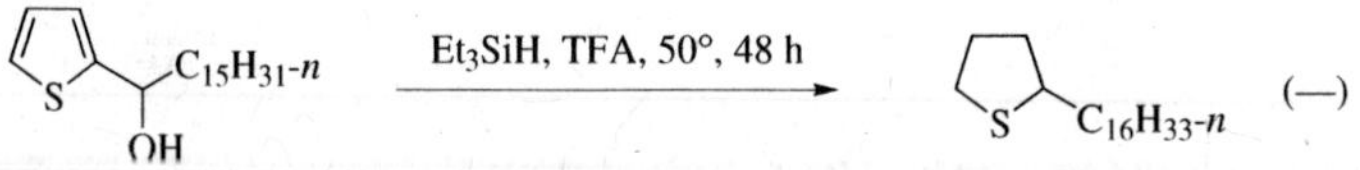

(Eq. 118)

Similar reactivity is realized with 2-acetylthiophene using triethylsilane with aluminum chloride.[259] Treatment of the ethylene glycol acetal of 2-thiophenecarbaldehyde with Et_3SiH/TFA results in reduction of the ring and oxidation of the side chain to the silylated carboxylic acid (Eq. 119),[260] whereas similar treatment of 2-thiophenecarbaldehyde gives 2-methyltetrahydrothiophene and 2-

acetylthiophene gives 2-ethylthiophene.[257] Some thiophenes are reduced to a mixture of tetrahydrothiophenes and 2,5-dihydrothiophenes.[210,259,261]

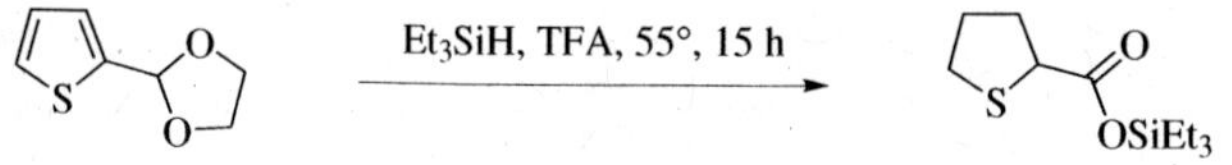
(45%) (Eq. 119)

Partial reduction of polyarenes has been reported. Use of boron trifluoride hydrate ($BF_3{\bullet}OH_2$) as the acid in conjunction with triethylsilane causes the reduction of certain activated aromatic systems.[217,262] Thus, treatment of anthracene with a 4–6 molar excess of $BF_3{\bullet}OH_2$ and a 30% molar excess of triethylsilane gives 9,10-dihydroanthracene in 89% yield after 1 hour at room temperature (Eq. 120). Naphthacene gives the analogously reduced product in 88% yield under the same conditions. These conditions also result in the formation of tetralin from 1-hydroxynaphthalene (52%, 4 hours), 2-hydroxynaphthalene (37%, 7 hours), 1-methoxynaphthalene (37%, 10 hours), 2-methoxynaphthalene (26%, 10 hours), and 1-naphthalenethiol (13%, 6 hours). Naphthalene, phenanthrene, 1-methylnaphthalene, 2-naphthalenethiol, phenol, anisole, toluene, and benzene all resist reduction under these conditions.[217] Use of deuterated triethylsilane to reduce 1-methoxynaphthalene gives tetralin-1,1,3-d_3 as product, thus yielding information on the mechanism of these reductions.[262] 2-Mercaptonaphthalenes are reduced to 2,3,4,5-tetrahydronaphthalenes in poor to modest yields.[217,263]

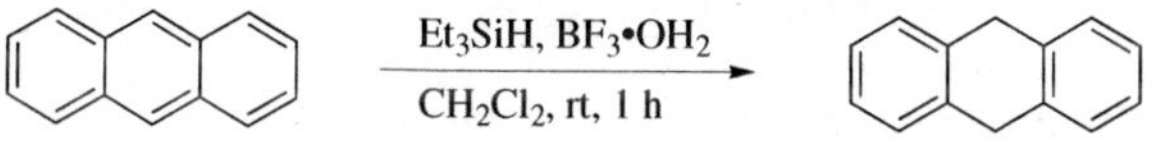
(89%) (Eq. 120)

The combination of $PhMeSiH_2$ (or Ph_2SiH_2) and Cp_2TiMe_2 (10 mol%) reduces pyridines to N-silylated-di- or tetrahydropyridines or the N-silylated piperidines.[264,265] With quinoline, only the pyridine ring is reduced preferentially to the benzene ring (Eq. 121).

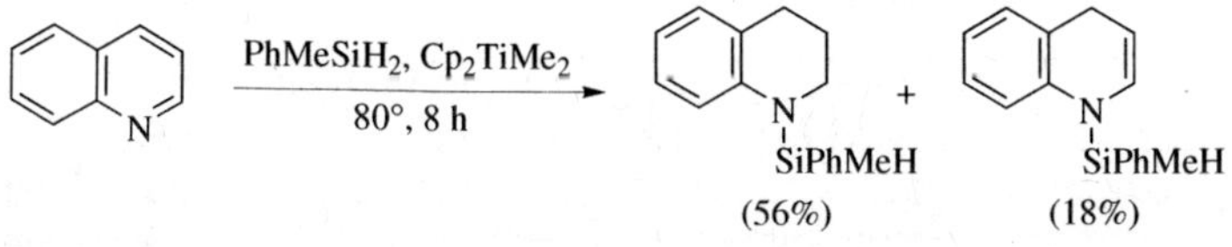
(Eq. 121)

Miscellaneous Unsaturated Substrates. Exposure of 1,1′ bis(*trans*-2-cyanovinyl)ferrocene to a mixture of two equivalents of triethylsilane and 320 equivalents of trifluoroacetic acid at 50° for three hours gives a product with the carbon-carbon double bonds reduced in 83% yield, but leaving the nitrile groups intact (Eq. 122).[179]

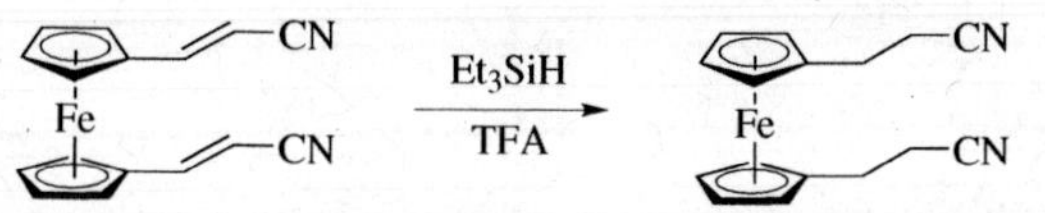
(Eq. 122)

Treatment of a chloroform or dichloromethane solution of 1-bromo-2,2-diphenylethene or 1-bromo-2,2-bis(4′-methoxyphenyl)ethene with a slight excess of triethylsilane and a 9- to 10-fold excess of TFA gives the corresponding ethanes in 62% and 88% yields, respectively, after one hour at 0° (Eq. 123).[184]

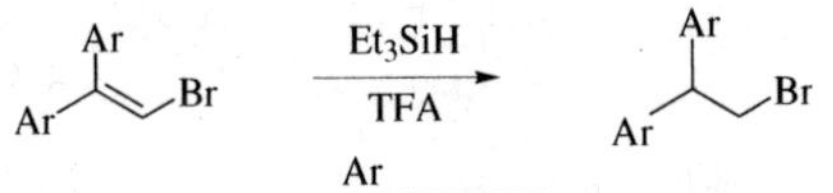

(Eq. 123)

Ar	
Ph	(62%)
4-$MeOC_6H_4$	(88%)

The carbonyl groups of 1,3-indanediones are generally resistant to the action of combinations of acid and silanes at room temperature.[266] Accordingly, treatment of a variety of 2-benzylidene-1,3-indanediones with Et_3SiH/TFA (ratio of 1 : 5 : 10) in CCl_4 at 55° for 7–20 hours gives the corresponding substituted 2-benzyl-1,3-indanediones in 54-78% yields (Eq. 124).[266] Use of a 27-fold excess of trifluoroacetic acid in the absence of a cosolvent reportedly leads to reduction of the carbonyl groups to give a mixture of products.[267]

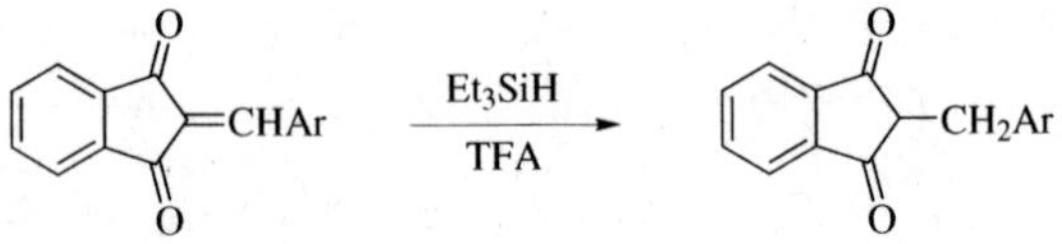

(Eq. 124)

An interesting hydroiodination reaction occurs when a mixture of cyclohexene and triethylsilane in dichloromethane is treated with a mixture of bis(pyridine) iodonium tetrafluoroborate and tetrafluoroboric acid in diethyl ether (Eq. 125). A 50% yield of iodocyclohexane is produced after one hour at 20°.[268]

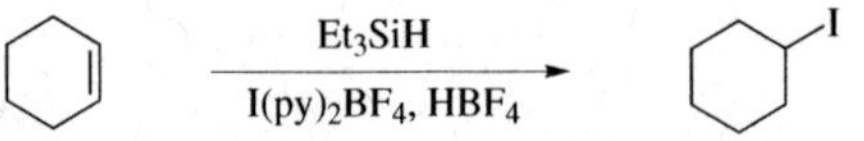

(Eq. 125)

Reduction of Ethers

Because of the high stability of the triphenylmethyl carbocation, the reductive ether cleavage of trityl ethers with Et_3SiH/trimethylsilyl triflate (TMSOTf) is highly successful. This reaction even occurs in the presence of highly reactive sugar ketals, leaving the ketals intact (Eq. 126).[269]

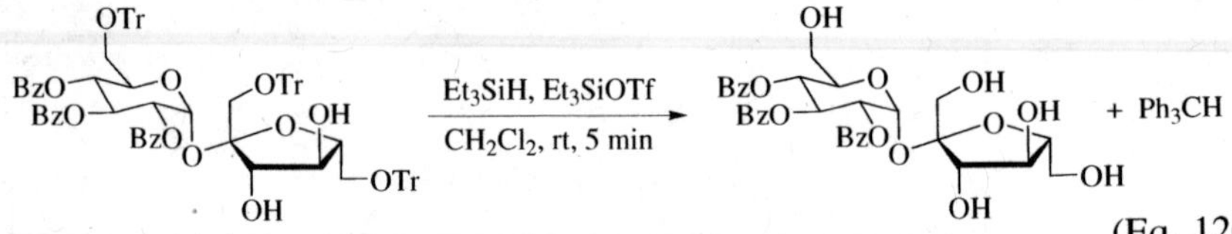

(87%)

(Eq. 126)

The combination of PMHS and $Pd(PPh_3)_4$ reduces allyl ethers to propene and alcohols.[270] The best combination for the reductive cleavage of ethers appears to be $Et_3SiH/(C_6F_5)_3B$, which gives excellent yields of the alcohol (via the silyl ether) and alkane (Eq. 127).[145]

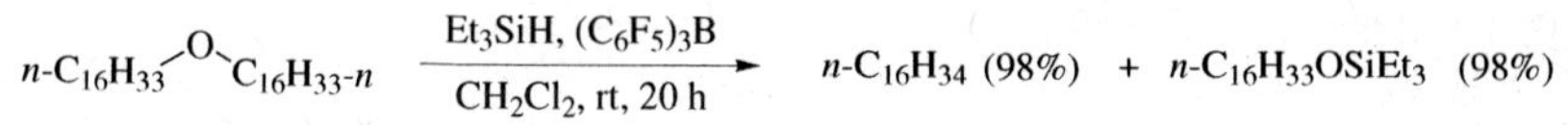

(Eq. 127)

Dialkyl ethers are reduced with the combination of Et_3SiH/TFA, although the yields vary.[144,271] *tert*-Butyl triphenylcyclopropenyl ether is reduced to the corresponding cyclopropene (Eq. 128),[272] and a dibenzyl-like ferrocene-derived ether is reduced to the corresponding alkane (Eq. 129).[179]

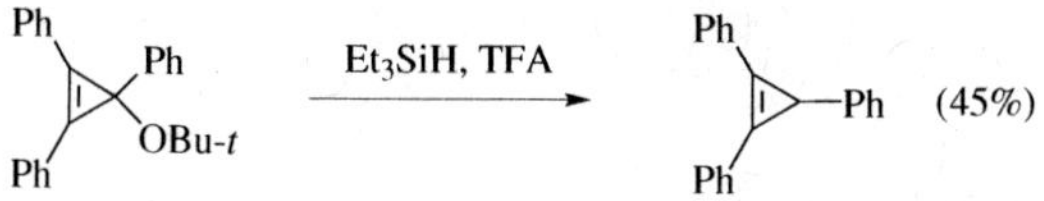

(Eq. 128)

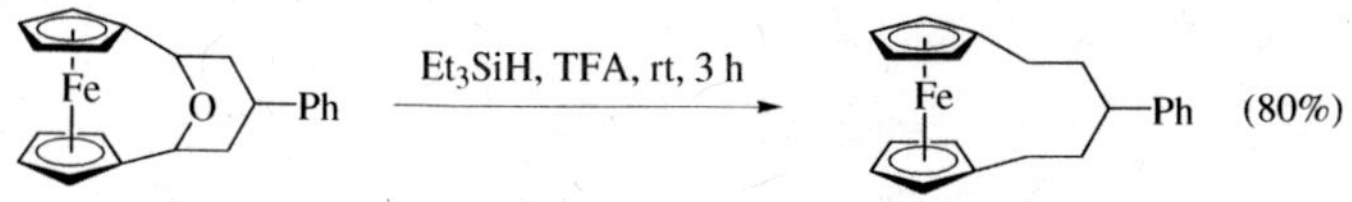

(Eq. 129)

Reduction of Allyl Acetates

Allyl acetates are reduced to the corresponding olefins with PMHS/$Pd(PPh_3)_4$ or Ph_2SiH_2/$Pd(PPh_3)_4$.[196,273] Unfortunately, double bond migration occurs in many of these reactions (Eqs. 130 and 131).[196,273] The combinations of Ph_2SiH_2/$Pd(P(Tol\text{-}p)_3)_4$/$ZnCl_2$[274] and Et_3SiH/TFA[275] are also employed in this transformation.

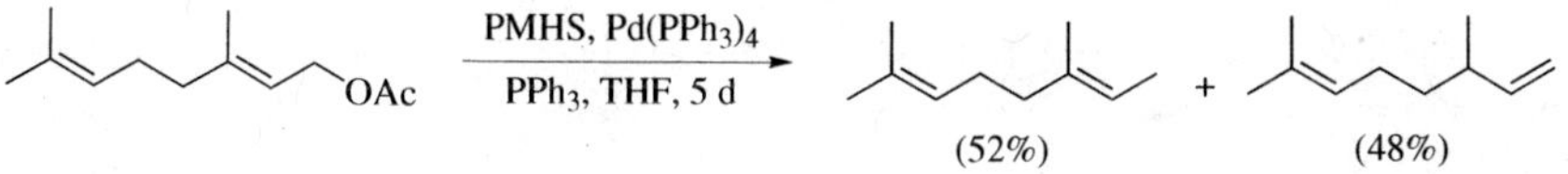

(Eq. 130)

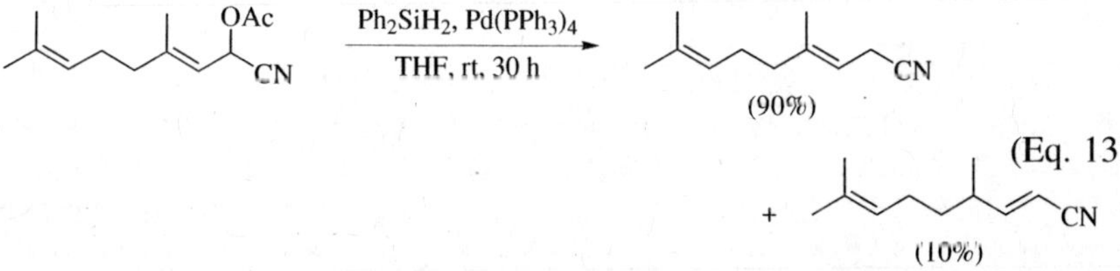

(Eq. 131)

The Et_3SiH/TFA reduction of a 3-acetoxy enol ether is reported. The diastereoselectivity is high for the Z isomer, but much lower for the E isomer (Eq. 132).[276]

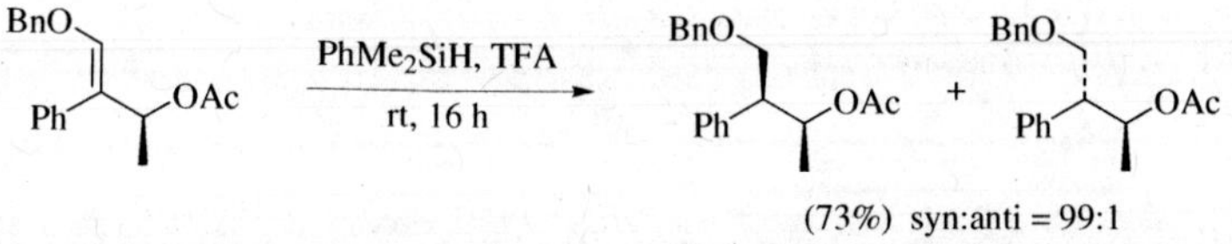

(Eq. 132)

Reduction of Carboxylic Acids

Aromatic and aliphatic carboxylic acids are reduced to the trifluoroacetates of the alcohol with Et_3SiH/TFA.[277] Use of an excess of the triethylsilane can give further reduction to the methyl group. The combination of PMHS/TBAF (tetra-*n*-butylammonium fluoride) reduces benzoic acids to the benzyl alcohols in good yields.[278] Comparable yields of this useful transformation can be realized through the use of PMHS/Ti(OPr-*i*)$_4$ (or PMHS/Ti(OEt)$_4$).[279] Both aromatic and aliphatic carboxylic acids can be reduced with $EtMe_2SiH$ and the ruthenium-based catalyst **57** (Eq. 133).[280] The latter reagent/catalyst combination also reduces esters to alcohols in high yield.

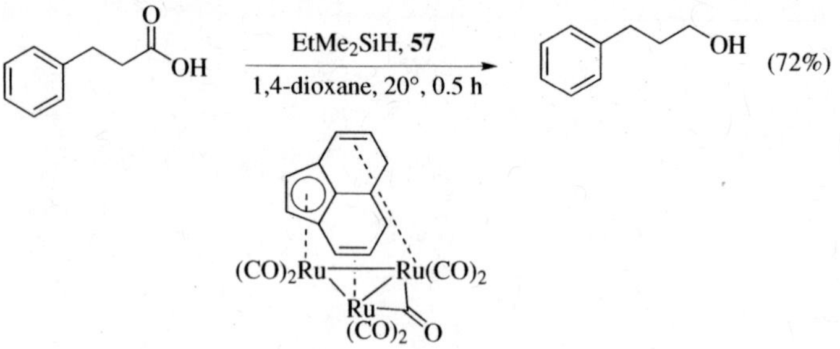

57

(Eq. 133)

The highly reactive reagent combination of $Et_3SiH/(C_6F_5)_3B$ reduces carboxylic acids to methyl groups (Eq. 134).[281,282] Isolation of the intermediate silyl ether is also possible.[282]

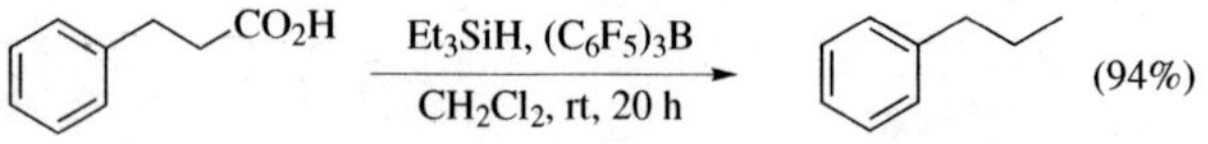

(Eq. 134)

Benzoic acids with electron-donating groups on the ring are reduced to toluene derivatives with the reagent combination Et_3SiH/TFA/TFAA.[283] *p*-Anisic acid gives 4-methylanisole in 97% yield under these conditions (Eq. 135). Formation of the corresponding benzyl trifluoroacetates occurs for substrates without activating groups. *p*-Nitrobenzoic acid is unreactive under these conditions, as are dibasic acids such as phthalic or succinic acid.[283] The same conditions reduce alkyl carboxylic acids to trifluoroacetates.[277] Use of the silane **58** or **59** provides cinnamaldehyde in fair yield from cinnamic acid (Eq. 136).[284]

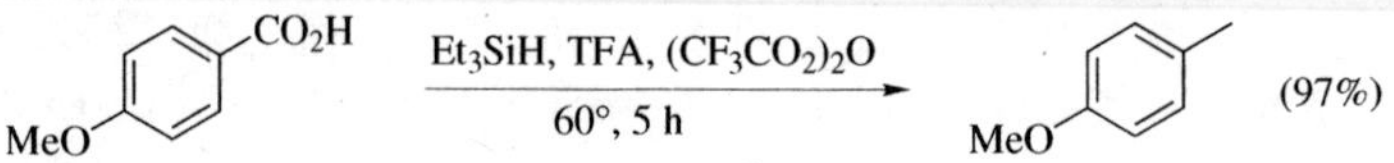

(Eq. 135)

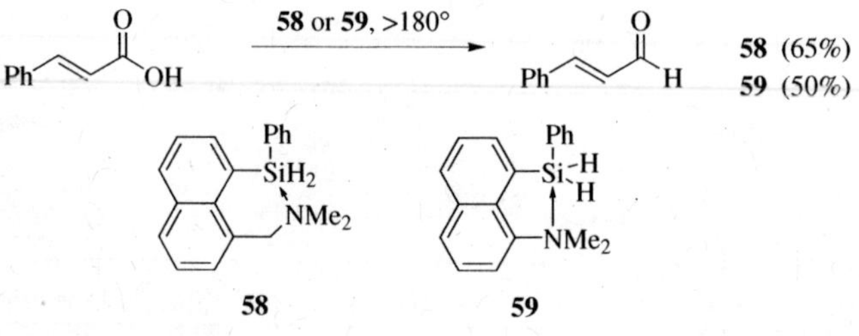

58 **59**

(Eq. 136)

Reduction of Acid Halides and Acid Anhydrides

The organosilane reduction of acid chlorides to aldehydes has been accomplished in high yields with the use of the pentacoordinated organosilane **60** (Eq. 137).[107] This transformation has been reported to occur with tribenzylsilane and triethylsilane, but yields were not reported.[285,286]

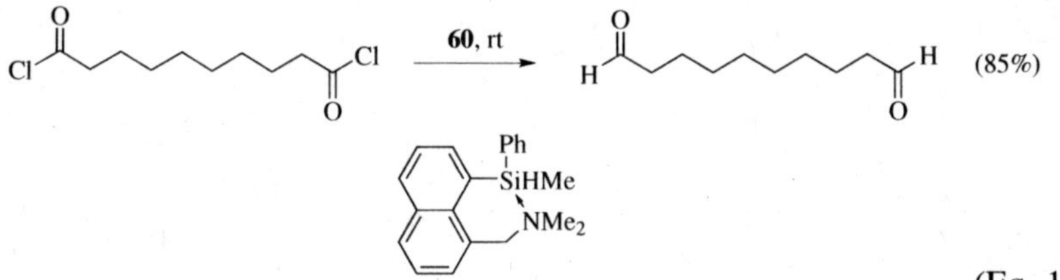

60

(Eq. 137)

The combination $Et_3SiH/(C_6F_5)_3B$ reduces acid chlorides to methyl groups (Eq. 138).[281,282] If a smaller amount of triethylsilane is used, the same combination reduces aryl acid chlorides to the trimethylsilyl ethers of the benzyl alcohols.[281,282]

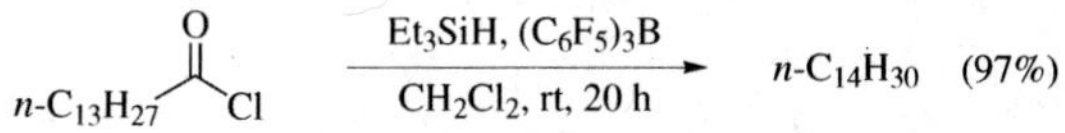

(Eq. 138)

One study of the Et_3SiH/TFA reduction of acid anhydrides reports the formation of one equivalent each of the alcohol and the trifluoroacetate ester of the acid (Eq. 139).[287]

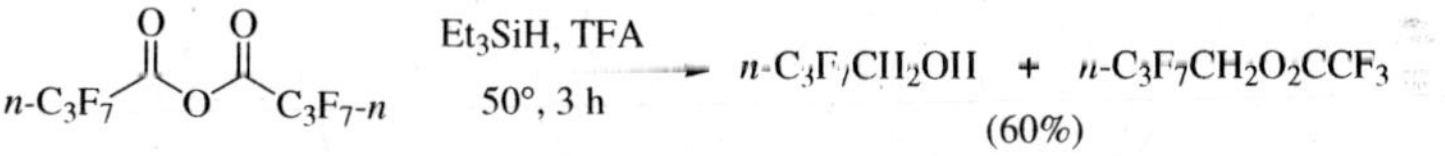

(Eq. 139)

Reduction of Esters and Lactones

The combination of $(EtO)_3SiH$/CsF (or KF) provides a convenient reagent for the reduction of esters to alcohols.[76,80,83] The yields are in the 70% range. Potassium tetraethoxyhydridosilicate also reduces esters in moderate yields.[288] The combination of PMHS/Cp_2TiCl_2/n-BuLi reduces esters in high yields even in the presence of an epoxide and a trisubstituted olefin (Eq. 140).[289] The reagent combination can reduce a methyl ester in the presence of a *tert*-butyl ester (Eq. 141).[290]

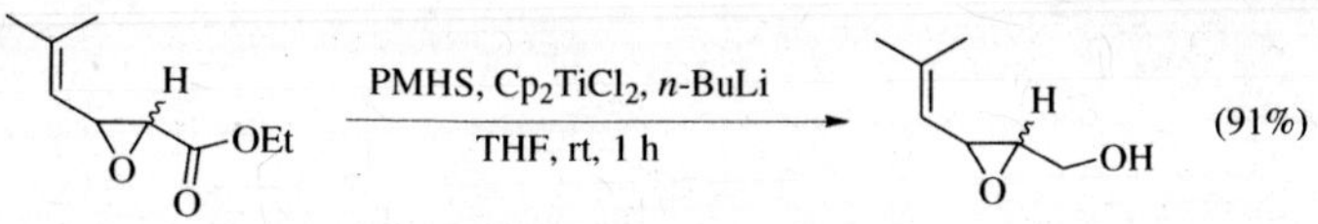

(Eq. 140)

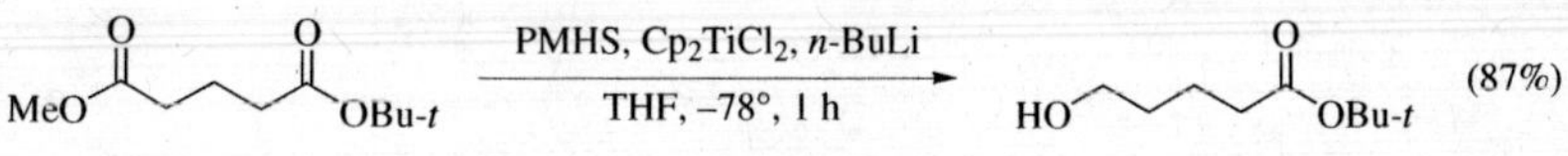

(Eq. 141)

Ester reductions with $(EtO)_3SiH/(EBTHI)TiCl_2/n\text{-}BuLi$ (EBTHI = ethylene-bis(η^5-tetrahydroindenyl)titanium) result in good yields of the corresponding alcohols.[290] Excellent yields of alcohols result from the reduction of esters with the PMHS/Ti(OPr-i)$_4$ system.[279,291,292] The reaction catalyzed by $(EBTHI)TiCl_2/$ n-BuLi occurs in lower yields.[289] Methyl cinnamate is reduced with PMHS/TBAF in good yield.[278] The ruthenium complex **57** (Eq. 135) catalyzes the $EtMe_2SiH$ reduction of esters to alcohols, although a mixture of the alcohol and the ether are often obtained (Eq. 142).[280]

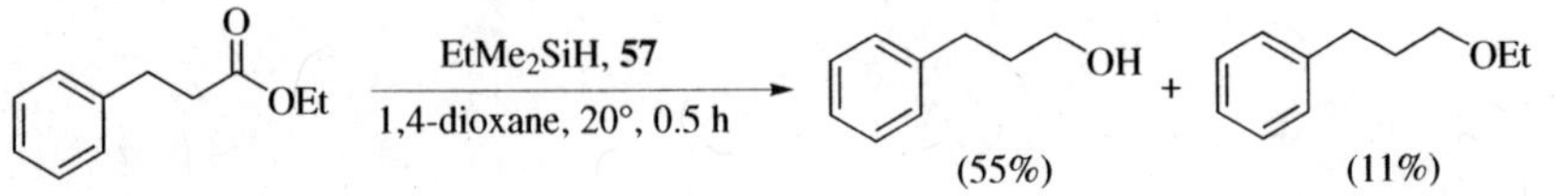

(Eq. 142)

High yields in the reduction of esters with $Ph_2SiH_2/[RhCl(cod)]_2$ are reported.[293] The combination of $(MeO)_3SiH/LiOMe$ is reported to reduce esters to the alcohols, although the advantages of this system over others does not seem to warrant working with the highly hazardous trimethoxysilane.[294] The reduction of the carbonyl group of an ester or lactone is possible. This results in the formation of the corresponding ether. This reaction can be carried out employing $PhSiH_3/(PPh_3)(CO)_4MnC(O)Me$,[295] $PhSiH_3/Mn(CO)_5Br$,[295] Cl_3SiH/γ-irradiation,[296] or $Et_3SiH/TiCl_4/TMSOTf$.[297] The reduction of lactones to cyclic ethers is nicely accomplished with the $EtMe_2SiH$/ruthenium catalyst system **57** (Eq. 143).[280] The same transformation can be carried out with $Et_3SiH/TiCl_4$[297] or $PhSiH_3/Mn(CO)_5Br$.[295] The reductive etherification of esters occurs by treating an ester with $Et_3SiH/ZnCl_2$ (Eq. 144).[298]

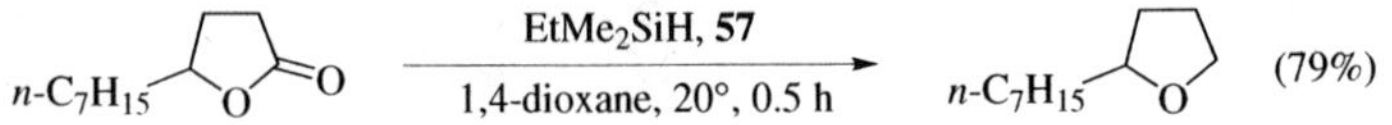

(Eq. 143)

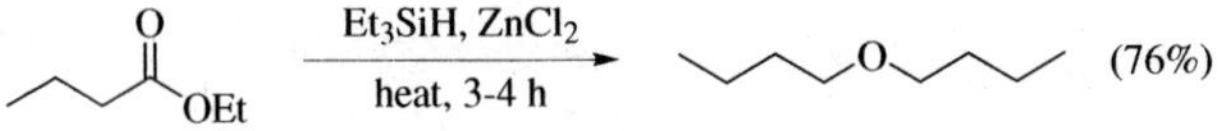

(Eq. 144)

The reduction of esters to aldehydes is a useful transformation and can be accomplished in good yields with $Et_3SiH/[RuCl_2(CO)_3]_2$[299] or with $Ph_3SiH/$ $(C_6F_5)_3B$.[116] Thio esters are reduced to aldehydes in good yields with $Et_3SiH/Pd/C$.[300] Lactones can be reduced to hemiacetals with $PMHS/Cp_2TiF_2$ or $PMHS/Cp_2Ti(OC_6H_4Cl\text{-}4)_2$ (Eq. 145).[301,302]

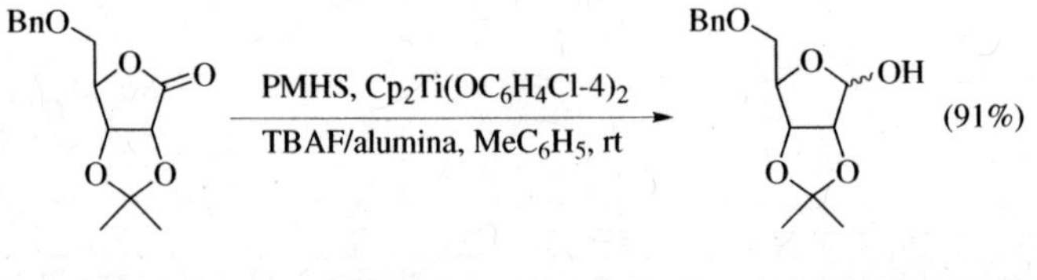

(Eq. 145)

The reduction of an ester to the silylated acetal occurs with $Et_3SiH/Et_2NH/[RuCl_2(CO)_3]_2$ (and other Ru catalysts) (Eq. 146),[299] $Et_3SiH/(C_6F_5)_3B$,[281,282] or $Ph_3SiH/(C_6F_5)_3B$,[115] and with $Ph_2MeSiH/Mn(CO)_5C(O)Me$.[295] The latter system reduces methyl benzoate to toluene. An intramolecular version of the ester to silylated acetal transformation is effected with TBAF (Eq. 147).[303,304]

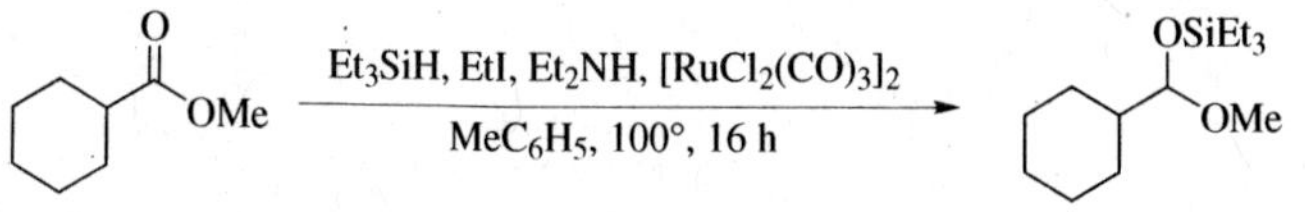

(92%) (Eq. 146)

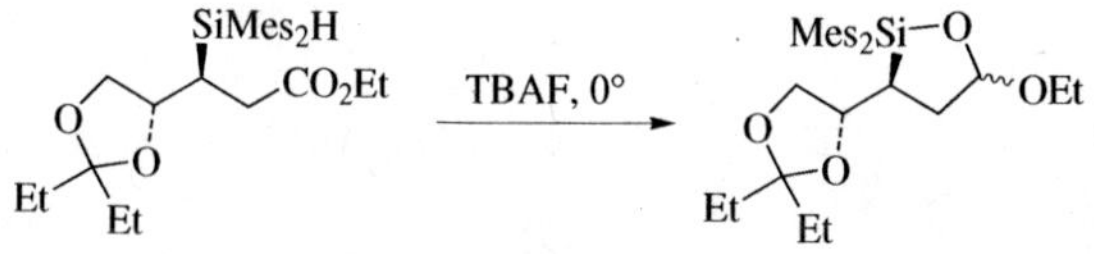

(91%) dr = 98:2 (Eq. 147)

The reaction of lactones of benzyl alcohols with Et_3SiH/TFA results in complete reduction of the alcohol part of the lactone to the methylene group while preserving the carboxylate function (Eq. 148).[305]

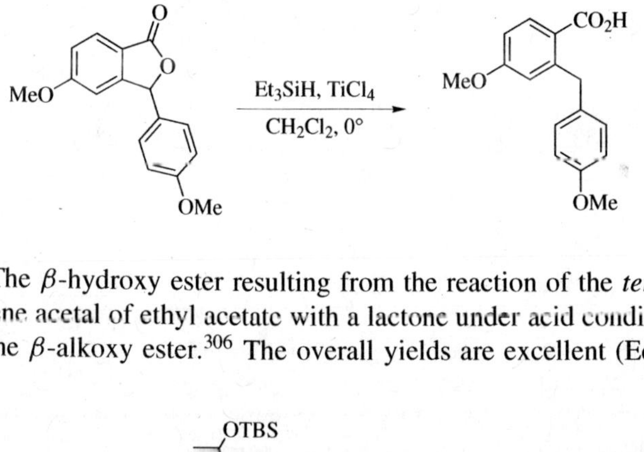

(87%) (Eq. 148)

The β-hydroxy ester resulting from the reaction of the *tert*-butyldimethylsilyl ketene acetal of ethyl acetate with a lactone under acid conditions can be reduced to the β-alkoxy ester.[306] The overall yields are excellent (Eq. 149).

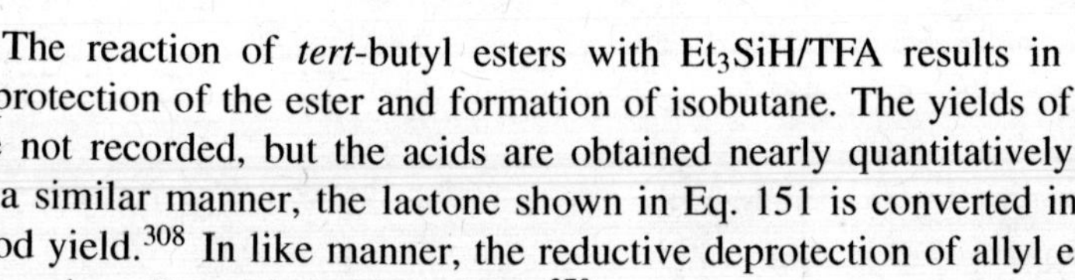

(Eq. 149)

(91%) cis:trans >99:1

The reaction of *tert*-butyl esters with Et_3SiH/TFA results in the reductive deprotection of the ester and formation of isobutane. The yields of the isobutane are not recorded, but the acids are obtained nearly quantitatively (Eq. 150).[307] In a similar manner, the lactone shown in Eq. 151 is converted into the acid in good yield.[308] In like manner, the reductive deprotection of allyl esters provides the carboxylic acids in high yields.[270]

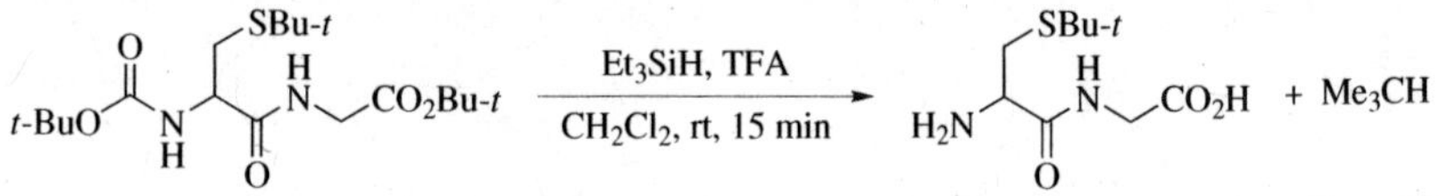

(100%) (Eq. 150)

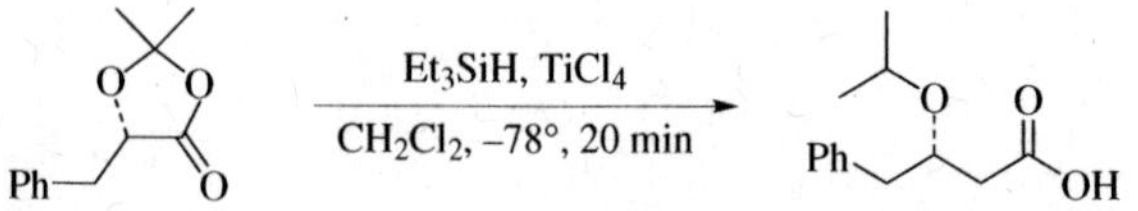

(91%) (Eq. 151)

The reduction of trifluoroacetates to alkanes occurs with the trifluoroacetates of benzylic and tertiary alcohols. This transformation is reported to occur with reagent combinations such as Ph_3SiH/nitrobenzene[193] and $EtCl_2SiH/AlBr_3$.[192] Secondary trifluoroacetates give more modest yields.[192]

Reduction of Aldehydes

Reduction to Alcohols. Aldehydes do not normally react spontaneously with organosilicon hydrides to form alcohols. Exceptional behavior is displayed with organocobalt cluster complex carbonyl compounds, which form the corresponding alcohols (R = H, Me, Ph, etc.) after treatment with one equivalent of triethylsilane in refluxing benzene under a carbon monoxide atmosphere and acid workup (Eq. 152).[309,310] Aside from these specific examples of anomalous behavior, the generally observed lack of reactivity is due to the combination of the relatively weak electrophilicity of the aldehyde carbonyl carbon center and the extremely feeble nucleophilicity of most tetravalent silyl hydrides. The reduction of aldehyde carbonyl groups by organosilicon hydrides can be promoted by several means. One way is by the introduction of acidic or electrophilic substances that coordinate with the carbonyl oxygen and thereby enhance the electrophilicity of the carbonyl carbon toward receiving a weakly nucleophilic silyl hydride. As mentioned previously, a second way is through the introduction into the reaction medium of substances possessing high nucleophilicity toward silicon centers. Such substances are thought to activate the silyl hydride by forming valence-expanded silicon species with enhanced hydride-donating properties capable of attacking even weakly electrophilic centers such as the carbonyl groups of common aldehydes. Both means of promotion can be synthetically useful.

$$(CO)_9Co_3CCOR \xrightarrow[\text{3. } H_2O]{\text{1. } Et_3SiH,\ C_6H_6,\ CO \quad \text{2. } H_2SO_4} (CO)_9Co_3CCH(OH)R \qquad \text{(Eq. 152)}$$

R = H (46%)
R = Me (88%)
R = Ph (68%)

Promotion by Acid. In principle, the reduction of aldehydes to alcohols and alcohol derivatives by organosilicon hydrides should occur upon exposure to either Lewis or Brønsted acids, as represented in Eq. 2. In practice, although organosilicon hydride reductions of either aliphatic or aromatic aldehydes do occur rapidly under acid conditions, they are frequently complicated by the formation of other products. The reductions rarely give clean yields of alcohols when conducted under anhydrous conditions. The reaction of a mixture of 1-butanal and triethylsilane that occurs upon addition of excess trifluoroacetic acid is a typical example. Analysis of the reaction mixture immediately following the addition of acid shows the formation of 37% of di-*n*-butyl ether along with *n*-butyl alcohol and *n*-butyl trifluoroacetate in a combined yield of 58% (Eq. 153).[311] No unreacted aldehyde remains. The same process transforms benzaldehyde into dibenzyl ether in 80% yield (Eq. 154).[311,312] In both reactions, the silicon-containing products are triethylsilyl trifluoroacetate and hexaethyldisiloxane. The Et_3SiH/TFA combination can also lead to the trifluoroacetate and toluene derivatives when used with some aryl aldehydes.[69]

$$n\text{-}C_3H_7CHO \xrightarrow[\text{TFA}]{Et_3SiH} \underset{(37\%)}{(n\text{-}C_4H_9)_2O} + \underset{(58\%)}{n\text{-}C_4H_9OH + n\text{-}C_4H_9O_2CCF_3} \qquad \text{(Eq. 153)}$$

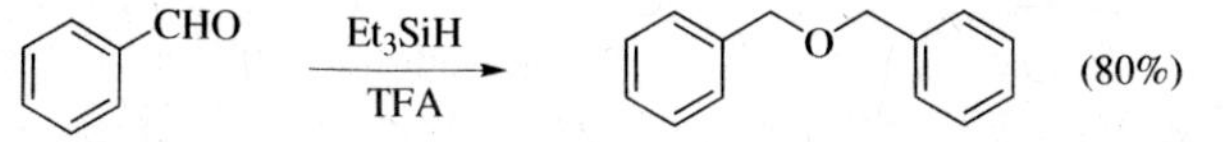

(Eq. 154)

The reduction of aldehydes with the combination $Et_3SiH/BF_3{\bullet}OEt_2$ gives both the alcohol and the symmetrical ether,[70] as do the Et_3SiH/TFA (and other acids) combinations.[313] Addition of boron trifluoride etherate to a mixture of 1-octanal and triethylsilane leads to the formation of di-*n*-octyl ether in 66% yield and *n*-octyl alcohol in 34% yield (Eq. 155).[74]

$$n\text{-}C_7H_{15}CHO \xrightarrow[BF_3{\bullet}OEt_2]{Et_3SiH} (n\text{-}C_8H_{17})_2O\ (34\%) + n\text{-}C_8H_{17}OH\ (66\%) \qquad \text{(Eq. 155)}$$

The addition of water and a non-hydrogen-bonding solvent to the reduction medium causes the reactions to shift toward the formation of alcohol products.[313] For example, triethylsilane in a mixture of concentrated hydrochloric acid and acetonitrile (5 : 4) reduces 1-heptanal to 1-heptanol in quantitative yield after 3 hours at room temperature. In a mixture of triethylsilane in sulfuric acid, water, and acetonitrile (2 : 2 : 5), *n*-heptanal gives a 97% yield of the same alcohol after 1.25 hours (Eq. 156).[313]

$$n\text{-}C_6H_{13}CHO \xrightarrow[H_2SO_4 \text{ or } HCl]{Et_3SiH,\ H_2O,\ MeCN} n\text{-}C_7H_{15}OH\ (97\%) \qquad \text{(Eq. 156)}$$

Triethylsilane reduces benzaldehyde to benzyl alcohol in 98% yield after 32 hours in a reaction medium containing sulfuric acid, water, and sulfolane (1 : 2 : 5) (Eq. 157). Neither benzene nor dimethylformamide is effective as an interfacing solvent for producing alcohol products under these conditions.[313]

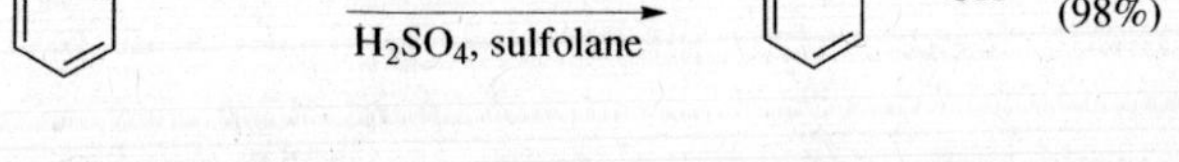

(Eq. 157)

In contrast to the propensity of Brønsted and some Lewis acids such as boron trifluoride etherate to promote the organosilicon hydride reduction of aldehydes to ethers under anhydrous conditions, uncomplexed boron trifluoride used with triethylsilane in dichloromethane solvent leads to the formation of primary alcohols in good yields from aliphatic aldehydes and from aromatic aldehydes containing electron-withdrawing groups.[1] The success of this method depends on the absence of significant quantities of Brønsted acids in the reaction medium and requires that the boron trifluoride gas be scrubbed of hydrogen fluoride prior to introduction. Using this method, 1-undecanal gives a 92% isolated yield of 1-undecanol after only 10 minutes at 0° (Eq. 158).

$$n\text{-}C_{10}H_{21}CHO \xrightarrow[BF_3,\ CH_2Cl_2]{Et_3SiH} n\text{-}C_{11}H_{23}OH \quad (92\%) \qquad \text{(Eq. 158)}$$

The same technique causes the transformation of *p*-anisaldehyde (R = MeO) and *p*-cyanobenzaldehyde (R = CN) into the corresponding substituted benzyl alcohols in quantitative yields within 10 minutes at 0° (Eq. 159).[1] The reduction of aryl aldehydes to benzyl alcohols without over-reduction to the arylmethanes also occurs with the reagent combinations PMHS/TBAF,[278] PMHS/Triton® B,[278] and $Ph_3SiH/(C_6F_5)_3B$.[116] The $Ph_3SiH/(C_6F_5)_3B$ combination can be used to isolate the benzyl silyl ethers.[282] Treatment of *p*-nitrobenzaldehyde ($R = NO_2$) with a catalytic amount of the Lewis acid trimethylsilyl iodide (TMSI, generated in situ from trimethylsilyl chloride and sodium iodide) and tetramethyldisiloxane gives the benzyl alcohol in 91% isolated yield.[314]

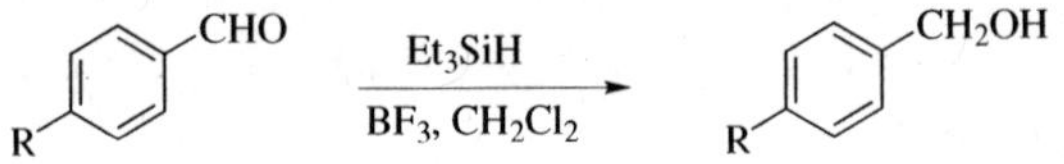

(Eq. 159)

R	
MeO	(100%)
CN	(100%)

The reagent combinations $PMHS/ZnCl_2$,[315] $PMHS/[Bu_2(AcO)Sn]_2O$,[316] and $PMHS/HCuPPh_3$[317] all promote reduction of aldehydes to the corresponding alcohols in good yields. Trichlorosilane in dimethylformamide reduces aldehydes to alcohols in high yields.[318]

Promotion by Valence Expansion. Addition of nucleophilic substances to mixtures of aldehydes and organosilicon hydrides promotes the reduction of the carbonyl group as depicted previously in Eq. 6. The reductions can occur under homogeneous[83] or heterogeneous[79,80,319] conditions, both with[83,320] and without solvent.[83,319] When the reactions occur under anhydrous conditions with catalytic amounts of nucleophile, the first-formed product is frequently a silyl ether. This ether can be regarded as an intermediate that normally undergoes facile acid- or base-catalyzed hydrolysis to give a final alcohol product (Eq. 160).[80] The silicon-containing products are usually silanols and/or disiloxanes produced by hydrolysis of the intermediate silyl ethers. These reductions are normally quite chemoselective and tolerate many other functional groups.

$$R_3SiH + R'CHO \xrightarrow{Nu^-} R_3SiOCH_2R \xrightarrow{H_2O} (R_3Si)_2O + R_3SiOH + R'CH_2OH \quad \text{(Eq. 160)}$$

R = H, alkyl, aryl
R' = alkyl, alkenyl, aryl

Fluoride ion is effective in promoting the reduction of aldehydes by organosilicon hydrides (Eq. 161). The source of fluoride ion is important to the efficiency of reduction. Triethylsilane reduces benzaldehyde to triethylbenzyloxysilane in 36% yield within 10–12 hours in anhydrous acetonitrile solvent at room temperature when tetraethylammonium fluoride (TEAF) is used as the fluoride ion source and in 96% yield when cesium fluoride is used.[83] The carbonyl functions of both *p*-anisaldehyde and cinnamaldehyde are reduced under similar conditions. Potassium bromide or chloride, or tetramethylammonium bromide or chloride are not effective at promoting similar behavior under these reaction conditions.[83] Moderate yields of alcohols are obtained by the KF-catalyzed PMHS, $(EtO)_3SiH$, or $Me(EtO)_2SiH$ reduction of aldehydes.[80,83,79]

$$ArCHO + Et_3SiH \xrightarrow[MeCN]{F^-} ArCH_2OSiEt_3 \quad Ar = Ph\ (36\text{-}96\%) \quad \text{(Eq. 161)}$$

Diphenylsilane reacts with two equivalents of neat *n*-heptanal in the presence of anhydrous cesium fluoride within three minutes at room temperature to form di-*n*-heptoxydiphenylsilane quantitatively (Eq. 162).[319] Potassium fluoride and potassium phthalate are considerably less effective promoters, even at temperatures up to 140°.[319]

$$2\ n\text{-}C_6H_{13}CHO + Ph_2SiH_2 \xrightarrow{CsF} (n\text{-}C_7H_{15}O)_2SiPh_2\ (100\%) \quad \text{(Eq. 162)}$$

Alkoxy-substituted organosilicon hydrides are more reactive toward carbonyl functions in the presence of nucleophiles than are organosilicon hydrides that have only alkyl or aryl substituents at the silicon center. The order of reactivity of the silanes used is $(EtO)_3SiH > (EtO)_2SiMeH$, and that of the fluoride salts is CsF > KF.[83] The use of these silane/fluoride salt pairs can lead to some very chemoselective transformations.[79,80,319] For example, after hydrolytic workup, an equimolar mixture of benzaldehyde, diethoxymethylsilane, and cesium fluoride gives an 80% yield of benzyl alcohol after only 10 minutes at room temperature under heterogeneous conditions.[80] The use of triethoxysilane and potassium fluoride gives a 90% yield of benzyl alcohol after six hours at room temperature. The same combination of reagents converts 1-heptanal into 1-heptanol in 70% yield within four hours without affecting benzophenone or 1,3-diphenylpropan-2-one when either is added to the same reaction mixture (Eq. 163).[83]

$$n\text{-}C_6H_{13}CHO + (EtO)_3SiH \xrightarrow[2.\ H_2O]{1.\ KF} n\text{-}C_7H_{15}OH \quad (70\%) \quad \text{(Eq. 163)}$$

These reaction conditions also permit the chemoselective quantitative reduction of benzaldehyde to benzyl alcohol without any concomitant reduction of either acetophenone or 3,3-dimethylbutan-2-one present in the same reaction mixture.[83] Additionally, this useful method permits the reduction of aldehyde functions in polyfunctional compounds without affecting amide, anhydride, ethylenic, bromo, chloro, or nitro groups.[79,80,319]

An improved variation of this reduction method involves the use of potassium fluoride (either anhydrous or as the dihydrate) or potassium formate in a polar aprotic solvent such as dimethylformamide or dimethyl sulfoxide in conjunction with either diethoxymethylsilane or PMHS. The intermediate silyl ethers are worked up by acidic hydrolysis when diethoxymethylsilane is used and by methanolysis when PMHS is the reducing agent.[82] A high chemoselectivity among carbonyl group reductions may be accomplished using this method by adjusting the reaction conditions. Aldehydes are especially easy to reduce this way.

The combination of diethoxymethylsilane and KF in dimethylformamide produces a 90% yield of benzyl alcohol from benzaldehyde in 0.25 hour at 20° following workup.[82] In a similar way, 1-heptanal forms 1-heptanol in 85% yield within 1.75 hours at 10°. Other combinations of the salts, solvents, and organosilicon hydrides give useful, if somewhat lower, yields of products. Potassium fluoride dihydrate, although a more active catalyst than the anhydrous salt, requires use of an excess of organosilicon hydride because the water that is present destroys some of the silane. Use of potassium formate and PMHS in dimethylformamide permits facile selective reduction of both alkyl and aryl aldehydes in the presence of ketones and esters.[82] The system of $Me(EtO)_2SiH$ and KO_2CH is very selective toward the reduction of aldehydes in the presence of ketones.[82] In a similar approach aldehydes are reduced with $[HSi(OEt)_4]K$.[288]

Fluoride ion catalyzes the hydrosilylation of both alkyl and aryl aldehydes to silyl ethers that can be easily hydrolyzed to the free alcohols by treatment with 1 M hydrogen chloride in methanol.[320] The most effective sources of fluoride are TBAF and tris(diethylamino)sulfonium difluorotrimethylsilicate (TASF). Somewhat less effective are CsF and KF. Solvent effects are marked. The reactions are facilitated in polar, aprotic solvents such as hexamethylphosphortriamide (HMPA) or 1,3-dimethyl-3,4,5,6-tetrahydro-2(1*H*)-pyrimidinone (DMPU), go moderately well in dimethylformamide, but do not proceed well in either tetrahydrofuran or dichloromethane. The solvent effects are dramatically illustrated in the reaction of undecanal and dimethylphenylsilane to produce undecyloxyphenyldimethylsilane. After one hour at room temperature with TBAF as the source of fluoride and a 10 mol% excess of silane, yields of 91% in HMPA, 89% in DMPU, 56% in dimethylformamide, 9% in tetrahydrofuran, and only 1% in dichloromethane are obtained (Eq. 164).[320]

$$n\text{-}C_{10}H_{21}CHO + PhMe_2SiH \xrightarrow[\text{rt}]{\text{TBAF}} n\text{-}C_{11}H_{23}OSiMe_2Ph \quad (1\text{-}91\%) \qquad \text{(Eq. 164)}$$

The reduction of aldehydes to alcohols takes place under mild conditions upon treatment with a mixture of trimethoxysilane and lithium methoxide (20 mol%

excess of each) in diethyl ether at room temperature (Eq. 165). The reaction occurs with both alkyl and aryl aldehydes and can be used to reduce aldehydes in the presence of ketones, esters, and nitriles. Workup is by treatment with 1 M aqueous hydrochloric acid.[91] For example, benzaldehyde forms benzyl alcohol in 85% isolated yield within 20 hours under these conditions, whereas *o*-nitrobenzaldehyde and *p*-anisaldehyde give the corresponding alcohols in yields of 55 and 86%, respectively. 1-Octanal yields 1-octanol in 80% yield after just six hours.[91] Triethoxysilane and diethoxymethylsilane are not as effective as reducing agents as trimethoxysilane. Sodium methoxide, alkali metal ethoxides, and, especially, potassium methoxide also are effective nucleophilic promoters. Lithium and sodium pinacolates are strong promoters that cause the reduction of both aldehydes and ketones.[91]

$$RCHO + (MeO)_3SiH \xrightarrow[Et_2O,\ rt]{MeOLi} RCH_2OH \quad \begin{array}{l} R = n\text{-}C_7H_{15}\ (80\%) \\ R = Ph\ (85\%) \\ R = 2\text{-}O_2NC_6H_4\ (55\%) \\ R = 4\text{-}MeOC_6H_4\ (86\%) \end{array} \qquad \text{(Eq. 165)}$$

There seems little doubt that the active reducing agents in these kinds of reductions are pentavalent hydridosilicates. In fact, it is possible to produce the stable potassium salts of these species in high yield by reacting equivalent amounts of the appropriate trialkoxysilanes and potassium alkoxides in large amounts of tetrahydrofuran or 1,2-dimethoxyethane (DME) at room temperature (Eq. 166).[107] A variety of alkoxy groups may be used (R = Et, *i*-Pr, Ph), but neither lithium nor sodium alkoxides are effective in this reaction.[107] Potassium tetraethoxyhydrosilicate shows high reducing properties toward both aldehydes and ketones without the need for added catalysts (Eq. 167).[288] It reduces benzaldehyde to benzyl alcohol in 90% yield and 1-pentanal to 1-pentanol in 80% yield following aqueous acid workup.[288]

$$\underset{R = Et,\ i\text{-}Pr,\ Ph}{(RO)_3SiH + ROK} \xrightarrow[\text{or DME}]{THF} [(RO)_4SiH]^- K^+ \quad (80\text{-}90\%) \qquad \text{(Eq. 166)}$$

$$[(EtO)_4SiH]^- K^+ + RCHO \xrightarrow[2.\ H_3O^+]{1.\ THF,\ 0^\circ\ \text{or rt}} RCH_2OH \quad \begin{array}{l} R = n\text{-}Bu\ (80\%) \\ R = Ph\ (90\%) \end{array} \qquad \text{(Eq. 167)}$$

A similar reducing system is created by combining dilithium catecholate and trichlorosilane at −78° in tetrahydrofuran. It is speculated that the relatively unstable pentacoordinate bis(1,2-benzenediolato)hydridosilicate (**61**) is formed in situ and that it is this species that can reduce aldehydes and ketones, but not esters, to alcohols when they are added to the reaction mixture at 0° (Eq. 168).[93] In a like manner, the dilithium salt of 2,2′-dihydroxybiphenyl, which forms a pentacoordinate intermediate that is stable enough to react at room temperature, can also be used to promote the reduction reaction. The alkoxides of aliphatic diols such as 1,2-ethanediol and pinacol are not very effective as ligand promoters in this system and those of simple alcohols are without effect. Use of the dilithium catecholate/trichlorosilane combination gives benzyl alcohol from benzaldehyde in 96% yield within two hours. Substitution of 2,2′-dihydroxybiphenyl for catechol provides a 92% yield of 2,2-dimethylpropanol from 2,2-dimethylpropanal within five hours at room temperature.[93]

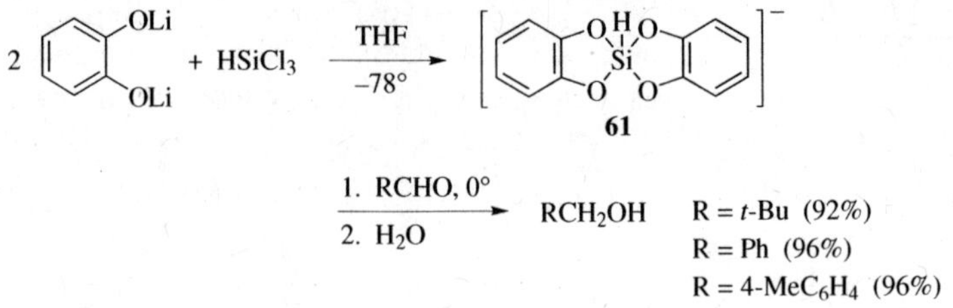

(Eq. 168)

Chemoselectivity between aldehydes and ketones is demonstrated by this method in the competitive reduction of a mixture of pentanal and cyclohexanone. The ratios of primary and secondary alcohols are 75 : 25 when catechol is used at 0° and 79 : 21 when 2,2′-dihydroxybiphenyl is used at room temperature. These regents are not as chemoselective as other reducing agents such as $LiAlH(OBu\text{-}t)_3$ (87 : 13) and $LiAlH(OCEt_3)_3$ (94 : 6) at 0°.[93]

Several types of organosilicon hydrides are effective reducing agents toward carbonyl functions because of valence expansion produced by intramolecular effects. Aryl silyl hydrides with amine functions are especially prone to having the proper configuration to permit such intramolecular valence expansion.[321,322] The valence expanded silicon hydrides compounds **58**–**60** react spontaneously with both *p*-nitrobenzaldehyde and *p*-anisaldehyde to give, within 0.5 to 3 days, the respective benzyl alcohols in quantitative yields following aqueous acidic workup.[321] Under the same conditions, a mixture of α-naphthylphenylsilane and *N*,*N*-dimethylbenzylamine fails to react even after 17 days.[321] It is of interest to note that the silyl hydrides **58**, **59**, and **60** (Eqs. 136 and 137) all have trigonal bipyramidal structures in which the active hydrogens occupy equatorial positions. Compound **58** is such an effective carbonyl reducing agent that it reduces carbon dioxide to formaldehyde via a stable silylformate intermediate.[323]

The 10-*Si*-5-hydridosiliconate ion **62** is known in association with lithium,[323] tetrabutylammonium,[101] and bis(phosphoranyl)iminium[93] cations. It is synthesized by hydride addition to the 8-*Si*-4-silane **63**, which is derived from hexafluoroacetone.[101] Benzaldehyde and related aryl aldehydes are reduced by solutions of **62** in dichloromethane at room temperature[101] or in tetrahydrofuran at 0°[96] within two hours. The alkyl aldehyde, 1-nonanal, is also reduced by **62** in tetrahydrofuran at 0°.[96] Good to excellent yields of the respective alcohols are obtained following hydrolytic workup. The reactions are not accelerated by addition of excess lithium chloride,[96] but neutral **63** catalyzes the reaction, apparently through complexation of its silicon center with the carbonyl oxygen prior to delivery of hydride from **62**.[101]

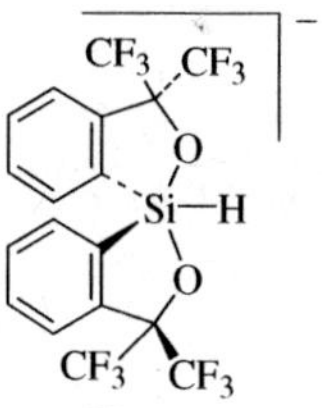

62

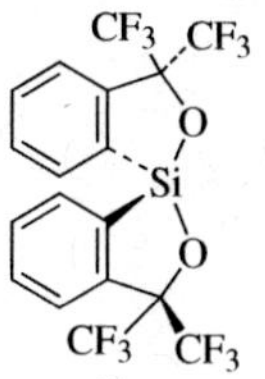

63

The solid bases CaO and hydroxyapatite catalyze the hydrosilylation of benzaldehyde by triethoxysilane at 90° in yields of 59% and 72% within one and two hours, respectively.[323,324] These reductions also very likely involve activation by valence expansion of the silicon hydride reagent.

Reductive Amidation of Aldehydes. The reductive amidation of aldehydes using an organosilane as the reducing agent has been realized. Benzaldehyde reacts over a 74-hour period with triethylsilane and acetonitrile in 75% aqueous sulfuric acid at room temperature to produce an 80% isolated yield of *N*-benzylacetamide (Eq. 169).[313] Octanal fails to react under the same conditions.[313] Reductive amidation of aldehydes also occurs with the reagent combination Et_3SiH/TFA/primary amide (Eq. 170).[326]

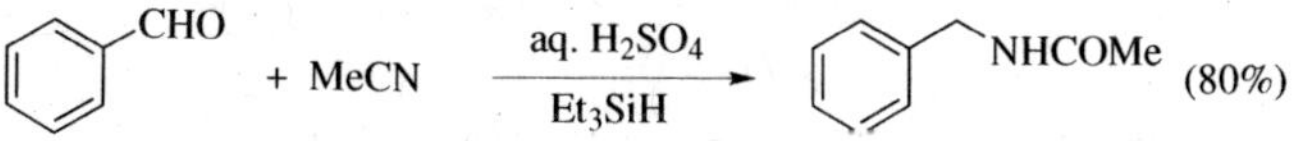

(Eq. 169)

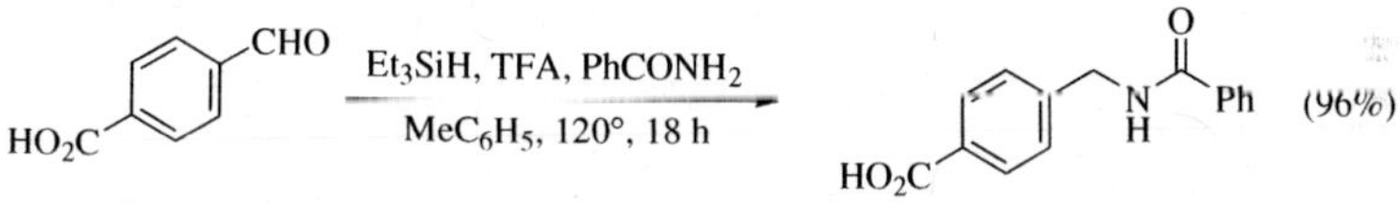

(Eq. 170)

Reductive Esterification. Aldehydes can give ester products when treated with combinations of organosilicon hydrides and carboxylic acids that have appreciable basicity. Benzaldehyde gives a product mixture consisting of 12% dibenzyl ether and 88% of benzyl formate when it is treated for 8 hours at room temperature with a slight excess of triethylsilane in formic acid.[313] *p*-Nitrobenzaldehyde produces 33% bis(*p*-nitrobenzyl) ether and 66% of *p*-nitrobenzyl trifluoroacetate when it reacts with Et_3SiH/TFA for 5 hours at room temperature. Other aldehydes give small, variable amounts of esters under similar reaction conditions. Although this general approach to the synthesis of esters from aldehydes is an attractive one, it appears not yet to be optimized for maximum synthetic utility because of the frequent formation of considerable amounts of ether products (Eq. 171).[313]

$$RCHO + R'CO_2H \xrightarrow{Et_3SiH} (RCH_2)_2O + RCH_2O_2CR' \quad \text{(variable)} \qquad \text{(Eq. 171)}$$

R = alkyl, aryl

R' = H, CF_3, Cl_2CH

Reductive Etherification. As indicated earlier, aldehydes as well as ketones often give very good yields of ethers when they are treated with Brønsted acids or other electrophilic species in the presence of organosilicon hydrides (Eq. 172). In the absence of added alcohols, symmetrical ethers are obtained.

$$\mathrm{RCHO} \xrightarrow[\mathrm{HX}]{\mathrm{R'_3SiH}} \mathrm{(RCH_2)_2O} \qquad \text{(Eq. 172)}$$

When alcohols are added to the reaction mixture, unsymmetrical ether products may be obtained. Starting with a mixture of aldehydes can also give rise to the formation of unsymmetrical ethers. These ether products are formed under conditions different from those used in the formation of ethers directly from alcohols. Thus, it is postulated that the reaction sequence that leads from the carbonyl substrate to the ether involves the intermediate formation of hemiacetals, acetals, or their protonated forms and alkoxycarbenium ions, which are intercepted and reduced to the final ether products by the organosilicon hydrides present in the reaction mix. The probable mechanistic scheme that is followed when Brønsted acids are present is outlined in Scheme 2.[311,327,328]

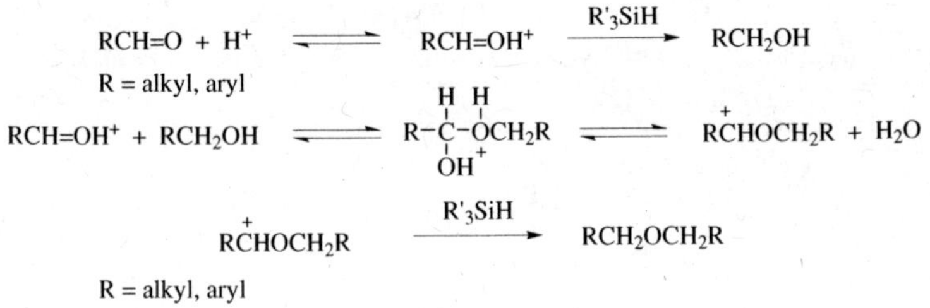

Scheme 2

Reduction of aldehydes to symmetrical ethers can be accomplished in good to excellent yields with $Et_3SiH/Ph_3C^+ ClO_4^-$,[329] $Et_3SiH/ZnCl_2$,[330] $Me_2ClSiH/In(OH)_3$,[331] $Et_3SiH/BiCl_3$,[332] $(HMe_2Si)_2O$/TMSOTf (or TMSCl/NaI),[314] Et_3SiH (or $PhMe_2SiH$)/Bu_4NClO_4,[333] Et_3SiH/TMSOTf,[334] Et_3SiH/H_2SO_4,[328] and Et_3SiH/TFA.[313] The reaction of 1.4 equivalents of triethylsilane with two equivalents of trifluoroacetic acid rapidly reduces benzaldehyde to dibenzyl ether in 80% yield at temperatures below 40°.[311] Similar treatment of *n*-butanal with two equivalents of triethylsilane and three equivalents of trifluoroacetic acid produces di-*n*-butyl ether in a more modest 37% yield.[311] Variations of these simple reaction conditions permit greater yields of desired ether products to be obtained. For example, 1-heptanal reacts with a 10 mol% excess of triethylsilane to give a 90% yield of di-*n*-heptyl ether within 45 minutes at room temperature when the reaction is run in a twenty-fold excess of trifluoroacetic acid acting as solvent.[313] With the exception of *p*-nitrobenzaldehyde, which gives only a 33% yield of the symmetrical ether (the remainder is converted into *p*-nitrobenzyl trifluoroacetate), other representative aryl aldehydes normally give yields of symmetrical ethers on the order of 80% or greater.[313]

Unsymmetrical ethers may be produced from the acid-promoted reactions of aldehydes and organosilicon hydrides when alcohols are introduced into the reaction medium (Eq. 173).[327,328] An orthoester can be used in place of the alcohol in this transformation.[327,335] A cyclic version of this conversion is reported.[336] Treatment of a mixture of benzaldehyde and a 10 mol% excess of triethylsilane with methanol and sulfuric, trifluoroacetic, or trichloroacetic acid produces benzyl methyl ether in 85–87% yields.[328] Changing the alcohol to ethanol, 1-propanol, 2-propanol, or 1-heptanol gives the corresponding unsymmetrical benzyl alkyl ethers in 45–87% yield with little or no side products.[328] A notable exception is the tertiary alcohol 2-methyl-2-propanol, which requires 24 hours.[328] 1-Heptanal gives an 87% yield of *n*-heptyl methyl ether with added methanol and a 49% yield of benzyl *n*-heptyl ether with added benzyl alcohol under similar conditions.[328]

$$\mathrm{RCHO} + \mathrm{R''OH} \xrightarrow[\mathrm{HX}]{\mathrm{R'_3SiH}} \mathrm{RCH_2OR''} \quad (45\text{-}87\%) \qquad \text{(Eq. 173)}$$

R = alkyl, aryl; R' = aryl
R'' = alkyl

The yield of ethyl *n*-pentyl ether formed from the reduction of 1-pentanal by Et_3SiH/TFA in ethanol is 57% after 6–8 hours at 50–60°.[327] The yield of product increases to 72% when one equivalent of ethyl orthoformate and some anhydrous hydrogen chloride are added to the reaction medium.[327] Presumably, this reduces the amount of free water in the reaction medium.

An interesting and effective variation of this general synthetic approach uses electrogenerated acid (EG acid) to assist in the formation of ethers from aldehydes.[333] This method permits the synthesis of both symmetrical and unsymmetrical ethers. The experiments are conducted using platinum electrodes in a simple undivided cell. A mixture of aldehyde and a 20 mol% excess of either triethylsilane or dimethylphenylsilane in dichloromethane solvent containing lithium perchlorate and tetra-*n*-butylammonium perchlorate is electrolyzed by the passage of small amounts of current (0.04–0.45 Faradays/mol) to give symmetrical ethers (Eq. 174). In this way, both dibenzyl and dialkyl ethers may be produced in excellent yields (86–96%).[333] Unsymmetrical ethers are produced in 50–99% yields when alkoxytrimethylsilanes are added to the reaction mixture (Eq. 175).[333] The alkoxy groups can include allyl, propargyl, and 3-phenylpropyl moieties. Phenol trimethylsilyl ether is ineffective in producing phenyl ethers.[333]

$$\mathrm{RCHO} \xrightarrow[\text{EG acid}]{\mathrm{R'_3SiH}} \mathrm{RCH_2OCH_2R} \quad (86\text{-}96\%) \qquad \text{(Eq. 174)}$$

R = alkyl, Ph; R' = Me, Et, Ph

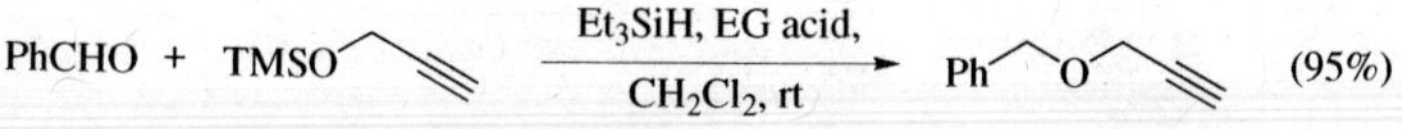

(Eq. 175)

Various chemical species with Lewis acid properties are also effective in promoting the direct conversion of aldehydes into ethers by organosilicon hydrides.

They offer the advantage that reductions can be effected under conditions that permit the conversion of substrates that may be adversely sensitive to the presence of strong Brønsted acids. For example, in the presence of a 10% excess of triethylsilane, addition of one-half equivalent of boron trifluoride etherate to octanal results, within one hour, in the formation of a 66% yield of dioctyl ether after a basic hydrolytic workup. Benzaldehyde provides a 75% yield of dibenzyl ether under the same reaction conditions. The remainder of the mass is found as the respective alcohol.[70] Zinc chloride is also capable of catalyzing this reaction. With its use, simple alkyl aldehydes are converted into the symmetrical ethers in about 50% yields.[330]

Superior yields of ethers from aldehydes are obtained by the use of several other electrophilic species. The addition of 5 mol% of trityl perchlorate to a mixture of triethylsilane and 3-phenylpropanal in dichloromethane at 0° produces an 83% yield of bis-(3-phenylpropyl) ether within 10 minutes (Eq. 176).[329] Reductive polycondensation of isophthalaldehyde occurs with two equivalents of triethylsilane in the presence of 10 mol% of trityl perchlorate to give 40–72% yields of polyether with average molecular weights ranging from 6,500 to 11,400 daltons (Eq. 177).[337] Addition of one equivalent of an alkoxytrimethylsilane to the reaction mixture produces unsymmetrical ethers in good to excellent yields. Thus, a mixture of (*E*)-cinnamaldehyde, 3-phenylpropoxytrimethylsilane, and triethylsilane in dichloromethane reacts under the influence of a catalytic amount of trityl perchlorate to give the unsymmetrical ether in 88% yield (Eq. 178).[329]

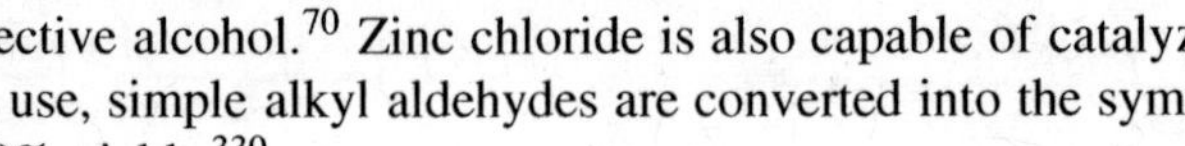

(Eq. 176)

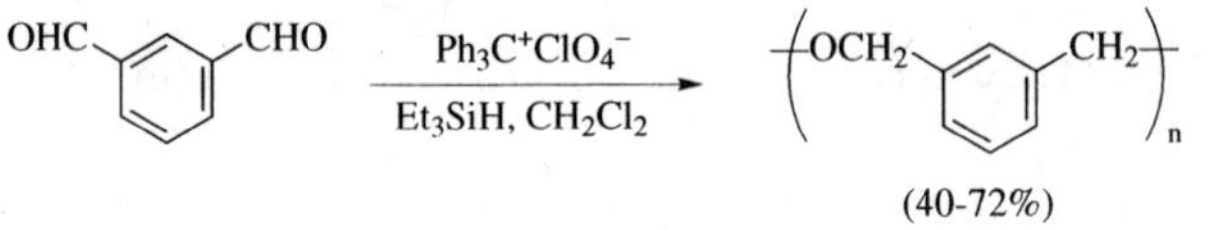

(Eq. 177)

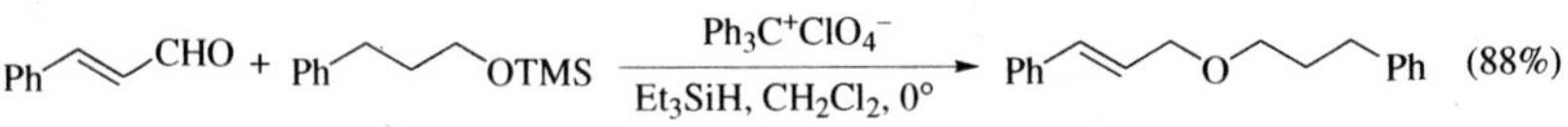

(Eq. 178)

The use of trimethylsilyl-based electrophilic catalysts with organosilicon hydrides also promotes the conversion of aldehydes into ethers and avoids the need to employ the potentially hazardous trityl perchlorate salt.[314,334,338] One reagent pair that is particularly effective in the reductive conversion of aldehydes into symmetrical ethers is a catalytic amount of trimethylsilyl triflate combined with either trimethylsilane, triethylsilane, PMHS,[334] or 1,1,3,3-tetramethyldisiloxane (TMDO, **64**) as the reducing agent (Eq. 179).[314] Either

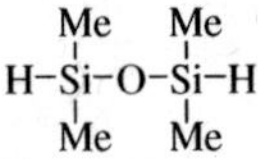

64

dichloromethane or benzene can be used as the solvent. The reactions occur at temperatures ranging from 0° to 80°. These conditions produce symmetrical ethers from both aromatic and aliphatic aldehydes in yields frequently exceeding 90%. Aromatic aldehydes tend to give minor amounts of benzyl alcohols as by-products.[334] The synthesis of cyclic ethers from dialdehydes or keto aldehydes is also possible (Eqs. 180 and 181).[339]

$$\underset{\text{R = alkyl, aryl}}{\text{RCHO}} \xrightarrow[\text{TMSOTf or TMSI}]{(\text{HMe}_2\text{Si})_2\text{O}} (\text{RCH}_2)_2\text{O} \qquad \text{(Eq. 179)}$$

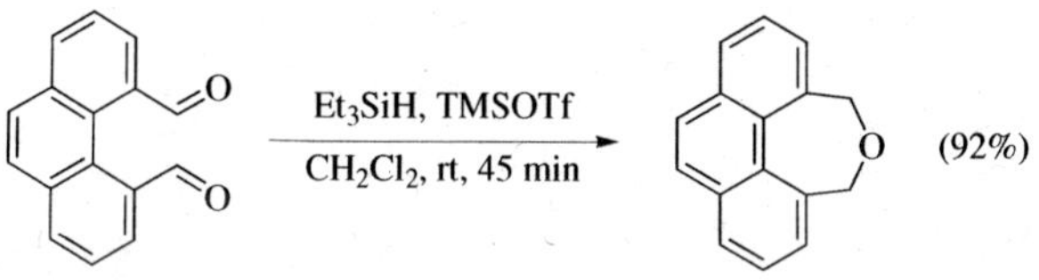

(Eq. 180)

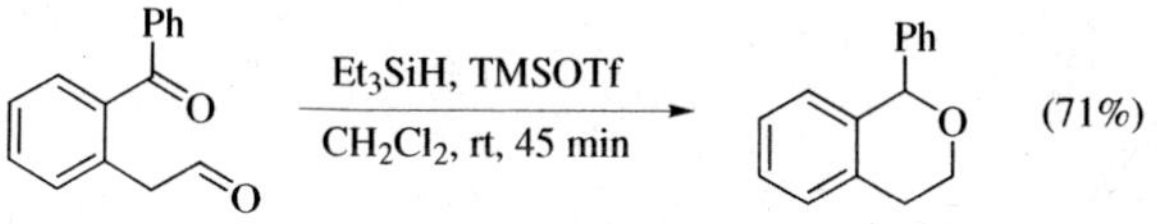

(Eq. 181)

The formation of unsymmetrical ethers from the reduction of aldehydes in the presence of tetrahydropyran (THP) ethers is reported (Eq. 182).[340]

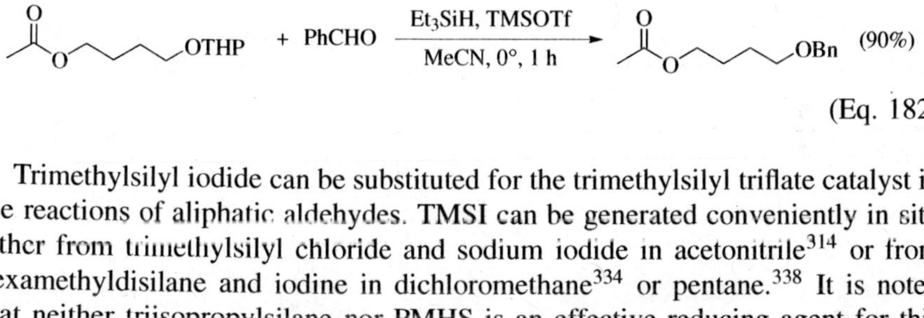

(Eq. 182)

Trimethylsilyl iodide can be substituted for the trimethylsilyl triflate catalyst in the reactions of aliphatic aldehydes. TMSI can be generated conveniently in situ either from trimethylsilyl chloride and sodium iodide in acetonitrile[314] or from hexamethyldisilane and iodine in dichloromethane[334] or pentane.[338] It is noted that neither triisopropylsilane nor PMHS is an effective reducing agent for this purpose when used with TMSI under these conditions.[314,334]

Equivalent amounts of aldehydes and alkoxytrimethylsilanes react to form unsymmetrical ethers in near quantitative yields in the presence of either trimethylsilane or triethylsilane and catalytic amounts (ca. 10 mol%) of TMSI in dichloromethane.[329,333,334,341] The procedure is particularly convenient experimentally when trimethylsilane is used with TMSI because the catalyst provides its own color indicator for the reduction step (color change from deep violet to vivid red-gold) and the only silicon-containing product following aqueous workup is the volatile hexamethyldisiloxane (bp 99–100°). It is possible to introduce trimethylsilane (bp 7°) either as a previously prepared solution in dichloromethane or by bubbling it directly into the reaction mixture. Cyclohexyloxytrimethylsilane and *n*-butanal react by this method to give a 93% isolated yield of *n*-butyl cyclohexyl ether (Eq. 183).[334]

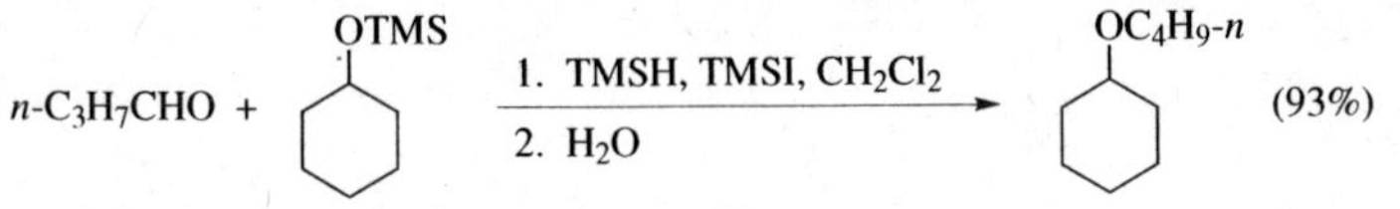

(Eq. 183)

Trimethylsilane in pentane is a particularly good system for the TMSI-catalyzed reductive coupling of tertiary alkoxytrimethylsilanes with aldehydes to form sterically crowded tertiary-primary ethers.[337] In this way, 1-(*tert*-butoxymethyl)-3-methylbenzene is formed in 87% yield (Eq. 184).[338] Reaction of terephthaldehyde with two equivalents of the trimethylsilyl ether of 1-adamantanol under these conditions leads to a good yield of the diadamantyl ether of 1,4-benzenedimethanol (Eq. 185).[338]

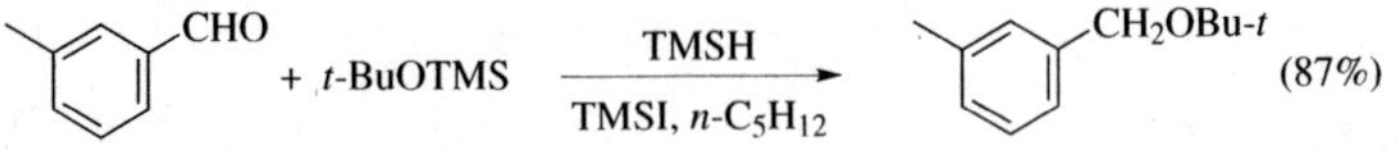

(Eq. 184)

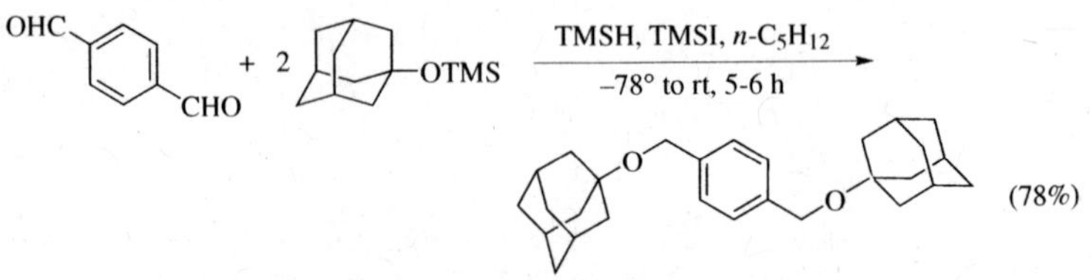

(Eq. 185)

Cyclic ethers can also be formed in a fashion similar to that of the reactions described previously (Eq. 186),[306,342] and also result from the reductive etherification of bis(trimethylsilylated) diols and dialdehydes (Eq. 187).[343]

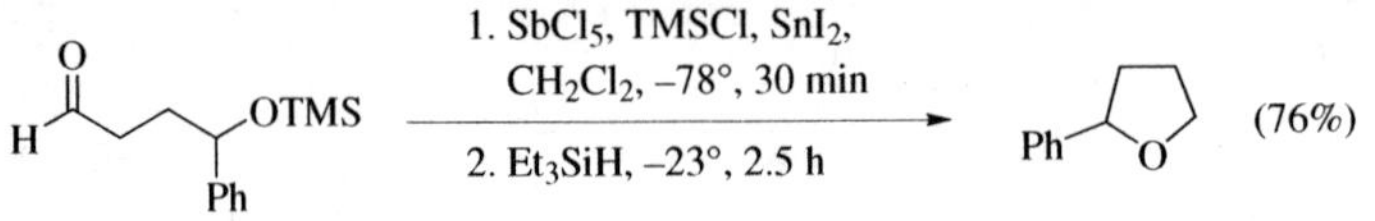

(Eq. 186)

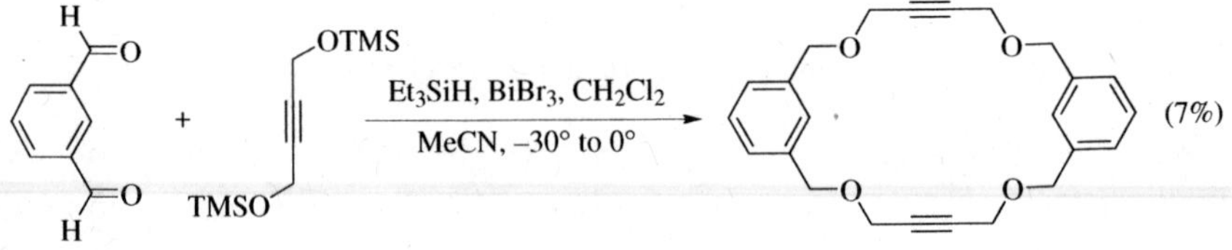

(Eq. 187)

The reductive silylation of aldehydes provides a one-step route to silyl ethers. This is accomplished with the reagent combinations $PhMe_2SiH/CuH(PPh_3)$,[317] $Et_3SiH/ZnCl_2$ (or $SnCl_2$ or $NiCl_2$),[343] Ph_2SiH_2/CsF (or KF, $BnMe_3$ NF, KO_2CH),[75,319] $Et_3SiH/(C_6F_5)_3B$,[115,281] $PhMe_2SiH/CsF$,[320,345–347] and Et_3SiH/TBAF.[76] Montmorillonite clay that has been subjected to ion exchange with ferric ion catalyzes the hydrosilylation of benzaldehyde with triethylsilane to give benzyl triethylsilyl ether in 79% yield.[324,325]

Various non-conjugated diene aldehydes react with $Et_3SiH/Ni(cod)_2/PPh_3$ to give O-triethylsilylated cycloalkanols in low to high yields. Acyclic dienes can lead to the silylated cycloalkanols in moderate yields with the proper catalyst (Eq. 188).[348] Bicyclic systems are also generated by this methodology (Eq. 189).[349]

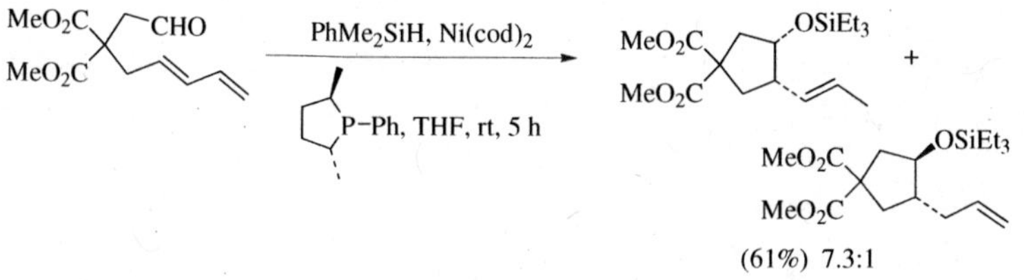

(Eq. 188)

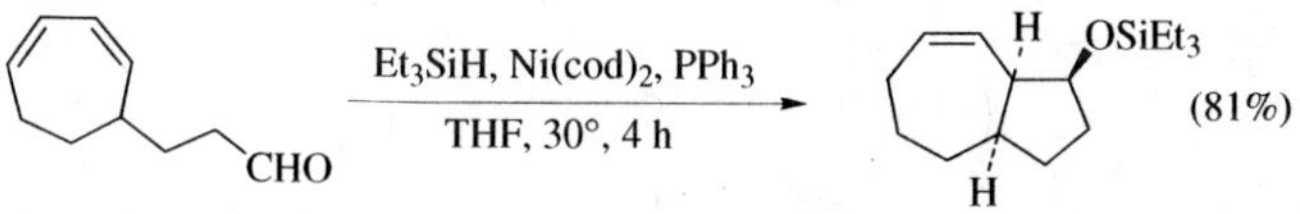

(Eq. 189)

$Et_3SiH/Ni(cod)_2$ brings about the reaction of an aldehyde and an alkyne to provide the silylated allyl alcohol (Eq. 190).[350] The reaction also occurs in an intramolecular mode.

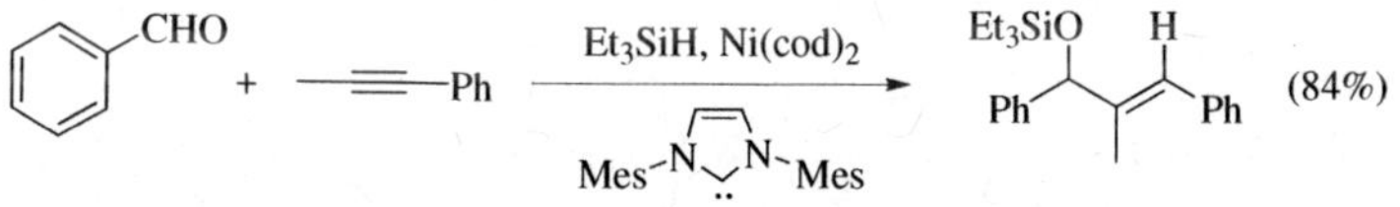

(Eq. 190)

Reduction to Alkanes. Carbonyl groups can be reductively deoxygenated to methylene functions if both of the two steps represented by Eqs. 1 and 2 proceed to completion. With aldehydes, this process leads to the transformation of the CHO group into a CH_3 group.

The relative instability of primary alkyl carbenium ions in the condensed phase and the weak intrinsic nucleophilicity of organosilicon hydrides are the reasons that primary alkyl alcohols are not reduced to hydrocarbons. For these same reasons, aliphatic aldehydes do not undergo complete deoxygenation to methyl-terminated hydrocarbons when treated with acids and organosilicon hydrides under usual laboratory conditions. In contrast, many aryl aldehydes can be transformed into methylarenes by this method. Since the organosilane reduction of benzyl alcohols to the corresponding toluene derivatives is known, it is not surprising that the reduction of an aryl aldehyde to a toluene is possible. This transformation has been carried out with Et_3SiH/TFA,[69,351,352] $(EtO)_3SiH$, and Et_3SiH with various catalysts,[353] $Et_3SiH/(C_6F_5)_3B$,[281] PMHS/Pd/C,[316] and PMHS/$(C_6F_5)_3B$.[354] The last combination also reduces alkyl aldehydes to the corresponding alkanes (Eq. 191).[281,282,354]

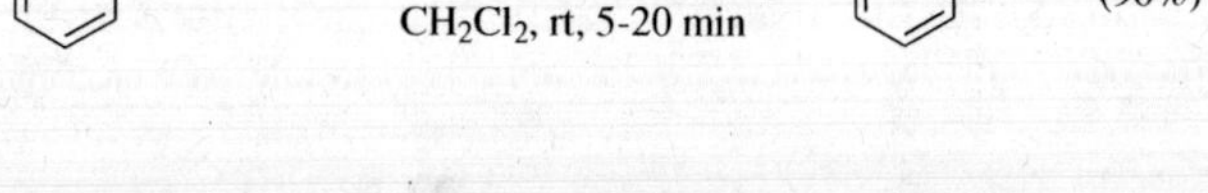

(Eq. 191)

Trifluoroacetic acid solutions of benzaldehydes having electron-donating ring substituents form the corresponding methyl arenes when at least two equivalents of an organosilicon hydride are added to the solution at room temperature. The reaction conditions permit preservation of the integrity of functions such as halogen, alkoxy, carboxylate, cyano, and nitro. There is little difference in the reducing abilities of triethylsilane, tri-*n*-propylsilane, and tri-*n*-hexylsilane in these reactions. Thus, the silane reducing agent can be chosen that best suits purification of the desired product. Basic aqueous workup converts the silicon reaction products derived from the organosilicon hydride into the corresponding silanols and disiloxanes, which may be removed from the desired reduction products by simple distillation.[69]

Benzaldehyde itself forms no toluene; only dibenzyl ether and benzyl trifluoroacetate are formed. Triethylsilane (2.2 equivalents) causes the transformation of *p*-anisaldehyde into *p*-methylanisole in 76% yield after only 30 minutes. Use of a three-fold excess of dimethylphenylsilane in place of the triethylsilane results in a slight improvement in yield to 83% after 45 minutes.[69]

Similar treatment of a trifluoroacetic acid solution of *p*-tolualdehyde with triethylsilane gives only a 20% yield of *p*-xylene after 11 hours reaction time followed by basic workup. Use of 2.5 equivalents of dimethylphenylsilane enhances the yield to 52% after only 15 minutes. This reaction proceeds stepwise through the formation of a mixture of the trifluoroacetate and the symmetrical ether. These intermediates slowly form the desired *p*-xylene product along with Friedel-Crafts side products under the reaction conditions (Eq. 192).[73] Addition of co-solvents such as carbon tetrachloride or nitromethane helps reduce the amount of the Friedel-Crafts side products.[73]

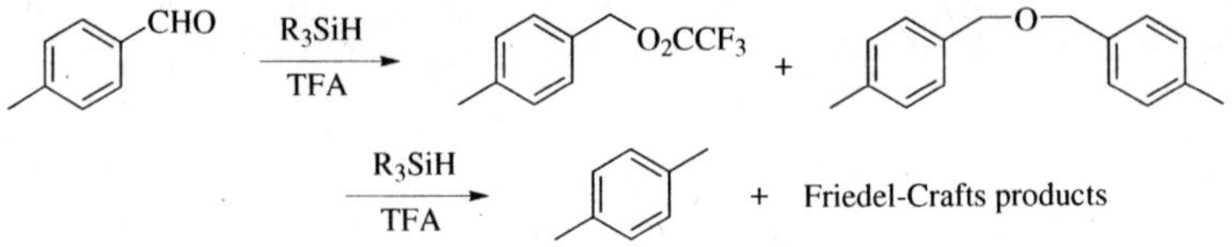

(Eq. 192)

Treatment of a polyfunctional chromium-tricarbonyl-complexed hydroxy aldehyde with an excess of Et_3SiH/TFA for 4.5 hours gives an 82% yield of fully reduced product with both the formyl and hydroxy groups completely and selectively reduced (Eq. 193).[352]

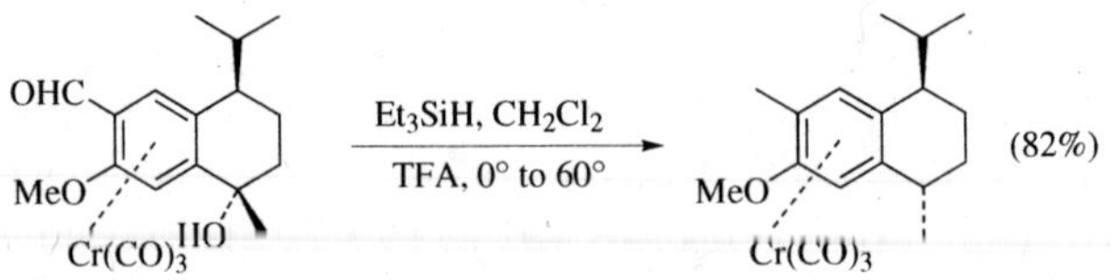

(Eq. 193)

The sequence of reagent and substrate addition can be quite important in these reactions. For example, a trifluoroacetic acid solution of 2,4,6-trimethylbenzaldehyde forms isodurene in 98% yield within 15 minutes when 2.2 equivalents of triethylsilane are added to the reaction mixture at room temperature.[69] In contrast, when trifluoroacetic acid is added to a stirred solution of triethylsilane and 2,4,6-trimethylbenzaldehyde, isodurene is formed in only 70% yield after basic aqueous workup. Minor side products under these reaction conditions are mesitylene (formed via acid-catalyzed decarbonylation of the aldehyde) and the Friedel-Crafts product 2,4,6,2′,3′,4′,6′-heptamethyldiphenylmethane (Eq. 194).[311]

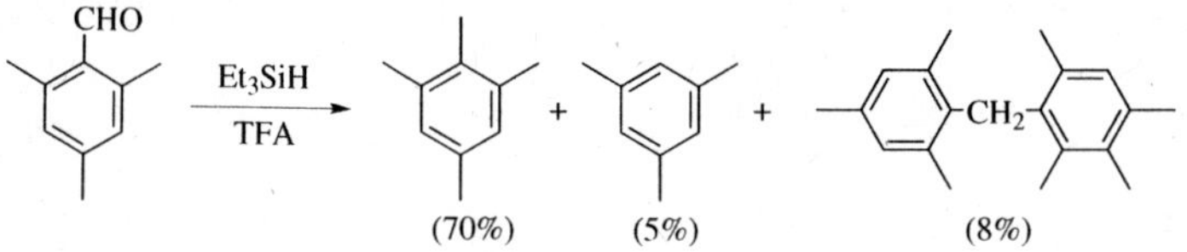

(Eq. 194)

The $Et_3SiH/BF_3{\cdot}OEt_2$ combination fails to cause complete deoxidative reduction of aldehydes, forming instead mixtures of primary alcohols and symmetrical ethers.[74] By contrast, aryl aldehydes lacking electron-withdrawing ring substituents, when reacted in dichloromethane with at least two equivalents of triethylsilane and gaseous boron trifluoride at 0°, form the corresponding methylarenes within a few minutes (Eq. 195).[1] Even benzaldehyde produces a 52% yield of toluene by this method when 18 equivalents of triethylsilane are added to suppress formation of Friedel-Crafts oligomers. The method offers the advantage that fluorotriethylsilane is formed, which is volatile and is easily separated from the desired organic products.[1]

$$\text{ArCHO} \xrightarrow[\text{BF}_3,\ 0°]{\text{Et}_3\text{SiH, CH}_2\text{Cl}_2} \text{ArMe} + \text{Et}_3\text{SiF} \qquad \text{(Eq. 195)}$$

Ar = Ph (52%)
Ar = 4-MeC_6H_4 (45%)
Ar = 4-ClC_6H_4 (68%)
Ar = 4-$MeOC_6H_4$ (100%)

If Friedel-Crafts products are desired, a clever method exists for the direct conversion of aryl aldehydes into diarylmethanes. Reaction of a mixture of an aromatic aldehyde and a catalytic amount of trimethylsilyl trifluoromethanesulfonate and excess polymethylhydrosiloxane in either benzene or toluene at reflux results in the formation of the respective arylphenyl or tolylmethanes in reasonably good yields within 1–3 hours (Eq. 196).[314] Thus, benzaldehyde reacts in refluxing benzene containing a few drops of TMSOTf and excess PMHS to give diphenylmethane in 92% yield within two hours and in refluxing toluene within one hour to give a 95% yield of a mixture of phenyl-*p*-tolylmethane and phenyl-*o*-tolylmethane in a 70 : 30 ratio. *p*-Tolualdehyde gives a 60% yield of phenyl-*p*-tolylmethane when heated at reflux in benzene for 2.5 hours and an 80% yield of di-*p*-tolylmethane and *p*-tolyl-*o*-tolylmethane in a 90 : 10 ratio when heated at reflux in toluene for 30 minutes. *o*-Chlorobenzaldehyde gives a mixture of 25% phenyl-*o*-chlorophenylmethane and 55% of bis(*o*-chlorophenyl)ether, and *p*-chlorobenzaldehyde gives a 65% yield of a mixture of phenyl-*p*-chlorophenylmethane and bis(*p*-chlorophenyl)ether in a 75 : 25 ratio when heated at reflux in benzene for 3 hours.[314]

$$ArCHO + Ar'H \xrightarrow[TMSOTf]{PMHS} ArCH_2Ar' \quad (25\text{-}95\%) \qquad \text{(Eq. 196)}$$

Ar = Ph, 4-MeC_6H_4, 4-ClC_6H_4, 2-ClC_6H_4
Ar' = Ph, 4-MeC_6H_4

The TFA-catalyzed triethylsilane reductive condensation of an aldehyde with indoles provides a convenient route to 3-substituted indoles in modest to good yields (Eq. 197).[355]

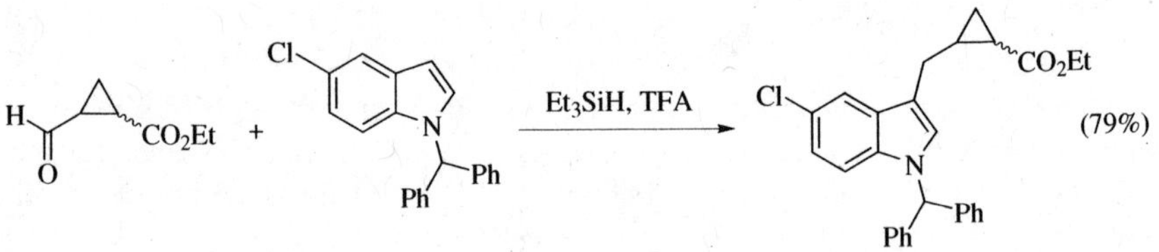

(Eq. 197)

Reduction to Methylene Halides. Treatment of aryl aldehydes with selected organosilicon hydrides and an appropriate trimethylsilyl halide produces benzyl chlorides, bromides, and iodides directly in good to excellent yields by reductive halogenation (Eq. 198). This protocol offers the advantage of being simple and leading only to the monohalo derivatives. The method is specific for the carbonyl group of both aryl ketones and aldehydes and preserves the integrity of many other groups (e.g. ring halogen, alkyl, alkoxy, cyano, nitro, hydroxy, ester) that may be found in polyfunctional compounds. Alkyl aldehydes form symmetrical ethers instead of halides under these reaction conditions.[314,356,357] Several variations of this general method exist. In the most straightforward approach for synthesizing iodides, the addition of an external trimethylsilyl reagent is not required. Aromatic aldehydes normally react within minutes at room temperature with iodine and TMDO in dichloromethane solution to produce benzyl iodides in high yields (66–87%) (Eq. 199). A reactive silyl iodide is believed to be formed in situ from tetramethyldisiloxane and iodine under these conditions. The reaction is not limited to aryl rings with only electron-donating groups; chloro-, hydroxy-, alkyl-, alkoxy-, cyano-, and carboalkoxy-substituted rings all undergo the transformation.[357] The same transformation can be carried out with diiodosilane.[358]

$$ArCHO \xrightarrow[TMSX]{TMDO} ArCH_2X \quad (40\text{-}97\%) \qquad \text{(Eq. 198)}$$

Ar = various substituted aryl

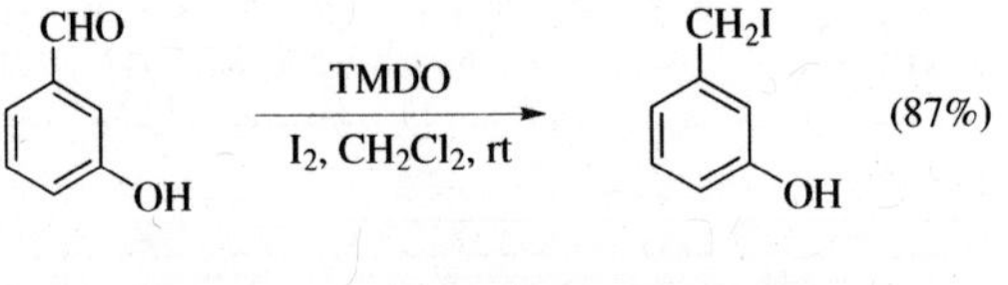

(Eq. 199)

Another variation of this method involves the treatment of an acetonitrile solution of the aryl aldehyde, trimethylsilyl chloride, and either sodium iodide, if iodide products are desired, or lithium bromide, if bromide products are desired, with TMDO. After an appropriate reaction time (5–195 minutes) at a temperature in the range of −70° to 80°, the upper siloxane layer is removed and the benzyl iodide or bromide product is isolated from the remaining lower portion after precipitation of the inorganic salts by addition of dichloromethane. For example, *p*-anisaldehyde reacts to form *p*-methoxybenzyl bromide in 84% isolated yield under these conditions (Eq. 200).[314,356]

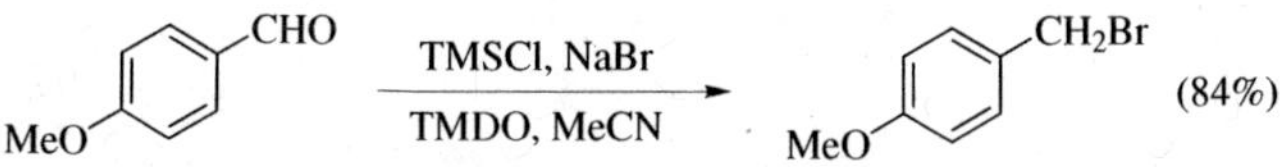

(Eq. 200)

In the preparation of iodides, but not bromides, PMHS may be substituted for the TMDO. Chlorides can be obtained if thionyl chloride and zinc iodide are added to suppress the formation of symmetrical ethers.[314] An example of this type of reductive chlorination is shown by the TMDO-mediated conversion of *p*-tolualdehyde into *p*-methylbenzyl chloride (Eq. 201).[313] To obtain chlorides from aldehydes having electron-withdrawing groups such as nitro or carbomethoxy, the initial reaction is first carried out at −70° and the mixture is then heated to reflux in order to reduce the formation of symmetrical ether by-products. Zinc chloride is substituted for zinc iodide for the synthesis of chlorides of substrates with electron-donating groups such as methoxy and hydroxy.[314]

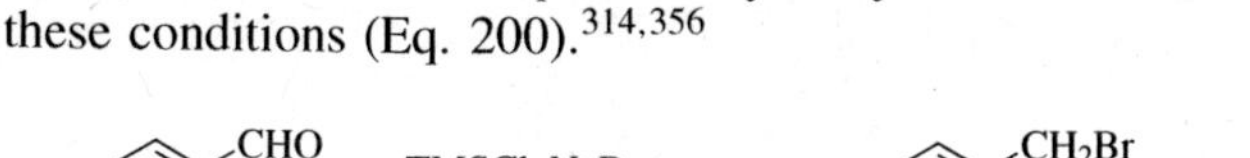

(Eq. 201)

Reductive Amination. Reaction of an aminohydrodimethylsilane with aldehydes in the presence of a Lewis acid catalyst gives the corresponding amine in good to high yields (Eq. 202).[359] The use of an Et_3SiH/TFA/amine reagent combination also leads to the reductive amination of aldehydes (Eq. 203).[360] Comparable reductive aminations of aldehydes are possible in moderate yields with $PhSiH_3/Bu_2SnCl_2$/amine,[361,362] and in good yields with $PMHS/TiCl_4$/amine[363] or with amine/Cl_3SiH.[364]

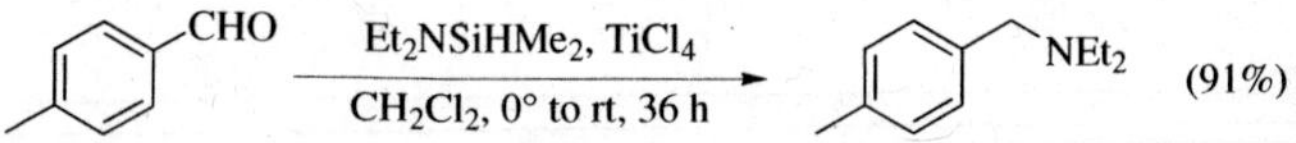

(Eq. 202)

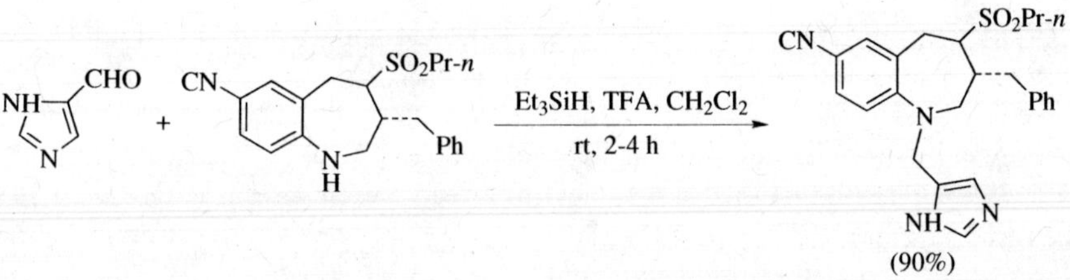

(Eq. 203)

Reductive Thiolation. Treatment of aldehydes with triethylsilane, thiols, and boron trifluoride monohydrate[6,217] yields sulfides in a one-flask process. For example, this method gives a 97% yield of benzyl isopropyl sulfide from benzaldehyde and 2-propanethiol (Eq. 204).[365]

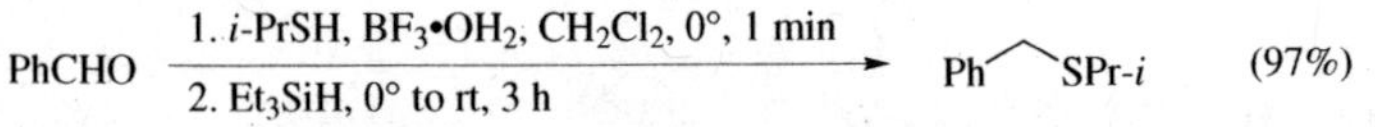

(97%) (Eq. 204)

Reduction of Ketones

The selective organosilane reduction of ketone functions can be effected in the presence of a number of other functional groups including epoxides,[320,366] ketals,[86,367] thioketals,[368] other ketones,[369,370] β-lactams,[371] alkynes,[372] esters,[79,80,83,84,87,320,373,374] α-bromides,[76,80,83] amides[80,83,84,86,276,320,375] ureas,[84,276] trifluoroacetamides,[83,376] sulfonamides,[83,86] and nitro groups.[80]

Reduction to Alcohols. The organosilane-mediated reduction of ketones to secondary alcohols has been shown to occur under a wide variety of conditions. Only those reactions that are of high yield and of a more practical nature are mentioned here. As with aldehydes, ketones do not normally react spontaneously with organosilicon hydrides to form alcohols. The exceptional behavior of some organocobalt cluster complex carbonyl compounds was noted previously. Introduction of acids or other electrophilic species that are capable of coordination with the carbonyl oxygen enables reduction to occur by transfer of silyl hydride to the polarized carbonyl carbon (Eq. 2). This permits facile, chemoselective reduction of many ketones to alcohols.

Certain catalysts promote the reduction of ketones with organosilanes. The reduction of acetophenone with Et_3SiH is catalyzed by the diphosphine **65** and gives only a small amount of overreduction to ethylbenzene.[377] Aryl alkyl enones and ynones are reduced to the corresponding alcohols with triethoxysilane and the titanium-based catalyst **66**.[378] Trichlorosilane reduces acetophenone in 90% yield with *N*-formylpyrrolidine catalysis.[379]

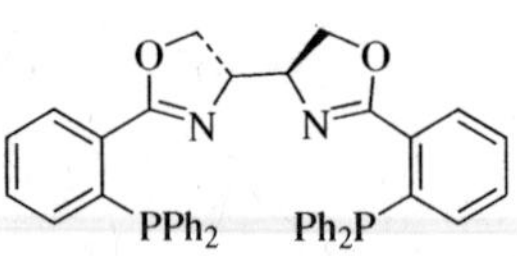

65

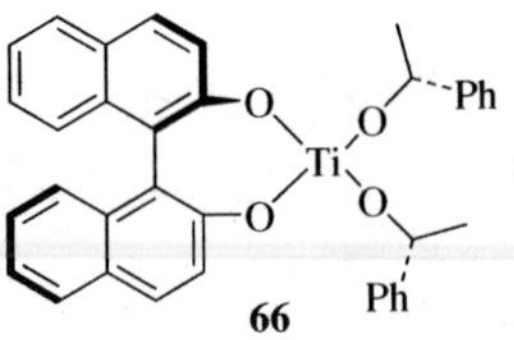

(dimer)

66

Promotion by Acid. The same range of Lewis and Brønsted acids that promote the silane reduction of aldehydes can be used for the reduction of ketones. These acid-catalyzed reductions appear to proceed by direct hydride transfer rather than by a single-electron transfer mechanism.[380] Similar to the case of aldehydes, the silane reductions of ketones promoted by Brønsted acids rarely give clean yields of alcohols when conducted under anhydrous conditions. Instead, mixtures of alcohols, esters, and silyl ethers often result.[313,381] The reagent combination of Et_3SiH/HCl (or H_2SO_4) gives good yields of the alcohol, although by-products of the sym-ether among others can complicate the reduction.[313] Use of zinc chloride to promote organosilicon hydride reduction of ketones to siloxanes that can be hydrolyzed to alcohols is well known. It is one of the first Lewis acid catalysts reported to be useful for this purpose,[382,383] although others are known.[384] The combination of $Et_3SiH/TFA/NH_4F$ provides a good yield of the alcohol with some ether formation.[135] High yields of the alcohol from both aryl and alkyl ketones are realized by the $HMDS/(AcOBu_2Sn)_2O$[316] and $HMDS/Sn(OTf)_2$[385] reagent combinations. The $Et_3SiH/(C_6F_5)_3B$ combination cleanly reduces aryl ketones to the substituted benzyl alcohols.[116]

Under certain conditions, the trifluoroacetic acid catalyzed reduction of ketones can result in reductive esterification to form the trifluoroacetate of the alcohol. These reactions are usually accompanied by the formation of side products, which can include the alcohol, alkenes resulting from dehydration, ethers, and methylene compounds from over-reduction.[68,70,207,208,313,386] These mixtures may be converted into alcohol products if hydrolysis is employed as part of the reaction workup. An example is the reduction of cyclohexanone to cyclohexanol in 74% yield when treated with a two-fold excess of both trifluoroacetic acid and triethylsilane for 24 hours at 55° and followed by hydrolytic workup (Eq. 205).[203]

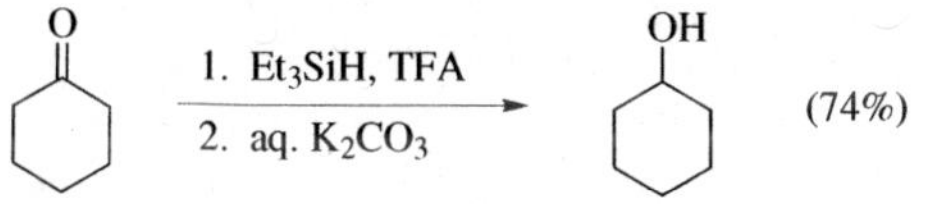

(Eq. 205)

Promotion by Valence Expansion. As in reactions of aldehydes, addition of nucleophilic substances to mixtures of ketones and organosilicon hydrides promotes reduction of the carbonyl group as depicted in Eq. 6. Ketones are conveniently reduced in high yields with reagent combinations of $(EtO)_3SiH$ or $Me(EtO)_2SiH$ and KF (or CsF).[80] The pentacoordinate silane **67** itself reduces ketones in high yields (Eq. 206).[81] In a somewhat similar approach, the lithium salt of silicate **68** is a good reducing agent for ketones (Eq. 207).[96] Other hydridosilicates are known to similarly reduce ketones.[93]

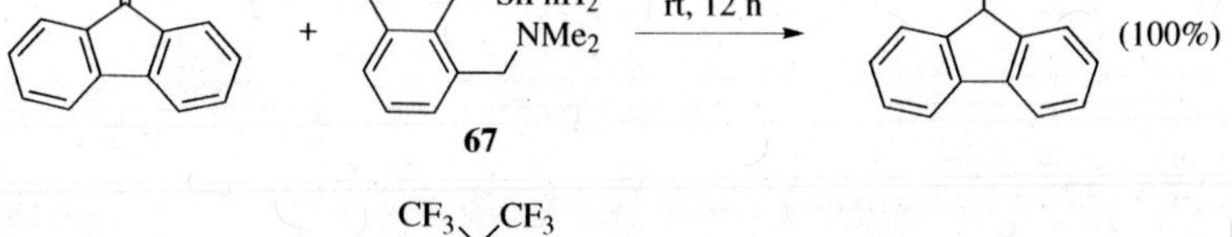

67

(Eq. 206)

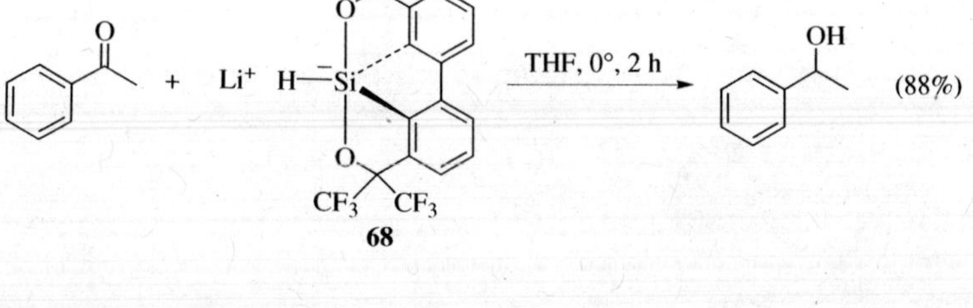

68

(Eq. 207)

The PMHS/TBAF system provides both an excellent and practical approach to the reduction of aryl ketones to the benzyl alcohols.[278] Similarly, the PMHS/Triton®B combination gives high yields of the benzyl alcohols.[278]

Diastereoselective Reductions. The diastereoselectivity of organosilane reductions of ketones has been the topic of a number of studies. Extensive studies of the Brønsted acid promoted reductions of alkyl-substituted cyclohexanones by mono-, di-, and trialkylsilanes show that chain branching and other steric features of both the silane and the carbonyl substrate can be important factors in determining the isomeric compositions of reduction product mixtures.[68,381,384,386] In general, the reduction of various substituted cyclohexanones does not show effective diastereoselectivity even when very sterically hindered silanes such as Ph_3SiH[387,388] or $(t\text{-Bu})_3SiH$[386] are employed. A quite good system is a silane under Triton®B or TBAF catalysis.[278] The best system for the trans-selective reduction of 4-*tert*-butylcyclohexanone is the sterically encumbered $(TMSO)_3SiH$/TBAF (Eq. 208).[278] This system is not successful in the stereoselective reduction of 2-methylcyclohexanone, giving a cis:trans selectivity of 18 : 82, although 3-methylcyclohexanone gives a cis:trans ratio of 7 : 93 and a high (>90%) yield of 3-methylcyclohexanol. The combination of PMHD/dibutylacetoxytin oxide (DBATO) reduces 4-*tert*-butylcyclohexanone exclusively to *trans*-4-*tert*-butylcyclohexanol.[316] The active reducing agent in this system is likely a tin hydride species. A system of Ph_2MeSiH (or Ph_3SiH)/TBAF/HMPA reduces 2-methylcyclohexanone in a cis:trans ratio of 95 : 5.[320] Only *cis*-2-methylcyclohexanol is isolated from the reduction of 2-methylcyclohexanone with $Ph_3SiH/(C_6F_5)_3B$.[116] Other systems give only moderate selectivities in the reduction of substituted cyclohexanones.[70,79,93,116,278,313,367,381,382,384,386,389–392]

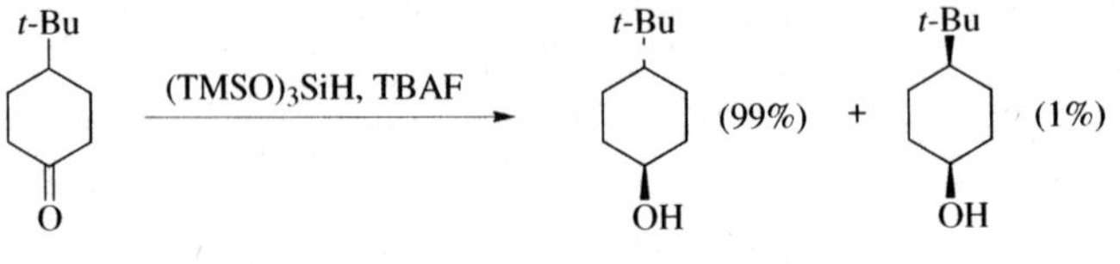

(Eq. 208)

Naphthoquinone is reduced to 1,2,3,4-tetrahydronaphthalene with Et_3SiH/TFA in 60% yield.[393] Quinones can be reduced to hydroquinones in good yields with hydridosiloxanes such as TMDO with iodide present (Eq. 209).[314,316,357] The reductive dehydration of a 1,3-diketone leads to an enone (Eq. 210).[374]

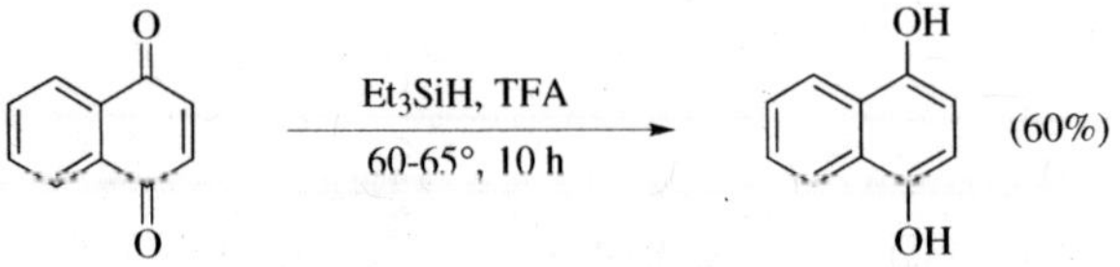

(Eq. 209)

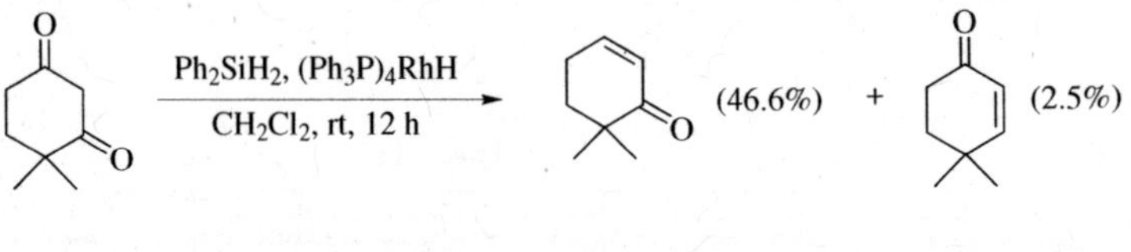

(Eq. 210)

Reduction of the ketone carbonyl of *cis*-1,2,3,4,4a,9b-hexahydro-8-hydroxydibenzofuran-3-one with trifluoroacetic acid and triethylsilane at 0° produces a mixture of the α- and β-isomers of the C3 alcohol with an $\alpha:\beta$ ratio of 1:4 (Eq. 211).[394] This result can be compared with the isomer ratio of 100:1 that results when sodium borohydride is used as the reducing agent.[394] The same cis pair of alcohol isomers is formed in 77% combined yield, but in a reversed ratio of $\alpha:\beta = 4:1$, when the less saturated tetrahydrodibenzofuran analog is used as the substrate (Eq. 212).[394]

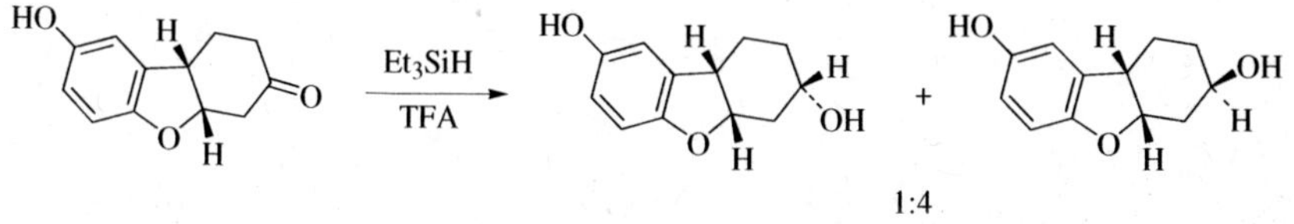

(Eq. 211)

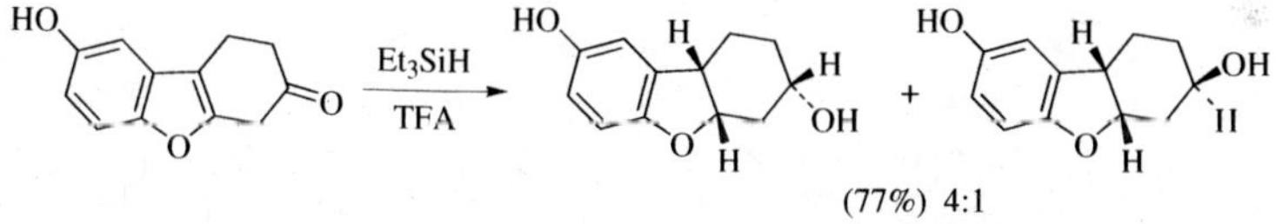

(Eq. 212)

Treatment of a pentacyclic 1α,11-(2-oxethano) thioketal steroid with excess Et_3SiH/TFA causes reduction of the carbon-carbon double bonds as well as the 17-carbonyl group to give a single reaction product (Eq. 213).[368] Other work utilizes trifluoroacetic acid, triethylsilane, and anisole in the presence of a catalytic amount of boron trifluoride etherate to reduce the acetyl carbonyl of a 3-acetyl-2-azetidinone derivative with a dr of 8:1 (Eq. 214).[395]

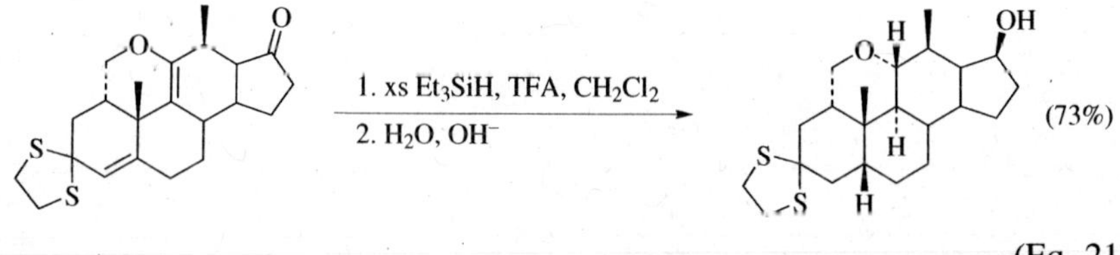

(Eq. 213)

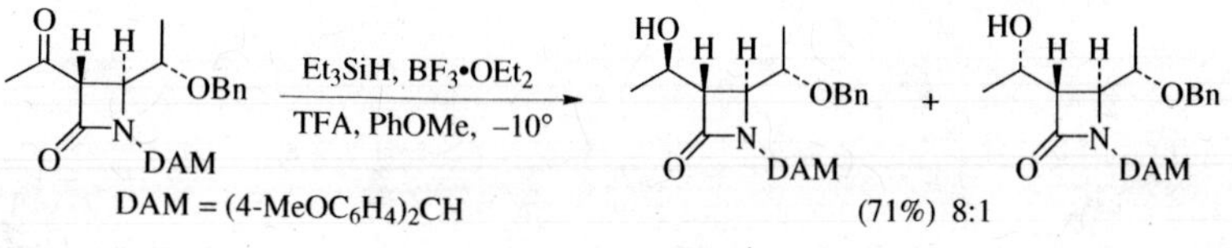

(Eq. 214)

In a similar way, a mixture consisting of 2% boron trifluoride etherate in trifluoroacetic acid and triethylsilane brings about the regioselective reduction of the acyclic carbonyl group of the diketovinyl chloride shown in Eq. 215 in high yield (>94%), but with formation of approximately equal amounts of the two possible diastereomers formed from the creation of a new chiral center.[396]

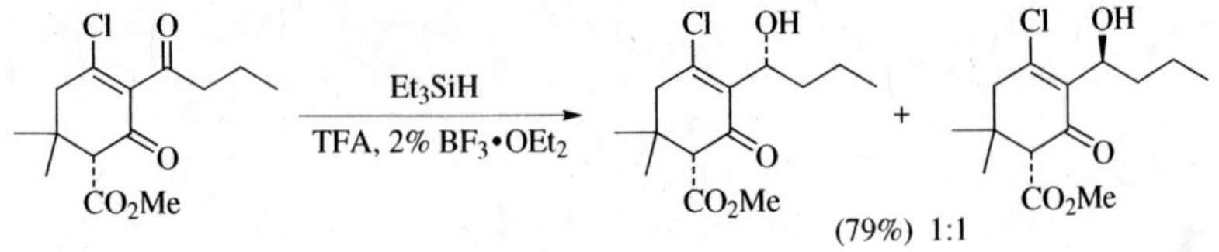

(Eq. 215)

The stereoselective reduction of the carbonyl group of β-hydroxy ketones can be accomplished by a silylation-intramolecular reduction sequence. The best results are obtained when diisopropylchlorosilane is employed.[397] The clever use of an intramolecular hydride transfer from a pre-anchored silyl hydride site allows β-hydroxy ketones to be reduced to 1,3-diols with a very high degree of diastereoselectivity.[397–399] A specific example is the reaction of the hydroxy ketone shown in Eq. 216, first with chlorodiisopropylsilane to form the acyclic siloxane and then with a Lewis acid catalyst to cause intramolecular hydride transfer with formation of a pair of cyclic trans- and cis-disiloxanes. The respective anti- and syn-diols are obtained after fluoride ion catalyzed hydrolysis. The trans:cis ratio of the disiloxane intermediates varies with the Lewis acid employed: $TiCl_4$ (30 : 1), $MgBr_2{\bullet}OEt_2$ (60 : 1), $SnCl_4$ (120 : 1), and $BF_3{\bullet}OEt_2$ (320 : 1).[397–399] A similar reduction with chlorodimethylsilane also gives good results (Eq. 217).[400]

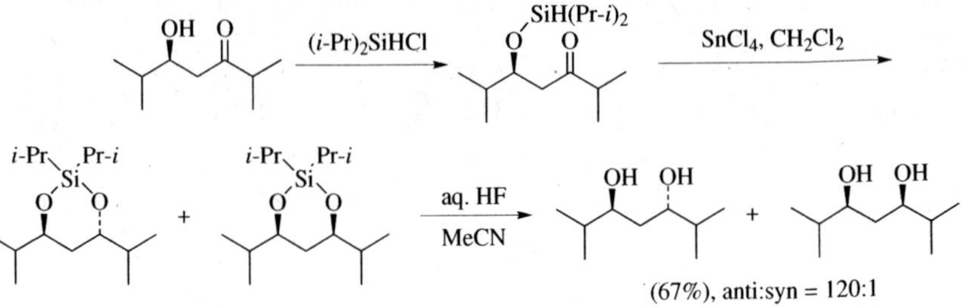

(Eq. 216)

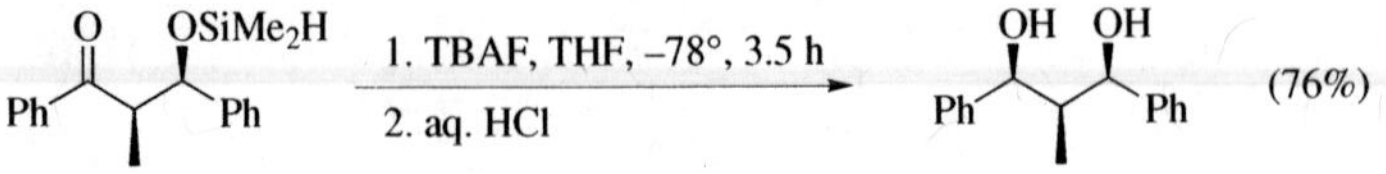

(Eq. 217)

The diastereoselectivity of the reduction of α-substituted ketones has been the subject of much investigation. The reagent combination of trifluoroacetic acid and dimethylphenylsilane is an effective method for the synthesis of erythro isomers of 2-amino alcohols, 1,2-diols, and 3-hydroxyalkanoic acid derivatives.[86,87,276,375] Quite often the selectivity for formation of the erythro isomer over the threo isomer of a given pair is >99 : 1. Examples where high erythro preference is found in the products are shown below (Eqs. 218–220).[276] Similar but complementary results are obtained with R_3SiH/TBAF, where the threo isomer product predominates (Eqs. 221 and 222).[86,87,320] The threo isomer also predominates with the PMHS/TBAF system (Eq. 223).[401]

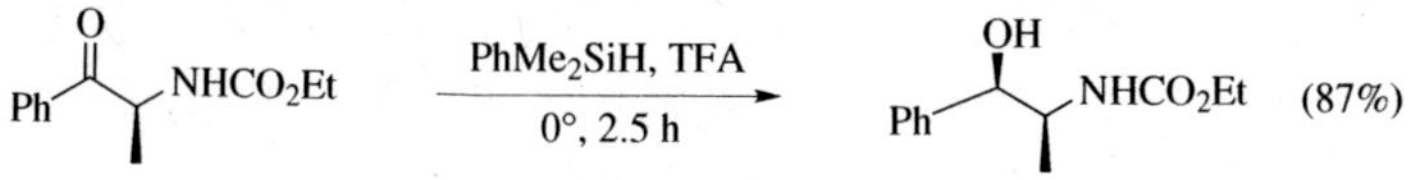

(Eq. 218)

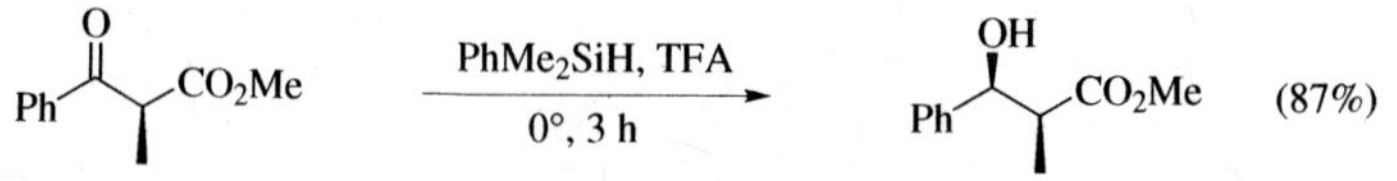

(Eq. 219)

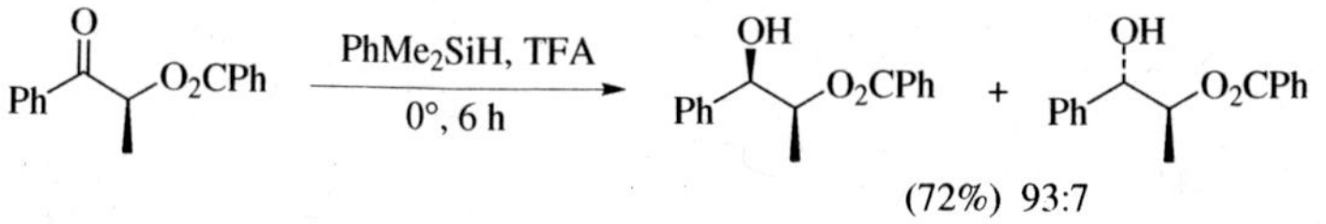

(Eq. 220)

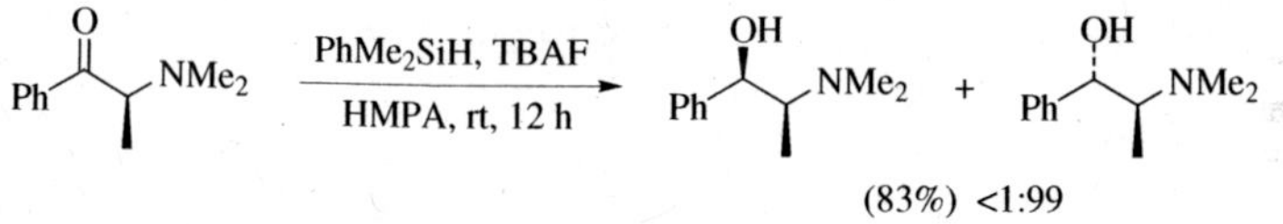

(Eq. 221)

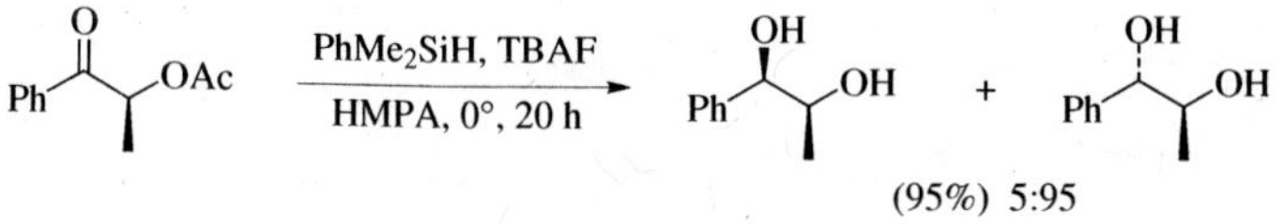

(Eq. 222)

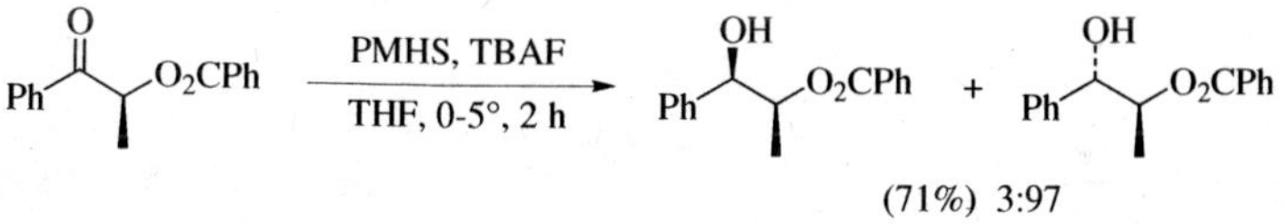

(Eq. 223

α Fluoroketo esters are reduced with high stereoselectivity but in only moderate yields with the combination Ph_3SiH (or $PhMe_2SiH$)/$AlCl_3$ (Eq. 224).[90] The use of $PhMe_2SiH$/TBAF does not give comparable selectivities.[90]

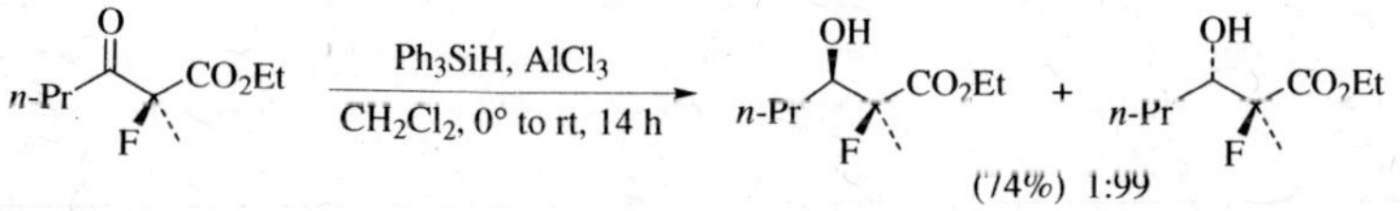

(Eq. 224)

Other systems that are highly stereoselective in the reduction of 2-substituted ketones include PMHS/Triton®-B (erythro:threo = 95 : 5),[278] $(TMSO)_3SiH$/Triton®-B (erythro:threo = 95 : 5),[278] and $PhMe_2SiH$/TASF/HMPA (erythro: threo = 93 : 7).[320]

Reductive Amidation. The Et_3SiH reduction of ketones in the presence of acid and acetonitrile results in the reductive amidation of the ketone (Eq. 225).[313] Ethers and carbinols may be by-products.

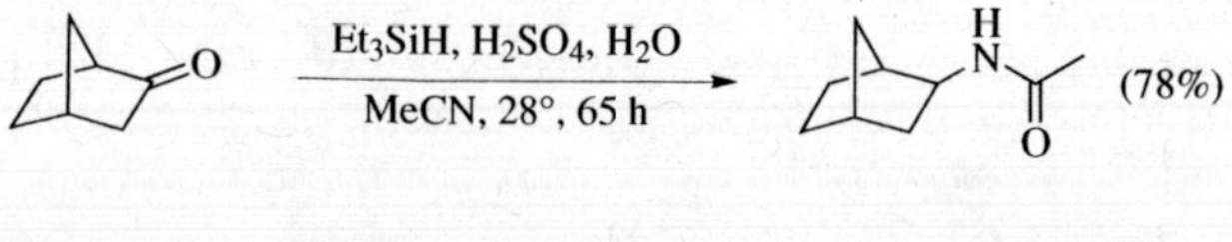

(Eq. 225)

Reductive Esterification. Organosilane reductions of γ- or δ-keto acids and esters provide the corresponding lactones as the final products (Eqs. 226 and 227).[69,79,402]

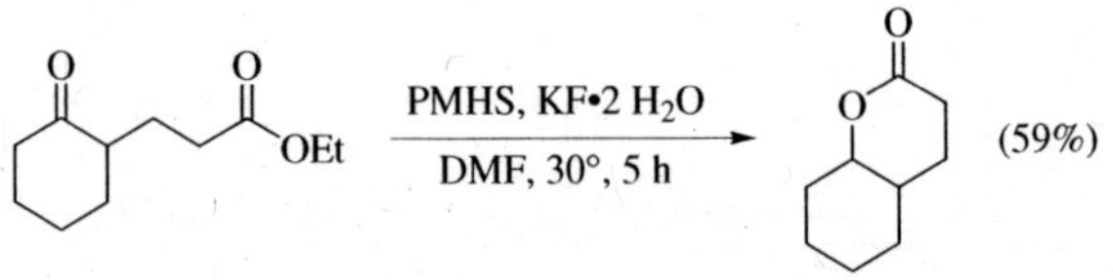

(Eq. 226)

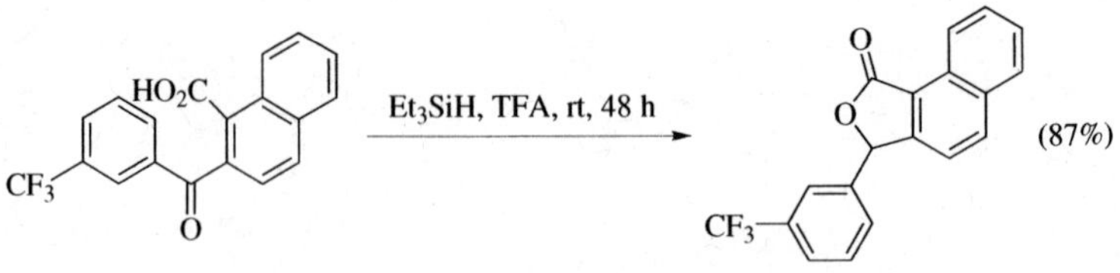

(Eq. 227)

Reductive Etherification. The organosilane reduction of ketones can result in the direct formation of symmetrical ethers. The treatment of ketones with $Et_3SiH/BiBr_3$ (or $BiCl_3$) gives good yields of the symmetrical ethers. The reaction is much better with aldehydes than with ketones, with acetophenone and benzophenone giving only traces of the ether.[332,343] Benzophenone is converted into bis(diphenylmethyl)ether in good yield with $(HMe_2Si)_2O$/TMSOTf.[314] The treatment of various ketones with Et_3SiH/TMSOTf (or TMSI) leads to the symmetrical ethers in excellent yields.[334] Treatment of cyclohexanone with $(n\text{-Bu})_3SiH$/TFA gives dicyclohexyl ether and cyclohexyl trifluoroacetate with formation of the ether favored at lower temperatures.[313] Other combinations of triethylsilane and various catalysts all result in the formation of mixtures of the ether and the silyl ether of the corresponding alcohol.[70,313,353] The PEHS/TFA combination also converts cyclohexanone into a mixture of the ether and the trifluoroacetate.[207]

It is possible to form cyclic ethers from diketones as seen by the cyclic hemiketal formed by the reduction of 1,5-cyclooctanedione (Eq. 228) and the conversion of 2,5-hexanedione into *cis*- and *trans*-2,5-dimethyltetrahydrofuran (Eq. 229).[392] A naphthopyran can be formed in a similar manner.[339]

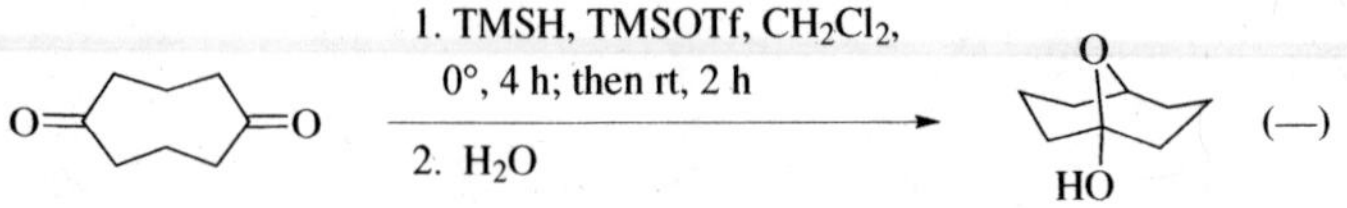

(Eq. 228)

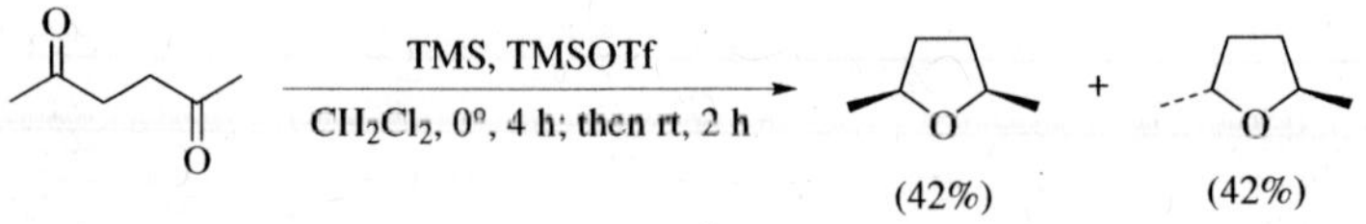

(Eq. 229)

The organosilane reduction of ketones in the presence of alcohols provides an excellent route to unsymmetrical ethers. The reaction of cyclohexanone with ethanol and Et_3SiH/TFA gives cyclohexyl ethyl ether in good yield.[327,328] The same conversion can be accomplished with $Et_3SiH/HC(OEt)_3/HCl$.[327] Adamantanone provides adamantyl methyl ether when treated with $Et_3SiH/HC(OMe)_3$/Nafion®-H,[335] and adamantyl cyclohexyl ether with Et_3SiH/TMSI/c-C_6H_{11}OTMS, but diadamantyl ether with Et_3SiH/TMSOTf.[334]

The employment of the trimethylsilyl ether of an alcohol as the partner to ketones in the synthesis of unsymmetrical ethers is highly useful. Ketones react with trimethylsilyl ethers in an electrolytic process with Et_3SiH or $PhMe_2SiH$ as the reducing agent.[333] Benzyloxytrimethylsilane with $Et_3SiH/BiBr_3$ provides unsymmetrical benzyl ethers in high yields from ketones and aldehydes with aldehydes being more reactive.[343] The reaction has been carried out with both the trimethylsilyl ethers of secondary alcohols and hindered ketones.[334] The combination of Et_3SiH/TMSOTf/ROTMS gives good yields of the desired ethers (Eq. 230).[341]

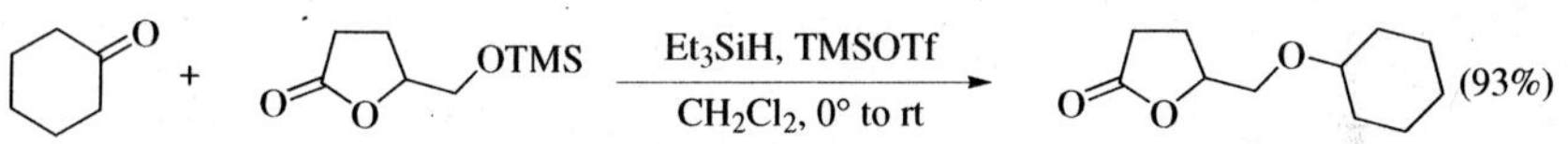

(Eq. 230)

The cyclization of γ-hydroxy ketones is useful for the formation of pyrans,[306,403] both directly and via rearrangement, as illustrated in Eq. 231.[153] As with their acyclic counterparts, these cyclizations also occur with the silyl ethers of the hydroxy ketones where $Et_3SiH/BiBr_3$ is used with the TBS and TES ethers.[342,404] A methyl thiomethyl ether is also capable of undergoing the reductive cyclization.[405] In like manner, 1,4-diols and ε-hydroxy ketones provide oxepanes, with Et_3SiH or $PhMe_2SiH$/TMSOTf being especially effective (Eqs. 232 and 233).[336,406] The trimethylsilyl ether of the alcohol also provides the oxepane.[306]

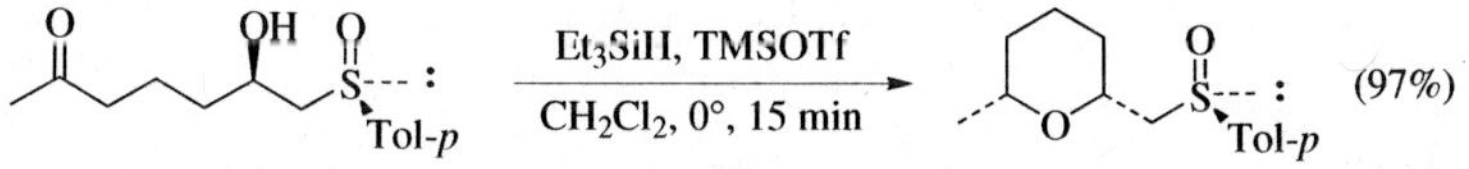

(Eq. 231)

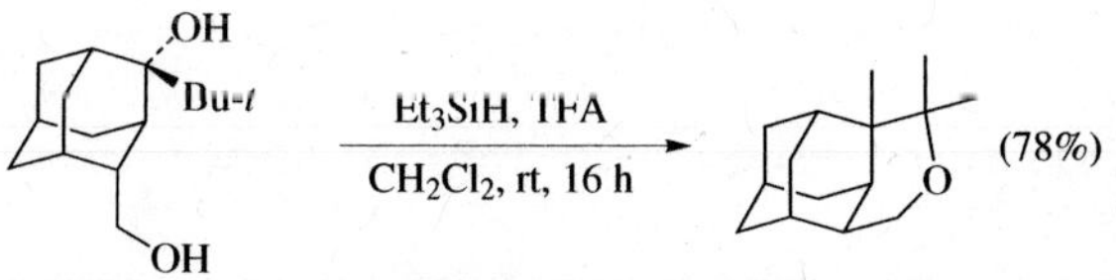

(Eq. 232)

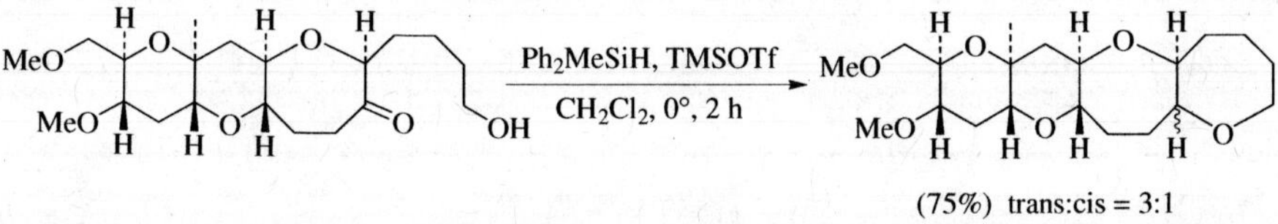

(Eq. 233)

The reductive ether formation from keto epoxides is an acid-catalyzed process (Eqs. 234[407] and 235[408]).

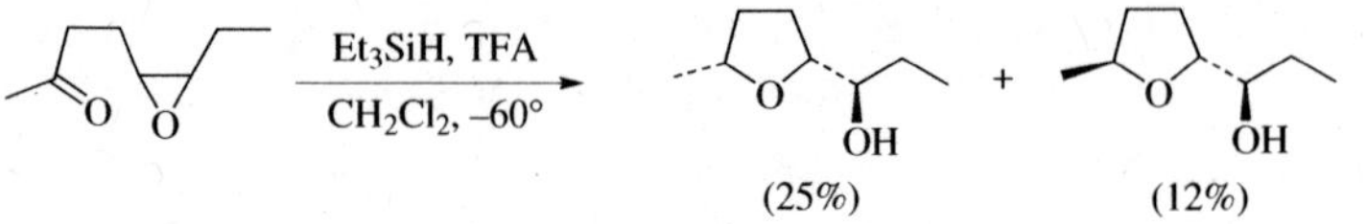

(Eq. 234)

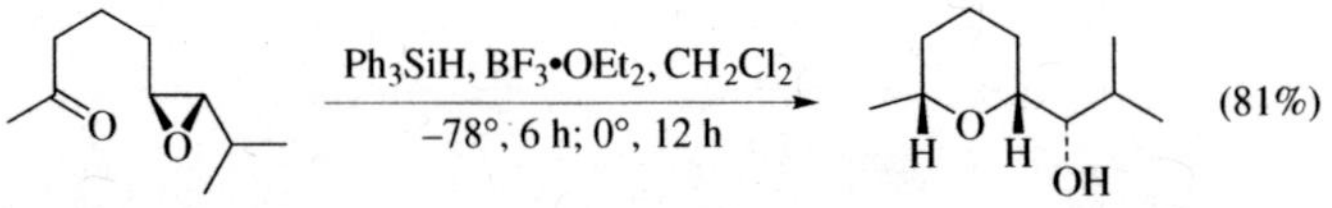

(Eq. 235)

Diketones are reductively cyclized in a TFA-catalyzed reaction. The cyclization of the cage structure shown in Eq. 236 illustrates this ring closure in the formation of an acetal of trifluoroacetaldehyde.[409] The organosilane reduction of triketone **69** followed by Jones oxidation gives the cyclic ketoether in fair yield (Eq. 237).[410]

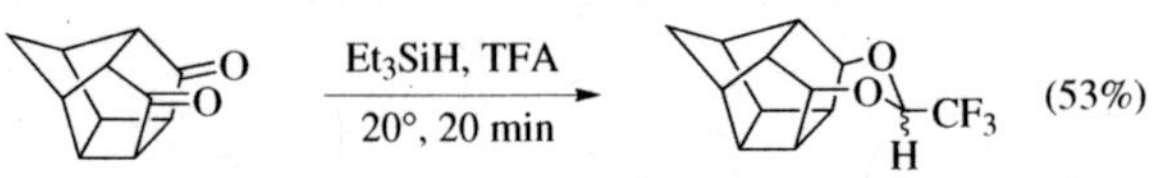

(Eq. 236)

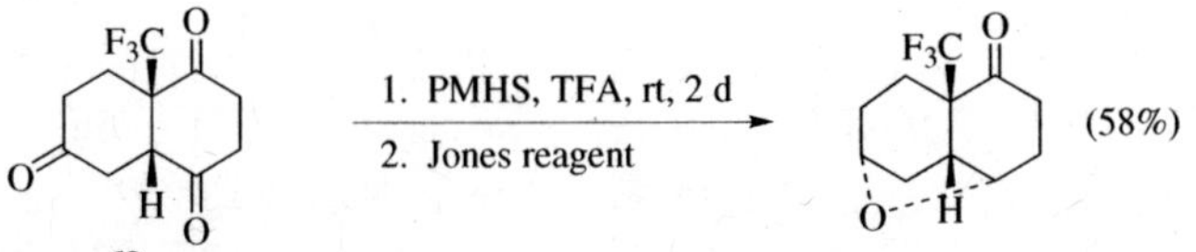

69

(Eq. 237)

Reductive Silylation. The reductive silylation of ketones and subsequent deprotection of the silylated alcohol to the alcohol can be accomplished in a variety of ways. Phenyldimethylsilane in the presence of KF, CsF, or RbF gives the (1-phenylethoxy)phenyldimethylsilane from acetophenone in high yields.[345–347] The reagent combination of $Ph_3SiH/(C_6F_5)_3B$ gives good yields of the silyl ether with aryl ketones.[115] The combination of $Et_3SiH/(C_6F_5)_3B$ gives an excellent yield of silyl ether with threo selectivity in the reduction of α-substituted ketones (Eq. 238).[372]

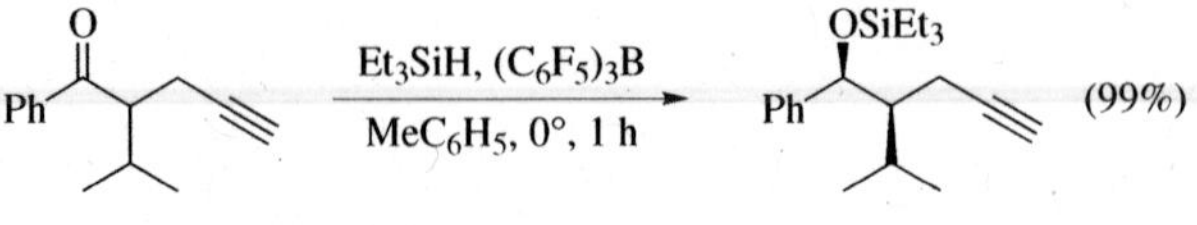

(Eq. 238)

The reduction of ketones or aldehydes with Ph_2SiH_2/KF produces either the mono- or dialkoxydiphenylsilane depending on the stoichiometry of the reaction.[75,319] The dendrimeric catalysts **70**, **71**, or **72** work with Et_3SiH to give the silyl ether of acetophenone in excellent yield (Eq. 239).[117]

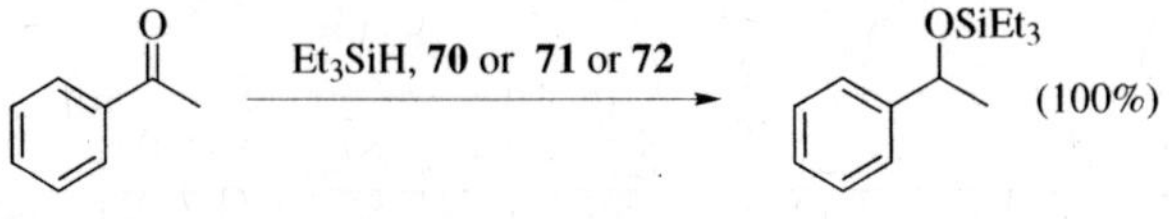

(Eq. 239)

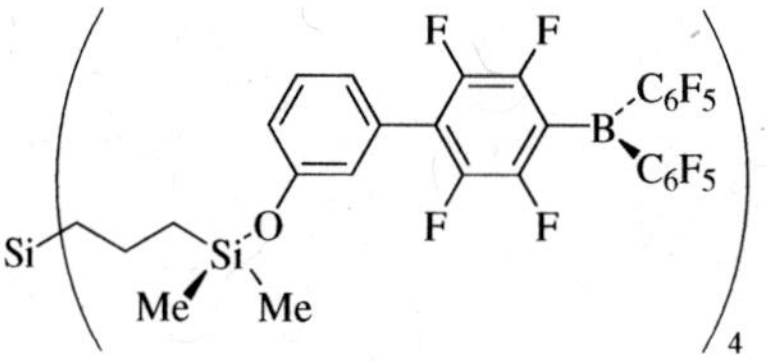

70

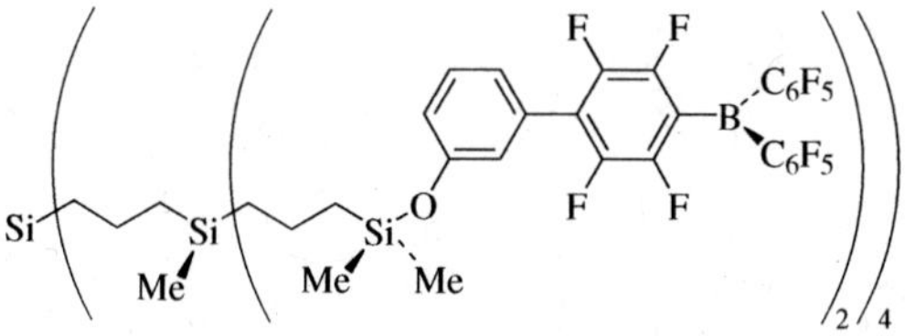

71

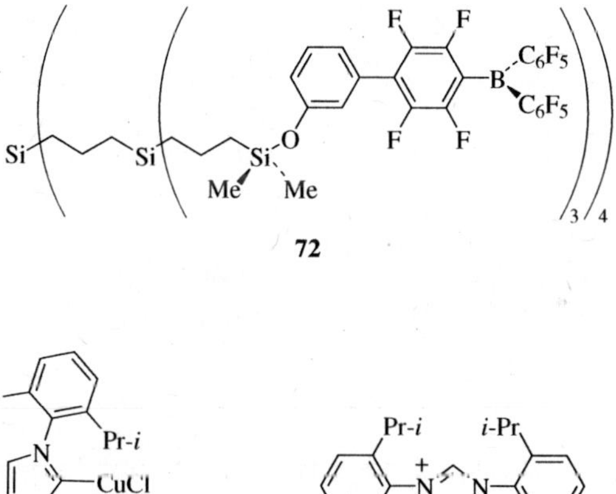

72

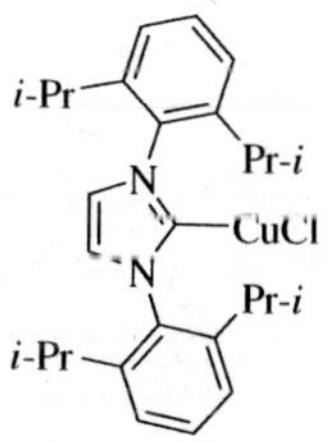

73

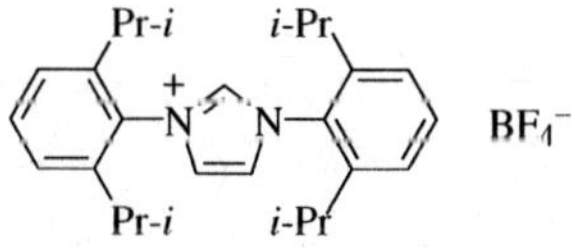

74

Benzil is reductively triethylsilylated to the bis(silyl) ether in 83% yield.[411] The combination of $Et_3SiH/ZnCl_2$ reductively triethylsilylates ketones in good yield.[382] Excellent yields of triethylsilyl ethers from ketones are accomplished with the use of triethylsilane and catalyst **73** or **74**.[412] *tert*-Butyldimethylsilyl ethers can be synthesized by the reaction of TBSH/TBSOTf with a ketone (Eq. 240).[392]

(Eq. 240)

Reduction to Halocarbons. The best conditions for the reductive chlorination of ketones use the reagent combination $Me_2ClSiH/In(OH)_3$ (Eq. 241).[331] Examples include conversions of aryl ketones to benzyl chlorides, ethynyl ketones to propargyl chlorides, and alkyl ketones to alkyl chlorides (Eq. 242).[331] Addition of lithium iodide to the reaction mixture yields the corresponding iodide product. The combination of $TMDO/I_2$ reductively iodinates aryl ketones and aldehydes in good yields (Eq. 243).[357]

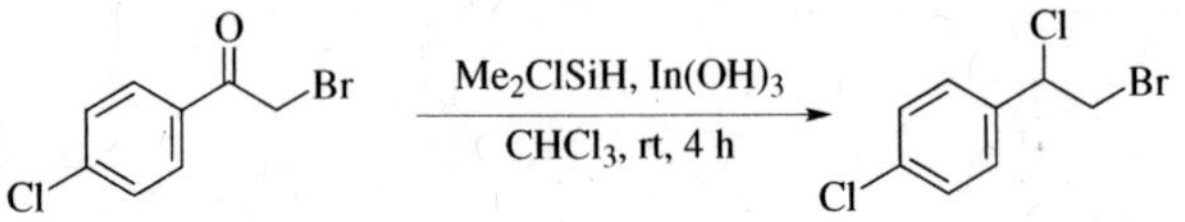

(99%) (Eq. 241)

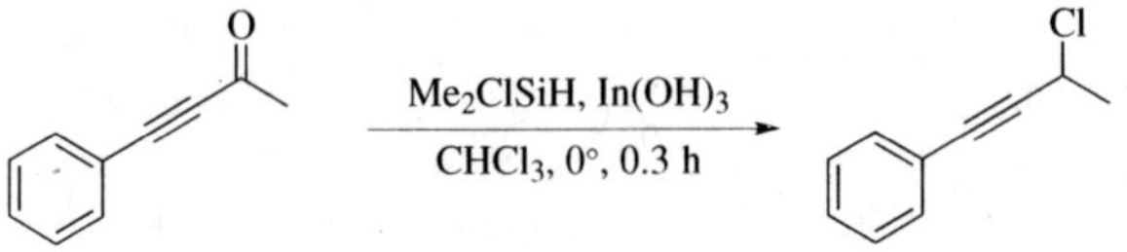

(78%) (Eq. 242)

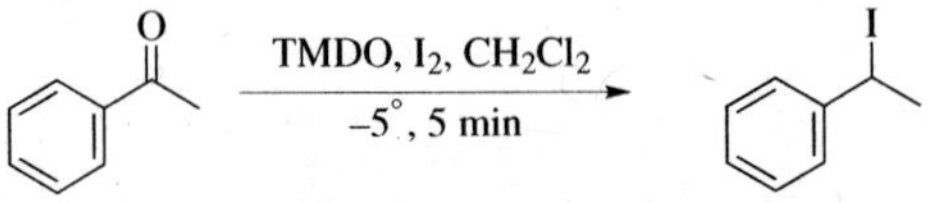

(76%) (Eq. 243)

Reduction to Alkanes. Carbonyl groups can be reductively deoxygenated to methylene functions if both of the two steps represented by Eqs. 1 and 2 are followed to completion. The direct reduction of a carbonyl function to a methylene group is an important reaction of organosilanes that is applicable to a variety of ketones. A number of reagents have been employed in the direct reduction of aromatic ketones such as acetophenone to the corresponding methylene derivatives. These include $Et_3SiH/BF_3{\bullet}OEt_2$,[210,217,375,413] Et_3SiH/TFA,[73,135,180,207,210,257,376,414–419] Et_3SiH/HCO_2H,[208] $Et_3SiH/HClO_4$,[214] $Et_3SiH/AlCl_3/HCl$,[136] Et_3SiH/CF_3SO_3H,[420] $ClMe_2SiH/In(OH)_3$,[331] PMHS/$(C_6F_5)_3B$,[354] $Et_3SiH/TFA/NH_4F$,[135] and Et_3SiH (or $PhMe_2SiH$)/$TiCl_4$.[421] Diaryl ketones are reduced with Et_3SiH/PPHF (pyridinium poly(hydrogen fluoride)) in excellent yields.[135] The triketone **75** is reduced to a diketone in good yield using Et_3SiH/TFA at 0° (Eq. 244).[418] On the other hand, 2-methyl-2*H*-indene-1,3-dione is reduced completely to 2-methyl-2,3-dihydro-1*H*-indene at elevated temperature (Eq. 245).[422]

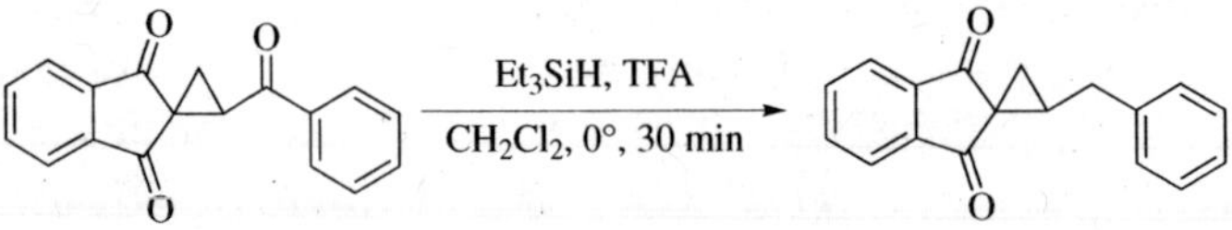

75

(81%) (Eq. 244)

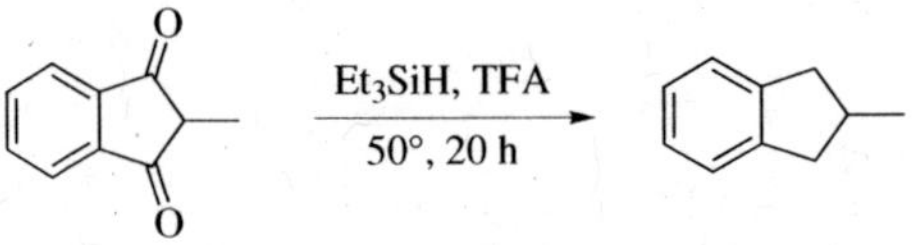

(100%) (Eq. 245)

The reduction of 2-acetylfuran to 2-ethylfuran is possible with Et_3SiH/TFA/BF_3•OEt_2 (80%).[211] Both $Et_3SiH/AlCl_3$[259] and Et_3SiH/TFA[257] reduce 2-acetylthiophene to 2-ethylthiophene without any reduction of the thiophene group. A series of strong acid catalysts and various silanes was used to study the reduction of ketones, with aliphatic ketones leading to the corresponding ethers and aromatic ketones providing methylene compounds.[353] The acylpyrone **76** is also reduced to the methylene derivative in good yield (Eq. 246).[423]

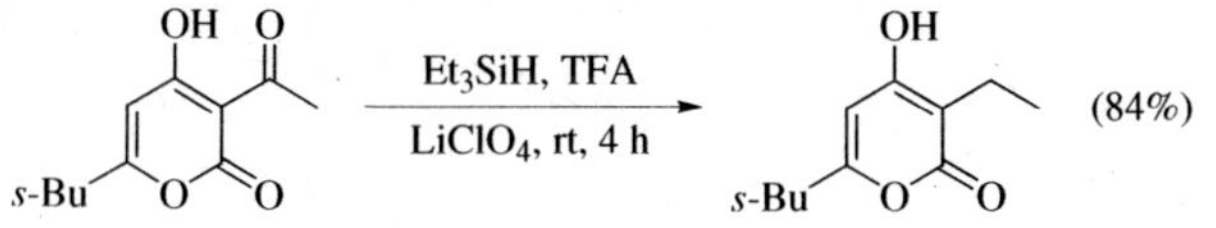

76

(Eq. 246)

The triflic acid catalyzed triethylsilane reduction of cyclohexanone gives cyclohexane, albeit in low yield.[420] 2-Methylcyclohexanone is reduced to methylcyclohexane and adamantanone to adamantane with $EtMe_2SiH/BF_3$ in good yield.[1] Adamantanone is also reduced to adamantane with Et_3SiH/BF_3•OEt_2[217] and TMSH/NaTFPB.[424] The reagent PMHS/$(C_6F_5)_3B$ effects the reduction of alkyl ketones to methylenes.[354] Thus, 1-phenyl-3-heptanone is reduced to 1-phenylheptane in excellent yield under these conditions (Eq. 247).

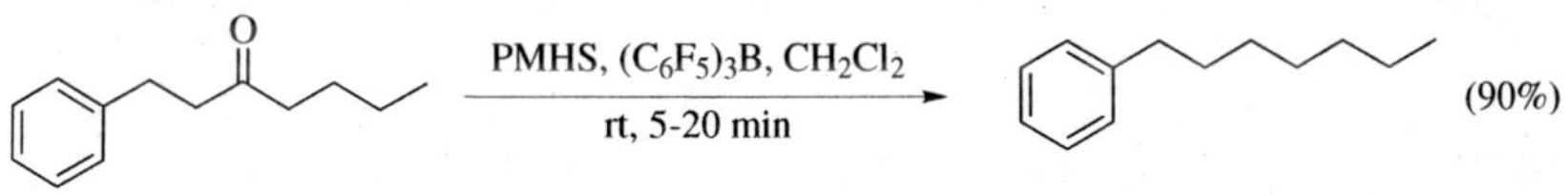

(Eq. 247)

Acyl cobalt complexes are reduced to their alkyl counterparts in good yields with Et_3SiH/TFA (Eq. 248).[183,310,425] Acyl ferrocene derivatives are reduced to the respective methylene compounds with Et_3SiH/TFA (Eqs. 249[180] and 250).[179] Acylcyclopentadienylmanganese tricarbonyl is similarly reduced in good yield.[351]

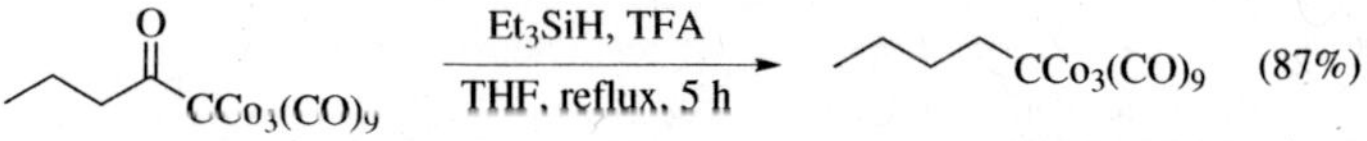

(Eq. 248)

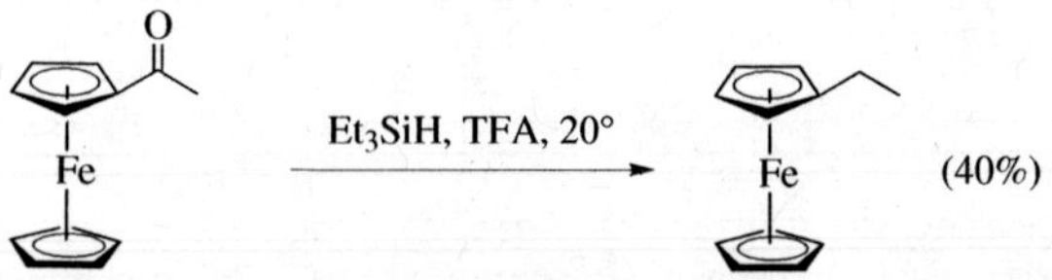

(Eq. 249)

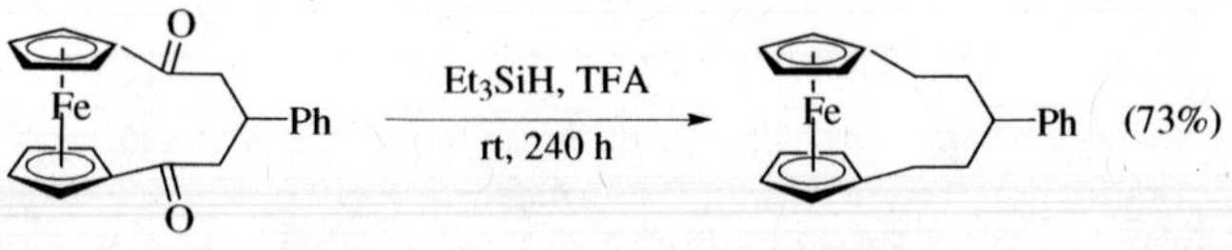

(Eq. 250)

Reductive Amination. An excellent application of organosilane reductions is found in the reductive amination of ketones. This method is a useful alternative to the cyanoborohydride approach to this transformation. In one version the ketone is reacted with *N*,*N*-diethylaminodimethylsilane, thus incorporating both the amine and the hydride in a single organosilane reagent (Eq. 251).[359] Other aminosilanes are also useful in this sequence. Another version of this transformation makes use of the inexpensive PMHS reducing agent to react a ketone directly with an amine in a two-step, one-pot, high-yield reaction.[363] The $PhSiH_3/(n\text{-}Bu)_2SnCl_2$ combination also reductively aminates ketones in good yields.[361]

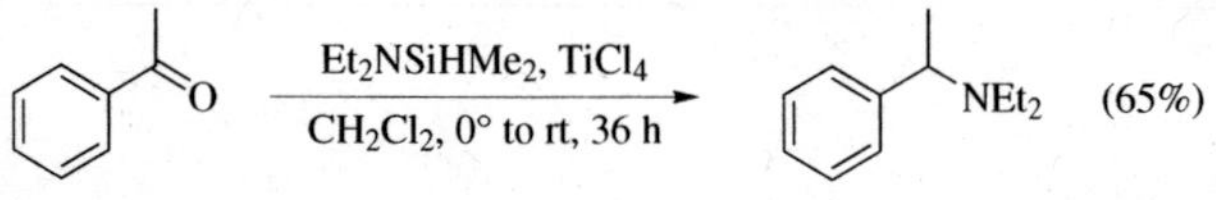

(Eq. 251)

Reductive Thiolation. Ketones are reductively thiolated when treated first with a thiol under acidic conditions followed by addition of a silane (Eq. 252).[365,426]

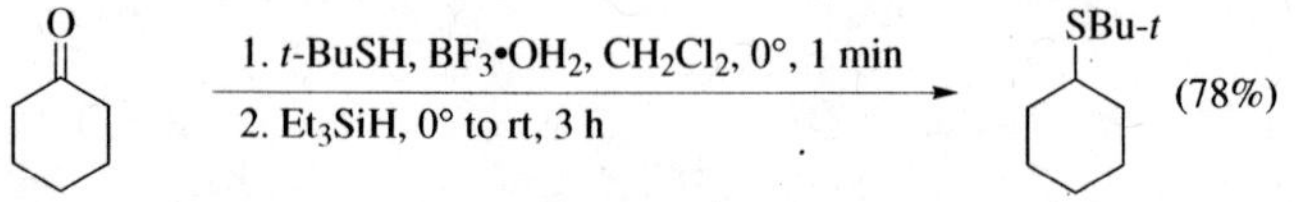

(Eq. 252)

Miscellaneous Ketone Reductions. The reductive allylation of aromatic ketones occurs with the reagent combination of Me_2ClSiH/allyltrimethylsilane/$InCl_3$ (Eq. 253).[427]

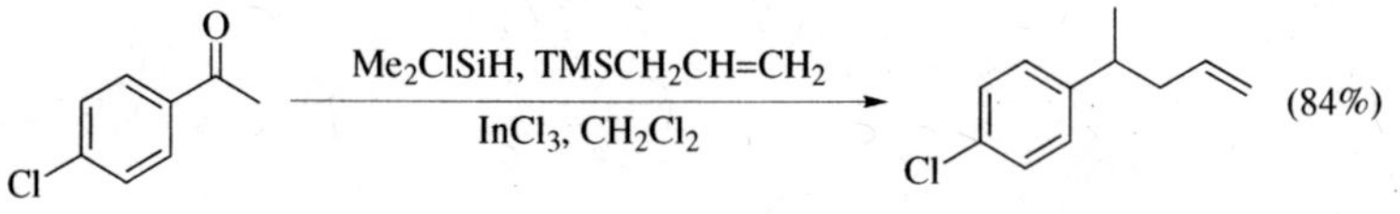

(Eq. 253)

Cyclododecanone is reduced to a mixture of *cis*- and *trans*-cyclododecene in high yield with trimethylsilane and tetrakis-3,5-bis(trifluoromethylphenyl)borate (TFPB, **77**) (Eq. 254).[424]

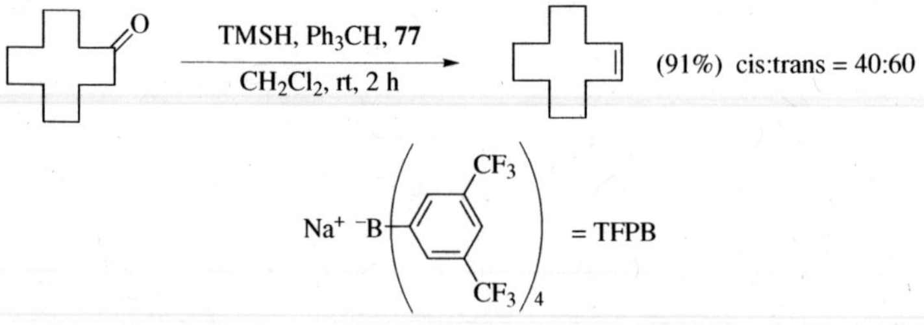

(Eq. 254)

77

Because the organosilane reduction of ketones passes through a positively charged intermediate via the complexation or protonation of the carbonyl oxygen, the presence of suitably placed C=C functions can lead to cyclizations with the hydride of the silane adding to the C=C group. This strategy applies to a number of conversions of δ,ε-unsaturated ketones into tetrahydrofuran derivatives (Eq. 255).[428,429] In these systems, the cyclization is clearly favored over the simple carbonyl reduction pathway. The phenylsilane reduction of 6-methylhept-5-ene-2-one results in the straightforward 1,2-reduction of the carbonyl group to the alcohol.[79,320,358] Under different conditions involving sulfuric acid, cyclization and hydration is preferred over reduction (Eq. 256).[243]

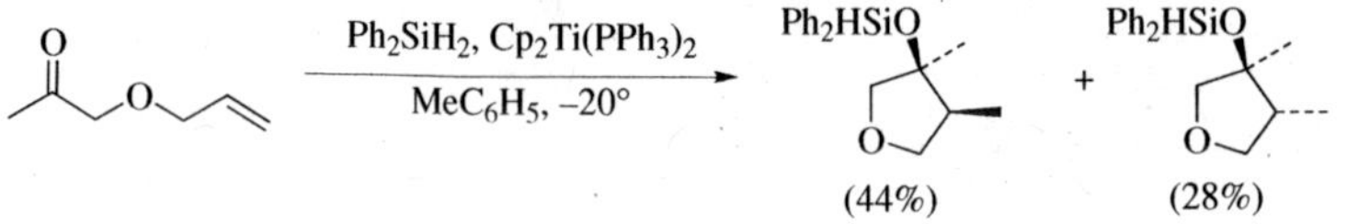

(Eq. 255)

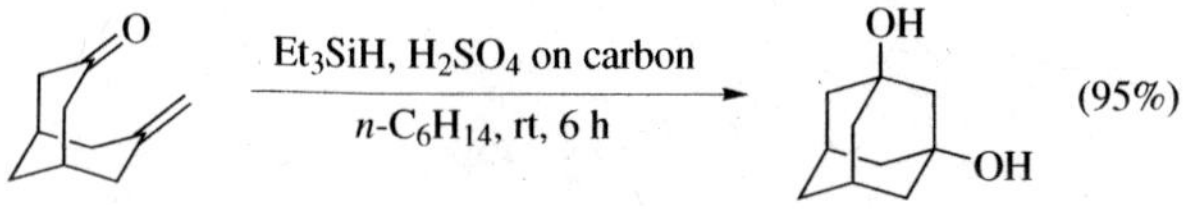

(Eq. 256)

Reduction of Amides

The organosilane/Cp_2TiF_2 reduction of N,N-diethylbenzamide illustrates the various possible pathways of reductive coupling, carbonyl reduction, and aldehyde formation (Eq. 257).[430] Conditions have been developed for all of these pathways. The combination of $PhMeSiH_2/Cp_2TiF_2$ gives good to excellent yields of the 1,2-diamine in nearly equal molar ratios of meso to racemic diastereomers (Eq. 258).[430] On the other hand, conditions for the reduction of amides to the amines include the use of $Ph_2SiH_2/HRh(CO)(PPh_3)_3$,[431] various silanes with ruthenium catalysts,[432] the $EtMe_2SiH$/Ru-complex,[280] and Et_3SiH/various Mn, Ru, Re, Os, Rh, Ir, Pd, and Pt catalysts.[432] All of these reducing systems give high yields of the amine (Eq. 259). The reduction of an amide to the aldehyde is best accomplished with $Ph_2SiH_2/Ti(OPr\text{-}i)_4$ wherein the yields range from 65% to 90% (Eq. 260).[433]

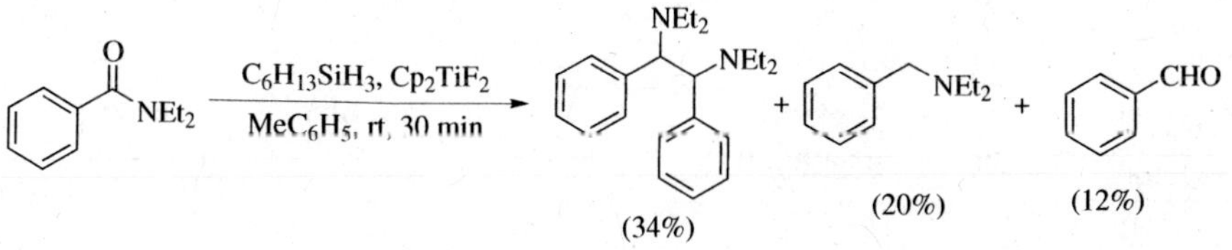

(Eq. 257)

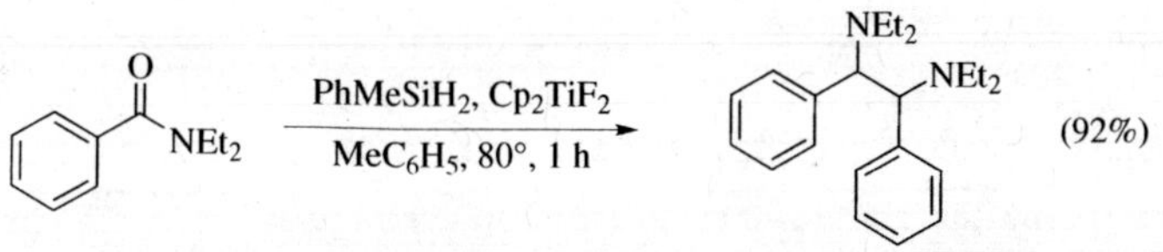

(Eq. 258)

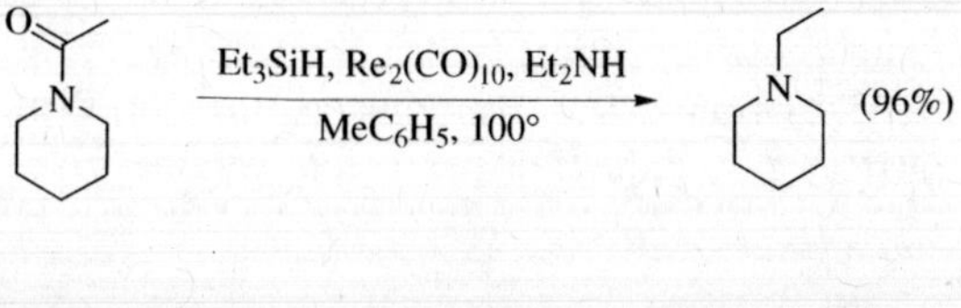

(Eq. 259)

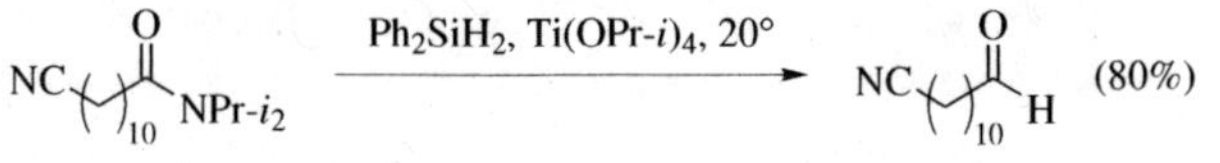

(Eq. 260)

Reduction of α,β-Unsaturated Aldehydes

When carried out under standard conditions with Et_3SiH/TFA, reduction of acrolein leads to a mixture of allyl alcohol, 1-propanol, and di-*n*-propyl ether in addition to allyl trifluoroacetate and *n*-propyl trifluoroacetate.[434] The 1,2-reduction of cinnamaldehyde with triethoxysilane in the presence of fluoride ion provides the corresponding allyl alcohol in good yields (Eq. 261).

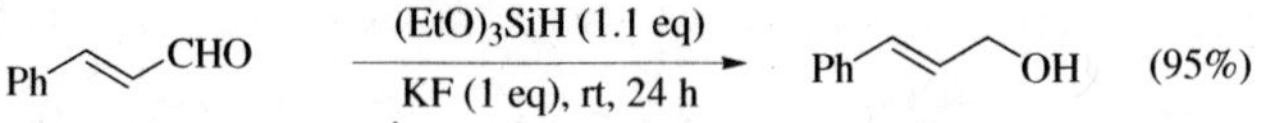

(Eq. 261)

Diphenylsilane catalyzed by various salts promotes the 1,2-reduction of cinnamaldehyde.[318] Cesium fluoride catalysis is particularly effective.[320] It is possible to stop these reactions at the silyl ether stage.[73,320] The 1,2-reduction of citral is accomplished in high yield with diphenylsilane and Wilkinson's catalyst (Eq. 262).[435] Interestingly, the trialkylsilanes, ethyldimethylsilane and triethylsilane, give high yields of the 1,4-reduction product whereas diisopropylsilane gives a 1 : 1 mixture of 1,2- and 1,4-reduction (Eq. 263).[435]

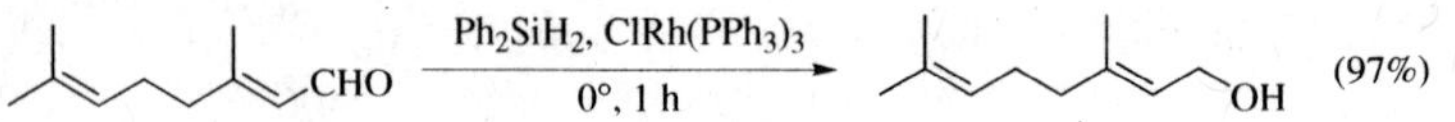

(Eq. 262)

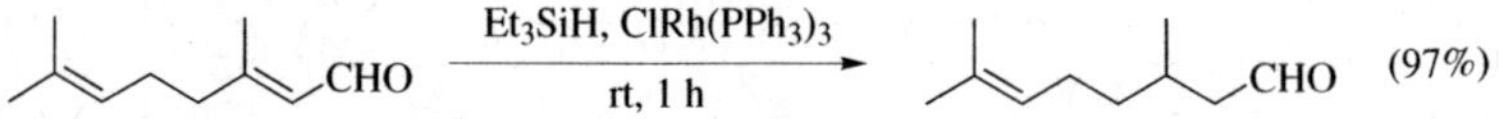

(Eq. 263)

The 1,4-reduction of α,β-unsaturated aldehydes is best carried out with diphenylsilane in the presence of zinc chloride and tetrakis(triphenylphosphine) palladium[436] or a combination of triethylsilane and tris(triphenylphosphine) chlororhodium.[437] Other practical approaches use phenylsilane with nickel (0) and triphenylphosphine[438] and diphenylsilane with cesium fluoride.[83] It is possible to isolate the initial silyl enol ether intermediate from the 1,4-hydrosilylation of α,β-unsaturated aldehydes (Eq. 264).[73,411] The silyl enol ethers are produced as a mixture of E and Z isomers.

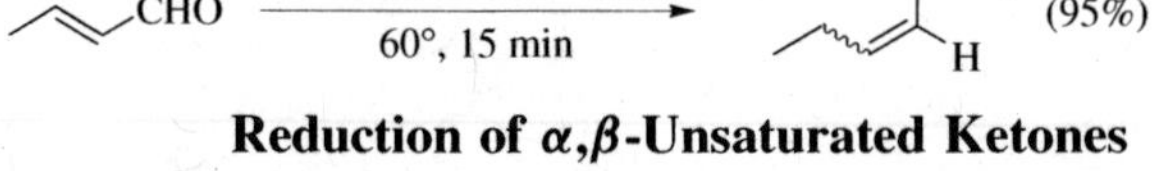

(Eq. 264)

Reduction of α,β-Unsaturated Ketones

The 1,2-hydrosilylation of α,β-unsaturated ketones is possible and provides a convenient route to allyl alcohols. The standard conditions of Et_3SiH/TFA lead to overreduction to the saturated alcohol with mesityl oxide.[434,439] The combination of $Et_3SiH/AlCl_3/HCl$ with mesityl oxide gives a mixture of the 1,2-reduction product 4-methylbut-3-ene-2-ol and the fully reduced product, 2-methylpent-2-ene.[136] The $Ph_2SiH_2/RhH(PPh_3)_4$ reduction of cyclohexenone gives reaction at both the C=O and the C=C bonds, resulting in a mixture of cyclohexenol and cyclohexanol, with cyclohexenol predominating.[320] Diethylsilane, diphenylsilane, or phenylsilane with Wilkinson's catalyst effectively reduce the carbonyl group of α,β-unsaturated ketones (Eq. 265).[435] Under similar conditions, triethylsilane gives predominantly reduction of the C=C bond (Eq. 266).[435]

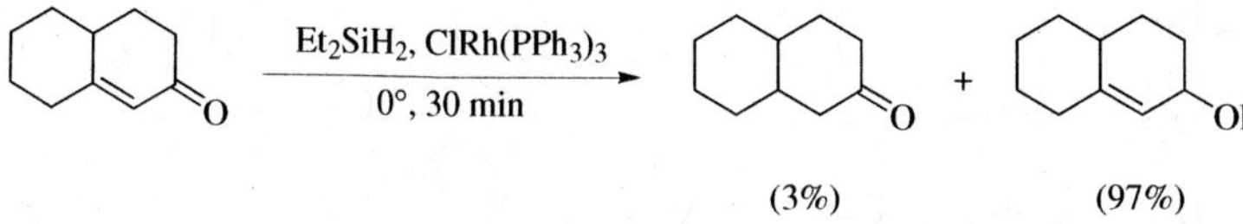

(Eq. 265)

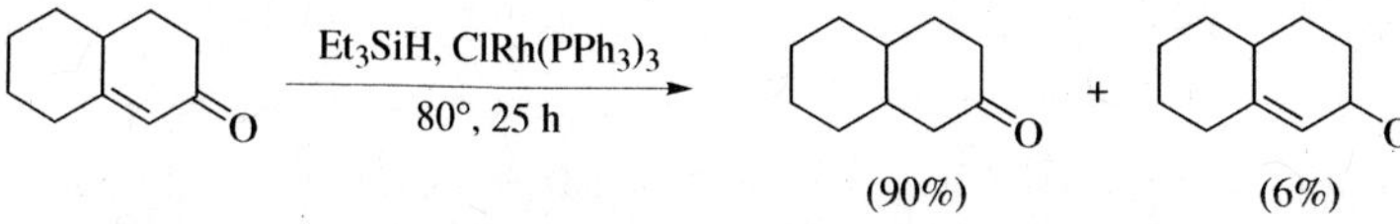

(Eq. 266)

The triethylsilane reduction of the alkylidene-1,3-dione **78** occurs in a 1,2-fashion at the acyclic carbonyl group (Eq. 267).[396]

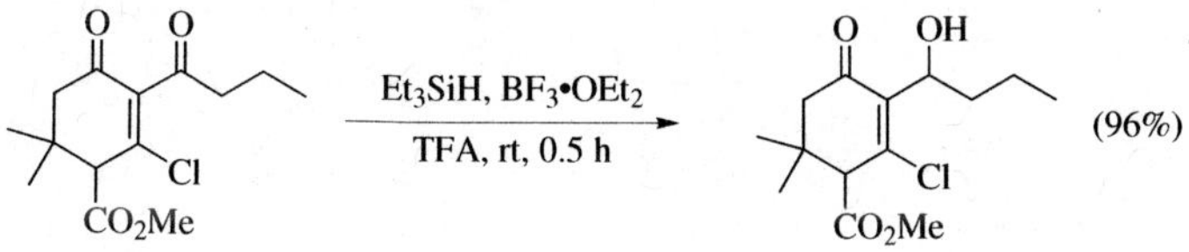

78

(Eq. 267)

The 1,2-reduction of enones is also accomplished with the combination of $PhMe_2SiH$/TBAF, conditions that show good anti stereoselectivity in the reduction of α'-substituted-α,β-unsaturated ketones (Eq. 268).[86,440] Similar behavior is seen with the reagent combinations Ph_2SiH_2/CsF,[320] Ph_2SiH_2/RhH$(PPh_3)_4$,[374] and $(EtO)_3SiH$/CsF.[80] Cyclohexenone is reduced to cyclohexenol with hydridosilicates such as **79**.[90,93] The triethylsilane/TFA reduction of dibenzylidene ketone causes reduction of the carbonyl group along with both double bonds to give 1,5-diphenyl-3-pentanol in 85% yield.[434,439]

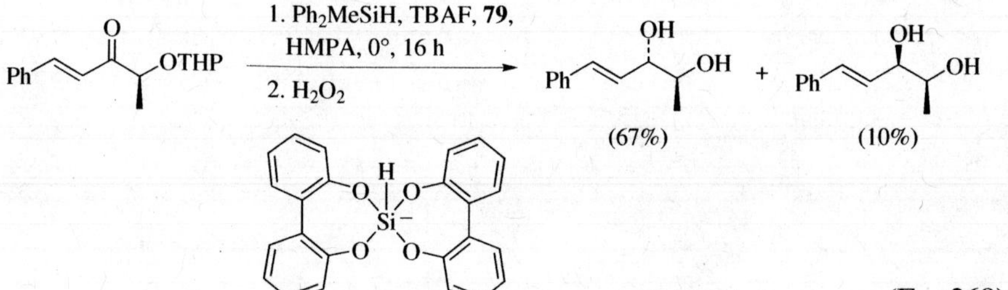

79

(Eq. 268)

The full reduction of a carbonyl to a methylene group occurs with certain β-aminoalkylidene-1,3-diones (Eq. 269).[395,441] A similar reduction is seen with the β-methoxy- and β-chloro analogs.[396] The γ-(carboxyethyl)cyclohexenone **80** is fully reduced to the methylene derivative (Eq. 270).[434,439,442]

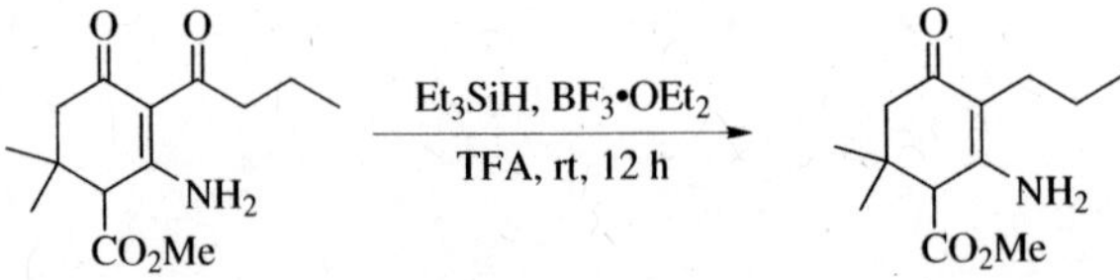

(97%) (Eq. 269)

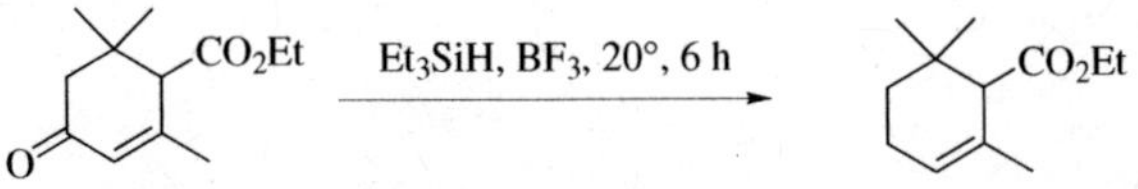

(64%) (Eq. 270)

80

α,β-Unsaturated α-aryl ketones can be reduced to the methylene compounds without concomitant reduction of the C=C bond. Enone **81** is reduced in a 1,2-fashion to the corresponding methylene compound (Eq. 271).[443]

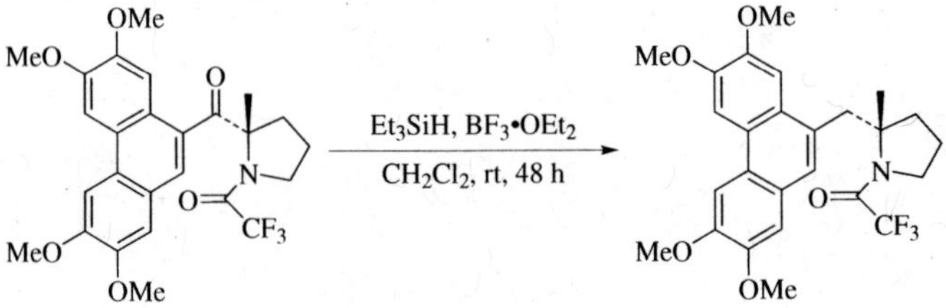

(64%) (Eq. 271)

81

The 1,4-hydrosilylation of enones can be accomplished under a variety of conditions using various reagent combinations. Among the useful ones are: Et_3SiH/TFA,[266,434,439] $Ph_2SiH_2/RhH(PPh_3)_4$,[374] $PhMe_2SiH$/CuF (or CuCl)$(PPh_3)_3$,[444–446] PMHS/Pd-nanocomposite,[219] $Ph_2SiH_2/ZnCl_2/Pd(PPh_3)_4$,[436] $Et_3SiH/RhCl(PPh_3)_3$,[435] $Ph_2SiH_2/RhCl(PPh_3)_3$,[435] $PhSiH_3/RhCl(PPh_3)_3$,[435] $EtMe_2SiH/RhCl(PPh_3)_3$,[435] $PhSiH_3/[CuH(PPh_3)]_6$,[447] Ph_2SiH_2/CsF,[83] $PhSiH_3/Mn(dpm)_3$,[448] $Et_3SiH/TiCl_4$,[435,449] $Ph_3SiH/Mo(CO)_6$,[450] $PhMe_2SiH$/CuF/$(PPh_3)_3$,[444,445] $Cl_3SiH/Ni/CoCl_2$,[451] and PMHS/Pd/C.[316] All of these methods show good to excellent selectivity for 1,4- over 1,2-hydrosilylation. The $PhMe_2SiH$/CuCl system reduces the unsaturated diketone in Eq. 272 with a high degree of cis-selectivity.[445]

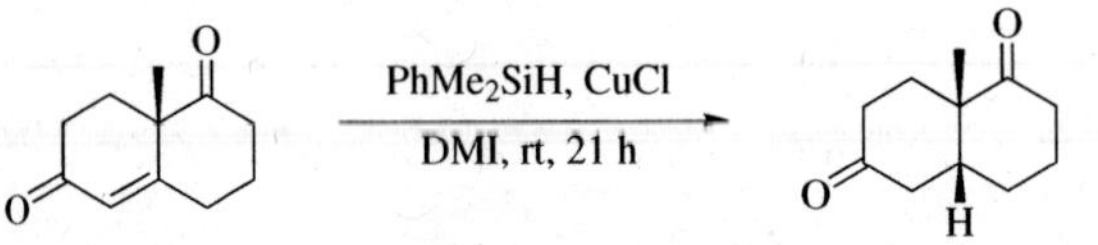

(81%) cis:trans = 98:2 (Eq. 272)

The $PhSiH_3/Mn(dpm)_3$ combination nicely reduces the conjugated C=C bond in a polyfunctional ketone without affecting the trityl group or causing reaction at the α-alkoxyketone function (Eq. 273).[448]

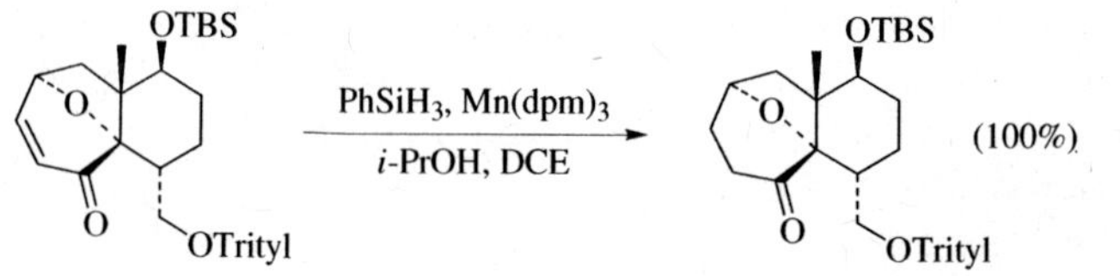

(Eq. 273)

$\alpha,\beta,\gamma,\delta$-Unsaturated ketones are reduced at the α,β-double bond with the combination $PhMe_2SiH/RhCl(PPh_3)_3$ (Eq. 274).[437,452] Employment of diethylsilane or diphenylsilane results in only the 1,2-hydrosilylation products.

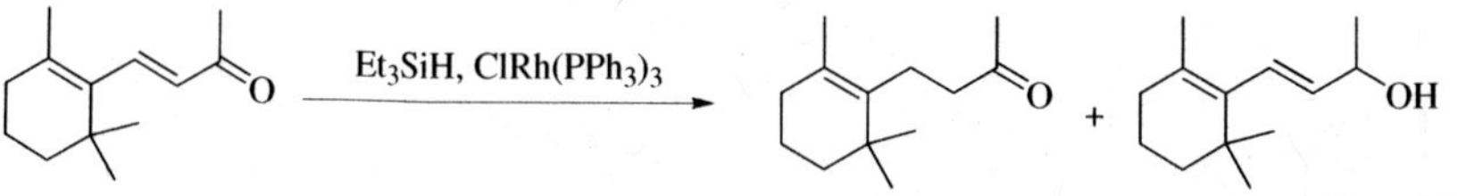

(Eq. 274)

The 1,4-reduction of an enone in the presence of a ketal is shown in Eq. 275.[436] The Et_3SiH/TFA system reduces the polyfunctional cyclic α,β-enone in Eq. 276 without affecting the α-hydroxy carboxyl or α-chloro keto groups.[453]

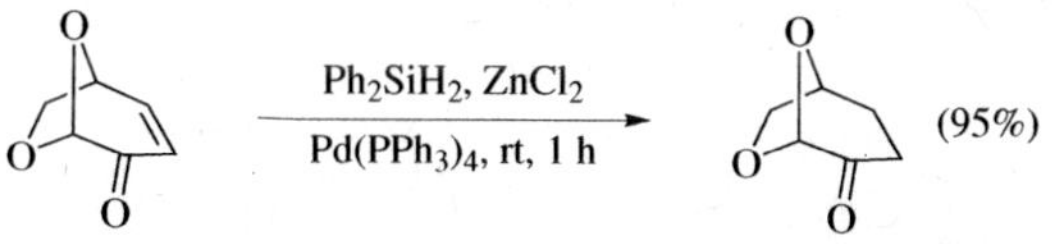

(Eq. 275)

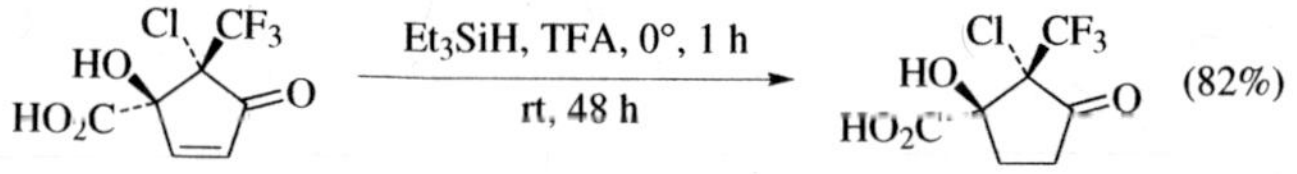

(Eq. 276)

The use of PMHS/TFA and a base leads to the 1,4-reduction of enones (Eq. 277).[454]

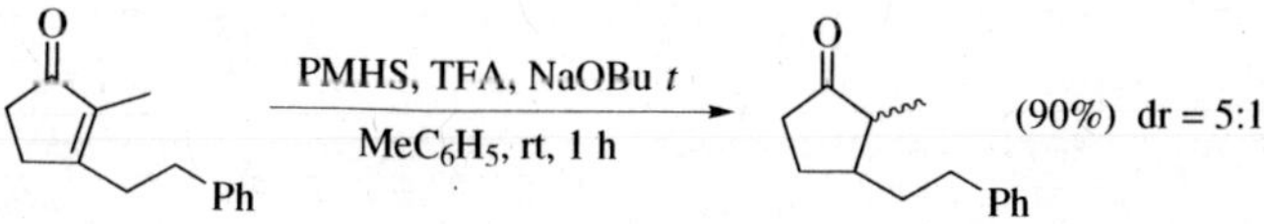

(Eq. 277)

The 1,4-hydrosilylation of enones can be used as a method for the introduction of the silyl enol ether functionality, and may be accomplished with the combination of $PhMe_2SiH/RhH(PPh_3)_4$ (Eq. 278),[374] $Et_3SiH/RhCl(PPh_3)_3$,[411] $(HMe_2Si)_2O/[CuH(PPh_3)]$,[455] Et_3SiH/Pt-complex,[456] chloromethyldimethylchlorosilane/$RhH(PPh_3)_4$,[374] or $Ph_3SiH/RhCl(PPh_3)_3$.[411]

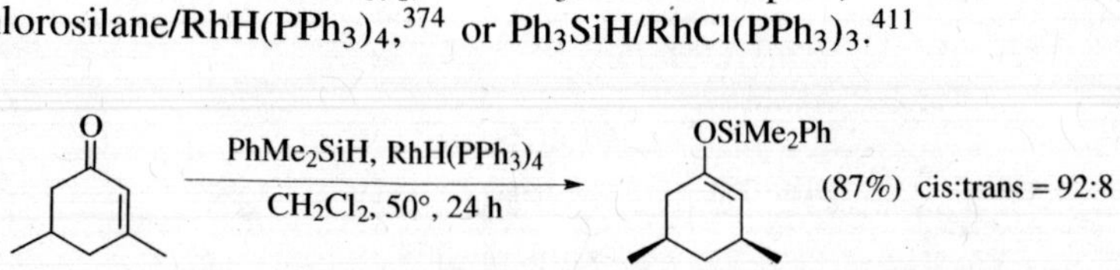

(Eq. 278)

Several steroidal α,β-unsaturated ketones have been subjected to silane reductions. In particular, enones of ring A have been reduced in a 1,4-manner to give the saturated ketone with a preference for trans AB ring fusion. This and comparable transformations are nicely accomplished with the $Ph_2SiH_2/ZnCl_2/Pd(PPh_3)_4$ system (Eq. 279)[457] used in the 1,4-reduction of non-steroidal enones. Other conditions for this transformation use $Et_3SiH/TiCl_4$[449] and Et_3SiH/TFA,[243] but these lead to mixed stereochemistry at the AB ring fusion. The $PhSiH_3/Mo(dpm)_3$ combination is also used for this reduction.[448]

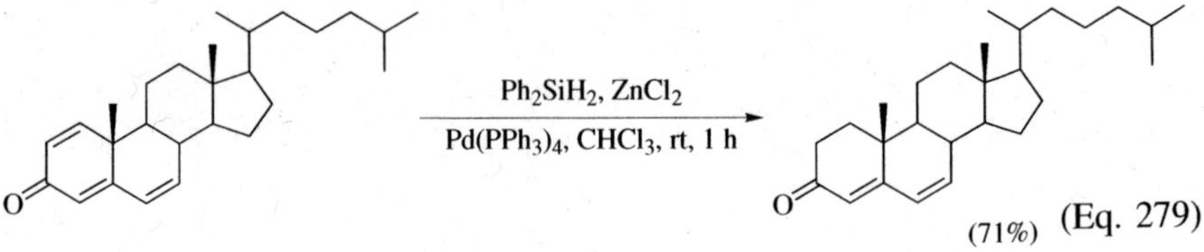

(Eq. 279)

The conjugated dienone **82** reacts with $Et_3SiH/TiCl_4$ to yield the nonconjugated product **83** shown in Eq. 280.[449] Other analogous dienones behave similarly with triethylsilane, but TMDO gives the best yields.[458]

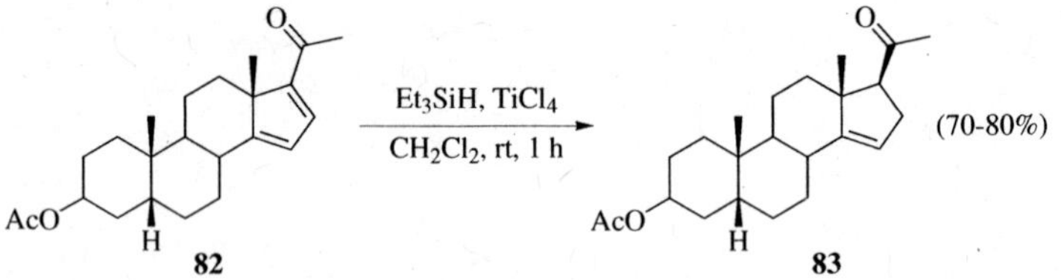

82 **83**

(Eq. 280)

The intermediate enolate or enol ether from the initial reduction of an enone may be alkylated in situ (Eq. 281).[455] β-Substituted cyclopentenones may be asymmetrically reduced and alkylated[459] (see section on asymmetric reductions of enones). Enolates may also be trapped with an aldehyde in a reductive aldol condensation of an enone with an aldehyde,[455] permitting a regioselective aldol condensation to be carried out as shown in Eq. 282.[455] This class of reductive aldol condensation reactions can also occur in a cyclic manner (Eq. 283).[460]

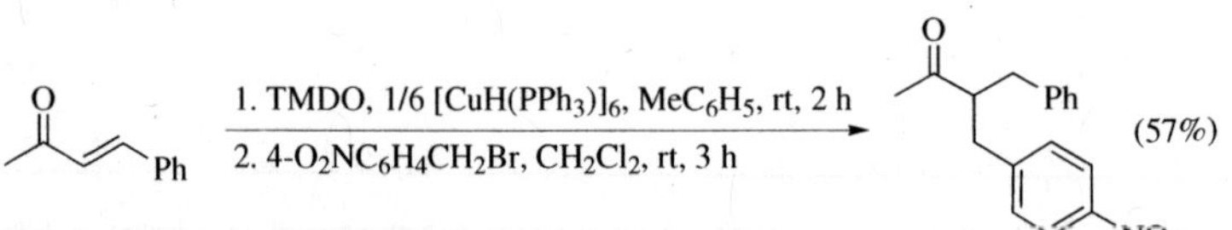

(Eq. 281)

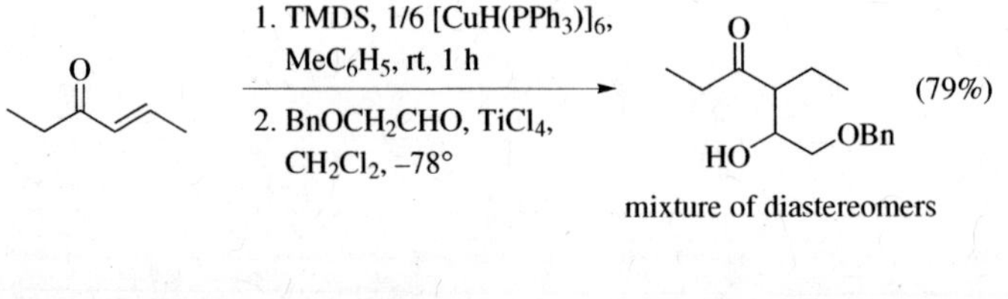

mixture of diastereomers

(Eq. 282)

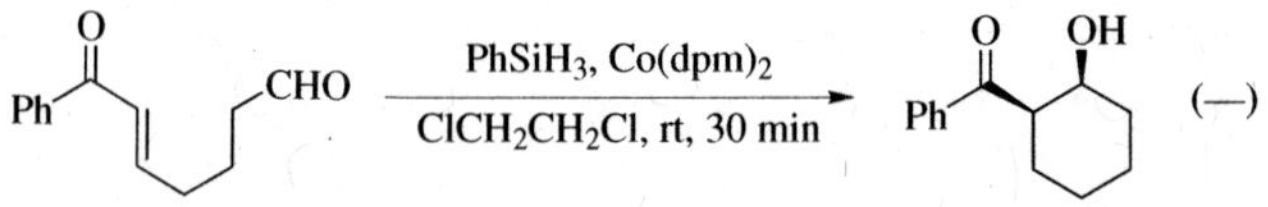

(Eq. 283)

A tandem 1,4-hydrosilylation/Michael addition of suitably arranged bis enones is reported employing the systems $PhSiH_3/Co(dpm)_2$ (Eq. 284),[460,461] various silanes/$Co(dpm)_2$,[462] $PhMe_2SiH/Co(dpm)_2$,[460] $Et_3SiH/AlCl_3$,[461] and $Et_2MeSiH/Rh_4(CO)_{12}$.[464] The competing 2 + 2 cycloaddition reaction can be a complication (Eq. 285).[460]

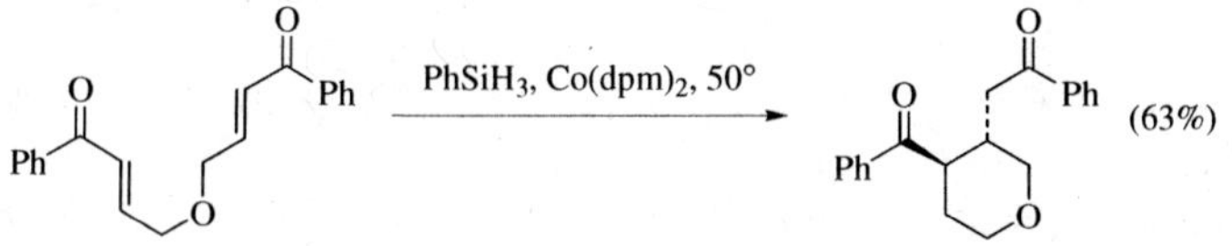

(Eq. 284)

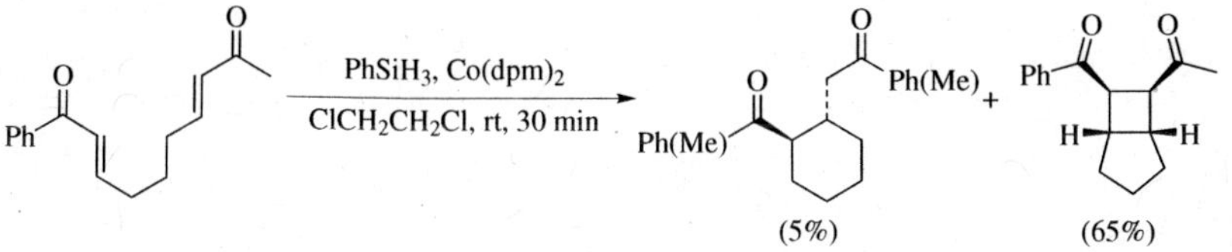

(Eq. 285)

A further example of the trapping of the in situ generated silyl enol ether from the reduction of an enone is the conversion of an enone into an α-hydroxy ketone via oxidation of the silyl enol ether (Eq. 286).[465]

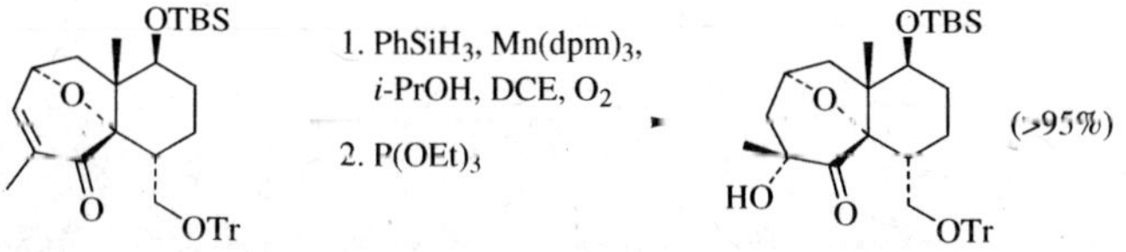

(Eq. 286)

Reduction of α,β-Unsaturated Esters

The reduction of the C=C bond of α,β-unsaturated esters has been carried out with various silane/catalyst combinations. The combination of $Et_3SiH/RhCl(PPh_3)_3$ gives the silyl ketene acetal along with the silylated hemiacetal, the result of the 1,2-reduction of the carbonyl group (Eq. 287).[466] The use of other silanes in this transformation gives similar results. Acrylates give hydrosilylation of the C=C bond, leading to β-silyl esters and their silyl ketene acetals as the major products (Eq. 288).[466,467]

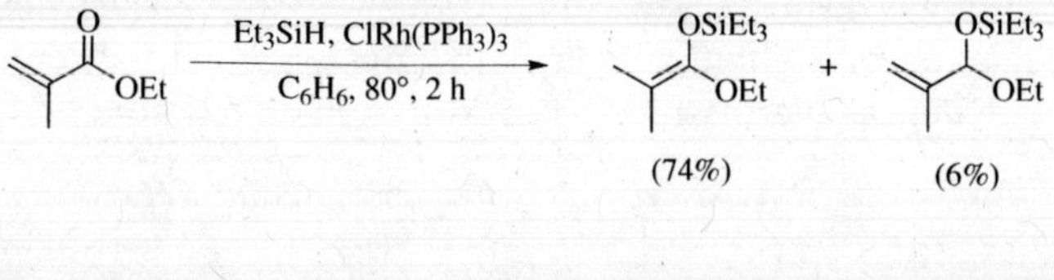

(Eq. 287)

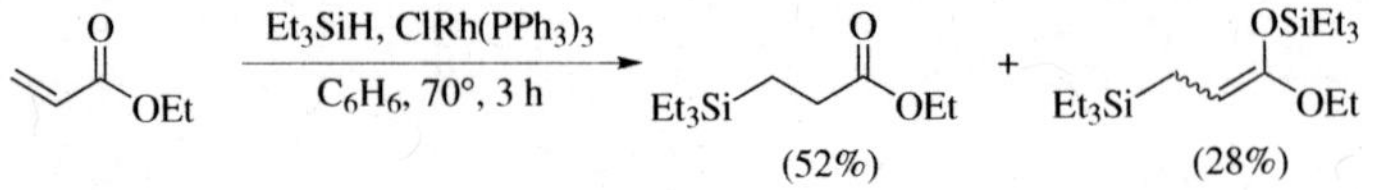

(Eq. 288)

Triethoxysilane and $RhH(PPh_3)_4$ produce the silyl ketene acetal in high yield.[374] Methyl acrylate is reduced to methyl propionate with $Cl_3SiH/CoCl_2$[448] or PMHS/Pd-nanocomposite catalyst,[219] and to a mixture of methyl propionate, the methyl β-silylpropionate, and the methyl α-silylpropionate with R_3SiH/H_2PtCl_6 or $R_3SiH/RhCl(PPh_3)_3$ where R = Me, Et, or *n*-Pr.[467] The reduction of ethyl acrylate shows similar behavior.[451,466] The combination of $Et_3SiH/Mo(CO)_6$ works well for the conjugate reduction of α,β-unsaturated esters.[450] This method appears to be free of some of the side reactions mentioned above. The use of metallic nickel, triphenylphosphine, and phenylsilane reduces α,β-unsaturated esters to the saturated esters in moderate yields.[438] Dimethylphenylsilane carries out the same transformation in the presence of CuCl.[444,445] Triethylsilane and trifluoroacetic acid can reduce α,β-unsaturated esters in good yields[468] as can Et_3SiH/TMSOTf.[469] Excellent yields of the silane reduction of α,β-unsaturated esters are obtained by the reaction of PMHS and sodium *tert*-butoxide in the presence of catalyst **73** (Eq. 289).[454]

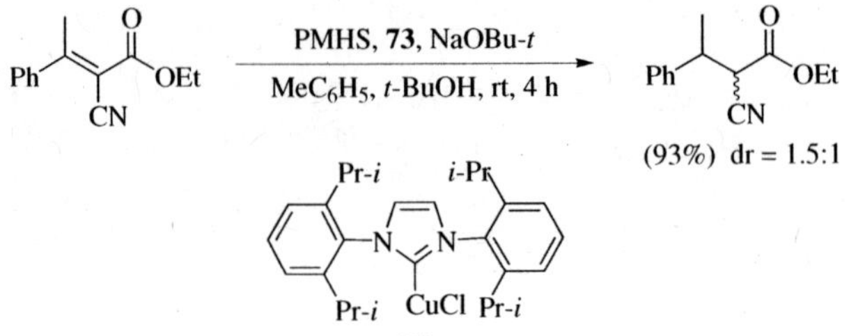

73

(Eq. 289)

The $Ph_2SiH_2/AlCl_3$ reduction of β-sulfonyl-α,β-unsaturated esters results in the formation of the β-sulfonyl ester. Good yields are obtained and $AlCl_3$ is the best Lewis acid catalyst for this reaction (Eq. 290).[373]

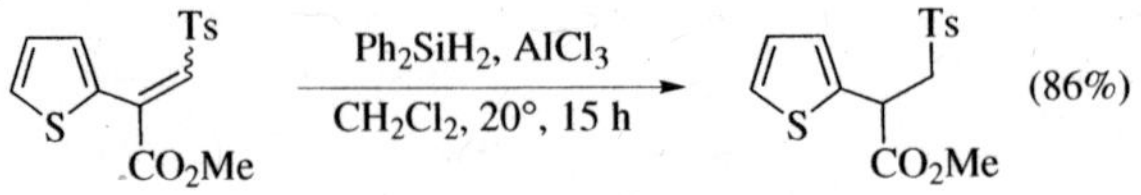

(Eq. 290)

In the reductive aldol condensation of an α,β-unsaturated ester and an aldehyde shown in Eq. 291, the initial step is believed to be the addition of an in situ formed rhodium hydride to the α,β-unsaturated ester, followed by reaction of the resulting rhodium enolate with the aldehyde.[470] The reaction has been carried out both inter-[470] and intramolecularly[471,472] as well as in an asymmetric fashion (Eq. 291).

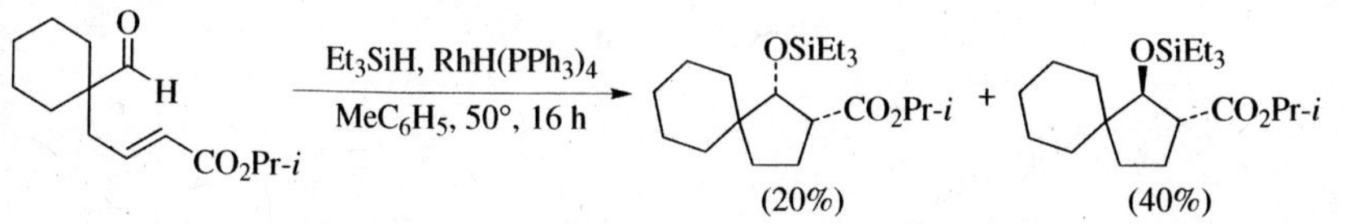

(Eq. 291)

Similarly, methyl methacrylate reacts with ketones and TMSH/$RuCl_3$•$3H_2O$ to give β-trimethylsiloxy-2,2-dimethyl methyl esters in good yields. A lactone example is shown in Eq. 292.[473] Methyl acrylate, trans methyl (*E*)-cinnamate, and 3,4-dehydro-δ-lactone react in an analogous manner, albeit in lower yields.[473]

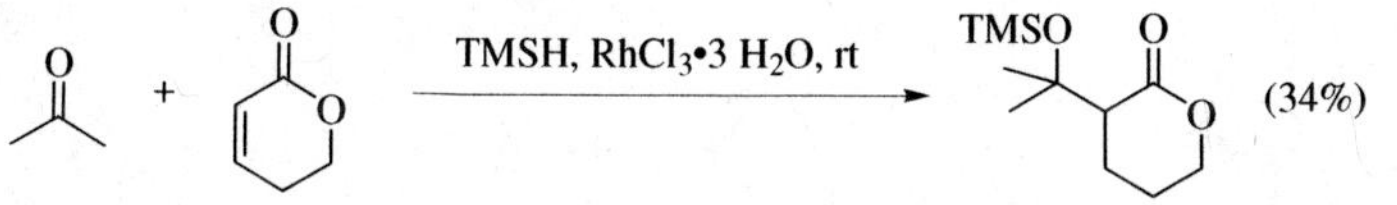

(Eq. 292)

Allyl acrylates have been reacted with the combination of $ClMe_2SiH$/ $[(cod)RhCl]_2$/Me-DuPHOS (1,2-bis(2,5-dimethylphospholano)benzene) to bring about reduction of the α,β-unsaturated ester followed by a Claisen rearrangement to the γ,δ-unsaturated carboxylic acid (Eq. 293).[474] Other silanes did not perform as well in this sequence.

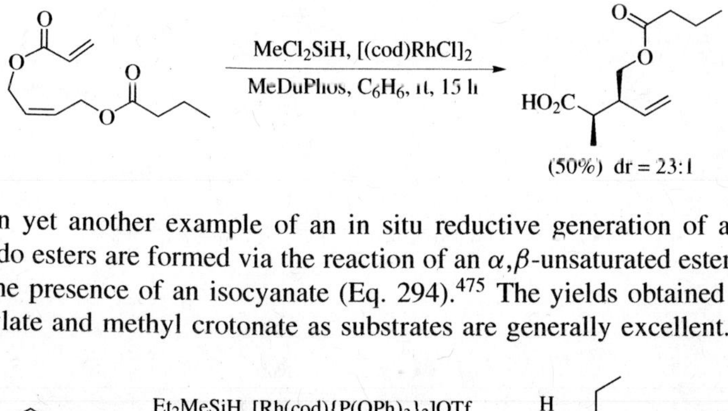

(Eq. 293)

In yet another example of an in situ reductive generation of an enolate, β-amido esters are formed via the reaction of an α,β-unsaturated ester with a silane in the presence of an isocyanate (Eq. 294).[475] The yields obtained using methyl acrylate and methyl crotonate as substrates are generally excellent.

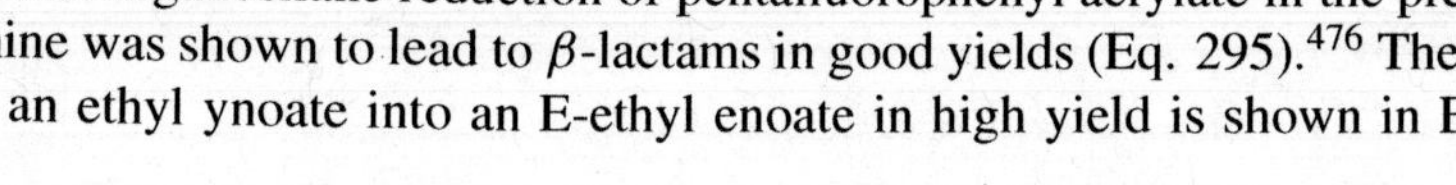

(Eq. 294)

The organosilane reduction of pentafluorophenyl acrylate in the presence of an imine was shown to lead to β-lactams in good yields (Eq. 295).[476] The conversion of an ethyl ynoate into an E-ethyl enoate in high yield is shown in Eq. 296.[477]

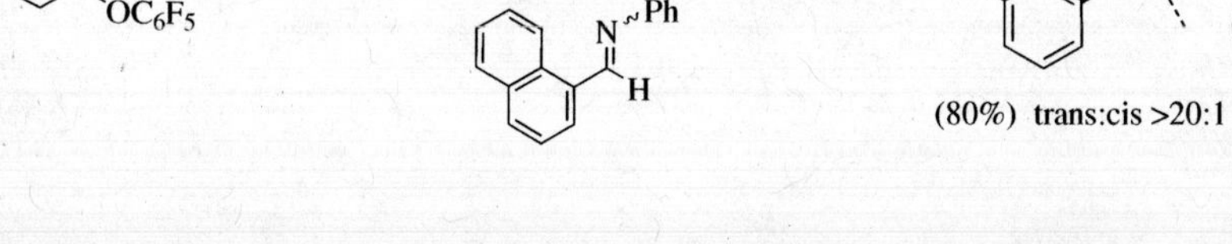

(Eq. 295)

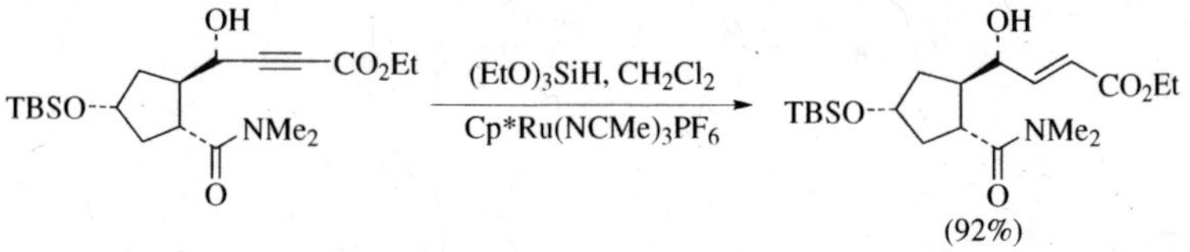

(Eq. 296)

Reduction of α,β-Unsaturated Amides

The conjugate hydrosilylation of α,β-unsaturated amides can be carried out in high yields with $PhSiH_3/Mo(CO)_6$ (Eq. 297)[450] or $Ph_2SiH_2/ZnCl_2/Pd(PPh_3)_4$.[436] Primary, secondary, and tertiary amides are equally reactive.[450] The reduction of a β-tributylstannyl-α,β-unsaturated tosylamide is also reported.[469]

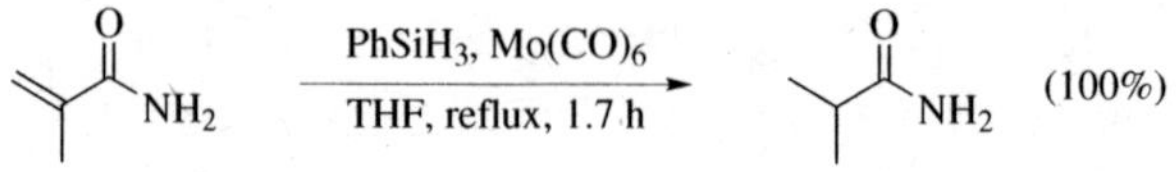

(Eq. 297)

Reduction of α,β-Unsaturated Nitriles

The reaction of α,β-unsaturated nitriles with organosilanes in the presence of Wilkinson's catalyst gives the α-silyl nitriles in good yields (Eq. 298).[466] The $PhSiH_3/Mo(CO)_6$ combination reduces α,β-unsaturated nitriles to the nitriles in good yields, although the β,β-disubstituted 3-methylcrotonitrile does not react.[450] Cinnamonitrile is reduced to the saturated nitrile in good yield with PMHS/Pd-nanocomposite[219] and in only modest yield with $Ph_2SiH_2/ZnCl_2/Pd(PPh_3)_4$.[436] The ferrocenyl unsaturated nitrile shown in Eq. 299 is reduced with Et_3SiH/TFA in good yield.[179] The $PhSiH_3$/Co(dpm) reductive aldolizations of an α,β-unsaturated nitrile and an aldehyde also occur in good yields (Eq. 300).[478]

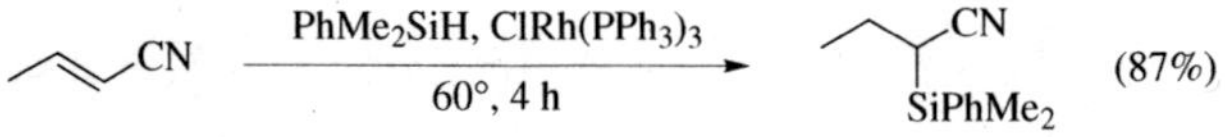

(Eq. 298)

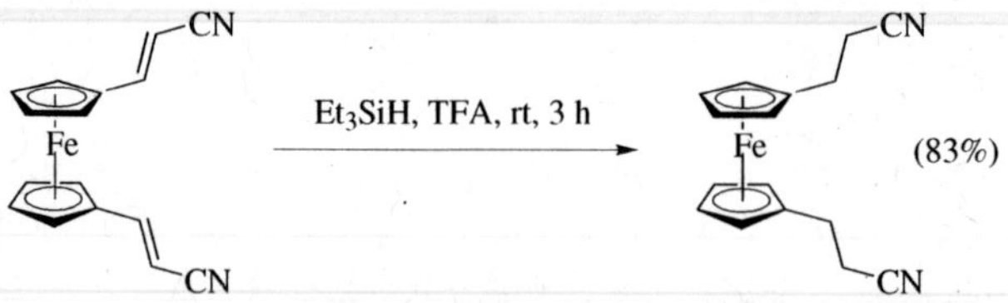

(Eq. 299)

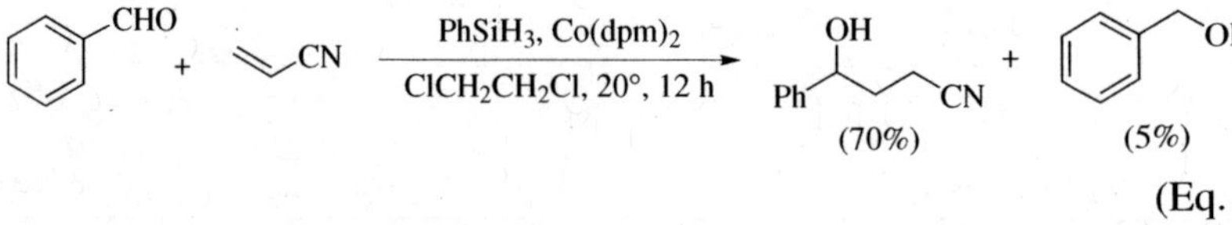

(Eq. 300)

Reduction of Acetals, Ketals, Hemiacetals, Hemiketals, and Orthoesters

The ketals of both aryl and alkyl ketones can be reduced to the methylene derivatives in good yields (Eqs. 301 and 302).[479]

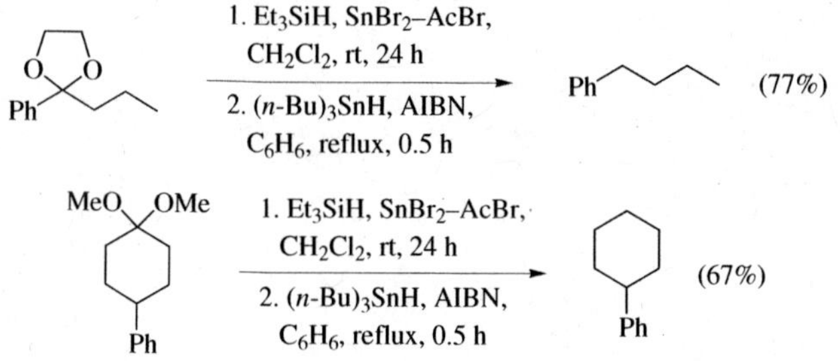

(Eq. 301)

(Eq. 302)

Simple dimethylacetals and ketals are reduced with the Et_3SiH/acid combination where the acid catalyst can be TMSOTf,[339,480–482] $BF_3{\cdot}OEt_2$,[483–485] TFA,[327,486] Nafion®-H,[335] tin-montmorillonite,[353] FSO_3H/BSA,[487] $TiCl_4$,[488–492] $AlCl_3$/HCl,[146] and FSO_3H/BSU.[487] Of these, Nafion®-H, $BF_3{\cdot}OEt_2$, BSA, and TMSOTf all give excellent isolated yields of the corresponding ether. The 1, 3-propanediol acetals of acetophenone and benzaldehyde are reduced to mono-ortho-benzyldiols with $PhSiH_3/RhCl(PPh_3)_3$ (Eq. 303),[493] Et_3SiH/TFA,[494] $Et_3SiH/BF_3{\cdot}OEt_2$,[495] or PMHS/$AlCl_3$ (Eq. 304).[496]

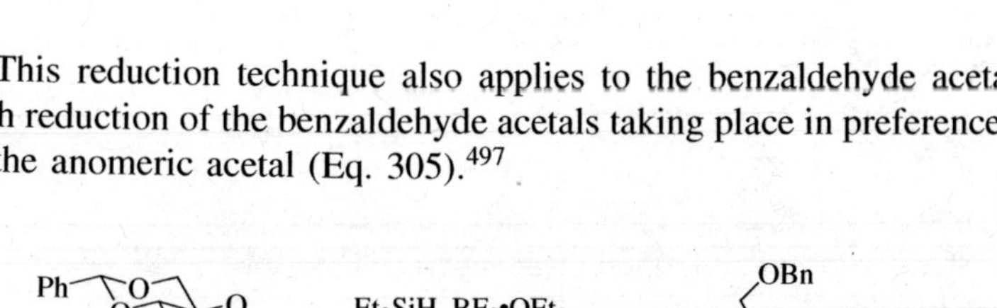

(Eq. 303)

(Eq. 304)

This reduction technique also applies to the benzaldehyde acetals of sugars with reduction of the benzaldehyde acetals taking place in preference to reduction of the anomeric acetal (Eq. 305).[497]

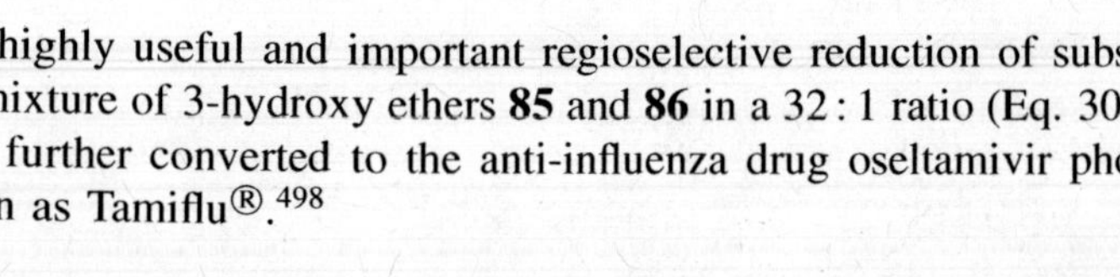

(Eq. 305)

A highly useful and important regioselective reduction of substrate **84** leads to a mixture of 3-hydroxy ethers **85** and **86** in a 32 : 1 ratio (Eq. 306). Compound **85** is further converted to the anti-influenza drug oseltamivir phosphate, better known as Tamiflu®.[498]

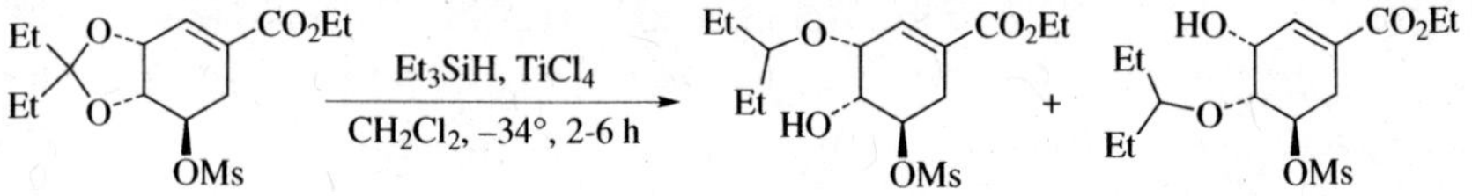

84 85 (56%) 32:1 86

(Eq. 306)

The triethylsilane reduction of the peroxy ethyl ether shown in Eq. 307 takes place at the C–O bond of the methyl ether without reduction of either the iodide or the peroxide functionalities (Eq. 307).[499] In contrast, a bridged peroxy ether undergoes reduction of both C–O bonds of the peroxide linkage rather than at the ether bridge (Eq. 308).[499]

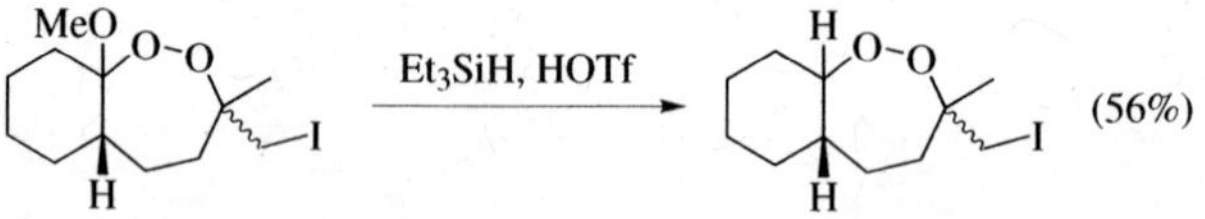

(Eq. 307)

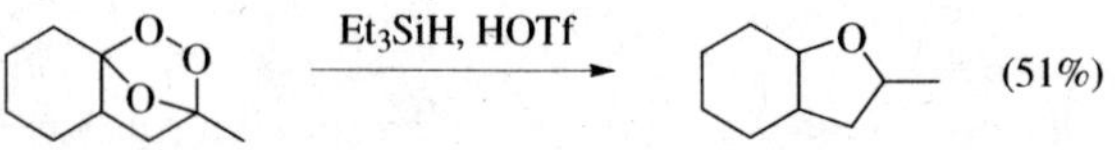

(Eq. 308)

single unidentified diastereomer

The Et_3SiH/tetracyanoethylene combination reduces acetals and ketals to the corresponding ethers but the yields are mixed.[500] The full reduction of benzaldehyde acetals to the toluene derivatives is realized by the initial reduction with $Et_3SiH/SnBr_2$–AcBr followed by Bu_3SnH/AIBN (azobis(isobutyronitrile)) or $LiAlH_4$.[479] The overall yields are excellent.

Diphenylmethylsilyl-protected hemiacetals are reduced upon treatment with $Ph_2SiH_2/Mn(CO)_3Ac$.[295] $Et_3SiH/TiCl_4$ reduces *tert*-butyldimethylsilyl ketals.[306] The combination of TBSH/$Sn(OTf)_2$ and a silyl ether converts ethylene glycol acetals and ketals into ethers (Eq. 309).[501]

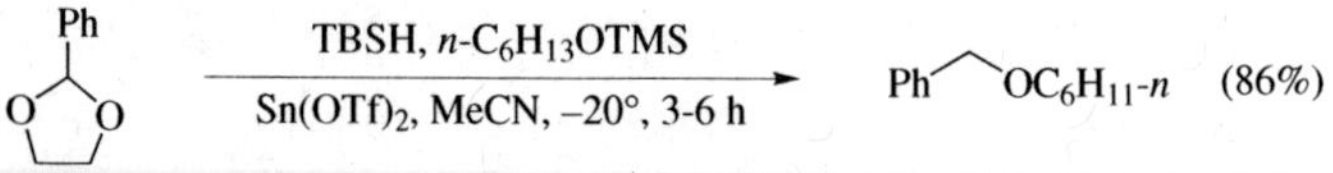

(Eq. 309)

α-Trimethylsilyloxythianes are reduced to the respective thianes with Et_3SiH/TMSCl/$InCl_3$ (Eq. 310).[426,502] Trimethylsilane with TMSI or TMSOTf effects this conversion as well.[392]

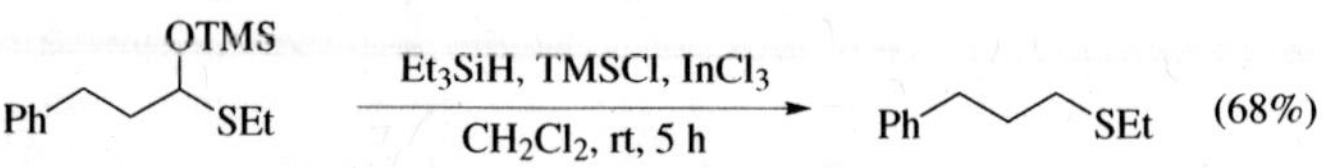

(Eq. 310)

A number of conditions are available for the reduction of the anomeric acetals to the reduced sugars. Among these are the combinations $Et_3SiH/BF_3{\bullet}OEt_2$/TFA[483] and Et_3SiH/TMSOTf,[503,504] although some isomerization is found to occur with this latter system. The polysaccharide shown in Eq. 311 is converted into the the deoxysugar in good yield.[483]

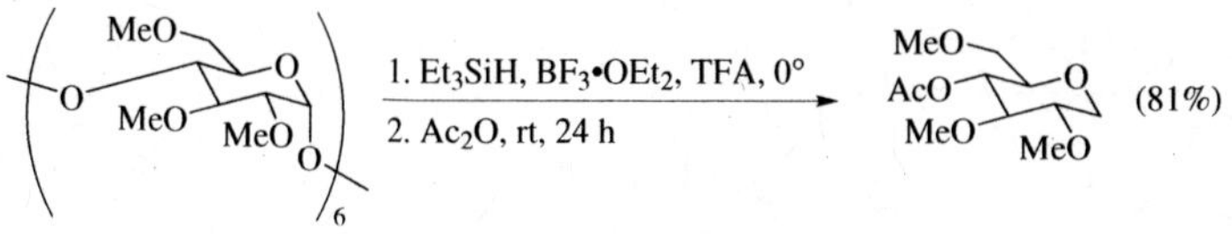
(Eq. 311)

Diiodosilane reduces acetals to alkyl iodides in a reductive iodination reaction (Eq. 312).[358,505] Alkyl bromides are formed from the reductive bromination of benzaldehyde acetals with the combination $Et_3SiH/SnBr_2$.[506]

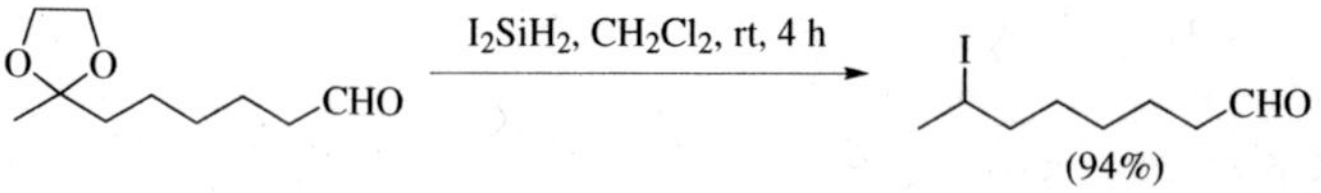
(Eq. 312)

There is a report of a reductive amination of an acetal. Thus, $(Et_2N)Me_2SiH/TiCl_4$ reacts with the dimethylacetal of benzaldehyde to form benzyldiethylamine in good yield (Eq. 313).[359]

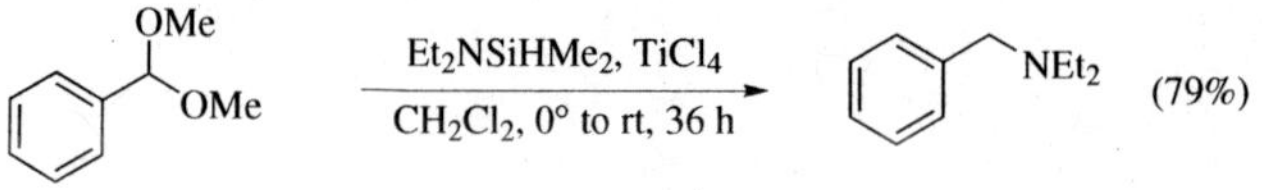
(Eq. 313)

Hemiacetals and hemiketals also undergo reduction to ethers with organosilanes under acid catalysis. These reductions generally occur in good yield. They are carried out with $Et_3SiH/BF_3{\bullet}OEt_2$,[497,507–512] Et_3SiH/TFA,[162,513–516] $PhMe_2SiH/TMSOTf$,[517] and $Et_3SiH/TMSOTf$,[518,519] which prove especially useful for sugar systems. The combination of $Ph_3SiH/TiCl_4$ reduces hemiketals in the presence of a dithioketal (Eq. 314).[520]

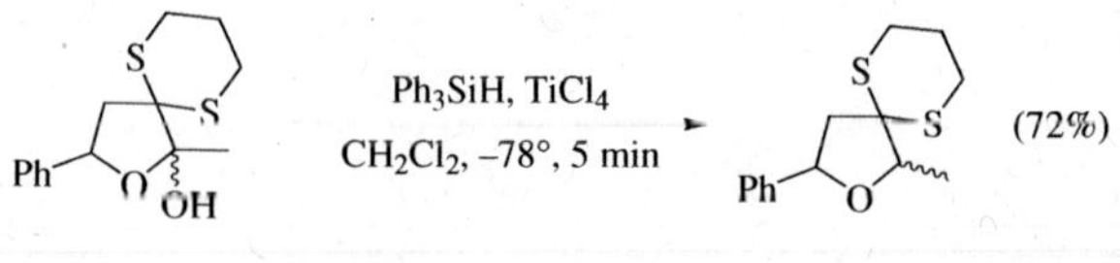
(Eq. 314)

The $PhSiH_3/RhCl(PPh_3)_3$ combination reduces trimethyl orthobenzoate to the dimethyl acetal of benzaldehyde with small amounts of methyl benzyl ether product (Eq. 315).[493]

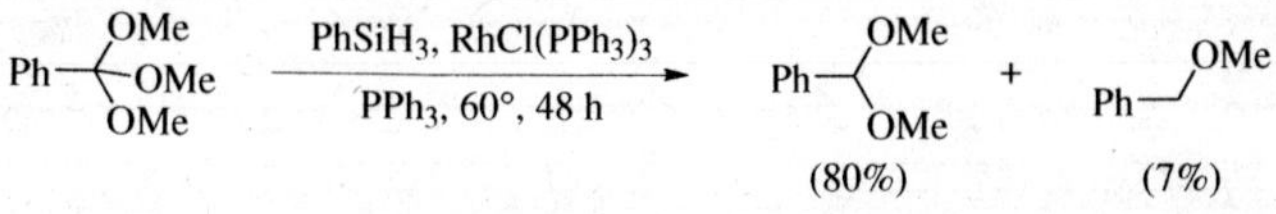
(Eq. 315)

Reduction of Aminals and Hemiaminals

As demonstrated with acetals and ketals, aminals are also readily reduced with silanes under acid catalysis. The $Et_3SiH/BF_3{\bullet}OEt_2$ combination reduces aminals to the amines in excellent yields (Eq. 316).[521–523] A similar reduction with Et_3SiH/TFA occurs in equally high yield (Eq. 317).[524] Hemiaminals are reduced with Et_3SiH/TFA (Eq. 318).[525]

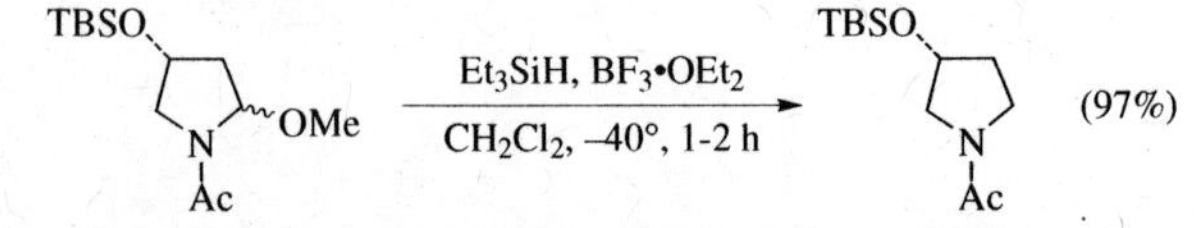

(Eq. 316)

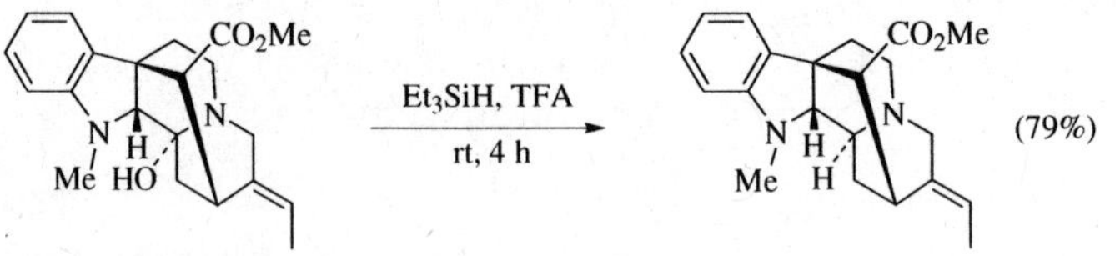

(Eq. 317)

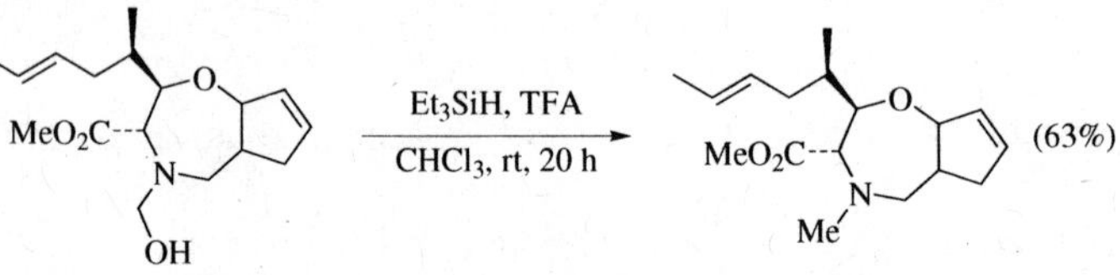

(Eq. 318)

Amidoaminals and amidohemiaminals are reduced to the amides with organosilanes and an acid catalyst. Best among the reported reagent combinations are $Et_3SiH/TiCl_4$,[524] and $Et_3SiH/BF_3{\bullet}OEt_2$.[521] Et_3SiH/TFA[526] (Eq. 319) and $Et_3SiH/BF_3{\bullet}OEt_2$[527] are effective in reducing amidoaminals in high yields.

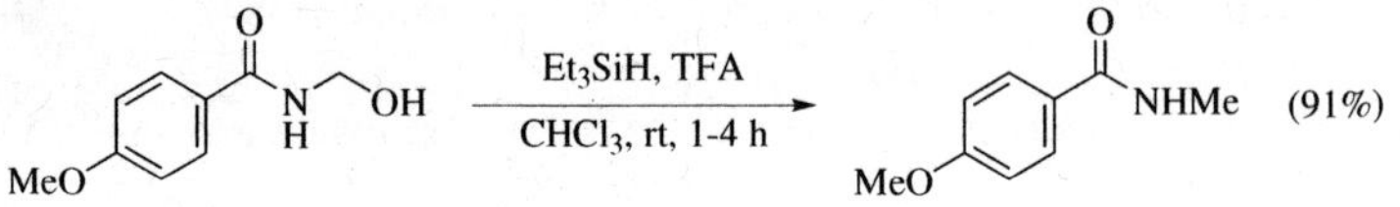

(Eq. 319)

N-Trimethylsilyloxymethylimines can be reduced to the corresponding *N*-alkylimines with $Et_3SiH/BF_3{\bullet}OEt_2$ (Eq. 320).[522]

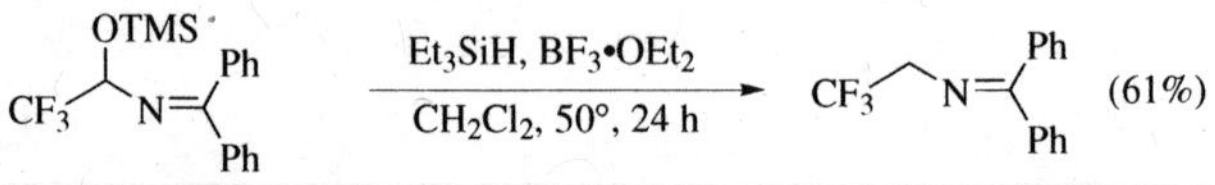

(Eq. 320)

O-Aminomethyl lactones are reduced to amino acids with the Et_3SiH/TFA combination (Eq. 321).[528]

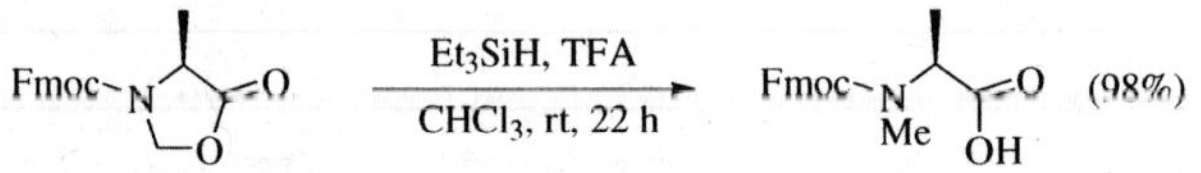

(Eq. 321)

Reduction of Enamines

Enamines are reduced to amines in good yields with Et_3SiH/TFA.[529–533] This reagent combination causes a variety of indoles to undergo stereoselective cis reduction of the indole ring while other potentially reducible functional groups in the molecule are unaffected (Eq. 322).[534] The organosilane reduction of enamides proceeds in excellent yields with the Et_3SiH/TFA reagent system (Eq. 323).[524,535]

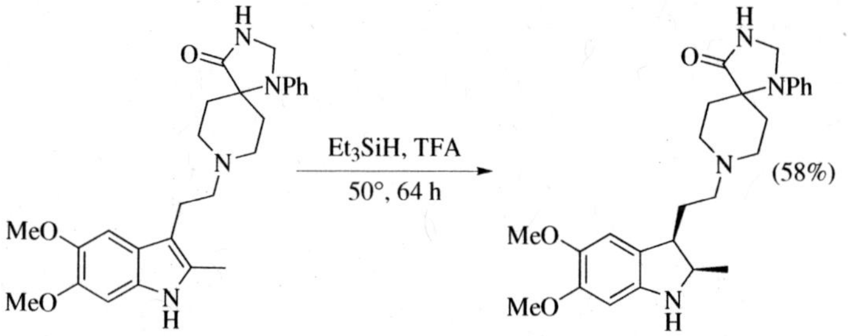

(Eq. 322)

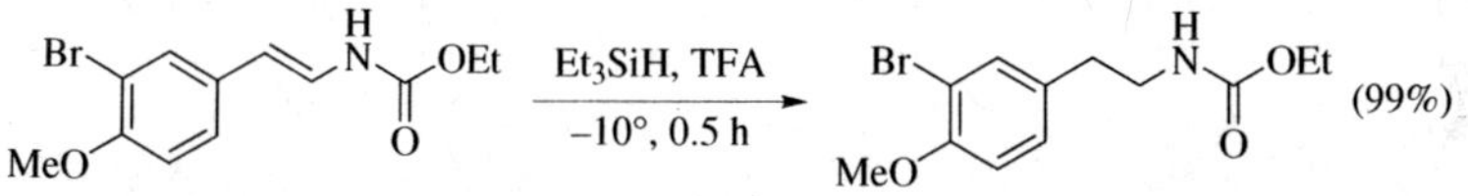

(Eq. 323)

The enamide double bond is reduced in preference over that of the enone moiety in each of the two examples shown below (Eqs. 324 and 325).[536,537] Other enamides are reduced under similar conditions.[235,537,538]

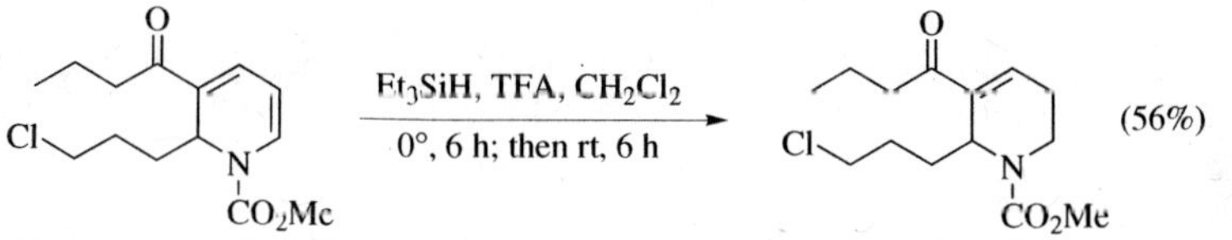

(Eq. 324)

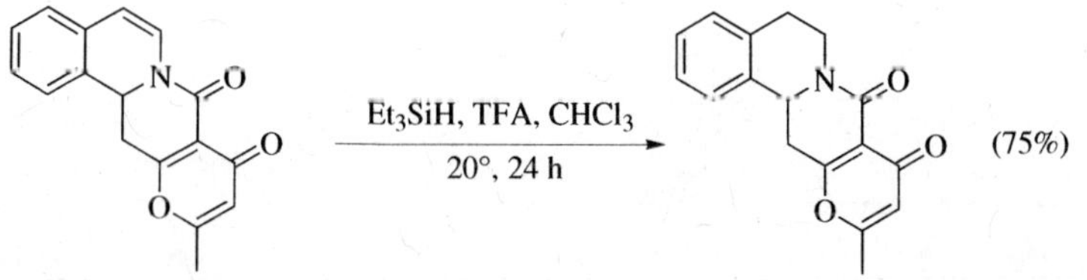

(Eq. 325)

Reduction of Imines

The reduction of imines with organosilanes is reported to take place with the reagent combinations PMHS/$ZnCl_2$,[539] Et_3SiH/TFA,[540,541] Cl_3SiH,[318,542] PMHS/butyltin trioctanoate,[543] $PhSiH_3$,[544] Et_3SiH/HCO_2H,[208] $PhMe_2SiH$/TFA,[276] $Cl_3SiH/BF_3{\cdot}OEt_2$,[545] Et_3SiH (and related silanes)/$RhCl(PPh_3)_3$,[546] and Cl_2SiH_2.[545] Imines are reduced in high yields with various silanes in the presence of Wilkinson's catalyst or $PdCl_2$.[120] Triethylsilane and diethylsilane are the most effective reducing agents. Diimine **87** undergoes reduction of the imine functions with diethylsilane without reduction of the amide functionality, even though the amide carbonyl is in the more reactive α-heteroaryl position (Eq. 326).[547]

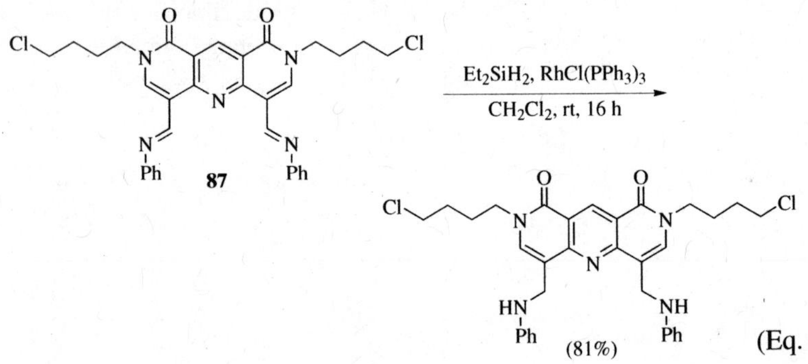

(Eq. 326)

Reduction of Oximes

Oximes can be reductively converted into the Boc-protected primary amines in good yields with the combination of PMHS/$(Boc)_2O$/Pd/C (Eq. 327).[548] *O*-Benzyloxyimines[549] are reported to be reduced in good to excellent yields by this method.

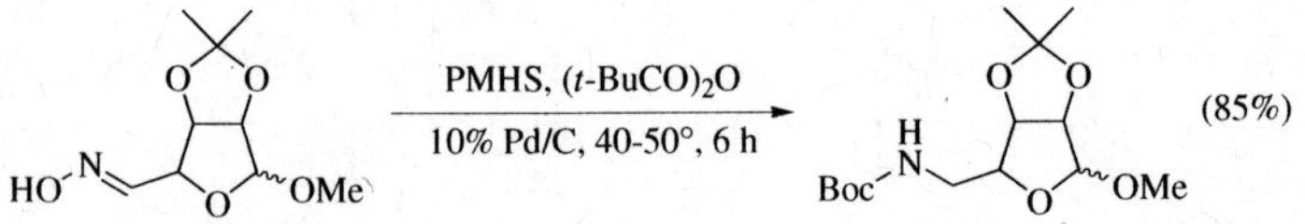

(Eq. 327)

O-Benzoyloximes are nicely reduced to *O*-benzoylhydroxylamines with Et_3SiH/TFA (Eq. 328).[550,551] *O*-Acetyloximes are reduced with Et_3SiH/TMSOTf in moderate to high yields.[552] The diethylphosphatoimine **88** is reduced to the hydroxylamine derivative.[553]

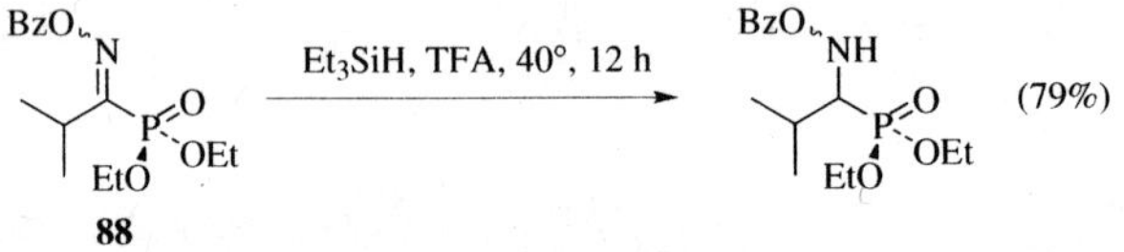

(Eq. 328)

Reduction of Nitroalkanes

The combination PMHS/Pd/C reduces nitrobenzenes to anilines in high yields (Eq. 329),[316] as does $Et_3SiH/RhCl(PPh_3)_3$ (Eq. 330).[554] This latter combination can also reduce both nitro and enone functionalities.[554]

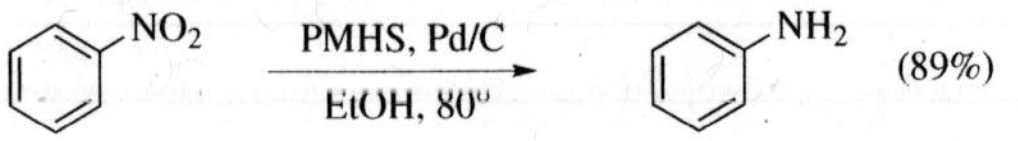

(Eq. 329)

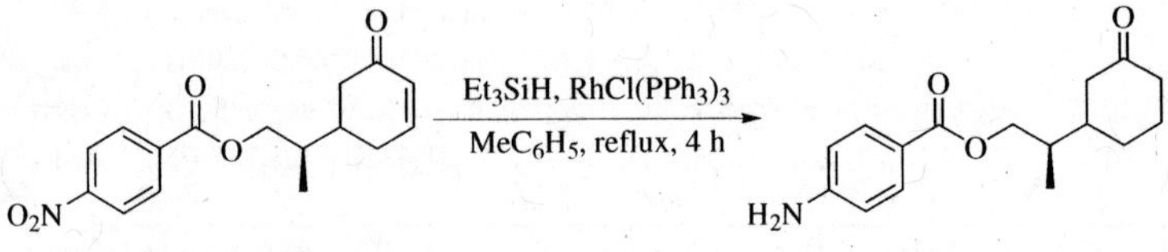

(Eq. 330)

Tertiary nitro alkanes or activated nitro groups are reduced to the alkane with loss of the nitro group by a combination of $PhSiH_3$ and catalytic amounts of $(n\text{-Bu})_3SnH$ and 1,1′-azobis(cyclohexanecarbonitrile) (ACHN), even in the presence of other potentially reducible functional groups.[555] The examples shown in Eqs. 331 and 332 illustrate this behavior.[555]

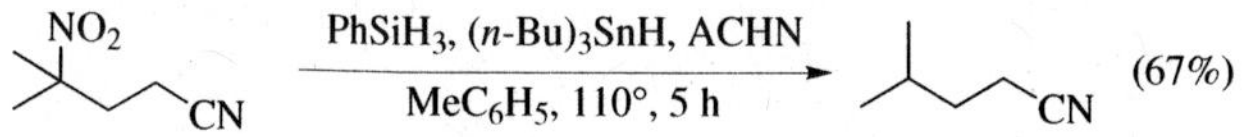

(Eq. 331)

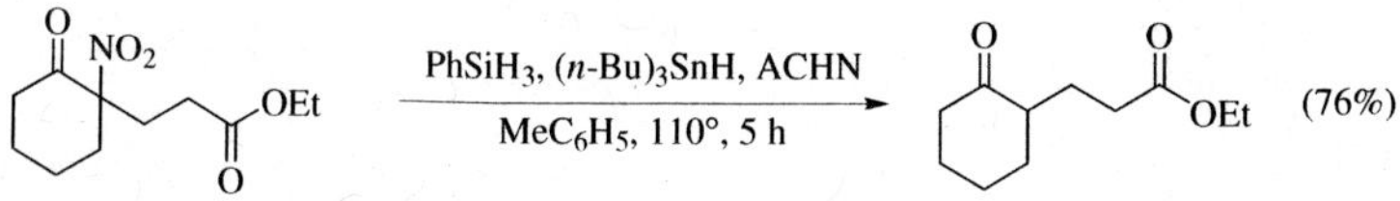

(Eq. 332)

Reduction of Miscellaneous Nitrogen-Containing Compounds

The combination of $PhSiH_3$ with a catalytic amount of bis(tri-*n*-butyltin) oxide reduces azides to primary amines in excellent yields (Eq. 333).[556] The reducing agent is $(n\text{-Bu})_3SnH$ formed in situ by the silane. Azides are converted into Boc-protected primary amines with the PMHS/Pd/C reagent (Eq. 334).[557,558]

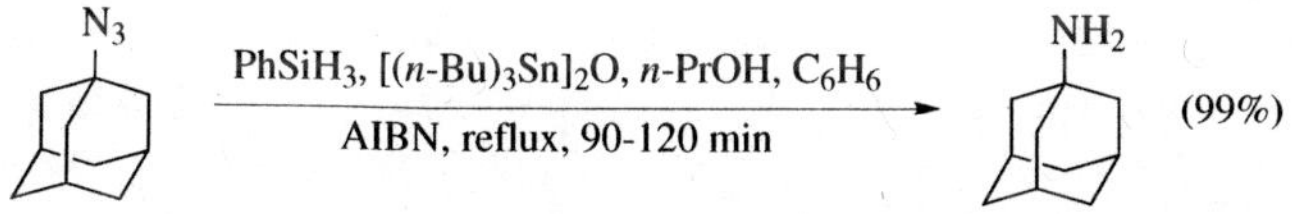

(Eq. 333)

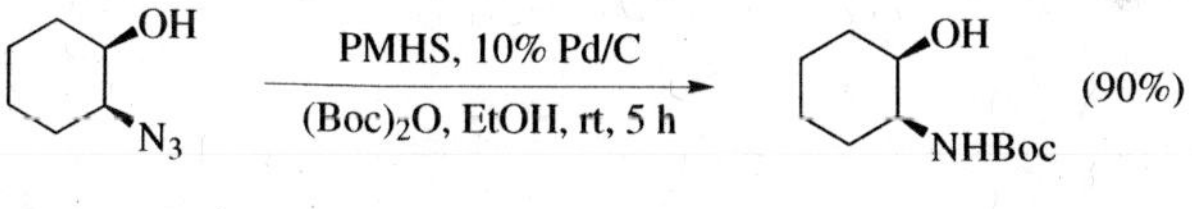

(Eq. 334)

The combination of $Et_3SiH/Co_2(CO)_8$ reduces nitriles to the *N*,*N*-bis(trimethylsilyl)amine (Eq. 335).[559] These derivatives can be readily hydrolyzed to the primary amines.

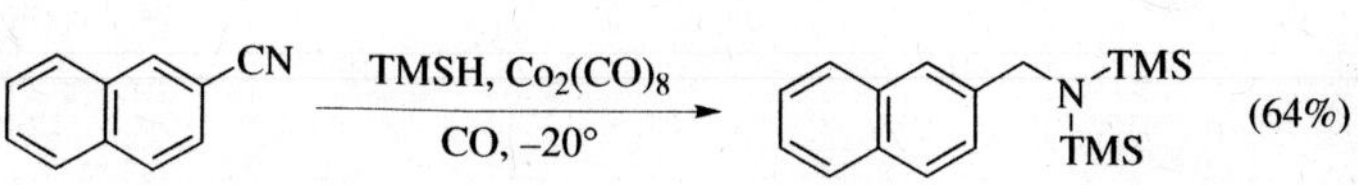

(Eq. 335)

The organosilane reduction of hydrazones to hydrazines is readily accomplished in good yields with Et_3SiH/TFA (Eq. 336).[560,561] *N*-Tosylimines[294] are reduced to their *N*-Boc tosylamino counterparts,[294] and are also reduced with $(MeO)_3SiH$/LiOMe in good yields.[294] Benzyl-protected hydroxylamines are reduced with $PhMe_2SiH$/TFA.[551]

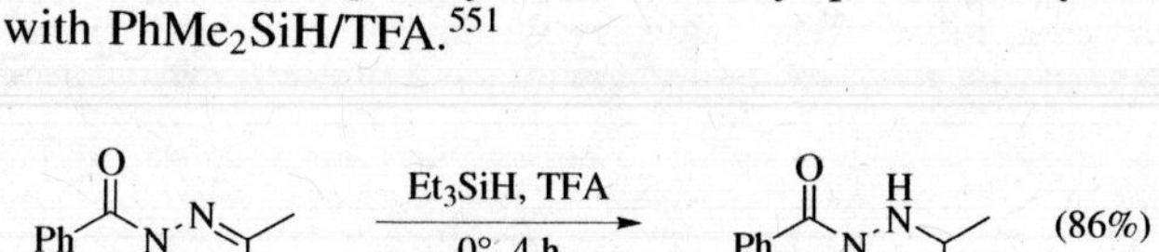

(Eq. 336)

In an analogous fashion to the reductive deprotection of allyl alcohols and allyl esters, the deallylation of allylamines is also possible (Eq. 337).[270]

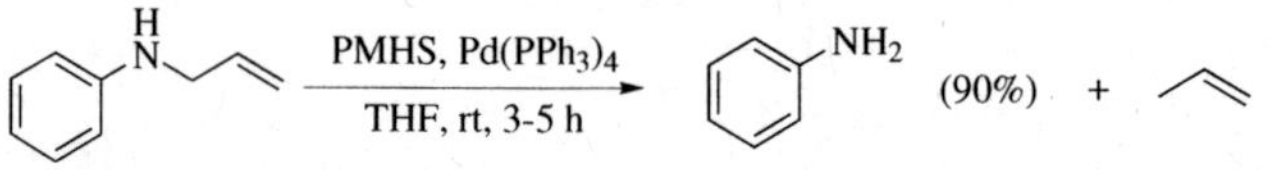

(Eq. 337)

The tetrachloroferrate or tetrafluoroborate salts of alkylated alkyl- or arylnitriles (nitrilium ions) are reduced to imines with triethylsilane. Subsequent hydrolysis of the intermediate imines leads to aldehydes in good yields, thus providing an excellent overall route to aldehydes from nitriles (Eq. 338).[28,562]

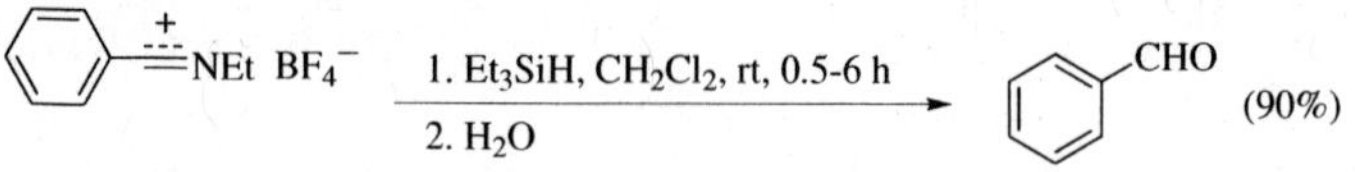

(Eq. 338)

Aryldiazonium salts are reduced to the benzene derivatives in good yields (Eq. 339).[563]

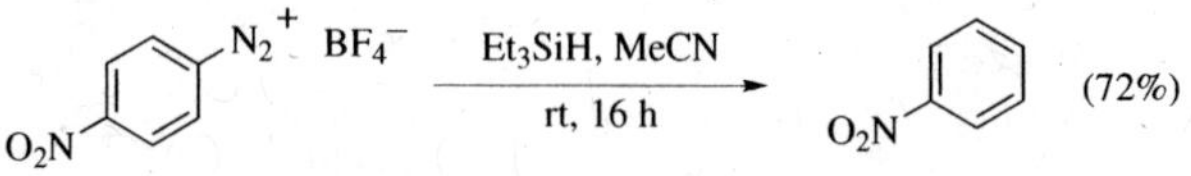

(Eq. 339)

Reduction of Miscellaneous Sulfur-Containing Compounds

Et_3SiH/TFA reduces disulfides to the corresponding mercaptans in modest yields (Eq. 340).[564] Naphthyl thio ethers are reduced in rather poor yields to tetrahydronaphthalene with the combination $Et_3SiH/BF_3{\cdot}OH_2$ (Eq. 341).[263] There is one report of the reduction of a diaryl sulfide to the hydrocarbon but the yield is low (Eq. 342).[217]

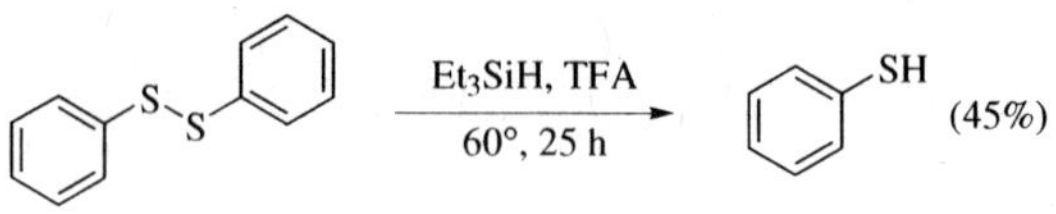

(Eq. 340)

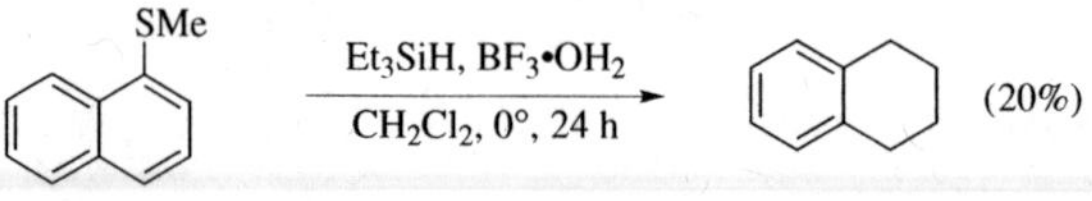

(Eq. 341)

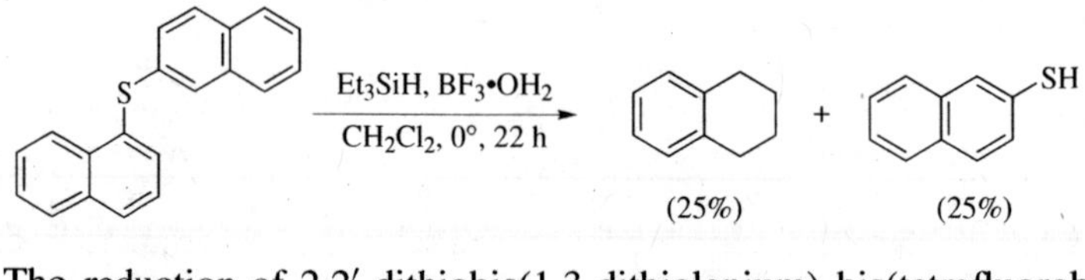

(Eq. 342)

The reduction of 2,2′-dithiobis(1,3-dithiolanium) bis(tetrafluoroborate) to the thiocarbonate compound by triethylsilane takes place quantitatively (Eq. 343).[565]

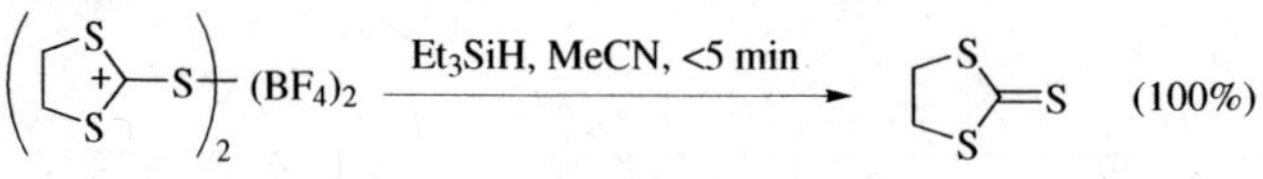

(Eq. 343)

Reduction of Small-Ring Heterocycles

The reduction of epoxides occurs with the reagent combinations Et_3SiH/BF_3[566] and $Et_3SiH/BF_3{\cdot}OEt_2$.[407] The reductive iodination of epoxides is carried out with the combination of TMDO and iodine (Eq. 344).[357] The yields are good to excellent. The reductive bromination of epoxides is accomplished in a similar manner. Trimethylchlorosilane is found to enhance these reactions.[567]

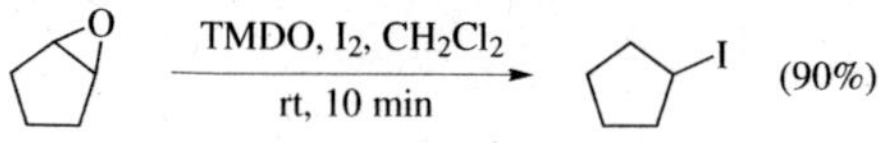

(Eq. 344)

N-Tosylaziridines are reduced with the combination PMHS/Pd/C in ethanol to yield the ring-opened *N*-tosyl primary amines (Eq. 345).[568]

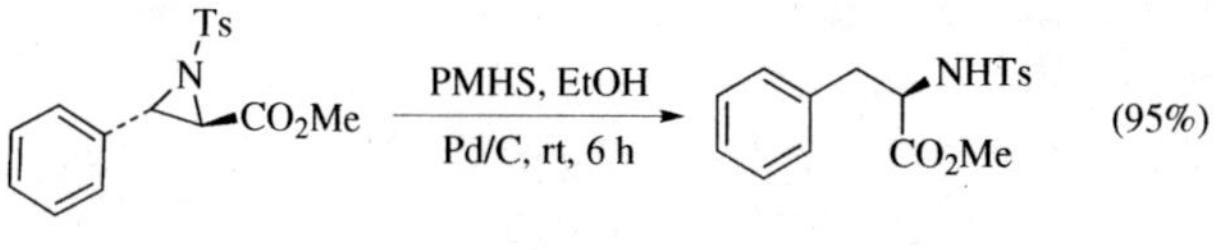

(Eq. 345)

Asymmetric Reduction of Ketones

The asymmetric organosilane reduction of prochiral ketones has been studied as an alternative to the asymmetric hydrogenation approach. A wide variety of chiral ligand systems in combination with transition metals can be employed for this purpose. The majority of these result in good to excellent chemical yields of the corresponding alcohols along with a trend for better ee results with aryl alkyl ketones than with prochiral dialkyl ketones.

Oxazolinylferrocenylphosphines complexed in situ with $RuCl_2PPh_3$ is one system used to catalyze the diphenylsilane reduction of prochiral ketones.[569–571] The best results are obtained when AgOTf or $Cu(OTf)_2$ is employed as an additive. Although the chemical yields are modest, the ee values range from 43% (methyl cyclohexyl ketone) to 97% (propiophenone). Chiral bis(oxazolinyl)bipyridine (bipymox) ligands in combination with rhodium(II) chloride, $AgBF_4$, and diphenylsilane reduce prochiral ketones in good to excellent yields and 24 to 90% ee.[572] The complexes formed with rhodium(I) are not as effective and some silyl enol ethers are formed from enolizable ketones. Rhodium(I) complexes with (+)- or (−)-DIOP [(2,2-dimethyl-1,3-dioxolane-4,5-diyl)bis(methylene)bis(diphenylphosphine)] ligands catalyze the diphenylsilane reduction of 17-keto steroids with only a moderate enhancement for the α-hydroxy steroid over the results obtained from the corresponding achiral reductions.[573]

The combination of $[Rh(Cl(NBD)]_2$ and ligands **89**, **90**, **91**, or **92** with diphenylsilane asymmetrically reduces aryl alkyl ketones, including acetophenones, in excellent yields and in 81 to 90% ee (Eq. 346).[574] The best results are with ferrocene **91** and acetophenone in toluene.[575] Other phosphine-substituted ferrocenes do not give comparable results. Rhodium(I) complexes of TADDOL-derived

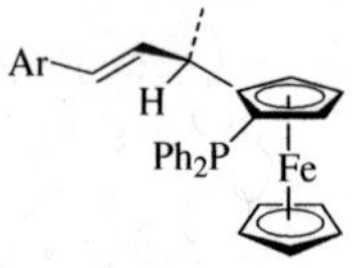

89 Ar = Ph
90 Ar = 3-$CF_3C_6H_4$
91 Ar = 4-$CF_3C_6H_4$
92 Ar = C_6F_5

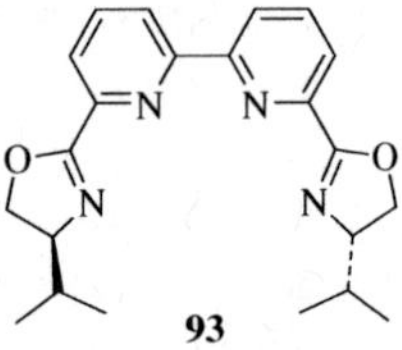

93

phosphates containing a chiral dihydrooxazole unit such as **93** are also used to catalyze the diphenylsilane reduction of prochiral ketones.

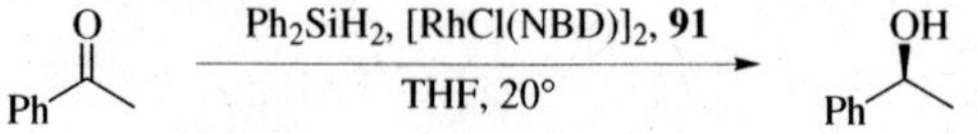

(94%) 89% ee (Eq. 346)

The combination of 1-naphthylphenylsilane and a rhodium(I) catalyst derived from Evans' mixed phosphine/sulfur ligands, for example **94**, reduces acetophenone in 95% ee. Other aryl alkyl ketones are reduced with similarly excellent ee values.[576] Under similar conditions, dialkyl ketones and β-keto esters are reduced in good yields and moderate to excellent ee. The asymmetric reduction of aryl alkyl ketones with diphenylsilane is accomplished with excellent ee values through the use of a rhodium(I) complex with TRAP (2,2′-bis[(dialkylphosphino) methyl]-1,1′-biferrocene) ligands.[577,578] These systems are also successful for the 1,2 asymmetric reduction of α,β-unsaturated ketones. The (*S*,*S*)-Phos-Biox ligand **95**, when complexed with rhodium(I), provides a catalyst that can be used in the diphenylsilane reduction of acetophenones to give the corresponding R-configured alcohols in high yields and >90% ee.[579] The complex **96** is highly successful in the diphenylsilane reduction of aryl alkyl ketones with high yields and >90% ee values.[580,581] Dialkyl ketones are reduced with more mixed results, with 2-octanone showing 63% ee and ethyl 4-oxopentanoate 95% ee. The same system can be used to reduce 2-phenylcyclohexanone with no cis/trans selectivity, but excellent enantioselectivity (Eq. 347).[390]

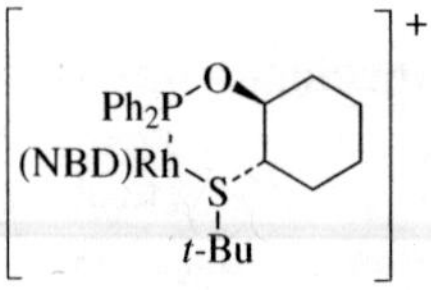

94

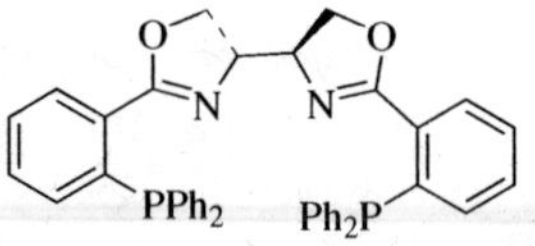

95

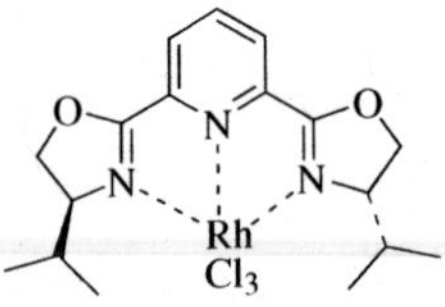

96

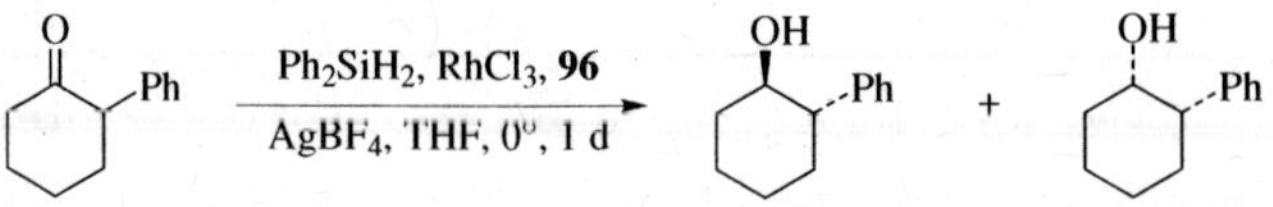

(46%) 1S,2R 99% ee   (46%) 1S,2S 96% ee (Eq. 347)

A chiral oxazolinoferrocene ligand with iridium(I) is used for the diphenylsilane reduction of aryl alkyl ketones in nearly quantitative yields and >83% ee values.[582] The dialkyl ketone, 2-octanone, is reduced with a poor 19% ee under these conditions. A catalyst prepared by the alkylation of [1,2-bis(tetrahydroindenyl)ethane]titanium(IV) 1,1′-binaphth-2,2′-diolate with methyllithium or *n*-butyllithium can be employed in the methyldiethoxysilane reduction of acetophenone with 99% ee.[583,584] Other ketones do not show nearly the same ee values. Methylsilane is the actual proposed reducing agent in this system. The phosphinophenyloxazoline **97** is an effective ligand for the asymmetric rhodium(I) diphenylsilane reduction of aryl alkyl ketones, even with propiophenone, which has proven difficult with other systems, showing 91% ee and 91% yield (Eq. 348).[585] A more general system involving mesitylphenylsilane and catalyst **98** permits the reduction of aryl alkyl ketones in very high chemical yields and >96% ee.[586] The reduction of dialkyl ketones ranges from 72% ee for 2-octanone to 96% ee for the more stereo-differentiated adamantly methyl ketone.

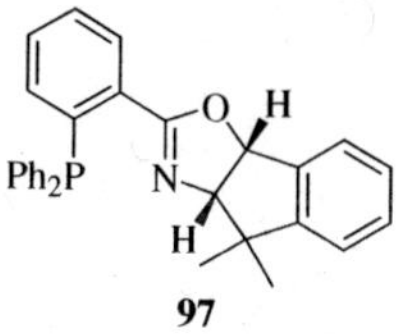

97

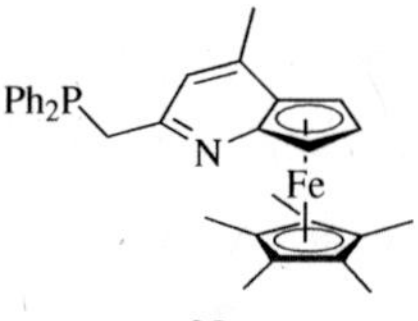

98

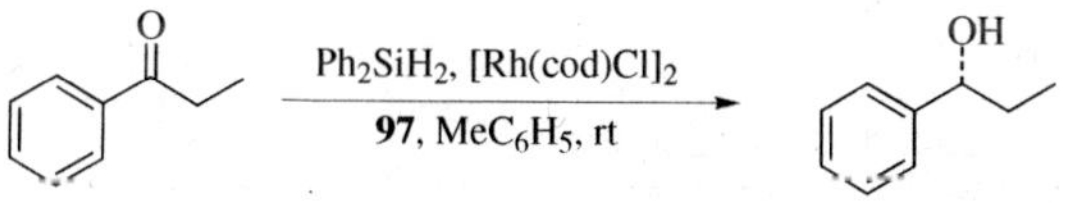

(Eq. 348)

Alkylated (*R*,*R*)-tetrahydroindenyltitanium difluoride and phenylsilane serve to asymmetrically reduce a variety of ketones, especially aryl alkyl ketones, in excellent chemical yields and >96% ee.[587] The use of the easier to handle and less expensive PMHS is also highly effective in these transformations. In a related study using the (*R*,*R*)-tetrahydroindenyltitanium 1,1′-binaphth-2,2′-diolate precursor to the active catalyst, similarly impressive results are obtained.[588]

The in situ generation of CuH from organosilanes in the presence of either a BIPHEP (**99**) or a SEGPHOS (**100**) type ligand represents a general method for the asymmetric hydrosilylation of aryl alkyl ketones at low temperatures.

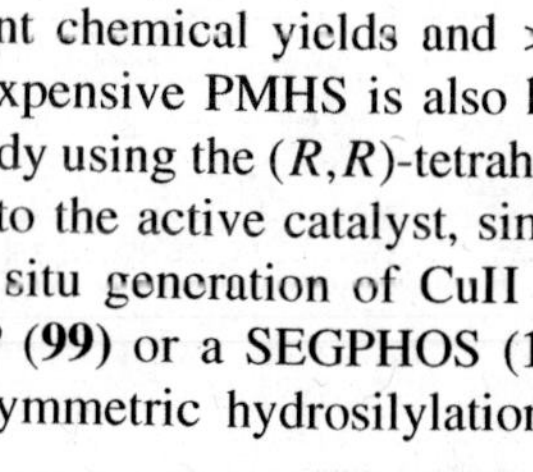

99

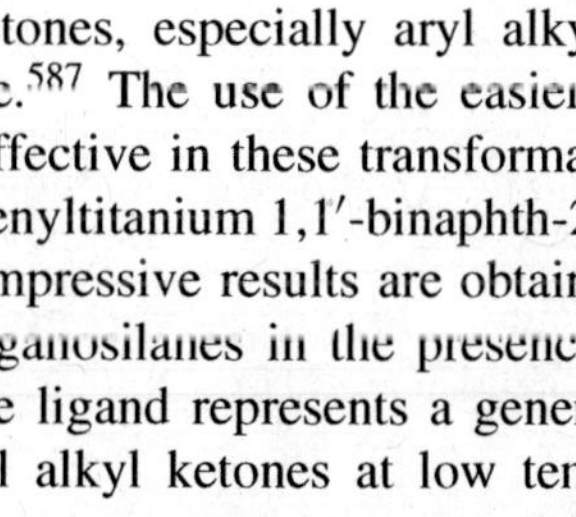

100

These excellent catalyst systems show high reactivity, substrate-to-ligand ratios of >100,000 : 1, high chemical yields, the ability to employ the inexpensive PMHS as the hydride donor, and typical ee values in the >90% range.[589,590] The most promising ligand found to date is DTBM-SEGPHOS (5,5′-bis[di(3,5-di-*tert*-butyl-4-methoxyphenyl)phosphino]-4,4′-bi-1,3-benzodioxole) **100** (Eq. 349). The BIPHEP ligand with copper(I) is also capable of asymmetrically reducing aryl alkyl ketones under similar conditions and with comparable results.[589,591]

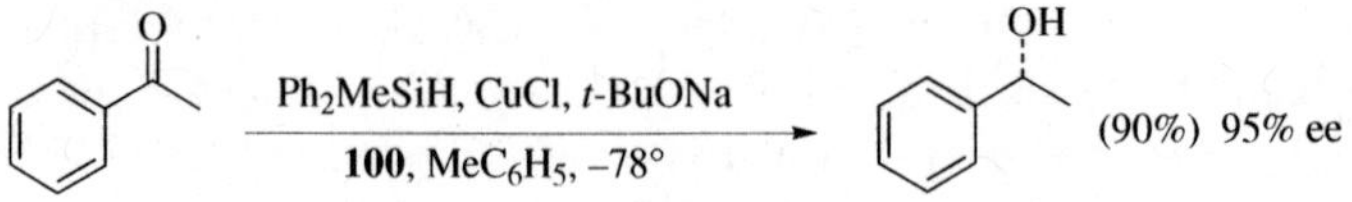

(Eq. 349)

Asymmetric Reduction of α,β-Unsaturated Ketones

The enantioselective hydrosilylation of 2-pentylcyclopentenone is effected with PMHS and an active catalyst derived from (*R*,*R*)-ethylenebis(tetrahydro-indenyl)titanium difluoride and phenylsilane (EBTHI)Ti (Eq. 350).[587] The use of diphenylsilane, a rhodium catalyst, and (*R*,*R*)-(*S*,*S*)-BuTRAP as the chiral ligand gives similar results.[576] Other related approaches give greatly inferior enantioselectivies.[592–594]

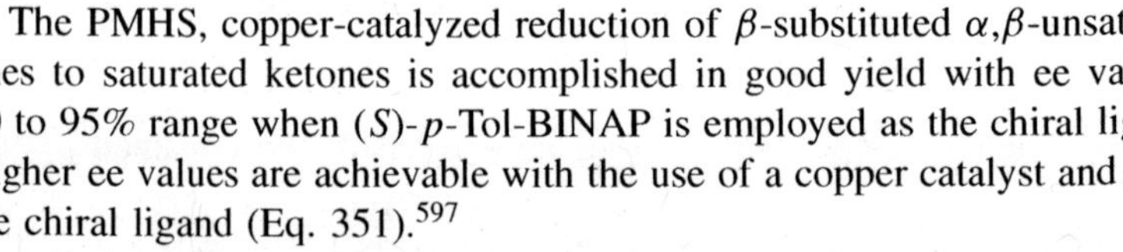

(Eq. 350)

The PMHS, copper-catalyzed reduction of β-substituted α,β-unsaturated ketones to saturated ketones is accomplished in good yield with ee values in the 90 to 95% range when (*S*)-*p*-Tol-BINAP is employed as the chiral ligand.[595,596] Higher ee values are achievable with the use of a copper catalyst and (*R*)-**100** as the chiral ligand (Eq. 351).[597]

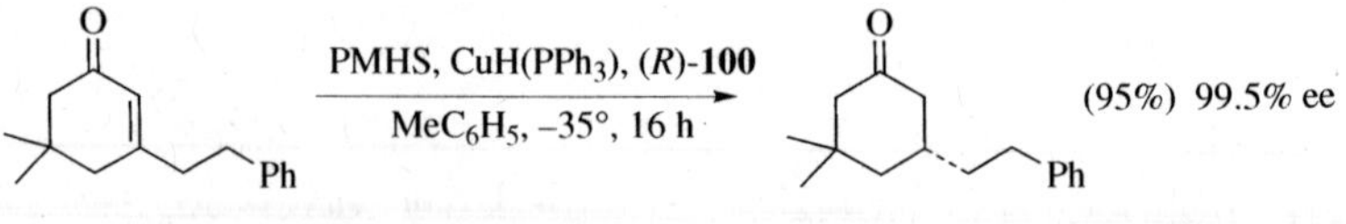

(Eq. 351)

The sequence of chiral 1,4-reduction of a β-substituted cyclopentenone followed by electrophilic trapping of the intermediate enolate provides an efficient route to chiral 2,3-disubstituted cyclopentanones that generates two chiral centers in the process (Eq. 352).[459]

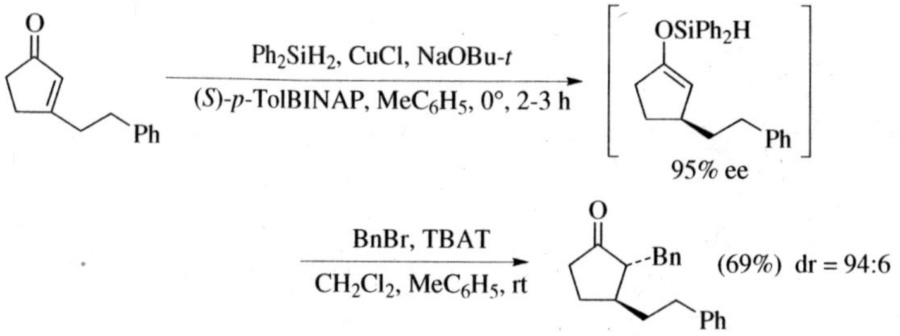

(Eq. 352)

Asymmetric Reduction of α,β-Unsaturated Esters and Lactones

The chiral hydrosilylation of β-substituted α,β-unsaturated esters to their saturated counterparts is the subject of reports by two groups. The combination of triphenylphosphinecopper hydride and (*R*)-DTBM-SEGPHOS is reported to give excellent yields of the β-substituted esters (Eq. 353).[598] Comparable yields, but with lower ee values, are reported for this transformation.[599,600]

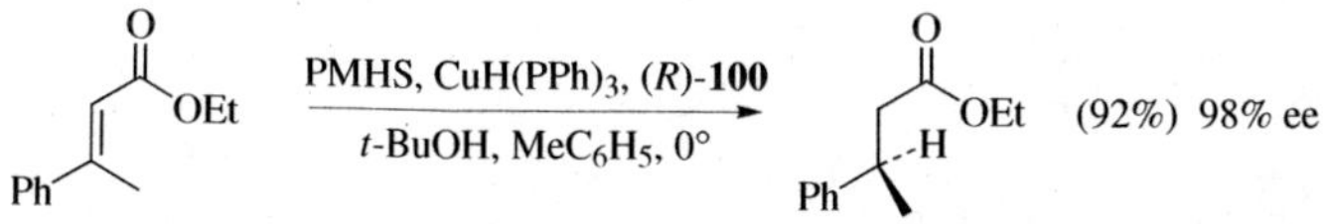

(Eq. 353)

The copper-catalyzed chiral reduction of β-substituted α,β-unsaturated lactones with PMHS and (*S*)-*p*-Tol-BINAP in the presence of a hindered alcohol can be carried out in moderate to good yields with moderate ee values.[599] The reaction is useful for both butenolides and pentenolides. Inferior results are realized with diphenylsilane as the reducing agent. Excellent results employing PMHS and the DTBM-SEGPHOS ligand are possible (Eq. 354).[598]

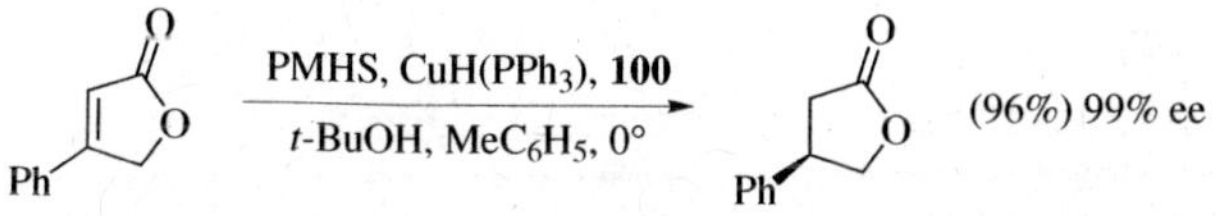

(Eq. 354)

The asymmetric reductive rhodium- and iridium-catalyzed aldol reaction of α,β-unsaturated esters with aldehydes[470,601,602] is proposed to involve a rhodium(I) or iridium(I) hydride as the active catalyst, which adds to the α,β-unsaturated ester, followed by reaction of the intermediate rhodium or iridium enolate with an aldehyde.[470] These transformations provide an excellent entry into α-alkyl-β-hydroxy esters in both high yields and high enantiomeric ratios. The reactions were first carried out with moderate success using the BINAP (2,2′-bis(diphenylphosphino)-1,1′-binaphthyl) ligand to introduce the asymmetry.[470,603] The use of various pybox ligands improves the yields of these transformations.[604] The best results are obtained from the ligand indane-pybox (**101**) and an iridium catalyst,[604] which was applied to an enantioselective synthesis of the macrocycle borrelidin (Eq. 355).[601,602]

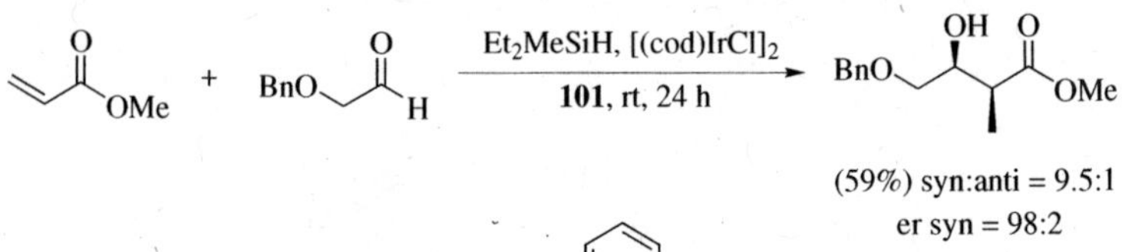

(Eq. 355)

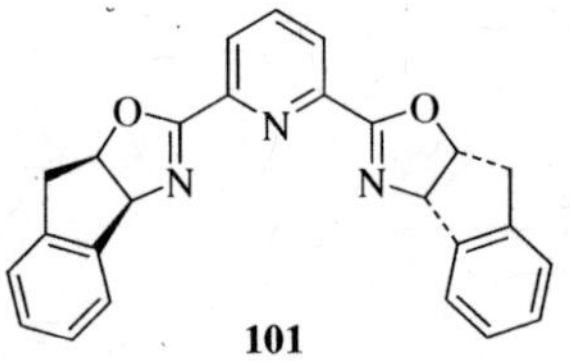

101

Asymmetric Reduction of α,β-Unsaturated Lactams

The chiral reduction of β-substituted α,β-unsaturated lactams with PMHS in the presence of (*S*)-*p*-Tol-BINAP as the chiral ligand with a copper catalyst results in β-substituted lactams in excellent yield and with greater than 90% ee.[599] This method has been applied in an efficient enantioselective synthesis of the antidepressant (−)-paroxetine (Eq. 356).

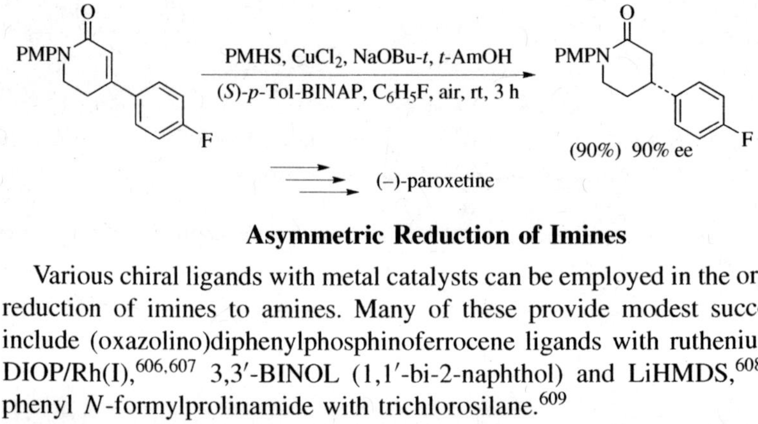

(Eq. 356)

Asymmetric Reduction of Imines

Various chiral ligands with metal catalysts can be employed in the organosilane reduction of imines to amines. Many of these provide modest success. These include (oxazolino)diphenylphosphinoferrocene ligands with ruthenium,[605] (−)-DIOP/Rh(I),[606,607] 3,3′-BINOL (1,1′-bi-2-naphthol) and LiHMDS,[608] and (*S*)-phenyl *N*-formylprolinamide with trichlorosilane.[609]

Some excellent findings in the asymmetric organosilane reduction of both aryl alkyl and dialkyl imines have resulted in the development of practical, scaleable methodologies for this key transformation. The reduction of imines with the ethylenebis(η^5-tetrahydroindenyl)titanium (EBTHI)–TiF_2-derived catalyst **102** with either phenylsilane or PMHS as the reducing agent gives high chemical yields of the corresponding amine and ee values well in excess of 90% with most at 99% (Eq. 357).[610–613] Straight-chain dialkyl imines are not as successful; for example, 2-(*N*-benzylimino)octane gives a 96% yield of (*S*)-(2-benzylamino)octane with 69% ee.[612] The CuH approach employed so successfully for the asymmetric organosilane reduction of ketones can be applied with equal success to the reduction of phosphoryl imines, thus providing a route to the asymmetric reduction of imines to primary amines via the hydrolysis of the resulting aminophosphorane.[598,614]

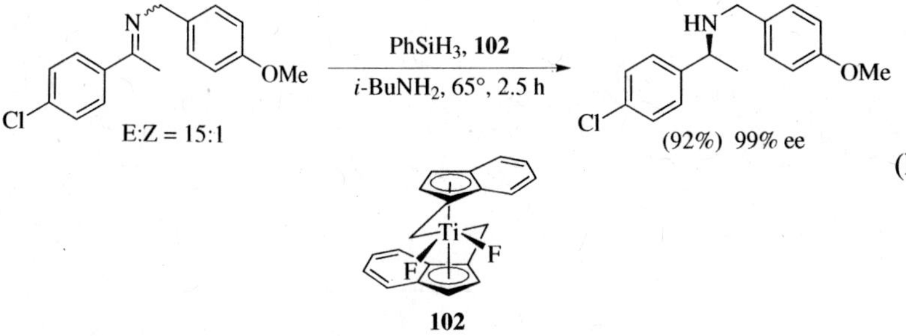

(Eq. 357)

COMPARISON WITH OTHER METHODS

Many different methods are known and used for the reduction of organic functional groups. These have been reviewed many times over the years and are too numerous to repeat here. The sequence of hydrosilylation of a multiple bond followed by removal of the silyl group is tantamount to the addition of hydrogen. Coupled with the keen current interest in asymmetric reductions, the use of hydrogen in asymmetric reductions and related reactions is highlighted here. The numerous asymmetric hydrogenations and asymmetric reductions with metal hydrides, including lithium aluminum hydride, sodium borohydride, and borane, coordinated or reacted with chiral diols, amino alcohols, diamines, and variations of these have been extensively reviewed.[615–631] In view of the very large number of methods for the reduction of organic functional groups and the high interest in asymmetric reductions, the choice of competitive examples is limited to those that are representative of asymmetric hydrogenations.

Asymmetric Hydrogenation of Olefins

EBTHI–Ti, when treated with n-BuLi, catalyzes the hydrogenation of trisubstituted olefins in good yields and excellent enantioselectivity though undetermined configuration (ee = 83% to >99%) (Eq. 358)[632] A zirconium version of this approach is also successful in the asymmetric hydrogenation of terminal olefins, although the enantioselectivities are not high.[633] On the other hand, this system gives excellent ee values when applied to the hydrogenation of disubstituted cyclic olefins (Eq. 359)[634]

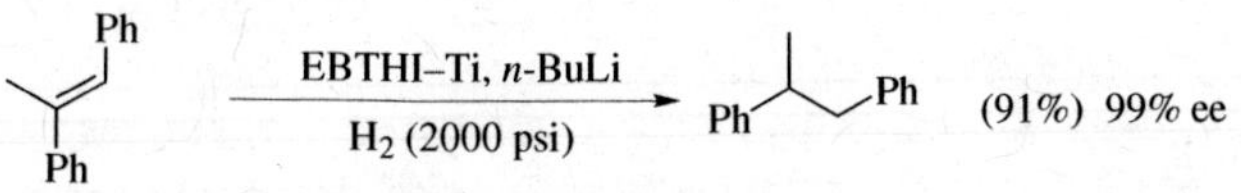

(Eq. 358)

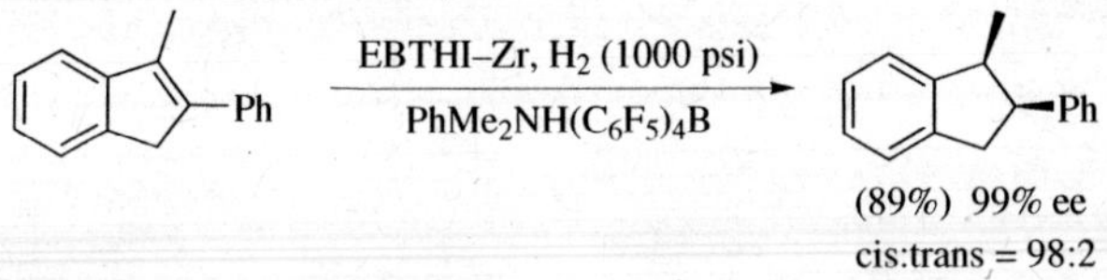

(Eq. 359)

Chiral phosphinodihydrooxazole iridium ligands are used to hydrogenate trisubstituted olefins in moderate yields and high enantioselectivity albeit of undetermined configuration.[635] In a similar fashion and with equally impressive results, phosphine-oxazoline complexes of iridium **104**, derived from $[Ir(cod)Cl]_2$ and **103**, are able to catalyze the hydrogenation of stilbenes (Eq. 360) and β-methyl cinnamic esters with both excellent conversion and enantioselectivity of undetermined configuration.[636] The complex **105** also gives excellent results (Eq. 361).[637]

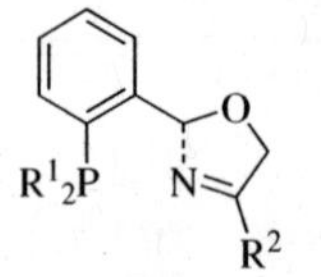

103
R^1, R^2 = alkyl, aryl

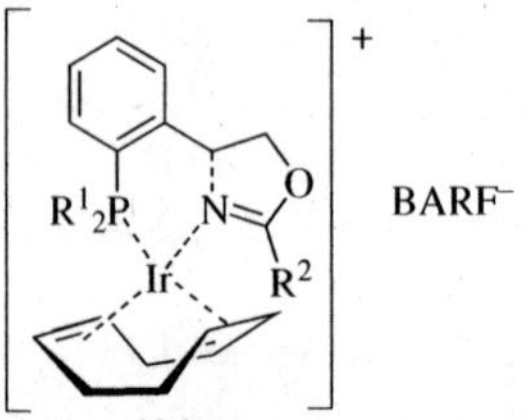

104
R^1 = *c*-C_6H_{11}, Ph
R^2 = *t*-Bu, adamantyl,
3,5-(*t*-Bu)$_2C_6H_3$, $CHPh_2$

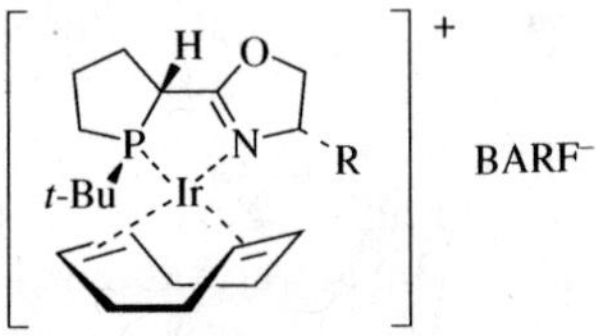

105
R = *i*-Pr, *t*-Bu, *i*-Bu, Ph, Bn

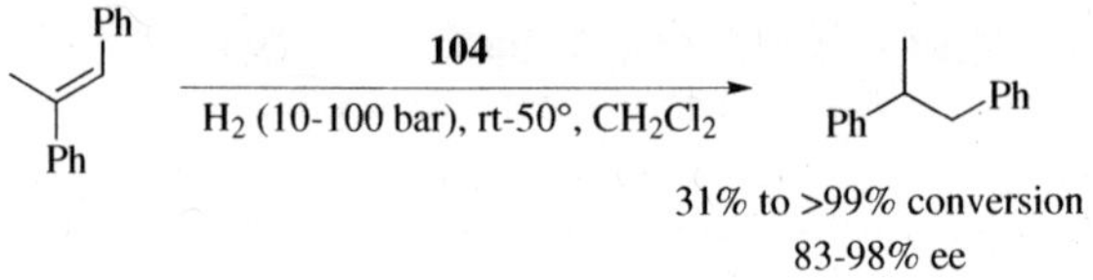

(Eq. 360)

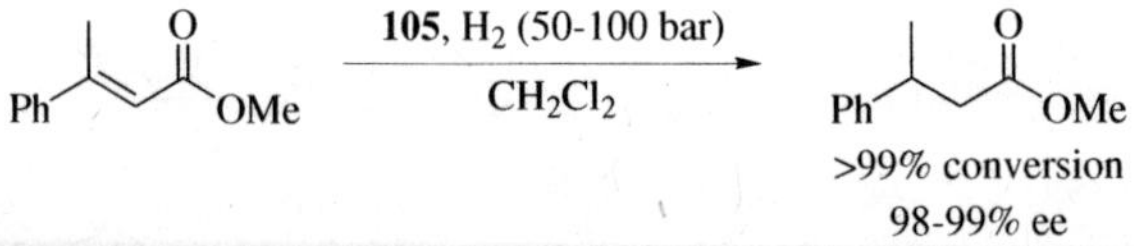

(Eq. 361)

The asymmetric hydrogenation of trisubstituted olefins with iridium complexes of chiral phosphinite-oxazoline ligands of the general structure **106** also provides excellent results with ee values in the 85–99% range.[638] The asymmetric hydrogenation of imines with these systems gives only moderate results. A similar fused phosphinite oxazoline iridium catalyst, **107**, gives good results with 1,1-disubstituted and trisubstituted olefins with ee values of >97%, although ethyl β-methylcinnamate gives poor results.[639]

Asymmetric Hydrogenation of Ketones

A number of asymmetric hydrogenations of prochiral ketones to highly enantiomerically enriched alcohols are available. A select few are highlighted here.

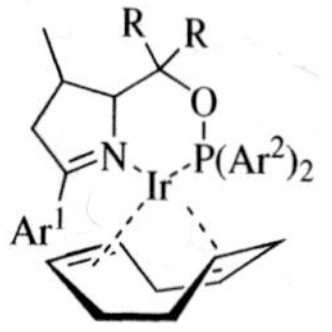

106

R	Ar^1	Ar^2
Bn	Ph	Ph
Bn	Ph	$2\text{-}MeC_6H_4$
Me	Ph	Ph
Bn	$3,5\text{-}Me_2C_6H_3$	Ph
Bn	$3,5\text{-}(t\text{-}Bu)_2C_6H_3$	Ph

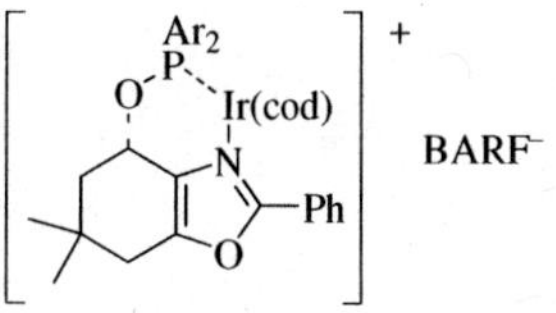

107

Ar = Ph, $2\text{-}MeC_6H_4$

The PennPhos ligands, for example **108**, complexed with rhodium, provide an excellent system for the hydrogenation of aryl alkyl ketones with ee values in the range of 94–96% (Eq. 362). Phenyl isopropyl ketone shows only a 72% ee under similar conditions. Dialkyl ketones exhibit ee values in the range of 73–94% with this system (Eq. 363).[640]

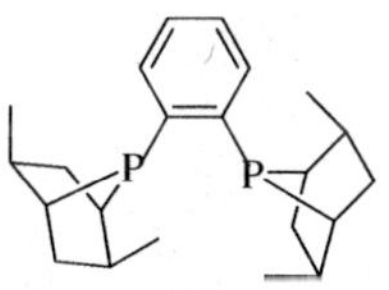

108

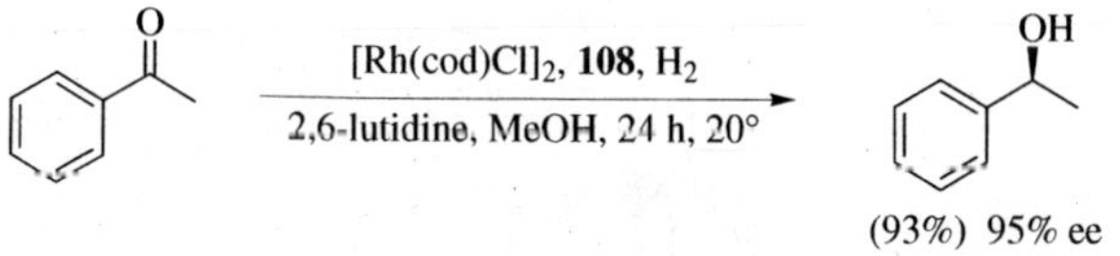

(Eq. 362)

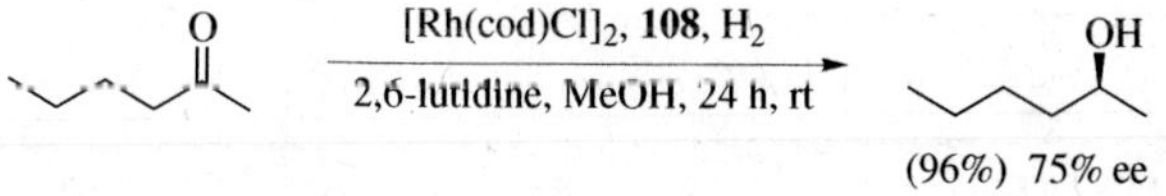

(Eq. 363)

Enantioselectivities in the range of 97.7–99.9%, with the majority in the range of 98.4–99.1%, are obtained in the asymmetric hydrogenation of aryl alkyl ketones with ruthenium catalyst **109**.[641] The same systems can hydrogenate β-keto esters (95.2–98.6% ee) and α,β-unsaturated acids (96.2% in a single example).[642]

Asymmetric transfer hydrogenation can be employed in the asymmetric hydrogenation of prochiral ketones with a ruthenium complex of bis(oxazolinylmethyl) amine ligand **110**. Enantioselectivities are greater than 95%.[643]

The BINAP system of general structure **111** can be used in asymmetric hydrogenations; the compound in which Ar = $3,5\text{-}Me_2C_6H_3$, $R^1 = R^2 = 4\text{-}MeOC_6H_4$,

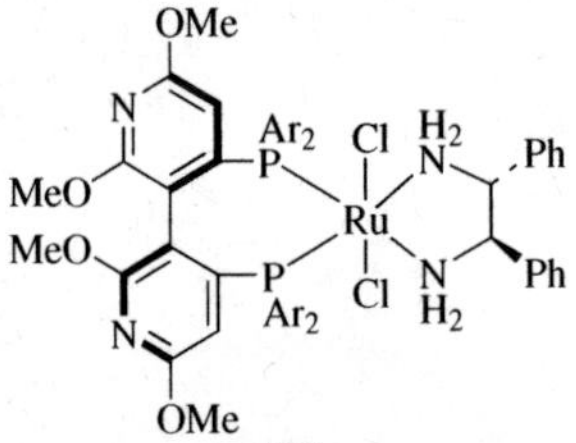

109

Ar = Ph, 4-MeC_6H_5, 3,5-$Me_2C_6H_3$

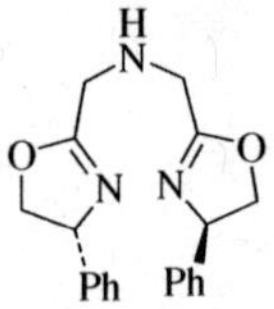

110

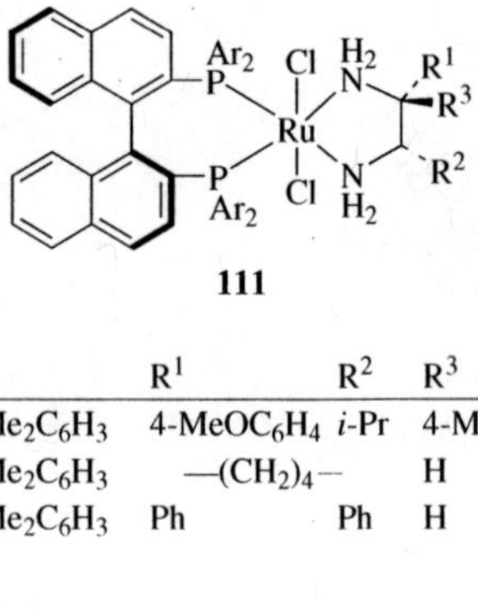

111

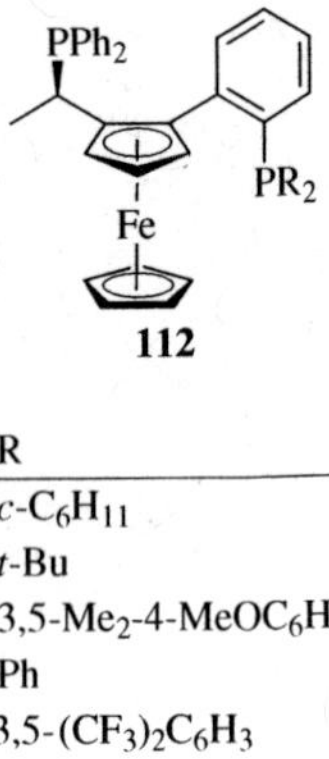

112

Ar	R^1	R^2	R^3
3,5-$Me_2C_6H_3$	4-$MeOC_6H_4$	*i*-Pr	4-$MeOC_6H$
3,5-$Me_2C_6H_3$	—$(CH_2)_4$—		H
3,5-$Me_2C_6H_3$	Ph	Ph	H

R
c-C_6H_{11}
t-Bu
3,5-Me_2-4-$MeOC_6H_2$
Ph
3,5-$(CF_3)_2C_6H_3$

and $R^3 = i$-Pr effects the asymmetric hydrogenation of cyclopropyl methyl ketone (95% ee), cyclopropyl phenyl ketone (96% ee), and other aryl alkyl ketones (94 to 100% ee), and is also useful for the 1,2-reduction of enones (>90% ee).[644] 2,4-Pentanedione is hydrogenated to the 2-(*R*)-4-(*R*)-2,4-pentanediol in 97% ee with ligand **112** and $[RuI_2(p\text{-cymene})]_2$. This system gives a wider range of enantioselectivity with prochiral ketones (22–97% ee) and α,β-unsaturated acids and esters (8–95% ee).[645]

o-BINAPO ligands of the type **113** complexed with ruthenium give good enantioselectivity in the hydrogenation of β-keto esters with the more hindered ortho-substituted aryl substituents giving the best results.[646] The selectivities range from 87–99% ee. These same systems hydrogenate the double bond of β-amido acrylates in >90% ee.

The TunaPhos ligands of general structure **114**, when complexed with $[RuPhCl_2]_2$, bring about the asymmetric hydrogenation of β-keto esters with

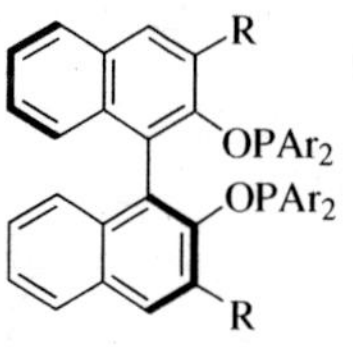

113

R	Ar
H	Ph
Me	Ph
Ph	Ph
3,5-$Me_2C_6H_3$	Ph
Ph	3,5-$Me_2C_6H_3$

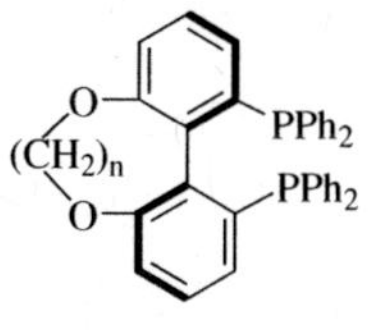

114 n = 1-6

high ee values. The results compare very favorably with those obtained with (*R*)-BINAP and (*R*)-BIPHEP. The best results are found where n = 4, which gives a dihedral angle of the phosphines of 88 degrees.[647]

(*R*)-BINAP-$RuBr_2$ can be successfully applied to the enantioselective hydrogenation of β-keto esters in the synthesis of (+)-(2*R*,3*R*)-corynomycolic acid **115**. (*S*)-MeO-BIPHEP-$RuBr_2$ was used in a similar manner in the synthesis of (*R*)-fluoxetine (**116**, Prozac®) and (*S*)-duloxetine (**117**).[648]

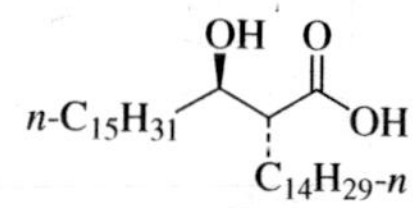

115

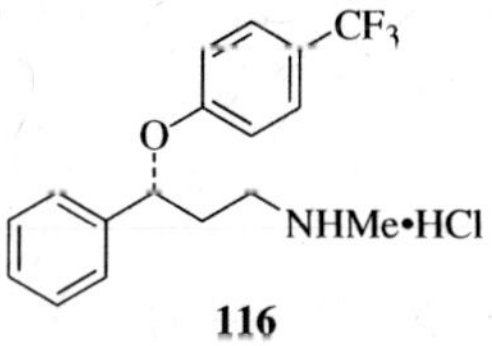

116

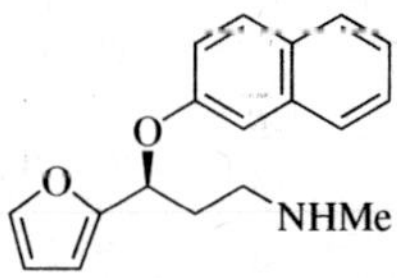

117

The highly enantioselective reductive amination of α-keto acids as a route to amino acids is possible with ligand **118** [(3*R*,4*R*)-1-(*N*-benzyl)-3,4-bis(diphenylphosphanyl)pyrrolidine, DEGUPHOS] and $[Rh(cod)_2]BF_4$.[649] (*R*,*R*) NORPHOS (2-exo-3-endo-bis(diphenylphosphino)bicyclo[2.2.1]heptene) and (2*S*,3*S*)-CHIRAPHOS (bis(diphenylphosphino)butane) are also good ligands for this transformation. Arylpyruvic acids give the best results (>95% ee).

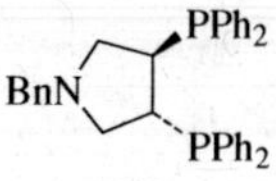

118

The industrially important cis-(+)-methyl jasmonate **119** is conveniently prepared by the hydrogenation of enone **120** with Me-DuPHOS and [Ru(1,2 : 5,6-η-cod)(η^3-methallyl)$_2$.[650]

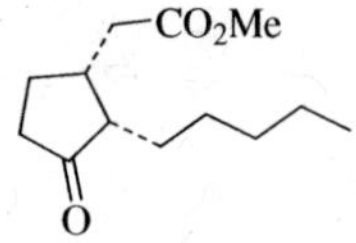

119

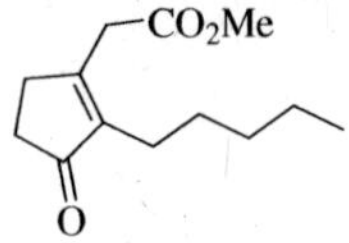

120

(S)-C_3-TunePhos (1,13-bis(diphenylphosphino)-7,8-dihydro-6*H*-dibenzo[f,h][1,5]dioxonin) ruthenium catalyzes the hydrogenation of α-phthalimido ketones with enantioselectivities of >94%,[651] leading to a highly enantioselective route to β-aminoethanols.

Asymmetric Hydrogenation of Enol Acetates

The diphosphine ligand **108** is useful in the asymmetric hydrogenation of enol acetates to chiral acetates, with 80.9% to >99% ee values being realized.[652] The ruthenium TunaPhos complexes from ligand **114** catalyze the asymmetric hydrogenation of enol acetates with high enantiomeric excesses (Eq. 364).[653] High yields and high ee values are obtained via hydrogenation of enol acetates with an achiral ruthenium catalyst and a lipase.[654] This same system is used to convert prochiral ketones into chiral acetates with high enantiomeric excess.

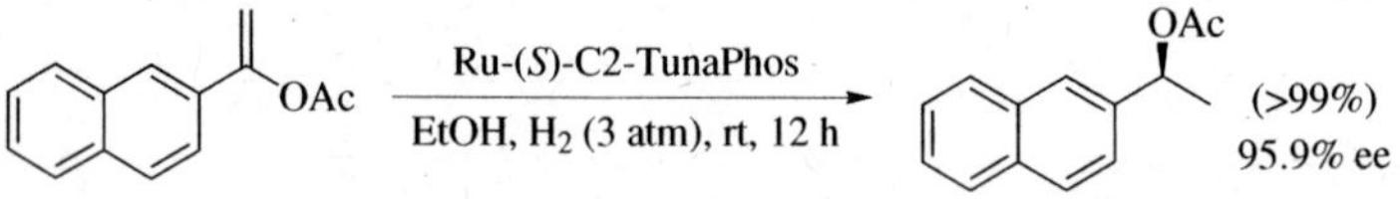

(Eq. 364)

Asymmetric Hydrogenation of α,β-Unsaturated Acids

A study of various diphosphine ligands with rhodium catalyst systems for the hydrogenation of 2-methylenesuccinamic acid favors the DuPHOS (substituted 1,2-bis(phospholano)benzene) Rh(I) and Et-ferroTANE® (1,1′-bis-2,4-diethylphosphotano)ferrocene) Rh(I) systems, with the former being slightly better than the latter (Eq. 365).[655] The conversions are high for both systems. BINAP-ruthenium complexes are successful in the asymmetric hydrogenation of α,β-unsaturated acids, with catalyst **121** showing the best results of those complexes studied.[656] The chiral diaminoferrocenediphosphine ligand **122** catalyzes the reduction of trisubstituted acrylic acids with ee values of >92% (Eq. 366).[657]

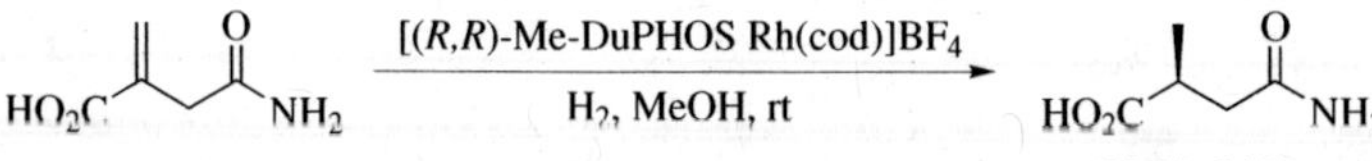

(Eq. 365)

Itaconic acids are reduced in very high enantiomeric excesses (>97%) with Rh-TangPhos catalysts.[658] Itaconic acid is reduced in 99.5% ee with the sugar-derived ferrocenyl phosphine **123**.[659]

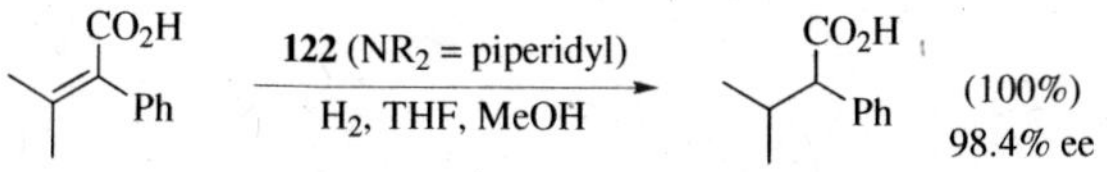

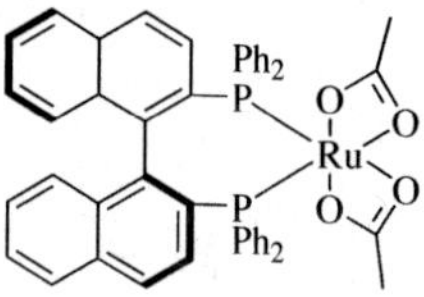

121

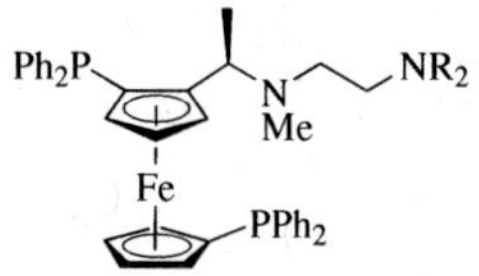

122

(Eq. 366)

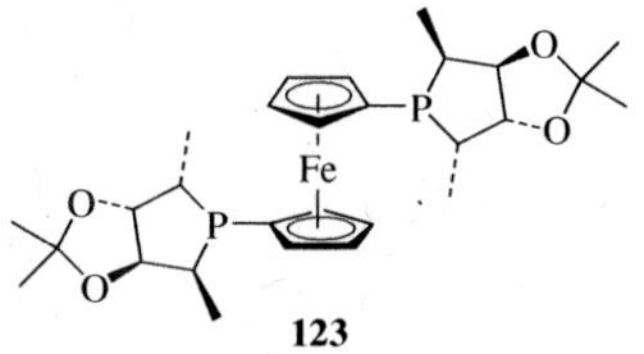

123

Asymmetric Hydrogenation of Acetamidoacrylates

The understandably strong interest in the synthesis of highly enantiomerically enriched α- and β-amino acids has made the asymmetric hydrogenation of α- and β-acetamidoacrylates an active area of investigation. The catalyst Rh-TangPhos catalyze the reduction of β-aryl-α-acetamidoacrylates with high enantioselectivity (>99% and >97%, respectively).[660] Chiral norbornadienyl diphosphoryl rhodium(I) complexes of the type **124** catalyze the asymmetric hydrogenation of α-acetamidoacrylates with high ee values (Eq. 367).[661] Rhodium(I) trap complexes catalyze the hydrogenation of the α-acetamidoacrylates with ee values in the 80-88% range.[662]

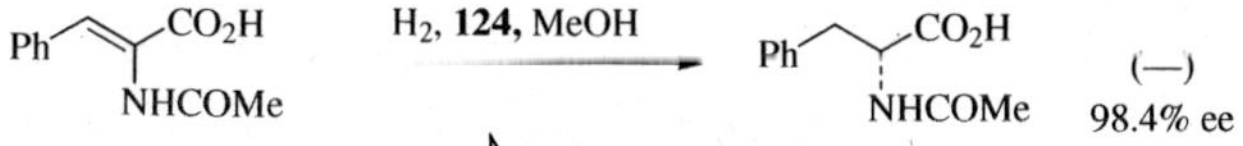

(Eq. 367)

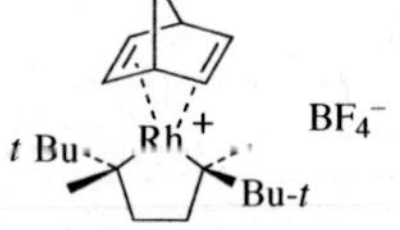

124

The rhodium(I) complexes with hydroxyphospholane ligand **125**[663] or **126**[660] catalyze the asymmetric hydrogenation of α-acetamidoacrylates with ee values in excess of 98%. System **125** is also very effective in the asymmetric hydrogenation of β-acetamidoacrylates (up to 99.6% ee).[664] The planar-chiral heterocyclic ligand **127** complexed with rhodium(I) catalyzes the hydrogenation of α-acetamidoacrylates in excellent yields and ee values from 79–96% under mild conditions.[665]

Other systems that prove successful in the highly enantioselective hydrogenation of α-acetamidoacrylates include the spirophosphinites **128** (94.2–97.2% ee)[666] and the Josiphos ligands **129** with rhodium(I) (84–96% ee). Excellent

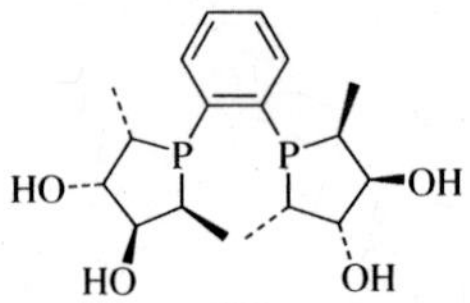

125

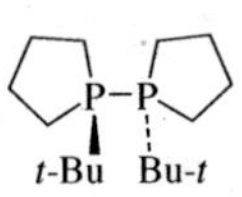

126

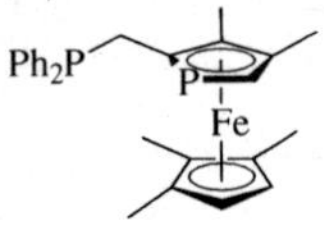

127

results are also obtained with dimethyl itaconate and styrenes.[667] The bis (phospholanes) of type **130**, again with rhodium(I), catalyze the hydrogenation of α-acetamidoacrylates in 92.6–99.1% ee.[668] The ligands **131**,[659] **132**,[669] **133**,[670] and **134**[670] all show good results in the asymmetric hydrogenation of α-acetamidoacrylates, with **131** being especially effective, often rendering ee values of 0>99.9%.

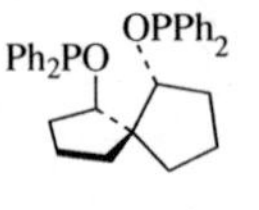

128

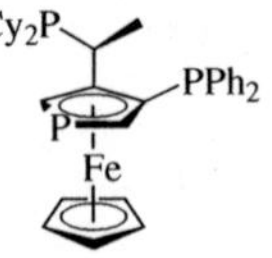

129

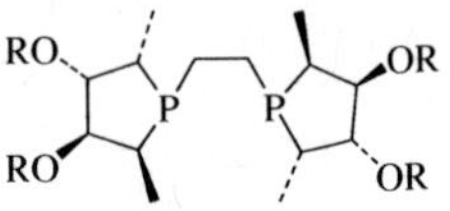

130 R = Bn, *t*-Bu

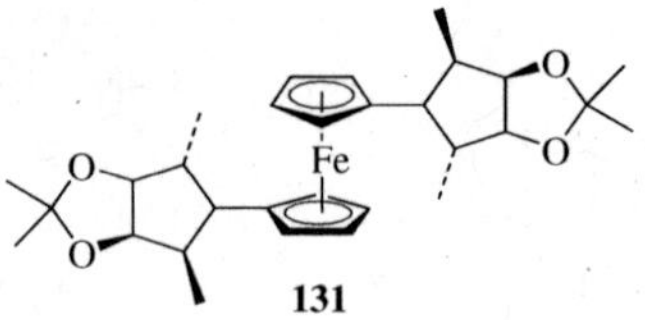

131

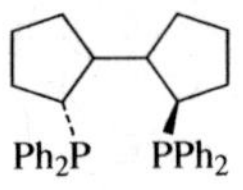

132

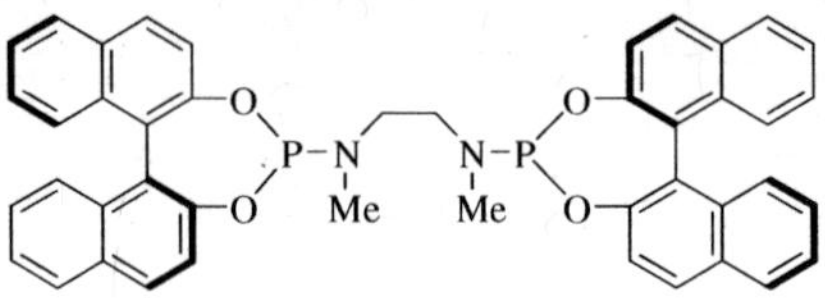

133

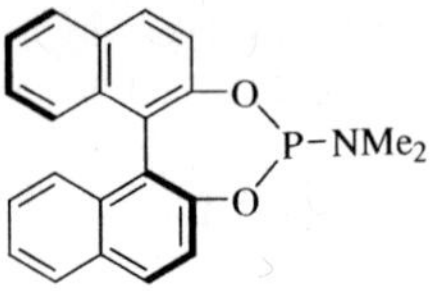

134

The various TunaPhos ligands with ruthenium(0) all catalyze the asymmetric hydrogenation of 2-acetylaminocyclopent-1-enecarboxylic acid ethyl ester in >99% ee.[671] The larger ring sizes give lower ee values. Good results are obtained in the asymmetric hydrogenation of β-acetamidoacrylates with the Et-ferroTANE® rhodium(I) complex.[672] The Rh-TangPhos catalyst system brings about the hydrogenation of α-aryl-β-substituted enamides with high enantioselectivity.[660]

Asymmetric Hydrogenation of Enamides

Enamides, in addition to the acrylates shown above, are also asymmetrically hydrogenated with many of the same systems that prove useful for the acetamidoacrylate reductions. The Rh(I)/BICP (2(*R*)-2′(i)-bis(dipenylphosphino)-1(*R*),1′(*R*)-dicyclopentane) **132** and Rh(I)/DuPHOS systems work well (>90% ee) for the asymmetric hydrogenation of β-acetamidovinyl methoxymethyl ethers in an approach to enantiomerically enriched β-aminoethanols.[673] The Rh-binaphane system **138** catalyzes the reduction of aryl alkyl enamides in up to 99.6% ee.[674] Cyclic enamides are reduced in 37–99% enantiomeric excess with the Rh(I)/**135** system (Eq. 368).[675]

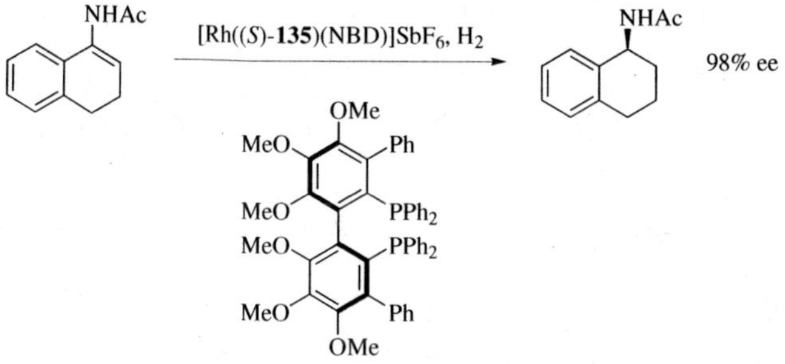

135

(Eq. 368)

The Rh(I)/**136** or Rh(I)/**137** combination can be used in the asymmetric hydrogenation of 1-arylenamides in 90–99% ee, with Rh(I)/**137** being the better of the two.[676] Me-DuPHOS and related ligands with rhodium(I) reduce 1-aryl-2-alkylenamides in >90% ee[677] whereas the Rh(I)/DIOP combination carries this out in 97.3–99% ee selectivity.[678] Finally, the Rh(I)/**138** system reduces β-substituted-α-arylenamides in 95–99% ee, and α-substituted acetamidoethylenes in 75.7–90% ee.[674]

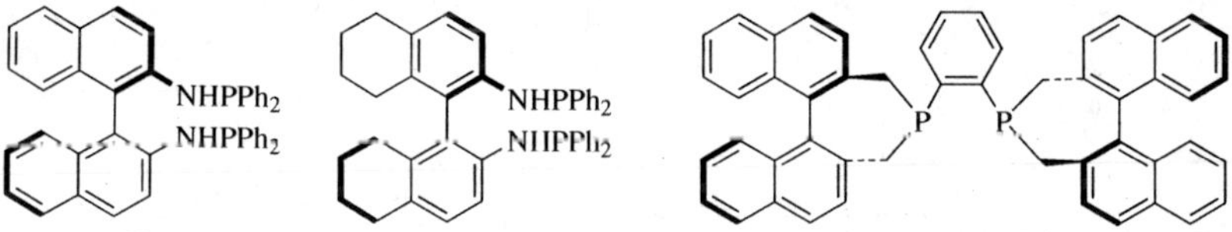

136 **137** **138**

Asymmetric Hydrogenation of Imines

As an extension of the asymmetric hydrogenation of prochiral ketones to enantiomerically enriched alcohols, the reduction of imines has been a topic of interest in obtaining chiral amines of high enantiomeric purity. Several entries to enantiomerically enriched amines based on the approaches outlined above are available. These asymmetric hydrogenations have proved to be more difficult than those for prochiral ketones, but nevertheless show good promise.

Iridium(III) hydride forms complexes with DIOP, BDPP (2,4-bis(diphenylphosphino)pentane), NORPHOS, and BINAP ligands to produce amines in 11–80% ee.[679] Similar modest results are obtained in the reduction of *N*-arylketimines with an iridium(III) complex with (2*S*,3*S*)-CHIRAPHOS as the chiral ligand.[680] The indium complexes with chiral phosphinodihydrooxazoles catalyze the enantioselective hydrogenation of imines in supercritical carbon dioxide with up to 80% ee, but generally lower ee values are observed in dichoromethane. The Rh(I)/chiral phosphine-catalyzed hydrogenation of imines is reported to give the chiral amines in up to 60% ee.[681] This work presents a crystal structure of an intermediate rhodium(diphos)imine complex. The iridium(III) complex with the diphosphine ligand **138** gives amines in up to 99% ee and in excellent yields (Eq. 369).[682] Cyclic imines undergo asymmetrical reduction via transfer hydrogenation using the catalyst EBTHI-Ti **102** (Eq. 370).[683]

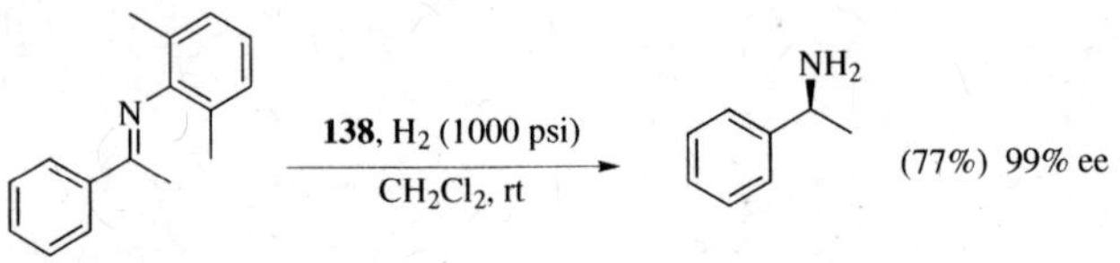

(Eq. 369)

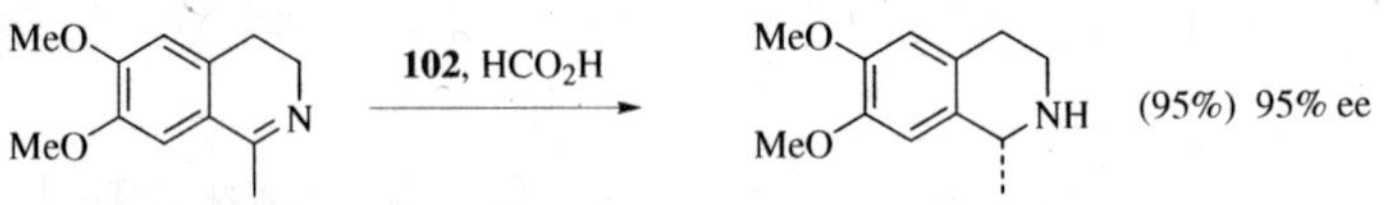

(Eq. 370)

The asymmetric hydrogenation of acyclic imines with the *ansa*-titanocene catalyst **102** gives the chiral amines in up to 92% ee.[684,685] This same system applied to cyclic imines produces the chiral amines with >97% ee values.[684,685] The mechanism of these reductions has been studied.[686]

EXPERIMENTAL CONDITIONS

Normal precautions to protect laboratory workers from exposure to chemical reagents should be followed. Strong acids such as trifluoroacetic acid and trifluoromethanesulfonic acid are often used in the preparations and should be handled with extreme care. The physical properties of organosilicon hydrides are similar to those of the analogous hydrocarbons, after taking account of the differences in molecular weights. They are generally lipophilic in nature. As previously mentioned, the chemical properties of organosilicon hydrides are considerably more benign than those of many metal-based reducing agents. However, organosilicon hydrides do react with strong bases and acids to produce hydrolysis products and hydrogen gas. This reaction occurs more rapidly with bases than with acids. Also, some of the lower molecular weight organosilicon hydrides, especially the parent compound SiH_4, are pyrophoric. A few of the organosilicon hydrides, such as trimethoxysilane and triethoxysilane, are toxic and have the ability to cause corneal damage.

Many of the synthetically useful reactions of organosilicon hydrides are conducted in solution using solvents such as CH_2Cl_2, $CHCl_3$, CCl_4, MeCN, or THF. In general, it is important that anhydrous reaction conditions be used and that normal purification procedures be followed to ensure that the solvents used are pure and anhydrous.

Finally, it must be mentioned that there are advantages in synthetic methods using polymeric organosilicon hydride reagents, such as PMHS, which are both relatively inexpensive and give high molecular weight products that are reasonably easy to separate from the desired organic products.

EXPERIMENTAL PROCEDURES

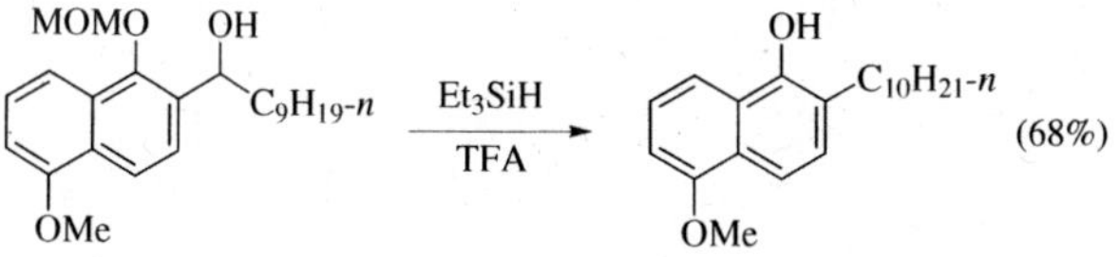

2-Decyl-5-methoxy-1-naphthol [Reduction of a Secondary Benzylic Alcohol to a Methylene Group with Concomitant Loss of a MOM Protecting Group].[167] To a solution of 2-(1-hydroxydecyl)-5-methoxy-1-methoxymethyleneoxynaphthalene (0.525 g, 1.4 mmol) and Et_3SiH (1.628 g, 14 mmol) in CH_2Cl_2 (8 mL) was added TFA (2.16 mL, 28 mmol) in CH_2Cl_2 (3 mL) at room temperature and under an atmosphere of argon. The reaction mixture was stirred for 2 hours at room temperature, and then was poured into a saturated aqueous $NaHCO_3$ solution (20 mL) and extracted with CH_2Cl_2 (3 × 15 mL). The extract was washed with saturated aqueous $NaHCO_3$ solution (15 mL), brine (15 mL), dried with $MgSO_4$, and evaporated. The crude product was purified by chromatography (SiO_2, benzene as eluent) to afford 2-decyl-5-methoxy-1-naphthol as needles: 0.319 g (68%); mp 61–62° (hexane); IR (CCl_4) 3600, 1600, 1505, 1254, 1055, 880 cm^{-1}; ^1H NMR (100 MHz, $CDCl_3$) δ 7.60 (d, $J = 9$ Hz, 1H), 7.45 (dd, $J = 2, 8$ Hz, 1H), 7.15 (t, $J = 8$ Hz, 1H), 7.02 (d, $J = 9$ Hz, 1H), 6.56 (dd, $J = 2, 8$ Hz, 1H), 4.90 (br s, 1H), 3.94 (s, 3H), 2.68 (t, $J = 7$ Hz, 2H), 1.68 (m, 2H), 1.26 (m, 14H), 0.90 (t, $J = 8$ Hz, 3H).

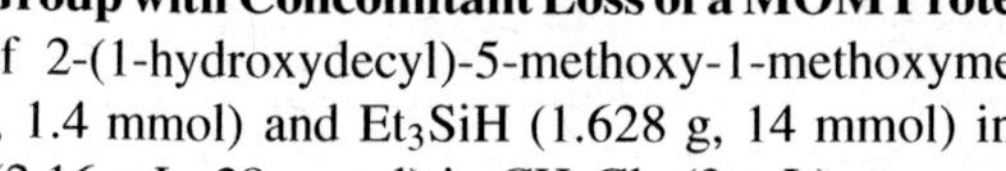

Cyclohexane [Aluminum Chloride Catalyzed Reduction of a Dichloroalkane to a Hydrocarbon].[189] After a solution of *cis*-1,2-dichlorocyclohexane (0.1582 g, 1.033 mmol) in CH_2Cl_2 (3 mL) was cooled to 0°, Et_3SiH (0.299 g, 2.57 mmol) and $AlCl_3$ (0.0345 g, 0.173 mmol) were added. The mixture was stirred for 30 minutes and then quenched with water (10 mL). Heptane (23.1 mg, 0.231 mmol) was added as an internal standard and the aqueous layer was separated and extracted with CH_2Cl_2. The combined organic layer was dried ($MgSO_4$) and analyzed by GLC: 0.064 g (74%).

$$n\text{-}C_{11}H_{23}CO_2Et \xrightarrow[\text{2. aq. HCl, 30 min}]{\text{1. }(EtO)_3SiH\text{, CsF, }N_2\text{, 30 min, 60°}} n\text{-}C_{12}H_{25}OH \quad (90\%)$$

1-Dodecanol [Fluoride-Promoted Reduction of an Ester to an Alcohol].[83] A mixture of ethyl dodecanoate (2.18 g, 10.0 mmol) and triethoxysilane (3.77 g, 23.0 mmol) was added to CsF (1.52 g, 10.0 mmol) under nitrogen. The reaction was followed by IR spectroscopy. After 30 minutes at 60°, 12 N HCl (1 mL) in acetone (5 mL) was added. After 30 minutes, the mixture was extracted with ether (2 × 150 mL). The combined extracts were dried with $MgSO_4$ and the solvents were removed. The residue was distilled under vacuum to give 1-dodecanol: 1.8 g (90%); bp 145°/15 Torr. The GLC retention time was identical with that of an authentic sample.

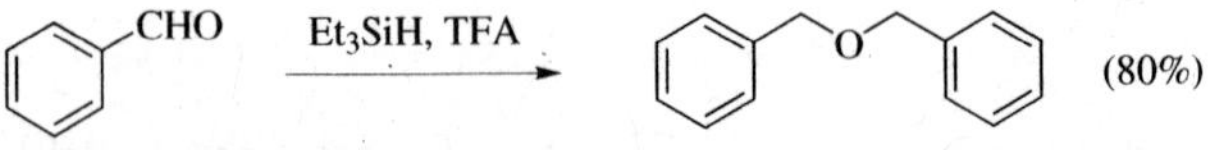

Dibenzyl Ether [Brønsted Acid Promoted Reduction of an Aldehyde to a Symmetrical Ether].[311] To a stirred solution of benzaldehyde (5.4 g, 0.05 mol) and TFA (11.4 g, 0.1 mol) under argon was added dropwise, with cooling, Et_3SiH (8.1 g, 0.07 mol) at a rate such that the temperature of the reaction mixture did not exceed 40°. The solution turned a crimson color that gradually disappeared. Analysis by GLC showed the complete absence of the aldehyde immediately after addition of all of the silane. The products were separated by vacuum distillation at 20 Torr, collecting the fractions up to 125°. Dibenzyl ether was obtained from the residue by freezing out: 4 g (0.02 mol, 80%); mp 3–6°; n_D^{25} 1.5608.

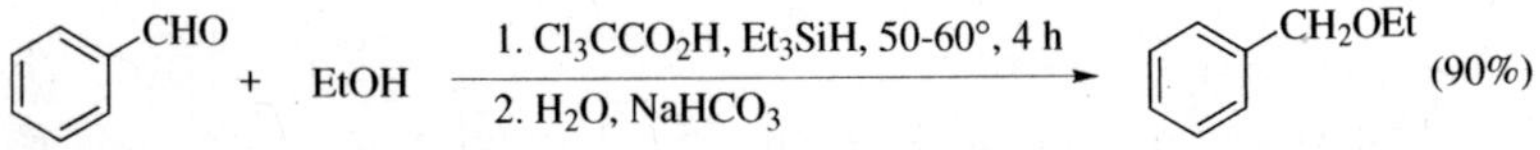

Ethyl Benzyl Ether [Brønsted Acid Promoted Reduction of an Aldehyde to an Unsymmetrical Ether].[327] To a cooled mixture of benzaldehyde (4.3 g, 41 mmol) and absolute ethanol (3.7 g, 80 mmol) was added trichloroacetic acid (18.2 g, 111 mmol). Et_3SiH (6.96 g, 60 mmol) was then added dropwise with stirring while the mixture was maintained at 50–60°. After 4 hours, the reaction mixture was diluted with water, neutralized with aqueous $NaHCO_3$ solution, and extracted with Et_2O. The dried ether extract was distilled and the 170–190° fraction was collected. Distillation from sodium gave ethyl benzyl ether: 4.8 g (90%); bp 187–189°.

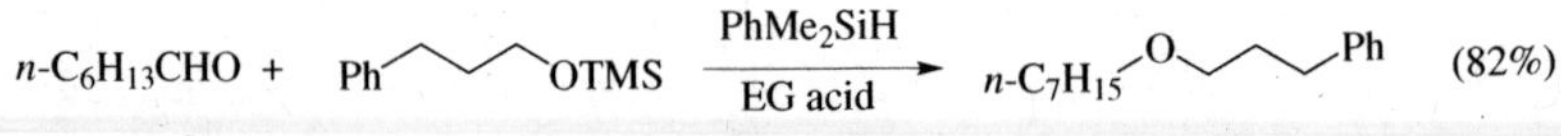

1-Heptyl 3-Phenylpropyl Ether [Electrogenerated Acid-Promoted Reduction of an Aldehyde to an Unsymmetrical Ether].[333] A mixture of 1-heptanal (1.0 mmol), 3-phenylpropoxytrimethylsilane (1.2 mmol), tetra-*n*-butylammonium perchlorate (0.1 mmol), and lithium perchlorate (0.1 mmol) was dissolved in CH_2Cl_2 (3 mL) in an undivided cell. The mixture was electrolyzed under constant current (1.67 mA cm^{-2}) with platinum electrodes at ambient temperature. After 5 minutes, dimethylphenylsilane (1.2 mmol) was added dropwise and the electrolysis was continued (0.06 Faraday/mol). After completion of the reaction, one drop of Et_3N was added and the solution was concentrated. The residue was chromatographed on SiO_2 to give 1-heptyl 3-phenylpropyl ether: 0.82 mmol (82%); bp 80–83°/1.0–2.0 Torr; IR (neat) 2955, 2925, 1605, 1110 cm^{-1}; ^{1}H NMR ($CDCl_3$) δ 7.22 (s, 5H), 3.37 (t, $J = 6$ Hz, 2H), 2.86–2.47 (m, 2H), 1.50–2.10 (m, 2H), 1.31 (br m, 10H), 0.87 (m, 3H).

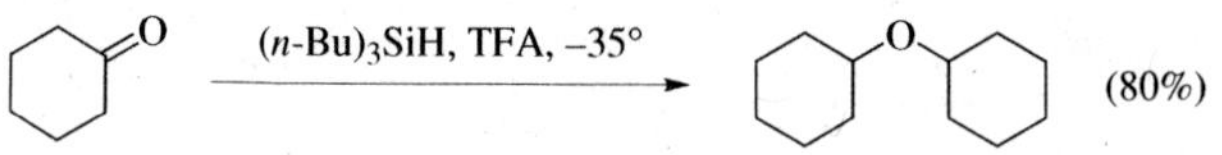

Dicyclohexyl Ether [Brønsted Acid Promoted Reduction of a Ketone to a Symmetrical Ether].[313] Cyclohexanone (3.92 g, 40 mmol) and tri(n-butyl) silane (1.78 g, 20 mmol) were placed in a round-bottomed flask. TFA (75 mmol) was added slowly over a one-hour period to the reaction mixture, which was held at −35°. After complete addition, the reaction flask was placed in a freezer at −15° for 70 hours. Direct distillation gave dicyclohexyl ether: 2.91 g (16 mmol, 80%); bp 119–121°/18 Torr.

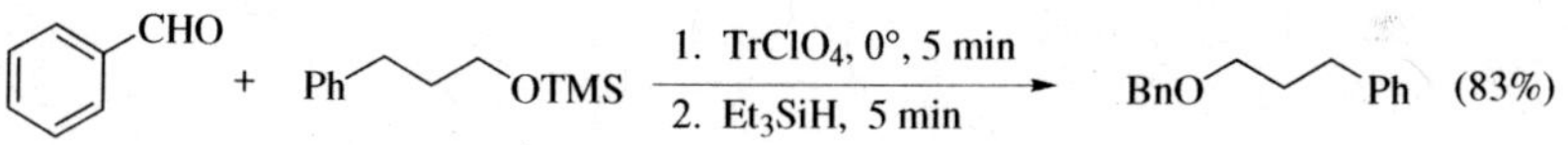

Benzyl 3-Phenylpropyl Ether [Trityl Perchlorate Catalyzed Reduction of an Aldehyde to an Unsymmetrical Ether].[329] Under an argon atmosphere, a CH_2Cl_2 (2 mL) solution of benzaldehyde (53 mg, 0.5 mmol) and 3-phenylpropoxytrimethylsilane (0.5 mmol) was added to trityl perchlorate (9 mg, 0.026 mmol), and the solution was stirred for 5 minutes at 0°. A CH_2Cl_2 (1 mL) solution of Et_3SiH (59 mg, 0.5 mmol) was added and stirring was continued for another 5 minutes. Then phosphate buffer was added, and the organic materials were extracted with Et_2O and dried over $MgSO_4$. After removal of the solvents under reduced pressure, isolation by TLC on SiO_2 provided 94 mg (83%) of benzyl 3-phenylpropyl ether: IR (NaCl) 1100 cm^{-1}; ^{1}H NMR ($CDCl_3$) δ 7.10 (s, 5H), 7.00 (s, 5H), 4.30 (s, 2H), 3.30 (t, $J = 6$ Hz, 2H), 2.8-2.4 (m, 2H), 2.1-1.5 (m, 2H).

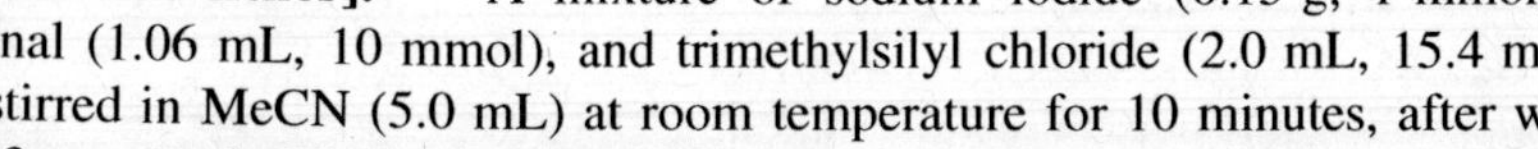

Di-n-pentyl Ether [TMSI-Catalyzed Reduction of an Aldehyde to a Symmetrical Ether].[314] A mixture of sodium iodide (0.15 g, 1 mmol), 1-pentanal (1.06 mL, 10 mmol), and trimethylsilyl chloride (2.0 mL, 15.4 mmol) was stirred in MeCN (5.0 mL) at room temperature for 10 minutes, after which 1,1,3,3-tetramethyldisiloxane (TMDO, 1.79 mL, 10 mmol) was added. When the exothermic reaction had ended (30 minutes), a solution of 2.5 N HF in MeOH (30 mL) was added to the reaction mixture, which was then refluxed for 5 minutes. Work-up was carried out by diluting the solution with CH_2Cl_2 (40 mL), washing with water (30 mL) and saturated aqueous $NaHCO_3$ solution (20 mL), drying, and evaporating the solvents. Crude di-n-pentyl ether was purified by distillation: 0.65 g (84%); bp 185–189°/760 Torr.

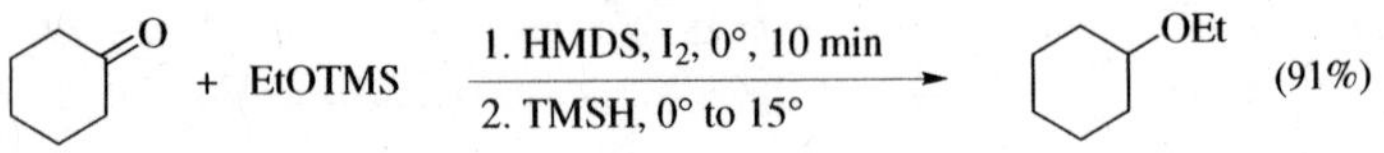

Cyclohexyl Ethyl Ether [TMSI-Catalyzed Reduction of a Ketone to an Unsymmetrical Ether].[334] In a 100-mL, three-necked flask equipped with a rubber septum, thermometer, magnetic stirring bar, and nitrogen inlet were placed finely powdered iodine (0.13 g, 0.50 mmol) and hexamethyldisilane (0.079 g, 0.54 mmol) in CH_2Cl_2 (14 mL). The violet solution was stirred 10 minutes at room temperature, cooled to 0°, and a solution of cyclohexanone (1.04 g, 10 mmol) and ethoxytrimethylsilane (1.10 g, 10 mmol) in 10 mL of CH_2Cl_2 was introduced via syringe. The reaction mixture was stirred for 10 minutes at 0°, after which TMSH was added directly from a gas cylinder by means of Tygon tubing attached to a hypodermic needle inserted through the rubber septum. The gas was allowed to slowly bubble through the solution until the color changed from violet to red-gold. During this time the internal temperature rose from 0° to 15°. The cold bath was removed and stirring was continued at room temperature for 2 hours. The mixture was washed with 10% aqueous $Na_2S_2O_3$ solution (4 × 30 mL) and water (4 × 30 mL), and dried over $MgSO_4$. The volatiles were removed under reduced pressure on a steam bath to obtain pure cyclohexyl ethyl ether: 1.37 g (91%); bp 141–144°, ^{13}C NMR ($CDCl_3$) δ 76.8, 63.0, 32.6, 26.4, 24.2, 15.9.

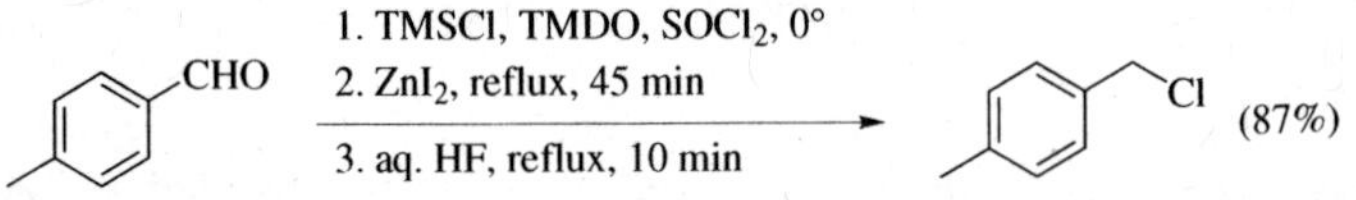

4-Methylbenzyl Chloride [Reductive Halogenation of an Aldehyde to a Benzyl Chloride].[314] A mixture of 4-methylbenzaldehyde (1.18 g, 10 mmol), chlorotrimethylsilane (2.0 mL, 15.7 mmol), 1,1,3,3-tetramethyldisiloxane (TMDO, 1.79 mL, 10 mmol), and thionyl chloride (1.0 mL, 13.7 mmol) was cooled at 0°. Then ZnI_2 (0.02 g) was added and a very exothermic reaction took place. When the spontaneous heating had ended, the mixture was heated at reflux with stirring for 45 minutes, and a 2.5 M solution of HF in MeOH (10 mL) was added. After being heated at reflux for 10 minutes, the solution was cooled to 0°, filtered, and taken up in CH_2Cl_2 (30 mL)/water (40 mL). The aqueous layer was extracted with CH_2Cl_2 (2 × 10 mL). The combined organic phases were dried (Na_2SO_4) and the solvents were evaporated to afford crude 4-methylbenzyl chloride, which was purified by distillation: 1.22 g, 87%; bp 190–195°/760 Torr.

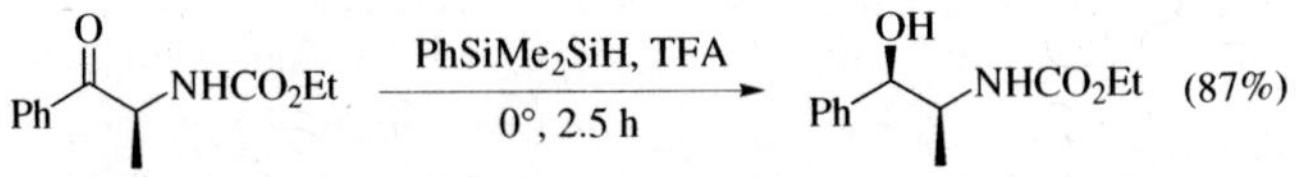

(1*R*,2*S*)-2-[(Ethoxycarbonyl)amino]-1-phenyl-1-propanol [Brønsted Acid Promoted Reduction of an α-Amino Ketone to an Erythro α-Hydroxy

Amine].[276] Dimethylphenylsilane (0.184 mL, 1.20 mmol) was added slowly to a TFA (1 mL) solution of (*S*)-2-[(ethoxycarbonyl)amino]-1-phenyl-1-propanone (221 mg, 1.00 mmol) at 0°, and the solution was stirred for 2.5 hours at 0°. Saturated aqueous $NaHCO_3$ solution (20 mL) was added, and the resulting mixture was extracted with CH_2Cl_2 (10 mL). The extract was dried over $MgSO_4$, filtered, and concentrated under reduced pressure to give the crude product, whose ^1H NMR spectrum showed exclusive formation (>99% selectivity) of (1*R*, 2*S*)-2-[(ethoxycarbonyl)amino]-1-phenyl-1-propanol. Purification by preparative TLC (SiO_2, AcOEt/hexane, 1/1) afforded (1*R*, 2*S*)-2-[(ethoxycarbonyl)amino]-1-phenyl-1-propanol (194 mg, 87%) as colorless crystals: mp 71°; $[\alpha]^{20}{}_D - 40°$ (*c* 0.245, CH_2Cl_2); IR (KBr) 3350, 1694, 1552, 1273, 1043, 1028, 708 cm^{-1}; ^1H NMR ($CDCl_3$) δ 7.34 (s, 5H), 4.9 (br s, 1H), 4.84 (d, $J = 3$ Hz, 1H), 4.10 (q, $J = 7.2$ Hz, 2H), 4.2-3.8 (m, 1H), 2.83 (br s, 1H), 1.24 (t, $J = 7$ Hz, 3H), 0.99 (d, $J = 7$ Hz, 3H); MS (70 eV) m/z (relative intensity): M^+ 223 (trace), 117 (18), 116 (66), 107 (11), 88 (21), 79 (15), 77 (14), 72 (11), 51 (5), 44 (100), 29 (23), 27 (7), 18 (5). Anal. Calcd for $C_{12}H_{17}NO_3$: C 64.55, H 7.67, N 6.27. Found: C 64.35, H 7.53, N 6.25.

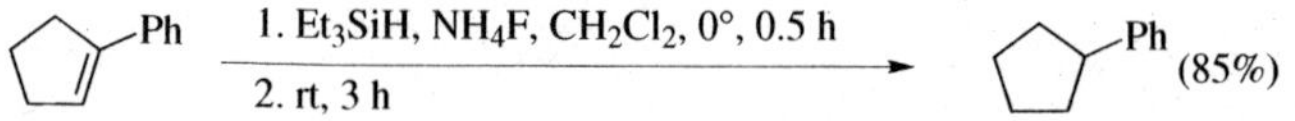

Phenylcyclopentane [Brønsted Acid Catalyzed Reduction of an Alkene to an Alkane].[135] To a stirred solution of 1-phenylcyclopentene (1.44 g, 10 mmol), NH_4F (0.48 g, 13 mmol), and Et_3SiH (1.5 g, 13 mmol) was added TFA (5.1 g, 50 mmol) at 0° over a 10-minute period. The reaction mixture was then stirred for 20 minutes at 0° and at room temperature for 3 hours. The reaction mixture was quenched with ice water and extracted with CH_2Cl_2. The organic extract was washed with 10% aqueous saturated $NaHCO_3$ solution, dried ($CaCl_2$), and concentrated. Distillation provided phenylcyclopentane: 1.22 g (85%); bp 108–111°/20 Torr.

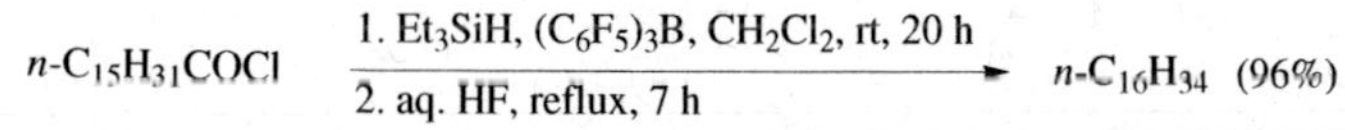

***n*-Hexadecane [Tris(pentafluorophenyl)boron-Catalyzed Reduction of an Acid Chloride to an Alkane].**[282] Et_3SiH (20 mmol) was added to a stirred solution of hexadecanoyl chloride (5 mmol) and tris(pentafluorophenyl)boron (5 mol%) in CH_2Cl_2. The reaction mixture was stirred for 20 hours at room temperature, quenched with Et_3N (0.25 g), filtered through Celite, and concentrated. The residue was mixed with 40% HF (5 mL) in EtOH (30 mL) and heated at reflux for 7 hours. Water (60 mL) was added, and the crude product was extracted with pentane (3 × 30 mL). The combined pentane solutions were washed with water and dried over $MgSO_4$. After the solvent and triethylfluorosilane were removed under vacuum, the product was purified by flash chromatography to give a 96% yield of *n*-hexadecane.

$n\text{-}C_{11}H_{23}CO_2H \xrightarrow[\text{2. aq. HF, reflux, 7 h}]{\text{1. } Et_3SiH,\ (C_6F_5)_3B,\ CH_2Cl_2,\ \text{rt, 20 h}} n\text{-}C_{12}H_{26}$ (91%)

***n*-Dodecane [Tris(pentafluorophenyl)boron-Catalyzed Reduction of a Carboxylic Acid to an Alkane].**[282] Dodecanoic acid (5 mmol) in CH_2Cl_2 was added to a CH_2Cl_2 solution of tris(pentafluorophenyl)boron (5 mol%) and Et_3SiH (30 mmol). The reaction mixture was stirred for 20 hours at room temperature, quenched with Et_3N (0.25 g), filtered through Celite, and concentrated. The residue was mixed with 40% HF (5 mL) in EtOH (30 mL) and heated at reflux for 7 hours. Water (60 mL) was added and the crude product was extracted with pentane (3 × 30 mL). The combined pentane solutions were washed with water and dried over $MgSO_4$. After the solvent and triethylfluorosilane were removed under vacuum, the product was purified by flash chromatography to give a 91% yield of *n*-dodecane.

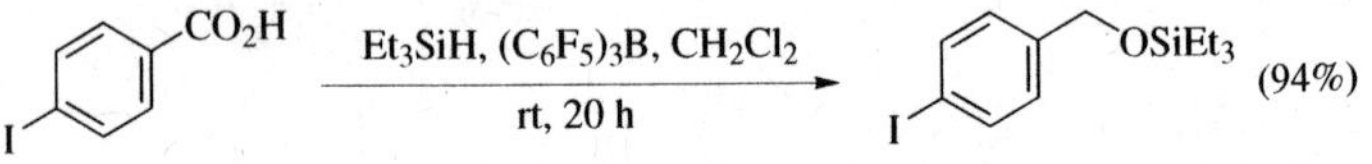

4-Iodobenzyloxytriethylsilane [Tris(pentafluorophenyl)boron-Catalyzed Reduction of a Carboxylic Acid to a Benzyl Triethylsilyl Ether].[282] Et_3SiH (16.5 mmol) was added to a stirred solution of 4-iodobenzoic acid (5 mmol) and tris(pentafluorophenyl)boron (5 mol%) in CH_2Cl_2. The reaction mixture was stirred for 20 hours at room temperature, quenched with Et_3N (0.25 g), filtered through Celite, and concentrated. The solvent was removed under vacuum and the product was purified by flash chromatography to give a 94% yield of 4-iodobenzyloxytriethylsilane. ^1H NMR ($CDCl_3$, 500 MHz) δ 7.48 (d, $J = 8.4$ Hz, 2H), 7.24 (d, $J = 8.4$ Hz, 2H), 4.71 (s, 2H), 1.02, 0.68 (q, $J = 8.0$ Hz, 6H); ^{13}C NMR ($CDCl_3$, 126 MHz) δ 140.8, 131.7, 128.3, 121.07, 64.4, 7.2, 4.9; GC–MS m/z (% relative intensity, ion): 300 (1, M^+), 271 (77, M–Et), 169 (100).

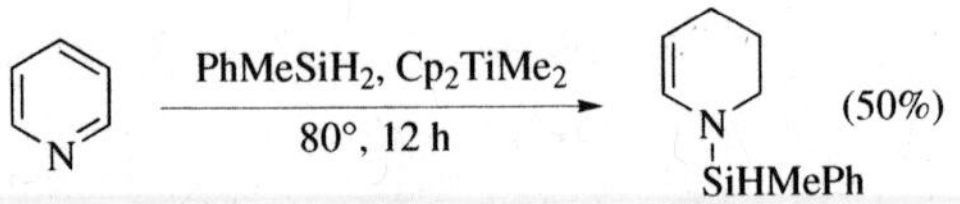

***N*-(Phenylmethylsilyl)-1,2,3,4-tetrahydropyridine [Reduction of a Pyridine].**[264] Phenylmethylsilane (3.5 mL, 25.6 mmol) and pyridine (1.0 mL, 12.5 mmol) were added to Cp_2TiMe_2 (0.13 g, 0.7 mmol, 6 mol%). The solution color changed to dark blue, then purple, accompanied by gas evolution. The mixture was stirred for 12 hours at 80°. The ^1H-NMR spectrum showed that >95% of the pyridine had reacted to give a yield of ca. 80% of the crude product, which was distilled under vacuum to give 1.29 g (50%) of the title compound as a colorless liquid: bp 57°/0.12 Torr; ^1H NMR (300 MHz, C_6D_6) δ 7.5 and 7.2 (2m, 5H), 5.01 (q, $J = 3.3$ Hz, 1H), 4.62 (m, 1H), 2.99 (m, 2H), 2.00 (m, 2H), 1.57 (m, 2H), 0.28 (d, $J = 3.3$ Hz, 3H); ^{29}Si NMR (59.9 MHz, C_6D_6) δ −10.5;

EIMS m/z (% relative intensity, ion): 203 (100, M^+), 188 (18.5, M^+-CH_3), 121 (76.6, $M^+-C_5H_8N$).

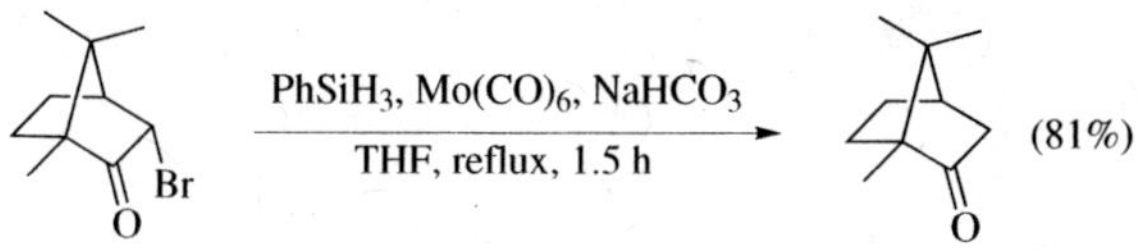

Camphor [Reduction of an α-Bromo Ketone to a Ketone].[197] A mixture of α-bromocamphor (2.24 g, 9.69 mmol), $Mo(CO)_6$ (0.11 g, 0.53 mmol), phenylsilane (1.30 g, 12 mmol), and $NaHCO_3$ (0.88 g, 10.5 mmol) in THF (6 mL) was heated at reflux for 1.5 hours. The mixture was cooled to room temperature, water (0.15 mL) was added, and the solvent was removed under reduced pressure. Distillation afforded camphor in 81% yield.

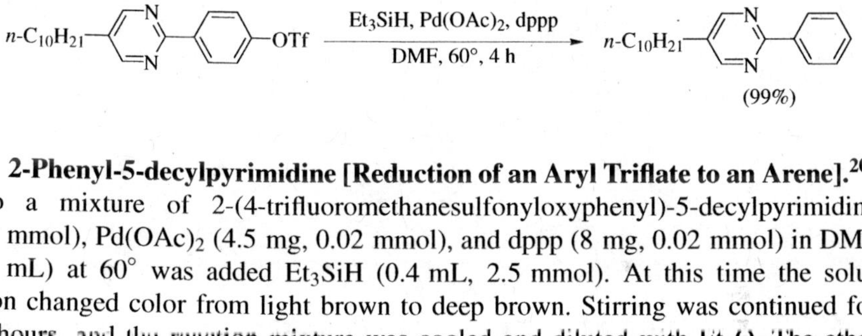

2-Phenyl-5-decylpyrimidine [Reduction of an Aryl Triflate to an Arene].[201] To a mixture of 2-(4-trifluoromethanesulfonyloxyphenyl)-5-decylpyrimidine (1 mmol), $Pd(OAc)_2$ (4.5 mg, 0.02 mmol), and dppp (8 mg, 0.02 mmol) in DMF (5 mL) at 60° was added Et_3SiH (0.4 mL, 2.5 mmol). At this time the solution changed color from light brown to deep brown. Stirring was continued for 4 hours, and the reaction mixture was cooled and diluted with Et_2O. The ether phase was washed with water, and with aqueous saturated solutions of $NaHCO_3$ and NaCl. The ether layer was dried (Na_2SO_4) and concentrated. The crude product was purified (99% yield) by chromatography: IR (KBr) 1549, 1435, 745, 691, 654 cm^{-1}; ^1H NMR: δ 8.62 (s, 2H), 8.41 (m, 2H), 7.46 (m, 3H), 2.62 (t, $J = 7.7$ Hz, 2H), 1.8-1.4 (m, 2H), 1.27 (br s, 14H), 0.88 (t, $J = 6.2$ Hz, 3H).

$$(n\text{-}C_{16}H_{33})_2O \xrightarrow[CH_2Cl_2,\ rt,\ 20\ h]{Et_3SiH,\ (C_6F_5)_3B} \underset{(98\%)}{n\text{-}C_{16}H_{34}} + \underset{(98\%)}{n\text{-}C_{16}H_{33}OSiEt_3}$$

***n*-Hexadecane and 1-(Triethylsiloxy)hexadecane [Reduction of a Symmetrical Ether].**[145] Et_3SiH (1.1 mmol) was added to tris(pentafluorophenyl)boron (5 mol%) and bis-(*n*-hexadecyl) ether (1 mmol) in 1 mL of CH_2Cl_2. The reaction mixture was stirred for 20 hours at room temperature and then quenched with Et_3N (0.05 mL), filtered through Celite, and concentrated. GLC analysis with an internal standard showed the presence of *n*-hexadecane (98%) and 1-(triethylsiloxy)hexadecane (98%).

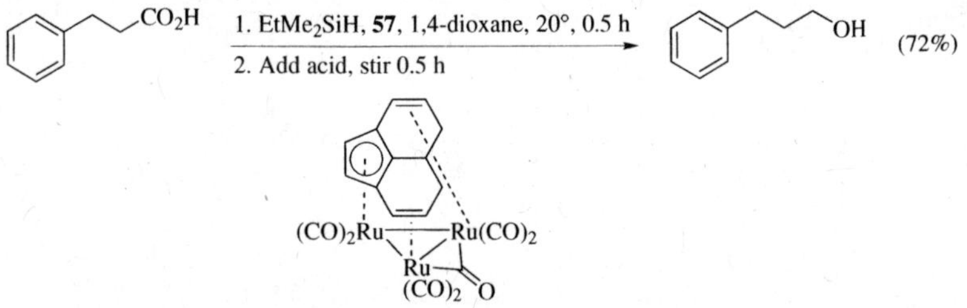

57

3-Phenyl-1-propanol [Reduction of a Carboxylic Acid to an Alcohol].[280] To a solution of ruthenium catalyst **57** (616.5 mg, 25.2 μmol) in dioxane (0.45 mL) was added dimethylethylsilane (0.84 mL, 6.3 mmol). After the mixture had been stirred for 30 minutes at room temperature, hydrocinnamic acid (380 mg, 2.55 mmol) was added, and the stirring was continued for 30 minutes. Vigorous gas evolution occurred. The reaction was quenched with aqueous HCl, and the mixture was extracted with Et_2O. The organic phase was washed with aqueous saturated $NaHCO_3$ and aqueous saturated NaCl solutions, and then dried with $MgSO_4$. The solvent was removed under vacuum and the product was purified by chromatography (EtOAc/hexane 1/9) to give 3-phenylpropyl alcohol: 248 mg, 72%.

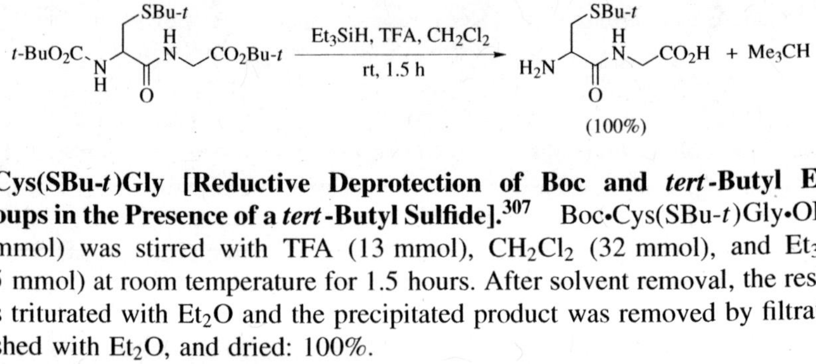

Cys(SBu-*t*)Gly [Reductive Deprotection of Boc and *tert*-Butyl Ester Groups in the Presence of a *tert*-Butyl Sulfide].[307] Boc•Cys(SBu-*t*)Gly•OBu-*t* (1 mmol) was stirred with TFA (13 mmol), CH_2Cl_2 (32 mmol), and Et_3SiH (2.5 mmol) at room temperature for 1.5 hours. After solvent removal, the residue was triturated with Et_2O and the precipitated product was removed by filtration, washed with Et_2O, and dried: 100%.

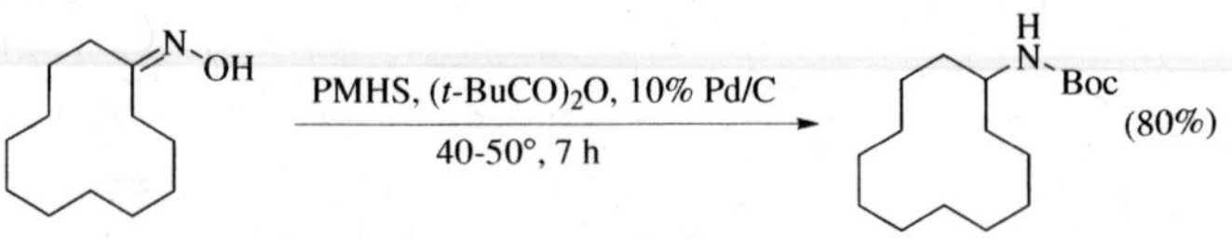

***N*-Boc-cyclododecylamine [Reductive Boc-protection of an Oxime].[549]** To a stirred solution of cyclododecanone oxime (1 mmol) in EtOH (10 mL) were added PMHS (180 mg, 3.0 mmol), di-*tert*-butyl dicarbonate (240 mg, 1.1 mmol), and 10% Pd/C (10 mg). The reaction mixture was stirred at 40–50° for 7 hours, after which time it was filtered and the filtrate concentrated under vacuum. The crude product was purified by column chromatography to give *N*-Boc-cyclododecylamine: 80%.

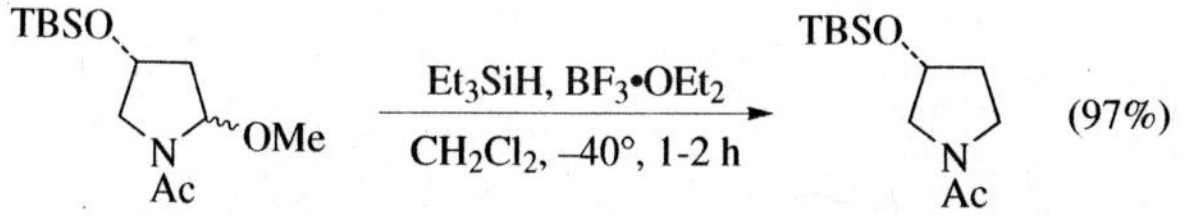

(3*R*)-*N*-Acetyl-3-(*tert*-butyldimethylsiloxy)pyrrolidine [Reduction of an Aminal to an Amine].[521] A solution of *N*-acetyl-2-methoxy-4-*tert*-butyldimethylsiloxy)pyrrolidine (2 mmol) and Et_3SiH (4 mmol) in dry CH_2Cl_2 (3 mL) was treated with $BF_3{\bullet}OEt_2$ (4 mmol) at −40°. The reaction was monitored by TLC (1−2 hours). The mixture was diluted with $CHCl_3$ and washed with aqueous saturated $NaHCO_3$ and aqueous saturated NaCl solution. The organic layer was dried ($MgSO_4$) and evaporated under reduced pressure. Purification of the residue by chromatography on SiO_2 (hexane/EtOAc 10 : 1) gave the title compound as a colorless syrup: 97%; $[\alpha]^{24.5}{}_D$ −23.4° (*c* 1.22, $CHCl_3$); IR (film) 3350, 1650 cm^{-1}; ^1H NMR ($CDCl_3$) δ 4.58 – 4.25 (m, 1H), 3.80-3.10 (m, 4H), 2.10−1.60 (m, 2H), 1.98, 1.95 (2s, 3H), 0.80 (s, 9H), 0.01 (s, 6H); MS *m/z*: 228 (M^+-CH_3); Anal. Calcd for $C_{12}H_{25}NO_2Si$: C 59.21; H 10.35; N 5.75; Si 11.54. Found: C 58.85; H 10.38; N 5.80; Si 11.18.

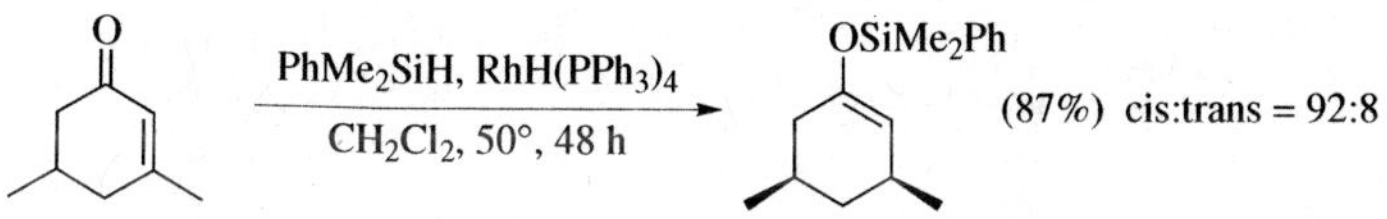

3,5-Dimethyl-l-cyclohexen-l-yl Dimethylphenylsilyl Ether [Reductive 1,4-Hydrosilylation of an Enone].[374] 3,5-Dimethyl-2-cyclohexen-1-one (0.124 g, 1.0 mmol) and $RhH(PPh_3)_4$ (6 mg, 0.0052 mmol) were treated with $PhMe_2SiH$ (0.177 g, 1.3 mmol) at 50° for 48 hours. After hexane (1 mL) was added, the reaction mixture was filtered and concentrated under vacuum. The crude product was purified by Kugelrohr distillation (90°/ 1 Torr) to give a colorless liquid: 225 mg, 87%: IR 2952, 2918, 2913, 2902, 2870, 1666, 1428, 1369, 1253, 1197, 1182, 1121, 1079, 826, 787, 699 cm^{-1}; ^1H NMR (200 MHz, $CDCl_3$) δ 7.58 (m, 2H), 7.37 (m, 3H), 4.66 (s, 1H), (s, 1H), 2.21 (br s, 1H), 1.93 (m, 1H), 1.63 (m, 4H), 0.90 (d, J = 7.9 Hz, 3H), 0.86 (d, J = 7.0 Hz, 3H), 0.43 (s, 6H); ^{13}C NMR δ 149.7, 138.0, 133.4, 129.5, 127.7, 111.3, 40.8, 38.4, 30.9, 29.6, 22.6, 22.0, −0.9, −1.0; EIMS m/z: M^+ calcd for $C_{16}H_{24}OSi$, 260.1596; found 260.1584. Anal. Calcd for $C_{16}H_{24}OSi$: C 73.80; H 9.30; found: C 73.78; H 9.25.

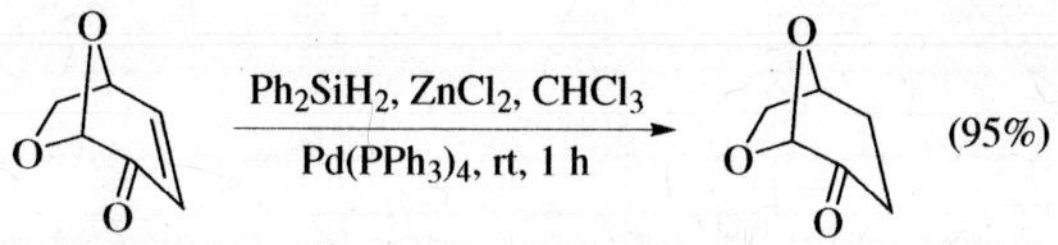

6,8-Dioxabicyclo[3.2.1]octan-4-one [1,2-Reduction of an Enone in the Presence of an Acetal].[436] 6,8-Dioxabicyclo[3.2.1]oct-2-ene-4-one (19 mg, 0.15 mmol) was dissolved in 3 mL of $CHCl_3$ along with Ph_2SiH_2 (55 mg,

0.30 mmol) and $ZnCl_2$ (20 mg, 0.15 mmol). $Pd(PPh_3)_4$ (17 mg, 0.015 mmol) was added and the mixture was stirred at room temperature until the reaction was complete as determined by GLC. The reaction mixture was filtered through a short SiO_2 column and purified by Kugelrohr distillation to give 18 mg (95%) of 6,8-dioxabicyclo[3.2.1]octan-4-one, having physical properties identical with literature values.

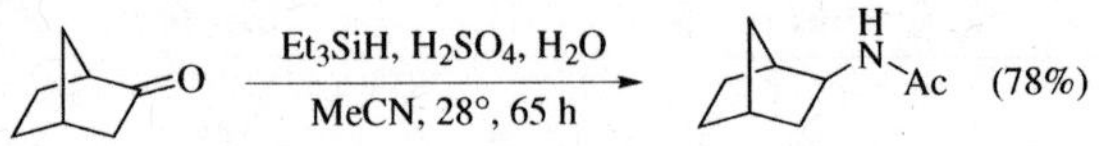

***N*-(*exo*-2-Norbornyl)acetamide [Reductive Amidation of a Ketone].**[313] To an acetonitrile (15 mL) solution of norcamphor (6.6 g, 60 mmol) and triethylsilane (7.7 g, 66 mmol) was added water (3.0 mL) followed by 9.0 mL of concentrated H_2SO_4 (9.0 mL) at ice-bath temperature. The heterogeneous reaction mixture was stirred rapidly at room temperature for 65 hours. The mixture was then quenched by addition of 50% aqueous NaOH solution (30 mL) and the aqueous solution was extracted with CH_2Cl_2 (3 × 50 mL). The combined extracts were passed through anhydrous $MgSO_4$ and the CH_2Cl_2 was removed under reduced pressure. The residue was washed three times with pentane to remove hexamethyldisiloxane and other soluble reaction products. The crude product was crystallized from Et_2O to give *N*-(*exo*-2-norbornyl)acetamide (6.5 g, 78%) whose ^1H-NMR spectrum and melting point were in accord with literature values.

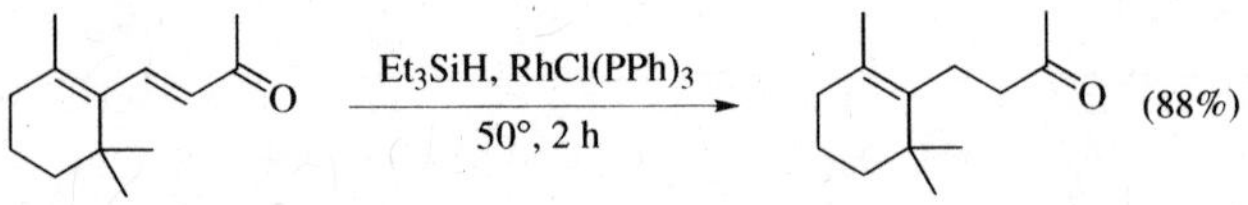

Dihydro-β-Ionone [1,4-Reduction of an α,β-Unsaturated Ketone].[435] A mixture of β-ionone (1.91 g, 10 mmol), Et_3SiH (1.27 g, 11 mmol), and $RhCl(PPh_3)_3$ (9 mg, 0.01 mmol) was stirred at 50° for 2 hours under nitrogen. The NMR spectrum of the reaction mixture showed the exclusive formation of the 1,4-addition (silyl enol ether) product: ^1H NMR (CCl_4) δ 5.23 (m, 1H), 4.28 (t, $J = 7$ Hz, 1H), 2.25-1.75 (m, 4H), 1.72 (br s, 3H), 1.63 (br s, 3H), 1.75-1.20 (m, 3H), 0.4–1.20 (m, 21H). The silyl enol ether was readily desilylated by treatment with K_2CO_3 (10 mg)/MeOH (10 mL) with stirring for 1 hour at room temperature. After solvent removal, the crude product was distilled under reduced pressure to give dihydro-β-ionone: 1.70 g, 88.1% bp 88°/2.5 Torr.

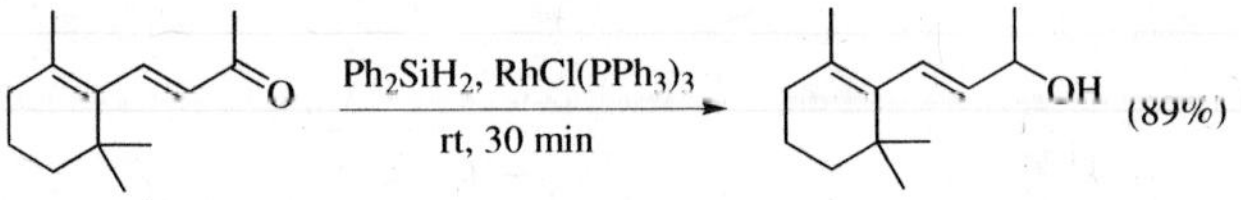

β-Ionol [1,2-Reduction of an α,β-Unsaturated Ketone].[435] A mixture of β-ionone (1.93 g, 10 mmol), Ph_2SiH_2 (2.02 g, 11 mmol), and $RhCl(PPh_3)_3$ (9 mg, 0.01 mmol) was stirred at room temperature under nitrogen. An exothermic reaction took place and the reaction was complete in 30 minutes. The NMR spectrum of the reaction mixture indicated the 1,2-reduction (silyl ether) product was formed exclusively: ^{1}H NMR (CCl_4) δ 7.80-7.10 (m, 10H), 5.55-5.18 (m, 3H), 5.40 (s, 1H), 4.39 (m, 1H), 2.30-1.00 (m, 5H), 1.53 (m, 3H), 1.27 (d, $J = 6$ Hz, 3H), 1.0 (m, 5H), 0.86 (s, 3H), 0.77 (s, 3H). To the reaction mixture was added *n*-hexane (50 mL), and the precipitated catalyst was removed by filtration. Then MeOH (10 mL) and K_2CO_3 (10 mg) were added to the filtrate. Methanolysis was complete within 1 hour at room temperature. The ratio of dihydro-β-ionone/β-ionol was 0 : 100 based on GLC and NMR analyses. After solvent evaporation, the residue was distilled to give β-ionol: 1.74 g, 89%; bp 99°/2 Torr.

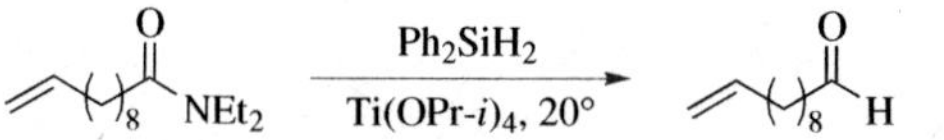

(90%)

Undec-10-enal [Reduction of an Amide to an Aldehyde].[433] To a dry flask containing neat *N*,*N*-diethylundec-10-enamide (0.155 mL, 0.65 mmol) under argon was added Ph_2SiH_2 (0.135 mL, 0.72 mmol) and Ti(OPr-*i*)$_4$ (0.196 mL, 0.65 mmol). Initial effervescence was observed [CAUTION!] and the reaction mixture was stirred at room temperature until TLC analysis showed complete consumption of the starting material (ca. 5 hours). The mixture was diluted with THF (20 mL), treated with 1 M HCl (10 mL), stirred for 1 hour, and poured onto Et_2O (80 mL). The organic layer was washed with 1 M HCl (3 × 10 mL), saturated aqueous $NaHCO_3$ solution (2 × 10 mL), and saturated aqueous NaCl solution (10 mL), and then dried ($MgSO_4$) and concentrated under vacuum. Flash column chromatography on SiO_2 (hexane:Et_2O 15 : 85) afforded undecenal (99 mg, 90%).

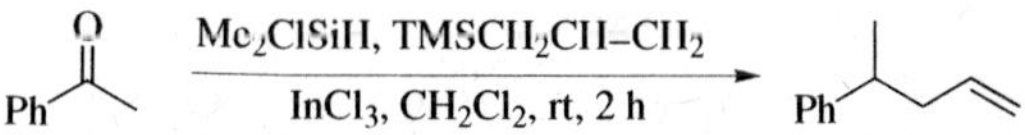

(86%)

4-Phenylpent-1-ene [Reductive Allylation of an Aryl Ketone].[427] Acetophenone (2.0 mmol) was added to a mixture of $InCl_3$ (0.1 mmol), $ClMe_2SiH$ (2.2 mmol), and allyltrimethylsilane (2.2 mmol) in CH_2Cl_2 (10 mL) at room temperature. The reaction mixture was stirred for 2 hours, quenched with water (20 mL), and extracted with Et_2O (3 × 20 mL). After the organic layer was dried ($MgSO_4$) and concentrated under vacuum, the residue was purified by chromatography (hexane) on SiO_2 to give 4-phenylpent-1-ene (86%).

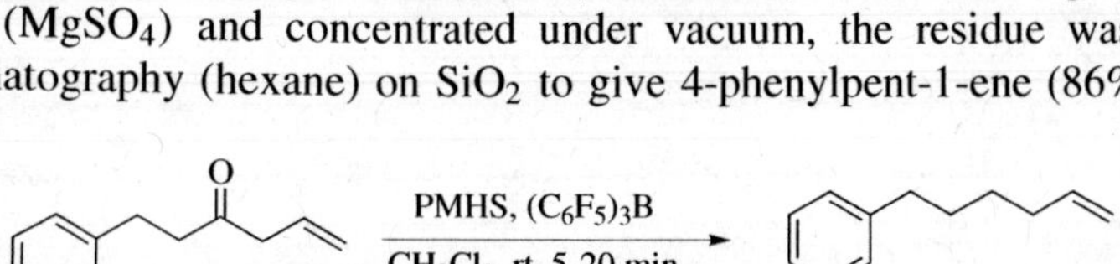

(88%)

6-Phenylhex-1-ene [Reduction of an Aliphatic Ketone Function to a Methylene Function].[354] To a solution of 6-phenylhex-1-ene-4-one (1 mmol) in dry CH_2Cl_2 (5 mL) and tris(pentafluorophenyl)boron (5 mol%) was slowly added PMHS (3 mmol) at room temperature. After 20 minutes, a vigorous effervescence was observed. The solvent was evaporated and the reaction mixture was dissolved in hexane and then filtered through a SiO_2 pad using hexane. Evaporation of the volatiles afforded the 6-phenylhex-1-ene (88%) in pure form.

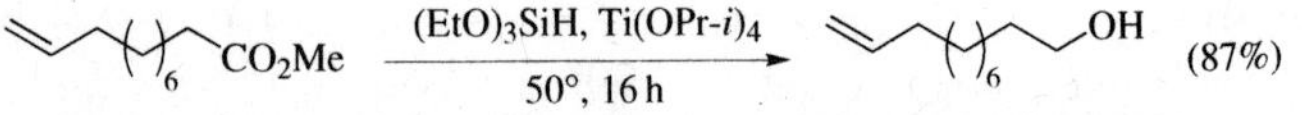

10-Undecen-1-ol [Reduction of an Ester to an Alcohol].[291] Triethoxysilane (1.7 mL, 9 mmol) and methyl 10-undecenoate (594 mg, 3 mmol) were added to a test tube. $Ti(OPr\text{-}i)_4$ (45 μL, 0.15 mmol) was added, and the test tube was fitted with a drying tube packed with Drierite to exclude excess moisture. The contents of the vessel were then stirred while being heated in an oil bath at 50° for 16 hours. The reaction mixture was washed into a 100-mL round-bottomed flask with THF (10 mL). Then 1 N NaOH (20 mL) was added slowly with stirring. NOTE: CAUTION: vigorous bubbling was observed. After 4 hours, the mixture was added to Et_2O (50 mL) and water (50 mL). After shaking, the layers were separated, and the aqueous layer was extracted with an additional 50 mL of Et_2O. The combined organic extracts were washed with 1 M HCl (2 × 50 mL), dried ($MgSO_4$), filtered, and concentrated under vacuum to afford 10-undecen-1-ol as a clear oil: 443 mg, 87%. The product was >95% pure as determined by GLC and ^1H NMR analyses.

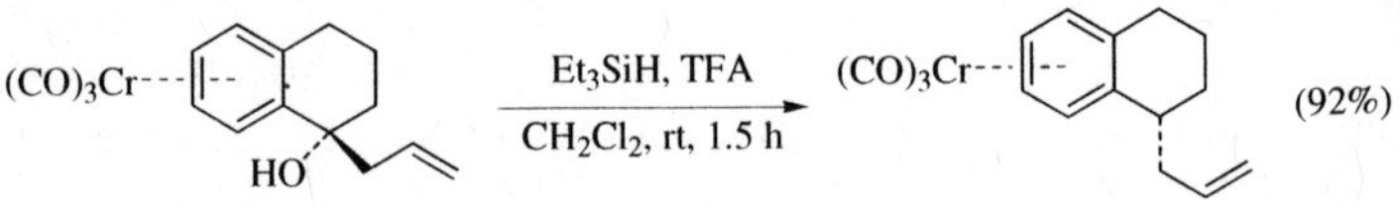

Tricarbonyl(1-*endo*-allyltetralin)chromium [Stereoselective Reduction of an Alcohol to a Hydrocarbon].[182] A solution of tricarbonyl(1-*exo*-allyl-1-*endo*-tetralol)chromium (150 mg, 0.46 mmol), Et_3SiH (0.22 mL, 1.4 mmol), and CH_2Cl_2 (1 mL) was stirred at room temperature for 1.5 hours under nitrogen. The mixture was poured into water (10 mL) and extracted with Et_2O (2 × 10 mL). Evaporation of the organic layer under reduced pressure and purification by silica gel chromatography on SiO_2 (1 : 8 Et_2O/petroleum ether) afforded tricarbonyl(1-*endo*-allyltetralin)chromium as yellow crystals: 131 mg, 92%; mp 88–89°; IR ($CHCl_3$)1960, 1880, 1635 cm^{-1}; ^1H NMR ($CDCl_3$) δ 6.15-4.80 (m, 7H), 2.95-2.10 (m, 5H), 2.05-1.35 (m, 4H).

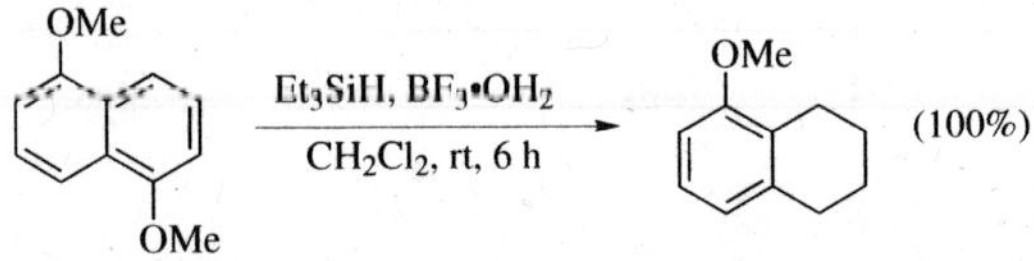

5-Methoxytetralin [Partial Reduction of a Substituted Naphthalene to a Tetralin].[262] 1,5-Dimethoxynaphthalene (300 mg, 1.0 mmol) dissolved in

CH_2Cl_2 (3–4 mL) was added dropwise to a flask containing $BF_3{\bullet}OH_2$ (1.1 g, 13 mmol) at 0°. After the addition was completed, the mixture was stirred for 10–15 minutes and allowed to warm to room temperature, and Et_3SiH (742 mg, 6.4 mmol) was added dropwise. The reaction mixture was stirred at room temperature for an additional 5–6 hours, neutralized with cold saturated aqueous Na_2CO_3 solution, and extracted with CH_2Cl_2 (3 × 15 mL). The combined organic extracts were washed with water (2 × 10 mL), dried ($MgSO_4$), and evaporated to leave ca. 270 mg of brownish oil, which according to NMR and GC analyses contained ca. 90% of 5-methoxytetralin. Purification of the crude product was accomplished by column chromatography on SiO_2 (CH_2Cl_2/Et_2O).

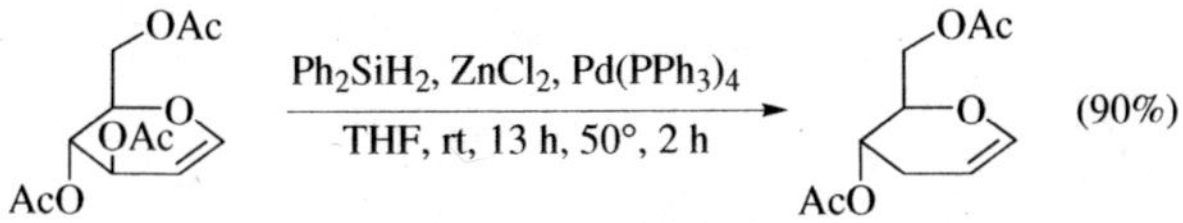

1,2,3-Trideoxy-D-*ribo*-hex-1-enopyranose Diacetate [Reduction of an Allyl Ester].[196] To a THF (10 mL) solution containing tri-*O*-acetylglucal (349 mg, 1.28 mmol), diphenylsilane (489 mg, 403 mmol), and $ZnCl_2$ (541 mg, 4.0 mmol) was added $Pd(PPh_3)_4$ (70 mg, 0.06 mmol, 12 mol%). The solution was stirred at room temperature for 13 hours, then at 50° for 2 hours, and then mixed with Et_2O and washed several times with water. The ether layer was dried over Na_2SO_4 and the solvent was evaporated. The yield of 1,2,3-trideoxy-D-ribo-hex-1 enopyranose diacetate was 90% as determined by NMR using bibenzyl as an internal standard. The product was partially decomposed upon chromatographic purification over SiO_2, yielding 120 mg (53%).

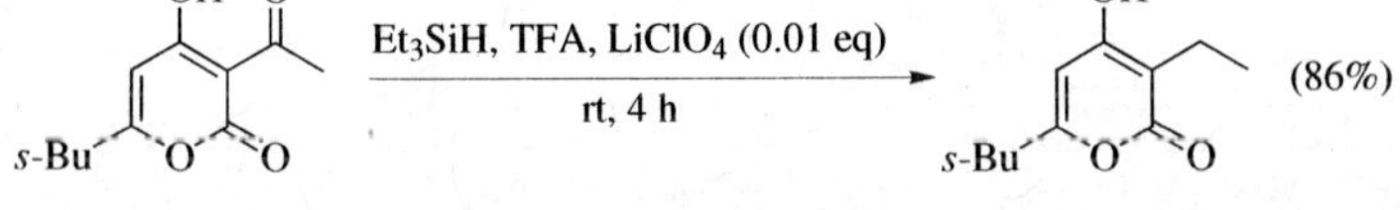

6-(2-Butyl)-4-hydroxy-3-ethyl-2-pyrone (Germicidin) [Reduction of a Ketone Carbonyl to a Methylene Group in a Multifunctional Compound].[423] A TFA (15 mL) solution containing 3-acetyl-6-(2-butyl)-4-hydroxy-2-pyrone (2 mmol), Et_3SiH (1.29 mL, 8.0 mmol), and $LiClO_4$ (2 mg, 0.02 mmol) was stirred at room temperature for about four hours while being monitored by TLC. The solvent was evaporated under vacuum and the residue was purified by column chromatography on silica gel ($CHCl_3$ eluent) to yield racemic 6-(2-butyl)-4-hydroxy-3-ethyl-2-pyrone: 337 mg, 86%; mp 95–97° (Et_2O/hexane); IR (KBr) 1160, 1285, 1430, 1595, 1680, 2885, 2945, 2980 cm^{-1}; ^1H NMR ($CDCl_3$) δ 6.22 (s, 1H), 2.48 (*q*, $J = 7.4$ Hz, 2H) and (m, 1H), 1.75-1.24 (m, 2H), 1.20 (d, $J = 6.7$ Hz, 3H), 1.11 (t, $J = 7.5$ Hz, 3H), 0.89 (t, $J = 7.5$ Hz, 3H); ^{13}C NMR ($CDCl_3$) δ 69.6, 168.8, 168.0, 105.0, 100.9, 39.8, 27.5, 17.7, 16.4, 12.4, 11.6.

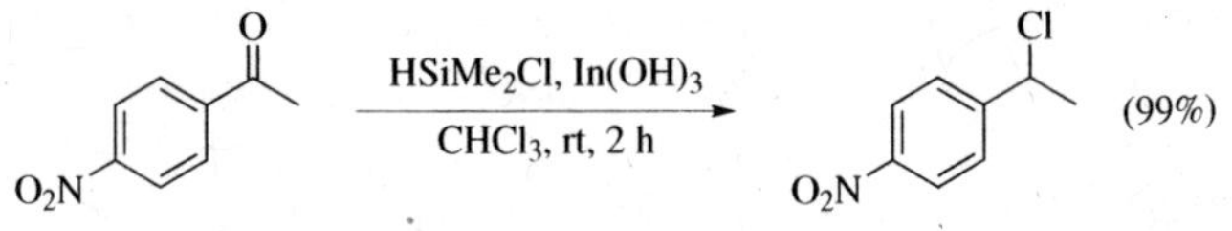

1-(1-Chloroethyl)-4-nitrobenzene [Deoxygenative Chlorination of a Ketone].[331] To a mixture of $In(OH)_3$ (0.1 mmol) and *p*-nitroacetophenone (2.0 mmol) in $CHCl_3$ (4.0 mL) was added $ClSiMe_2H$ (2.4 mmol) under nitrogen. The reaction mixture was stirred for 2 hours at room temperature, and then was poured into EtOAc (50 mL) and washed with saturated aqueous $NaHCO_3$ solution (50 mL). The organic layer was dried over $MgSO_4$ and concentrated under vacuum to yield 99% of 1-(1-chloroethyl)-4-nitrobenzene. The physical and spectral data of the product were in excellent accord with known values.

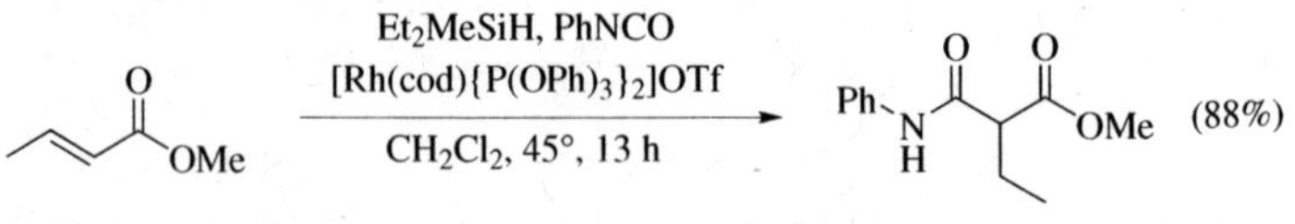

Methyl 2-(Phenylcarbamoyl)butanoate [Hydrocarbamoylation of an α,β-Unsaturated Ester].[475] To a solution of $[Rh(cod)\{P(OPh)_3\}_2]OTf$ (9.9 mg, 0.01 mmol) in CH_2Cl_2 (4 mL) was added a mixture of phenyl isocyanate (121 mg, 1.0 mmol), methyl crotonate (210 mg, 2.1 mmol), and Et_2MeSiH (205 mg, 2.0 mmol) in CH_2Cl_2 (2 mL). The resulting mixture was heated at reflux for 13 hours under a nitrogen atmosphere. The solvent was removed under reduced pressure and the residue was purified by flash column chromatography on SiO_2 (4 : 1 hexane/EtOAc) to afford methyl 2-(phenylcarbamoyl)butanoate: 194 mg, 88%; mp 76.0–77.0° (hexane/EtOAc); IR (CCl_4) 3371, 1722, 1689 cm^{-1}; ^1H NMR ($CDCl_3$) δ 8.74 (br s, 1H), 7.54 (d, $J = 8.4$ Hz, 2H), 7.30 (dd, $J = 8.4$ and 7.5 Hz, 2H), 7.10 (t, $J = 7.5$ Hz, 1H), 3.76 (s, 3H), 3.32 (t, $J = 7.3$ Hz, 1H), 2.03 (dq, $J = 7.4$ and 7.3 Hz, 2H), 0.99 (t, $J = 7.4$ Hz, 3H); ^{13}C NMR ($CDCl_3$) δ 172.7, 166.3, 137.4, 128.8, 124.3, 119.9, 54.9, 52.5, 24.9, 11.8. Anal. Calcd for $C_{12}H_{15}NO_3$: C 65.14; H 6.83; N 6.33. Found: C 65.30; H 6.57; N 6.28.

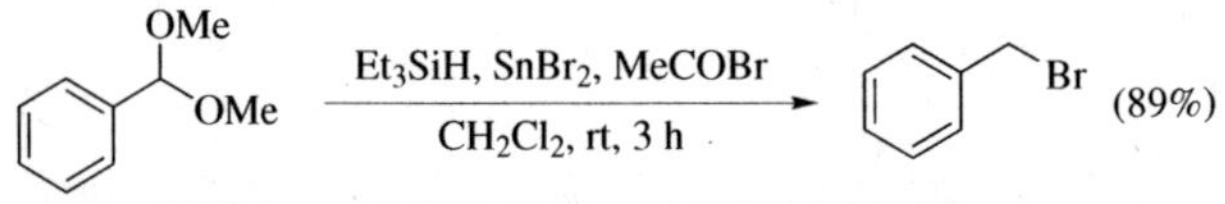

Benzyl Bromide [Reductive Bromination of an Acetal].[506] To a suspension of tin(II) bromide (5.1 mg, 0.02 mmol) and benzaldehyde dimethyl acetal (54.8 mg, 0.36 mmol) in CH_2Cl_2 (2.5 mL) were added successively Et_3SiH (65.0 mg, 0.56 mmol) and acetyl bromide (96.8 mg, 0.79 mmol) in CH_2Cl_2 (1 mL) at room temperature under an argon atmosphere. The mixture was stirred for 3 hours at room temperature and quenched with a phosphate buffer (pH 7).

The organic materials were extracted with CH_2Cl_2 and the extract was dried over Na_2SO_4. Benzyl bromide (54.7 mg, 89%) was isolated by TLC on SiO_2.

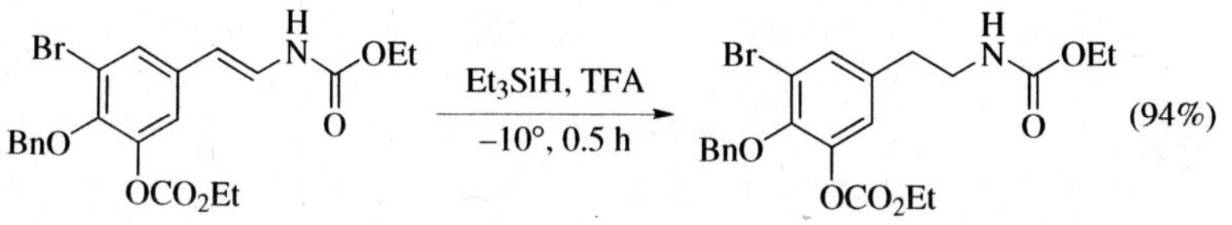

2-(Benzyloxy)-3-bromo-5-[(2-ethoxycarbonyl)ethyl]phenyl Ethyl Carbamate [Reduction of an Enamide to an Amide].[535] A mixture of Et_3SiH (170 μL, 1.07 mmol) and (*E*)-2-(benzyloxy)-3-bromo-5-(2-ethoxycarbonyl) vinyl)phenyl ethyl carbamate (32 mg, 0.69 mmol) was cooled to −10° in an ice-salt bath under nitrogen, and treated with pre-cooled TFA (1.0 mL) in one portion. The two-phase mixture was *rapidly* stirred at −10° for 0.5 hour, poured into ice-cold saturated aqueous $NaHCO_3$ solution, and worked up by extraction with CH_2Cl_2. The extracts were dried (Na_2SO_4) and concentrated to give essentially pure title product: 30 mg, 94%; ^{1}H NMR ($CDCl_3$) δ 7.5-7.3 (m, 5H), 7.29 (d, $J = 2.0$ Hz, 1H), 6.96 (d, $J = 2.0$ Hz, 1H), 4.99 (s, 2H), 4.74 (br t, $J = 5.6$ Hz, 1H), 4.21 (q, $J = 7.2$ Hz, 2H), 4.10 (q, $J = 7.2$ Hz, 2H), 3.38 (dt, $J = 5.6$ and 6.8 Hz, 2H), 2.74 (t, $J = 6.8$ Hz, 2H), 1.29 (t, $J = 7.2$ Hz, 3H), 1.22 (t, $J = 7.2$ Hz, 3H); ^{13}C NMR ($CDCl_3$) δ 156.5, 152.9, 146.8, 145.1, 136.6, 136.4, 131.1, 128.4, 128.3, 128.2, 122.6, 118.1, 75.5, 65.2, 60.8, 41.7, 35.2, 14.6, 14.1; HRMS (CI) m/z: $[M + H]^+$ calcd for $C_{21}H_{25}BrNO_6$, 466.0865; found, 466.0864.

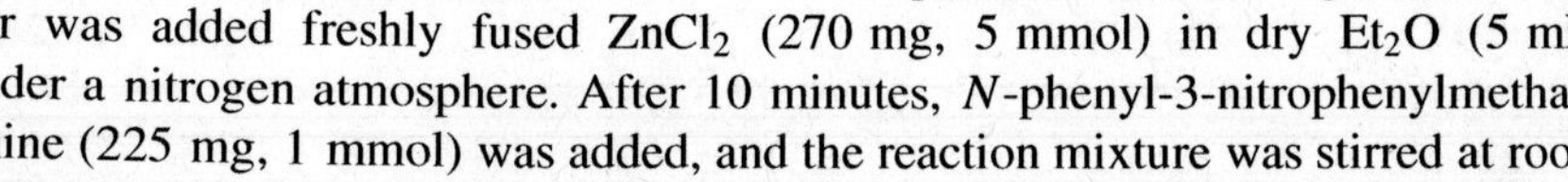

3-Nitrobenzylamine [Reduction of an Imine to an Amine].[539] To PMHS (300 mg) in a 25-mL flask fitted with a septum inlet and magnetic stirring bar was added freshly fused $ZnCl_2$ (270 mg, 5 mmol) in dry Et_2O (5 mL) under a nitrogen atmosphere. After 10 minutes, *N*-phenyl-3-nitrophenylmethanimine (225 mg, 1 mmol) was added, and the reaction mixture was stirred at room temperature for 12 hours and extracted with 1 M HCl (2 × 15 mL). The aqueous layer was washed with CH_2Cl_2 (15 mL) to remove non-amine impurities. The purified aqueous layer was made basic (pH ~ 10) with 1 N NaOH and extracted with EtOAc (3 × 15 mL). The combined organic layers were washed with water (15 mL) and brine (15 mL), and dried (Na_2SO_4). The volatiles were removed and the residue was purified by column chromatography to yield 170 mg (75%) of the title product: ^{1}H NMR ($CDCl_3$) δ 7.20 (d, $J = 7.5$ Hz, 2H), 7.70-6.50 (m, 7H), 3.50 (s, 2H); MS m/z: M^+ 228, 136, 106, 91, 77.

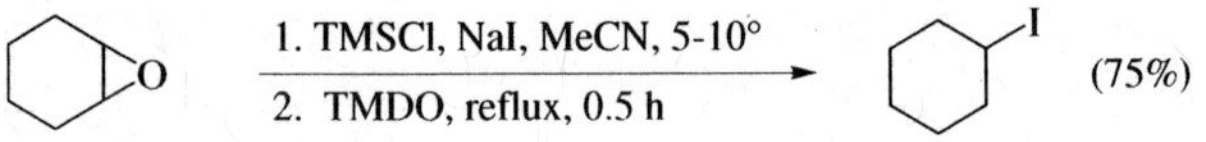

Cyclohexyl Iodide [Iodoreduction of an Oxirane to an Iodoalkane].[357] A mixture of cyclohexene oxide (1.01 mL, 10 mmol), NaI (2.00 g, 13.3 mmol), and TMSCl (1.92 mL, 15 mmol) in anhydrous MeCN (10 mL) was stirred at 5–10° for 2–3 minutes. Then TMDO (1.79 mL, 10 mmol) was added and the mixture was heated at reflux for 0.5 hour. The remaining siloxane products were destroyed by adding 45% aqueous HF (2.0 mL) and heating at reflux for 5 minutes. The reaction mixture was taken up in CH_2Cl_2 (30 mL), and the organic layer was washed with water (20 mL), 1 N $NaHSO_3$ (10 mL), and water (20 mL) again. Drying (Na_2SO_4) and evaporation of the solvents afforded crude cyclohexyl iodide, which was purified by Kugelrohr distillation to give pure product, 1.58 g (75%); bp 180–183°; ^{1}H NMR (CCl_4) δ 4.14 (m, 1H), 1.92 (m, 4H), 1.41 (m, 6H).

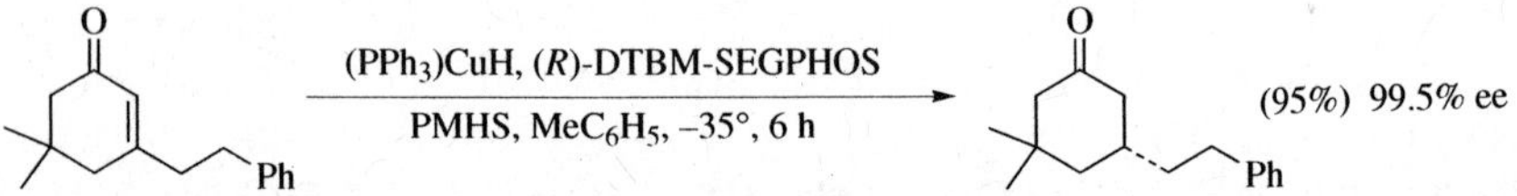

(*R*)-3,3-Dimethyl-5-(2-phenylethyl)cyclohexanone [Asymmetric 1,4-Reduction of an Enone].[597] To a 5-mL round-bottomed flask, flame-dried and purged with argon, was added $CuHPPh_3$ (2.3 mg, 7.0 μmol) and (*R*)-DTBM-SEGPHOS (1.9 mg, 1.6 μmol). Toluene (0.60 mL) was added and the solution was cooled to −35°. After PMHS (215 μL, 3.3 mmol) was introduced by syringe, 3-(2-phenylethyl)-5,5-dimethylcyclohexenone (192 mg, 0.84 mmol) was added. The mixture was stirred at −35° for 12 hours until the reaction was complete as determined by TLC (20% Et_2O/ligroin) and was then quenched by pouring into 3 N NaOH. Ether and water were added, and the mixture was stirred for 2 hours at room temperature. The aqueous layer was extracted with Et_2O (2 x), and the combined organic layers were washed with brine, dried over $MgSO_4$, filtered, and concentrated. The residue was purified by flash chromatography (10% Et_2O/ligroin) to afford the title ketone as a clear oil: 185 mg (95%), chiral GC (ketal from (*R*,*R*)-2,3-butanediol, Chiraldex B-DM 140) showed 99.5% ee. ^{1}H NMR ($CDCl_3$, 400 MHz) δ 7.31-7.17 (m, 5H), 2.64 (t, J = 8.2 Hz, 2H), 2.44 (ddd, J = 9.2, 2.0, 2.0 Hz, 1H), 2.21 (d, J = 13.2 Hz, 1H), 2.10 (ddd, J = 13.2, 2.2, 2.2 Hz, 1H), 1.35 (t, J = 12.4 Hz, 1H), 1.09 (s, 3H), 0.88 (s, 3H), 1.99-1.91 (m, 2H), 1.73-1.62 (m, 3H); ^{13}C NMR ($CDCl_3$, 125 MHz) δ 25.3, 32.3, 33.2, 34.6, 35.4, 39.3, 45.3, 47.6, 54.7, 126.0, 128.4, 128.6, 142.2, 212.0; HRMS calcd for $C_{16}H_{22}O$ 230.1671; found 230.1663.

TABULAR SURVEY

A thorough coverage of the literature through 2004 has been carried out based on the search of certain silanes. A small number of additional pertinent articles, particularly regarding the asymmetric silane reductions that were published later, are included.

Tables are organized by the functional group classes undergoing change in the substrates. Table entries are ordered by increasing carbon count of the starting substrate. Protecting groups are included in the carbon count. Unspecified yields are denoted by (—).

The following abbreviations are used in the tables:

10-C-6	10-crown-6
18-C-6	18-crown-6
Ac	acetyl
acac	acetylacetone
ACHN	1,1′-azobis(cyclohexylnitrile)
Ad	adamantyl
AIBN	azobis(isobutyronitrile)
An	anthracenyl
BARF	tetrakis[3,5-bis(trifluoromethyl)phenyl]borate
BDE	bond dissociation energy
BINAP	2,2′-bis(diphenylphosphino)-1,1′-binaphthyl
BINOL	1,1′-bi-2-naphthol
BIPHEMP	2,2′-bis(diphenylphosphino)-6,6′-dimethyl-1,1′-biphenyl
BIPHEP	2,2′-bis(diphenylphosphino)-1,1′-biphenyl
bipy	bipyridyl
Bipymox	6,6′-bis(oxazolinyl)-2,2′-bipyridine
Bmpp	benzylmethylphenylphosphine
Bn	benzyl
Boc	*tert*-butyloxycarbonyl
BOM	benzyloxymethyl
BOX	bis(oxazolino)
BPPFA	1′,2-bis[diphenylphosphinoferrocenyl]ethyldimethylamine
BSA	bis(trimethylsilyl)acetamide
BTAF	benzyltrimethylammonium fluoride
Bz	benzoyl
Cbz	carbobenzyloxy
cod	1,5-cyclooctadiene
coe	cyclooctene
cot	cyclooctatetraene
Cp	cyclopentadienyl
Cp*	pentamethylcyclopentadienyl
CSA	camphorsulfonic acid
DAST	(diethylamino)sulfur trifluoride
dba	dibenzylidene acetone
DBATO	dibutylacetoxytin oxide
dbpp	di-(*tert*-butylphenyl)phosphite
DBU	1,8-diazabicyclo[5.4.0]undece-7-ene

DCE	1,2-dichloroethane
de	diastereomeric excess
dea	diethanolamine
dee	diethoxyethane
DIBALH	diisobutylaluminum hydride
DIPOF	[2-(4,5-diphenyl-4,5-dihydro-1,3-oxazolin-2-yl)-ferrocenyl]diphenylphosphine
DIOP	(2*S*,3*S*)-*O*-isopropylidene-2,3-dihydroxy-1,4-bis-(diphenylphosphino)butane
diphos	1,2-bis(diphenylphosphino)ethane (see dppe)
DME	1,2-dimethoxyethane
DMF	*N*,*N*-dimethylformamide
DMI	1,3-dimethyl-2-imidazolidinone
DMPU	1,3-dimethyl-3,4,5,6-tetrahydro-2-(1*H*)-pyrimidinone
DMSO	dimethyl sulfoxide
DMTS	dimethylthexylsilyl
dpm	2,2,6,6-tetramethylheptane-3,5-dionate ("dipivaloylmethanato")
dppb	1,4-bis(diphenylphosphino)butane
dppbz	1,2-bis(diphenylphosphino)benzene
dppe	1,2-bis(diphenylphosphino)ethane
dppf	bis(diphenyllphosphino)ferrocene
dppfc	diphenylphosphinoferrocene
dppm	bis(diphenylphosphino)methane
dppp	1,3-bis(diphenylphosphino)propane
dr	diastereomeric ratio
ds	diastereoselectivity
dma	dimethylacetamide
ebpe	*N*,*N*′-ethylenebis(1-phenylethylamine)
EBTHI	ethylenebis(η^5-tetrahydroindenyl)titanium
ee	enantiomeric excess
EE	ethoxyethyl
EG	electrogenerated
EH	2-ethylhexanoate
er	enantiomeric ratio
Fc	ferrocene
Fmoc	fluorenylmethoxycarbonyl
HAp	hydroxyapatite
HMDS	hexamethyldisilazane
H-MOP	2-diphenylphosphino-1,1′-binaphthyl
HMPA/HMPT	hexamethylphosphoric triamide
KU-1	phenol-formaldehyde sulfocationite
LAH	lithium aluminum hydride
Me-DuPHOS	1,2-bis-(2,5-dimethylphospholano)benzene
MEM	methoxyethoxymethyl

MOM	methoxymethyl
Ms	methanesulfonyl
mont	montmorillonite
MPM	methoxyphenylmethyl
NBD	norbornadiene
Nf	nonaflate
nmdpp	(*S*)-neomenthyldiphenylphosphine
NMP	*N*-methylpyrrolidone
NORPHOS	(−)-(*R*,*R*)-2-exo-3-endo-bis(diphenylphosphino)-bicyclo[2.2.1]hept-5-ene
Np	naphthyl
NR	no reaction
Ns	4-nitrobenzenesulfonate
o-dppb	1,2-bis(diphenylphosphino)benzene
OTFA	trifluoroacetate
PE	phosphatidylethanolamine
Piv	pivaloyl
PEHS	polyethylhydrogensiloxane
PMB	*p*-methoxybenzyl
PMHS	polymethylhydrogensiloxane
PNB	*p*-nitrobenzoate
PPA	polyphosphoric acid
PPFA	[*N*,*N*-dimethyl-1[2-(diphenylphosphino)-ferrocenyl]ethylamine
ppf	2-[diphenylphosphinoferrocenyl]ethyldimethylamine
PPHF	pyridinium poly(hydrogen fluoride)
PPTS	pyridinium *p*-toluenesulfonate
PTC	phase-transfer catalysis
$P(tm\text{-}tp)_3$	tris(2,2″,6,6″-tetramethyl-*m*-terphenyl-5′-yl)phosphane
$P(tp)_3$	tris(*m*-terphenyl-5′-yl)phosphane
p-Tol-BINAP	2,2′-bis(*p*-tolylphosphino)-1,1′-binaphthyl
PTSA	*p*-toluenesulfonic acid
pybox	pyridinebis(oxazoline)
pymox	pyridinemono(oxazoline)
rt	room temperature
SEGPHOS	(4,4′-bi-1,3-benzodioxole)-5,5′-diylbis(diarylphosphine)
TADDOL	$\alpha,\alpha,\alpha',\alpha'$-tetraaryl-1,2-dioxolane-4,5-dimethanol
TASF	tris(diethylamino)sulfonium difluorotrimethylsilicate
TBAF	tetra-*n*-butylammonium fluoride
TBAT	tetrabutylammonium triphenyldifluorosilicate
TBDPS	*tert*-butyldiphenylsilyl
TBS	*tert*-butyldimethylsilyl
TEA	triethylamine
TEAF	tetraethylammonium fluoride
TES	triethylsilyl

Tf	trifluoromethanesulfonyl
TFA	trifluoroacetic acid
TFAA	trifluoroacetic anhydride
TFPB	tetrakis-3,5-bis(trifluoromethylphenyl)borate
THEATi(OPr-*i*)	[tris(hydroxyethyl)amino]titanium(IV)isopropoxide
THF	tetrahydrofuran
THP	tetrahydropyran
TIPS	triisopropylsilyl
TMDO	1,1,3,3-tetramethyldisiloxane
TMEDA	*N*,*N*,*N*′,*N*′-tetramethylethylenediamine
TMS	trimethylsilyl
TMSBr	trimethylsilyl bromide
TMSCl	trimethylsilyl chloride
TMSI	trimethylsilyl iodide
Tol	tolyl
Tr	triphenylmethyl
TRAP	2,2′-bis[(dialkylphosphino)methyl]-1,1′-biferrocene
TRISPHOS	2,2′,2″-tris(2,4,8,10)-tetra-*tert*-butyldibenzo[*d*,*f*]-[1,3,2]dioxaphosphepin-6-yl-6-oxy)tri-2-propylamine
Trityl	triphenylmethyl
Ts	*p*-toluenesulfonyl
Vi	vinyl

CHART 1. LIGAND AND CATALYST STRUCTURES USED IN TABLES

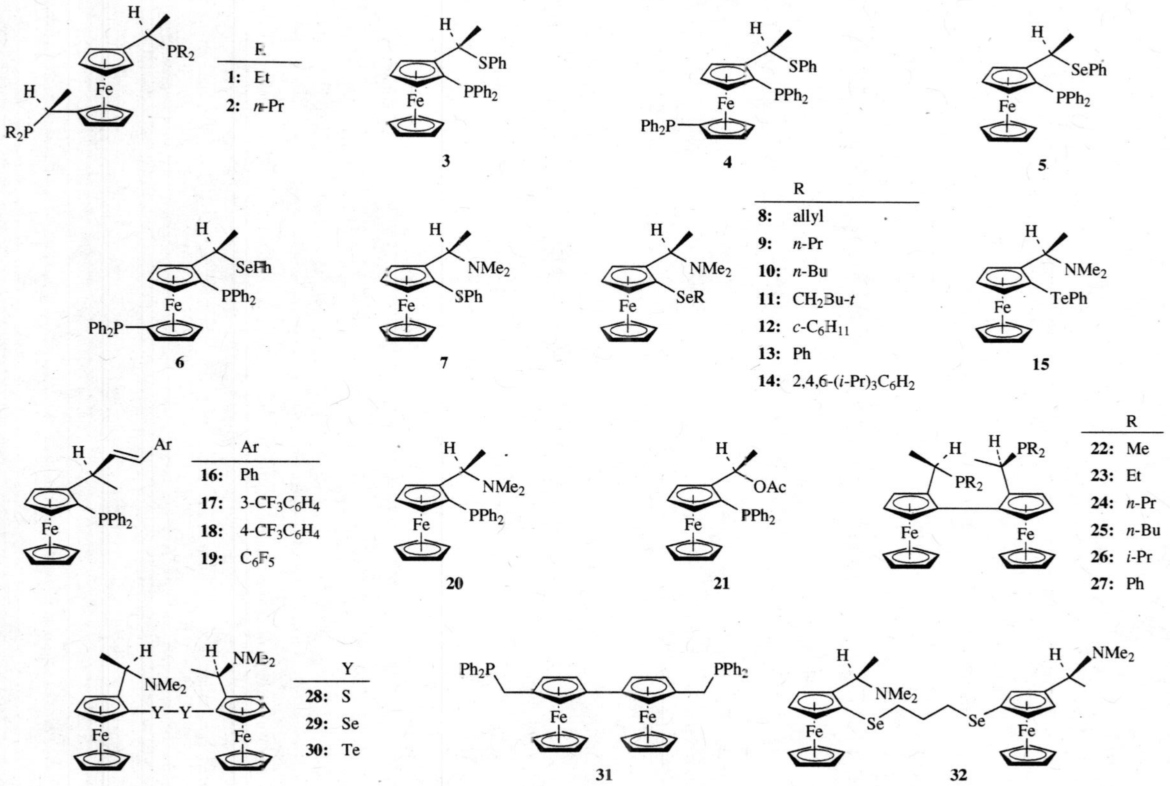

CHART 1. LIGAND AND CATALYST STRUCTURES USED IN TABLES (*Continued*)

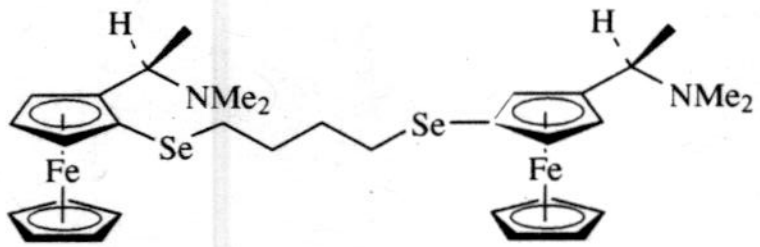

33

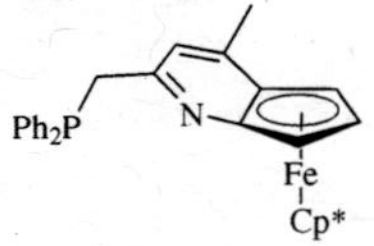

34

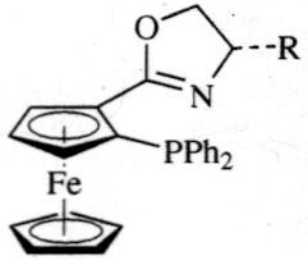

	R
35:	*i*-Pr
36:	*t*-Bu
37:	Ph

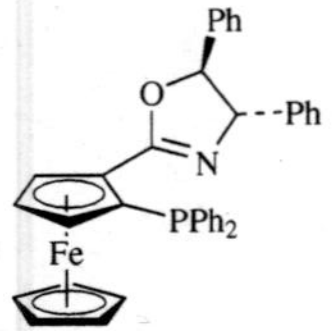

38

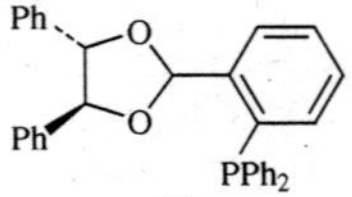

39

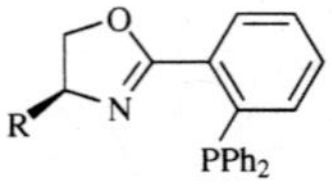

	R
40:	Me
41:	*i*-Pr
42:	*t*-Bu

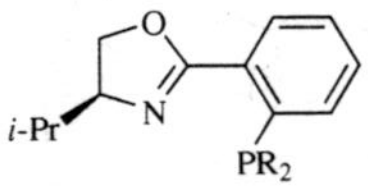

	R
43:	$c\text{-}C_6H_{11}$
44:	1-naphthyl

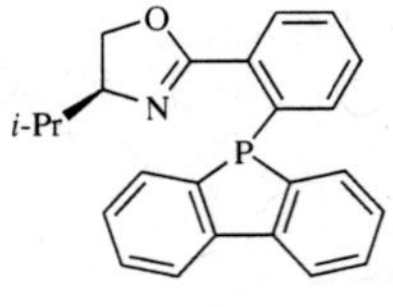

45

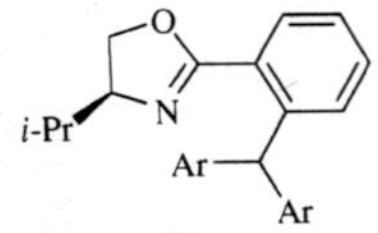

Ar = 3,5-$(CF_3)_2C_6H_3$

46

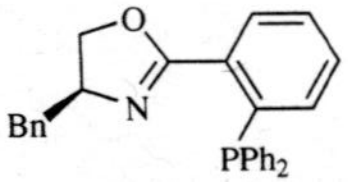

47

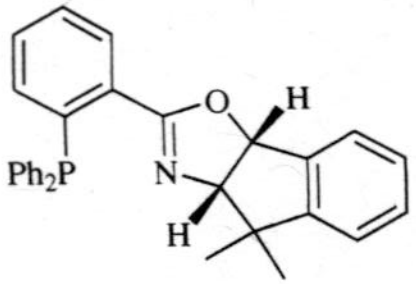

48

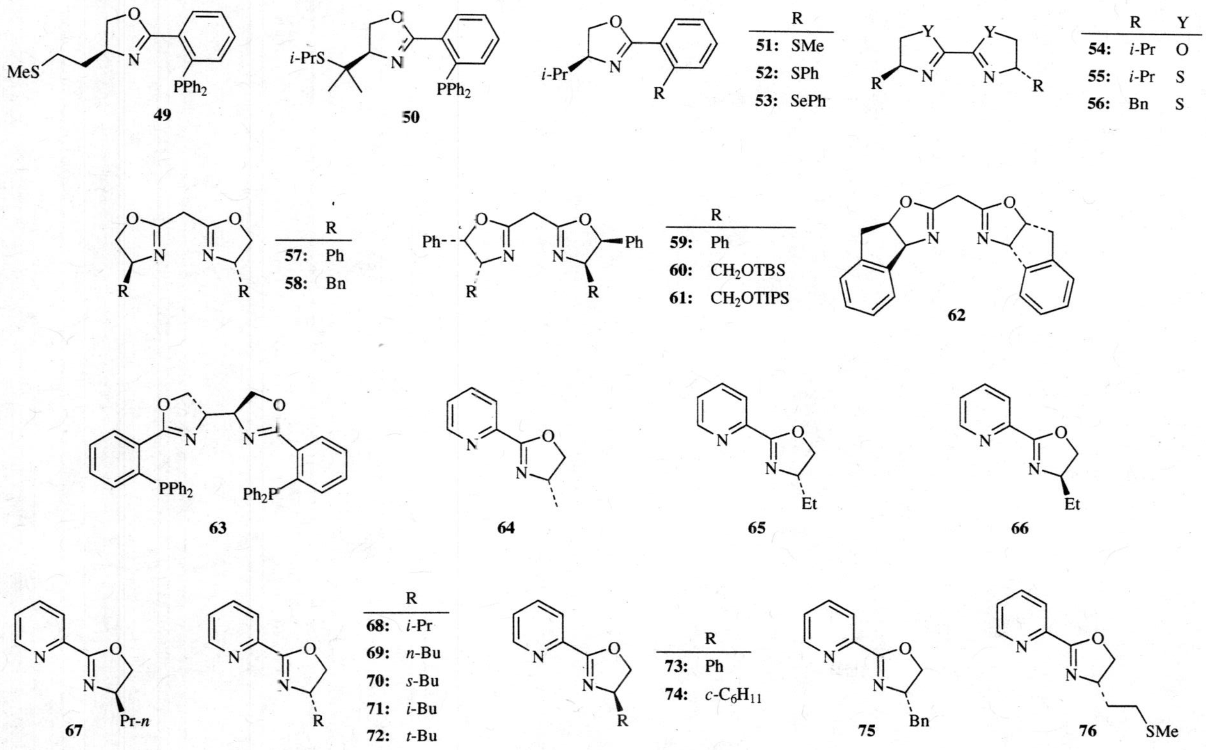
MeS
PPh2
49
i-PrS
PPh2
50
i-Pr
R
R: 51: SMe; 52: SPh; 53: SePh
R / Y: 54: i-Pr O; 55: i-Pr S; 56: Bn S
R: 57: Ph; 58: Bn
Ph
R: 59: Ph; 60: CH2OTBS; 61: CH2OTIPS
62
PPh2
Ph2P
63
64
Et
65
Et
66
Pr-n
67
R: 68: i-Pr; 69: n-Bu; 70: s-Bu; 71: i-Bu; 72: t-Bu
R: 73: Ph; 74: c-C6H11
Bn
75
SMe
76

CHART 1. LIGAND AND CATALYST STRUCTURES USED IN TABLES (*Continued*)

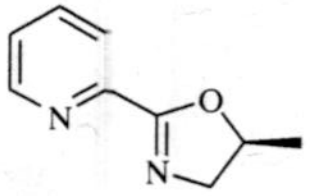

77

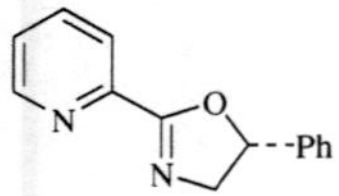

78

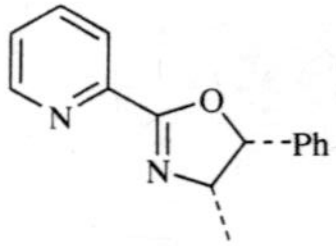

79

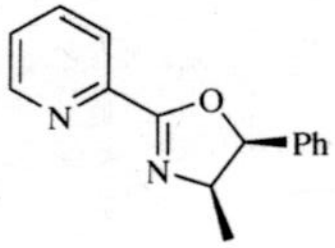

80

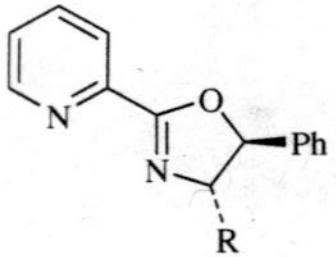

	R
81:	Me
82:	CH_2OH
83:	CH_2OMe

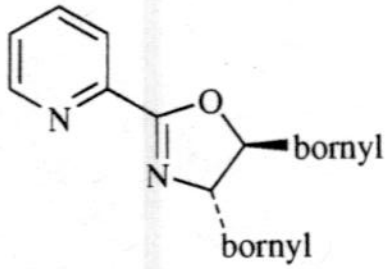

84

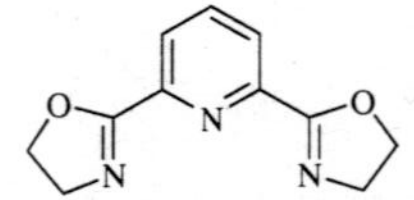

85

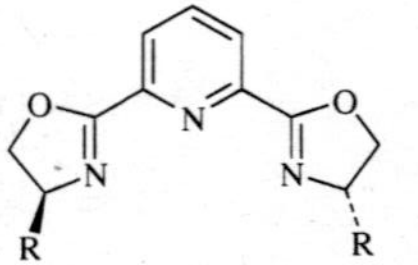

	R
86:	Me
87:	*i*-Pr
88:	*t*-Bu
89:	*s*-Bu
90:	Ph

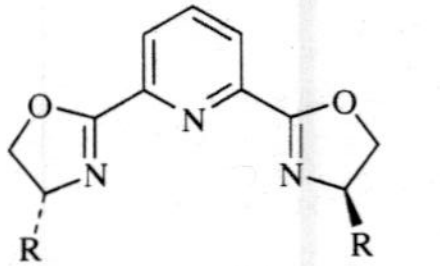

	R
91:	Me
92:	Et
93:	Ph

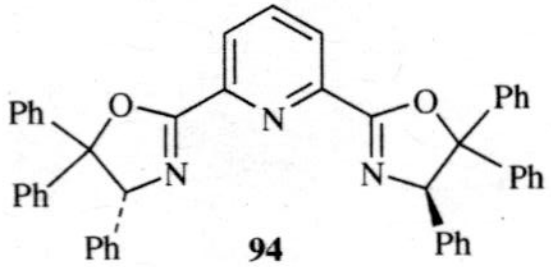

94

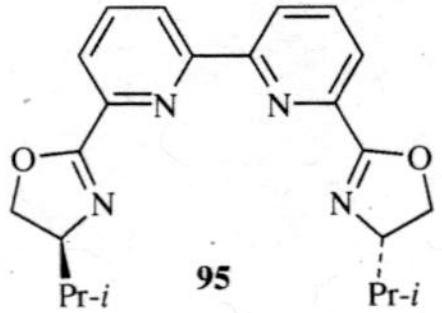

95

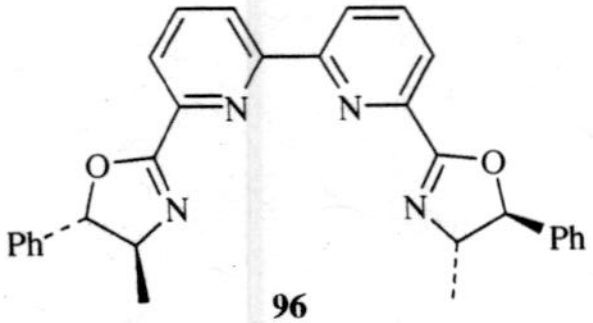

96

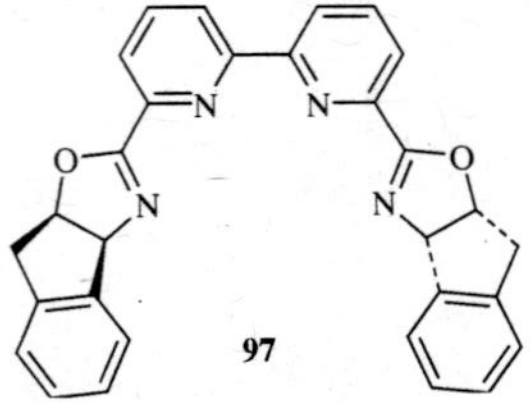

97

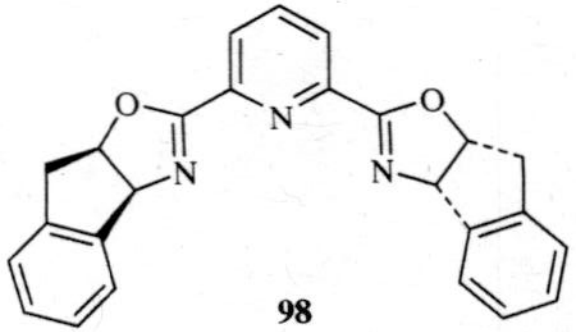

98

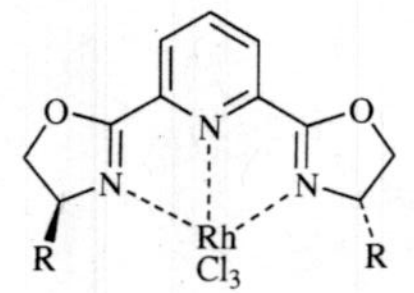

R	
99:	*i*-Pr
100:	*s*-Bu
101:	*i*-Bu
102:	Me

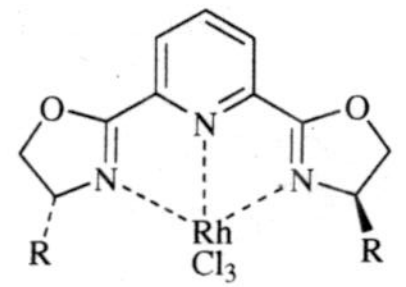

R	
103:	Et
104:	Ph

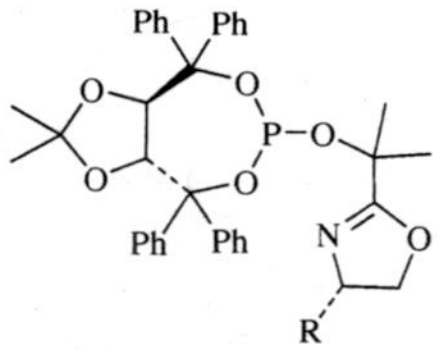

R	
105:	Ph
106:	*i*-Pr

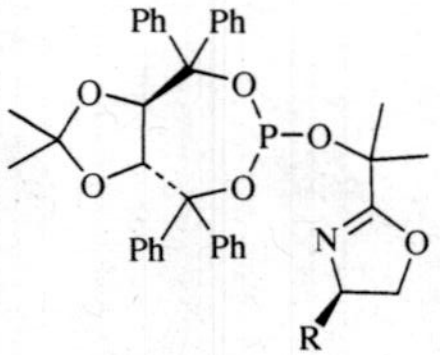

R	
107:	Ph
108:	*i*-Pr

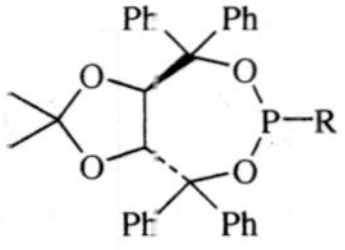

R	
109:	Ph
110:	4-MeC_6H_4
111:	2,4,6-$Me_3C_6H_2$
112:	2-naphthyl

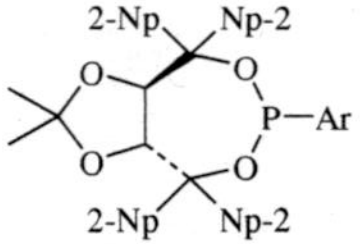

Ar	
113:	Ph
114:	4-MeC_6H_4
115:	2-naphthyl

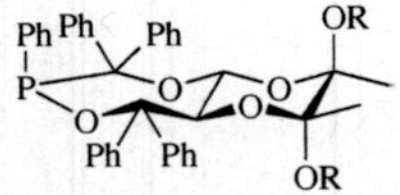

R	
116:	Me
117:	Et

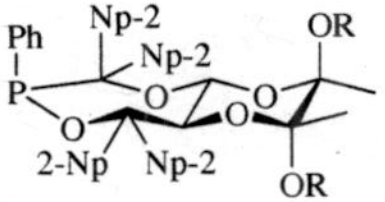

R	
118:	Me
119:	Et

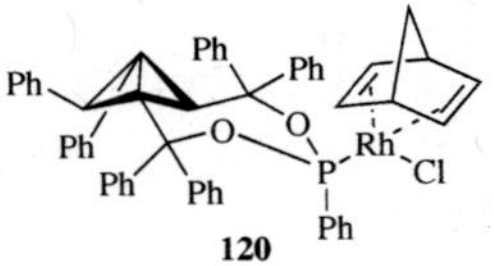

120

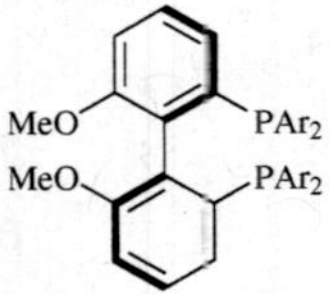

Ar	
121:	3,5-$Me_2C_6H_3$
122:	3,5-Me_2-4-$MeOC_6H_2$

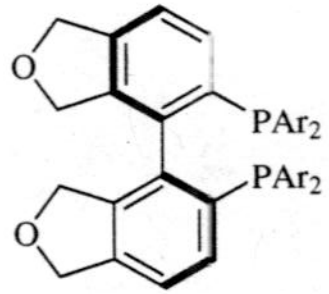

Ar	
123:	3,5-$Me_2C_6H_3$
124:	3,5-Me_2-4-$MeOC_6H_2$

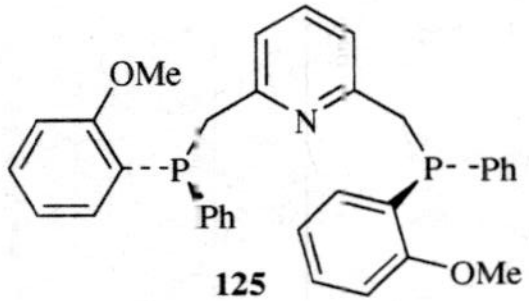

125

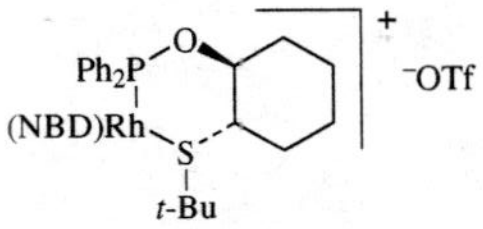

126

CHART 1. LIGAND AND CATALYST STRUCTURES USED IN TABLES (*Continued*)

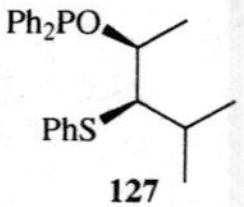

127

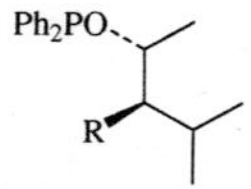

	R
128:	PhS
129:	PMPS

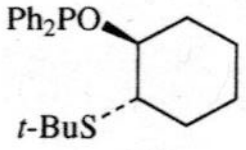

130

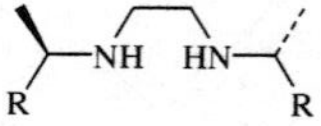

	R
131:	Ph
132:	4-ClC_6H_4
133:	*c*-C_6H_{11}
134:	1-naphthyl

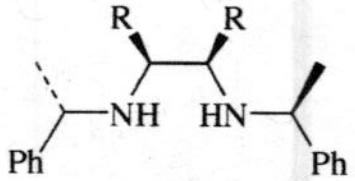

	R
135:	H
136:	*n*-Bu
137:	Ph

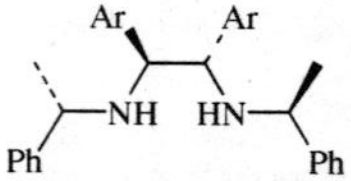

	Ar
138:	2-$MeOC_6H_4$
139:	Ph

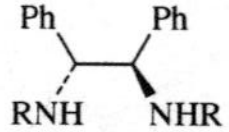

	R
140:	H
141:	$(CH_2)_2OMe$
142:	CH_2naphthyl-1
143:	3-$MeOC_6H_4$
144:	4-$MeOC_6H_4$
145:	2,4,6-$Me_3C_6H_2$
146:	Bn

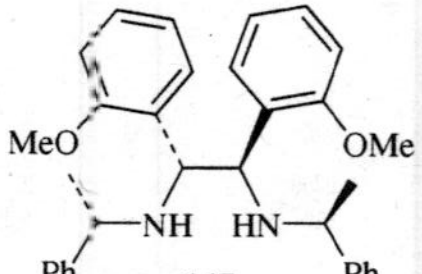

147

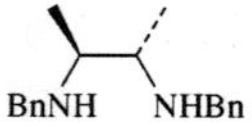

148

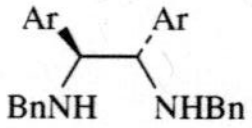

	Ar
149:	2-$MeOC_6H_4$
150:	2,5-$(MeO)_2C_6H_3$

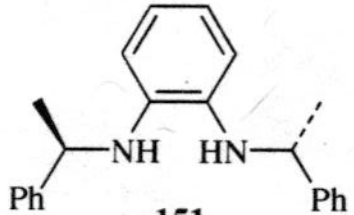

151

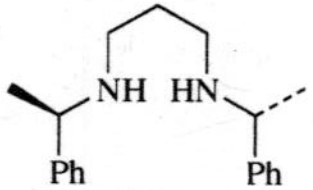

152

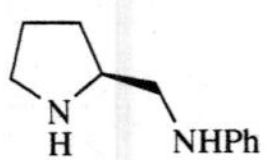

153

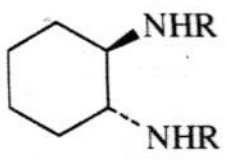

	R
154:	H
155:	*i*-Bu
156:	*neo*-C_6H_{13}
157:	*c*-C_6H_{11}
158:	CH_2naphthyl-1

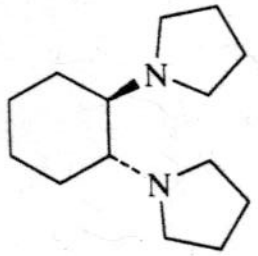

159

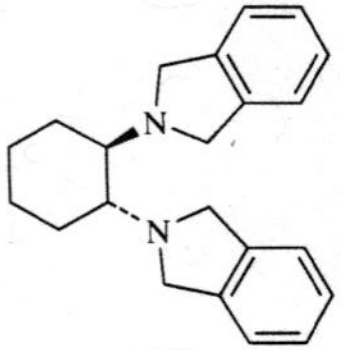

160

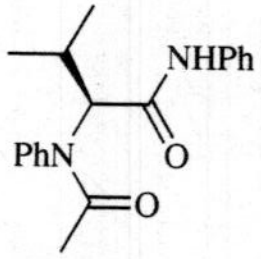

161

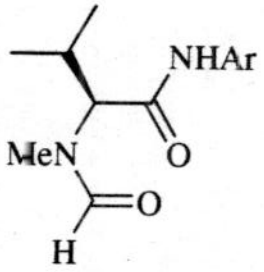

	Ar
162:	Ph
163:	4-MeOC_6H_4
164:	3,5-$(MeO)_2C_6H_4$
165:	3,5-$Me_2C_6H_4$
166:	3,5-$(CF_3)_2C_6H_4$
167:	3,5-$Cl_2C_6H_4$

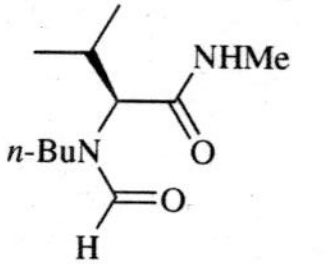

168

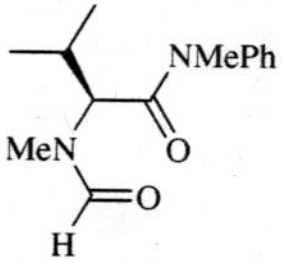

169

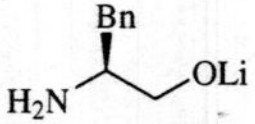

170

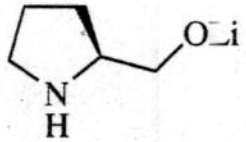

171

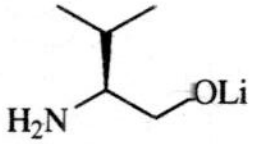

172

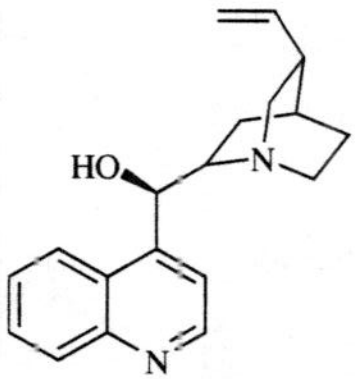

173

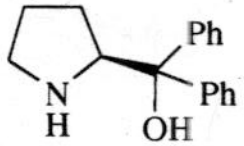

174

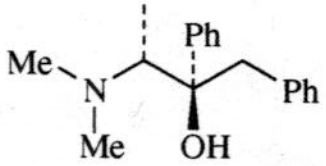

175

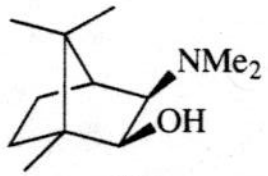

176

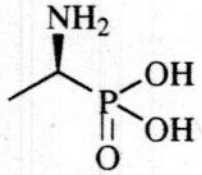

177

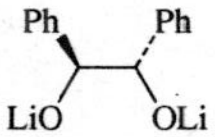

178

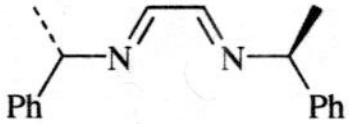

179

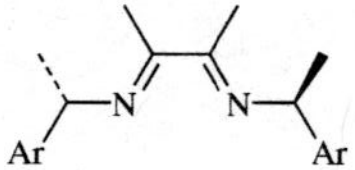

	Ar
180:	Ph
181:	1-naphthyl

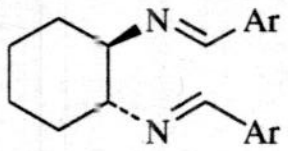

	Ar
182:	1-naphthyl
183:	2-naphthyl
184:	2,4,6-$Me_3C_6H_2$
185:	2,4,6-$(MeO)_3C_6H_2$
186:	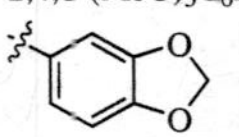

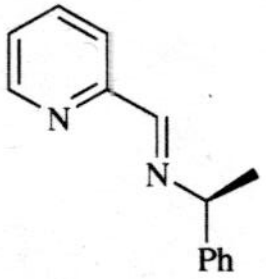

187

CHART 1. LIGAND AND CATALYST STRUCTURES USED IN TABLES (*Continued*)

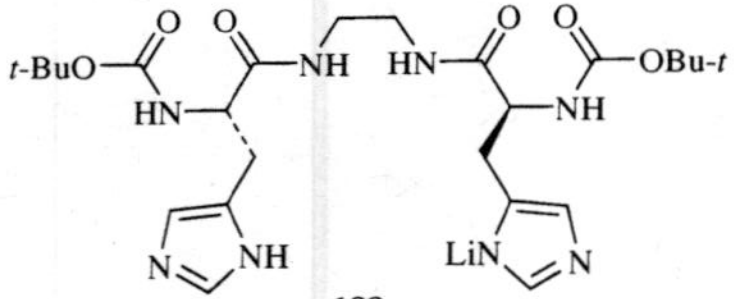
188

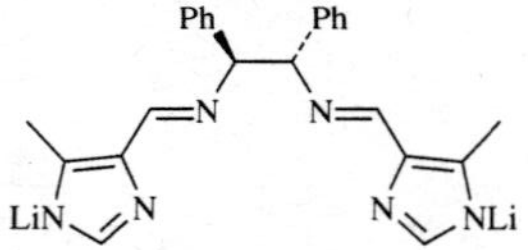
189

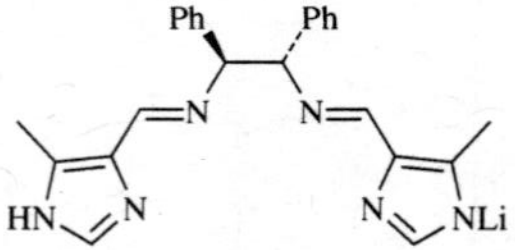
190

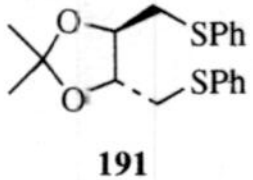
191

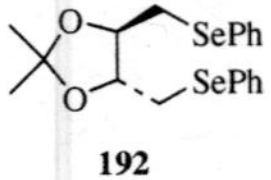
192

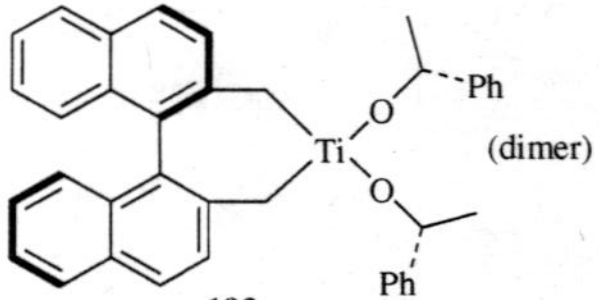
193

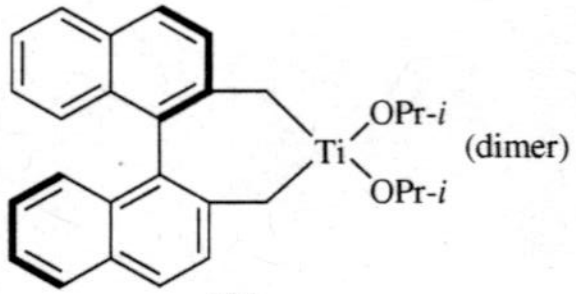
194

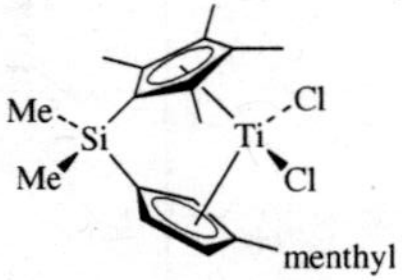
195

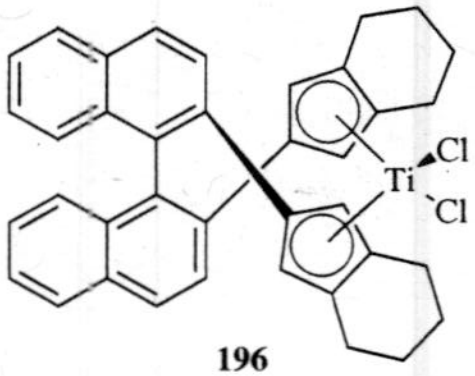
196

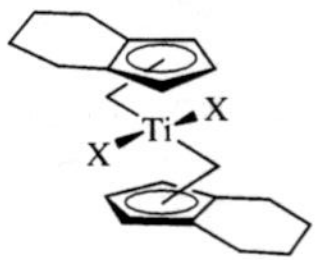
197

X_2 = 1,1'-binaphth-2,2'-diolate

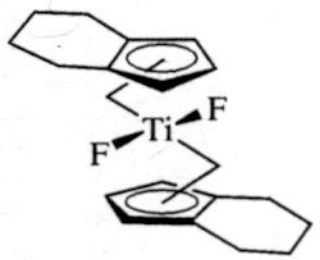
198

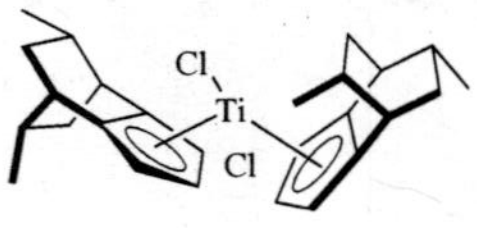
199

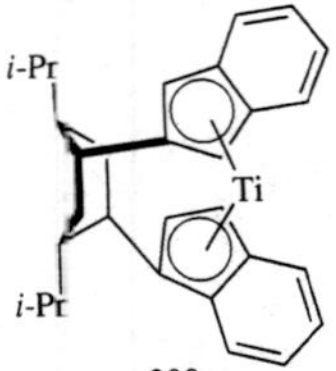
200

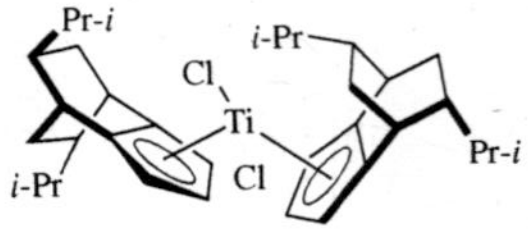
201

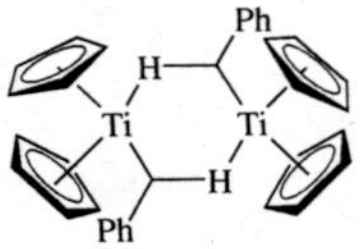
202

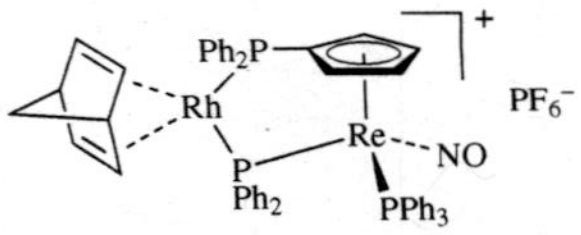
203

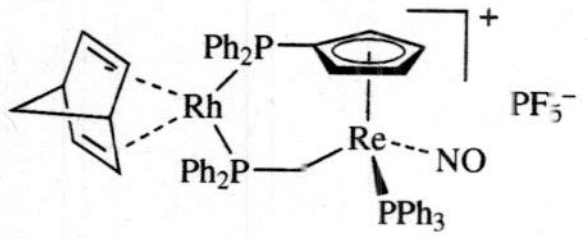

204

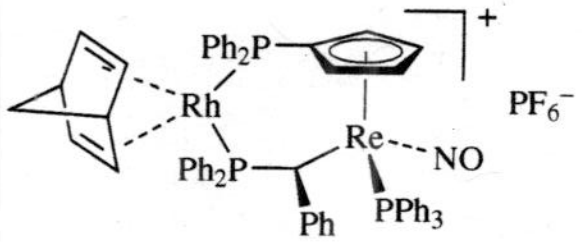

205 ($S_{Re}S_C$)

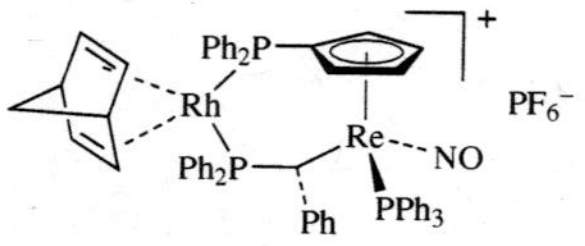

206 ($S_{Re}S_C$)

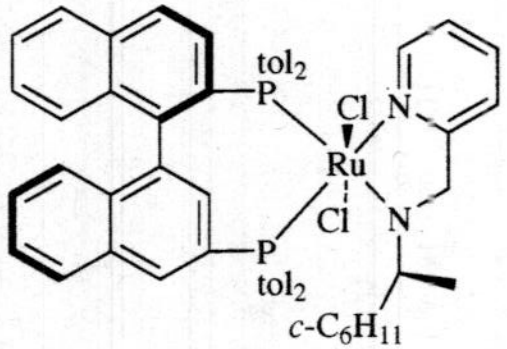

207

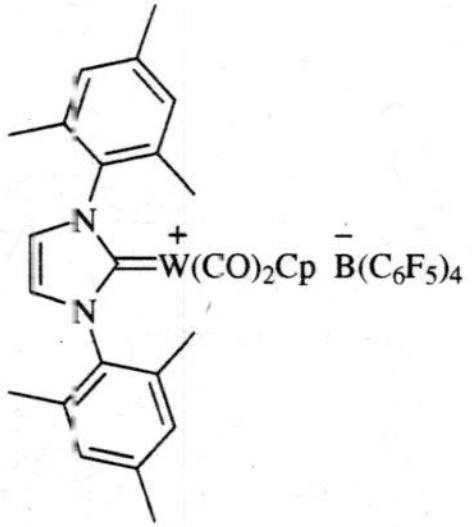

208

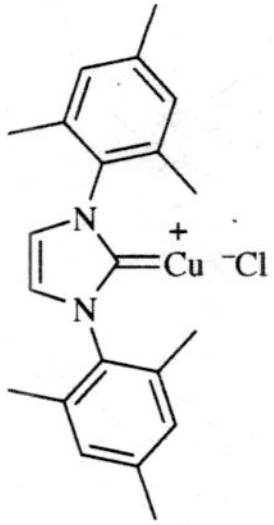

209

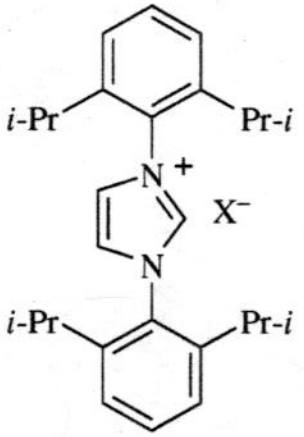

210: X = BF_4
211: X = Cl

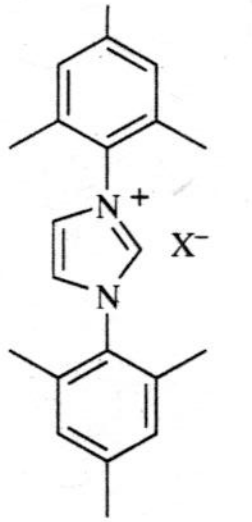

212: X = BF_4
213: X = Cl

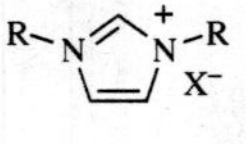

	R	X
214:	Me	BF_4
215:	*t*-Bu	Cl
216:	2-Ad	BF_4

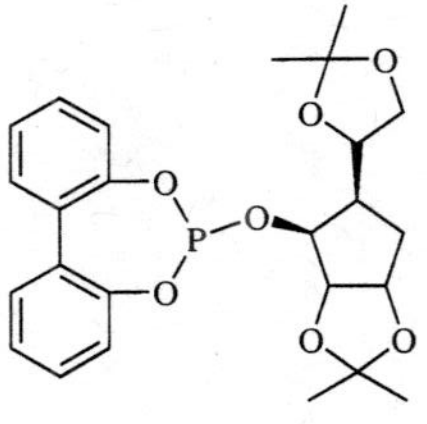

217

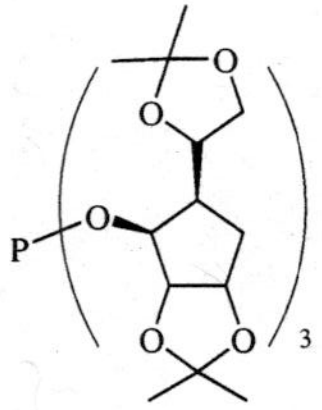

218

CHART 1. LIGAND AND CATALYST STRUCTURES USED IN TABLES (*Continued*)

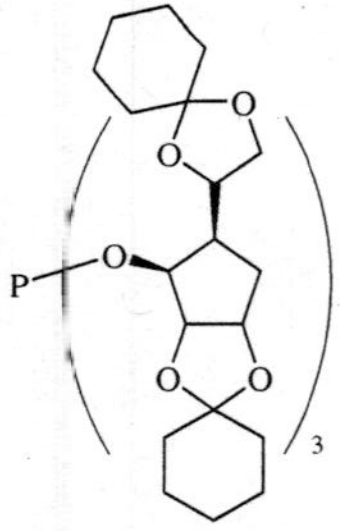

219

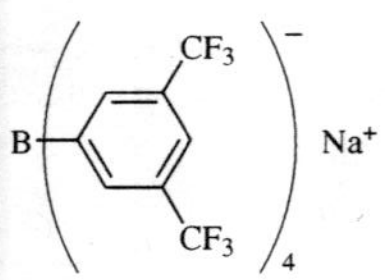

220

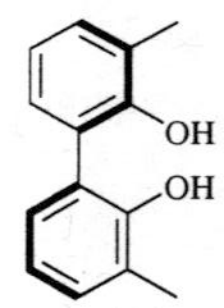

221

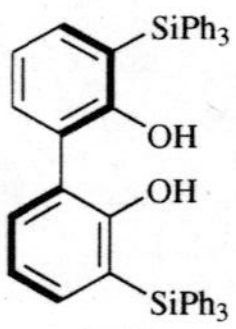

222

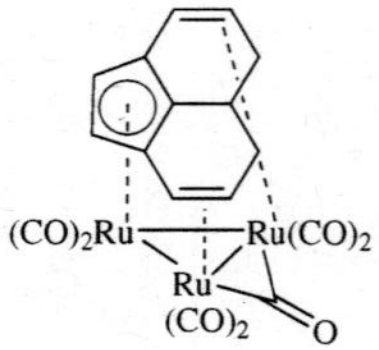

223

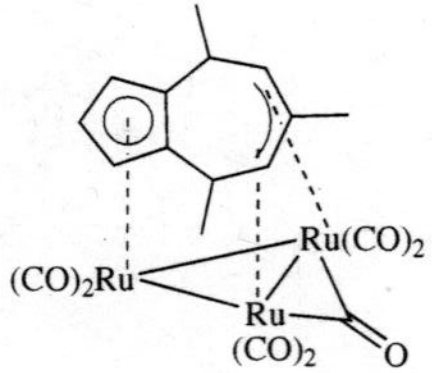

224

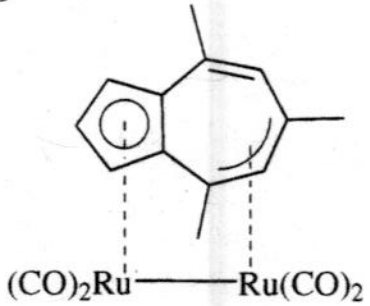

225

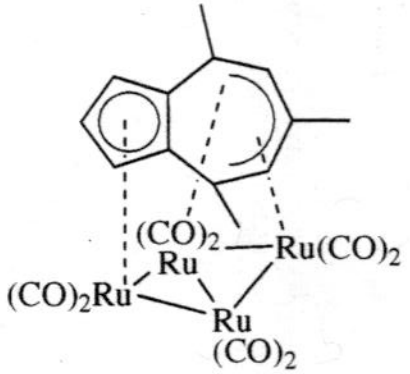

226

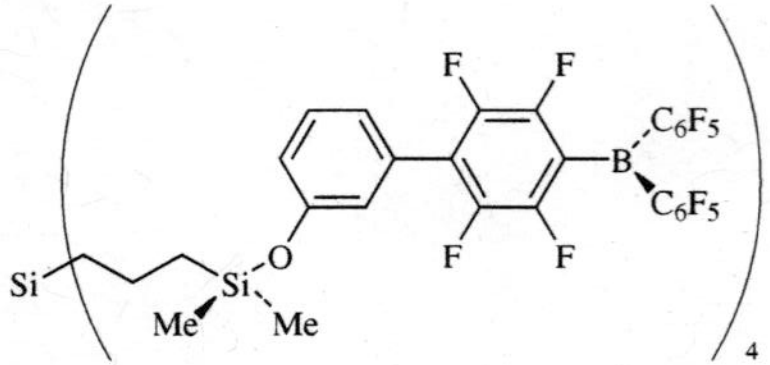

227

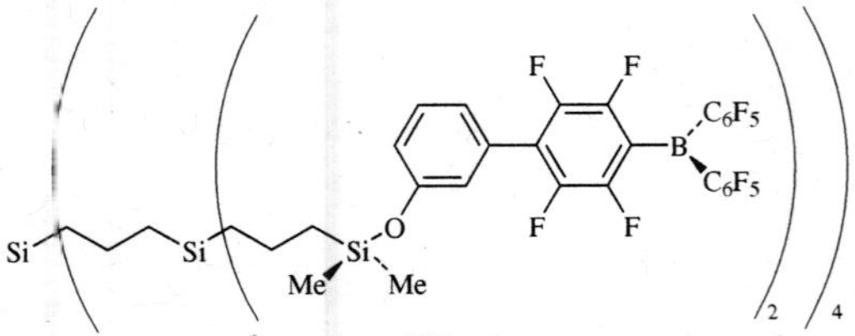

228

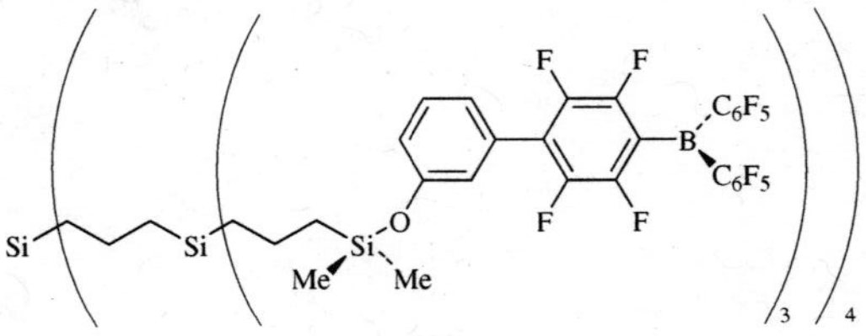

229

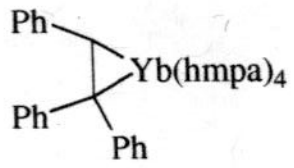

230

CHART 2. ORGANOSILANE COMPOUND DESIGNATIONS USED IN TABLES

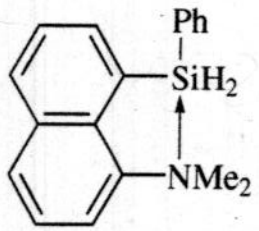

A

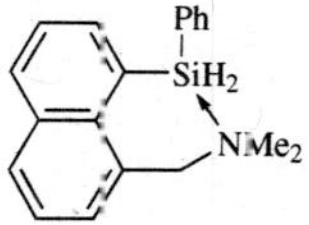

B

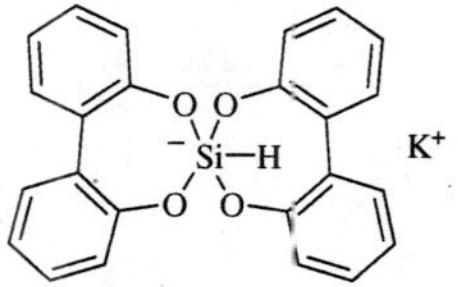

C

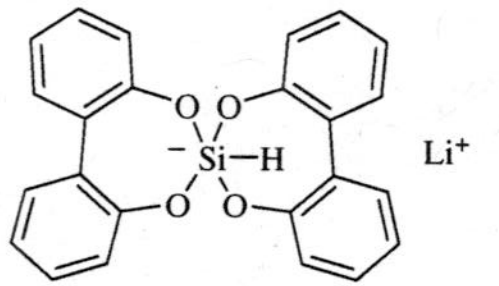

D

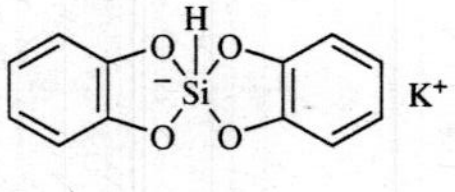

E

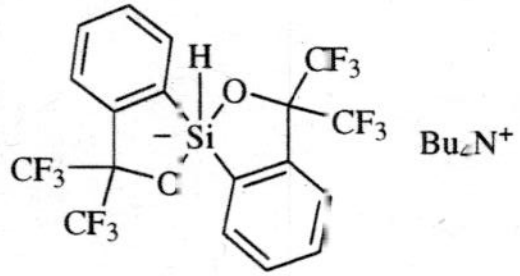

F

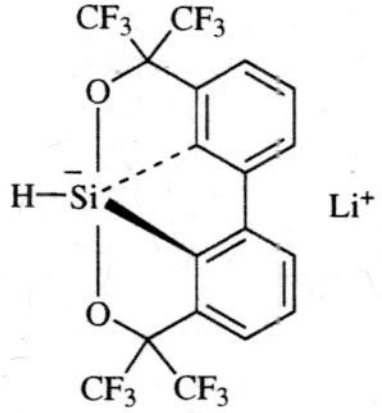

G

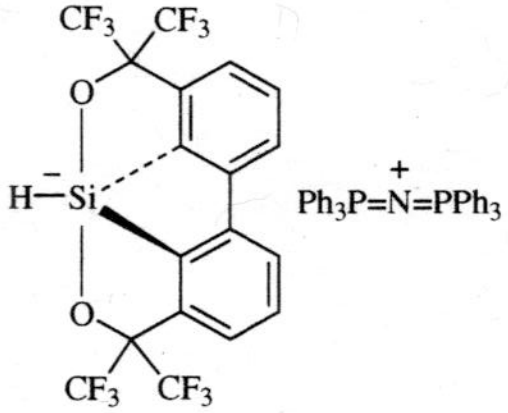

H

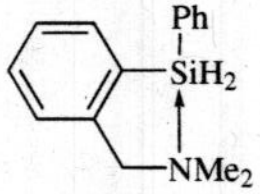

I

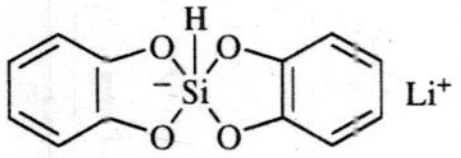

J

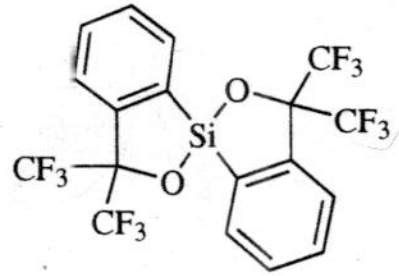

K

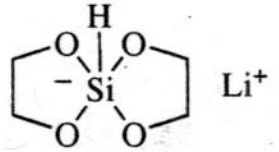

L

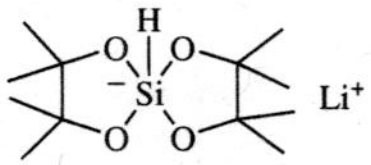

M

TABLE 1. ORGANOSILANE REDUCTION OF ALKENES

Alkene	Conditions	Product(s) and Yield(s) (%)	Refs.
C_{4-6}			
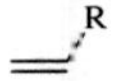	Et_3SiH (x eq), TFA, 50°		234
R	x / Time		
OBu-*n*	1 / 10 h	(80)	
OAc	2 / 75 h	(13)	
O_2CCF_3	2 / 75 h	(0)	
SC_4H_9-*n*	2 / 20 h	(15)	
C_5 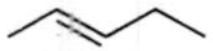 or 	$PhSiH_3$, **202**	 (—)	687
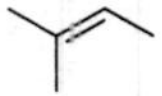	Et_3SiH (1 eq), TFA (2 eq), 50°, 10 h	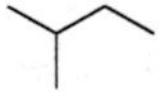 (—)	203
	Et_3SiD (1 eq), TFA (2 eq), 50°, 24 h	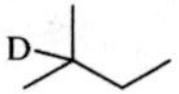 (90)	221
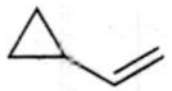	Et_3SiH (1 eq), TFA (2 eq), rt, 30 min	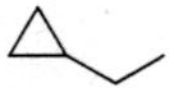 (—)	232
	Et_3SiH, TFA (x eq)	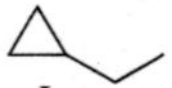 **I** + 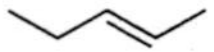 **II**	233

x	Time	Temp	I	II
2.5	0.1 h	0°	(—)	(—)
4	0.5 h	20°	(41)	(6.3)
4	1 h	20°	(40)	(10.8)
2.5	1.5 h	20°	(35)	(2.8)
2.5	4 h	20°	(39)	(6)
2.5	15 h	20°	(31)	(10)

$C_{5\text{-}16}$

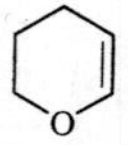

Et_3SiH (2 eq), $AlCl_3$ (1.0 eq), HCl, CH_2Cl_2, rt

(92) — 136

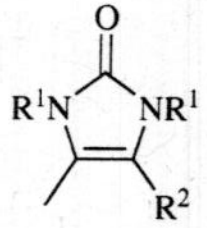

R_3SiH (1 eq), TFA, 50°, 20 h

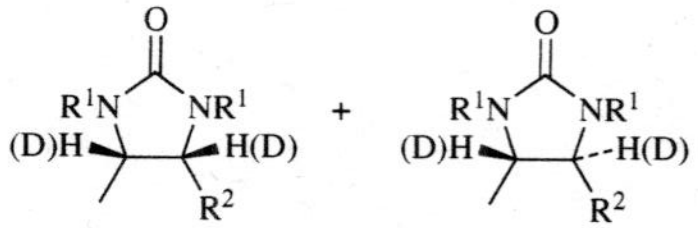

I + II — 688

R^1	R^2	R_3SiH	Time	Temp	I + II	% I
H	$(CH_2)_5CO_2Et$	Et_3SiH or Et_3SiD	20 h	50°	(70)	50
Ac	$(CH_2)_5CO_2Et$	Et_3SiH or Et_3SiD	60 h	rt	(70)	95
Ac	$(CH_2)_5CO_2Et$	Ph_3SiH	60 h	rt	(30)	96
Ac	$(CH_2)_5CO_2Et$	Ph_2SiD_2	60 h	rt	(23)	94
Ac	$(CH_2)_5CO_2Et$	Ph_3GeD	20 h	rt	(27)	98
H	Me	Et_3SiH	20 h	50°	(65)	50
Ac	Me	Et_3SiH	20 h	rt	(78)	92

C_6

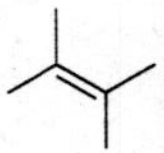

Et_3SiH (1.2 eq), 65% $HClO_4$, CH_2Cl_2, 20°, 30 min

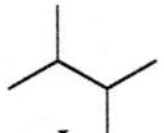

I (83) — 214

Et_3SiH (1.1-1.2 eq), HCO_2H, HOAc, KU-1, 55°, 5 h — I (42) — 208

Et_3SiH (1.1-1.2 eq), HCO_2H, KU-1, 55°, 5 h — I (92) — 208

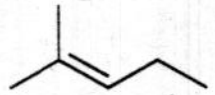

Et_3SiH, TMSOTf, CD_2Cl_2, −75°, 5 min

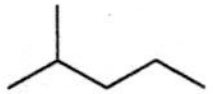

(98)[a] — 216

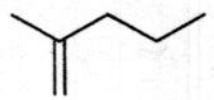

Et_3SiH (1 eq), TFA (2 eq), 50°, 10 h

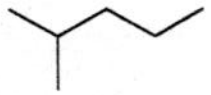

(—) — 203

TABLE 1. ORGANOSILANE REDUCTION OF ALKENES (*Continued*)

Alkene	Conditions	Product(s) and Yield(s) (%)	Refs.
C_6 	Et_3SiH (1 eq), TFA (2 eq)	 **I**	222
	Temp / Time		
	20° / 400 h	(75)	
	50° / 3 h	(48)	
	50° / 6 h	(67)	
	50° / 24 h	(87)	
	80° / 24 h	(80)	
	Et_3SiH, TMSOTf, CD_2Cl_2, –75°, 5 min	**I** (93)[a]	216
	Et_3SiD (1 eq), TFA (2 eq), 50°, 24 h	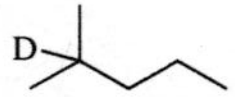 (90)	221
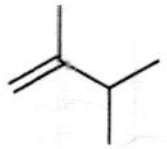	Et_3SiH, TMSOTf, CD_2Cl_2, –75°, 5 min	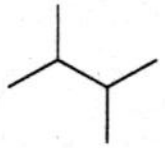 (96)[a]	216
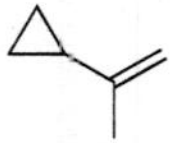	Et_3SiH (1 eq), TFA (2 eq), rt, 72 h	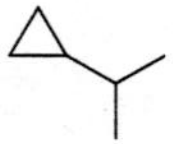 (20) + 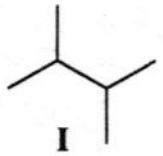 **I** + 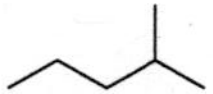 **II**; **I** + **II** (80)	232
	Et_3SiH, TFA (x eq)	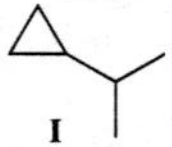 **I** + 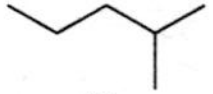 **II**	233
	x / Time / Temp	**I** / **II**	
	2 / 0.5 h / –10°	(78) / (0)	
	3 / 0.5 h / –10°	(59) / (0.5)	
	2 / 3 h / 20°	(52) / (2)	
	2 / 50 h / 20°	(40) / (—)	
	Et_3SiD (1 eq), TFA (2 eq), –10°, 0.5 h	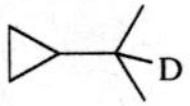 (80)	221

	Et_3SiH (2 eq), TFA (4 eq), 20°, 90 h	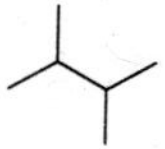 (65)	233
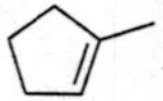	Et_3SiH (1 eq), TFA (2 eq), 50°, 10 h	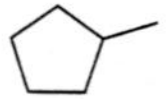 (—)	203
	Et_3SiD (1 eq), TFA (2 eq), 50°, 50 h	 (60)	221
	$EtCl_2SiH$ (2 eq), $AlBr_3$ (1 eq), PTSA (1 eq), 40°, 2 h	 **I** (23)	192
	$EtCl_2SiH$ (2 eq), $AlCl_3$ (1 eq), PTSA (1 eq), 40°, 2 h	**I** (17)	192
	PMHS, Pd/C, EtOH 80°	**I** (100)	316
	$PhSiH_3$, **202**	**I** (—)	687
	Et_3SiH (10.2 eq), TFA (13.2 eq), 20°, 24 h	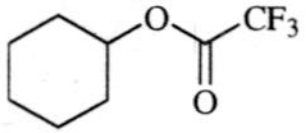 (65)	230
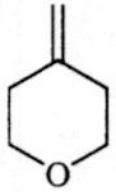	Et_3SiH (x eq), $AlCl_3$ (3 eq), HCl, CH_2Cl_2, rt	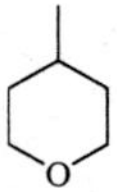 **I**	136
	x 3, Time 3.5 h	(23)	
	x 4, Time 1 h	(40)	
	x 5, Time 3 h	(41)	
	Et_3SiH (3 eq), $AlCl_3$ (2 eq), HCl, 20°, 1 h	**I** (40)	146
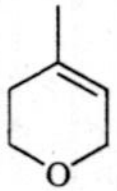	Et_3SiH (5 eq), $AlCl_3$ (2.5 eq), HCl, CH_2Cl_2, rt, 3 h	**I** (35)	136

TABLE 1. ORGANOSILANE REDUCTION OF ALKENES (*Continued*)

Alkene	Conditions	Product(s) and Yield(s) (%)	Refs.
C_7 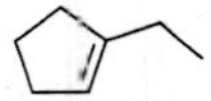	Et_3SiH (1.1-1.2 eq), HCO_2H, KU-1, 55°, 5 h	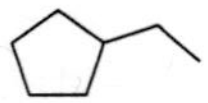 (100)	208
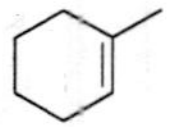	Et_3SiH (0.87 eq), TFA, rt, 120 h	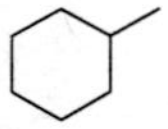 **I** (78)	202
	Et_3SiH (1 eq), TFA (2 eq), 50°, 10 h	**I** (—)	203
	Et_3SiD (1 eq), TFA (2 eq), 50°, 24 h	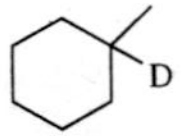 (80)	221
	Et_3SiH (1 eq), TFA (x eq)	**I**	203
	x / Temp / Time		
	1.5 / 50° / 10 h	(40)	
	2 / 50° / 10 h	(67)	
	2 / 20° / 120 h	(73)	
	2.5 / 50° / 10 h	(72)	
	10 / 20° / 120 h	(78)	
	Et_3SiH (1 eq), TFA (8 eq), $O_2NC_6H_5$, 50°, 3 h	**I** (92)	207
	$Et_3SiO(EtHSiO)_nSiEt_3$ (0.1 eq), TFA (8 eq), $O_2NC_6H_5$, 50°, 3 h	**I** (94)	207
	Et_3SiH (1.2 eq), 65% $HClO_4$, CH_2Cl_2, 20°, 30 min	**I** (97)	214
	Et_3SiH (1.4 eq), HCl (xs), $AlCl_3$ (0.2 eq), CH_2Cl_2, rt	**I** (100)	213

Ph_2SiH_2 (1.2 eq), H_2SO_4 (30 eq) **I** 209

Solvent	Time	Temp	96% H_2SO_4	90% H_2SO_4
EtOH	1.5 h	60°	(100)	(—)
n-BuOH	2 h	rt	(72)	(62)
n-$C_8H_{17}OH$	2 h	rt	(65)	(63)
n-$C_{12}H_{25}OH$	2 h	60°	(73)	(—)
n-$C_{12}H_{25}OH$	3.5 h	60°	(—)	(65)
(*n*-Bu)$_2$O	2.5 h	rt	(—)	(61)
(*n*-C_8H_{17})$_2$O	2.5 h	rt	(—)	(69)
(*n*-C_8H_{17})$_2$O	2 h	60°	(68)	(70)
n-$C_4H_9OC_{12}H_{25}$-*n*	2 h	60°	(93)	(89)

Et_3SiH (1.1-1.2 eq), HCO_2H, KU-1, 55°, 5 h **I** 208

Solvent	
HCO_2H	(96)
HCO_2H, HOAc	(77)
HOAc	(37)
MeOH	(5)
CH_2Cl_2	(0)

Ph_2SiH_2 (1 eq), $O_2NC_6H_5$, acid, 130°, 10 h **I** 134

Acid	
CF_3CO_2H	(50)
ClF_2CCO_2H	(25)
Cl_3CCO_2H	(5)
Cl_2HCCO_2H	(0)
CH_3CO_2H	(0)
4-Me$C_6H_4SO_3H$	(35)

TABLE 1. ORGANOSILANE REDUCTION OF ALKENES (*Continued*)

Alkene	Conditions	Product(s) and Yield(s) (%)	Refs.
C_7 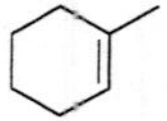	Ph_2SiH_2, $O_2NC_6H_5$, 130°	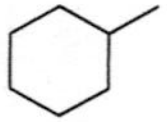 (98) **I**	193
	$Et_3SiO(EtHSiO)_nSiEt_3$ (0.25 mol%), TFA, 50°, 20 h	**I** (100)	207
	$Et_3SiO(EtHSiO)_nSiEt_3$ (0.1 mol%), TFA (x eq), solvent, 20° x Solvent Time 3 $CHCl_3$ 5 h 5 4-$O_2NC_6H_5$ 5 h	**I** (52) (89)	207
	$EtCl_2SiH$ (2 eq), $AlCl_3$ (1 eq), PTSA (1 eq), 40°, 2 h	**I** (65)	192
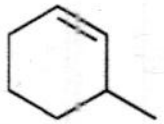 or 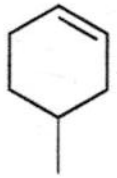	$PhSiH_3$, **202**	**I** (—)	687
	Et_3SiH (5 eq), TFA (10 eq), 20°, 24 h	 (3.5) + 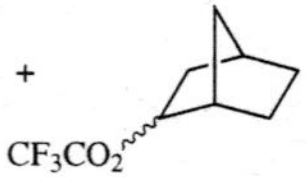 (—)	230
	Et_3SiH (5 eq), TFA (10 eq), $BF_3•OEt_2$ (0.4 eq), 20°, 24 h	**I** (22)	230
	$PhSiH_3$, **202**	**I** (—)	687
	Et_3SiH (10.2 eq), TFA (20.3 eq), 20°, 24 h	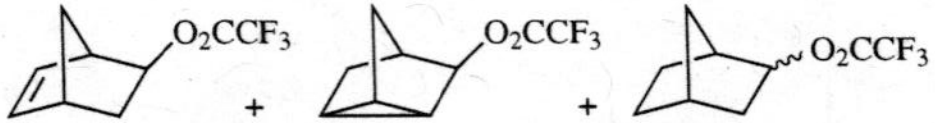 **I** + **II** + **III** **I** + **II** + **III** (—), **I**:**II**:**III** = 62:20:17	230
C_8 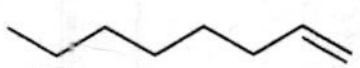	PMHS, Pd/C, EtOH, 80°	 (88)	316

Reactant	Conditions	Product(s) and Yield(s) (%)	Refs.
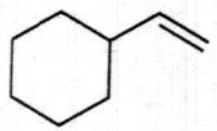	Et_3SiH (5.1 eq), TFA (13.1 eq), 20°, 24 h	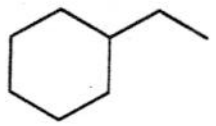 (20) + 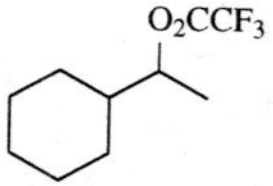	230
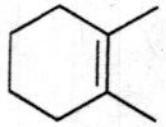	Et_3SiH (1 eq), TFA (2 eq), 20°	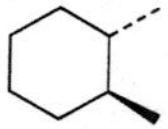 **I** + 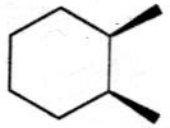 **II** Time / I + II / I:II 2 h (13) 1.5 24 h (58) 1.3 144 h (72) 1.5 240 h (74) 1.5	229
	Et_3SiH (5.1 eq), TFA (13.1 eq), 20°, 24 h	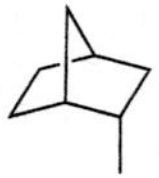 (45)	230
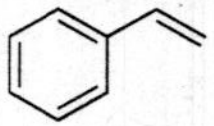	Et_3SiH (1 eq), TFA (2 eq), 50°, 10 h	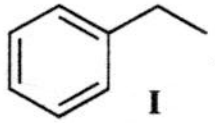 **I** (—)	203
	$PhSiH_3$, **202**	**I** (—)	687
	PMHS-Pd nanocomposite, C_6H_6, rt, 4 h	**I** (95)	219
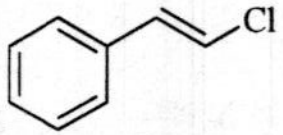	PMHS-Pd nanocomposite, C_6H_6, rt, 6 h	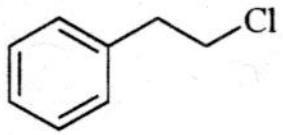 (85)	219
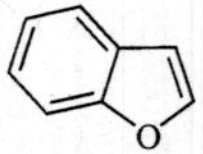	Et_3SiH (1.5 eq), TFA (x eq), 3 h	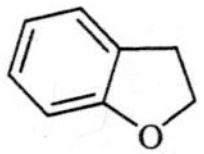 x / Temp 2.5 70° (54) 3.5 60° (48) 5 60° (66)	251
	Et_3SiH (1.5 eq), CF_3CO_2D (2.5 eq), 60°, 7 h	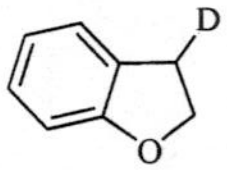 (47)	251
C_{8-9}			
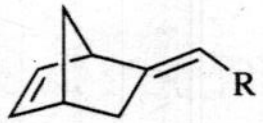	Et_3SiH (6.5 eq), TFA (12.1 eq), $BF_3•OEt_2$ (1 eq), 20°, 24 h	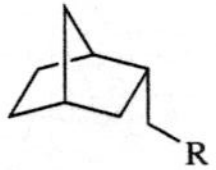 R H (85) Me (80)	230

TABLE 1. ORGANOSILANE REDUCTION OF ALKENES (*Continued*)

Alkene	Conditions	Product(s) and Yield(s) (%)	Refs.
C_{8-18} 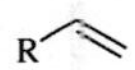	Et_3SiH (2 eq), $PdCl_2$ (10 mol%), EtOH		220
R	Time		
n-C_6H_{13}	1 d	(100)	
n-C_8H_{17}	1 d	(90)	
n-$C_{10}H_{21}$	1.5 d	(97)	
n-$C_{12}H_{25}$	1 d	(73)	
n-$C_{14}H_{29}$	1.5 d	(97)	
n-$C_{16}H_{33}$	1 d	(96)	
C_9 	PMHS, Pd/C, EtOH, 80°	 (25)	316
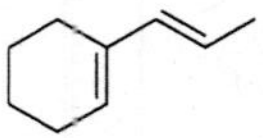	Et_3SiH, TFA	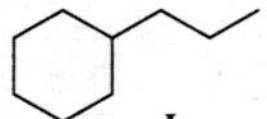 **I** (—)	233
	Et_3SiH (2 eq), TFA (3 eq)	**I** (70)	231
	Et_3SiH (6.5 eq), TFA (12.1 eq), BF_3•OEt_2 (1 eq), 20°, 24 h	**I** (75)	230
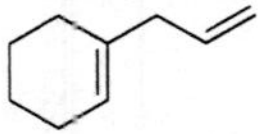	Et_3SiH (2 eq), TFA (3 eq)	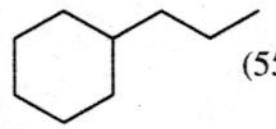 **I** (55) + 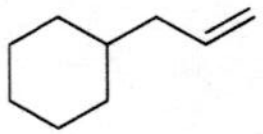 **II** (15)	231
	Et_3SiH (6.5 eq), TFA (12.1 eq), BF_3•OEt_2 (1 eq), 20°, 24 h	**I** (25) + **II** (—)	230
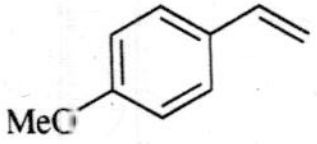	PMHS-Pd nanocomposite, C_6H_6, rt, 3 h	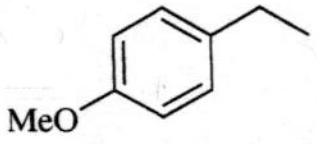 (95)	219
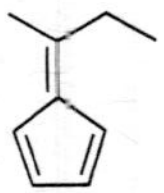	Et_3SiH (1 eq), TFA (8 eq), 60°, 10 h	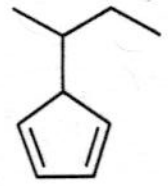 (73)	689

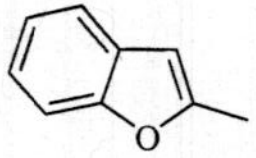	Et_3SiH (1.5 eq), TFA (2.5 eq), 60°	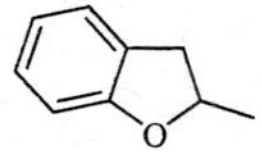	Time: 0.5 h (84); 2 h (91)	251
	Et_3SiH (1.5 eq), CF_3CO_2D (2.5 eq), 60°, 2 h	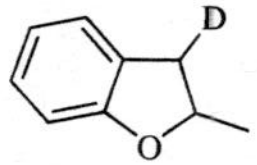	(86)	251
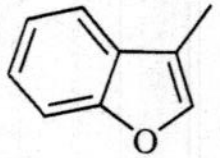	Et_3SiH (1.5 eq), TFA (2.5 eq), 60°	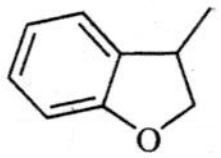	Time: 0.2 h (84); 7 h (91)	251
	Et_3SiH (1.5 eq), CF_3CO_2D (2.5 eq), 60°, 7 h	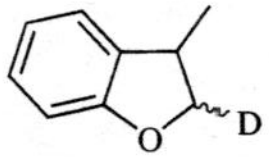	(60)	251
C_{10} 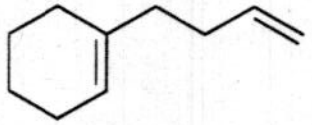	Et_3SiH (2 eq), TFA (3 eq)	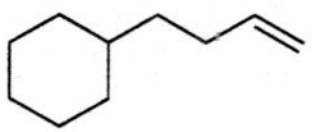	(65)	231
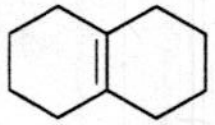	R_3SiH (x eq), TFA (5 eq), rt, <5 d	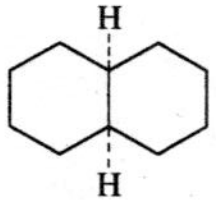 **I** + 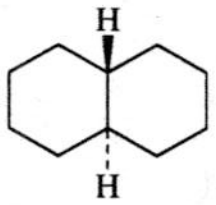 **II**		204

R_3Si	x	rel % I	I:II
n-BuH_2Si	0.36	37	0.59
Et_2HSi	0.60	40	0.67
Et_3Si	1.1	42	0.72
(*i*-$C_5H_{11})_3Si$	1.1	35	0.54
(*c*-$C_5H_9)_3Si$	1.1	54	1.17
Ph_3Si	1.1	58	1.38
(*s*-Bu$)_3Si$	1.1	72	2.57
(*t*-Bu$)_2HSi$	0.55	77	3.35
(*t*-Bu$)_2MeSi$	1.1	83	4.88
(*t*-Bu$)_3Si$	1.1	93	13.3

TABLE 1. ORGANOSILANE REDUCTION OF ALKENES (*Continued*)

Alkene	Conditions	Product(s) and Yield(s) (%)	Refs.
C_{10} 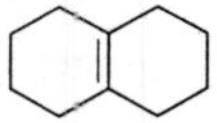	R_3SiH (1.5 eq), acid	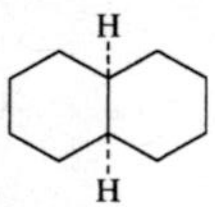 **I** + 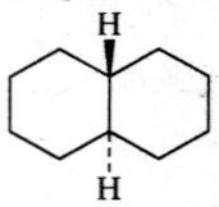 **II**	212
	R_3Si — Acid Et_3Si — TFA Et_3Si — n-$C_3F_7CO_2H$ Ph_2HSi — TFA Ph_2MeSi — TFA Ph_3Si — TFA $(t\text{-}Bu)_3Si$ — TFA	**I:II** 1:2.9 1:4 1:1.5 1:1.3 1.8:1 13.3:1	
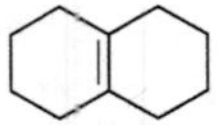 92% + 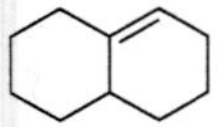 8%	Ph_3SiH (1.2 eq), TFA (6 eq), CH_2Cl_2, rt, 24 h	**I** (57) + **II** (32)	127
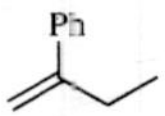	(*R*)-1-NpPhMeSiH, (96.2% opt. purity, 1 eq), TFA, CH_2Cl_2, rt, 20 min	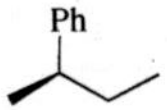 (19) 2.05% ee	142
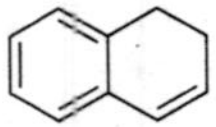	Et_3SiH (1.6 eq) TFA, 50-60°, 1 h	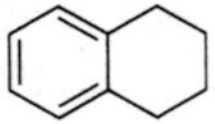 **I** (90)	224
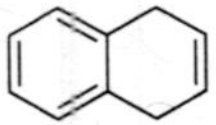	Et_3SiH (1.6 eq) TFA, 50-60°, 70 h	**I** (5)	224
C_{11} 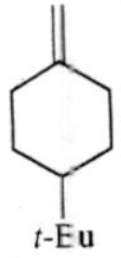	R_3SiH (x eq), TFA (5 eq), rt, <5 d	 **I** + 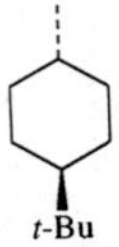 **II**	204

R$_3$Si	x	I	I:II
n-BuH$_2$Si	1.1	(4)	0.04
Et$_2$HSi	0.60	(8)	0.09
Et$_3$Si	1.1	(10)	0.11
(*i*-C$_5$H$_{11}$)$_3$Si	1.1	(9)	0.10
(*s*-Bu)$_3$Si	1.1	(16)	0.19
(*t*-Bu)$_2$HSi	0.55	(16)	0.19

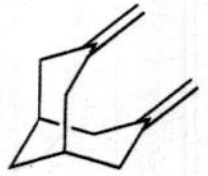

Et$_3$SiH (1.5 eq), H$_2$SO$_4$ (1.25 eq), CH$_2$Cl$_2$, rt, 6 h

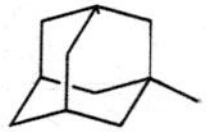

I (62) +

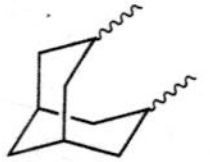

II (14)

I + **II** (76), **I**:**II** = 81:19

243

Et$_3$SiH (1.5 eq), TFA (1.25 eq), CH$_2$Cl$_2$, rt, 6 h

I (29) +

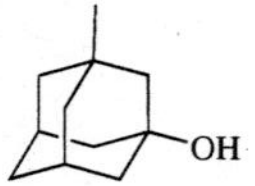

III (45)

243

Et$_3$SiH (1.5 eq), H$_2$SO$_4$/carbon, *n*-C$_6$H$_{14}$, rt, 6 h

I (25) + **III** (56)

243

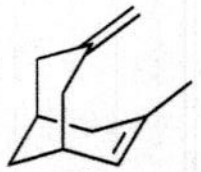

Et$_3$SiH (1.5 eq), H$_2$SO$_4$ (1.25 eq), CH$_2$Cl$_2$, rt, 6 h

I (65) + **III** (15)

243

Et$_3$SiH (1.5 eq), H$_2$SO$_4$/carbon, *n*-C$_6$H$_{14}$, rt, 6 h

I (11) + **III** (47)

243

Et$_3$SiH (1.5 eq), H$_2$SO$_4$ (1.25 eq), CH$_2$Cl$_2$, rt, 6 h

I + **II** + **III** (21), **I**:**II**:**III** = 29:66:5

243

Et$_3$SiH (1.5 eq), H$_2$SO$_4$/carbon, *n*-C$_6$H$_{14}$, rt, 6 h

I (9) +

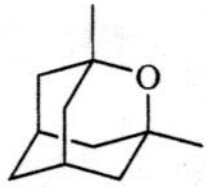

(12)

243

TABLE 1. ORGANOSILANE REDUCTION OF ALKENES (*Continued*)

Alkene	Conditions	Product(s) and Yield(s) (%)	Refs.
C_{11} 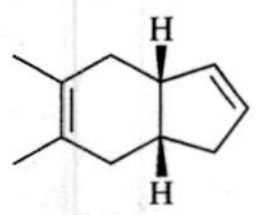	Et_3SiH (x eq), TFA (y eq)	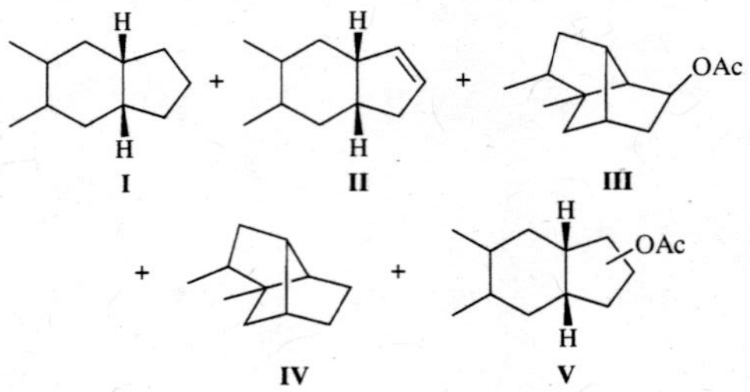	690

x	y	Temp	Time	I	II	III	IV
5.5	1.3	rt	0.5 h	(2)	(18)	(5)	(65)
5.5	1.3	rt	6 h	(4)	(14)	(9)	(68)
5.5	1.3	rt	64 h	(15)	(1)	(6)	(67)
5.5	1.3	80°	1 h	(3)	(1)	(1)	(81)
5.5	1.3	80°	2 h	(12)	(—)	(2)	(69)
5.5	1.3	80°	9 h	(17)	(—)	(3)	(60)
13	5	80°	0.5 h	(2)	(5)	(3)	(76)
13	5	80°	17 h	(22)	(—)	(3)	(22)
13	5	80°	34 h	(41)	(—)	(7)	(17)

Alkene	Conditions	Product(s) and Yield(s) (%)	Refs.
C_{12-14} 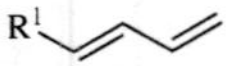	Et_3SiH (5 eq), $Ni(cod)_2$ (20 mol%), PPh_3 (40 mol%), MeC_6H_5, R^2CHO (1 eq), 50°	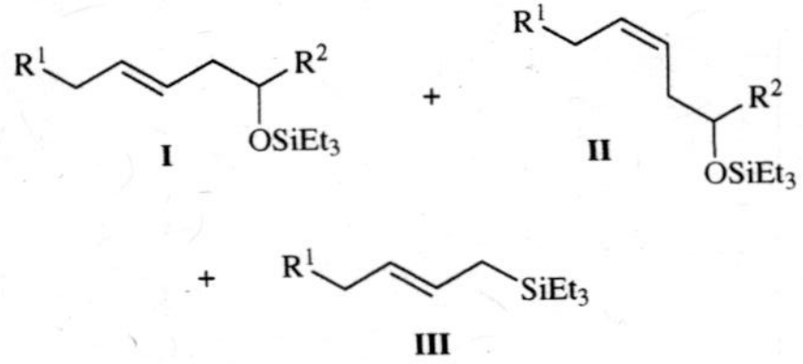	691

R^1	R^2	Time	I:II	I + II	III
$BnCH_2$	4-$MeOC_6H_4$	20 h	100:0	(72)	(0)
$BnCH_2$	4-$MeO_2CC_6H_4$	19 h	100:0	(66)	(0)
$MOMOCH_2C_6H_4$	4-$MeOC_6H_4$	18 h	100:0	(76)	(0)
$MOMOCH_2C_6H_4$	4-$MeO_2CC_6H_4$	44 h	100:0	(59)	(0)
$MOMOCH_2C_6H_4$	*n*-Pr	27 h	3:1	(71)	(0)
$MOMOCH_2C_6H_4$	*i*-Pr	24 h	2:1	(81)	(0)
$BnO(CH_2)_3$	Ph	18 h	100:0	(63)	(7)
$BnO(CH_2)_3$	4-$MeOC_6H_4$	19 h	100:0	(49)	(31)
$BnO(CH_2)_3$	4-$MeO_2CC_6H_4$	44 h	100:0	(54)	(0)
$BnO(CH_3)_3$	*i*-Pr	18 h	100:0	(63)	(11)

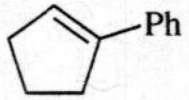

Et_3SiH (1.3 eq), NH_4F (1.3 eq), TFA (5 eq), 0°, 0.5 h, rt, 3 h

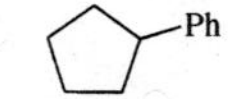

(85) 135

C_{12}

Et_3SiH (1.2 eq), 65% $HClO_4$, CH_2Cl_2, 20°, 30 min

(100) 214

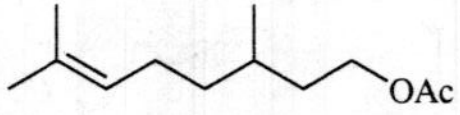

Et_3SiH (2 eq), TFA (2.5 eq), $LiClO_4$ (x eq), *i*-$PrNO_2$

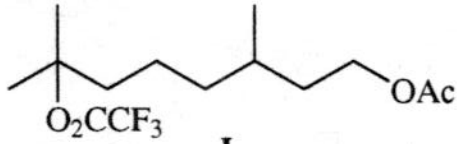

I

\+

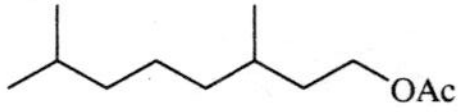

II

205

x	Temp	Time	I	II
0	80°	22 h	(3)	(75)
0	50°	22 h	(54)	(24)
0.1	50°	22 h	(2)	(80)
0.2	20°	22 h	(29)	(60)
1	20°	22 h	(1)	(90)
1	20°	3 h	(45)	(42)

Et_3SiH (2 eq), Cl_2CHCO_2H, $LiClO_4$ (1 eq), *i*-$PrNO_2$, 20°, 48 h

I (20) + **II** (65) 205

TABLE 1. ORGANOSILANE REDUCTION OF ALKENES (*Continued*)

Alkene	Conditions	Product(s) and Yield(s) (%)	Refs.
C_{12}			
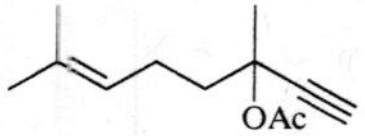	Et_3SiH (2 eq), TFA (2.5 eq), $LiClO_4$ (1 eq), *i*-$PrNO_2$, 20°, 20 h	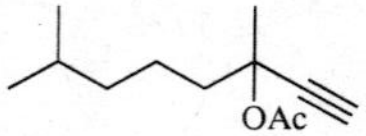 (20)	205
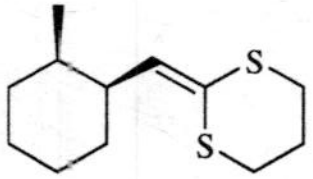	Et_3SiH, TFA	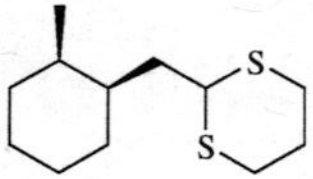 (97)	223
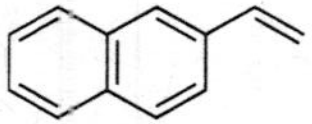	PMHS-Pd nanocomposite, C_6H_6, rt, 4 h	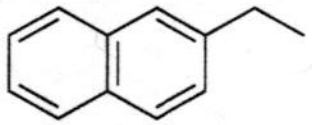 (96)	219
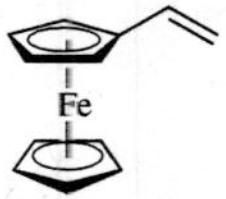	PMHS-Pd nanocomposite, C_6H_6, rt, 7 h	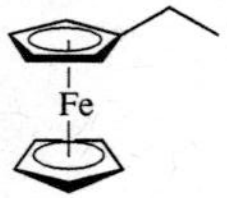 (85)	219
C_{13}			
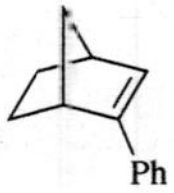	Et_3SiH (1.2 eq), TFA, CH_2Cl_2, rt, 30 min	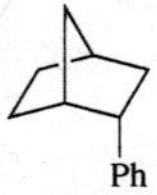 (—)	164
C_{14}			
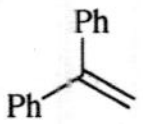	Et_3SiH (1.3 eq), NH_4F (1.3 eq), TFA (5 eq), 0°, 0.5 h; rt, 3 h	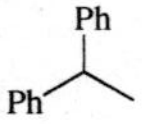 (93)	135
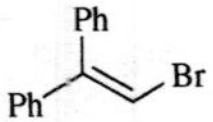	Et_3SiH (1.25 eq), TFA (9 eq), $CHCl_3$, 0°, 1 h	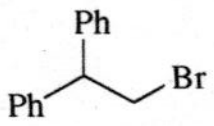 (62)	184
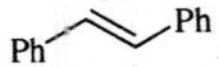	PMHS-Pd nanocomposite, C_6H_6, rt, 4 h	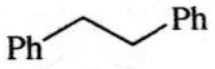 (94)	219

C_{14} 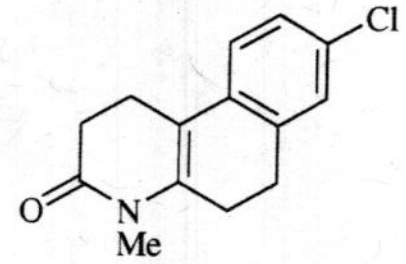	Et_3SiH, TFA	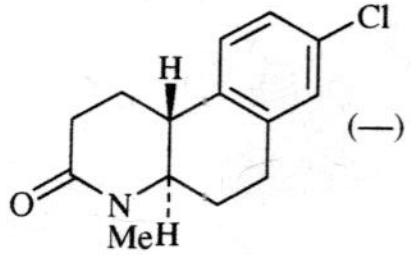 (—) + 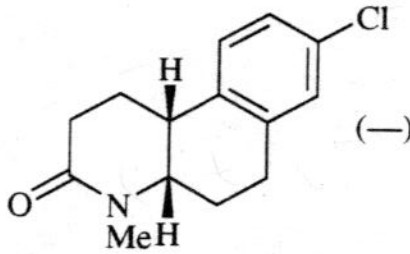 (—)	692
C_{15} 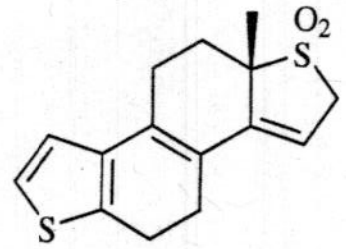	Et_3SiH, TFA, rt, 2 h	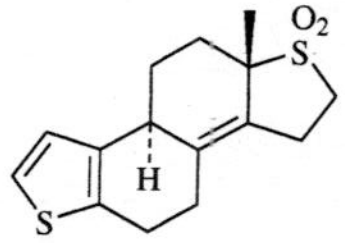 (68)	693
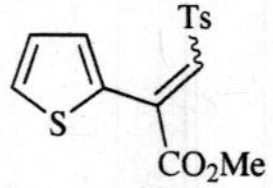	Ph_2SiH_2 (1.2 eq), $AlCl_3$ (1 eq), CH_2Cl_2, 20°, 15 h	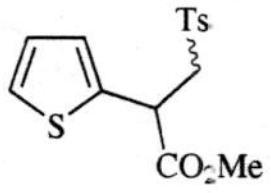 **I** (86)	373
	Et_3SiH (1.2 eq), $AlCl_3$ (1 eq), CH_2Cl_2, 20°, 15 h	**I** (86)	373
C_{15-16} 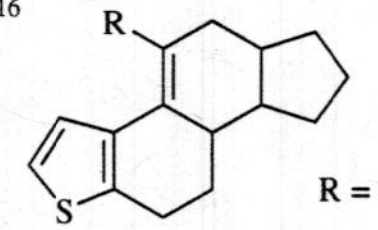 R = H, Me	Et_3SiH (1.1 eq), TFA (1.1 eq)	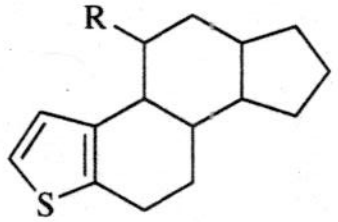 (—)	237
C_{16} 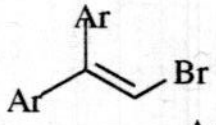 Ar = 4-$MeOC_6H_4$	Et_3SiH (1.5 eq), TFA (3 eq), $CHCl_3$, 0°, 1 h	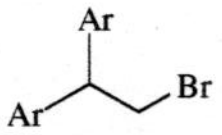 (88)	184
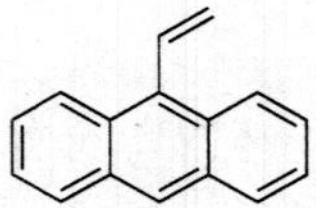	PMHS-Pd nanocomposite, C_6H_6, rt, 6 h	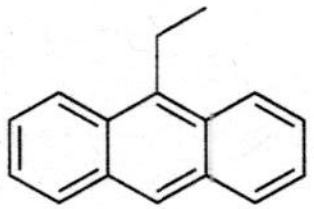 (92)	219

TABLE 1. ORGANOSILANE REDUCTION OF ALKENES (*Continued*)

Alkene	Conditions	Product(s) and Yield(s) (%)	Refs.
C_{16-18}			
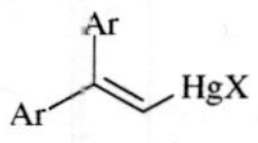 Ar = 4-MeOC$_6$H$_4$; X = Cl, Br, CF$_3$CO$_2$	Et$_3$SiH (1.5 eq), TFA (3 eq), CHCl$_3$, 0°, 1 h	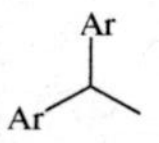 (—) + 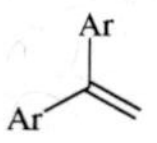 (—)	184
C_{17}			
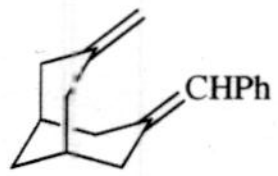	Et$_3$SiH (1.5 eq), H$_2$SO$_4$ (1.25 eq), CH$_2$Cl$_2$, rt, 6 h	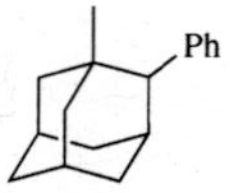 **I** (71) + 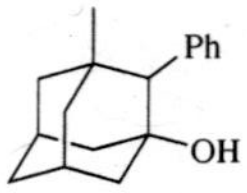 **II** (23)	243
	Et$_3$SiH (1.5 eq), H$_2$SO$_4$/carbon, n-C$_6$H$_{14}$, rt, 6 h	**I** (4) + **II** (47)	243
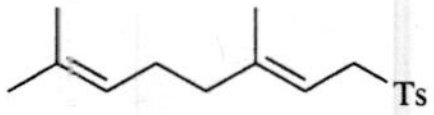	Et$_3$SiH (2 eq), TFA (2.5 eq), LiClO$_4$ (1 eq), i-PrNO$_2$, 20°, 20 h	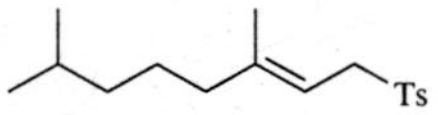 (7) + 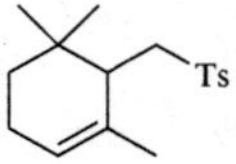 (93)	205
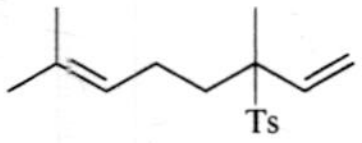	Et$_3$SiH (2 eq), TFA (2.5 eq), LiClO$_4$ (1 eq), i-PrNO$_2$, 20°, 20 h	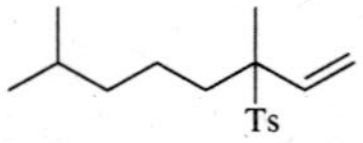 (87)	205
C_{18}			
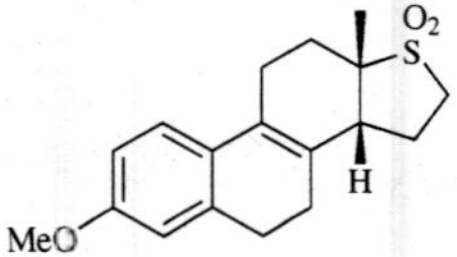	Et$_3$SiH, TFA, rt, 0°	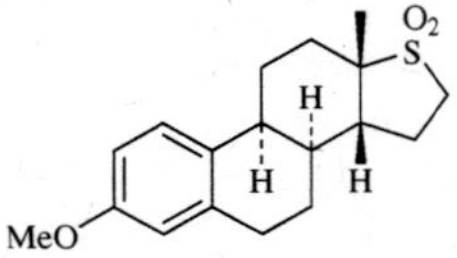 (75)	694
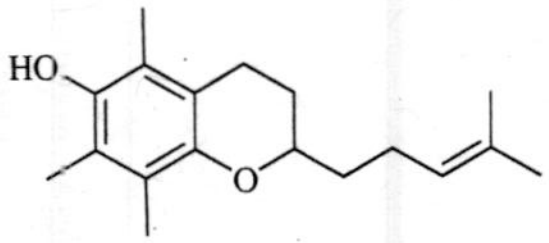	Et$_3$SiH (2 eq), TFA (2.5 eq), LiClO$_4$ (1 eq), i-PrNO$_2$, 20°, 20 h	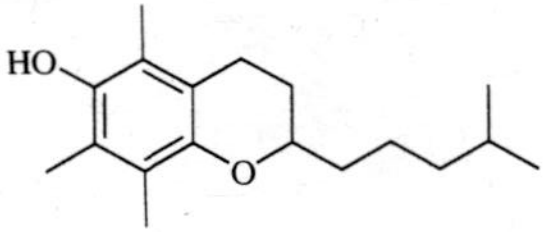 (85)	205

C_{18-21}			
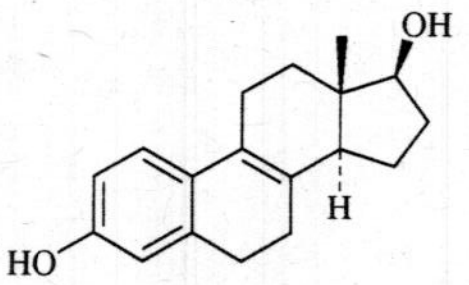	1. Et_3SiH, TFA 2. OH^-, H_2O	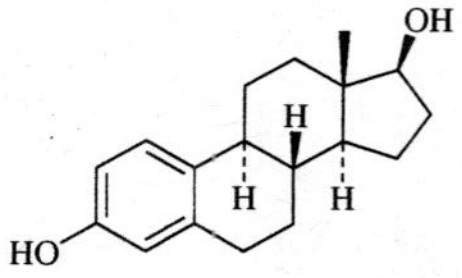 (96)	239
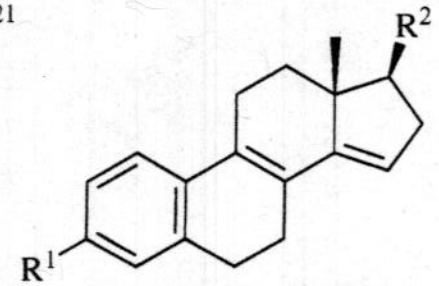	Et_3SiH, TFA, solvent	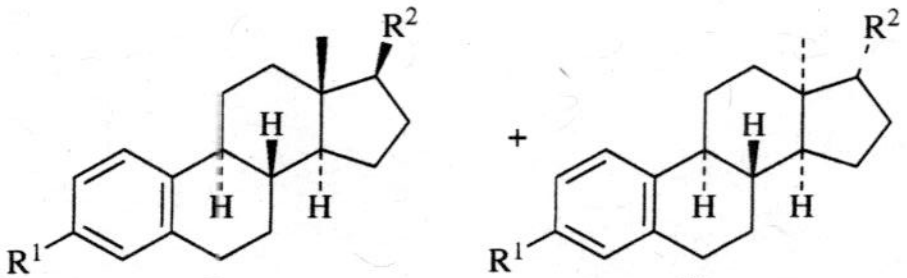 I + II	695
R^1 / R^2	Solvent	I / II	
OH / OH	C_6H_6	(31) / (62)	
OMe / OH	C_6H_6	(22) / (78)	
OMe / OH	CH_2Cl_2	(34) / (66)	
OMe / OAc	C_6H_6	(20) / (40)	
OMe / OAc	CH_2Cl_2	(64) / (22)	
C_{19}			
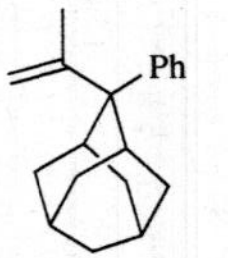	Et_3SiH (10 eq), $MeSO_3H$ (1 eq), CH_2Cl_2, –78°, 12 h; 0°, 10 h	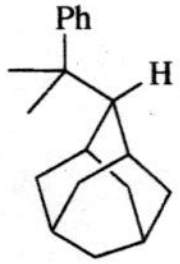 (95)	140
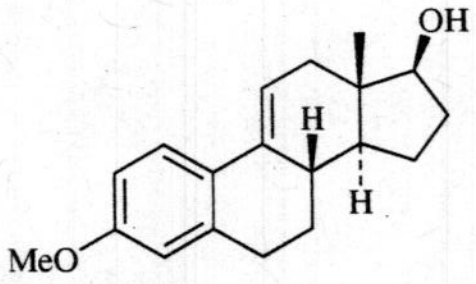	Et_3SiH, TFA	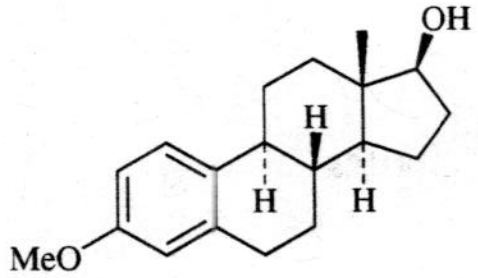 (60)	239

TABLE 1. ORGANOSILANE REDUCTION OF ALKENES (*Continued*)

Alkene	Conditions	Product(s) and Yield(s) (%)	Refs.
C_{19}			
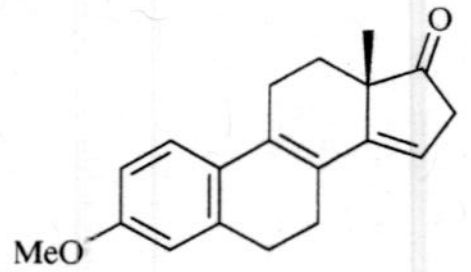	Et_3SiH, TFA	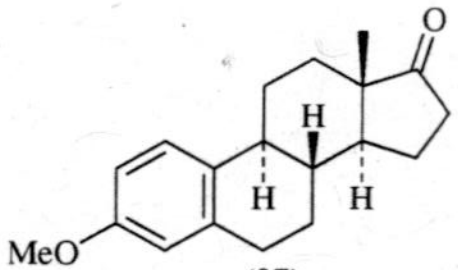 (27) + 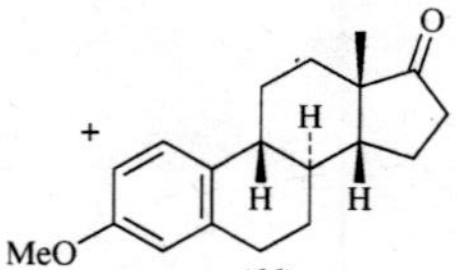 (66)	241
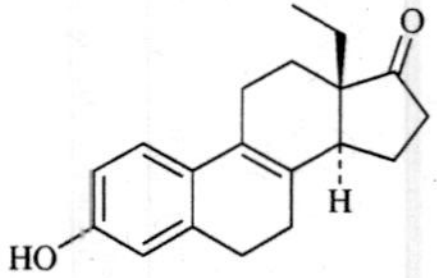	Et_3SiH, TFA	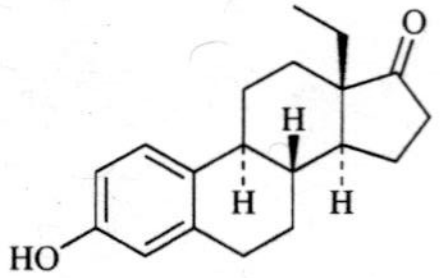 (70)	226
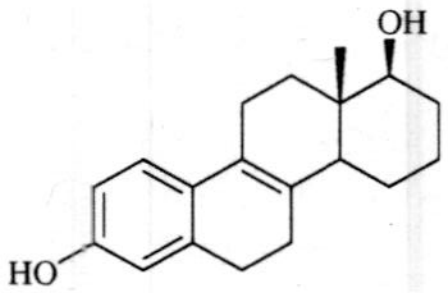	Et_3SiH, TFA	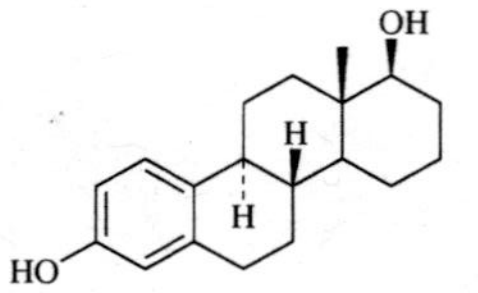 (—)	240
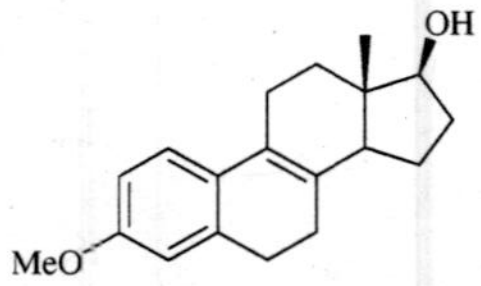	Et_3SiH, TFA	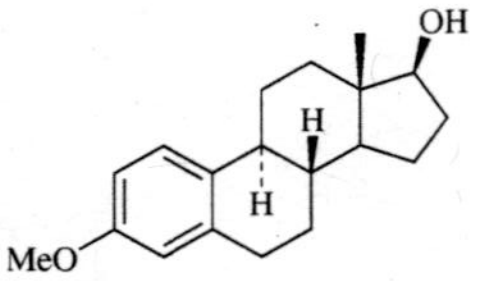 (60)	226
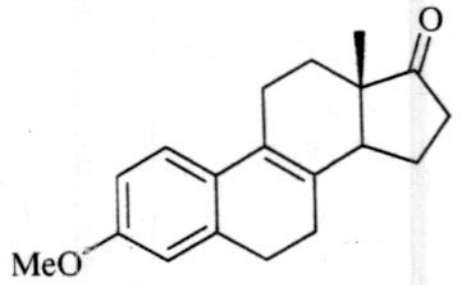	Et_3SiH, TFA	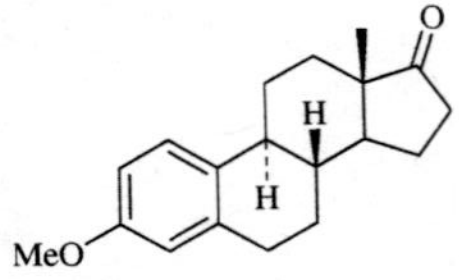 (60)	226

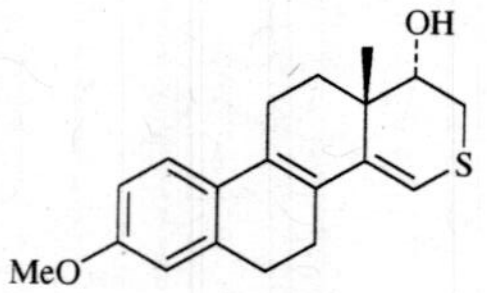	Et_3SiH (5 eq), TFA, CH_2Cl_2, rt, 70 h	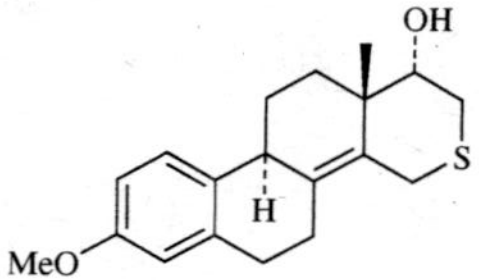 (60.3)	696
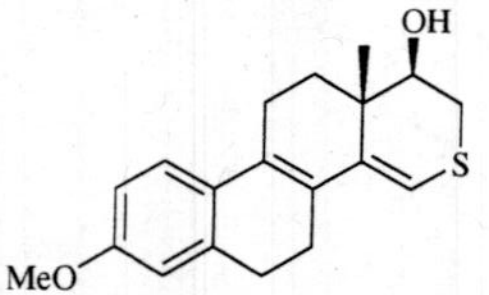	Et_3SiH (5 eq), TFA, CH_2Cl_2, rt, 40 h	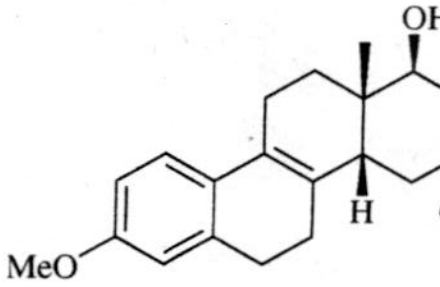 (30) + 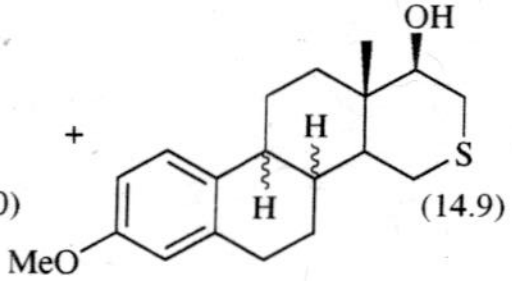 (14.9)	696
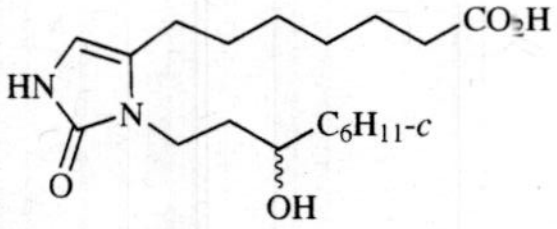	Et_3SiH (1.5 eq), TFA, 50°, 11 h	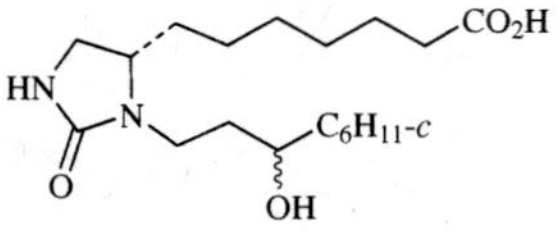 (89)	697
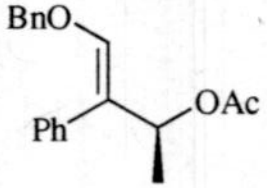	$PhMe_2SiH$ (1.2 eq), TFA, rt, 16 h	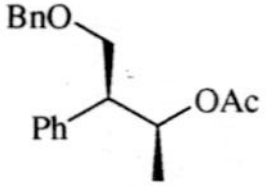 **I** + 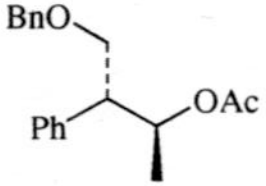 **II** **I** + **II** (73), **I**:**II** = 99:1	276
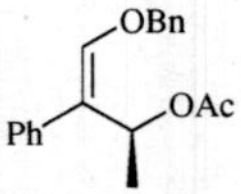	$PhMe_2SiH$ (1.2 eq), TFA, rt, 16 h	**I** + **II** (77), **I**:**II** = 24:76	276
C_{20} 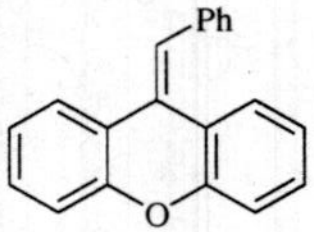	Et_3SiH, TFA, CH_2Cl_2, rt, 15 min	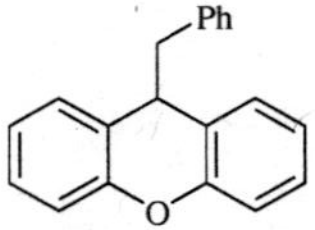 (94)	64

TABLE 1. ORGANOSILANE REDUCTION OF ALKENES (*Continued*)

Alkene	Conditions	Product(s) and Yield(s) (%)	Refs.
C_{20}			
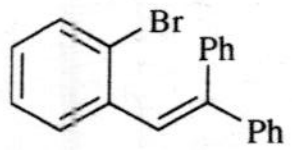	Et_3SiH, TFA	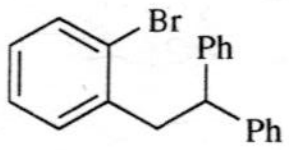 (—)	698
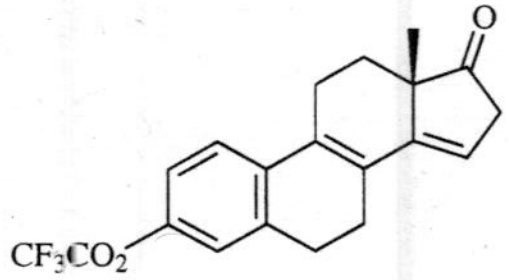	Et_3SiH, TFA	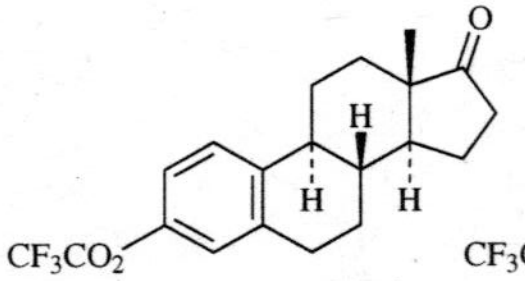 (11) + 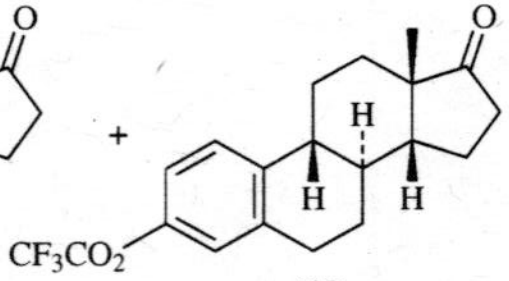 (66)	241
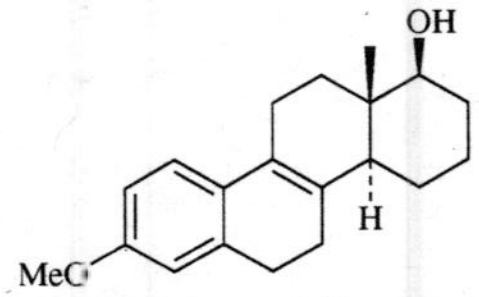	Et_3SiH, TFA, CH_2Cl_2	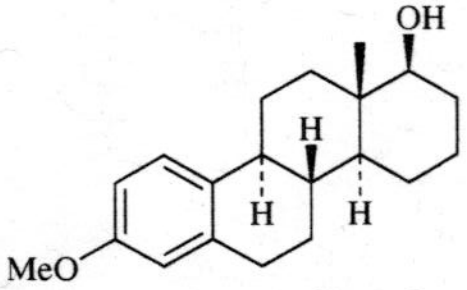 (55)	696
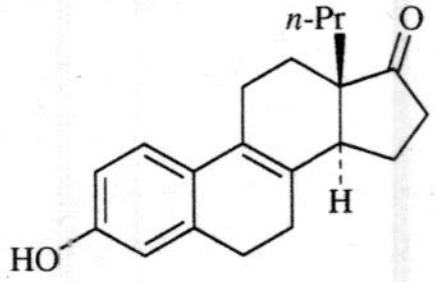	Et_3SiH, TFA	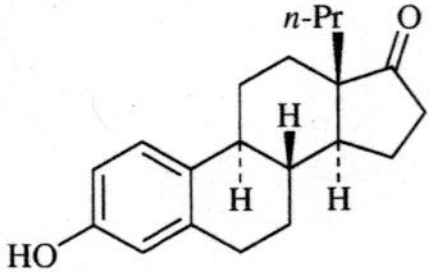 (70)	226
C_{21}			
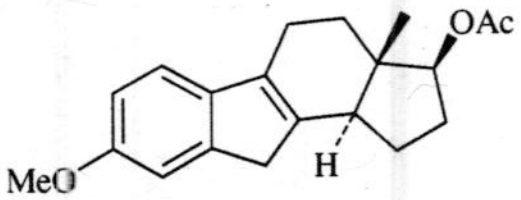	Et_3SiH, TFA	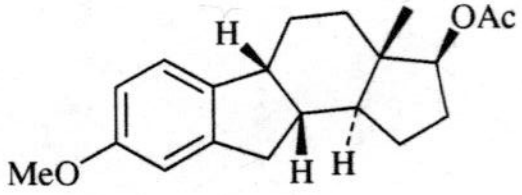 (65)	699
C_{22}			
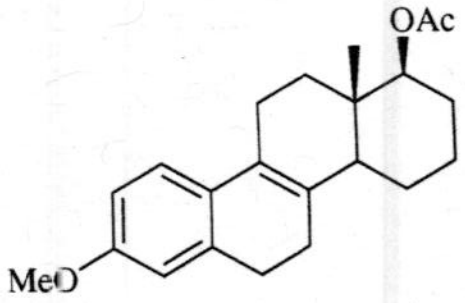	1. Et_3SiH, TFA 2. $NaBH_4$	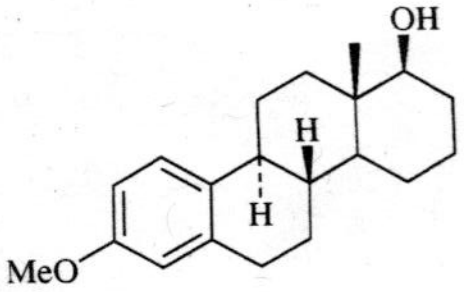 (—)	240

C_{23} 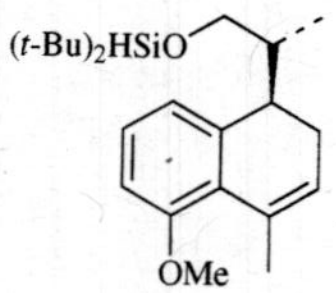	TBAF, TFA, high dilution	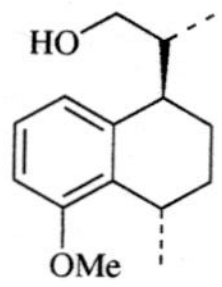 (65-75)	225
	Et_3SiH, TBAF	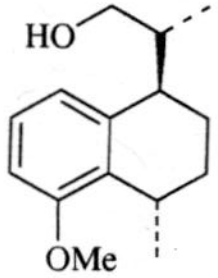 **I** (—) + 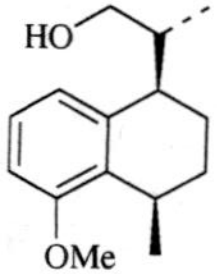 **II** (—) **I**:**II** = 3:2	225
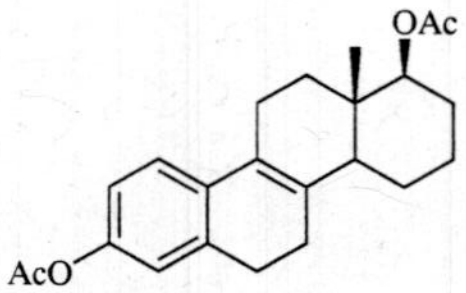	1. Et_3SiH TFA 2. $NaBH_4$	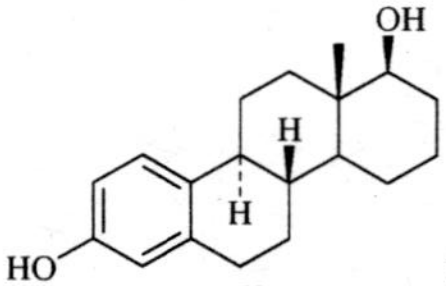 (60) + 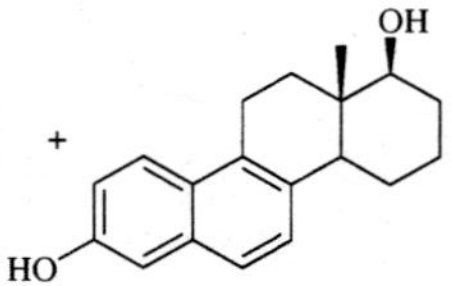 (—)	240
C_{24} 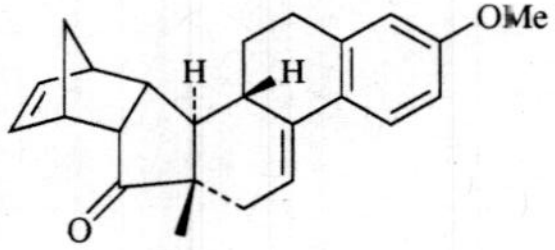	Et_3SiH (5 eq), TFA (10 eq), CH_2Cl_2, rt, 20 h	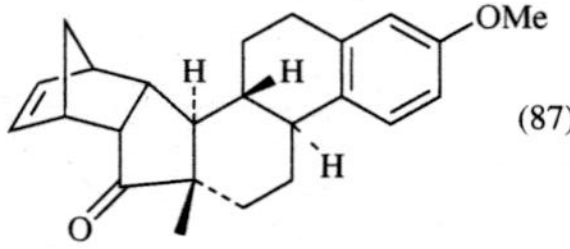 (87)	227
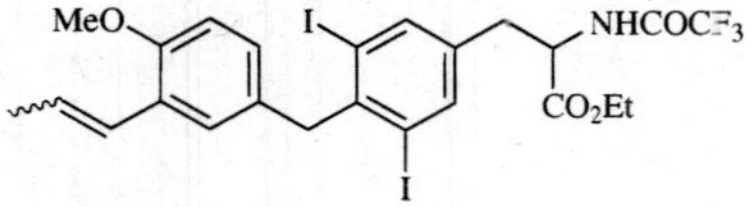	Et_3SiH, TFA, CH_2Cl_2, 50°, 18 h	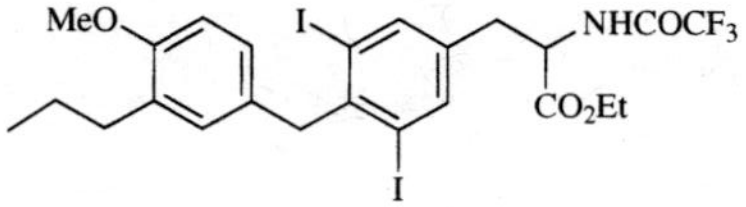 (27)	700

TABLE 1. ORGANOSILANE REDUCTION OF ALKENES (*Continued*)

Alkene	Conditions	Product(s) and Yield(s) (%)	Refs.
C_{25} 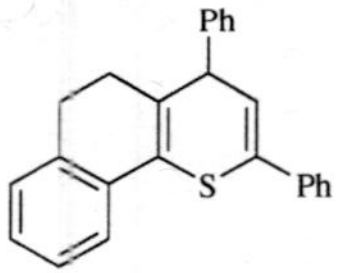	Et_3SiH (2 eq), TFA, rt, 7 min	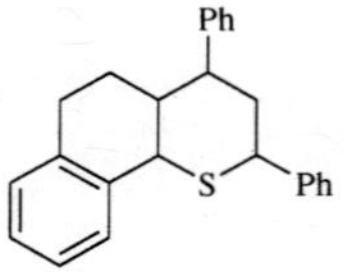 (99)[b]	228
C_{28} 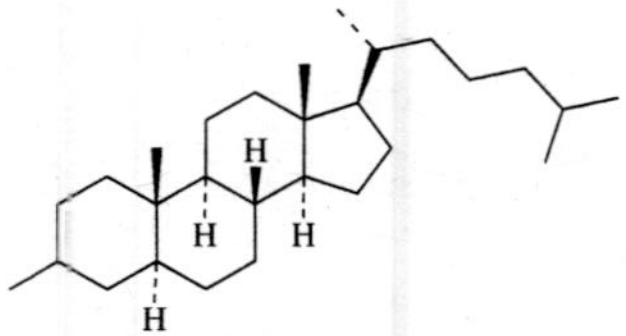	Ph_3SiH (1.2 eq), TFA (6.5 eq), CH_2Cl_2, rt, 24 h	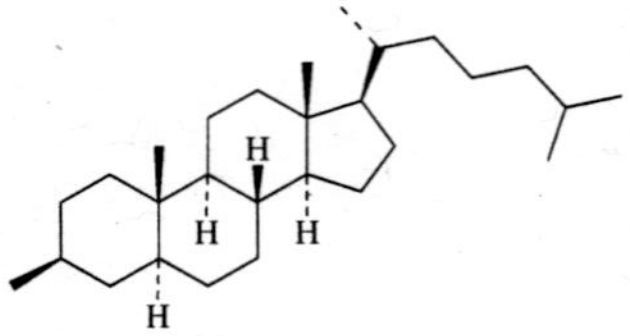 (66)	127
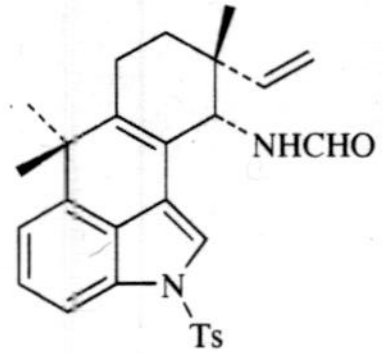	Et_3SiH, TFA	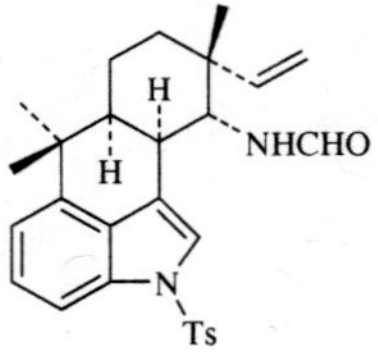 (2) + 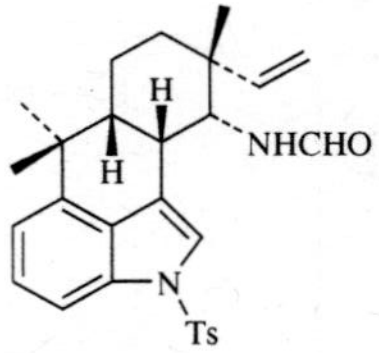 (58) + 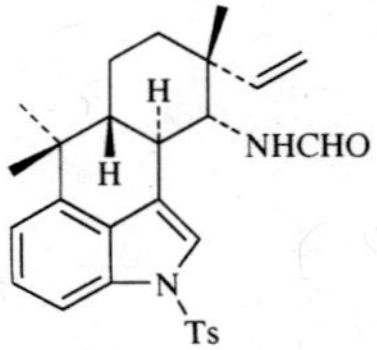 (19)	701

[a] The yield was determined by NMR spectroscopy.
[b] The product was a single diastereomer of unknown stereochemistry.

TABLE 2. ORGANOSILANE REDUCTION OF ALKYNES

Alkyne	Conditions	Product(s) and Yield(s) (%)	Refs.
C_5 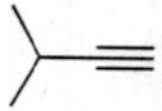	Et_3SiH (2 eq), TFA (3 eq), rt, 120 h	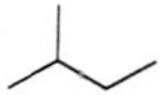 (5)	244
C_6 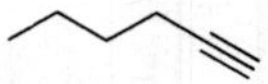	Et_3SiH (10 eq), $AlCl_3$ (0.5 eq), HCl, CH_2Cl_2, 20°, 30 min	 (44) + 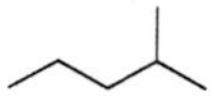 (7)	146
C_8 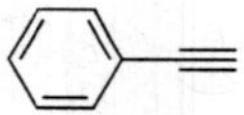	Et_3SiH (5 eq), $AlCl_3$ (0.2 eq), HCl, CH_2Cl_2, 20°, 90 min	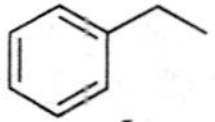 **I** (40–50)	146
	Et_3SiH (5 eq), $AlCl_3$ (0.2 eq), HCl, CH_2Cl_2, –40°, 300 min	**I** (55)	146
	Et_3SiH (2 eq), TFA (3 eq), rt, 120 h	**I** (30)	244
C_9 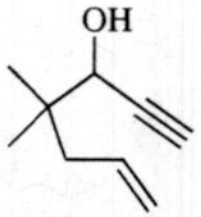	Et_3SiH (1.2 eq), $Pd(dppe)Cl_2$ (5 mol%), HCO_2H (1.2 eq), MeC_6H_5, 70°, 2 h	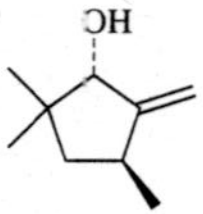 (52) + 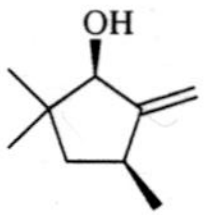 (26)	246
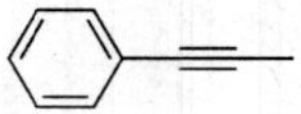	Et_3SiH (2 eq), TFA (3 eq), rt, 120 h	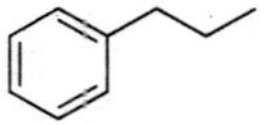 (7)	244
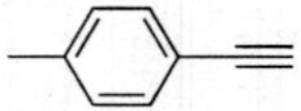	Et_3SiH (2 eq), TFA (3 eq), rt, 120 h	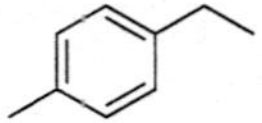 (21)	244
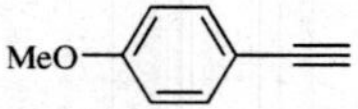	Et_3SiH (2 eq), TFA (3 eq), rt, 120 h	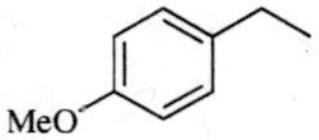 (20)	244

TABLE 2. ORGANOSILANE REDUCTION OF ALKYNES (*Continued*)

Alkyne	Conditions	Product(s) and Yield(s) (%)	Refs.
C_{10} 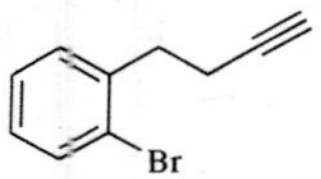	Et_3SiH (2 eq), $(Ph_3P)_4Pd$ (3 mol%), Cs_2CO_3 (2 eq), DMF, 60°, 20 h	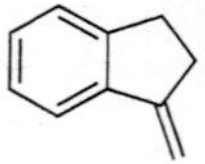 (62)	249
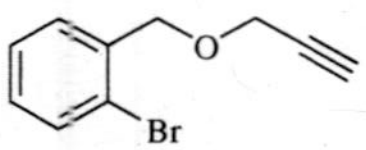	Et_3SiH (2 eq), $(Ph_3P)_4Pd$ (3 mol%), Cs_2CO_3 (2 eq), DMF, 80°, 4 h	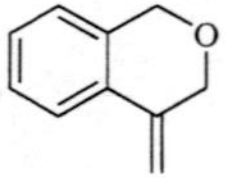 (54)	249
C_{11} 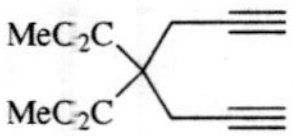	Et_3SiH (10 eq), $Pd_2(dba)_3$•$CHCl_3$ (2.5 mol%), (*o*-tol)$_3$P (5 mol%), HOAc, rt, 25 min	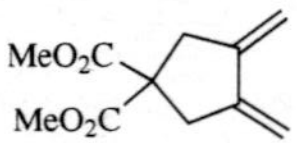 (34)	248
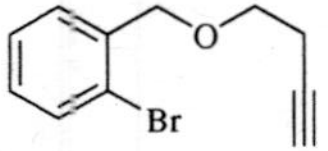	Et_3SiH (2 eq), $(Ph_3P)_4Pd$ (3 mol%), Cs_2CO_3 (2 eq), DMF, 80°, 3 h	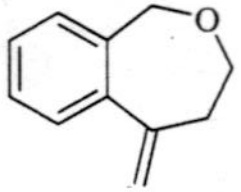 (48)	249
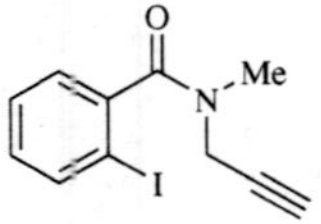	Et_3SiH (2 eq), $(Ph_3P)_4Pd$ (3 mol%), Cs_2CO_3 (2 eq), DMF, 100°, 12 h	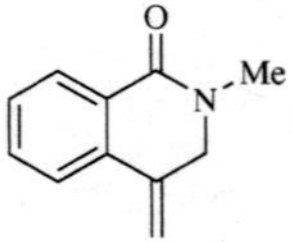 (82)	249
C_{12} 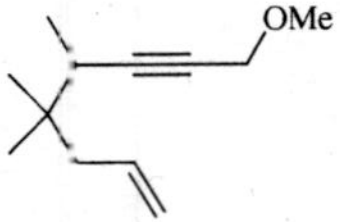	PMHS (10 eq), $Pd_2(bpa)_3$•$CHCl_3$ (2.5 mol%), (*o*-tol)$_3$P (5 mol%), HOAc (1 eq), $ClCH_2CH_2Cl$, rt	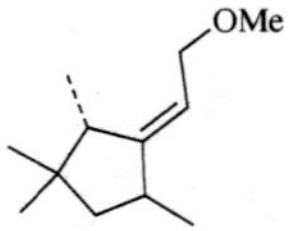 (73)	245
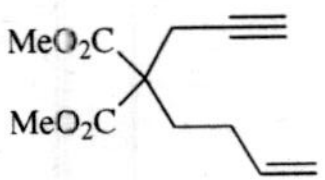	PMHS (10 eq), $Pd_2(bpa)_3$•$CHCl_3$ (5 mol%), HOAc (1 eq), C_6H_6, rt	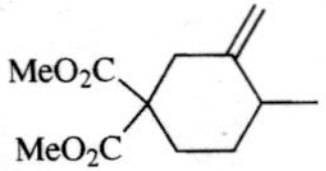 (79)	245

C13			
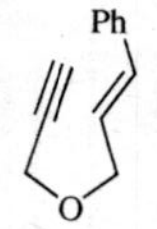	Et_3SiH (1.2 eq), $Pd(dppe)Cl_2$ (5 mol%), HCO_2H (1.2 eq), MeC_6H_5	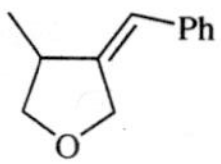 (64)	246
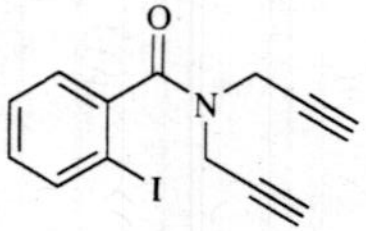	Et_3SiH (2 eq), $(Ph_3P)_4Pd$ (3 mol%), Cs_2CO_3 (2 eq), DMF, 80°, 12 h	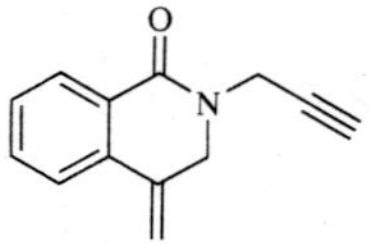 (80)	249
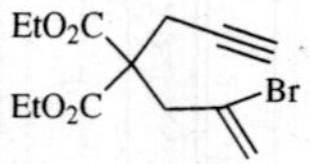	Et_3SiH (2 eq), $(Ph_3P)_4Pd$ (3 mol%), Cs_2CO_3 (2 eq), DMF, 80°, 4 h	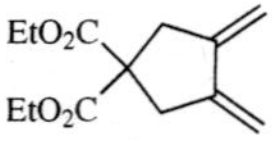 (72)	249
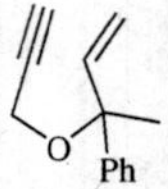	Et_3SiH (1.2 eq), $Pd(dppe)Cl_2$ (5 mol%), HCO_2H (1.2 eq), 1,4-dioxane, 70°, 10 h	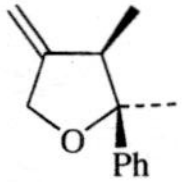 (73) + 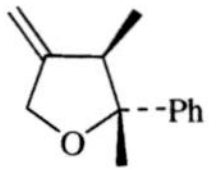 (8)	246
C14			
	Et_3SiH (2 eq), TFA (3 eq), rt, 120 h	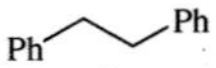 **I** (3)	244
	Et_3SiH (12 eq), $AlCl_3$ (1.2 eq), HCl, CH_2Cl_2, 20°, 195 min	**I** (35)	146
	Et_3SiH (97 eq), $AlCl_3$ (2.7 eq), HCl, CH_2Cl_2, 20°, 170 min	**I** (50)	146
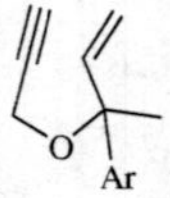	Et_3SiH (1.2 eq), $Pd(dppe)Cl_2$ (5 mol%), HCO_2H (1.2 eq), 1,4-dioxane, 70°, 10 h	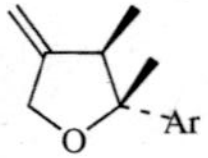	246

Ar	Solvent	Temp	Time	Yield	Ref.
4-$MeOC_6H_4$	1,4-dioxane	70°	10 h	(60)	246
2-MeC_6H_4	MeC_6H_5	80°	3 h	(88)	

TABLE 2. ORGANOSILANE REDUCTION OF ALKYNES (*Continued*)

Alkyne	Conditions	Product(s) and Yield(s) (%)	Refs.
C_{14} 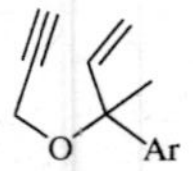 Ar = 4-MeC_6H_4	Et_3SiH (10 eq), $Pd_2(dba)_3$ (10 mol%), PPh_3 (10 mol%), HOAc (2 eq), MeC_6H_5, 50°, 2 h	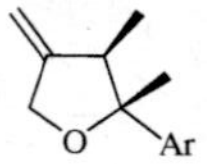 (70) + 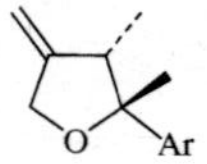 (8)	247
C_{15} 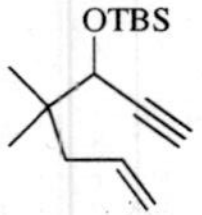	Et_3SiH (1.2 eq), Pd(dppe)Cl_2 (5 mol%), HCO_2H (1.2 eq), MeC_6H_5, 60°, 2 h	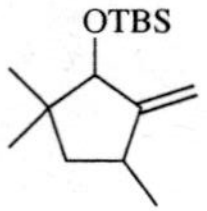 (0)	246
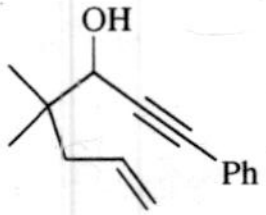	Et_3SiH (1.2 eq), Pd(dppe)Cl_2 (5 mol%), HCO_2H (1.2 eq), MeC_6H_5, 70°, 2 h	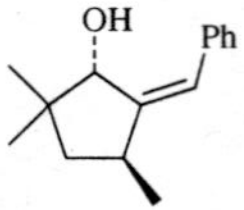 (50) + 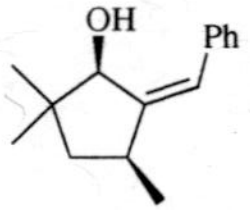 (5)	246
 Ar = 4-MeC_6H_4	Et_3SiH (10 eq), $Pd_2(dba)_3$ (10 mol%), PPh_3 (10 mol%), HOAc (2 eq), MeC_6H_5, 50°, 2 h	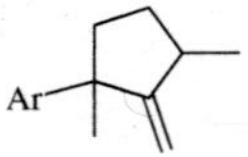 (5) + 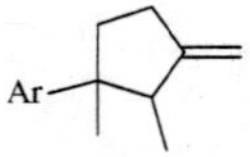 (47)	247
C_{16-18} 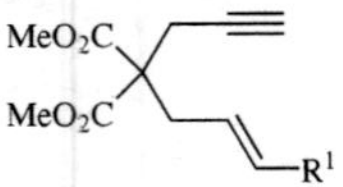	Silane (10 eq), $Pd_2(bpa)_3$•$CHCl_3$ (2.5 mol%), $(o\text{-tol})_3P$ (5 mol%), HOAc (1 eq), C_6H_6, rt	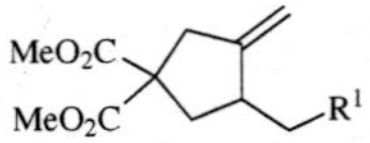	245

R^1	Silane	
C_6H_{11}	PMHS	(96)
CH_2OTBS	PMHS	(88)
CH_2OTBS	Et_3SiH	(59)
$CH(OAc)_2$	PMHS	(75)
$CH(OAc)_2$	Et_3SiH	(83)

C_{17}				
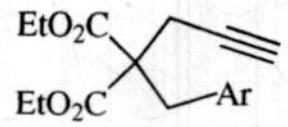 Ar = 2-BrC$_6$H$_4$	Et_3SiH (2 eq), $(Ph_3P)_4Pd$ (3 mol%), Cs_2CO_3 (2 eq), DMF, 80°, 4 h	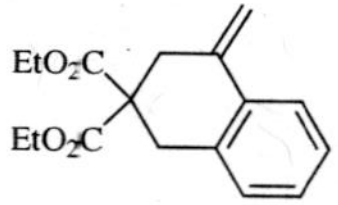	(85)	249
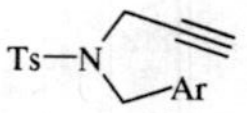 Ar = 2-BrC$_6$H$_4$	Et_3SiH (2 eq), $(Ph_3P)_4Pd$ (3 mol%), Cs_2CO_3 (2 eq), DMF, 100°, 12 h	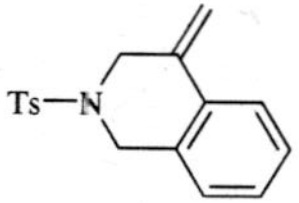	(65)	249
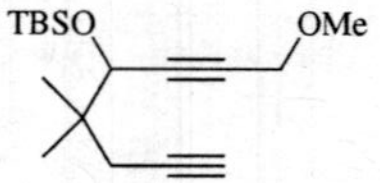	Et_3SiH (1.0 eq), $Pd_2(dba)_3$•$CHCl_3$ (2.5 mol%), (o-tol)$_3$P (5 mol%), HOAc, rt, 60 min	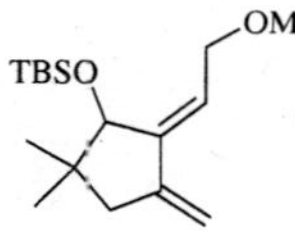	(70)	248
C_{18}				
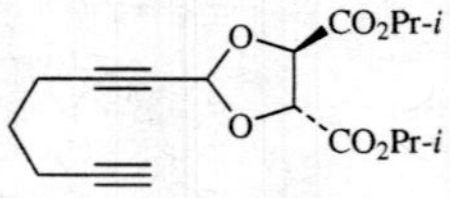	Et_3SiH (1.0 eq), $Pd_2(dba)_3$•$CHCl_3$ (2.5 mol%), (o-tol)$_3$P (5 mol%), HOAc, rt, 25 min	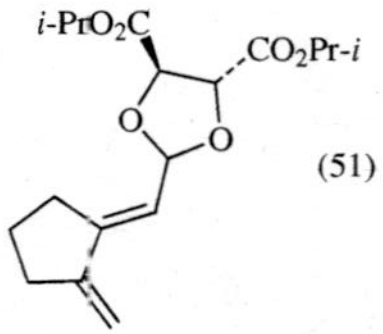	(51)	248
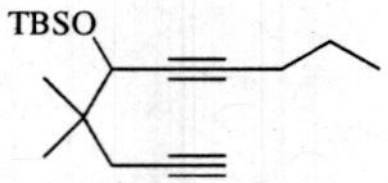	Et_3SiH (1.0 eq), $Pd_2(dba)_3$•$CHCl_3$ (2.5 mol%), (o-tol)$_3$P (5 mol%), HOAc, rt, 15 min	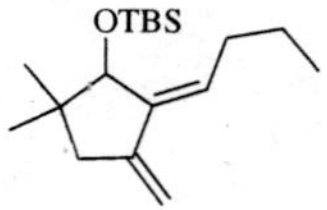	(60)	248
C_{19}				
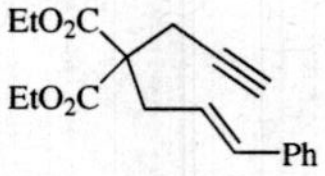	Et_3SiH (1.2 eq), Pd(dppe)Cl_2 (5 mol%), HCO_2H (1.2 eq), 1,4-dioxane, 60°, 2 h	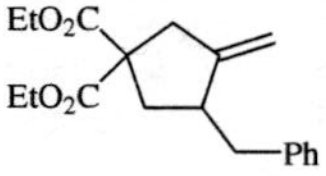	(85)	246

TABLE 2. ORGANOSILANE REDUCTION OF ALKYNES (*Continued*)

Alkyne	Conditions	Product(s) and Yield(s) (%)	Refs.
C_{19} 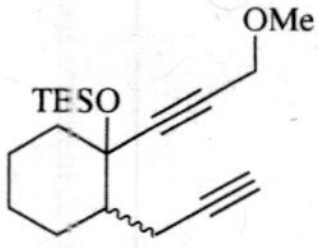	Et_3SiH (10 eq), $Pd_2(dba)_3$•$CHCl_3$ (2.5 mol%), (*o*-tol)$_3$P (5 mol%), HOAc, rt, 15 min	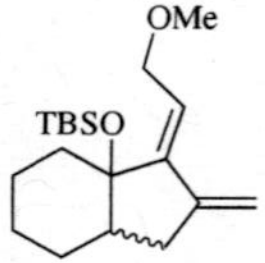 (86)	248
C_{20} 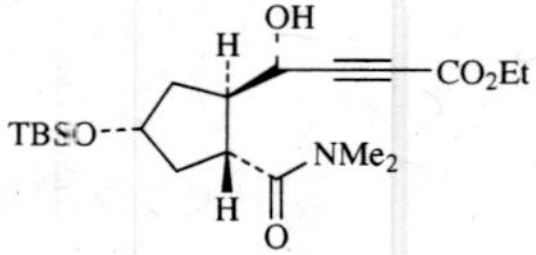	$(EtO)_3SiH$, $Cp^*Ru(NCMe)_3PF_6$, CH_2Cl_2	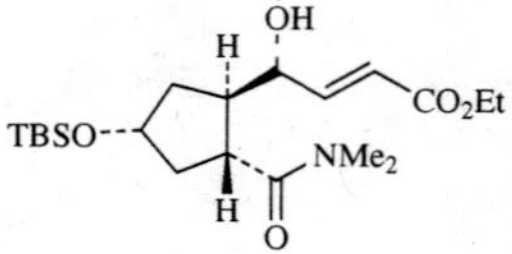 (92)	477
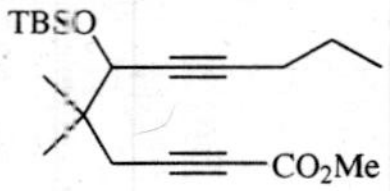	Et_3SiH (10 eq), $Pd_2(dba)_3$•$CHCl_3$ (2.5 mol%), (*o*-tol)$_3$P (5 mol%), HOAc, rt, 20 min	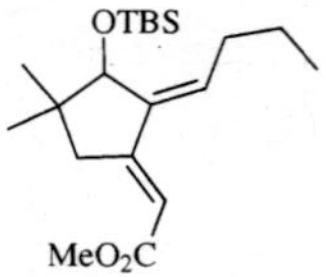 (95)	248
C_{20} 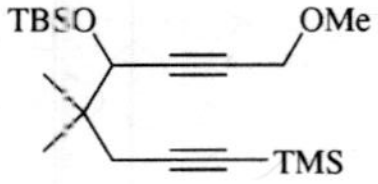	Et_3SiH (10 eq), $Pd_2(dba)_3$•$CHCl_3$ (2.5 mol%), (*o*-tol)$_3$P (5 mol%), HOAc, rt, 3 min	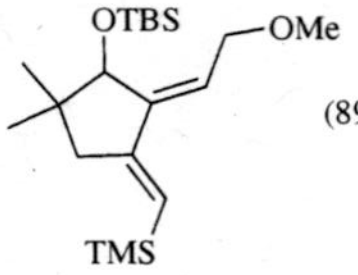 (89)	248
C_{21} 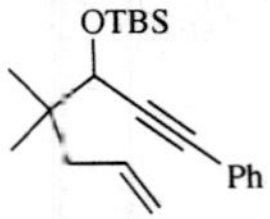	Et_3SiH (1.2 eq), $Pd(dppe)Cl_2$ (5 mol%), HCO_2H (1.2 eq), MeC_6H_5, 60°, 2 h	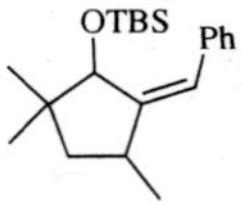 (0)	246

C_{23}	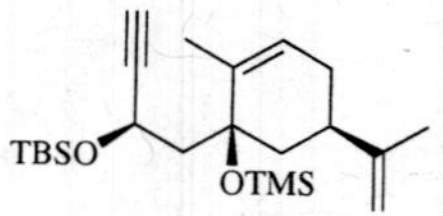	PMHS (10 eq), $Pd_2(dbpa)_3 \cdot CHCl_3$ (2.5 mol%), $(o\text{-tol})_3P$ (5 mol%), HOAc (1 eq), $ClCH_2CH_2Cl$, rt	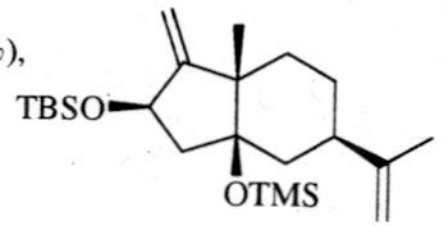	(90)	241
C_{25}	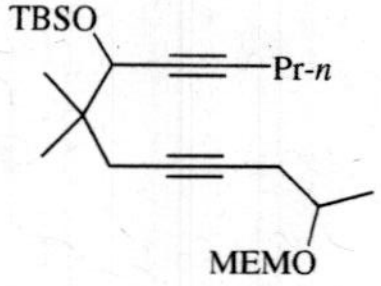	Et_3SiH (10 eq), $Pd_2(dba)_3 \cdot CHCl_3$ (2.5 mol%), $(o\text{-tol})_3P$ (5 mol%), HOAc, rt, 30 min	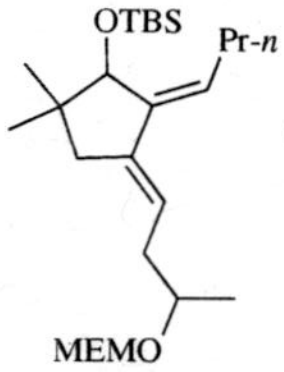	(83)	248
C_{30}	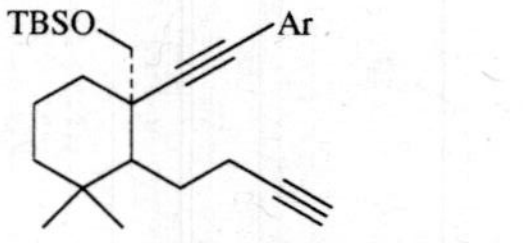 Ar = 2,5-$(MeO)_2$-4-MeC_6H_2	Et_3SiH (2 eq), $(dba)_3PdCl_2 \cdot CHCl_3$ (2.5 mol %), tri(2-furyl)phosphine (10 mol%), HCO_2H (2 eq), MeC_6H_5, 80°	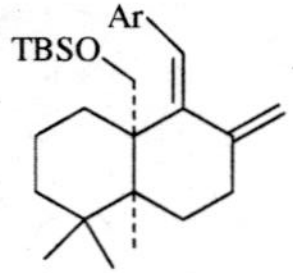	(79)	702
C_{40}	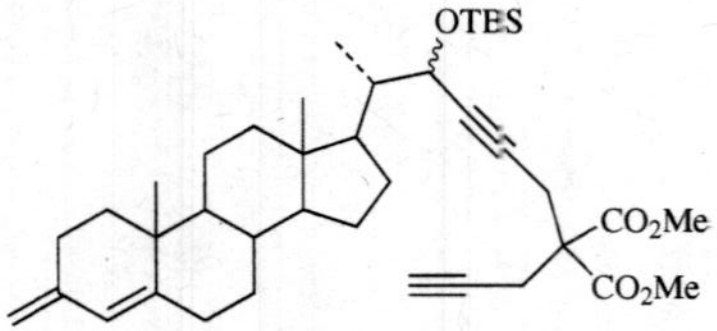	Et_3SiH (10 eq), $Pd_2(dba)_3 \cdot CHCl_3$ (2.5 mol%), $(o\text{-tol})_3P$ (5 mol%), HOAc, rt, 90 min	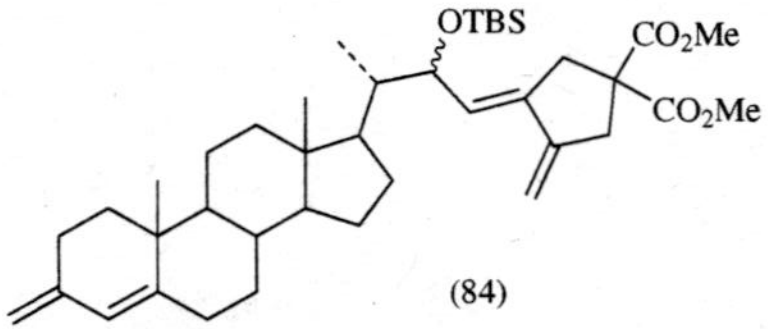	(84)	248

TABLE 3. ORGANOSILANE REDUCTION OF AROMATIC HYDROCARBONS

Aromatic Hydrocarbon	Conditions	Product(s) and Yield(s) (%)	Refs.
C_4 	Et_3SiH (3 eq), $AlCl_3$ (1.3 eq), CH_2Cl_2, 20°, 1.5 h	 (—) +  (—) I II	259
	Et_3SiH (3 eq), TFA (30 eq), 20°, time	I + II	261
	Time	I II	
	36 min	(10) (17)	
	1026 min	(23) (42)	
	3993 min	(28) (48)	
	Et_3SiH (3 eq), TFA (30 eq), $LiClO_4$ (1%), 20°, time	I + II	261
	Time	I II	
	11 min	(20) (31)	
	66 min	(30) (49)	
	232 min	(31) (53)	
	Et_3SiH (3 eq), BF_3•OEt_2 (0.057 mol%), 2 h	I (30) + II (50)	210
	R_3SiH (x eq), HCl (xs), $AlCl_3$ (0.3 eq), CH_2Cl_2, rt	I	213
	R_3Si x Time		
	Et_3Si 5 15 h	(66)	
	Ph_2HSi 9.4 1 h	(39)	
	PhH_2Si 9.4 15 h	(43)	
	$EtCl_2Si$ 1.7 15 h	(0)	
C_{4-28} 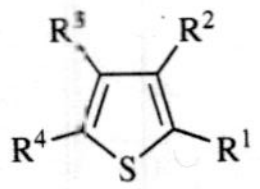	Et_3SiH (x eq), TFA (y eq), 50°	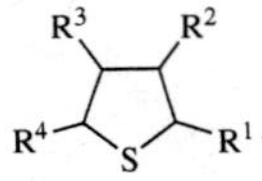	257

R^1	R^2	R^3	R^4	x	y	Time	
H	H	H	H	2	9	80 h	(60)
Cl	H	H	Cl	2	9	50 h	(—)
Me	H	H	H	2	7	20 h	(80)
Me	Me	Br	H	2	9	50 h	(—)
Et	H	H	Cl	3	8	100 h	(—)
CO_2H	H	H	Me	3	8	65 h	(—)
Me	Me	H	Me	2	7	20 h	(80)
$(CH_2)_4CO_2H$	H	H	H	2	7	30 h	(65)
H	Ph	H	H	3	13	20 h	(70)
$(CH_2)_5CO_2Me$	H	H	H	2	9	30 h	(65)
t-Bu	H	H	*t*-Bu	3	8	100 h	(15)
H	Ph	Ph	H	5	20	50 h	(70)
Ph	H	H	Ph	5	20	50 h	(—)
Ph	Ph	Ph	Ph	6	27	50 h	(—)

C_5

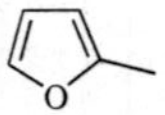

Et_3SiH (x eq), TFA (y eq), BF_3•OEt_2 (z eq), 20°, (addition of TFA/BF_3•OEt_2 to mixture of the furan and Et_3SiH)

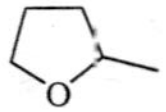

I

x	y	z	Time	**I**
3	5	0	60 min	(4)
4	8	0.2	4 min	(18)
4	17	0.4	4 min	(31)

211

Et_3SiH (x eq), TFA (y eq), BF_3•OEt_2 (z eq), 20°, 4 min

I

x	y	z	**I**
4	17	0.4	(50)
10	50	0	(72)
10	50	0.4	(70)

211

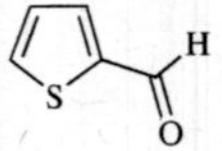

Et_3SiH (4 eq), TFA (8 eq), 50°, 30 h

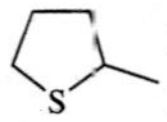

(65) 257

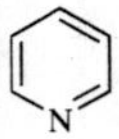

Ph_2SiH_2(2 eq), Cp_2TiMe_2 (10 mol%), 80°, 8 h

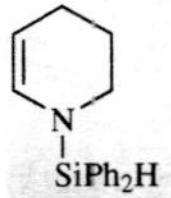

(94) 265

TABLE 3. ORGANOSILANE REDUCTION OF AROMATIC HYDROCARBONS (*Continued*)

Aromatic Hydrocarbon	Conditions	Product(s) and Yield(s) (%)	Refs.
C_6 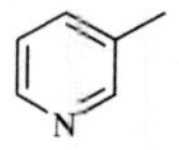	$PhMeSiH_2$ (1.5 eq), Cp_2TiMe_2 (10 mol%), 6 d	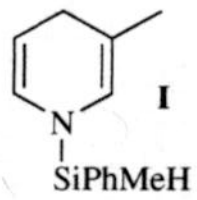 **I** + 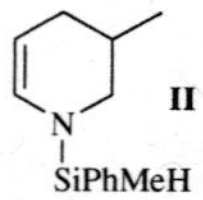 **II** + 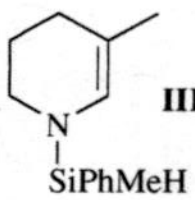 **III**	264
	Temp	**I** / **II** / **III**	
	0°	(88) / (0) / (12)	
	24°	(85) / (0) / (15)	
	50°	(27) / (17.2) / (56)	
	60°	(19) / (27) / (49)	
	80°	(3) / (57) / (39)	
	$PhMeSiH_2$ (1.5 eq), Cp_2TiMe_2 (10 mol%), 80°, 6 d	**III** (30)	264
	$PhMeSiH_2$ (2 eq), Cp_2TiMe_2 (10 mol%), 80°, 8 h	**II** (51) + **III** (34)	265
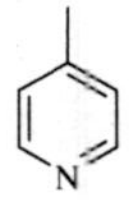	$PhMeSiH_2$ (1.5 eq), Cp_2TiMe_2 (10 mol%), 80°, 6 d	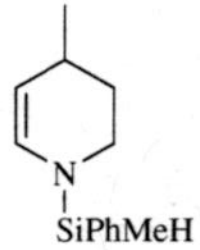 **I** (53)	264
	Ph_2SiH_2 (2 eq), Cp_2TiMe_2 (10 mol%), 80°, 8 h	**I** (83)	265
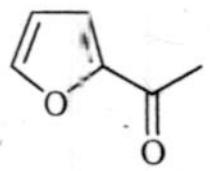	Et_3SiH (10 eq), TFA (x eq), BF_3•OEt_2 (y eq), 20°	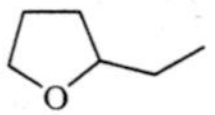	211
	x / y / Time		
	30 / 0 / 20 h	(65)	
	50 / 0 / 2 h	(79)	
	28 / 0.6 / 5 min	(80)	

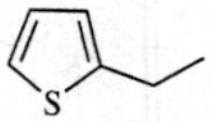	Et_3SiH (3 eq), TFA (30 eq), 20°, 40 min	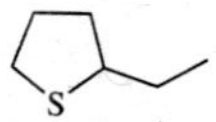 **I** (90)	261
	Et_3SiH (3 eq), TFA (10 eq), *p*-TsOH (10 eq), 20°, 40 min	**I** (80)	261
	Et_3SiH (3 eq), TFA (10 eq), *p*-TsOLi (x eq), $LiClO_4$ (y eq), 20°, 40 min	**I** (80) x y 1 0 (55) 2 0 (60) 3 0 (65) 0 0.95 (78) 0 1 (82)	261
	Et_3SiH (1.1–1.2 eq), HCO_2H, KU-1, 55°, 5 h	**I** (93)	208
	Et_3SiH (1.1–1.2 eq), HCO_2H, HOAc, KU-1, 55°, 5 h	**I** (67)	208
	Et_3SiH (3 eq), BF_3•OEt_2 (0.057 mol%), 50°, 4 h	**I** (80)	210
	Et_3SiH (3 eq), $AlCl_3$ (0.3 eq), CH_2Cl_2, 20°, 10 min	**I** (80)	259
	Et_3SiH (3 eq), HCl (xs), $AlCl_3$ (0.3 eq), CH_2Cl_2, rt, 10 min	**I** (80)	213
	$EtCl_2SiH$ (2 eq), $AlBr_3$ (3 eq), PTSA (3 eq), 80°, 10 h	**I** (75)	192
	Et_3SiH (5 eq), TFA (10 eq), 50°, 50 h	**I** (55)	257
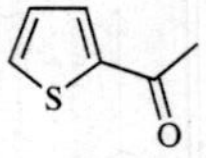	Et_3SiH (3 eq), $AlCl_3$ (1.3 eq), CH_2Cl_2, 20°, 1.5 h	**I** (70)	259
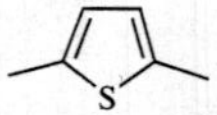	Et_3SiH (3 eq), HCl (xs), $AlCl_3$ (0.3 eq), CH_2Cl_2, rt, 90 min	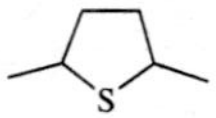 (76)	213

TABLE 3. ORGANOSILANE REDUCTION OF AROMATIC HYDROCARBONS (*Continued*)

Aromatic Hydrocarbon	Conditions	Product(s) and Yield(s) (%)	Refs.
C_7			
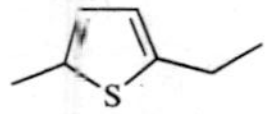	Et_3SiH (3 eq), $AlCl_3$ (0.3 eq), CH_2Cl_2, 20°, 1.5 h	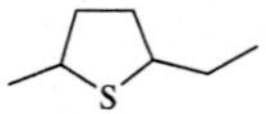 (75)	259
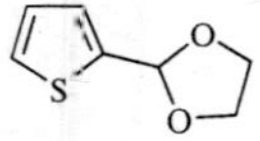	Et_3SiH (2–5 eq), TFA (7–9 eq), 55°, 15 h	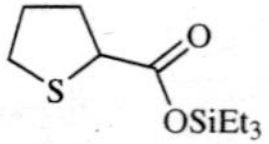 (45)	260
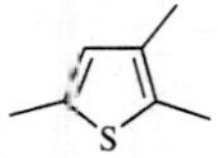	R_3SiH (x eq), HCl (xs), $AlCl_3$ (0.3 eq), CH_2Cl_2, rt	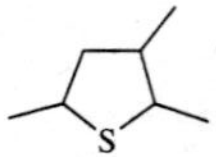	213
	R_3 — Time — x		
	Et_3 — 1.5 h — 2.7	(61)	
	Ph_2H — 15 h — 9.4	(33)	
	$EtCl_2$ — 15 h — 4.3	(0)	
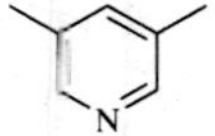	$PhMeSiH_2$ (1.5 eq), Cp_2TiMe_2 (10 mol%), 80°, 8 d	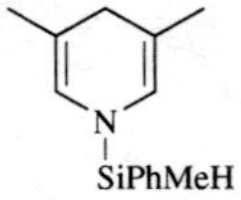 (60)	264
	Ph_2SiH_2 (2 eq), Cp_2TiMe_2 (10 mol%), 80°, 8 h	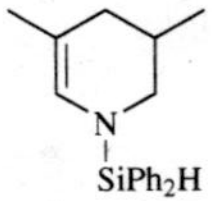 (70)	265
C_8			
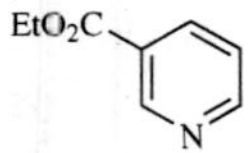	$PhMeSiH_2$ (1.5 eq), Cp_2TiMe_2 (10 mol%), 80°, 3 h	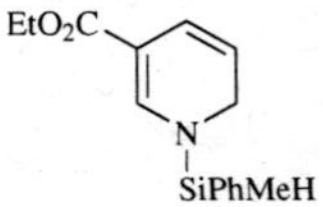 (90)	264
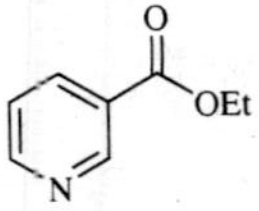	Ph_2SiH_2 (2 eq), Cp_2TiMe_2 (10 mol%), 80°, 8 h	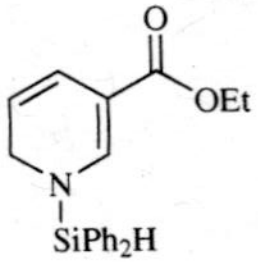 (100)	265

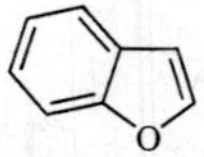	Et_3SiH (1.5 eq), TFA (x eq)	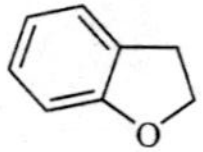	251
	x / Temp / Time		
	2.5 / 70° / 3 h	(54)	
	1.5 / 60° / 3 h	(14)	
	3.5 / 60° / 3 h	(48)	
	5 / 60° / 3 h	(66)	
	Et_3SiH (1.5 eq), CF_3CO_2D (2.5 eq), 60°, 7 h	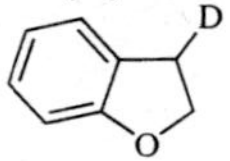 (47)	251
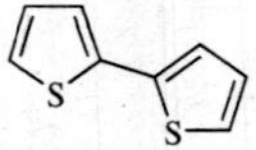	Et_3SiH (5 eq), TFA (15 eq), 20–50°, 1.5h	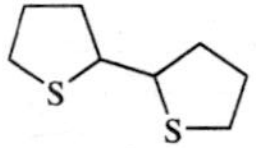 (—)	703
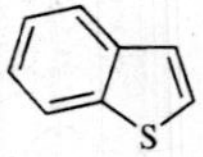	Et_3SiH (3 eq), TFA (9 eq), 50°, 125 h	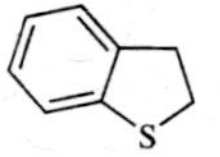 (55)	257
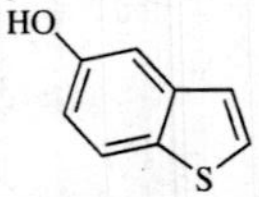	1. Et_3SiH (2.5 eq), TFA, 55°, 24h 2. Et_3SiH (0.033 eq), 55°, 24 h	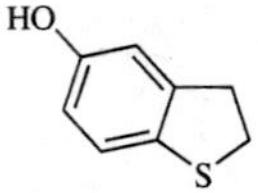 (30)	704
C_9 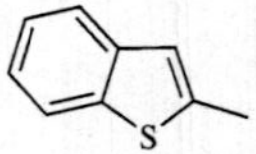	Et_3SiH (1-1.1 eq), TFA (7 eq), 50°, 20 h	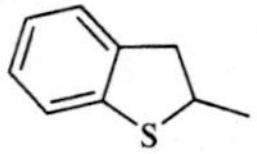 (80)	253, 257
	Ph_2SiD_2 (4 eq), TFA (11 eq), C_6H_6, 50°	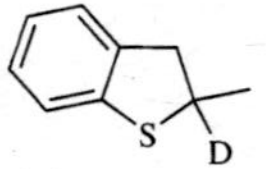 (—)	253
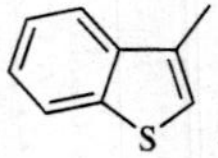	Et_3SiH (1-1.1 eq), TFA (7 eq), 50°, 20 h	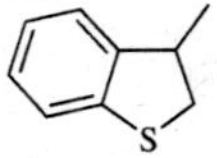 (90)	253, 257
	Ph_2SiD_2 (4 eq), TFA (11 eq), C_6H_6, 50°	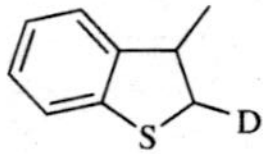 (—)	253

TABLE 3. ORGANOSILANE REDUCTION OF AROMATIC HYDROCARBONS (*Continued*)

Aromatic Hydrocarbon	Conditions	Product(s) and Yield(s) (%)	Refs.
C_9 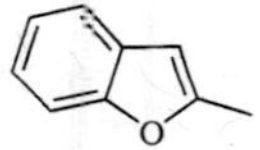	Et_3SiH (1.5 eq), TFA (2.5 eq), 60°, 2 h	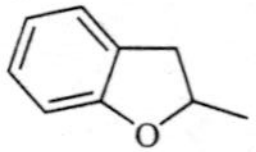 (91)	251
	Et_3SiH (1.5 eq), CF_3CO_2D (2.5 eq), 60°, 2 h	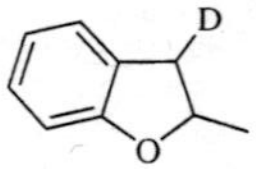 (86)	251
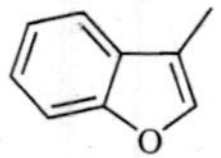	Et_3SiH (1.5 eq), TFA (2.5 eq), 60°, 2 h	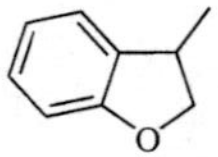 (62)	251
	Et_3SiH (1.5 eq), CF_3CO_2D (2.5 eq), 60°, 7 h	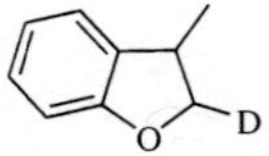 (60)	251
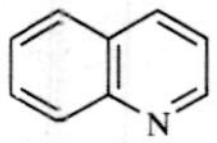	$PhMeSiH_2$ (1.5 eq), Cp_2TiMe_2 (10 mol%), 80°, 36 h	**I** 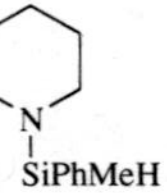 + **II** 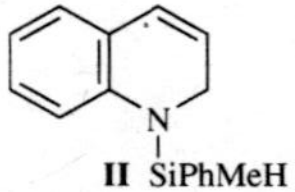 **I** + **II** (—), **I**:**II** = 3:1	264
	Ph_2SiH_2 (2 eq), Cp_2TiMe_2 (10 mol%), 80°, 8 h	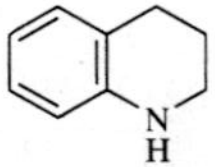 (55) + 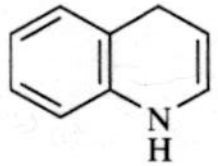 (19)	265
	$PhMeSiH_2$ (1.5 eq), Cp_2TiMe_2 (10 mol%), 200 psi H_2, 80°, 48 h	**I** (54)	264

C_{10}			
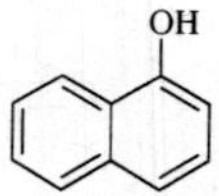	Et_3SiH (1.3 eq), BF_3•OH_2 (4-6 eq), CH_2Cl_2, rt, 4 h	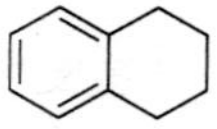 **I** (52)	217
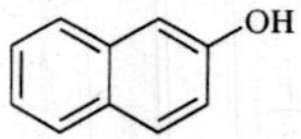	Et_3SiH (1.3 eq), BF_3•OH_2 (4-6 eq), CH_2Cl_2, rt, 10 h	**I** (37)	217
C_{10-16}			
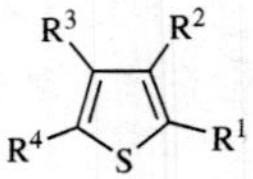	Et_3SiH (3 eq), TFA, $CHCl_3$, 50°, 50 h	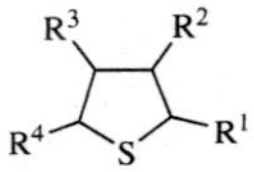	254

R^1	R^2	R^3	R^4	
H	Ph	H	H	(80)
Ph	Ph	H	H	(75)
Ph	H	Ph	H	(75)
Ph	H	H	Ph	(75)
H	Ph	Ph	H	(75)

C_{11}			
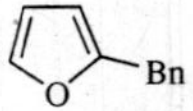	Et_3SiH (10 eq), TFA (50 eq)	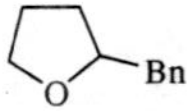 (80)	211
	Et_3SiH (10 eq), TFA (28 eq), BF_3•OEt_2 (0.6 eq), 20°, 5 min	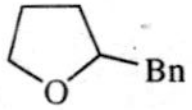 (89)	211
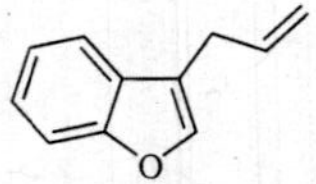	Et_3SiH, TFA, 20°, 1.5 h	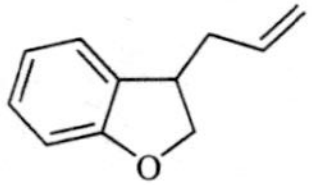 (95)	252
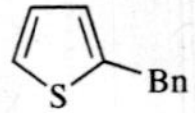	Et_3SiH (5 eq), TFA (9 eq), 50°, 30 h	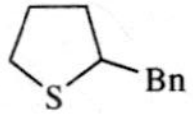 (75)	257
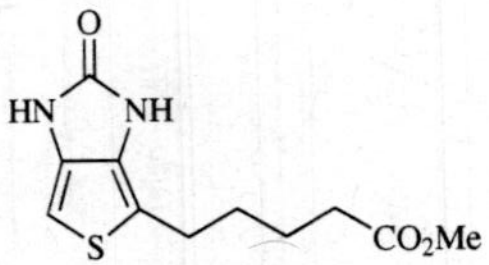	1. Et_3SiH, TFA, BF_3•OEt_2, Al_2O_3, 45-50°, 30 h 2. KOH, EtOH, 20°, 12 h	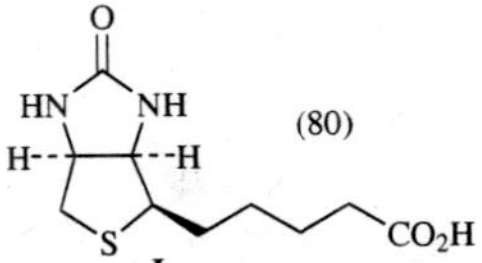 **I** (80)	255
	1. Et_3SiH TFA, Al_2O_3, 45-50°, 30 h 2. KOH, EtOH, 20°, 12 h	**I** (5-8)	255

TABLE 3. ORGANOSILANE REDUCTION OF AROMATIC HYDROCARBONS (*Continued*)

Aromatic Hydrocarbon	Conditions	Product(s) and Yield(s) (%)	Refs.
C_{11}			
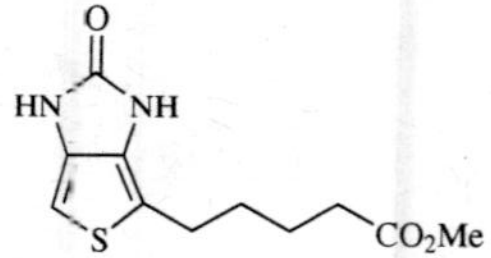	1. Et_3SiH, 50°, 30 h 2. KOH, EtOH, 20°, 12 h	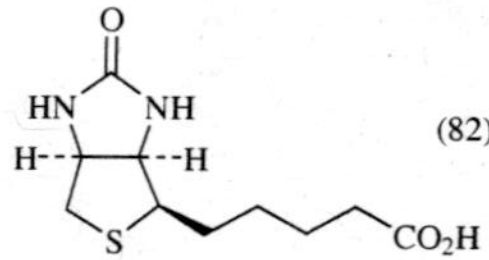 (82)	255, 256
C_{12}			
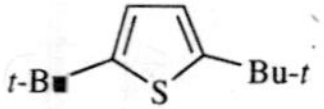	Et_3SiH (3 eq), TFA, $CHCl_3$, 50°, 50 h	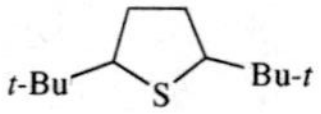 (15)	254
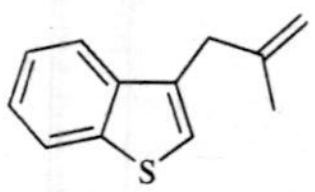	Et_3SiH, TFA, 20°, 1.5 h	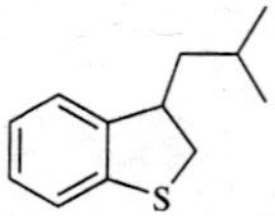 (60)	252
C_{13}			
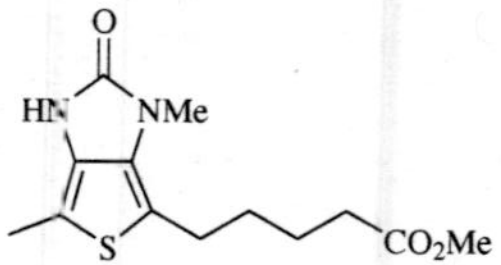	1. Et_3SiH, 100°, 30 h 2. KOH, EtOH, 20°, 12 h	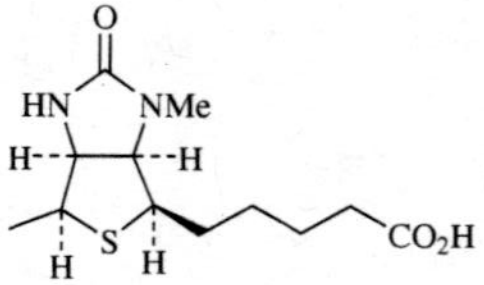 (72)	255
C_{14}			
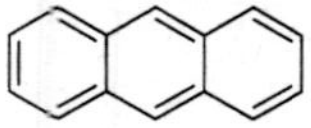	Et_3SiH (1.3 eq), $BF_3{\bullet}OH_2$ (10 eq), CH_2Cl_2, rt, 1 h	 (89)	217
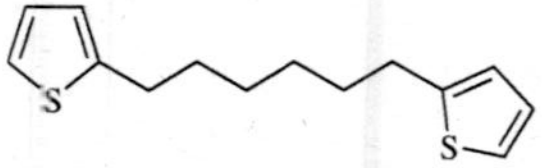	Et_3SiH (4 eq), TFA (8 eq), 50°, 30 h	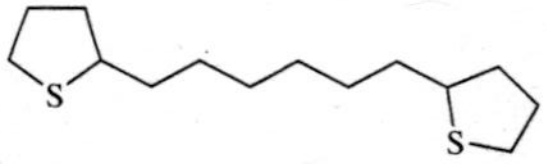 (70)	257
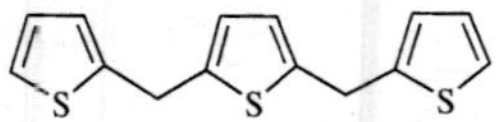	Et_3SiH (7 eq), TFA (13 eq), 50°, 50 h	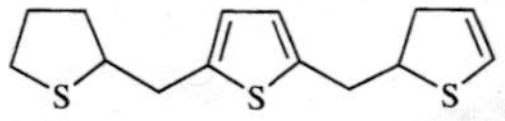 (60)	257

C_{15-16} 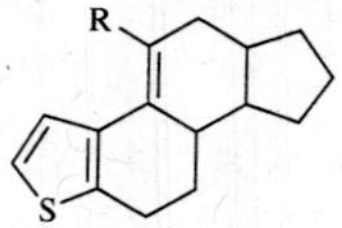 R = H, Me	Et_3SiH (xs), TFA (xs)	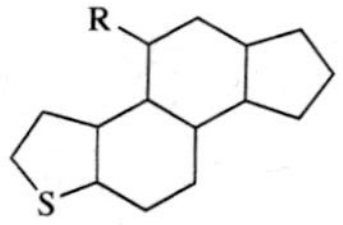 (—)	237
C_{16} 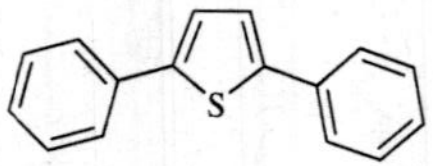	Et_3SiH (3 eq), HCl (xs), $AlCl_3$ (2 eq), CH_2Cl_2, rt	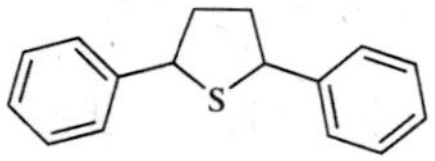 (66)	213
C_{18} 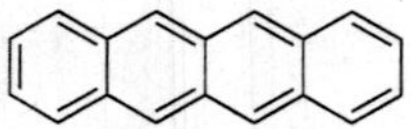	Et_3SiH (1.3 eq), BF_3•OH_2 (10 eq), CH_2Cl_2, rt, 4 h	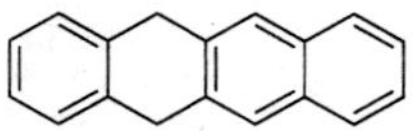 (88)	217
C_{18-21} 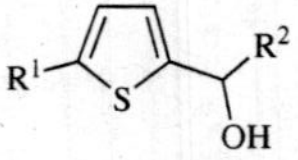 R^1 / R^2: H / n-$C_{15}H_{31}$; H / n-$C_{16}H_{33}$; Me / n-$C_{12}H_{25}$	Et_3SiH, TFA, 50°, 48 h	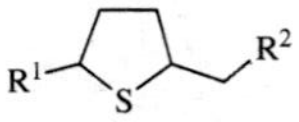 (—)	258
C_{19-23} 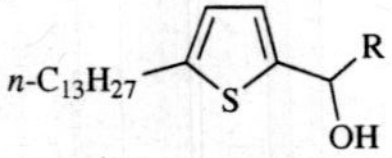 R: Me; Et; n-Pr; n-Bu; n-C_5H_{11}	Et_3SiH, TFA, 50°, 48 h	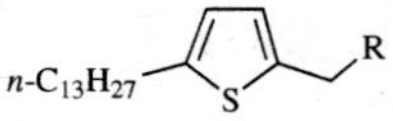 (—) + 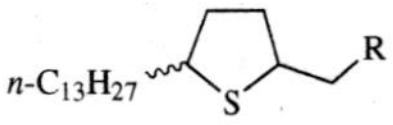 (—)	258

TABLE 4. ORGANOSILANE REDUCTION OF HALOCARBONS

Halocarbon	Conditions	Product(s) and Yield(s) (%)	Refs.
C_3			
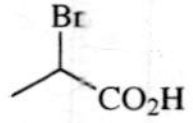	PMHS, $(Ph_3P)_4Pd$ (5 mol%), MeCN, DMSO, Bn_3N (1.4 eq), 110°, 3 h	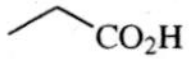 (35)	199
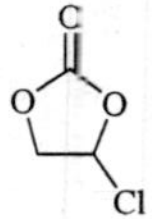	Et_3SiH (1.4 eq), $PdCl_2$ (5-10 mol%), rt, 1 h	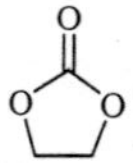 (81)	195
C_4			
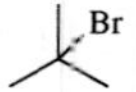	Ph_3SiH, CH_2Cl_2, BF_3•OEt_2, rt	 (—)	59
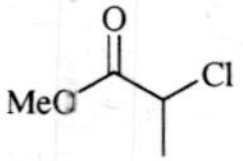	Et_3SiH (1.4 eq), $PdCl_2$ (5-10 mol%), rt, 1 h	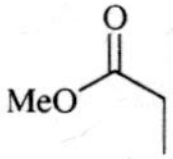 (>95)	195
C_5			
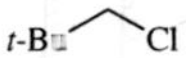	Et_3SiH (0.89 eq), $AlCl_3$, rt, 15 h	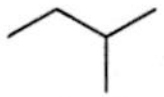 (48.6)	185
	Et_3SiH (5 eq), $AlCl_3$ (0.1 eq), HCl	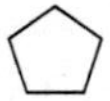 **I** (71)	146
	Et_3SiH (12 eq), $AlCl_3$ (0.4 eq), C_5H_{12}, 40°, <1 h	**I** (71)	186
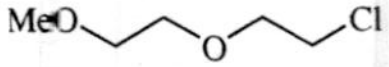	Et_3SiH (1.4 eq), $PdCl_2$ (5-10 mol%), rt, 10 min	 (>95)	195
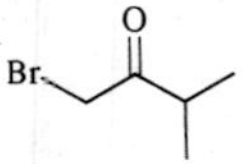	$PhSiH_3$ (180 mol%), $Mo(CO)_6$ (5 mol%), PPh_3 (19 mol%), $NaHCO_3$ (80 mol%), THF, 65°, 1.0 h	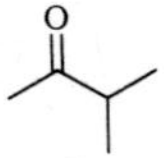 **I** (>95)	197

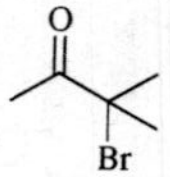	$PhSiH_3$ (160 mol%), $Mo(CO)_6$ (4 mol%), PPh_3 (16 mol%), $NaHCO_3$ (80 mol%), diglyme, 75°, 1.2 h	**I** (100)	197
C_5			
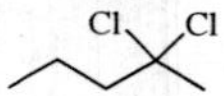	Et_3SiH (2.5 eq), $AlCl_3$ (0.25 eq), 20°, 30 min	 (67)	189
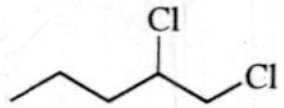	Et_3SiH (2.5 eq), $AlCl_3$ (0.25 eq), 20°, 30 min	**I** (70) + 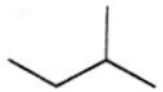 (8)	189
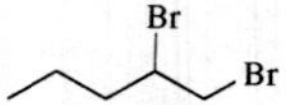	Et_3SiH (2.5 eq), $AlCl_3$ (0.25 eq), 20°, 30 min	**I** (72)	189
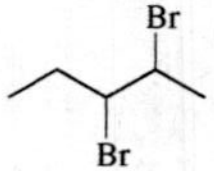	Et_3SiH (2.5 eq), $AlCl_3$ (0.25 eq), 20°, 30 min	**I** (72)	189
	Et_3SiH (2.5 eq), $AlCl_3$ (0.25 eq), 20°, 30 min	**I** (85)	189
C_{5-11}			
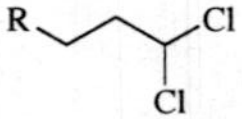	Et_3SiH (2.5 eq), $AlCl_3$ (0.25 eq), 20°, 30 min	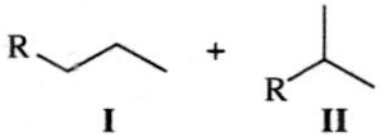	189
C_6			
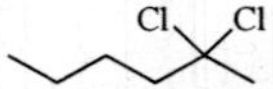	Et_3SiH (2.5 eq), $AlCl_3$ (0.25 eq), 20°, 30 min	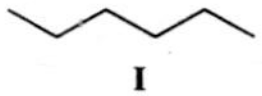 (72)	189
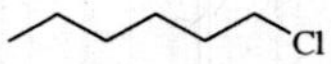	Et_3SiH (0.89 eq), $AlCl_3$, rt, 15 h	**I** (57)	185
	Et_3SiH (5 eq), $AlCl_3$ (0.1 eq)	**I** (~55%)	146

R	I	II
Et	(64)	(5)
n-Pr	(72)	(7)
n-C_4H_9	(68)	(7)
n-C_8H_{17}	(58)	(—)

TABLE 4. ORGANOSILANE REDUCTION OF HALOCARBONS (*Continued*)

Halocarbon	Conditions	Product(s) and Yield(s) (%)	Refs.
C_6 	Et_3SiH (5 eq), $AlCl_3$ (0.1 eq), HCl	**I** (64)	146
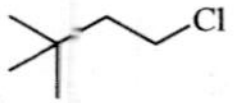	Et_3SiH (0.89 eq), $AlCl_3$, rt, 15 h	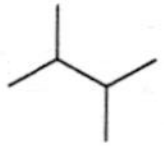 (50)	185
	Et_3SiH (5 eq), $AlCl_3$ (0.1 eq)	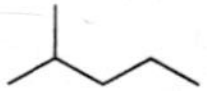 (~55%)	146
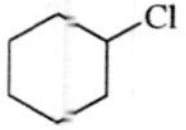	Et_3SiH (5 eq), $AlCl_3$ (0.1 eq), HCl	 (71) **I**	146
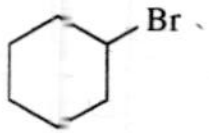	Et_3SiH (5 eq), $AlCl_3$ (0.1 eq), HCl	**I** (90)	146
	R_3SiH (1.3-1.5 eq), CH_2Cl_2, $AlCl_3$ (25 mol%)	**I** (95) R_3SiH = Et_3SiH, *n*-$BuSiH_3$	188
	Et_3SiH (1.25 eq), $AlCl_3$ (0.25 eq), CH_2Cl_2, 18-20°, 15 min	**I** (95)	187
	$H(Ph_2Si)_4H$ (1.0 eq), $AlCl_3$ (1.0 eq), CH_2Cl_2, 18-20°, 2 h	**I** (5)	187
	Et_3SiH (12 eq), $AlCl_3$ (0.4 eq), C_5H_{12}, 40°, <1 h	**I** (90)	186
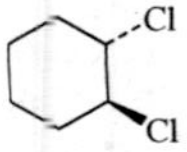	Et_3SiH (2.5 eq), $AlCl_3$ (0.25 eq), 20°, 30 min	**I** (67)	189
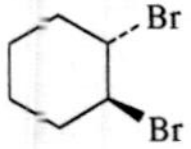	Et_3SiH (2.5 eq), $AlCl_3$ (0.25 eq), 20°, 30 min	**I** (74)	189

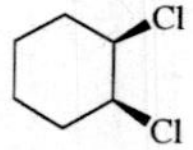	Et_3SiH (2.5 eq), $AlCl_3$ (0.25 eq), 20°, 30 min	**I** (74)	189
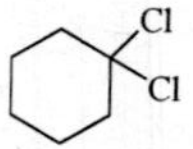	Et_3SiH (2.5 eq), $AlCl_3$ (0.25 eq), 20°, 30 min	**I** (88)	189
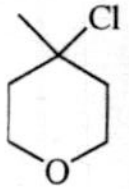	Et_3SiH (x eq), $AlCl_3$ (y eq), HCl, CH_2Cl_2, rt	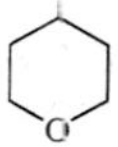 x / y / Time: 3 / 2 / 0.5 h (92); 2 / 2 / 0.5 h (75); 3 / 1.5 / 1.5 h (37); 3 / 0.75 / 20 h (0)	136
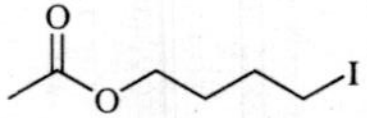	Et_3SiH (1.4 eq), $PdCl_2$ (5-10 mol%), rt, 5 min	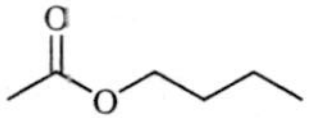 (>95)	195
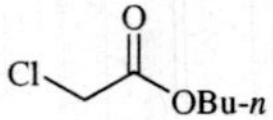	$PhSiH_3$ (170 mol%), $Mo(CO)_6$ (6 mol%), PPh_3 (26 mol%), $NaHCO_3$ (110 mol%), C_6H_6, 80°, 24 h	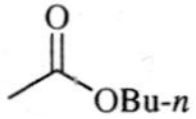 (75)	197
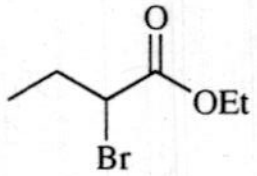	$PhSiH_3$ (190 mol%), $Mo(CO)_6$ (5 mol%), PPh_3 (17 mol%), $NaHCO_3$ (100 mol%), THF, 65°, 5.5 h	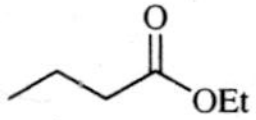 (75)	197
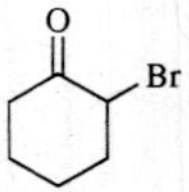	$PhSiH_3$ (200 mol%), $Mo(CO)_6$ (5 mol%), PPh_3 (23 mol%), $NaHCO_3$ (100 mol%), THF, 65°, 1.33 h	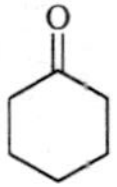 (80)	197
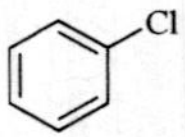	Et_3SiH (1.4 eq), $PdCl_2$ (5-10 mol%), rt, 1 h	 **I** (>95)	195
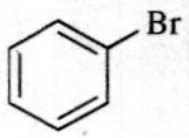	Et_3SiH (1.4 eq), $PdCl_2$ (5-10 mol%), rt, 40 min	**I** (>95)	195

TABLE 4. ORGANOSILANE REDUCTION OF HALOCARBONS (*Continued*)

Halocarbon	Conditions	Product(s) and Yield(s) (%)	Refs.
C_{6-12}			
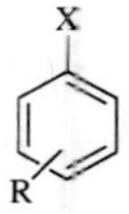	PMHS, $(Ph_3P)_4Pd$ (5 mol%), MeCN, DMSO, Bn_3N (1.4 eq)	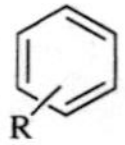 R / X / Temp / Time H / I / 60° / 3 h / (53) H / Br / 110° / 18 h / (75) H / Cl / 110° / 18 h / NR 4-Cl / Br / 110° / 18 h / (65) 4-Ac / Br / 110° / 3 h / (96) 4-O_2N / Br / 110° / 3 h / (93) 4-BrC_6H_4 / Br / 110° / 5 h / (72) 4-CHO / Br / 110° / 3 h / (93) 4-CO_2H / Br / 110° / 18 h / (77)	199
C_7			
	Et_3SiH (5 eq), $AlCl_3$ (0.1 eq), HCl	 (73) **I**	146
	Et_3SiH (12 eq), $AlCl_3$ (0.4 eq), C_5H_{12}, 40°, <1 h	**I** (64)	186
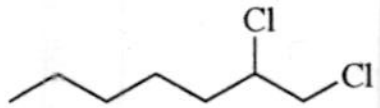	Et_3SiH (2.5 eq), $AlCl_3$ (0.25 eq), 20°, 30 min	**I** (76) + 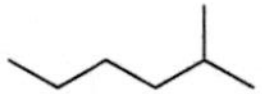 **II** (11) + 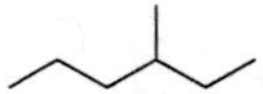 **III** (7)	189
	Et_3SiH (6 eq), $AlCl_3$ (0.55 eq), 20°, 30 min	**I** (83) + **II** (3) + **III** (3)	189
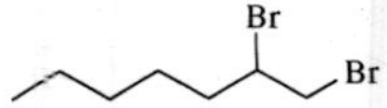	Et_3SiH (6 eq), $AlCl_3$ (0.55 eq), 20°, 30 min	**I** (76)	189
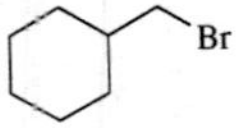	R_3SiH (1.3 eq), CH_2Cl_2, $AlCl_3$ (25 mol%), R_3 = Et_3, $EtMe_2$, Ph_2Et, $(C_6H_{11})_2Et$	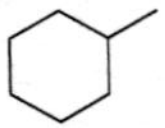 (90-95) **I**	188

	Et_3SiD (1.4 eq), $PdCl_2$ (5-10 mol%), rt	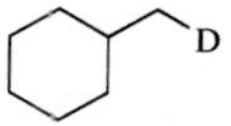 (—)	195
	Et_3SiH (12 eq), $AlCl_3$ (0.4 eq), C_5H_{12}, 40°, <1 h	**I** (52)	186
	Et_3SiD (12 eq), $AlCl_3$ (0.4 eq), C_5H_{12}, 40°, <1 h	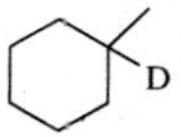 (—)	186
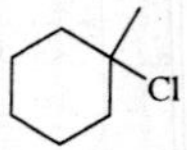	R_3SiH (x eq), $AlCl_3$ (y eq), CH_2Cl_2, 18-20°, 15 min	**I** — R_3Si / x / y: Et_3Si 1.3 0.25 (95); $H(SiPh_2)_4H$ 0.6 0.25 (96)	187
	$EtCl_2SiH$ (2 eq), $AlBr_3$ (0.25 eq), 20°, 2 h	**I** (92)	192
	R_3SiH (1.3 eq), CH_2Cl_2, $AlCl_3$ (25 mol%) $R_3 = Et_3$, n-BuH_2, Ph_2H, $(c\text{-}C_6H_{11})_2Et$	**I** (93-96)	188
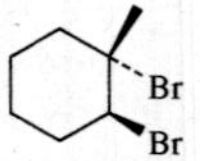	Et_3SiH (2.5 eq), $AlCl_3$ (0.25 eq), 20°, 30 min	**I** (87)	189
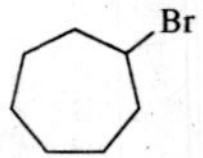	Et_3SiH (1.3 eq), CH_2Cl_2, $AlCl_3$ (25 mol%)	 **I** (85)	188
	Et_3SiH (1.4 eq), $PdCl_2$ (5-10 mol%), rt	**I** (—)	195
	R_3SiH (1.3 eq), CH_2Cl_2, $AlCl_3$ (25 mol%) $R_3 = Et_3$, n-BuH_2, Ph_2H, $(c\text{-}C_6H_{11})_2Et$	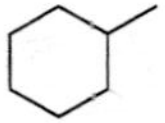 **II** (3-7)	188
	Et_3SiH (5 eq), $AlCl_3$ (0.1 eq), HCl	**I** (39) + **II** (26)	146
	n-$BuSiH_3$ (5 eq), $AlCl_3$ (0.1 eq), HCl	**I** (1) + **II** (65)	146

TABLE 4. ORGANOSILANE REDUCTION OF HALOCARBONS (*Continued*)

Halocarbon	Conditions	Product(s) and Yield(s) (%)	Refs.
C_7 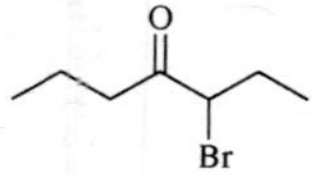	Ph_2SiH_2 (1.1 eq), $(Ph_3P)_4Pd$ (7 mol%), $ZnCl_2$ (0.33 eq), THF, rt, 12 h	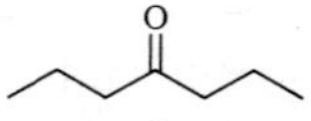 **I** (50)	197
	Ph_2SiH_2 (1.4 eq), $(Ph_3P)_4Pd$ (8 mol%), rt, THF, 24 h	**I** (60)	197
	Ph_2SiH_2 (0.9 eq), $(Ph_3P)_4Pd$ (3.5 mol%), NaOAc (1.6 eq), THF, rt, 4 h	**I** (60)	197
	Ph_2SiH_2 (1.1 eq), $(Ph_3P)_4Pd$ (7 mol%), K_2CO_3 (1.0 eq), THF, rt, 6 h	**I** (70)	197
	$PhSiH_3$ (w mol%), $Mo(CO)_6$ (x mol%), PPh_3 (y mol%), $NaHCO_3$ (z mol%)	**I**	197

w	x	y	z	Solvent	Temp	Time	I
200	20	0	0	THF	65°	4.0 h	(75)
180	49	0	0	THF	65°	2.5 h	(100)
230	23	0	260	THF	65°	1.75 h	(100)
150	5	17	100	THF	65°	2.25 h	(100)
150	6	23	110	C_6H_6	80°	2.75 h	(97)

Halocarbon	Conditions	Product(s) and Yield(s) (%)	Refs.
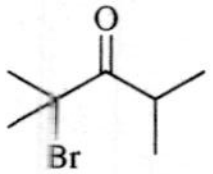	Ph_2SiH_2 (3 eq), $(Ph_3P)_4Pd$ (10 mol%), K_2CO_3 (1.8 eq), THF, rt, 12 h	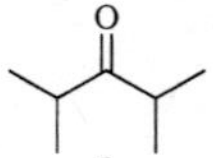 **I** (25)	197
	$PhSiH_3$ (190 mol%), $Mo(CO)_6$ (5 mol%), PPh_3 (19 mol%), $NaHCO_3$ (110 mol%), C_6H_6, 80°, 2.5 h	**I** (70)	197
	Et_3SiH (12 eq), $AlCl_3$ (0.4 eq), C_5H_{12}, 40°, <1 h	 **I** (96)	186
	Et_3SiH (5 eq), $AlCl_3$ (0.1 eq), HCl	**I** (96)	146

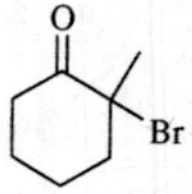	PhSiH₃ (150 mol%), $Mo(CO)_6$ (6 mol%), PPh_3 (20 mol%), $NaHCO_3$ (80 mol%), THF, 65°, 1.0 h	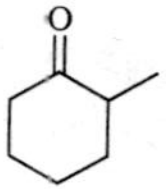 (100)	197
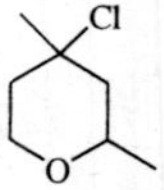	Et_3SiH (3 eq), $AlCl_3$ (x eq), HCl, CH_2Cl_2, rt x / Time 0.75 / 20 h 2 / 1.5 h	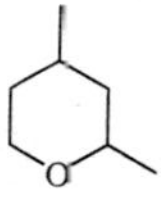 + 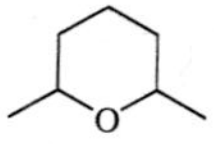 (23) (6) (30) (28)	136
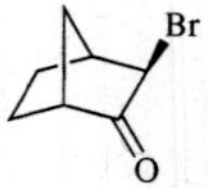	$PhSiH_3$ (160 mol%), $Mo(CO)_6$ (6 mol%), PPh_3 (25 mol%), $NaHCO_3$ (80 mol%), MeC_6H_5, 75°, 5.5 h	 (90)	197
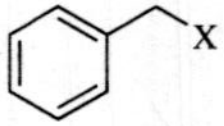 X = Cl, Br	Et_3SiH (1.4 eq), $PdCl_2$ (5-10 mol%), rt, 2 min	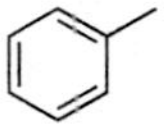 (>95)	195
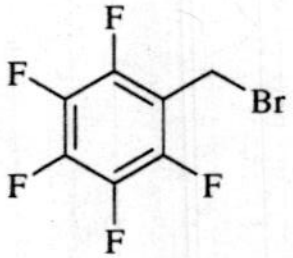	Et_3SiH (1 4 eq), $PdCl_2$ (5-10 mol%), rt, 2 min	 (>95)	195
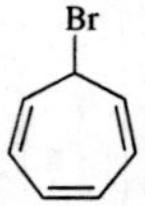	R_3SiH (x eq), $AlCl_3$ (0.25 eq), CH_2Cl_2, 18 to 20° R_3Si / x / y / Time Et_3Si / 1.5 / 0.25 / 30 min $H(SiPh_2)_4Si$ / 0.6 / 0.6 / 1 h	 (80) (46)	187

TABLE 4. ORGANOSILANE REDUCTION OF HALOCARBONS (*Continued*)

Halocarbon	Conditions	Product(s) and Yield(s) (%)	Refs.
C_7 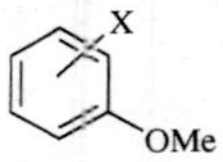	Et_3SiH (1.4 eq), $PdCl_2$ (5-10 mol%), rt, 2 min	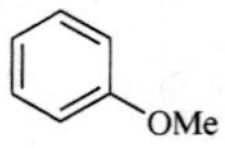 (>95)	195

X	Time
2-Br	2 min
2-I	10 min
4-I	25 min

Halocarbon	Conditions	Product(s) and Yield(s) (%)	Refs.
$C_{7\text{-}19}$ RX	$[HSi(OEt)_4]K$, THF	RH (I) + RR (II)	288

R	X	Temp	Time	I	II
n-$C_{12}H_{25}$	Cl	45°	48 h	(33)	(—)
n-$C_{12}H_{25}$	Br	rt	15 h	(47)	(—)
Bn	Cl	rt	40 h	(16)	(14)
Bn	Br	rt	20 h	(37)	(34)
Ph_2CH	Cl	rt	15 h	(12)	(40)
Ph_3C	Br	rt	15 h	(47)	(22)

Halocarbon	Conditions	Product(s) and Yield(s) (%)	Refs.
$C_{7\text{-}11}$ ArCTf	Et_3SiH (x eq), $Pd(OAc)_2$, ligand, DMF	ArH	201

Ar	x	Ligand	Temp	Time	ArH
4-$O_2NC_6H_4$	2.5	dppp	100°	7 d	(67)
4-$O_2NC_6H_4$	2.5	dppf	60°	20 min	(100)
4-BrC_6H_4	2.5	dppp	43°	1 h	(90)
4-$(OHC)C_6H_4$	2.5	dppp	60°	1 h	(99)
4-$MeO_2CC_6H_4$	2.5	dppp	60°	14 h	(100)
1-$C_{10}H_7$	2.5	dppp	60°	5 h	(100)
2-(CF_3SO_3)-4-t-BuC_6H_3	5	dppp	60°	22 h	(>99)

Halocarbon	Conditions	Product(s) and Yield(s) (%)	Refs.
C_8 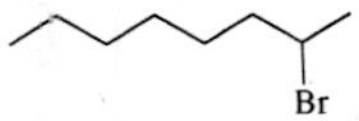	R_3SiH (1.3 eq), CH_2Cl_2, $AlCl_3$ (25 mol%)		188

R_3Si	Yield
Et_3Si	(85-90)
Et_2MeSi	(85-90)
Ph_2EtSi	(7-8)
$(c$-$C_6H_{11})_2EtSi$	(7-8)

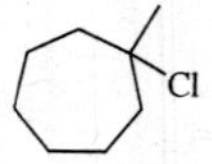	R_3SiH (1.2-1.5 eq), CH_2Cl_2, $AlCl_3$ (25 mol%) $R_3Si = Et_3Si$, Et_2MeSi, n-BuH_2Si, Ph_2HSi Ph_3Si, $(C_6H_{11})_2EtSi$	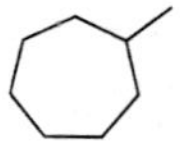	(80-95)	188
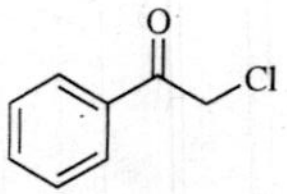	$PhSiH_3$ (140 mol%), $Mo(CO)_6$ (4 mol%), PPh_3 (18 mol%), $NaHCO_3$ (70 mol%), THF, 65°, 4.5 h	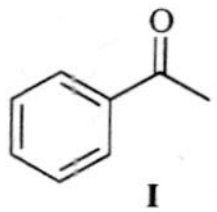 **I**	(90)	197
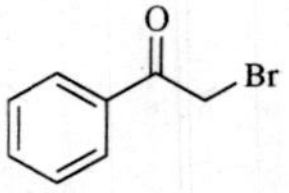	Ph_2SiH_2 (1.3 eq), $ZnCl_2$ (2.8 eq), $(Ph_3P)_4Pd$ (7 mol%), $CHCl_3$, rt, 1 h	**I** (70)		197
	Ph_2SiH_2 (1.5 eq), $(Ph_3P)_4Pd$ (8 mol%), THF, rt, 0.5 h	**I** (73)		197
	Et_3SiH (1.5 eq), $(Ph_3P)_4Pd$ (8 mol%), THF, rt, 20 h	**I** (5)		197
	$PhSiH_3$ (1.5 eq), $(Ph_3P)_4Pd$ (8 mol%), THF, rt, 0.1 h	**I** (60)		197
	Ph_2SiH_2 (1.4 eq), $Pd(OAc)_2$ (8 mol%), PPh_3 (0.16 eq), THF, rt, 24 h	**I** (70)		197
	$PhSiH_3$ (120 mol%), $Mo(CO)_6$ (6 mol%), PPh_3 (27 mol%), $NaHCO_3$ (110 mol%), C_6H_6, 80°, 3.25 h	**I** (95)		197
	PMHS, $(Ph_3P)_4Pd$ (5 mol%), MeCN, DMSO, Bn_3N (1.4 eq), 110 °, 3 h	**I** (80)		199

TABLE 4. ORGANOSILANE REDUCTION OF HALOCARBONS (*Continued*)

Halocarbon	Conditions	Product(s) and Yield(s) (%)	Refs.
C_8			
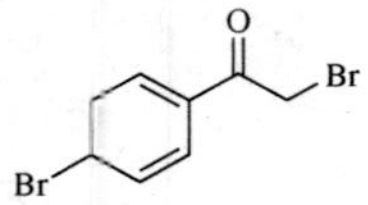	$PhSiH_3$ (210 mol%), $Mo(CO)_6$ (5 mol%), PPh_3 (20 mol%), $NaHCO_3$ (150 mol%), THF, 65°, 4.5 h	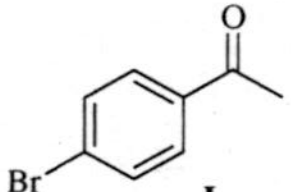 **I** (100)	197
	$PhSiH_3$ (150 mol%), $Mo(CO)_6$ (8 mol%), PPh_3 (33 mol%), $NaHCO_3$ (130 mol%), C_6H_6, 80°, 2.0 h	**I** (>95)	197
	$PhSiH_3$ (200 mol%), $Mo(CO)_6$ (5 mol%), PPh_3 (19 mol%), $NaHCO_3$ (140 mol%), MeC_6H_5, 95°, 1.0 h	**I** (>95)	197
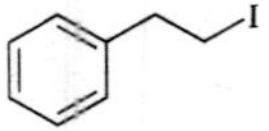	Et_3SiH (1.4 eq), $PdCl_2$ (5-10 mol%), rt, 10 min	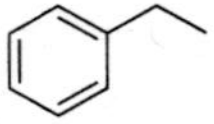 (>95)	195
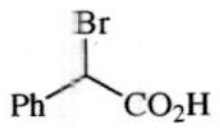	PMHS, $(Ph_3P)_4Pd$ (5 mol %), MeCN, DMSO, Bn_3N (1.4 eq), 110°, 18 h	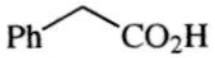 (55)	199
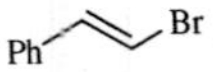	PMHS, $(Ph_3P)_4Pd$ (5 mol %), MeCN, DMSO, Bn_3N (1.4 eq), 60°, 3 h	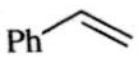 (37)	199

C_9 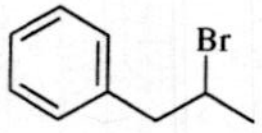	Et_3SiH (12 eq), $AlCl_3$ (0.4 eq), C_5H_{12}, 40°, <1 h	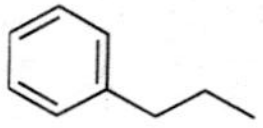 I (43)	186
	Et_3SiH (5 eq), $AlCl_3$ (0.1 eq), HCl	I (43)	
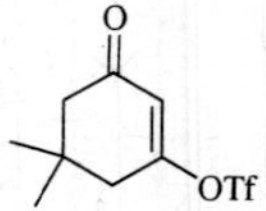	Et_3SiH(2.5 eq), $Pd(OAc)_2$, dppp, Et_3N (2 eq), DMF, rt, 40 min	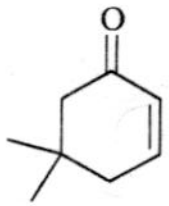 (68)	146, 201
C_{10} 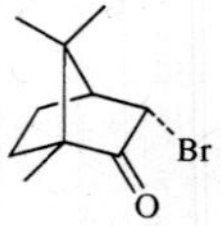	$PhSiH_3$ (160 mol%), $Mo(CO)_6$ (11 mol%), THF, 65°, 2.5 h	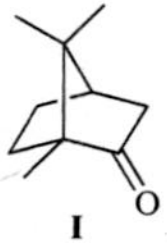 I (100)	197
	$PhSiH_3$ (280 mol%), $Mo(CO)_6$ (65 mol%), PPh_3 (24 mol%), $NaHCO_3$ (130 mol%), MeC_6H_5, 75°, 12.0 h	I (45)	197
	Ph_2SiH_2 (3 eq), $(Ph_3P)_4Pd$ (10 mol%), K_2CO_3 (1.8 eq), THF, rt, 12 h	I (55)	197
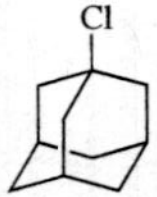	Et_3SiH (1.3 eq), $AlCl_3$ (0.25 eq), CH_2Cl_2, 18-20°, 15 min	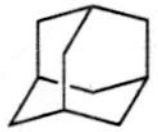 I (88)	187
	$H(Ph_2Si)_4H$ (0.6 eq), $AlCl_3$ (0.25 eq), CH_2Cl_2, 18-20°, 15 min	I (88)	187
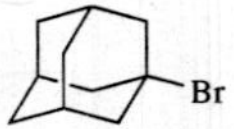	Et_3SiH (1.4 eq), $PdCl_2$ (5-10 mol%), 80°, 4 h	I (84)	195

TABLE 4. ORGANOSILANE REDUCTION OF HALOCARBONS (*Continued*)

Halocarbon	Conditions	Product(s) and Yield(s) (%)	Refs.
C_{10} 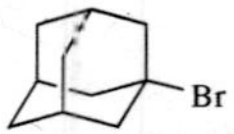	Et_3SiH (5 eq), $AlCl_3$ (0.1 eq), HCl	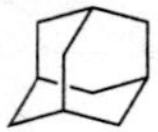 **I** (84)	146
	R_3SiH (1.3-1.5 eq), CH_2Cl_2, $AlCl_3$ (25 mol%) R_3Si = Et_3Si, Et_2MeSi, Ph_2HSi, $(c\text{-}C_6H_{11})_2EtSi$	**I** (90-96)	188
	Et_3SiH (12 eq), $AlCl_3$ (0.4 eq), C_5H_{12}, 40°, <1 h	**I** (79)	186
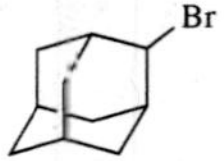	Et_3SiH (12 eq), $AlCl_3$ (0.4 eq), C_5H_{12}, 40°, <1 h	**I** (84)	186
	Et_3SiH (1.4 eq), $PdCl_2$ (5-10 mol%), 80°, 6 h	**I** (80)	195
	Et_3SiH (1.4 eq), $PdCl_2$ (5-10 mol%), rt, 2 h	**I** (78)	195
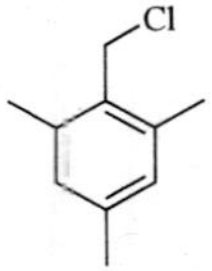	Et_3SiH (1.46 eq), TFA (3 eq), CH_2Cl_2, rt, 2.5 h	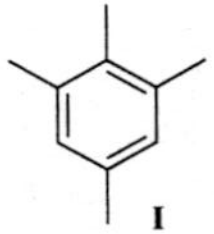 **I** (79) + 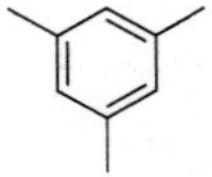 (9)	128
	Et_3SiH (1.5 eq), CH_2Cl_2, $AlCl_3$ (25 mol%)	**I** (80)	188
	Ph_2EtSiH (1.5 eq), CH_2Cl_2, $AlCl_3$ (25 mol%)	**I** (45)	188
	$(c\text{-}C_6H_{11})_2EtSi$ (1.5 eq), CH_2Cl_2, $AlCl_3$ (25 mol%)	**I** (35)	188

C_{10} 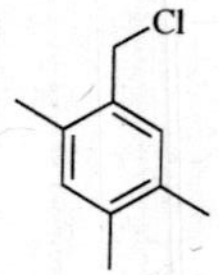	R_3SiH (x eq), $AlCl_3$ (0.25 eq), CH_2Cl_2, 18-20°, 1 h R_3SiH / x Et_3SiH / 1.3 $H(SiPh_2)_4H$ / 0.6	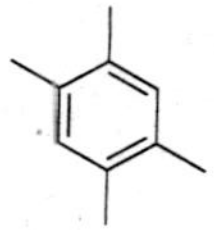 (78) (56)	187
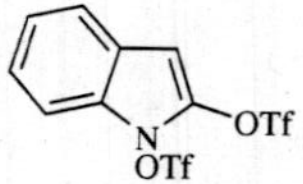	Et_3SiH (2.5 eq), $Pd(OAc)_2$, dppp, DMF, rt, 10 h	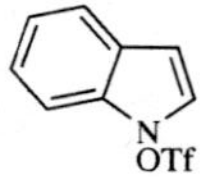 (83)	201
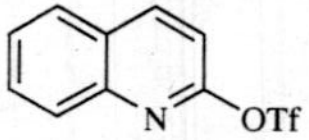	Et_3SiH (2.5 eq), $Pd(OAc)_2$, dppp, DMF, 60°, 4 h	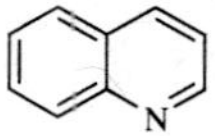 (97)	201
C_{11} 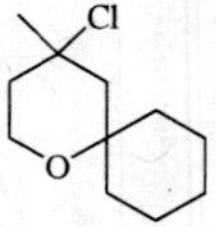	Et_3SiH (3 eq), $AlCl_3$ (1 eq), HCl, CH_2Cl_2, rt, 1 h	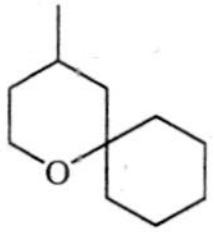	136
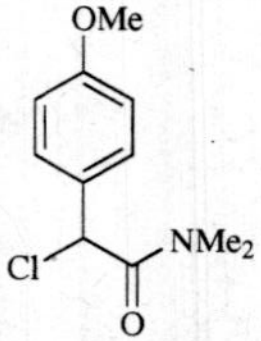	Et_3SiH (1.5 eq), TFA (5 eq), CH_2Cl_2, rt, 18 h	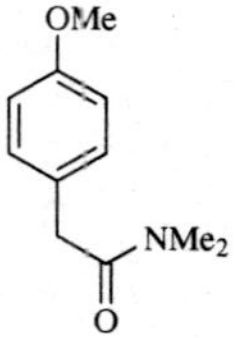 (54)	198
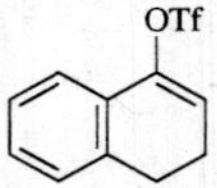	Et_3SiH (2.5 eq), $Pd(OAc)_2$, dppp, DMF, 60°, 4 h	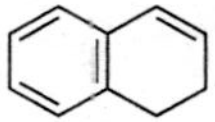 (88)	201
C_{12} 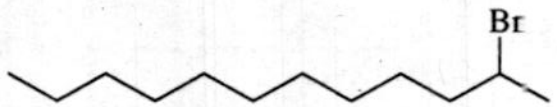	n-$BuSiH_3$ (5 eq), $AlCl_3$ (0.1 eq), HCl	 (82)	146

TABLE 4. ORGANOSILANE REDUCTION OF HALOCARBONS (*Continued*)

Halocarbon	Conditions	Product(s) and Yield(s) (%)	Refs.
C_{12}			
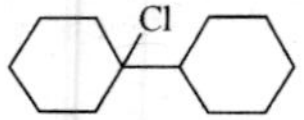	Et_3SiH (1.3 eq), $AlCl_3$ (0.25 eq), CH_2Cl_2, 18-20°, 15 min	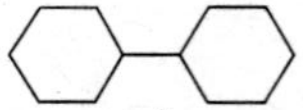 **I** (95)	187
	$H(Ph_2Si)_4H$ (0.6 eq), $AlCl_3$ (0.25 eq), CH_2Cl_2, 18-20°, 15 min	**I** (96)	187
C_{13}			
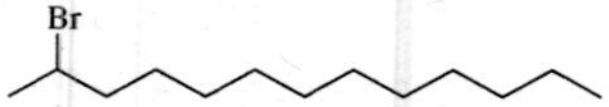	n-$BuSiH_3$, $AlCl_3$ (0.4 eq), C_5H_{12}, 40°, <1 h	 **I** (82)	186
	n-Bu_3SiH, $AlCl_3$ (0.4 eq), C_5H_{12}, 40°, <1 h	**I** (87)	186
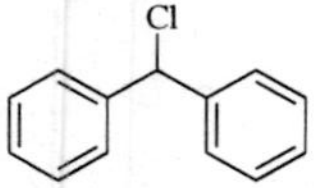	Et_3SiH (5 eq), $AlCl_3$ (0.1 eq), HCl	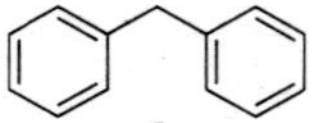 **I** (100)	146
	Et_3SiH (12 eq), $AlCl_3$ (0.4 eq), C_5H_{12}, 40°, <1 h	**I** (100)	186
	Et_3SiH (1.5 eq), $AlCl_3$ (0.5 eq), CH_2Cl_2, 18-20°, 15 min	**I** (95)	187
	$H(Ph_2Si)_4H$ (0.6 eq), $AlCl_3$ (0.5 eq), CH_2Cl_2, 18-20°, 15 min	**I** (95)	187
	R_3SiH (1.3-1.5 eq), CH_2Cl_2, $AlCl_3$ (25 mol%) R_3Si = Et_3Si, Et_2MeSi, Ph_2HSi, $(c\text{-}C_6H_{11})_2EtSi$	**I** (80-88)	188
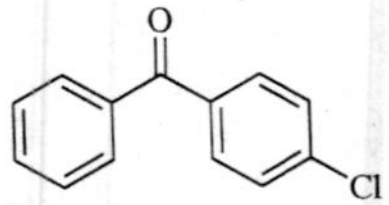	$HMe_2SiOSiMe_2H$ (2 eq), 10% Ni/C, PPh_3 (40%), dioxane, reflux, 15 h	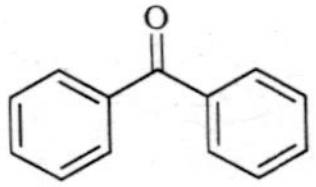 (96)	200

C_{14}				
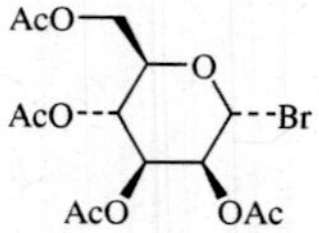	Et_3SiH (1.4 eq), $PdCl_2$ (5-10 mol%), rt, 40 min	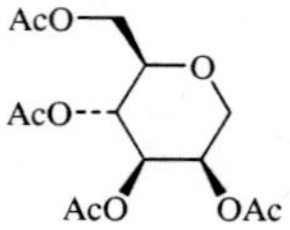	(92)	195
C_{16}				
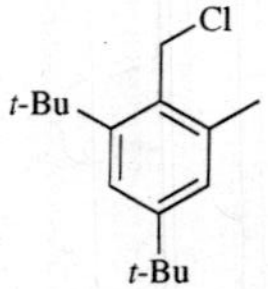	Et_3SiH (2 eq), TFA (2 eq), CH_2Cl_2, 40°, 40 min, rt, 15 h	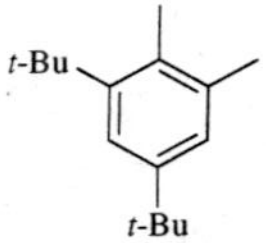	(50)	128
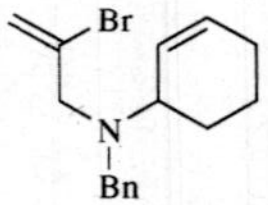	1. $Ni(cod)_2$ (1.23 eq), Et_3N (3 eq), MeCN 2. Et_3SiH (4 eq)	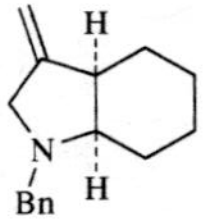	(77)	706, 705
C_{17}				
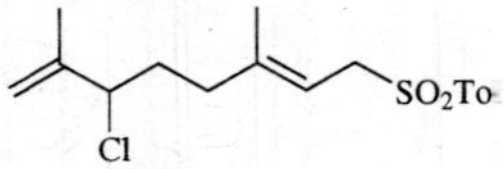	Ph_2SiH_2 (2.9 eq), $ZnCl_2$ (3-4 eq), $(Ph_3P)_4Pd$ (5.0 mol%), THF, rt. 12 h; 50° 6 h	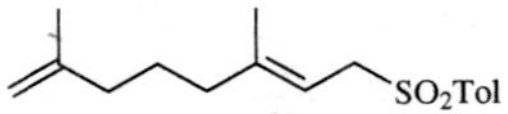	(58)	196
C_{19}				
Ph_3CX	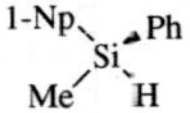 , C_6H_6, reflux, 4 h	Ph_3CH (—) + 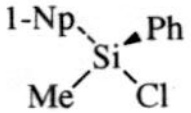	(71)	60
	R_3SiH, rt	Ph_3CH		59

X	R_3Si	Solvent	Time	
Cl	Ph_3Si	CH_2Cl_2	10 min	(100)
Cl	Ph_3Si	C_6H_6	24 h	(97)
Cl	Et_3Si	CH_2Cl_2	10 min	(100)
Br	Ph_3Si	CH_2Cl_2	10 min	(100)
Br	Ph_3Si	C_6H_6	24 h	(100)
Br	Et_3Si	CH_2Cl_2	10 min	(100)

TABLE 4. ORGANOSILANE REDUCTION OF HALOCARBONS (*Continued*)

Halocarbon	Conditions	Product(s) and Yield(s) (%)	Refs.
C_{19} 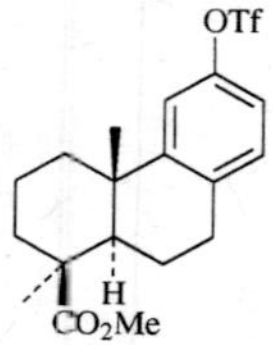	Et_3SiH (3 eq), $Pd(OAc)_2$, dppp, DMF, 60°, 22 h	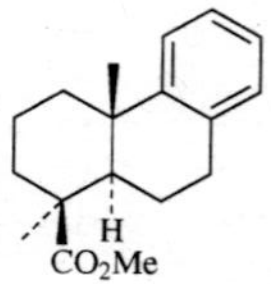 (92)	201
C_{21} 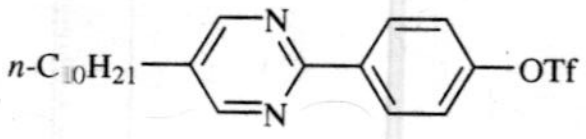	Et_3SiH (2.5 eq), $Pd(OAc)_2$, dppp, DMF, 60°, 4 h	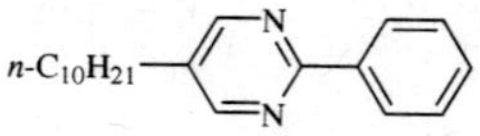 **I** (99)	201
	Et_3SiH (2.5 eq), $Pd(OAc)_2$, dppp, DMF, 80°, 9 d	**I** (41)	201
	Et_3SiH (2.5 eq), $Pd(OAc)_2$, dpp, DMF, 60°, 10 min	**I** (100)	201
C_{25} 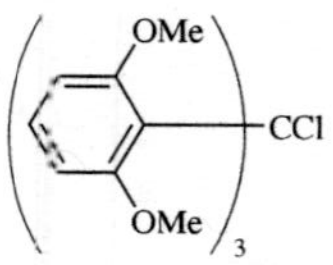	Et_3SiH (4 eq), HOAc, rt, 1 h	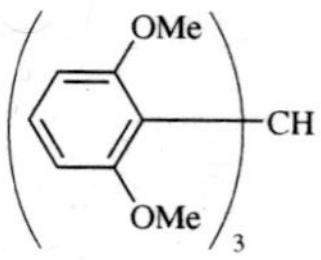 (95)	29

TABLE 5. ORGANOSILANE REDUCTION OF ALCOHOLS

Alcohol	Conditions	Product(s) and Yield(s) (%)	Refs.
C_6			
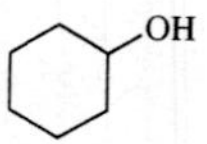	Et_3SiH (4 eq), $AlCl_3$ (1.5 eq), HCl, CH_2Cl_2, rt, 3.5 h	 **I** (70) + 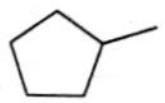 **II** (7)	136
	Et_3SiH (4 eq), $AlCl_3$ (1.5 eq), CH_2Cl_2, rt, 3.5 h	**I** (53) + **II** (6)	136
C_6-C_{10}			
R-OH	Et_3SiH (x eq), $(C_6F_5)_3B$ (10 mol %), C_6H_{14} or CH_2Cl_2, rt, 20 h	R-H **I** + $ROSiEt_3$ **II**	144

R	x	I	II
c-C_6H_{11}	6.0	(—)	(>95)
Ph	1.1	(0)	(>99)
c-$C_6H_{11}CH_2$	6.0	(91)	(—)
Ph(Me)CH	6.0	(—)	(>95)
$BnCH_2$	3.0	(>95)	(—)
n-C_9H_{19}	1.1	(—)	(>95)
n-C_9H_{19}	3.0	(>99)	(0)
$Bn(CH_2)_2$	3.0	(>95)	(—)
2-C_9H_{17}	3.0	(—)	(>95)
$Bn(Me)_2C$	3.0	(—)	(86)

TABLE 5. ORGANOSILANE REDUCTION OF ALCOHOLS (*Continued*)

Alcohol	Conditions	Product(s) and Yield(s) (%)	Refs.
C_7 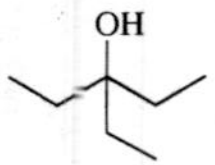	Ph_3SiH (1.2 eq), TFA (6.5 eq), CH_2Cl_2, rt, 24 h	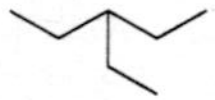 (78)	127
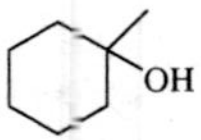	Et_3SiH (2 eq), $AlCl_3$ (1.1 eq), HCl, CH_2Cl_2, rt, 45 min	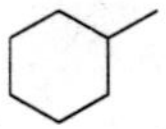 **I** (91)	136
	Et_3SiH (3 eq), $AlCl_3$ (1.5 eq), CH_2Cl_2, rt	**I** (94)	136
C_8 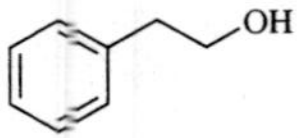	Et_3SiH (3.0 eq), $(C_6F_5)_3B$ (5 mol %), C_6H_{14} or CH_2Cl_2, rt, 20 h	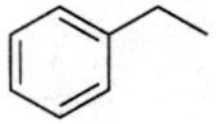 **I** (>95)	145
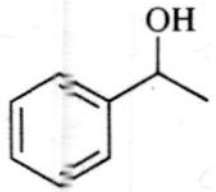	Et_3SiH (1 eq), acid, rt, 96 h	**I** Acid: H_2SO_4/HOAc (24); TFA (80)	134
	Et_3SiH (x mol%), TFA (y mol%), 50°, 15 h	**I** Solvent / x / y: $CHCl_3$ 0.5 9 (94); 4-$O_2NC_6H_5$ 0.2 12 (56)	134
	Ph_2ClSiH (2 eq), $InCl_3$ (0.05 eq), CH_2Cl_2, rt, 1 h	**I** (88)	172
	Et_3SiH (x eq), Et_3SiD (x eq), CH_2Cl_2, 0°, 0.5 h	**I** + 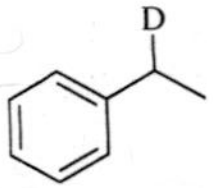 x / k_H/k_D: 1.67 1.05; 1.0 0.99; 0.71 1.01; 0.5 0.98	133

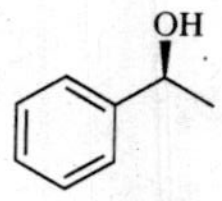	Et_3SiD, CH_2Cl_2, 0°, 0.5 h	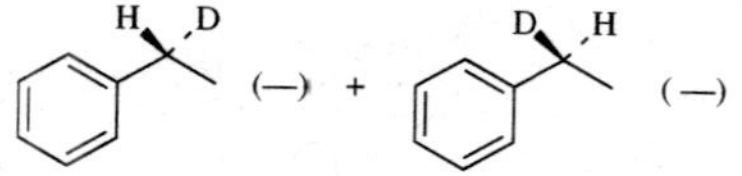 (—) + (—)	133
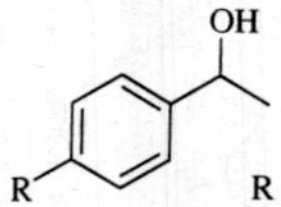	Ph_2ClSiH (2 eq), $InCl_3$ (0.05 eq)	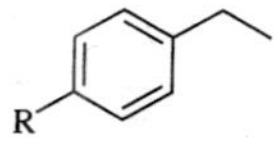	172
R	**Solvent / Temp / Time**		
Cl	CH_2Cl_2 / rt / 2 h	(70)	
Br	$ClCH_2CH_2Cl$ / 80° / 3 h	(83)	
NO_2	$ClCH_2CH_2Cl$ / 80° / 3 h	(95)	
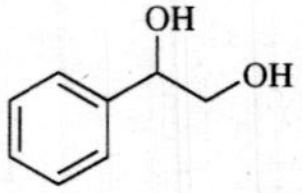	Et_3SiH (2 eq), $BF_3{\bullet}OEt_2$ (2 eq), CH_2Cl_2, rt, 20 h	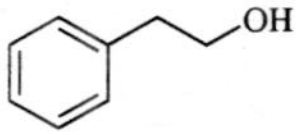 (64)	133
C_9 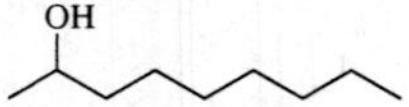	Ph(t-$BuCH_2$)MeSiH (2-10 mol%), BF_3, CH_2Cl_2, rt, 10 min	 (50)	126
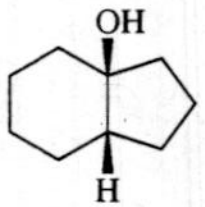	Ph_3SiH, TFA, CH_2Cl_2	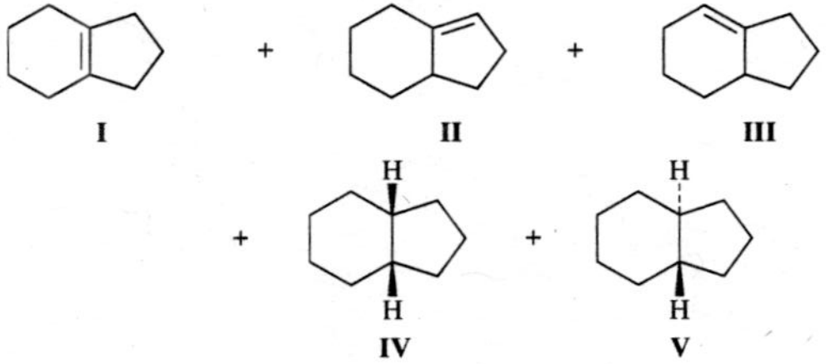 **I** + **II** + **III** + **IV** + **V** (100), **I**:**II**:**III**:**IV**:**V** = 95:5:0:0:0	131
	Et_3SiH, TFA, CH_2Cl_2	**I** + **II** + **III** + **IV** + **V** (100), **I**:**II**:**III**:**IV**:**V** = 83:7:0:6:4	131

TABLE 5. ORGANOSILANE REDUCTION OF ALCOHOLS (*Continued*)

Alcohol	Conditions	Product(s) and Yield(s) (%)	Refs.
C_9 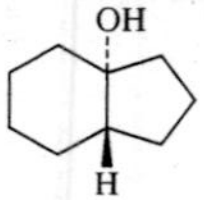	Ph_3SiH, TFA, CH_2Cl_2	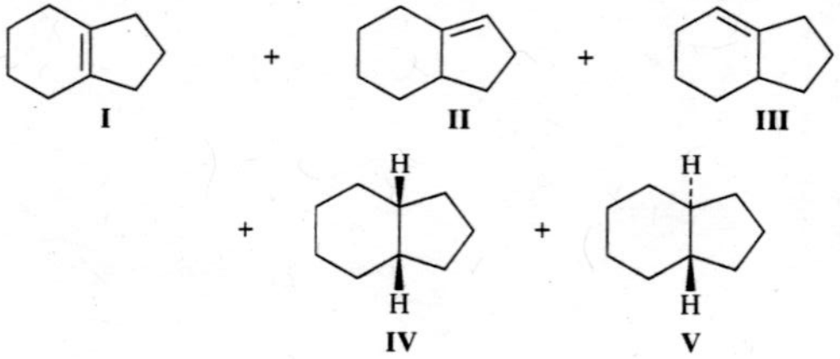 **I** + **II** + **III** + **IV** + **V** (100), **I**:**II**:**III**:**IV**:**V** = 90:6:3:0:1	131
	Et_3SiH, TFA, CH_2Cl_2	**I** + **II** + **III** + **IV** + **V** (100), **I**:**II**:**III**:**IV**:**V** = 62:13:7:3:15	131
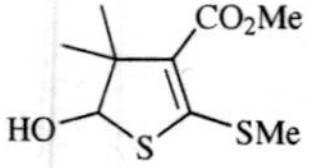	Et_3SiH, $BF_3•OEt_2$, CH_2Cl_2, 0°, 1 h; 20°, 3 h	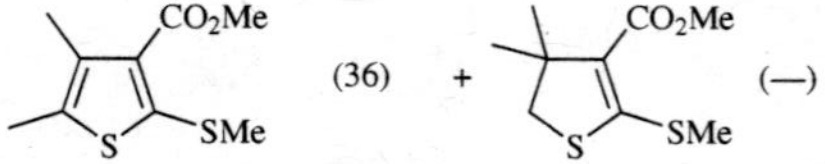 (36) + (—)	708
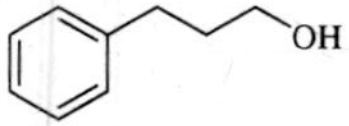	Et_3SiH (3.0 eq), $(C_6F_5)_3B$ (5 mol%), C_6H_{14} or CH_2Cl_2, rt , 20 h	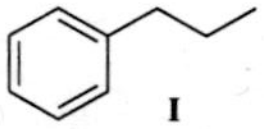 **I** (>95)	145
	Ph_2ClSiH (2 eq), $InCl_3$ (5 mol%), $ClCH_2CH_2Cl$, 80°, 5 h	**I** (0)	172
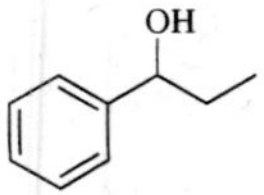	Ph_2ClSiH (2 eq), $InCl_3$ (5 mol%), CH_2Cl_2, rt, 1 h	**I** (90)	172
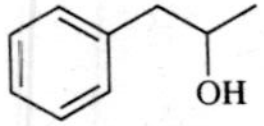	Ph_2ClSiH (2 eq), $InCl_3$ (5 mol%), $ClCH_2CH_2Cl$, 80°, 6 h	**I** (54)	172

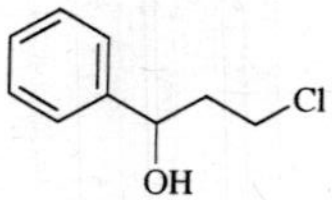

Ph_2ClSiH (2 eq), $InCl_3$ (0.05 eq), $ClCH_2CH_2Cl$, 80°, 0.5 h

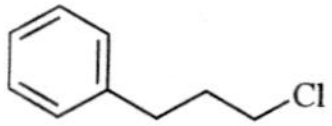

(95)

172

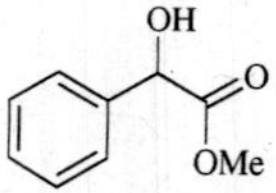

Et_3SiH (2 eq), BF_3•OEt_2 (2 eq), CH_2Cl_2, 0°, 30 min

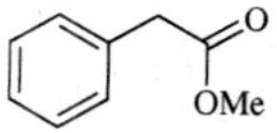

(0)

133

C_{10}

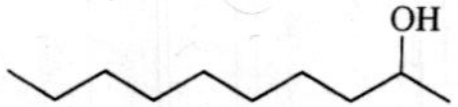

R_3SiH (2 eq), catalyst (5 mol%), 4 h

I

172

R_3Si	Catalyst	Solvent	Temp	
Ph_2ClSi	$InCl_3$	DCE	80°	(76)
Ph_2ClSi	none	DCE	80°	(5)
Ph_2ClSi	$AlCl_3$	DCE	80°	(23)
Ph_2ClSi	BF_3•OEt_2	DCE	80°	(tr)
Ph_2HSi	$InCl_3$	DCE	80°	(19)
Et_3Si	$InCl_3$	DCE	80°	(0)
Me_2ClSi	$InCl_3$	DCE	rt	(0)
Ph_2ClSi	$InCl_3$	C_6H_{14}	70°	(33)
Ph_2ClSi	$InCl_3$	C_6H_6	80°	(20)
Ph_2ClSi	$InCl_3$	THF	63°	(0)
Ph_2ClSi	$InCl_3$	MeCN	80°	(0)

Ph_2ClSiH (2 eq), $InCl_3$ (0.05 eq), $ClCH_2CH_2Cl$, 80°, 4 h

I (76)

172

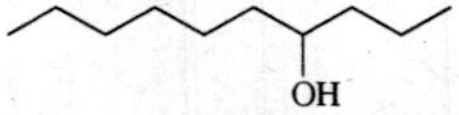

Ph_2ClSiH (2 eq), $InCl_3$ (0.05 eq), $ClCH_2CH_2Cl$, 80°, 4 h

I (74)

172

TABLE 5. ORGANOSILANE REDUCTION OF ALCOHOLS (*Continued*)

Alcohol	Conditions	Product(s) and Yield(s) (%)	Refs.
C_{10}			
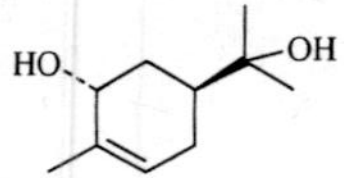	Et_3SiH (3 eq), $LiClO_4$, Et_2O, rt, 16 h	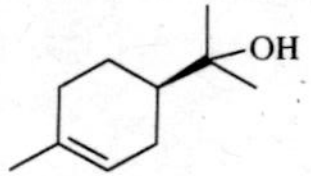 (52)	173
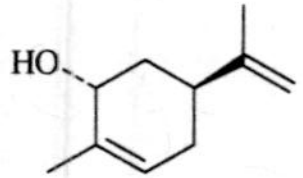	Et_3SiH (3 eq), $LiClO_4$, Et_2O, rt, 16 h	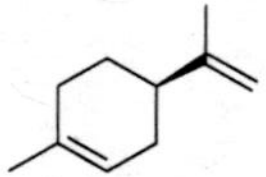 (68)	173
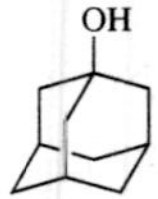	Et_3SiH (1.3 eq), NH_4F (1.3 eq), TFA (5 eq), 0°, 0.5 h; rt, 3 h	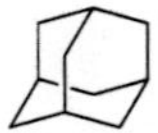 **I** (97)	135
	Ph_2ClSiH (2 eq), $InCl_3$ (0.05 eq), CH_2Cl_2, 80°, 3 h	**I** (99)	126, 172
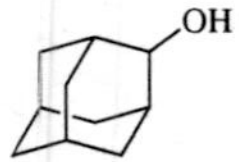	Et_3SiH (3.3 eq), BF_3, CH_2Cl_2, rt, 30 min	**I** (86)	126
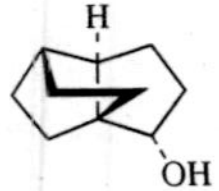	Et_3SiH, BF_3, CH_2Cl_2, rt, 30 min	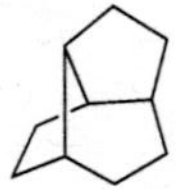 (100)	132
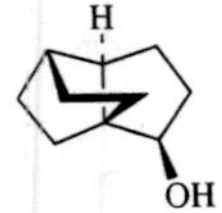	Et_3SiH, BF_3, CH_2Cl_2, rt, 30 min	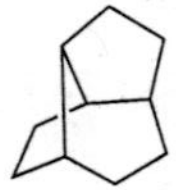 (100)	132

	Et_3SiH, BF_3, CH_2Cl_2, 5-10 min	 **I** (80-85) +  **II** (5)	503
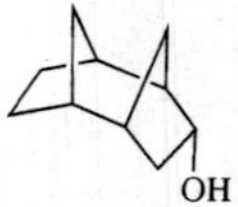	Et_3SiH, BF_3, CH_2Cl_2, 5-10 min	**I** (55-60)	503
	Et_3SiH, BF_3, CH_2Cl_2, 60 min	**I** (85) + **II** (<5)	503
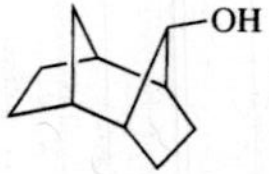	Et_3SiH, BF_3, CH_2Cl_2, 5-10 min	**II** (50-55) + 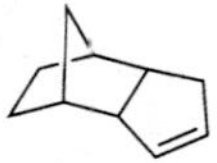 **III** (10) + 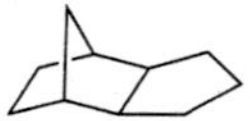 **IV** (30-35)	503
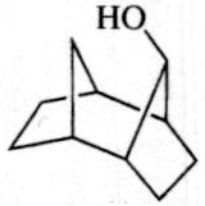	Et_3SiH, BF_3, CH_2Cl_2, 5-10 min	**II** (20-25) + **III** (5) + **IV** (40-45) +  **V** (15-25)	503
	Et_3SiH, BF_3, CH_2Cl_2, 5-10 min	**I** (95) + **II** (5)	503
	Et_3SiH, BF_3, CH_2Cl_2, 5-10 min	**I** (5)	503
	Et_3SiH, BF_3, CH_2Cl_2, 60 min	**I** (15) + **II** (<5)	503
	Et_3SiH, BF_3, CH_2Cl_2, 5-10 min	**II** (40-45) + **IV** (40) +  **VI** (15-20)	503
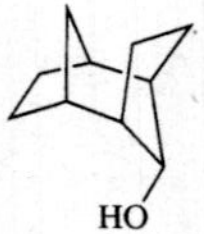	Et_3SiH, BF_3, CH_2Cl_2, 5-10 min	**II** (35-40) + **IV** (35-40) + **VI** (20-25)	503

TABLE 5. ORGANOSILANE REDUCTION OF ALCOHOLS (*Continued*)

Alcohol	Conditions	Product(s) and Yield(s) (%)	Refs.
C_{10} 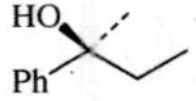 I +/or 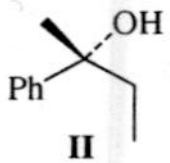 II	R_3SiH (1 eq), TFA, CH_2Cl_2, rt, 20 min	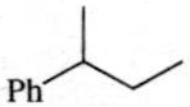	141, 142
Substrate	R_3Si / Silane (%) Opt. Purity	( ) / Conf.	
I + II	(*R*)-1-NpPhMeSi / 88.8	(82) / R	
II	(*R*)-1-NpPhMeSi / 96.2	(81) / R	
I	(*R*)-1-NpPhMeSi / 96.2	(75) / R	
II	(*S*)-1-NpPhMeSi / 89.0	(84) / S	
I	(*S*)-1-NpPhMeSi / 89.0	(75) / S	
II	Et_3Si / —	(85) / —	
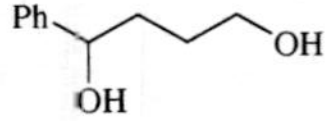	Ph_2ClSiH (2 eq), $InCl_3$ (0.05 eq), $ClCH_2CH_2Cl$, 80°, 3 h	 (53)	172
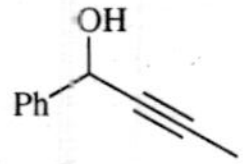	Et_3SiH, TFA	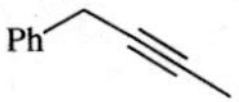 (—)	708
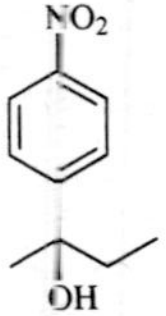	Ph(*t*-$BuCH_2$)MeSiH (2-10 mol%), BF_3, CH_2Cl_2, 0°, 10 min	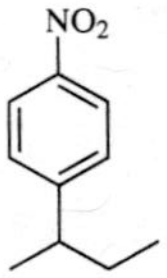 (100)	126
C_{11} 	Ph_3SiH (1.1 eq), TFA, CH_2Cl_2, rt	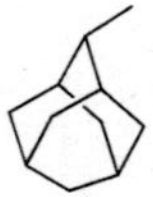 (41)	127

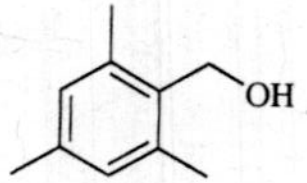	Ph_3SiH (3 eq), TFA, rt, 43 h	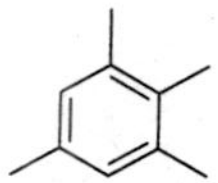 (41, 89[a])	26
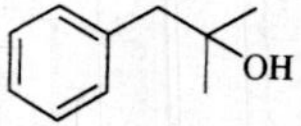	Et_3SiH (1 eq), acid, rt Acid / Time H_2SO_4-HOAc / 96 h 4-$MeC_6H_4CO_2H$/HOAc / 672 h TFA / 96 h	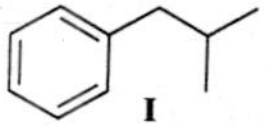 **I** (10) (60) (100)	134
	Ph_2ClSiH (2 eq), $InCl_3$ (5 mol%), $ClCH_2CH_2Cl$, 80°, 3 h	**I** (99)	172
C_{11} 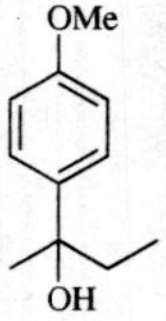	Ph(t-$BuCH_2$)MeSiH (2-10 mol%), BF_3, CH_2Cl_2, 0°, 2 min	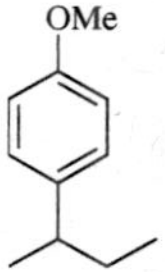 **I** (100)	126
	Ph(t-$BuCH_2$)MeSiH (2-10 mol%), TFA, CH_2Cl_2, 0°, 2 min	**I** (100)	126
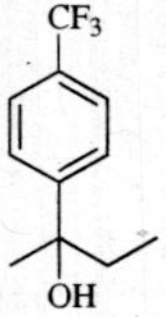	Ph(t-$BuCH_2$)MeSiH (2-10 mol%), BF_3, CH_2Cl_2, 0°, 2 min	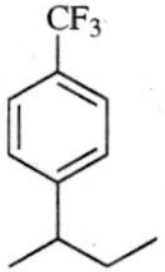 (80)	126

TABLE 5. ORGANOSILANE REDUCTION OF ALCOHOLS (*Continued*)

Alcohol	Conditions	Product(s) and Yield(s) (%)	Refs.
C_{11} 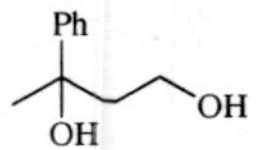	Et_3SiH (2 eq), BF_3•OEt_2 (2 eq), CH_2Cl_2, 0°, 30 min	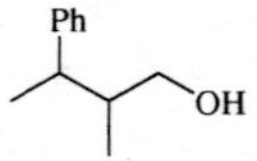 (71)	133
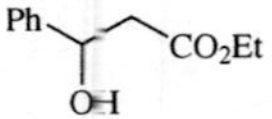	Ph_2ClSiH (2 eq), $InCl_3$ (0.05 eq), $ClCH_2CH_2Cl$, 80°, 0.5 h	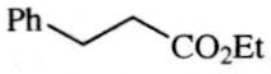 (68)	172
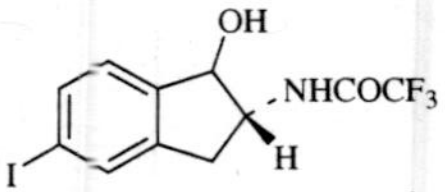	Et_3SiH, BF_3•OEt_2, 70°	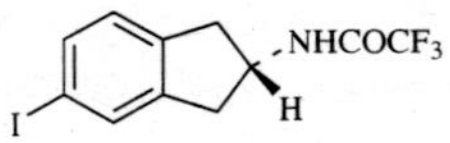 (72)	709
C_{11-17} 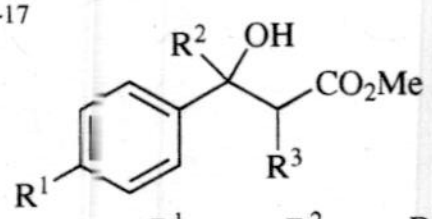	Et_3SiH (1 eq), BF_3•OEt_2 (1 eq), CH_2Cl_2, 0°, 1 h	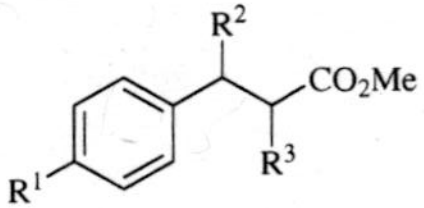	170
R^1 R^2 R^3			
H Me H		(85)	
H H Me		(100)	
H Me Me		(100)	
H H Ph		(94)	
H Me Ph		(89)	
OMe Me H		(94)	
C_{12} 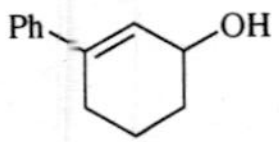	Et_3SiH (3 eq), $LiClO_4$, Et_2O, rt, 16 h	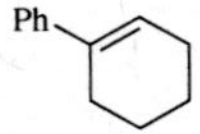 (92)	173
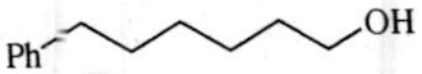	Et_3SiH (3.0 eq), $(C_6F_5)_3B$ (5 mol%), C_6H_{14} or CH_2Cl_2, rt, 20 h	 (>99)	145
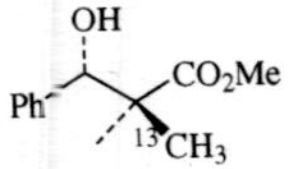	$EtMe_2SiH$, HBF_4	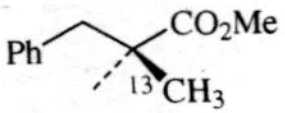 (Đ)	604

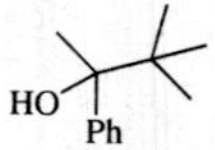	1-NpPhMeSiH (1.1 eq), CH_2Cl_2, BF_3, –60°, 6 min	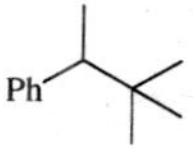 (86.5) + 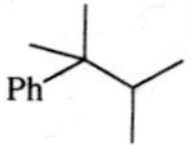 (13.5)	126
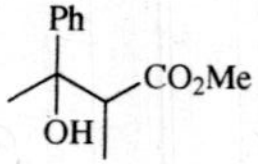	Et_3SiH (2 eq), $BF_3{\bullet}OEt_2$ (2 eq), CH_2Cl_2, rt, 20 h	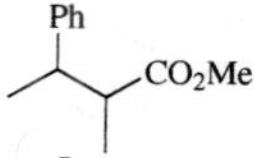 I + 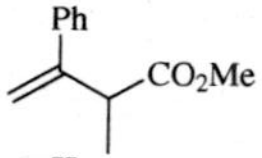 II **I + II** (87), **I**:**II** = 92:8	133
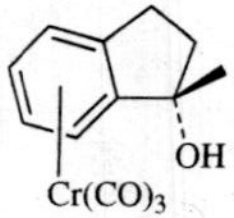	Et_3SiH, TFA	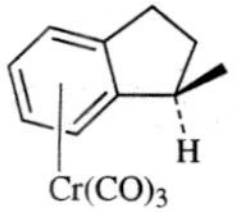 (95)	710
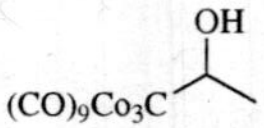	Et_3SiH, TFA	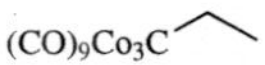 (72)	309
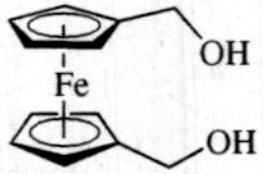	Et_3SiH (2 eq), TFA, rt, 20 min	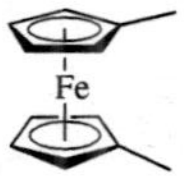 (88)	175
C_{12-14}			
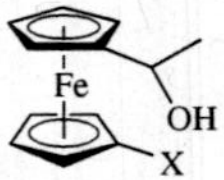	Et_3SiH (5 eq), TFA (x eq), HOAc, 20°, time	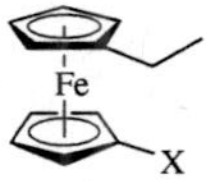	178

X	x	Time	
H	3.5	1 h	(95)
Cl	3.5	20 h	(70)
Cl	10	2 h	(90)
CO_2Me	10	3 h	(85)
CN	10	20 h	(55)

TABLE 5. ORGANOSILANE REDUCTION OF ALCOHOLS (*Continued*)

Alcohol	Conditions	Product(s) and Yield(s) (%)	Refs.
C_{12-20} 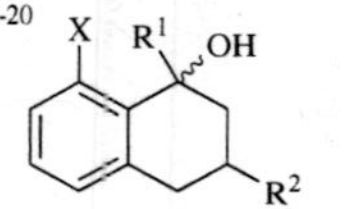	Et_3SiH (3 eq), TFA (3 eq), CH_2Cl_2, –78° to 0°	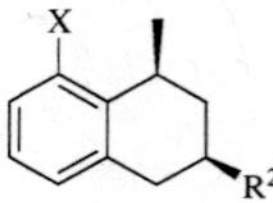 **I** + 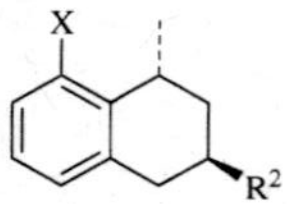 **II**	162
X / R^1 / R^2		**I + II** / **I:II**	
CH / Me / Me		(93) / 100:0	
H / Me / *i*-Pr		(70) / 100:0	
H / Ph / Me		(86) / 10:1	
H / Ph / *i*-Pr		(75) / 100:0	
H / Me / Me		(88) / 100:0	
C_{13} 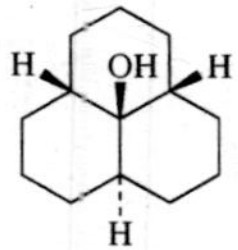	Ph_3SiH (1.1 eq), TFA, CH_2Cl_2, rt, 48 h	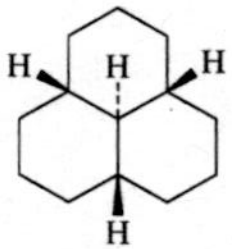 (92)	127
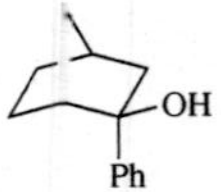	R_3SiH (1.2 eq), TFA, CH_2Cl_2, rt, 30 min	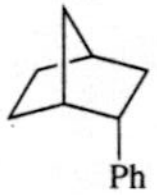 R_3Si: Et_3Si (100)[a]; Ph_3Si (65); PhH_2Si (100)[a]	164
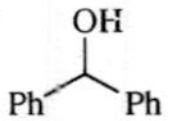	Et_3SiH (3.0 eq), $(C_6F_5)_3B$ (5 mol%), C_6H_{14} or CH_2Cl_2, rt, 20 h	 **I** (98)	145
	Et_3SiH (1 eq), acid, rt	**I** Acid / Time: H_2SO_4/HOAc, 30 min (60); 4-$MeC_6H_4CO_2H$/HOAc, 30 min (55); TFA, 2 h (70)	134
	Ph_3SiH (3 eq), TFA, rt, 48 h	**I** (70)	26
	Ph_2ClSiH (2 eq), $InCl_3$ (0.05 eq), CH_2Cl_2, rt, 2 h	**I** (87)	172

Reactant		Conditions	Product(s) and Yield(s) (%)	Refs.
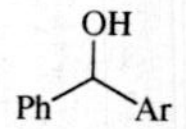		HMe$_2$SiOSiMe$_2$H (1.0 eq), TMSOTf (cat.), C_6H_6, 80°, 240 min	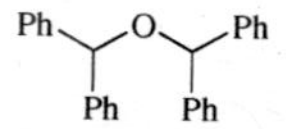 (91)	314
		Cl_3SiH (3 eq), (n-Pr)$_3$N (1 eq), 55-75°, 1 h	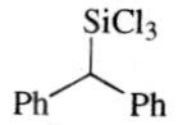 (95)	711
	Ar = 3-BrC_6H_4	Et_3SiH, TFA	 I (—)	712
	Ar = 4-$O_2NC_6H_4$	Ph_2ClSiH (2 eq), $InCl_3$ (5 mol%), CH_2Cl_2, rt, 3 h	I (95)	172
C_{13-15} 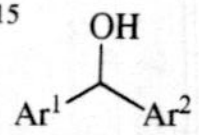		Et_3SiH, HOTf, $CHCl_3$	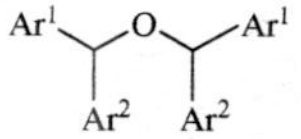 I +  II	713

Ar[1]	Ar[2]	I	II
Ph	Ph	(30)	(70)
Ph	4-ClC_6H_4	(4)	(29)
Ph	4-BrC_6H_4	(11)	(71)
4-$MeOC_6H_4$	4-$MeOC_6H_4$	(73)	(17)
4-$MeOC_6H_4$	4-$MeOC_6H_4$	(22)	(78)
4-$MeOC_6H_4$	4-$MeOC_6H_4$	(22)	(78)
4-$MeOC_6H_4$	4-$MeOC_6H_4$	(5)	(95)
4-$MeOC_6H_4$	4-$MeOC_6H_4$	(0)	(100)
2,5-$(MeO)_2C_6H_3$	4-ClC_6H_4	(29)	(0)
2,5-$(MeO)_2C_6H_3$	4-ClC_6H_4	(0)	(84)
2,5-$(MeO)_2C_6H_3$	4-BrC_6H_4	(0)	(90)
2,5-$(MeO)_2C_6H_3$	4-IC_6H_4	(0)	(92)

Reactant	Conditions	Product(s) and Yield(s) (%)	Refs.
C_{13-18} 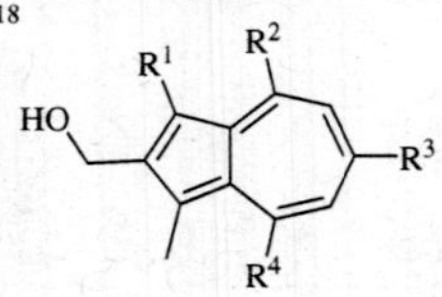	Et_3SiH (xs), TFA, 50°, 19 h	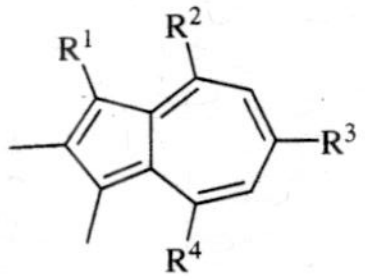	163

R[1]	R[2]	R[3]	R[4]	
H	H	Me	H	(100)
H	H	Me	Me	(99)
H	Me	Me	H	(100)
H	Me	Me	Me	(100)
Me	H	Me	Me	(96)
H	Me	t-Bu	Me	(100)

TABLE 5. ORGANOSILANE REDUCTION OF ALCOHOLS (*Continued*)

Alcohol	Conditions	Product(s) and Yield(s) (%)	Refs.
C_{14} 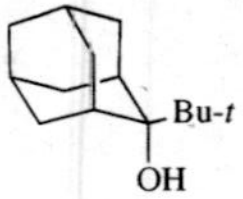	$(n\text{-}C_6H_{13})_3SiH$, TFA, CH_2Cl_2, rt, 10 min	 (90)	151, 152
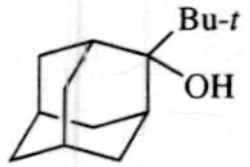	Et_3SiH, TFA, CH_2Cl_2, rt, 16 h	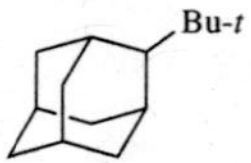 (82)	153
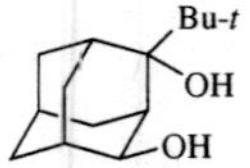	Et_3SiH, TFA, CH_2Cl_2, rt, 16 h	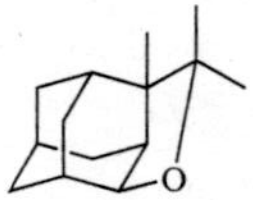 **I** (81)	153
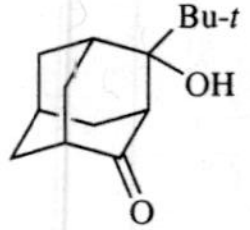	Et_3SiH, TFA, CH_2Cl_2, rt, 16 h	**I** (79)	153
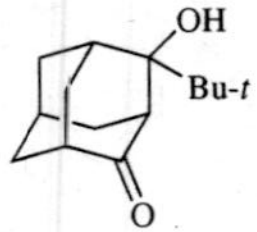	Et_3SiH, TFA, CH_2Cl_2, rt, 16 h	**I** (77)	153
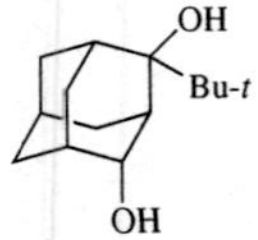	Et_3SiH, TFA, CH_2Cl_2, rt, 16 h	**I** (78)	153
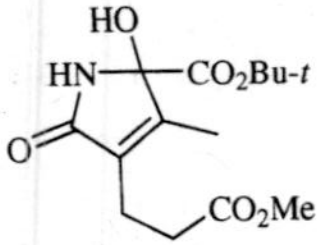	Et_3SiH (xs), TFA, 18°, 18 h; 45°, 3 h	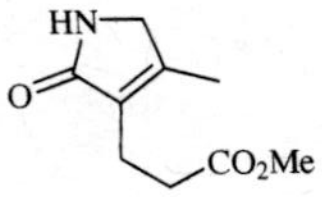 (87)	714

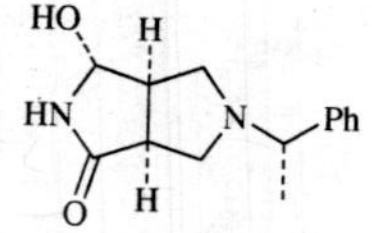	Et_3SiH, TFA, rt, 16 h	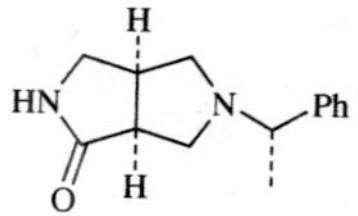 (31)	715
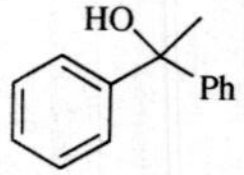	Et_3SiH (1.3 eq), NH_4F (1.3 eq), TFA (5 eq), 0°: 0.5 h, rt, 4 h	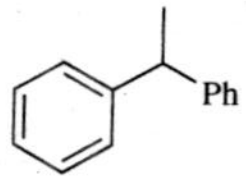 (96)	135
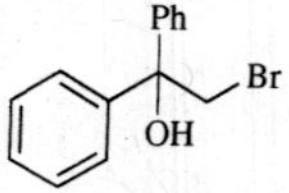	Et_3SiH (1.4 eq), TFA (1 eq), $CHCl_3$, 0°, 1 h	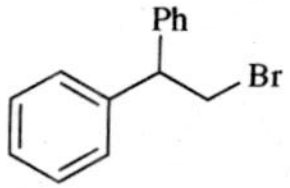 (66)	184
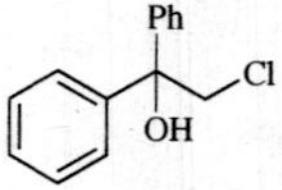	Et_3SiH (1.25 eq), TFA (9 eq), $CHCl_3$, –15°, 1 h	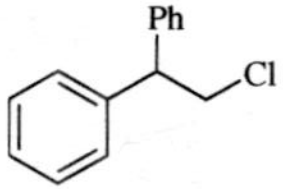 (77)	184
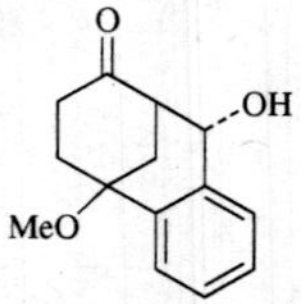	Et_3SiH, TFA	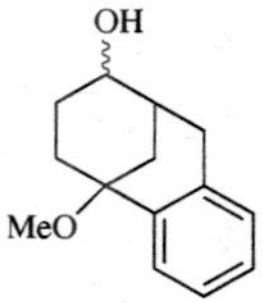 (75)	716
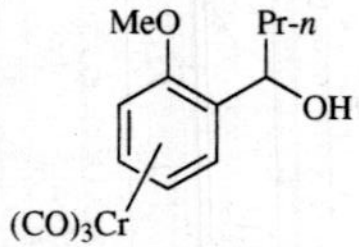	Et_3SiH, TFA	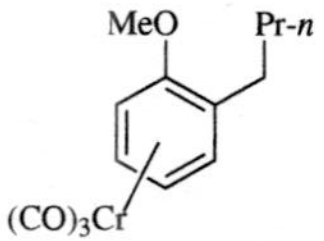 (86)	717
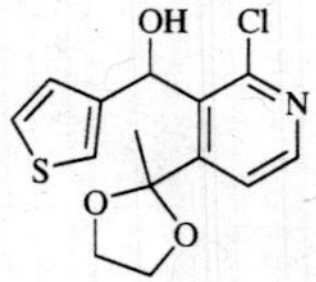	Et_3SiH (1.1 eq), TFA, rt, 18 h	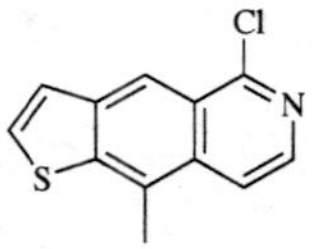 (60)	486

TABLE 5. ORGANOSILANE REDUCTION OF ALCOHOLS (*Continued*)

Alcohol	Conditions	Product(s) and Yield(s) (%)	Refs.
C_{14}			
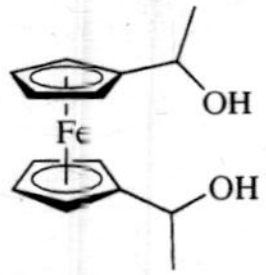	Et_3SiH (2 eq), TFA, rt, 20 min	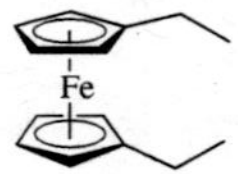 (88)	179
C_{14-15}			
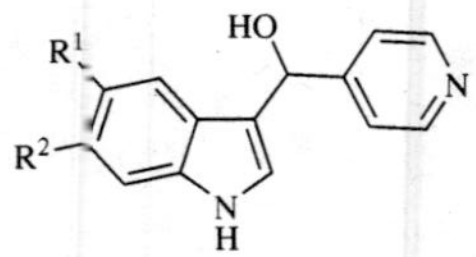	Et_3SiH, TFA, CH_2Cl_2, rt	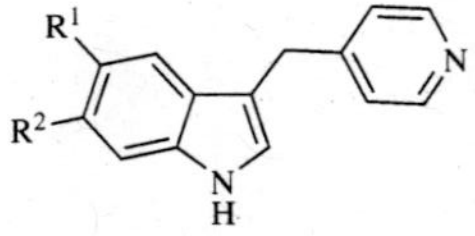 R[1] / R[2] / Time: H / H / 16 h (66) F / H / 18 h (80) H / F / 18 h (68) NO_2 / H / 2 h (78) CN / H / 4 h (73)	718
C_{15}			
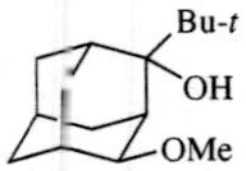	Et_3SiH, TFA, CH_2Cl_2, rt, 16 h	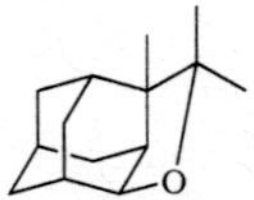 (8) + 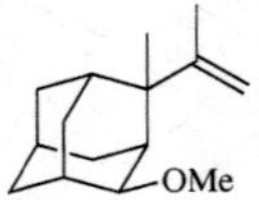 (53)	153
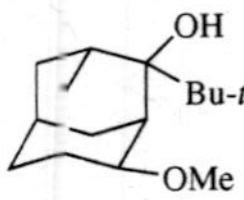	Et_3SiH, TFA, CH_2Cl_2, rt, 16 h	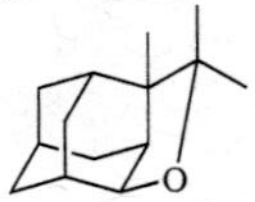 (8) + 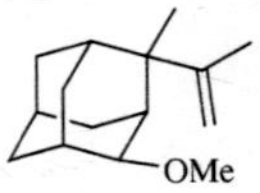 (56)	153
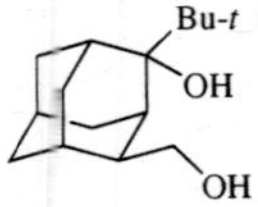	Et_3SiH, TFA, CH_2Cl_2, rt, 16 h	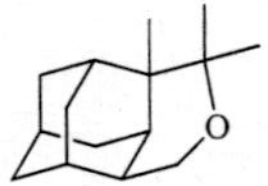 (83)	153
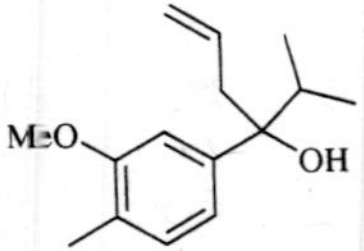	Et_3SiH, TFA	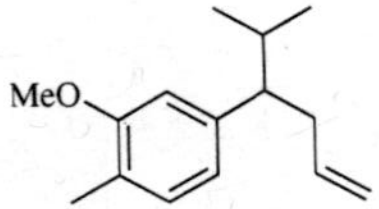 (—)	719

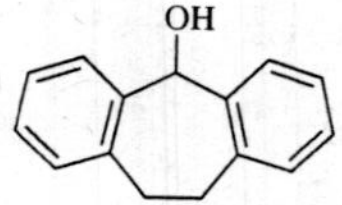	Et_3SiH (1.3 eq), NH_4F (1.3 eq), TFA (5 eq), 0°, 0.5 h; rt, 4 h	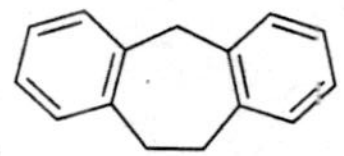 (99)	135
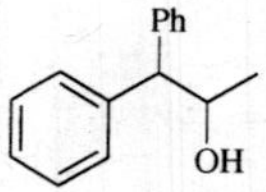	Ph_2ClSiH (2 eq) $InCl_3$ (0.05 eq), $ClCH_2CH_2Cl$, 80°, 1 h	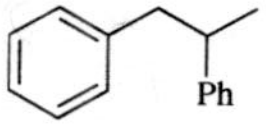 (92)	172
C_{16}			
n-$C_{16}H_{33}OH$	Et_3SiH (3.0 eq), $(C_6F_5)_3B$ (5 mol%), C_6H_{14} or CH_2Cl_2, rt, 20 h	n-$C_{16}H_{34}$ (95)	145
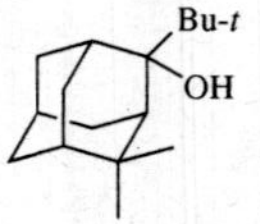	Et_3SiH, TFA, CH_2Cl_2, rt, 16 h	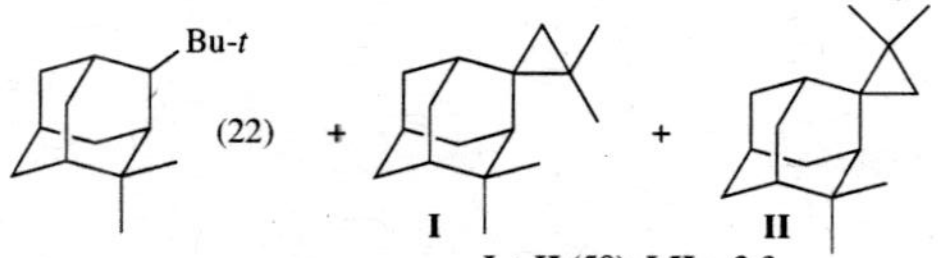 (22) + **I** + **II** **I** + **II** (50), **I**:**II** = 2:3	153
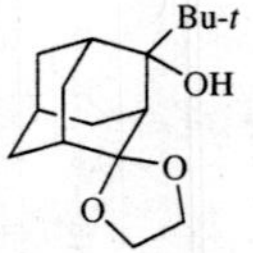	Et_3SiH, TFA, CH_2Cl_2, rt, 16 h	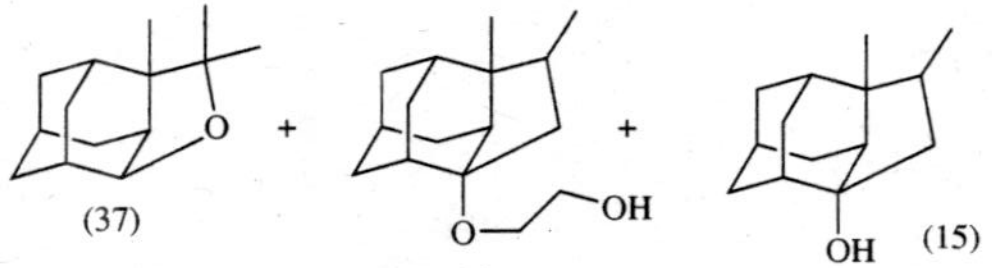 (37) + (26) + (15)	153
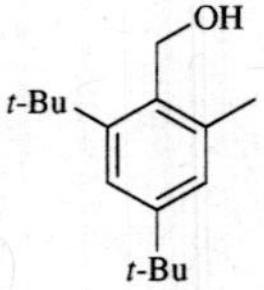	Et_3SiH (3.1 eq), TFA (6.4 eq), CH_2Cl_2, rt, 5 h	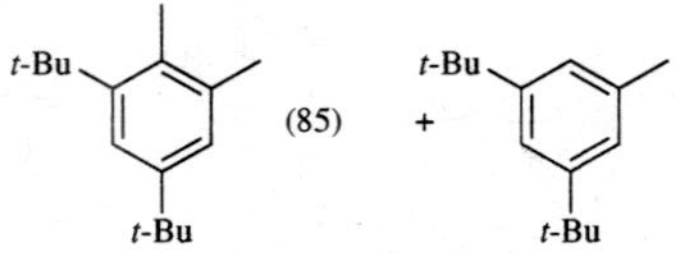 (85) + (15)	128

Alcohol	Conditions	Product(s) and Yield(s) (%)	Refs.
C_{16}			
 I or 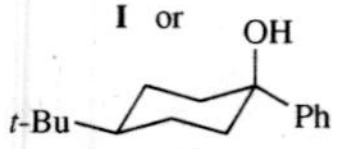 II	R_3SiH (1.3 eq), TFA (2.1 eq), CH_2Cl_2, rt, 0.5 h	 III +  IV + 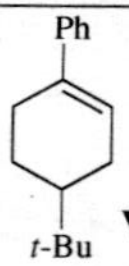 V	26

R_3Si	Alcohol	III	IV	V	IV:III
Et_3Si	I	(20)	(39)	(41)	1.8
$(n\text{-}C_8H_{17})_3Si$	I	(15)	(21)	(64)	1.4
Ph_3Si	I	(10)	(18)	(72)	1.8
Et_2HSi	I	(12)	(54)	(34)	4.5
Ph_2HSi	I	(11)	(42)	(47)	3.8
PhH_2Si	I	(3)	(13)	(84)	4.3
Et_3Si	II	(28)	(50)	(22)	1.8
$(n\text{-}C_8H_{17})_3Si$	II	(24)	(36)	(40)	1.5
Ph_3Si	II	(20)	(29)	(51)	1.5
Ph_2HSi	II	(13)	(49)	(38)	3.8

Alcohol	Conditions	Product(s) and Yield(s) (%)	Refs.
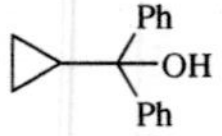	Ph_3SiH, TFA, CH_2Cl_2, rt, 0.5 h	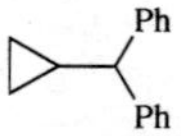 (16) + 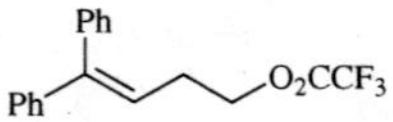 (84)	24
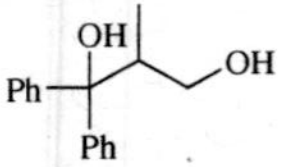	Et_3SiH, CH_2Cl_2, 0°, 0.5 h	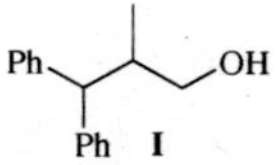 I (—) + 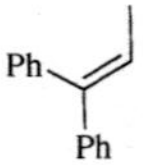 (—) + HCHO	133
	Et_3SiH (2 eq), BF_3•OEt_2 (2 eq), CH_2Cl_2, –40°, 30 min	I + 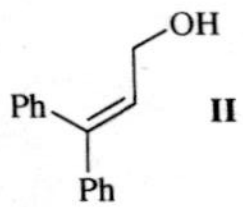 II I + II (92), I:II = >95:5	137

Reactant	Conditions	Product(s) and Yield(s) (%)	Refs.
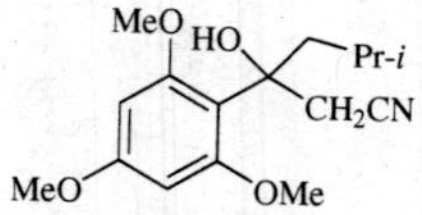	Et_3SiH (2.7 eq), TFA (7.8 eq), 80°, 2 h	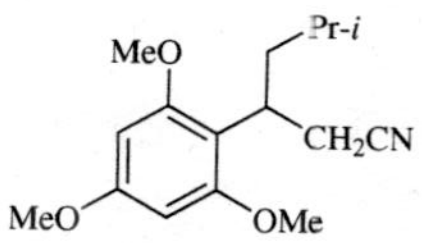 (92)	720
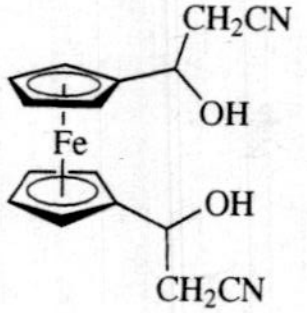	Et_3SiH (2 eq), TFA, rt, 3 h	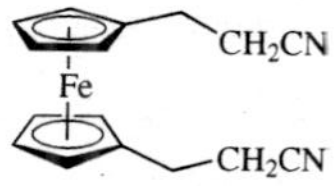 (83)	179
C_{16}			
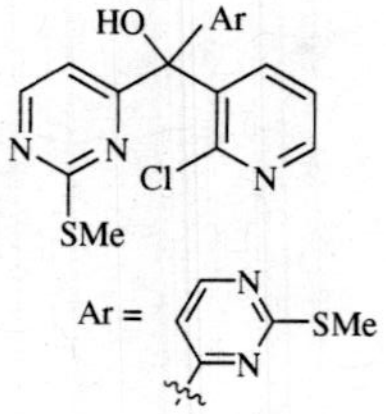	Et_3SiH (x eq), TFA (y eq) x y 2 8.3 4 8.2 8 8.2 8.1 4.3 7.8 2.0	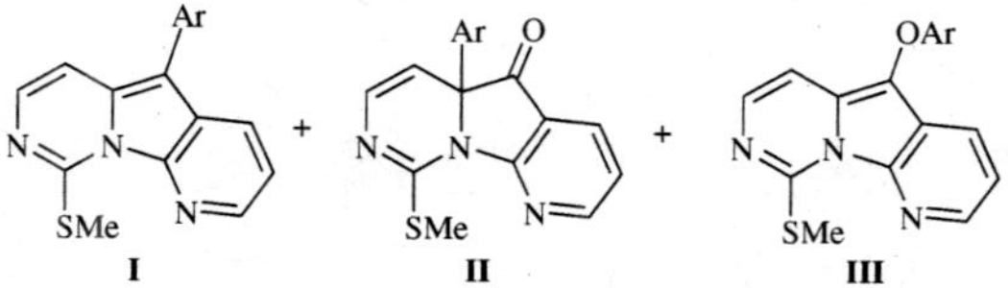 I (15) (18) (22) (34) (33) + 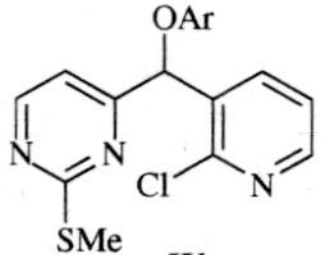 IV II + III + IV (—)	721
C_{17}			
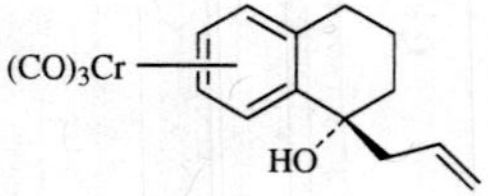	Et_3SiH (3 eq), TFA (3 eq), CH_2Cl_2, rt, 1.5 h	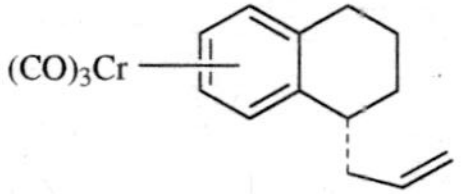 (92)	182
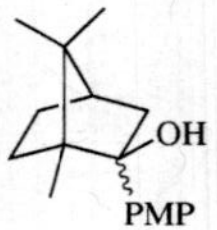	Ph_3SiH, TFA, CH_2Cl_2, rt, 0.5 h	 (—) only isomer	24

TABLE 5. ORGANOSILANE REDUCTION OF ALCOHOLS (*Continued*)

Alcohol	Conditions	Product(s) and Yield(s) (%)	Refs.
C_{17} 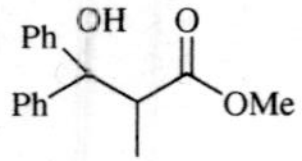	Et_3SiH (2 eq), $BF_3{\bullet}OEt_2$ (2 eq), CH_2Cl_2, 0°, 30 min	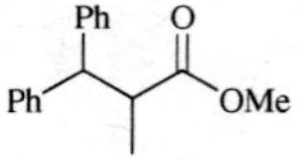 (89)	137
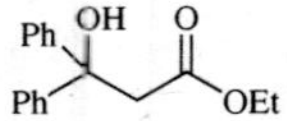	Et_3SiH (2 eq), $BF_3{\bullet}OEt_2$ (2 eq), CH_2Cl_2, 0°, 30 min	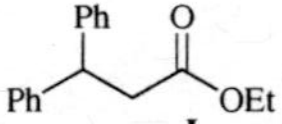 **I** + 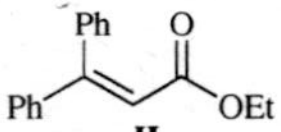 **II** **I** + **II** (68), **I**:**II** = 73:27	137
C_{18} 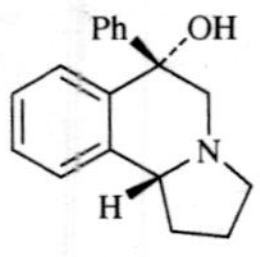	Et_3SiH, TFA, 0-5°, 24 h	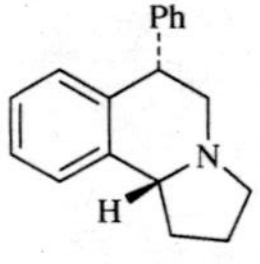 (trace)	722
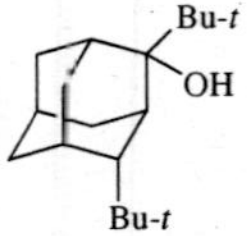	Et_3SiH, TFA, CH_2Cl_2, rt, 16 h	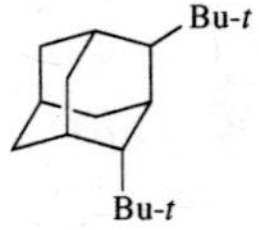 (31) + 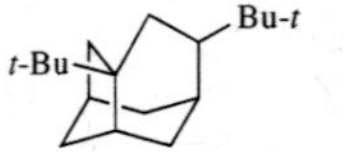 (38)	153
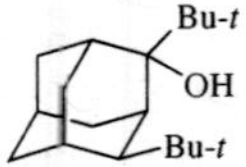	Et_3SiH, TFA, CH_2Cl_2, rt, 16 h	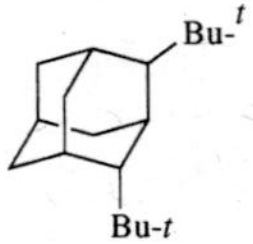 (80)	153
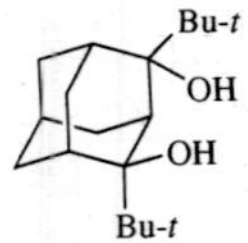	Et_3SiH, TFA, CH_2Cl_2, rt, 16 h	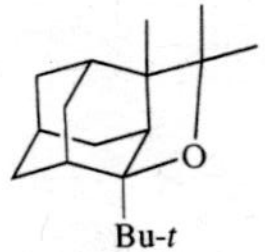 (78)	153

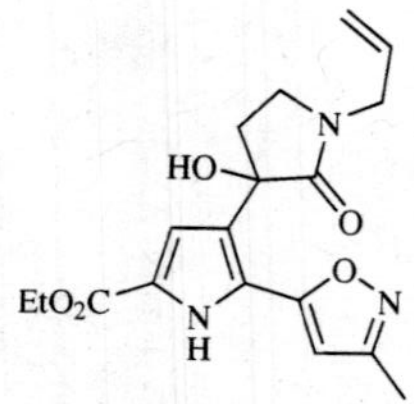	Et_3SiH (xs), TFA, rt, 2 h	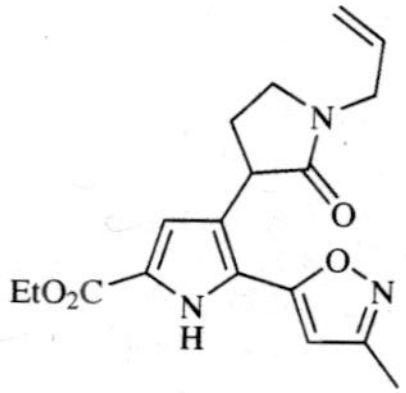 (86)	157
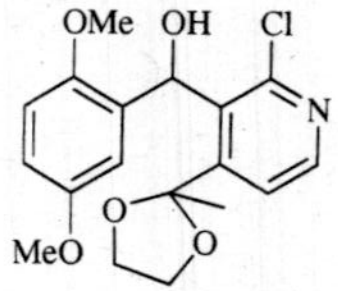	Et_3SiH (1.1 eq), TFA, rt, 18 h	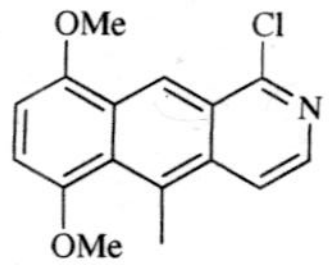 (65)	486
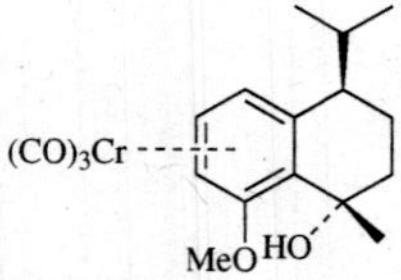	Et_3SiH (xs), TFA	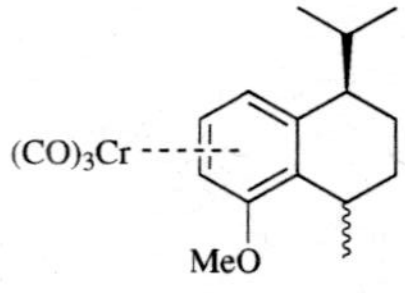 (75)	723
C_{18-21}			
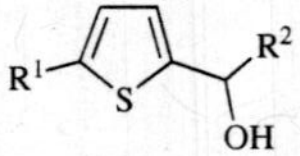	Et_3SiH, TFA, 50°, 48 h	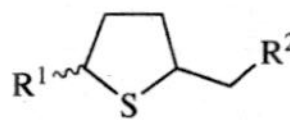 (—) R^1 / R^2: H / n-$C_{15}H_{31}$; H / n-$C_{16}H_{33}$; Me / n-$C_{12}H_{25}$	258
C_{19}			
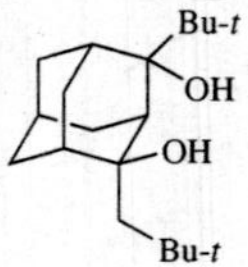	Et_3SiH, TFA, CH_2Cl_2, rt, 16 h	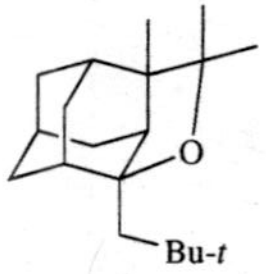 (81)	153
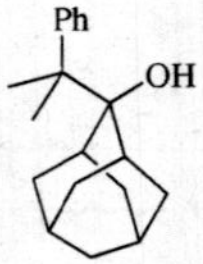	Et_3SiH (10 eq), $MeSO_3H$ (1 eq), CH_2Cl_2, −78°, 12 h; 0°, 10 h	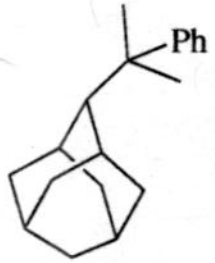 (95)	140

TABLE 5. ORGANOSILANE REDUCTION OF ALCOHOLS (*Continued*)

Alcohol	Conditions	Product(s) and Yield(s) (%)	Refs.
C_{19} 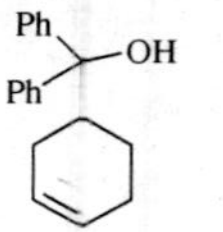	Et_3SiH (1 eq), TFA (17 eq), CH_2Cl_2, rt, 20 h	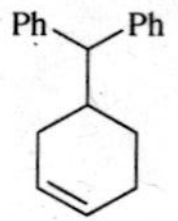 (—)	724
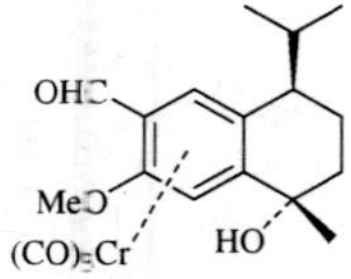	Et_3SiH (6 eq), TFA (9 eq), 60°, 4.5 h	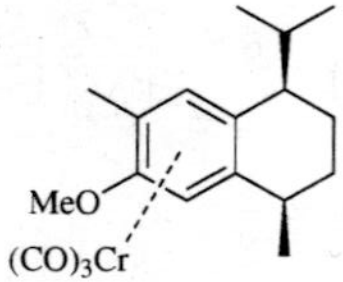 (82)	352, 725
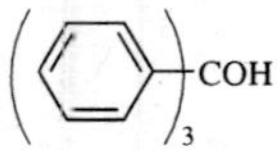	Et_3SiH (3.0 eq), $(C_6F_5)_3B$ (5 mol%), C_6H_{14} or CH_2Cl_2, rt, 20 h	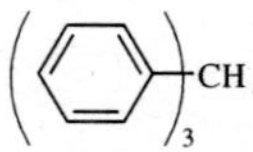 **I** (98)	145
	Et_3SiH (1 eq), acid, rt, 30 min	**I** Acid: H_2SO_4/HOAc (88); 4-$MeC_6H_4CO_2H$/HOAc (80); TFA (59)	134
	Ph_3SiH (3 eq), TFA, rt, 48 h	**I** (74)	26
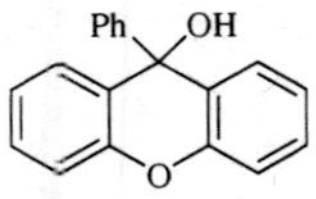	Et_3SiH (3 eq), HOAc, rt, 48 h	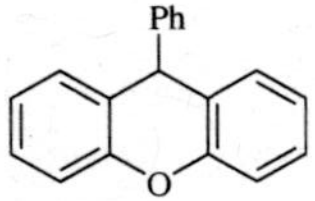 **I** (95)	26
	Ph_3SiH (3 eq), TFA, rt, 48 h	**I** (80)	26
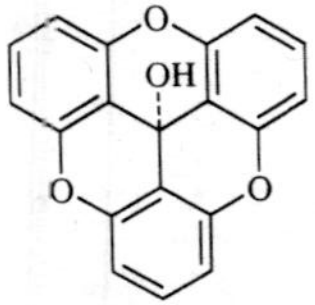	Et_3SiH (4 eq), HOAc, reflux, 40 h	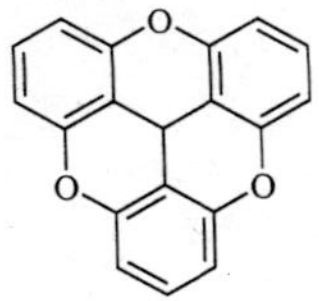 (89)	29

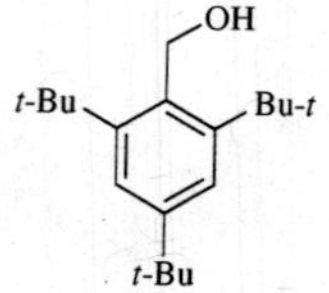	Et_3SiH (3.2 eq), TFA (5.9 eq), CD_2Cl_2, rt, 1 h	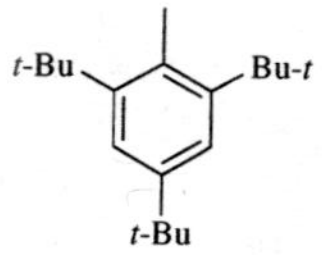 (90)	128
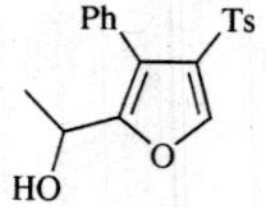	Et_3SiH, TFA, rt	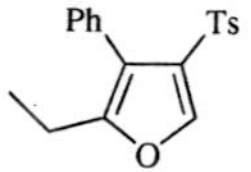 (95)	726
C_{19-23}			
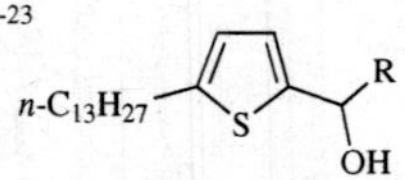 R = Me, Et, *n*-Pr, *n*-Bu, *n*-C_5H_{11}	Et_3SiH, TFA, 50°, 48 h	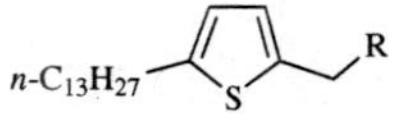 (—) + 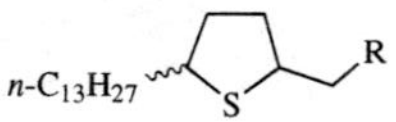 (—)	258
C_{20}			
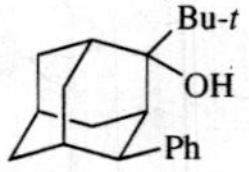	Et_3SiH, TFA, CH_2Cl_2, rt, 16 h	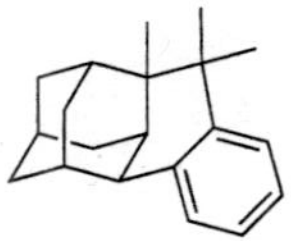 (78)	153
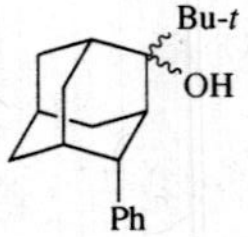 1:1	Et_3SiH, TFA, CH_2Cl_2, rt, 16 h	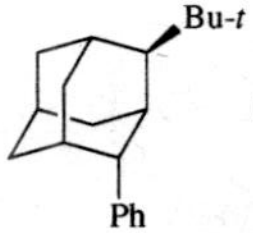 **I** + 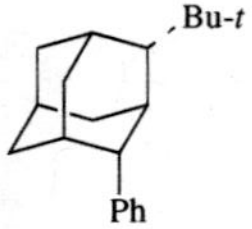 **II** **I** + **II** (73), **I**:**II** = 3:2	153
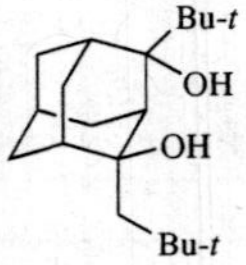	Et_3SiH, TFA, CH_2Cl_2, rt, 16 h	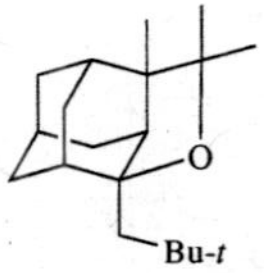 (81)	153

TABLE 5. ORGANOSILANE REDUCTION OF ALCOHOLS (*Continued*)

Alcohol	Conditions	Product(s) and Yield(s) (%)	Refs.
C_{20} 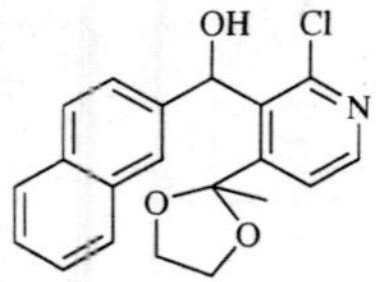	Et_3SiH (1.1 eq), TFA, rt, 18 h	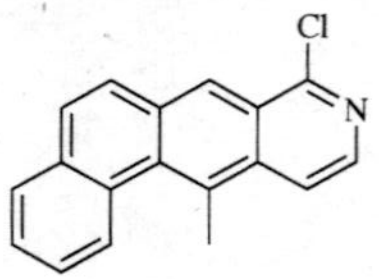 (37)	486
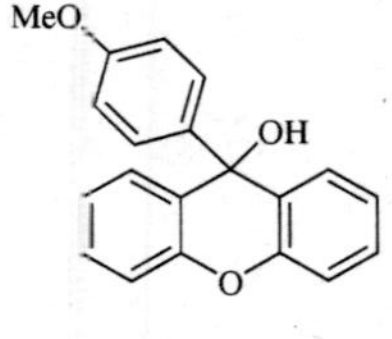	Et_3SiH (3 eq), TFA, rt, 48 h	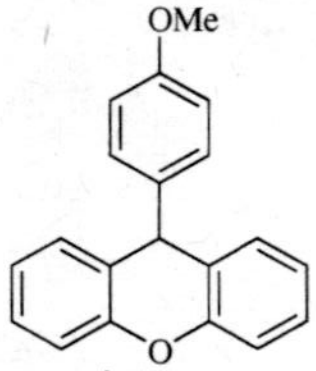 **I** (68)	26
	Ph_3SiH (3 eq), TFA, rt, 48 h	**I** (94)	26
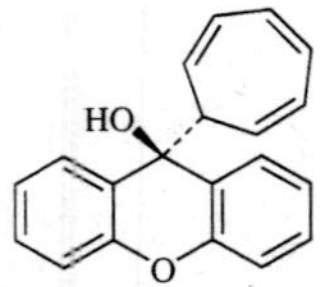	Et_3SiH, TFA, CH_2Cl_2, rt, 15 min	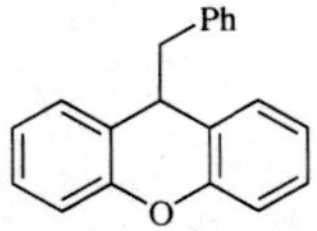 (88)	64
	Et_3SiD, TFA, CH_2Cl_2, rt, 15 min	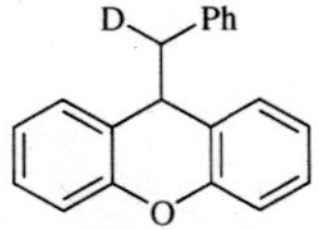 (75)	64
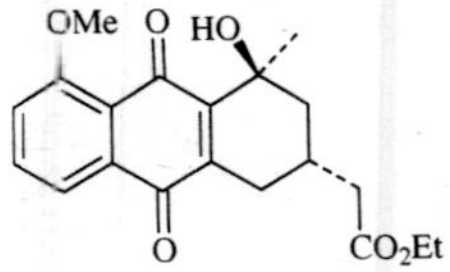	Et_3SiH (3 eq), TFA (3 eq), CH_2Cl_2, –78° to 0°	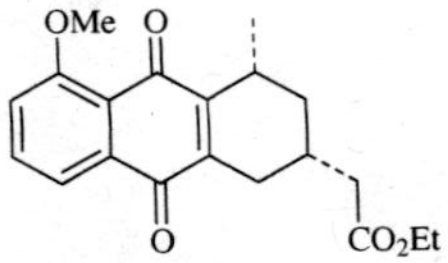 (95)	162

	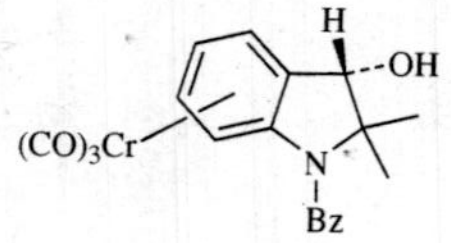	Et_3SiH, HPF_6, CH_2Cl_2, −30°, 1 h	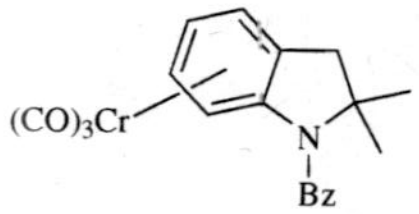 (52)	727
	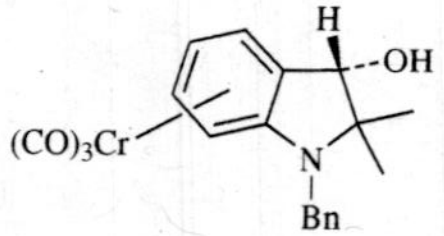	Et_3SiH, HPF_6, CH_2Cl_2, −30°, 0.25 h	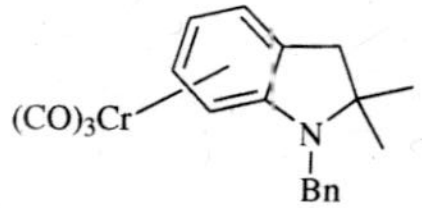 (89)	727
C_{21}	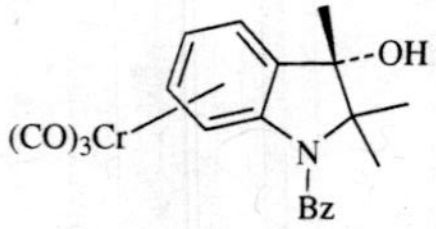	Et_3SiH, HPF_6, CH_2Cl_2, −30°, 1 h	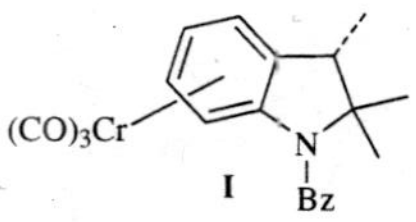 **I** (65)	727
		Et_3SiH, TFA, CH_2Cl_2, rt, 1 h	**I** (60)	727
	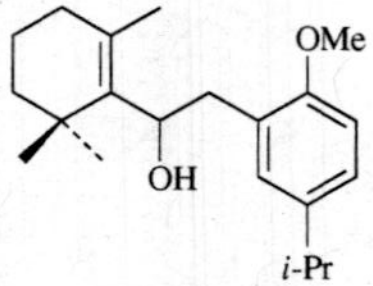	Et_3SiH, BF_3•OEt_2, CH_2Cl_2, −10°, 1 h	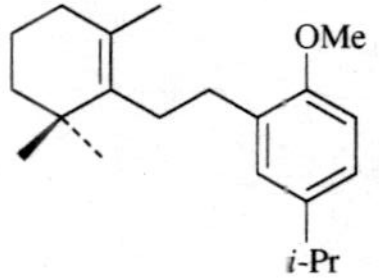 (69)	174
	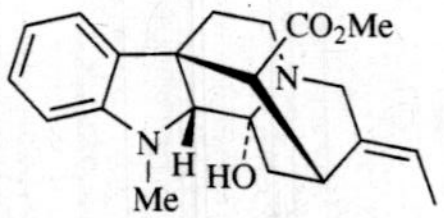	Et_3SiH, TFA, rt, 4 h	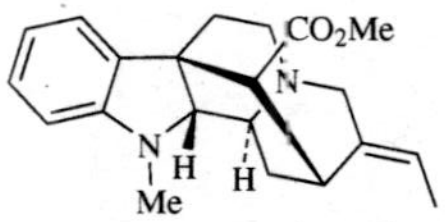 (79)	728
C_{22}	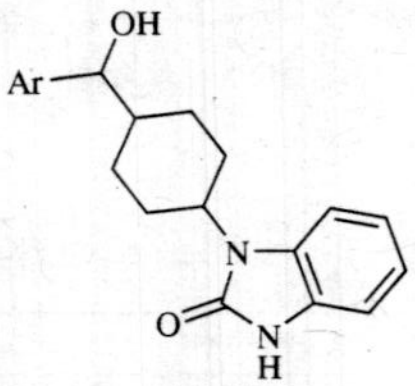 Ar = 3,5-$(MeO)_2C_6H_3$	Et_3SiH (3 eq), TFA, rt, 24 h	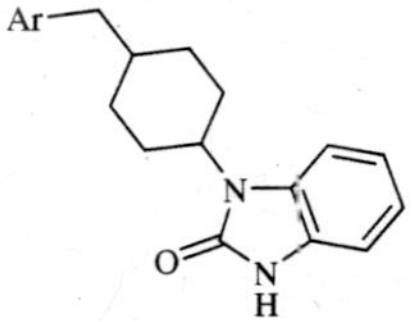 (55.5)	729

TABLE 5. ORGANOSILANE REDUCTION OF ALCOHOLS (*Continued*)

	Alcohol	Conditions	Product(s) and Yield(s) (%)	Refs.
C_{22}	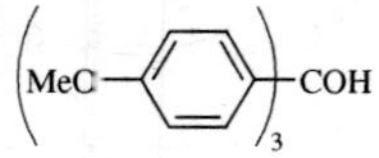	Ph_3SiH (3 eq), TFA, rt, 48 h	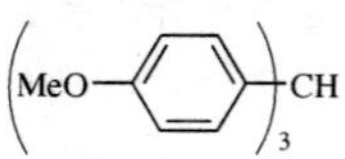 (87)	26
	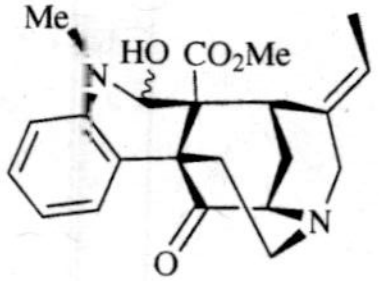	Et_3SiH, TFA, CH_2Cl_2, rt	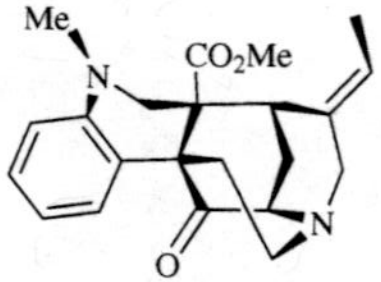 (—)	730
	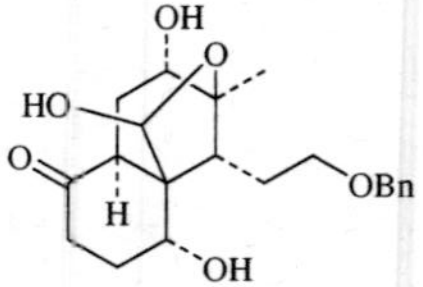	1. *i*-Bu_2AlH, hexanes, MeC_6H_5, −78° 2. Et_3SiH (1.5 eq), $BF_3•OEt_2$ (1.1 eq), −20°, 24 h	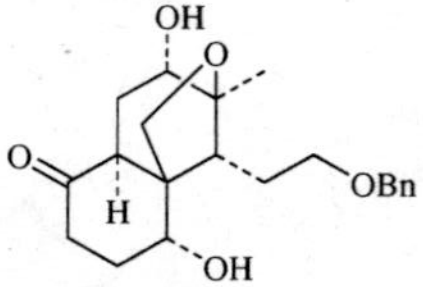 (65)	510
C_{23}	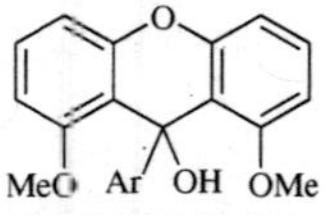 Ar = 2,6-$(MeO)_2C_6H_3$	Et_3SiH (4 eq), HOAc, reflux, 24 h	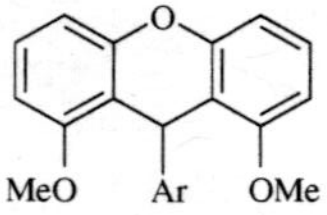 (83)	29
	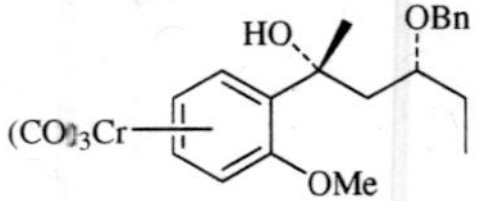	Et_3SiH, $BF_3•OEt_2$, CH_2Cl_2, −78° to 0°	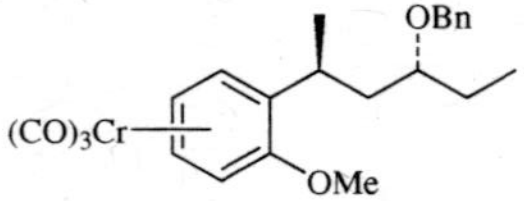 (75)	181
C_{24}	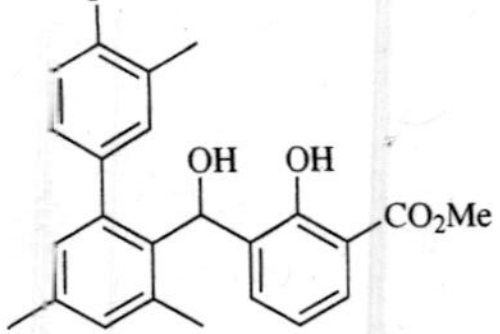	Et_3SiH, TFA, rt, 15 h	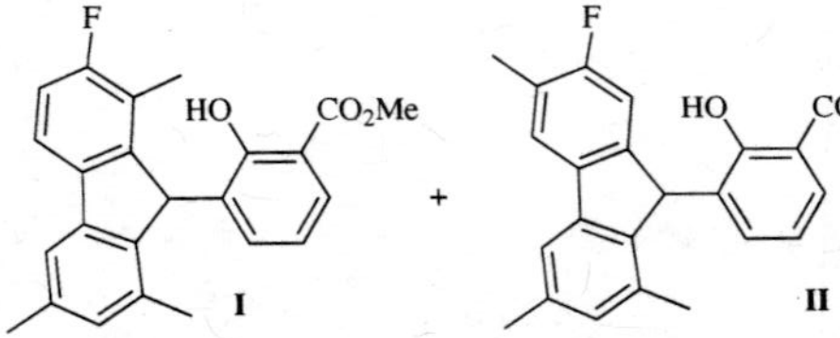 **I** + **II** **I + II** (45)	171

C_{25}			
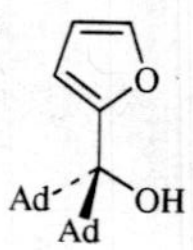	Et_3SiH (2.3 eq), TFA, CH_2Cl_2, rt, 3 h	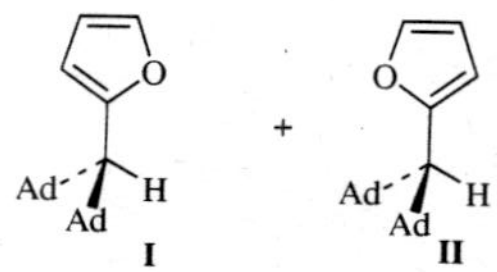 **I** + **II** = (86), **I**:**II** = 2:1	160
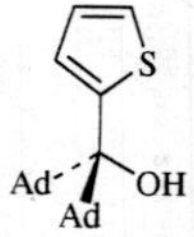	Et_3SiH (2.4 eq), TFA, CH_2Cl_2, rt, 4 h	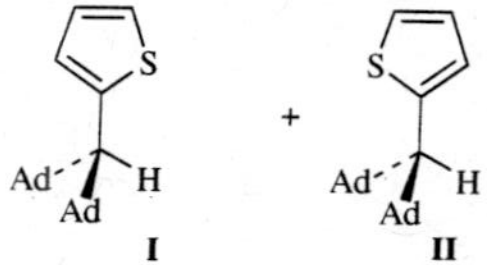 **I** + **II** = (81), **I**:**II** = 1:1	160
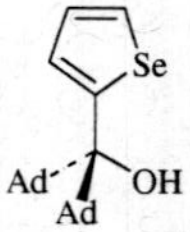	Et_3SiH (2.4 eq), TFA, CH_2Cl_2, rt, 4 h	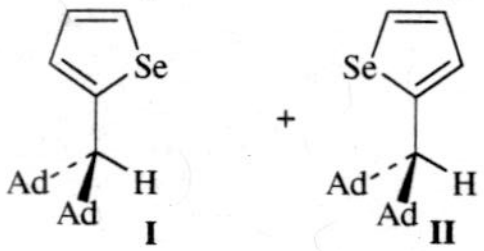 **I** + **II** = (53), **I**:**II** = 1:1.8	160
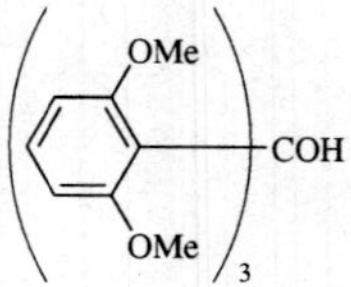	Et_3SiH, HOAc	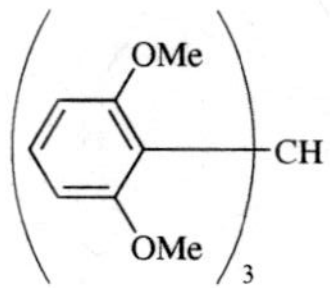 **I** (100)	65
	1. HOAc, rt, 24 h 2. Et_3SiH (3 eq), rt, 1 min	**I** (95)	26
C_{26}			
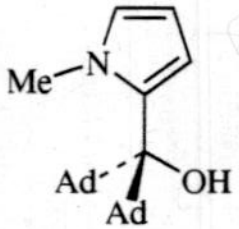	Et_3SiH, TFA, CH_2Cl_2, –10° to rt, 16 h	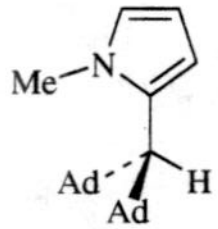 (18) + 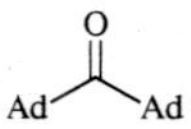 (78)	731
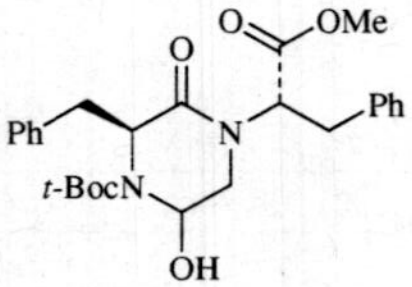	Et_3SiH, TFA (15 eq), CH_2Cl_2	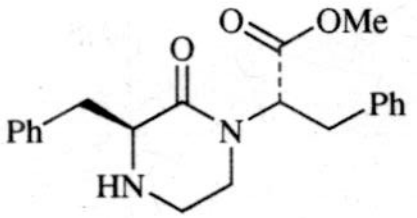 (86)	732

TABLE 5. ORGANOSILANE REDUCTION OF ALCOHOLS (*Continued*)

Alcohol	Conditions	Product(s) and Yield(s) (%)	Refs.
C_{26} 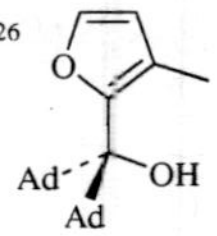	Et_3SiH (3.0 eq), TFA, CH_2Cl_2, rt, 20 h	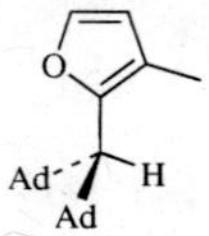 **I** + 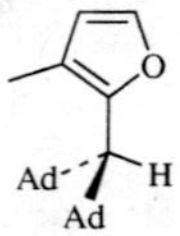 **II** **I** + **II** (30), **I**:**II** = 3 to 1:1	160
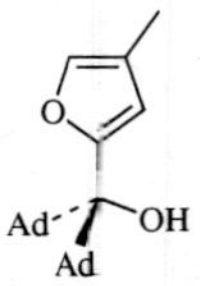	Et_3SiH (2.4 eq), TFA, CH_2Cl_2, rt, 2 h	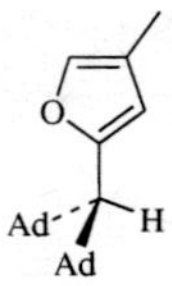 **I** + 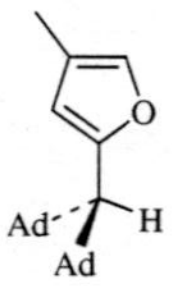 **II** **I** + **II** (58), **I**:**II** = 1:2	160
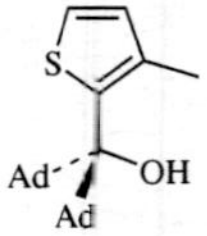	Et_3SiH (2.5 eq), TFA, CH_2Cl_2, rt, 16 h	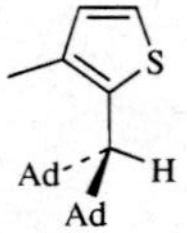 **I** (56) + 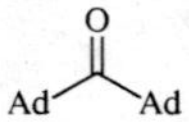 (27)	160
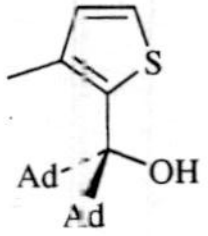	Et_3SiH (2.5 eq), TFA, CH_2Cl_2, rt, 2 h	**I** (93)	160
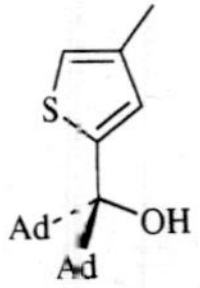	Et_3SiH (2.4 eq), TFA, CH_2Cl_2, rt, 4 h	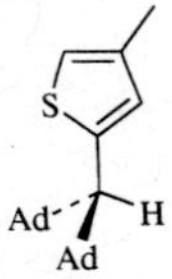 **I** + 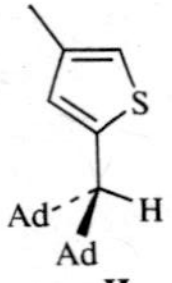 **II** **I** + **II** (82), **I**:**II** = 1:1	160

Reactant	Conditions	Product(s) and Yield(s) (%)	Refs.
C_{27} 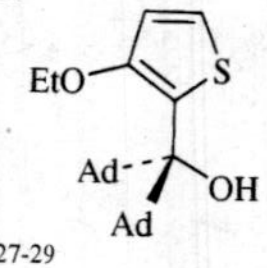	Et_3SiH or $PhMe_2SiH$, TFA, CH_2Cl_2	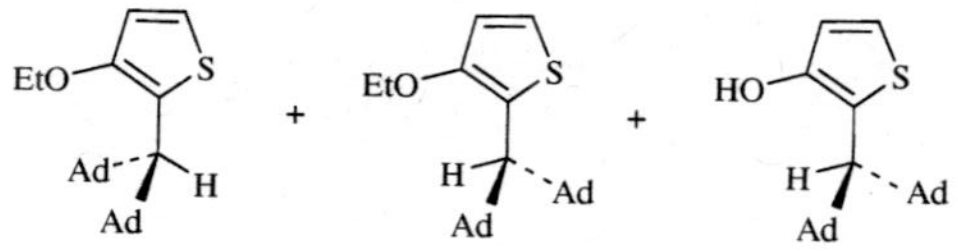 (—)[b]	161
C_{27-29} 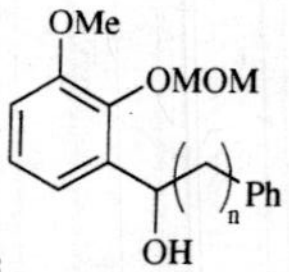	Et_3SiH, TFA, rt	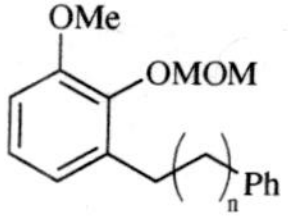 n = 11 (—) n = 13 (30)	733
C_{28} 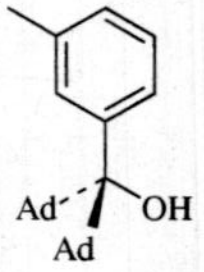	Ph_2SiH_2, TFA, CH_2Cl_2, 3 h	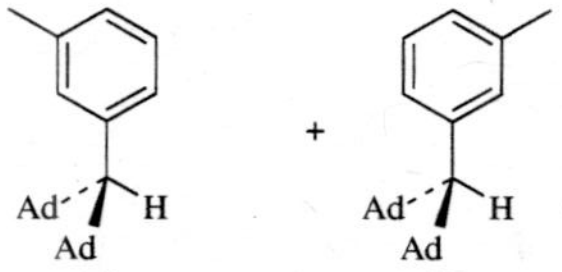 I + II **I** + **II** (—), **I**:**II** = 0.53	159
	Ph_3SiH, TFA, CH_2Cl_2, 3 h	**I** + **II** (—), **I**:**II** = 0.32	159
	i-Pr_3SiH, TFA, CH_2Cl_2, 3 h	**I** + **II** (—), **I**:**II** = 0.72	159
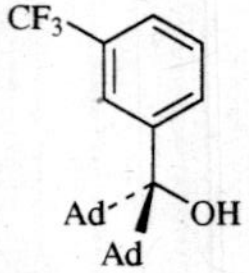	R_3SiH, TFA, CH_2Cl_2, 3 h	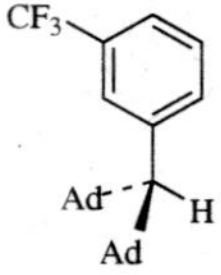 **I** (—) + 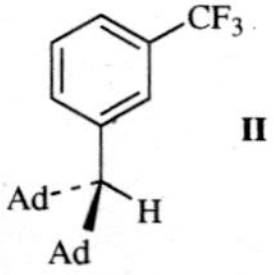 **II** (—)	159

R_3Si	I:II
n-$C_6H_{13}H_2Si$	0.52
Ph_2HSi	0.52
Ph_3Si	0.49
(*i*-Pr)$_3$Si	0.60
Et_3Si	0.41
Me_2PhSi	0.39
(TMS)$_3$Si	5.4

TABLE 5. ORGANOSILANE REDUCTION OF ALCOHOLS (*Continued*)

Alcohol	Conditions	Product(s) and Yield(s) (%)	Refs.
C_{28} 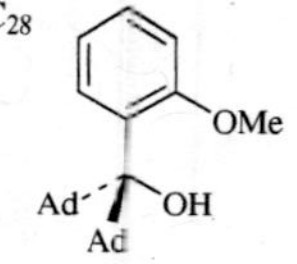	R_3SiH, TFA, CH_2Cl_2, –10° to rt, 16 h	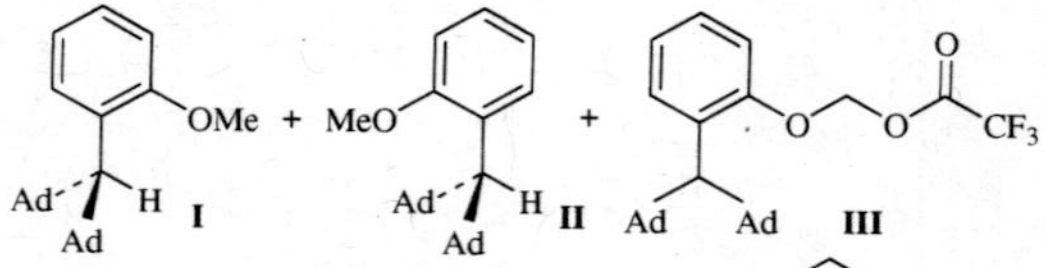 I + II + III + 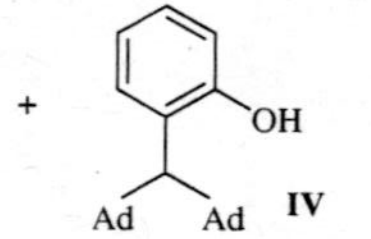 IV	158
C_{28-32} 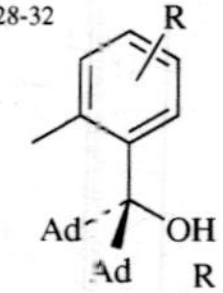	Et_3SiH, TFA, CH_2Cl_2, 3 h	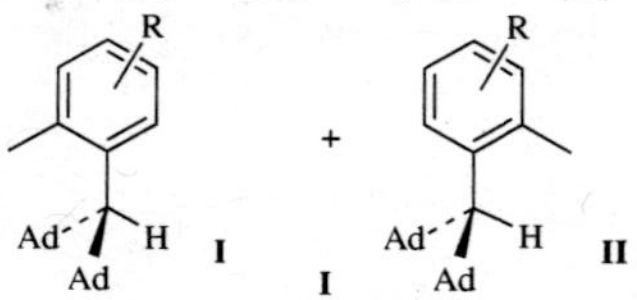 I + II I:II = —	159

R_3Si	I	II	III	IV
None	(—)	(—)	(91)	(9)
n-$C_6H_{13}H_2Si$	(12)	(47)	(36)	(5)
Ph_2HSi	(9)	(45)	(38)	(8)
Ph_3Si	(0)	(37)	(49)	(15)
(i-Pr)$_3$Si	(0)	(28)	(58)	(14)
Et_3Si	(8)	(41)	(40)	(11)
Me_2PhSi	(27)	(34)	(31)	(9)
$(Me_3Si)_3Si$	(21)	(9)	(54)	(17)

R	I
H	(—)
4-Me	(74)
5-Me	(79)
5-t-Bu	(57)
4-Cl	(94)
5-Cl	(83)
4-F	(66)
5-F	(77)
4-MeO	(71)
3,4,5-Me_3	(85)

C_{29}			
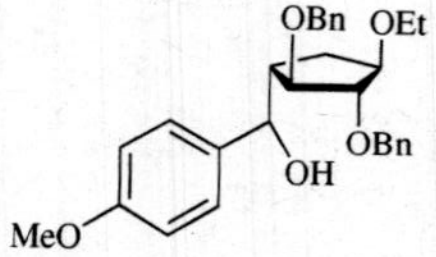	Et_3SiH, TFA, CH_2Cl_2, 0°, 2 h	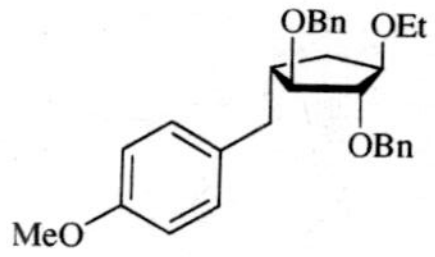 (76)	168
C_{29-33}			
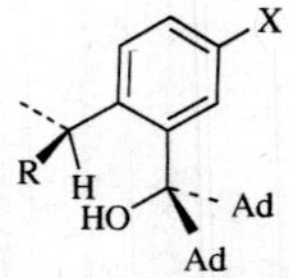 R, X = H, H; Me, *i*-Pr	Et_3SiH (large xs), TFA	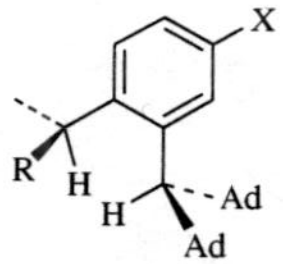 (—)	734

TABLE 5. ORGANOSILANE REDUCTION OF ALCOHOLS (*Continued*)

Alcohol	Conditions	Product(s) and Yield(s) (%)	Refs.
C_{31} 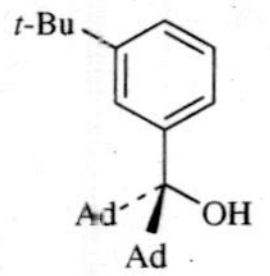	R_3SiH, TFA, CH_2Cl_2, 3 h	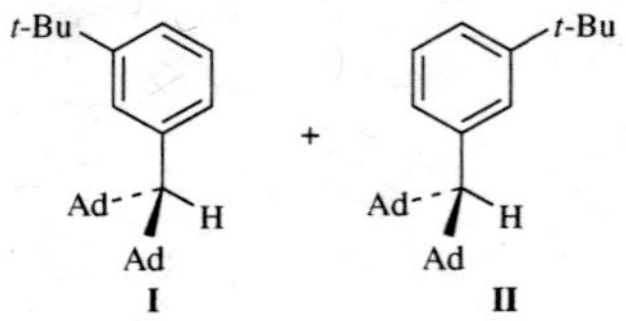 (—)	159
	R_3Si	I:II	
	n-$C_6H_{13}H_2Si$	0.74	
	Ph_2HSi	1.3	
	Ph_3Si	2.5	
	$(i\text{-}Pr)_3Si$	21	
	Et_3Si	3.4	
	Me_2PhSi	4.1	
	$(TMS)_3Si$	23	
C_{32} 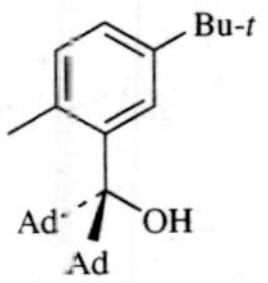	R_3SiH, TFA, CH_2Cl_2, 24 h	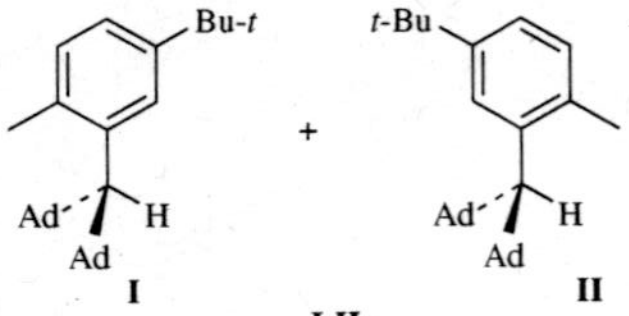 (—)	159
	R_3Si	I:II	
	n-$C_6H_{13}H_2Si$	2.9	
	Ph_2HSi	5.8	
	Ph_3Si	1.6	
	$(i\text{-}Pr)_3Si$	NR	
	Et_3Si	6.2	
	Me_2PhSi	2.5	
	$(TMS)_3Si$	1.2	

C$_{57}$

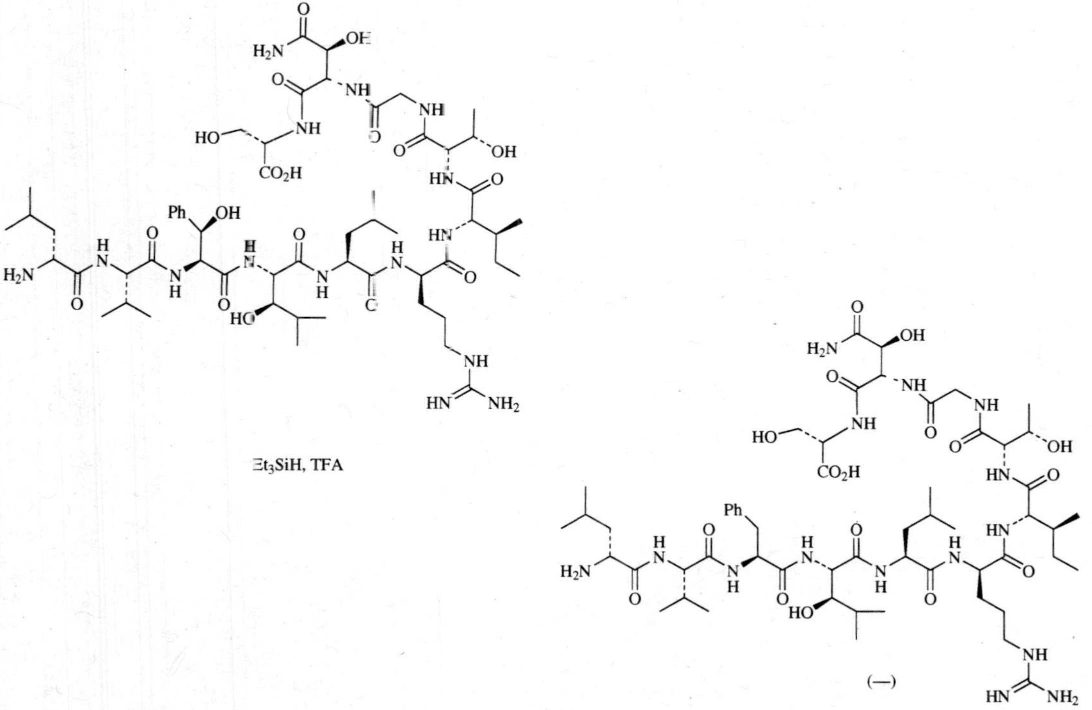

(—)

169

[a] The yield was determined by gas chromatography.

[b] Various product ratios were observed for this reaction.

TABLE 6. ORGANOSILANE REDUCTION OF ETHERS

Ether	Conditions	Product(s) and Yield(s) (%)	Refs.
C_5 	Et_3SiH (3.0 eq), $(C_6F_5)_3B$ (5 mol%), CH_2Cl_2, rt, 20 h	 (97)	144, 145
	Et_3SiH (1.1 eq), $(C_6F_5)_3B$ (10 mol%), CH_2Cl_2, rt, 20 h	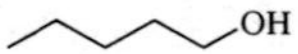 (88)	144, 145
C_6 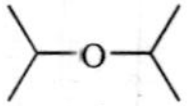	Et_3SiH (1.1 eq), $(C_6F_5)_3B$ (5 mol%), CH_2Cl_2, rt, 20 h	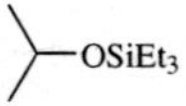 (>95) + C_3H_8	144, 145
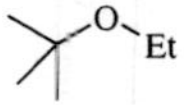	Et_3SiH (2 eq), TFA (10 eq), reflux, 24 h	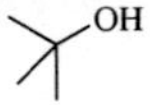 (60)	271
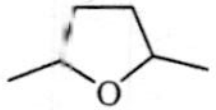	Et_3SiH (3.0 eq), $(C_6F_5)_3B$ (10 mol%), CH_2Cl_2, rt, 20 h; then TBAF	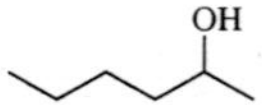 (91)	144
C_7 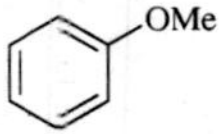	Et_3SiH (1.1 eq), $(C_6F_5)_3B$ (5 mol%), CH_2Cl_2, rt, 20 h	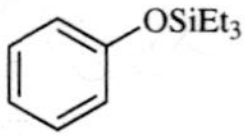 (>99) + CH_4	145
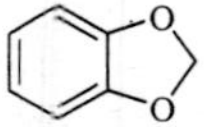	Et_3SiH (3.0 eq), $(C_6F_5)_3B$ (5 mol%), CH_2Cl_2, rt, 20 h; then TBAF	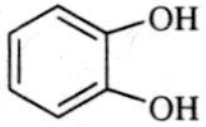 (79)	145
C_8 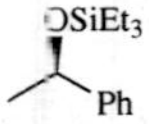	Et_3SiD (1.5 eq), $(C_6F_5)_3B$ (5 mol%), C_6H_{14}, rt, 20 h	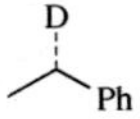 (—) 41% ee	145
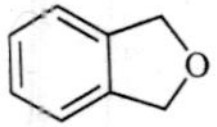	Et_3SiH (3.0 eq), $(C_6F_5)_3B$ (5 mol%), CH_2Cl_2, rt, 20 h	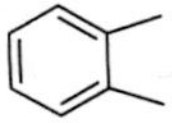 (78)	145
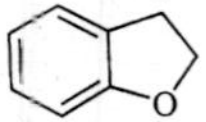	Et_3SiH (1.1 eq), $(C_6F_5)_3B$ (5 mol%), CH_2Cl_2, rt, 20 h	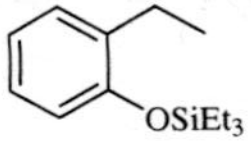 (>99)	145

C_9	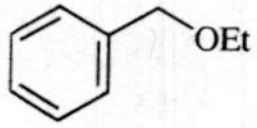	Et_3SiH (2 eq), TFA (10 eq), reflux, 24 h	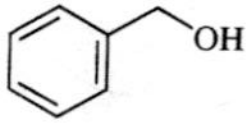 (7)	271
	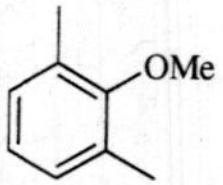	Et_3SiH (1.1 eq), $(C_6F_5)_3B$ (5 mol%), CH_2Cl_2, rt, 20 h	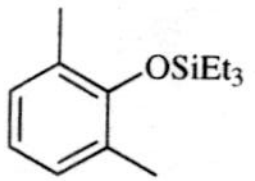 (>99) + CH_4	145
C_{9-16}		PMHS (2 eq), $(Ph_3P)_4Pd$ (0.01 eq), THF, rt, 3-5 h	ROH + 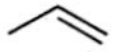	270

R	
$Ph(CH_2)_3$	(94)
PhMeCH	(87)
Ph_2CH	(85)
2-*i*-Pr-4-Me-*c*-C_6H_9	(89)
2-$C_{10}H_7$	(90)
4-FC_6H_4	(92)
4-$MeOC_6H_4$	(90)
4-$NO_2C_6H_4$	(88)
PrenylO$(CH_2)_4$	(92)
BnO$(CH_2)_4$	(90)
PMBO$(CH_2)_4$	(87)
TsO$(CH_2)_4$	(92)
THPO$(CH_2)_4$	(85)
AcO$(CH_2)_4$	(89)
MOMO$(CH_2)_4$	(85)
TBSO$(CH_2)_4$	(90)

C_{10}	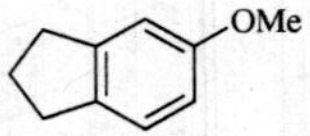	Et_3SiH (1.1 eq), $(C_6F_5)_3B$ (5 mol%), CH_2Cl_2, rt, 20 h	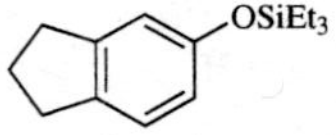 (96) + CH_4	145
C_{11}		PMHS (2 eq), $(Ph_3P)_4Pd$ (0.01 eq), THF, rt, 3-5 h	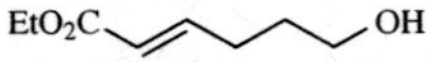 (86)	270

TABLE 6. ORGANOSILANE REDUCTION OF ETHERS (*Continued*)

Ether	Conditions	Product(s) and Yield(s) (%)	Refs.
C_{11} 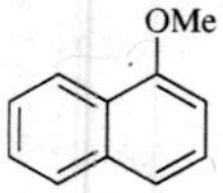	Et_3SiH (1.3 eq), $BF_3{\bullet}OH_2$ (4-6 eq), CH_2Cl_2, rt, 10 h	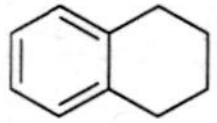 **I** (37)	217
	Et_3SiD (6.4 eq), $BF_3{\bullet}OH_2$ (12.8 eq), CH_2Cl_2, rt, 24 h	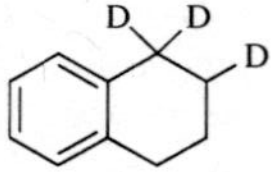 (100)	262
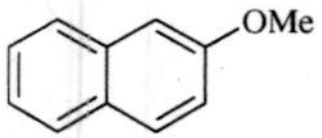	Et_3SiH (1.3 eq), $BF_3{\bullet}OH_2$ (4-6 eq), CH_2Cl_2, rt, 10 h	**I** (26)	217
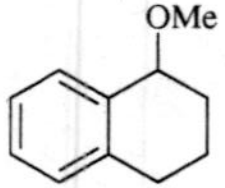	Et_3SiD (6.4 eq), $BF_3{\bullet}OH_2$ (12.8 eq), CH_2Cl_2, rt, 24 h	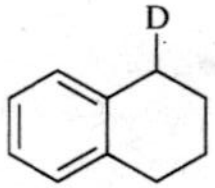 (100)	262
C_{12} 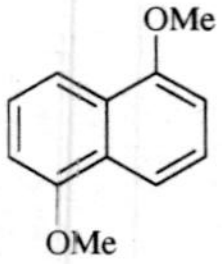	Et_3SiH (6.4 eq), $BF_3{\bullet}OH_2$ (12.8 eq), CH_2Cl_2, rt, 6 h	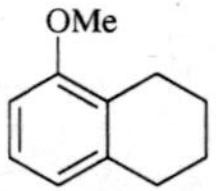 (100)	262
	Et_3SiD (6.4 eq), $BF_3{\bullet}OH_2$ (12.8 eq), CH_2Cl_2, rt, 24 h	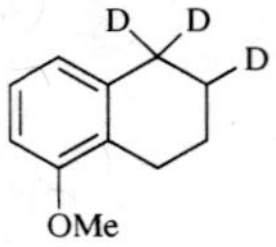 (100)	262
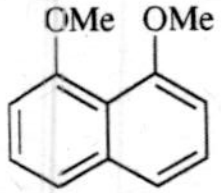	Et_3SiH (6.4 eq), $BF_3{\bullet}OH_2$ (12.8 eq), CH_2Cl_2, rt, 6 h	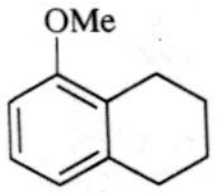 (100)	262

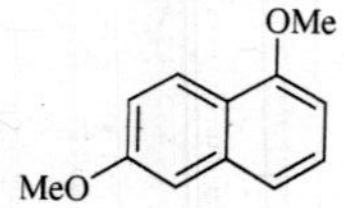	Et$_3$SiH (6.4 eq), BF$_3$•OH$_2$ (12.8 eq), CH$_2$Cl$_2$, rt, 24 h	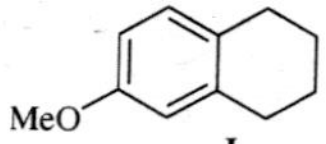 **I** (65)	262
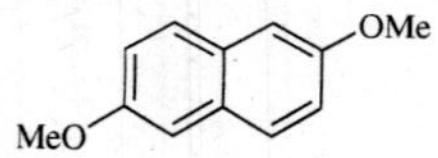	Et$_3$SiH (€.4 eq), BF$_3$•OH$_2$ (12.8 eq), CH$_2$Cl$_2$, rt, 24 h	**I** (95)	262
	Et$_3$SiD (6 4 eq), BF$_3$•OH$_2$ (12.8 eq), CH$_2$Cl$_2$, rt, 37 h	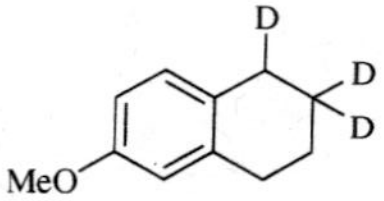 (60)	262
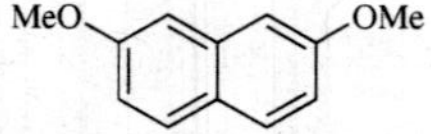	Et$_3$SiH (6.4 eq), BF$_3$•OH$_2$ (12.8 eq), CH$_2$Cl$_2$, rt, 6 h	**I** (50)	262
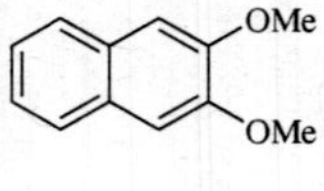	Et$_3$SiH (6.4 eq), BF$_3$•OH$_2$ (12.8 eq), CH$_2$Cl$_2$, rt, 48 h	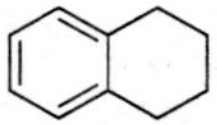 (80)	262
	Et$_3$SiD (6.4 eq), BF$_3$•OH$_2$ (12.8 eq), CH$_2$Cl$_2$, rt, 144 h	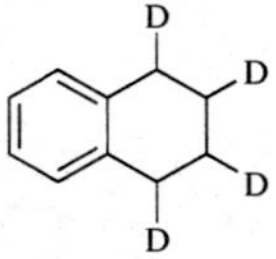 (24)	262
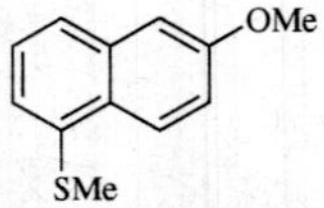	Et$_3$SiH (6.4 eq), BF$_3$•OH$_2$ (12.8 eq), CH$_2$Cl$_2$, rt, 24 h	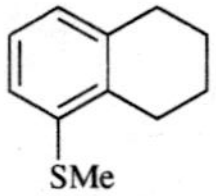 **I** (100)	262
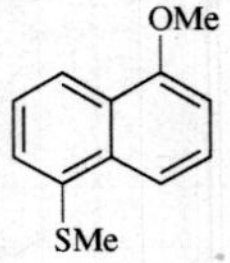	Et$_3$SiH (3-5 eq), BF$_3$•OH$_2$ (12.8 eq), CH$_2$Cl$_2$, C°, 24 h	**I** (100)	263

TABLE 6. ORGANOSILANE REDUCTION OF ETHERS (*Continued*)

Ether	Conditions	Product(s) and Yield(s) (%)	Refs.
C_{12}			
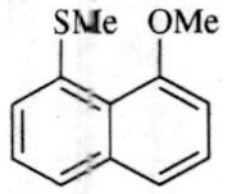	Et_3SiH (3-5 eq), BF_3•OEt_2, CH_2Cl_2, 0°, 24 h	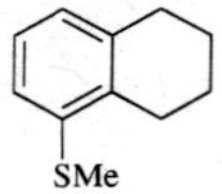 (70)	263
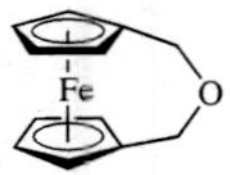	Et_3SiH (2 eq), TFA, rt, 20 min	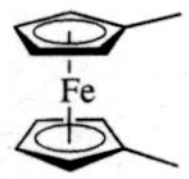 (88)	179
C_{13}			
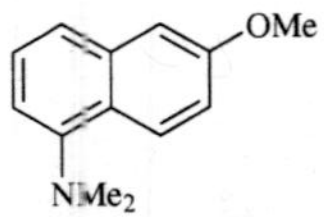	Et_3SiH (6.4 eq), BF_3•OEt_2, CH_2Cl_2, rt, 24 h	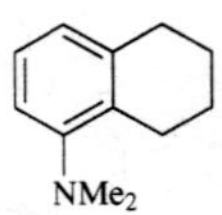 (100)	262
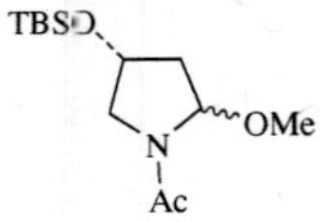	Et_3SiH (2 eq), BF_3•OEt_2 (2 eq), CH_2Cl_2, –40°, 1-2 h	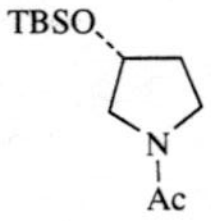 (97)	521
C_{14}			
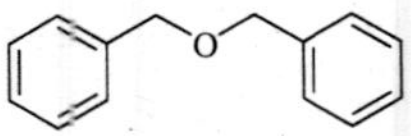	Et_3SiH (1.1 eq), $(C_6F_5)_3B$ (5 mol%), CH_2Cl_2, rt, 20 h	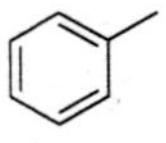 **I** (77) + 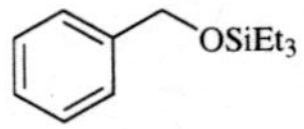 (93)	145
	Et_3SiH (3.0 eq), $(C_6F_5)_3B$ (5 mol%), CH_2Cl_2, rt, 20 h	**I** (86)	145
	Et_3SiH (2 eq), TFA (10 eq), reflux, 24 h	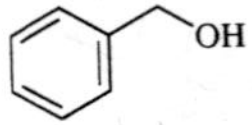 (3)	271

C_{15}			
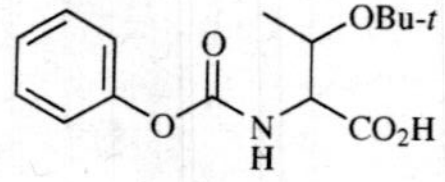	Et_3SiH (2.5 eq), TFA, CH_2Cl_2, rt, 45 min	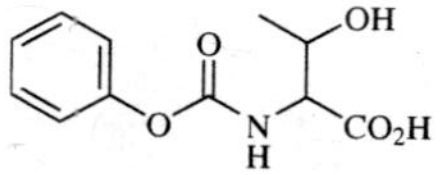 (96) + Me_3CH	307
C_{16}			
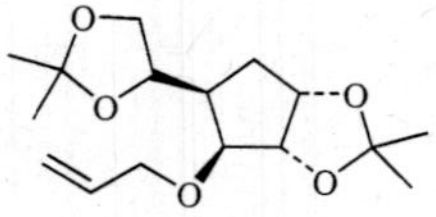	PMHS (2 eq), $(Ph_3P)_4Pd$ (0.01 eq), THF, rt, 3-5 h	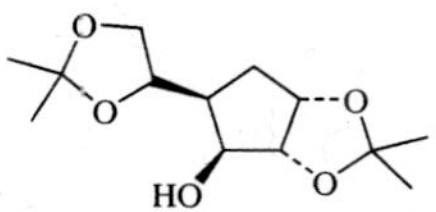 (85)	270
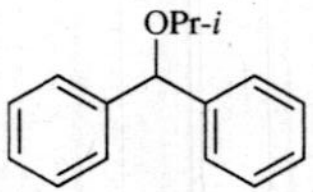	Et_3SiH (2 eq), TFA (10 eq), reflux, 24 h	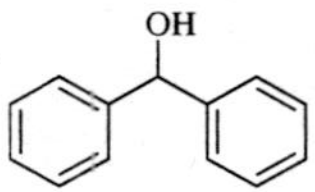 (60)	271
C_{20}			
n-$C_{10}H_{21}$-O-$C_{10}H_{21}$-n	Et_3SiH (1.1 eq), $(C_6F_5)_3B$ (10 mol%), CH_2Cl_2, rt, 20 h	n-$C_{10}H_{22}$ (93) + n-$C_{10}H_{21}OSiEt_3$ (95) **I**	144
	Et_3SiH (3.0 eq), $(C_6F_5)_3B$ (10 mol%), CH_2Cl_2, rt, 20 h	**I** (>99)	144
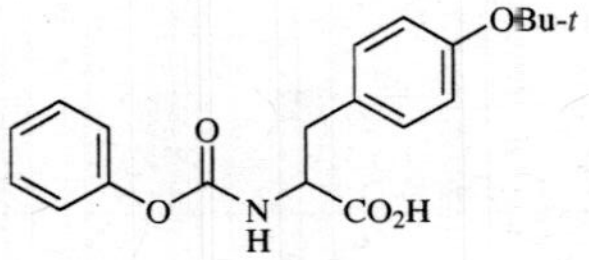	Et_3SiH (2.5 eq), TFA, CH_2Cl_2, rt, 35 min	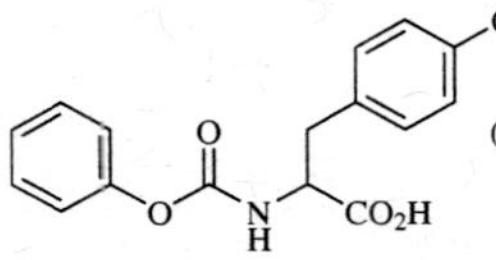 (100) + Me_3CH	307

TABLE 6. ORGANOSILANE REDUCTION OF ETHERS (*Continued*)

Ether	Conditions	Product(s) and Yield(s) (%)	Refs.
C_{20}			
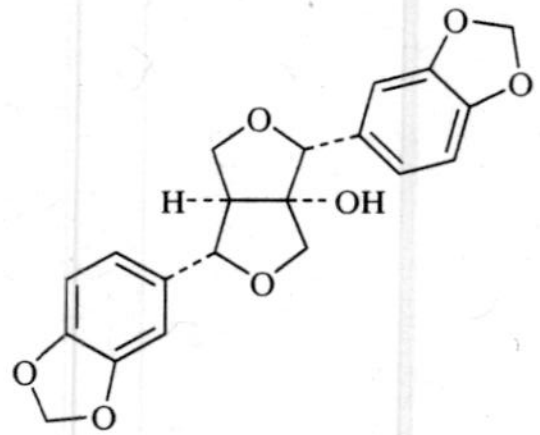	1. BF_3•OEt_2 (2.25 eq), CH_2Cl_2, 0°, 1 h 2. Et_3SiH (13.9 eq), rt, 16 h	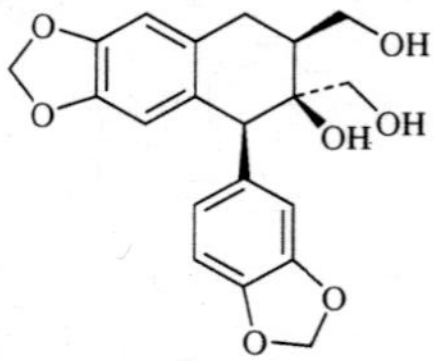 (36)	735
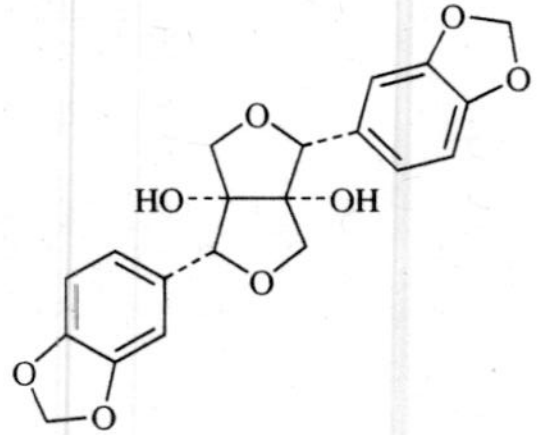	1. BF_3•OEt_2 (2.25 eq), CH_2Cl_2, 0°, 30 min 2. Et_3SiH (23 eq), rt, 16 h	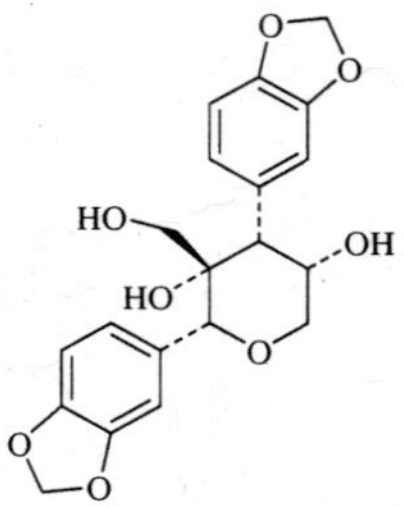 (44) + 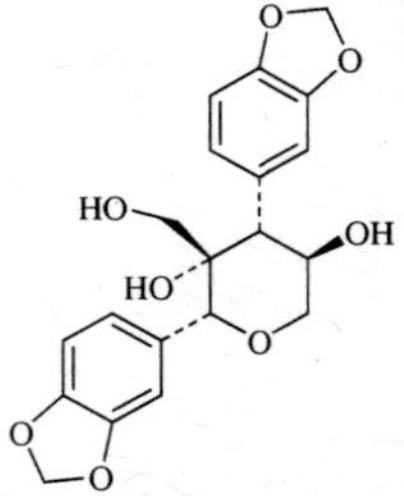 (44)	735
C_{21}			
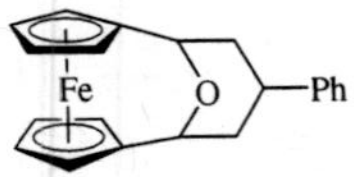	Et_3SiH (4 eq), TFA, rt, 3 h	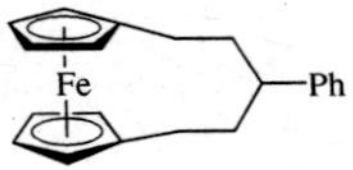 (80)	179

Reactant	Conditions	Product(s) and Yield(s) (%)	Refs.
C_{22}			
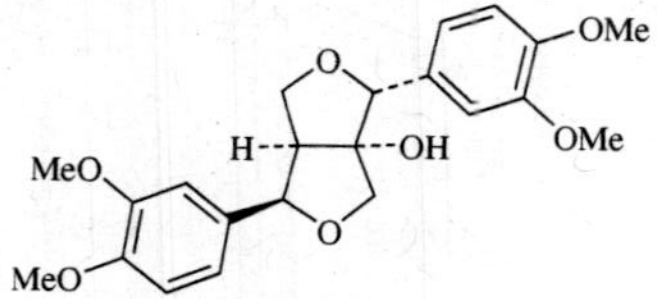	Et_3SiH (4.5 eq), $BF_3{\bullet}OEt_2$ (1.5 eq), CH_2Cl_2, 0°, 1 h; rt, 16 h	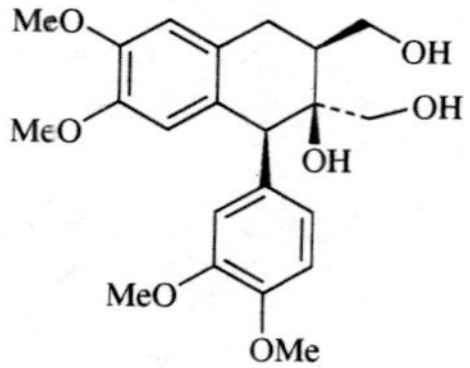 (66) + 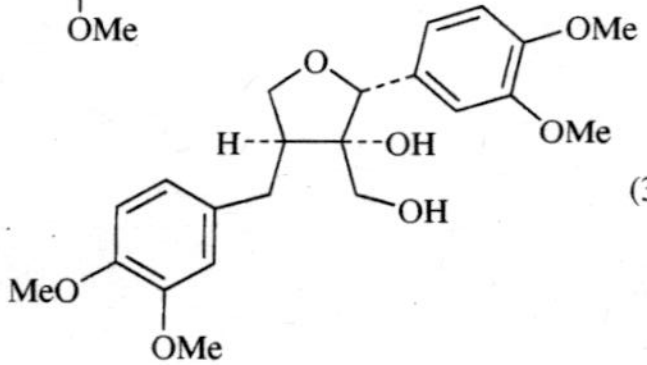 (3)	484, 736
C_{23}			
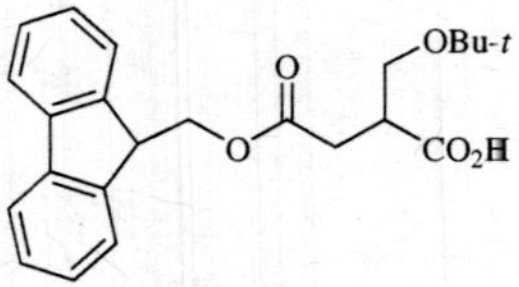	Et_3SiH (2.5 eq), TFA, CH_2Cl_2, rt, 15 min	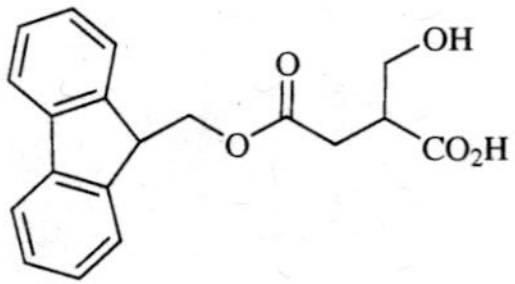 (99) + Me_3CH	307
C_{25}			
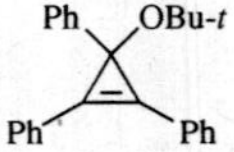	Et_3SiH (3 eq), HOAc, rt, 48 h	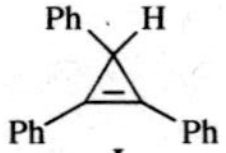 I (80)	26
	Et_3SiH, TFA	**I** (45)	272

TABLE 6. ORGANOSILANE REDUCTION OF ETHERS (*Continued*)

Ether	Conditions	Product(s) and Yield(s) (%)	Refs.
C_{30-44}			
	Et_3SiH (1.2 eq), Et_3SiOTf (x eq), CH_2Cl_2, rt	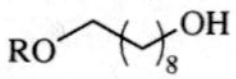 + Ph_3CH	269
R	x / Time		
Ac	0.005 / 4 min, 30 sec	(96)	
Piv	0.02 / 2 min	(99)	
Bz	0.01 / 3 min, 30 sec	(95)	
Bn	0.01 / 1 min, 20 sec	(93)	
MPM	0.01 / 1 min, 16 sec	(93)	
TBDPS	0.01 / 4 min	(89)	
MOM	0.01 / 24 min	(93)	
C_{32}			
n-$C_{15}H_{33}$–O–$C_{16}H_{33}$-n	Et_3SiH (1.1 eq), $(C_6F_5)_3B$ (5 mol%), CH_2Cl_2, rt, 20 h	n-$C_{16}H_{34}$ (98) + n-$C_{16}H_{33}OSiEt_3$ (98) **I**	145
	Et_3SiH (3.0 eq), $(C_6F_5)_3B$ (5 mol%), CH_2Cl_2, rt, 20 h	**I** (95)	145
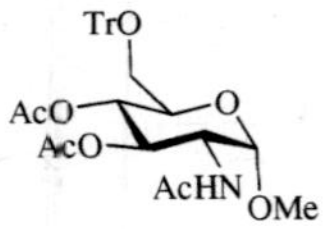	Et_3SiH (3.6 eq), TMSOTf (0.01 eq), CH_2Cl_2, rt, 32 min	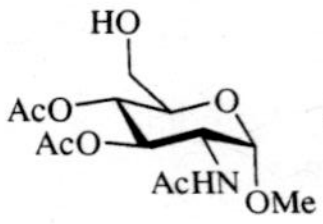 (86) + Ph_3CH	269

C_{42}	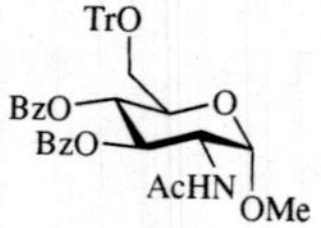	Et_3SiH (1.2 eq), TMSOTf (0.01 eq), CH_2Cl_2, rt, 32 min	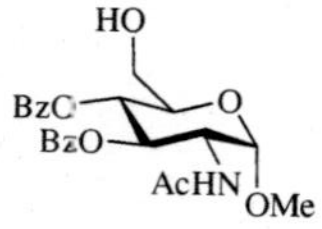 (99) + Ph_3CH **I**	269
		Et_3SiH (1.2 eq), Et_3SiOTf (0.01 eq), CH_2Cl_2, rt, 65 min	**I** (92) + Ph_3CH	269
C_{47}	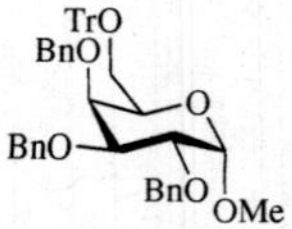	Et_3SiH (1.2 eq), TMSOTf (0.01 eq), CH_2Cl_2, rt, 5 min	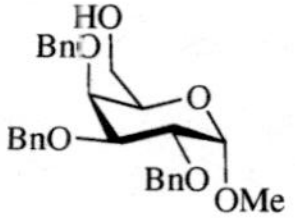 (89) + Ph_3CH	269
	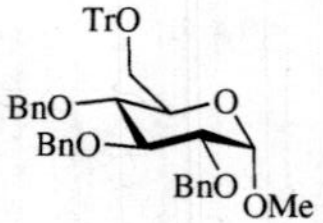	Et_3SiH (1.2 eq), TMSOTf (0.01 eq), CH_2Cl_2, rt, 11 min	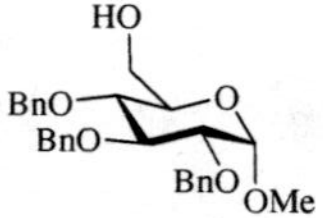 (88) + Ph_3CH	269
C_{90}	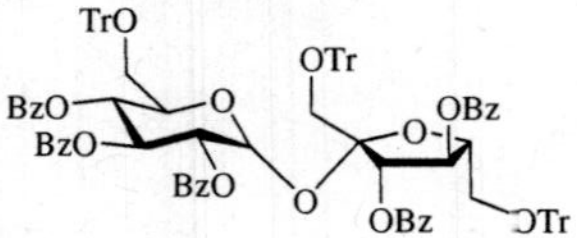	Et_3SiH (1.2 eq), Et_3SiOTf (0.01 eq), CH_2Cl_2, rt, 5 min	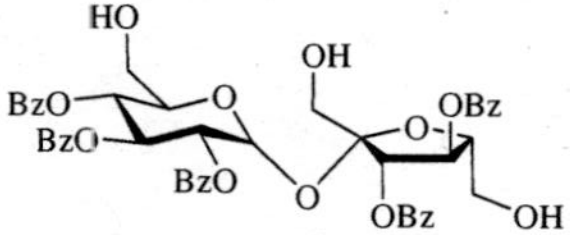 (86) + Ph_3CH	269
C_{92}	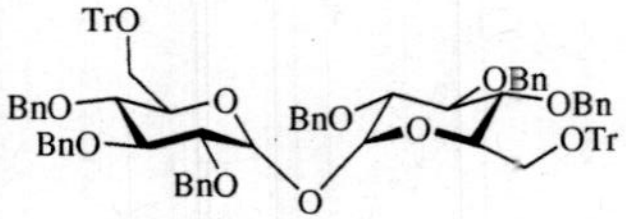	Et_3SiH (2.5 eq), Et_3SiOTf (0.01 eq), CH_2Cl_2, rt, 40 min	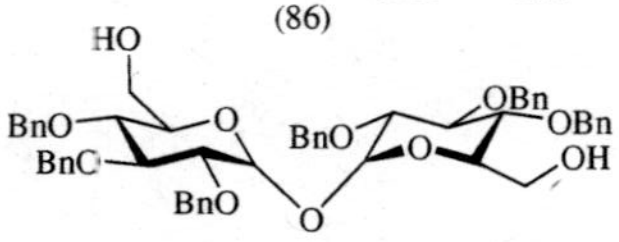 (96) + Ph_3CH	269

TABLE 7. ORGANOSILANE REDUCTION OF ALLYL ESTERS

Allyl Ester	Conditions	Product(s) and Yield(s) (%)	Refs.
C_{10}			
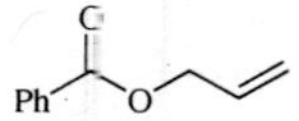	PMHS (2 eq), $(Ph_3P)_4Pd$ (0.01 eq), THF, rt, 3-5 h	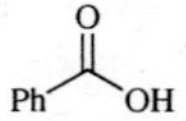 (87) + 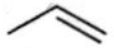 (—)	270
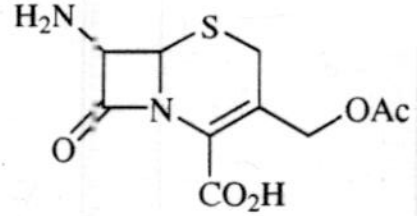	Et_3SiH, TFA, CH_2Cl_2, BF_3•OEt_2	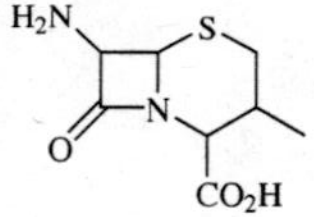 (—)	275
C_{11}			
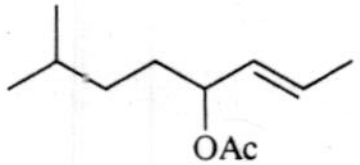	PMHS (2.0 eq), $(Ph_3P)_4Pd$ (4.0 mol%), THF, rt, 4 d	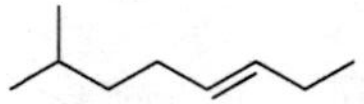 **I** + 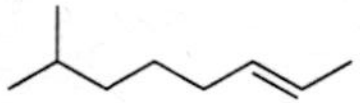 **II**; **I** + **II** (98), **I**:**II** = 1:1	196, 273
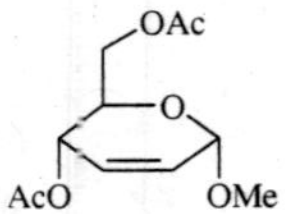	Ph_2SiH_2 (2.5 eq), $ZnCl_2$ (2.1 eq), $[(p\text{-tol})_3P]_4Pd$, 2 h	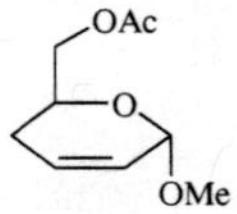 **I** + 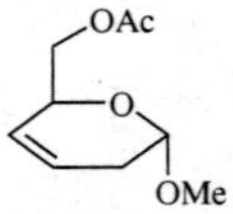 **II**; **I** + **II** (75), **I**:**II** = 33:67	273
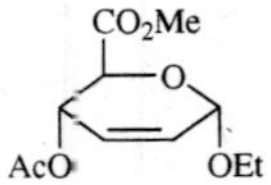	Ph_2SiH_2 (2.1 eq), $ZnCl_2$ (1.8 eq), $[(p\text{-tol})_3P]_4Pd$, 3 h	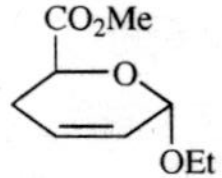 **I** + 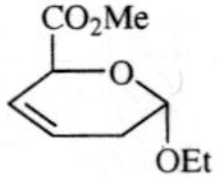 **II**; **I** + **II** (92), **I**:**II** = 44:56	273
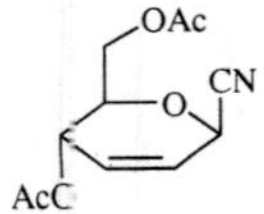	Ph_2SiH_2 (2 eq), $ZnCl_2$ (2.4 eq), $[(p\text{-tol})_3P]_4Pd$, 4 h	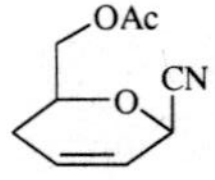 **I** + 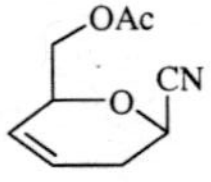 **II**; **I** + **II** (68), **I**:**II** = 50:50	273
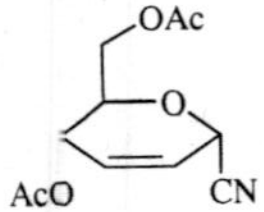	Ph_2SiH_2 (2 eq), $ZnCl_2$ (1.9 eq), $[(p\text{-tol})_3P]_4Pd$, 5 h	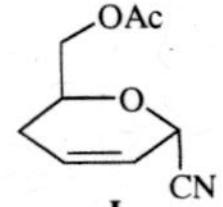 **I** + 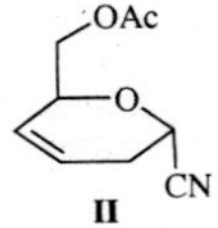 **II**; **I** + **II** (63), **I**:**II** = 75:25	273

Reactant	Conditions	Product(s) and Yield(s) (%)	Refs.
C_{12} 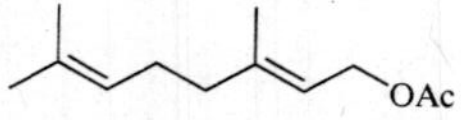	PMHS (1.75 eq), $(Ph_3P)_4Pd$ (6 mol%), PPh_3 (20 mol%), THF, 5 d	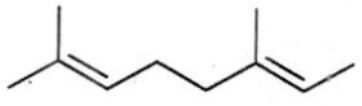 **I** (52) + 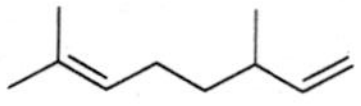 **II** (48)	273
	PMHS (1.7 eq), $(Ph_3P)_4Pd$ (5.0 mol%), THF, rt, 5 d	**I** + **II** (99), **I**:**II** = 1:1	196
	Ph_2SiH_2 (2.0 eq), $ZnCl_2$ (3-4 eq), $(Ph_3P)_4Pd$ (7.2 mol%), THF, rt, 30 h	**I** + **II** (99), **I**:**II** = 1:1	196
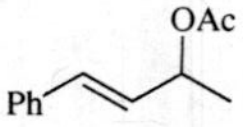	PMHS (2.1 eq), $(Ph_3P)_4Pd$ (4.0 mol%), THF, rt, 24 h	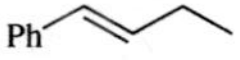 **I** + 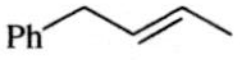 **II** **I** + **II** (79), **I**:**II** = 6:1	196
	Ph_2SiH_2 (2.0 eq), $(Ph_3P)_4Pd$ (10 mol%), THF, 0°, 24 h	**I** + **II** (100), **I**:**II** = 6:4	196
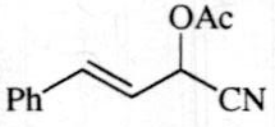	PMHS (2.2 eq), $(Ph_3P)_4Pd$ (4.6 mol%), THF, 20 min	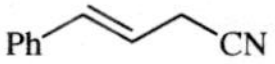 (93)	273
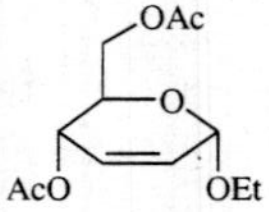	Ph_2SiH_2 (1.7 eq), $ZnCl_2$ (2 eq), $[(p\text{-tol})_3P]_4Pd$, 3 h	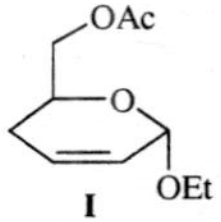 **I** + 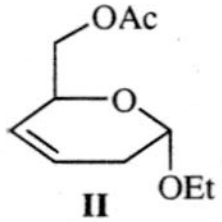 **II** **I** + **II** (55), **I**:**II** = 25:75	273
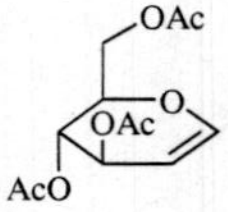	Ph_2SiH_2 (4 eq), $ZnCl_2$ (2.2 eq), $[(p\text{-tol})_3P]_4Pd$, 24 h	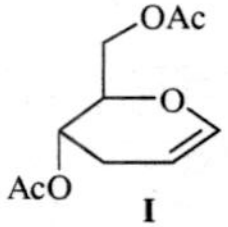 **I** (58)	273
	Ph_2SiH_2 (3.5 eq), $ZnCl_2$ (3-4 eq), $(Ph_3P)_4Pd$ (12.0 mol%), THF, rt, 13 h; 50°, 2 h	**I** (90)	196

TABLE 7. ORGANOSILANE REDUCTION OF ALLYL ESTERS (*Continued*)

Allyl Ester	Conditions	Product(s) and Yield(s) (%)	Refs.
C_{12} 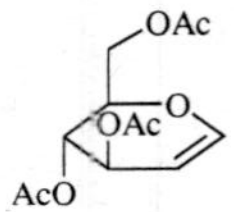	Ph_2SiD_2 (2.6 eq), $ZnCl_2$ (5.4 eq), $[(p\text{-tol})_3P]_4Pd$, 24 h	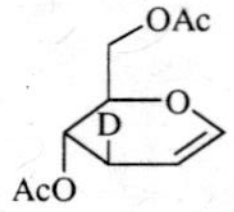 (59)	273
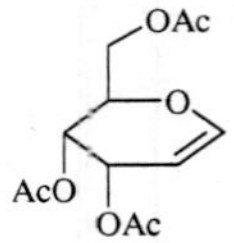	Ph_2SiD_2 (4 eq), $ZnCl_2$ (8 eq), $[(p\text{-tol})_3P]_4Pd$, 24 h	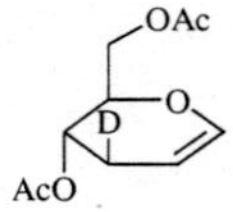 (42)	273
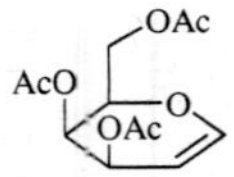	Ph_2SiH_2 (2 eq), $ZnCl_2$ (4 eq), $[(p\text{-tol})_3P]_4Pd$, 120 h	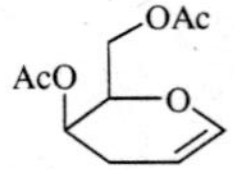 (26)	273
	Ph_2SiD_2 (2.7 eq), $ZnCl_2$ (8.4 eq), $[(p\text{-tol})_3P]_4Pd$, 80 h	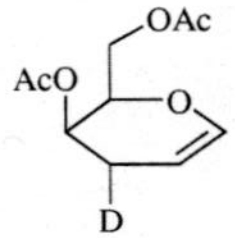 (10)	273
C_{13} 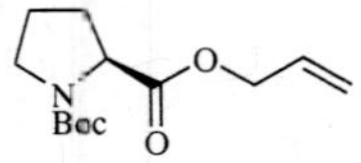	PMHS (2 eq), $(Ph_3P)_4Pd$ (0.01 eq), THF, rt, 3-5 h	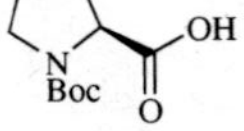 (85) + 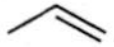 (—)	270
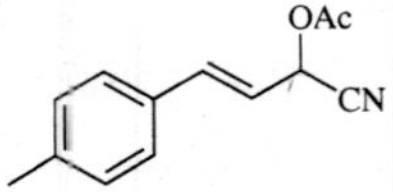	PMHS (1.8 eq), $(Ph_3P)_4Pd$ (3 mol%), THF, 1.5 h	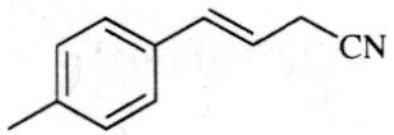 (100)	273
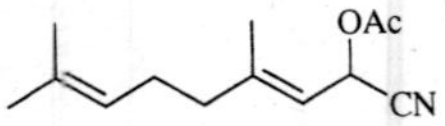	PMHS (2.5 eq), $(Ph_3P)_4Pd$ (3.5 mol%), THF, 48 h	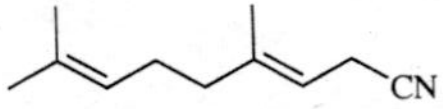 **I** + 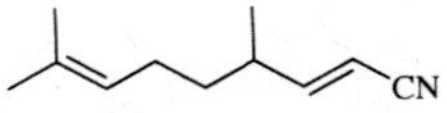 **II** **I** + **II** (75), **I**:**II** = 1:10	273

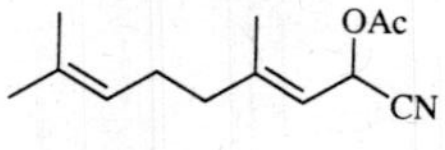	PMHS (2.5 eq) $(Ph_3P)_4Pd$ (3.5 mol%), THF, rt, 48 h	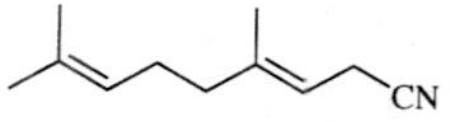 + **II** **I** **I** + **II** (75), **I**:**II** = 10:1	196
	Ph_2SiH_2 (2.2 eq), $(Ph_3P)_4Pd$ (7.2 mol%), THF, rt, 30 h	**I** (90) + **II** (10%)	196
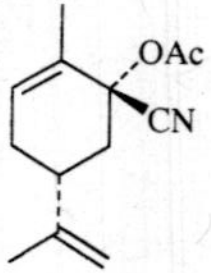	Ph_2SiH_2 (3.5 eq), $ZnCl_2$ (3-4 eq), $(Ph_3P)_4Pd$ (12.0 mol%), THF, rt, 4 h	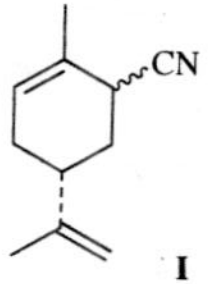 + 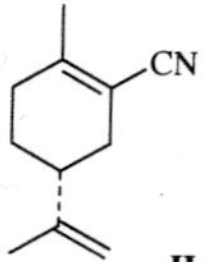 **I** **II** **I** + **II** (73), **I**:**II** = 1:1	196
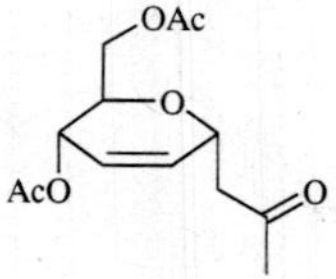	Ph_2SiH_2 (2 eq), $ZnCl_2$ (1.9 eq), $[(p\text{-tol})_3P]_4Pd$, 4 h	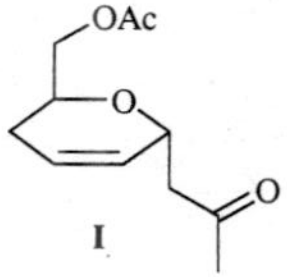 + 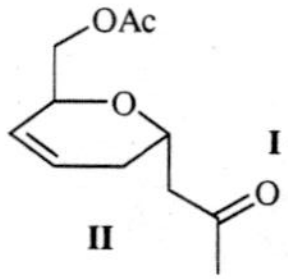 **I** **II** **I** + **II** (65), **I**:**II** = 50:50	273
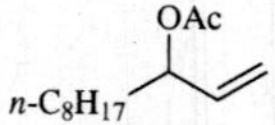	PMHS (3.7 eq), $(Ph_3P)_4Pd$ (8.0 mol%), THF, rt, 1.5 h	 (74)	196, 273
C_{14}			
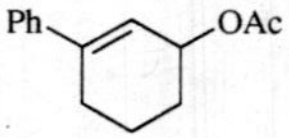	Et_3SiH (3 eq), $LiClO_4$, Et_2O, rt, 16 h	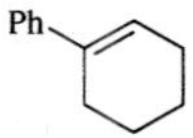 (92)	173
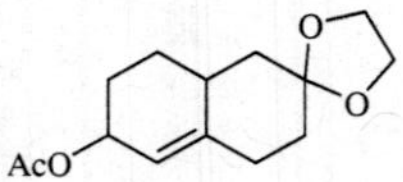	Et_3SiH (3 eq), $LiClO_4$, Et_2O, rt, 16 h	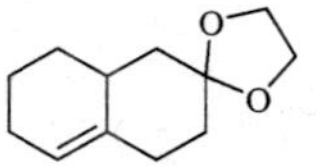 (48) + 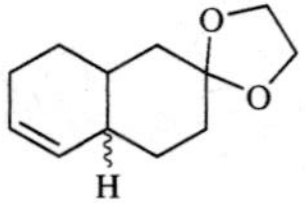 (12)	173
C_{15}			
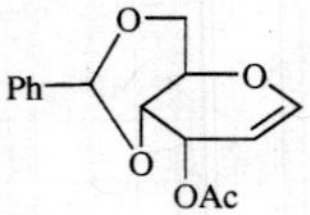	Ph_2SiH_2 (3.3 eq), $ZnCl_2$ (5 eq), $[(p\text{-tol})_3P]_4Pd$, 10 h	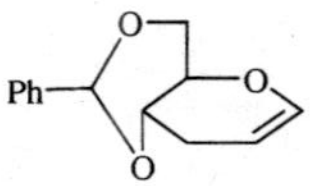 (63)	273

TABLE 7. ORGANOSILANE REDUCTION OF ALLYL ESTERS (*Continued*)

	Allyl Ester	Conditions	Product(s) and Yield(s) (%)	Refs.
C_{15}	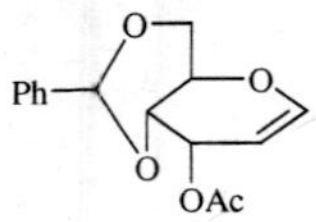	Ph_2SiD_2 (3.7 eq), $ZnCl_2$ (16.9 eq), $[(p\text{-tol})_3P]_4Pd$, 10 h	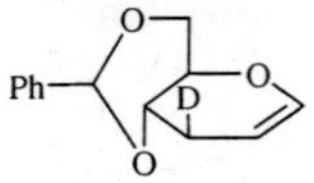 (55)	273
C_{16}	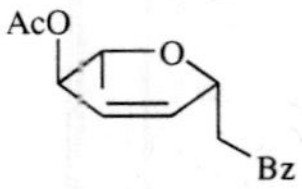	Ph_2SiH_2 (2.9 eq), $ZnCl_2$ (2.4 eq), $[(p\text{-tol})_3P]_4Pd$, 4 h	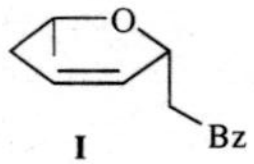 **I** + 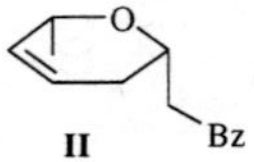 **II**; **I** + **II** (79), **I**:**II** = 50:50	273
	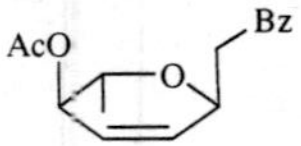	Ph_2SiH_2 (2.9 eq), $ZnCl_2$ (2.4 eq), $[(p\text{-tol})_3P]_4Pd$, 4 h	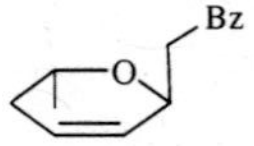 **I** + 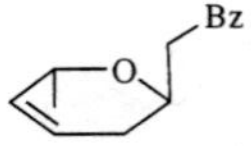 **II**; **I** + **II** (79), **I**:**II** = 50:50	273
C_{17}	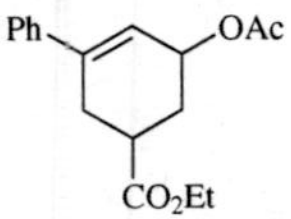	Et_3SiH (3 eq), $LiClO_4$, Et_2O, rt, 16 h	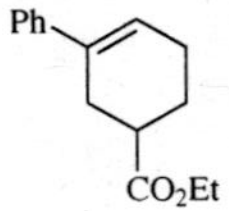 (89) + 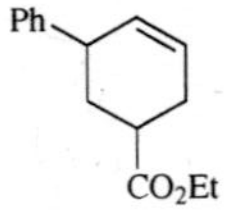 (5)	173
	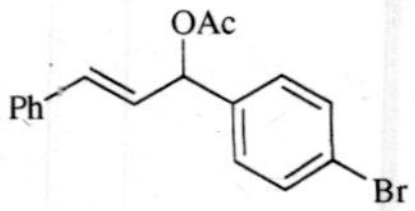	PMHS (1.8 eq), $(Ph_3P)_4Pd$ (3 mol%), THF, 1.5 h	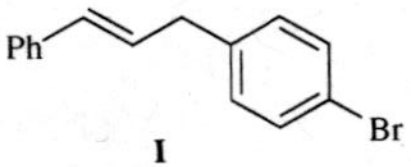 **I** + 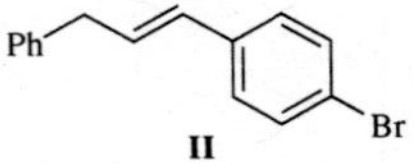 **II**; **I** + **II** (89), **I**:**II** = 1:1	196, 273
		Ph_2SiH_2 (1.9 eq), $(Ph_3P)_4Pd$ (5.7 mol%), THF, rt, 1 h	**I** + **II** (90), **I**:**II** = 1:1	196

C_{18}	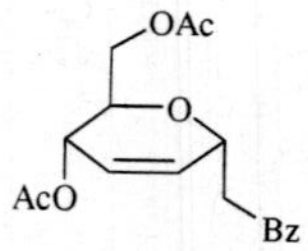	Ph_2SiH_2 (1.4 eq), $ZnCl_2$ (2.4 eq), $[(p\text{-tol})_3P]_4Pd$, 4 h	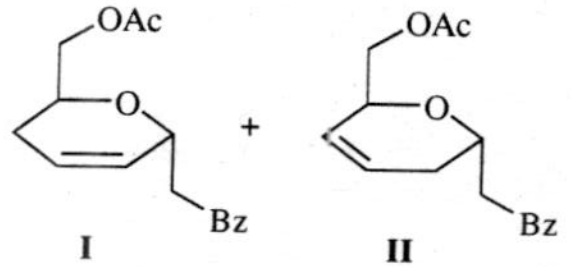 **I** + **II** (94), **I**:**II** = 50:50	273
	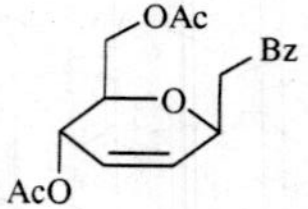	Ph_2SiH_2 (1.6 eq), $ZnCl_2$ (2.4 eq), $[(p\text{-tol})_3P]_4Pd$, 6 h	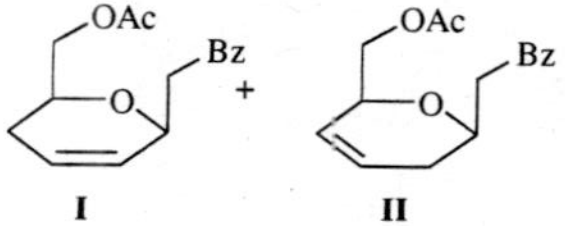 **I** + **II** (91), **I**:**II** = 50:50	273
C_{20}	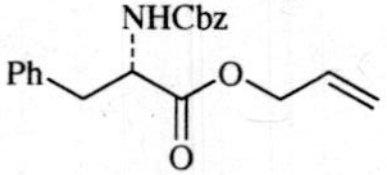	PMHS (2 eq), $(Ph_3P)_4Pd$ (0.01 eq), THF, rt, 3-5 h	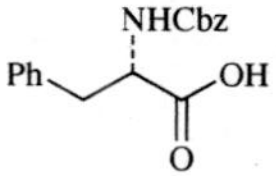 (85) + 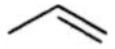 (—)	270
C_{26}	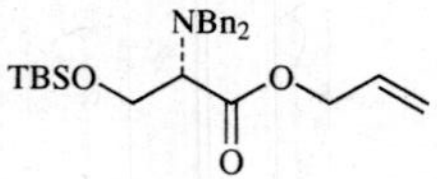	PMHS (2 eq), $(Ph_3P)_4Pd$ (0.01 eq), THF, rt, 3-5 h	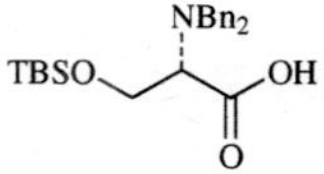 (87) + 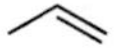 (—)	270
C_{40}	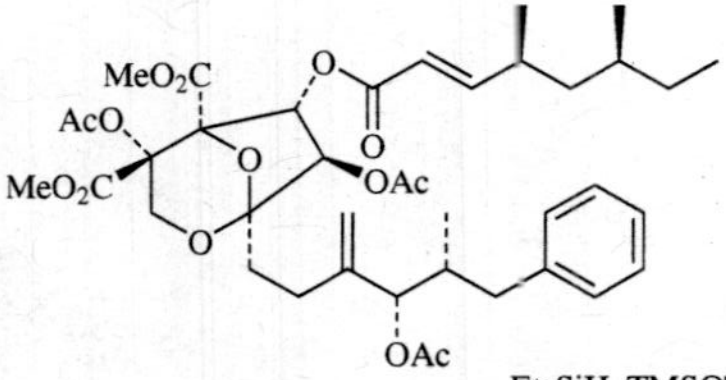	Et_3SiH, TMSOTf, CH_2Cl_2, 0-20°, 23 h	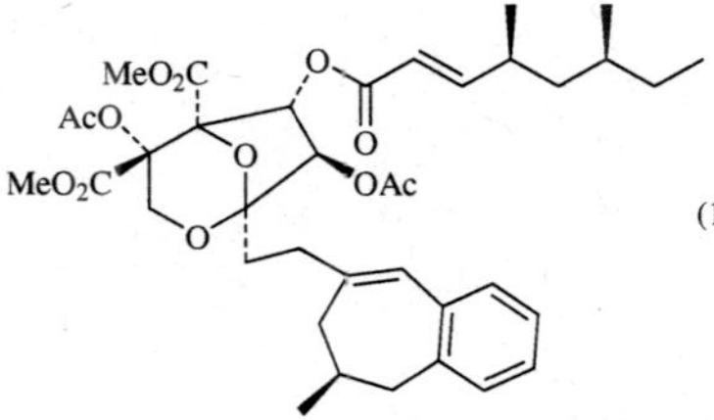 (16)	737

TABLE 8. ORGANOSILANE REDUCTION OF ACIDS

Carboxylic Acid	Conditions	Product(s) and Yield(s) (%)	Refs.
C_5 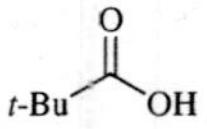	Et_3SiH (2 eq), TFA (10 eq), TFAA 60°, 5 h	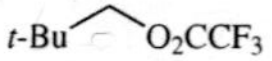 **I** (70)	277
	Ph_2SiH_2 (2 eq), TFA (10 eq), TFAA, 60°, 5 h	**I** (80)	277
C_7 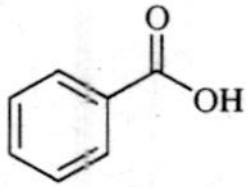	1. $EtMe_2SiH$ (4 eq), **201** (1 mol%), 1,4-dioxane, 20°, 0.5 h 2. Add acid, 18 h	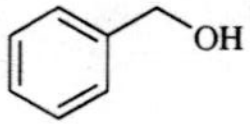 (46)	280
	Et_3SiH (xs), TFA, TFAA, 60°, 5 h	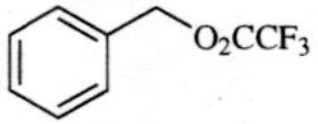 (30)	277
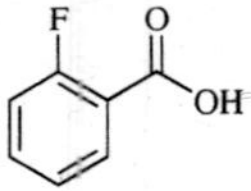	**B**, >180°	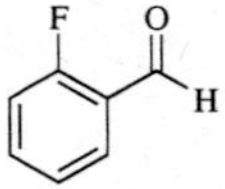 **I** (72)	284
	A, 150-170°	**I** (60)	284
C_{7-10} 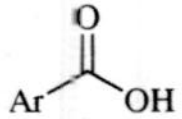	PMHS (3 eq), TBAF (2 mol%), THF, rt	 Ar: Ph (82); 4-BrC_6H_4 (74); 4-ClC_6H_4 (67); 2,4-$Cl_2C_6H_3$ (69); 3,5-$Cl_2C_6H_3$ (75); 3-MeC_6H_4 (78); 4-MeC_6H_4 (70); 2,4-$Me_2C_6H_3$ (72); 3,4,5-$(MeO)_3C_6H_2$ (79)	278

$C_{7\text{-}11}$ 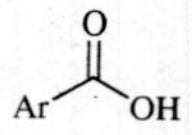	Et_3SiH (3.3 eq), $(C_6F_5)_3B$ (5 mol%), CH_2Cl_2, rt, 20 h	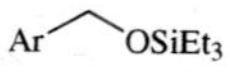	Ar: Ph (95); 4-FC_6H_4 (93); 4-BrC_6H_4 (93); 4-IC_6H_4 (94); 4-MeC_6H_4 (91); 2-$C_{10}H_7$ (96)	281, 282
$C_{7\text{-}18}$ 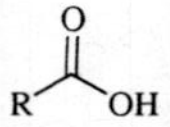	Et_3SiH (6 eq), $(C_6F_5)_3B$ (5 mol%), CH_2Cl_2, rt, 20 h	RMe	R: Ph (85); $BnCH_2$ (91); $Bn(CH_2)_2$ (37); $Bn(CH_2)_3$ (86); $Bn(CH_2)_4$ (88); n-$C_{11}H_{23}$ (94); n-$C_{17}H_{35}$ (91)	281
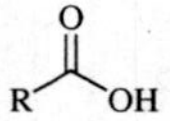	PMHS (10 eq), $Ti(OEt)_4$ (100 mol%), THF, 16 h	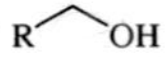	R: Ph (92); 4-$O_2NC_6H_4$ (78); 4-HOC_6H_4 (81); Bn (81); n-$C_{17}H_{35}$ (86)	279

TABLE 8. ORGANOSILANE REDUCTION OF ACIDS (*Continued*)

Carboxylic Acid	Conditions	Product(s) and Yield(s) (%)	Refs.
C_{7-18} 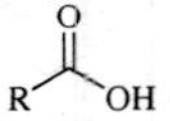	PMHS (10 eq), Ti(OPr-*i*)$_4$ (14 mol%), THF	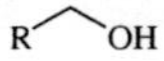 R / Time: Ph, 16 h (63); 4-$O_2NC_6H_4$, 16 h (86); 4-MeC_6H_4, 30 h (69); 4-$MeOC_6H_4$, 48 h (76); 2-$MeOC_6H_4$, 30 h (70); *n*-$C_{17}H_{35}$, 16 h (88)	279
C_8 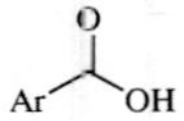	Et_3SiH (xs), TFA, TFAA, 60°, 5 h	ArMe Ar: 4-$MeOC_6H_4$ (97); 4-MeC_6H_4 (45)	283
C_9 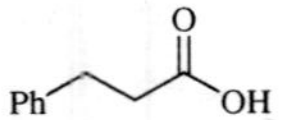	Et_3SiH (6.0 eq), $(C_6F_5)_3B$ (5 mol%), CH_2Cl_2, rt, 20 h	 (94)	282
	1. $EtMe_2SiH$ (4 eq), **223** (1 mol%), 1,4-dioxane, 20°, 0.5 h 2. Add acid, 0.5 h	 (72)	280
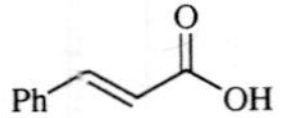	**B**, >180°	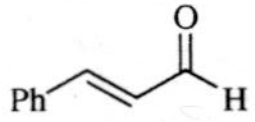 **I** (50)	284
	A, 150-170°	**I** (65)	284
C_{10} 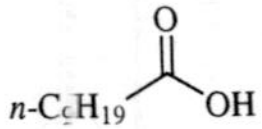	1. $EtMe_2SiH$ (4 eq), **223** (1 mol%), 1,4-dioxane, 20°, 0.5 h 2. Add acid, 6 h	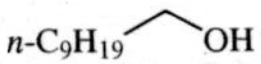 (80)	280

	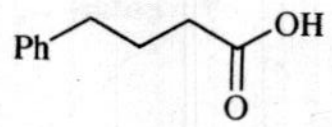	Et_3SiH (6.0 eq), $(C_6F_5)_3B$ (5 mol%), CH_2Cl_2, rt, 20 h	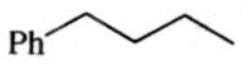 I + 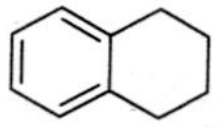 II; I + II (—), I:II = 37:63	282
C_{11}	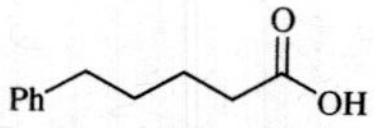	Et_3SiH (6.0 eq), $(C_6F_5)_3B$ (5 mol%), CH_2Cl_2, rt, 20 h	 (93)	282
	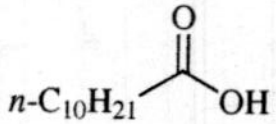	Et_3SiH (6.0 eq), $(C_6F_5)_3B$ (5 mol%), CH_2Cl_2, rt, 20 h	n-$C_{11}H_{24}$ (91)	282
C_{12}	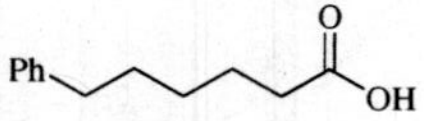	Et_3SiH (6.0 eq), $(C_6F_5)_3B$ (5 mol%), CH_2Cl_2, rt, 20 h	 (93)	282
C_{17}	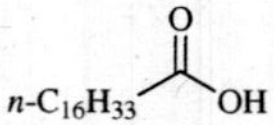	Et_3SiH (6.0 eq), $(C_6F_5)_3B$ (5 mol%), CH_2Cl_2, rt, 20 h	n-$C_{17}H_{36}$ (94)	282

TABLE 9. ORGANOSILANE REDUCTION OF ACYL HALIDES

Acyl Halide	Conditions	Product(s) and Yield(s) (%)	Refs.
$C_{2\text{-}12}$			
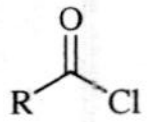	C, rt	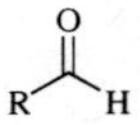 R: Me (—); t-Bu (—); Ph (90); $BnCH_2$ (87); PhCH=CH (86); 4-$O_2NC_6H_4$ (90); 4-$MeOC_6H_4$ (91); 2-C_4H_3O (87); 2-C_4H_3S (86); $Br(CH_2)_3$ (90); $Cl(CH_2)_3$ (82); $EtO_2C(CH_2)_8$ (80)	109
C_3			
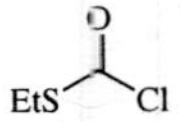	A, rt	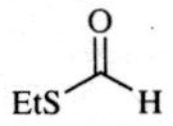 (>95)[a]	109
C_7			
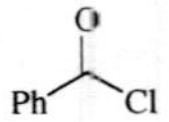	Bn_3SiH, Et_2O, reflux	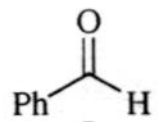 I (—)	286
	Et_3SiH, $AlCl_3$, Et_2O, reflux	I (—)	286
	Et_3SiH (6 eq), $(C_6F_5)_3B$ (5 mol%), CH_2Cl_2, rt, 20 h	PhMe (84)	281
	Cl_3SiH (3 eq), $(n\text{-}Pr)_3N$ (1 eq), MeCN, rt, 1 h; 85°, 0.5 h	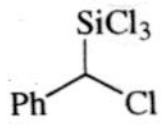 (91)	711
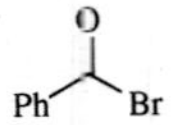	Bn_3SiH, Et_2O, reflux	I (—)	286
	Et_3SiH, Et_2O, reflux	I (—)	286
	A, rt	I (45)[a]	109

C_7			
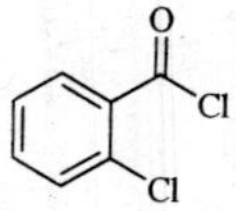	Et_3SiH, $AlCl_3$, Et_2O, reflux	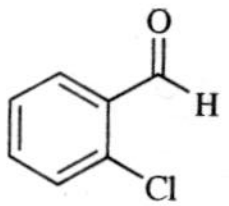 (—)	286
C_8			
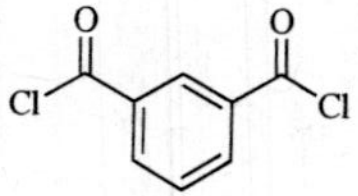	**C**, rt	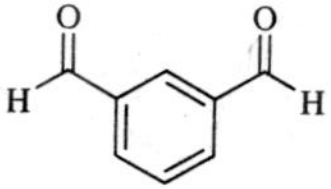 (89)	109
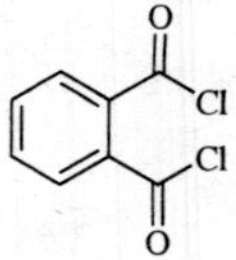	**C**, rt	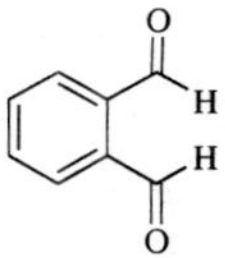 **I** (15)[a]	109
	A, rt	**I** (89)	109
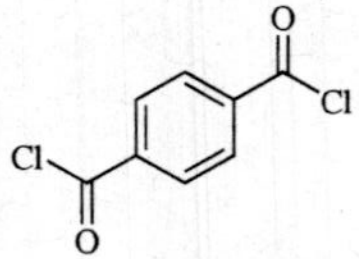	**C**, rt	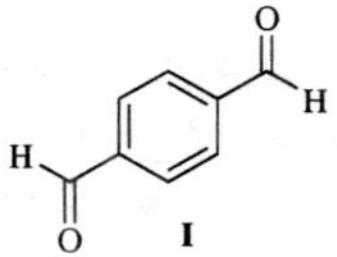 **I** (85)	109
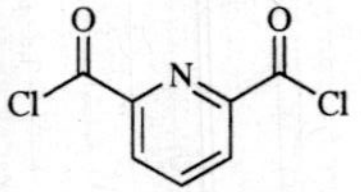	**C**, rt	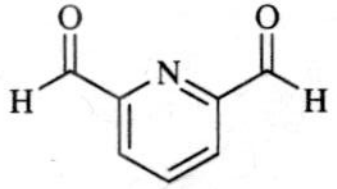 (>95)[a]	109
C_9			
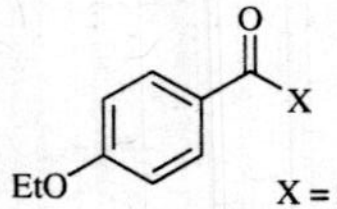 X = Cl, Br	Bn_3SiH, Et_2O, reflux	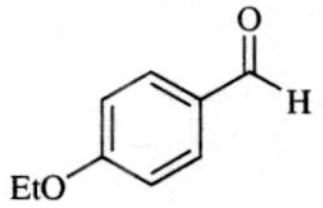 **I** (—)	286

TABLE 9. ORGANOSILANE REDUCTION OF ACYL HALIDES (*Continued*)

Acyl Halide	Conditions	Product(s) and Yield(s) (%)	Refs.
C_{10}			
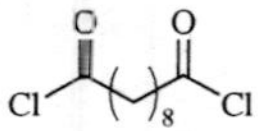	C, rt	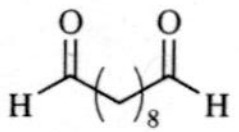 (85)	109
C_{11}			
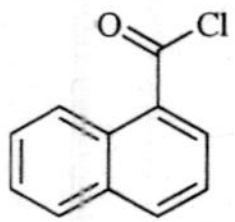	Et_3SiH (2.2 eq), $(C_6F_5)_3B$ (5 mol%), CH_2Cl_2, rt, 20 h	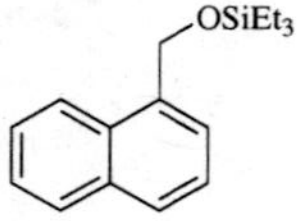 (95)	281, 282
C_{14}			
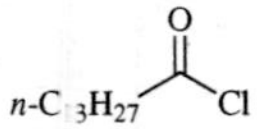	Et_3SiH (4.0 eq), $(C_6F_5)_3B$ (5 mol%), CH_2Cl_2, rt, 20 h	n-$C_{14}H_{30}$ (97)	282
C_{16}			
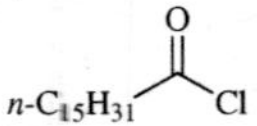	Et_3SiH (4 eq), $(C_6F_5)_3B$ (5 mol%), CH_2Cl_2, rt, 20 h	n-$C_{16}H_{34}$ (95)	281
C_{22}			
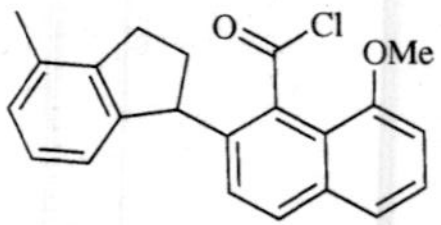	Et_3SiH, TFA, CCl_4, 15 min	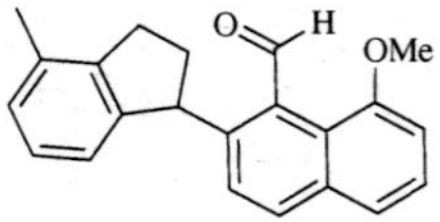 (—)	285

[a] The yield was determined by NMR analysis.

TABLE 10. ORGANOSILANE REDUCTION OF ESTERS AND LACTONES

Ester	Conditions	Product(s) and Yield(s) (%)	Refs.
C_4			
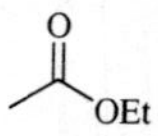	Ph_2MeSiH (1.2 eq), $Mn(CO)_5Ac$ (3.0 mol%), C_6D_6, rt, 1.75 h	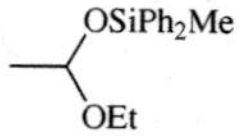 (84)	295
	$PhSiH_3$ (1.2 eq), $Mn(CO)_4PPh_3Ac$ (1.5-3.0 mol%), C_6D_6, rt, 15 min	Et_2O (85) + $PhSiH(OEt)_2$ (—)	295
	$PhSiH_3$ (1.2 eq), $Mn(CO)_5Ac$ (1.5-3.0 mol%), C_6D_6, rt, 1.5 h	Et_2O (100)	295
	$PhSiH_3$ (1.2 eq), $Mn(CO)_5Br$ (1.5-3.0 mol%), C_6D_6, rt, 4 h	Et_2O (55)	295
	Ph_2SiH_2 (1.2 eq), $Mn(CO)_4PPh_3Ac$ (1.5-3.0 mol%), C_6D_6, rt, 4 h	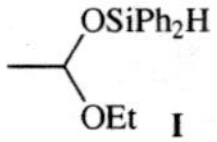 **I** (95) + $Ph_2SiH(OEt)$ (5)	295
	Ph_2SiH_2 (1.2 eq), $Mn(CO)_5Ac$ (1.5-3.0 mol%), C_6D_6, rt, 4 h	**I** + 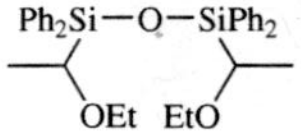 **II** **I** + **II** (—), **I**:**II** = 2:1	295
	Ph_2SiH_2 (1.2 eq), $Mn(CO)_5SiMe_2Ph$ (3.0 mol%), C_6D_6, *hv*, 20°, 35 min	Et_2O (100)	295
	Et_3SiH (2.2 eq), **201** (0.2 mol%), rt, 26 h	$EtOSiEt_3$ (94) + EtOEt (<1)	377
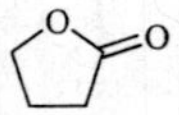	$PhSiH_3$ (1.2 eq), $Mn(CO)_5Br$ (1.5-3.0 mol%), C_6D_6, rt, 25 min	 (35)	295

TABLE 10. ORGANOSILANE REDUCTION OF ESTERS AND LACTONES (*Continued*)

Ester	Conditions	Product(s) and Yield(s) (%)	Refs.
C_4 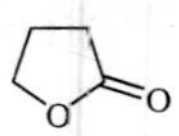	1. 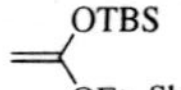 OEt, $SbCl_5$, TMSCl, SnI_2, CH_2Cl_2, −78°, 30 min 2. Et_3SiH (1.5 eq), −23°, 2.5 h	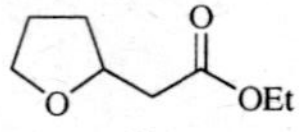 **I** (75)	306
	1. 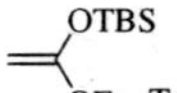 OEt, $TrSbCl_6$ (5-30 mol%), CH_2Cl_2, −78°, 30 min 2. Et_3SiH (1.5 eq), −23°, 2.5 h	**I** (72)	306
C_{4-5} 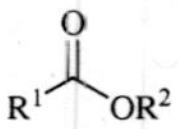	Cl_3SiH, γ-irradiation	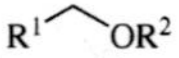 R^1 / R^2: H / *n*-Pr (77) Me / Et (85) Me / *n*-Pr (99) Et / Me (100) Et / Et (100) Et / *n*-Pr (96)	296
C_{4-6} 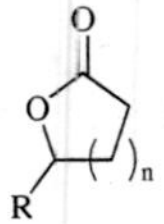	$(MeO)_3SiH$ (3.0 eq), LiOMe (8 mol%), THF, rt, 0.5 h	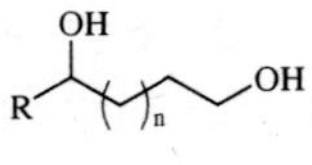 n / R: 0 / Me (48) 1 / Me (56) 2 / H (100) 3 / H (95)	294
C_{4-9} 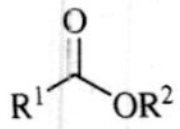	Et_3SiH (1.5 eq), Ru catalyst (1 mol%), MeC_6H_5, 100°, 16 h	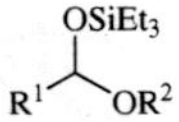	299

R[1]	R[2]	Catalyst	
Et	Me	$Ru_3(CO)_{12}$	(57)
Et	Me	$[RuCl_2(CO)_3]_2$, EtI, Et_2NH (5 eq)	(57)
Me	*i*-Pr	$Ru_3(CO)_{12}$	(74)
Me	*i*-Pr	$[RuCl_2(CO)_3]_2$, EtI, Et_2NH (5 eq)	(91)
Me	*t*-Bu	$Ru_3(CO)_{12}$	(45)
t-Bu	Me	$[RuCl_2(CO)_3]_2$, EtI, Et_2NH (5 eq)	(66)
c-C_6H_{11}	Me	$Ru_3(CO)_{12}$	(61)
c-C_6H_{11}	Me	$[RuCl_2(CO)_3]_2$, EtI, Et_2NH (5 eq)	(97)
Me	Ph	$Ru_3(CO)_{12}$	(94)
Me	Ph	$[RuCl_2(CO)_3]_2$, EtI, Et_2NH (5 eq)	(98)
Ph	Me	$Ru_3(CO)_{12}$	(57)[a]
Ph	Me	$[RuCl_2(CO)_3]_2$, EtI, Et_2NH (5 eq)	(63)[a]
Bn	Me	$Ru_3(CO)_{12}$	(84)
Bn	Me	$[RuCl_2(CO)_3]_2$, EtI, Et_2NH (5 eq)	(90)

C_5 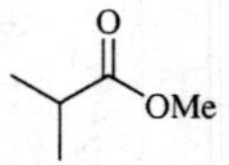	R_3SiH, EtI, Et_2NH, $[RuCl_2(CO)_3]_2$, MeC_6H_5, 100°, 16 h	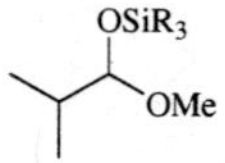 (—) R_3 = Et_3, *t*-$BuMe_2$, Me_2Ph	299
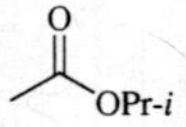	$PhSiH_3$ (1.2 eq), $Mn(CO)_4PPh_3Ac$, (3.0 mol%), C_6D_6, rt, 30 min	EtOPr-*i* (95)	295
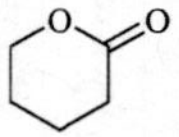	$PhSiH_3$ (1.2 eq), $Mn(CO)_4PPh_3Ac$, (3.0 mol%), C_6D_6, rt, 30 min	 (40)	295
	1. 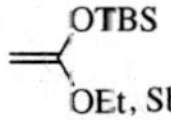 OEt, $SbCl_5$, TMSCl, SrI_2, CH_2Cl_2, –78°, 30 min 2. Et_3SiH (1.5 eq), –23°, 2.5 h	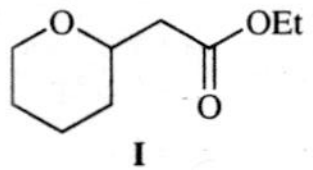 I (91)	306
	1. 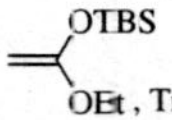 OEt, $TrSbCl_6$ (5-30 mol%), CH_2Cl_2, –78°, 30 min 2. Et_3SiH (1.5 eq), –23°, 2.5 h	I (95)	306

Ester	Conditions	Product(s) and Yield(s) (%)	Refs.
C_5 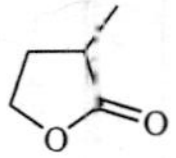	1. 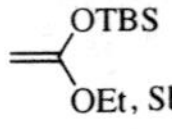 OEt, $SbCl_5$, TMSCl, SnI_2, CH_2Cl_2, –78°, 30 min 2. Et_3SiH (1.5 eq), –23°, 2.5 h	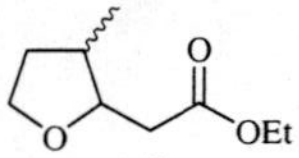 **I** (71) cis:trans = 36:64	306
	1. 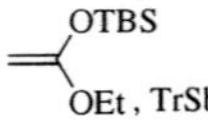 OEt, $TrSbCl_6$ (5-30 mol%), CH_2Cl_2, –78°, 30 min 2. Et_3SiH (1.5 eq), –23°, 2.5 h	**I** (56) cis:trans = 31:69	306
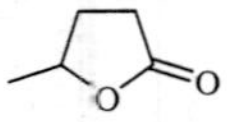	1. 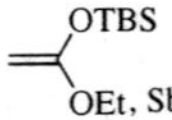 OEt, $SbCl_5$, TMSCl, SnI_2, CH_2Cl_2, –78°, 30 min 2. Et_3SiH (1.5 eq), –23°, 2.5 h	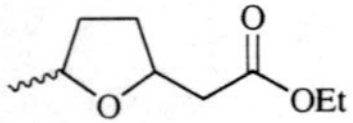 **I** (56) cis:trans = 31:69	306
	1. 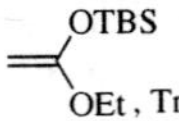 OEt, $TrSbCl_6$ (5-30 mol%), CH_2Cl_2, –78°, 30 min 2. Et_3SiH (1.5 eq), –23°, 2.5 h	**I** (56) cis:trans = 31:69	306
C_{5-20} 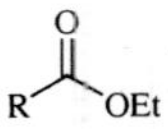	Et_3SiH (1 eq), $ZnCl_2$, heat, 3-4 h	RCH_2OCH_2R R = Et (70)	298
		n-Pr (76)	
		i-Bu (82)	
		n-Bu (72)	
		n-C_5H_{11} (69)	
		n-$C_{15}H_{31}$ (64)	
		n-$C_{17}H_{35}$ (70)	
		Ph (60)	
		4-BrC_6H_4 (56)	

C_6 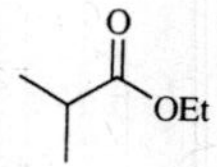	R_3SiH (1.5 eq), catalyst, MeC_6H_5, 100°, 15 h	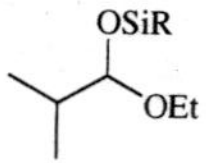	299
	R_3 — Catalyst		
	Et_3 — $Ru_3(CO)_{12}$	(78)	
	Et_3 — $[RuCl_2(CO)_3]_2$	(86)	
	t-$BuMe_2$ — $Ru_3(CO)_{12}$	(79)	
	t-$BuMe_2$ — $[RuCl_2(CO)_3]_2$	(87)	
	$PhMe_2$ — $Ru_3(CO)_{12}$	(87)	
	$PhMe_2$ — $[RuCl_2(CO)_3]_2$	(72)	
	$PhSiH_3$ (1.2 eq), $Mn(CO)_5Br$ (1.5-3.0 mol%), C_6D_6, rt, 30 min	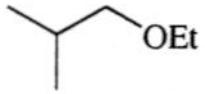 (61)	295
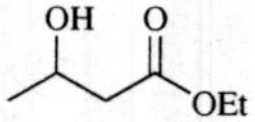	$(MeO)_3SiH$ (3.0 eq), LiOMe (6 mol%), THF, 65°, 0.5 h	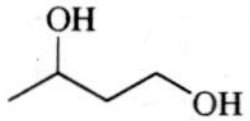 (100)	294
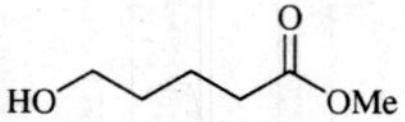	$(MeO)_3SiH$ (3.5 eq), LiOMe (6 mol%), THF, 65°, 0.5 h	 (80)	294
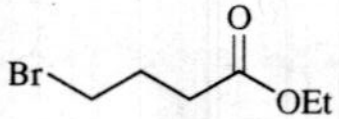	$PhSiH_3$ (1.2 eq), $Mn(CO)_5Br$ (1.5-3.0 mol%), C_6D_6, rt, 35 min	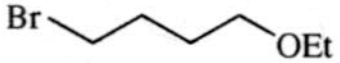 (72)	295
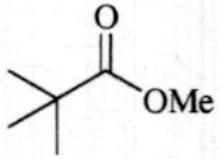	$PhSiH_3$ (1.2 eq), $Mn(CO)_5Br$ (1.5-3.0 mol%), C_6D_6, rt, 30 min	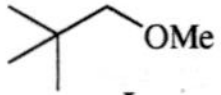 **I** (10)	295
	$PhSiH_3$ (1.2 eq), $Mn(CO)_5Br$ (1.5-3.0 mol%), C_6D_6, rt, 12 h	**I** (10)	295
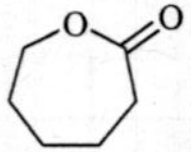	$PhSiH_3$ (1.2 eq), $Mn(CO)_5Br$ (1.5-3.0 mol%), C_6D_6, rt, 30 min	 (65)	295

TABLE 10. ORGANOSILANE REDUCTION OF ESTERS AND LACTONES (*Continued*)

Ester	Conditions	Product(s) and Yield(s) (%)	Refs.
C_6 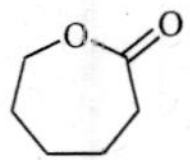	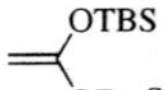 1. OEt, $SbCl_5$, TMSCl, SnI_2, CH_2Cl_2, –78°, 30 min 2. Et_3SiH (1.5 eq), –23°, 2.5 h	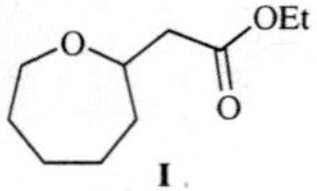 **I** (87)	306
	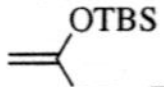 1. OEt, $TrSbCl_6$(5-30 mol%), CH_2Cl_2, –78°, 30 min 2. Et_3SiH (1.5 eq), –23°, 2.5 h	**I** (90)	306
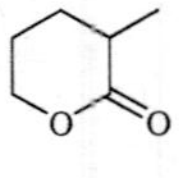	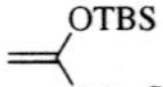 1. OEt, $SbCl_5$, TMSCl, SnI_2, CH_2Cl_2, –78°, 30 min 2. Et_3SiH (1.5 eq), –23°, 2.5 h	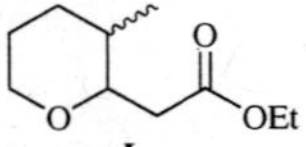 **I** (84) cis:trans = 12:88	306
	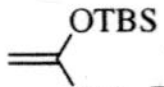 1. OEt, $TrSbCl_6$(5-30 mol%), CH_2Cl_2, –78°, 30 min 2. Et_3SiH (1.5 eq), –23°, 2.5 h	**I** (83) cis:trans = 10:90	306
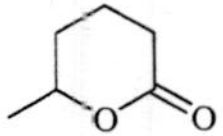	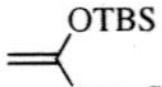 1. OEt, $SbCl_5$, TMSCl, SnI_2, CH_2Cl_2, –78°, 30 min 2. Et_3SiH (1.5 eq), –23°, 2.5 h	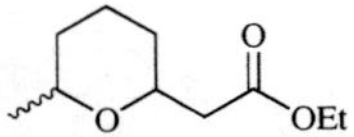 **I** (79) cis:trans > 99:1	306
	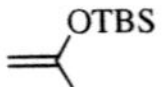 1. OEt, $TrSbCl_6$(5-30 mol%), CH_2Cl_2, –78°, 30 min 2. Et_3SiH (1.5 eq), –23°, 2.5 h	**I** (82) cis:trans > 99:1	306

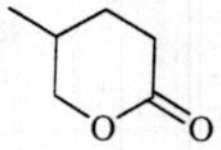	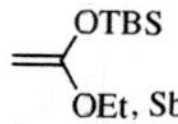 1. OEt, $SbCl_5$, TMSCl, SnI_2, CH_2Cl_2, −78°, 30 min 2. Et_3SiH (1.5 eq), −23°, 2.5 h	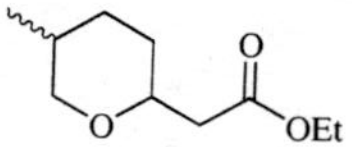 **I** (87) cis:trans = 7:93	306
	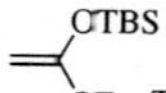 1. OEt , $TrSbCl_6$(5-30 mol%), CH_2Cl_2, −78°, 30 min 2. Et_3SiH (1.5 eq), −23°, 2.5 h	**I** (82) cis:trans = 4:96	306
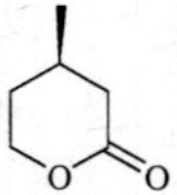	Et_3SiH, 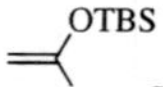 OEt , $SbCl_3$, TMSCl, SnI_2, CH_2Cl_2	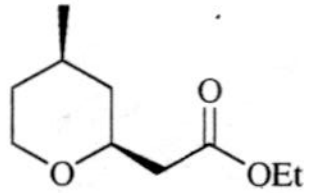 (84)	306
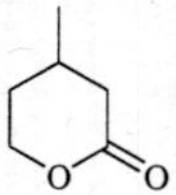	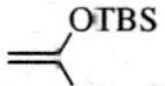 1. OEt, $SbCl_5$, TMSCl, SnI_2, CH_2Cl_2, −78°, 30 min 2. Et_3SiH (1.5 eq), −23°, 2.5 h	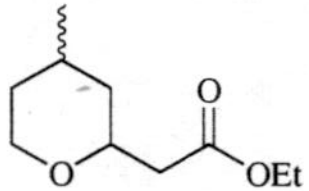 **I** (82) cis:trans > 99:1	306
	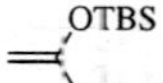 1. OEt , $TrSbCl_6$(5-30 mol%), CH_2Cl_2, −78°, 30 min 2. Et_3SiH (1.5 eq), −23°, 2.5 h	**I** (89) cis:trans > 99:1	306
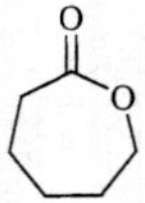	$(EtO)_3SiH$ (2.2 eq), CsF (1 eq), rt, 1 min	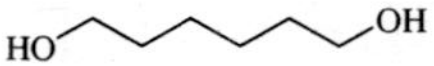 (50)	81
C_7			
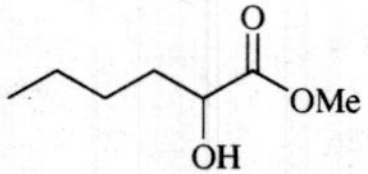	$(MeO)_3SiH$ (3.5 eq) LiOMe (6 mol%), THF, 65°, 9.5 h	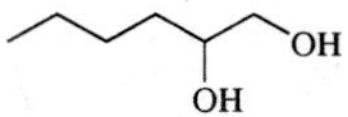 (100)	294
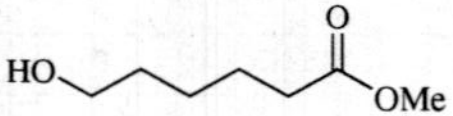	$(MeO)_3SiH$ (3.5 eq) LiOMe (6 mol%), THF, 65°, 0.5 h	 (63)	294

TABLE 10. ORGANOSILANE REDUCTION OF ESTERS AND LACTONES (*Continued*)

	Ester	Conditions	Product(s) and Yield(s) (%)	Refs.
C_7	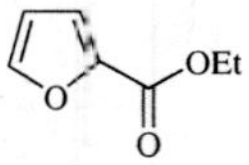	1. Cp_2TiCl_2 (5 mol%), *n*-BuLi (10 mol%), THF, –15°, 15 min 2. $(EtO)_3SiH$ (2 eq), ester addition, rt, 0.5-2 h	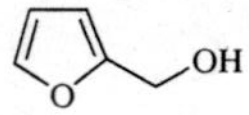 (75)	290
	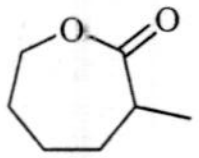	1. 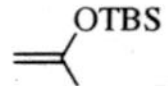 OEt, $SbCl_5$, TMSCl, SnI_2, CH_2Cl_2, –78°, 30 min 2. Et_3SiH (1.5 eq), –23°, 2.5 h	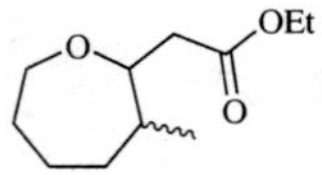 **I** (91) cis:trans > 99:1	306
		1. 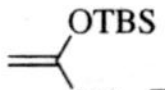 OEt, $TrSbCl_6$ (5-30 mol%), CH_2Cl_2, –78°, 30 min 2. Et_3SiH (1.5 eq), –23°, 2.5 h	**I** (86) cis:trans > 99:1	306
	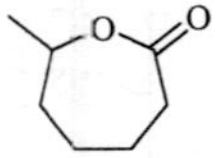	1. 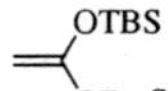 OEt, $SbCl_5$, TMSCl, SnI_2, CH_2Cl_2, –78°, 30 min 2. Et_3SiH (1.5 eq), –23°, 2.5 h	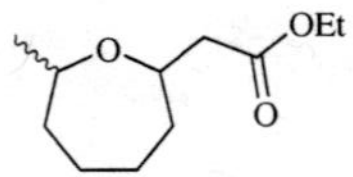 **I** (85) cis:trans > 99:1	306
		1. 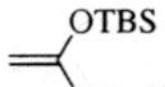 OEt, $TrSbCl_6$ (5-30 mol%), CH_2Cl_2, –78°, 30 min 2. Et_3SiH (1.5 eq), –23°, 2.5 h	**I** (81) cis:trans > 99:1	306
	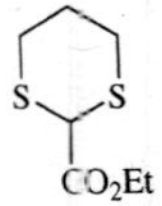	Ph_3SiH, $(C_6F_5)_3B$ (1 mol%), MeC_6H_5, rt	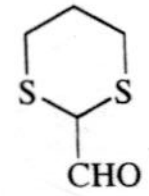 (45)	116

$C_{7\text{-}9}$

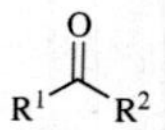

$(EtO)_3SiH$ (1.5 eq), HAp, rt

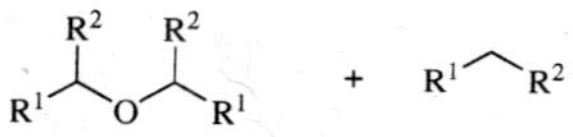

I + II

353

R^1	R^2	Time	
n-C_5H_{11}	MeO	24 h	(44)
Ph	EtO	20 h	(97)

$C_{7\text{-}28}$

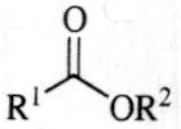

R_3SiH (x eq), CsF (1 eq)

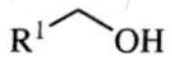

81

R^1	R^2	R_3Si	x	Temp	Time	
n-C_4H_9	Et	Ph_2HSi	1.1	140°	4 h	(90)
n-C_4H_9	Et	$(EtO)_2MeSi$	2.2	120°	3 h	(90)
n-C_4H_9	Et	$(EtO)_3Si$	2.2	rt	1 min	(90)
Ph	Et	Ph_2HSi	1.1	140°	4 h	(65)
Ph	Et	$(EtO)_3Si$	2.2	100°	3 h	(65)
$MeO_2C(CH_2)_7$	Me	Ph_2HSi	1.1	140°	4 h	(65)
$MeO_2C(CH_2)_7$	Me	$(EtO)_2MeSi$	2.2	120°	2 h	(65)
$MeO_2C(CH_2)_7$	Me	$(EtO)_3Si$	2.2	rt	12 h	(65)
n-$C_{11}H_{23}$	Et	$(EtO)_2MeSi$	2.2	100°	20 h	(90)
n-$C_{11}H_{23}$	Et	$(EtO)_3Si$	2.2	60°	0.5 h	(90)
$H_2C{=}CH(CH_2)_8$	Me	Ph_2HSi	1.1	140°	10 h	(70)
$H_2C{=}CH(CH_2)_8$	Me	$(EtO)_2MeSi$	2.2	80°	48 h	(65)
$H_2C{=}CH(CH_2)_8$	Me	$(EtO)_3Si$	2.2	rt	20 h	(65)
$H_2C{=}CH(CH_2)_8$	Me	$(EtO)_3Si$	2.2	60°	0.5 h	(70)
n-$C_8H_{17}CH{=}CH(CH_2)_7$	Me	Ph_2HSi	1.1	140°	4 h	(70)
n-$C_8H_{17}CH{=}CH(CH_2)_7$	Me	$(EtO)_2MeSi$	2.2	120°	3 h	(85)
n-$C_8H_{17}CH{=}CH(CH_2)_7$	Me	$(EtO)_3Si$	2.2	rt	4 h	(90)
$H_2C{=}CH(CH_2)_8$	menthyl	$(EtO)_3Si$	2.2	rt	72 h	(75)
$H_2C{=}CH(CH_2)_8$	menthyl	$(EtO)_3Si$	2.2	60°	9 h	(75)
n-$C_8H_{17}CH{=}CH(CH_2)_7$	menthyl	$(EtO)_3Si$	2.2	rt	72 h	(80)

TABLE 10. ORGANOSILANE REDUCTION OF ESTERS AND LACTONES (*Continued*)

Ester	Conditions	Product(s) and Yield(s) (%)	Refs.
C_8 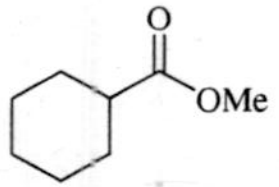	Et_3SiH, EtI, Et_2NH, $[RuCl_2(CO)_3]_2$, MeC_6H_5, 100°, 16 h	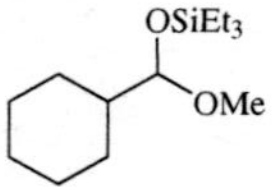 (92)	299
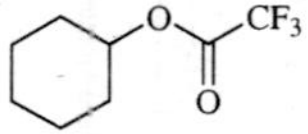	$EtCl_2SiH$ (2 eq), $AlBr_3$ (1 eq), 20°, 2 h	 (69)	192
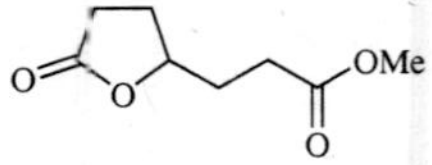	PMHS (5 eq), Cp_2TiF_2 (2 mol%), TBAF/alumina (1 mol%), MeC_6H_5, rt	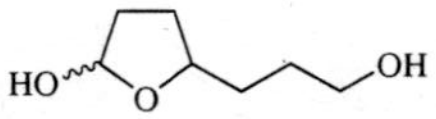 (88)	301
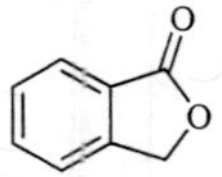	PMHS (5 eq), $Cp_2Ti(OC_6H_4Cl\text{-}4)_2$ (2 mol%), TBAF/alumina (1 mol%), MeC_6H_5, rt	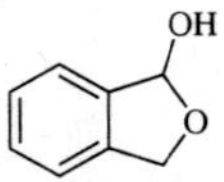 (75) + 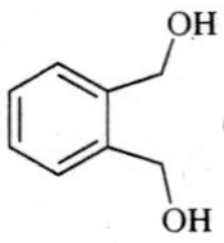 (14)	302
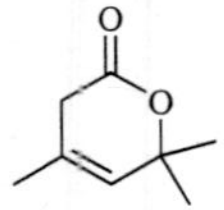	PMHS (5 eq), $Cp_2Ti(OC_6H_4Cl\text{-}4)_2$ (2 mol%), TBAF/alumina (1 mol%), MeC_6H_5, rt	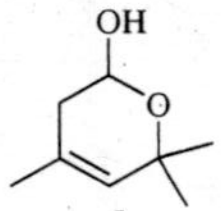 **I** (89)	301, 302
	PMHS (5 eq), Cp_2TiF_2 (2 mol%), TBAF/alumina (1 mol%), MeC_6H_5, rt	**I** (76)	301
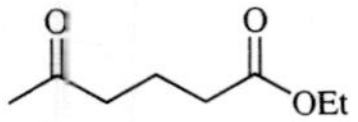	1. Cp_2TiCl_2 (5 mol%), *n*-BuLi (10 mol%), THF, −15°, 15 min 2. $(EtO)_3SiH$ (3.3 eq), ester addition, rt, 0.5-2 h	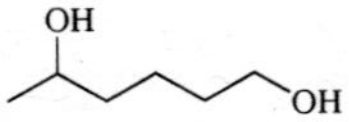 (78)	290
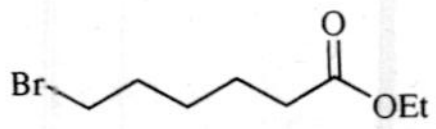	1. Cp_2TiCl_2 (5 mol%), *n*-BuLi (10 mol%), THF, −15°, 15 min 2. $(EtO)_3SiH$ (2 eq), ester addition, −20° to rt, 8 h	 **I** (78)	290

	PMHS (5 eq), Cp_2TiCl_2 (15 mol%), EtMgBr (2 eq), THF, rt, 3 h	**I** (53) + 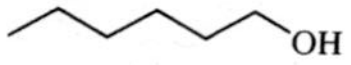 (6)	289
	$(EtO)_3SiH$ (2.5 eq), Ti(OPr-i)$_4$, (5 mol%), 40-55°, 6 h	**I** (75)	291
	Ph_3SiH, $(C_6F_5)_3B$ (1 mol%), MeC_6H_5, rt	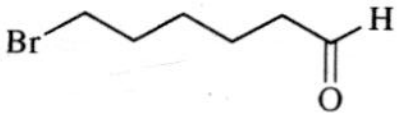 (70)	116
	1. $EtMe_2SiH$ (4 eq), **223** (1 mol%), 1,4-dioxane, 20°, 0.5 h 2. Add ester, 2 h	**I** (47) +  (40)	280
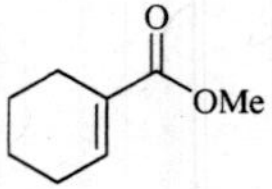	PMHS (2.5 eq), Cp_2TiCl_2, *n*-BuLi (2 eq, 5 mol%), THF, rt, 1 h	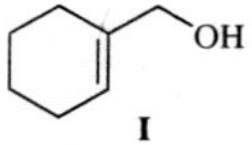 **I** (65)	289
	1. Cp_2TiCl_2 (5 mol%), *n*-BuLi (10 mol%), THF, –15°, 15 min 2. $(EtO)_3SiH$ (2 eq), ester addition, rt, 0.5-2 h	**I** (71)	290
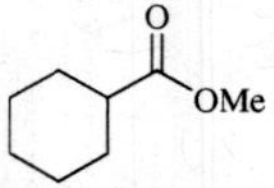	$(EtO)_3SiH$ (2.5 eq), Ti(OPr-*i*)$_4$ (5 mol%), 40-55°, 16 h	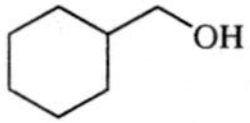 (88)	291
	Et_3SiH (1.5 eq), $[RuCl_2(CO)_3]_2$, EtI, Et_2NH, MeC_6H_5, 100°, 16 h	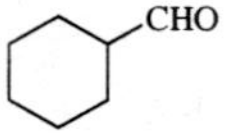 (90)	299

TABLE 10. ORGANOSILANE REDUCTION OF ESTERS AND LACTONES (*Continued*)

Ester	Conditions	Product(s) and Yield(s) (%)	Refs.
C_8			
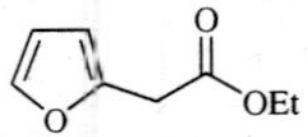	PMHS (2.5 eq), Cp_2TiCl_2 (5 mol%), EtMgBr (2 eq), THF, rt, 17.5 h	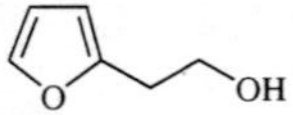 (75)	289
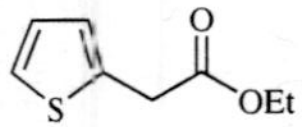	$(EtO)_3SiH$ (2.5 eq), $Ti(OPr\text{-}i)_4$, (5 mol%), 40-55°, 22 h	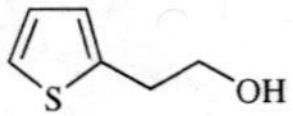 **I** (93)	291
	PMHS (2.5 eq), $Ti(OPr\text{-}i)_4$ (25 mol%), 65°, 24 h	**I** (8) + 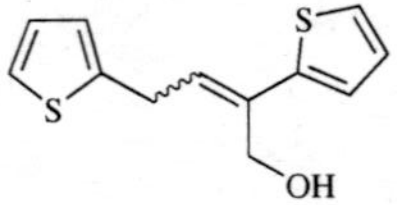 (20)	292
	1. Cp_2TiCl_2 (5 mol%), *n*-BuLi (10 mol%), THF, –15°, 15 min 2. $(EtO)_3SiH$ (2 eq), ester addition, rt, 0.5-2 h	**I** (88)	290
	PMHS (2.5 eq), Cp_2TiCl_2 (5 mol%), EtMgBr (2 eq), THF, rt, 17.5 h	**I** (92)	289
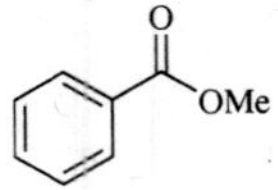	1. Cp_2TiCl_2 (5 mol%), *n*-BuLi (10 mol%), THF, –15°, 15 min 2. $(EtO)_3SiH$ (2 eq), ester addition, rt, 0.5-2 h	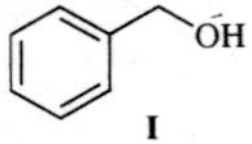 **I** (93)	290
	$(EtO)_3SiH$ (2.5 eq), $PhSiH_3$ (1.4 eq), $Ti(OPr\text{-}i)_4$ (5 mol%), 40-55°, 16 h	**I** (75)	291
	PMHS (2.5 eq), Cp_2TiCl_2 (2 mol%), EtMgBr (2 eq), THF, rt, 1.5 h	**I** (94)	289
	1. $EtMe_2SiH$ (4 eq), **223** (1 mol%), 1,4-dioxane, 20°, 0.5 h 2. Add ester, 24 h	**I** (52)	280

	PMHS (xs), zinc catalyst (0.2 eq), additive, $(i\text{-Pr})_2O$, 70°, 4 h	I		738

Zinc catalyst	Additive	
$Zn(EH)_2$	$NaBH_4$ (1eq)	(100)
$Zn(O_2CC_6H_5)_2$	$NaBH_4$ (1 eq)	(96)
$Zn(OAc)_2$	$NaBH_4$ (1 eq)	(2)
$Zn(EH)_2$	MeMgCl (1 eq)	(100)
$Zn(EH)_2$	$AlEt_3$ (2 eq)	(8)
$Zn(EH)_2$	BH_3 (1 eq)	(62)
$Zn(EH)_2$	$AlH(Bu\text{-}t)_2$ (1 eq)	(66)
ZnI_2	LiH	(78)
$ZnCl_2$	$LiAlH_4$	(13)
$ZnEt_2$	$Me_2N(CH_2)_2OH$ (1 eq)	(13)
$ZnEt_2$	$H_2N(CH_2)_2NH_2$ (1 eq)	(84)
$ZnEt_2$	$Me_2N(CH_2)_2NMe_2$ (1 eq)	(84)

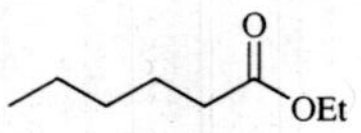	$PhSiH_3$ (1.2 eq), $Mn(CO)_5Br$ (1.5-3.0 mol%), C_6D_6, rt, 30 min		(81)	295
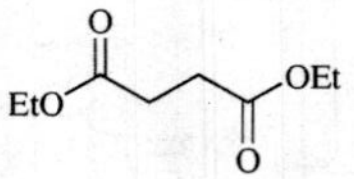	$PhSiH_3$ (1.2 eq), $Mn(CO)_5Br$ (1.5-3.0 mol%), C_6D_6, rt, 20 min	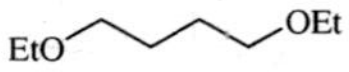	(68)	295
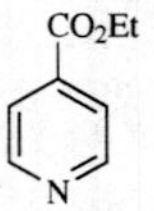	$PhMeSiH_2$ (1.5 eq), Cp_2TiMe_2 (10 mol%), 80°, 39 h	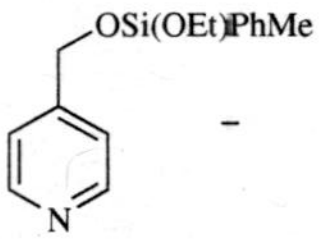 I + 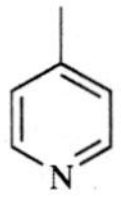 II	**I + II** (90), **I:II** = 4:1	264

TABLE 10. ORGANOSILANE REDUCTION OF ESTERS AND LACTONES (*Continued*)

Ester	Conditions	Product(s) and Yield(s) (%)	Refs.
C_{8-11}			
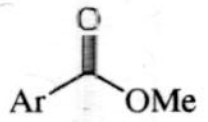	PMHS (3 eq), TBAF (2 mol%), THF, rt		278
Ar			
Ph		(95)	
4-BrC_6H_4		(96)	
4-ClC_6H_4		(97)	
2,4-$Cl_2C_6H_3$		(94)	
3,5-$Cl_2C_6H_3$		(83)	
3-MeC_6H_4		(90)	
4-MeC_6H_4		(93)	
2,4-$Me_2C_6H_3$		(93)	
3,4,5-$(MeO)_3C_6H_2$		(81)	
C_{8-14}			
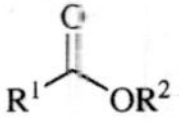	PMHS (2.5 eq), Ti(OPr-*i*)$_4$ (x mol%)	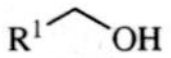	292
R^1 / R^2	x / Temp / Time		
$Br(CH_2)_5$ / Et	100 / 23 / 24 h	(87)	
$I(CH_2)_5$ / Et	100 / 23 / 24 h	(79)	
2-FC_6H_4 / Me	100 / 65 / 2 h	(80)	
2-BrC_6H_4 / Et	100 / 65 / 1.25 h	(87)	
4-$MeOC_6H_4$ / Me	25 / 65 / 24 h	(89)	
n-C_9H_{19} / Et	25 / 65 / 24 h	(93)	
CH_2=$CH(CH_2)_8$ / Me	25 / 70 / 24 h	(78)	
CH≡$C(CH_2)_8$ / Me	100 / 40 / 22 h	(63)	
1-adamantyl / Me	25 / 70 / 24 h	(>99)	
t-$BuO_2C(CH_2)_7$ / Me	100 / 40 / 5.5 h	(56)	

Reactant	Conditions	Product(s) and Yield(s) (%)	Refs.
C_{8-19} 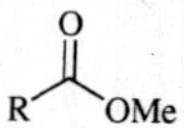	PMHS (x eq), Ti(OPr-i)$_4$ (y eq), THF	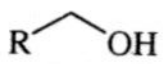	279

R	x	y	Time	Yield
Ph	10	1	5 h	(86)
Bn	10	1	8 h	(76)
$BnCH_2$	10	1	16 h	(65)
(E)-PhCH=CH	10	1	5 h	(82)
4-$O_2NC_6H_4$	10	1	16 h	(84)
4-HOC_6H_4	10	1	16 h	(81)
4-$MeOC_6H_4$	10	1	16 h	(88)
n-$C_{11}H_{23}$	10	1	16 h	(89)
n-$C_{17}H_{35}$	10	1	16 h	(98)
n-$C_{17}H_{35}$	5	0.5	16 h	(93)
n-$C_{17}H_{35}$	2	1	16 h	(92)
n-$C_{17}H_{35}$	10	0.5	80 h	(87)

Reactant	Conditions	Product(s) and Yield(s) (%)	Refs.
C_9 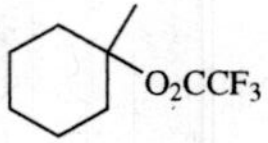	$EtCl_2SiH$ (2 eq), $AlBr_3$ (0.25 eq), 40°, 2 h	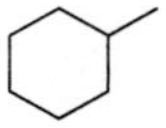 I (88)	192
	$EtCl_2SiH$ (2 eq), $AlCl_3$ (0.5 eq), 40°, 2 h	I (84)	192
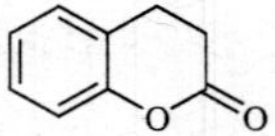	Et_3SiH (6.0 eq), $(C_6F_5)_3B$ (5 mol%), CH_2Cl_2, rt, 20 h	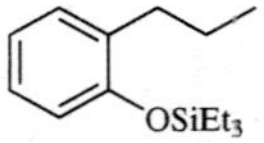 (>99)	282
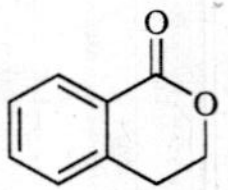	PMHS (5 eq), $Cp_2Ti(OC_6H_4Cl\text{-}4)_2$ (2 mol%), TBAF/alumina (1 mol%), MeC_6H_5, rt	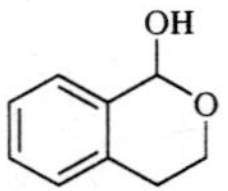 (87) + 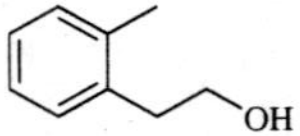 (2)	301, 302
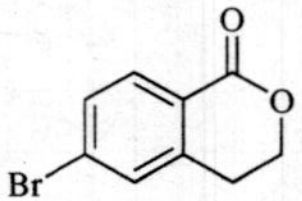	PMHS (5 eq), Cp_2TiF_2 (5 mol%), TBAF/alumina (1 mol%), MeC_6H_5, rt	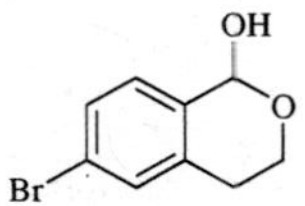 (69)	301, 302

Ester	Conditions	Product(s) and Yield(s) (%)	Refs.
C_9 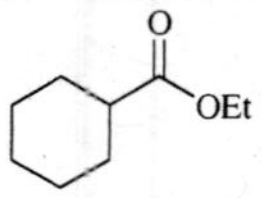	PMHS (2.5 eq), Cp_2TiCl_2 (5 mol%), EtMgBr (2 eq), THF, rt, 5 h	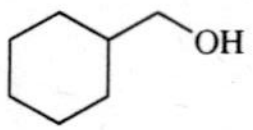 (88)	289
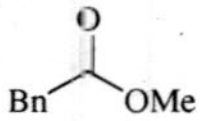	PMHS (3 eq), TBAF (2 mol%), THF, rt	 (76)	278
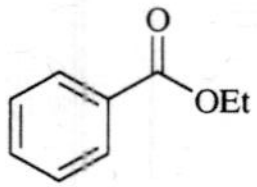	PMHS (1.2 eq), KF•$2H_2O$ (1.3 eq), DMSO, 80°, 6.5 h	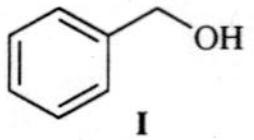 **I** (81)	82, 83
	$[HSi(OEt)_4]K$, THF, rt, 8-15 h	**I** (86)	288
	$(EtO)_3SiH$ (2.3 eq), CsF (1 eq), 60°, 30 min	**I** (90)	83
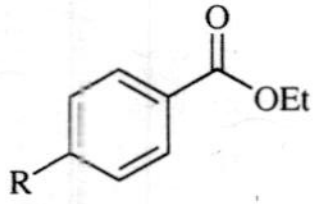	1. Cp_2TiCl_2 (5 mol%), *n*-BuLi (10 mol%), THF, −15°, 15 min 2. $(EtO)_3SiH$ (3.3 eq), ester addition, rt, 0.5-2 h	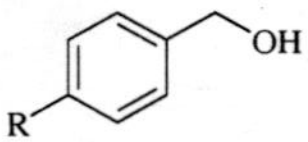 (88) R: OH (88); NH_2 (81)	290
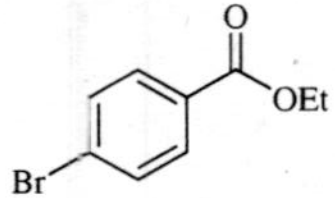	PMHS (2.5 eq), Cp_2TiCl_2 (15 mol%), EtMgBr (2 eq), THF, rt, 3 h	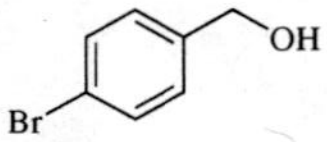 (40) + 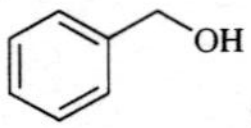 (9)	289
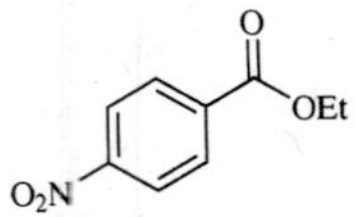	$(EtO)_3SiH$ (1.3 eq), $PhSiH_3$ (1.4 eq), $Ti(OPr\text{-}i)_4$ (5 mol%), 40-55°, 20 h	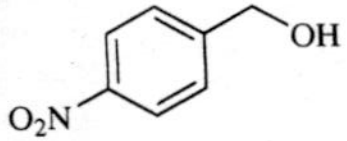 (75)	291

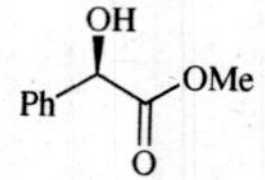	PMHS (3 eq), TBAF (2 mol%), THF, rt	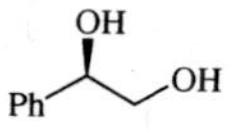 (90) >95% ee	278
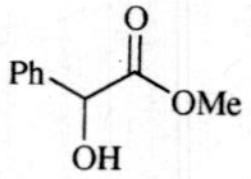	$(MeO)_3SiH$ (3.5 eq), LiOMe (6 mol%), THF, 65°, 9.5 h	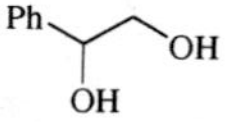 **I** (94)	294
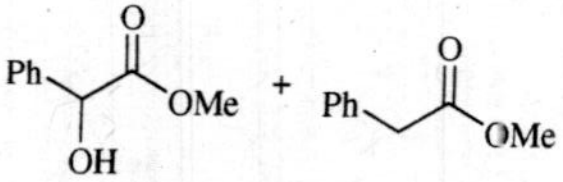	$(MeO)_3SiH$ (3.5 eq), LiOMe (6 mol%), THF, 65°, 9.5 h	**I** (78) +  (—)	294
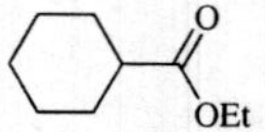	$PhSiH_3$ (1.2 eq), $Mn(CO)_5Br$ (1.5-3.0 mol%), C_6D_6, rt, 25 min	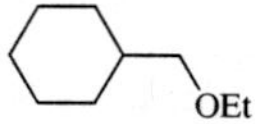 (70)	295
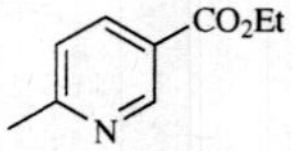	$PhMeSiH_2$ (1.5 eq), Cp_2TiMe_2 (10 mol%), rt, 24 h	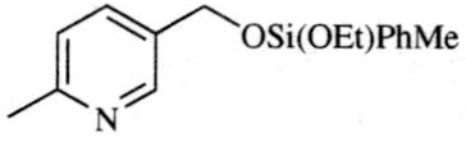 (80) + 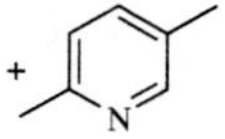 (20)	264
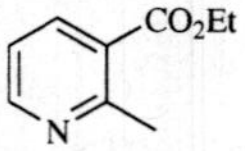	$PhMeSiH_2$ (1.5 eq), Cp_2TiMe_2 (10 mol%), 6 d	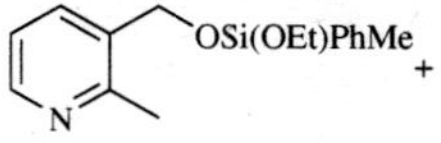 **I** + 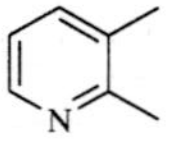 **II** **I** + **II** (100), **I**:**II** = 3:2	264
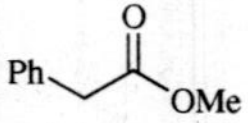	$PhSiH_3$ (1.2 eq), $Mn(CO)_4PPh_3Ac$ (3.0 mol%), C_6D_6, rt, 30 min	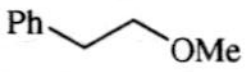 (83)	295
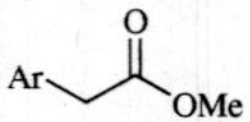	Et_3SiH (5.0 eq), $TiCl_4$ (1.5 eq), TMSOTf (3.1 eq), CH_2Cl_2, rt, 20 h	 Ar: Ph (40); 4-BrC_6H_4 (80); 4-$O_2NC_6H_4$ (72)	297
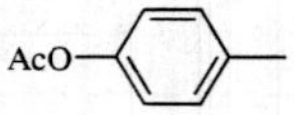	$PhSiH_3$ (1.2 eq), $Mn(CO)_4PPh_3Ac$ (3.0 mol%), C_6D_6, rt, 12 h	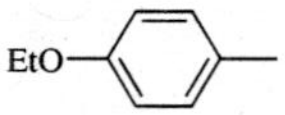 (12)	295

TABLE 10. ORGANOSILANE REDUCTION OF ESTERS AND LACTONES (*Continued*)

Ester	Conditions	Product(s) and Yield(s) (%)	Refs.
C_9 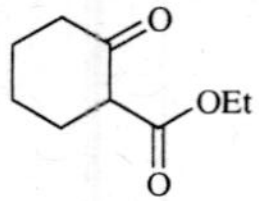	Ph_3SiH, $(C_6F_5)_3B$ (2 mol%), MeC_6H_5, rt	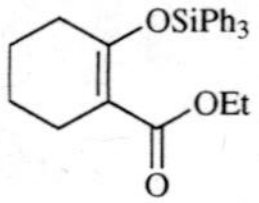 $(78)^b$	116
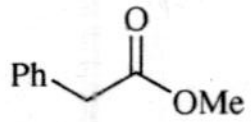	Et_3SiH (1.5 eq), $[RuCl_2(CO)_3]_2$, EtI, Et_2NH, MeC_6H_5, 100°, 16 h	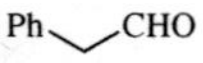 (65)	299
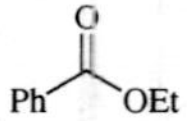	Ph_3SiH, $(C_6F_5)_3B$ (2 mol%), MeC_6H_5, rt	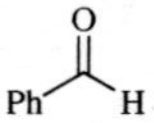 $(80)^c$	116
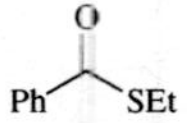	Et_3SiH (2-3 eq), Pd/C (2-5 mol%), acetone, rt, 30-60 min	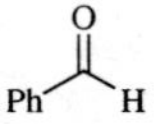 (91)	300
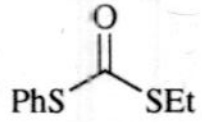	Et_3SiH (2-3 eq), Pd/C (2-5 mol%), acetone, rt, 30-60 min	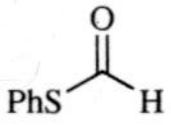 (75)	300
$C_{9\text{-}10}$ 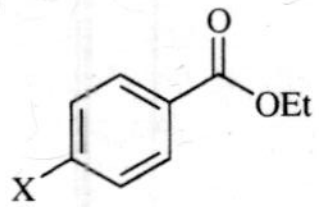	Ph_3SiH (1 eq), $(C_6F_5)_3B$ (2 mol%)	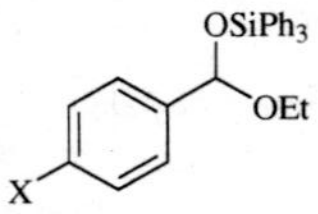 X = H (80) X = Me, Cl, NO_2 (—)	115
$C_{9\text{-}15}$ 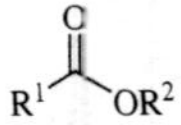	Ph_2SiH_2 (3 eq), catalyst (1.25 mol%), PPh_3 (5 eq), THF, rt	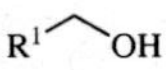	293

R^1	R^2	Catalyst	Time	
Ph	Et	$[RhCl(cod)]_2$	144 h	(70)
Ph	Et	$RhCl(PPh_3)_3$	24 h	(56)
$Br(CH_2)_6$	Et	$[RhCl(cod)]_2$	72 h	(92)
$Br(CH_2)_6$	Et	$RhCl(PPh_3)_3$	24 h	(92)
Bn	Et	$[RhCl(cod)]_2$	72 h	(92)
Me	$n\text{-}C_{10}H_{21}$	$[RhCl(cod)]_2$	72 h	(94)
$n\text{-}C_9H_{19}$	Et	$[RhCl(cod)]_2$	72 h	(98)
$n\text{-}C_9H_{19}$	*i*-Bu	$[RhCl(cod)]_2$	72 h	(92)
$n\text{-}C_9H_{19}$	*i*-Bu	$RhCl(PPh_3)_3$	6 h	(95)
$n\text{-}C_{11}H_{23}$	*i*-Pr	$[RhCl(cod)]_2$	72 h	(66)
$n\text{-}C_{11}H_{23}$	*i*-Pr	$RhCl(PPh_3)_3$	6 h	(80)
$n\text{-}C_7H_{15}$ 	*i*-Pr	$[RhCl(cod)]_2$	72 h	(63)

C_{10}

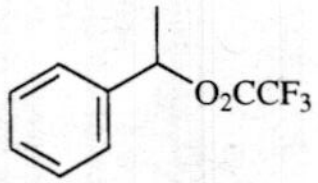	Ph_3SiH, $O_2NC_6H_5$, 80°, 13 h	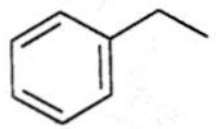 (90)	193
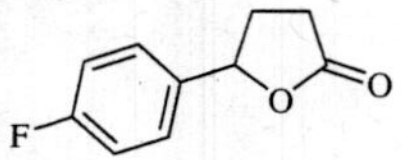	PMHS (5 eq), $Cp_2Ti(OC_6H_4Cl\text{-}4)_2$ (2 mol%), TBAF/alumina (1 mol%), MeC_6H_5, rt	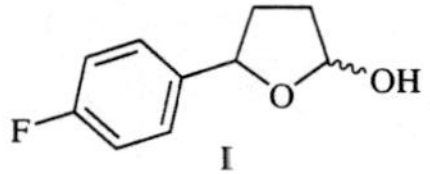 I (96)	301, 302
	PMHS (5 eq), Cp_2TiF_2 (2 mol%), TBAF/alumina (1 mol%), MeC_6H_5, rt	**I** (97)	301
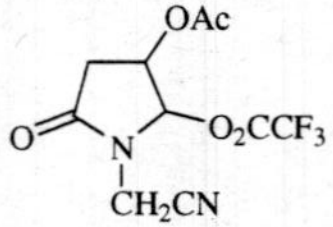	Et_3SiH (1.2 eq), TFA (13 eq), rt, 1 h	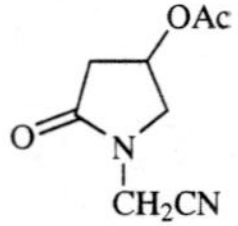 (81)	739
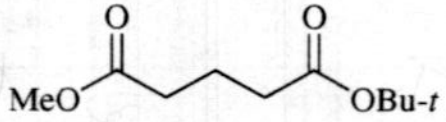	1. Cp_2TiCl_2 (5 mol%), *n*-BuLi (10 mol%), THF, –15°, 15 min 2. $(EtO)_3SiH$ (2 eq), ester addition, –20° to rt, 8 h	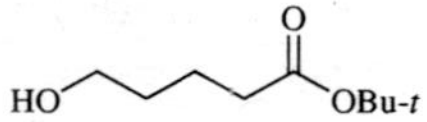 (87)	290

Ester	Conditions	Product(s) and Yield(s) (%)	Refs.
C_{10} 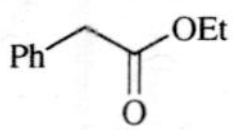	$(EtO)_3SiH$ (2.5-3.0 eq), $Ti(OPr\text{-}i)_4$, (5 mol%), 40-55°, 10 h	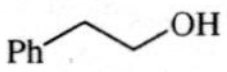 (89)	291
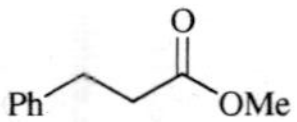	1. $EtMe_2SiH$ (2.4 eq), **223** (1 mol%), 1,4-dioxane, 20°, 0.5 h 2. Add ester, 0.5 h	 (97)	280
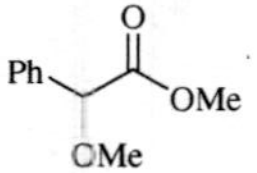	$(MeO)_3SiH$ (3.5 eq), LiOMe (6 mol%), THF, 65°, 9.5 h	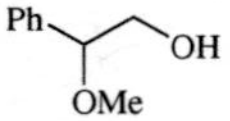 (76)	294
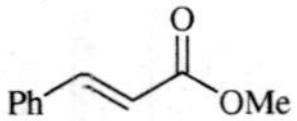	PMHS (3 eq), TBAF (2 mol%), THF, rt	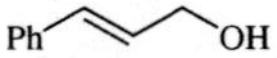 (84)	278
$C_{10\text{-}11}$ 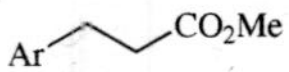	Et_3SiH (5.0 eq), $TiCl_4$ (1.5 eq), TMSOTf (3.1 eq), CH_2Cl_2, rt, 20 h	 Ar: Ph (75); 4-$MeOC_6H_4$ (89)	297
C_{11} 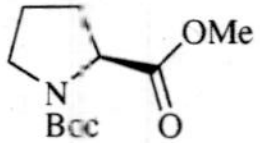	PMHS (2.5 eq), Cp_2TiCl_2 (5 mol%), EtMgBr (2 eq), THF, rt, 23 h	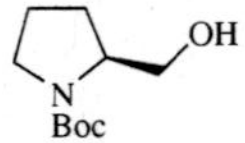 (70)	289
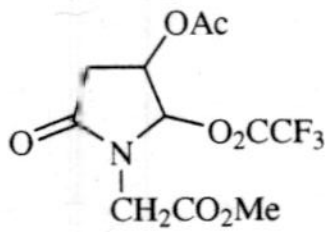	Et_3SiH (1.2 eq), TFA (13 eq), rt, 1 h	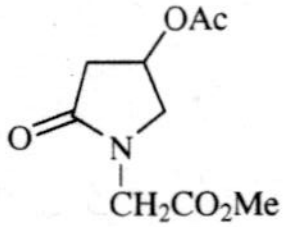 (80)	739
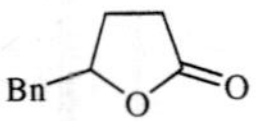	PMHS (5 eq), $Cp_2Ti(OC_6H_4Cl\text{-}4)_2$ (2 mol%), TBAF/alumina (1 mol%), MeC_6H_5, rt	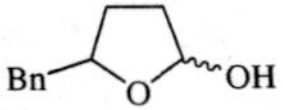 **I** (93)	301, 302
	PMHS (5 eq), Cp_2TiF_2 (2 mol%), TBAF/alumina (1 mol%), MeC_6H_5, rt	**I** (87)	301

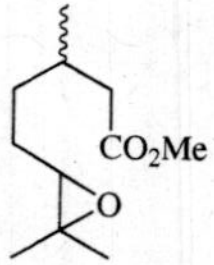	PMHS (2.5 eq), Cp_2TiCl_2 (5 mol%), EtMgBr (2 eq), THF, –78° to rt, 1.5 h	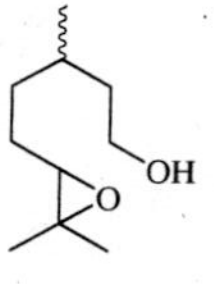 (44) + 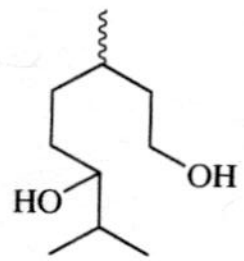 (28)	289
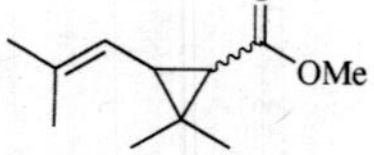	PMHS (2.5 eq), $Ti(OPr\text{-}i)_4$ (100 mol%), 65°, 8 h	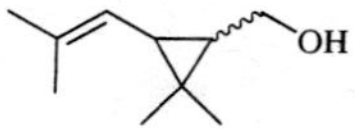 (58)	292
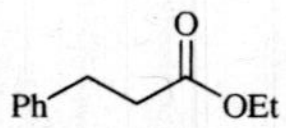	1. Cp_2TiCl_2 (5 mol%), *n*-BuLi (10 mol%), THF, –15°, 15 min 2. $(EtO)_3SiH$ (1.2 eq), ester addition, –20° to rt, 8 h	 **I** (82)	290
	PMHS (2.5 eq), Cp_2TiCl_2 (5 mol%), EtMgBr (2 eq), THF, rt, 23 h	**I** (94)	289
	1. $EtMe_2SiH$ (2.4 eq), **2** (1 mol%), 1,4-dioxane, 20°, 0.5 h 2. Add ester, 1 h	**I** (55) +  **II** (11)	280

TABLE 10. ORGANOSILANE REDUCTION OF ESTERS AND LACTONES (*Continued*)

Ester	Conditions	Product(s) and Yield(s) (%)	Refs.
C_{11} 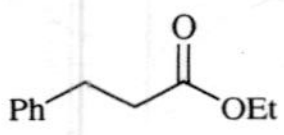	1. R_3SiH (2.4 eq), **223** (1 mol%), solvent 2. Add ester	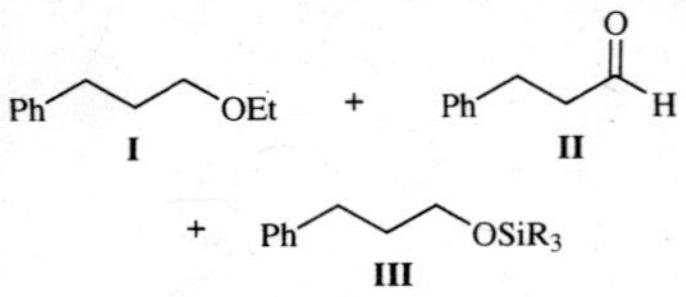	280

R_3Si	Solvent	Temp	Time	% Conversion	I:II:III
$EtMe_2Si$	1,4-dioxane	20°	1 h	(100)	20:0:80
$HMe_2Si(CH_2)_2(Me)_2Si$	1,4-dioxane	20°	2 h	(100)	16:0:84
Et_2MeSi	1,4-dioxane	20°	5 h	(100)	76:0:24
$PhMe_2Si$	1,4-dioxane	20°	21 h	(95)	17:0:83
Ph_2MeSi	1,4-dioxane	20°	21 h	(74)	16:0:84
Et_3Si	1,4-dioxane	20°	21 h	(7)	0:100:0
$(i\text{-}Pr)_3Si$	1,4-dioxane	20°	24 h	(0)	0:0:0
$EtMe_2Si$	C_6H_6	20°	1 h	(100)	42:0:58
$EtMe_2Si$	Et_2O	20°	2 h	(100)	45:16:39
$EtMe_2Si$	tetrahydropyran	20°	2 h	(94)	53:0:47
$EtMe_2Si$	oxepane	20°	8 h	(69)	61:0:39
$EtMe_2Si$	C_5H_5N	20°	22 h	(0)	0:0:0
$EtMe_2Si$	1,4-dioxane	10°	1 h	(100)	28:0:72

Ester	Conditions	Product(s) and Yield(s) (%)	Refs.
	1. $EtMe_2SiH$ (2.4 eq), **224** (1 mol%) 2. Add ester	**I** (56)	280

	1. $EtMe_2SiH$ (2.4 eq), **225** (1 mol%) 2. Add ester	**I** (34) + **II** (0.3)	280
	1. $EtMe_2SiH$ (2.4 eq), **226** (1 mol%) 2. Add ester	**I** (40) + **II** (0.8)	280
	1. Et_2MeSiH (2.4 eq), **223** (1 mol%), THP, 20°, 0.5 h 2. Add ester, 1 h	**I** (87)	280
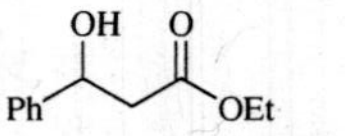	$(MeO)_3SiH$ (3.0 eq), LiOMe (6 mol%), THF, 65°, 0.5 h	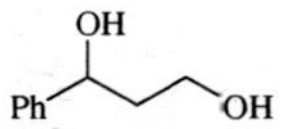 (92)	294
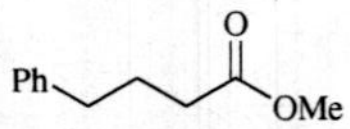	Et_3SiH (x eq), $TiCl_4$ (y eq), co-catalyst (z eq), CH_2Cl_2, rt, 20 h		297

x	y	Co-catalyst	z	
5.0	3.0	None	—	(27)
5.0	1.5	TMSOTf	3.0	(81)
5.0	—	TMSOTf	3.0	(—)
5.0	1.5	AgOTf	3.0	(63)
5.0	1.5	AgOTf, TMSCl	3.0, 3.0	(76)

TABLE 10. ORGANOSILANE REDUCTION OF ESTERS AND LACTONES (*Continued*)

Ester	Conditions	Product(s) and Yield(s) (%)	Refs.
C_{11} 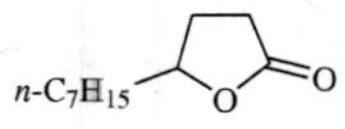	1. $EtMe_2SiH$ (4 eq), **223** (1 mol%), 1,4-dioxane, 20°, 0.5 h 2. Add ester, 3 h	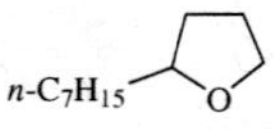 (79)	280
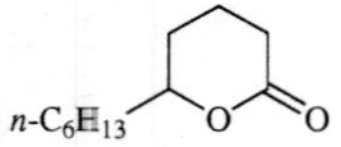	1. $EtMe_2SiH$ (4 eq), **223** (1 mol%), 1,4-dioxane, 20°, 0.5 h 2. Add ester, 2 h	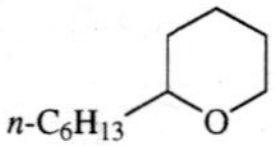 (73)	280
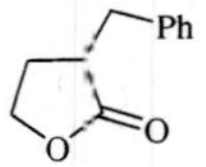	Et_3SiH (5.0 eq), $TiCl_4$ (1.5 eq), TMSOTf (3.1 eq), CH_2Cl_2, rt, 20 h	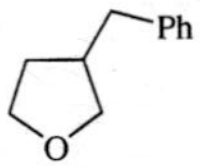 (67)	297
C_{11-16} 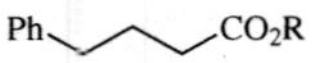	Et_3SiH (5.0 eq), $TiCl_4$ (1.5 eq), TMSOTf (3.1 eq), CH_2Cl_2, rt, 20 h	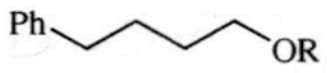 R — Me (81) Et (78) *i*-Pr (39) *n*-Bu (79) *c*-C_6H_{11} (34) Ph (67)	297
C_{12} 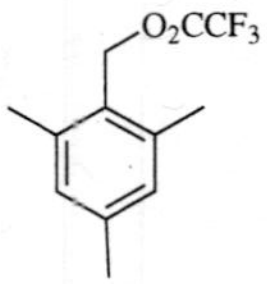	Ph_2SiH_2, $O_2NC_6H_5$, 80°, 7 h	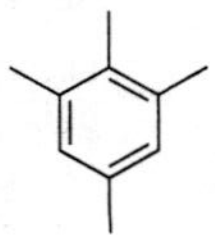 (90)	193

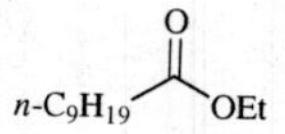

Ph$_2$SiH$_2$ (x eq), [Rh(cod)]$_2$ (y mol%), PPh$_3$ (2y eq), THF, rt

I

293

x	y	Time	I
4	5.0	72 h	(96)
4	2.5	72 h	(92)
4	1.0	72 h	(40)
3	2.5	72 h	(98)
2	2.5	72 h	(72)
3	2.5	6 h	(96)
3	1.0	6 h	(51)
3	1.0	24 h	(56)

Silane (2.5 eq), catalyst (x mol%) — I — 292

Silane	Catalyst	x	Temp	Time	I
PMHS	Ti(OPr-*i*)$_4$	25	65	24 h	(100)
PMHS	Ti(OPr-*i*)$_4$	100	23	25 h	(100)
PMHS	Ti(OPr-*i*)$_4$	25	65	2 h	(100)
Ph$_2$SiH$_2$	Ti(OPr-*i*)$_4$	25	65	25 h	(100)
Cl$_3$SiH	Ti(OPr-*i*)$_4$	100	23	24 h	(0)
PMHS	Ti(OBu-*n*)$_4$	25	65	24 h	(75)
PMHS	Ti(OBu-*n*)$_4$	100	65	2 h	(90)
Ph$_2$SiH$_2$	THEATi(OPr-*i*)	100	23	96 h	(0)

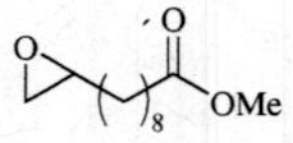

PMHS (2.5 eq), Ti(OPr-*i*)$_4$ (100 mol%), 65°, 2.5 h

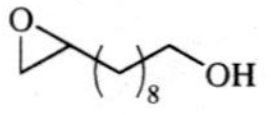

I (81) — 292

1. (EBTHI)$_2$TiCl$_2$ (5 mol%), *n*-BuLi (10 mol%), THF, –15°, 15 min
2. (EtO)$_3$SiH (2 eq), ester addition, –20° to rt, 8 h

I (67) — 290

TABLE 10. ORGANOSILANE REDUCTION OF ESTERS AND LACTONES (*Continued*)

Ester	Conditions	Product(s) and Yield(s) (%)	Refs.
C_{12} 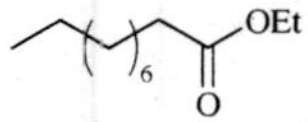	$(EtO)_3SiH$ (2.5-3.0 eq), $Ti(OPr\text{-}i)_4$ (5 mol%), 40-55°, 10 h	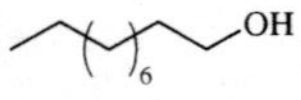 (95) **I**	291
	Ph_3SiH, $(C_6F_5)_3B$ (1 mol%), MeC_6H_5, rt	**I** (69)	116
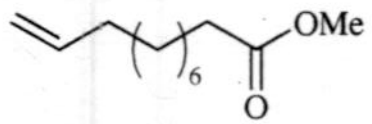	$(EtO)_3SiH$ (2.5-3.0 eq), $Ti(OPr\text{-}i)_4$ (5 mol%), 40-55°, 16 h	 (87) **I**	291
	PMHS (3 eq), TBAF (2 mol%), THF, rt	**I** (95)	278
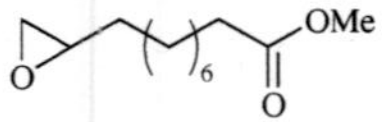	$(EtO)_3SiH$ (2.5-3.0 eq), $Ti(OPr\text{-}i)_4$ (5 mol%), 40-55°, 21 h	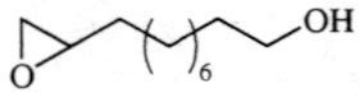 (83)	291
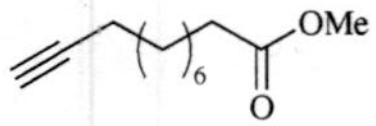	$(EtO)_3SiH$ (2.5-3.0 eq), $Ti(OPr\text{-}i)_4$ (5 mol%), 40-55°, 18 h	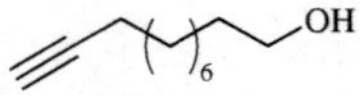 (70)	291
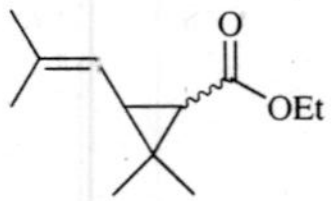	$(EtO)_3SiH$ (2.5 eq), $PhSiH_3$ (1.4 eq), $Ti(OPr\text{-}i)_4$ (5 mol%), 40-55°, 22 h	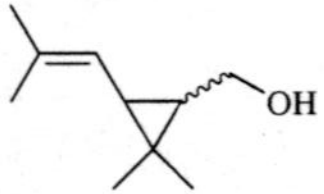 (80) **I**	291
	1. $(EBTHI)_2TiCl_2$ (5 mol%), *n*-BuLi (10 mol%), THF, –15°, 15 min 2. $(EtO)_3SiH$ (2 eq), ester addition, –20° to rt, 8 h	**I** (83)	290
	PMHS (2.5 eq), Cp_2TiCl_2 (5 mol%), EtMgBr (2 eq), THF, rt, 1 h	**I** (91)	289
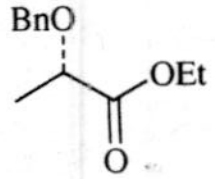	PMHS (10 eq), $Ti(OPr\text{-}i)_4$ (1 eq), THF	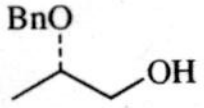 (89)	279

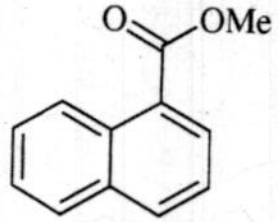	Et_3SiH (3.3 eq), $(C_6F_5)_3B$ (5 mol%), CH_2Cl_2, rt, 20 h	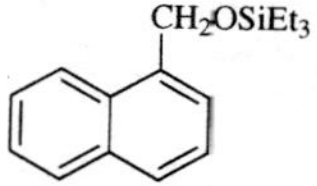 (92)	282
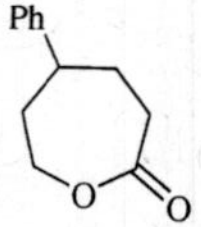	Et_3SiH (5.0 eq), $TiCl_4$ (1.5 eq), TMSOTf (3.1 eq), CH_2Cl_2, rt, 20 h	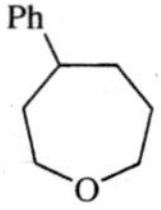 (40)	297
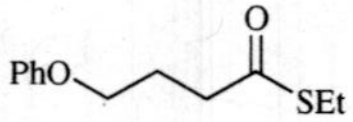	Et_3SiH (2-3 eq), Pd/C (2-5 mol%), acetone, rt, 30-60 min	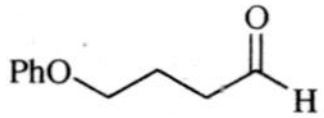 (94)	300
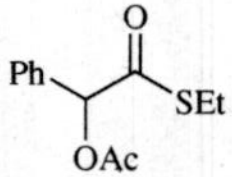	Et_3SiH (2-3 eq), Pd/C (2-5 mol%), acetone, rt, 30-60 min	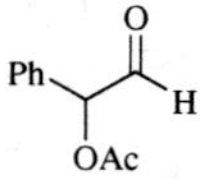 (84)	300
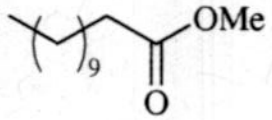	$(EtO)_3SiH$, CsF, 60°, 0.5 h	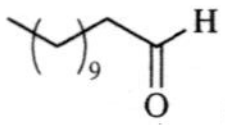 **I** (70)	83
	PMHS, KF•$2H_2O$, DMSO, 80°, 6 h	**I** (70)	83
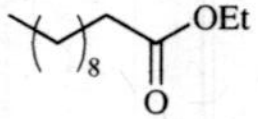	$[HSi(OEt)_4]K$, THF, rt, 8-15 h	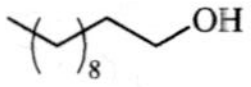 (65)	288
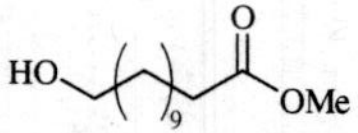	$(EtO)_3SiH$ (2.5-3.0 eq), $Ti(OPr\text{-}i)_4$ (5 mol%), 40-55°, 10 h	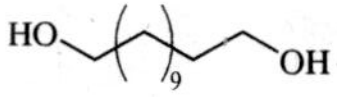 (88)	291

TABLE 10. ORGANOSILANE REDUCTION OF ESTERS AND LACTONES (*Continued*)

Ester	Conditions	Product(s) and Yield(s) (%)	Refs.
C_{13} 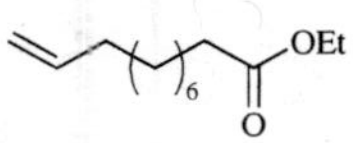	$(EtO)_3SiH$ (2.3 eq), CsF (1 eq), 60°, 30 min	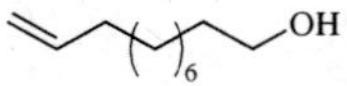 **I** (70)	83
	PMHS (2.3 eq), $KF{\cdot}2H_2O$ (1 eq), DMSO, 80°, 6 h	**I** (70)	83
	1. $(EBTHI)_2TiCl_2$ (5 mol%), *n*-BuLi (10 mol%), THF, –15°, 15 min 2. $(EtO)_3SiH$ (2 eq), ester addition, –20° to rt, 8 h	**I** (62)	290
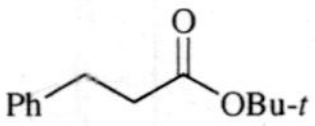	1. $EtMe_2SiH$ (4 eq), **223** (1 mol%), 1,4-dioxane, 20°, 0.5 h 2. Add ester, 12 h	 (80) +  (7)	280
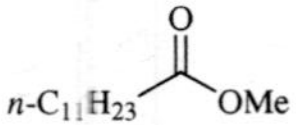	1. $EtMe_2SiH$ (4 eq), **223** (1 mol%), 1,4-dioxane, 20°, 0.5 h 2. Add ester, 3 h	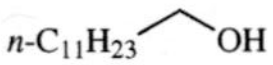 (69) + 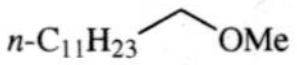 (17)	280
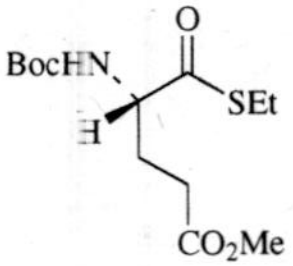	1. Et_3SiH (2-3 eq), Pd/C (2-5 mol%), acetone, rt, 30-60 min 2. Camphorsulfonic acid (0.1 eq), MeOH, rt	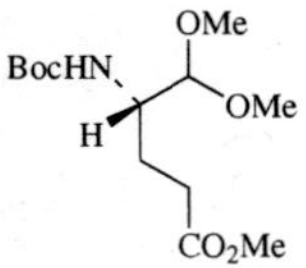 (95)	300
$C_{13\text{-}16}$ 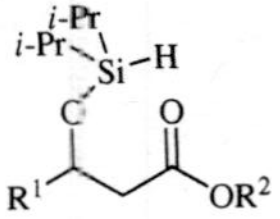	TBAF (0.5 eq), CH_2Cl_2, rt, 15 min	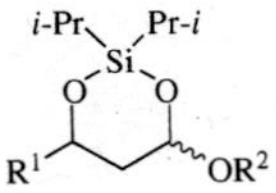 R^1 / R^2 / cis:trans *n*-Pr / Me / 1:1 (97) *n*-Pr / Me / 2:1 (95) *n*-Pr / Et / 1:1 (98) *n*-Bu / Me / 1:1 (98) Ph / Me / 2:1 (96)	303

C_{14}			
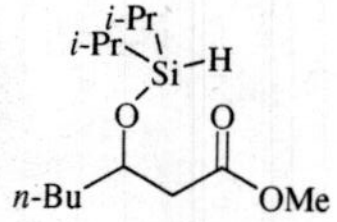	TBAF (0.5 eq), AcOEt, rt, 15 min	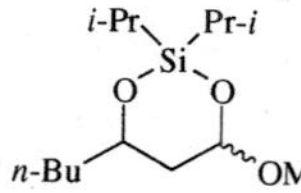 (81)	303
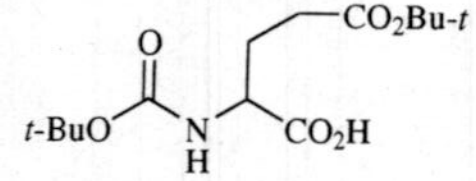	Et_3SiH (2.5 eq), TFA, CH_2Cl_2 rt, 65 min	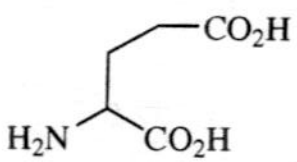 (96) + Me_3CH (—)	307
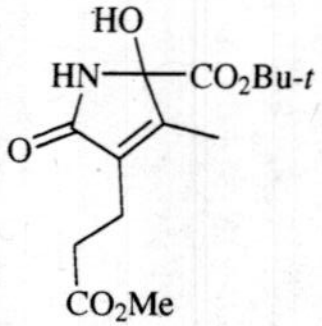	Et_3SiH (xs), TFA, 18°, 18 h; 45°, 3 h	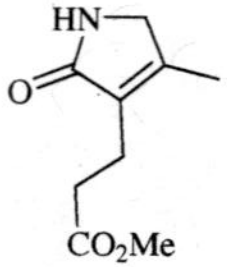 (87)	714
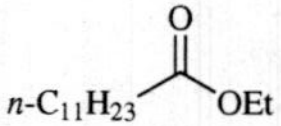	1. Et_2MeSiH (4 eq), **223** (1 mol%), THP, 20°, 0.5 h 2. Add ester, 22 h	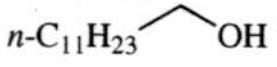 (17) + 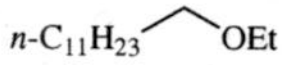 (75)	280
C_{14-15}			
	Et_3SiH (5.0 eq), $TiCl_4$ (1.5 eq), TMSOTf (3.1 eq), CH_2Cl_2, rt, 20 h	 R: *i*-Pr (72); *t*-Bu (65)	297
C_{15}			
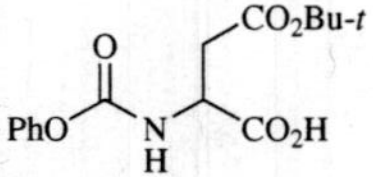	Et_3SiH (2.5 eq), TFA, CH_2Cl_2, rt, 45 min	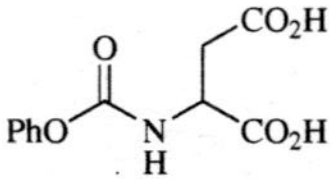 (100) + Me_3CH (—)	307
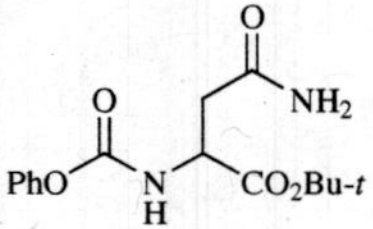	Et_3SiH (2.5 eq), TFA, CH_2Cl_2, rt, 35 min	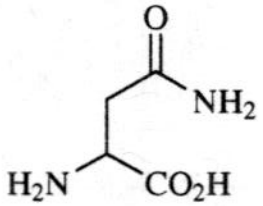 (98) + Me_3CH (—)	307

TABLE 10. ORGANOSILANE REDUCTION OF ESTERS AND LACTONES (*Continued*)

	Ester	Conditions	Product(s) and Yield(s) (%)	Refs.
C_{15}	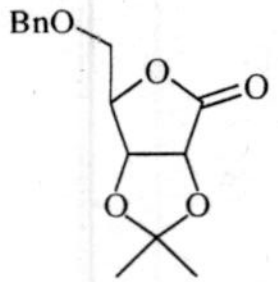	PMHS (5 eq), $Cp_2Ti(OC_6H_4Cl\text{-}4)_2$ (2 mol%), TBAF/alumina (1 mol%), MeC_6H_5, rt	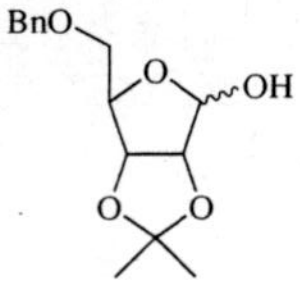 (91)	302
	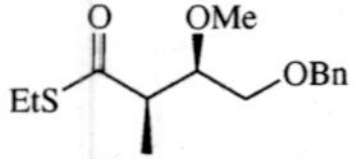	Et_3SiH, Pd/C	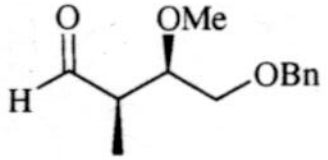 (—)	740
C_{16}	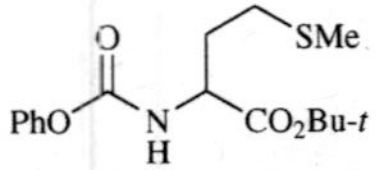	Et_3SiH (2.5 eq), TFA, CH_2Cl_2, rt, 25 min	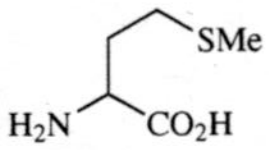 (95) + Me_3CH (—)	307
	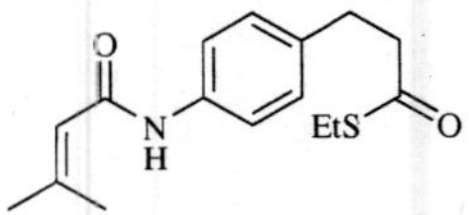	Et_3SiH (2-3 eq), Pd/C (2-5 mol%), acetone, rt, 30-60 min	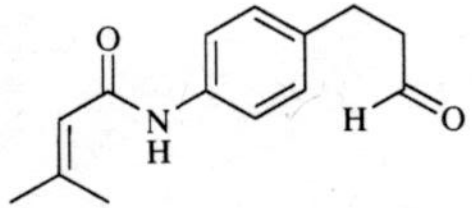 (97)	300
C_{17}	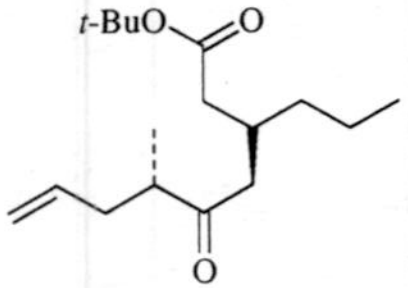	Et_3SiH, TFA, CH_2Cl_2	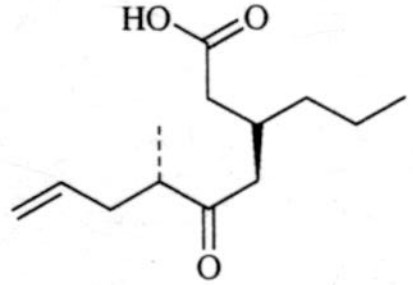 (100) + Me_3CH (—)	741
	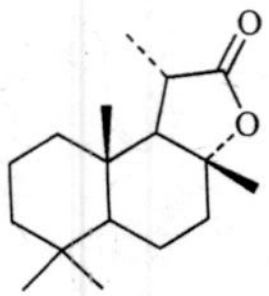	PMHS (5 eq), $Cp_2Ti(OC_6H_4Cl\text{-}4)_2$ (3 mol%), TBAF/alumina (1 mol%), MeC_6H_5, rt	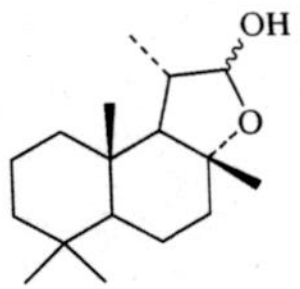 (94)	302

Substrate	Conditions	Product(s) and Yield(s) (%)	Refs.
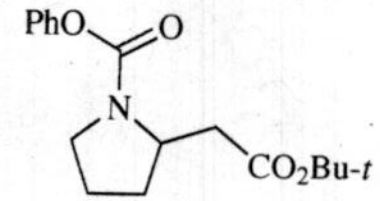	Et_3SiH (2.5 eq), TFA, CH_2Cl_2, rt, 25 min	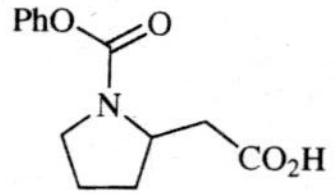 (90) + Me_3CH (—)	307
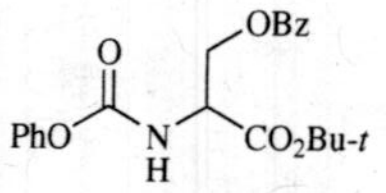	Et_3SiH (2.5 eq), TFA, CH_2Cl_2, rt, 45 min	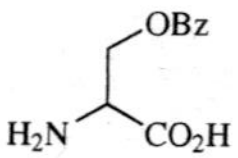 (100) + Me_3CH (—)	307
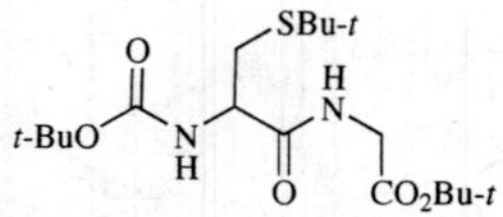	Et_3SiH (2.5 eq), TFA, CH_2Cl_2, rt, 15 min	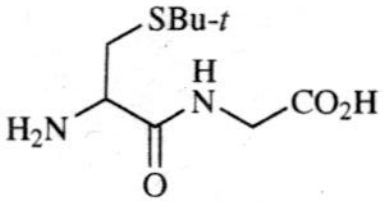 (100) + Me_3CH (—)	307
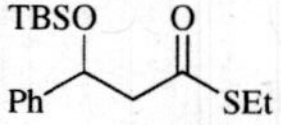	Et_3SiH (2-3 eq), Pd/C (2-5 mol%), acetone, rt, 30-60 min	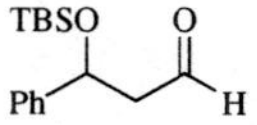 (88)	300
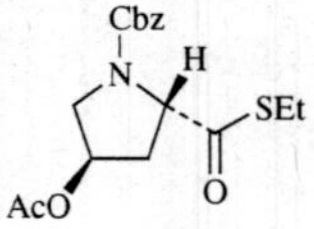	Et_3SiH (2-3 eq), Pd/C (2-5 mol%), acetone, rt, 30-60 min	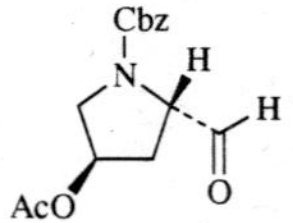 (88)	300
C_{17-18}			
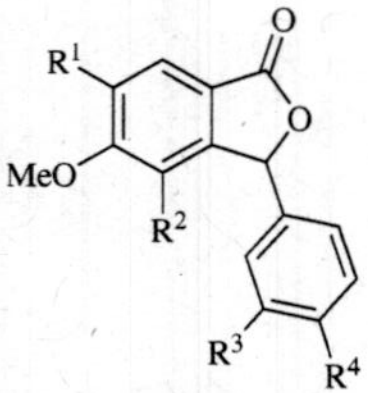	Et_3SiH (3 eq), TFA (5 eq)	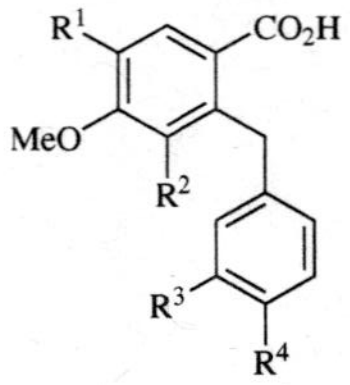	305

Time	R^1	R^2	R^3	R^4	
24 h	MeO	H	—OCH_2O—		(63)
18 h	H	MeO	H	MeO	(80)
24 h	H	MeO	MeO	MeO	(54)
24 h	MeO	H	MeO	MeO	(85)

TABLE 10. ORGANOSILANE REDUCTION OF ESTERS AND LACTONES (*Continued*)

Ester	Conditions	Product(s) and Yield(s) (%)	Refs.
C_{17-19} 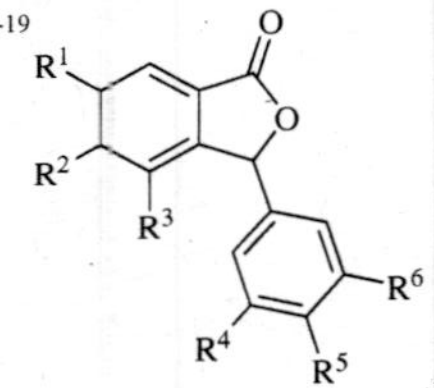	Et_3SiH (3 eq), $TiCl_4$, CH_2Cl_2, 0°	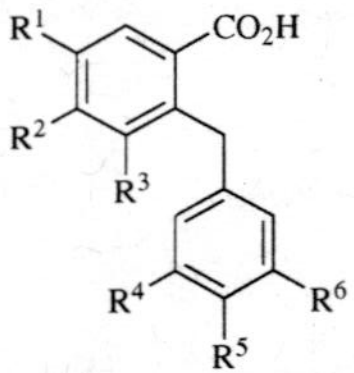 (see table below)	305
C_{18} 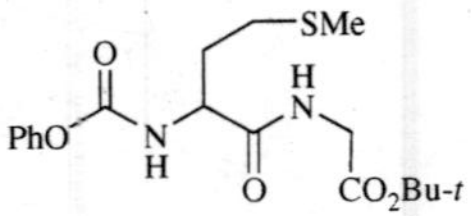	Et_3SiH (2.5 eq), TFA, CH_2Cl_2, rt, 45 min	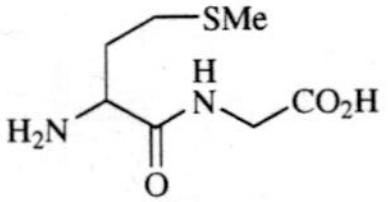 (100) + Me_3CH (—)	307
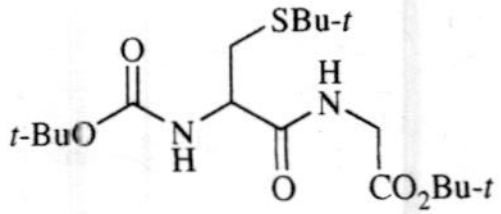	Et_3SiH (2.5 eq), TFA, CH_2Cl_2, rt, 90 min	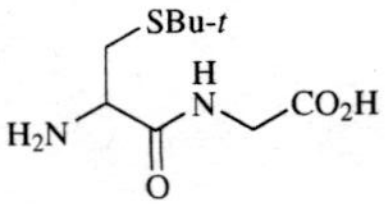 (100) + Me_3CH (—)	307
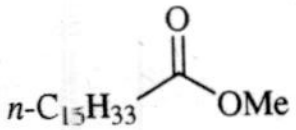	Et_3SiH (4.0 eq), $(C_6F_5)_3B$ (5 mol%), CH_2Cl_2, rt, 20 h	n-$C_{17}H_{36}$ (96)	282

R^1	R^2	R^3	R^4	R^5	R^6	Time	
MeO	MeO	H	H	MeO	H	3 h	(87)
MeO	MeO	H	—OCH_2O—		H	3 h	(83)
H	MeO	MeO	H	MeO	H	2 h	(80)
MeO	MeO	H	MeO	MeO	H	3 h	(82)
—OCH_2O—		H	MeO	MeO	MeO	6 h	(73)
H	MeO	MeO	MeO	MeO	H	4 h	(81)
MeO	MeO	H	MeO	MeO	MeO	5 h	(78)
H	MeO	MeO	MeO	MeO	MeO	5 h	(73)

C_{19}	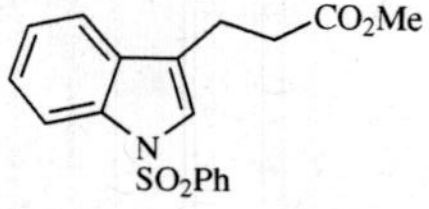	Et_3SiH (5.0 eq), $TiCl_4$ (1.5 eq), TMSOTf (3.1 eq), CH_2Cl_2, rt, 20 h	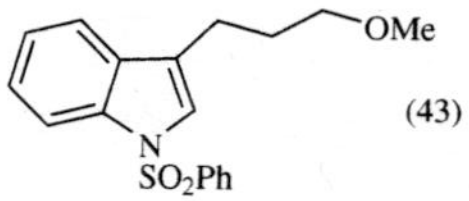 (43)	297
	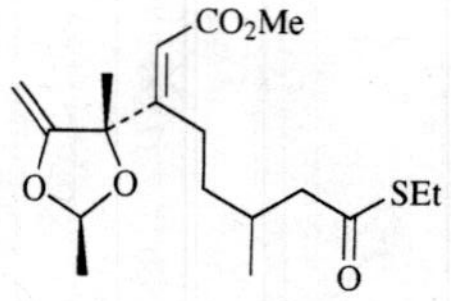	Et_3SiH (2-3 eq), Pd/C (2-5 mol%), acetone, rt, 30-60 min	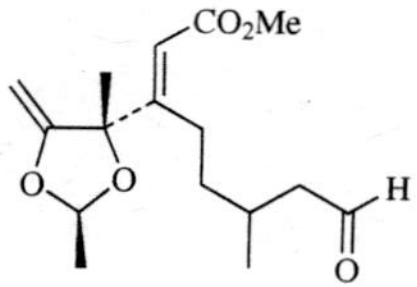 (92)	300
	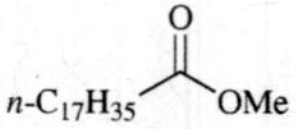	Et_3SiH (6 eq), $(C_6F_5)_3B$ (5 mol%), CH_2Cl_2, rt, 20 h	n-$C_{18}H_{38}$ (98)	281
		1. $EtMe_2SiH$ (4 eq), **223** (1 mol%), 1,4-dioxane, 20°, 0.5 h 2. Add ester, 6 h	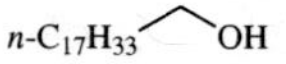 (32) + 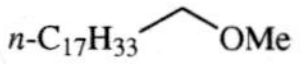 (22)	280
	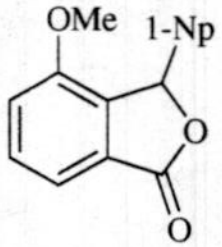	Et_3SiH (xs), TFA	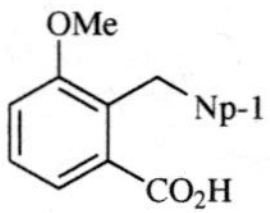 (90)	304
	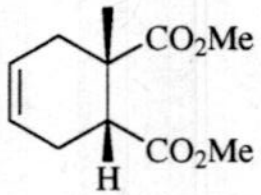	PMHS (10 eq), $Ti(OPr$-$i)_4$ (1 eq), THF	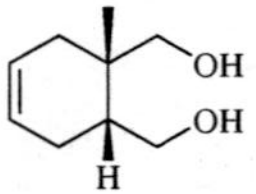 (78)	279
	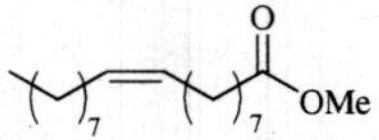	PMHS (2.5 eq), $Ti(OPr$-$i)_4$ (100 mol%), 65°, 24 h	 **I** (89)	292
		PMHS (2.5 eq), Cp_2TiCl_2 (2 mol%), n-BuLi (2 eq), THF, rt, 1.5 h	**I** (82)	289

TABLE 10. ORGANOSILANE REDUCTION OF ESTERS AND LACTONES (*Continued*)

Ester	Conditions	Product(s) and Yield(s) (%)	Refs.
C_{19}			
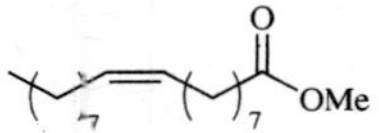	$(EtO)_3SiH$ (2.5-3.0 eq), $Ti(OPr\text{-}i)_4$ (5 mol%), 40 to 55°, 10h	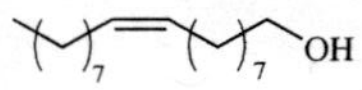 (92)	291
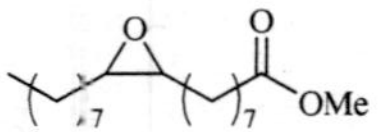	PMHS (2.5 eq), $Ti(OPr\text{-}i)_4$ (100 mol%), 65°, 5 h	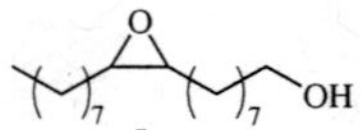 **I** (85)	292
	PMHS (2.5 eq), (EBTHI)$TiCl_2$ (5 mol%), *n*-BuLi (2 eq), THF, rt, 1 h	**I** (64)	289
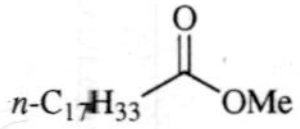	1. $EtMe_2SiH$ (4 eq), **223** (1 mol%), 1,4-dioxane, 20°, 0.5 h 2. Add ester, 6 h	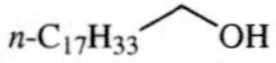 (32) +  (22)	280
C_{20}			
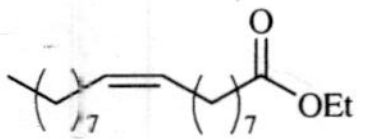	1. $(EBTHI)_2TiCl_2$ (5 mol%), *n*-BuLi (10 mol%), THF, –15°, 15 min 2. $(EtO)_3SiH$ (2 eq), ester addition, –20° to rt, 8 h	 (90)	290
C_{21}			
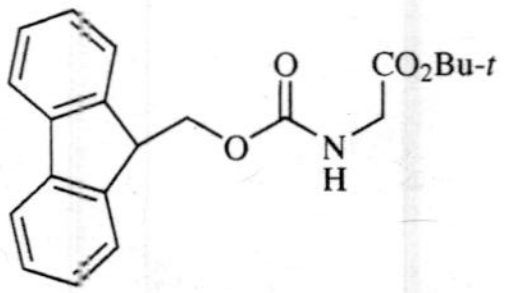	Et_3SiH (2.5 eq), TFA, CH_2Cl_2, rt, 20 min	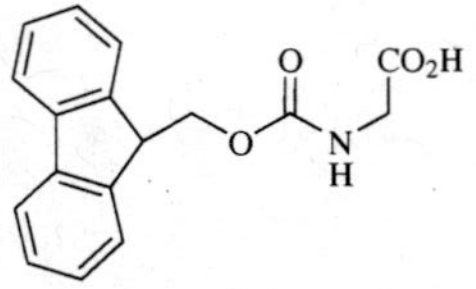 (85) + Me_3CH (—)	307
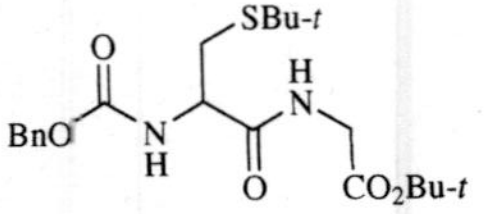	Et_3SiH (2.5 eq), TFA, CH_2Cl_2, rt, 40 min	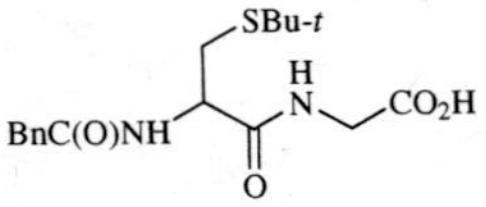 (78) + Me_3CH (—)	307

C_{21}	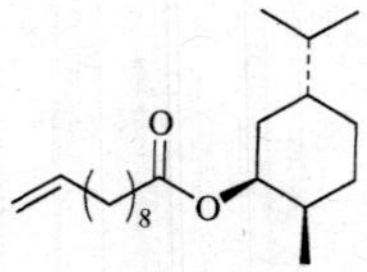	$(EtO)_3SiH$ (1.1 eq), CsF (1 eq), 60°, 9 h	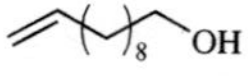 (75)	79
C_{22}	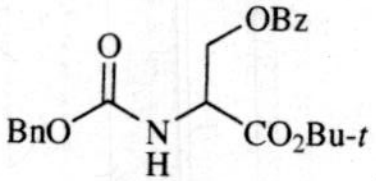	Et_3SiH (2.5 eq), TFA, CH_2Cl_2, rt, 30 min	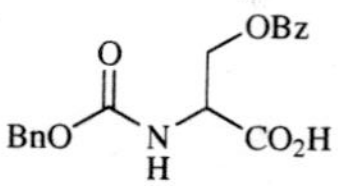 (98) + Me_3CH (—)	307
	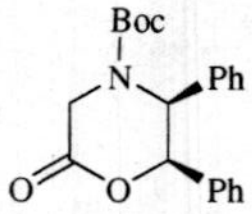	PMHS (5 eq), $Cp_2Ti(OC_6H_4Cl\text{-}4)_2$ (5 mol%), TBAF/alumina (1 mol%), MeC_6H_5, rt	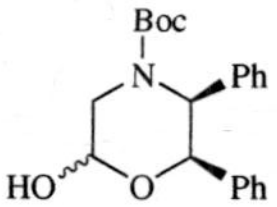 (92)	302
C_{25}	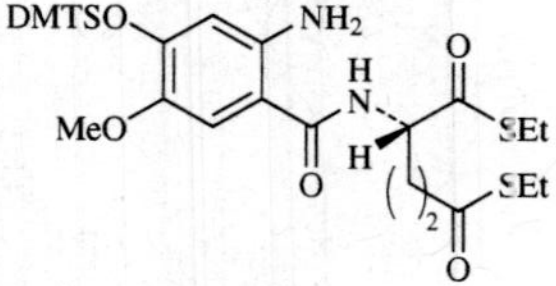	1. Et_3SiH (5 eq), Pd/C (15 mol%), CH_2Cl_2, rt, 40 min 2. Camphorsulfonic acid (0.1 eq), MeOH 3. TBAF, AcOH, MeOH	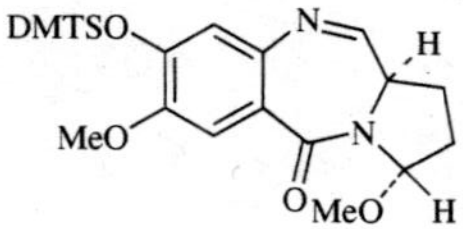 (65)	300
C_{26}	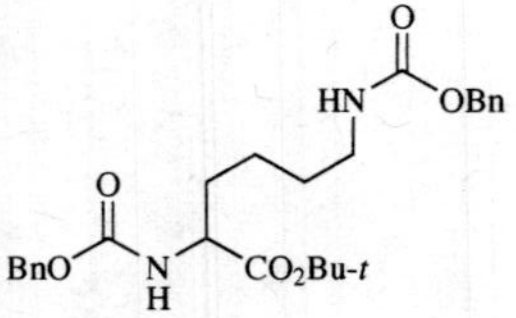	Et_3SiH (2.5 eq), TFA, CH_2Cl_2, rt, 35 min	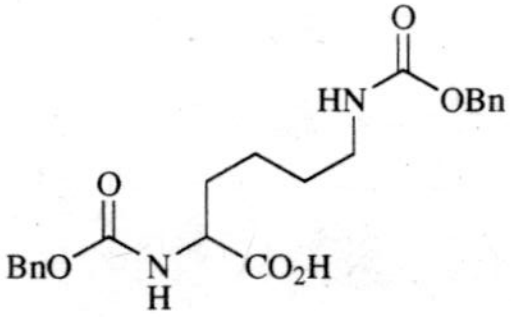 (83) + Me_3CH (—)	307

TABLE 10. ORGANOSILANE REDUCTION OF ESTERS AND LACTONES (*Continued*)

	Ester	Conditions	Product(s) and Yield(s) (%)	Refs.
C_{27}	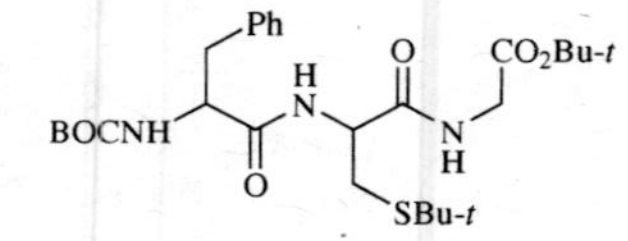	Et_3SiH (2.5 eq), TFA, CH_2Cl_2, rt, 75 min	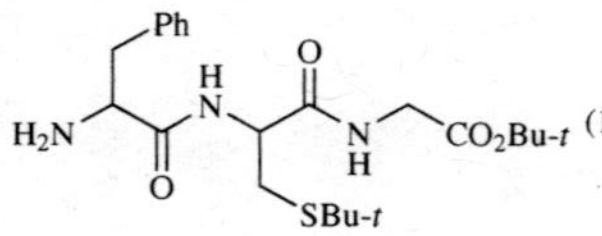 (100) + Me_3CH (—)	307
	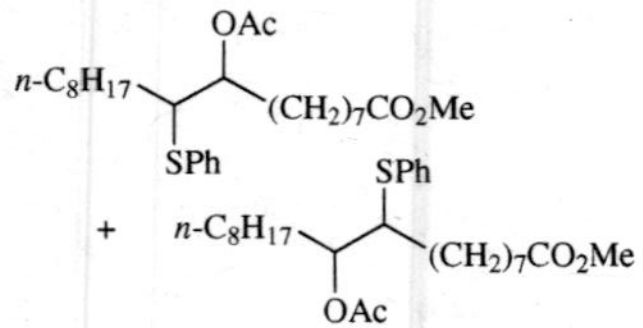	$BnMe_2SiH$, TFA, rt, 20 h	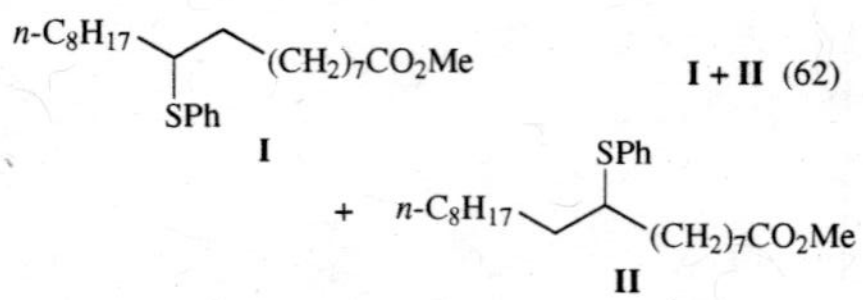 **I** + **II** (62)	742
C_{28}	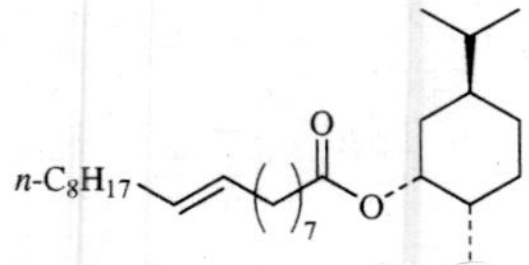	$(EtO)_3SiH$, CsF, rt, 72 h	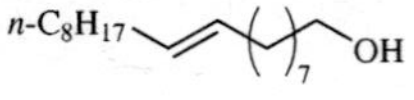 (80)	83
	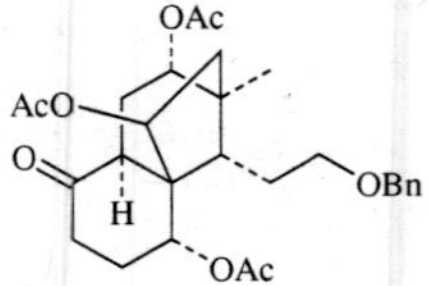	Et_3SiH (1.5 eq), $BF_3•OEt_2$ (1.1 eq), –20°, 24 h	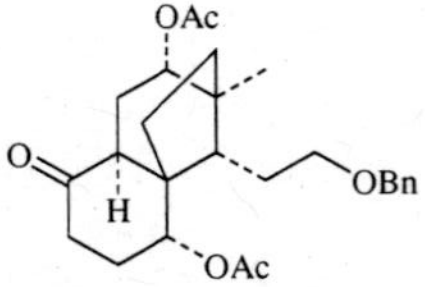 (65)	510
	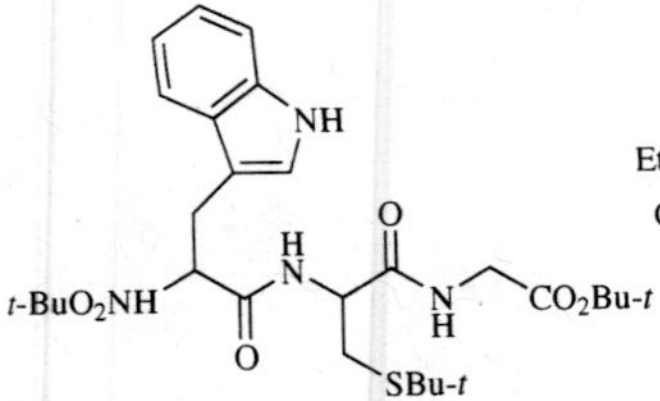	Et_3SiH (2.5 eq), TFA, CH_2Cl_2, rt, 95 min	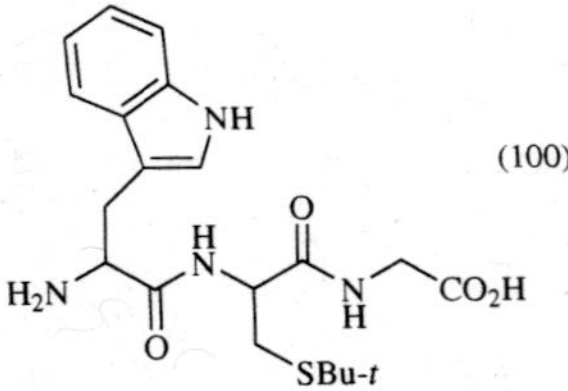 (100) + Me_3CH (—)	307

C_{29}	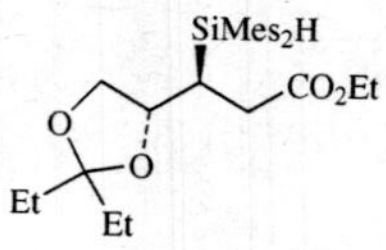	TBAF (10 mol%), 0°	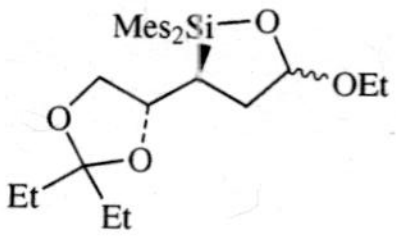	(91) dr = 98:2	743
C_{33}	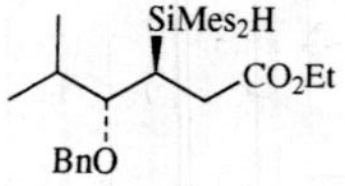	TBAF (10 mol%), 0°	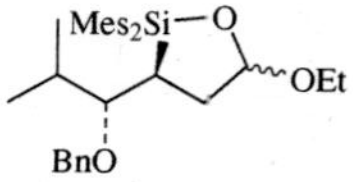	(—) dr = 96:4	743
C_{35}	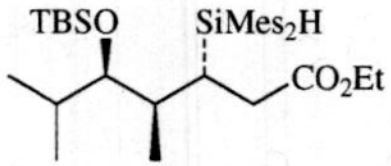	TBAF (10 mol%), 0°	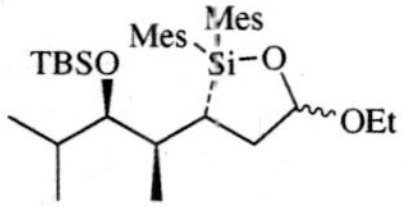	(—) dr = 96:4	743
C_{36}	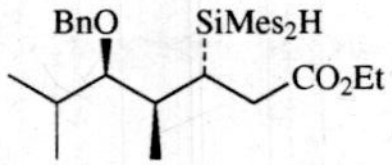	TBAF (10 mol%), 0°	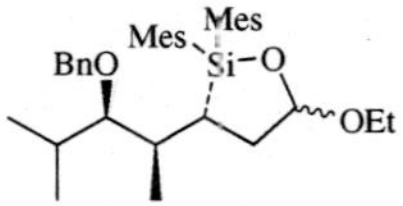	(82) dr = 85:15	743
C_{39}	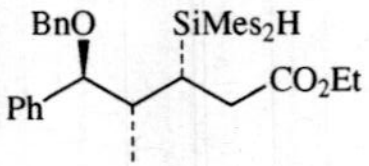	TBAF (10 mol%), 0°	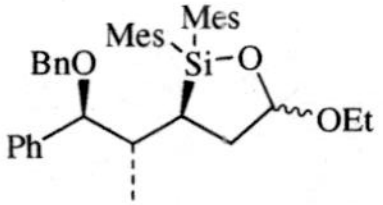	(79) dr = 93:7	743

[a] The yield was determined by NMR spectroscopy.

[b] No reduction occurred in this reaction.

[c] The product was isolated as the phenylhydrazone.

TABLE 11. ORGANOSILANE REDUCTION OF ALDEHYDES

Aldehyde	Conditions	Product(s) and Yield(s) (%)	Refs.
C_2 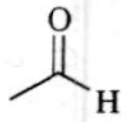	Me_3SiH (1.25 eq), TMSI (0.3 eq), 1-AdOTMS (1.25 eq), $n\text{-}C_5H_{12}$, –78° to rt, 5-6 h	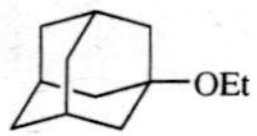 (69)	338
$C_{2\text{-}7}$ 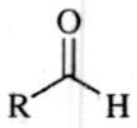	Et_3SiH (3 eq), TFA (2.9 eq), $PhCONH_2$, solvent, 18 h	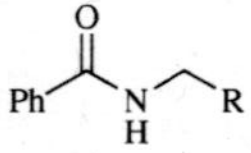 R | Solvent | Temp Me | MeCN | 22° | (95) *t*-Bu | MeC_6H_5 | 120° | (90) $c\text{-}C_6H_{11}$ | MeC_6H_5 | 120° | (92)	326
$C_{2\text{-}7}$ 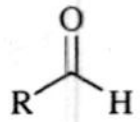 R = Me, Et, *i*-Pr, *t*-Bu, Ph, $4\text{-}ClC_6H_4$, $4\text{-}MeC_6H_4$, $4\text{-}NCC_6H_4$	Et_3SiH, $HClO_4$, MeCN, rt	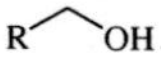 (—)	380
C_3 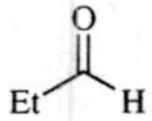	Et_3SiH (2 eq), TMSOTf (0.1 eq), CH_2Cl_2, rt, 2 h	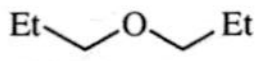 (100)[a]	334
	Et_3SiH, TMSOTf, $c\text{-}C_6H_{11}OTMS_3$, CH_2Cl_2, 0°, 2 h	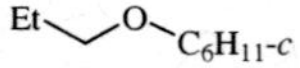 (100)[a]	334
$C_{4\text{-}7}$ 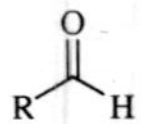 R = *n*-Pr, Ph	$Ph(Np\text{-}1)_2SiH_2$, $ClRh(PPh_3)_3$ (5×10^{-4} M)	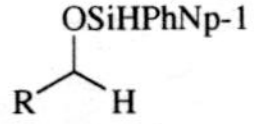 (—)	744

C$_4$

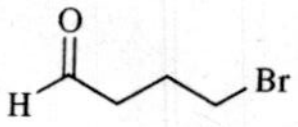

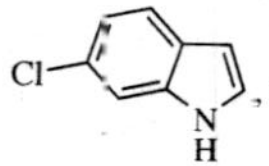, Et_3SiH (3 eq), TFA

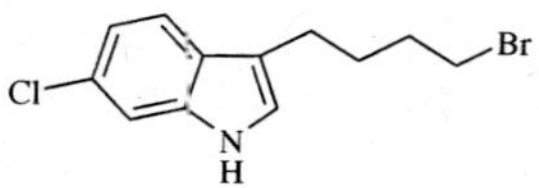

(48) 355

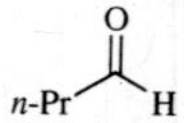

See table.

Conditions	Temp	Time	R		
Et_3SiH, TMSOTf, c-$C_6H_{11}OTMS$, CH_2Cl_2	0°	2 h	c-C_6H_{11}	(100)[a]	334
Et_3SiH, TFA	30-40°	—	CH_2Pr-n	(37)	311
Et_3SiH, TFA, EtOH	50-60°	8 h	Et	(78)	327
TMSH, TMSI, c-$C_6H_{11}OTMS$, CH_2Cl_2	0°	2 h	c-C_6H_{11}	(100)	334
Et_3SiH, TMSI, n-$C_8H_{17}OTMS$, CH_2Cl_2	0°	2 h	n-C_8H_{17}	(100)	334

C$_{4\text{-}9}$

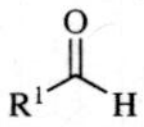

Et_3SiH (1.2 eq), R^2OTMS (1.2 eq), $BiBr_3$ (10-30 mol%), rt

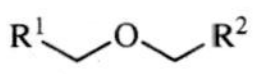

343

R^1	R^2	Time	
(E)-MeCH=CH	Bn	2 h	(76)
t-Bu	Bn	<5 min	(58)
Ph	Bn	<5 min	(84)
Ph	Ph	1 h	(0)
Ph	2-octyl	1 h	(81)
n-C_7H_{15}	Bn	<5 min	(86)
4-$MeO_2CC_6H_4$	Bn	1 h	(83)

C$_{4\text{-}7}$

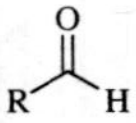

Et_3SiH (1 eq), EtOH (x eq), TFA (y eq), 50-60°

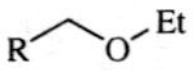

R	x	y		
n-C_3H_7	2	4	(78)	327
n-C_4H_9	2	4	(57)	
Ph	3	5	(89)	

C$_5$

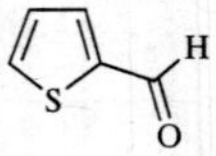

Et_3SiH (4 eq), TFA (8 eq), 50°, 30 h

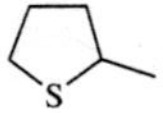

(65) 257

TABLE 11. ORGANOSILANE REDUCTION OF ALDEHYDES (*Continued*)

	Aldehyde	Conditions	Product(s) and Yield(s) (%)	Refs.
C_5	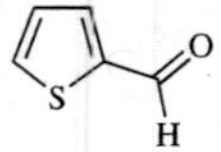	$PhMe_2SiH$ (1.1 eq), CsF (10 mol%), 18-C-6 (5 mol%), CH_2Cl_2, rt, 3 h	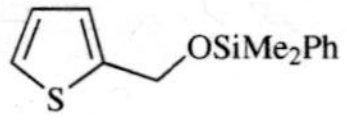 (54)	346, 347
	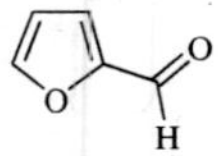	$(EtO)_3SiH$ (1.1 eq), KF (1 eq), rt, 1 h	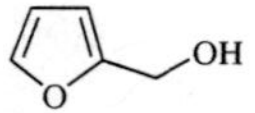 **I** (85)	79
		$Me(EtO)_2SiH$ (1.1 eq), KF (1 eq), rt, 18 h	**I** (90)	79
		**H**, 0°, 2 h	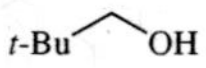 (92)	93
	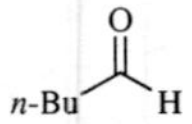	Et_3SiH (1 eq), $HC(OEt)_3$ (1 eq), EtOH, HCl, TFA (3 eq), 50-60°, 6-8 h	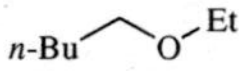 (72)	327
		Et_3SiH, TFA, 30-40°	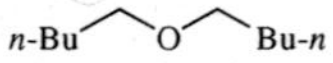 **I** (50)	311
		$(HMe_2Si)_2O$, TMSOTf, C_6H_6, 80°, 30 min	**I** (90)	314
		$(HMe_2Si)_2O$, TMSCl, NaI, MeCN, rt, 30 min	**I** (84)	314, 356
		$(HMe_2Si)_2O$, TFA, EtOH, 50-60°, 8 h	**I** (57)	327
		$(HMe_2Si)_2O$, TFA, HCl, EtOH, $HC(OEt)_3$, 50-60°, 8 h	**I** (72)	329
	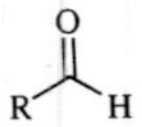 R = *i*-Bu, *s*-Bu	Et_3SiH, $ZnCl_2$	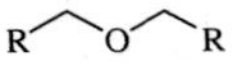 (52)	330

Reactant	Conditions	Product(s) and Yield(s) (%)	Refs.
C_{5-9} 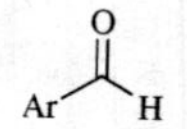	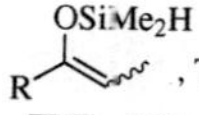, TBAF (6 mol%), THF, −78°, 3.5 h	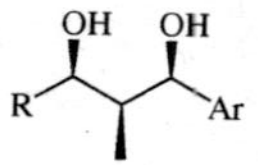 I + 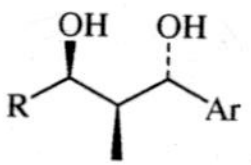 II	400

Ar	R	Z:E	I + II	I:II
2-C_4H_3O	Ph	>98:2	(73)	91:9
Ph	Et	74:26	(64)	68:32
Ph	*i*-Pr	75:25	(64)	88:12
Ph	*t*-Bu	>98:2	(65)	99:1
4-$O_2NC_6H_4$	Ph	>98:2	(84)	85:15
4-MeC_6H_4	Ph	>98:2	(87)	96:4
4-$MeOC_6H_4$	Ph	>98:2	(74)	94:6
$BnCH_2$	Ph	>98:2	(43)	82:18

Reactant	Conditions	Product(s) and Yield(s) (%)	Refs.
C_{5-11} 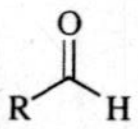	Cl_3SiH (1.5 eq), CH_2Cl_2/DMF (4:1), 0°, 4-6 h	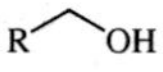	318

R	
2-C_4H_3O	(81)
2-C_4H_3S	(99)
Ph	(98)
4-ClC_6H_4	(95)
4-$O_2NC_6H_4$	(98)
4-$MeOC_6H_4$	(100)
$BnCH_2$	(98)
(*E*)-PhCH=CH	(87)
PhC≡C	(72)
H_2C=CH$(CH_2)_8$	(80)

Reactant	Conditions	Product(s) and Yield(s) (%)	Refs.
C_5 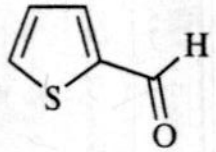	$PhMe_2SiH$ (1.1 eq), CsF (0.1 eq), 18-C-6 (0.05 eq), CH_2Cl_2, rt, 11 h	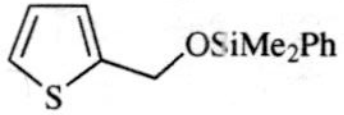 (64)	347
C_6 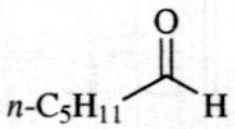	Ph_2SiH_2 (1.1 eq), TBAF (5 eq), HMPA, rt, 2.5 h	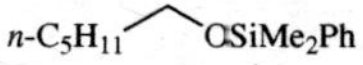 (62)	320

TABLE 11. ORGANOSILANE REDUCTION OF ALDEHYDES (Continued)

Aldehyde	Conditions	Product(s) and Yield(s) (%)	Refs.
C_6			
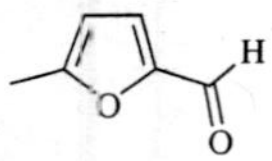	$PhMe_2SiH$ (1.1 eq), CsF (10 mol%), 18-C-6 (5 mol%), CH_2Cl_2, rt, 0.5 h	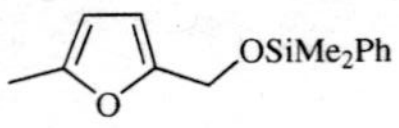 (58)	346, 347
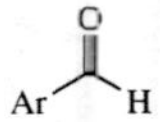	$PhMe_2SiH$ (1.1 eq), CsF (10 mol%), 18-C-6 (0.5 mol%), CH_2Cl_2, rt	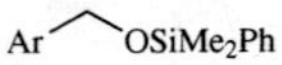	345
Ar	Time		
2-pyridinyl	9 h	(67)	
3-pyridinyl	20 h	(54)	
4-pyridinyl	4 h	(60)	
C_7			
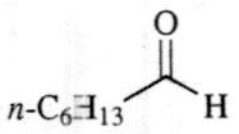	Ph_2SiH_2 (0.5 eq), 1,2-$C_6H_4(CO_2K)_2$, 100°, 24 h	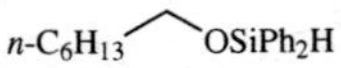 **I** (60)	319
	Ph_2SiH_2 (0.5 eq), KF, 100°, 6 h	**I** (50)	319
	Ph_2SiH_2 (0.5 eq), CsF, rt, 3 min	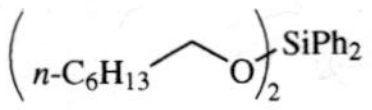 (100)	83, 319
	$(EtO)_3SiH$ (1.1 eq), KF (1 eq), rt, 4 h	 **II** (70)	79, 83
	Et_3SiH (1.1 eq), HCl, 28°, 3 h	**II** (100)	313
	Et_3SiH (1.25 eq), H_2SO_4, H_2O (11 eq), 28°, 1.25 h	**II** (97)	313
	$Me(EtO)_2SiH$ (1.5 eq), KF (1 eq), DMF, 10°, 1.75 h	**I** (85)	83
	1. Cyclohexanone, $Me(EtO)_2SiH$ (1.2 eq), KO_2CH (1 eq), DMF, 45°, 19 h 2. H_3O^+	**II** (72) + c-$C_6H_{10}O$ (recovered)	82

	$HMe_2SiOSiMe_2H$ (1.0 eq), TMSOTf (cat.), C_6H_6, 80°, 30 min	n-$C_6H_{13}CH_2OCH_2C_6H_{13}$-$n$ **III** (96)	314
	Et_3SiH, TFA, rt, 45 min	**III** (90)	313
	$PhMe_2SiH$, EG acid, CH_2Cl_2, rt,	**III** (92)	333
	Et_3SiH, $ZnCl_2$	**III** (48)	330
	$(HMe_2Si)_2O$, TMSCl, C_6H_6, rt, 30 min	**III** (96)	330
	$(HMe_2Si)_2O$, TMSCl, NaI, C_6H_6, rt, 30 min	**III** (90)	314
	Et_3SiH, H_2SO_4, MeOH, rt, 1 h	**III** (87)	328
	TMSH, TMSI, TMSOEt, CH_2Cl_2, 0°, 2 h	**III** (90)	334
	Et_3SiH, TFA, BnOH, rt, 1 h	n-$C_6H_{13}CH_2OBn$ (49)	328
	$PhMe_2SiH$, EG acid	n-$C_6H_{13}CH_2O(CH_2)_3Ph$ (82)	333
	TMSH (1.1 eq), TMSI (5 mol%), EtOTMS (1 eq), CH_2Cl_2, 0-15°; rt, 2 h	n-$C_6H_{13}CH_2OEt$ (90)	334
+ 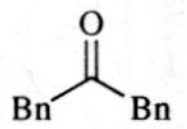	$(EtO)_3SiH$ (2.3 eq), KF (1 eq), rt, 20 h	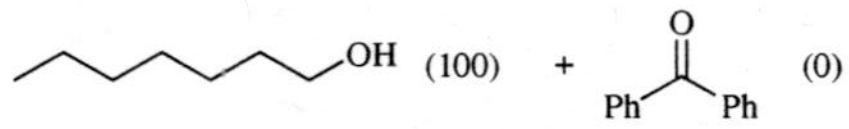	83
	TMSH (1.25 eq), TMSI (0.3 eq), n-C_5H_{12}, –78° to rt, 5-6 h, t-BuOTMS (1.25 eq)	n-$C_6H_{13}CH_2OBu$-t (63)	338

TABLE 11. ORGANOSILANE REDUCTION OF ALDEHYDES (*Continued*)

Aldehyde	Conditions	Product(s) and Yield(s) (%)	Refs.
C_7 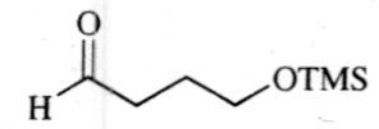	1. $SbCl_5$, TMSCl, SnI_2, CH_2Cl_2, –78°, 30 min 2. Et_3SiH (1.5 eq), –23°, 2.5 h	 **I** (89)	306
	1. $TrSbCl_6$ (5-30 mol%), CH_2Cl_2, –78°, 30 min 2. Et_3SiH (1.5 eq), –23°, 2.5 h	**I** (91)	306
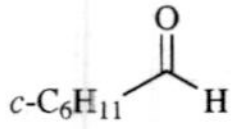	$PhSiH_3$ (0.4 eq), $Mn(dpm)_3$ (3 mol%), *i*-PrOH, DCE, O_2, rt	 (92)	367
	t-BuOTMS (1.25 eq), TMSH (1.25 eq), TMSI (0.3 eq), *n*-C_5H_{12}, –78° to rt, 5-6 h	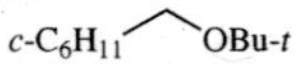 (84)	338
	, Et_3SiH (1 eq), TMSOTf (0.1 eq), CH_2Cl_2, –30°	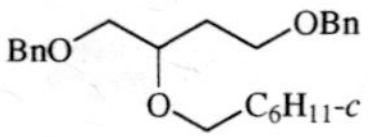 (84)	341
	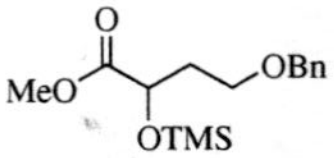, Et_3SiH (1 eq), TMSOTf (0.1 eq), CH_2Cl_2, –30°	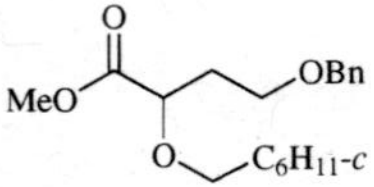 (36)	341
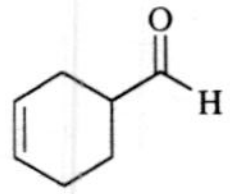	R_3SiH (x eq), $(Ph_3P)CuH$ (3 mol%), MeC_6H_5	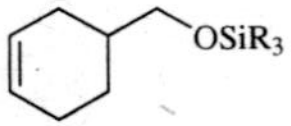 R_3Si / x / Time: $PhMe_2Si$ 1.3 2 h (95); Ph_2MeSi 2.5 — (98)	317
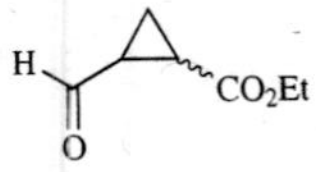	5-Chloro-*N*-benzhydrylindole, Et_3SiH (3 eq), TFA	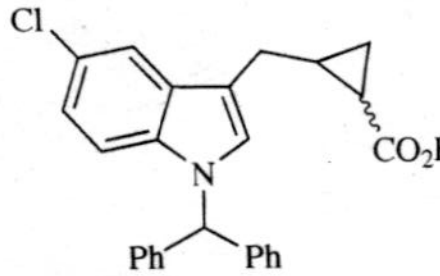 (79)	355

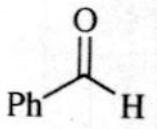	Et_3SiH (1.1 eq), Fe-mont, rt, 0.3 h	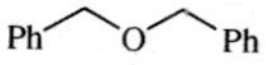 I (80)	353
	$(EtO)_3SiH$ (1.1 eq), HAp, rt, 0.3 h	**I** (72)	353
	$(EtO)_3SiH$ (1.1 eq), CaO, rt 0.3 h	**I** (59)	353
	Et_3SiH (1 eq), $(Ph_3P)_3RhCl$ (0.1 mol%), rt, 5 min	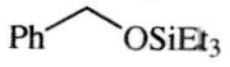 (95)	411
	PMHS (10% xs), $[Bu_2(AcO)Sn]_2O$ (2 mol%), EtOH, reflux	 **II** (100)	316
	PMHS, Pd/C, EtOH, 80°	PhMe (84)	316
	PMHS, TMSOTf (cat.), C_6H_5, reflux, 2 h	 (92)	314
	PMHS, TMSOTf (cat.), MeC_6H_5, reflux, 1 h	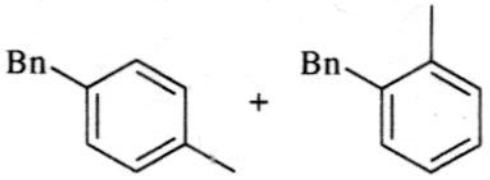 **III** + **IV** **III + IV** (95)	314
	$(EtO)_3SiH$ (2.3 eq), KF (1 eq), rt, 6 h	**II** (90)	83
	$Me(EtO)_2SiH$ (2.3 eq), KF (1 eq), DMF, 20°, 0.25 h	**II** (90)	83
	$Me(EtO)_2SiH$, KF, 20°, 0.25 h	**II** (90)	83
	Silane, salt, solvent	**II**	82

Silane	Salt	Solvent	Temp	Time	
$Me(EtO)_2SiH$	KF	DMF	20°	0.25 h	(90)
PMHS	$KF•2H_2O$	DMF	35°	1 h	(76)
PMHS	$KF•2H_2O$	DMSO	rt	1 h	(67)
PMHS	KO_2CH	DMF	80°	4 h	(64)
$Me(EtO)_2SiH$	KO_2CH	DMF	80°	11 h	(75)

TABLE 11. ORGANOSILANE REDUCTION OF ALDEHYDES (*Continued*)

Aldehyde	Conditions	Product(s) and Yield(s) (%)	Refs.
C_7			
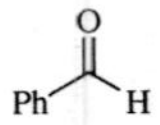	PMHS (1 eq), $ZnCl_2$ (1 eq), Et_2O, rt, 24 h	 (72) **II**	315
	Et_3SiH, TBAF, MeCN, 12 h	**II** (36)	77
	Et_3SiH, CsF, MeCN, 12 h	**II** (96)	77
	G, 0°, 2 h	**II** (96)	93
	$PhSiH_3$ (0.4 eq), $Mn(dpm)_3$ (3 mol%), *i*-PrOH, DCE, O_2, rt	**II** (68)	367
	K, CH_2Cl_2, rt, 2 h	**II** (95)	101
	J, THF, 0°, 2 h	**II** (95)	96
	L, THF, 0°, 2 h	**II** (61)	96
	M, LiCl (4 eq), THF, 0°, 2 h	**II** (69)	96
	Et_3SiH (1.1 eq), BF_3•OEt_2 (2.0 eq), rt	**I** (75) + **II** (25)	74
	Et_3SiH, TFA, 30-40°	**I** (80)	311
	$Et_3SiO(EtHSiO)_nSiEt_3$ (0.5 mol%), TFA (8 eq), $CHCl_3$, 50°, 1 h	**I** (92)	207
	Et_3SiH (2 eq), TMSOTf (0.1 eq), CH_2Cl_2, rt, 2 h	**I** (major) + **II** (minor)	334

$(HMe_2Si)_2O$, TFA, EtOH, 30-40°	**I** (80)	313
Et_3SiH, TFA, rt, 30 min	**I** (87)	313
Et_3SiH, TFA, $CHCl_3$, rt, 15 min	**I** (93)	313
Et_3SiH, TFA, CCl_4, rt, 15 min	**I** (89)	313
Et_3SiH, TFA, MeCN, rt, 15 min	**I** (96)	313
Et_3SiH, Cl_2CHCO_2H, rt, 30 min	**I** (93)	313
Et_3SiH, BF_3•OEt_2, rt, 1 h	**I** (75)	333
Et_3SiH, EG acid, CH_2Cl_2, rt	**I** (91)	333
$PhMe_2SiH$, EG acid, CH_2Cl_2, rt	**I** (96)	74
Et_3SiH, $TrClO_4$, CH_2Cl_2, 0°, 10 min	**I** (84)	329
Et_3SiH, $TrClO_4$, BnOTMS, CH_2Cl_2, 0°, 10 min	**I** (72)	329
$(HMe_2Si)_2O$, TMSOTf, C_6H_6, 80°, 20 min	**I** (97)	314
Cl_3SiH (3 eq), $(n\text{-}Pr)_3N$ (1 eq), MeCN, 57-58°, 1 h	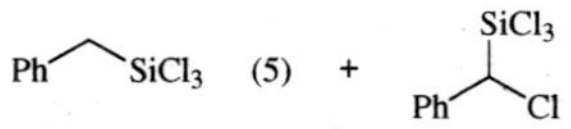 (5) + (42)	711

TABLE 11. ORGANOSILANE REDUCTION OF ALDEHYDES (*Continued*)

Aldehyde	Conditions	Product(s) and Yield(s) (%)	Refs.
C_7 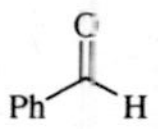	Et_3SiH (1.4 eq), TBAF (1.4 eq), MeCN, rt, 12 h	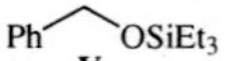 **V** (36)	76
	Et_3SiH (1.5 eq), CsF (1.5 eq), MeCN, rt, 10 h	**V** (96)	76
	$PhMe_2SiH$ (1.2 eq), TBAF (2 eq), HMPA, rt, 1.5 h	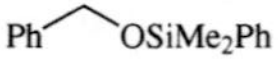 **VI** (91)	320
	$PhMe_2SiH$ (1.1 eq), CsF (0.1 eq), 18-C-6 (0.05 eq), CH_2Cl_2, rt, 11 h	**VI** (57)	347
	R_3SiH (1.5 eq), catalyst, MeC_6H_5	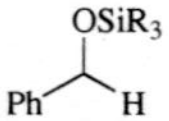	324

R_3Si	Catalyst	Temp	Time	
Et_3Si	Fe-mont	rt	1 h	(86)
$PhMe_2Si$	Fe-mont	rt	3 h	(94)
$(EtO)_3Si$	Fe-mont	40°	3 h	(—)
Et_3Si	Cu-mont	rt	3 h	(77)
Et_3Si	Na-mont	rt	47 h	(0)
Et_3Si	silica	rt	21 h	(0)
Et_3Si	Ca-Y-zeolite	rt	7 h	(0)

Aldehyde	Conditions	Product(s) and Yield(s) (%)	Refs.
	Et_3SiH (1.1 eq), ROH, acid, 28°		328

R	Acid	
Me	H_2SO_4	(87)
Me	CF_3CO_2H	(87)
Me	CCl_3CO_2H	(85)
Et	H_2SO_4	(87)
n-Pr	H_2SO_4	(68)
i-Pr	H_2SO_4	(69)
n-C_7H_{15}	CF_3CO_2H	(81)
t-Bu	CF_3CO_2H	(26)
t-Bu	H_2SO_4	(45)

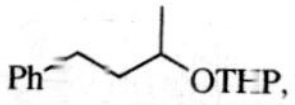, Et_3SiH (2.2 eq), TMSOTf (10 mol%), MeCN, 0°, 1 h	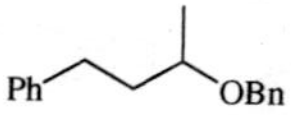 (95)	340
, Et_3SiH (2.2 eq), TMSOTf (10 mol%), MeCN, 0°, 1 h	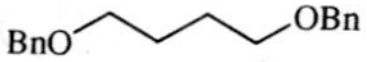 (95)	340
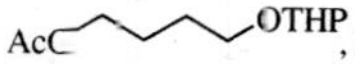, Et_3SiH (2.2 eq), TMSOTf (10 mol%), MeCN, 0°, 1 h	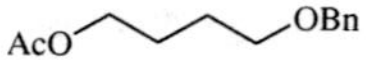 (98)	340
, Et_3SiH (2.2 eq), TMSOTf (10 mol%), MeCN, 0°, 1 h	 (98) Ar = 4-$MeOC_6H_4$	340
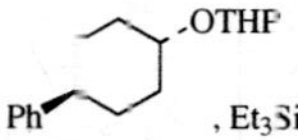, Et_3SiH (2.2 eq), TMSOTf (10 mol%), MeCN, 0°, 1 h	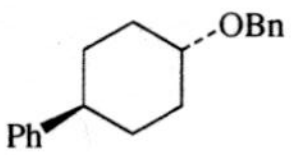 (97)	340
1-AdOTMS (1.25 eq), TMSH (1.25 eq), TMSI (0.3 eq), n-C_5H_{12}, –78° to rt, 5-6 h	 (75)	338
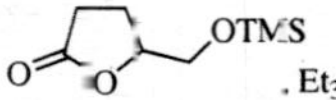, Et_3SiH (1 eq), TMSOTf (0.1 eq), CH_2Cl_2, 0° to rt	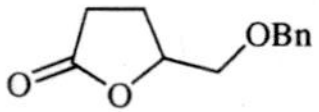 (89)	341
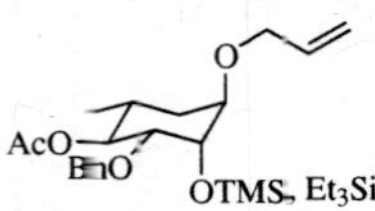, Et_3SiH (1 eq), TMSOTf (2.0 eq), CH_2Cl_2, –78° to –30°	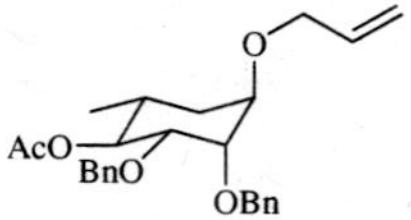 (96)	341
TMSH, TMSI, c-$C_6H_{11}OTMS$, CH_2Cl_2, 0°, 2 h	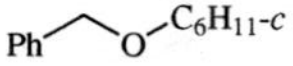 (98)	334

TABLE 11. ORGANOSILANE REDUCTION OF ALDEHYDES (*Continued*)

Aldehyde	Conditions	Product(s) and Yield(s) (%)	Refs.
C_7 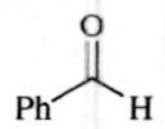	Et_3SiH, TFA, n-$C_7H_{15}OH$, rt, 1 h	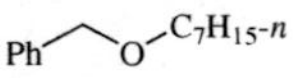 (81)	328
	Et_3SiH, Ph_3CClO_4, n-$C_6H_{13}CH(Me)OTMS$, CH_2Cl_2, 0°, 10 min	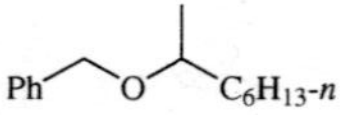 (—)	329
	$PhMe_2SiH$, EG acid, $Ph(CH_2)_3OTMS$, CH_2Cl_2, rt,	 (99)	333
	Et_3SiH, Ph_3CClO_4, $Ph(CH_2)_3OTMS$, CH_2Cl_2, 0°, 10 min	 (83)	329
	$PhMe_2SiH$, EG acid, $TMSO(CH_2)_2OTMS$, CH_2Cl_2, rt,	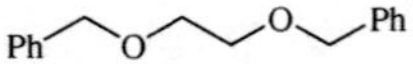 (63)	333
	Ph(CH₂)₃OTHP, Et_3SiH (2.2 eq), catalyst (10 mol%), MeCN, 0°, 1 h		340
	Catalyst		
	$Sn(OTf)_2$	(96)	
	$BF_3 \cdot OEt_2$	(20)	
	TMSI	(88)	
	TMSOTf	(98)	
	Et_3SiH, TFA, MeOH, rt, 1 h	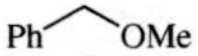 **VII** (87)	328
	Et_3SiH, Cl_3CCO_2H, MeOH, rt, 1 h	**VII** (85)	328
	Et_3SiH, H_2SO_4, EtOH, rt, 1 h	 (87)	328
	Et_3SiH, EG acid, $TMSOCH_2C{\equiv}CH$, CH_2Cl_2, rt,	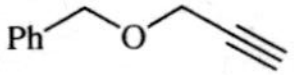 (95)	333

Reactant	Conditions	Product(s) and Yield(s) (%)	Refs.
	Et_3SiH, EG acid, $TMSOCH_2CH{=}CH_2$, CH_2Cl_2, rt,	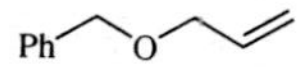 (71)	333
	Et_3SiH, H_2SO_4, *n*-PrOH, rt, 1 h	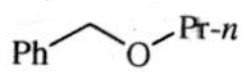 (68)	328
	Et_3SiH, H_2SO_4, *i*-PrOH, rt, 1 h	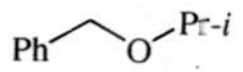 (69)	328
	Et_3SiH, H_2SO_4, *t*-BuOH, rt, 1 h	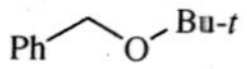 (45)	328
	1. $HC(OMe)_3$, CH_2Cl_2, rt, 2 h 2. Et_3SiH, Nafion®-H, CH_2Cl_2, reflux, 3 h	**VII** (94.6)	335
	Et_3SiH (1 eq), MeOH (2 eq), TFA (4 eq), 50 to 60°	**VII** (83)	327
	Et_3SiH (1 eq), TMSOTf (0.1 eq), ROTMS (0.83 eq), CH_2Cl_2	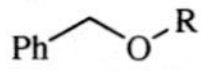 (see table below)	341

R	Temp	
Bn	−78° to −30°	(96)
$H_2C{=}CHCH_2$	−30° to 0°	(99)
i-Bu	0° to rt	(81)
i-Pr	0° to rt	(92)
c-C_6H_{11}	−78° to −30°	(99)
t-Bu	0° to rt	(16)
Ph	0° to rt	(0)

Reactant	Conditions	Product(s) and Yield(s) (%)	Refs.
+ 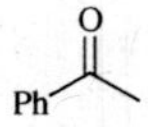	$(EtO)_3SiH$ (2.3 eq), KF (1 eq), rt, 36 h	**II** (100)	83
	PDMS (2.3 eq), K_2CO_3 (1 eq), 60°, 2.5 h	**II** (100)	83
	1. 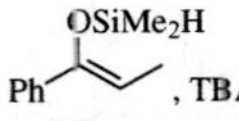, TBAF (6 mol%), THF, −78°, 3.5 h 2. H_3O^-	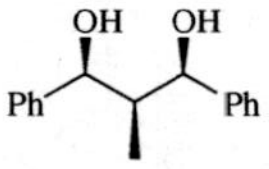 **VIII** + 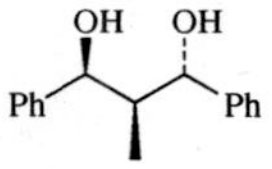 **IX** **VIII** + **IX** (91), **VIII**:**IX** = 97:3	400

Aldehyde	Conditions	Product(s) and Yield(s) (%)	Refs.
C_7 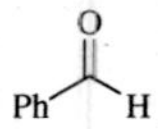	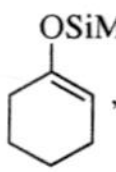 , TBAF (6 mol%), –78°, 15 h	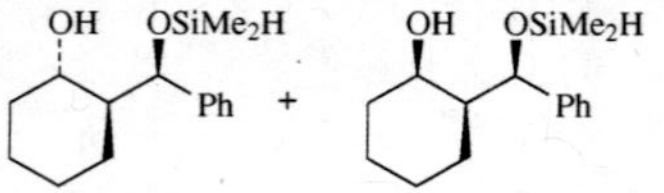 X, XI 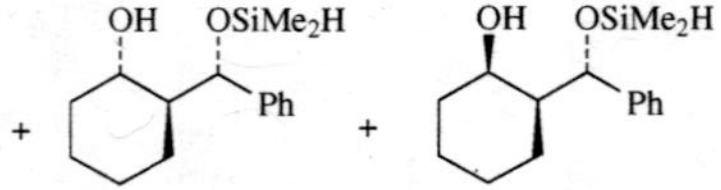 XII, XIII **X + XI + XII + XIII** (75), **X**:**XI**:**XII**:**XIII** = 27:6:53:14	400
	Et_3SiH (3 eq), TFA (2-3 eq), $RCONH_2$, PhCHO (x eq), solvent	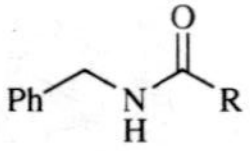	326

R	Solvent	Temp	Time	x = 0.33	x = 3
Bn	MeC_6H_5	120°	36 h	(92)	(68)
Ph	MeC_6H_5	120°	18 h	(91)	(94)
4-FC_6H_4	MeC_6H_5	120°	18 h	(92)	(93)
4-$MeOC_6H_4$	MeC_6H_5	120°	18 h	(95)	(91)
PhO	MeCN	22°	18 h	(90)	(85)
BnO	MeCN	22°	18 h	(95)	(92)
t-BuO	MeCN	22°	18 h	(92)	(81)
PhNH	MeC_6H_5	22°	18 h	(92)	(97)
BnNH	MeC_6H_5	22°	18 h	(89)	(88)

Aldehyde	Conditions	Product(s) and Yield(s) (%)	Refs.
	Me_2ClSiH (1.2 eq), $In(OH)_3$ (5 mol%), $CHCl_3$, rt, 5.5 h	 **XIV** (90)	331
	Me_2ClSiH (1.2 eq), $InCl_3$ (5 mol%), $CHCl_3$, rt, 5.5 h	**XIV** (60)	331

Substrate	Conditions	Product(s) and Yield(s) (%)	Refs.
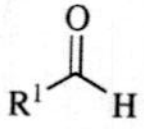	1. R^2SH, $BF_3•OH_2$, CH_2Cl_2, 0°, 1 min 2. Et_3SiH, 0° to rt, 3 h	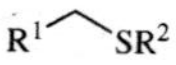 R^1 / R^2 Ph / Ph (91) Ph / *i*-Pr (97) Ph / *t*-Bu (94) Ph / *s*-Bu (92) *n*-C_6H_{13} / *i*-Pr (86)	365
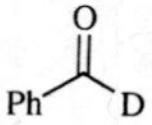	$MesPhSiH_2$, $[Rh(cod)Cl]_2$ (2.5 mol%), (–)-**4**, THF, rt	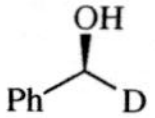 (74) 95% ee	586
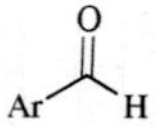	PMHS, TMSOTf (cat.), C_6H_6, reflux, 3 h	ArBn **I** + 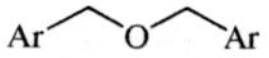 **II** Ar / **I** / **II** 2-ClC_6H_4 / (25) / (55) 4-ClC_6H_4 / **I** + **II** (65) **I**:**II** = 75:25	314
	Et_3SiH, TFA, 0°, 10 h	**II** Ar = 4-ClC_6H_5 (80)	313
	$PhMe_2SiH$, EG acid, CH_2Cl_2, rt,	**II** Ar = 4-ClC_6H_5 (86)	333
	$(HMe_2Si)_2O$, TMSOTf, C_6H_6, 80°, 20 min	**II** Ar = 4-ClC_6H_5 (90)	314
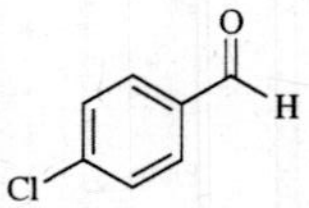	Et_3SiH (1.2 eq), TMSCl (1.2 eq), $In(OH)_3$ (5 mol %), $CHCl_3$, rt, 2 h	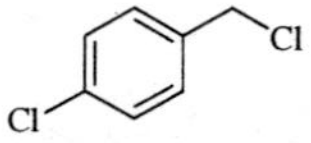 **I** (34)	331
	Me_2ClSiH (1.2 eq), $In(OH)_3$ (5 mol%), $CHCl_3$, rt, 2 h	**I** (34)	331
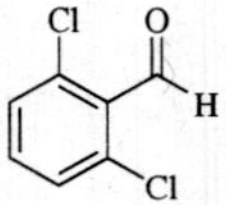	Cl_3SiH (3 eq), (*n*-Pr)$_3$N (1 eq), MeCN, 51-64°, 1 h	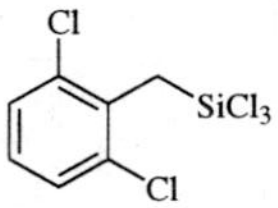 (61)	711
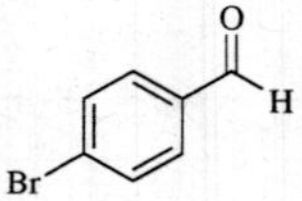	$(EtO)_3SiH$ (1.1 eq), KF (1 eq), 100°, 1 h	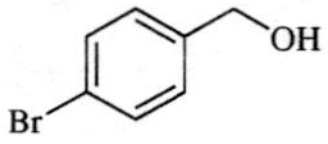 (90)	79, 80

TABLE 11. ORGANOSILANE REDUCTION OF ALDEHYDES (*Continued*)

Aldehyde	Conditions	Product(s) and Yield(s) (%)	Refs.
C_7 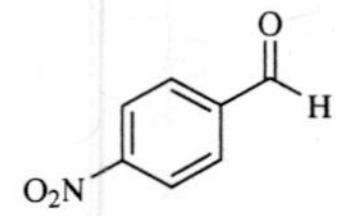	Et_3SiH, TFA, rt, 5 h	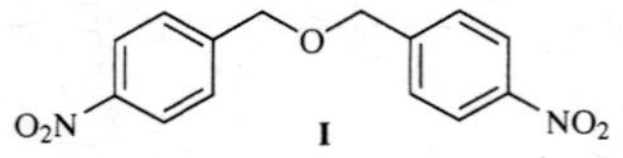 **I** (33)	313
	$(HMe_2Si)_2O$, TMSOTf, CH_2Cl_2, 30°, 2 h	**I** (55)	314
	C (1 eq), rt, 12 h	**I** (100)	84
	$(EtO)_3SiH$ (2.3 eq), KF (1 eq), 100°, 2 h	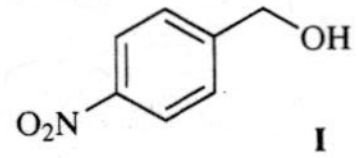 **I** (80)	83, 80, 79
	PMHS (1 eq), $ZnCl_2$ (1 eq), Et_2O, rt, 24 h	**I** (65)	315
	$PhSiH_3$ (0.4 eq), $Mn(dpm)_3$ (3 mol%), *i*-PrOH, DCE, O_2, rt	**I** (80)	367
	B (1 eq), rt, 12 h	**I** (100)	84
	$PhMe_2SiH$ (1.3 eq), $(Ph_3P)CuH$ (3 mol%), MeC_6H_5, 2 h	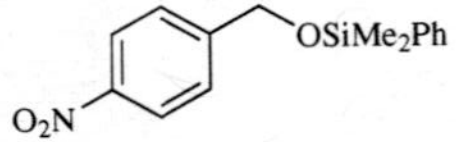 (89)	317
	Et_3SiH (1.0 eq), TMSOTf (0.1 eq), ROTMS (0.83 eq), CH_2Cl_2, 0° to rt	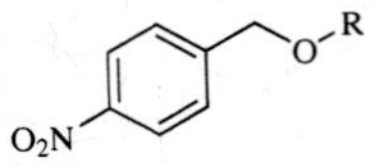 R: *n*-C_6H_{13} (88); *i*-Pr (73); *t*-Bu (63)	341

Reactant	Conditions	Product(s) and Yield(s) (%)	Refs.
	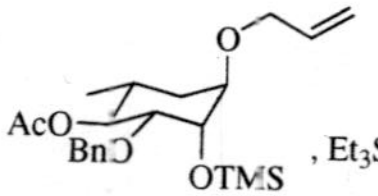 , Et_3SiH (1 eq), TMSOTf (2.0 eq), CH_2Cl_2, –78° to –30°	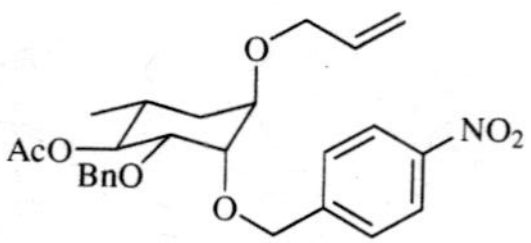 (52)	341
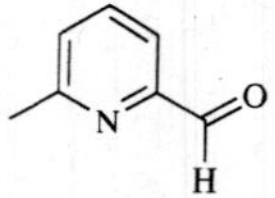	$PhMe_2SiH$ (1.1 eq), CsF (10 mol%), 18-C-6 (0.5 mol%), CH_2Cl_2, rt, 1 h	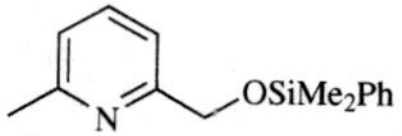 (58)	345
C_{7-8}			
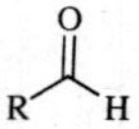	$(MeO)_3SiH$ (1.3 eq), LiOMe (1.2 eq), Et_2O, rt	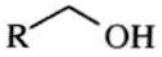	91

R	Time	
Ph	20 h	(85)
$2-O_2NC_6H_4$	15 h	(55)
$4-MeOC_6H_4$	15 h	(86)
$n-C_7H_{15}$	6 h	(80)

Reactant	Conditions	Product(s) and Yield(s) (%)	Refs.
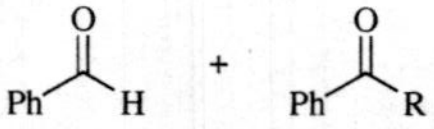	Ph_3SiH (1 eq), $(C_6F_5)_3B$ (2 mol%), MeC_6H_5, rt	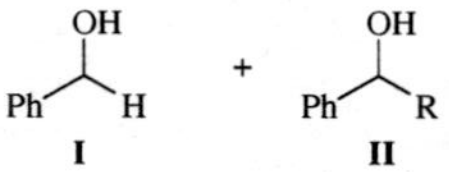	116

R	I:II
Me	6:1
OEt	>99:1

Reactant	Conditions	Product(s) and Yield(s) (%)	Refs.
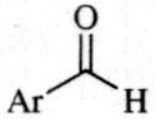	$HMe_2SiOSiMe_2H$ (1.0 eq), TMSOTf (cat.), solvent	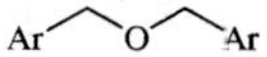	314

Ar	Solvent	Temp	Time	
Ph	C_6H_6	80°	20 min	(97)
$2-ClC_6H_4$	C_6H_6	80°	20 min	(94)
$4-ClC_6H_4$	C_6F_6	80°	20 min	(90)
$4-ClC_6H_4$	TFA	rt	30 min	(94)
$4-O_2NC_6H_4$	CH_2Cl_2	–78°	70 min	(9)
$4-O_2NC_6H_4$	CH_2Cl_2	30°	120 min	(55)
$4-HO_2CC_6H_4$	C_6F_6	80°	120 min	(97)
$4-HO_2CC_6H_4$	TFA	rt	240 min	(46)

TABLE 11. ORGANOSILANE REDUCTION OF ALDEHYDES (*Continued*)

Aldehyde	Conditions	Product(s) and Yield(s) (%)	Refs.
C_{7-8} 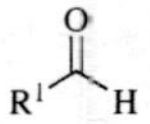	$R^2{}_3SiH$ (1.2 eq), *n*-Bu_4NClO_4 (0.1 eq), $LiClO_4$ (0.1 eq), CH_2Cl_2, electrolysis	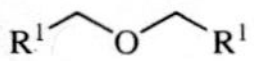	333
R^1	$R^2{}_3Si$		
Ph	Et_3Si	(91.0)	
Ph	$PhMe_2Si$	(95.7)	
4-MeC_6H_4	$PhMe_2Si$	(94.5)	
4-ClC_6H_4	$PhMe_2Si$	(86.3)	
n-C_6H_{13}	$PhMe_2Si$	(91.8)	
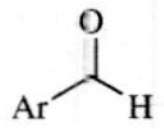	Me_3SiH (1.25 eq), TMSI (0.3 eq), *t*-BuOTMS (1.25 eq), *n*-C_5H_{12}, –78° to rt, 5-6 h	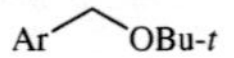 Ar: Ph (67); 4-FC_6H_4 (77); 4-MeC_6H_4 (71); 3-MeC_6H_4 (87); 2-MeC_6H_4 (21); 2-$MeOC_6H_4$ (64)	338
	, Et_3SiH (2.2 eq), TMSOTf (10 mol%), MeCN, 0°, 1 h	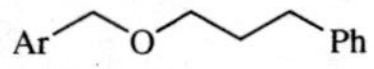 Ar: Ph (98); 4-ClC_6H_4 (95); 4-$O_2NC_6H_4$ (90); 4-$MeOC_6H_4$ (90); 4-MeC_6H_4 (98)	340
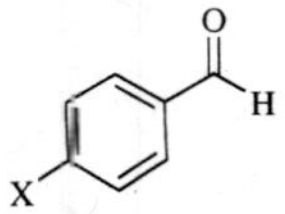	Ph_3SiH (1 eq), $(C_6F_5)_3B$ (2 mol%)	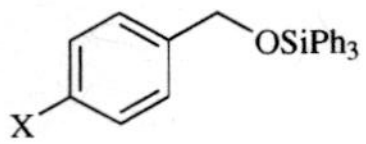 X: H (81); Me (82); Cl (81); NO_2 (96)	115

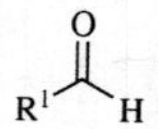

Et_3SiH (1.2 eq), TMSCl (50 mol%), $InCl_3$ (20 mol%), R^2STMS (1.2 eq), CH_2Cl_2, rt, 5 h

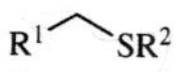

R^1	R^2	
Ph	Et	(87)
Ph	*i*-Pr	(83)
Ph	Ph	(82)
$BnCH_2$	Et	(81)
n-C_8H_{17}	Et	(78)

426

C_{7-9}

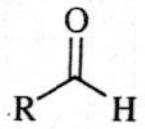

1. Et_3SiH, BF_3, CH_2Cl_2
2. $BF_3•OEt_2$, rt

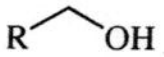

I + RMe **II**

R	Time	I	II
Ph	11 min	(—)	(52)
4-$MeOC_6H_4$	10 min	(0)	(100)
4-NCC_6H_4	10 min	(100)	(0)
4-$O_2NC_6H_4$	5 min	(100)	(0)

1

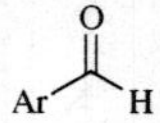

Ph_3SiH, $(C_6F_5)_3B$ (2 mol%), MeC_6H_5, rt

Ar	
Ph	(87)
4-MeC_6H_4	(82)
4-ClC_6H_4	(81)
4-$O_2NC_6H_4$	(96)

116

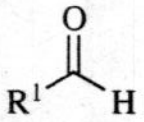

$R^3{}_3SiH$ (1.2 eq), R^2OTMS, $(n\text{-}Bu)_4NClO_4$ (0.1 eq), $LiClO_4$ (0.1 eq), CH_2Cl_2, electrolysis

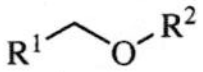

333

R^1	R^2	$R^3{}_3Si$	
n-C_6H_{13}	$Bn(CH_2)_2$	$PhMe_2Si$	(81.9)
Ph	$Bn(CH_2)_2$	$PhMe_2Si$	(98.8)
Ph	$H_2C{=}CHCH_2$	Et_3Si	(71.2)
Ph	$HC{\equiv}CCH_2$	Et_3Si	(95.0)
Ph	$(CH_2)_2OTMS$	$PhMe_2Si$	(63.2)[b]
(*E*)-PhCH=CH	$H_2C{=}CHCH_2$	Et_3Si	(51.1)
(*E*)-PhCH=CH	$HC{\equiv}CCH_2$	Et_3Si	(49.8)

TABLE 11. ORGANOSILANE REDUCTION OF ALDEHYDES (*Continued*)

Aldehyde	Conditions	Product(s) and Yield(s) (%)	Refs.
C_{7-9}			
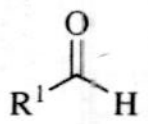	Et_3SiH (1.3 eq), $BiCl_3$ (0.11 eq), R^2OH, CH_2Cl_2, rt	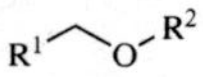	332
R^1	R^2		
c-C_6H_{11}	$BnCH_2CH_2$	(77)	
c-C_6H_{11}	$BnCH_2CHMe$	(70)	
Ph	$HO_2C(CH_2)_{11}$	(38)	
Ph	Et	(74)	
Ph	*n*-C_9H_{19}	(95)	
Ph	Bn	(80)	
Ph	$BnCH_2CH_2$	(93)	
Ph	$BnCH_2CHMe$	(82)	
Bn	$BnCH_2CH_2$	(61)	
$BnCH_2$	$BnCH_2CH_2$	(73)	
$BnCH_2$	$BnCH_2CHMe$	(69)	
n-C_7H_{15}	$BnCH_2CH_2$	(63)	
n-C_7H_{15}	$BnCH_2CHMe$	(64)	

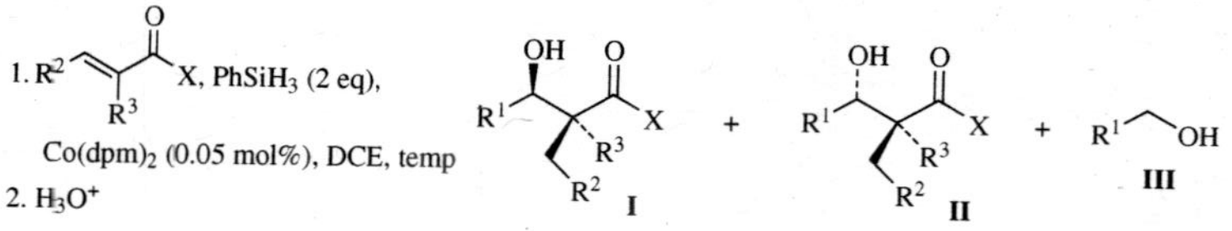

478

R^1	R^2	R^3	X	Temp	Time	I + II	I:II	III
Ph	H	H	NMe_2	20°	2 h	(95)	80:20	(3)
(*E*)-PhCH=CH	H	H	NMe_2	20°	4 h	(96)	72:28	(trace)
$BnCH_2$	H	H	NMe_2	20°	6 h	(90)	70:30	(trace)
Ph	Me	H	NMe_2	50°	3 h	(72)	70:30	(10)
(*E*)-PhCH=CH	Me	H	NMe_2	20°	4 h	(68)	72:28	(12)
Ph	H	Me	NMe_2	20°	5 h	(50)	—	(31)
(*E*)-PhCH=CH	H	Me	NMe_2	20°	4 h	(70)	—	(14)
(*E*)-PhCH=CH	H	H	OMe	20°	20 h	(80)	50:50	(10)
$BnCH_2$	H	H	OMe	20°	20 h	(62)	50:50	(14)

C_{7-8}

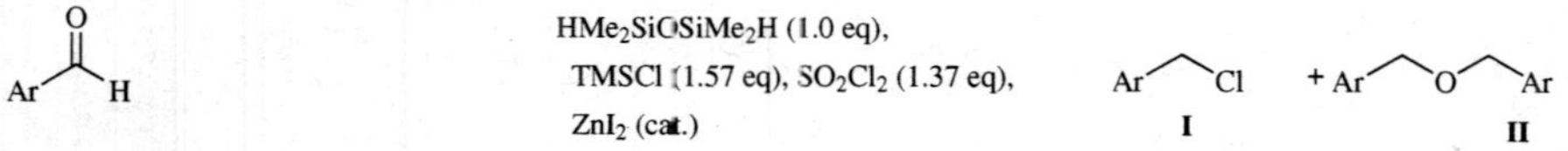

314

Ar	Time	Temp	I	II
Ph	20 min	70°	(91)	(—)
2-ClC_6H_4	60 min	70°	(55)	(45)
4-ClC_6H_4	45 min	70°	(85)	(—)
4-$O_2NC_6H_4$	180 min	–70°	(80)	(20)
3-HOC_6H_4	195 min	–70°	(40)	(—)
4-MeC_6H_4	15 min	70°	(87)	(—)
4-$MeOC_6H_4$	150 min	–70°	(60)	(40)
4-$MeO_2CC_6H_4$	180 min	–70°	(60)	(40)

TABLE 11. ORGANOSILANE REDUCTION OF ALDEHYDES (*Continued*)

Aldehyde	Conditions	Product(s) and Yield(s) (%)	Refs.
C_{7-9} 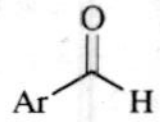	$HMe_2SiOSiMe_2H$ (1.0 eq), TMSCl (1.5 eq), LiBr (1.7 eq), MeCN, 80°	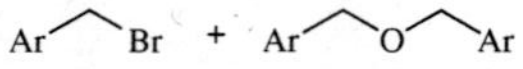 I + II	314, 356
Ar	**Time**	**I** / **II**	
Ph	30 min	(94) / (—)	
2-ClC_6H_4	15 min	(64) / (—)	
4-ClC_6H_4	20 min	(90) / (—)	
3-HOC_6H_4	45 min	(77) / (23)	
4-MeC_6H_4	20 min	(97) / (—)	
4-$MeO_2CC_6H_4$	80 min	(80) / (20)	
4-$MeOC_6H_4$	15 min	(84) / (—)	
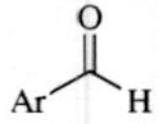	$HMe_2SiOSiMe_2H$ (1.0 eq), I_2 (0.6 eq), CH_2Cl_2, –5°	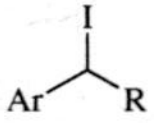	357
Ar	**Time**		
Ph	10 min	(80)	
2-ClC_6H_4	10 min	(72)	
3-HOC_6H_4	5 min	(87)	
4-MeC_6H_4	10 min	(70)	
4-$MeOC_6H_4$	10 min	(66)	
4-NCC_6H_4	15 min	(75)	
4-$MeO_2CC_6H_4$	45 min	(75)	
2,4-$Me_2C_6H_3$	8 min	(83)	
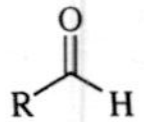	Et_3SiH (1.2 eq), $BiBr_3$, (10-30 mol%), rt	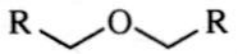	343

R	Time	
Ph	<5 min	(84)
4-ClC_6H_4	<5 min	(93)
4-BrC_6H_4	<5 min	(89)
2-$MeOC_6H_4$	8 h	(83)
4-$MeO_2CC_6H_4$	<5 min	(85)
4-$Me_2NC_6H_4$	20 h	(0)
n-C_7H_{15}	2 h	(88)

C_{7-10}

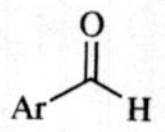

R_3SiH, $SnCl_2$ (cat.)

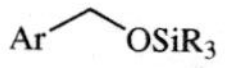

344

Ar	R_3Si	Time	
Ph	Et_3Si	180 min	(12.1)
Ph	$(n\text{-}Bu)_3Si$	120 min	(27.5)
Ph	$Et(i\text{-}C_5H_{11})_2Si$	120 min	(28.3)
4-(TMS)C_6H_4	Et_3Si	180 min	(11.5)
4-(TMS)C_6H_4	$(n\text{-}Bu)_3Si$	120 min	(19.8)
4-*t*-BuC_6H_4	Et_3Si	180 min	(15.1)
4-*t*-BuC_6H_4	$(n\text{-}Bu)_3Si$	50 min	(71.8)

R_3SiH (x eq), $NiCl_2$ (cat.), 90 min

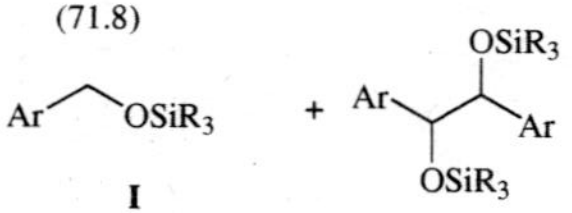

I + II

344

Ar	R_3Si	x	I	II
Ph	Et_3Si	1.0	(23.4)	(58.2)
Ph	$(n\text{-}Bu)_3Si$	1.0	(36.5)	(36.0)
Ph	$Et(i\text{-}C_5H_{11})_2Si$	1.0	(37.6)	(39.2)
4-(TMS)C_6H_4	Et_3Si	2.0	(10.1)	(80.2)
4-(TMS)C_6H_4	$(n\text{-}Bu)_3Si$	2.0	(20.1)	(40.3)
4-*t*-BuC_6H_4	Et_3Si	2.0	(52.9)	(—)
4-*t*-BuC_6H_4	$(n\text{-}Bu)_3Si$	2.0	(53.1)	(21.6)
4-*t*-BuC_6H_4	Et_3Si	1.0	(21.7)	(47.3)

PMHS (3 eq), Triton-B (2 mol%), THF, rt

278

Ar	
Ph	(99)
4-NCC_6H_4	(84)
4-MeC_6H_4	(92)
3,4,5-$(MeO)_3C_6H_2$	(83)

TABLE 11. ORGANOSILANE REDUCTION OF ALDEHYDES (*Continued*)

Aldehyde	Conditions	Product(s) and Yield(s) (%)	Refs.
C_{7-10} 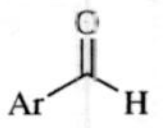	PMHS (3 eq), TBAF (2 mol%), THF, rt	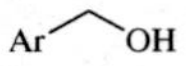 Ar: Ph (92); 4-ClC_6H_4 (88); 4-$O_2NC_6H_4$ (91); 2-$OHCC_6H_4$ (79); 3,4-[OCH_2O]C_6H_3 (95); 4-MeC_6H_4 (98); 3,4,5-$(MeO)_3C_6H_2$ (96)	278
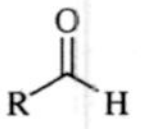	$(EtO)_3SiH$ (1.1 eq), MF (1 eq)	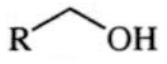	79

R	MF	Temp	Time	Yield (%)
4-$O_2NC_6H_4$	KF	100°	2 h	(80)
4-$MeCONHC_6H_4$	CsF	80°	8 h	(80)
(*E*)-PhCH=CH	KF	rt	24 h	(95)
$Me_2C=CH(CH_2)_2CH(Me)CH_2$	KF	rt	1 h	(80)

Aldehyde	Conditions	Product(s) and Yield(s) (%)	Refs.
C_{7-11} 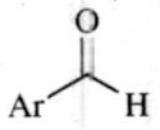	Et_3SiH, $ZnCl_2$ (cat.), >100°, 90 min	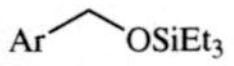 Ar: Ph (71.9); 4-*t*-BuC_6H_4 (72.5); 4-(TMS)C_6H_4 (56.5)	344
C_{7-12} 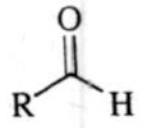	Et_3SiH (x eq), acid (y eq), solvent, 28°	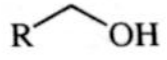 **I** + 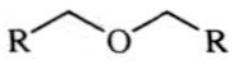 **II** + Other **III**	313

R	x	Acid	y	Solvent	Time	I	II	III
n-C_6H_{13}	1.1	TFA	20	none	0.75 h	(90)	(10)	(0)
Ph	1.2	TFA	1.5	none	0.5 h	(87)	(13)	(0)
Ph	1.1	TFA	10	$CHCl_3$	0.25 h	(93)	(7)	(0)
Ph	1.1	TFA	10	CCl_4	0.25 h	(89)	(11)	(0)
Ph	1.1	TFA	2.0	MeCN	0.25 h	(96)	(4)	(0)
Ph	1.1	Cl_2CHCO_2H	6.0	none	0.5 h	(93)	(7)	(0)
Ph	1.1	HCO_2H	14	none	8 h	(12)	(88)	(0)
Ph	1.1	$MeCO_2H$	9.2	none	5 h	(8)	(27)	(65)
Ph	1.1	HCl	0	MeCN	4 h	(9)	(3)	(88)
Ph	1.2	H_2SO_4	0	MeCN	1 h	(0)	(50)	(50)
Ph	2.5	TFA	20	none	0.25 h	(5)	(80)	(15)
Ph	2.5	TFA	20	none	44 h	(0)	(12)	(88)
4-$O_2NC_6H_4$	2.2	TFA	11	none	5 h	(33)	(67)	(0)
4-ClC_6H_4	2.2	TFA	11	none	10 h	(80)	(20)	(0)
4-MeC_6H_4	6.0	TFA	2.5	CH_2Cl_2	6 h	(81)	(17)	(2)
4-MeC_6H_4	6.0	TFA	2.5	none	118 h	(70)	(22)	(8)
1-$C_{10}H_7$	6.0	TFA	2.4	none	1.5 h	(67)	(33)	(0)
2-$C_{10}H_7$	6.0	TFA	2.4	none	18 h	(84)	(16)	(0)
n-$C_{11}H_{23}$	1.0	TFA	2.5	none	16 h	(93)	(7)	(0)

C_{7-11}

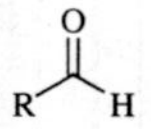

1. BF_3 (gas), CH_2Cl_2, 0°
2. Et_3SiH (x eq), 0°

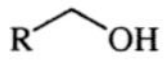

I + RMe II

1

R	x	Time	I	II
Ph	18	11 min	(—)	(52)
4-ClC_6H_4	10	10 min	(—)	(68)
4-$O_2NC_6H_4$	1.5	5 min	(100)	(—)
4-MeC_6H_4	10	10 min	(—)	(45)
4-$MeOC_6H_4$	2	10 min	(—)	(100)
4-NCC_6H_4	3	10 min	(100)	(—)
n-$C_{10}H_{21}$	1.5	10 min	(92)	(—)

TABLE 11. ORGANOSILANE REDUCTION OF ALDEHYDES (*Continued*)

Aldehyde	Conditions	Product(s) and Yield(s) (%)	Refs.
C_{7-8} 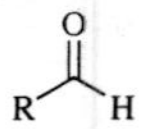	$[HSi(OEt)_4]K$, THF, 0°	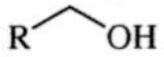 R / Time: Ph, 3 h (90); 4-BrC_6H_4, 4 h (84); n-C_7H_{15}, 4 h (80)	288
C_{7-14} 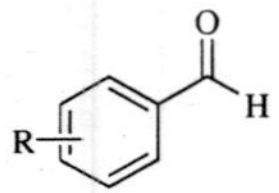	Ph_2MeSiH (2.5 eq), $(Ph_3P)CuH$ (3 mol%), MeC_6H_5	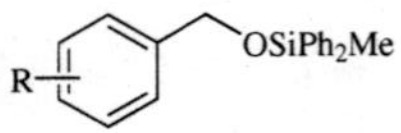 R / Time: 2-Br, 30 min (95); 4-Br, 45 min (98); 4-O_2N, 30 min (89); 4-MeO, 120 min (92); 4-TsO, 60 min (89); 3-BnO, 120 min (97)	317
C_{7-23} 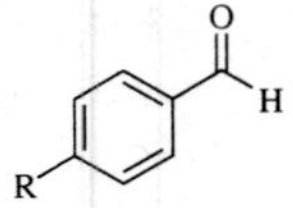	Et_3SiH (3 eq), TFA (2.9 eq), $PhCONH_2$ (3 eq), MeC_6H_5, 120°, 18 h	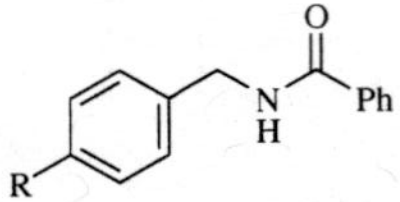 R: F (96); CO_2H (96); CO_2Me (91); OAc (75); OBn (90); OTBS (46); OTBDPS (91)	326
C_8 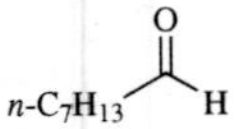	Et_3SiH (1.1 eq), BF_3•OEt_2 (2.0 eq), rt	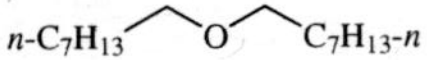 (66) + 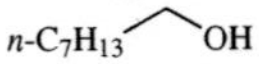 **I** (34)	74
	L, THF, 0°, 2 h	**I** (81)	96
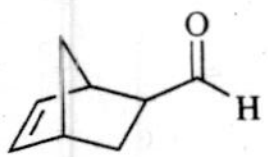	$(EtO)_3SiH$ (1.1 eq), KF (1 eq), rt, 1 h	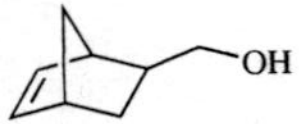 (70)	80

Reactant	Conditions	Product(s) and Yield(s) (%)	Refs.
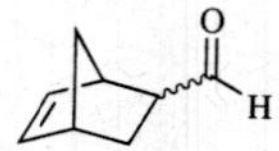	$PhMe_2SiH$ (1.3 eq), $(Ph_3P)CuH$ (3 mol%), MeC_6H_5, 2 h	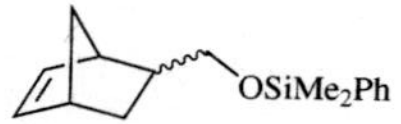 (98)	317
	Et_3SiH (5 eq), $Ni(cod)_2$ (0.2 eq), PPh_3 (0.4 eq), THF, 55°, 17 h	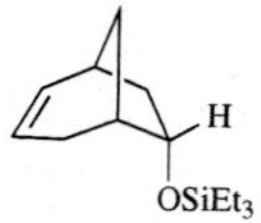 (13)	349
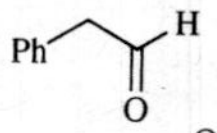	Me_2ClSiH (1.2 eq), $In(OH)_3$ (5 mol%), $CHCl_3$, 60°, 5.5 h	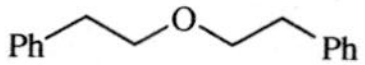 (70)	331
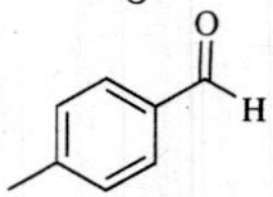	R_3SiH (x eq), TFA (y eq), solvent	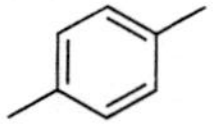	73

R_3Si	x	y	Solvent	Time	
Et_3Si	2.2	8.0	TFA	11 h	(20)
$PhMe_2Si$	2.2	8.0	TFA	13 h	(46)
$PhMe_2Si$	2.5	5.0	TFA	15 min	(52)
Et_3Si	6.0	5.0	MeCN	170 h	(50)
Et_3Si	2.2	6.7	MeCN	180 h	(43)
Et_3Si	6.0	5.0	$MeNO_2$	54 h	(61)
Et_3Si	2.2	5.0	CCl_4	100 h	(50)
Et_3Si	6.0	5.0	CCl_4	336 h	(66)

Reactant	Conditions	Product(s) and Yield(s) (%)	Refs.
	PMHS, TMSOTf (cat.), C_6H_6, reflux, 2.5 h	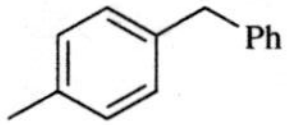 (60)	314
	PMHS, TMSOTf (cat.), MeC_6H_5, reflux, 0.5 h	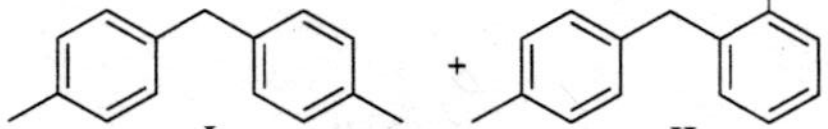 I + II; **I + II** (80), **I**:**II** = 90:10	314
	F, 0°, 2 h	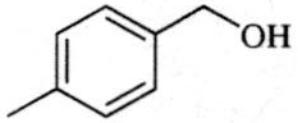 (96)	93

TABLE 11. ORGANOSILANE REDUCTION OF ALDEHYDES (*Continued*)

Aldehyde	Conditions	Product(s) and Yield(s) (%)	Refs.
C_8 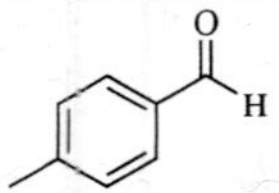	Et_3SiH (6 eq), TFA (5 eq), CCl_4, rt	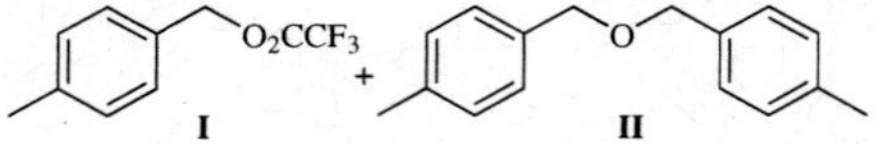 I + II + 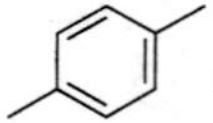 III (see time table below)	73
	$[Cp^*(Me_3P)Ir(SiMes_2)(H)][OTf]$ (1 eq), CD_2Cl_2	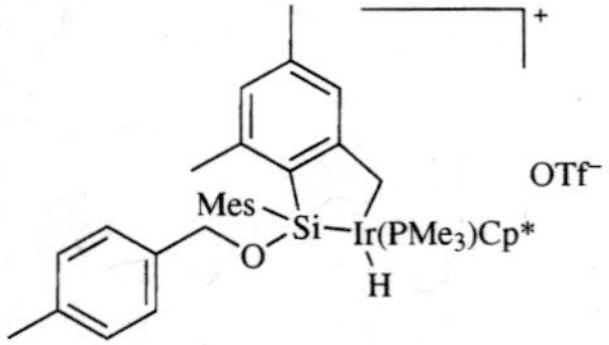 OTf^- (85)	745
	$[Cp^*(Me_3P)Ir(\eta_2\text{-}SiPh_2C_6H_4)(H)]^+$ $[(C_6F_5)_4B]^-$, CH_2Cl_2, rt, 10 min	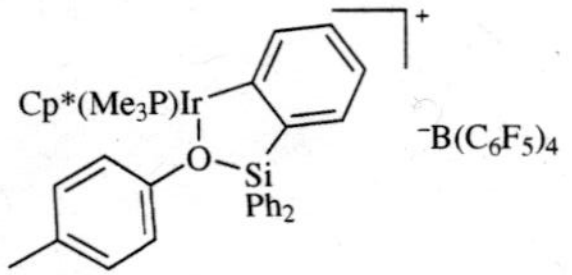 $^-B(C_6F_5)_4$ (48)	745
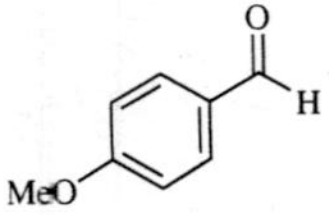	PMHS (3 eq), $(C_6F_5)_3B$ (5 mol%), CH_2Cl_2, rt, 5-20 min	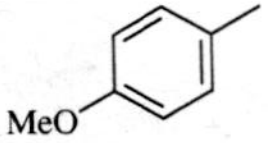 I (65)	354

Time	I	II	III	Other
16 h	(31)	(56)	(13)	(0)
22 h	(31)	(46)	(23)	(0)
45 h	(42)	(30)	(26)	(2)
89 h	(38)	(12)	(39)	(11)
209 h	(9)	(0)	(61)	(30)
336 h	(0)	(0)	(66)	(34)

	R_3SiH (x eq), TFA (5.0 eq), rt	**I**	R₃Si / x / Time: Et_3Si, 2.5, 30 min (76) $PhMe_2Si$, 6.0, 90 min (80) $PhMe_2Si$, 6.0, 30 min (81) Et_3Si, 6.0, 45 min (83)	73
	Et_3SiH (1.4 eq), TBAF (1.4 eq), MeCN, rt, 12 h	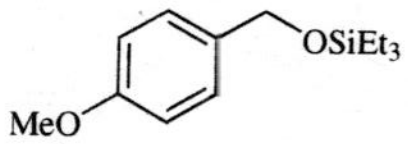 (19)		76
	PMHS (2.5 eq), $(Ph_3P)CuH$ (3 mol%), MeC_6H_5, 3 h	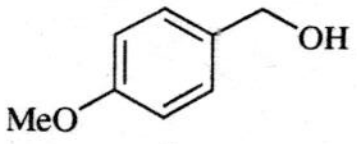 **I** (94)		317
	$PhSiH_3$ (0.4 eq), $Mn(dpm)_3$ (3 mol%), *i*-PrOH, DCE, O_2, rt	**I** (56)		367
	PMHS (1 eq), $ZnCl_2$ (1 eq), Et_2O, rt, 24 h	**I** (68)		315
	C (1 eq), rt, 12 h	**I** (100)		84
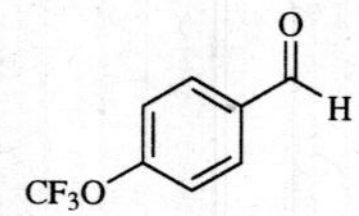	Et_3SiH, TFA	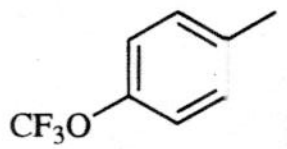 (—)		414

For R_3Si = Et_3Si and $PhMe_2Si$ in the first row, the printed column headers are R_3Si, x and Time.

TABLE 11. ORGANOSILANE REDUCTION OF ALDEHYDES (*Continued*)

Aldehyde	Conditions	Product(s) and Yield(s) (%)	Refs.
C_8			
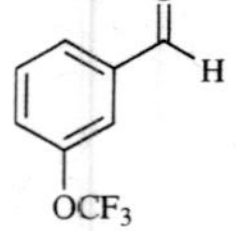	Et_3SiH, TFA	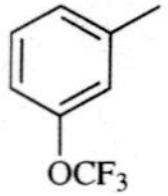 (—)	414
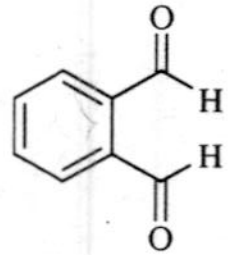	TMSH (4 eq), TMSOTf (1.6 mol%), CH_2Cl_2, 0°, 4 h; rt, 2 h	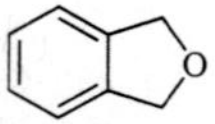 (99)	392
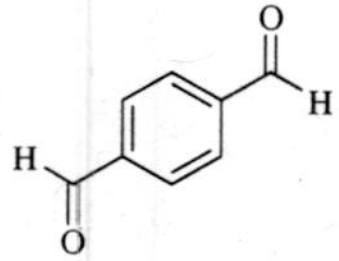	TMSH (2.5 eq), TMSI (0.3 eq), *t*-BuOTMS (2.5 eq), *n*-C_5H_{12}, –78° to rt, 5-6 h	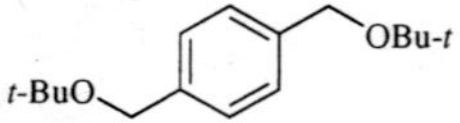 (70)	338
	1-AdOTMS (2.5 eq), TMSH (2.5 eq), TMSI (0.3 eq), *n*-C_5H_{12}, –78° to rt, 5-6 h	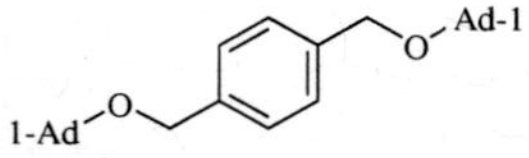 (78)	338
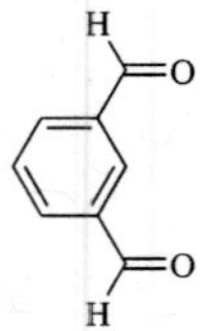	(*Z*)-$TMSOCH_2CH{=}CHCH_2OTMS$, Et_3SiH (2.4 eq), $BiBr_3$ (10 mol%), CH_2Cl_2/MeCN, –30° to 0°	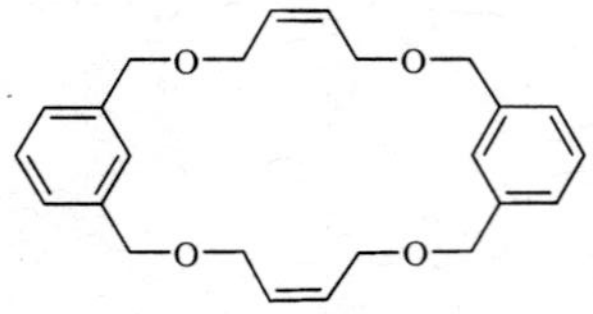 **I** (<1)	343
	(*Z*)-$TMSOCH_2CH{=}CHCH_2OTMS$, Et_3SiH (1.2 eq), TMSOTf (10 mol%), CH_2Cl_2, –30° to 0°,	**I** (8) + others	343

Reactant	Conditions	Product(s) and Yield(s) (%)	Refs.
	$TMSOCH_2C{\equiv}CCH_2OTMS$, Et_3SiH (2.4 eq), $BiBr_3$ (10 mol%), CH_2Cl_2/MeCN, –30° to 0°	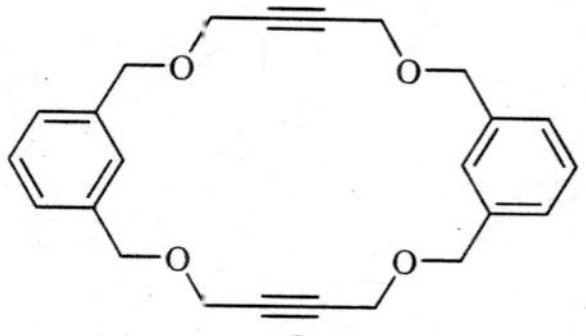 **I** (7) + others	343
	$TMSOCH_2C{\equiv}CCH_2OTMS$, Et_3SiH (2.4 eq), TMSOTf (10 mol%), CH_2Cl_2, –30° to 0°	**I** (10) + others	343
	Et_3SiH (2 eq), $TrClO_4$ (10 mol%), CH_2Cl_2	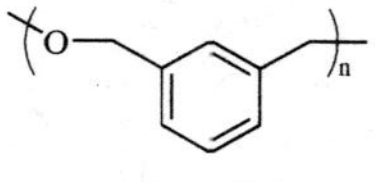 **I**	337
	R_3SiH (2 eq), Ph_3CClO_4 (10 mol %), CH_2Cl_2, –20°, 1 h	**I**	337
C_{8-17} 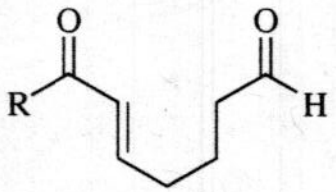	$PhSiH_3$ (1.2 eq), $Co(dpm)_2$ (5 mol%), rt	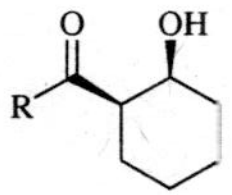	461
C_9 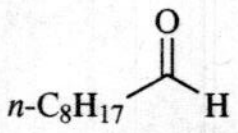	Et_3SiH (1 eq), $TrClO_4$ (5 mol%), CH_2Cl_2, 0°, 5 min	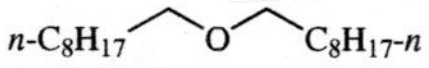 (88)	329

Temp	Time	M_n	
0°	0.25 h	6500	(52)
0°	6 h	10,400	(40)
–20°	0.25 h	8800	(58)
–50°	24 h	11,400	(54)

R_3Si	M_n	
Et_3Si	8600	(63)
$(n\text{-}C_6H_{13})_3Si$	5600	(24)
$PhMe_2Si$	9200	(63)
Ph_3Si	9900	(23)

R	
Me	(38)
2-C_4H_3O	(75)
2-C_4H_3S	(73)
Ph	(87)
4-$F_3CC_6H_5$	(72)
2-$C_{10}H_7$	(68)

TABLE 11. ORGANOSILANE REDUCTION OF ALDEHYDES (*Continued*)

Aldehyde	Conditions	Product(s) and Yield(s) (%)	Refs.
C_9			
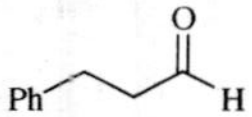	Et_3SiH (5 eq), $Ni(cod)_2$ (0.2 eq), PPh_3 (0.4 eq), THF, 30°, 10 h	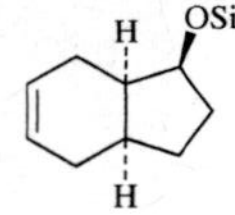 (44) + 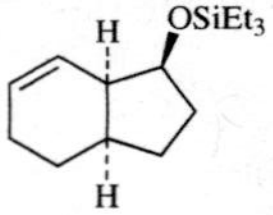 (22)	349
	PMHS (3 eq), $(C_6F_5)_3B$ (5 mol%), CH_2Cl_2, rt, 5-20 min	 (90)	354
	PMHS (1 eq), $ZnCl_2$ (1 eq), Et_2O, rt, 24 h	 **I** (72)	315
+ 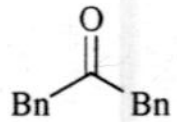	$Me(EtO)_2SiH$ (1.2 eq), KO_2CH (1 eq), DMF, 80°, 15 h	**I** + 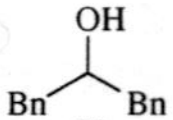 **II** **I** + **II** (83), **I**:**II** = 100:0	82
	Et_3SiH (1 eq), $TrClO_4$ (5 mol%), CH_2Cl_2, 0°, 5 min	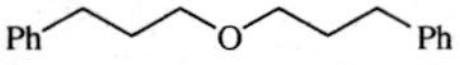 (83)	329
	$PhMe_2SiH$ (1.3 eq), $(Ph_3P)CuH$ (3 mol%), MeC_6H_5, 2 h	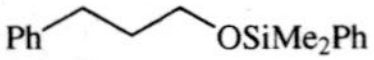 (90)	317
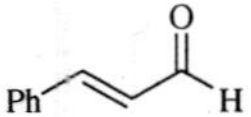	Et_3SiH (0.72 eq), H_2PtCl_6	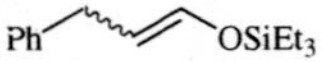 (—)	76
	Et_3SiH (1.5 eq), CsF (1.5 eq), MeCN, rt, 10 h	 **I** (—)	76
	$PhMe_2SiH$ (1.2 eq), TBAF (2 eq), HMPA, rt, 0.5 h	 (100)	320
	EtO_3SiH (1.5 eq), HAp, rt, 2 h	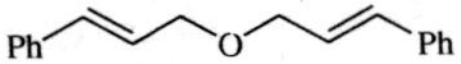 **I** + 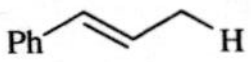 **II** **I** + **II** (70)	353

	I_2SiH_2 (1.2 eq), $CDCl_3$, rt Time: 6 min; 28 min; 126 min; 4 min	 I : II = (70) : (30); (35) : (65); (17) : (83); (100) : (0)	358
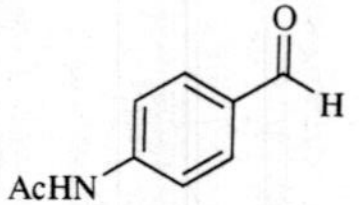	$(EtO)_3SiH$ (1.1 eq), KF (1 eq), 100°, 1 h	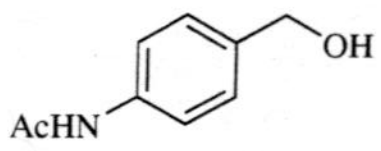 (80)	79, 80
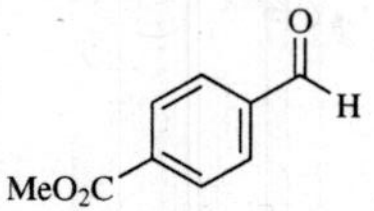	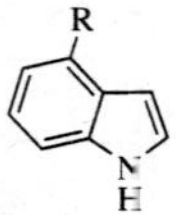, Et_3SiH (3 eq), TFA	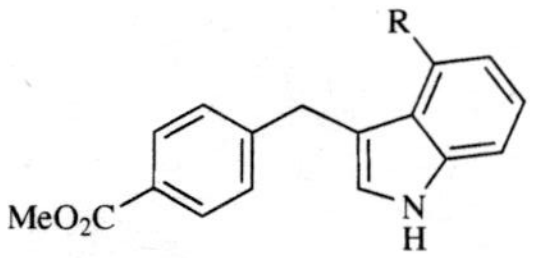 R = H (55); F (90)	355
	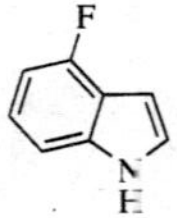, Et_3SiH (3 eq), BF_3•OEt_2	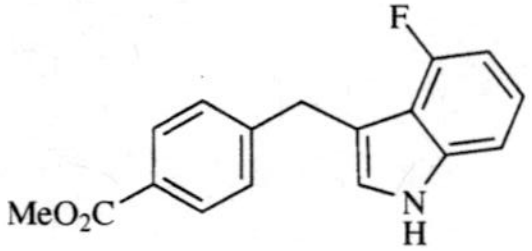 (71)	355
	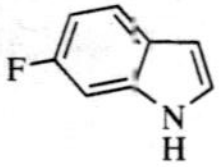, Et_3SiH (3 eq), TFA	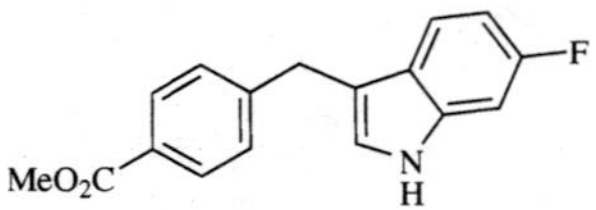 (64)	355
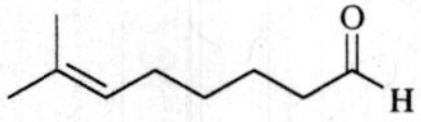	$(EtO)_3SiH$ (1.1 eq), KF (1 eq), rt, 1 h	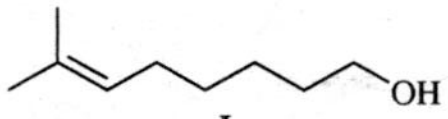 I (80)	80
	$Me(EtO)_2SiH$ (1.1 eq), KF (1 eq), rt, 18 h	I (80)	80

TABLE 11. ORGANOSILANE REDUCTION OF ALDEHYDES (*Continued*)

Aldehyde	Conditions	Product(s) and Yield(s) (%)	Refs.
C_9			
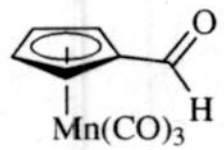	Et_3SiH (3 eq), TFA (5 eq), 40-50°, 6 h	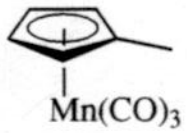 (73)	351
C_{10}			
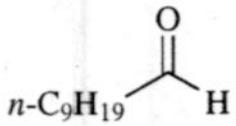	Et_3SiH, $ZnCl_2$	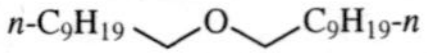 (50)	330
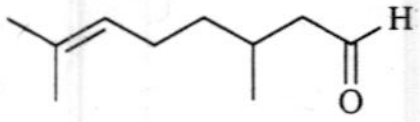	$PhMe_2SiH$ (1.3 eq), $(Ph_3P)CuH$ (3 mol%), MeC_6H_5, 2 h	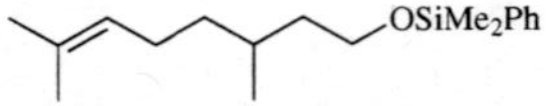 (93)	317
	$Me(EtO)_2SiH$ (1.5 eq), KF (1 eq), DMF, 35°, 2.5 h	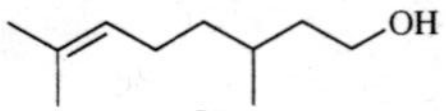 **I** (68)	82
	PMHS (2.5 eq), KF (1.25 eq), DMF, 30°, 0.75 h	**I** (70)	82
	Ph_2SiD_2 (1.1 eq), $(Ph_3P)_3RhCl$	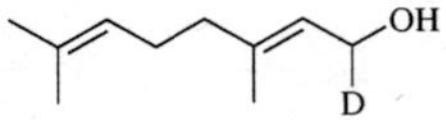 (92)	435
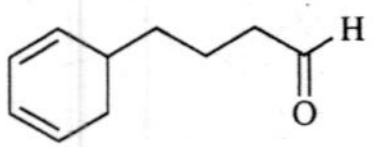	Et_3SiH (5 eq), $Ni(cod)_2$ (0.2 eq), PPh_3 (0.4 eq), THF, 30°, 3 h	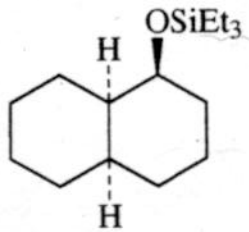 (60)	349
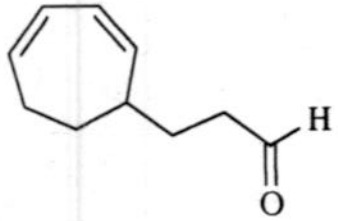	Et_3SiH (5 eq), $Ni(cod)_2$ (0.2 eq), PPh_3 (0.4 eq), THF, 30°, 4 h	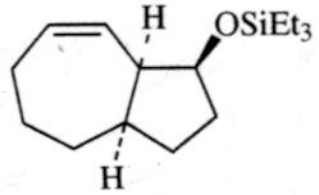 (81)	349
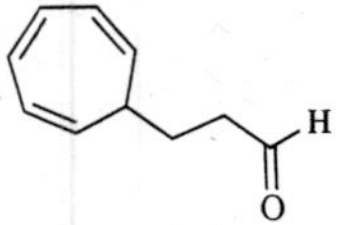	Et_3SiH (5 eq), $Ni(cod)_2$ (0.2 eq), PPh_3 (0.4 eq), THF, 30°, 2 h	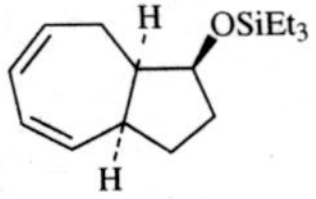 (47)	349

Reactant	Conditions	Product(s) and Yield(s) (%)	Refs.
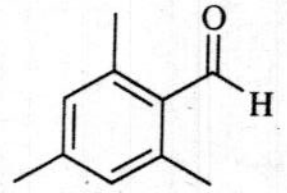	Et_3SiH (2.2 eq), TFA (10.0 eq), rt, 15 min	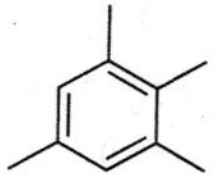 (98)	73
	Et_3SiH, TFA, 30-40°	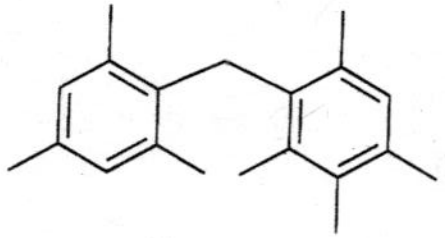 (8)	311
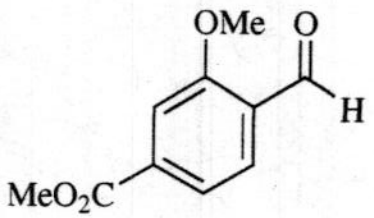	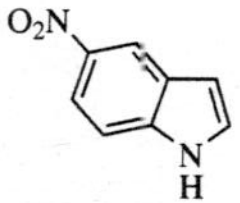, Et_3SiH (3 eq), $BF_3•OEt_2$	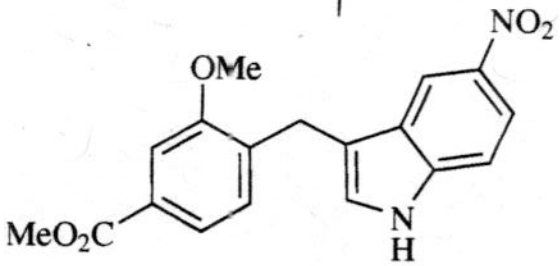 (52)	355
	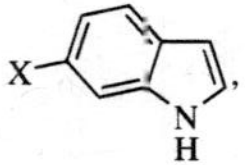, Et_3SiH (3 eq), acid	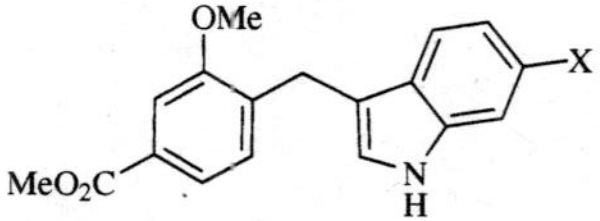	355
	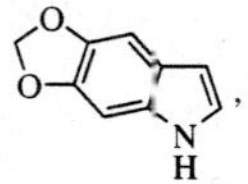, Et_3SiH (3 eq) TFA	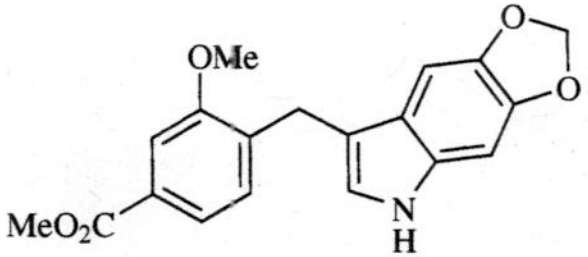 (20)	355
C_{16-18} 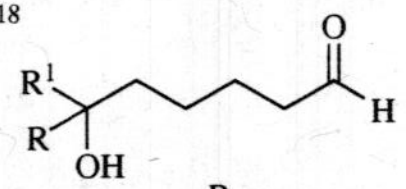	Et_3SiH (2 eq), $BF_3•OEt_2$, CH_2Cl_2, –78°, 1 h; 20°, 16 h	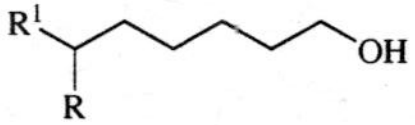	746

X	Acid	
Cl	TFA	(61)
Cl	$BF_3•OEt_2$	(43)
F	$BF_3•OEt_2$	(47)

R	R^1	
Et	Et	(53)
—$(CH_2)_4$—		(—)
—$(CH_2)_5$—		(51)
Ph	Me	(97)
Ph	Et	(—)
Ph	Ph	(—)

Aldehyde	Conditions	Product(s) and Yield(s) (%)	Refs.
C_{11} 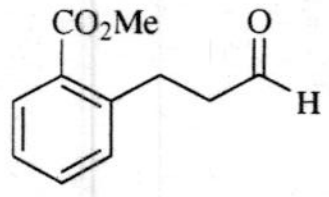	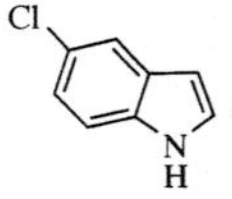, Et_3SiH (3 eq), TFA	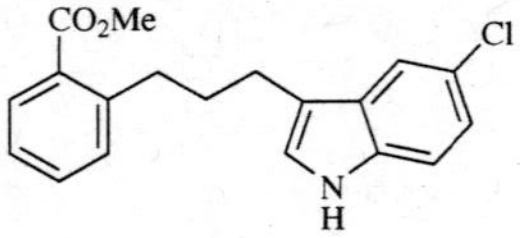 (55)	355
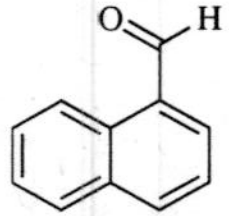	Et_3SiH (1.1 eq), $(C_6F_5)_3B$ (5 mol%), CH_2Cl_2, rt, 20 h	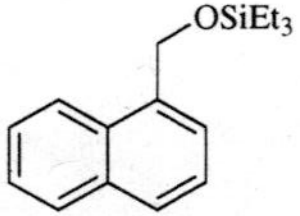 (96)	282
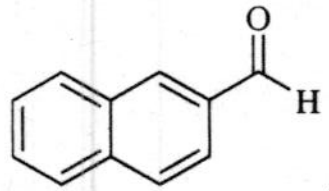	Me_2ClSiH (1.2 eq), $In(OH)_3$ (5 mol%), $CHCl_3$, rt, 2.5 h	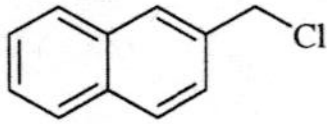 (92)	331
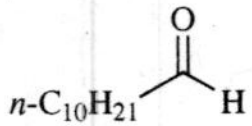	Et_3SiH (3.0 eq), $(C_6F_5)_3B$ (5 mol%), CH_2Cl_2, rt, 20 h	 (96)	282
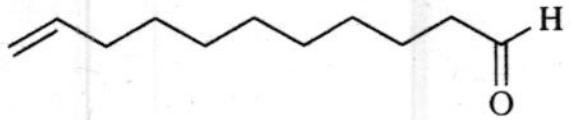	$PhMe_2SiH$ (1.3 eq), $(Ph_3P)CuH$ (3 mol%), MeC_6H_5, 2 h	 (95)	317
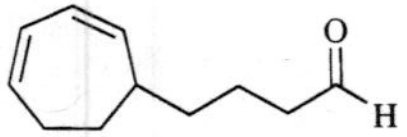	Et_3SiH (5 eq), $Ni(cod)_2$ (0.2 eq), PPh_3 (0.4 eq), THF, 35°, 8 h	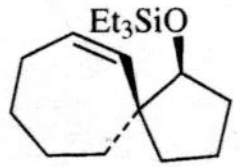 (36)	349
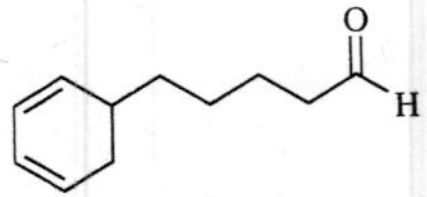	Et_3SiH (5 eq), $Ni(cod)_2$ (0.2 eq), PPh_3 (0.4 eq), THF, 55°, 10 h	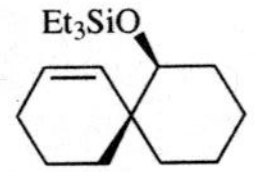 (30)	349

Substrate	Conditions	Product(s) and Yield(s) (%)	Refs.
C_{12}			
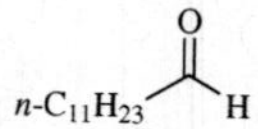	Et_3SiH (3 eq), $(C_6F_5)_3B$ (5 mol%), CH_2Cl_2, rt, 20 h	 (95)	281
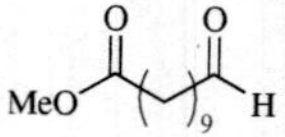	PMHS (1 eq), $ZnCl_2$ (1 eq), Et_2O, rt, 24 h	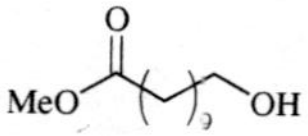 (70)	315
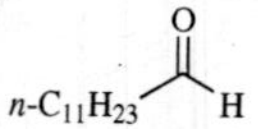	Et_3SiH, $ZnCl_2$	n-$C_{11}H_{23}$–O–$C_{11}H_{23}$-n (55)	330
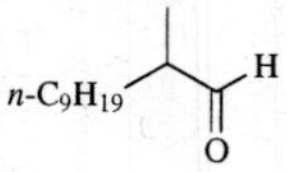	$PhMe_2SiH$ (1.3 eq), $(Ph_3P)CuH$ (3 mol%), MeC_6H_5, 2 h	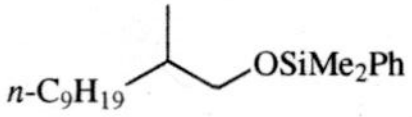 (92)	317
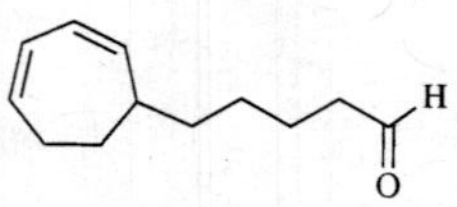	Et_3SiH (5 eq), $Ni(cod)_2$ (0.2 eq), PPh_3 (0.4 eq), THF, 45°, 34 h	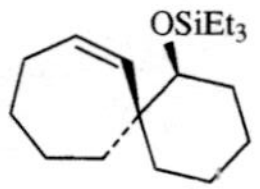 (22)	349
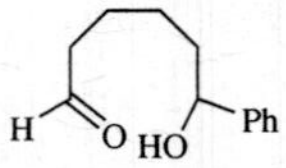	Ph_2MeSiH (2 eq), TMSOTf (1 eq), CH_2Cl_2, 0°, 2 h	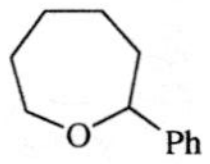 (90)	336
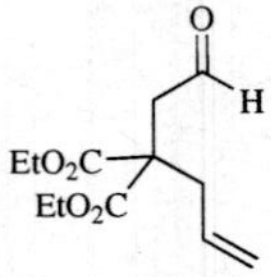	Ph_2SiH_2, $Cp_2Ti(PMe_3)_2$ (10 mol%), PMe_3 (60 mol%), MeC_6H_5, –20°	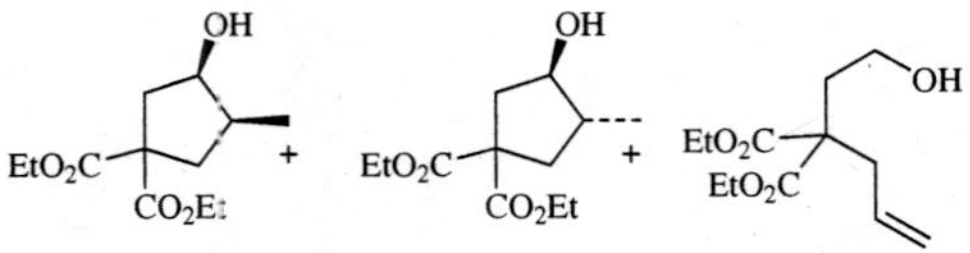 I + II + III; **I** + **II**:**III** = 99:1, **I**:**II** = 99:1, **I** (65)	428, 429

TABLE 11. ORGANOSILANE REDUCTION OF ALDEHYDES (*Continued*)

Aldehyde	Conditions	Product(s) and Yield(s) (%)	Refs.
C_{12} 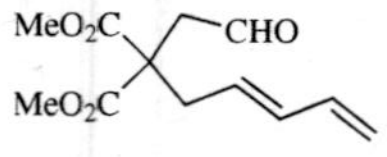	Et_3SiH (5 eq), $Ni(cod)_2$ (20 mol%), ligand (40 mol%), THF, rt, 12 h	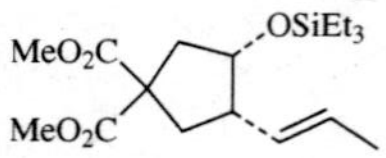	348
	R_3SiH (5 eq), **41** (20 mol%), $Ni(cod)_2$ (10 mol%), solvent, 5 h	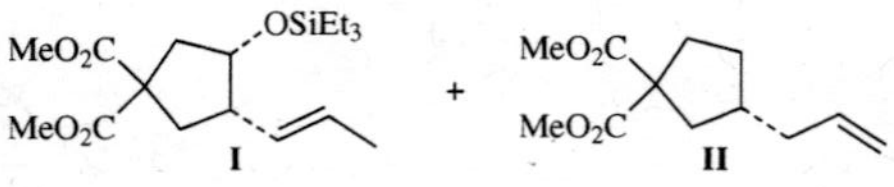	348

Ligand	Time		% ee
PPh_3	12 h	(84)	0
(*S*)-BINAP	62 h	(2)	0
(*R*)-H-MOP	40 h	(2)	16
(*S*,*R*)-PPFA	76 h	(3)	10
NMDPP	27 h	(2)	(—)
118	50 h	(13)	3

R_3Si	Solvent	Temp	Time	**I** + **II**	**I**:**II**	% ee **I**	% ee **II**
Et_3Si	THF	rt	5 h	(84)	4.3:1	2	47
t-$BuMe_2Si$	THF	rt	8 h	(83)	>50:1	16	—
$(EtO)_3Si$	THF	0°	5 h	(60)	12:1	46	33
Ph_3Si	THF	0°	2 h	(80)	1.7:1	47	53
Ph_2MeSi	THF	0°	2 h	(83)	1.2:1	27	78
$PhMe_2Si$	THF	0°	7 h	(82)	1.9:1	21	72
Ph_2MeSi	MeC_6H_5	0°	3 h	(88)	1.7:1	17	77
Ph_2MeSi	DMF	0°	3 h	(87)	1.1:1	39	81
Ph_2MeSi	DMF	–20°	28 h	(73)	1.1:1.2	44	86
Ph_2MeSi	MeCN	–20°	24 h	(83)	1:1	40	85

Reactant	Conditions	Product(s) and Yield(s) (%)	Refs.
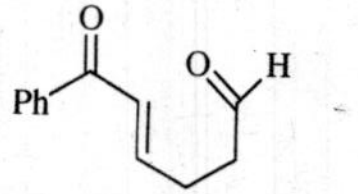	$PhSiH_3$ (1.2 eq), $Co(cpm)_2$ (5 mol%), rt	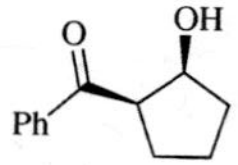 (70)	461
C_{13}			
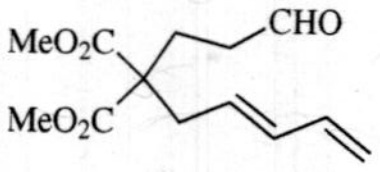	R_3SiH (5 eq), $Ni(cod)_2$ (10 mol%), **41** (20 mol%), DMF	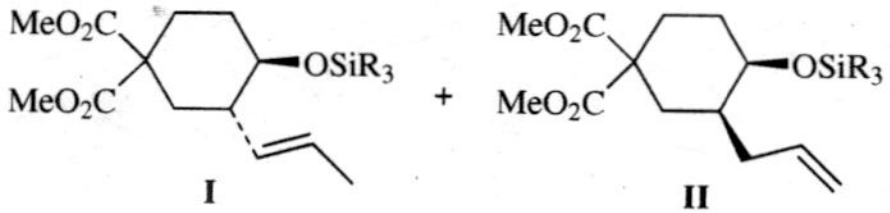 I + II	348

R_3Si	Time
$(EtO)_3Si$	24 h
Ph_2MeSi	9 h

I + II	I:II	% ee I:II
(51)	1:2.6	42:66
(61)	7.3:1	61:66

Reactant	Conditions	Product(s) and Yield(s) (%)	Refs.
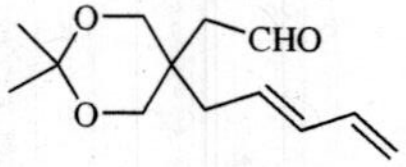	$(EtO)_3SiH$ (5 eq), $Ni(cod)_2$ (10 mol%), **41** (20 mol%), MeCN, –30°, 10 h	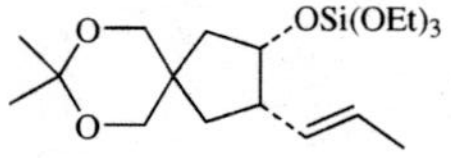 (80) 64% ee	348
C_{13-14}			
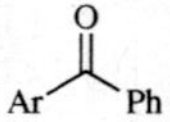	Cl_3SiH (3 eq), $(n\text{-}Pr)_3N$ (x eq), 1 h	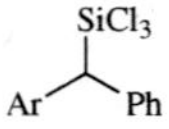	711

Ar	x	Temp	
$4\text{-}ClC_6H_4$	1	55-72°	(73)
$4\text{-}MeC_6H_4$	1.2	60-67°	(88)

TABLE 11. ORGANOSILANE REDUCTION OF ALDEHYDES (*Continued*)

Aldehyde	Conditions	Product(s) and Yield(s) (%)	Refs.
C_{14} 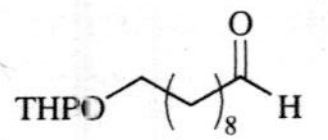	PMHS (1 eq), $ZnCl_2$ (1 eq), Et_2O, rt, 24 h	 (80)	315
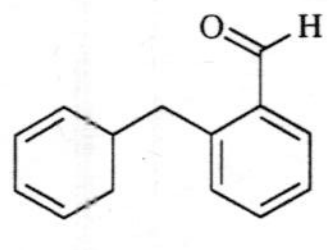	Et_3SiH (5 eq), $Ni(cod)_2$ (0.2 eq), PPh_3 (0.4 eq), THF, 55°, 8 h	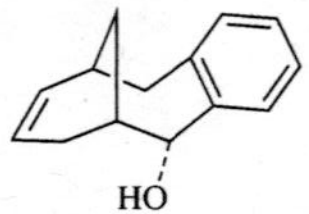 (34) + 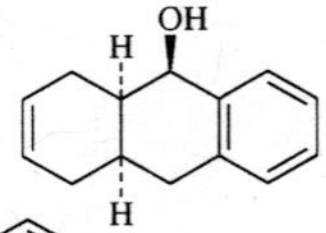 (17) + 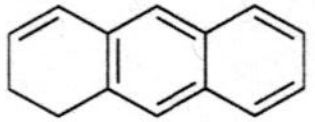 (17)	349
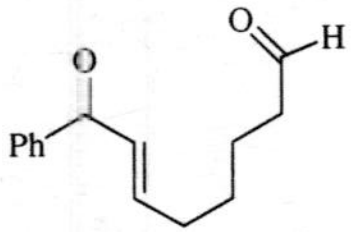	$PhSiH_3$ (1.2 eq), $Co(dpm)_2$ (5 mol%), 35°	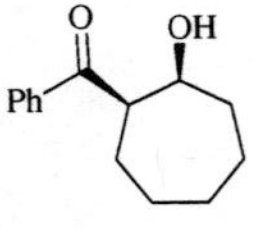 (35)	461
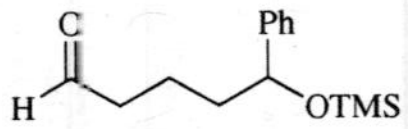	1. $SbCl_5$, TMSCl, SnI_2, CH_2Cl_2, −78°, 30 min 2. Et_3SiH (1.5 eq), −23°, 2.5 h	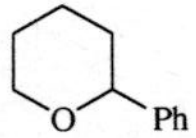 **I** (76)	306
	1. $TrSbCl_6$ (5-30 mol%), CH_2Cl_2, −78°, 30 min 2. Et_3SiH (1.5 eq), −23°, 2.5 h	**I** (90)	306
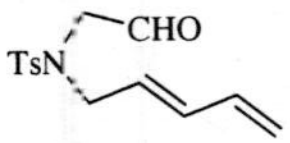	R_3SiH (5 eq), $Ni(cod)_2$ (10 mol%), **41** (20 mol %), solvent	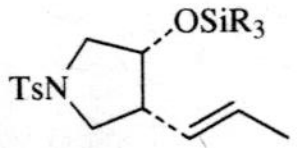 **I** + 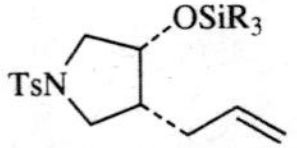 **II**	348

R_3Si	Solvent	Time	**I** + **II**	**I**:**II**	% ee **I**:**II**
$(EtO)_3Si$	MeCN	6 h	(60)	4.6:1	48:41
Ph_2MeSi	DMF	42 h	(22)	1:2.4	10:67

	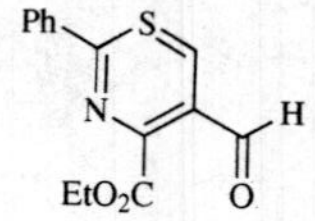	Et_3SiH, TFA	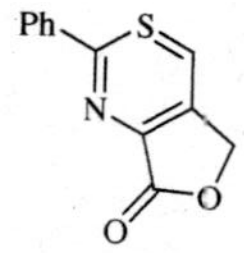 (40)	747
C_{15}	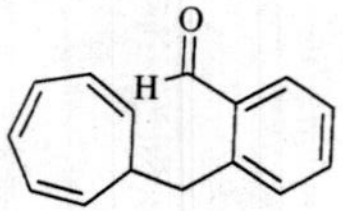	Et_3SiH (5 eq), $Ni(cod)_2$ (0.2 eq), PPh_3 (0.4 eq), THF, 35°, 16 h	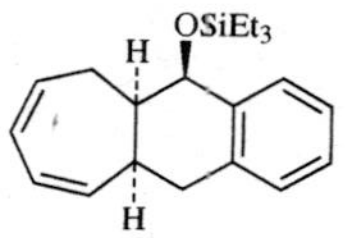 (15) + 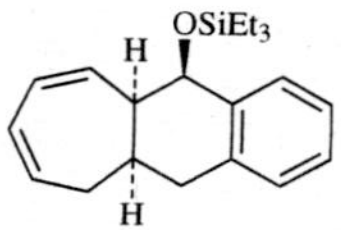 (13)	349
	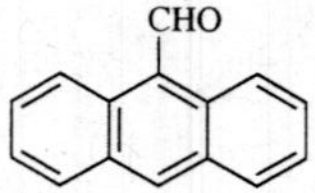	$PhMe_2SiH$ (1.3 eq), $(Ph_3P)CuH$ (3 mol%), MeC_6H_5, 2 h	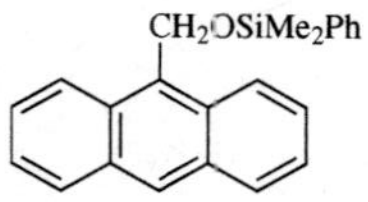 (96)	317
C_{16}	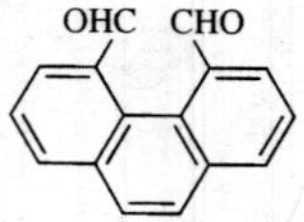	Et_3SiH (8 eq), TMSOTf (1 eq), CH_2Cl_2, rt, 45 min	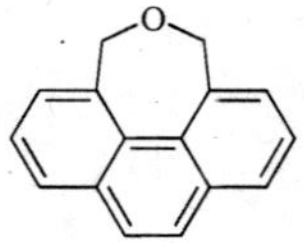 (92)	339
C_{19}	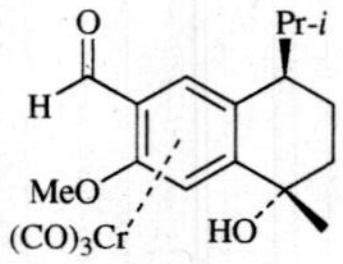	Et_3SiH (6 eq), TFA (9 eq), 60°, 4.5 h	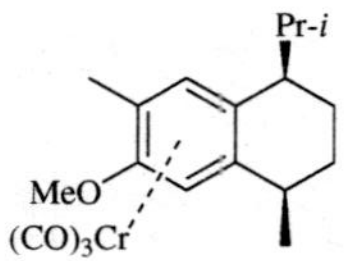 (82)	352

TABLE 11. ORGANOSILANE REDUCTION OF ALDEHYDES (*Continued*)

Aldehyde	Conditions	Product(s) and Yield(s) (%)	Refs.
C_{21}			
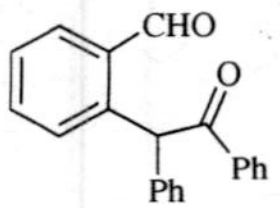	Et_3SiH (8 eq), TMSOTf (1 eq), CH_2Cl_2, rt, 45 min	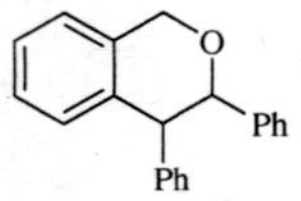 (59)	339
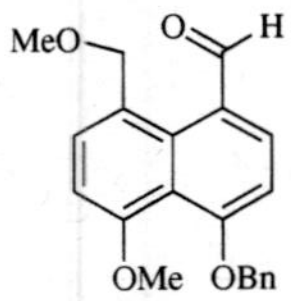	PMHS (2.5 eq), $(Ph_3P)CuH$ (3 mol%), MeC_6H_5, 1.5 h	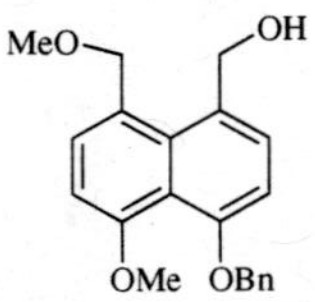 (85)	317

[a] The yield was determined by NMR spectroscopy.

[b] The product associated with this yield is $(PhCH_2OCH_2)_2$.

TABLE 12. ORGANOSILANE REDUCTION OF KETONES

	Ketone	Conditions	Product(s) and Yield(s) (%)	Refs.
C_3	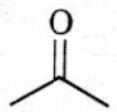	Et_3SiH (1.1-1.2 eq), HCO_2H, HOAc, KU-1, 55°, 5 h	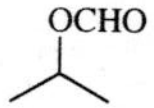 (95)	208
		Et_3SiH (1 eq), $ZnCl_2$, 100°, 24 h	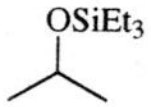 (78)	382
		Ph_2SiH_2 (1 eq), $[Cp^*(Me_3P)Ir(\eta^2\text{-}SiPh_2C_6H_4)(H)]^+$ $[(C_6F_5)_4B]^-$ (5 mol%), CD_2Cl_2, rt, 3h	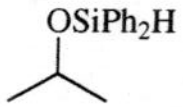 (33) + $Ph_2Si(OPr\text{-}i)_2$ (40)	745
		(*E*)-CH_3CH=$CHCO_2Me$, TMSH (1.3 eq), $RhCl_3$•$3H_2O$ (0.009 mol%), rt	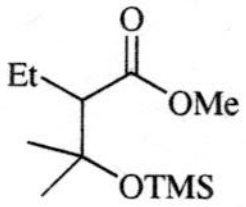 (89)	473
		CH_2=$CHCO_2Me$, TMSH (1.3 eq), $RhCl_3$•3 H_2O (0.009 mol%), rt	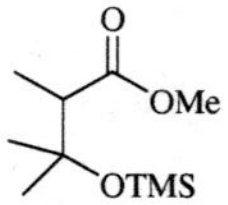 (24)	473
		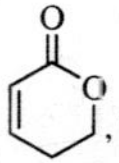, TMSH (1.3 eq), $RhCl_3$•3 H_2O (0.009 mol%), rt	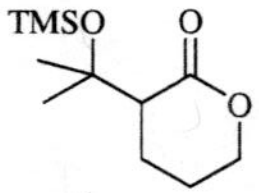 (34)	473
		Ph_2SiH_2, $(Ph_3P)_3RuCl_2$ or $(Ph_3P)_3RhCl$ (5x10^{-4} M)	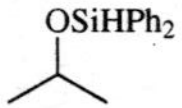 (—)	744
	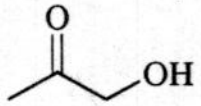	Ph_2SiH_2 (1.3 eq), $(Ph_3P)_4RhH$ (0.4 mol%), CH_2Cl_2, rt, 12 h	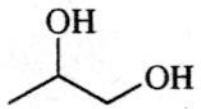 (80)	374

TABLE 12. ORGANOSILANE REDUCTION OF KETONES (*Continued*)

Ketone	Conditions	Product(s) and Yield(s) (%)	Refs.
C_{3-6} 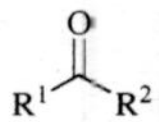	Et_3SiH (x eq), acid, H_2O (y eq), MeCN, 28°	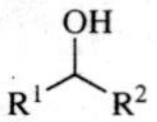 (100)	313
C_{3-8}	Et_3SiH (x eq), TFA (y eq), solvent, 28°	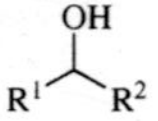 I + 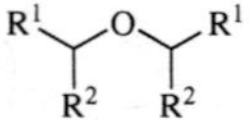 II + other III	313
C_{3-10}	Et_3SiH, $HClO_4$, MeCN, rt	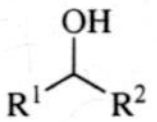 (—)	380

C_{3-6}:

R^1	R^2	x	Acid	y	Time
Me	Me	1.1	HCl	0	2.5 h
t-Bu	Me	1.25	H_2SO_4	11	1.5 h
—$(CH_2)_5$—		1.1	HCl	0	2.5 h
—$(CH_2)_5$—		1.25	H_2SO_4	11	1.5 h

C_{3-8}:

R^1	R^2	x	y	Solvent	Time	I	II	III
Me	Me	1.1	20	none	1.3 h	(32)	(68)	(0)
—$(CH_2)_5$—		1.1	20	none	1.5 h	(25)	(75)	(0)
—$(CH_2)_5$—		1.0	3.0	none	1.0 h	(42)	(58)	(0)
Ph	CH_2Br	2.5	20	none	0.25 h	(5)	(80)	(15)
Ph	CH_2Br	2.5	20	none	44 h	(0)	(12)	(88)
4-$O_2NC_6H_4$	Me	6.0	10	CCl_4	22 h	(7)	(93)	(0)

C_{3-10}:

R^1	R^2
Me	Me
Ph	Me
4-ClC_6H_4	Me
Ph	Et
4-MeC_6H_4	Me
4-NCC_6H_4	Me
Ph	n-Pr

C_{3-9}

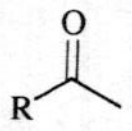

I_2SiH_2 (150 mol%) solvent, 22°

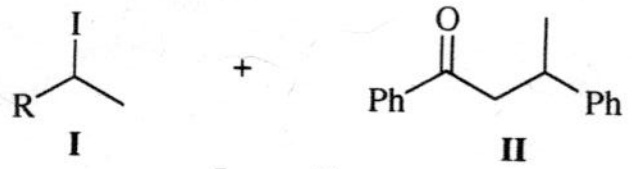

358

R	Solvent	Time	I	II
Me	$CDCl_3$	43 h	(40)	(0)
Me	$CDCl_3$	3 h	(100)	(0)
Ph	CCl_4	20 h	(38)	(12)
Bn	$CDCl_3$	48 h	(15)	(0)
Bn	$CDCl_3$	3 h	(68)	(0)

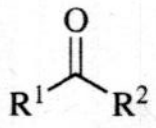

1. R^3SH, $BF_3{\bullet}OH_2$, CH_2Cl_2, 0°, 1 min
2. Et_3SiH, 0° to rt, 3 h

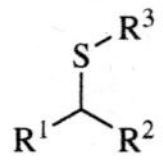

365

R^1	R^2	R^3	
Me	Me	*i*-Pr	(68)
—$(CH_2)_5$—		Ph	(76)
—$(CH_2)_5$—		*t*-Bu	(78)
Ph	Me	Ph	(88)
4-MeC_6H_4	Me	Ph	(90)
4-MeC_6H_4	Me	*i*-Pr	(83)

C_{3-13}

R_3SiH (1 eq), $(Ph_3P)_3RhCl$ (x mol%)

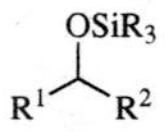

411

R^1	R^2	R_3SiH	x	Temp	Time	
Me	Me	Et_3SiH	0.5	rt	10 min	(95)
t-Bu	Me	Ph_2SiH_2	0.1	0°	10 min	(97)
—$(CH_2)_5$—		Et_3SiH	0.5	rt	5 min	(98)
—$(CH_2)_6$—		Ph_2SiH_2	0.1	0°	10 min	(98)
Ph	Me	Et_3SiH	1.0	60°	15 min	(97)
Ph	*t*-Bu	$EtMe_2SiH$	0.1	40°	300 min	(96)
Ph	Ph	Et_2SiH_2	0.1	rt	5 min	(95)

TABLE 12. ORGANOSILANE REDUCTION OF KETONES (*Continued*)

Ketone	Conditions	Product(s) and Yield(s) (%)	Refs.
$C_{3\text{-}13}$ 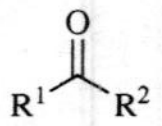	PMHS (10% xs), $(Bu_2(AcO)Sn)_2O$ (2 mol%), EtOH, reflux	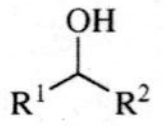	316
R^1 / R^2			
Me / Me		(100)	
$H_2C{=}CH$ / Me		(65)	
Ph / Me		(81)	
Ph / Ph		(80)	
C_4 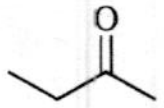	Et_3SiH (1 eq), EtOH (2 eq), TFA (4 eq), 50-60°	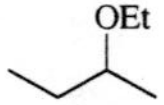 **I** (20)	327
	Et_3SiH (1 eq), $HC(OEt)_3$ (1 eq), EtOH, HCl, TFA (3 eq), 50-60°, 6-8 h	**I** (60)	327
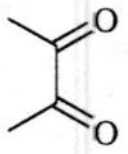	Et_3SiH (2.5 eq), $(Ph)P_3RhCl$ (0.5 mol%), 70°, 3 h	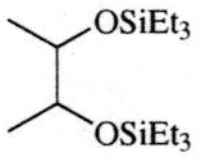 (85)	411
$C_{4\text{-}10}$ 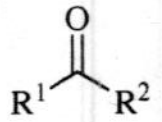	Et_3SiH (1.2 eq), R^3OTMS (1.2 eq), $BiBr_3$ (10-30 mol%), rt	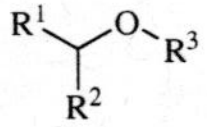	343
R^1 / R^2	R^3 / Time		
$HOCH_2$ / Et	Bn / 1 h	(—)	
$H_2C{=}CH$ / Me	Bn / 20 h	(trace)	
—$(CH_2)_5$—	Bn / <5 min	(89)	
$n\text{-}C_6H_{13}$ / Me	Bn / 1 h	(96)	
Ph / Me	Bn / 20 h	(trace)	
Bn / Et	2-octyl / 1 h	(88)	

C_{4-8}

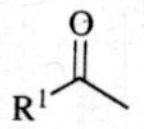

R₃SiH (1.2 eq), $BnCH_2CH_2OTMS$, $(n\text{-}Bu)_4NClO_4$ (0.1 eq), $LiClO_4$ (0.1 eq), CH_2Cl_2, electrolysis

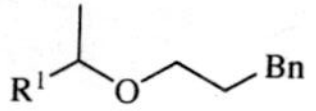

333

R^1	R_3Si	
Et	$PhMe_2Si$	(95)
MeO_2CCH_2	Et_3Si	(87)
Ph	$PhMe_2Si$	(80)

C_{4-10}

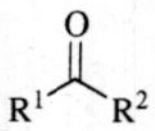

Et_3SiH (5 eq), **209** (3 mol%), CuCl (3 mol%), NaOBu-*t* (20 mol%), MeC_6H_5, rt

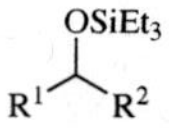

412

R^1	R^2	Time	
Me	Et	1 h	(93)
Et	Et	1 h	(82)
Me	n-C_6H_{13}	0.75 h	(96)
—$(CH_2)_5$—		1 h	(73)
—$(CH_2)_5$—		3 h	(83)
—$(CH_2)_4CH(t\text{-}Bu)$—		20 h	(35)
—$(CH_2)_4CHMe$—		1 h	(95)
Ph	Me	3 h	(94)
Ph	Et	3 h	(95)
$BnCH_2$	Me	1 h	(73)
$BnCH_2$	Me	2 h	(100)

TABLE 12. ORGANOSILANE REDUCTION OF KETONES (*Continued*)

Ketone		Conditions	Product(s) and Yield(s) (%)	Refs.
C_{4-12}				
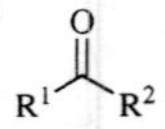		Et_3SiH (5 eq), **210** (3 mol%), CuCl (3 mol%), NaOBu-*t* (20 mol%), MeC_6H_5, rt	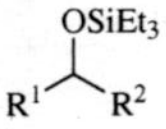	412
R^1	R^2			
Me	Et		(80)	
Et	Et		(91)	
n-C_6H_{13}	Me		(82)	
n-C_9H_{19}	Me		(86)	
—$(CH_2)_5$—			(93)	
—$(CH_2)_4$CHMe—			(91)	
c-C_3H_5	c-C_3H_5		(87)	
Ph	Me		(94)	
$BnCH_2$	Me		(85)	
Ph	Et		(96)	
1-$C_{10}H_7$	Me		(93)	
C_5				
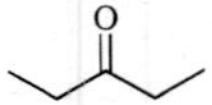		Et_3SiH (1.2 eq), **208** (0.2 mol%), rt, 15 min	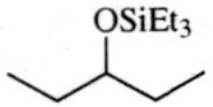 (93) + 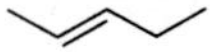 (5)	377
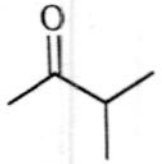		TMSH (1.1 eq), TMSI (5 mol%), CH_2Cl_2, 0°	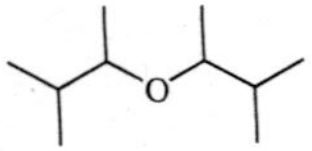 (—), dr = 100:0 (R,R:S,S)	334

Substrate	Conditions	Product(s) and Yield(s) (%)	Refs.
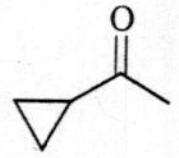	Et_3SiH (2.2 eq), **203** (0.2 mol%), rt, 168 h	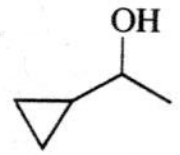 (77) + 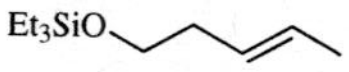 (14)	377
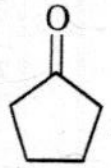	Et_3SiH (2 eq), TMSOTf (0.1 eq), CH_2Cl_2, rt, 2 h	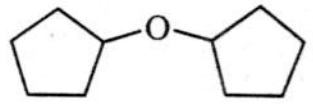 (100)[a]	334
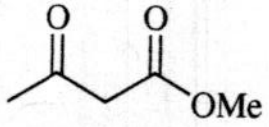	Ph_2SiH_2 (1.3 eq), $(Ph_3P)_4RhH$ (0.3 mol%), CH_2Cl_2, rt, 4 h	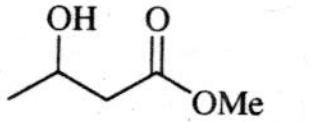 (86)	374
	Et_3SiH (1 eq), $(Ph_3P)_3RhCl$	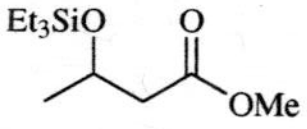 I + 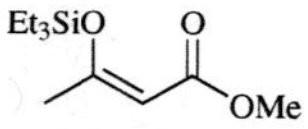 II; **I** + **II** (80), **I**:**II** = 81:19	411
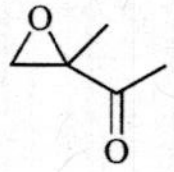	$PhMe_2SiH$ (1.2 eq), TBAF (5 mol%), HMPA, 0°, 12 h	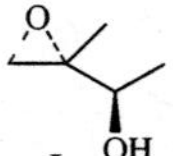 I + 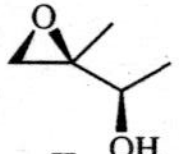 II; **I** + **II** (95), **I**:**II** = 71:29	320
C_{5-7} 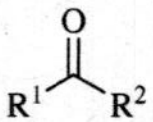	R_3SiH (1 eq), Cp_2TiPh_2 (1.6 mol%)	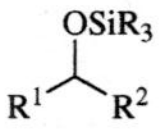	748

R^1	R^2	R_3Si	Temp	Time	Yield
Et	Et	Ph_2HSi	120°	11 h	(91)
Et	Et	PhMeHSi	90°	8 h	(66)
—$(CH_2)_4$—		Ph_2HSi	120°	11 h	(76)
Me	n-C_5H_9	PhH_2Si	120°	5 h	(60)
Me	n-C_5H_9	Ph_2HSi	120°	11 h	(91)
Me	n-C_5H_9	Ph_2HSi	120°	20 h	(88)

TABLE 12. ORGANOSILANE REDUCTION OF KETONES (*Continued*)

Ketone		Conditions	Product(s) and Yield(s) (%)	Refs.
C_{5-17} 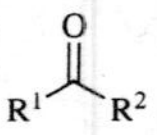		PMHS, $Sn(OTf)_2$ (10 mol%), L-histidine (10 mol%), MeOH, rt, 12-14 h	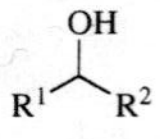	385
R^1	R^2			
CO_2Et	CF_3		(75)	
2-C_4H_3S	CO_2Et		(98)	
Ph	Me		(98)	
Ph	CO_2Me		(99)	
4-$O_2NC_6H_4$	CO_2Me		(98)	
Ph	CO_2Et		(99)	
4-$MeOC_6H_4$	CO_2Me		(99)	
4-MeC_6H_4	CO_2Me		(96)	
1-$C_{10}H_7$	CO_2Me		(98)	
2-$C_{10}H_7$	CO_2Me		(96)	
Ph	CH_2-*N*-phthalimide		(89)	
4-ClC_6H_4	CH_2-*N*-phthalimide		(50)	
4-$O_2NC_6H_4$	CH_2-*N*-phthalimide		(98)	
4-MeC_6H_4	CH_2-*N*-phthalimide		(88)	
C_6 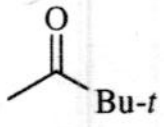		TMSH (1.1 eq), TMSI (5 mol%), CH_2Cl_2, 0°	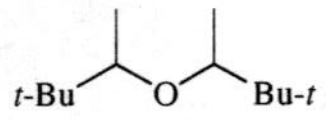 (—), dr = 100:0 (R,S:S,R)	334
		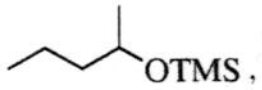, TMSH (1.1 eq), TMSI (5 mol%), CH_2Cl_2, 0°	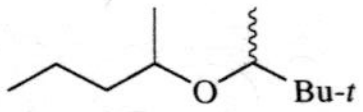 (—), dr = 54:45	334
		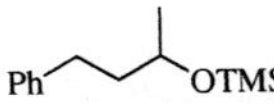, TMSH (1.1 eq), TMSI (5 mol%), CH_2Cl_2, 0°	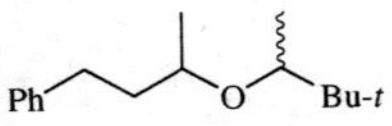 (—), dr = —	334

	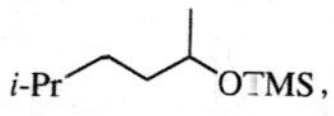, TMSH (1.1 eq), TMSI (5 mol%), CH_2Cl_2, 0°	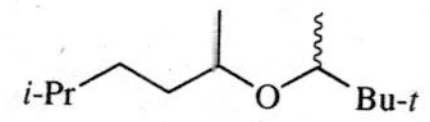 (—), dr = 50:50	334
	$(EtO)_3SiH$ (1 eq), CsF, rt, 1 h	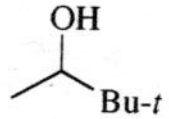 **I** (80)	80
	$(EtO)_3SiH$ (1 eq), KF, 100°, 24 h,	**I** (0)	80
	H, 0°, 5 h	**I** (92)	93
	Et_3SiH (1.1 eq), H_2SO_4, H_2O, MeCN, 28°, 65 h	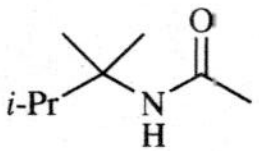 (67)	313
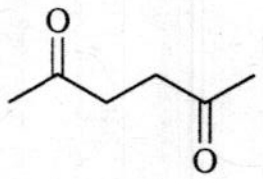	TMSH (4 eq), TMSOTf (1.6 mol%), CH_2Cl_2, 0°, 4 h; rt, 2 h	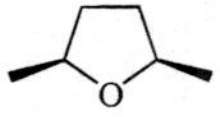 **I** + 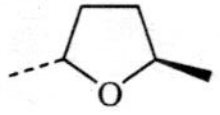 **II** **I + II** (85), **I**:**II** = 50:50	390
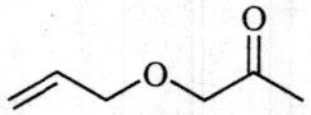	Ph_2SiH_2 (1.0 eq), $Cp_2Ti(PPh_3)_2$, MeC_6H_5, –20°	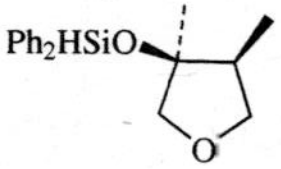 **I** + 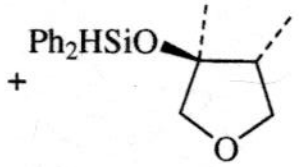 **II** + 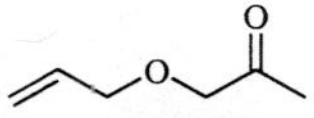 **III** **I + II**:**III** = 80:1; **I**:**II** = 3:2, **I+II** (72)	428
	Ph_2SiH_2, $Cp_2Ti(PMe_3)_2$ (10 mol%), PMe_3 (50 mol%), MeC_6H_5, –20°	**I + II**:**III** = 80:1; **I**:**II** = 3:2, **I** (72)	429
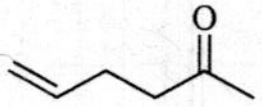	Et_3SiH (1.2 eq), **208** (0.2 mol%), rt, 19 h	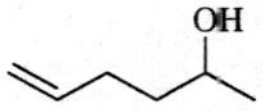 **I** (98)	377

TABLE 12. ORGANOSILANE REDUCTION OF KETONES (*Continued*)

Ketone	Conditions	Product(s) and Yield(s) (%)	Refs.
C_6 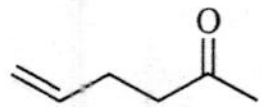	$PhMe_2SiH$ (1.2 eq), TBAF (5 eq), HMPA, rt, 12 h	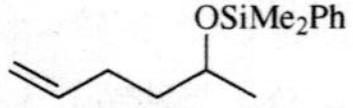 (57)	320
	Et_3SiH (3 eq), **209** (3 mol%), NaOBu-*t* (20 mol%), MeC_6H_5, rt, 1 h	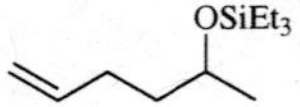 (97)	412
	$Me(EtO)_2SiH$ (1.5 eq), KF (1 eq), DMF, 60°, 5 h	**I** (85)	82, 83
	PMHS (1.5 eq), $KF•2H_2O$ (1 eq), DMSO, 60°, 1.5 h	**I** (65)	82
	$(EtO)_3SiH$ (2.3 eq), CsF (1 eq), 0°, 15 min	**I** (90)	83
	$Me(EtO)_2SiH$ (2.3 eq), KF (1 eq), DMF, 60°, 5 h	**I** (85)	82, 83
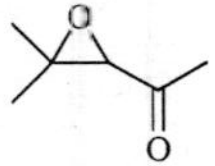	$PhMe_2SiH$ (1.2 eq), TBAF (5 mol%), HMPA, 0°, 12 h	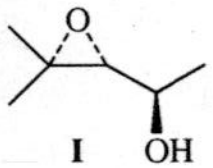 **I** + 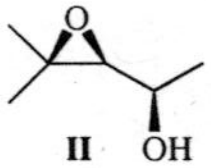 **II** (100), **I**:**II** = 65:35	320
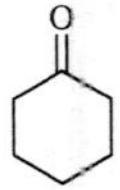	Et_3SiH (3 eq), $AlCl_3$ (0.5 eq), HCl, CH_2Cl_2, rt, 5 h	 **I** (60) + 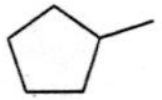 (8)	136
	Et_3SiH (1.5 eq), CF_3SO_3H (4 eq), CH_2Cl_2, 0° to rt, 2 h	**I** (22)[b] + 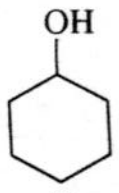 **II** (75)[b]	420
	1. Et_3SiH (2 eq), BF_3, CH_2Cl_2, 0°, 1.5 min 2. H_2O	**I** (82) + **II** (0)	1

1. Et_3SiH (4 eq), BF_3, CH_2Cl_2, 0°, 30 min 2. H_2O	**I** (0) + **II** (90)	1
$(EtO)_3SiH$ (1 eq), KF, 100°, 12 h	**II** (75)	80
$(i\text{-}Pr)_2Si(OPr\text{-}i)H$, $SnCl_4$, CH_2Cl_2, 0°, 3 h	**II** (92)	399
Silane (1.5 eq), salt (1 eq), DMF	**II**	82

Silane	Salt	Temp	Time	
$Me(EtO)_2SiH$	KF	80°	1 h	(77)
PMHS	KF	60°	1.5 h	(70)
PMHS	$KF{\cdot}2H_2O$	30°	1 h	(64)
$Me(EtO)_2SiH$	KO_2CH	80°	5 h	(75)
PMHS	KO_2CH	80°	23 h	(—)

R_3SiH (1.2 eq), $[Rh(C_2H_4)_2]_2$ (0.5 mol%), ligand (2 mol%), C_6H_6, rt	**II**	389

R_3SiH	Ligand	Time	
Et_3SiH	$P(tm\text{-}tp)_3$	21 h	(81)
Et_3SiH	$P(tp)_3$	21 h	(36)
Et_3SiH	PPh_3	21 h	(13)
$PhMe_2SiH$	$P(tm\text{-}tp)_3$	20 h	(97)
$PhMe_2SIH$	$P(tp)_3$	20 h	(37)
$PhMe_2SiH$	PPh_3	20 h	(27)
$PhMe_2SiH$	$P(tm\text{-}tp)_3$	6 h	(95)
$PhMe_2SiH$	$P(tp)_3$	6 h	(65)
$PhMe_2SiH$	PPh_3	6 h	(9)
$PhMe_2SiH$	$P(2\text{-}C_4H_3O)_3$	3 h	(22)
$PhMe_2SiH$	$P(o\text{-}tol)_3$	3 h	(32)
$PhMe_2SiH$	$PMes_3$	3 h	(25)
$PhMe_2SiH$	PEt_3	3 h	(<2)
$PhMe_2SiH$	$P(C_6H_{11}\text{-}c)_3$	3 h	(<2)
$PhMe_2SiH$	$P(Bu\text{-}t)_3$	3 h	(31)

TABLE 12. ORGANOSILANE REDUCTION OF KETONES (*Continued*)

Ketone	Conditions	Product(s) and Yield(s) (%)	Refs.
C_6 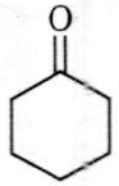	**L**, THF, 0°, 2 h	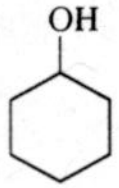 **II** (80)	96
	H, 0°, 5 h	**II** (100)	93
	G, rt, 2 h	**II** (95)	93
	Et_3SiH (1.3 eq), NH_4F (1.3 eq), TFA (5 eq), −30°, 2 h	**II** + 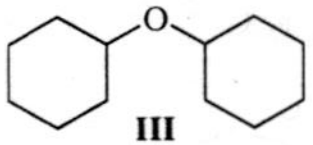 **III** **II** + **III** (—), **II**:**III** = 95:5	135
	Et_3SiH (1.1 eq), BF_3•OEt_2 (2.0 eq), rt	**II** (94) + **III** (6)	74
	Et_3SiH (2 eq), TMSOTf (0.1 eq), CH_2Cl_2, rt, 2 h	**III** (100)[a]	334
	TMSH (1.1 eq), TMSI (5 mol%), CH_2Cl_2, 0-15°; rt, 2 h	**III** (88)	334
	$Et_3SiO(EtHSiO)_nSiEt_3$ (0.2 mol%), TFA (4 eq), $CHCl_3$, 50°, 10 h	**III** (20) + 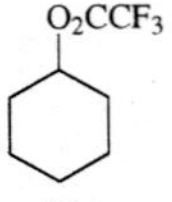 **IV** (55)	136
	$Et_3SiO(EtHSiO)_nSiEt_3$ (0.2 mol%), TFA (4 eq), $CHCl_3$, 50°, 10 h	**III** (20) + **IV** (55)	136
	R_3SiH, acid	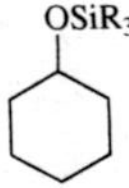 **V** + **III**	353

R_3Si	Acid	Amount[c]	Time	V	III
Et_3Si	Fe-mont	0.2 g	0.5 h	(7)	(81)
Et_3Si	CF_3SO_3H	0.1 mmol	0.5 h	(4)	(80)
Et_3Si	CF_3SO_3H	1.1 mmol	0.3 h	(67)	(4)
$PhMe_2Si$	Fe-mont	0.2 g	1.5 h	(0)	(84)
$(EtO)_3Si$	Fe-mont	0.2 g	20 h	(0)	(0)
Et_3Si	Na-mont	0.2 g	47 h	(0)	(0)
Et_3Si	SiO_2	0.2 g	21 h	(0)	(0)

n-$BuSiH_3$ (0.5 eq), TFA (x eq) **III** + **IV** 313

x	Temp	Time	III	IV
6.0	55°	3.5 h	(38)	(62)
2.5	55°	1.5 h	(47)	(53)
3.0	rt	24 h	(49)	(51)
6.0	0°	3.5 h	(60)	(40)
3.0	0°	24 h	(67)	(33)
3.0	−15°	25 h	(82)	(18)
2.5	−15°	72 h	(90)	(10)

n-$BuSiH_3$ (0.5 eq), TFA (2.5 eq), −15°, 72 h **III** (90) + **IV** (10) 313

TABLE 12. ORGANOSILANE REDUCTION OF KETONES (*Continued*)

Ketone	Conditions	Product(s) and Yield(s) (%)	Refs.
C_6 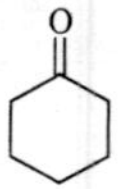	Et_3SiH (5 eq), CuCl (3 mol%), NaOBu-*t* (20 mol%), ligand (3 mol%), MeC_6H_5, rt	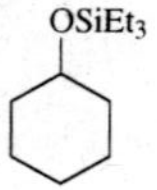 **V**	412
	Ligand / Time		
	210 / 2 h	(99)	
	211 / 2 h	(80)	
	213 / 8 h	(92)	
	212 / 8 h	(35)	
	216 / 2 h	(99)	
	215 / 8 h	(38)	
	214 / 24 h	(3)	
	none / 24 h	(0)	
	Et_3SiH (1 eq), $ZnCl_2$, 95°, 48 h	**V** (87)	382
	TBSH (3 eq), TBSOTf (0.01 eq), rt, 2 h	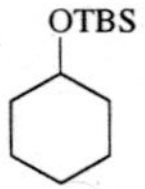 (90)	392
	R_3SiH (1.5 eq), Fe-mont, CH_2Cl_2	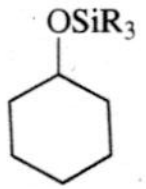	324
	R_3Si / Temp / Time		
	Et_3Si / rt / 1 h	(79)	
	$PhMe_2Si$ / 40° / 6 h	(—)	
	$(EtO)_3Si$ / 40° / 11 h	(—)	

Et_3SiH (1.1 eq), MeOH, TFA, 28°	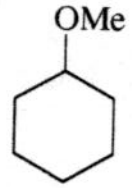	(88)	328
Et_3SiH (1.1 eq), EtOH, TFA, 28°	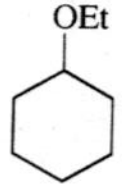 **I**	(68)	328
Et_3SiH (1.1 eq), EtOH, H_2SO_4, 28°	**I** (44)		328
Et_3SiH (1 eq), EtOH (3 eq), TFA (5 eq), 50-60°	**I** (16)		327
Et_3SiH (1 eq), $HC(OEt)_3$ (1 eq), EtOH, HCl, TFA (3 eq), 50-60°, 6-8 h	**I** (72)		327
TMSH (1.1 eq), TMSI (5 mol%), EtOTMS (1 eq), CH_2Cl_2, 0-15°; rt, 2 h	**I** (91)		334
Et_3SiH (1 eq), $HC(OEt)_3$ (1 eq), EtOH, TFA, 50-60°, 6-8 h	**I** (55)		327
Et_3SiH (1 eq), TMSOTf (0.1 eq), ROTMS (0.83 eq), CH_2Cl_2, −78° to −30°	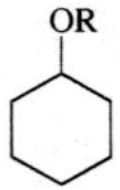		341
R			
Bn	(100)		
c-C_6H_{11}	(95)		
t-Bu	(35)		

TABLE 12. ORGANOSILANE REDUCTION OF KETONES (*Continued*)

Ketone	Conditions	Product(s) and Yield(s) (%)	Refs.
C_6 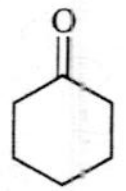	Et_3SiH, TMSOTf, TMSI, n-$C_8H_{17}OTMS$, CH_2Cl_2, 0°, 2 h	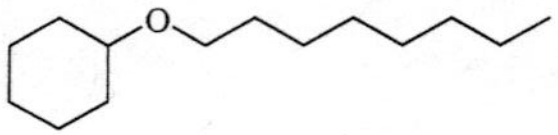 (100)[a]	334
	TMSH (1.1 eq), TMSI (5 mol%), c-$C_6H_{11}OTMS$ (1 eq), CH_2Cl_2, 0-15°; rt, 2 h	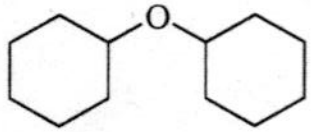 (93)	334
	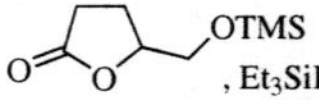 , Et_3SiH (1 eq), TMSOTf (0.1 eq), CH_2Cl_2, 0° to rt	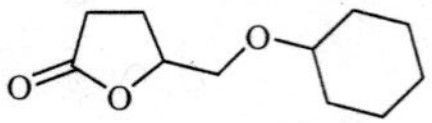 (93)	341
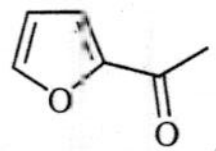	Et_3SiH (10 eq), TFA (x eq), BF_3•OEt_2 (y eq), 20°	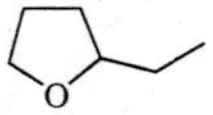 **I**	211
	x / y / Time		
	30 / 0 / 5 min	(17)	
	30 / 0 / 30 min	(31)	
	30 / 0 / 1 h	(46)	
	30 / 0 / 2 h	(52)	
	30 / 0 / 20 h	(65)	
	50 / 0 / 5 min	(34)	
	50 / 0 / 30 min	(67)	
	50 / 0 / 1 h	(70)	
	50 / 0 / 2 h	(79)	
	28 / 0.6 / 5 min	(80)	
	28 / 0.6 / 1 h	(80)	
	$PhMe_2SiH$ (1.1 eq), CsF (13.4 mol%), 18-C-6 (6.7 mol%), CH_2Cl_2, rt, 10 h	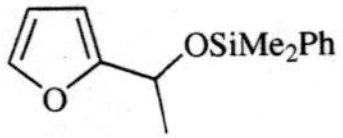 (61)	346, 347

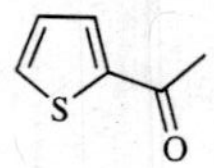	Et_3SiH (5 eq), TFA (10 eq), 50°, 50 h	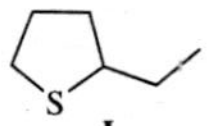 **I** (55)	257
	Et_3SiH (3 eq), $AlCl_3$ (1.3 eq), CH_2Cl_2, 20°, 1.5 h	**I** (70)	259
	$PhMe_2SiH$ (1.1 eq), CsF (5 mol%), 18-C-6 (2.5 mol%), CH_2Cl_2, rt, 11 h	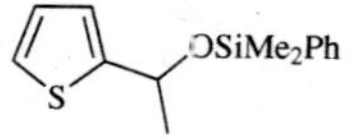 (45)	346
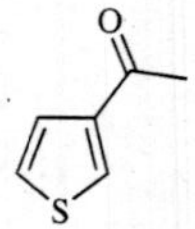	$PhMe_2SiH$ (1.1 eq), CsF (5 mol%), 18-C-6 (5 mol%), CH_2Cl_2, rt, 10 h	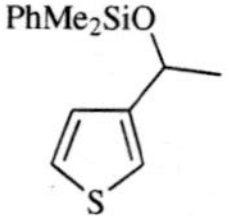 (53)	346
C_{6-9} 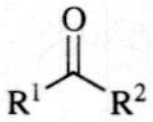	Et_3SiH (1.2 eq), TMSCl (50 mol%), $InCl_3$ (20 mol%), R^3STMS (1.2 eq), CH_2Cl_2, rt, 5 h	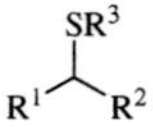	426
R^1 / R^2	R^3		
—$(CH_2)_5$—	Et	(93)	
Ph / H	Et	(87)	
Ph / H	*i*-Pr	(83)	
Ph / H	Ph	(82)	
$BnCH_2$ / H	Et	(81)	
Ph / H	Et	(98)	
n-C_8H_{17} / Me	Et	(78)	
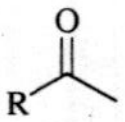	Et_3SiH (1.3 eq), $BiCl_3$ (0.11 eq), CH_2Cl_2, rt, 3 h	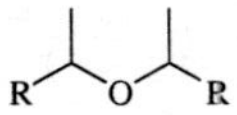	332
R			
n-Bu		(29)	
Ph		(trace)	
$BnCH_2$		(50)	

TABLE 12. ORGANOSILANE REDUCTION OF KETONES (*Continued*)

Ketone	Conditions	Product(s) and Yield(s) (%)	Refs.
C_{6-13} 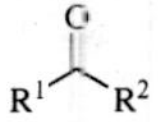	R_3SiH (1.5 eq), catalyst, n-C_7H_{16}, 90°	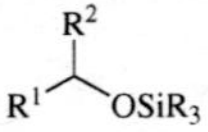	353

R^1	R^2	R_3Si	Catalyst	Time	Yield (%)
—$(CH_2)_5$—		$(EtO)_3Si$	HAp	1 h	(99)
—$(CH_2)_5$—		$(EtO)_3Si$	CaO	2 h	(61)
—$(CH_2)_5$—		$PhMe_2Si$	CaO	8 h	(0)
—$(CH_2)_5$—		Et_3Si	CaO	8 h	(0)
—$(CH_2)_5$—		$(EtO)_3Si$	CaO	2 h	(24)
n-C_6H_{13}	Me	$(EtO)_3Si$	HAp	1 h	(93)
n-C_6H_{13}	Me	$(EtO)_3Si$	CaO	2 h	(71)
Ph	Me	$(EtO)_3Si$	HAp	5 h	(72)
Ph	Me	$(EtO)_3Si$	HAp	20 h	(0)
Ph	Me	$PhMe_2Si$	HAp	23 h	(0)
Ph	Me	Et_3Si	CaO	7 h	(70)
Ph	Me	$(EtO)_3Si$	$Ca(OH)_2$	21 h	(0)
Ph	Me	$(EtO)_3Si$	CaF_2	21 h	(0)
PhCH=CH	Me	$(EtO)_3Si$	HAp	3 h	(94)
Ph	Ph	$(EtO)_3Si$	HAp	1 h	(82)
Ph	Ph	$(EtO)_3Si$	CaO	1 h	(72)

Ketone	Conditions	Product(s) and Yield(s) (%)	Refs.
C_{6-10} 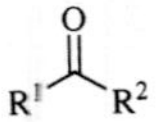	1. BF_3 (gas), CH_2Cl_2, 0° 2. Et_3SiH (x eq), 0°	 I + II	1

R^1	R^2	x	Time	I	II
—$(CH_2)_5$—		2	1.5 min	(82)	(—)
—$(CH_2)_5$—		4	30 min	(—)	(90)
—$(CH_2)_4CHMe$—		2.2	60 min	(—)	(88)
4-$O_2NC_6H_4$	Me	4	3 min	(100)	(—)
4-NCC_6H_4	Me	3	10 min	(100)	(—)
n-C_8H_{17}	Me	3.3	60 min	(—)	(80)
adamantanone		2.2	60 min	(—)	(100)

C_{6-10}

Et_3SiH (1.3 eq), $BiCl_3$ (0.11 eq), R^3OH, CH_2Cl_2, rt

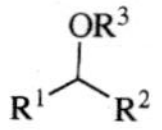

332

R^1	R^2	R^3	
—$(CH_2)_5$—		$BnCH_2CH_2$	(75)
Ph	Me	$BnCH_2CH_2$	(29)
$BnCH_2$	Me	Et	(24)

C_{6-11}

R_3SiH (1.1 eq), CsF (1 eq)

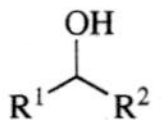

79

R^1	R^2	R_3Si	Temp	Time	
$H_2C{=}C(CH_2)_2$	Me	$(EtO)_3Si$	0°	15 min	(90)
Ph	MeCHBr	$(EtO)_3Si$	70°	30 min	(70)
Me	$PhNHCOCH_2$	$(EtO)_3Si$	rt	10 h	(90)
Ph	$MeO_2C(CH_2)_2$	$Me(EtO)_2Si$	rt	2.5 min	(85)
i-$PrO_2C(CH_2)_2$	*i*-PrO_2C	$(EtO)_3Si$	0°	30 min	(70)

C_{6-13}

Et_3SiH (1.2 eq), $BiBr_3$ (10-30 mol%), rt

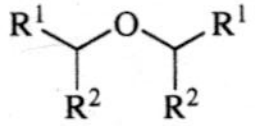

343

R^1	R^2	Time	
—$(CH_2)_5$—		2 h	(72)
Ph	Me	20 h	(trace)
$BnCH_2$	Et	2 h	(61)
Ph	Ph	24 h	(trace)

TABLE 12. ORGANOSILANE REDUCTION OF KETONES (*Continued*)

Ketone	Conditions	Product(s) and Yield(s) (%)	Refs.
C_{6-13} 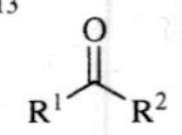	Ph_2SiH_2 (0.5 eq), salt	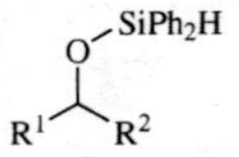 I + 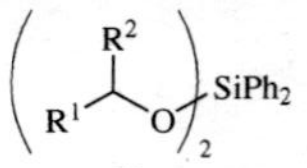 II	319

R^1	R^2	Salt	Temp	Time	I	II
—$(CH_2)_5$—		HCO_2K	100°	24 h	(70)	(0)
Ph	Me	HCO_2K	180°	4 h	(25)	(0)
Ph	Me	KF	100°	4 h	(40)	(0)
Ph	Ph	KF	100°	24 h	(30)	(0)
Ph	Ph	KF	150°	13 h	(0)	(30)
—$(CH_2)_5$—		KF	100°	15 h	(70)	(0)
Ph	Me	CsF	rt	5 min	(0)	(85)
Ph	Ph	CsF	rt	30 min	(0)	(100)
—$(CH_2)_5$—		CsF	rt	15 min	(0)	(99)
Ph	Me	$BnMe_3NF$	rt	5 min	(0)	(90)
Ph	Ph	$BnMe_3NF$	rt	5 min	(0)	(95)
—$(CH_2)_5$—		$BnMe_3NF$	rt	10 min	(0)	(85)

$(MeO)_3SiH$ (1.3 eq), $LiOCMe_2CMe_2OLi$ (2.4 eq), Et_2O, rt — 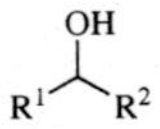 — Refs. 91

R^1	R^2	Time	Yield
$2-C_4H_3O$	Me	15 h	(68)
$n-C_6H_{13}$	Me	15 h	(81)
—$(CH_2)_5$—		15 h	(83)
—$(CH_2)_6$—		15 h	(76)
Ph	Me	20 h	(97)
$4-BrC_6H_4$	Me	15 h	(94)
Ph	Et	15 h	(93)
Ph	Ph	15 h	(79)
2-tetralone		15 h	(86)

C_{6-13}

$(EtO)_3SiH$, catalyst, solvent, 90°

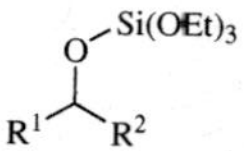

R^1	R^2	Catalyst	Solvent	Time	Product	Refs.
—$(CH_2)_5$—		HAp	n-C_7H_{16}	1 h	(99)	324
Ph	Ph	HAp	n-C_7H_{16}	1 h	(82)	
Me	Ph	HAp	n-C_7H_{16}	5 h	(72)	
Ph	EtO	HAp	n-C_7H_{16}	20 h	(97)	
n-C_5H_{11}	MeO	HAp	n-C_7H_{16}	24 h	(44)	
Me	PhCH=CH	HAp	n-C_7H_{16}	3 h	(94)	
—$(CH_2)_5$—		CaO	n-C_7H_{16}	2 h	(61)	
—$(CH_2)_5$—		CaO	MeC_6H_5	3 h	(21)	
—$(CH_2)_5$—		CaO	CH_2Cl_2	2 h	(0)	
Ph	Ph	CaO	n-C_7H_{16}	1 h	(72)	
Ph	Ph	CaO	MeC_6H_5	14 h	(47)	
Me	Ph	CaO	n-C_7H_{16}	17 h	(70)	
Me	Ph	CaO	MeC_6H_5	13 h	(46)	
Ph	Ph	MgO	n-C_7H_{16}	15 h	(35)	
Me	Ph	$Mg(OH)_2$	n-C_7H_{16}	20 h	(0)	
Me	Ph	$Ca(OH)_2$	n-C_7H_{16}	21 h	(0)	
Me	Ph	CaF_2	n-C_7H_{16}	21 h	(0)	

Et_3SiH (x eq), catalyst, rt

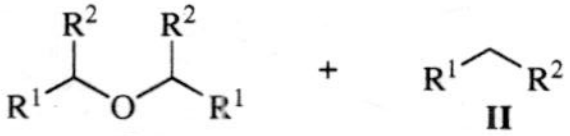

I + II

R^1	R^2	Catalyst	x	Time	I	II	Refs.
—$(CH_2)_5$—		Fe-mont	1.1	0.5 h	(81)	(0)	353
—$(CH_2)_5$—		CF_3SO_3H	1.1	0.5 h	(80)	(0)	
Ph	Me	Fe-mont	2.2	4 h	(0)	(72)	
Ph	Ph	Fe-mont	2.2	60 h	(0)	(98)	
Ph	Ph	Sn-mont	2.2	2 h	(0)	(98)	
Ph	Ph	CF_3SO_3H	2.2	24 h	(0)	(94)	
Ph	Ph	CF_3SO_3H	2.2	24 h	(0)	(94)	
Ph	Ph	Sn-mont/Et_3N	2.2	2 h	(0)	(1)	

TABLE 12. ORGANOSILANE REDUCTION OF KETONES (*Continued*)

Ketone	Conditions	Product(s) and Yield(s) (%)	Refs.
C_7 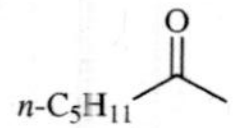	**H**, 0°, 5 h	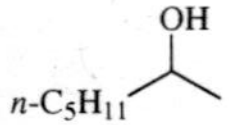 (90)	93
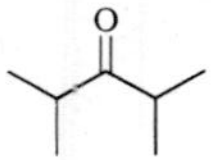	**H**, 0°, 30 h	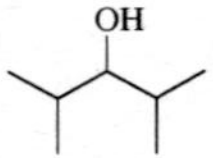 (50)	93
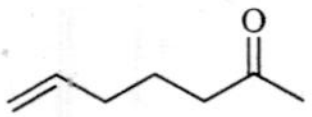	Ph_2SiH_2 (1.0 eq), $Cp_2Ti(PPh_3)_2$, MeC_6H_5, –20°	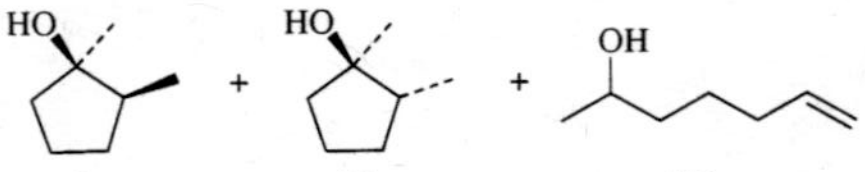 **I** + **II** + **III**; **I + II**:**III** = 17:1; **I**:**II** = 43:1, **I + II** (64)	428, 429
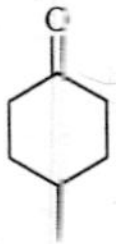	R_3SiH (1.1 eq), acid catalyst (1 eq), 0° to rt, 24 h	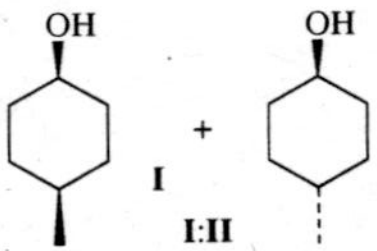 **I** + **II**	72, 381, 384

R_3Si	Acid Catalyst	**I**:**II**
n-BuH_2Si	$BF_3•OEt_2$	19:81
Et_2HSi	$BF_3•OEt_2$	39:61
Et_3Si	$BF_3•OEt_2$	60:40
n-BuH_2Si	H_2SO_4	18:82
Et_2HSi	H_2SO_4	26:74
Et_3Si	H_2SO_4	35:65
Et_3Si	TFA (6.8 eq)	36:64
$(n$-$Bu)_3Si$	H_2SO_4	25:75
$(i$-$C_5H_{11})_3Si$	TFA (6.7 eq)	33:67
$(c$-$C_5H_9)_3Si$	TFA (7 eq)	48:52
$(i$-$Bu)_3Si$	TFA (7 eq)	56:44
$(s$-$Bu)_3Si$	TFA (7 eq)	58:42
$(t$-$Bu)_3Si$	TFA (7.2 eq)	67:33

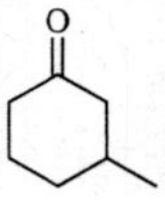

Et_3SiH (1.1 eq), acid, rt	**I + II** (100)		74

Acid	I:II
$BF_3•OEt_2$	60:40
TFA	36:64
H_2SO_4	35:65

Silane, catalyst — **I + II** (>90) — 278

Silane	Catalyst	I:II
PMHS	Triton-B	18:82
PMHS	TBAF	16:84
$(TMSO)_3SiH$	Triton-B	16:84
$(TMSO)_3SiH$	TBAF	14:86

$PhSiH_3$ (0.4 eq), $Mn(dpm)_3$ (3 mol%), *i*-PrOH, DCE, O_2, rt — **I + II** (98), **I**:**II** 13:1 — 367

$(t\text{-Bu})_2SiH_2$ (1 eq), TFA (6.6 eq), rt, 20h — 386

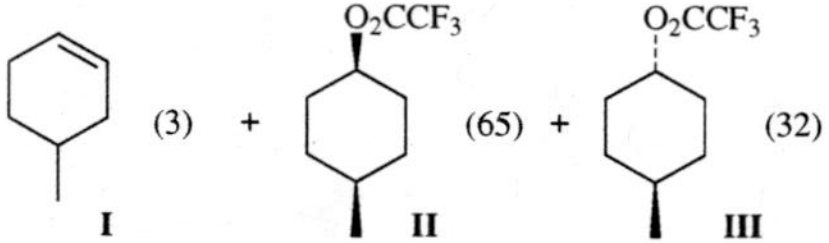

I (3) + **II** (65) + **III** (32)

$(t\text{-Bu})_2MeSiH$ (1 eq), TFA (6.6 eq), rt, 20 h — **I** (5) + **II** (62) + **III** (33) — 386

Silane, catalyst — 278

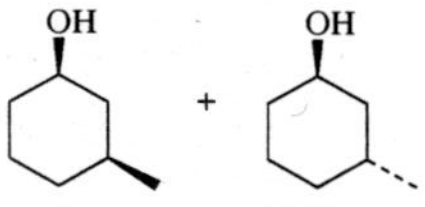

I + **II** — **I + II** (>90)

Silane	Catalyst	I:II
PMHS	Triton-B	33:67
PMHS	TBAF	30:70
$(TMSO)_3SiH$	Triton-B	15:85
$(TMSO)_3SiH$	TBAF	7:93

TABLE 12. ORGANOSILANE REDUCTION OF KETONES (*Continued*)

Ketone	Conditions	Product(s) and Yield(s) (%)	Refs.
C_7 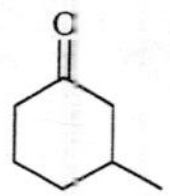	R_3SiH (1.1 eq), acid catalyst (1 eq), 0° to rt, 24 h	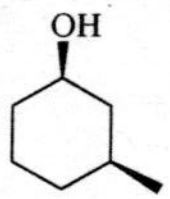 **I** + 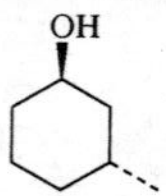 **II**	72, 381, 384

R_3Si	Acid Catalyst	I:II
n-BuH_2Si	BF_3•OEt_2	76:24
Et_2HSi	BF_3•OEt_2	52:48
Et_3Si	BF_3•OEt_2	33:67
n-BuH_2Si	H_2SO_4	81:19
Et_2HSi	H_2SO_4	71:29
Et_3Si	H_2SO_4	61:39
Et_3Si	TFA (6.8 eq)	58:42
$(n\text{-}Bu)_3Si$	H_2SO_4	65:35
$(n\text{-}C_6H_{13})_3Si$	H_2SO_4	68:32
$(i\text{-}C_5H_{11})_3Si$	TFA (6.7 eq)	62:38
$(c\text{-}C_5H_9)_3Si$	TFA (7 eq)	49:51
$(i\text{-}Bu)_3Si$	TFA (7 eq)	39:61
$(s\text{-}Bu)_3Si$	TFA (7 eq)	33:67
$(t\text{-}Bu)_3Si$	TFA (7.2 eq)	27:73

Ketone	Conditions	Product(s) and Yield(s) (%)	Refs.
	Et_3SiH (1.1 eq), acid, rt	**I** + **II** (100)	74

Acid	I:II
BF_3•OEt_2	33:67
TFA	58:42
H_2SO_4	61:39

Ketone	Conditions	Product(s) and Yield(s) (%)	Refs.
	$PhSiH_3$ (0.4 eq), $Mn(dpm)_3$ (3 mol%), i-PrOH, DCE, O_2, rt	**I** + **II** (81), **I**:**II** 1:9	367

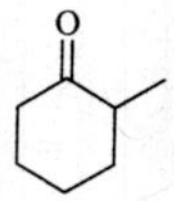

R_3SiH (1.1 eq),
acid catalyst, 0° to rt, 24 h

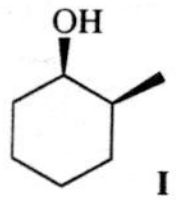
I
+
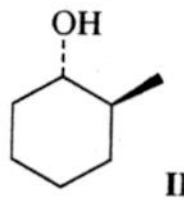
II

381, 384

R_3Si	Acid Catalyst	I:II
n-BuH_2Si	BF_3•OEt_2 (1 eq)	34:66
Et_2HSi	BF_3•OEt_2 (1 eq)	52:48
Et_3Si	BF_3•OEt_2 (1 eq)	64:36
n-BuH_2Si	H_2SO_4 (1 eq)	31:69
Et_2HSi	H_2SO_4 (1 eq)	41:59
Et_3Si	H_2SO_4 (1 eq)	54:46
Et_3Si	TFA (6.8 eq)	52:48
(*n*-Bu)$_3$Si	H_2SO_4 (1 eq)	46:54
n-PrH_2Si	H_2SO_4 (1 eq)	49:51
(*n*-C_6H_{13})$_3$Si	TFA (1 eq)	56:44
(*i*-C_5H_{11})$_3$Si	TFA (6.7 eq)	52:48
(*c*-C_5H_9)$_3$Si	TFA (7 eq)	41:59
(*i*-Bu)$_3$Si	TFA (7 eq)	38:62
(*s*-Bu)$_3$Si	TFA (2.3 eq)	36:64
(*s*-Bu)$_3$Si	TFA (7 eq)	36:64
(*t*-Bu)$_3$Si	TFA (7.2 eq)	+33:67

Silane, catalyst

I + II (>90)

278

Silane	Catalyst	I:II
PMHS	Triton-B	33:67
PMHS	TBAF	13:87
(TMSO)$_3$SiH	Triton-B	30:70
(TMSO)$_3$SiH	TBAF	18:82

TABLE 12. ORGANOSILANE REDUCTION OF KETONES (*Continued*)

Ketone	Conditions	Product(s) and Yield(s) (%)	Refs.
C_7 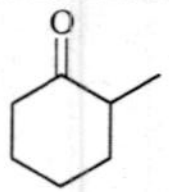	R_3SiH (x eq), TBAF (5 eq), HMPA	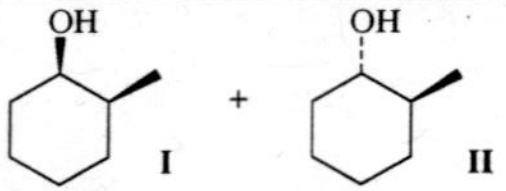	320
	R_3Si / x / Temp / Time: $PhMe_2Si$ 1.2 0° 24 h; Ph_2HSi 1.0 0° 5 h; Ph_2MeSi 1.2 0° 24 h; Ph_3Si 1.2 rt 12 h	**I + II** / **I:II**: (99) 76:24; (74) 86:14; (81) 94:6; (40) 95:5	
	Et_3SiH (1.1 eq), acid, rt	**I + II** (100)	74
	Acid: BF_3•OEt_2; TFA; H_2SO_4	**I:II**: 64:36; 48:52; 54:46	
	$PhSiH_3$ (0.4 eq), $Mn(dpm)_3$ (3 mol%), *i*-PrOH, DCE, O_2, rt	**I + II** (87), **I:II** = 1:2.1	367
	Et_3SiH (1 eq), $ZnCl_2$, reflux, 24 h	**I + II** (50), **I:II** = —	382
	Ph_3SiH, $(C_6F_5)_3B$ (2 mol%), MeC_6H_5, rt	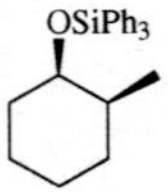 (75)	116
	TBSH (3 eq), TBSOTf (0.01 eq), rt, 2 h	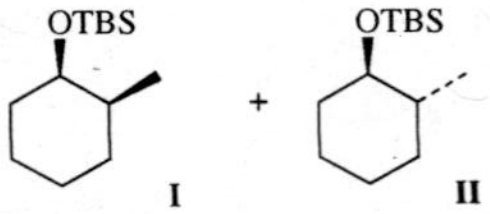 **I + II** (85), **I:II** = 77:23	392
	TBSH (1.2 eq), CuCl (0.5 mol%), NaOMe (3 mol%), **123** (0.1 mol%), Et_2O, rt, <3 h	**I + II** (99), **I:II** = 1.4:1	749

	Et_3SiH (5 eq), CuCl (3 mol%), NaOBu-*t* (3 mol%), *o*-dppb (0.05 mol%), MeC_6H_5	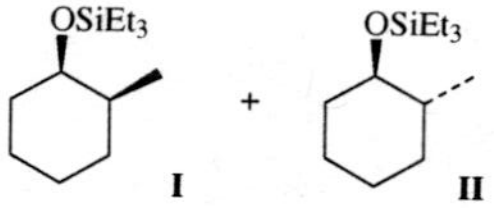 **I** + **II** (93), **I**:**II** 1:4	749
	$(t\text{-}Bu)_2MeSiH$ (1 eq), TFA (6.6 eq), rt, 20 h	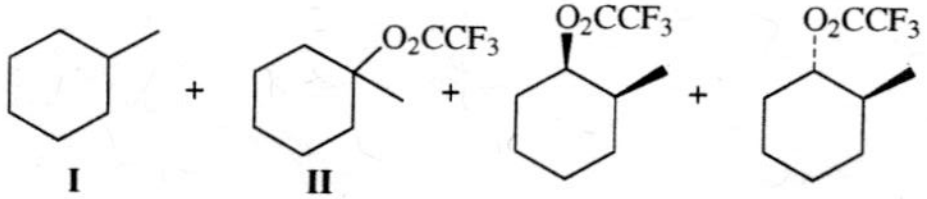 **I** + **II** (60) **III** (14) **IV** (26)	386
	$(t\text{-}Bu)_2SiH_2$ (1 eq), TFA (6.5 eq), rt, 20 h	**I** (30) + **III** (46) + **IV** (24)	386
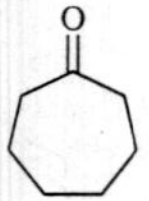	Et_3SiH (2 eq), TMSOTf (0.1 eq), CH_2Cl_2, rt, 2 h	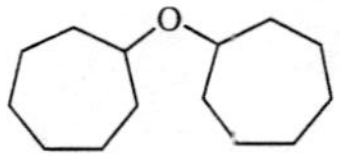 (100)[a]	334
	$PhSiH_3$ (0.4 eq), $Mn(dpm)_3$ (3 mol%), *i*-PrOH, DCE, O_2, rt	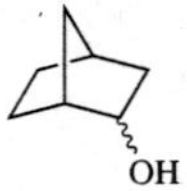 (54) endo:exo = 14:1	367
	Et_3SiH (5 eq), CuCl (3 mol%), NaOBu-*t* (20 mol%), **210** (3 mol%), MeC_6H_5, rt	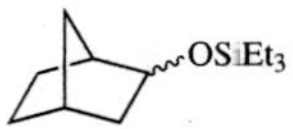 (88)	412
	Et_3SiH (1.1 eq), H_2SO_4, H_2O, MeCN, 28°, 65 h	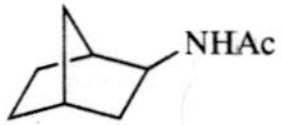 (78)	313

TABLE 12. ORGANOSILANE REDUCTION OF KETONES (*Continued*)

Ketone	Conditions	Product(s) and Yield(s) (%)	Refs.
C_7 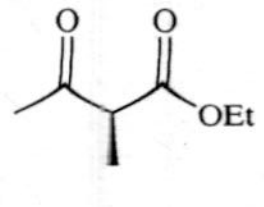	*$PhMe_2SiH$ (1.2 eq),* TFA, 0°, 10 h	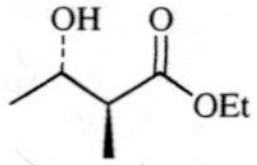 I + 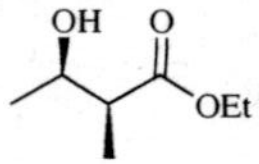 II **I + II** (90), **I**:**II** = 1:1	87, 276
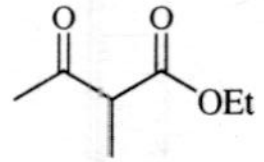	TMSH (1.1 eq), TMSI (5 mol%), n-$C_5H_{11}OTMS$ (1 eq), CH_2Cl_2, 0-15°; rt, 2 h	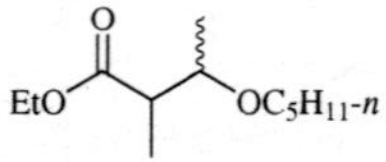 (90)	334
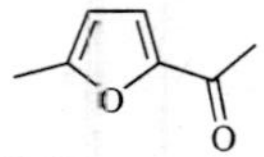	$PhMe_2SiH$ (1.1 eq), CsF (5 mol%), 18-C-6 (2.5 mol%), CH_2Cl_2, rt, 1 h	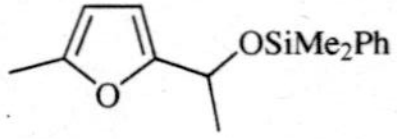 (85)	346
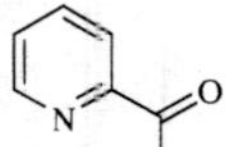	$PhMe_2SiH$ (1.1 eq), CsF (10 mol%), 18-C-6 (0.5 mol%), CH_2Cl_2, rt, 5.5 h	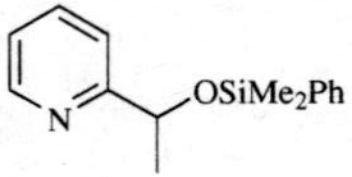 (67)	345
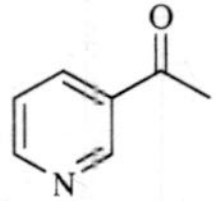	$PhMe_2SiH$ (1.1 eq), CsF (10 mol%), 18-C-6 (0.5 mol%), CH_2Cl_2, rt, 15 h	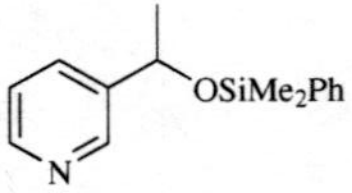 (60)	345
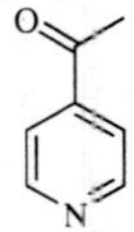	$PhMe_2SiH$ (1.1 eq), CsF (10 mol%), 18-C-6 (0.5 mol%), CH_2Cl_2, rt, 3 h	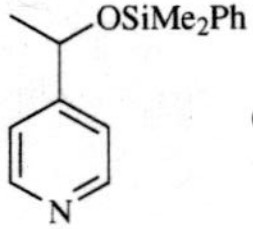 (71)	345
C_{7-10} 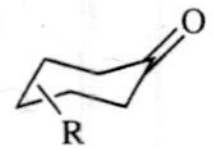	Et_3SiH (0.8 eq), TFA (6.5 eq), rt, 24 h	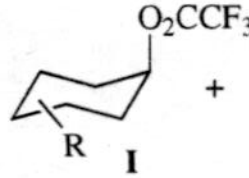 I + 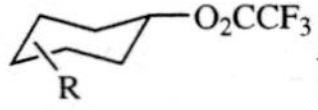 II + 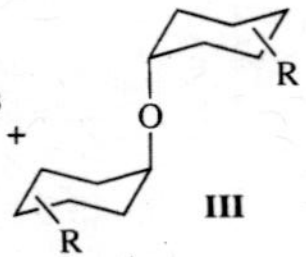 III + 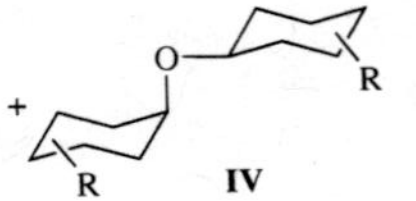 IV	72

R	I + II	I:II	III + IV	III:IV
4-Me	(83)	0.56	(17)	—
3-Me	(87)	0.72	(13)	—
2-Me	(>99)	0.92	(<1)	—
3,3,5-Me_3	(84)	8.1	(16)	3.65
4-*t*-Bu	(83)	0.47	(17)	1.04

$C_{7\text{-}11}$

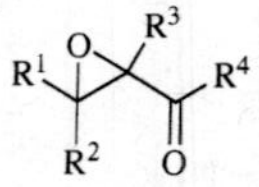

$(MeO)_3SiH$ (1.2 eq), LiOMe (0.04 eq), solvent

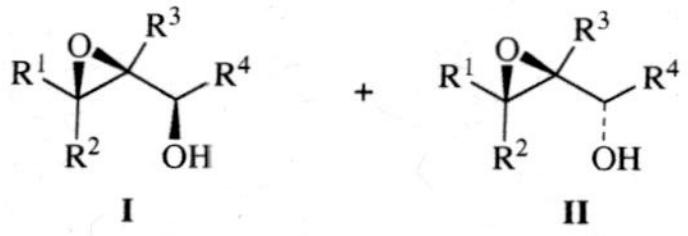

I + II

366

R^1	R^2	R^3	R^4	Solvent	Temp	Time	I + II	I:II
H	H	H	*n*-Bu	Et_2O	−20°	2 h	(78)	11:89
H	H	H	*n*-Bu	HMPA	0°	23 h	(80)	81:19
H	H	Me	*n*-Bu	Et_2O	−20°	2 h	(84)	11:89
H	H	Me	*n*-Bu	HMPA	0°	23 h	(90)	44:56
H	H	H	Ph	Et_2O	−20°	9 h	(100)	8:92
H	H	H	Ph	HMPA	0°	22 h	(98)	90:10
H	H	Me	Ph	Et_2O	−20°	18 h	(91)	34:66
H	H	Me	Ph	HMPA	0°	22 h	(91)	72:28
Me	H	H	Ph	Et_2O	−20°	6 h	(99)	9:91
Me	H	H	Ph	HMPA	0°	22 h	(100)	93:7
Me	Me	H	Ph	Et_2O	−20°	2 h	(88)	0:100
Me	Me	H	Ph	HMPA	0°	22 h	(100)	60:40

TABLE 12. ORGANOSILANE REDUCTION OF KETONES (*Continued*)

Ketone	Conditions	Product(s) and Yield(s) (%)	Refs.
C_{7-12}			
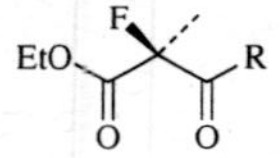	$PhMe_2SiH$ (1.25 eq), $AlCl_3$, CH_2Cl_2, rt	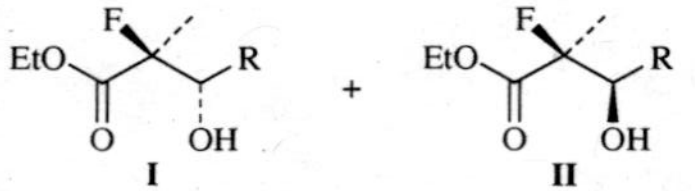 I + II	90
R	Temp Time	I + II I:II	
Me	0° 2 h	(45) 98:2	
Et	rt 2 h	(65) 96:4	
n-Pr	rt 3 h	(74) 99:1	
i-Pr	rt 12 h	(45) 98:2	
n-Bu	rt 3 h	(71) 97:3	
Ph	rt 4 h	(65) 99:1	
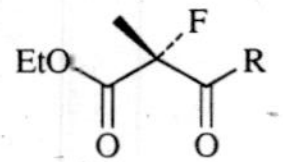	$PhMe_2SiH$ (1.25 eq), TBAF (0.05 eq), DMF	 I + II	90
R	Temp Time	I + II I:II	
Me	0° 10 h	(51) 21:79	
Et	rt 21 h	(67) 14:86	
n-Pr	rt 24 h	(74) 17:83	
Ph	rt 40 h	(49) 4:96	
C_{7-17}			
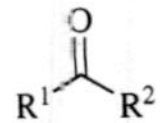	R_3SiH (1 eq), Et_2O, $ZnCl_2$, 35°	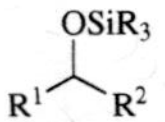	383

R^1	R^2	R_3Si	Time	Yield (%)
—$(CH_2)_5$—		$(n\text{-}Bu)_3Si$	2 h	(25)
—$(CH_2)_5$—		Et_3Si	2 h	(26)
n-Pr	n-Pr	$(n\text{-}Bu)_3Si$	3 h	(31)
Et	n-Bu	$(n\text{-}Bu)_3Si$	3 h	(34)
n-Bu	n-Pr	Et_3Si	2 h	(79)
n-Bu	n-Bu	Et_3Si	3 h	(57)
Me	$n\text{-}C_7H_{15}$	Et_3Si	4 h	(22)
Et	Bn	Et_3Si	3 h	(29)
$n\text{-}C_5H_{11}$	$n\text{-}C_5H_{11}$	Et_3Si	5 h	(59)
$n\text{-}C_8H_{17}$	$n\text{-}C_8H_{17}$	Et_3Si	4 h	(35)

C_8	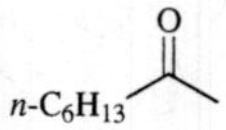	$PhMe_2SiH$ (1.2 eq), $[Rh(C_2H_4)_2]_2$ (0.5 mol%), P(tm-tp)$_3$ (2 mol%), C_6H_6, rt, 6 h	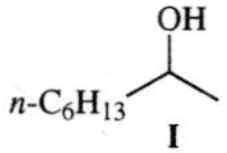 **I** (91)	389
		$PhMe_2SiH$ (1.2 eq), $[Rh(C_2H_4)_2]_2$ (0.5 mol%), P(tp)$_3$ (2 mol%), C_6H_6, rt, 6 h	**I** (41)	389
		$PhMe_2SiH$ (1.2 eq), $[Rh(C_2H_4)_2]_2$ (0.5 mol%), PPh_3 (2 mol%), C_6H_6, rt, 6 h	**I** (31)	389
	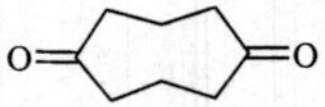	TMSH (4 eq), TMSOTf (1.6 mol%), CH_2Cl_2, 0°, 4 h; rt, 2 h	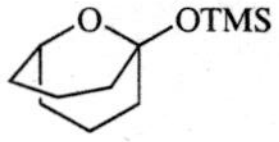 (—)	392
	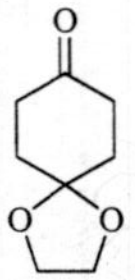	$PhSiH_3$ (0.4 eq), $Mn(dpm)_3$ (3 mol%), i-PrOH, DCE, O_2, rt	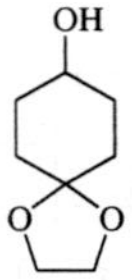 (92)	367

TABLE 12. ORGANOSILANE REDUCTION OF KETONES (*Continued*)

Ketone	Conditions	Product(s) and Yield(s) (%)	Refs.
C_8 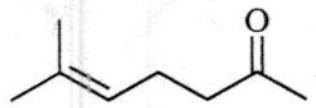	PMHS (2.5 eq), $(Ph_3P)CuH$ (3 mol%), MeC_6H_5, 10 h	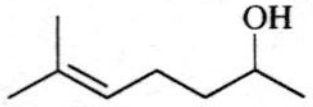 **I** (97)	317
	$Me(EtO)_2SiH$ (1.5 eq), KF (1 eq), DMF, 80°, 3 h	**I** (80)	82
	Et_3SiH (1 eq), $ZnCl_2$, 105°, 16 h	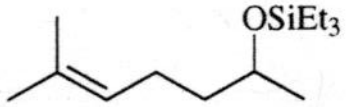 (70)	382
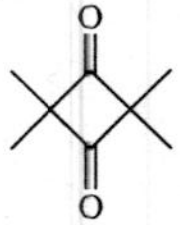	R_3SiH (2 eq), $(Ph_3P)_3RhCl$, C_6H_6, rt,	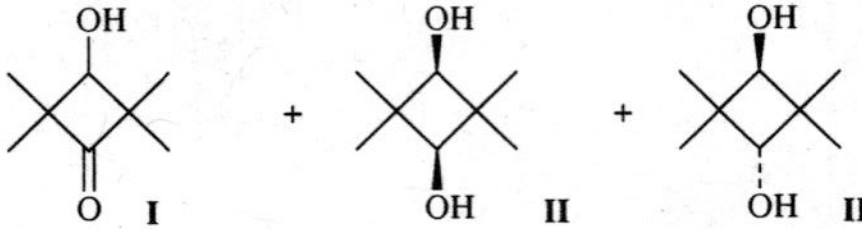 **I** + **II** + **III**	370

R_3Si	Time	% Conversion	I	II + III	II:III
n-$C_5H_{11}H_2Si$	2 h	100	(0)	(100)	52:48
n-BuMeHSi	2 h	35	(100)	(0)	—
$(n\text{-}Pr)_2HSi$	1 h	60	(100)	(0)	—
$(n\text{-}Pr)_2HSi$	5 h	100	(100)	(0)	—
Ph_2HSi	2 h	100	(5)	(95)	65:35
Ph_2HSi (1 eq)	2 h	51	(—)	(0)	—
Et_3Si	24 h	0	(—)	(—)	—
Et_3Si (60°)	24 h	0	(100)	(—)	—
Et_2MeSi	24 h	100	(100)	(0)	—
Et_2MeSi, then Ph_2HSi	24 h, +3 h	100	(35)	(65)	46:54

Reactant	Conditions	Product(s) and Yield(s) (%)	Refs.
	R_3SiH, $[Rh(cod)Cl]_2$, ligand, C_6H_6, rt, 2 h	**I + II + III**	370

R_3Si	Ligand	I	II + III	II:III
Ph_2HSi	PPh_3	(—)	(100)	63:37
n-$C_5H_{11}H_2Si$	PPh_3	(—)	(100)	53:47
Ph_2HSi	diphenylenephenylphosphine	(—)	(100)	57:43
n-$C_5H_{11}H_2Si$	diphenylenephenylphosphine	(—)	(100)	55:45
Ph_2HSi	dppe	(80)	(20)	69:31

Reactant	Conditions	Product(s) and Yield(s) (%)	Refs.
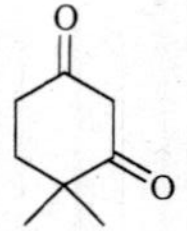	Ph_2SiH_2 (1.3 eq), $(Ph_3P)_4RhH$ (0.5 mol%), CH_2Cl_2, rt, 12 h	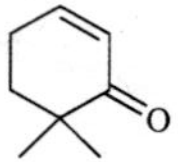 (46.6) + 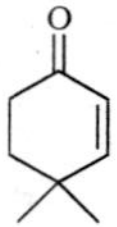 (2.5)	374
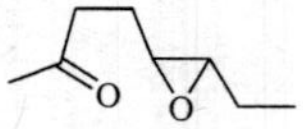	Et_3SiH (1.1 eq), TFA (1.1 eq), CH_2Cl_2, –60°	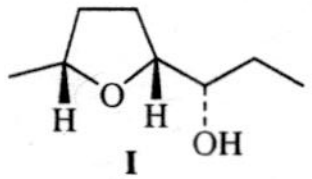 **I** + 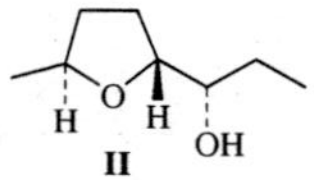 **II** **I + II** (68), **I:II** = 2:1	408
	Ph_3SiH (1.1 eq), BF_3•OEt_2 (1.1 eq), CH_2Cl_2, –60°	**I + II** (51), **I:II** = 5:1	408
	Ph_3SiH (1.1 eq), TIPSOTr (1.1 eq), CH_2Cl_2, –60°	**I + II** (50), **I:II** = 4:1	408
	$PhMe_2SiH$ (1.1 eq), BF_3•OEt_2 (1.1 eq), CH_2Cl_2, –60°	**I + II** (—), **I:II** = 2:1	408
	$MePh_2SiH$ (1.1 eq), BF_3•OEt_2 (1.1 eq), CH_2Cl_2, –60°	**I + II** (—), **I:II** = 4:1	408
	t-$BuPh_2SiH$ (1.1 eq), BF_3•OEt_2 (1.1 eq), CH_2Cl_2, –60°	**I + II** (—), **I:II** = 4:1	408

TABLE 12. ORGANOSILANE REDUCTION OF KETONES (*Continued*)

Ketone	Conditions	Product(s) and Yield(s) (%)	Refs.
C_8 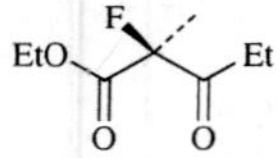	$PhMe_2SiH$ (1.25 eq), catalyst (x eq), CH_2Cl_2, rt	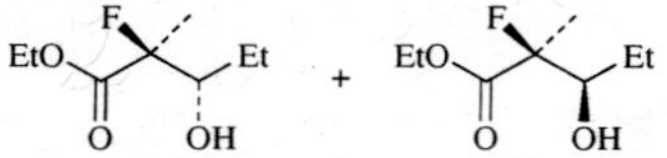 I + II	90
	Catalyst / x / Time	I + II / I:II	
	$AlCl_3$ 1.5 2 h	(65) 96:4	
	$AlCl_3$ 1.5 3 h	(80) 96:4	
	$AlCl_3$ 1.5 4 h	(73) 98:2	
	$AlCl_3$ 1.0 4 h	(51) 98:2	
	$EtCl_2Al$ 1.5 3 h	(61) 97:3	
	$EtCl_2Al$ 1.1 4 h	(54) 98:2	
	$EtCl_2Al$ 1.1 10 h	(68) 94:6	
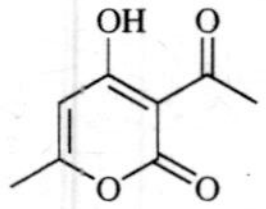	Et_3SiH, TFA, $LiClO_4$ (0.01 eq), rt, 4 h	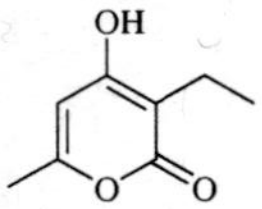 (85)	423
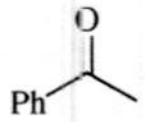	Et_3SiH (2.5 eq), TFA (5 eq), 20°	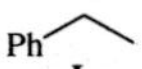 I (70)	180
	$Et_3SiO(EtHSiO)_nSiEt_3$ (0.5 mol%), TFA (9 eq), $CHCl_3$, 50°, 15 h	I (94)	207
	Et_3SiH (2 mol%), TFA (9 eq), $CHCl_3$, 50°, 3 h	I (100)	207
	Et_3SiH (3 eq), TFA (5 eq), 50°, 15 h	I (80)	210
	Et_3SiH (3 eq), BF_3•OEt_2 (14.6 mol%), 50°, 1 h	I (80)	210
	Et_3SiH (1.1-1.2 eq), HCO_2H, HOAc, KU-1, 55°, 5 h	I (81)	208

Et_3SiH (1.2 eq), 65% $HClO_4$, CH_2Cl_2, 20°, 30 min	**I** (74)		214
Et_3SiH (2.5 eq), $AlCl_3$ (0.5 eq), HCl, CH_2Cl_2, rt, 15 min	**I** (95)		136
Et_3SiH (1.5 eq), CF_3SO_3H (4 eq), CH_2Cl_2, 0° to rt, 2 h	**I** (80)[b]		420
Cl_3SiH (1.5 eq), CH_2Cl_2/DMF (4:1), 0°, 12 h	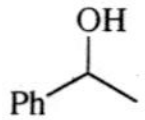 **II**	(54)	318
$(MeO)_3SiH$ (1.5 eq), $LiOCMe_2CMe_2OLi$ (0.08 eq), THF, 0°	**II** (98)		92
PMHS (200 mol%), $Sn(OTf)_2$ (10 mol%), TMEDA (10 mol%), MeOH, rt	**II** (98)		385
$(EtO)_2MeSiH$ (2.3 eq), CsF (1 eq), 100°, 2.5 h	**II** (70)		83
$(EtO)_3SiH$ (2.3 eq), CsF (1 eq), 0°, 0.5 h	**II** (80)		83
PMHS (2.3 eq), KF (1 eq), DMF, 0°, 0.5 h	**II** (80)		83
$Me(EtO)_2SiH$ (1.5 eq), KF (1 eq), DMF, 80°, 1.5 h	**II** (78)		82
PMHS (1.5 eq), KF (1 eq), DMF, 30°, 2.5 h	**II** (78)		82, 83
R_3SiH (1 eq), MF	**II**		80, 83

R_3Si	MF	Temp	Time	
$Me(EtO)_2Si$	CsF	100[c]	2.5 h	(70)
$Me(EtO)_2Si$	KF	100[c]	24 h	(50)
$(EtO)_3Si$	KF	100[c]	1 h	(50)
$(EtO)_3Si$	CsF	0°	0.5 h	(80)

TABLE 12. ORGANOSILANE REDUCTION OF KETONES (*Continued*)

Ketone	Conditions	Product(s) and Yield(s) (%)	Refs.
C_8 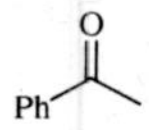	PMHS (1 eq), $ZnCl_2$ (1 eq), Et_2O, rt, 24 h	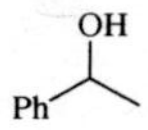 **II** (52)	315
	D, 0°, 12 h	**II** (50)	93
	E, rt, 12 h	**II** (50)	93
	H, rt, 12 h	**II** (97)	93
	G, 0°, 2 h	**II** (98)	93
	L, THF, 0°, 2 h	**II** (88)	96
	$PhSiH_3$ (0.4 eq), $Mn(dpm)_3$ (3 mol%), *i*-PrOH, DCE, O_2, rt	**II** (24)	367
	Et_3SiH (1.2 eq), **208** (0.2 mol%), rt, 23 h	**I** (2) + **II** (89)	377
+ 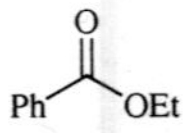	PMHS (1.2 eq), KF (1.3 eq), DMSO, 40°, 4 h	**II** (85) + 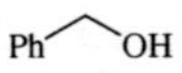 **III** (0)	82, 83
+ 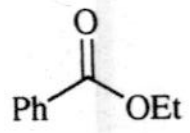	$(EtO)_3SiH$ (2.3 eq), CsF (1 eq), rt, 1 min	**II** (100) + **III** (0)	83

+ 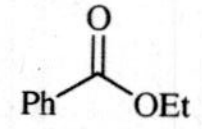	$Me(EtO)_2SiH$ (2.3 eq), CsF (1 eq), DMF, 40°, 4 h	**II** (100) + **III** (0)		83
+ 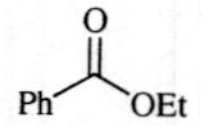	Ph_3SiH (1 eq), $(C_6F_5)_3B$ (2 mol%), MeC_6H_5, rt	**II** (>99) + **III** (<1)		116
	Cl_3SiH (1.5 eq), activator (x eq), CH_2Cl_2, 0° to rt	**II**		379

Activator	x	Time	
DMF	1.5	12 h	(52)
DMF	0.1	24 h	(28)
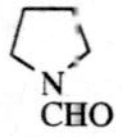	1.5	12 h	(90)
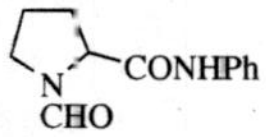	1.5	12 h	(95)
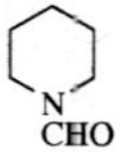	1.5	12 h	(72)
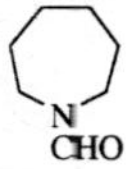	0.1	24 h	(27)
MeNHCHO	1.5	12 h	(26)
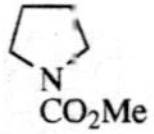	1.5	12 h	(0)
Me_2NAc	1.5	12 h	(0)

Et_3SiH (2 eq), TMSOTf (0.1 eq), CH_2Cl_2, rt, 2 h	**II** (minor) +	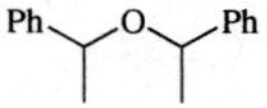 (major)	334

TABLE 12. ORGANOSILANE REDUCTION OF KETONES (*Continued*)

Ketone	Conditions	Product(s) and Yield(s) (%)	Refs.
C_8 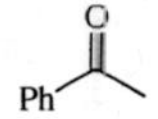	1. $HC(OMe)_3$, CH_2Cl_2, rt, 2 h 2. Et_3SiH, Nafion®-H, CH_2Cl_2, reflux, 3 h	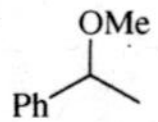 (91)	335
	Et_3SiH (1 eq), TMSOTf (0.1 eq), ROTMS (0.83 eq), CH_2Cl_2	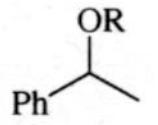 R / Temp allyl, −30° to 0° (85) Bn, −78° to −30° (100) *c*-C_6H_{11}, 0° to rt (96) *t*-Bu, 0° to rt (28)	341
	Et_3SiH (1 eq), $ZnCl_2$, 95°, 150 h	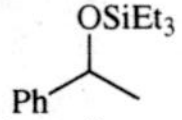 **I** (40)	382
	Et_3SiH, **227** or **228** or **229** (5 mol%)	**I** (100)	117
	Ph_2SiH_2 (1 eq), $[Cp^*(Me_3P)Ir(\eta_2\text{-}SiPh_2C_6H_4)(H)]^+$ $[(C_6F_5)_4B]^-$ (5 mol%), CD_2Cl_2, rt, 18h	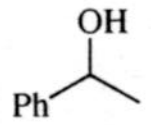 (20) + 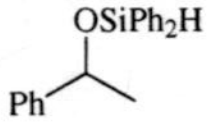 (54)	745
	Ph(Np-1)SiH_2, $(Ph_3P)_3RhCl$ (5 x 10^{-4} M)	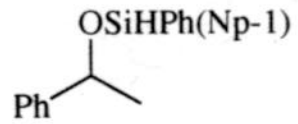 (80) + 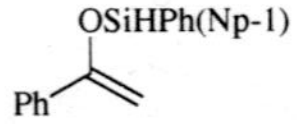 (20)	744

$PhMe_2SiH$ (1.2 eq), TBAF (2 eq), HMPA, rt, 0 5 h	**II** (40) + 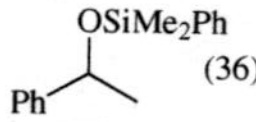 **III** (36)	320
$PhMe_2SiH$ (1.1 eq), CsF (0.[illegible] eq), 18-C-6 (0.05 eq), CH_2Cl_2, rt, 7 h	**III** (54)	346, 347
$PhMe_2SiH$ (1.1 eq), catalyst (0.1 eq), 18-C-6 (0.05 eq), solvent, rt	**III**	346, 347

Catalyst	Solvent	Time	
CsF	CH_2Cl_2	5 h	(100)
CsF	THF	14 h	(100)
CsF	C_6H_6	29 h	(92)
RbF	THF	21 h	(76)
RbF	CH_2Cl_2	21 h	(100)
KF	CH_2Cl_2	75 h	(52)
KF	THF	25 h	(46)
KF	C_6H_6	35 h	(73)
NaF	THF	19 h	(0)
LiF	THF	75 h	(0)
KF-Al_2O_3	THF	24 h	(1)

TABLE 12. ORGANOSILANE REDUCTION OF KETONES (*Continued*)

	Ketone	Conditions	Product(s) and Yield(s) (%)	Refs.
C_8	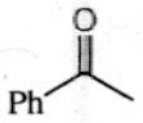	Me_2ClSiH, $TMSCH_2CH{=}CH_2$, catalyst (5 mol%), solvent, rt, 2 h	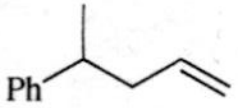	427
		Catalyst — **Solvent**		
		$InCl_3$ — CH_2Cl_2	(86)	
		$InBr_3$ — CH_2Cl_2	(60)	
		InI_3 — CH_2Cl_2	(85)	
		$AlCl_3$ — CH_2Cl_2	(1)	
		$Sc(OTf)_3$ — CH_2Cl_2	(<0.2)	
		$InCl_3$ — $ClCH_2CH_2Cl$	(63)	
		$InCl_3$ — C_6H_{12}	(0)	
		$InCl_3$ — DMF	(0)	
		$InCl_3$ — THF	(0)	
		$InCl_3$ — MeCN	(35)	
		$InCl_3$ — $CHCl_3$	(16)	
		$InCl_3$ — CCl_4	(0)	
		Me_2ClSiH (1.2 eq), $In(OH)_3$ (5 mol%), $CHCl_3$, rt, 0.5 h	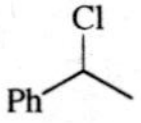 (78)	331
	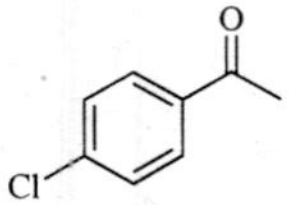	Me_2ClSiH (1.2 eq), $In(OH)_3$ (5 mol%), $CHCl_3$, rt, 0.5 h	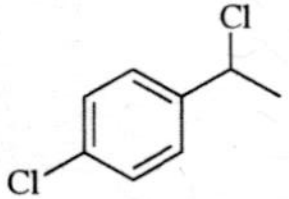 (76)	331
	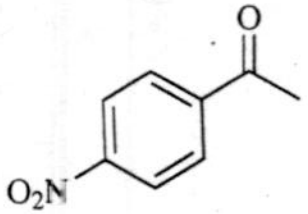	1. Et_3SiH, BF_3, CH_2Cl_2 2. $BF_3{\bullet}OEt_2$, rt, 30 min	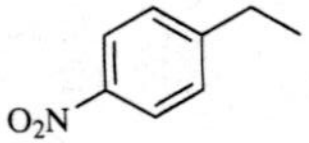 **I** (100) + 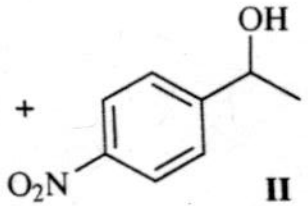 **II** (0)	1
		1. Et_3SiH, BF_3, CH_2Cl_2 2. $BF_3{\bullet}OEt_2$, rt, 3 min	**I** (0) + **II** (100)	1

	$ClMe_2SiH$ (1.2 eq), catalyst (5 mol%), $CHCl_3$, rt, 2 h	**I + II**	331

Catalyst	I	II
none	(0)	(0)
$In(OH)_3$	(0)	(99)
$Sc(OTf)_3$	(0)	(69)
$TiCl_4$	(7)	(0)
$(C_6F_5)_3B$	(81)	(0)

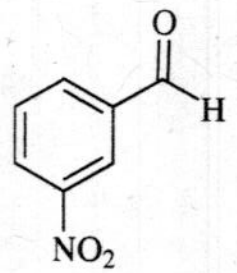	$(EtO)_3SiH$ (1 eq), CsF, rt, 1 h	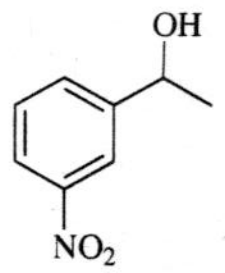 (80)	80
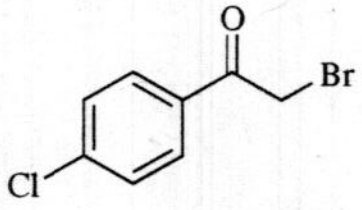	Me_2ClSiH (1.2 eq), $In(OH)_3$ (5 mol%), $CHCl_3$, rt, 4 h	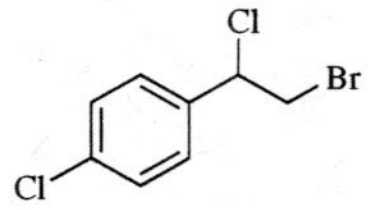 (99)	331
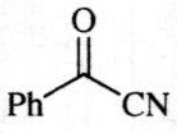	Et_3SiH (1 eq), $(Ph_3P)_3RhCl$	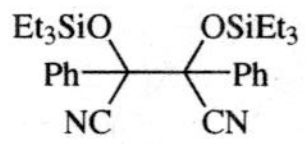 (50)	411
	Et_3SiH (1 eq), $PdCl_2$	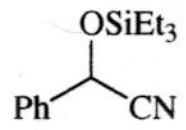 (45)	411
C_{8-9} 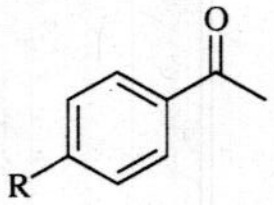	Ph_3SiH (1 eq), $(C_6F_5)_3B$ (2 mol%)	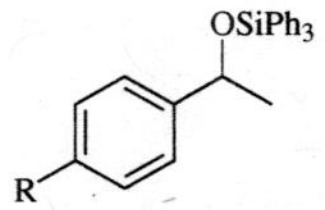	115

R	
H	(76)
Me	(84)
Cl	(80)
NO_2	(91)

TABLE 12. ORGANOSILANE REDUCTION OF KETONES (*Continued*)

Ketone	Conditions	Product(s) and Yield(s) (%)				Refs.
C_{8-9} 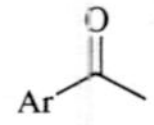	Ph_3SiH, $(C_6F_5)_3B$ (2 mol%), MeC_6H_5, rt	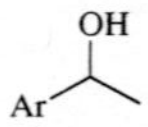	Ar			116
			Ph		(76)	
			4-MeC_6H_4		(84)	
			4-ClC_6H_4		(80)	
			4-$O_2NC_6H_4$		(91)	
	$HMe_2SiOSiMe_2H$ (1.0 eq), I_2 (0.6 eq), CH_2Cl_2, –5°	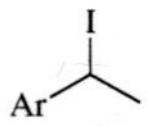	Ar	Time		357
			Ph	5 min	(76)	
			4-MeC_6H_4	20 min	(65)	
C_{8-10} 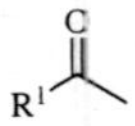	Et_3SiH (1 eq), R^2OTMS (1 eq), $TrClO_4$ (5 mol%), CH_2Cl_2, 0°, 5 min	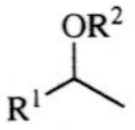	R^1	R^2		329
			Ph	$BnCH_2CH_2$	(69)	
			Ph	Bn	(76)	
			$BnCH_2$	$BnCH_2CH_2$	(68)	
C_{8-11} 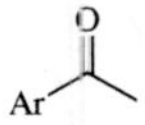	PMHS (3 eq), TBAF (2 mol%), THF, rt	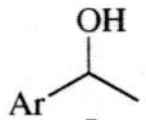 **I**	Ar			278
			Ph		(93)	
			4-FC_6H_4		(86)	
			4-ClC_6H_4		(96)	
			4-$O_2NC_6H_4$		(92)	
			3,4-$[OCH_2O]C_6H_3$		(93)	
			4-MeC_6H_4		(94)	
			3,4,5-$(MeO)_3C_6H_2$		(89)	
			4-NCC_6H_4		(86)	
	PMHS (3 eq), Triton-B (2 mol%), THF, rt	**I**	Ar			278
			4-FC_6H_4		(97)	
			4-ClC_6H_4		(88)	
			4-MeC_6H_4		(88)	
			3,4,5-$(MeO)_3C_6H_2$		(93)	

C_{8-12}

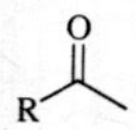

$(EtO)_3SiH$, **194** (10 mol%), Et_2O

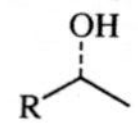

378

R	Temp	Time	
n-C_6H_{13}	40°	12 h	(84)
c-C_6H_{11}	40°	12 h	(77)
n-$C_4H_9C{\equiv}C$	40°	12 h	(50)
(E)-n-$C_5H_{11}CH{=}CH$	40°	12 h	(>98)
1-$C_{10}H_7$	50°	5 h	(>98)
Ph	50°	5 h	(>98)
4-$MeOC_6H_4$	50°	5 h	(>98)
4-ClC_6H_4	50°	5 h	(>98)
4-$O_2NC_6H_4$	50°	5 h	(>98)
3-$O_2NC_6H_4$	50°	5 h	(>98)

C_{8-13}

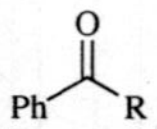

$[H_2Si(OSiPr\text{-}i)_3]K$ (x eq), rt, 4 h

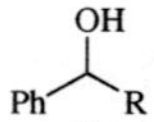

I

R		x
Me	(36)	0.5
Me	(80)	1.0
Ph	(50)	0.5

750

$[HSi(OEt)_4]K$, THF, rt

I

R	Time	
Me	15 h	(53)
Ph	4 h	(73)

288

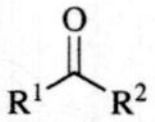

Ph_2SiH_2 (1 eq), KF, 100°

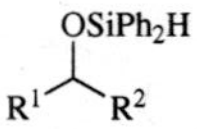

R^1	R^2	
—$(CH_2)_5$—		(70)
Ph	Me	(40)
Ph	Ph	(30)

78

TABLE 12. ORGANOSILANE REDUCTION OF KETONES (*Continued*)

Ketone	Conditions	Product(s) and Yield(s) (%)	Refs.
C_{8-13}			
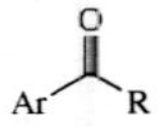	Me_2ClSiH, $TMSCH_2CH{=}CH_2$, $InCl_3$ (5 mol%), CH_2Cl_2, rt	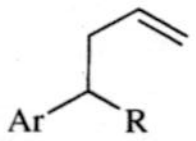	427
Ar: *p*-ClC_6H_4; R: Me	2 h	(84)	
Ar: Ph; R: Et	2 h	(67)	
Ar: Ph; R: $ClCH_2CH_2$	1 h; 60°, 3 h	(70)	
Ar: $PhCH_2CH_2$; R: Me	2 h	mixture	
Ar: *p*-$EtO_2CC_6H_4$; R: Me	1 h; 60°, 3 h	(44)	
Ar: Ph; R: Ph	2 h	(99)	
Ar: *p*-$NO_2C_6H_4$; R: Ph	1 h; 60°, 2 h	(75)	
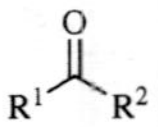	Et_3SiH (1.1 eq), H_2SO_4, H_2O, MeCN, 28°	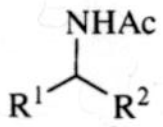	313
		R^1 / R^2 / Time	
		—$(CH_2)_5$— / 72 h (30)	
		Ph / Me / 72 h (85)	
		Ph / Ph / 48 h (63)	
C_{8-17}			
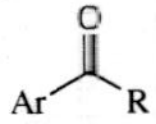	Et_3SiH (x eq), TFA (y eq), solvent, rt		73

Ar	R	x	y	Solvent	Time	
Ph	Me	2.2	10.0	TFA	15 min	(100)
4-ClC_6H_4	Me	2.2	6.7	TFA	200 h	(100)
Ph	$BrCH_2$	4.4	7.0	TFA	44 h	(93)
Ph	Et	2.5	5.4	TFA	15 min	(100)
Ph	*n*-Pr	2.2	10.0	TFA	15 min	(100)
Ph	*i*-Pr	2.5	5.4	TFA	15 min	(100)
Ph	*c*-C_3H_5	2.6	9.5	H_2O-TFA	7 h	(48)
Ph	*n*-Bu	2.2	10.0	TFA	15 min	(100)
Ph	$HO_2C(CH_2)_3$	2.2	6.5	TFA	48 h	(100)
Ph	*c*-C_4H_7	2.2	9.9	H_2O-TFA	6 h	(25)
Ph	*c*-C_4H_7	5.0	5.0	CCl_4-TFA	7.5 h	(26)
Ph	$HO_2C(CH_2)_4$	2.4	7.0	TFA	5.5 h	(100)
Ph	Ph	2.2	15.0	TFA	15 min	(100)
4-$O_2NC_6H_4$	Ph	2.4	7.7	TFA	47 h	(100)
4-$HO_2CC_6H_4$	Ph	3.0	6.7	CCl_4-TFA	120 h	(100)
Ph	*n*-$C_{10}H_{21}$	2.2	10.0	TFA	15 min	(100)

C_{8-18}

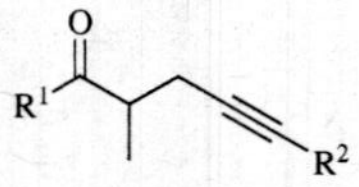

Et_3SiH, $(C_6F_5)_3B$ (2 mol%), MeC_6H_5, 0°, 1 h

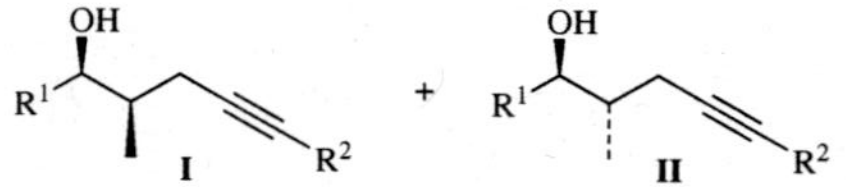

372

R^1	R^2	I + II	I:II
Et	H	(100)	4.4:1
t-Bu	H	(100)	>30:1
c-C_6H_{11}	H	(93)	5.0:1
Ph	H	(90)	7.0:1
Ph	Me	(100)	5.0:1
Ph	Ph	(100)	3.0:1
Ph	TMS	(100)	7.7:1
2-MeC_6H_4	H	(94)	15:1

TABLE 12. ORGANOSILANE REDUCTION OF KETONES (*Continued*)

Ketone	Conditions	Product(s) and Yield(s) (%)	Refs.
C_9 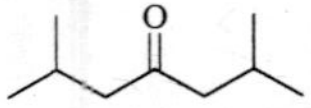	$(EtO)_3SiH$ (1 eq), CsF, rt, 1 h	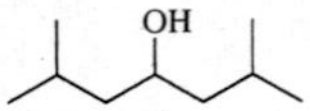 **I** (95)	80
	$Me(EtO)_2SiH$ (1.5 eq), KF (1 eq), DMF, 60°, 5.5 h	**I** (75)	82
	PMHS (1.5 eq), $KF•2H_2O$ (1 eq), DMSO, 60°, 7 h	**I** (85)	82
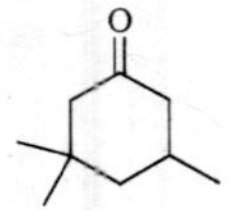	1. R_3SiH (1.1 eq), acid catalyst (1 eq), 0° to rt, 24 h 2. Aq. OH^-	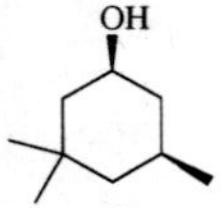 **I** + 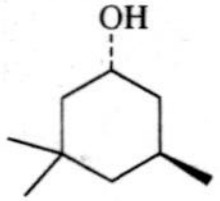 **II** **I** + **II** (—)	381, 384

R_3Si	Acid Catalyst	I:II
n-BuH_2Si	$BF_3•OEt_2$	13:87
Et_2HSi	$BF_3•OEt_2$	9:91
Et_3Si	$BF_3•OEt_2$	5:95
Et_3Si	TFA (6.8 eq)	16:84
n-BuH_2Si	H_2SO_4	26:74
Et_2HSi	H_2SO_4	15:85
Et_3Si	H_2SO_4	10:90
(*n*-Bu)$_3$Si	H_2SO_4	11:89

Ketone	Conditions	Product(s) and Yield(s) (%)	Refs.
	Et_3SiH (1.1 eq), acid, rt	**I** + **II** (100)	74
	(*t*-Bu)$_2SiH_2$ (1 eq), TFA (6.6 eq), rt, 20 h	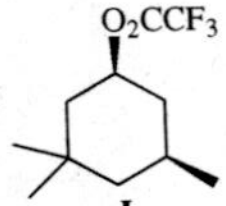 **I** (11) + 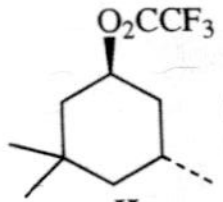 **II** (89)	386

Acid	I:II
$BF_3•OEt_2$	95:5
TFA	84:16
H_2SO_4	90:10

	$(t\text{-Bu})_2$MeSiH (1 eq), TFA (6.6 eq), rt, 20 h	**I** (<1) + **II** (60) + 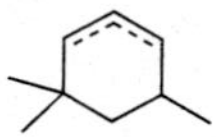 (40)	386
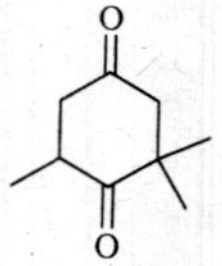	$PhSiH_3$ (0.4 eq), $Mn(dpm)_3$ (3 mol%), *i*-PrOH, DCE, O_2, rt	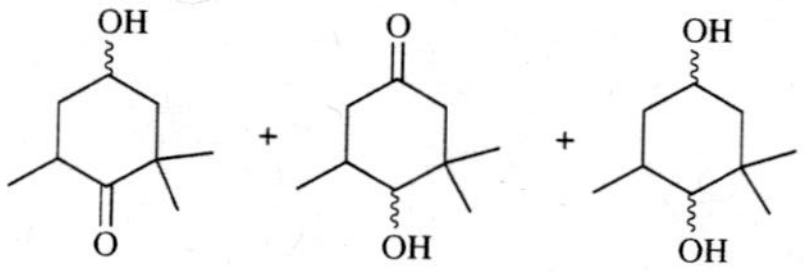 **I** **II** **III** **I** + **II** + **III** (85)	367
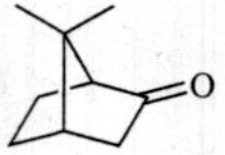	Ph_2SiH_2 (1.3 eq), $(Ph_3P)_4RhH$ (0.4 mol%), CH_2Cl_2, rt, 7 h	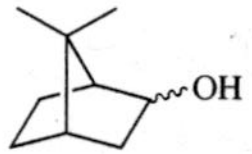 (85), exo:endo = 36:64	374
	$PhSiH_3$ (0.4 eq), $Mn(dpm)_3$ (3 mol%), *i*-PrOH, DCE, O_2, rt	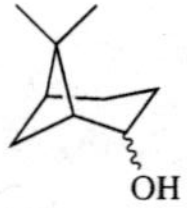 (53), endo:exo = 1:7	367
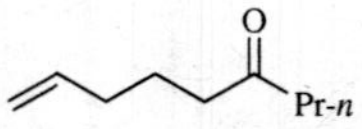	Ph_2SiH_2 (1.0 eq), $Cp_2Ti(PPh_3)_2$, MeC_6H_5, −20°	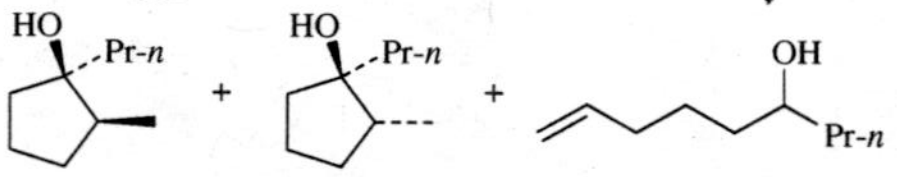 **I** **II** **III** **I** + **II** (64), (**I** + **II**):**III** = 23:1, **I**:**II** = 22:1	428, 429
	Ph_2SiH_2 (1.0 eq), $Cp_2Ti(PPh_3)_2$, MeC_6H_5, −20°	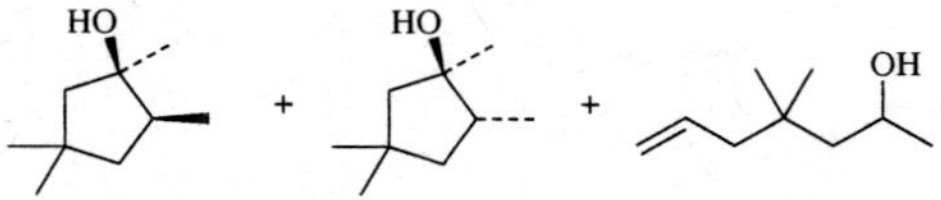 **I** **II** **III** **I** + **II** (56), (**I** + **II**):**III** = 16:1, **I**:**II** = 13:1	428, 429

TABLE 12. ORGANOSILANE REDUCTION OF KETONES (*Continued*)

Ketone	Conditions	Product(s) and Yield(s) (%)	Refs.
C_9 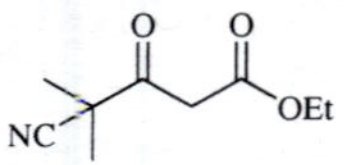	Ph_2SiH_2 (1.2 eq), $AlCl_3$ (1 eq), CH_2Cl_2, 20°, 19 h	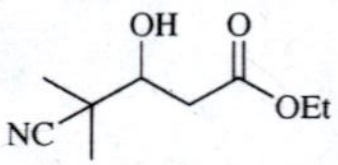 (90) + Ph_2SiHCl (—)	373
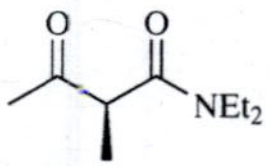	$PhMe_2SiH$ (1.2 eq), TFA, 0°, 3 h	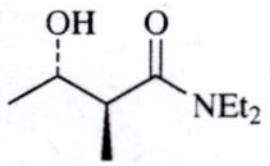 **I** + 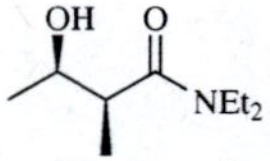 **II** **I** + **II** (94), **I**:**II** = 2:98	87, 276
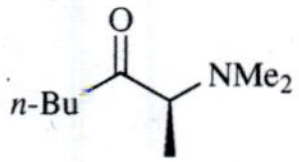	$PhMe_2SiH$ (1.1-1.2 eq), TBAF (5-10 mol%), HMPA, rt, 10 h	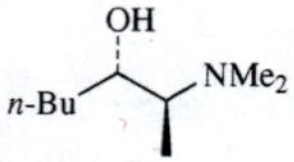 **I** + 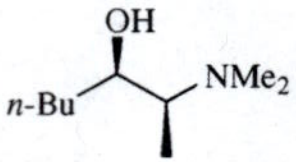 **II** **I** + **II** (83), **I**:**II** >99:1	320
	Ph_2SiH_2 (1.1-1.2 eq), TBAF (5-10 mol%), HMPA, rt, 12 h	**I** + **II** (97), **I**:**II** = 98:2	320
	Ph_3SiH (4 eq), catalyst (2.2 eq), CH_2Cl_2, −78°, 6 h; 0°, 12 h	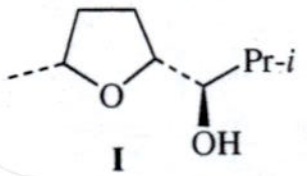 **I** + 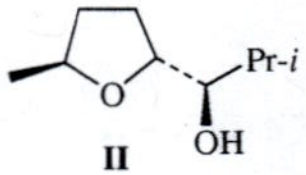 **II** + 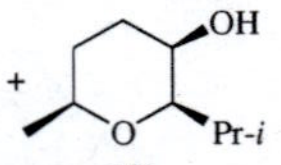 **III**	407

Catalyst	**I** + **II** + **III**	**I**:**II**:**III**
$BF_3•OEt_2$	(67)	6:1:2
TMSOTf	(81)	20:1:1.3

Ketone	Conditions	Product(s) and Yield(s) (%)	Refs.
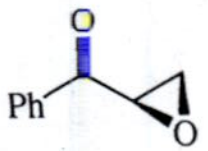	$(MeO)_3SiH$ (1.2 eq), LiOMe (0.04 eq), Et_2O, −20°, 1 d	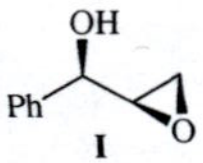 **I** + 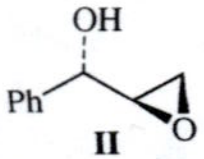 **II** **I** + **II** (97), **I**:**II** = 6:94	366
	$(MeO)_3SiH$ (1.2 eq), LiOMe (1.2 eq), Et_2O, −20°, 1 d	**I** + **II** (19), **I**:**II** = 68:32	366

	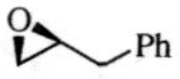 OLi (1.2 eq), (MeO)$_3$SiH (1.2 eq), Et$_2$O, −20°, 1 d	**I** + **II** (96), **I**:**II** = 4:96	366
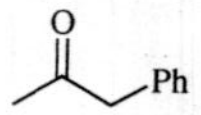	Et$_3$SiH (1.2 eq), **123** (0.05 mol%), CuCl (0.5 mol%), NaOBu-*t* (0.1 mol%), Et$_2$O, 4 h	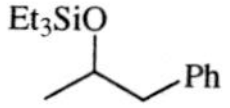 **I** (95)	749
	Et$_3$SiH (1.2 eq), CuCl (0.5 mol%), NaOBu-*t* (0.1 mol%), *o*-dppb (0.05 mol%), neat, microwave, 15 min	**I** (98)	749
	t-BuMe$_2$SiH (1.2 eq), CuCl (0.5 mol%), NaOMe (3 mol%), **123** (0.05 mol%), MeC$_6$H$_5$, Et$_2$O, or CH$_2$Cl$_2$, rt, <3 h	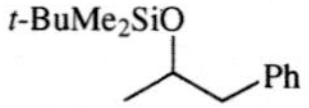 **I** (100)	749
	t-BuMe$_2$SiH (1.2 eq), CuCl (0.5 mol%), NaOMe (3 mol%), *o*-dppb (0.05 mol%), MeC$_6$F$_5$, rt, <5 h	**I** (99)	749
	t-BuMe$_2$SiH (1.2 eq), CuCl (0.5 mol%), NaOMe (3 mol%), *o*-dppb (0.05 mol%), neat, 150°, microwave, 15 min	**I** (96)	749
+ 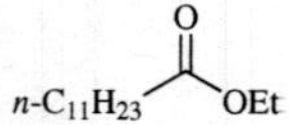	(EtO)$_2$MeSiH (2.3 eq), CsF (1 eq), rt, 5 h	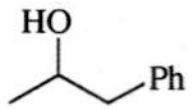 **II** (100) + 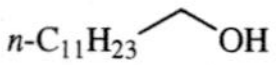 (0)	83
	PMHS (3 eq), (C$_6$F$_5$)$_3$B (5 mol%), CH$_2$Cl$_2$, rt, 5-20 min	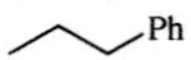 (86)	354
	(EtO)$_3$SiH (1 eq), CsF, rt, 1 h	**II** (70)	80
	PMHS (2.5 eq), (Ph$_3$P)CuH (3 mol%), MeC$_6$H$_5$, 22 h	**II** (78)	317

TABLE 12. ORGANOSILANE REDUCTION OF KETONES (*Continued*)

Aldehyde	Conditions	Product(s) and Yield(s) (%)	Refs.
C_9			
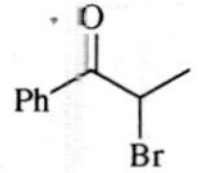	$(EtO)_3SiH$ (2.3 eq), CsF (1 eq), rt, 0.5 h	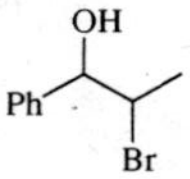 (70)	83
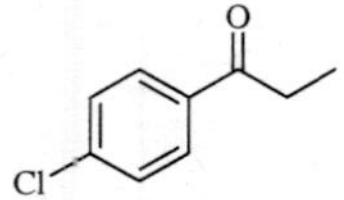	PMHS (3 eq), $(C_6F_5)_3B$ (5 mol%), CH_2Cl_2, rt, 5-20 min	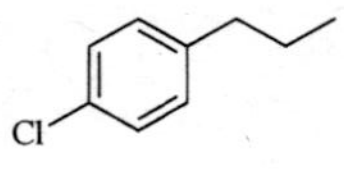 (83)	354
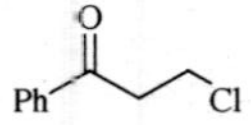	PMHS (3 eq), $(C_6F_5)_3B$ (5 mol%), CH_2Cl_2, rt, 5-20 min	 (84)	354
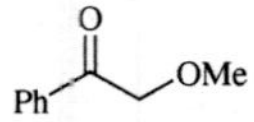	Me_2ClSiH (1.2 eq), $In(OH)_3$ (5 mol%), $CHCl_3$, rt, 3.5 h	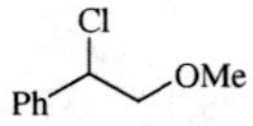 (46)	331
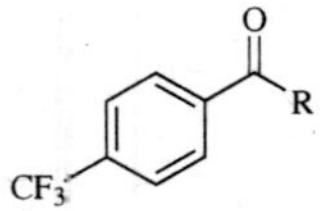 R = Me, CD_3	$Et_3SiH(D)$, $BF{\bullet}OEt_2$, CH_2Cl_2, 0°	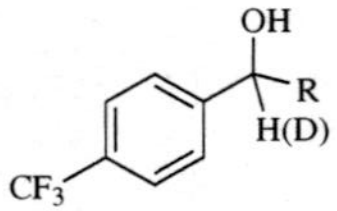 (—)	751
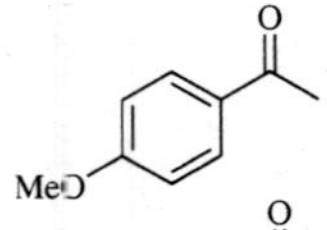	Me_2ClSiH (1.2 eq), $In(OH)_3$ (5 mol%), $CHCl_3$, rt, 3.5 h	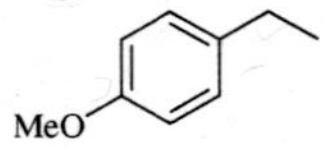 (38)	331
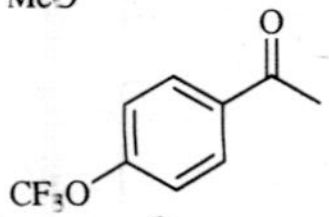	Et_3SiH, TFA	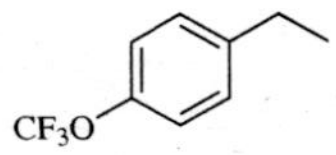 (—)	414
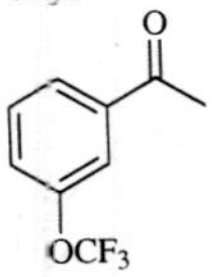	Et_3SiH, TFA	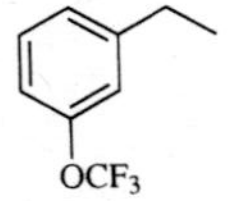 (—)	414

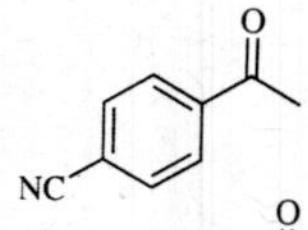	Me_2ClSiH (1.2 eq), $In(OH)_3$ (5 mol%), $CHCl_3$, rt, 3.5 h	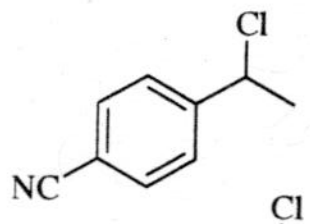 (68)	331
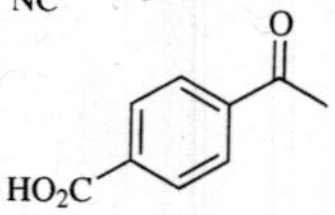	Me_2ClSiH (4 eq), $In(OH)_3$ (5 mol%), $CHCl_3$, 60°, 3.5 h	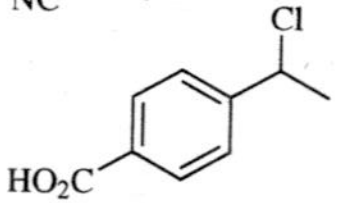 (70)	331
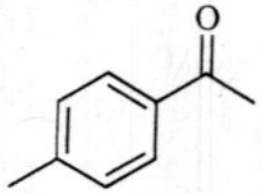	Me_2ClSiH (1.2 eq), $In(OH)_3$ (5 mol%), $CHCl_3$, rt, 3.5 h	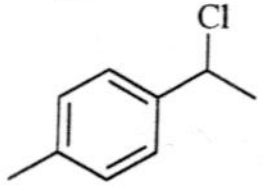 (94)	331
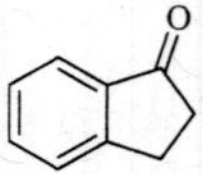	Et_3SiH (1.5 eq), CF_3SO_3H (4 eq), CH_2Cl_2, 0° to rt, 2 h	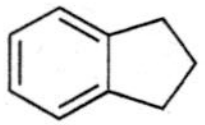 **I** (96)	420
	Et_3SiH (2.6 eq), PPHF, TFA, 0°, 10 min, rt, 3 h	**I** (87)	135
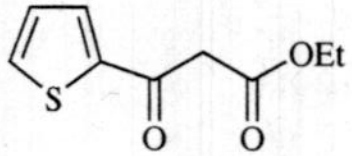	Ph_2SiH_2 (1.2 eq), $AlCl_3$ (1 eq), CH_2Cl_2, 20°, 18 h; 40°, 11 h	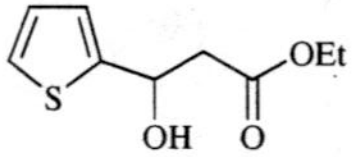 (64) + Ph_2HSiCl (—)	373
C_{9-12} 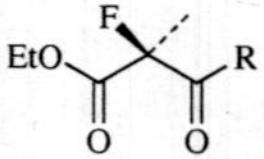	R_3SiH (1.25 eq), $AlCl_3$, CH_2Cl_2, rt	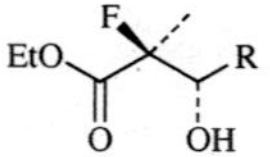 **I** + 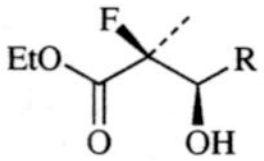 **II**	90

R	R_3Si	I + II	I:II
n-Pr	$PhMe_2Si$	(65)	99:1
n-Pr	Ph_3Si	(74)	99:1
n-Pr	Et_3Si	(69)	78:22
n-Pr	$Me(EtO)_2Si$	(22)	99:1
n-Pr	[—Si(H)(Me)O—]$_4$	(54)	94:6
Ph	$PhMe_2Si$	(49)	74:26
Ph	Ph_3Si	(65)	99:1

TABLE 12. ORGANOSILANE REDUCTION OF KETONES (*Continued*)

Ketone	Conditions	Product(s) and Yield(s) (%)	Refs.
C_{9-16} 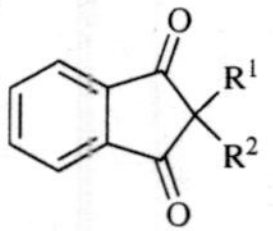	Et_3SiH (5 eq), TFA (20 eq), 50°, 20 h	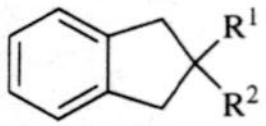	422

R^1	R^2	Yield (%)
H	H	(80)
H	Me	(100)
H	4-MeC_6H_4	(87)
H	4-BrC_6H_4	(85)
H	4-FC_6H_4	(93)
H	4-ClC_6H_4	(83)
H	4-IC_6H_4	(79)
Me	Me	(100)
Me	Ph	(20)
Me	Bn	(32)
Me	3-FC_6H_4	(21)
Me	3-ClC_6H_4	(20)
Me	3-BrC_6H_4	(20)
Et	3-BrC_6H_4	(15)

Reactant	Conditions	Product(s) and Yield(s) (%)			Refs.
C_{10-15}					
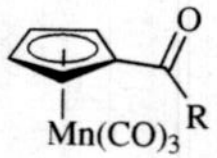	Et_3SiH (3 eq), TFA (5 eq), 40-50°	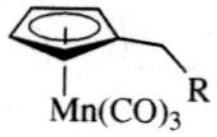	R	Time	351
			Me	15 h (84)	
			CF_3	15 h (95)	
			Ph	15 h (92)	
C_{10}					
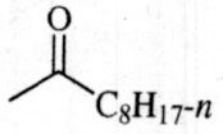	Et_3SiH (5 eq), **121** (0.05 mol%), CuCl (3 mol%), NaOBu-*t* (3 mol%), MeC_6H_5	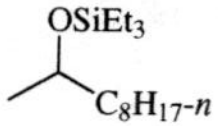 **I**	(87)		749
	Et_3SiH (5 eq), CuCl (3 mol%), NaOBu-*t* (3 mol%), **123** (0.05 mol%), MeC_6H_5	**I** (87)			749

TABLE 12. ORGANOSILANE REDUCTION OF KETONES (*Continued*)

Ketone	Conditions	Product(s) and Yield(s) (%)	Refs.
C_{10} 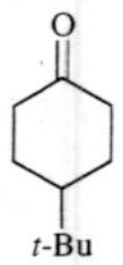	Et_3SiH (1.1 eq), acid, rt	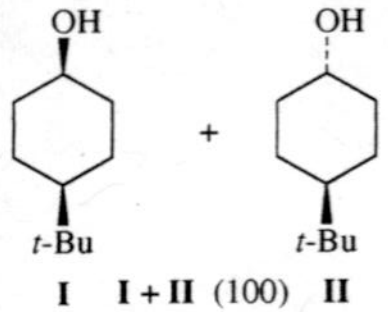 I + II (100) Acid / I:II BF_3•OEt_2 61:39 TFA 32:68 H_2SO_4 32:68	74
	PMHS (2.5 eq), KF•$2H_2O$ (1.25 eq), DMF, 60°, 3 h	**I** + **II** (72), **I**:**II** 18:82	82
	H, rt, 5 h	**I** + **II** (92), **I**:**II** 67:33	93
	G, 0°, 2 h	**I** (92), **I**:**II** 44:56	93
	R_3SiH (1.1 eq), acid catalyst (1 eq), 0° to rt, 24 h	**I** + **II** (~100)	72, 381

R_3Si	Acid Catalyst	I:II
n-BuH_2Si	BF_3•OEt_2	17:83
n-BuH_2Si	TFA (6.8 eq)	16:84
Et_2HSi	BF_3•OEt_2	36:64
Et_3Si	BF_3•OEt_2	61:39
Et_3Si	H_2SO_4	10:90
n-BuH_2Si	H_2SO_4	20:80
Et_2HSi	H_2SO_4	32:68
Et_3Si	TFA (6.8 eq)	32:68
$(n$-$Bu)_3Si$	H_2SO_4	22:78
$(n$-$C_6H_{13})_3Si$	H_2SO_4	21:79
n-PrH_2Si	TFA (2 eq)	20:80
$(i$-$C_5H_{11})_3Si$	TFA (6.7 eq)	30:70
$(c$-$C_5H_9)_3Si$	TFA (7 eq)	44:56
$(i$-$Bu)_3Si$	TFA (7 eq)	55:45
$(s$-$Bu)_3Si$	TFA (0.9 eq)	56:44
$(s$-$Bu)_3Si$	TFA (7 eq)	55:45
$(t$-$Bu)_3Si$	TFA (7.2 eq)	68:32

R_3SiH, Triton-B or TBAF	**I** + **II** (>90)	278
R_3Si — Catalyst	**I:II**	
PMHS — Triton-B	88:12	
$(TMSO)_3Si$ — Triton-B	91:9	
$(MeO)_3Si$ — Triton-B	79:21	
$(EtO)_3Si$ — Triton-B	73:27	
PMHS — TBAF	97:3	
$(TMSO)_3Si$ — TBAF	>99:1	
$(MeO)_3Si$ — TBAF	62:38	
$(EtO)_3Si$ — TBAF	79:21	
PMHS (10% xs), $[Bu_2(AcO)Sn]_2O$ (2 mol%), EtOH, reflux	**II** (65)	316
Ph_2SiH_2 (1.3 eq), **87** (5 mol%), $RhCl_3$ (1 mol%), $AgBF_4$ (2 mol%), THF, 0°, 1 d	**I** + **II** (92), **I:II** = 33:67	390

TABLE 12. ORGANOSILANE REDUCTION OF KETONES (*Continued*)

Ketone	Conditions	Product(s) and Yield(s) (%)	Refs.
C_{10} 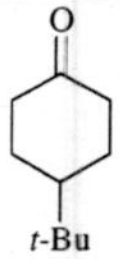	1. R_3SiH (1.5 eq), $(Ph_3P)_3RhCl$ (2 mol%), C_6H_6 2. TsOH, aq. MeOH, rt, 1 h	 **I** (—) + 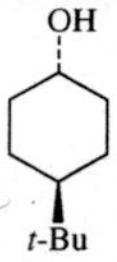 **II** (—)	391

R_3Si	Temp	Time	I:II
Et_3Si	rt	144 h	33:67
Et_3Si	45°	20 h	23:77
Et_3Si	80°	3 h	11:89
Et_3Si	110°	1.5 h	12:88
Ph_3Si	rt	64 h	19:81
Ph_3Si	80°	24 h	12:88
$PhMe_2Si$	80°	12 h	29:71
$(EtO)_3Si$	80°	20 h	29:71
Cl_3Si	80°	18 h	37:63
Et_2HSi	rt	0.25 h	46:54
Ph_2HSi	rt	0.25 h	57:43

Ketone	Conditions	Product(s) and Yield(s) (%)	Refs.
	1. R_3SiH (1.5 eq), $(Ph_3P)_3RuCl_2$ (2 mol%), C_6H_6, 80° 2. TsOH, aq. MeOH, rt, 1 h	**I** + **II** (60-90)	391

R_3Si	Time	I:II
Et_3Si	8 h	8:92
Et_3Si	20 h	5:95
Ph_3Si	24 h	7:93
$PhMe_2Si$	10 h	12:88
Et_2HSi	12 h	41:59
Ph_2HSi	12 h	49:51

Ph_2SiH_2 (1.3 eq), $(Ph_3P)_4RhH$ (0.3 mol%), CH_2Cl_2, rt, 5 h	**I + II** (81), **I:II** = 16:84	374
$PhSiH_3$ (0.4 eq), $Mn(dpm)_3$ (3 mol%), *i*-PrOH, DCE, O_2, rt	**I + II** (98), **I:II** = 1:14	367
Ph_2SiH_2 (1.5 eq), $(Ph_3P)_3RhCl$ (1 mol%), ligand (2 mol%), $AgBF_4$ (2 mol%), THF	**I + II**	572

Ligand	Temp	Time	I + II	I:II
95	rt	4 h	(87)	35:65
87	0°	1 d	(92)	33:67
72	5°	4 h	(74)	39:61

Et_3SiH (2 eq), TMSOTf (0.1 eq), CH_2Cl_2, rt, 2 h

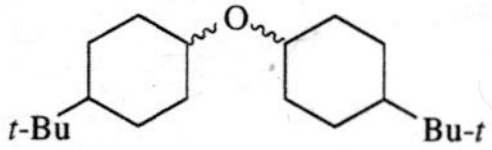

III (100)[a] 334

Et_3SiH (1.1 eq), H_2SO_4, H_2O, MeCN, 28°, 65 h

I + II + III (70) + 

(4) + 

(21) 313

TABLE 12. ORGANOSILANE REDUCTION OF KETONES (*Continued*)

Ketone	Conditions	Product(s) and Yield(s) (%)	Refs.
C_{10} 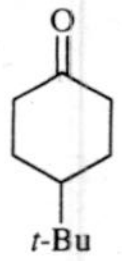	R_3SiH (x eq), TFA (y eq)	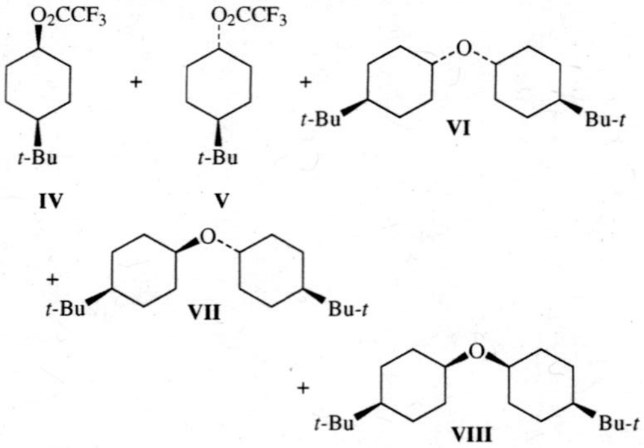	72, 74

R_3Si	x	y	IV + V	IV:V	VI + VII + VIII	trans, trans VI (%)	cis, trans VII (%)	cis, cis VIII (%)
n-BuH_2Si	1.0	2.5	(26)	0.031	(74)	67	29	4
n-PrH_2Si	1.05	2.1	(50)	0.25	(50)	53	40	7
PMHS	0.8	6.8	(66)	0.23	(34)	30	55	15
Et_3Si	1.0	7.5	(83)	0.49	(17)	25	48	27
(*c*-$C_5H_9)_3Si$	0.75	7.0	(89)	0.79	(11)	13	35	52
(*i*-$Bu)_3Si$	1.1	7.0	(86)	1.08	(14)	18	43	39
(*s*-$Bu)_3Si$	0.83	7.0	(94)	1.22	(6)	3	24	73

Ketone	Conditions	Product(s) and Yield(s) (%)	Refs.
	(*t*-$Bu)_2SiH_2$ (1 eq), TFA (6.6 eq), rt, 2 h	**IV** (66) + **V** (31) + 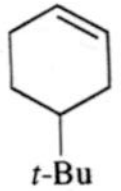 (3) **IX**	386

(t-Bu)$_3$SiH (1.2 eq), TFA (3 eq), rt

I + **IV** + **V** + **IX** +

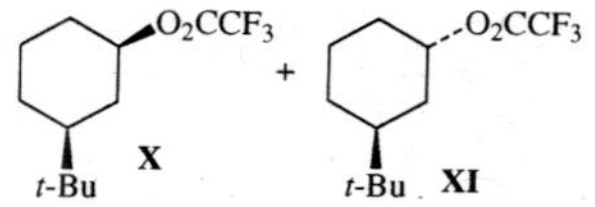

386

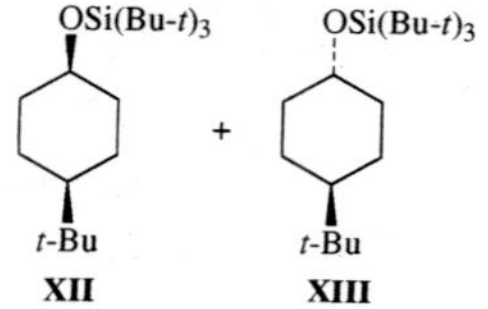

Time	% Conv.	I	IV	V	IX	X	XI	XII + XIII
21 h	50	(9)	(10)	(8)	(5.5)	(3)	(0.5)	(64)
27 h	53	(10)	(12)	(9.3)	(7.4)	(3.7)	(0.6)	(57)
93 h	85	(—)	(13)	(9)	(22)	(4.2)	(0.8)	(51)
264 h	94	(—)	(13)	(9.3)	(50)	(6.5)	(1.2)	(20)
720 h	97	(—)	(21)	(10)	(38)	(18)	(3)	(10)

(t-Bu)$_2$MeSiH (1.17 eq), TFA (3 eq), rt

I + **II** + **IV** + **V** + **IX**

386

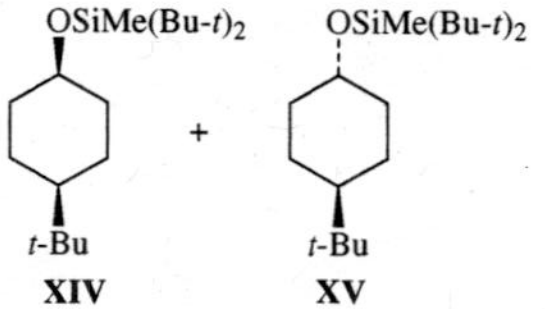

Time	IX	XIV	XV	I + IV	II + V
0.6 h	(0.7)	(70)	(27)	(1.8)	(0.5)
21 h	(6)	(41)	(18)	(27)	(7)
46 h	(8)	(28)	(10)	(38)	(16)
96 h	(8)	(27)	(8)	(39)	(20)

TABLE 12. ORGANOSILANE REDUCTION OF KETONES (*Continued*)

Ketone	Conditions	Product(s) and Yield(s) (%)	Refs.
C_{10} 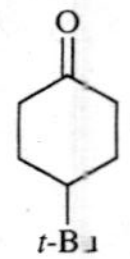	Et_3SiH (1.1 eq), $ZnCl_2$ (1.0 eq), rt	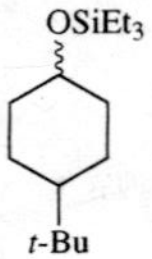 (100) cis:trans = 32:68 **XVI**	74
	Et_3SiH (1.1 eq), $ZnCl_2$ (0.1 eq), rt	**XVI** (100) cis:trans = 67:33	74
	Et_3SiH (1.1 eq), BF_3•OEt_2 (3.0 eq), rt	**XVI** (100) cis:trans = 61:39	74
	Et_3SiH (1.1 eq), $SnCl_2$ (0.1 eq), rt	**III** (50) + **XVI** (50) cis:trans = 42:58	74
	Et_3SiH (1.1 eq), $AlCl_3$ (0.1 eq), rt	**III** (55) + **XVI** (55) cis:trans = 60:40	74
	TBSH (3 eq), TBSOTf (0.01 eq), rt, 2 h	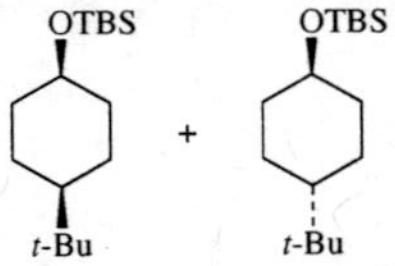 **XVII** **XVIII** **XVII + XVIII** (81), **XVII:XVIII** 66:34	392
	Et_3SiH (1.1 eq), HCO_2H (2.0 eq), rt	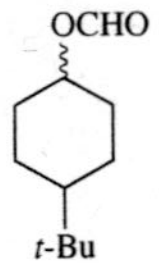 (100) cis:trans = 38:62	74
	Me_2ClSiH (1.2 eq), $In(OH)_3$ (5 mol%), $CHCl_3$, 60°, 2 h	 (76) cis:trans = 12:88	331

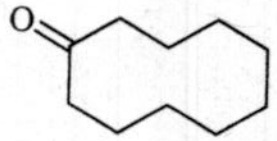	$PhSiH_3$ (0.4 eq), $Mn(dpm)_3$ (3 mol%), *i*-PrOH, DCE, O_2, rt	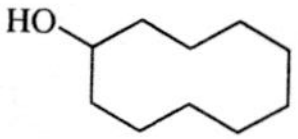 (22)	367
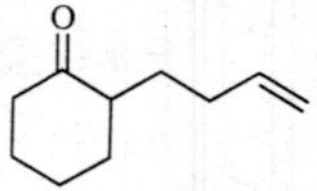	$PhMeSiH_2$ $Cp_2Ti(PMe_3)_2$ (10 mol%), PMe_3 (60 mol%), MeC_6H_5, –20°	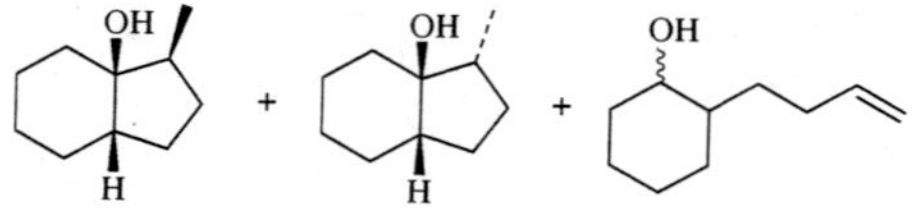 **I** + **II** + **III** **I** + **II**:**III** = 6:1, **I**:**II** = 70:1, **I** + **III** (72)	428, 429
	Et_3SiH (1.5 eq), H_2SO_4/carbon, *n*-C_6H_{14}, rt, 6 h	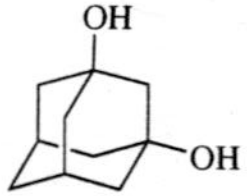 (95)	243
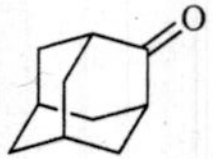	$PhSiH_3$ (0.4 eq), $Mn(dpm)_3$ (3 mol%), *i*-PrOH, DCE, O_2, rt	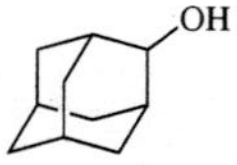 **I** (99)	367
	1. BF_3, CH_2Cl_2, 0° 2. $EtMe_2SiH$ (2.2 eq), 0°, 60 min	 **II** (100)	1
	Et_3SiH (1.5 eq), CF_3SO_3H (4 eq), CH_2Cl_2, 0° to rt, 2 h	**I** (79)[b] + **II** (20)[b]	420
	Et_3SiH (1.3 eq), BF_3•OEt_2 (4-6 eq), CH_2Cl_2, rt, 1 h	**II** (78)	217
	Et_3SiH (2 eq), TMSOTf (0.1 eq), CH_2Cl_2, rt, 2 h	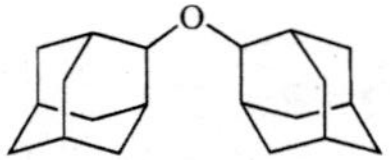 **III** (100)[a]	334

TABLE 12. ORGANOSILANE REDUCTION OF KETONES (*Continued*)

Ketone	Conditions	Product(s) and Yield(s) (%)	Refs.
C_{10} 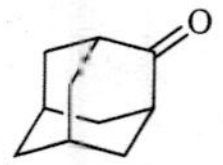	TMSH (xs), **220** (0.09 eq), Ph_3CCl (0.1 eq), CH_2Cl_2, rt, 2 h,	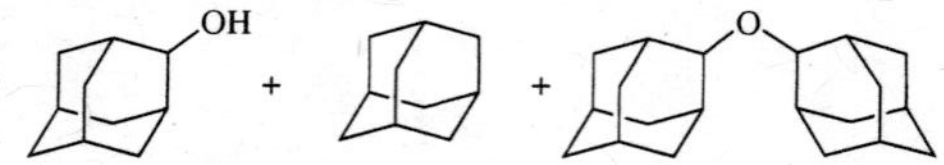 **I** + **II** + **III**; **I + II + III** (98), **I:II:III** = 8:15:75	424
	TMSH (xs), TMSCl (cat.), Ph_3CH (0.1 eq), **220** (0.09 eq), CH_2Cl_2, rt, 2 h	**I + II + III** (75), **I:II:III** = 26:40:29	424
	1. $HC(OMe)_3$, CH_2Cl_2, rt, 2 h 2. Et_3SiH, Nafion®-H, CH_2Cl_2, reflux, 3 h	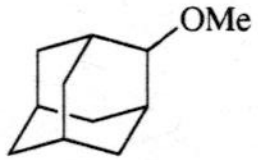 (93)	335
	Et_3SiH, TMSOTf, TMSI, c-$C_6H_{11}OTMS$, CH_2Cl_2, 0°, 2 h	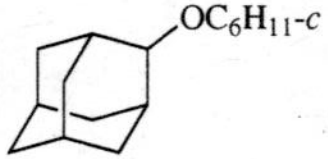 (100)[a]	334
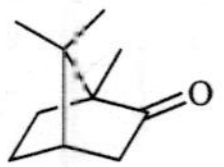	$(EtO)_3SiH$ (1 eq), CsF, rt, 1 min	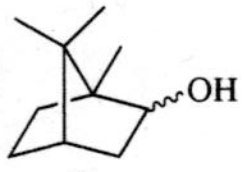 **I** (95)	80
	$PhSiH_3$ (0.4 eq), $Mn(dpm)_3$ (3 mol%), i-PrOH, DCE, O_2, rt	**I** (trace)	367
	Et_3SiH (1 eq), TMSOTf (1 eq), n-$C_6H_{13}OTMS$ (0.83 eq), CH_2Cl_2, 0° to rt	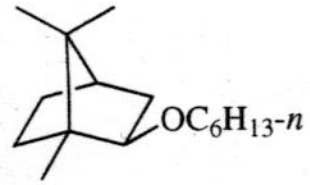 (88)	341

R_3SiH, $(Ph_3P)_3RhCl$, solvent, 0-80°

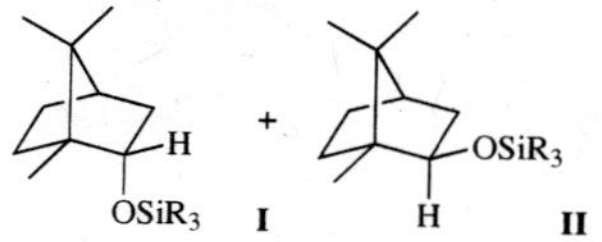

R_3Si	mmol Cat.	Solvent	I	II
Ph_3Si	0.1	none	(90)	(10)
Et_2HSi	0.1	none	(91)	(9)
PhMeHSi	0.1	none	(75)	(25)
Ph_2HSi	0.1	none	(73)	(27)
Et_3Si	1.0	hexane	(30)	(70)
$PhMe_2Si$	0.1	none	(0)	(0)

388

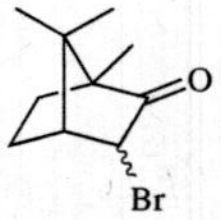

$(EtO)_3SiH$ (1.1 eq), CsF (1 eq), 70°, 3 h

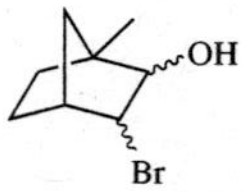

(60)

79, 80

$PhSiH_3$ (0.4 eq), $Mn(dpm)_3$ (3 mol%), *i*-PrOH, DCE, O_2, rt

(trace)

367

TABLE 12. ORGANOSILANE REDUCTION OF KETONES (*Continued*)

Ketone	Conditions	Product(s) and Yield(s) (%)	Refs.
C_{10} 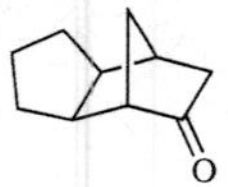	1. $HC(OMe)_3$, CH_2Cl_2, rt, 2 h 2. Et_3SiH, Nafion®-H, CH_2Cl_2 reflux, 3 h	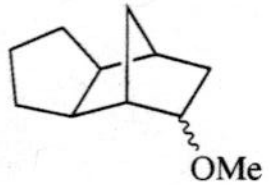 (88)	335
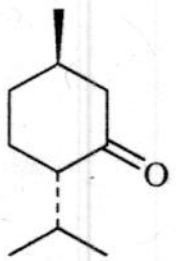	R_3SiH, $(Ph_3P)_3RhCl$, solvent, 0-80°	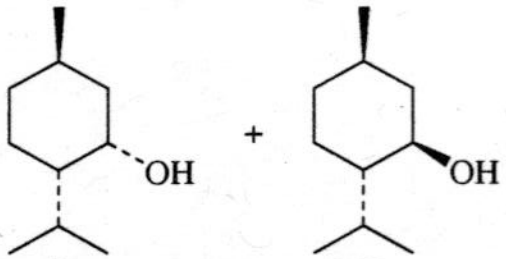 **I** + **II**	387, 388
	PhMe$_2$SiH (1.2 eq), [Rh(C_2H_4)$_2$]$_2$ (0.5 mol%), P(tm-tp)$_3$ (2 mol%), C_6H_6, rt, 20 h	**I + II** (92), **I:II** 45:55	389
	R_3SiH (1.1 eq), acid catalyst (1 eq), 0° to rt, 24 h	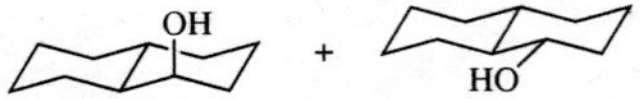 **I** + **II**; **I + II** (—)	384

R_3Si	mol% Cat.	Solvent	I	II
Ph_3Si	0.1	none	(90)	(10)
Et_2HSi	0.1	none	(83)	(17)
PhMeHSi	0.1	none	(86)	(14)
Ph_2HSi	0.1	none	(85)	(15)
Et_3Si	1.0	hexane	(64)	(36)
$PhMe_2Si$	0.1	none	(0)	(0)

R_3Si	Acid Catalyst	I:II
n-BuH$_2$Si	HOAc	18:82
Et_3Si	HOAc	39:61
Et_3Si	BF_3•OEt_2	64:36

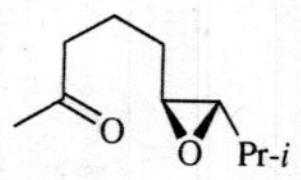	Ph_3SiH (4 eq), catalyst (2.2 eq), CH_2Cl_2, –78°, 6 h; 0°, 12 h Catalyst: BF_3•OEt_2; TMSOTf	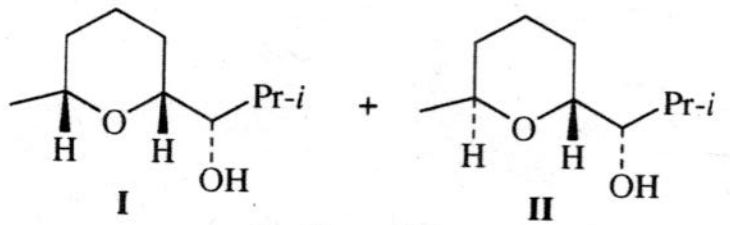 I + II: (81), I:II >99:1 (BF_3•OEt_2); (47), — (TMSOTf)	407
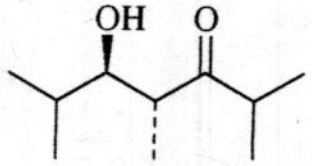	1. $(i\text{-}Pr)_2SiClH$, Et_3N, DMAP, hexane 2. $SnCl_4$, CH_2Cl_2, –80°	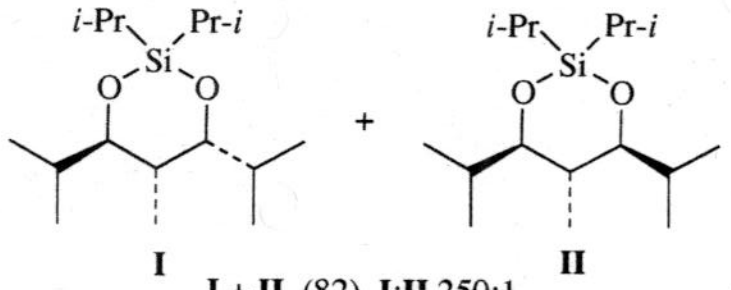 **I + II** (82), **I:II** 250:1	397
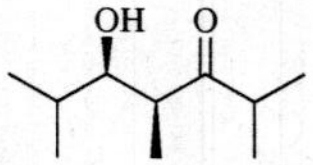	1. $(i\text{-}Pr)_2SiClH$, Et_3N, DMAP, hexane 2. $SnCl_4$, CH_2Cl_2, –80°	**I** + 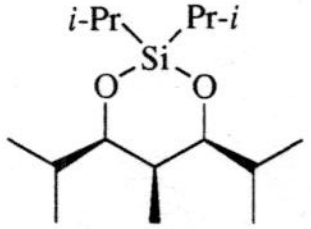 **III** **I + III** (75), **I:III** = 300:1	397
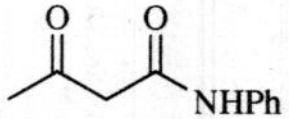	$(EtO)_3SiH$ (2.3 eq), CsF (1 eq), rt, 10 min	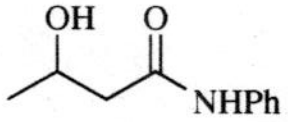 (90)	80, 83
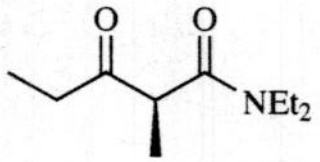	$PhMe_2SiH$ (1.2 eq), TFA, 0°, 6 h	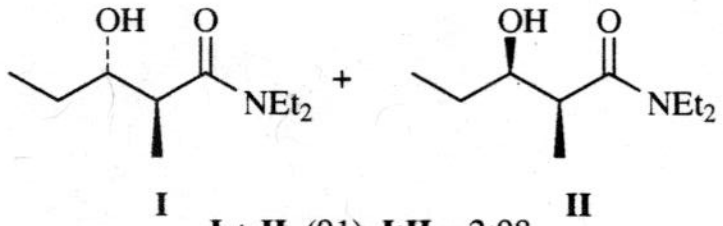 **I + II** (91), **I:II** = 2:98	87, 276
	$PhMe_2SiH$ (1.2 eq), TASF (10 mol%), DMPU 0°, 22 h	**I + II** (93), **I:II** 23:77	87, 276

TABLE 12. ORGANOSILANE REDUCTION OF KETONES (*Continued*)

Ketone	Conditions	Product(s) and Yield(s) (%)	Refs.
C_{10}			
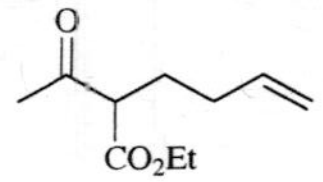	Ph_2SiH_2 (1.0 eq), $Cp_2Ti(PPh_3)_2$, MeC_6H_5, 21°	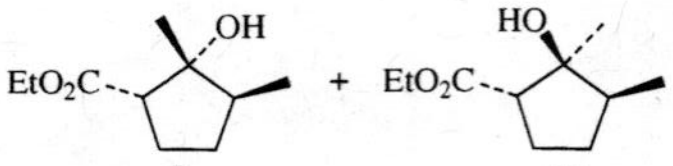 **I** + **II** + 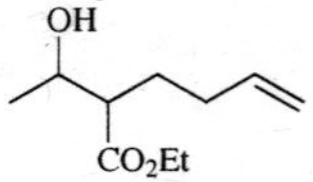 **III** **I** (78), **I** + **II**:**III** = 99:1, **I**:**II** = 90:1	428
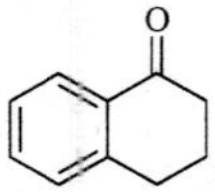	Et_3SiH (1.3 eq), $BF_3•OEt_2$ (4-6 eq), CH_2Cl_2, rt, 6 h	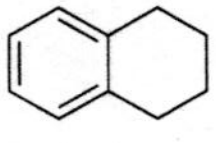 (67)	217
	$(MeO)_3SiH$ (1 eq), Et_2O:TMEDA (30:1), Li-(*R*)-BINOL (10 mol%), 0°, 24 h	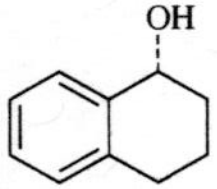 (39) 93% ee	592
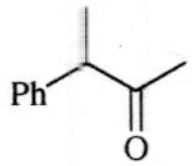	Silane, Triton-B	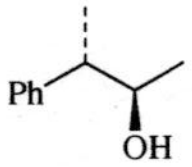 **I** + 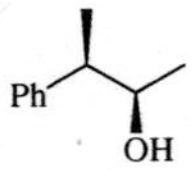 **II** **I** + **II** (>90)	278
	Silane	**I**:**II**	
	PMHS	95:5	
	$(TMSO)_3SiH$	95:5	
	$(MeO)_3SiH$	95:5	
	$(EtO)_3SiH$	73:27	
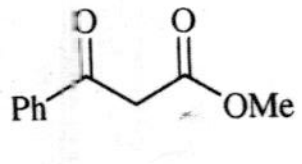	Ph_2SiH_2 (1.2 eq), $AlCl_3$ (1 eq), CH_2Cl_2, 20°, 20 h	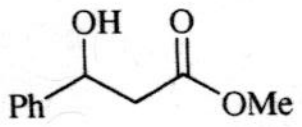 (90) + Ph_2HSiCl (—)	373

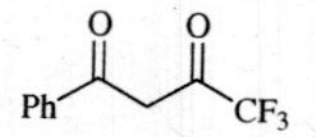	Ph_2SiH_2 (1.2 eq), $AlCl_3$ (1 eq), CH_2Cl_2, 20°, 22 h	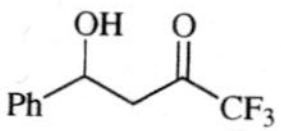 (30) + Ph_2SiHCl (—)	373
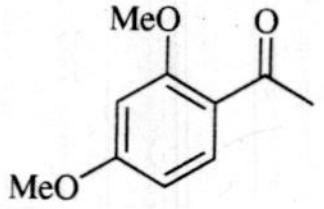	TBSH (1.2 eq), **123** (0.1 mol%), CuCl (0.5 mol%), NaOBu-*t* (3 mol%), MeC_6H_5, rt, 4 h	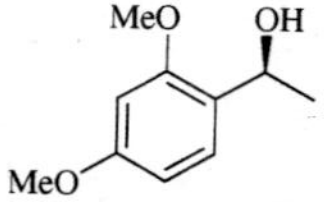 (90)[b] 80% ee	749
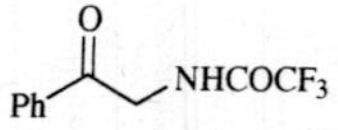	Et_3SiH (3.42 eq), TFA, reflux, 2 h; rt, 16 h	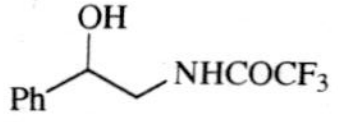 (50)	376
	Et_3SiH (4 eq), BF_3•OEt_2 (20 eq), reflux, 2 h; rt, 16 h	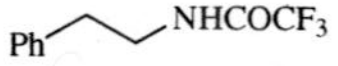 (78)	376
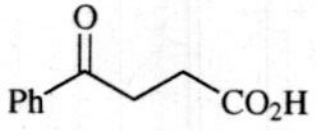	Et_3SiH (2.2 eq), TFA, rt	 (86) + 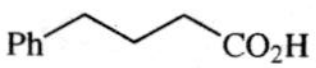 (14)	73
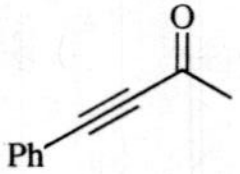	Me_2ClSiH (1.2 eq), $In(OH)_3$ (5 mol%), $CHCl_3$, 0°, 0.3 h	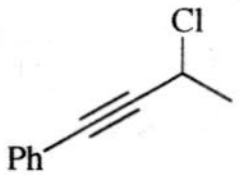 (78)	331
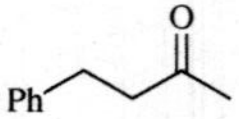	Me_2ClSiH (1.2 eq), $In(OH)_3$ (5 mol%), $CHCl_3$, 60°, 3 h	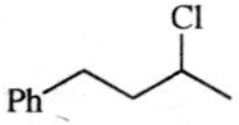 (92)	331
	Me_2ClSiH (1.2 eq), $In(OH)_3$ (5 mol%), LiI, $CHCl_3$, rt, 11 h	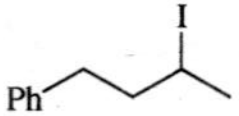 (74)	331

TABLE 12. ORGANOSILANE REDUCTION OF KETONES (*Continued*)

Ketone	Conditions	Product(s) and Yield(s) (%)	Refs.
C_{10-23} 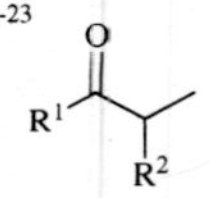	PMHS, (2-2.5 eq), TBAF (1 eq), THF, 0-5°, 2 h	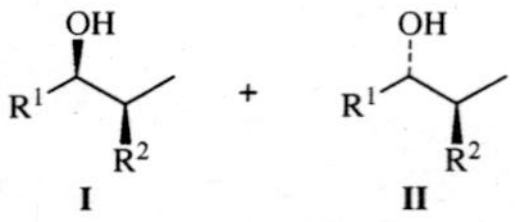	401

R^1	R^2	I + II	I:II
Ph	MeO	(77)	87:13
Me	BnO	(79)	73:27
Ph	AcO	(76)	97:3
Ph	Et_2N	(76)	100:0
Ph	*N*-piperidinyl	(80)	100:0
Ph	BnNH	(88)	26:74
Ph	BzO	(71)	97:3
Ph	BnMeN	(87)	100:0
Me	Bn_2N	(90)	95:5
Ph	Bn_2N	(95)	100:0

Ketone	Conditions	Product(s) and Yield(s) (%)	Refs.
C_{11} 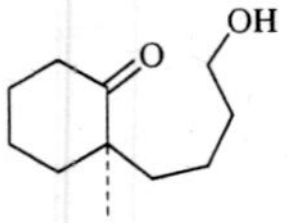	Ph_2MeSiH (2 eq), TMSOTf (1 eq), CH_2Cl_2, 0°, 2 h	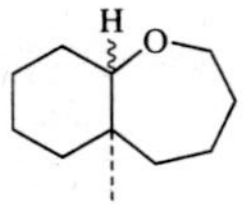 (50) trans:cis = 1:1	336
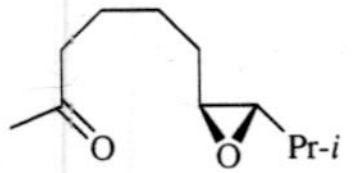	Ph_3SiH (4 eq), BF_3•OEt_2 (2.2 eq), CH_2Cl_2, −78° to 20°, 12 h	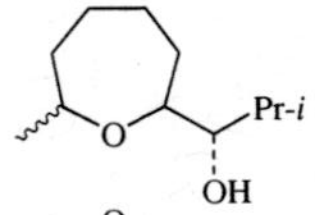 (81) cis:trans > 20:1	407
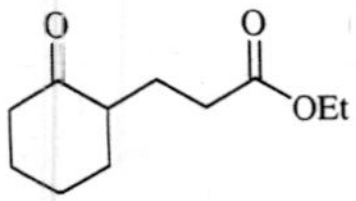	PMHS (1.2 eq), KF•$2H_2O$ (1.3 eq), DMF, 30°, 5 h	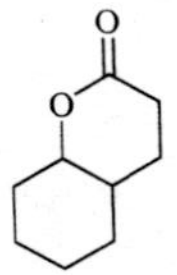 (59)	82

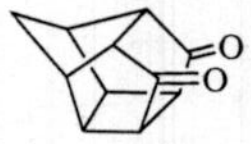	Et_3SiH (4 eq), TFA (36 eq), 20° 20 min	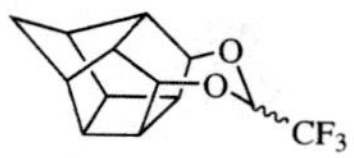 (53)	409
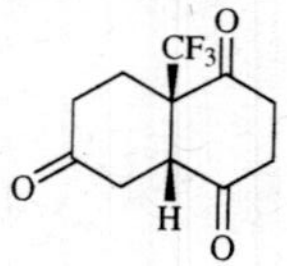	PMHS, TFA, rt, 2 d	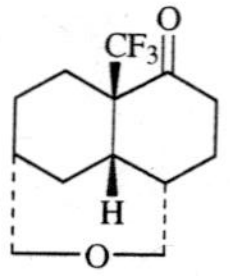 (58) + 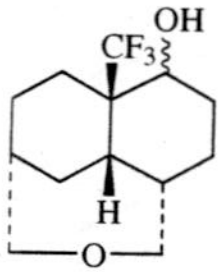 (—)	410
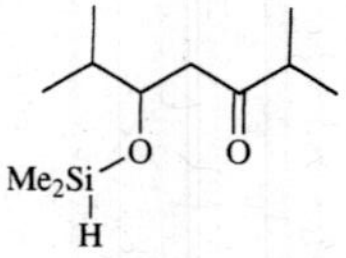	$SnCl_4$ (0.1 eq), –80°, 2 h	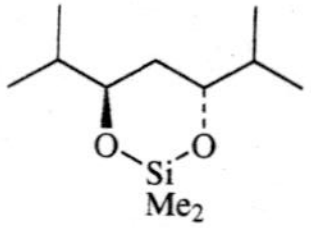 + 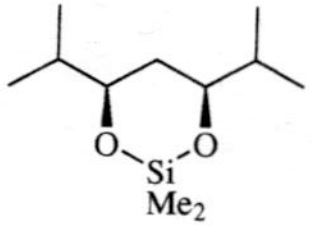 **I** **I + II** (37), **I**:**II** = — **II**	399
	$MgBr_2•OEt_2$ (0.1 eq), rt, 24 h	**I + II** (—), **I**:**II** = 14:1	399
	$BF_3•OEt_2$ (0.5 eq), –80°, 2 h	**I + II** (—), **I**:**II** = 23:1	399
	$ZnBr_2$ (0.12 eq), –80°, 8 h; rt, 16 h	**I + II** (—), **I**:**II** = 1:6	399
	$ZnCl_2$ (0.25 eq), –80°, 30 min; rt, 15 min	**I + II** (—), **I**:**II** = 1:4	399
	TFA (0.2 eq), –80°, 30 min	**I + II** (—), **I**:**II** = 2:1	399
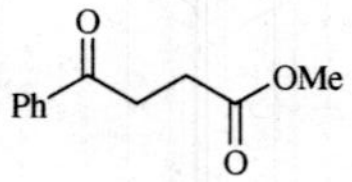	$Me(EtO)_2SiH$ (1 eq), CsF, rt, 2.5 h	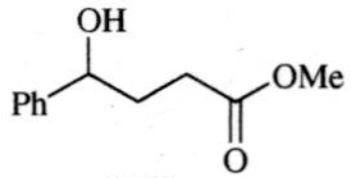 (85)	80, 83
	$Me(EtO)_2SiH$ (2.3 eq), CsF, rt, 2.5 h	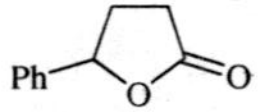 (85)	83
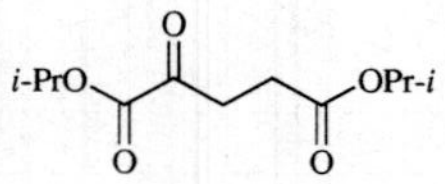	$(EtO)_3SiH$ (1 eq), CsF, rt, 0.5 h	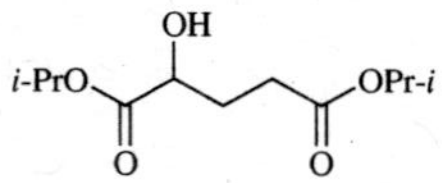 (70)	79, 80

TABLE 12. ORGANOSILANE REDUCTION OF KETONES (*Continued*)

Ketone	Conditions	Product(s) and Yield(s) (%)	Refs.
C_{11} 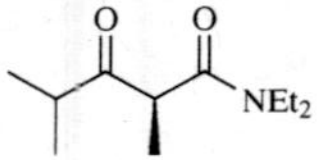	$PhMe_2SiH$ (1.2 eq), TFA, 0°, 20 h	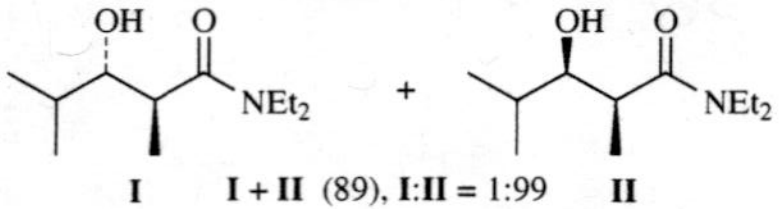 I + II (89), I:II = 1:99	87, 276
	$PhMe_2SiH$ (1.2 eq), TASF (10 mol%), DMPU, 0°, 24 h; rt, 72 h	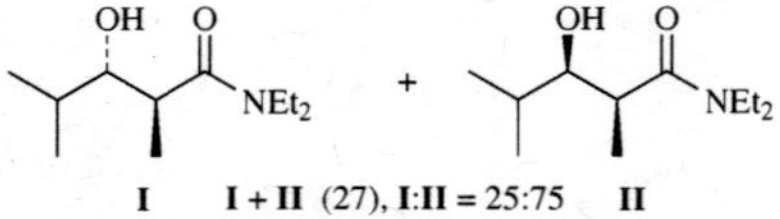 I + II (27), I:II = 25:75	87, 276
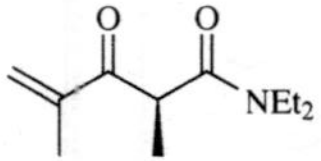	$PhMe_2SiH$ (1.2 eq), TFA, 0°, 16 h	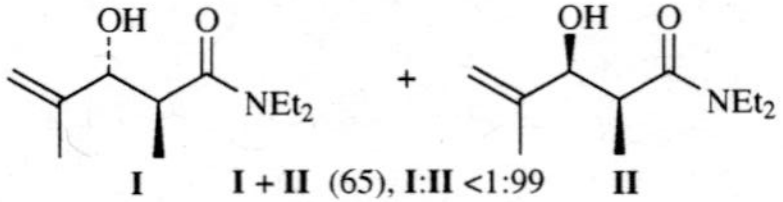 I + II (65), I:II <1:99	87, 276
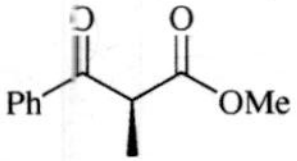	$PhMe_2SiH$ (1.2 eq), TFA, 0°, 3 h	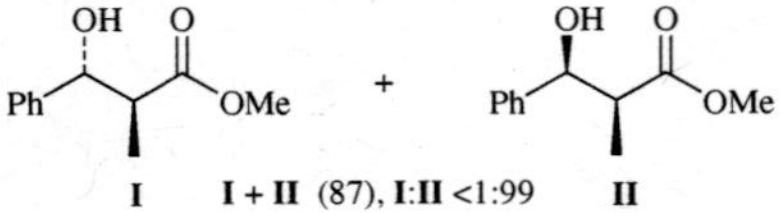 I + II (87), I:II <1:99	87
	$PhMe_2SiH$ (1.2 eq), TASF (10 mol%), DMPU, rt, 12 h	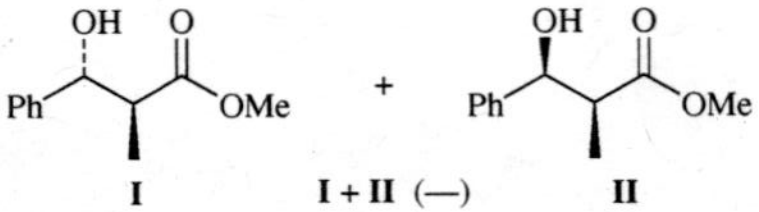 I + II (—)	87, 276
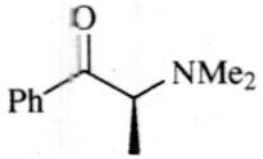	$PhMe_2SiH$ (1.1-1.2 eq), TBAF (5-10 mol%), HMPA, rt, 12 h	 I + II (83) threo:erythro >99:1	86
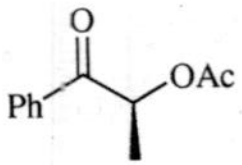	$PhMe_2SiH$ (1.1-1.2 eq), TBAF (5-10 mol%), HMPA, 0°, 20 h	 I + II (95) threo:erythro = 95:5	86, 320

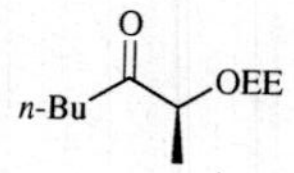	1. $PhMe_2SiH$ (1.2 eq), TBAF (5-10 mol%), HMPA, 0°, 12 h 2. H_2O	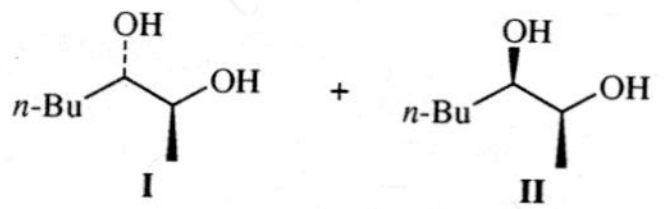 **I + II** (86) threo:erythro = 84:16	86, 320
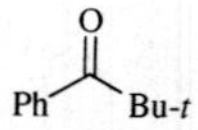	$(EtO)_3SiH$ (1 eq), CsF, rt, 2 h	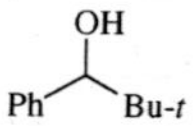 (75)	80
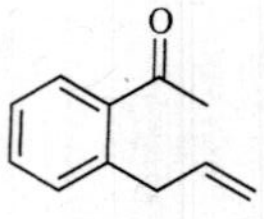	1. Ph_2SiH_2, $Cp_2Ti(PMe_3)_2$ (10 mol%), PMe_3 (60 mol%), MeC_6H_5, –20° 2. TBAF, THF, 15 min	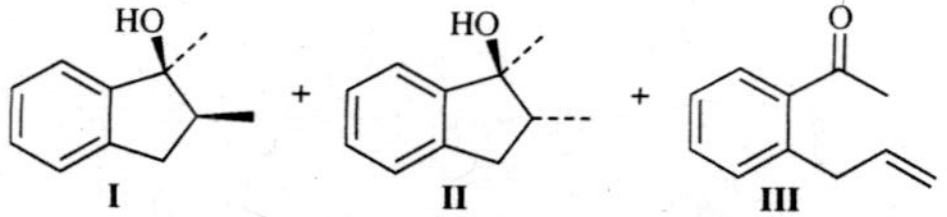 **I** (86), (**I** + **II**):**III** = 99:1, **I**:**II** = 99:1	428, 429
	1. Ph_2SiH_2, $Cp_2Ti(PMe_3)_2$ (10 mol%), PMe_3 (60 mol%), MeC_6H_5, –20° 2. HCl/acetone, 3 h	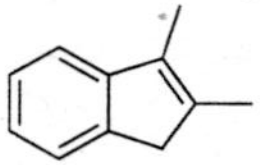 (71)	428, 429
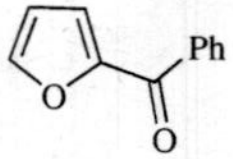	Et_3SiH (10 eq), TFA (x eq), $BF_3•OEt_2$ (y eq), 20°	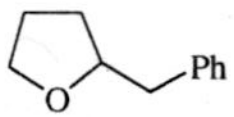	211
	x = 50, y = 0, Time = 5 min	(30)	
	x = 50, y = 0, Time = 30 min	(64)	
	x = 50, y = 0, Time = 2 h	(70)	
	x = 28, y = 0.6, Time = 5 min	(89)	
	x = 28, y = 0.6, Time = 1 h	(90)	
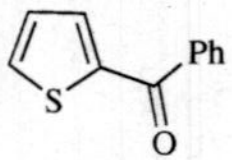	Et_3SiH (5 eq), TFA (9 eq), 50°, 30 h	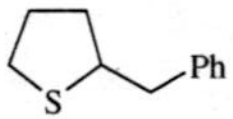 (75)	257
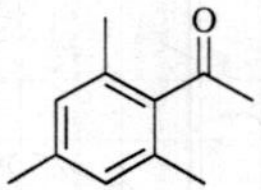	Et_3SiH (1.5 eq), CF_3SO_3H (4 eq), CH_2Cl_2, 0° to rt, 2 h	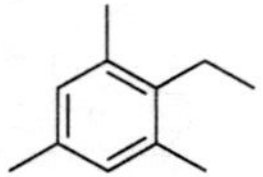 (84)	420

TABLE 12. ORGANOSILANE REDUCTION OF KETONES (*Continued*)

Ketone	Conditions	Product(s) and Yield(s) (%)	Refs.
C_{11}			
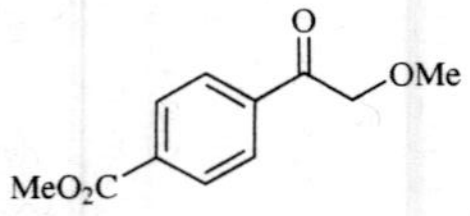	Me_2ClSiH (1.2 eq), $In(OH)_3$ (5 mol%), $CHCl_3$, rt, 3.5 h	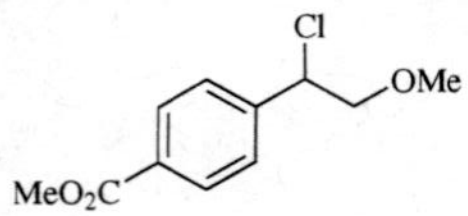 (46)	331
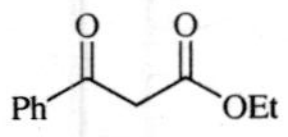	Me_2ClSiH (1.2 eq), $In(OH)_3$ (5 mol %), $CHCl_3$, rt, 2 h	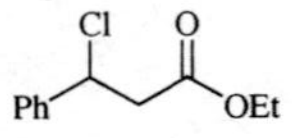 (64)	331
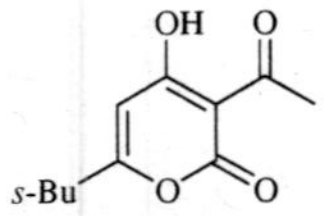	Et_3SiH, TFA, $LiClO_4$ (0.01 eq), rt, 4 h	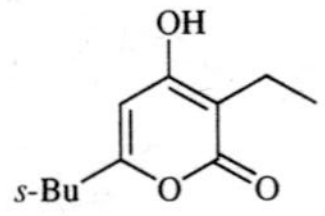 (84)	423
C_{11-13}			
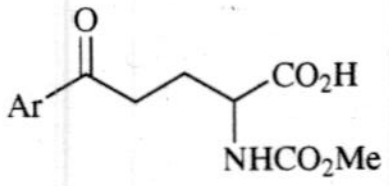	1. TMSCl, Et_3N 2. $PhMe_2SiH$, $TiCl_4$, CH_2Cl_2, –78° to rt	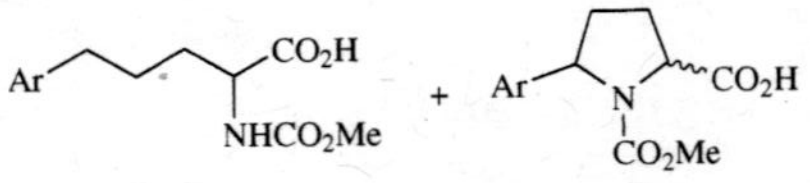 **I** + **II**	421

Ar	I + II	I:II
Ph	(97)	94:6
4-ClC_6H_4	(95)	98:2
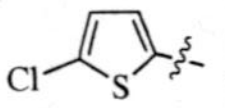	(95)	>99:1

C_{11-16}

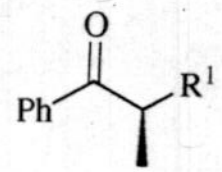

R_3SiH (1.1-1.2 eq),
TBAF (5-10 mol%), HMPA

I + II

320

R^1	R_3Si	Temp	Time	I + II	I:II
OAc	$PhMe_2Si$	0°	20 h	(95)	95:5
OAc	$PhMe_2Si$	0°	4 h	(93)	93.3:6.7
OAc	$(4\text{-}CF_3C_6H_4)Me_2Si$	0°	4 h	(72)	93.1:6.9
OAc	$(4\text{-}MeC_6H_4)Me_2Si$	0°	4 h	(83)	93.1:6.9
OAc	$(4\text{-}MeOC_6H_4)Me_2Si$	0°	4 h	(88)	93:7
OAc	$PhMe_2Si$	rt	14 h	(99)	93:7
OAc	Ph_2MeSi	rt	14 h	(89)	92:8
OAc	$(i\text{-}PrO)Ph_2Si$	rt	3 h	(91)	86:14
OAc	$(i\text{-}PrO)_3Si$	−50°	12 h	(71)	78:22
OBz	$PhMe_2Si$	0°	6 h	(82)	96:4
OEE	$PhMe_2Si$	0°	18 h	(55)	90:10

TABLE 12. ORGANOSILANE REDUCTION OF KETONES (*Continued*)

Ketone	Conditions	Product(s) and Yield(s) (%)	Refs.
C_{11-16} 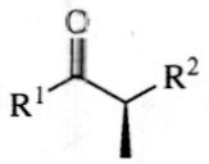	$PhMe_2SiH$ (1.1-1.2 eq), acid	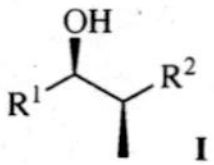 I + 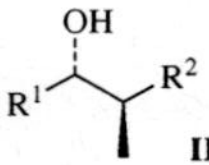 II	276

R^1	R^2	Acid	Temp	I + II	I:II
Ph	CO_2Me	TMSOTf (1 eq)	rt	(83)	79:21
Ph	CO_2Me	$AlCl_3$ (1 eq)	rt	(66)	78:22
Ph	NMe_2	TFA	rt	(0)	—
Ph	$NHCO_2Me$	TFA	0°	(87)	>99:1
Ph	$NHCO_2Et$	TFA	0°	(87)	>99:1
Ph	$NHCO_2Et$	$AlCl_3$ (1 eq)	0°	(64)	70:30
Ph	$NHCO_2Et$	TMSOTf (1 eq)	0°	(65)	71:29
2,5-$(MeO)_2C_6H_3$	$NHCO_2Me$	TFA	0°	(84)	>99:1
n-Bu	OBn	TFA	0°	(82)	47:53
Ph	$NHSO_2Ph$	TFA	0°	(66)	98:2
Ph	OBz	TFA	0°	(72)	93:7

C_{12}			
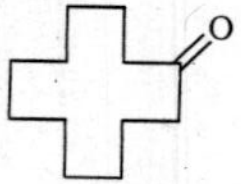	TMSH (xs), **220**-TFPB (0.09 eq), Ph_3CH (0.1 eq), CH_2Cl_2, rt, 2 h	 (91) cis:trans = 40:60	424
	$PhSiH_3$ (0.4 eq), $Mn(dpm)_3$ (3 mol%), *i*-PrOH, DCE, O_2, rt	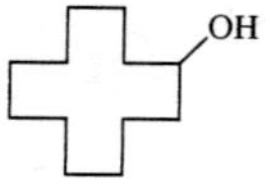 (13)	367
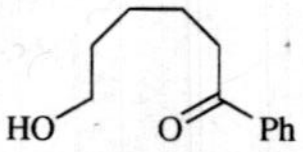	Ph_2MeSiH (2 eq), TMSOTf (1 eq), CH_2Cl_2, 0°, 2 h	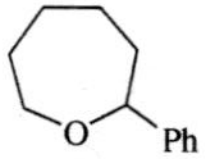 (79)	336
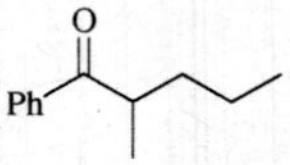	Et_3SiH, $B(C_6F_5)_3$ (2 mol%), MeC_6H_5, 0°, 1 h	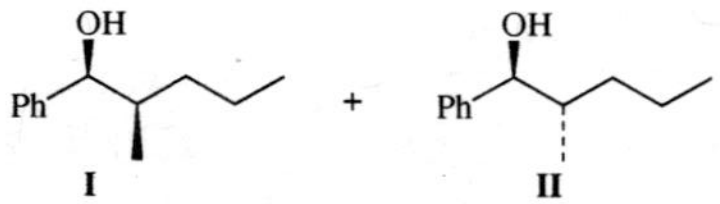 **I** + **II** (98), **I**:**II** = 1:1.5	372
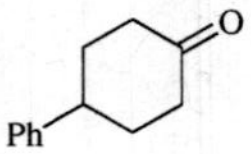	Cl_3SiH (1.5 eq), CH_2Cl_2:DMF (4:1), 0°, 12 h	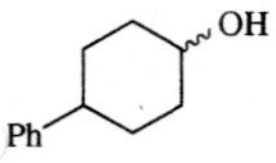 (85)	318
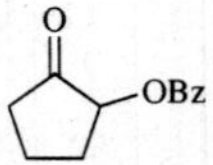	$PhMe_2SiH$ (1.1-1.2 eq), TBAF (5-10 mol%), HMPA, 0°, 4 h	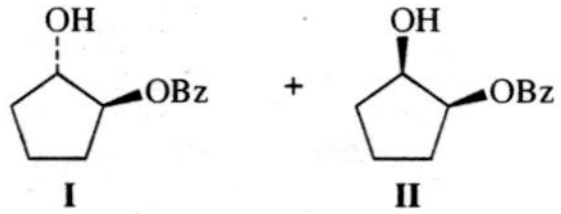 **I** + **II** (76), **I**:**II** = 33:67	320

TABLE 12. ORGANOSILANE REDUCTION OF KETONES (*Continued*)

Ketone	Conditions	Product(s) and Yield(s) (%)	Refs.
C_{12}			
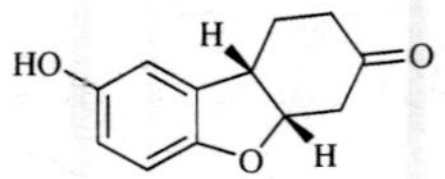	Et_3SiH (3.2 eq), TFA, 0°, 1 h	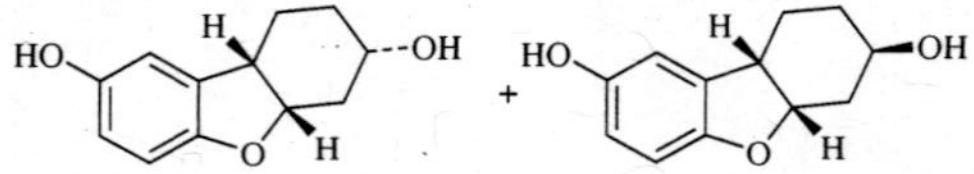 I II; **I** + **II** (56), **I**:**II** = 1:4	394
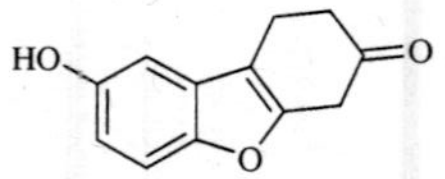	Et_3SiH (3.2 eq), TFA, rt, 24 h	**I** + **II** (77), **I**:**II** = 4:1	394
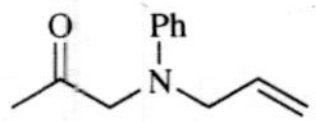	Ph_2SiH_2, $Cp_2Ti(PMe_3)_2$ (10 mol%), PMe_3 (60 mol%), MeC_6H_5, –20°	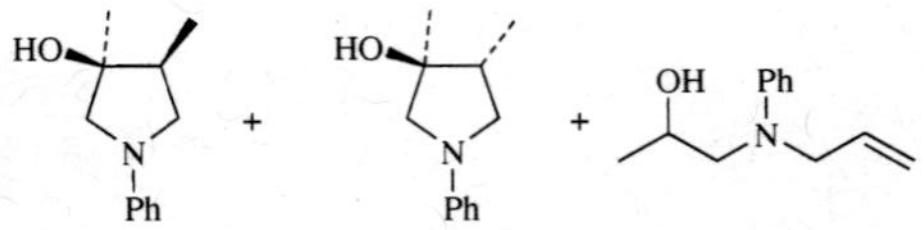 I II III; **I** + **II** + **III** (75), (**I** + **II**):**III** = 99:1, **I**:**II** = 1:1	428, 249
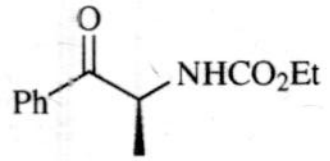	$PhMe_2SiH$ (1.2 eq), TFA, 0°, 2.5 h	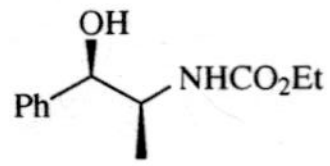 (87) threo:erythro > 99:1	86
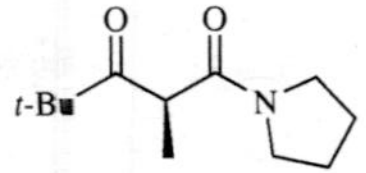	$PhMe_2SiH$ (1.2 eq), TASF (10 mol %), DMPU, 0°, 16 h	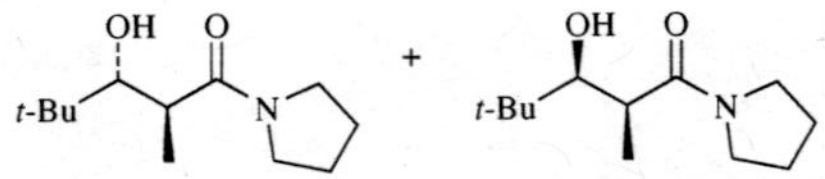 I II; **I** + **II** (90), **I**:**II** = 91:9	320

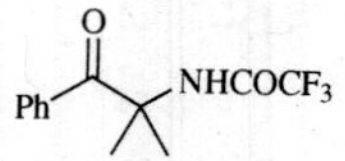	Et_3SiH (3.42 eq), TFA, reflux, 2 h; rt, 16 h	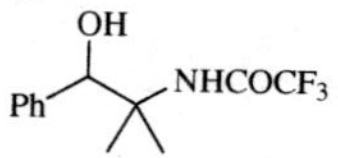 (72)	376
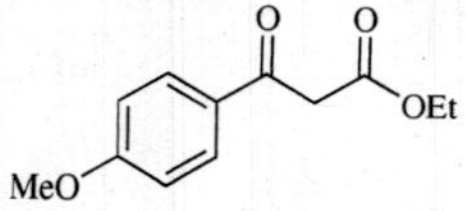	Ph_2SiH_2 (1.2 eq), $AlCl_3$ (1 eq), CH_2Cl_2, 20°, 18 h; 40°, 4 h	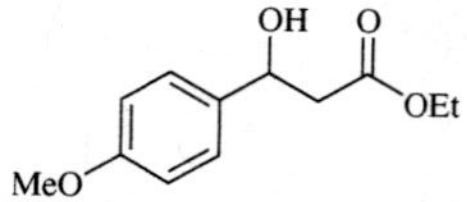 (78) + Ph_2HSiCl (—)	373
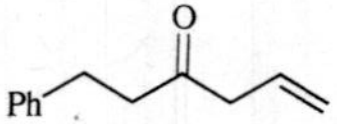	PMHS (3 eq), $(C_6F_5)_3B$ (5 mol%), CH_2Cl_2, rt, 5-20 min	 (88)	354
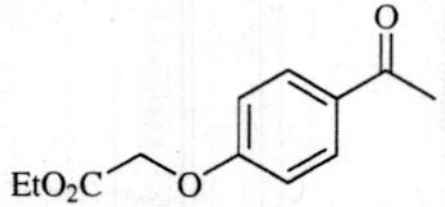	PMHS (3 eq), $(C_6F_5)_3B$ (5 mol%), CH_2Cl_2, rt, 5-20 min	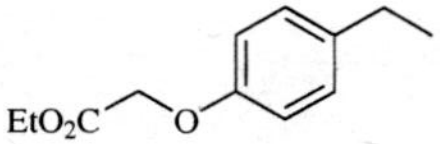 (82)	354
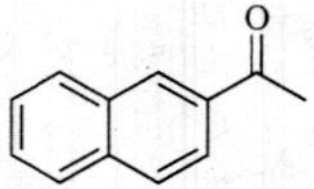	Et_3SiH (2.6 eq), $BF_3{\bullet}OH_2$ (4-6 eq), CH_2Cl_2, rt, 10 h	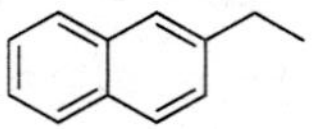 (70)	217
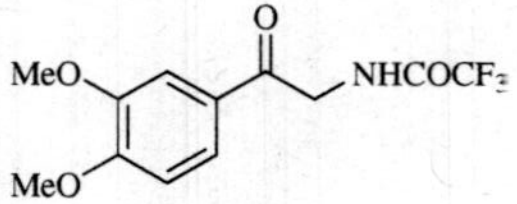	Et_3SiH (3.42 eq), TFA, reflux, 2 h, rt, 16 h	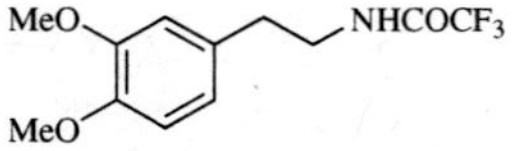 (74)	376
C_{12-13}			
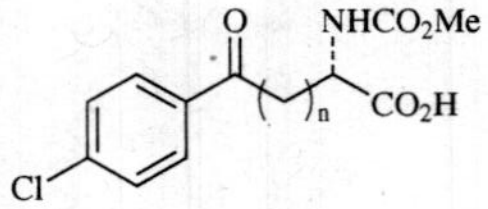	1. TMSCl, Et_3N 2. Et_3SiH, $TiCl_4$, CH_2Cl_2, 0°, rt	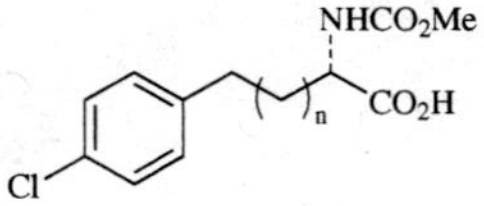 n = 1 (89) n = 2 (85)	421, 752

TABLE 12. ORGANOSILANE REDUCTION OF KETONES (*Continued*)

Ketone	Conditions	Product(s) and Yield(s) (%)	Refs.
C_{12-14} 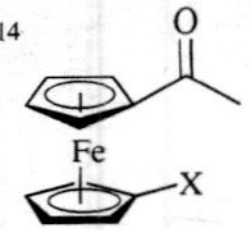	Et_3SiH (2.5 eq), TFA (5 eq), 20°	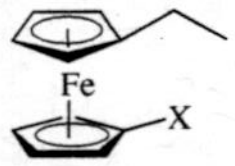 X: H (40); Cl (45); CO_2Me (65); CN (85)	180
C_{12-15} 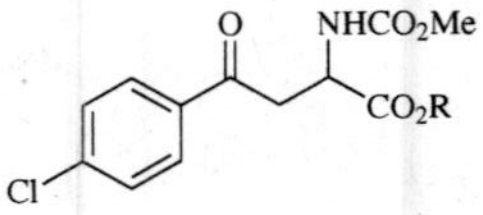	Et_3SiH, acid, solvent	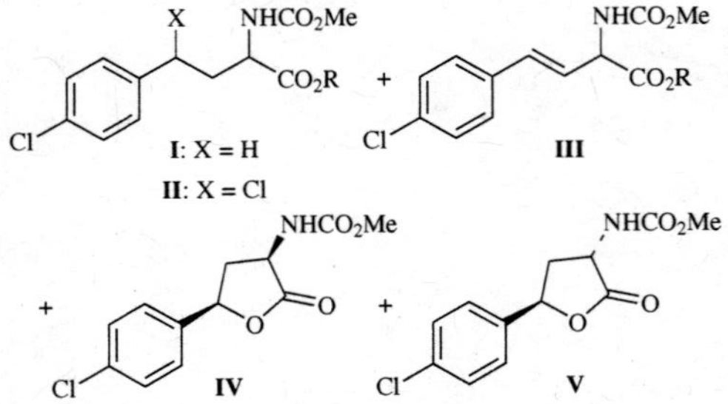 I: X = H; II: X = Cl; III; IV; V	421

R	Acid	Solvent	Temp	I	II	III	IV	V
H	TFA	none	reflux	(—)	(—)	(—)	(54)	(7)
Me	TFA	none	reflux	(25)	(—)	(—)	(49)	(10)
H	BF_3•OEt_2	none	70°	(—)	(—)	(—)	(65)	(6)
Me	BF_3•OEt_2	none	rt	(12)	(—)	(28)	(—)	(—)
H	$TiCl_4$	CH_2Cl_2	rt	(67)	(—)	(—)	(—)	(—)
Me	$TiCl_4$	CH_2Cl_2	rt	(25)	(38)	(—)	(—)	(—)
TMS	$TiCl_4$	CH_2Cl_2	rt	(88)	(—)	(—)	(—)	(—)
TMS	$AlCl_3$	CH_2Cl_2	rt	(45)	(—)	(23)	(—)	(—)
TMS	$SnCl_4$	CH_2Cl_2	rt	(—)	(—)	(—)	(—)	(—)

C_{12-18}

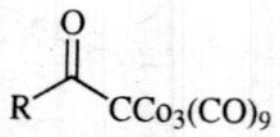

Et$_3$SiH (x eq), TFA (y eq), THF, reflux, 6 h

310, 425

R	x	y	Yield
Me	2.4	2.07	(90)
Et	2.33	2	(92)
n-Pr	2.33	2	(87)
i-Pr	2.67	3.0	(81)
n-Bu	2.33	2	(80)
n-C_6H_{13}	2.67	2.67	(80)
c-C_6H_{11}	2.5	3.0	(75)
Ph	2.5	2.5	(82)
4-MeC_6H_4	2.6	3.0	(78)
4-BrC_6H_4	2.4	2.07	(67)

Et$_3$SiH (x eq), CO, C_6H_6, reflux, 8 h

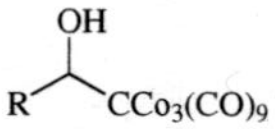

310

R	x	Yield
Me	1.1	(84)
Me	1.06	(90)
Et	1.07	(73)
n-Pr	1.07	(75)
i-Pr	1.0	(80)
n-Bu	1.02	(81)
n-C_6H_{13}	1.02	(81)
c-C_6H_{11}	1.02	(52)
Ph	1.37	(87)
4-MeC_6H_4	1.02	(70)
4-BrC_6H_4	1.02	(84)

TABLE 12. ORGANOSILANE REDUCTION OF KETONES (*Continued*)

Ketone	Conditions	Product(s) and Yield(s) (%)	Refs.
C_{13} 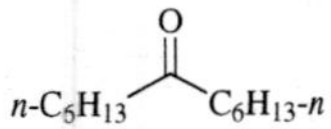	$PhSiH_3$ (0.4 eq), $Mn(dpm)_3$ (3 mol%), *i*-PrOH, DCE, O_2, rt	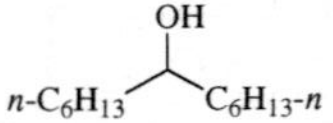 (43)	367
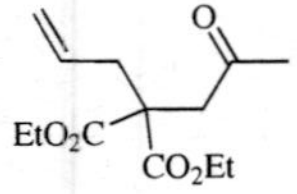	Ph_2SiH_2 (1.0 eq), $Cp_2Ti(PPh_3)_2$, MeC_6H_5, −20°	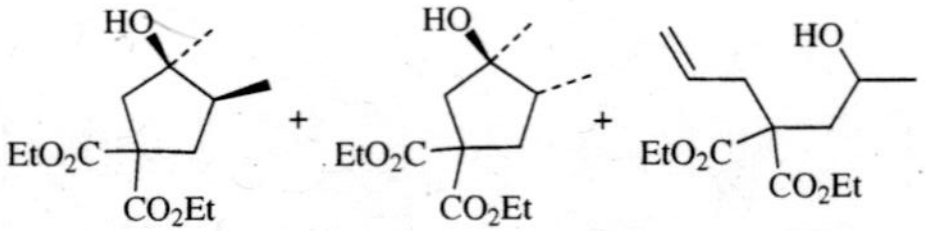 **I** **II** **III** **I** + **II** (68), **I** + **II**:**III** = 99:1, **I**:**II** = 10:1	428
	Ph_2SiH_2, $Cp_2Ti(PMe_3)_2$ (10 mol%), PMe_3 (60 mol%), MeC_6H_5, −20°	**I** (65), **I** + **II**:**III** = 9:1, **I**:**II** 99:1	429
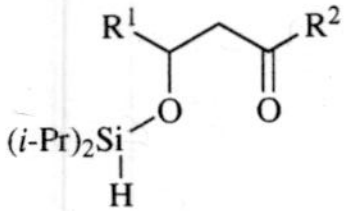	$SnCl_4$ (0.1 eq), −80°, 2 h	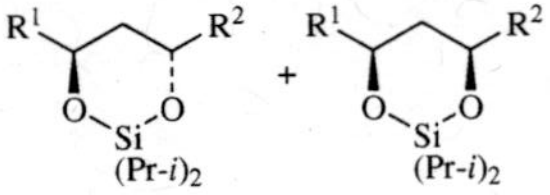 **I** **II**	399
C_{13-17}	Catalyst (x eq), CH_2Cl_2	**I** + **II**	398

R^1	R^2	I + II	I:II
i-Pr	Me	(67)	50:1
Me	*i*-Pr	(69)	40:1

R^1	R^2	Catalyst	x	Temp	Time	I + II	I:II
i-Pr	Me	$SnCl_4$	0.1	−80°	2 h	(61)	50:1
Me	*i*-Pr	$SnCl_4$	0.1	−80°	2 h	(68)	40:1
i-Pr	*i*-Pr	$SnCl_4$	0.1	−80°	2 h	(67)	120:1
i-Pr	*i*-Pr	$MgBr_2{\bullet}OEt_2$	0.1	rt	24 h	(—)	60:1
i-Pr	*i*-Pr	$TiCl_4$	0.1	−80°	30 min	(—)	30:1
i-Pr	*i*-Pr	$BF_3{\bullet}OEt_2$	0.5	−80°	2 h	(—)	320:1
n-Bu	*n*-Bu	$SnCl_4$	0.5	−80°	2 h	(44)	40:1

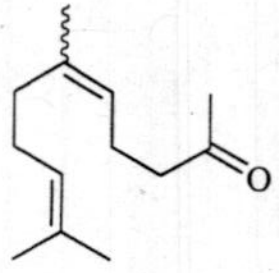	TBSH (1.2 eq), **123** (0.1 mol%), CuCl (3 mol%), NaOMe (3 mol %), Et_2O, rt, <3 h	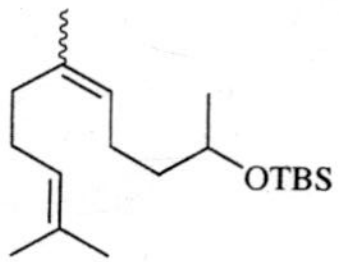	749
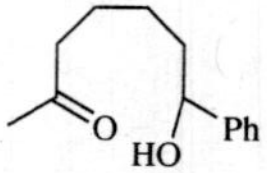	Et_3SiH (10 eq), TMSOTf (1 eq), CH_2Cl_2, 0°, 15 min	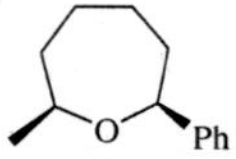 **I** (88)	336
	Ph_2MeSiH (2 eq), TMSOTf (1 eq), CH_2Cl_2, 0°, 2 h	**I** (85)	336
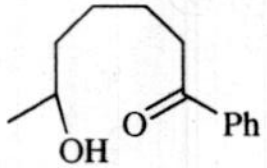	Et_3SiH (10 eq), TMSOTf (1 eq), CH_2Cl_2, 0°, 15 min	**I** (90)	336
	Ph_2MeSiH (2 eq), TMSOTf (1 eq), CH_2Cl_2, 0°, 2 h	**I** (83)	336
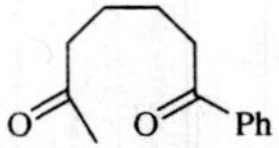	Ph_2MeSiH (2 eq), TMSOTf (1 eq), CH_2Cl_2, 0°, 2 h	**I** (85)	336
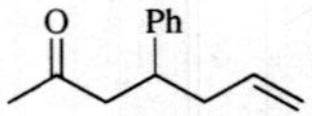	Ph_2SiH_2 (1.0 eq), $Cp_2Ti(PPh_3)_2$, MeC_6H_5	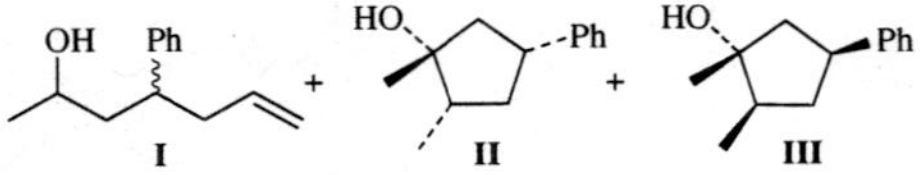 **I** + **II** + **III**	428, 429

Temp	II	(II + III):I	II:III
–20°	(63)	2.2:1	99:1
–5°	(59)	3:1	65:1
0°	(50)	3.5:1	46:1
21°	(63)	5:1	10:1
50°	(55)	6:1	10:1

TABLE 12. ORGANOSILANE REDUCTION OF KETONES (*Continued*)

Ketone	Conditions	Product(s) and Yield(s) (%)	Refs.
C_{13-14}			
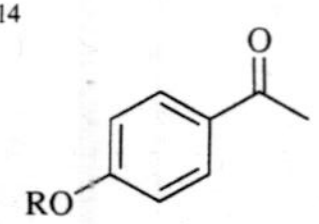	PMHS (3 eq), $(C_6F_5)_3B$ (5 mol%), CH_2Cl_2, rt, 5-20 min	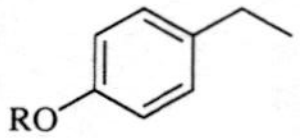 R: THP (84); TBS (90)	354
C_{13}			
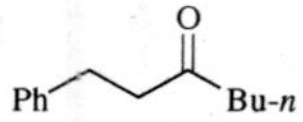	PMHS (3 eq), $(C_6F_5)_3B$ (5 mol%), CH_2Cl_2, rt, 5-20 min	 (90)	354
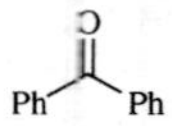	PMHS (3 eq), $(C_6F_5)_3B$ (5 mol%), CH_2Cl_2, rt, 5-20 min	 **I** (88)	354
	Et_3SiH (2.5 eq), TFA (5 eq), 20°	**I** (65)	180
	Et_3SiH (1.3 eq), NH_4F (1.3 eq), TFA (5 eq), 0°, 0.5 h; rt, 24 h	**I** (93)	135
	Et_3SiH (3.7 eq), **220** (0.09 eq), Ph_3CCl (0.1 eq), CH_2Cl_2, rt, 2 h	**I** (87)	424
	$(EtO)_3SiH$ (2.3 eq), CsF (1 eq), rt, 5 min	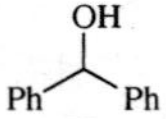 **II** (95)	83
	$Me(EtO)_2SiH$ (2.3 eq), KF (1 eq), DMF, 20°, 0.5 h	**II** (90)	83
	R_3SiH (1 eq), MF	**II**	80, 83

R_3Si	MF	Temp	Time	Yield (%)
$Me(EtO)_2Si$	CsF	rt	1.5 h	(75)
$(EtO)_3Si$	KF	100°	10 h	(95)
$(EtO)_3Si$	CsF	0°	5 min	(95)

	(EtO)$_3$SiH, CaO, n-C_7H_{16}, 90°, 2 h	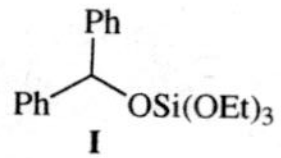 (72)	325
	(EtO)$_3$SiH, MgO n-C_7H_{16}, 90°, 15 h	**I** (35)	325
	Et_3SiH (1.1 eq) MeOH, H_2SO_4, 28°	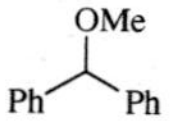 (43)	328
	$HMe_2SiOSiMe_2H$ (1.0 eq), TMSOTf (cat.), C_6H_6, 80°, 240 min	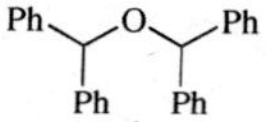 (95)	314
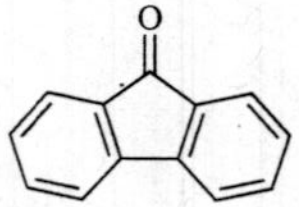	**C** (1 eq), rt, 12 h	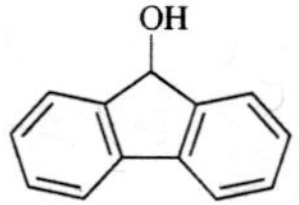 (100)	84
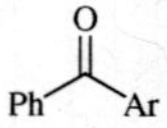	Et_3SiH (2.6 eq), P?HF, TFA, 0°, 10 min; rt		135
Ar	Time		
Ph	2.5 h	(96)	
4-BrC_6H_4	3.5 h	(96)	
4-IC_6H_4	5 h	(93)	
2-FC_6H_4	6 h	(94)	
4-$O_2NC_6H_4$	6 h	(91)	
4-HOC_6H_4	7 h	(91)	
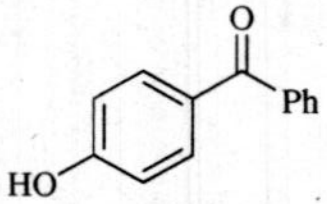	Et_3SiH (2.6 eq), NH_4F (1.3 eq), TFA (5 eq), 0°, 0.5 h; 60°, 16 h	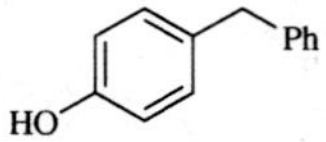 (78)	135

Ketone	Conditions	Product(s) and Yield(s) (%)	Refs.
C_{13}			
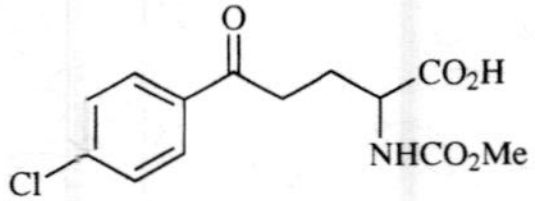	1. TMSCl, Et_3N 2. R_3SiH (x eq), $TiCl_4$, CH_2Cl_2, −78° to rt R_3Si / x: Et_3Si 1.5 Et_3Si 5.0 Ph_3Si 1.5 Ph_3Si 5.0 Ph_2MeSi 5.0 $PhMe_2Si$ 1.5 $PhMe_2Si$ 5.0	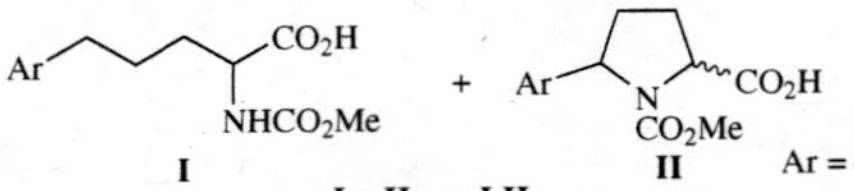 I + II Ar = 4-ClC_6H_4 I + II / I:II: (100) 88:12 (100) 85:15 (98) 75:25 (100) 93:7 (95) 95:5 (85) 94:6 (95) 98:2	421
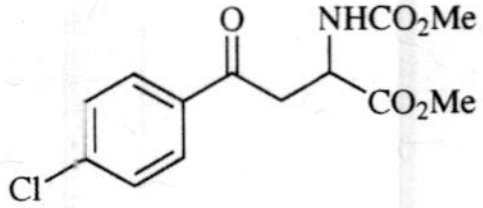	Et_3SiH (x eq), $TiCl_4$ (y eq), TMSOTf (z eq), CH_2Cl_2 x / y / z: 5.0 3.0 6.0 3.0 1.5 3.0 3.0 3.0 —	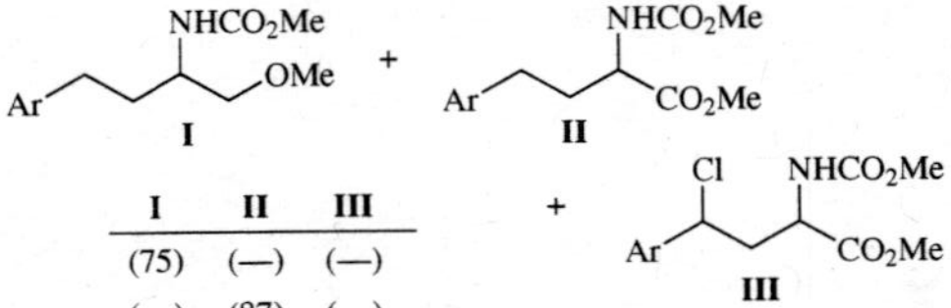 I + II + III Ar = 4-ClC_6H_4 I / II / III: (75) (—) (—) (—) (87) (—) (—) (25) (38)	297
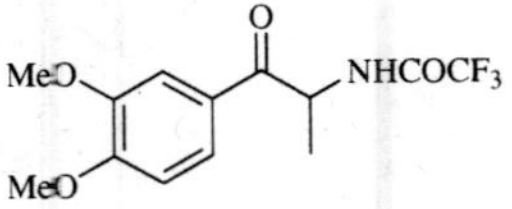	Et_3SiH (3.42 eq), TFA, reflux, 2 h; rt, 16 h	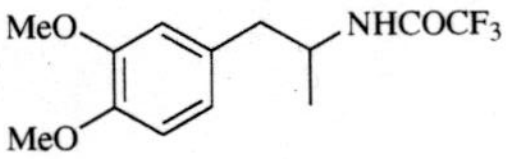 (88)	376, 419
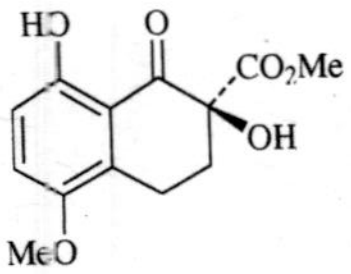	Et_3SiH (3.0 eq), TFA, 0° to rt, 16 h	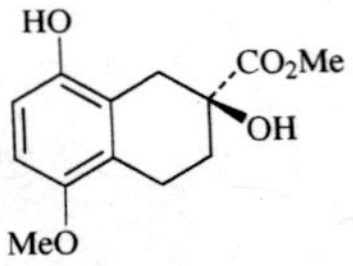 (90)	416

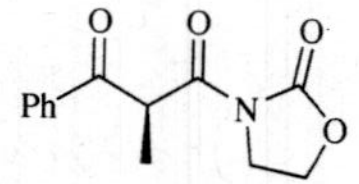

PhMe$_2$SiH (1.2 eq), TFA, 0°, 5 h

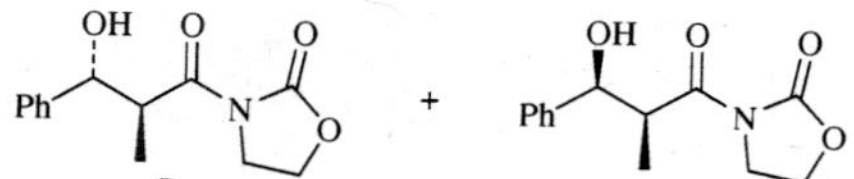

I + II

I + II (98), **I**:**II** < 1:99

87, 276

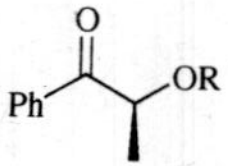

PhMe$_2$SiH (1.1-1.2 eq), TBAF (5-10 mol%), HMPA, 0°, 12 h

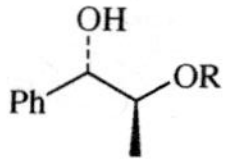

R	Time		threo:erythro
EE	12 h	(86)	84:16
THP	16 h	(77)	87:13
t-Bu	12 h	(90)	93:7

86

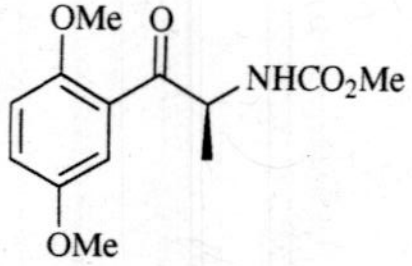

PhMe$_2$SiH (1.1-1.2 eq), TFA, 0°, 2.5 h

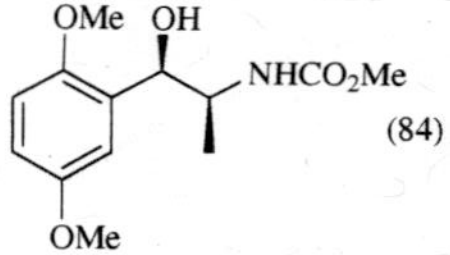

(84) threo:erythro < 1:99

86

C_{13-15}

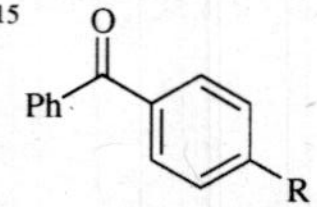

Et$_3$SiH (3 eq), BF$_3$•OEt$_2$ (3 eq), CH$_2$Cl$_2$, rt, 10-20 h

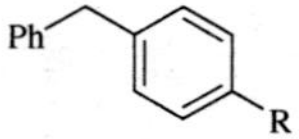

R	
H	(92)
NO$_2$	(95)
CF$_3$	(85)
CO$_2$Me	(97)

413

C_{13-19}

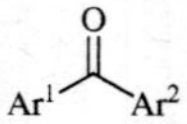

Et$_3$SiH (1.5 eq), CF$_3$SO$_3$H (4 eq), CH$_2$Cl$_2$, 0° to rt, 2 h

Ar[1]	Ar[2]	
Ph	Ph	(90)
Ph	4-MeC$_6$H$_4$	(92)
4-MeC$_6$H$_4$	4-MeC$_6$H$_4$	(85)
Ph	4-MeOC$_6$H$_4$	(79)
4-MeOC$_6$H$_4$	4-MeOC$_6$H$_4$	(0)
Ph	4-BrC$_6$H$_4$	(72)
Ph	4-O$_2$NC$_6$H$_4$	(95)
Ph	4-CF$_3$C$_6$H$_4$	(80)
Ph	4-FC$_6$H$_4$	(76)
Ph	2-FC$_6$H$_4$	(80)
Ph	4-HOC$_6$H$_4$	(78)
Ph	2,4,5-(Me)$_3$C$_6$H$_2$	(80)
2,5-(Me)$_2$C$_6$H$_3$	2,4,5-(Me)$_3$C$_6$H$_2$	(0)
2,4,5-(Me)$_3$C$_6$H$_2$	2,4,5-(Me)$_3$C$_6$H$_2$	(0)

420

TABLE 12. ORGANOSILANE REDUCTION OF KETONES (*Continued*)

Ketone	Conditions	Product(s) and Yield(s) (%)	Refs.
C_{14} 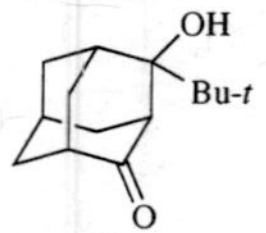	Et_3SiH, TFA, CH_2Cl_2, rt, 16 h	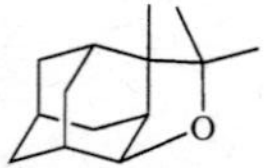 (77)	153
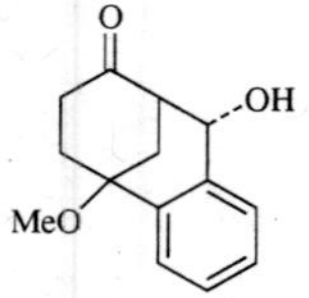	Et_3SiH, TFA	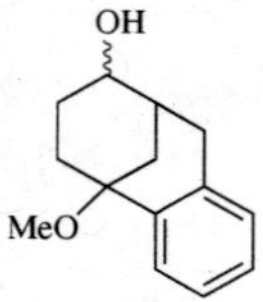 (75)	716
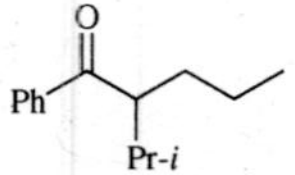	Et_3SiH, $(C_6F_5)_3B$ (2 mol%), MeC_6H_5, 0°, 1 h	 **I** + **II**; **I** + **II** (—), **I**:**II** = 98:2	372
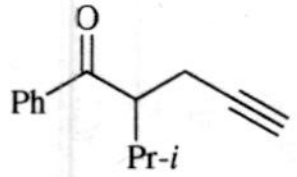	Et_3SiH, $(C_6F_5)_3B$ (2 mol%), MeC_6H_5, 0°, 1 h	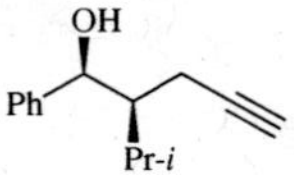 (99)	372
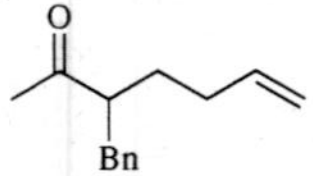	Ph_2SiH_2 (1.0 eq), $Cp_2Ti(PPh_3)_2$, PPh_3 (0.6 eq), MeC_6H_5	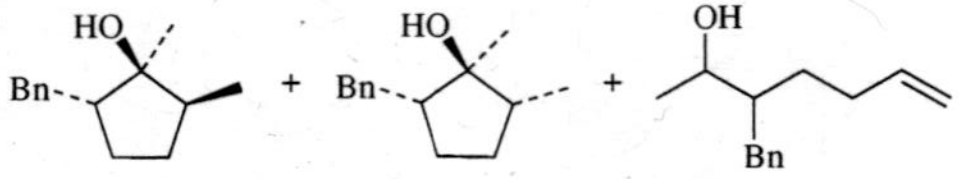 **I** + **II** + **III**; **I** + **II** (71), (**I** + **II**):**III** = 99:1, **I**:**II** = 20:1	428
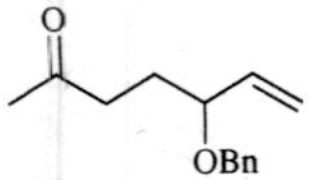	Ph_2SiH_2 (1.0 eq), $Cp_2Ti(PPh_3)_2$, PPh_3 (0.6 eq), MeC_6H_5, 21°	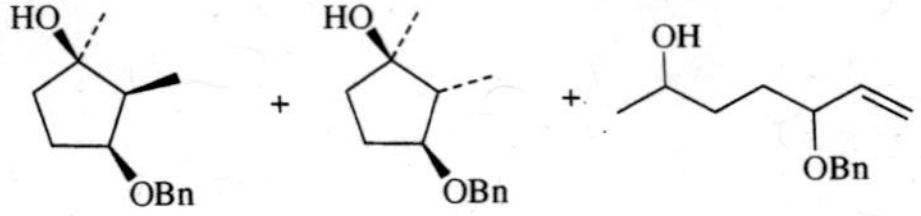 **I** + **II** + **III**; **I** + **II** + **III** (50), (**I** + **II**):**III** = 3:1, **I**:**II** = 3:2	428

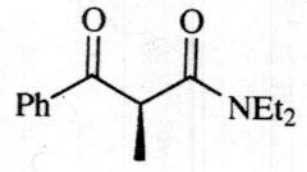	Ph_2SiH_2 (1.0 eq), $Cp_2Ti(PPh_3)_2$, PPh_3 (0.6 eq), MeC_6H_5, –20°	**I + II** (45), (**I + II**):**III** = 8.5:1, **I**:**II**:**III** = 4:1.6:1	428
	$PhMe_2SiH$ (1.2 eq), TASF (10 mol%) DMPU, 0°, 12 h	 **I** **I + II** (98), **I**:**II** > 99:1 **II**	87, 320
	$PhMe_2SiH$ (1.2 eq), TFA, 0°, 4 h	**I + II** (98), **I**:**II** < 1:99	87, 276
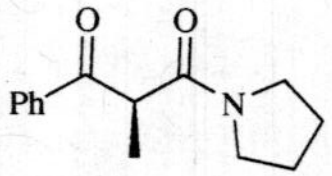	$PhMe_2SiH$ (1.2 eq) TASF (10 mol%), DMPU, 0°, 22 h	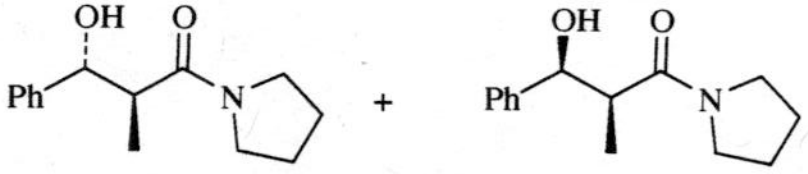 **I** **I + II** (98), **I**:**II** > 99:1 **II**	87, 320
	$PhMe_2SiH$ (1.2 eq) TFA, 0°, 3 h	**I + II** (99), **I**:**II** < 1:99	87, 276
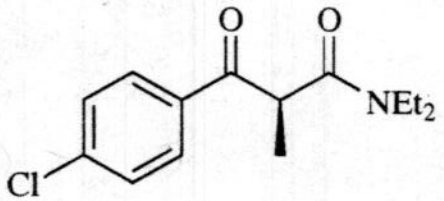	$PhMe_2SiH$ (1.2 eq) TASF (10 mol%), DMPU, 0°, 16 h	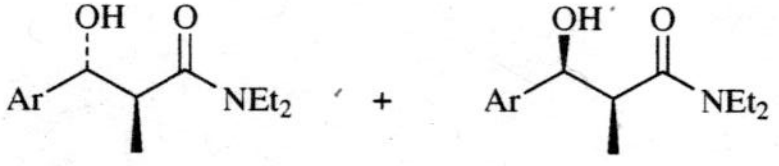 **I** **I + II** (86), **I**:**II** = 99:1 **II** Ar = 4-ClC_6H_4	87
	$PhMe_2SiH$ (1.2 eq) TFA, 0°, 4 h	**I + II** (99), **I**:**II** < 1:99	87, 276
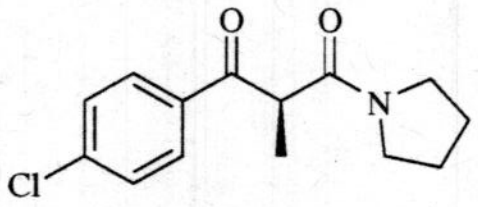	$PhMe_2SiH$ (1.2 eq) TASF (10 mol%), DMPU, 0°, 16 h	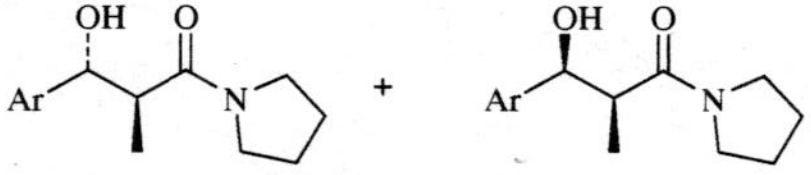 **I** **I + II** (98), **I**:**II** > 99:1 **II** Ar = 4-ClC_6H_4	320
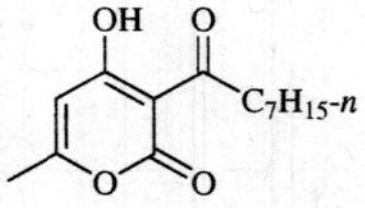	Et_3SiH, TFA, $LiClO_4$ (0.01 eq), rt, 4 h	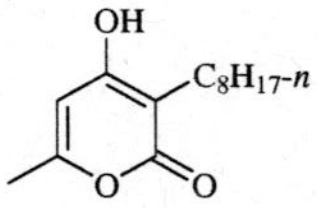 (88)	423

TABLE 12. ORGANOSILANE REDUCTION OF KETONES (*Continued*)

Ketone	Conditions	Product(s) and Yield(s) (%)	Refs.
C_{14} 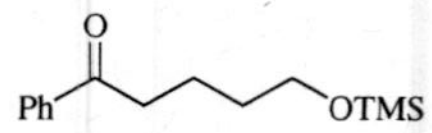	1. $SbCl_5$, TMSCl, SnI_2, CH_2Cl_2, –78°, 30 min 2. Et_3SiH (1.5 eq), –23°, 2.5 h	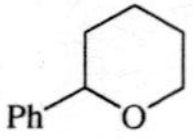 (90)	306
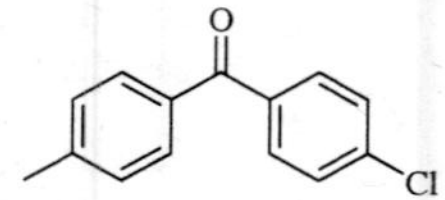	PMHS (3 eq), $(C_6F_5)_3B$ (5 mol%), CH_2Cl_2, rt, 5-20 min	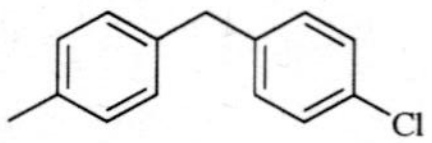 (85)	354
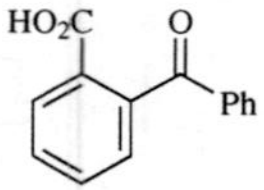	Et_3SiH (2.2 eq), TFA, rt	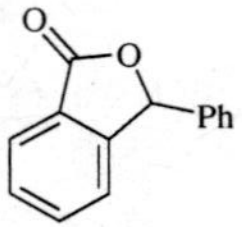 (100)	73
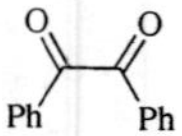	Et_3SiH (2.5 eq), $(Ph_3P)_3RhCl$ (0.5 mol%), C_6H_{14}, 70°, 4 h	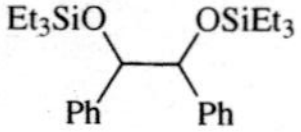 (83)	411
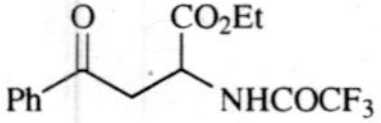	Et_3SiH (3.66 eq), $BF_3{\bullet}OEt_2$ (20 eq), rt, 48 h	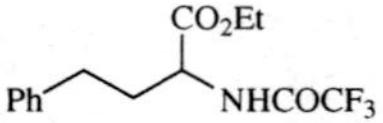 (22)	753
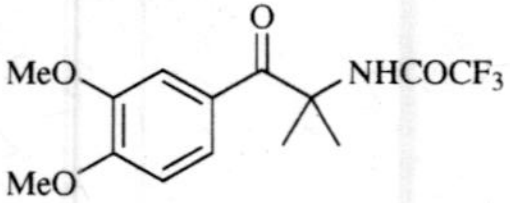	Et_3SiH (4 eq), $BF_3{\bullet}OEt_2$ (20 eq), reflux, 2 h; rt, 16 h	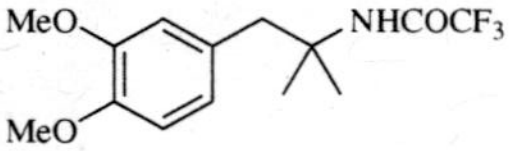 (78)	753
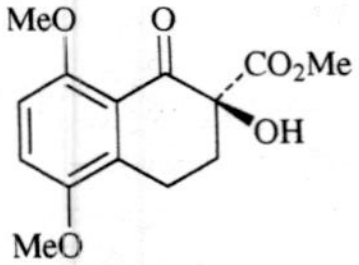	Et_3SiH (3.0 eq), TFA, 0° to rt, 16 h	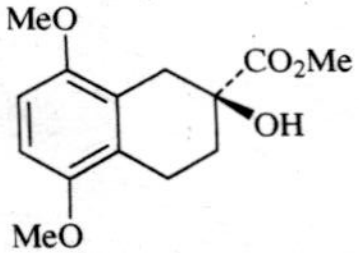 (75)	416

Reactant	Conditions	Product(s) and Yield(s) (%)	Refs.
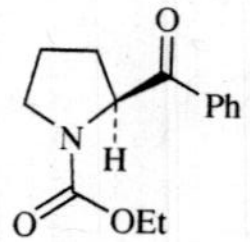	Et_3SiH, TFA	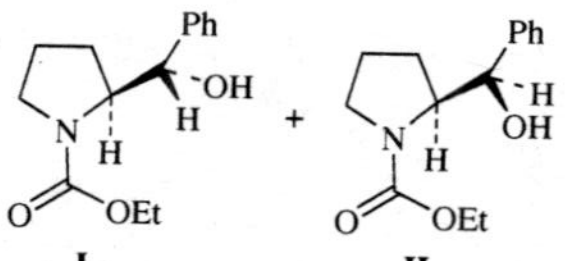 I + II; **I+II** = (—), **I:II** = 38:62	754
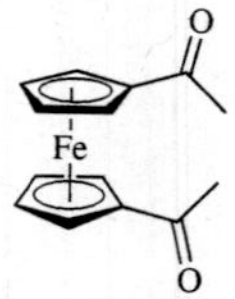	Et_3SiH (2 eq), TFA, rt, 120 h	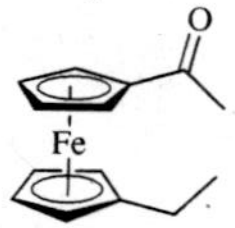 (55)	179
	Et_3SiH (20 eq), TFA, rt, 20 h	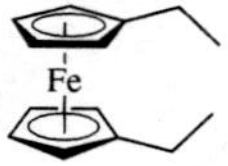 (80)	179
C_{15} 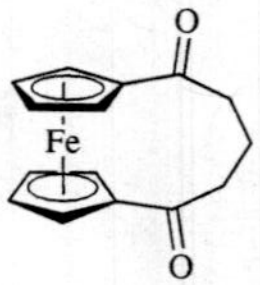	Et_3SiH (4 eq), TFA, rt, 4 h	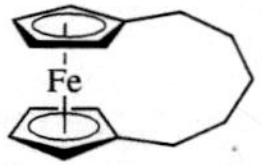 (86)	179
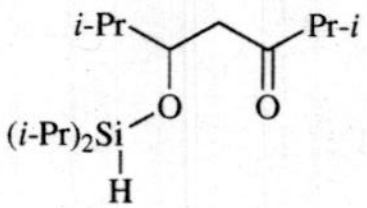	$SnCl_4$ (0.1 eq), –80°, 2 h	 I + II; **I + II** (—), **I:II** = 60:1	399
	$MgBr_2$•OEt_2 (0.1 eq), rt, 24 h	**I + II** (—), **I:II** = 60:1	
	$TiCl_4$ (0.5 eq), –80°, 30 min	**I + II** (—), **I:II** = 30:1	
	BF_3•OEt_2 (0.5 eq), –80°, 2 h	**I + II** (—), **I:II** = 320:1	
	$ZnBr_2$ (0.12 eq), –80°, 8 h; rt, 16 h	**I + II** (—), **I:II** = 2:1	
	$ZnCl_2$ (0.25 eq), –80°, 30 min; rt, 15 min	**I + II** (—), **I:II** = 1:2.5	
	TFA (0.2 eq), –80°, 30 min	**I + II** (—), **I:II** = 35:1	
	TBAF (0.2 eq), –80°, 30 min	**I + II** (—), **I:II** = 2:1	
	$(Ph_3P)_3RhCl$ (0.08 eq), C_6H_6, reflux, 12 h	**I + II** (—), **I:II** = 5:1	

TABLE 12. ORGANOSILANE REDUCTION OF KETONES (*Continued*)

	Ketone	Conditions	Product(s) and Yield(s) (%)	Refs.
C_{15}	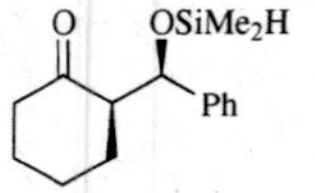	DMF, 100°, 24 h	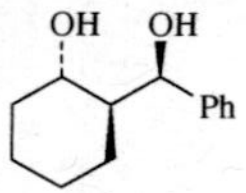 (67)	400
	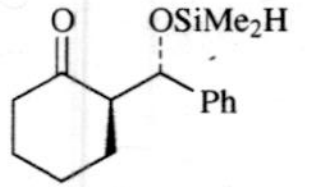	DMF, 100°, 24 h	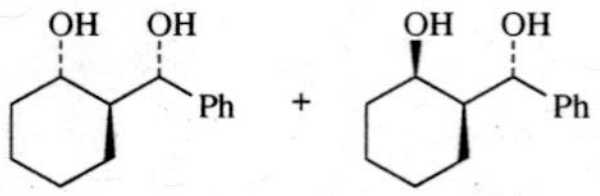 I + II **I + II** (26), **I**:**II** = 1:4	400
	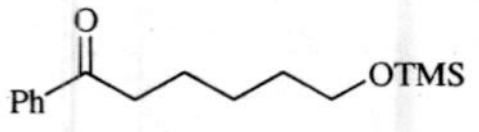	1. $SbCl_5$, TMSCl, SnI_2, CH_2Cl_2, –78°, 30 min 2. Et_3SiH (1.5 eq), –23°, 2.5 h	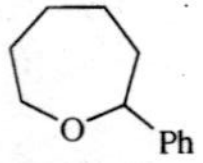 **I** (68)	306
		1. $TrSbCl_6$ (5-30 mol%), CH_2Cl_2, –78°, 30 min 2. Et_3SiH (1.5 eq), –23°, 2.5 h	**I** (77)	306
	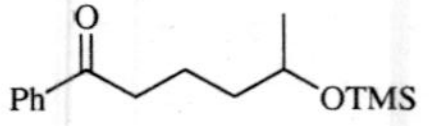	1. $SbCl_5$, TMSCl, SnI_2, CH_2Cl_2, –78°, 30 min 2. Et_3SiH (1.5 eq), –23°, 2.5 h	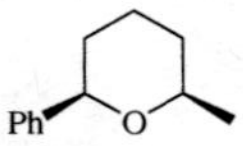 **I** (92)	306
		1. $TrSbCl_6$ (5-30 mol%), CH_2Cl_2, –78°, 30 min 2. Et_3SiH (1.5 eq), –23°, 2.5 h	**I** (96)	306
	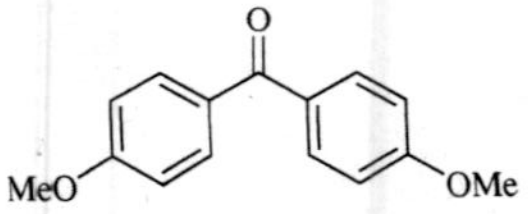	Et_3SiH (2.5 eq), TFA (5 eq), 20°	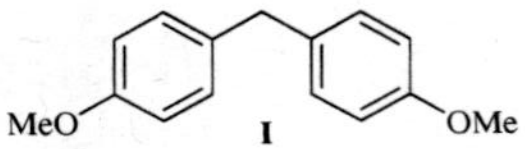 **I** (10)	180
		Et_3SiH (5 eq), Sn-mont, DCE, reflux, 20 h	**I** (75)	353
		Et_3SiH (5 eq), CF_3SO_3H (1 eq), DCE, reflux, 10 h	**I** (54)	353

Reactant	Conditions	Product(s) and Yield(s) (%)	Refs.
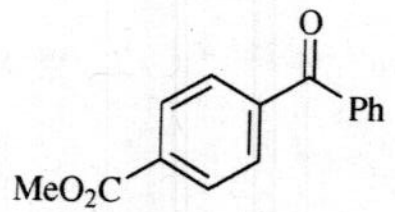	Et_3SiH, TFA, CH_2Cl_2	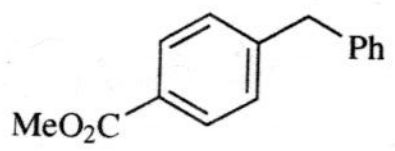 (96)	415
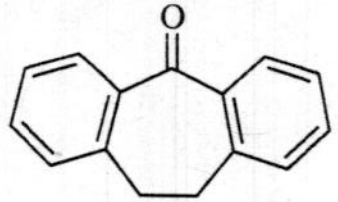	Et_3SiH (2.6 eq), PPHF, TFA, 0°, 10 min, rt, 3 h	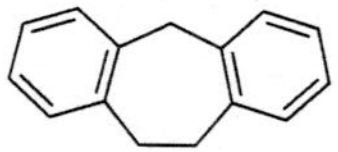 (98)	135
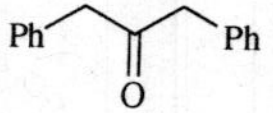	PMHS (3 eq), $(C_6F_5)_3B$ (5 mol%), CH_2Cl_2, rt, 5-20 min	 (88)	354
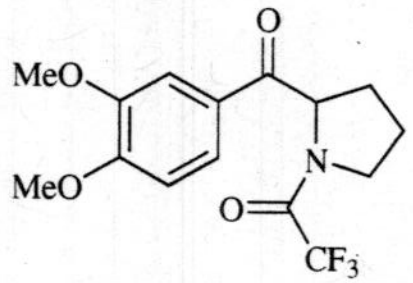	Et_3SiH (4 eq), BF_3•OEt_2 (20 eq), reflux, 2 h, rt, 16 h	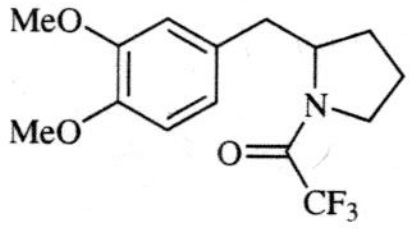 (88)	376
C_{15-16} 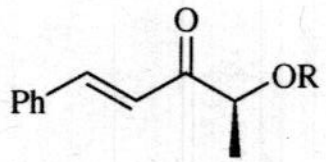	$PhMe_2SiH$ (1.2 eq), TBAF (5-10 mol%), HMPA, 0°	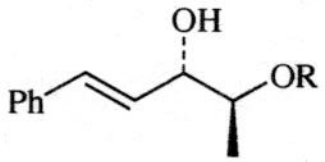 R / Time / / threo:erythro; *t*-Bu / 12 h / (95) / 95:5; THP / 16 h / (77) / 87:13	86
C_{15} 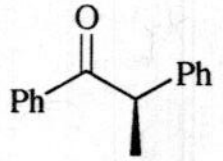	$PhMe_2SiH$ (1.1-1.2 eq), TASF (5-10 mol%), HMPA	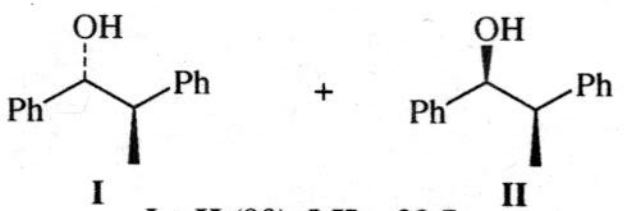 **I** + **II** (90), **I**:**II** = 93:7	320
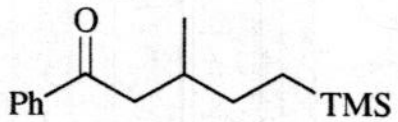	Et_3SiH, $(C_6F_5)_3B$ (2 mol%), MeC_6F_5, 0°, 1 h	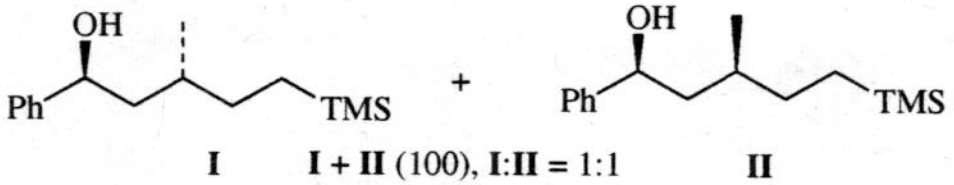 **I** + **II** (100), **I**:**II** = 1:1	372
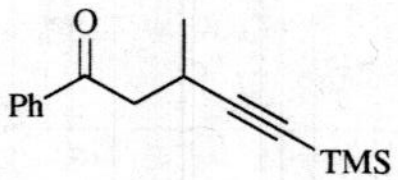	Et_3SiH, $(C_6F_5)_3B$ (2 mol%), MeC_6F_5, 0°, 1 h	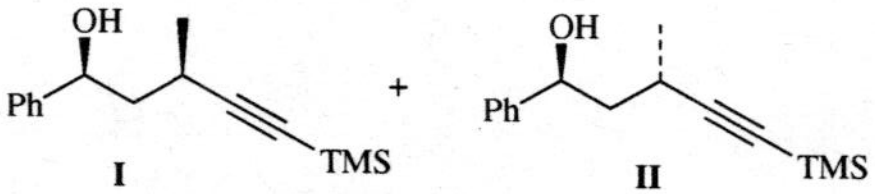 **I** + **II** (100), **I**:**II** = 6.5:1	372

TABLE 12. ORGANOSILANE REDUCTION OF KETONES (*Continued*)

Ketone	Conditions	Product(s) and Yield(s) (%)	Refs.
C_{15} 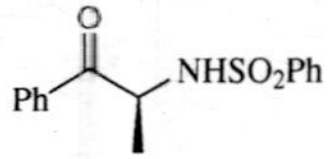	$PhMe_2SiH$ (1.1-1.2 eq), TFA, 0°, 20 h	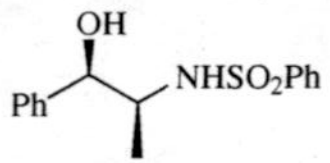 (66) threo:erythro = 2:98	86
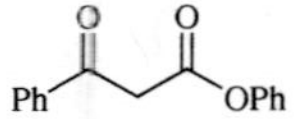	Ph_2SiH_2 (1.2 eq), $AlCl_3$ (1 eq), CH_2Cl_2, 20°, 24 h	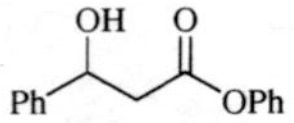 (77) + Ph_2SiHCl (—)	373
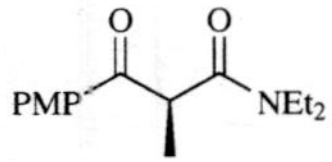	$PhMe_2SiH$ (1.2 eq), TASF (10 mol%), DMPU, 0°, 16 h	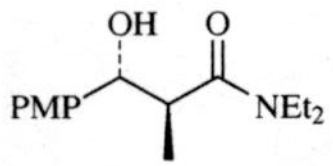 I + 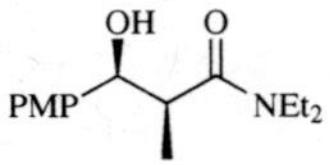 II **I + II** (92), **I:II** > 99:1	87
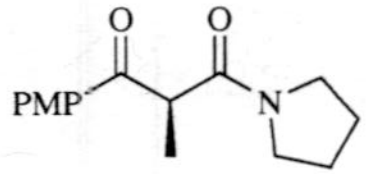	$PhMe_2SiH$ (1.2 eq), TASF (10 mol%), DMPU, 0°, 16 h	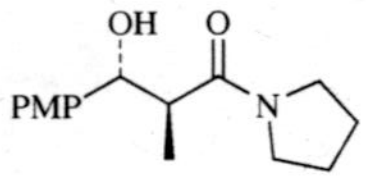 I + 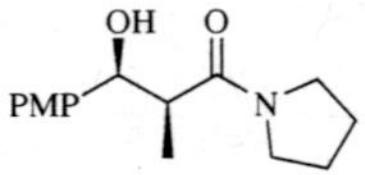 II **I + II** (92), **I:II** = 99:1	320
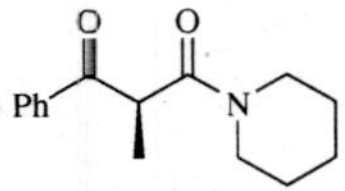	$PhMe_2SiH$ (1.2 eq), TFA, 0°, 5 h	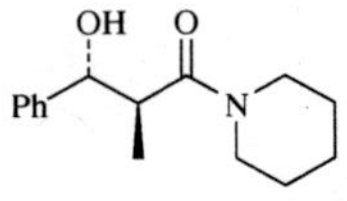 I + 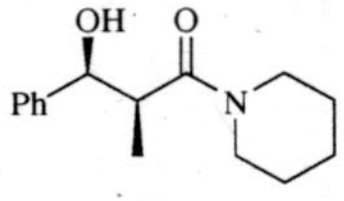 II **I + II** (93), **I:II** < 1:99	87, 320
	$PhMe_2SiH$ (1.2 eq), TASF, 0°	**I + II** (90), **I:II** = 99:1	87, 276
	Et_3SiH (1.1 eq), TFA, 0°, 4 h	**II** (90)	375
C_{16} 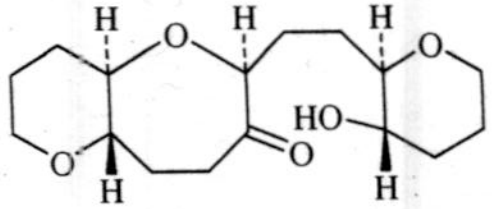	Et_3SiH (10 eq), TMSOTf (1 eq), CH_2Cl_2, 0°, 10 min	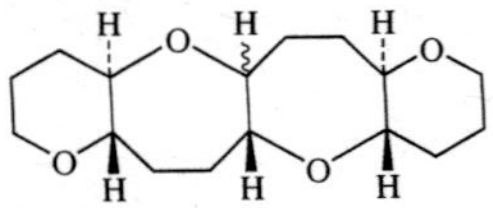 I (75) trans:cis = 3:1	336
	Ph_2MeSiH (10 eq), TMSOTf (1 eq), CH_2Cl_2, 0°, 10 min	**I** (88) trans:cis 4:1	336

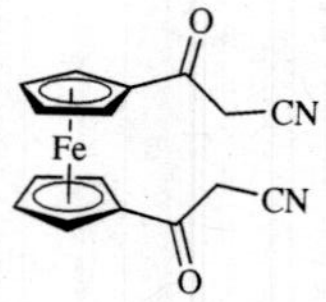	Et_3SiH (6 eq), TFA, rt, 5 h	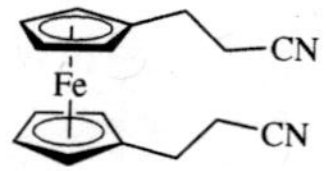 (41)	179
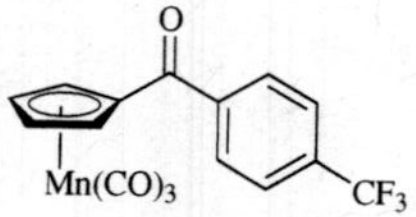	Et_3SiH, TFA, $CHCl_3$, 50-60°, 15 h	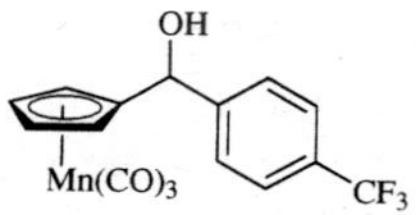 (95)	351
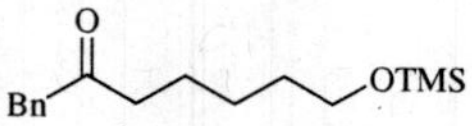	1. $SbCl_5$, TMSCl, SnI_2, CH_2Cl_2, −78°, 30 min 2. Et_3SiH (1.5 eq), −23°, 2.5 h	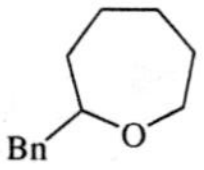 **I** (80)	306
	1. $TrSbCl_6$ (5-30 mol%), CH_2Cl_2, −78°, 30 min 2. Et_3SiH (1.5 eq), −23°, 2.5 h	**I** (95)	306
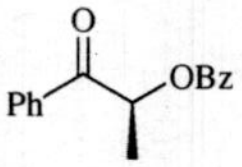	$PhMe_2SiH$ (1.2 eq), TBAF (5-10 mol%), HMPA, 0°, 12 h	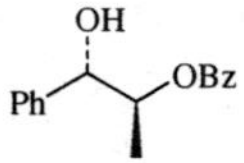 (82) threo:erythro = 96:4	86
	$PhMe_2SiH$ (1.2 eq), TFA, 0°, 6 h	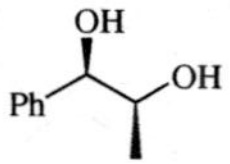 (72) threo:erythro = 7:93	86
C_{17} 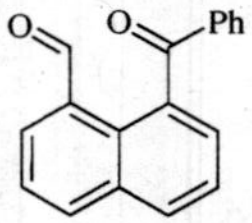	Et_3SiH (8 eq), TMSOTf (1 eq), CH_2Cl_2, rt, 45 min	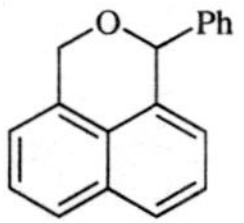 (98)	339

TABLE 12. ORGANOSILANE REDUCTION OF KETONES (*Continued*)

	Ketone	Conditions	Product(s) and Yield(s) (%)	Refs.
C_{17}	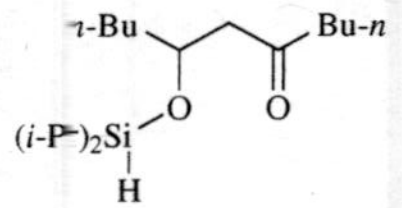	$SnCl_4$ (0.1 eq), –80°, 2 h	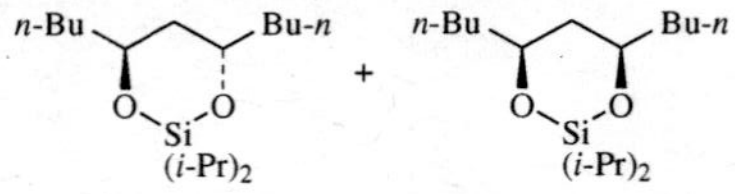 I I + II (60), I:II = 40:1 II	399
C_{18}	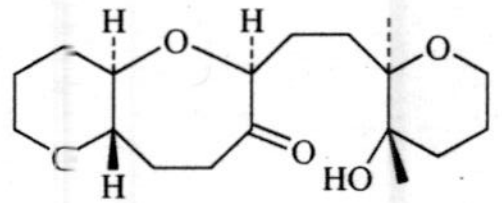	Ph_2MeSiH (2 eq), TMSOTf (1 eq), CH_2Cl_2, 0°, 2 h	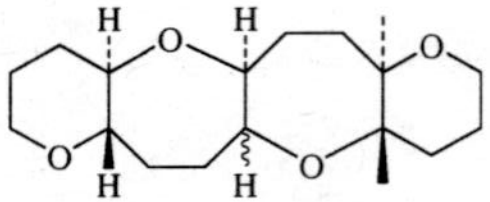 (81) trans:cis = 3:1	336
	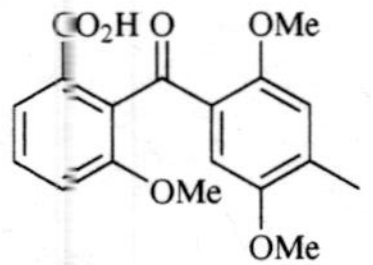	Et_3SiH, TFA	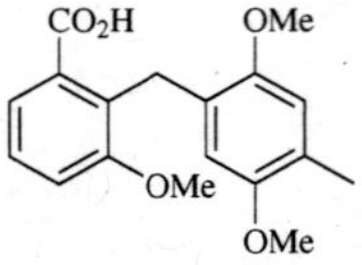 (80)	417
	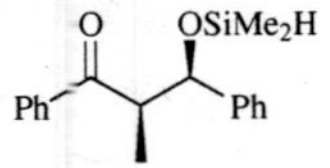	TBAF (6 mol%), THF, –78°, 3.5 h	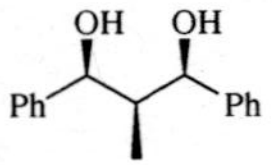 I (76)	400
		$BnOSiMe_2H$, TBAF (6 mol%), THF, –78°, 3.5 h	I (0)	400
	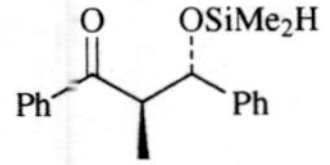	TBAF (6 mol%), THF, –78°, 3.5 h	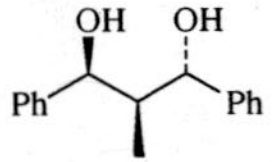 I (88)	400
		$BnOSiMe_2H$, TBAF (6 mol%), THF, –78°, 3.5 h	I + 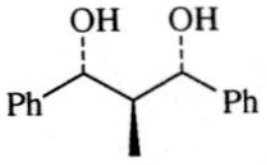 II I + II (46), I:II = 13:87	400

C_{18-19}

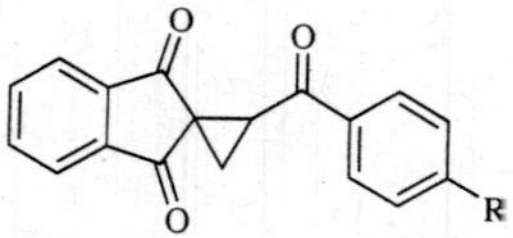

Et_3SiH (3 eq), TFA (10 eq), 50°

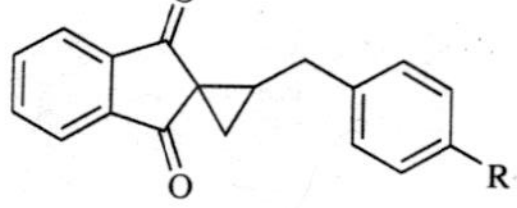

I

R	Time	
F	24 h	(57)
Cl	6 h	(78)
Br	6 h	(62)
OMe	2 h	(57)

418

C_{18-22}

R = H, Et, *n*-Bu — Et_3SiH (3 eq), TFA (5 eq), 50°, 6 h — **I** (70-90) — 755

C_{18-23}

R	Conditions	Product(s) and Yield(s) (%)	Refs.
H	Et_3SiH (3 eq), TFA (10 eq), CH_2Cl_2, 0°, 30 min	**I** (81)	756
Me		(76)	
Et		(78)	
n-Pr		(78)	
i-Pr		(74)	
n-Bu		(58)	
n-C_5H_{11}		(67)	

C_{19-23}

R = Me, *n*-Pr, *n*-C_5H_{11} — Et_3SiH (3 eq), TFA (5 eq), 6 h

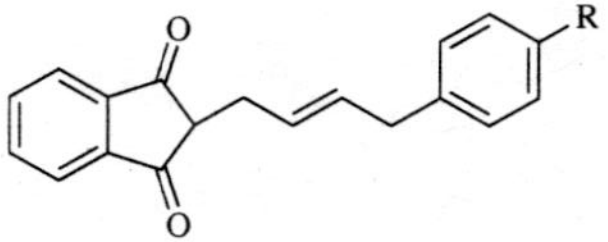

(70-90) — 755

C_{19}

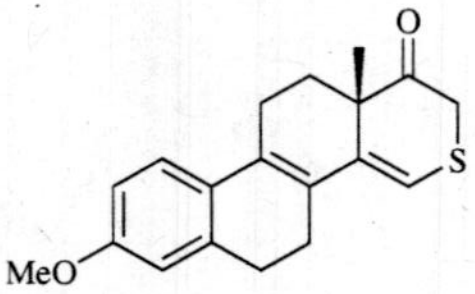

Et_3SiH (2.5 eq), TFA, CH_2Cl_2, rt, 40 h

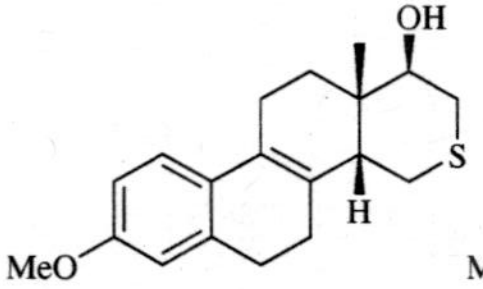

(22)

+

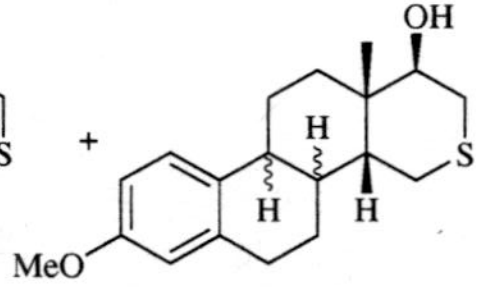

(11)

696

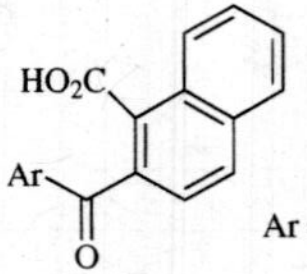

Ar = 3-$CF_3C_6H_4$

Et_3SiH (5.1 eq), TFA, rt, 48 h

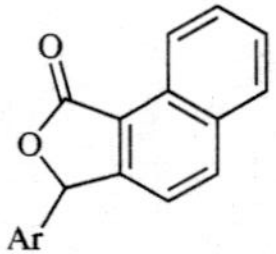

(87) — 402

Ketone	Conditions	Product(s) and Yield(s) (%)	Refs.
C_{19-26}			
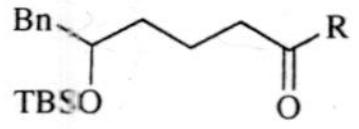	Et_3SiH (1.2 eq), $BiBr_3$ (5 mol%), MeCN, rt	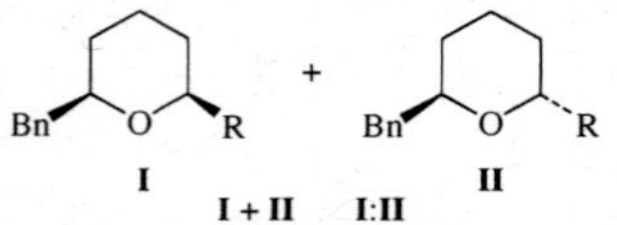	342
R		I + II / I:II	
Me		(90) >99:1	
i-Pr		(95) >99:1	
$HOCH_2$		(82) >99:1	
$H_2C{=}CHCH_2$		(95) >99:1	
EtO_2CCH_2		(97) >99:1	
$Br(CH_2)_6$		(96) >99:1	
Ph		(97) >99:1	
$BzOCH_2$		(93) >99:1	
$BnOCH_2$		(91) >99:1	
C_{20}			
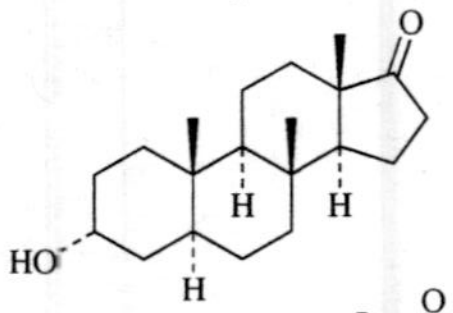	$PhSiH_3$ (0.4 eq), $Mn(dpm)_3$ (3 mol%), *i*-PrOH, DCE, O_2, rt	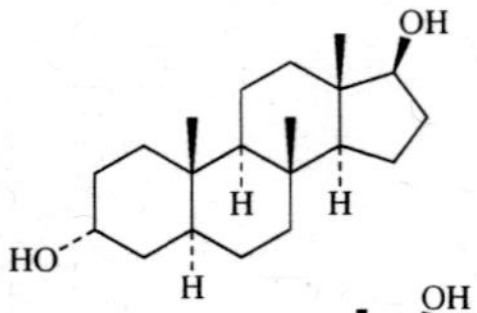 (47)	367
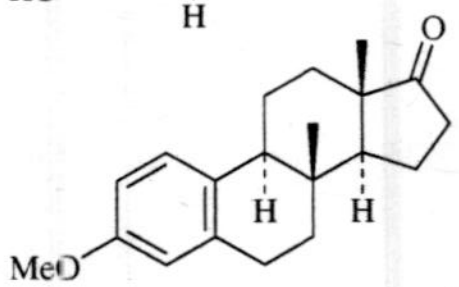	$PhSiH_3$ (0.4 eq), $Mn(dpm)_3$ (3 mol%), *i*-PrOH, DCE, O_2, rt	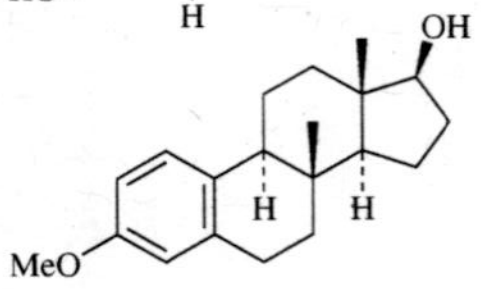 (61)	367
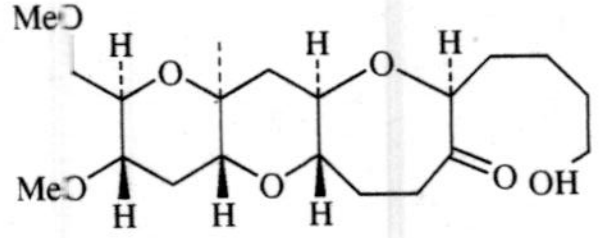	Ph_2MeSiH (2 eq), TMSOTf (1 eq), CH_2Cl_2, 0°, 2 h	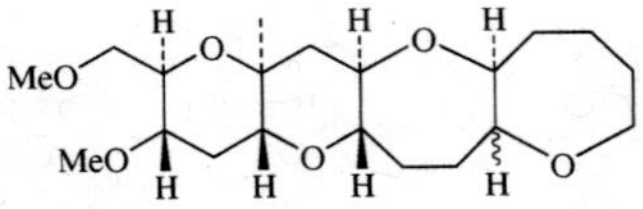 (75) trans:cis = 3:1	336

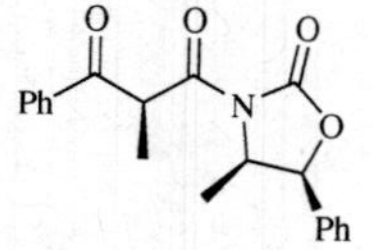	$PhMe_2SiH$ (1.2 eq), TFA, 0°, 4 h	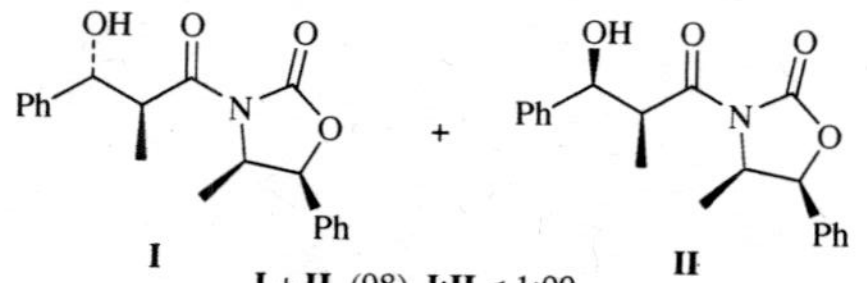 I + II (98), I:II < 1:99	87, 276
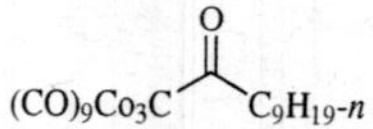	Et_3SiH, TFA	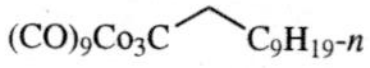 (88)	183
C_{20-23}			
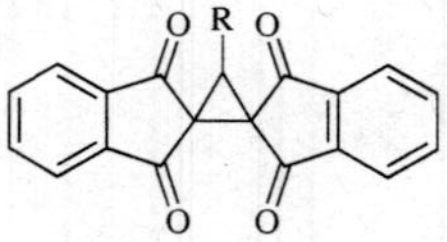	Et_3SiH (6 eq), TFA (20 eq), 20°	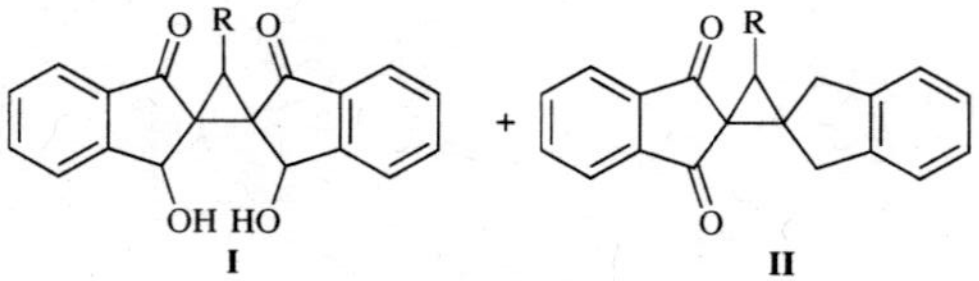	757

R	I	II
Me	(41)	(25)
Et	(47)	(20)
n-Pr	(43)	(30)
n-Bu	(43)	(20)

C_{21}			
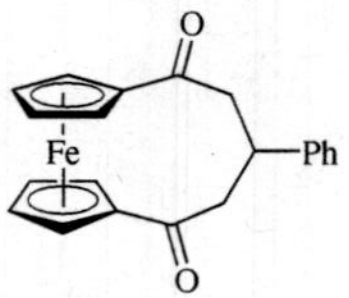	Et_3SiH (20 eq), TFA, rt, 240 h	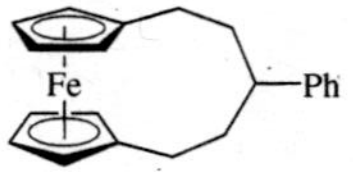 (73)	179
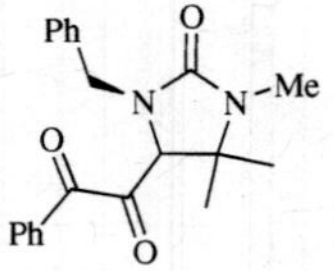	1. $PhMe_2SiH$ (4 eq), CsF (5 mol%), 18-C-6 (5 mol%), CH_2Cl_2, rt 2. HCl, acetone	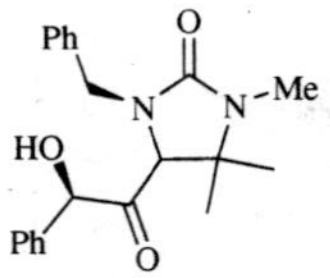 (40)	369

TABLE 12. ORGANOSILANE REDUCTION OF KETONES (*Continued*)

Ketone	Conditions	Product(s) and Yield(s) (%)	Refs.
C_{21} 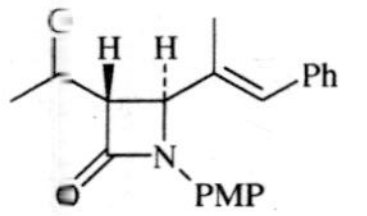	$PhMe_2SiH$, TASF, THF, rt, 70 h	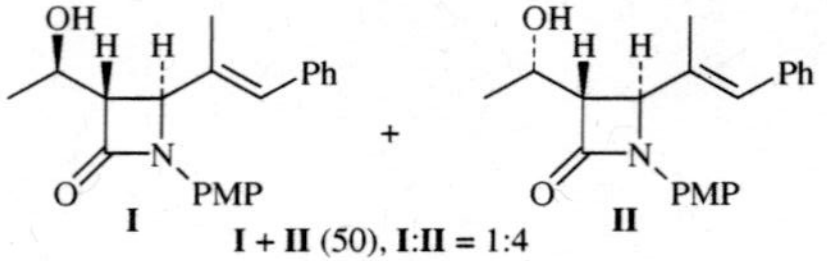 **I** + **II** (50), **I**:**II** = 1:4	371
	$PhMe_2SiH$, TASF, THF, HMPA, rt, 23 h	**I** + **II** (90), **I**:**II** = 1:8	371
	$PhMe_2SiH$, TFA, THF, 0°, 22 h	**I** + **II** (90), **I**:**II** = 1:1	371
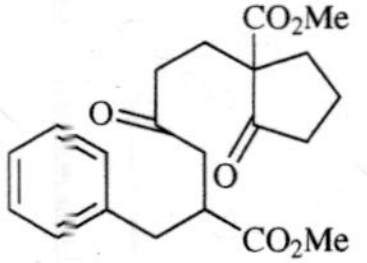	Et_3SiH, TFA, CH_2Cl_2, rt, 16 h	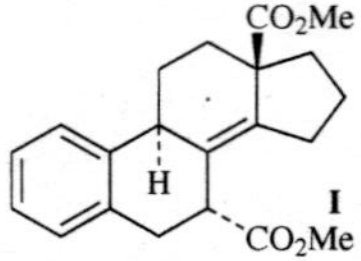 **I** (17) + 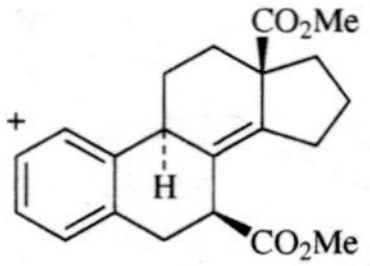 (18)	758
	1. TFA, 2 h 2. Et_3SiH, CH_2Cl_2	**I** (25)	758
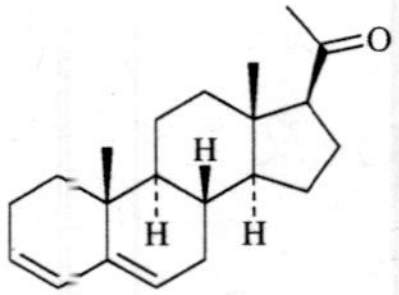	Ph_2SiH_2, Rh-(+)-diop, C_6H_6, 22°	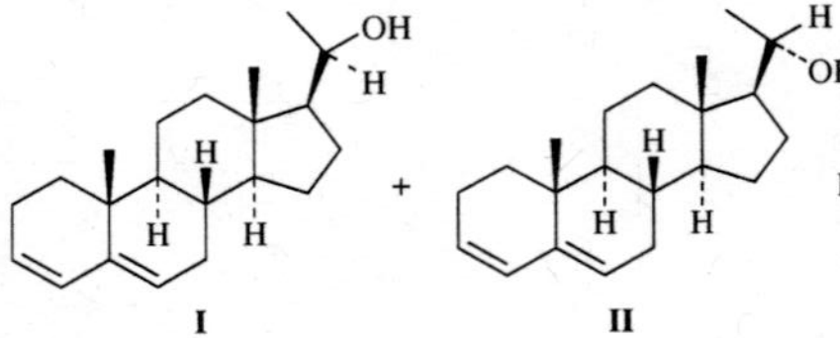 **I**:**II** 7:3	573
	Ph_2SiH_2, Rh-(–)-diop, C_6H_6, 22°	**I**:**II** = 7:3	573
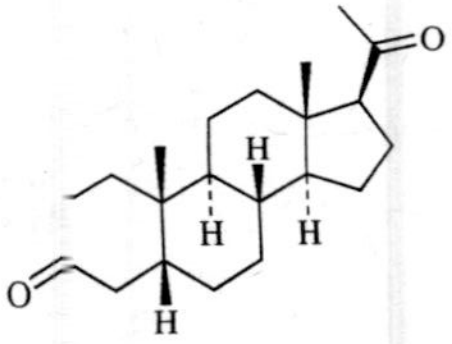	Ph_2SiH_2, Rh-(+)-diop, C_6H_6, 22°	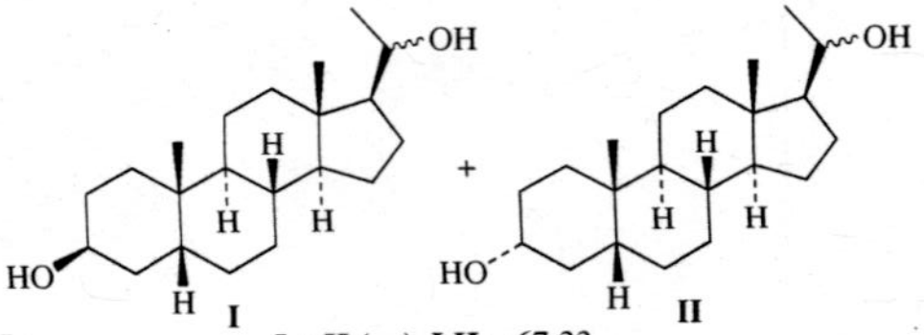 **I** + **II** (—), **I**:**II** = 67:33	573
	Ph_2SiH_2, Rh-(–)-diop, C_6H_6, 22°	**I** + **II** (—), **I**:**II** = 64:36	573

C_{23} 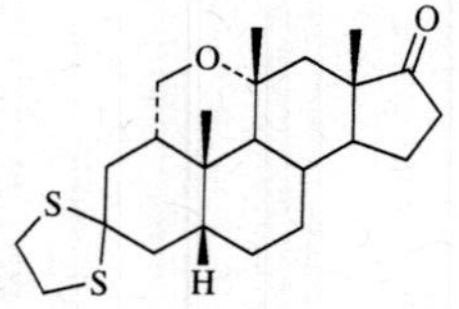	1. Et_3SiH (10 eq), TFA (50 eq), CH_2Cl_2, 50°, 20 h 2. KOH, MeOH, THF, rt	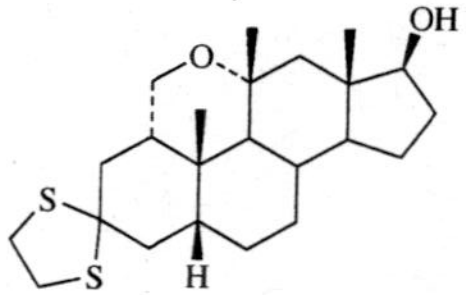 (73)	368
C_{24} 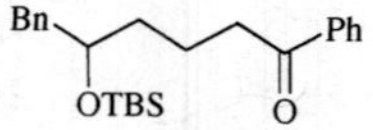	Et_3SiH (1.2 eq), $BiBr_3$ (5 mol%), MeCN, rt	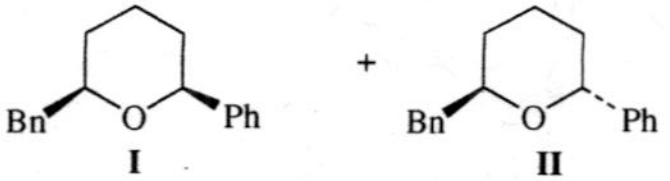 **I + II** (97) **I**:**II** > 99:1	342
	Et_3SiH (1.2 eq), TMSBr (15 mol%), MeCN, rt	**I + II** (99), **I**:**II** > 99:1	342
	Et_3SiH (1.2 eq), HBr (15 mol%), MeCN, rt	**I + II** (99), **I**:**II** > 99:1	342
C_{24-26} 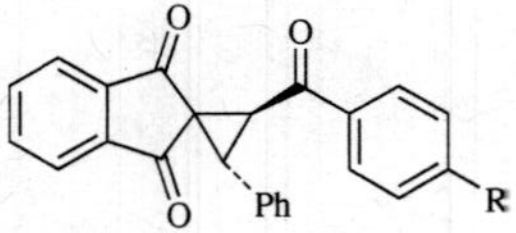	Et_3SiH (6 eq), TFA (20 eq), 50°, 6 h	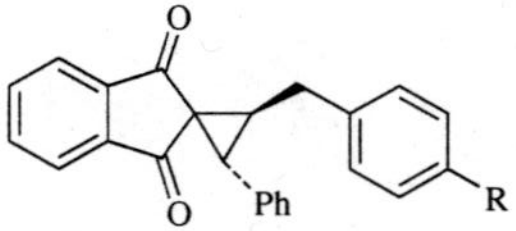 R = H, Me, Et (—)	757
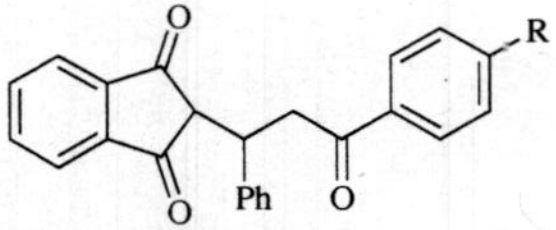	Et_3SiH (5 eq), TFA (10 eq), CCl_4, 50-55°, 8 h	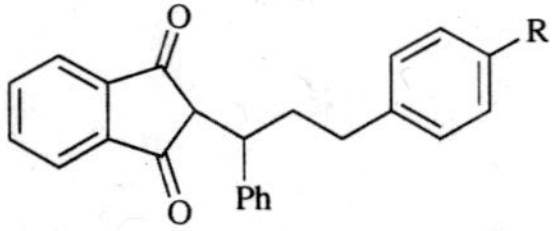 R: H (33); Me (49); Et (30)	266

	Ketone	Conditions	Product(s) and Yield(s) (%)	Refs.
C_{27}	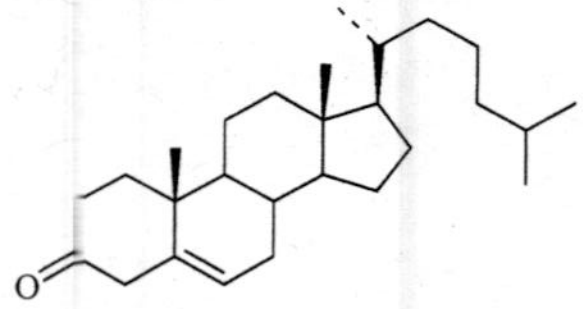	PMHS (3 eq), $(C_6F_5)_3B$ (5 mol%), CH_2Cl_2, rt, 5-20 min	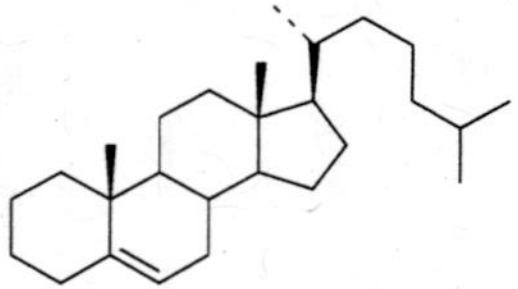 (—)	354
C_{28}	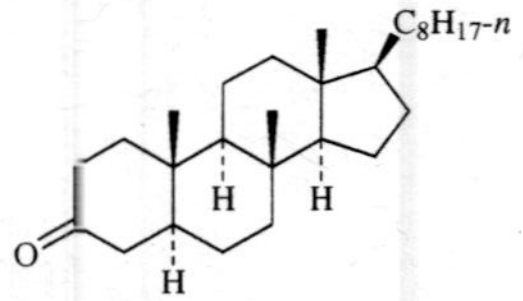	$PhSiH_3$ (0.4 eq), $Mn(dpm)_3$ (3 mol%), *i*-PrOH, DCE, O_2, rt	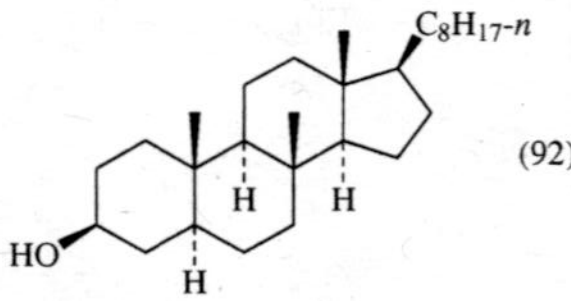 (92)	367
C_{31}	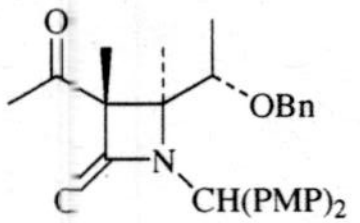	Et_3SiH (3 eq), TFA (4 eq), $BF_3•OEt_2$ (0.23 eq), $MeOC_6H_5$, –25°, 1.5 h	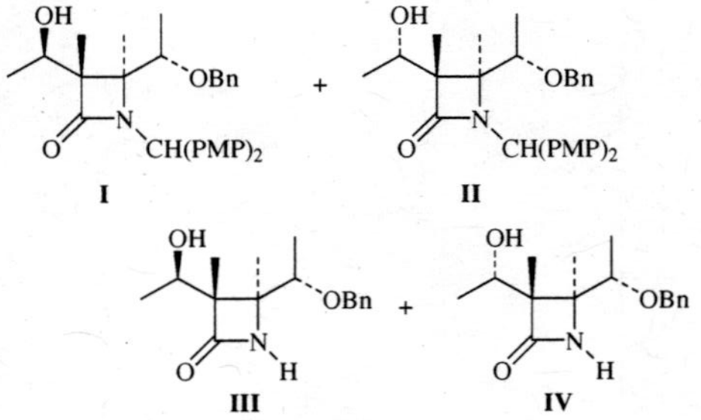 **I** + **II** (71), **I**:**II** = 8:1, **III** + **IV** (10)	395
	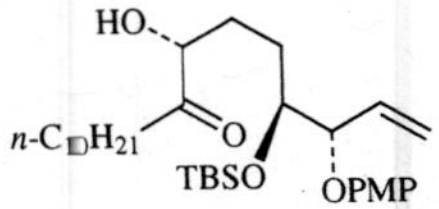	1. *t*-$BuMe_2SiH$, $BiBr_3$, MeCN, 0° 2. TBSOTf, 2,6-lutidine, 0°	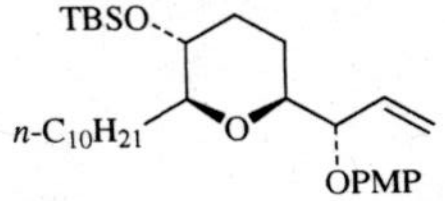 (93)	404

C_{37}	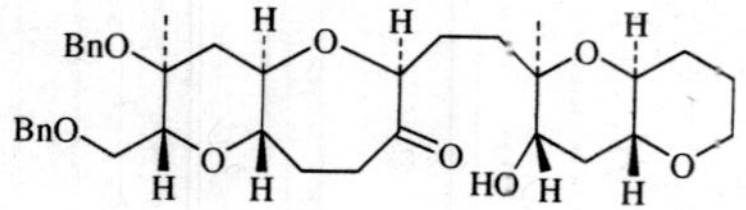	Ph_2MeSiH (2 eq), TMSOTf (1 eq), CH_2Cl_2, 0°, 2 h	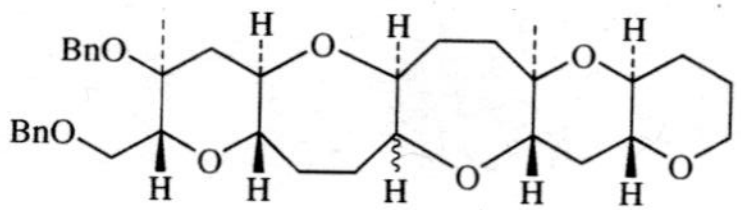 (55) trans:cis = 3:1	336
C_{38}	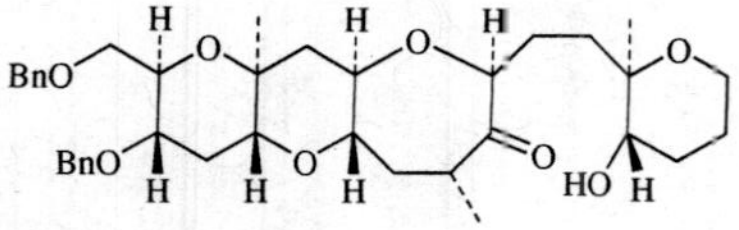	Ph_2MeSiH (2 eq), TMSOTf (1 eq), CH_2Cl_2, 0°, 2 h	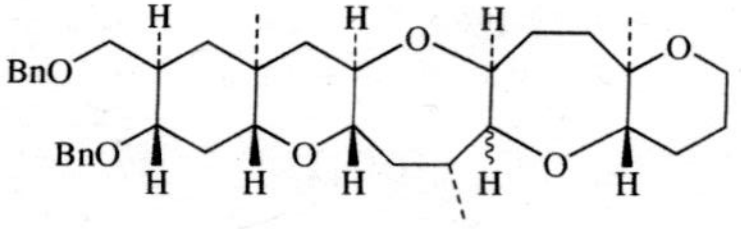 (62) trans:cis = 6:1)	336

[a] The yield was determined by NMR spectroscopy.

[b] The yield was determined by gas chromatography.

[c] This column gives the amount of acid used in a one mmol reaction.

TABLE 13. ORGANOSILANE REDUCTION OF AMIDES

Amide	Conditions	Product(s) and Yield(s) (%)	Refs.
C_{5-13}			
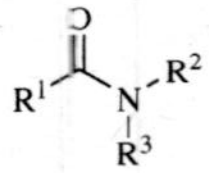	Et_3SiH (3-3.5 eq), cat-1 (1 mol%), cat-2 (5 mol%), cat-3 (5 mol%), MeC_6H_5, 100°	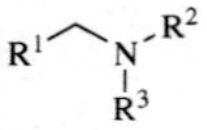	432

R^1	R^2	R^3	Cat-1	Cat-2	Cat-3	Yield (%)
H	—$(CH_2)_4$—		$(Ph_3P)_2RuCl_2(CO)_2$	EtI	none	(92.9)
H	—$(CH_2)_5$—		$(Ph_3P)_2RuCl_2(CO)_2$	EtI	none	(86.1)
H	—$(CH_2)_2O(CH_2)_2$—		$(Ph_3P)_2RuCl_2(CO)_2$	EtI	none	(87.0)
—$(CH_2)_3$—		Me	$Ru_3(CO)_{12}$	none	none	(87.7)
—$(CH_2)_4$—		Me	$Ru_3(CO)_{12}$	none	none	(86.6)
Me	—$(CH_2)_2O(CH_2)_2$—		$(Ph_3P)_2RuCl_2(CO)_2$	EtI	none	(92.3)
—$(CH_2)_5$—		Me	$Ru_3(CO)_{12}$	none	none	(99.2)
—$(CH_2)_6$—		H	$Ru_3(CO)_{12}$	none	none	(95.9)
4-MeC_6H_4	H	H	$[RuCl_2(CO)_3]_2$	EtI	Et_2NH	(50.3)
Ph	H	Me	$[RuCl_2(CO)_3]_2$	EtI	Et_2NH	(97.1)
n-C_7H_{15}	H	H	$Os_3(CO)_{12}$	none	Et_2NH	(56.0)
Me	H	Bn	$[RuCl_2(CO)_3]_2$	EtI	Et_2NH	(99.5)
Ph	Me	Me	$Os_3(CO)_{12}$	none	Et_2NH	(99.5)
H	Ph	Ph	$Ru_3(CO)_{12}$	none	none	(93.2)

Amide	Conditions	Product(s) and Yield(s) (%)	Refs.
C_5			
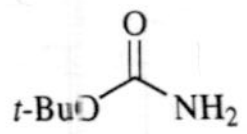	Et_3SiH (3 eq), TFA (2-3 eq), PhCHO (x eq), MeCN, 22°, 18 h	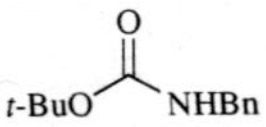	326

x	Yield (%)
0.33	(92)
3	(81)

C_7

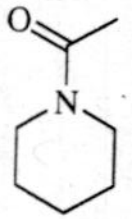

Et_3SiH (3-3.5 eq), cat-1 (1 mol%), cat-2 (5 mol%), cat-3 (5 mol%), MeC_6H_5, 100°

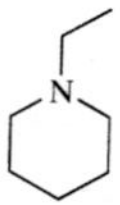

I

432

Cat-1	Cat-2	Cat-3	Time	
$Mn_2(CO)_{10}$	none	Et_2NH	16 h	(89.3)
$Re_2(CO)_{10}$	none	Et_2NH	16 h	(95.6)
$Ru_3(CO)_{12}$	none	none	16 h	(88.2)
$(Ph_3P)_2RuCl_2(CO)_2$	EtI	none	16 h	(96.1)
$(Ph_3P)_2RuCl_2(CO)_2$	MeI	none	16 h	(98.1)
$(Ph_3P)_2RuCl_2(CO)_2$	I_2	none	16 h	(94.1)
$(Ph_3P)_3RuH_2(CO)$	EtI	Et_2NH	40 h	(94.4)
$Ru(acac)_3$	EtI	Et_2NH	16 h	(88.0)
$Os_3(CO)_{12}$	none	Et_2NH	16 h	(99.8)
$Os_3(CO)_{12}$	none	pyridine	16 h	(99.3)
$(Ph_3P)_4RhH$	none	Et_2NH	16 h	(99.5)
$IrCl_3$	none	none	16 h	(94.4)
K_2IrCl_6	none	Et_2NH	16 h	(92.6)
$Pd(OH)_2/C$	none	Et_2NH	40 h	(77.7)
$PtCl_2$	none	Et_2NH	16 h	(78.5)

R_3SiH (3-3.5 eq), cat-1 (1 mol%), cat-2 (5 mol%), cat-3 (5 mol%), MeC_6H_5, 100°

I

432

R_3Si	Cat-1	Cat-2	Cat-3	Time	I
$PhMe_2Si$	$(Ph_3P)_2RuCl_2(CO)_2$	EtI	none	16 h	(90.2)
t-$BuMe_2Si$	$(Ph_3P)_2RuCl_2(CO)_2$	EtI	none	16 h	(70.4)
i-Pr_3Si	$[RuCl_2(CO)_3]_2$	EtI	Et_2NH	40 h	(50.6)
$ClMe_2Si$	$Os_3(CO)_{12}$	none	Et_2NH	16 h	(80.1)
$Me_2(EtO)Si$	$(Ph_3P)_2RuCl_2(CO)_2$	EtI	none	16 h	(93.1)
$Me(EtO)_2Si$	$(Ph_3P)_2RuCl_2(CO)_2$	EtI	none	16 h	(90.8)
$(EtO)_3Si$	$[RuCl_2(CO)_3]_2$	EtI	Et_2NH	16 h	(86.1)

TABLE 13. ORGANOSILANE REDUCTION OF AMIDES (*Continued*)

Amide	Conditions	Product(s) and Yield(s) (%)	Refs.
C_{7-8}			
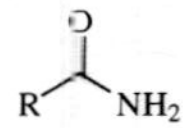	Et_3SiH (3 eq), TFA (2-3 eq), PhCHO (x eq)	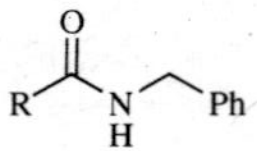	326
R	Solvent / Temp / Time	x = 0.33 / x = 3	
Ph	MeC_6H_5 120° 18 h	(91) (94)	
4-FC_6H_4	MeC_6H_5 120° 18 h	(92) (93)	
PhO	MeCN 22° 18 h	(90) (85)	
PhNH	MeC_6H_5 22° 18 h	(92) (97)	
Bn	MeC_6H_5 120° 36 h	(92) (68)	
4-$MeOC_6H_4$	MeC_6H_5 120° 18 h	(95) (91)	
BnO	MeCN 22° 18 h	(95) (92)	
BnNH	MeC_6H_5 22° 18 h	(89) (88)	
C_7 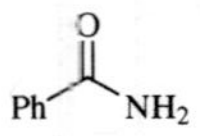	Et_3SiH (3 eq), TFA (2.9 eq), $EtCH(OEt)_2$ (0.33 eq), MeCN, 22°, 18 h	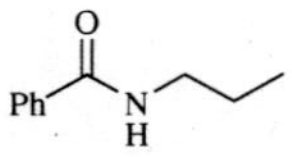 (65)	326
C_8 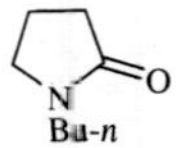	Ph_2SiH_2 (2.1 eq), $(Ph_3P)_3RhH(CO)$ (0.1 mol%), THF, rt, 0.5 h	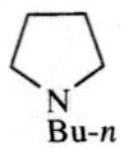 (66)	431
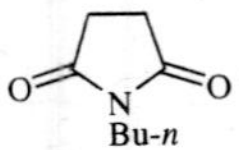	Ph_2SiH_2 (4.3 eq), $(Ph_3P)_3RhH(CO)$ (1 mol%), THF, rt, 3.5 h	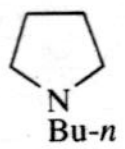 (70)	431

C_{9-14}

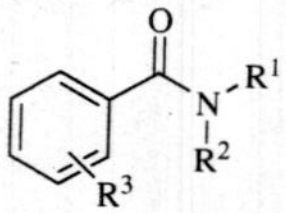

$PhMeSiH_2$ (2 eq), catalyst (1 mol%), MeC_6H_5, 80°, 1 h

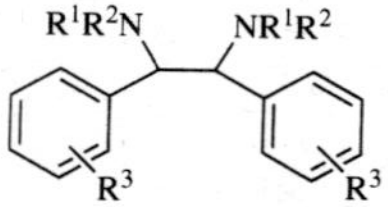

I

430

R^1	R^2	R^3	Catalyst		meso:rac
Me	Me	H	Cp_2TiF_2	(84)	52:48
Me	Me	H	Cp_2TiMe_2	(80)	52:48
Et	Et	H	Cp_2TiF_2	(92)	52:48
Et	Et	H	Cp_2TiMe_2	(88)	53:47
Et	Et	4-Cl	Cp_2TiF_2	(93)	53:47
Et	Et	4-Cl	Cp_2TiMe_2	(90)	53:47
Et	Et	4-Me	Cp_2TiF_2	(91)	62:38
Et	Et	4-CF_3	Cp_2TiF_2	(96)	53:47
Et	Et	4-MeO	Cp_2TiMe_2	(94)	56:44
Me	Ph	H	Cp_2TiF_2	(77)	50:50

TABLE 13. ORGANOSILANE REDUCTION OF AMIDES (*Continued*)

Amide	Conditions	Product(s) and Yield(s) (%)	Refs.
C_{9-14} 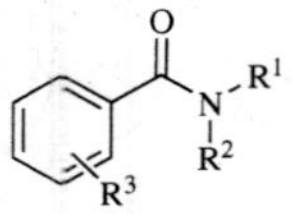	1. $PhMeSiH_2$ (3 mmol), Cp_2TiF_2 (0.1 mol), MeC_6H_5, 70°, 10 min 2. $PhMeSiH_2$ (2 mmol), amide (1 mmol), MeC_6H_5, 20°, 30 min	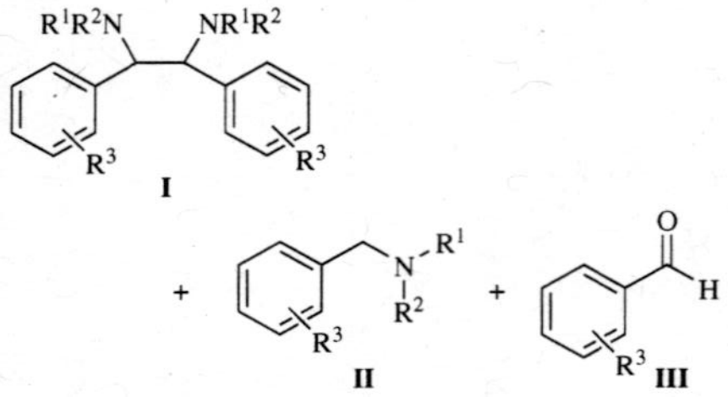 I + II + III	430

R^1	R^2	R^3	I	meso:rac	II	III
Me	Me	H	(81.3)	54:46	(2.0)	(16.7)
Et	Et	H	(97.7)	60:40	(0)	(2.3)
Et	Et	4-Cl	(44.8)	53:47	(29.9)	(25.3)
Et	Et	4-MeO	(100)	53:47	(0)	(0)
Et	Et	4-Me	(100)	50:50	(0)	(0)
Et	Et	4-CF_3	(55.9)	53:47	(20.6)	(23.5)
Me	Ph	H	(14.7)	55:45	(44.2)	(41.1)

Amide	Conditions	Product(s) and Yield(s) (%)	Refs.
C_{10} 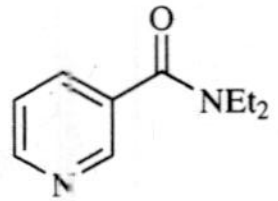	$PhMeSiH_2$ (2 eq), Cp_2TiF_2 (1 mol%), MeC_6H_5, 80°, 1 h	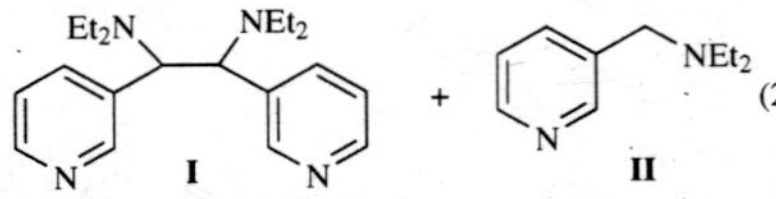 I (27) meso:rac = 57:43 + II (25)	430
	1. $PhMeSiH_2$ (3 mmol), Cp_2TiF_2 (0.1 mol), MeC_6H_5, 70°, 10 min 2. $PhMeSiH_2$ (2 mmol), amide (1 mmol), MeC_6H_5, 20°, 30 min	I (43.5) meso:rac = 56:44 + II (34.1) + 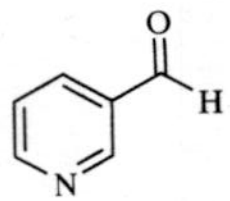 (22.4)	430

C_{10-19}

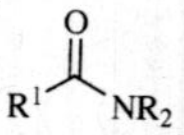

Ph_2SiH_2 (2.1 eq), $(Ph_3P)_3RhH(CO)$ (0.1 mol%), THF, rt

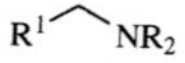

431

R^1	NR_2	Time	
$MeO_2CCH_2CH_2$	$N(CH_2)_5$	3 h	(98)
Ph	NEt_2	2 h	(85)
$4\text{-}BrC_6H_4$	NEt_2	20 h	(85)
i-Pr	NBu_2	24 h	(90)
t-Bu	NBu_2	48 h	(86)
$4\text{-}MeO_2CC_6H_4$	NEt_2	4.5 h	(70)
Ph	$NMe(C_5H_{11}\text{-}c)$	4 h	(91)
Ph	$N(C_6H_{11}\text{-}c)_2$	48 h	(83)

C_{11}

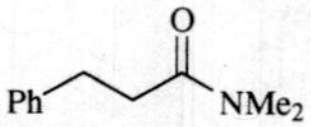

1. $EtMe_2SiH$ (2.5 eq), **2** (1 mol%), 1,4-dioxane, 20°, 0.5 h
2. Add acid, 0.5 h

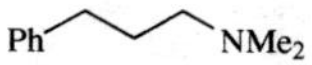

(75) 280

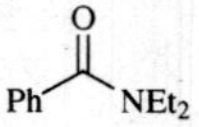

1. $EtMe_2SiH$ (2.5 eq), **2** (1 mol%), 1,4-dioxane, 20°, 0.5 h
2. Acid, 0.5 h

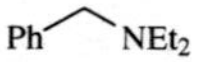

(56) 280

Silane (1 eq), Cp_2TiF_2 (0.1 mol), MeC_6H_5, 1 h

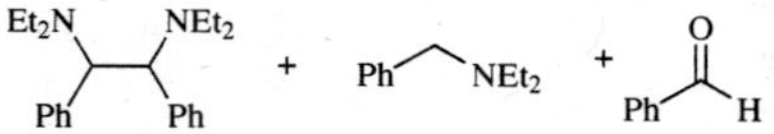

I + II + III

430

Silane	Temp	I:II:III	I meso:rac	
$PhSiH_3$	20°	92:8:0	54:46	(86)
$PhSiH_3$	80°	100:0:0	57:43	(83)
$n\text{-}C_6H_{13}SiH_3$	20°	51:30:19	55:45	(66)
$n\text{-}C_6H_{13}SiH_3$	80°	85:15:0	56:44	(90)
$[—(Me)(H)SiO—]_4$	20°	43:9:48	52:48	(40)
$[—(Me)(H)SiO—]_4$	80°	92:8:0	55:45	(81)
PMHS	20°	59:1:49	56:44	(37)
PMHS	80°	(—)	(—)	(0)

TABLE 13. ORGANOSILANE REDUCTION OF AMIDES (*Continued*)

Amide	Conditions	Product(s) and Yield(s) (%)	Refs.
C_{11}			
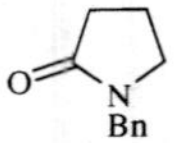	1. $EtMe_2SiH$ (2.5 eq), **225** (1 mol%), 1,4-dioxane, 20°, 0.5 h 2. Add acid, 2 h	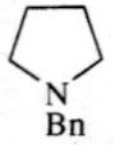 (45)	280
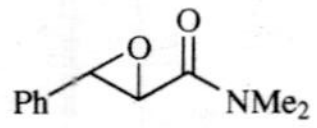	Ph_2SiH_2 (2.1 eq), $(Ph_3P)_3RhH(CO)$ (0.1 mol%), THF, rt, 3 h	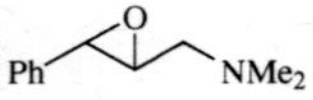 (65)	431
C_{12}			
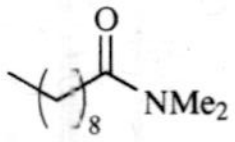	Ph_2SiH_2 (1.1 eq), $Ti(OPr\text{-}i)_4$ (1.0 eq), 20°	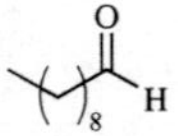 (74)	433
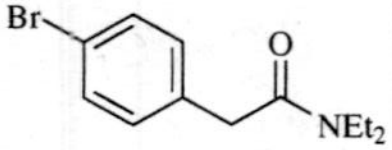	Ph_2SiH_2 (1.1 eq), $Ti(OPr\text{-}i)_4$ (1.0 eq), 20°	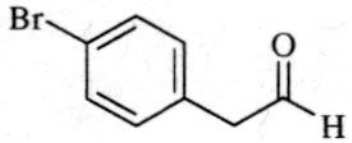 (80)	433
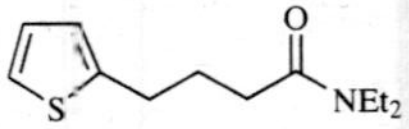	Ph_2SiH_2 (1.1 eq), $Ti(OPr\text{-}i)_4$ (1.0 eq), 20°	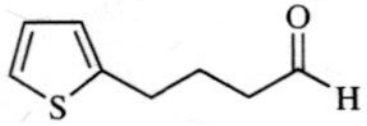 (65)	433
C_{14}			
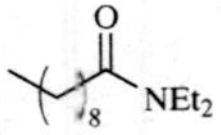	Ph_2SiH_2 (1.1 eq), $Ti(OPr\text{-}i)_4$ (1.0 eq), 20°	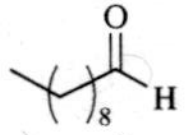 **I** (90)	433
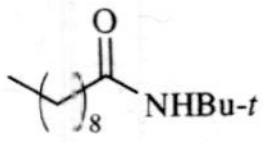	Ph_2SiH_2 (1.1 eq), $Ti(OPr\text{-}i)_4$ (1.0 eq), 20°	**I** (90)	433

Reactant	Conditions	Product(s) and Yield(s) (%)	Refs.
C_{15}			
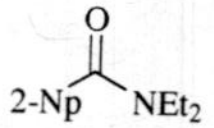	$PhMeSiH_2$ (2 eq), Cp_2TiF_2 (1 mol%), MeC_6H_5, 80°, 1 h	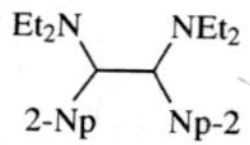 I (83) meso:rac= 54:46 + 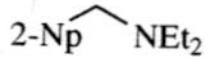 II (9)	430
	1. $PhMeSiH_2$ (3 mmol), Cp_2TiF_2 (0.1 mol), MeC_6H_5, 70°, 10 min 2. $PhMeSiH_2$ (2 mmol), amide (1 mmol), MeC_6H_5, 20°, 30 min	**I** (51.2) meso:rac = 48:52 + **II** (39.6) + 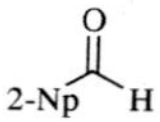 (9.2)	430
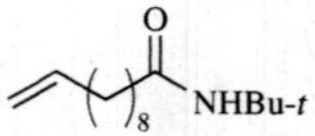	Ph_2SiH_2 (1.1 eq), $Ti(OPr\text{-}i)_4$ (1.0 eq), 20°	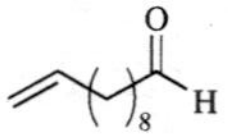 (90)	433
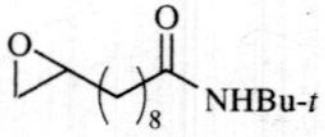	Ph_2SiH_2 (1.1 eq), $Ti(OPr\text{-}i)_4$ (1.0 eq), 20°	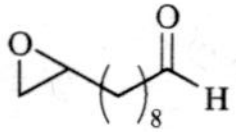 (50)	433
C_{16}			
	R_3SiH (2.1 eq), catalyst (0.1 mol%), THF, rt		431

R_3Si	Catalyst	Time	
Ph_2HSi	$(Ph_3P)_3RhH(CO)$	1 h	(94)
Ph_2HSi	$(Ph_3P)_4RhH$	0.5 h	(93)
Ph_2HSi	$[Rh(cod)_2]BF_4$	42 h	(86)
Ph_2HSi	$(Ph_3P)_3RhCl$	48 h	(93)
Ph_2HSi	$RhCl_3{\cdot}3H_2O$	48 h	(95)
Ph_3Si	$(Ph_3P)_3RhH(CO)$	—	(0)

TABLE 13. ORGANOSILANE REDUCTION OF AMIDES (*Continued*)

Amide	Conditions	Product(s) and Yield(s) (%)	Refs.
C_{16}			
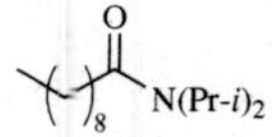	Ph_2SiH_2 (1.1 eq), $Ti(OPr\text{-}i)_4$ (1.0 eq), 20°	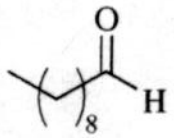 (78)	433
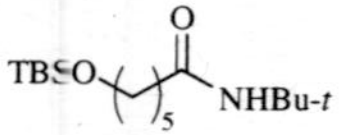	Ph_2SiH_2 (1.1 eq), $Ti(OPr\text{-}i)_4$ (1.0 eq), 20°	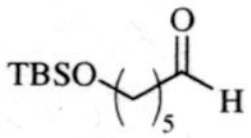 (87)	433
C_{17}			
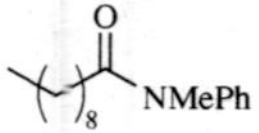	Ph_2SiH_2 (1.1 eq), $Ti(OPr\text{-}i)_4$ (1.0 eq), 20°	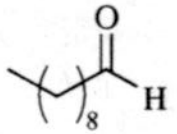 (83)	433
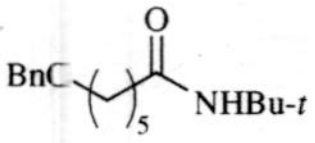	Ph_2SiH_2 (1.1 eq), $Ti(OPr\text{-}i)_4$ (1.0 eq), 20°	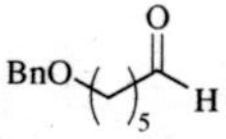 (81)	433
C_{18}			
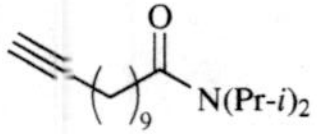	Ph_2SiH_2 (1.1 eq), $Ti(OPr\text{-}i)_4$ (1.0 eq), 20°	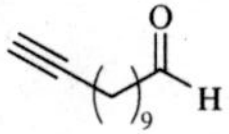 (80)	433
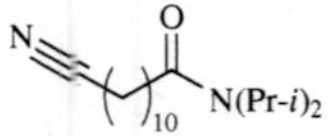	Ph_2SiH_2 (1.1 eq), $Ti(OPr\text{-}i)_4$ (1.0 eq), 20°	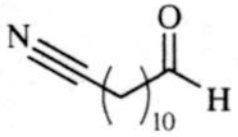 (80)	433
C_{20}			
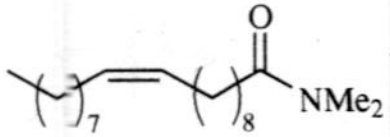	Ph_2SiH_2 (1.1 eq), $Ti(OPr\text{-}i)_4$ (1.0 eq), 20°	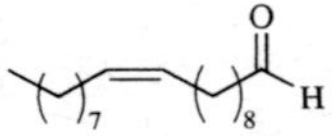 (83)	433

TABLE 14. ORGANOSILANE REDUCTIVE AMINATION OF ALDEHYDES AND KETONES

Aldehyde/Ketone	Amine	Conditions	Product(s) and Yield(s) (%)	Refs.
C_4				
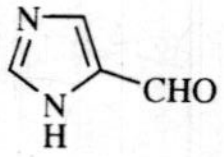	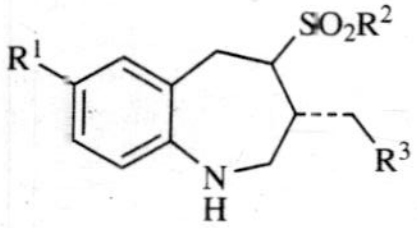	Et_3SiH (2 eq), TFA, CH_2Cl_2, rt	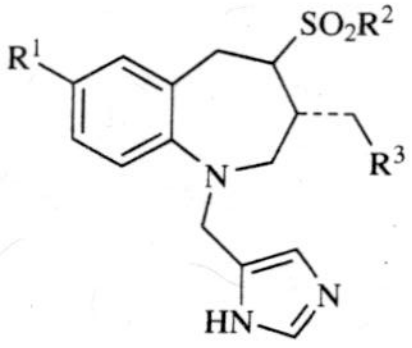	360
	R^1 / R^2 / R^3	Time		
	CN / *n*-Pr / Ph	2 h	(90)	
	CN / 4-$MeOC_6H_4$ / Ph	4 h	(94)	
	CN / 2-C_4H_3S / Ph	4 h	(93)	
	CN / Ph / 4-FC_6H_4	4 h	(95)	
	Ph / 2-C_4H_3S / 3-C_5H_4N	4 h	(93)	
	Ph / $(CH_2)_3NMe_2$ / Ph	4 h	(90)	
$C_{4\text{-}12}$				
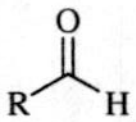	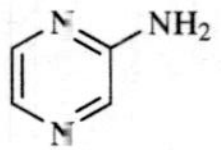	$PhSiH_3$ (1.1 eq), $(n\text{-}Bu)_2SnCl_2$ (0.02 eq), THF, 20° or reflux, 24-300 h	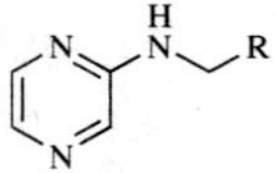 R: *n*-Pr (63); *i*-Pr (70); Ph (72); Bn (50); *n*-$C_{11}H_{23}$ (56)	362
	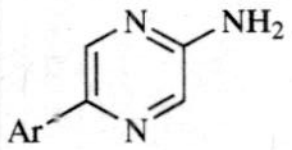 Ar = 4-$TBSOC_6H_4$	$PhSiH_3$ (1.1 eq), $(n\text{-}Bu)_2SnCl_2$ (0.02 eq), THF, TBAF, 20° or reflux, 24-300 h	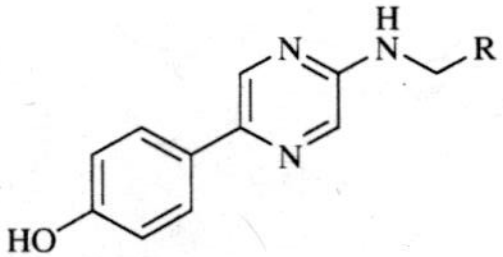 R: *n*-Pr (25); *i*-Pr (41); Ph (17); Bn (10); *n*-$C_{11}H_{23}$ (57)	362
C_5				
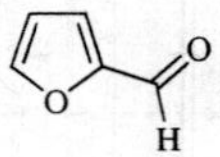	$Me_2SiH(NEt_2)$ (1.2 eq)	$TiCl_4$ (0.20 eq), CH_2Cl_2, 0° to rt, 36 h	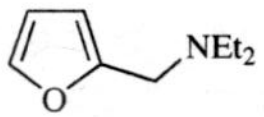 (65)	359

TABLE 14. ORGANOSILANE REDUCTIVE AMINATION OF ALDEHYDES AND KETONES (*Continued*)

Aldehyde/Ketone	Amine	Conditions	Product(s) and Yield(s) (%)	Refs.
C_6 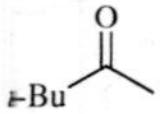	MeCN	Et_3SiH (1.1 eq), H_2SO_4, H_2O, 28°, 65 h	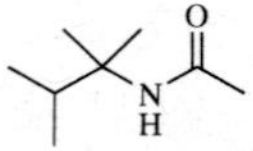 (67)	313
C_7 	MeCN	Et_3SiH (1.1 eq), H_2SO_4, H_2O, 28°, 65 h	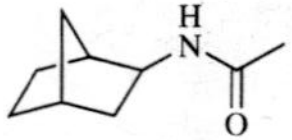 (78)	313
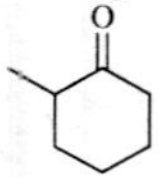	$BnNH_2$	1. Ti(OPr-*i*)$_4$, (1.25 eq), THF, rt, 1 h 2. PMHS (2 eq), THF, 4.5 h	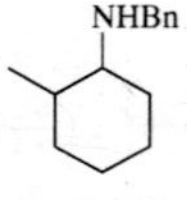 (88)	363
	$PhNH_2$	1. Ti(OPr-*i*)$_4$ (1.25 eq), THF, rt, 1 h 2. PMHS (2 eq), THF, 4.5 h	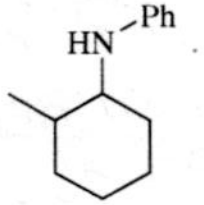 (87)	363
	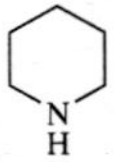	1. Ti(OPr-*i*)$_4$ (1.25 eq), THF, rt, 1 h 2. PMHS (2 eq), THF, 5 h	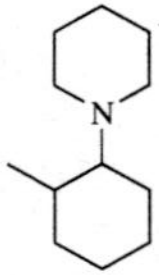 (85)	363
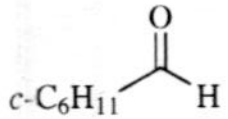	$Me_2SiH(NEt_2)$ (1.2 eq)	ZnI_2 (0.10 eq), CH_2Cl_2, 0° to rt, 36 h	 (81)	359
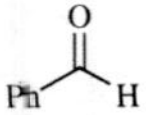	$BnNH_2$	1. Ti(OPr-*i*)$_4$ (1.25 eq), THF, rt, 1 h 2. PMHS (2 eq), THF, 5 h	 (90)	363
	$PhNH_2$	1. Ti(OPr-*i*)$_4$ (1.25 eq), THF, rt, 1 h 2. PMHS (2 eq), THF, 5 h	 (92)	363

	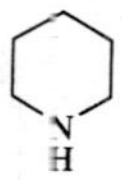	1. Ti(OPr-*i*)$_4$ (1.25 eq), THF, rt, 1 h 2. PMHS (2 eq), THF, 5.5 h	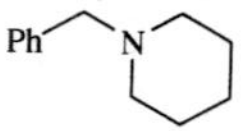 (85)	363
	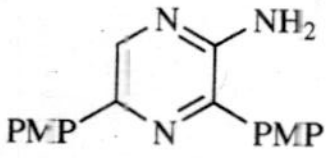	$PhSiH_3$ (1.1 eq), (n-Bu)$_2SnCl_2$ (0.02 eq), THF, 20°, 24 h	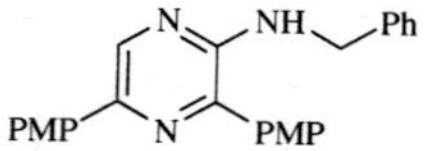 (37)	362
	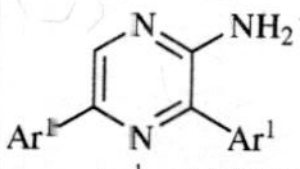 Ar^1 = 4-TBSOC_6H_4	$PhSiH_3$ (1.1 eq), (n-Bu)$_2SnCl_2$ (0.02 eq), THF, TBAF, 20°, 24 h	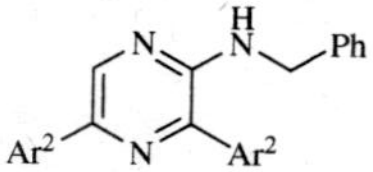 (78) Ar^2 = 4-HOC_6H_4	362
	$PhNH_2$	$PhSiH_3$ (1.1 eq), (n-Bu)$_2SnCl_2$ (0.02 eq), THF, rt, 2-24 h	 (80)	361
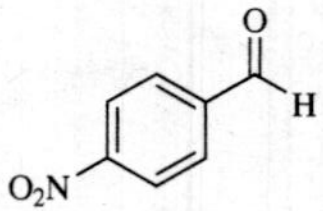	$BnNH_2$	1. Ti(OPr-*i*)$_4$ (1.25 eq), THF, rt, 1 h 2. PMHS (2 eq), THF, 4 h	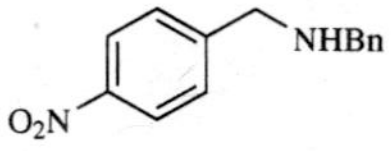 (85)	363
	$PhNH_2$	$PhSiH_3$ (1.1 eq), (n-Bu)$_2SnCl_2$ (0.02 eq), THF, rt, 2-24 h	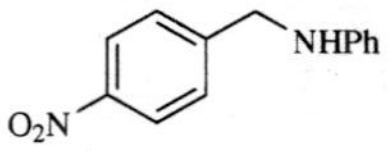 (85)	361
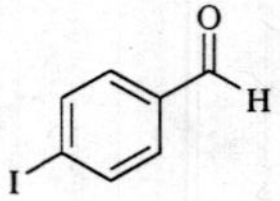	$PhNH_2$	$PhSiH_3$ (1.1 eq), (n-Bu)$_2SnCl_2$ (0.02 eq), THF, rt, 2-24 h	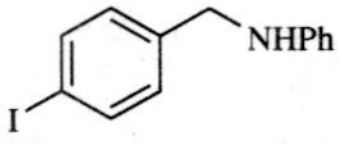 (86)	361
C_{7-9} 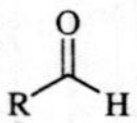	$BnNH_2$	1. $MgSO_4$, CH_2Cl_2, rt, 1 h 2. Cl_3SiH, CH_2Cl_2/DMF (4:1), 0°, 4 h	 R: c-C_6H_{11} (74); (E)-PhCH=CH (81); $BnCH_2$ (93)	318

TABLE 14. ORGANOSILANE REDUCTIVE AMINATION OF ALDEHYDES AND KETONES (*Continued*)

Aldehyde/Ketone		Amine		Conditions			Product(s) and Yield(s) (%)	Refs.
C_{7-10} 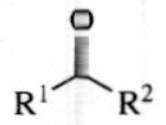		$R^3R^4NSiHMe_2$ (1.2 eq)		Catalyst (x eq), CH_2Cl_2, 0° to rt			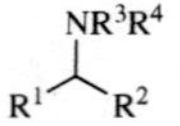	359
R^1	R^2	R^3	R^4	Catalyst	x	Time		
Ph	H	Me	Me	$TiCl_4$	0.2	36 h	(64)	
Ph	H	Me	Me	$TiCl_4$	0.4	36 h	(73)	
4-MeC_6H_4	H	Me	Me	$TiCl_4$	0.2	36 h	(67)	
4-MeC_6H_4	H	Me	Me	$TiCl_4$	0.4	36 h	(85)	
4-$O_2NC_6H_4$	H	Me	Me	$TiCl_4$	0.2	48 h	(94)	
Ph	Me	Me	Me	$TiCl_4$	0.4	36 h	(87)	
$BnCH_2$	H	Me	Me	ZnI_2	0.1	36 h	(72)	
$BnCH_2$	Me	Me	Me	$TiCl_4$	0.2	48 h	(67)	
Ph	H	—$(CH_2)_4$—		$TiCl_4$	0.3	36 h	(77)	
$BnCH_2$	H	—$(CH_2)_4$—		ZnI_2	0.1	36 h	(71)	
Ph	H	—$(CH_2)_2O(CH_2)_2$—		$TiCl_4$	0.2	36 h	(83)	
$BnCH_2$	H	—$(CH_2)_2O(CH_2)_2$—		ZnI_2	0.1	36 h	(66)	
Ph	H	H	CO_2Me	Ph_3CClO_4	0.05	2 h	(94)	
Ph	Me	H	CO_2Me	Ph_3CClO_4	0.05	1 h	(53)	
$BnCH_2$	H	H	CO_2Me	Ph_3CClO_4	0.05	3 h	(40)	
$BnCH_2$	H	H	CO_2Me	Ph_3CClO_4	0.05	1 h	(88)	
$BnCH_2$	Me	H	CO_2Me	Ph_3CClO_4	0.05	1 h	(86)	

C_{7-13}

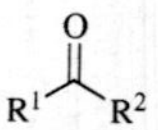

R^3NH_2

1. Activator (1.25 eq), solvent, rt, 1 h
2. PMHS (2 eq), THF

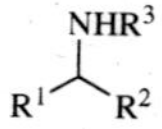

363

R^1	R^2	R^3	Activator	Solvent	Time (2)	
Ph	H	Bn	$AlCl_3$	MeC_6H_5	12 h	(40)
Ph	H	Bn	Pd-C	EtOH	10 h	(60)
Ph	H	Bn	Ti(OPr-*i*)$_4$	THF	5 h	(90)
Ph	H	Ph	$AlCl_3$	MeC_6H_5	12 h	(52)
Ph	H	Ph	Pd-C	EtOH	10 h	(62)
Ph	H	Ph	Ti(OPr-*i*)$_4$	THF	5 h	(92)
(*E*)-PhCH=CH	H	Bn	$AlCl_3$	MeC_6H_5	10 h	(48)
(*E*)-PhCH=CH	H	Bn	Pd-C	EtOH	8 h	(55)
(*E*)-PhCH=CH	H	Bn	Ti(OPr-*i*)$_4$	THF	4 h	(85)
(*E*)-PhCH=CH	H	Ph	$AlCl_3$	MeC_6H_5	12 h	(45)
(*E*)-PhCH=CH	H	Ph	Pd-C	EtOH	9 h	(56)
(*E*)-PhCH=CH	H	Ph	Ti(OPr-*i*)$_4$	THF	4.5 h	(88)
Ph	Me	Bn	$AlCl_3$	MeC_6H_5	13 h	(45)
Ph	Me	Bn	Pd-C	EtOH	10 h	(55)
Ph	Me	Bn	Ti(OPr-*i*)$_4$	THF	6 h	(90)
Ph	Me	Ph	$AlCl_3$	MeC_6H_5	13 h	(50)
Ph	Me	Ph	Pd-C	EtOH	10 h	(58)
Ph	Me	Ph	Ti(OPr-*i*)$_4$	THF	6 h	(90)

TABLE 14. ORGANOSILANE REDUCTIVE AMINATION OF ALDEHYDES AND KETONES (*Continued*)

Aldehyde/Ketone		Amine	Conditions	Product(s) and Yield(s) (%)	Refs.
C_{7-13}					
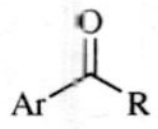		$Me_2SiH(NEt_2)$ (1.2 eq)	$TiCl_4$ (0.20 eq), CH_2Cl_2, 0° to rt, 36 h	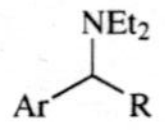	359
Ar	R				
Ph	H			(90)	
4-ClC_6H_4	H			(78)	
4-$O_2NC_6H_4$	H			(85)	
Ph	Me			(65)	
4-MeC_6H_4	H			(91)	
4-$MeOC_6H_4$	H			(92)	
4-NCC_6H_4	H			(84)	
Ph	Et			(39)	
4-$MeO_2CC_6H_4$	H			(84)	
Ph	Ph			(46)	
4-PhC_6H_4	H			(86)	
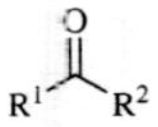		MeCN	Et_3SiH (1.1 eq), H_2SO_4, H_2O, 28°	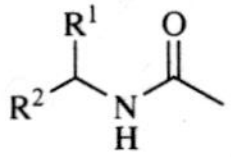	313
R^1	R^2		Time		
—$(CH_2)_5$—			72 h	(30)	
Ph	H		74 h	(80)	
Ph	Me		72 h	(85)	
n-C_7H_{15}	H		72 h	(0)	
Ph	Ph		48 h	(63)	

C_{7-23}

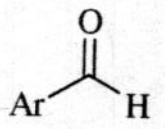

$PhCONH_2$ (3 eq)

Et_3SiH (3 eq), TFA (2.9 eq), MeC_6H_5, 120°, 18 h

326

Ar	
4-FC_6H_4	(96)
4-$MeO_2CC_6H_4$	(91)
4-$HO_2CC_6H_4$	(96)
4-$AcOC_6H_4$	(75)
4-$BnOC_6H_4$	(90)
4-$TBSOC_6H_4$	(46)
4-$TBDPSOC_6H_4$	(91)

C_8

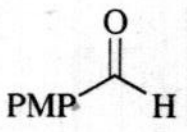

$PhNH_2$

R_3SiH (x eq), $(n\text{-}Bu)_2SnCl_2$ (0.02 eq), $PhNH_2$ (1.0 eq), THF, rt, 21 h

361

R_3Si	x	
Et_3Si	1.1	(0)
Ph_3Si	1.1	(0)
Ph_2HSi	1.1	(88)
Ph_2HSi	1.1	(—)
PhH_2Si	1.1	(82)
PhH_2Si	1.1	(83)
PhH_2Si	0.5	(83)
PhH_2Si	0.33	(83)
PhH_2Si	0.20	(—)
PMHS	2.0	(82)
PhH_2Si	1.1	(0)

NHMePh

$PhSiH_3$ (1.1 eq), $(n\text{-}Bu)_2SnCl_2$ (0.02 eq), THF, rt, 2-24 h

(83)

361

TABLE 14. ORGANOSILANE REDUCTIVE AMINATION OF ALDEHYDES AND KETONES (*Continued*)

Aldehyde/Ketone	Amine	Conditions	Product(s) and Yield(s) (%)	Refs.
C_8 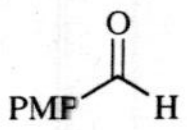	$ArNH_2$	$PhSiH_3$ (1.1 eq), $(n\text{-}Bu)_2SnCl_2$ (0.02 eq), THF, rt, 2-24 h	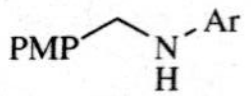	361
	Ar			
	$4\text{-}MeOC_6H_4$		(82)	
	$4\text{-}O_2NC_6H_4$		(85)	
	$4\text{-}t\text{-}BuC_6H_4$		(85)	
	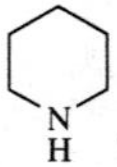	$PhSiH_3$ (1.1 eq), $(n\text{-}Bu)_2SnCl_2$ (0.02 eq), THF, rt, 16 h	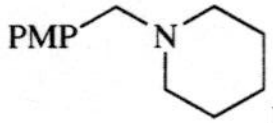 **I** (70)	361
		$PhSiH_3$ (1.1 eq), THF, rt, 16 h	**I** (13)	361
	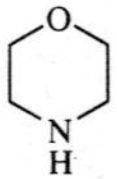	$PhSiH_3$ (1.1 eq), $(n\text{-}Bu)_2SnCl_2$ (0.02 eq), THF, rt, 16 h	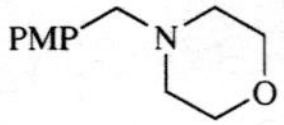 (78)	361
	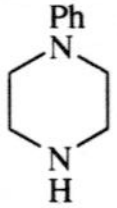	$PhSiH_3$ (1.1 eq), $(n\text{-}Bu)_2SnCl_2$ (0.02 eq), THF, rt, 16 h	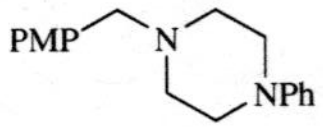 (67)	361
	Et_2NH	$PhSiH_3$ (1.1 eq), $(n\text{-}Bu)_2SnCl_2$ (0.02 eq), THF, rt, 16 h	 (49)	361
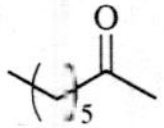	$PhNH_2$	$PhSiH_3$ (1.1 eq), $(n\text{-}Bu)_2SnCl_2$ (0.02 eq), THF, rt, 2-24 h	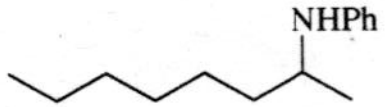 (85)	361
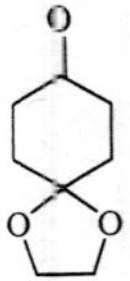	$PhNH_2$	$PhSiH_3$ (1.1 eq), $(n\text{-}Bu)_2SnCl_2$ (0.1 eq), THF, rt, 2-24 h	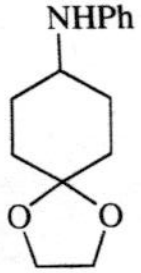 (91)	361

	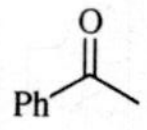	PhNH$_2$	PhSiH$_3$ (1.1 eq), $(n\text{-Bu})_2SnCl_2$ (0.1 eq), THF, rt, 2-24 h	 **I** (70)	361
		PhNH$_2$	1. Ti(OPr-*i*)$_4$ (1.25 eq), THF, rt, 1 h 2. PMHS (2 eq), THF, 6 h	**I** (90)	363
		BnNH$_2$	1. Ti(OPr-*i*)$_4$ (1.25 eq), THF, rt, 1 h 2. PMHS (2 eq), THF, 6 h	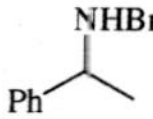 (90)	363
			1. Ti(OPr-*i*)$_4$ (1.25 eq), THF, rt, 1 h 2. PMHS (2 eq), THF, 6 h	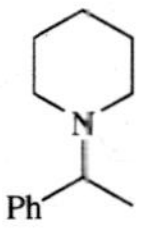 (88)	363
C$_9$	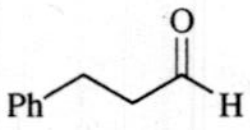	Me$_2$SiH(NEt$_2$) (1.2 eq)	ZnI$_2$ (0.10 eq), CH$_2$Cl$_2$, 0° to rt, 36 h	 (72)	359
	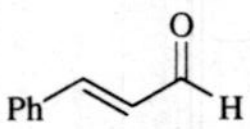	PhNH$_2$	PhSiH$_3$ (1.1 eq), $(n\text{-Bu})_2SnCl_2$ (0.02 eq), THF, rt, 2-24 h	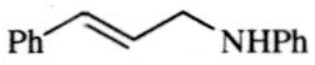 **I** (70)	361
			1. Ti(OPr-*i*)$_4$ (1.25 eq), THF, rt, 1 h 2. PMHS (2 eq), THF, 4.5 h	**I** (88)	363
		BnNH$_2$	1. Ti(OPr-*i*)$_4$ (1.25 eq), THF, rt, 1 h 2. PMHS (2 eq), THF, 4 h	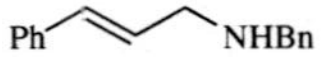 (85)	363
		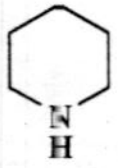	1. Ti(OPr-*i*)$_4$ (1.25 eq), THF, rt, 1 h 2. PMHS (2 eq), THF, 4.5 h	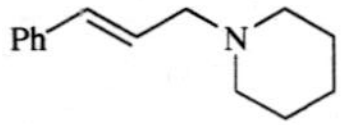 (85)	363

TABLE 14. ORGANOSILANE REDUCTIVE AMINATION OF ALDEHYDES AND KETONES (*Continued*)

Aldehyde/Ketone	Amine	Conditions	Product(s) and Yield(s) (%)	Refs.
C_{10} 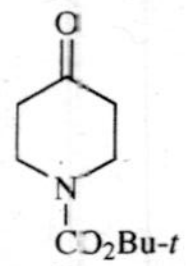	$PhNH_2$	$PhSiH_3$ (1.1 eq), (*n*-Bu)$_2SnCl_2$ (0.02 eq), THF, rt, 2-24 h	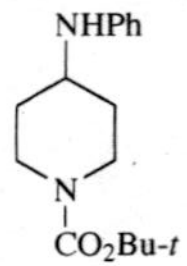 (86)	361
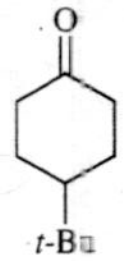	MeCN	Et_3SiH (1.1 eq), H_2SO_4, H_2O, 28°, 65 h	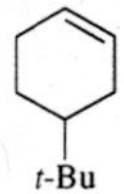 (4) + 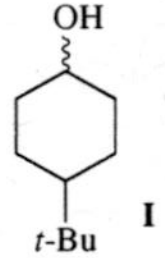 **I** + 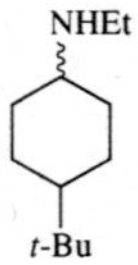 (21) + 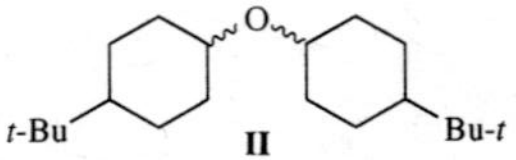 **II** **I** + **II** (70)	313
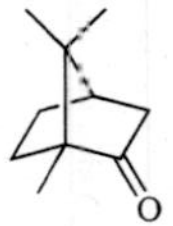	$BnNH_2$	1. Ti(OPr-*i*)$_4$ (1.25 eq), THF, rt, 1 h 2. PMHS (2 eq), THF, 4.5 h	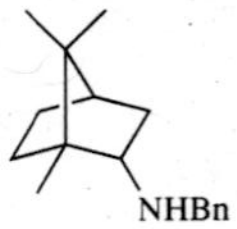 (88)	363
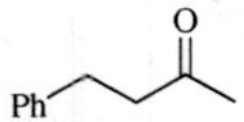	$Me_2SiH(NEt_2)$ (1.2 eq)	ZnI_2 (0.10 eq), CH_2Cl_2, 0° to rt, 36 h	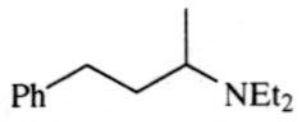 (46)	359
C_{13} 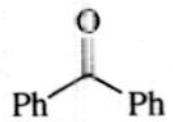	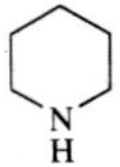	1. Ti(OPr-*i*)$_4$ (1.25 eq), THF, rt, 1 h 2. PMHS (2 eq), THF, 4.5 h	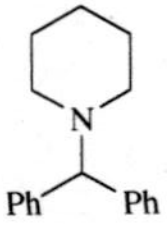 (80)	363
	$ArNH_2$ Ar: Ph, Bn	1. Ti(OPr-*i*)$_4$ (1.25 eq), THF, rt, 1 h 2. PMHS (2 eq), THF, 4.5 h	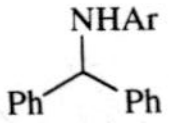 (85) (83)	363

C_{14}					
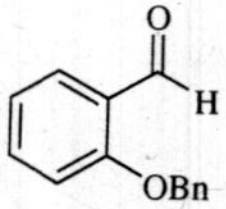	PhNH$_2$	PhSiH$_3$ (1.1 eq), $(n\text{-Bu})_2SnCl_2$ (0.02 eq), THF, rt, 2-24 h	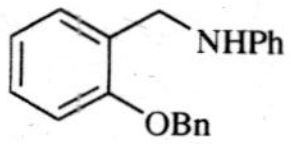	(82)	361

TABLE 15. ORGANOSILANE REDUCTION OF α,β-UNSATURATED ALDEHYDES

Unsaturated Aldehyde	Conditions	Product(s) and Yield(s) (%)	Refs.
C_3 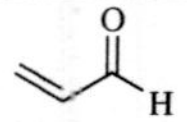	Et_3SiH (3 eq), TFA (6 eq), $CHCl_3$, 40°, 6 h	 (10) +  (12) + 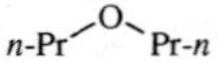 (9) + 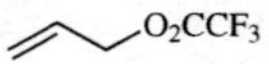 (—) + CF_3CO_2Pr-*n* (—)	434
C_4 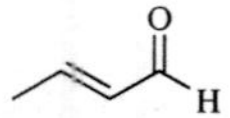	$PhSiH_3$ (1.01 eq), Ni, PPh_3 (0.02 eq)	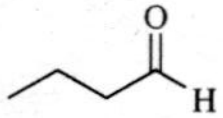 (100)	438
C_5 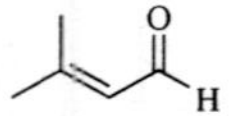	$PhSiH_3$ (1.01 eq), Ni, PPh_3 (0.02 eq)	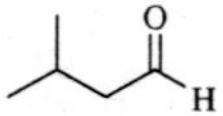 (100)	438
C_9 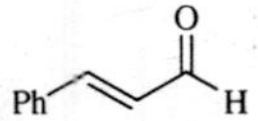	Ph_2SiH_2 (2.3 eq), CsF (1 eq), rt, 3 min	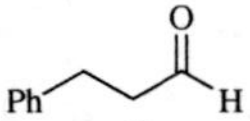 **I** (100)	83
	Et_3SiH (1.1 eq), $(Ph_3P)_3RhCl$, C_6H_6, 45°, 1 h	**I** (—)	435
	Ph_2SiH_2 (1.1 eq), $(Ph_3P)_3RhCl$, 0° to rt, 2 h	**I** (—)	435
	$PhSiH_3$ (1.01 eq), Ni, PPh_3 (0.02 eq)	**I** (98.6)	438
	$(EtO)_3SiH$ (1.1 eq), KF (1 eq), rt, 24 h	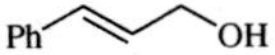 **I** (95)	79, 83
	$(EtO)_3SiH$ (1.1 eq), CsF (1 eq), rt, 1 h	**I** (80)	79
	$Me(EtO)_2SiH$ (1.1 eq), CsF (1 eq), rt, 2 h	**I** (95)	79, 83
	Ph_2SiH_2 (0.5 eq), salt	**I**	319
	Salt		
	HCO_2K	(85)	
	KF	(65)	
	$C_6H_4(CO_2K)_2$-1,2	(65)	
	CsF	(100)	

Substrate	Conditions	Product(s) and Yield(s) (%)	Refs.
	PMHS (1 eq), $ZnCl_2$ (1 eq), Et_2O, rt, 24 h	**I** (69)	315
	$Me(EtO)_2SiH$ (2.3 eq), CsF (1 eq), rt, 2 h	**I** (95)	83
	$(EtO)_3SiH$ (2.3 eq), KF (1 eq), rt, 24 h	**I** (95)	83
	$(EtO)_3SiH$, CsF, rt, 5 min	**I** (100)	83
	Et_3SiH (1.5 eq), CsF (1.5 eq), MeCN, rt, 10 h	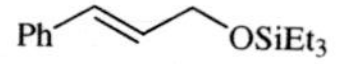 (—)	76
	$PhMe_2SiH$ (1.2 eq), TBAF (2 eq), HMPA, rt, 0.5 h	 (100)	320
	Et_3SiH (0.72 eq), H_2PtCl_6	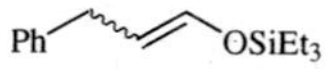 (—)	76
C_{10} 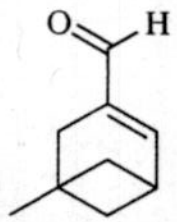	$PhSiH_3$ (1.5 eq), $[(Ph_3P)CuH]_6$ (5 mol%), MeC_6H_5, 5 h	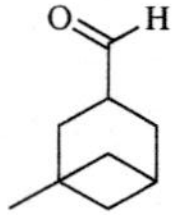 (80)	447
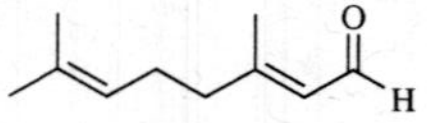	R_3SiH (1.1 eq), $(Ph_3P)_3RhCl$	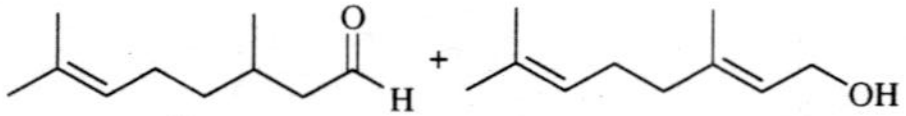 **I** + **II**	435
	R_3Si	**I**:**II** — **I** + **II**	
	$EtMe_2Si$	100:0 (92)	
	Et_3Si	100:0 (97)	
	$(i\text{-}Pr)_2HSi$	49:51 (94)	
	Et_2HSi	3:97 (95)	
	Ph_2SiH_2	0:100 (97)	
	Et_3SiD (1.1 eq), $(Ph_3P)_3RhCl$	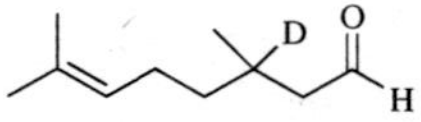 (100)	435
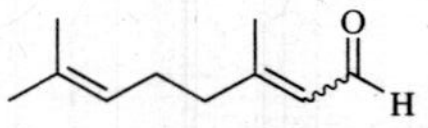	Ph_2SiH_2 (2.0 eq), $ZnCl_2$ (0.12 eq), $(Ph_3P)_4Pd$ (0.012 eq), $CHCl_3$, rt, 1.5 h	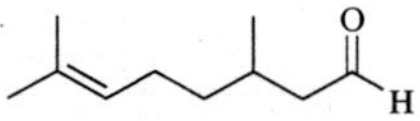 (96)	436

TABLE 16. ORGANOSILANE REDUCTION OF α,β–UNSATURATED KETONES

Unsaturated Ketone	Conditions	Product(s) and Yield(s) (%)	Refs.
C_4 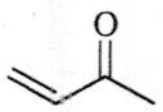	Et_3SiH (3 eq), TFA (10 eq), CCl_4, 50°, 6 h	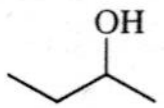 (94)	434, 439
	PMHS, Pd/C, EtOH, 80°	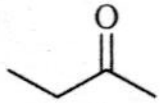 **I** (100)	316
	PMHS-Pd nanocomposite, C_6H_6, rt, 5 h	**I** (95)	219
	$PhMe_2SiH$ (1.1 eq), $(Ph_3P)_4RhH$ (0.5 mol%), CH_2Cl_2, 50°, 24 h	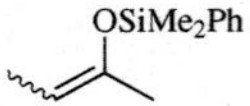 (73), E:Z = 33:67	374
	$ClCH_2Me_2SiH$ (1.1 eq), $(Ph_3P)_4RhH$ (0.2 mol%), CH_2Cl_2, rt,1.56 h	 (85), E:Z = 1:2	374
	$(EtO)_3SiH$ (1.1 eq), $(Ph_3P)_4RhH$ (0.3 mol%), CH_2Cl_2, rt, 20 h	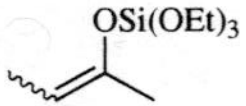 (75), E:Z = 56:44	374
	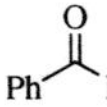 H, Et_2MeSiH, $Rh_4(CO)_{12}$ (0.5 mol %), solvent	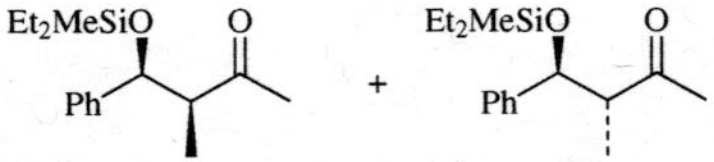 **I** + **II**	464

Solvent	Temp	Time	I + II	I:II
none	20°	1 h	(42)	50:50
hexane	20°	15 h	(97)	77:23
hexane	0°	3 h	(95)	86:14
hexane	−15°	3 h	(85)	88:12
C_6H_6	20°	1 h	(64)	81:19
toluene	−15°	7 h	(98)	87:13
CH_2Cl_2	−15°	9 h	(93)	85:15
THF	−15°	5 h	(53)	78:22

C_{4-10}

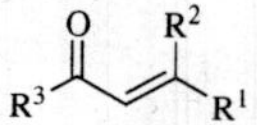

R^1	R^2	R^3
Me	H	H
Me	H	H
Me	Me	Me
Me	Me	Me
Ph	H	H
Me	H	Ph

R_3SiH (1 eq), $(Ph_3P)_3RhCl$ (0.5 mol%)

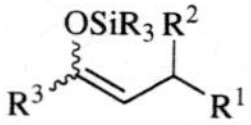

411

R_3Si	Temp	Time	
Et_3Si	60°	15 min	(95)
Ph_3Si	60°	30 min	(97)
Et_3Si	50°	15 min	(94)
$PhMe_2Si$	50°	30 min	(98)
Et_3Si	60°	15 min	(96)
Et_3Si	60°	30 min	(90)

C_5

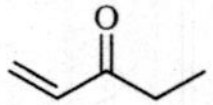

Et_2MeSiH, $Rh_4(CO)_{12}$ (0.5 mol%), ligand, RCHO, hexane

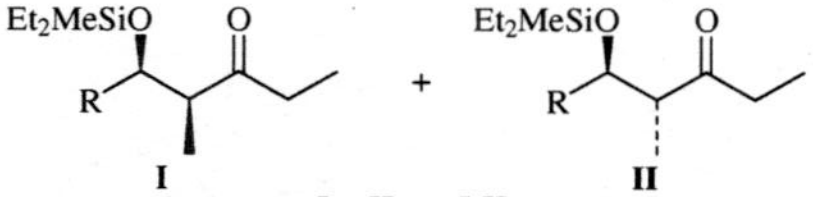

464

R	Ligand	Temp	Time	I + II	I:II
n-C_5H_9	none	0°	6.5 h	(21)	73:27
n-C_5H_9	$MePh_2P$	15°	6 h	(45)	74:26
Ph	$MePh_2P$	0°	20 h	(100)	84:16

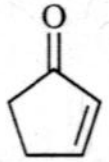

Cl_3SiH (2 eq), $CoCl_2$ (0.175 eq), neat; or CD_3CN, DMI (0.5 eq), rt, 5 h

I (92) 451

$PhMe_2SiH$ (2 eq), $CoCl_2$ (0.175 eq), DMI (0.5 eq), 80°, 10 h — **I** (65) 451

Ph_2SiH_2 (1.2 eq), $ZnCl_2$ (0.16 eq), $(Ph_3P)_4Pd$ (0.005 eq), $CHCl_3$, rt, 0.5 h — **I** (99) 436

$PhMe_2SiH$ (1.1 eq), $(Ph_3P)_4RhH$ (0.3 mol%), CH_2Cl_2, rt, 12 h

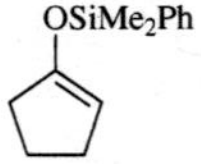

(83) 374

TABLE 16. ORGANOSILANE REDUCTION OF α,β–UNSATURATED KETONES (*Continued*)

Unsaturated Ketone	Conditions	Product(s) and Yield(s) (%)	Refs.
C_6 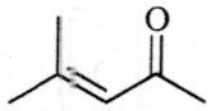	Et_3SiH (3 eq), TFA (6 eq), $CHCl_3$, 60°, 4 h	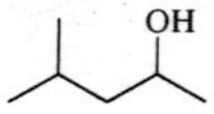 **I** (90)	434
	Et_3SiH (1.2 eq), $TiCl_4$ (2.4 eq), CH_2Cl_2, –5°, 1 h	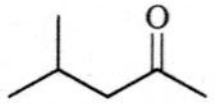 **II** (75)	449
	Ph_2SiH_2 (2.0 eq), $ZnCl_2$ (0.35 eq), $(Ph_3P)_4Pd$ (0.007 eq), $CHCl_3$, rt, 1 h	**II** (99)	436
	Et_3SiH (3 eq), $AlCl_3$ (1.2 eq), HCl, CH_2Cl_2, rt, 4 h	**I** (71) + **II** (10)	136
	R_3SiH (1.1 eq), $(Ph_3P)_3RhCl$	**I** + **II**	435
	R_3Si	**I** + **II** / **I**:**II**	
	$EtMe_2Si$	(99) 0:100	
	$PhMe_2Si$	(98) 1:99	
	Et_3Si	(95) 1:99	
	$(i\text{-}Pr)_2HSi$	(93) 13:87	
	Et_2HSi	(90) 57:43	
	$PhMeHSi$	(98) 61:39	
	Ph_2HSi	(99) 93:7	
	PhH_2Si	(95) 99:1	
	Ph_2SiH_2 (1.3-1.5 eq), $Mo(CO)_6$ (3-5 mol%), THF, reflux, 24 h	**II** (5)	450
	$PhSiH_3$ (1.3-1.5 eq), $Mo(CO)_6$ (3-5 mol%), THF, reflux, 23 h	**II** (95)	450

	$PhCHO$, Et_2MeSiH, $Rh_4(CO)_{12}$ (0.5 mol%), $MePh_2P$, C_6H_6, 20°, 7 h	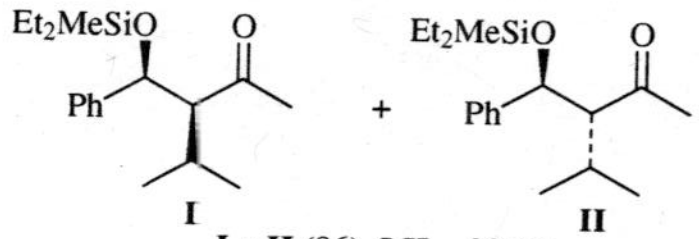 I + II (86), I:II = 80:20	464
	$PhCHO$, Et_2MeSiH, $Rh_4(CO)_{12}$ (0.5 mol%), C_6H_6, 20°, 7 h	I + II (0)	464
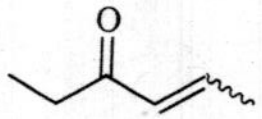	$PhMe_2SiH$ (1.1 eq), $(Ph_3P)_4RhH$ (0.2 mol%), CH_2Cl_2, rt, 6 h	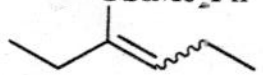 (82), E:Z = 77:23	374
	$ClCH_2Me_2SiH$ (1.1 eq), $(Ph_3P)_4RhH$ (0.3 mol%), CH_2Cl_2, rt, 10 h	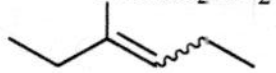 (83), E:Z = 4:1	374
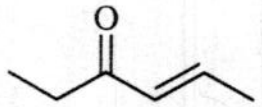	1. $HMe_2SiOSiMe_2H$ (1.5 eq), 1/6 $[(Ph_3P)CuH]_6$ (5 mol%), MeC_6H_5, rt, 1 h 2. $BnOCH_2CHO$, $TiCl_4$, CH_2Cl_2, –78°	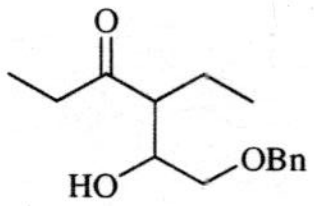 (79)	455
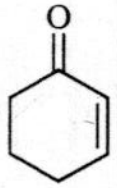	Ph_2SiH_2 (1.3 eq), $(Ph_3P)_4RhH$ (0.4 mol%), CH_2Cl_2, rt, 4 h	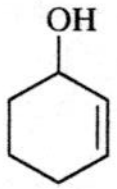 I (83)	374
	Ph_2SiH_2 (0.5 eq), o-$C_6H_4(CO_2K)_2$, 140°, 1.5 h	I + 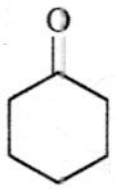 II + 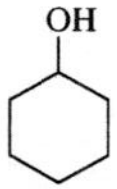 III I + II + III (70), I:II:III = 65:0:35	319

TABLE 16. ORGANOSILANE REDUCTION OF α,β–UNSATURATED KETONES (*Continued*)

Unsaturated Ketone	Conditions	Product(s) and Yield(s) (%)	Refs.
C_6 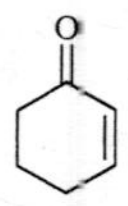	Ph_2SiH_2 (0.5 eq), CsF, 140°, 1.5 h	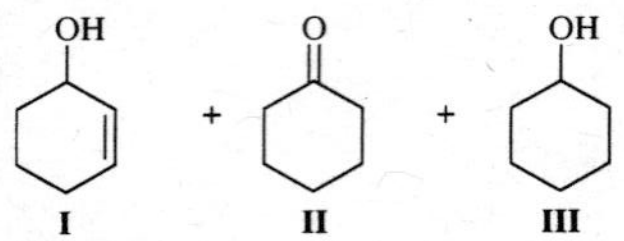 **I** + **II** + **III** (65), **I**:**II**:**III** = 60:10:30	319
	G, 0°, 2 h	**I** (85)	93
	H, 0°, 5 h	**I** + **II** (91), **I**:**II** = 97:3	93
	$PhMe_2SiH$ (4 eq), CuCl (2 eq), DMI, rt, 22 h	**III** (100)	445
	$PhMe_2SiH$ (2 eq), DMA, $(Ph_3P)_3CuF•2$ EtOH (1 eq)	**III** (91)	446
	1. $PhMe_2SiH$ (2 eq), $(Ph_3P)_3CuF•2$ EtOH, DMA, 0°, 0.5 h 2. Add enone, 0° to rt, 2 h	**III** (92)	444
	$PhMe_2SiH$ (1.2 eq), CuCl (50 mol%), PPh_3 (50 mol%), TBAF (25 mol%), DMA, 0° to rt, 17 h	**III** (67)	444
	$PhMe_2SiH$ (2 eq), $(Ph_3P)_3CuF•2$ EtOH (1.0 eq), DMA, 0° to rt, 2 h	**II** (92)	444
	$PhMe_2SiH$ (1.2 eq), CuCl (50 mol%), PPh_3 (50 mol%), TBAF (20 mol%), DMA, 0° to rt, 17 h	**II** (67)[a]	444
	Ph_2SiH_2 (1.2 eq), $ZnCl_2$ (0.16 eq), $(Ph_3P)_4Pd$ (0.005 eq), rt, 0.5 h	**II** (100)	436

$PhSiH_3$ (1.5 eq), $[(Ph_3P)CuH]_6$ (5 mol%), MeC_6H_5, 8 min	**II** (100)	447
$PhSiH_3$ (1.01 eq), Ni, PPh_3 (0.02 eq)	**II** (93.1)	438
$PhSiH_3$ (1.3-1.5 eq), $Mo(CO)_6$ (3-5 mol%), THF, reflux, 4 h	**II** (25)	450
Cl_3SiH (2 eq), $CoCl_2$ (0.175 eq), CD_3CN, DMI (0.5 eq), rt, 5 h	**II** (88)	451
Cl_3SiH (2 eq), $CoCl_2$ (0.05 eq), neat, or CD_3CN, DMPU (0.25 eq), rt, 5 h	**II** (88)	451
L-D CH_2Cl_2, rt, 2 h	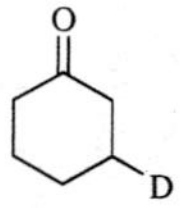	101
Et_3SiH (1 eq), TMSOTf (0.1 eq), BnOTMS (0.83 eq), CH_2Cl_2, −78° to −30°	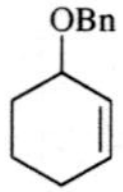 (50)	341
$ClCH_2Me_2SiH$ (1.1 eq), $(Ph_3P)_4RhH$ (0.1 mol%), CH_2Cl_2, rt, 0.5 h	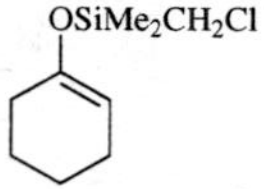 (83)	374
$PhMe_2SiH$ (1.1 eq), $(Ph_3P)_4RhH$ (0.3 mol%), CH_2Cl_2, rt, 12 h	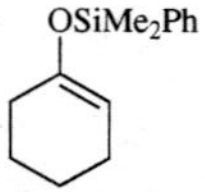 (84)	374

TABLE 16. ORGANOSILANE REDUCTION OF α,β-UNSATURATED KETONES (*Continued*)

Unsaturated Ketone	Conditions	Product(s) and Yield(s) (%)	Refs.
C_6 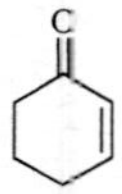	1. $PhMe_2SiH$, (1.5 eq), 1/6 $[CuH(PPh_3)]_6$ (5 mol%), MeC_6H_5, rt, 1 h 2. 4-$O_2NC_6H_4CHO$, BF_3•OEt_2, CH_2Cl_2, –78°	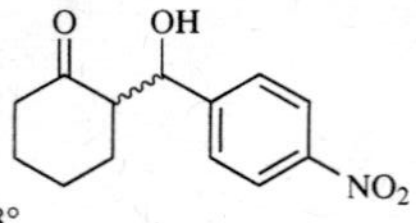 (73)	455
	Et_2MeSiH, PhCHO, $Rh_4(CO)_{12}$ (0.5 mol%), MeC_6H_5, 15°, 15 h	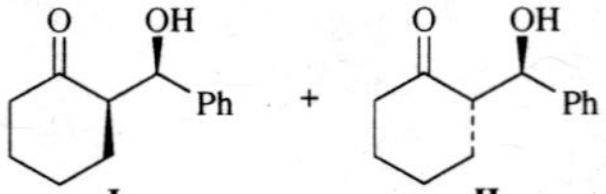 I + II I + II (57), I:II = 75:25	464
	Et_2MeSiH, PhCHO, $Rh_4(CO)_{12}$ (0.5 mol%), $MePh_2P$, MeC_6H_5, 0°, 5 h	I + II (75), I:II = 73:27	464
	1. $PhMe_2SiH$ (1.5 eq), 1/6 $[CuH(PPh_3)]_6$ (5 mol%), MeC_6H_5, rt 2. 2-Br-5-benzyloxybenzaldehyde, BF_3•OEt_2, CH_2Cl_2 3. TsOH (cat.), C_6H_6, reflux, 1.5 h	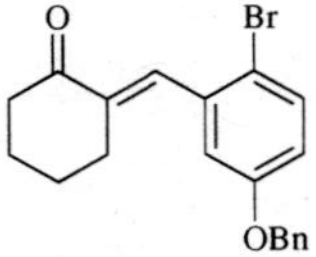 (77)	455
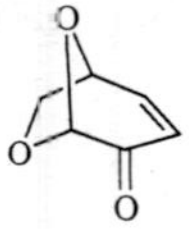	Ph_2SiH_2 (2.0 eq), $ZnCl_2$ (1.0 eq), $(Ph_3P)_4Pd$ (0.10 eq), $CHCl_3$, rt, 1 h	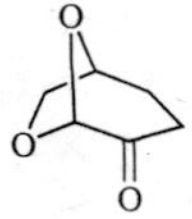 (95)	436

C_{6-12}

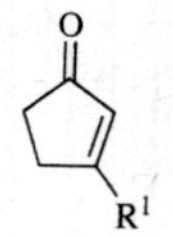

1. Ph_2SiH_2 (0.53 eq), CuCl (5%), NaOBu-*t* (5%), (*S*)-*p*-Tol-BINAP, MeC_6H_5, 0°, 2-3 h
2. R^2X, TBAT (1.2 eq), CH_2Cl_2/MeC_6H_5 (1:1), rt, 24 h

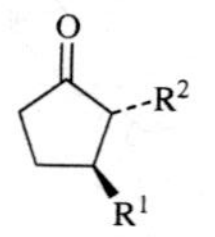

459

R^1	R^2X		dr
Me	BnBr	(62)	92:8
Me	$BrCH_2CO_2Et$	(52)	94:6
Me	1-Bromo-3-methyl-2-butene	(52)	80:20
$BnCH_2$	BnBr	(67)	94:6
$BnCH_2$	Allyl bromide	(64)	76:24
$BnCH_2$	MeI	(65)	73:27
$BnCH_2$	*n*-BuI	(42)	85:15

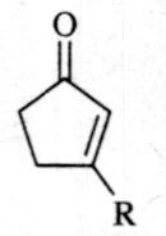

1. Ph_2SiH_2 (0.53 eq), CuCl (5%), NaOBu-*t* (5%), (*S*)-*p*-Tol-BINAP, MeC_6H_5, 0°, 2-3 h
2. BnBr, TBAT (1.2 eq), solvent

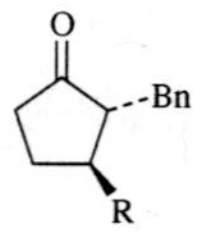

459

R	Solvent	Temp	
Me	CH_2Cl_2	rt	(62)
Me	THF	rt	(53)
Me	CH_2Cl_2:MeC_6H_5 (1:1)	rt	(62)
$BnCH_2$	CH_2Cl_2	50	(60)
$BnCH_2$	CH_2Cl_2:MeC_6H_5 (1:1)	rt	(64)
$BnCH_2$	CH_2Cl_2:MeC_6H_5 (1:1)	rt	(67)
$BnCH_2$	CH_2Cl_2:MeC_6H_5 (1:1)	rt	(69)

TABLE 16. ORGANOSILANE REDUCTION OF α,β–UNSATURATED KETONES (*Continued*)

Unsaturated Ketone	Conditions	Product(s) and Yield(s) (%)	Refs.
C_6 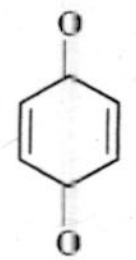	Et_3SiH (1.5 eq), TFA, (13 eq), 60-65°, 1 h	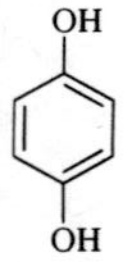 **I** (98)	393
	PMHS (10% xs), $(Bu_2(AcO)Sn)_2O$ (2 mol%), EtOH, reflux	**I** (81)	316
$C_{6\text{-}9}$ 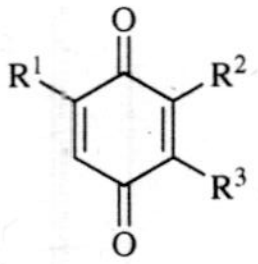	$HMe_2SiOSiMe_2H$ (1.5 eq), NaI (10 mol%), TMSCl (10 mol%), CH_2Cl_2, reflux, 30 min	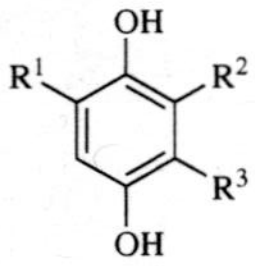	314, 357
R^1 = H, R^2 = H, R^3 = H		(85)	
R^1 = Cl, R^2 = H, R^3 = H		(96)	
R^1 = Me, R^2 = H, R^3 = H		(90)	
R^1 = MeO, R^2 = H, R^3 = H		(98)	
R^1 = Me, R^2 = Me, R^3 = Me		(90)	

$C_{6\text{-}15}$

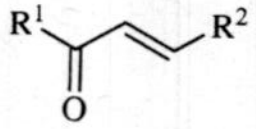

1.2 eq

Et_3SiH (1 eq), $InBr_3$ (10 mol %), R^3CHO (1 eq), EtCN, –78° to rt, 1h

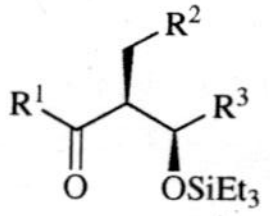

760

R^1	R^2	R^3		syn:anti
Et	Me	4-$MeOC_6H_4$	(61)	>99:1
Et	Me	Ph	(40)	>99:1
Me	Ph	4-$MeOC_6H_4$	(46)	>99:1
Ph	Me	4-$MeOC_6H_4$	(75)	90:10
Ph	Me	4-$O_2NC_6H_4$	(59)	>99:1
Ph	Me	Ph	(78)	92:8
Ph	Me	Bn	(40)	>99:1
Ph	Me	*t*-Bu	(73)	>99:1
Ph	Me	Et_2CH	(87)	90:10
Ph	Ph	4-$MeOC_6H_4$	(82)	>99:1
Ph	Ph	Ph	(65)	78:22

TABLE 16. ORGANOSILANE REDUCTION OF α,β-UNSATURATED KETONES (*Continued*)

Unsaturated Ketone	Conditions	Product(s) and Yield(s) (%)	Refs.
C_7 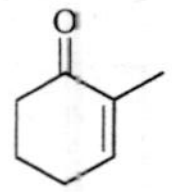	Et_3SiH (1.1 eq), $(Ph_3P)_3RhCl$ (10 mol%), 45°, 4 h	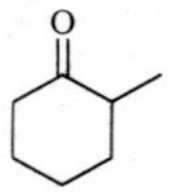 **I** + 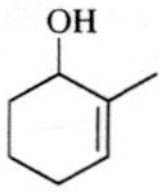 **II** **I** (90), **I**:**II** = 98:2	435
	TMSH (1.2 eq), $TiCl_4$ (2.4 eq), CH_2Cl_2, –5°, 1 h	**I** (73)	449
	Et_3SiH (1.2 eq), $TiCl_4$ (2.4 eq), CH_2Cl_2, –5°, 1 h	**I** (73)	449
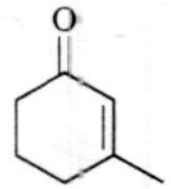	Ph_2SiH_2 (1.3 eq), $(Ph_3P)_4RhH$ (0.3 mol%), CH_2Cl_2, rt, 4 h	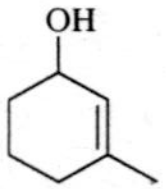 **I**	374
	Et_3SiH (1.1 eq), $(Ph_3P)_3RhCl$ (10 mol%), 50°, 30 h	**I** + 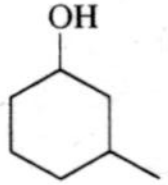 **II** **II** (97), **I**:**II** = 0:100	435
	Ph_2SiH_2 (1.1 eq), $(Ph_3P)_3RhCl$ (10 mol%), 0°, 340 min	**II** (97), **II**:**I** = 100:0	435
	$PhSiH_3$ (1.5 eq), $[(Ph_3P)CuH]_6$ (5 mol%), MeC_6H_5, 18 h	**II** (100)	447
	$PhSiH_3$ (1.01 eq), Ni, PPh_3 (0.02 eq)	**II** (100)	438
	Et_3SiH (1.2 eq), $TiCl_4$ (2.4 eq), CH_2Cl_2, –5°, 1 h	**II** (74)	449

	$HMe_2SiOSiMe_2H$ (1.5 eq), 1/6 $[(Ph_3P)CuH]_6$ (5 mol%), MeC_6H_5, rt, 24 h	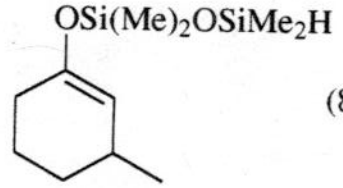 (85)	455
	$PhMe_2SiH$ (1.1 eq), $(Ph_3P)_4RhH$ (0.5 mol%), CH_2Cl_2, 50°, 24 h	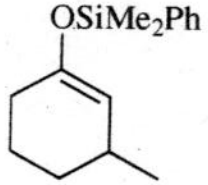 (89)	374
	1. $HMe_2SiOSiMe_2H$ (1.5 eq) 1/6 $[(Ph_3P)CuH]_6$ (5 mol%), MeC_6H_5, rt, 39 h 2. n-$C_5H_{11}CHO$, $TiCl_4$, CH_2Cl_2, –78°	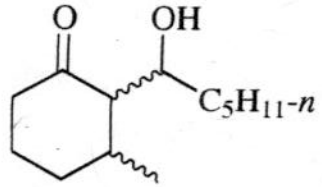 (72)	455
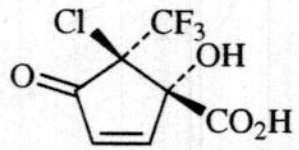	Et_3SiH (3 eq), TFA, rt, 48 h	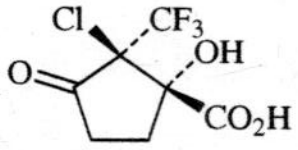 (82)	453
C_8 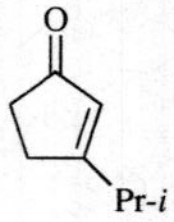	PMHS (1.6 eq), **209** (0.1 mol%), NaOBu-t (0.1 mol%), MeC_6H_5, rt, 1 h	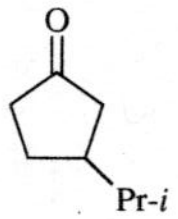 (81)	454
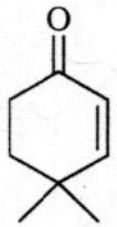	$PhSiH_3$ (1.01 eq), Ni, PPh_3 (0.02 eq)	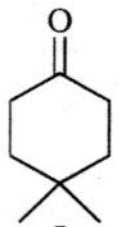 **I** (98)	438
	$PhSiH_3$ (1.5 eq), $[(Ph_3P)CuH]_6$ (5 mol%), MeC_6H_5, 8 min	**I** (85)	447

TABLE 16. ORGANOSILANE REDUCTION OF α,β-UNSATURATED KETONES (*Continued*)

Unsaturated Ketone	Conditions	Product(s) and Yield(s) (%)	Refs.
C_8 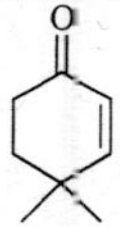	$PhMe_2SiH$ (1.1 eq), $(Ph_3P)_4RhH$ (0.4 mol%), CH_2Cl_2, rt, 12 h	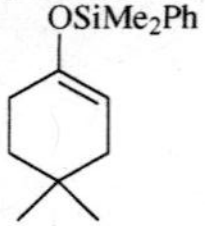 (96)	374
	$ClCH_2Me_2SiH$ (1.1 eq), $(Ph_3P)_4RhH$ (0.3 mol%), CH_2Cl_2, rt, 4 h	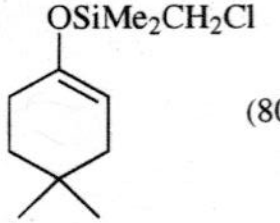 (80)	374
	1. $PhMe_2SiH$ (1.5 eq), 1/6 $[(Ph_3P)CuH]_6$ (5 mol%), MeC_6H_5, rt, 4 h 2. 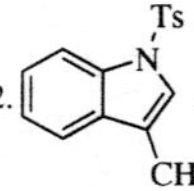, $TiCl_4$, CH_2Cl_2, –78°	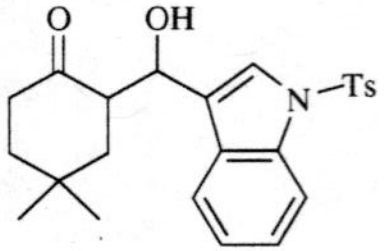 (82)	455
	1. $PhMe_2SiH$, (1.5 eq), $[(Ph_3P)CuH]_6$ (18 mol%), MeC_6H_5, rt, 4 h 2. Aldehyde, BF_3•OEt_2, CH_2Cl_2, –78°, 1 h	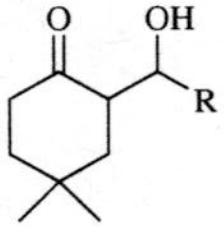 Aldehyde / R: cyclohex-3-ene-carbaldehyde / cyclohex-3-enyl (80) 4-$O_2NC_6H_4CHO$ / 4-$O_2NC_6H_4$ (68)	455
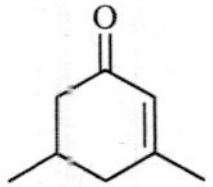	$PhMe_2SiH$ (4 eq), CuCl (2 eq), DMI, rt, 22 h	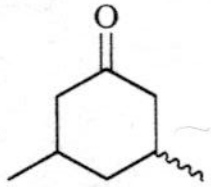 **I** (100) cis:trans = 97:3	445
	Cl_3SiH (2 eq), $CoCl_2$ (0.175 eq), CD_3CN, DMI (0.5 eq), rt, 12 h	**I** (50)	451
	$PhSiH_3$ (1.5 eq), $[(Ph_3P)CuH]_6$ (5 mol %), MeC_6H_5, 46 h	**I** (100), single isomer	447

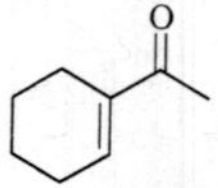	$PhMe_2SiH$ (1.1 eq), $(Ph_3P)_4RhH$ (0.5 mol%), CH_2Cl_2, 50°, 24 h	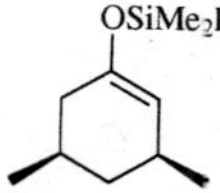 (87) cis:trans = 92:8	374
	Et_3SiH (1.1 eq), $(Ph_3P)_3RhCl$ (10 mol%), C_6H_6, 50°, 5 h	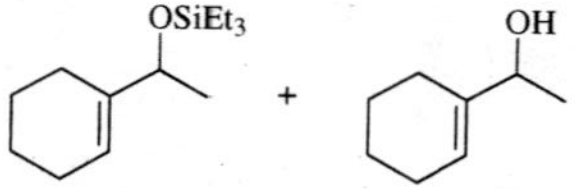 **I** + **II** — **I** (95), **I**:**II** = 87:13	435
	Cl_3SiH (2 eq), $CoCl_2$ (0.175 eq), CD_3CN, DMI (0.5 eq), rt, 2 h	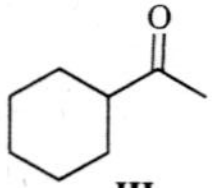 **III** (75)	451
	$PhSiH_3$ (1.3-1.5 eq), $Mo(CO)_6$ (3-5 mol%), THF, reflux, 4.5 h	**III** (100)	450
	Ph_2SiH_2 (1.8 eq), $ZnCl_2$ (0.38 eq), $(Ph_3P)_4Pd$ (0.013 eq), rt, 4 h	**III** (—)	436
	Ph_2SiH_2 (1.8 eq), $ZnCl_2$ (0.38 eq), $(Ph_3P)_4Pd$ (0.013 eq), $CHCl_3$, rt, 4 h	**III** (85)	436
	PMHS-Pd nanocomposite, C_6H_6, rt, 6 h	**III** (97)	219
	Ph_2SiH_2 (1.1 eq), $(Ph_3P)_3RhCl$ (10 mol%), C_6H_6, rt, 1 h	**I** + **II** (96), **I**:**II** = 97:3	435
	PhCHO, Et_2MeSiH, $Rh_4(CO)_{12}$ (0.5 mol%), $MePh_2P$, C_6H_6, 20°, 15 h,	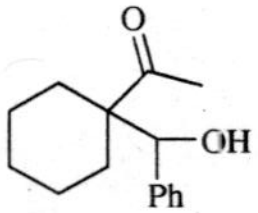 (69)	464

Unsaturated Ketone	Conditions	Product(s) and Yield(s) (%)	Refs.
C_{8-17} 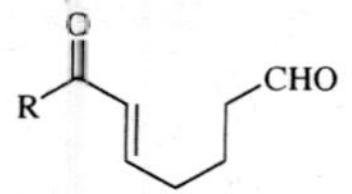	$PhSiH_3$ (1.2 eq), $Co(dpm)_2$ (5 mol%), rt	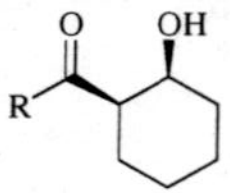	461
R			
Me		(38)	
Ph		(87)	
4-$CF_3C_6H_5$		(72)	
2-$C_{10}H_7$		(68)	
2-C_4H_3O		(75)	
2-C_4H_3S		(73)	
C_9 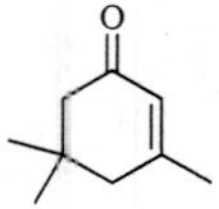	TMSH (1.2 eq), $TiCl_4$ (2.4 eq), CH_2Cl_2, –5°, 1 h	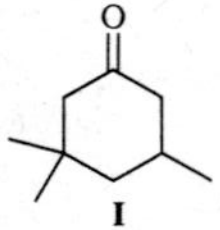 **I** (66)	449
	Et_3SiH (1.2 eq), $TiCl_4$ (2.4 eq), CH_2Cl_2, –5°, 1 h	**I** (53)	449
	$PhMe_2SiH$ (4 eq), CuCl (2 eq), DMI, rt, 20 h	**I** (13)	445
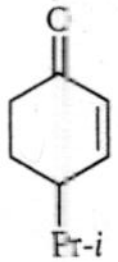	$PhSiH_3$ (1.5 eq), $[(Ph_3P)CuH]_6$ (0.5mol%), MeC_6H_5, 6 h	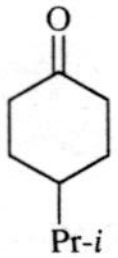 (99)	447
	1. $PhMe_2SiH$ (2.5 eq), 1/6 $[(Ph_3P)CuH]_6$ (5 mol%), MeC_6H_5, rt, 3 h 2. Me_2CHCH_2CHO, BF_3•OEt_2, CH_2Cl_2, –78°	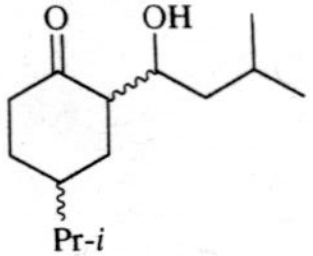 (89)	455

	1. $PhMe_2SiH$ (1.5 eq), 1/6 $[(Ph_3P)CuH]_6$ (5 mol%), MeC_6H_5, rt, 2 h 2. $ClCH_2CH_2OCH_2CHO$, $BF_3•OEt_2$, CH_2Cl_2, −78°	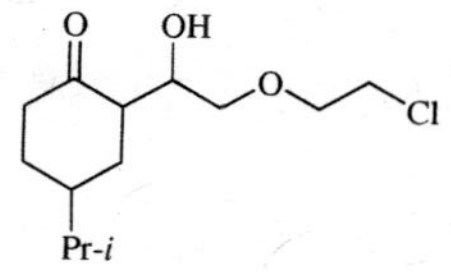 (87)	455
	1. $PhMe_2SiH$ (1.5 eq), 1/6 $[(Ph_3P)CuH]_6$ (5 mol%), MeC_6H_5, rt, 2 h 2. 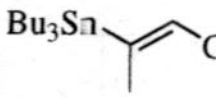 CHO, $BF_3•OEt_2$, CH_2Cl_2, −78°	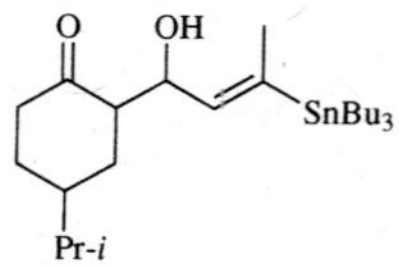 (76)	455
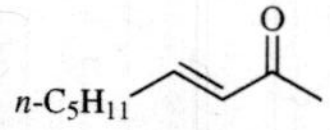	$PhMe_2SiH$ (2 eq), DMA, $(Ph_3P)_3CuF•2$ EtOH (1 eq)	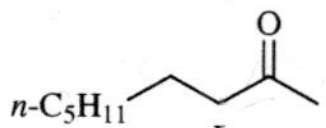 **I** (91)	446
	1. $PhMe_2SiH$ (2 eq), $(Ph_3P)_3CuF•$ 2 EtOH, DMA, 0°, 0.5 h 2. Add enone, 0° to rt, 2.5 h	**I** (91)	444
	$PhMe_2SiH$ (1.2 eq), CuCl (25 mol%), PPh_3 (25 mol%), TBAF (20 mol%), DMA, 0° to rt, 2 h	**I** (77)	444
	1. $PhMe_2SiD$ (2 eq), $(Ph_3P)_3CuF•2$ EtOH, DMA, 0°, 0.5 h 2. Add enone, 0° to rt, 2.5 h	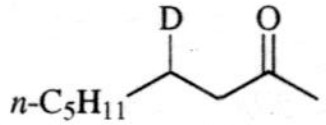 **II** (62) >99% D	444
	$PhMe_2SiH$, CuCl (25 mol%), PPh_3 (25 mol%), TBAF (20 mol%), DMA, 0° to rt, 2 h	**II** (77)	444

TABLE 16. ORGANOSILANE REDUCTION OF α,β–UNSATURATED KETONES (*Continued*)

Unsaturated Ketone	Conditions	Product(s) and Yield(s) (%)	Refs.
C_9 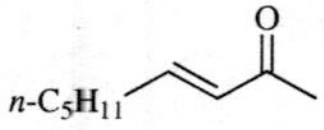	1. $PhMe_2SiH$ (2 eq), $(Ph_3P)_3CuF•2\ EtOH$, DMA, 0°, 0.5 h 2. Add enone, 0°, 45 min 3. D_2O	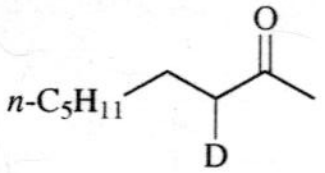 (67) 85% D	444
	1. $PhMe_2SiH$ (2 eq), $(Ph_3P)_3CuF•2\ EtOH$, DMA, 0°, 0.5 h 2. Add enone, 0°, 45 min 3. Allyl bromide	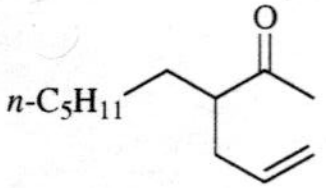 (45)	444
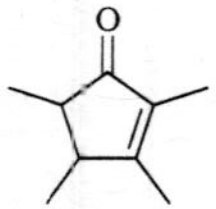	$PhSiH_3$ (1.01 eq), Ni, PPh_3 (0.02 eq)	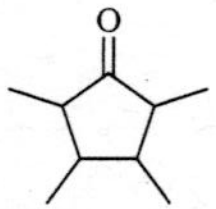 (81.9)	438
C_{10} 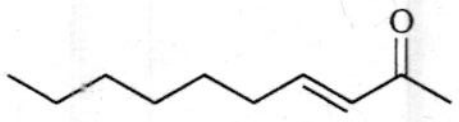	$PhSiH_3$ (1.3 eq), $Mn(dpm)_3$ (3 mol%), *i*-PrOH, DCE	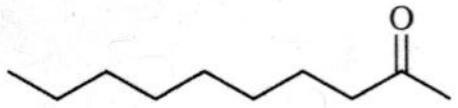 (74)	448
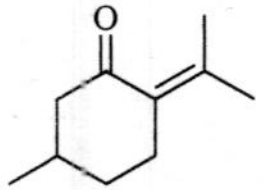	Et_2SiH_2, $(Ph_3P)_3RhCl$	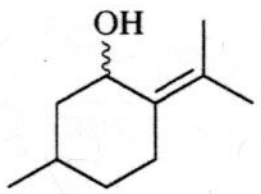 (—)	435
	$EtMe_2SiH$ (1.1 eq), $(Ph_3P)_3RhCl$ (10 mol%), C_6H_6, rt, 2 h	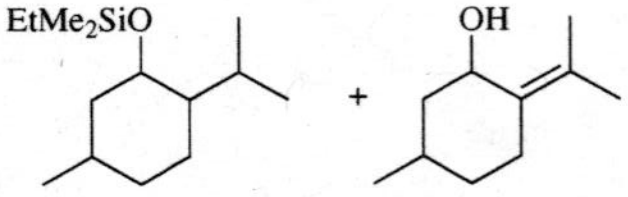 **I** + **II** **I** (97), **I**:**II** = 78:22	435
	Ph_2SiH_2 (1.1 eq), $(Ph_3P)_3RhCl$ (10 mol%), rt, 20 min	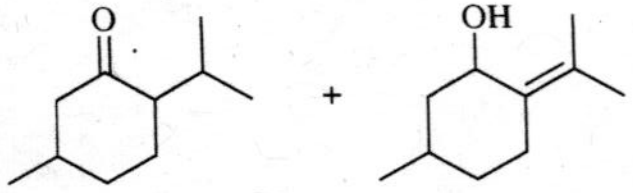 **I** + **II** **I** (98), **I**:**II** = 0:100	435

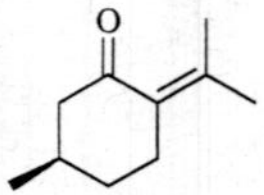

Ph$_2$SiH$_2$ (1.55 eq), ZnCl$_2$ (0.4 eq), (Ph$_3$P)$_4$Pd (0.018 eq), CHCl$_3$, rt, 3 h

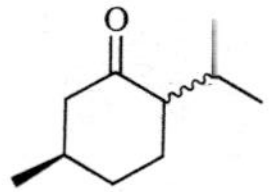

(90) cis:trans = 1:1 — 436

R$_3$SiH, (Ph$_3$P)$_3$RhCl — **I** + **II** (—) — 452

R$_3$Si	I:II
PhMe$_2$Si	75:25
Et$_3$Si	50:50
(n-Pr)$_3$Si	50:50
Et$_2$HSi	0:100
Ph$_2$HSi	0:100

R$_3$SiH (1.1 eq), (Ph$_3$P)$_3$RhCl (0.5 mol%), 50°, 2 h — **I** + **II** (—) — 437

R$_3$Si	I:II
Et$_3$Si	50:50
(n-Pr)$_3$Si	50:50
PhMe$_2$Si	75:25
Et$_2$HSi	0:100
Ph$_2$HSi	0:100

Et$_2$SiH$_2$, (Ph$_3$P)$_3$RhCl — **II** (—) — 435

Ph$_2$SiH$_2$ (1.3-1.5 eq), Mo(CO)$_6$ (3-5 mol%), THF, reflux, 15 h

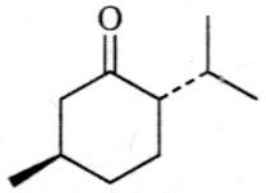

I

\+

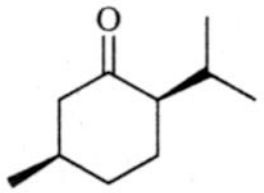

II

I + **II** (100) **I**:**II** = 4.5:1 — 450

PhSiH$_3$ (1.3-1.5 eq), Mo(CO)$_6$ (3-5 mol%), THF, reflux, 3 h — **I** + **II** (90), **I**:**II** = 4.5:1 — 450

TABLE 16. ORGANOSILANE REDUCTION OF α,β–UNSATURATED KETONES (*Continued*)

Unsaturated Ketone	Conditions	Product(s) and Yield(s) (%)	Refs.
C_{10} 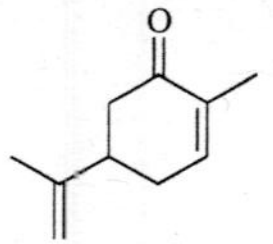	$PhMe_2SiH$ (4 eq), DMA, $(Ph_3P)_3CuF{\bullet}2$ EtOH (1 eq)	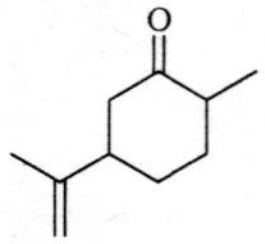 (85) dr = 8:1	446
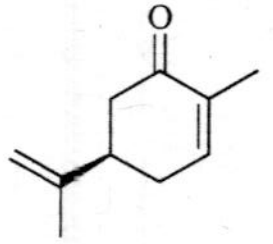	1. $PhMe_2SiH$ (2 eq), $(Ph_3P)_3CuF{\bullet}2$ EtOH, DMA, 0°, 0.5 h 2. Add enone, 0° to rt, 24 h	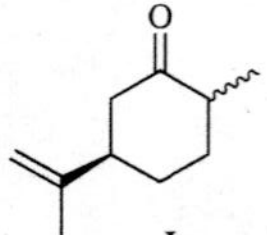 **I** (85)	444
	$PhMe_2SiH$, CuCl (160 mol%), PPh_3 (160 mol%), TBAF (150 mol%), DMA, 0° to rt, 4.5 h	**I** (87)	444
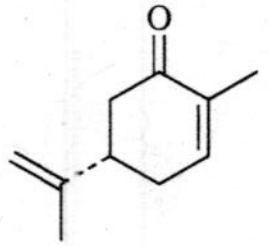	$PhMe_2SiH$ (4 eq), CuCl (2 eq), DMI, rt, 22 h	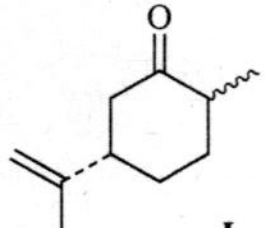 **I** (83)	445
	Ph_2SiH_2 (1.7 eq), $ZnCl_2$ (0.52 eq), $(Ph_3P)_4Pd$ (0.013 eq), $CHCl_3$, rt, 2 h	**I** (89) cis:trans = 1:4	436
	$PhSiH_3$ (2 eq), $Mn(dpm)_3$ (3 mol%), *i*-PrOH	**I** (50)	448
	$PhSiH_3$ (1.3-1.5 eq), $Mo(CO)_6$ (3-5 mol%), THF, reflux, 11.5 h	**I** (0)	450
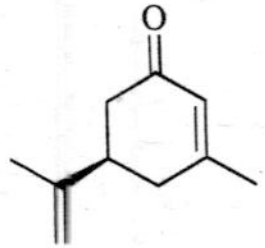	Ph_2SiH_2 (1.3 eq), $(Ph_3P)_4RhH$ (0.4 mol%), CH_2Cl_2, rt, 12 h	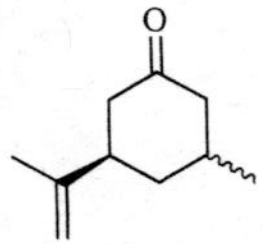 (84) cis:trans = 65:35	374

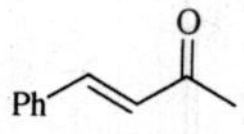

Conditions	Product(s) and Yield(s) (%)	Refs.
$(EtO)_3SiH$, CsF, rt, 30 min	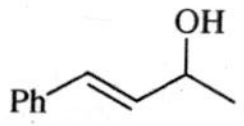 I (100)	83
Ph_2SiH_2 (0.5 eq), CsF, rt, 30 min	I (100)	319
R_3SiH (x eq), DMA, $(Ph_3P)_3CuF$•2 EtOH (y eq)	I + 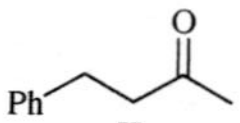 II	446

R_3Si	x	y	I	II
$PhMe_2Si$	2	1	(—)	(>99)
$PhMe_2Si$	1	1	(—)	(63)
$PhMe_2Si$	2	0.05	(trace)	(69)
Et_3Si	2	1	(—)	(95)
Ph_2HSi	2	1	(36)	(49)
$PhCl_2Si$	2	1	(—)	(—)

Conditions	Product(s) and Yield(s) (%)	Refs.
R_3SiH (2 eq), $(Ph_3P)_3CuF$•2 EtOH (1.0 eq), DMA, 0° to rt, 0.5 h	I + II	444

R_3Si	Time	I	II
$PhMe_2Si$	2 h	(—)	(>99)
Et_3Si	1.5 h	(—)	(96)
$(EtO)_3Si$	3.5 h	(34)	(58)
Ph_2HSi	18 h	(36)	(49)
$PhCl_2Si$	—	(0)	(0)

Conditions	Product(s) and Yield(s) (%)	Refs.
Et_3SiH (1.1 eq), $(Ph_3P)_3RhCl$, 45°, 15 h	I:II = 37:63	435
Ph_2SiH_2 (1.1 eq), $(Ph_3P)_3RhCl$, 0° to rt, 6 h	I:II = 60:40	435

TABLE 16. ORGANOSILANE REDUCTION OF α,β–UNSATURATED KETONES (*Continued*)

Unsaturated Ketone	Conditions	Product(s) and Yield(s) (%)	Refs.
C_{10} 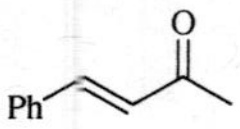	$PhMe_2SiH$ (3 eq), DMA	**I** + **II**	444

	Time	Temp	I	II
$CuF(PPh_3)_3$•2 EtOH	2 h	0° to rt	(—)	(>99)
$CuF(PPh_3)_3$	4 h	0° to rt	(—)	(55)
CuCl	21 h	0° to rt	(—)	(27)
$(CuOTf)_2$•C_6H_6	12 h	0° to rt	(trace)	(48)
Cu_2O	21 h	60°	(—)	(64)
CuCN	12 h	0° to rt	(trace)	(trace)

Unsaturated Ketone	Conditions	Product(s) and Yield(s) (%)	Refs.
	$PhMe_2SiH$ (2 eq), $(Ph_3P)_3CuF$•2 EtOH (1.0 eq), solvent, 0° to rt	**II**	444

Solvent	Time	II
DMA	2 h	(92)
DMF	2.5 h	(40)
NMP	6 h	(57)
THF	2.5 h	(42)
MeC_6H_5	12.5	(80)

Unsaturated Ketone	Conditions	Product(s) and Yield(s) (%)	Refs.
	Et_3SiH (1.2 eq), $TiCl_4$ (2.4 eq), CH_2Cl_2, 0°, 1 h	**II** (36)	449
	1. $HMe_2SiOSiMe_2H$, 1/6 $[(Ph_3P)CuH)]_6$ (5 mol%), MeC_6H_5, rt, 3 h 2. TBAF, MeOH, rt, 30 min	**II** (82)	455
	Ph_2SiH_2 (1.3 eq), $ZnCl_2$ (0.3 eq), $(Ph_3P)_4Pd$ (0.0009 eq), $CHCl_3$, rt, 2 h	**II** (100)	436
	Ph_2SiH_2 (2.3 eq), CsF (1 eq), rt, 30 min	**II** (100)	83
	Ph_2SiH_2 (1.3-1.5 eq), $Mo(CO)_6$ (3-5 mol%), THF, reflux, 7 h	**II** (95)	450

	$PhSiH_3$ (1.3-1.5 eq), $Mo(CO)_6$ (3-5 mol%), THF, reflux, 1.5 h	**II** (100)	450
	$PhSiH_3$ (1.01 eq), Ni, PPh_3 (0.02 eq)	**II** (96.9)	438
	1. $HMe_2SiOSiMe_2H$ (0.55 eq), 1/6 $[(Ph_3P)CuH]_6$ (2 mol%), MeC_6H_5, rt, 2 h 2. 4-$O_2NC_6H_4CH_2Br$, CH_2Cl_2, rt, 3 h	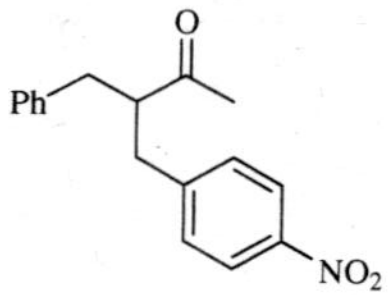 (57)	455
	1. PMHS (1.25 eq) 1/6 $[(PPh_3)CuH]_6$ (2 mol%), MeC_6H_5, rt 2. $BnOCH_2CHO$, $TiCl_4$, CH_2Cl_2, –78°	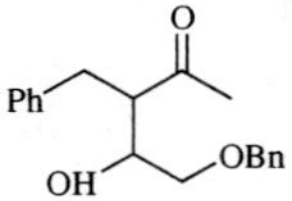 (83)	455
	PhCHO, Et_2MeSiH, $Rh_4(CO)_{12}$ (0.5 mol%), $MePh_2P$, C_6H_6, 20°, 3.5 h	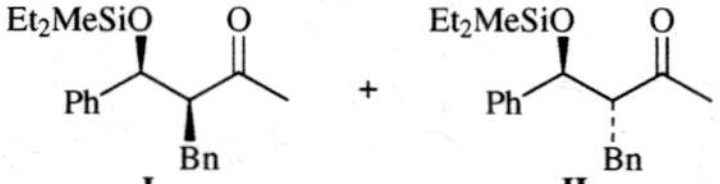 **I** + **II** (88), **I**:**II** = 80:20	464
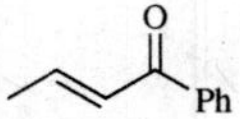	Et_3SiH (1.1 eq), $(Ph_3P)_3RhCl$, 45°, 15 h	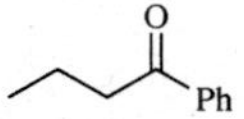 **I**	435
	Ph_2SiH_2 (1.1 eq), $(Ph_3P)_3RhCl$, 0° to rt, 6 h	**I** (—)	435
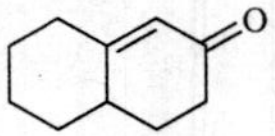	Et_3SiH (1.1 eq), $(Ph_3P)_3RhCl$ (10 mol%), 80°, 25 h	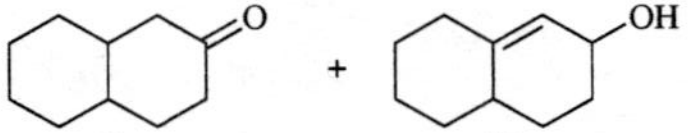 **I** (96), **I**:**II** = 94:6	435
	Et_2SiH_2 (1.1 eq), $(Ph_3P)_3RhCl$ (10 mol%), 0°, 30 min	**I** (98), **I**:**II** = 3:97	435

TABLE 16. ORGANOSILANE REDUCTION OF α,β-UNSATURATED KETONES (*Continued*)

Unsaturated Ketone	Conditions	Product(s) and Yield(s) (%)	Refs.
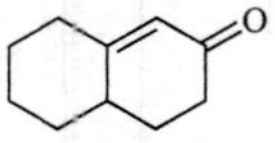	Et_3SiH (1 eq), $(Ph_3P)_3RhCl$ (0.5 mol%), 50°, 60 min	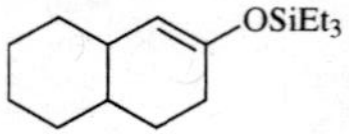 (92)	411
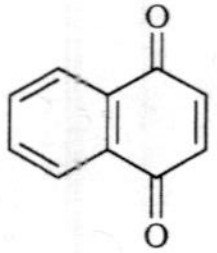	Et_3SiH (4 eq), TFA (20 eq), 60-65°, 8 h	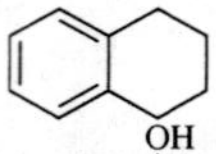 (50)	393
	Et_3SiH (5 eq), TFA (26 eq), 60-65°, 10 h	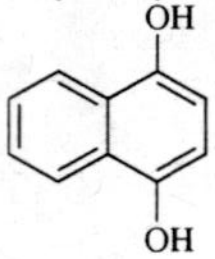 **I** (60)	393
	$HMe_2SiOSiMe_2H$ (1.5 eq), NaI (10 mol%), TMSCl (10 mol%), CH_2Cl_2, reflux, 30 min	**I** (75)	314
C_{11} 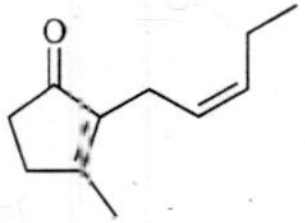	PMHS (1.6 eq), **209**, NaOBu-*t* (0.1 mol%), MeC_6H_5, rt, 1 h	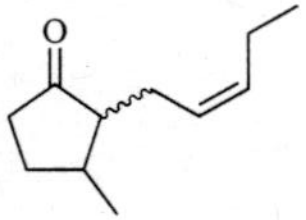 (85) dr = 4:1	454
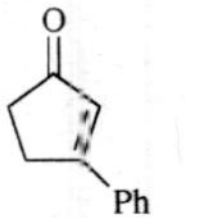	PMHS (1.6 eq), **209** (0.1 mol%), NaOBu-*t* (0.1 mol%), MeC_6H_5, rt, 1 h	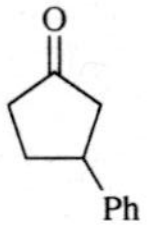 (95)	454
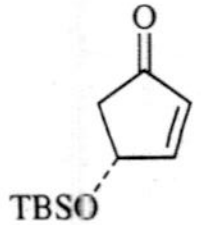	$PhSiH_3$ (1.5 eq), $[(Ph_3P)CuH]_6$ (5 mol%), MeC_6H_5, 10 min	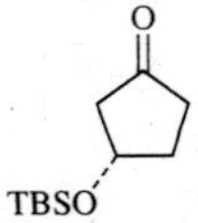 (70)	447
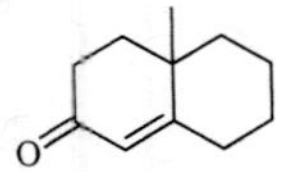	Et_3SiH (1.2 eq), $TiCl_4$ (2.4 eq), CH_2Cl_2, 1 h	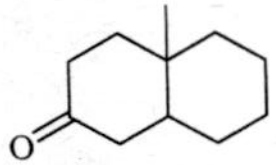 Temp: 0° (53) cis:trans = 9:1; –78° (50)	449

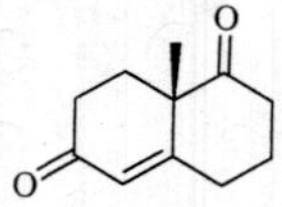	$PhMe_2SiH$ (4 eq), CuCl (2 eq), DMI, rt, 21 h	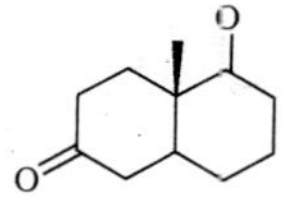 (81) cis:trans = 98:2	445
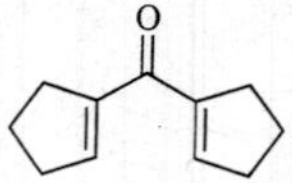	Et_3SiH (2 eq), $BF_3{\bullet}OEt_2$ (1.1 eq), CH_2Cl_2, −78°	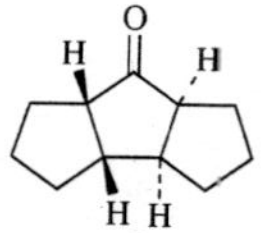 **I** (71) + 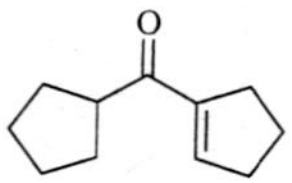 (14)	463
	Et_3SiH (2 eq), $SnCl_4$ (0.1 eq), CH_2Cl_2, −78°	**I** (61)	463
C_{12} 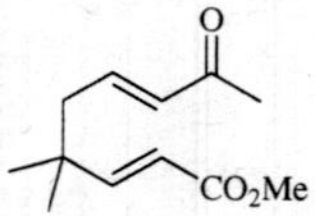	Et_3SiH (2.1 eq), $(Ph_3P)_3RhCl$ (1 mol%), MeC_6H_5, 50°, 16 h	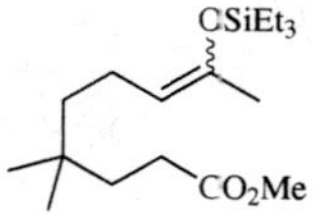 (91)	471
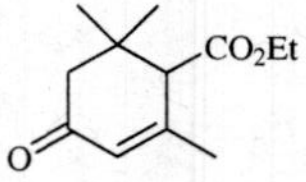	Et_3SiH (6 eq), BF_3 (xs), 20°, 6 h	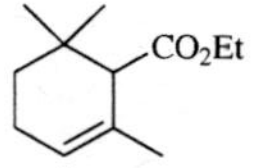 **I** (64)	442
	Et_3SiH (xs), $BF_3{\bullet}OEt_2$, 80-95°	**I** (66-74)	442
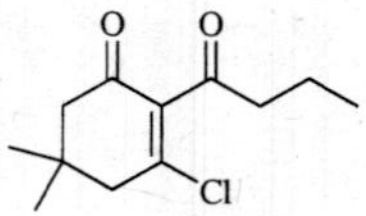	Et_3SiH or $EtMe_2SiH$ (4 eq), $BF_3{\bullet}OEt_2$ (2 mol%), TFA, rt, 0.5 h	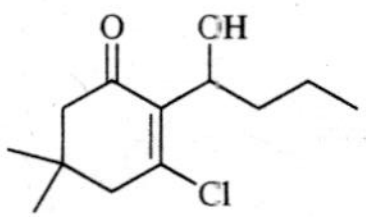 (96)	396
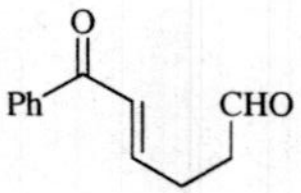	$PhSiH_3$ (1.2 eq), $Co(dpm)_2$ (5 mol%), rt	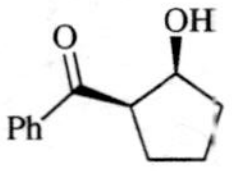 (70)	461
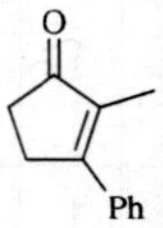	PMHS (1.6 eq), **209** (0.1 mol%), NaOBu-*t* (3 mol%), MeC_6H_5, rt, 3 h	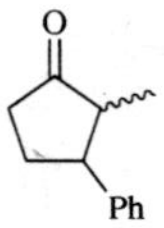 (94) dr = 5:1	454

TABLE 16. ORGANOSILANE REDUCTION OF α,β–UNSATURATED KETONES (*Continued*)

Unsaturated Ketone	Conditions	Product(s) and Yield(s) (%)	Refs.
C_{12} 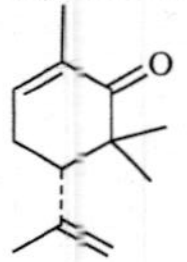	$PhSiH_3$ (1.3 eq), $Mn(dpm)_3$ (3 mol%), *i*-PrOH, DCE	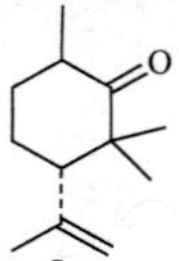 **I** (99)	448
	1. $PhSiH_3$ (1.3 eq), $Mn(dpm)_3$ (3 mol%), *i*-PrOH, DCE, O_2 2. $P(OEt)_3$ (1.1 eq)	**I** (59)	465
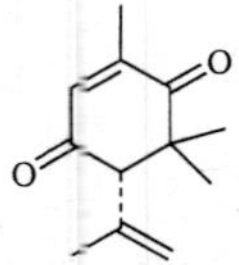	1. $PhSiH_3$ (1.3 eq), $Mn(dpm)_3$ (3 mol%), *i*-PrOH, DCE, O_2 2. $P(OEt)_3$ (1.1 eq)	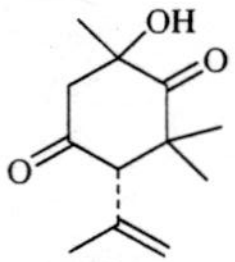 (87)	465
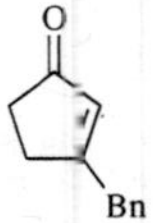	1. PMHS (0.53 eq), CuCl (10%), NaOBu-*t* (5%), (*S*)-*p*-Tol-BINAP, MeC_6H_5, 0°, 2-3 h 2. Allyl bromide, TBAT (1.2 eq)	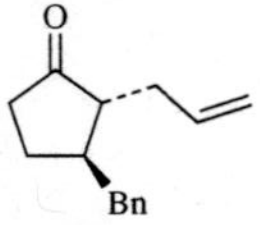 (60)	459
C_{13} 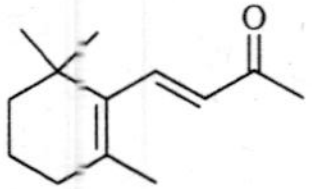	R_3SiH (1.1 eq), $(Ph_3P)_3RhCl$	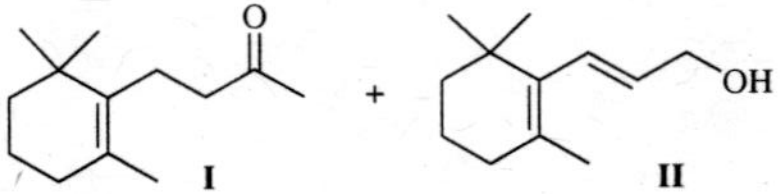 **I** + **II**	435
	R_3Si	**I + II** / **I:II**	
	$EtMe_2Si$	(94) / 98:2	
	$PhMe_2Si$	(95) / 85:15	
	Et_3Si	(95) / 78:22	
	Et_3SiH (1.2 eq), $TiCl_4$ (2.4 eq), CH_2Cl_2, 0°, 1 h	**I** (78)	449

Et_3SiH (1.2 eq), $(Ph_3P)_3RhCl$, (10-20 mol%), solvent, 50°	**I** (—) + 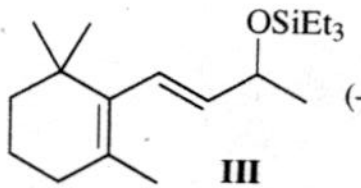 (—) **III**	435
Solvent / Time: none, 3 h; none, 3 h; C_6H_6 (0.2 M), 12 h	**I:III**: 64:36; 78:22; 86:14	
R_3SiH, $(Ph_3P)_3RhCl$	**I** + **II** (—)	437, 452
R_3Si: $PhMe_2Si$; Et_3Si; Et_2HSi; Ph_2HSi	**I:II**: 91:9; 44:56; 0:100; 0:100	
Ph_2SiH_2 (2.5 eq), $ZnCl_2$ (0.35 eq), $(Ph_3P)_4Pd$ (0.019 eq), $CHCl_3$, rt, 2 h	**I** (96)	436
Ph_2SiH_2 (1.1 eq), $(Ph_3P)_3RhCl$ (10 mol%), rt, 30 min	**I** (98), **I:II** = 100:0	435
$PhSiH_3$ (1.5 eq), $[(Ph_3P)CuH]_6$ (5 mol%), MeC_6H_5, 47 h	**I** (96)	447
$EtMe_2SiH$ (1.1 eq), $(Ph_3P)_3RhCl$ (10 mol%), C_6H_6, rt, 15 h	**I** (94), **I:II** = 98:2	435
Et_3SiH (1.1 eq), $(Ph_3P)_3RhCl$ (10 mol%), 50°, 2 h	**I** (96), **I:II** = 100:0	435
$PhSiH_3$ (1.3 eq), $Mn(dpm)_3$ (3 mol%), *i*-PrOH, DCE	**I** (25)	448
1. $PhSiH_3$ (1.3 eq) $Mn(dpm)_3$ (3 mol%), *i*-PrOH, DCE, O_2 2. $P(OEt)_3$ (1.1 eq)	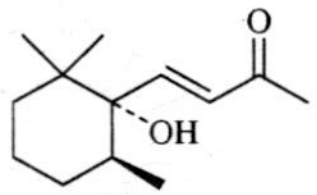 (51)	465

TABLE 16. ORGANOSILANE REDUCTION OF α,β-UNSATURATED KETONES (*Continued*)

Unsaturated Ketone	Conditions	Product(s) and Yield(s) (%)	Refs.
C_{13} 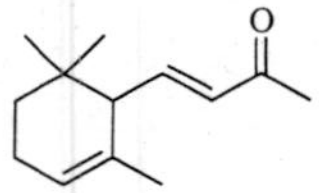	Et_3SiH, $(Ph_3P)_3RhCl$	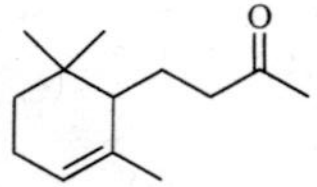 (96)	452
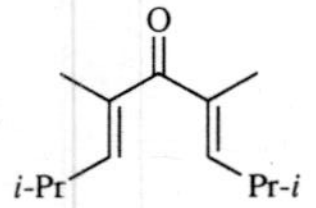	Et_3SiH (10 eq), BF_3•OEt_2 (1.1 eq), CH_2Cl_2, –78°	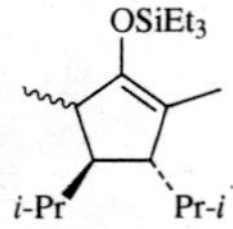 (24) 5.7:1 epimer ratio + 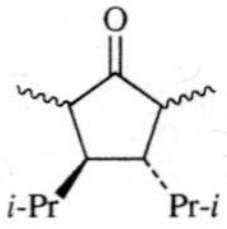 (69) 2.6:1 epimer ratio	463
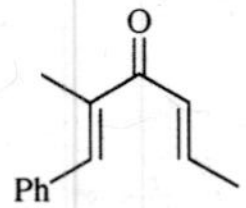	Et_3SiH (10 eq), BF_3•OEt_2 (1.1 eq), CH_2Cl_2, –78°	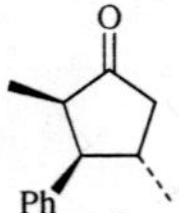 **I** (18) + 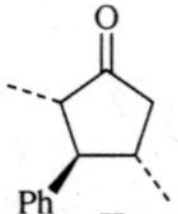 **II** (59)	463
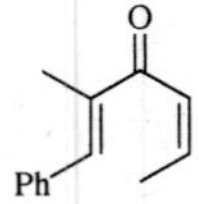	Et_3SiH (10 eq), BF_3•OEt_2 (1.1 eq), CH_2Cl_2, –78°	**I** (19) + **II** (62)	463
	Et_3SiH (2 eq), $SnCl_4$ (0.1 eq), CH_2Cl_2, –78°	**I** (53) + **II** (9) + 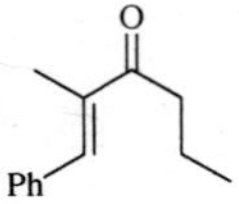 (12)	463
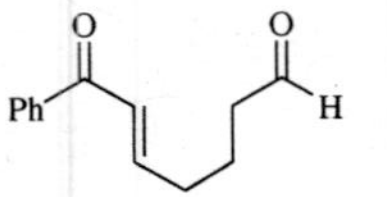	$PhSiH_3$ (120 mol%), $Co(dpm)_2$ (5 mol%), DCE, rt, 30 min	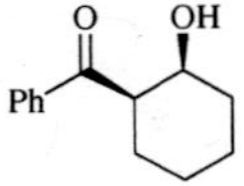 (—)	460
	$PhSiD_3$ (120 mol%), $Co(dpm)_2$ (5 mol%), DCE, rt, 30 min	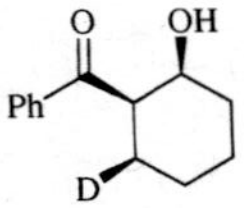 (—) + 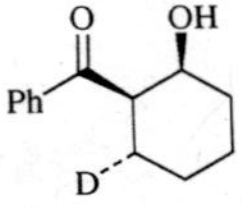 (—)	460

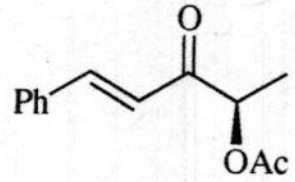

PhMe$_2$SiH, TBAF

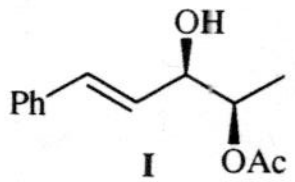

I

(95) 84% selectivity

440

Ph$_2$MeSiH, TBAF

I (—) 87% selectivity

440

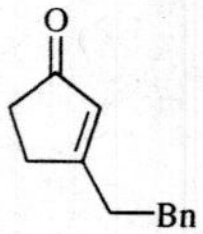

R$_3$SiH (1.3 eq), **209** (x mol%), NaOBu-*t* (0.1 mol%), solvent, rt

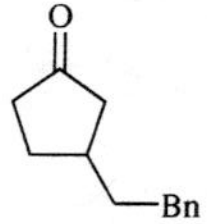

454

R$_3$SiH	Solvent	x	Time	
PMHS	MeC_6H_5	0.1	1 h	(88)
Ph_2SiH_2	MeC_6H_5	0.1	1 h	(87)
PMHS	THF	0.1	1 h	(86)
Ph_2SiH_2	THF	0.1	1 h	(88)
PMHS	MeC_6H_5	0.005	6 h	(86)

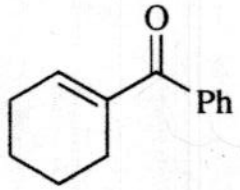

PhSiH$_3$ (1.3 eq), Mn(dpm)$_3$ (3 mol%), *i*-PrOH, DCE

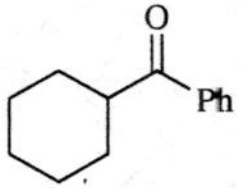

(50)

448

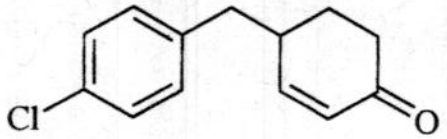

PhSiH$_3$ (1.5 eq), [(Ph$_3$P)CuH]$_6$ (5 mol%), MeC$_6$H$_5$, 6 h

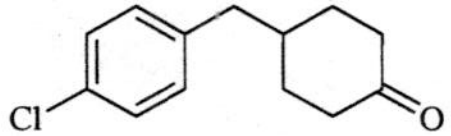

(97)

447

C_{13-15}

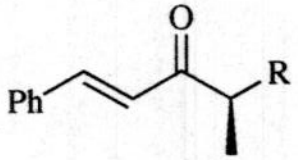

PhMe$_2$SiH (1.1-1.2 eq), TBAF (5-10 mol%), HMPA, 0°

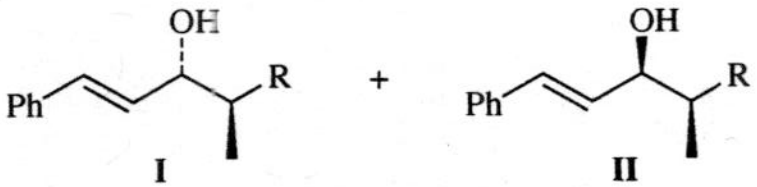

I + II

320

R	Time	I + II	I:II
OAc	12 h	(95)	84:16
OBu-*t*	10 h	(86)	91:9
OTHP	18 h	(77)	87:13

TABLE 16. ORGANOSILANE REDUCTION OF α,β-UNSATURATED KETONES (*Continued*)

	Unsaturated Ketone	Conditions	Product(s) and Yield(s) (%)	Refs.
C_{14}	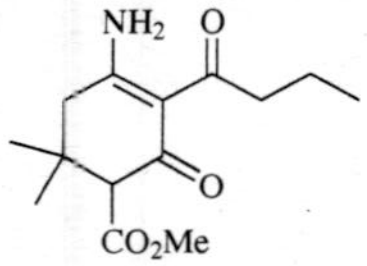	Et_3SiH or $EtMe_2SiH$ (4 eq), BF_3•OEt_2 (2 mol%), TFA, rt, 12 h	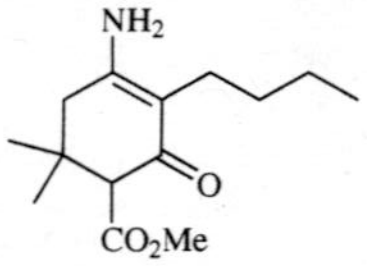 (98)	396
	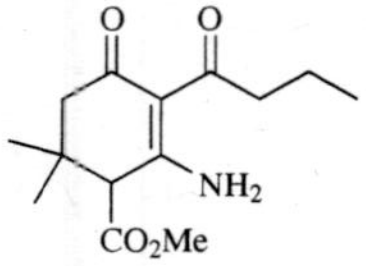	Et_3SiH or $EtMe_2SiH$ (4 eq), BF_3•OEt_2 (2 mol%), TFA, rt, 12 h	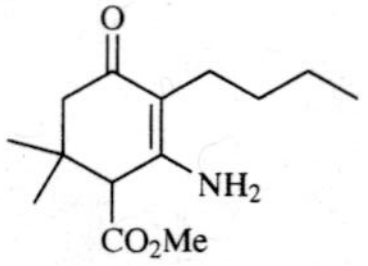 (97)	396
	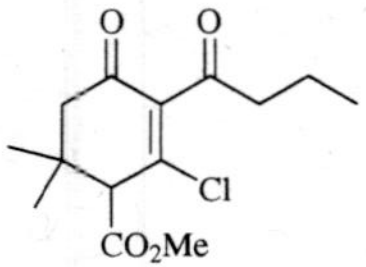	Et_3SiH or $EtMe_2SiH$ (4 eq), BF_3•OEt_2 (2 mol%), TFA, rt, 0.5 h	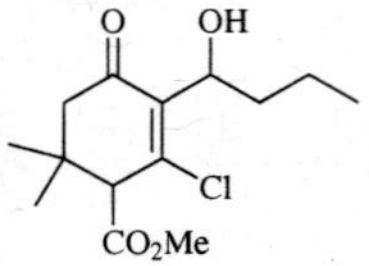 (96)	396
	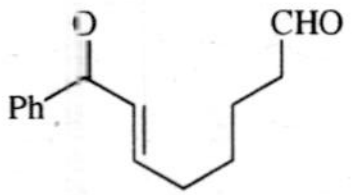	$PhSiH_3$ (1.2 eq), Co(dpm)$_2$ (5 mol%), DCE, 35°	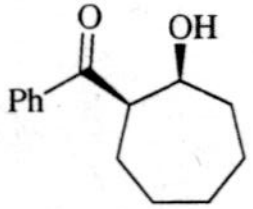 (35)	461
	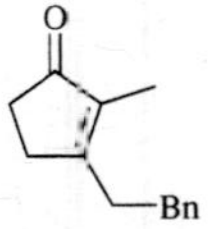	PMHS (1.6 eq), **209** (0.1 mol%), NaOBu-*t* (0.1 mol%), MeC_6H_5, rt, 1 h	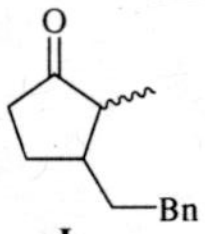 **I** (90) dr = 5:1	454
		PMHS (3 eq), **209**, $CuCl_2$•$2H_2O$ (1 mol%), NaOBu-*t* (6 mol%), MeC_6H_5, rt, 1 h	**I** (87) dr = 4:1	454

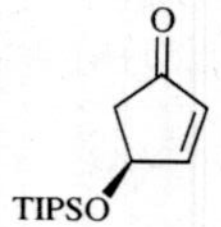	Et_3SiH, Pt[(vinylMe$_2$Si)$_2$O]$_2$	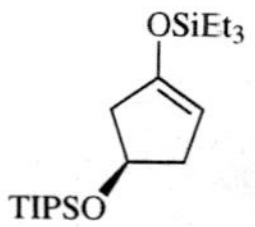 (96)	456
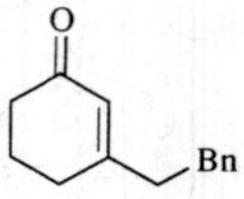	PMHS (1.6 eq), **209** (0.1 mol%), NaOBu-*t* (0.1 mol%), MeC_6H_5, rt, 1 h	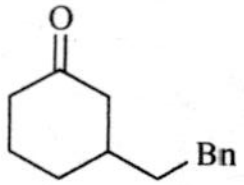 (93)	454
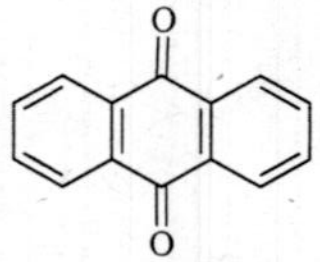	Et_3SiH (5 eq), TFA (20 eq), 60-65°, 20 h	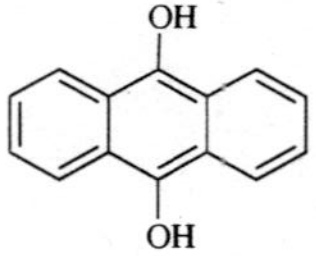 (54)	393
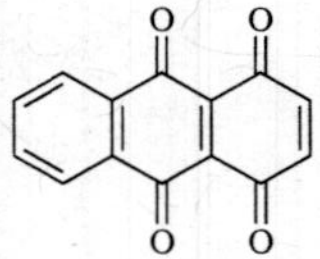	Et_3SiH (10 eq), TFA (20 eq), 60-65°, 0.5 h	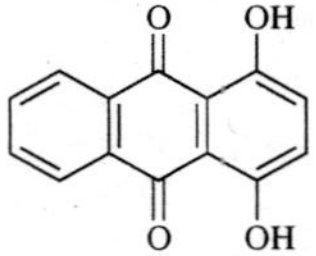 (99)	393
C_{15} 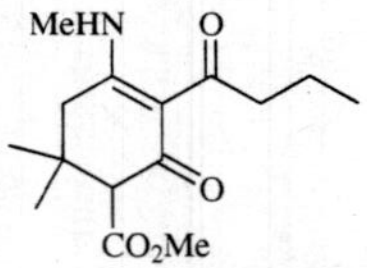	Et_3SiH or $EtMe_2SiH$ (4 eq), BF_3•OEt_2 (2 mol%), TFA, rt, 24 h	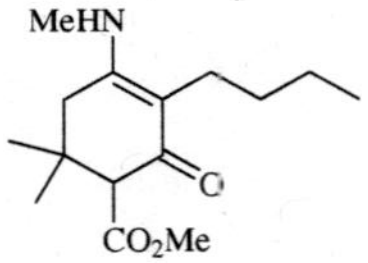 (98)	396
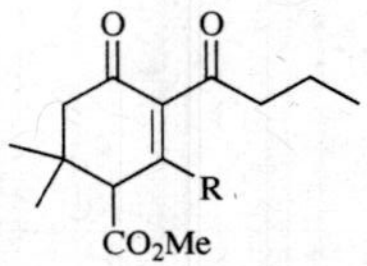	Et_3SiH or $EtMe_2SiH$ (4 eq), BF_3•OEt_2 (2 mol%), TFA, rt, 24 h	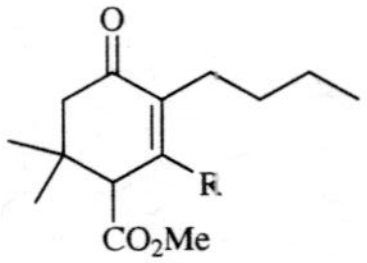 R: NHMe, Time: 24 h (98); R: OMe, Time: 8 h (98)	396

Unsaturated Ketone	Conditions	Product(s) and Yield(s) (%)	Refs.
C_{15}			
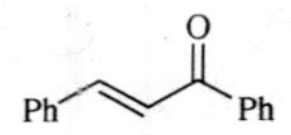	Ph_2SiH_2 (0.5 eq), CsF, rt, 30 min	 I I + II (95), I:II = 95:5 II	319
	Et_3SiH (3 eq), TFA (6 eq), $CHCl_3$, 60°, 7 h	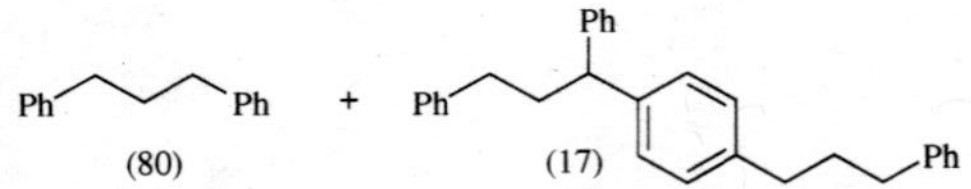 (80) (17)	434
	Et_3SiH (1.1 eq), $(Ph_3P)_3RhCl$, C_6H_6, rt, 15 h	II (—)	435
	Ph_2SiH_2 (1.1 eq), $(Ph_3P)_3RhCl$, 0° to rt, 6 h	II (—)	435
	$PhSiH_3$ (1.1 eq), $(Ph_3P)_3RhCl$, 0° to rt, 6 h	II (—)	435
	$PhMe_2SiH$ (2 eq), DMA, $(Ph_3P)_3CuF$•2 EtOH (1 eq)	II (98)	446
	$PhMe_2SiH$ (2 eq), $(Ph_3P)_3CuF$•2 EtOH (1.0 eq), DMA, 0° to rt, 3 h	II (98)	444
	$PhMe_2SiH$ (1.2 eq), CuCl (20 mol%), PPh_3 (20 mol%), TBAF (10 mol%), DMA, 0° to rt, 1 h	II (25)	444
	$PhMe_2SiH$ (1.2 eq), CuCl (110 mol%), PPh_3 (110 mol%), TBAF (100 mol%), DMA, 0° to rt, 2 h	II (94)	444
	Ph_2SiH_2 (2.3 eq), CsF (1 eq), 0°, 1 h	I (95) + II (5)	83

Substrate	Conditions	Product(s) and Yield(s) (%)	Refs.
	$PhSiH_3$ (1.5 eq), $[(Ph_3P)CuH]_6$ (5 mol%), MeC_6H_5, 108 min	**II** (96)	447
	PMHS-Pd nanocomposite, C_6H_6, rt, 6 h	**II** (95)	219
	1. $PhSiH_3$ (1.5 eq), 1/6 $[(Ph_3P)CuH]_6$ (5 mol%), MeC_6H_5, rt, 0.5 h 2. Me_2CHCH_2CHO, BF_3•OEt_2, CH_2Cl_2, –78°	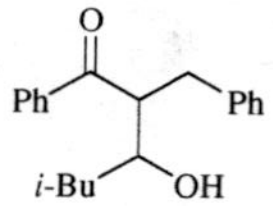 (73)	455
C15-16 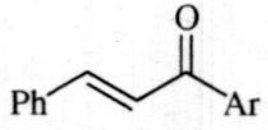 Ar = 4-BrC_6H_4	Ph_2SiH_2 (2.4 eq), $ZnCl_2$ (0.77 eq), $(Ph_3P)_4Pd$ (0.014 eq), $CHCl_3$, rt, 3 h	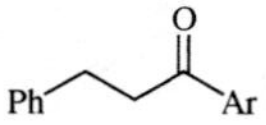 (98)	436
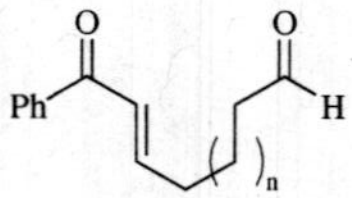	$PhSiH_3$ (120 mol%), $Co(dpm)_2$ (5 mol%), DCE, rt, 30 min	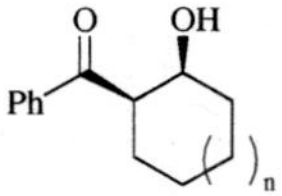 n = 1 (87) n = 2 (36)	460
C15-23 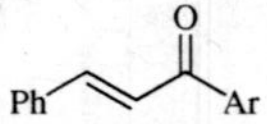	Et_3SiH (1 eq), TFA (10 eq), CCl_4, 50°, 6 h	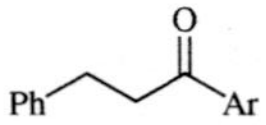 Ar: Ph (80); 4-FC_6H_4 (75); 4-MeC_6H_4 (75); 4-EtC_6H_4 (75); 4-*t*-BuC_6H_4 (76); 4-*n*-$C_6H_{13}C_6H_4$ (79); 4-PhC_6H_4 (79); 4-$BnCH_2C_6H_4$ (77)	439
C16 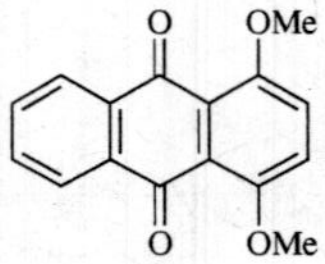	Et_3SiH (4 eq), TFA (10 eq), 60-65°, 20 h	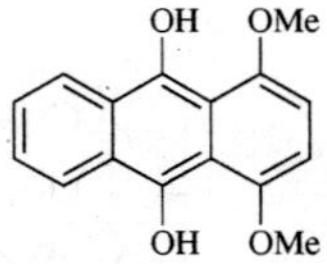 (50)	393

TABLE 16. ORGANOSILANE REDUCTION OF α,β–UNSATURATED KETONES (*Continued*)

Unsaturated Ketone	Conditions	Product(s) and Yield(s) (%)	Refs.
C_{16}			
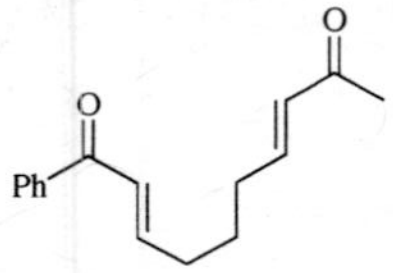	$PhSiH_3$ (2.4 eq), $Co(dpm)_2$ (5 mol%), 70°	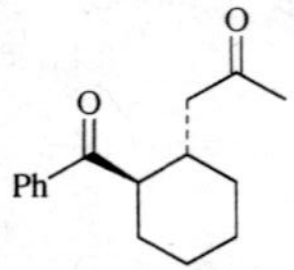 (47) + 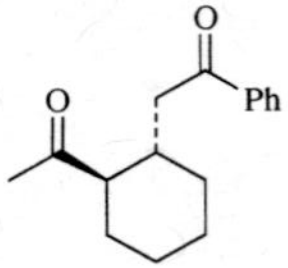 (15)	461
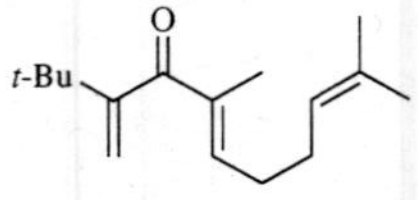	Et_3SiH (10 eq), BF_3•OEt_2 (1.1 eq), CH_2Cl_2, –78°	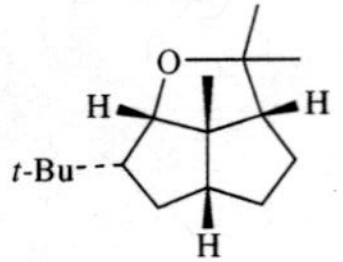 (81)	463
C_{16-17}			
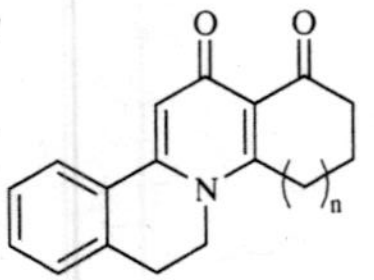	Et_3SiH, BF_3•OEt_2, TFA, rt, 24 h	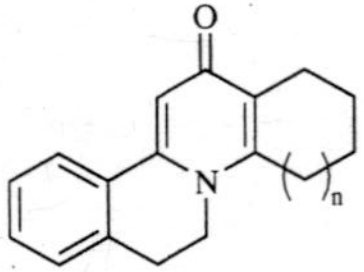 n = 0 (93); n = 1 (62)	441
C_{16-18}			
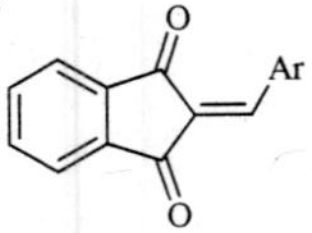	Et_3SiH (5 eq), TFA (10 eq), CCl_4, 50-55°	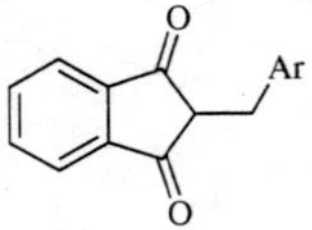	266

Ar	Time	Yield (%)
Ph	7 h	(54)
4-MeC_6H_4	8 h	(74)
4-EtC_6H_4	7 h	(53)
2-MeC_6H_4	20 h	(68)
4-$MeOC_6H_4$	9 h	(67)
4-ClC_6H_4	9 h	(78)

C_{16-19}

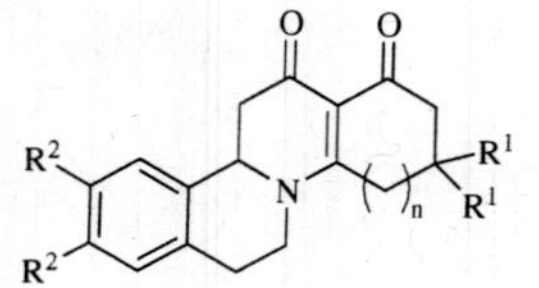

Et_3SiH, BF_3•OEt_2, TFA, rt, 24 h

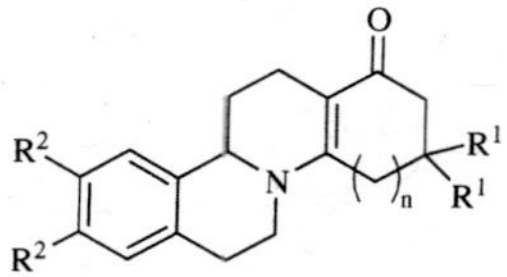

R^1	R^2	n	
H	H	0	(94)
H	H	1	(92)
Me	H	1	(92)
Me	OMe	1	(90)

441

C_{16-23}

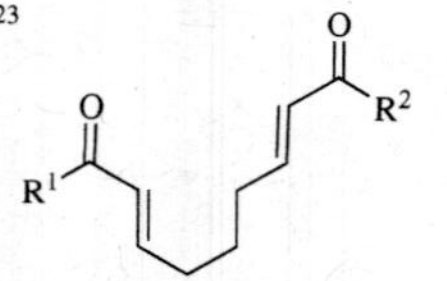

R^1	R^2
Ph	Me
4-$MeOC_6H_4$	4-$MeOC_6H_4$
Ph	4-$MeOC_6H_4$

$PhSiH_3$ (120 mol%), $Co(dpm)_2$ (5 mol%), DCE, rt, 30 min

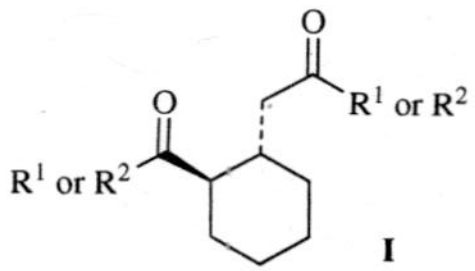
I

\+

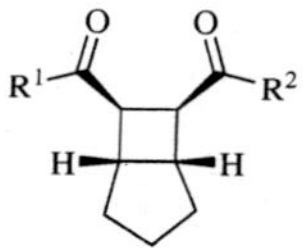
II

I	II
(5)	(65)
(50)	(0)
(46)	(40)

460

C_{17}

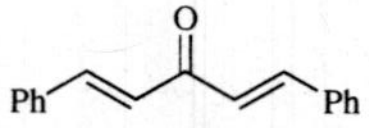

Et_3SiH (6 eq), TFA (10 eq), CCl_4, 50°, 6 h

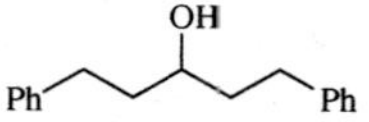
(85) 393

1. $PhMe_2SiH$ (2 eq), $(Ph_3P)_3CuF$•2 EtOH, DMA, 0°, 0.5 h
2. Enone, 0° to rt, 2 h

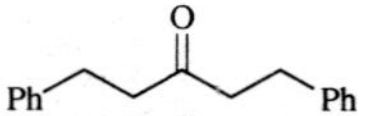
I
(52) 444

$PhMe_2SiH$, CuCl (25 mol%), PPh_3 (25 mol%), TBAF (20 mol%), DMA, 0° to rt, 1.5 h

I (64) 444

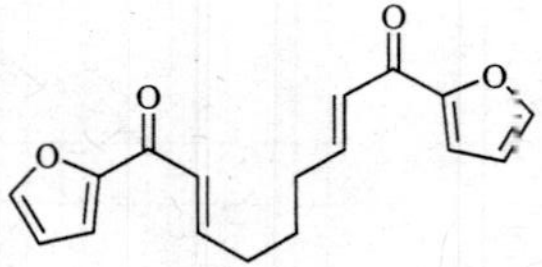

$PhSiH_3$ (2.4 eq), $Co(dpm)_2$ (5 mol%), 50°

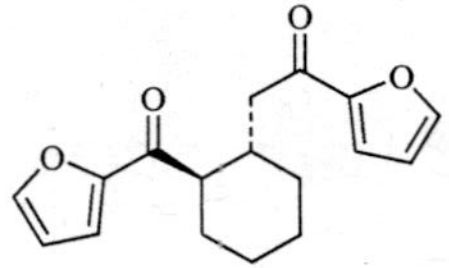
(52) 461

TABLE 16. ORGANOSILANE REDUCTION OF α,β-UNSATURATED KETONES (*Continued*)

Unsaturated Ketone	Conditions	Product(s) and Yield(s) (%)	Refs.
C_{17}			
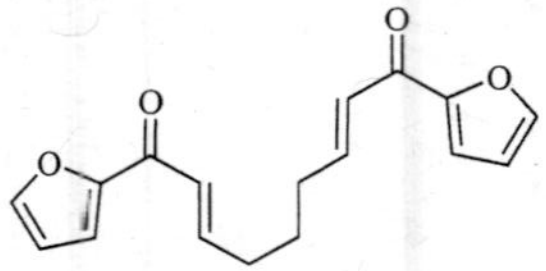	$PhSiH_3$ (240 mol%), $Co(dpm)_2$ (5 mol%), DCE, rt, 30 min	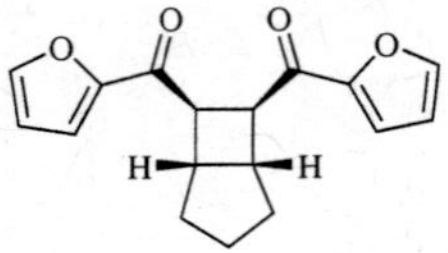 (65) + 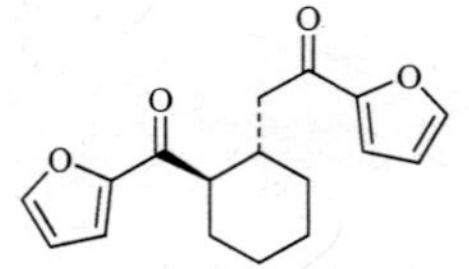 (2)	460
C_{18}			
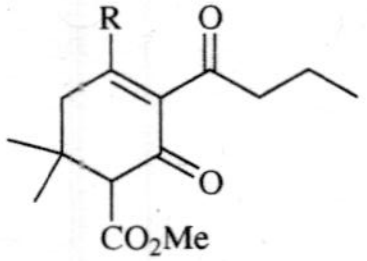	Et_3SiH or $EtMe_2SiH$ (4 eq), BF_3•OEt_2 (2 mol%), TFA, rt	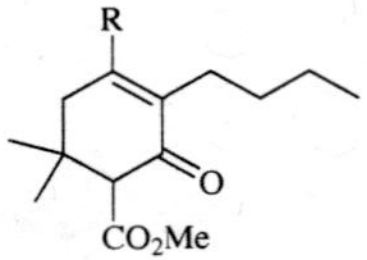 R = 1-pyrrolidinyl, Time 8 h (98); R = NEt_2, Time 24 h (94)	396
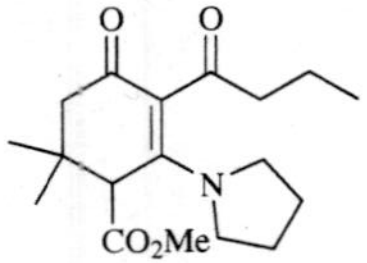	Et_3SiH or $EtMe_2SiH$ (4 eq), BF_3•OEt_2 (2 mol%), TFA, rt, 12 h	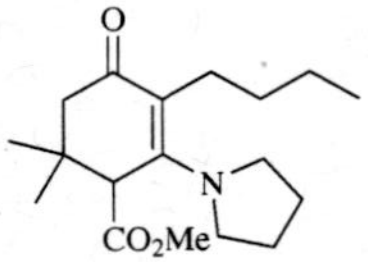 (99)	396
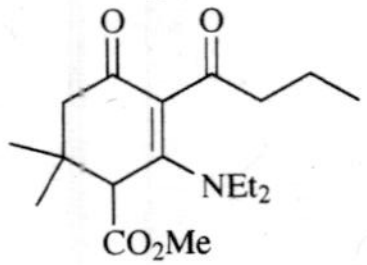	Et_3SiH or $EtMe_2SiH$ (4 eq), BF_3•OEt_2 (2 mol%), TFA, rt, 12 h	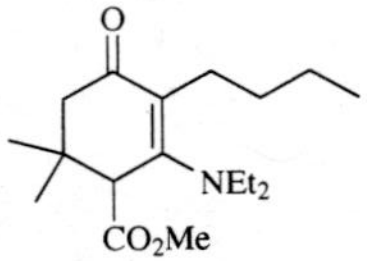 (98)	396
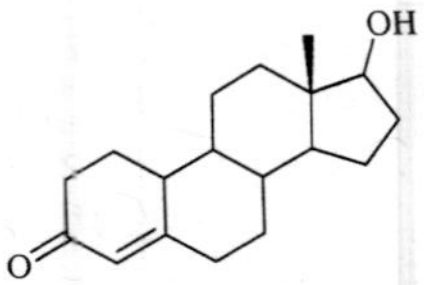	Et_3SiH, TFA, CH_2Cl_2, 42 h	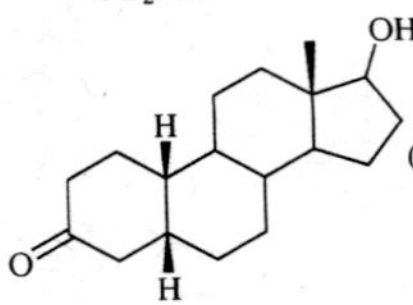 (60) + 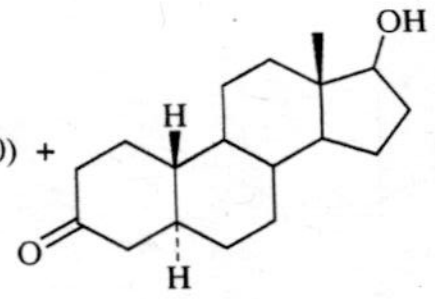 (40)	242

C_{19}

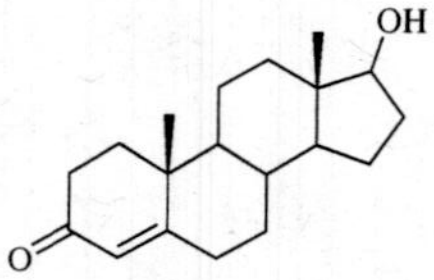	Et_3SiH, TFA, CH_2Cl_2, 44 h	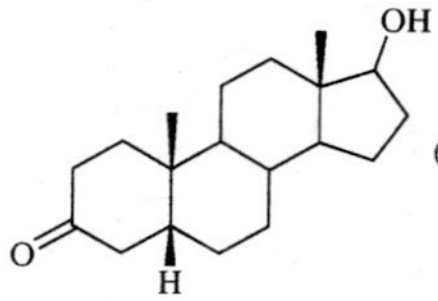 (86) + 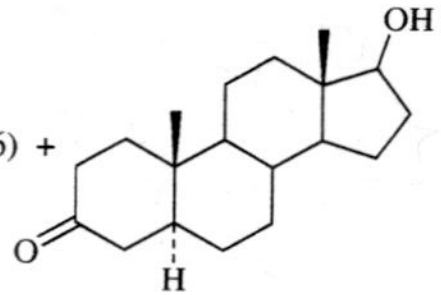 (14)	242
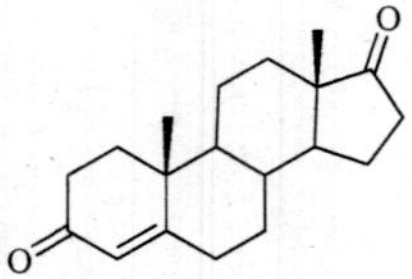	Et_3SiH, TFA, CH_2Cl_2, 60 h	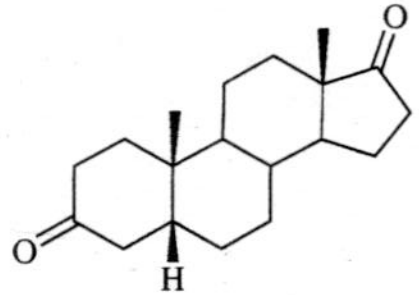 (83) + 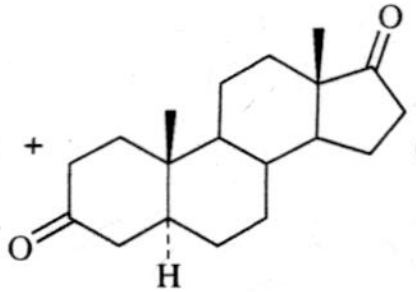 (17)	242
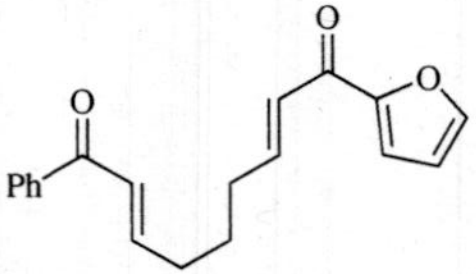	$PhSiH_3$ (2.4 eq), $Co(dpm)_2$ (5 mol%), 50°	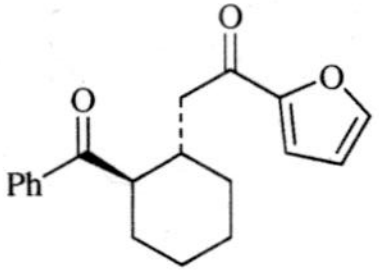 (27) + 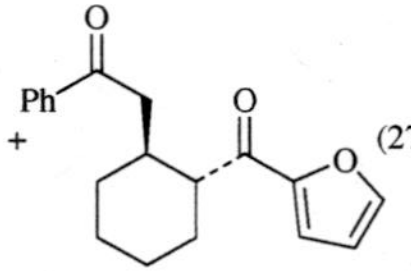 (27)	461
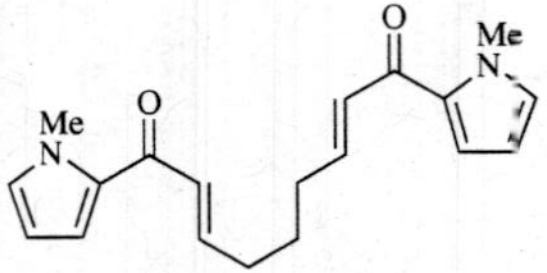	$PhMeSiH_2$ (400 mol%), $Co(dpm)_2$ (10 mol%), DCE, rt, 30 min	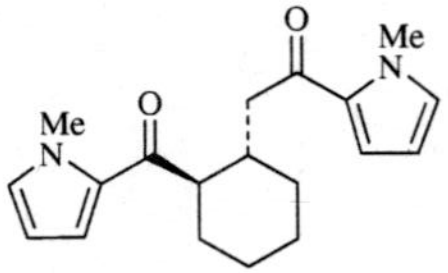 (75)	460
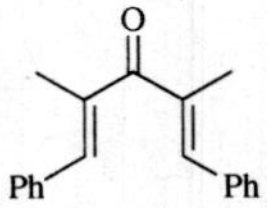	Et_3SiH (10 eq), Et_2AlCl (1.1 eq), CH_2Cl_2, −78°	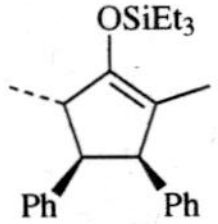 **I** (49) + 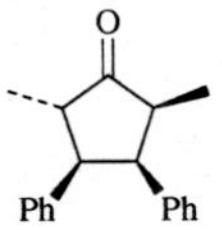 **II** (36) + 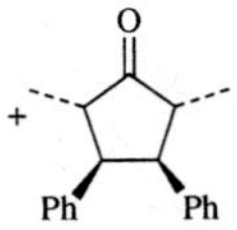 **III** (8)	463
	Et_3SiH (2 eq), $BF_3{\bullet}OEt_2$ (1.1 eq), CH_2Cl_2, −78°	**II** (88) + **III** (10)	463

TABLE 16. ORGANOSILANE REDUCTION OF α,β–UNSATURATED KETONES (*Continued*)

Unsaturated Ketone	Conditions	Product(s) and Yield(s) (%)	Refs.
C_{19} 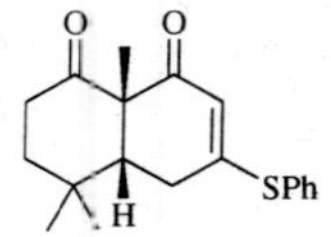	Ph_2SiH_2 (1.3 eq), $(Ph_3P)_4RhH$ (0.5 mol%), CH_2Cl_2, rt, 12 h	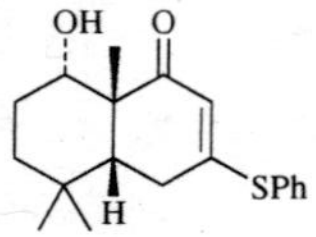 (56)	374
C_{20} 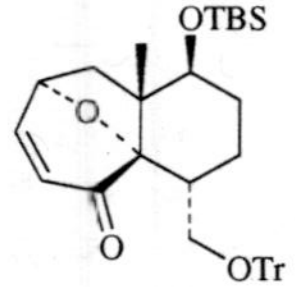	$PhSiH_3$ (1.3 eq), $Mn(dpm)_3$ (3 mol%), *i*-PrOH, DCE	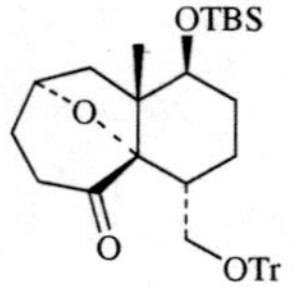 (100)	448
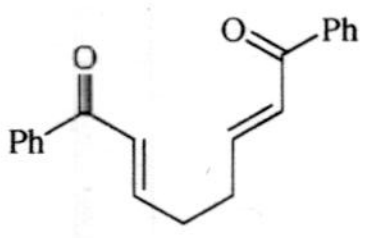	$PhSiH_3$ (2.4 eq), $Co(dpm)_2$ (5 mol%), 50°	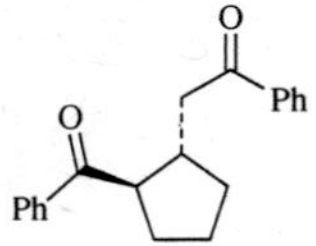 (62)	461
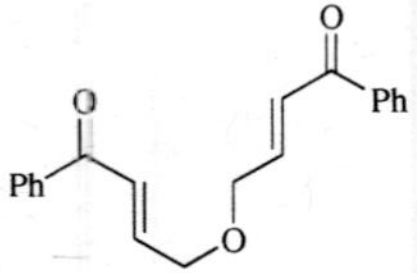	$PhSiH_3$ (2.4 eq), $Co(dpm)_2$ (5 mol%), 50°	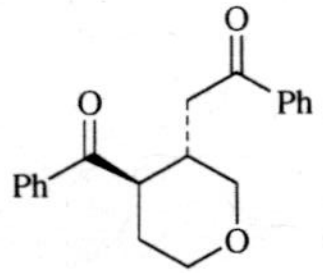 (63)	461
C_{20-21} 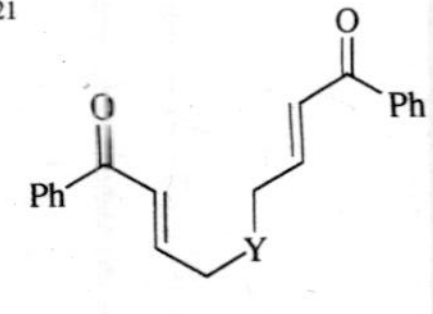	R_3SiH (x eq), $Co(dpm)_2$ (y eq), DDCE, 50°	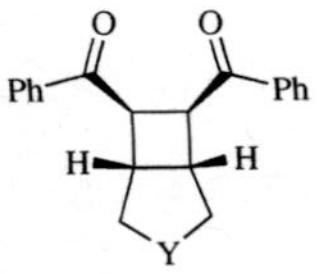 I + 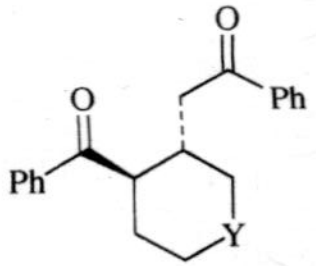 II	462

Y	R_3Si	x	y	I	II
O	PhH_2Si	1.2	0.05	(16)	(31)
O	PhH_2Si	2.4	0.05	(—)	(63)
O	PhMeHSi	4.0	0.10	(69)	(9)
CH_2	PhH_2Si	1.2	0.05	(38)	(41)
CH_2	PhH_2Si	2.4	0.05	(—)	(73)
CH_2	PhMeHSi	4.0	0.10	(72)	(11)

C_{21}

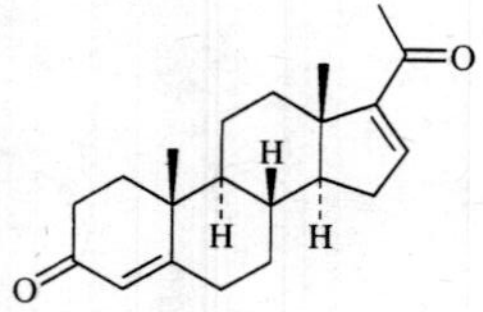

$PhSiH_3$ (1.3 eq), $Mn(dpm)_3$ (3 mol%), *i*-PrOH, DCE

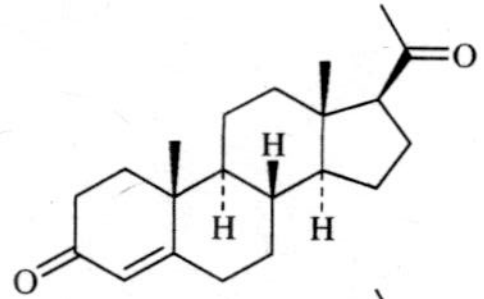

(99) 448

1. $PhSiH_3$ (1.3 eq), $Mn(dpm)_3$ (3 mol%), *i*-PrOH, DCE, O_2
2. $P(OEt)_3$ (1.1 eq)

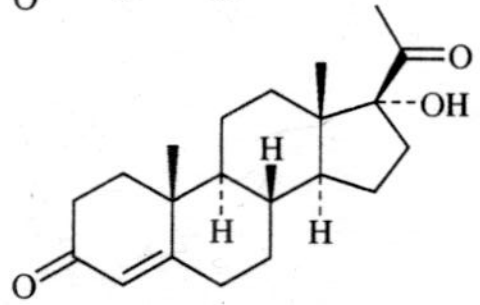

(85) 465

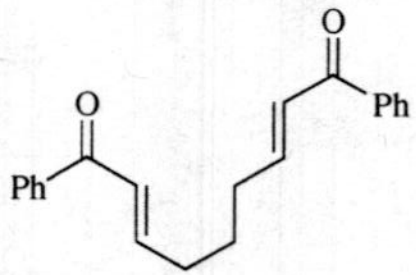

$PhSiH_3$ (240 mol%), $Co(dpm)_2$ (5 mol%), DCE, rt, 30 min

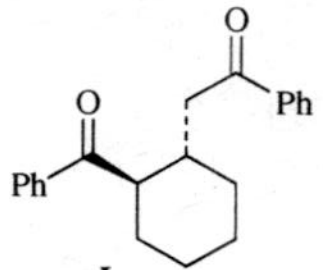

I

(70) 460

$PhMeSiH_2$ (400 mol%), $Co(dpm)_2$ (10 mol%), DCE, rt, 30 min

I (11) +

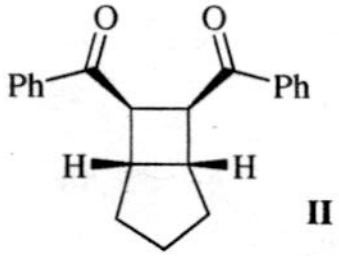

II

(72) 460

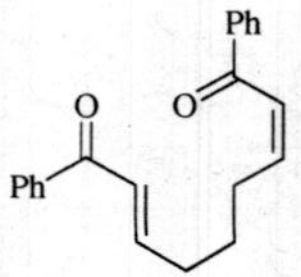

$PhSiH_3$ (120 mol%), $Co(dpm)_2$ (5 mol%), DCE rt, 30 min

II (—) 460

TABLE 16. ORGANOSILANE REDUCTION OF α,β-UNSATURATED KETONES (*Continued*)

Unsaturated Ketone	Conditions	Product(s) and Yield(s) (%)	Refs.
C_{23}			
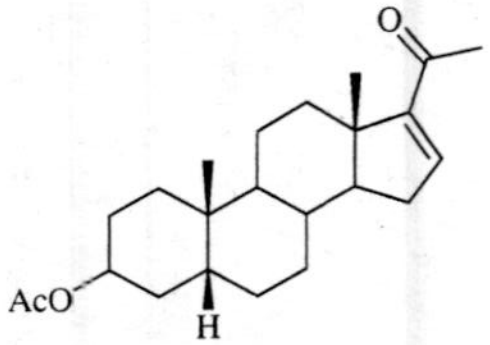	Et_3SiH (1.2 eq), $TiCl_4$ (2.4 eq), CH_2Cl_2, rt, 0.5 h	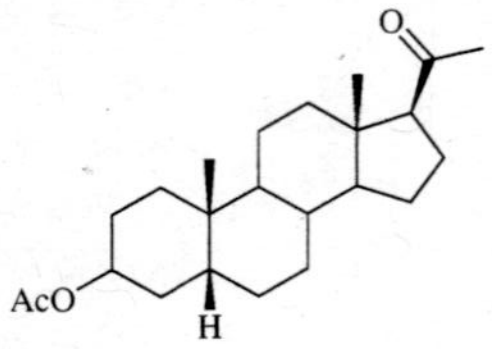 (80)	449
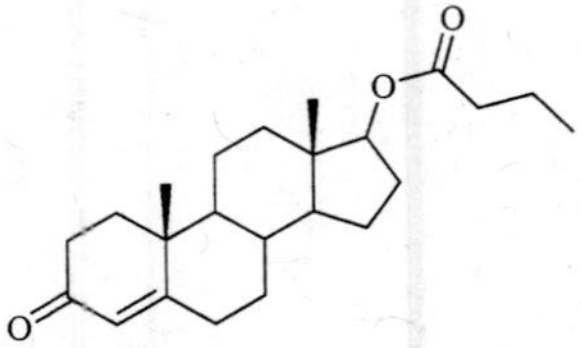	Et_3SiH, TFA, CH_2Cl_2, 48 h	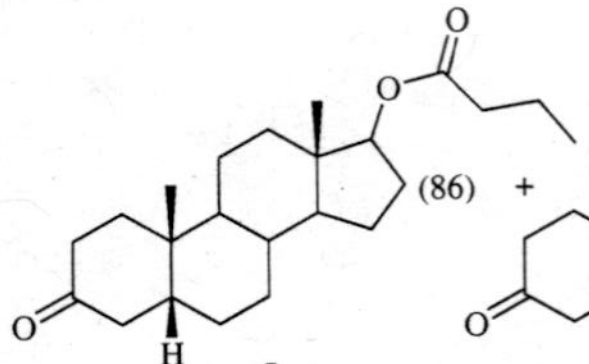 **I** (86) + 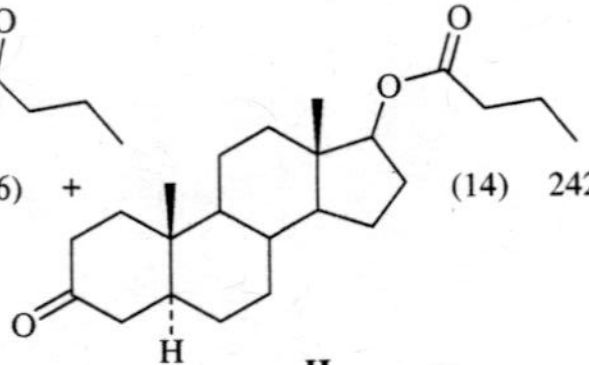 **II** (14)	242
	Et_3SiH, TFA, 30 h	**I** (96) + **II** (4)	242
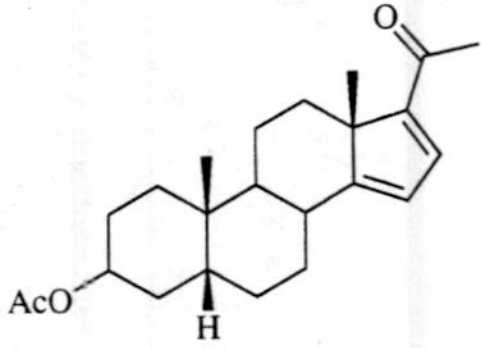	Et_3SiH (1.2 eq), $TiCl_4$ (2.4 eq), CH_2Cl_2, rt, 1 h	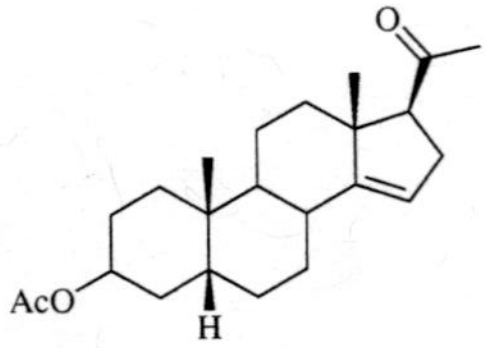 (70-80)	449
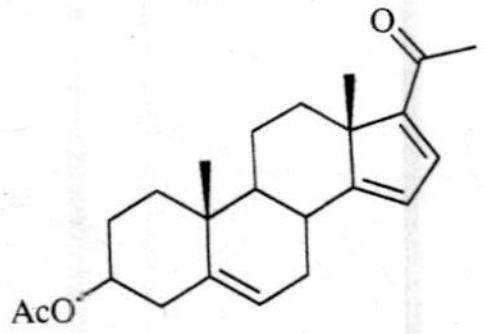	Et_3SiH (1.2 eq), $TiCl_4$ (2.4 eq), CH_2Cl_2, rt, 1 h	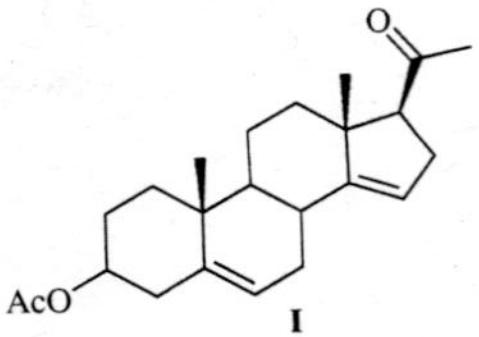 **I** (76)	449

R_3SiH, H_2PtCl_6, MeC_6H_5 **I** 458

R_3Si	Temp	Time	
Et_3Si	150°	5 h	(80)
$(n\text{-}Pr)_3Si$	160°	7 h	(70)
$MeCl_2Si$	60°	0.5 h	(50)
Me_2ClSi	60°	0.5 h	(45)
$Et_2(EtO)Si$	60°	0.6 h	(75)
$Me(EtO)_2Si$	rt	0.4 h	(40)
$HMe_2SiOSiMe_2$	rt	0.5 h	(75)
$[—(H)(Me)SiO—]_4$	60°	—	(50)

C_{26}

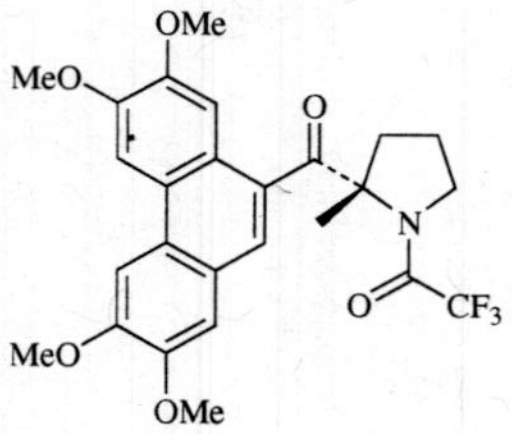

Et_3SiH (3.95 eq), $BF_3•OEt_2$ (19.8 eq), CH_2Cl_2 rt, 48 h

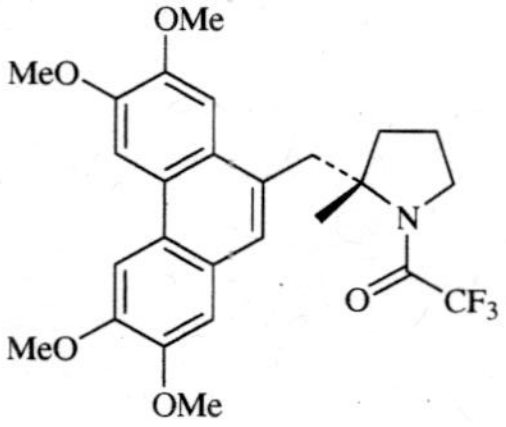

(64) 443

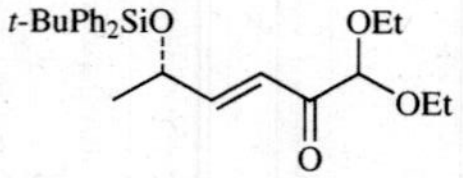

Et_3SiH, $(Ph_3P)_3RhCl$

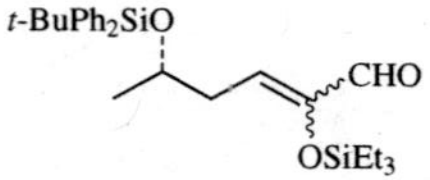

(32) 761

C_{27}

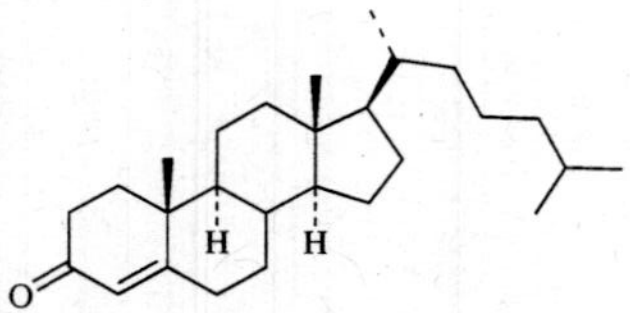

Ph_2SiH_2 (1.3 eq), $(Ph_3P)_4RhH$ (0.4 mol%), CH_2Cl_2, rt, 12 h

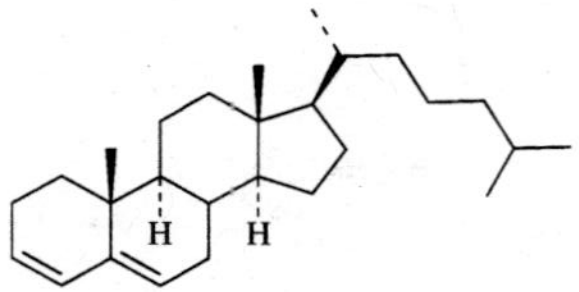

(76) 374

TABLE 16. ORGANOSILANE REDUCTION OF α,β-UNSATURATED KETONES (*Continued*)

Unsaturated Ketone	Conditions	Product(s) and Yield(s) (%)	Refs.
C_{27} 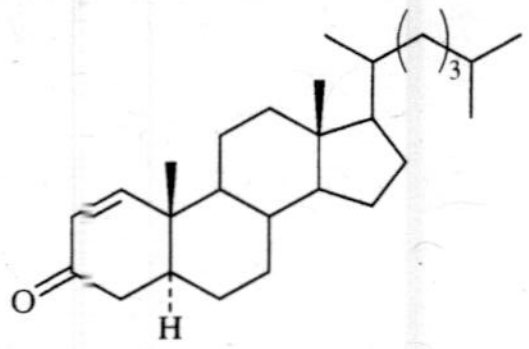	Ph_2SiH_2 (1.6 eq), $ZnCl_2$ (0.5 eq), $(Ph_3P)_4Pd$ (0.03 eq), $CHCl_3$, rt, 1 h	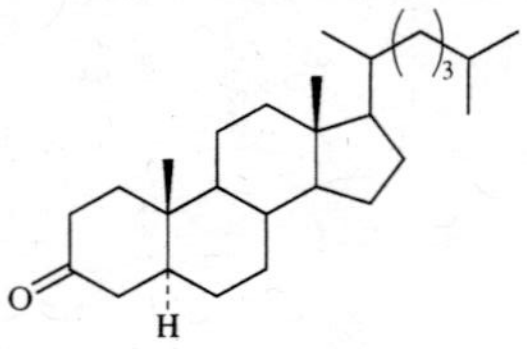 (100)	436, 457
	Ph_2SiD_2 (4 eq), $ZnCl_2$ (4 eq), $(Ph_3P)_4Pd$ (7.2 mol%), $CHCl_3$, rt, 1 h	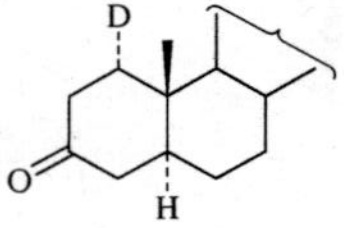 (92)	436
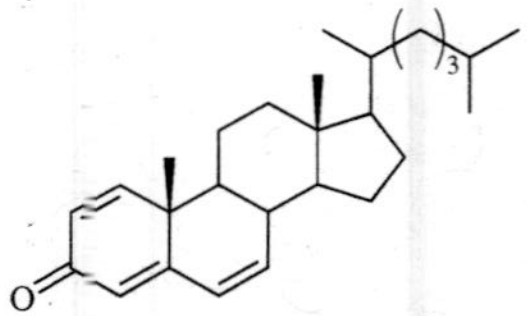	Ph_2SiH_2 (1.6 eq), $ZnCl_2$ (0.90 eq), $(Ph_3P)_4Pd$ (0.067 eq), $CHCl_3$, rt, 1 h	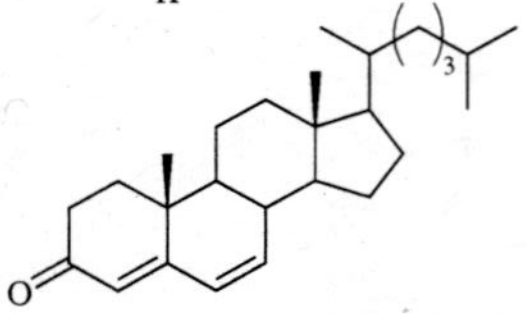 (71)	436
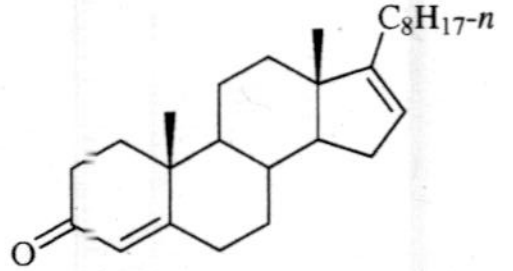	Et_3SiH (1.2 eq), $TiCl_4$ (2.4 eq), CH_2Cl_2, rt, 10 min	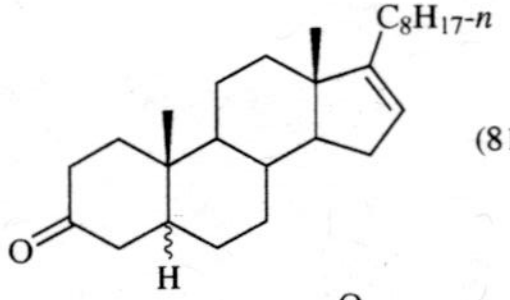 (81) 5α:5β = 3:1	449
C_{29} 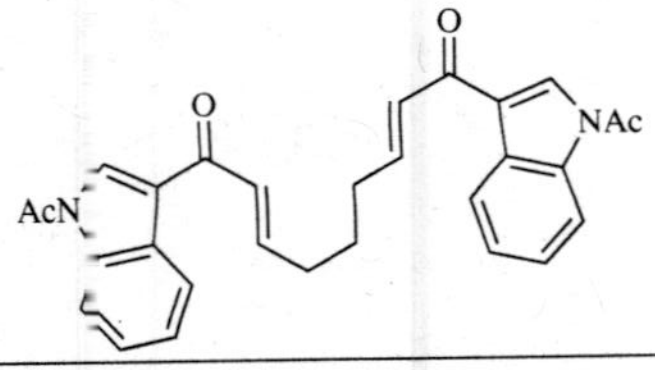	$PhSiH_3$ (2.4 eq), $Co(dpm)_2$ (5 mol%), 50°	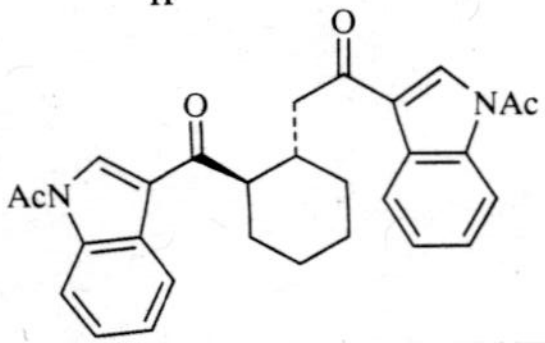 (68)	461

[a] The yield was determined by NMR analysis.

TABLE 17. ORGANOSILANE REDUCTION OF α,β–UNSATURATED ESTERS

Unsaturated Ester	Conditions	Product(s) and Yield(s) (%)	Refs.
C_4 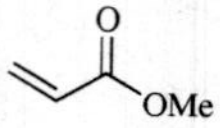	Cl_3SiH (2 eq), $CoCl_2$ (0.175 eq), CD_3CN, DMI (0.5 eq), 70°, 5 h	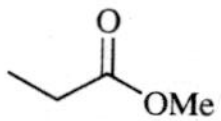 **I** (90)	451
	PMHS-Pd nanocomposite, C_6H_6, rt, 4 h	**I** (96)	219
	TMSH, H_2PtCl_6 (0.5 mol%)	**I** (22) + 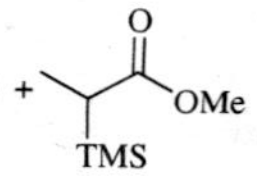 (33) + 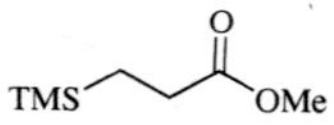 (4)	467
	Et_3SiH, H_2PtCl_6 (0.5 mol%)	**I** (42) + 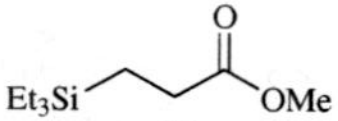 **II** (26)	467
	$(n\text{-}Pr)_3SiH$, $(Ph_3P)_3RhCl$ (0.5 mol%), 100°, 1 min	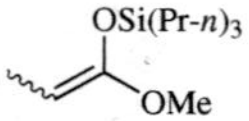 (31.6) + 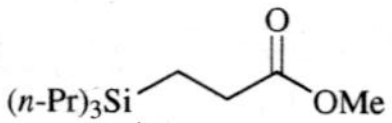 (38.9)	467
	Et_3SiH, $(Ph_3P)_3RhCl$ (0.1 mol%), C_6H_6, 50°, 1 h	**II** (54) + 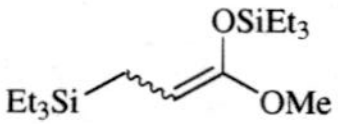 (23)	466
	$PhMe_2SiH$, $(Ph_3P)_3RhCl$ (0.1 mol%), C_6H_6, 50°, 12 h	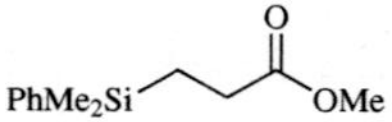 (80)	466
	$EtMe_2SiH$, $(Ph_3P)_3RhCl$ (0.1 mol%), C_6H_6, 30°, 1 h	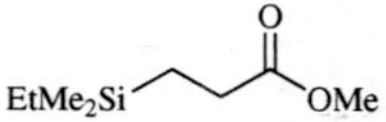 (78)	466
	Et_2MeSiH (1 eq), PhNCO (0.5 eq), [Rh(cod){P(OPh)$_3$}$_2$]OTf (1 mol%), CH_2Cl_2, 45°, 3 h	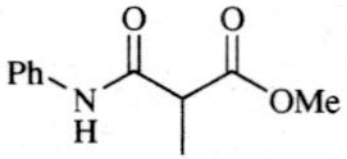 (90)	475

TABLE 17. ORGANOSILANE REDUCTION OF α,β–UNSATURATED ESTERS (*Continued*)

Unsaturated Ester	Conditions		Product(s) and Yield(s) (%)	Refs.
C_4 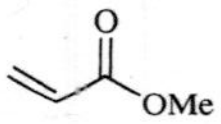	Et_2MeSiH (1 eq), catalyst (1 mol%), PhNCO (0.5 eq), CH_2Cl_2, rt		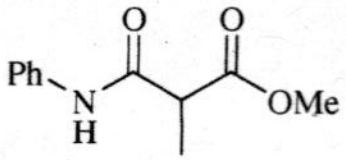	475
	Catalyst	Time		
	[Rh(cod){P(OPh)$_3$}$_2$]OTf	21 h	(86)	
	[Rh(cod){P(OPh)$_3$}$_2$]OTf	3 h	(90)	
	[Rh(cod)(PPh_3)$_2$]OTf	14 h	(95)	
	[Rh(cod)(PPh_2Me)$_2$]OTf	10 h	(89)	
	[Rh(cod)(dppb)]OTf	47 h	(68)	
	$Rh_4(CO)_{12}$	47 h	(79)	
	None	68 h	(0)	
	Et_2MeSiH (1 eq), RNCO (0.5 eq), [Rh(cod){P(OPh)$_3$}$_2$]OTf (1 mol%), CH_2Cl_2, 45°		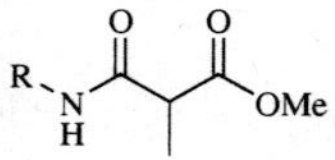	475
	R	Time		
	Ph	3 h	(90)	
	4-ClC_6H_4	3 h	(93)	
	4-BrC_6H_4	13 h	(96)	
	4-$MeOC_6H_4$	5 h	(73)	
	4-MeC_6H_4	15 h	(96)	
	3-MeC_6H_4	13 h	(96)	
	2-MeC_6H_4	13 h	(86)	
	2-ClC_6H_4	12 h	(94)	
	1-$C_{10}H_7$	6 h	(99)	
	2-C_4H_3O	4 h	(85)	
	p-Ts	12 h	(82)	
	Bz	14 h	(82)	
	Bn	15 h	(45)	
	c-C_3H_5	61 h	(43)	
	c-C_6H_{11}	24 h	(0)	

Cl_2MeSiH, $[Rh(cod)Cl]_2$ (2.5 mol%), Me-DuPHOS (5.5 mol%), *i*-PrCHO, CH_2Cl_2, rt, 16 h	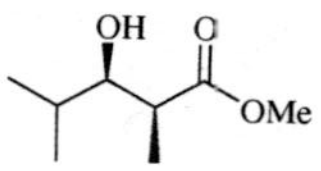	(15) syn:anti = 15:1	762
Cl_2MeSiH, $[Rh(cod)Cl]_2$ (2.5 mol%), Me-DuPHOS (5.5 mol%), PhCHO, CH_2Cl_2, rt, 16 h	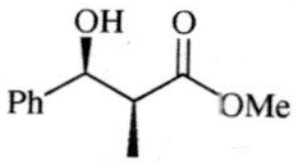	(69) syn:anti = 23:1	762
Cl_2MeSiH, $[Rh(cod)Cl]_2$ (2.5 mol%), Me-DuPHOS (5.5 mol%), (*E*)-PhCH=CHCHO, CH_2Cl_2, rt, 16 h	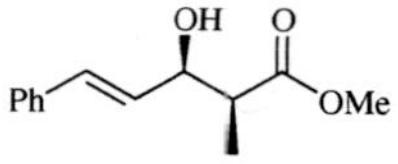	(41) syn:anti = >20:1	762
Cl_2MeSiH, $[Rh(cod)Cl]_2$ (2.5 mol%), Me-DuPHOS (5.5 mol%), 1-naphthaldehyde, CH_2Cl_2, rt, 16 h	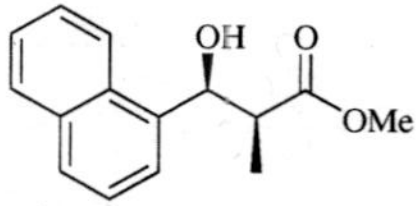	(82) syn:anti = 10:1	762
Et_2MeSiH, $[Ir(cod)Cl]_2$ (2.5 mol%), RCHO, **97** (7.5 mol%), rt, 24 h	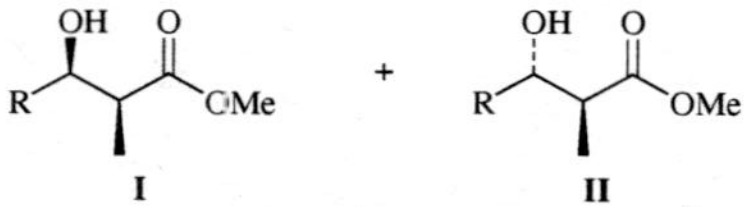		601

R	I + II	I:II	er of I
Ph	(68)	6.6:1	97:3
$BnOCH_2$	(49)	9.9:1	98:2
$TBSOCH_2$	(47)	8.2:1	98:2
Et	(<5)	—	—
$BnO(CH_2)_2$	(65)	2.7:1	91:9
(*E*)-PhCH=CH	(0)	—	—

TABLE 17. ORGANOSILANE REDUCTION OF α,β-UNSATURATED ESTERS (*Continued*)

Unsaturated Ester	Conditions	Product(s) and Yield(s) (%)	Refs.
C_4 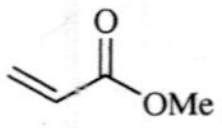	1. $PhSiH_3$ (2 eq), $Co(dpm)_2$ (0.05 mol%), RCHO, DCE, 20° 2. H_3O^+ R — Time (*E*)-PhCH=CH — 20 h $BnCH_2$ — 20 h	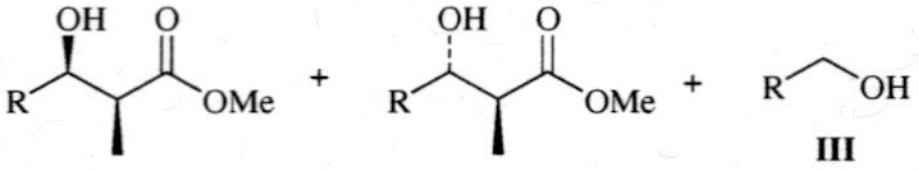 I + II + III I + II — I:II — III (80) — 50:50 — (10) (62) — 50:50 — (14)	475
	TMSH (1.3 eq), $RhCl_3$•$3H_2O$ (0.009 mol%), acetone, rt	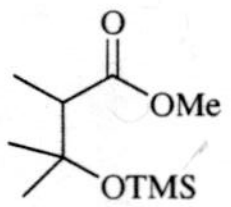 (24)	473
C_5 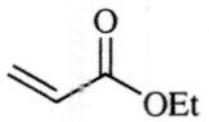	$PhMe_2SiH$ (2 eq), $CoCl_2$ (0.175 eq), CD_3CN, DMI (0.5 eq), 80°, 10 h	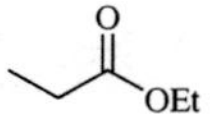 **I** (85)	451
	Cl_3SiH (2 eq), $CoCl_2$ (0.175 eq), CD_3CN, DMI (0.5 eq), 70°, 5 h	**I** (70)	451
	$PhMe_2SiH$, $(Ph_3P)_3RhCl$, 80°, 2 h	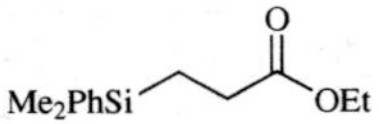 (85)	466
	$EtMe_2SiH$, $(Ph_3P)_3RhCl$, reflux, 12 h	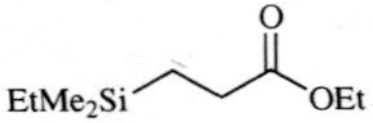 (83)	466
	$ClMe_2SiH$, $(Ph_3P)_3RhCl$, 70°, 12 h	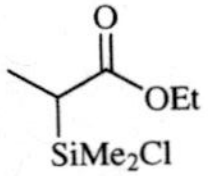 (65)	466

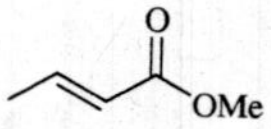	Ph_2SiD_2 (1.2 eq), $ZnCl_2$ (1.74 eq), $(Ph_3P)_4Pd$ (0.02 eq), PPh_3 (0.09 eq), rt, 72 h	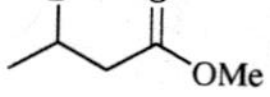 (90)	436
	1. TMSH, H_2PtCl_6 (0.5 mol%), 60°, 3 h 2. MeOH, rt	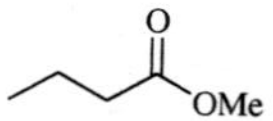 (28) + 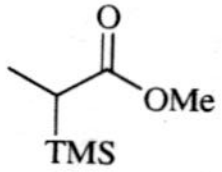 (17) + 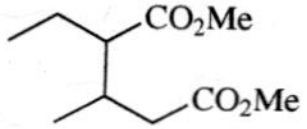 (54)	467
	TMSH (1.3 eq), $RhCl_3 \cdot 3\ H_2O$ (0.009 mol%), acetone, rt	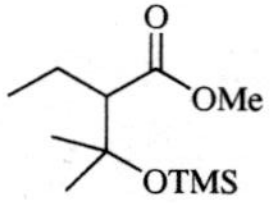 (89)	473

TABLE 17. ORGANOSILANE REDUCTION OF α,β-UNSATURATED ESTERS (*Continued*)

Unsaturated Ester	Conditions	Product(s) and Yield(s) (%)	Refs.
C_5 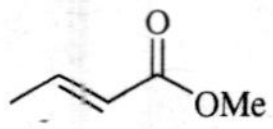	Et_2MeSiH (1 eq), ArNCO (0.5 eq), [Rh(cod){$P(OPh)_3$}$_2$]OTf (1 mol%), CH_2Cl_2, 45°	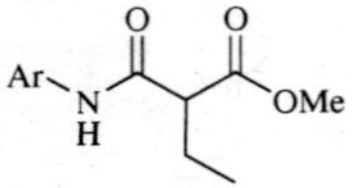	475

Ar	Time	Yield (%)
Ph	13 h	(88)
4-ClC_6H_4	4 h	(87)
4-BrC_6H_4	13 h	(80)
4-$MeOC_6H_4$	14 h	(91)
4-MeC_6H_4	12 h	(92)
3-MeC_6H_4	14 h	(95)
2-MeC_6H_4	14 h	(84)
2-ClC_6H_4	5 h	(95)
1-$C_{10}H_7$	15 h	(83)
2-C_4H_3O	4 h	(53)
p-Ts	12 h	(61)

Unsaturated Ester	Conditions	Product(s) and Yield(s) (%)	Refs.
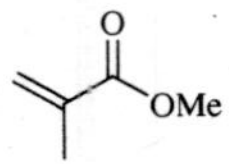	Et_2MeSiH (1 eq), PhNCO (0.5 eq), [Rh(cod){$P(OPh)_3$}$_2$]OTf (1 mol%), CH_2Cl_2, 80°, 24 h	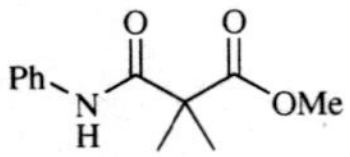 (26)	475
	$PhSiH_3$ (1.3-1.5 eq), $Mo(CO)_6$ (3-5 mol%), THF, reflux, 20 h	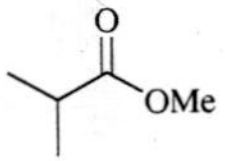 (90)	450
	TMSH (1.3 eq), $RhCl_3$•3 H_2O (0.009 mol%), 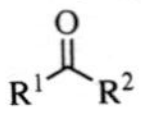 , rt	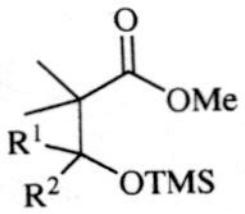	473

R^1	R^2	Yield (%)
Me	Me	(95)
Et	Me	(91)
—$CH_2CH(Me)(CH_2)_3$—		(66)
Me	H	(78)

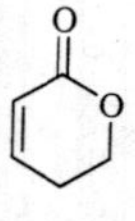	TMSH (1.3 eq), $RhCl_3 \cdot 3H_2O$ (0.009 mol%), acetone, rt	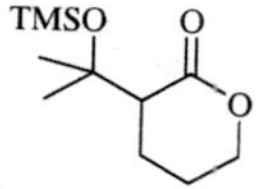 (34)	473
	$(EtO)_3SiH$ (1.1 eq), $(Ph_3P)_4RhH$ (0.3 mol%), CH_2Cl_2, rt, 12 h	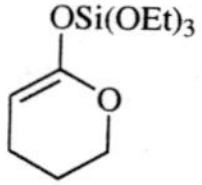 (85)	374
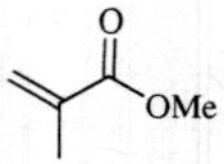	$(EtO)_3SiH$ (1.1 eq) $(Ph_3P)_4RhH$ (0.3 mol%), CH_2Cl_2, rt, 1.5 h	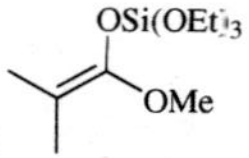 (91)	374
	Et_3SiH, $(Ph_3P)_3RhCl$, C_6H_6, 70°, 2 h	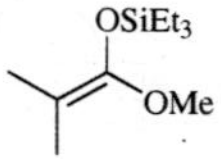 (74) + 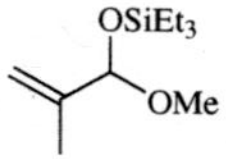 (6)	466
	$(n\text{-}Pr)_3SiH$, $(Ph_3P)_3RhCl$ (0.5 mol%), 100°, 1 min	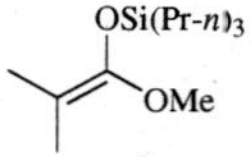 (70)	467
	$PhMe_2SiH$, $(Ph_3P)_3RhCl$, 60°, 2 h	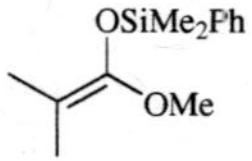 (71) + 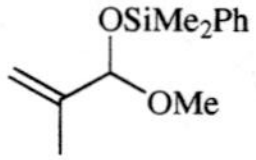 (24)	466
	$EtMe_2SiH$, $(Ph_3P)_3RhCl$, C_6H_6, 70°, 12 h	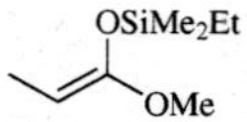 (86) + 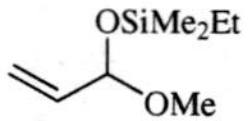 (6)	466

TABLE 17. ORGANOSILANE REDUCTION OF α,β–UNSATURATED ESTERS (*Continued*)

Unsaturated Ester	Conditions	Product(s) and Yield(s) (%)	Refs.
C_5			
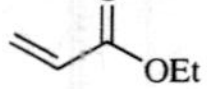	Et_3SiH, $(Ph_3P)_3RhCl$, C_6H_6, 70°, 3 h	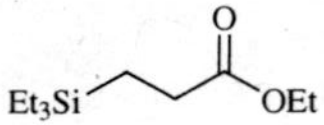 (52) +  (28)	466
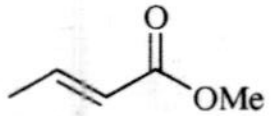	1. TMSH, $(Ph_3P)_3RhCl$ (0.5 mol%), 60°, 10 min 2. MeOH, rt	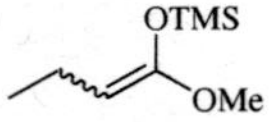 (77) + 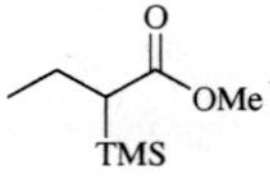 (8.6)	467
	1. Et_3SiH, $(Ph_3P)_3RhCl$ (0.5 mol%), 100°, 1 min 2. MeOH, rt	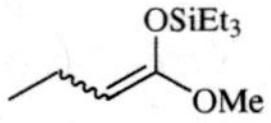 (85.6)	467
	1. $(n\text{-}Pr)_3SiH$, $(Ph_3P)_3RhCl$ (0.5 mol%), 100°, 1 min 2. MeOH, rt	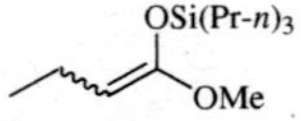 (73.5)	467
C_6			
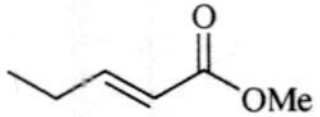	TMSH, H_2PtCl_6 (0.5 mol%)	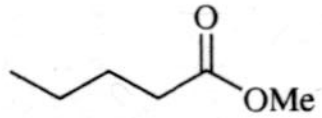 (56) + 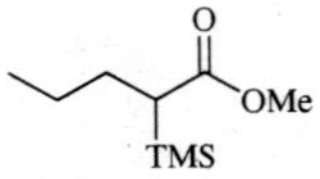 (3) + 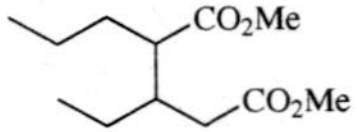 (25)	467
	$n\text{-}Pr_3SiH$, $(Ph_3P)_3RhCl$ (0.5 mol%), 100°, 1 min	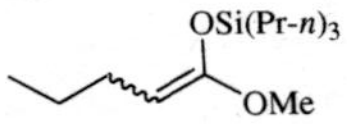 (75.4)	467

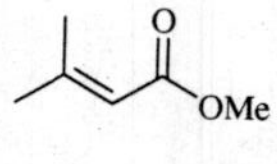	$PhSiH_3$ (1.01 eq), Ni, PPh_3 (0.02 eq)	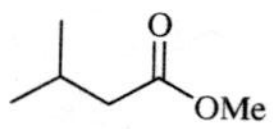 (7.7)	438
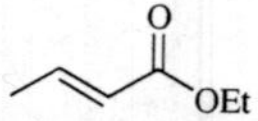	$PhSiH_3$ (1.3-1.5 eq), $Mo(CO)_6$ (3-5 mol%), THF, reflux, 11.5 h	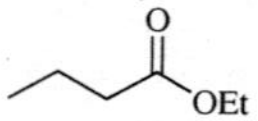 **I** (75)	450
	Ph_2SiH_2 (2.4 eq), $ZnCl_2$ (1.7 eq), $(Ph_3P)_4Pd$ (0.02 eq), PPh_3 (0.09 eq), $CHCl_3$, rt, 48 h	**I** (90)	436
	Et_3SiH, $(Ph_3P)_3RhCl$, 50°, 1 h	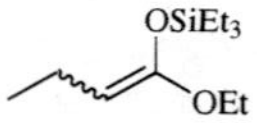 (95)	466
	$PhMe_2SiH$, $(Ph_3P)_3RhCl$, 50°, 1 h	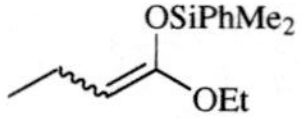 (95)	466
	$EtMe_2SiH$, $(Ph_3P)_3RhCl$, 80°, 1 h	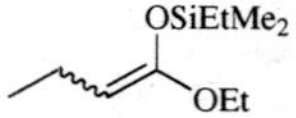 (82)	466
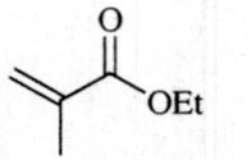	Et_3SiH, $(Ph_3P)_3RhCl$, C_6H_6, 80°, 2 h	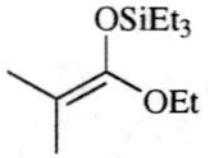 (74) + 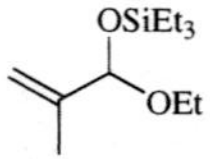 (6)	466
	$PhMe_2SiH$, $(Ph_3P)_3RhCl$, 65°, 1 h	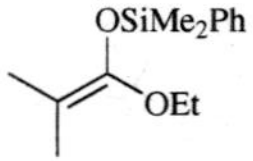 (70) + 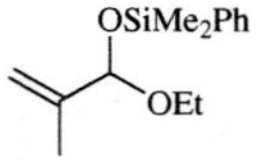 (17)	466
	$EtMe_2SiH$, $(Ph_3P)_3RhCl$, 40°, 16 h	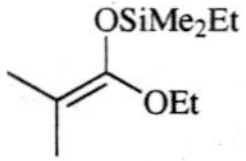 (86) + 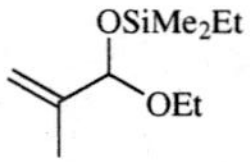 (5)	466
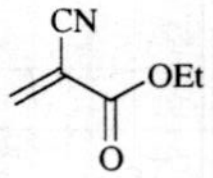	Et_3SiH (1 eq), TFA (3 eq), 0°, <5 min	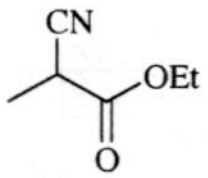 (85)	236

TABLE 17. ORGANOSILANE REDUCTION OF α,β–UNSATURATED ESTERS (*Continued*)

Unsaturated Ester	Conditions	Product(s) and Yield(s) (%)	Refs.
C_6 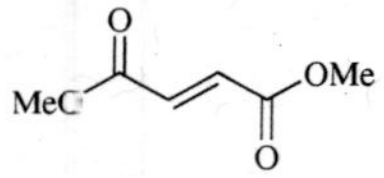	Et_2MeSiH (1 eq), PhNCO (0.5 eq), [Rh(cod){$P(OPh)_3$}$_2$]OTf (1 mol%), CH_2Cl_2, 45°, 13 h	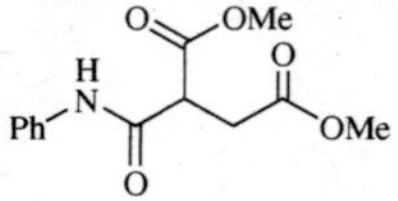 (40)	475
C_7 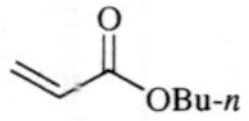	PMHS-Pd nanocomposite, C_6H_6, rt, 4 h	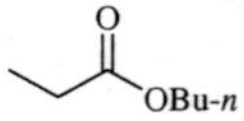 (94)	219
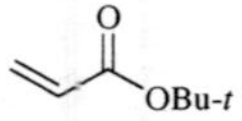	Cl_3SiH (1.3 eq), $(Ph_3P)_4Pd$ (5 mol%), RCHO, CH_2Cl_2, rt, 45 h	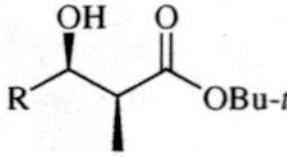 I + 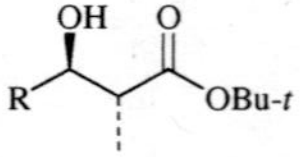 II	763
	R: Ph	I + II (37), I:II 72:28	
	R: 2-$C_{10}H_7$	I + II (32), I:II 73:27	
	R: 4-PhC_6H_4	I + II (36), I:II 62:38	
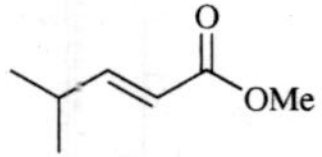	TMSH, H_2PtCl_6 (0.5 mol%)	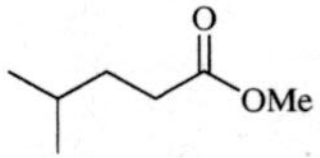 (60) + 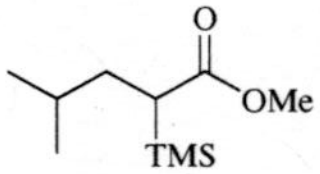 (3)	467
	$(n\text{-Pr})_3SiH$, $(Ph_3P)_3RhCl$ (0.5 mol%), 100°, 1 min	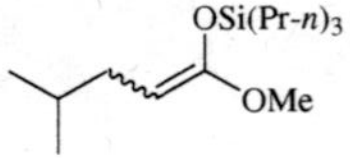 (76.5)	467
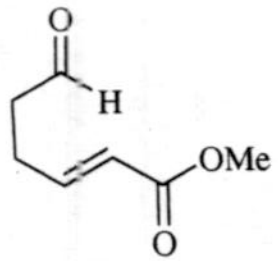	Et_3SiH (2.1 eq), [RhCl(C_8H_{14})$_2$]$_2$ (2.5 mol%), ligand (5 mol%), MeC_6H_5, 50°, 18 h	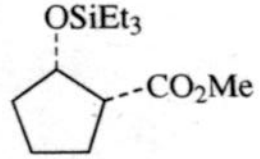 I + 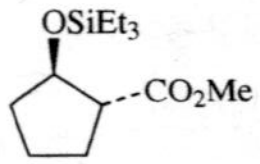 II	471

Ligand	I + II	I:II
$(c\text{-}C_6H_{11})_3P$	(79)	2.5:1
DIPHOS	(78)	3.3:1
$(p\text{-tol})_3P$	(27)	1::2
$(p\text{-tol})_3P$	(53)	2:1
$(PMF)_3P$	(61)	1:1.6
$(PMF)_3P$	(51)	2:1

R_3SiH (2.1 eq), Rh(I) catalyst (x mol%), ligand (4x eq), MeC_6H_5, 50°, 18 h

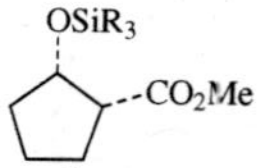

I

\+

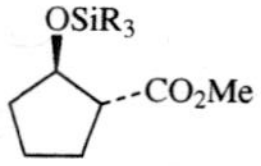

II

472

R_3Si	Catalyst	x	Ligand	I + II	I:II
Et_3Si	$(Ph_3P)_3RhCl$	1.0	none	(81)	3.0
$PhMe_2Si$	$(Ph_3P)_3RhCl$	1.0	none	(62)	2.4
Ph_2MeSi	$(Ph_3P)_3RhCl$	1.0	none	(49)	2.8
Ph_3Si	$(Ph_3P)_3RhCl$	1.0	none	(42)	1.5
Et_3Si	$[(c\text{-}C_8H_{14})_2RhCl]_2$	2.5	$(c\text{-}C_6H_{11})_3P$	(79)	2.5
Et_3Si	$[(c\text{-}C_8H_{14})_2RhCl]_2$	2.5	dppe	(78)	3.3
Et_3Si	$[(c\text{-}C_8H_{14})_2RhCl]_2$	2.5	$(p\text{-tol})_3P$	(27)	0.5
Et_3Si	$[(c\text{-}C_8H_{14})_2RhCl]_2$	2.5	$(PMP)_3P$	(61)	0.6
Et_3Si	$(Ph_3P)_4RhH$	1.0	none	(81)	0.5

R_3SiH (2.1 eq), $(Ph_3P)_3RhCl$ (1 mol%), MeC_6H_5, 50°, 18 h — I + II — 471

R_3Si	I + II	I:II
Et_3Si	(81)	2.0:1
$PhMe_2Si$	(62)	2.4:1
Ph_2MeSi	(49)	2.8:1
Ph_3Si	(42)	1.5:1

TABLE 17. ORGANOSILANE REDUCTION OF α,β–UNSATURATED ESTERS (*Continued*)

Unsaturated Ester	Conditions	Product(s) and Yield(s) (%)	Refs.
C_7 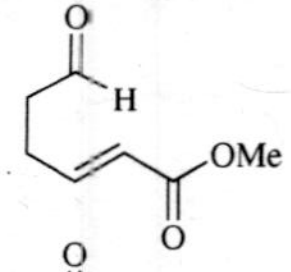	Et_3SiH (2.1 eq), $(Ph_3P)_4RhH$ (1 mol%), MeC_6H_5, 50°, 18 h	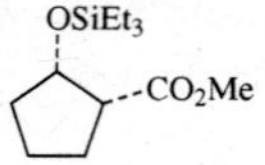 **I** + 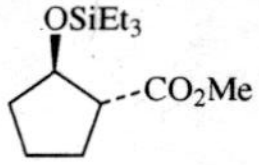 **II** **I** + **II** (81), **I**:**II** = 1:11	471
C_8 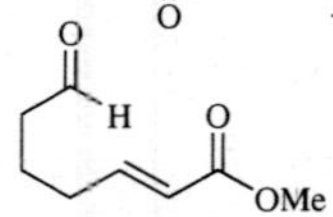	Et_3SiH (2.1 eq), $(Ph_3P)_3RhCl$ (1 mol%), MeC_6H_5, 50°, 16 h	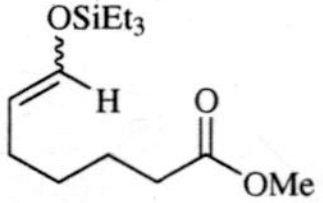 **I** + 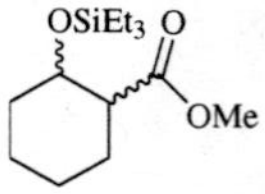 **II** **I** + **II** (66), **I**:**II** = 1:4.6	471
	Et_3SiH (2.1 eq), $(Ph_3P)_4RhH$ (1 mol%), MeC_6H_5, 50°, 16 h	**I** + **II** (65), **I**:**II** = 1:3	471
	Et_3SiH (2.1 eq), $(Ph_3P)_3RhCl$ (3 mol%), MeC_6H_5, 70°	**II** (29)	472
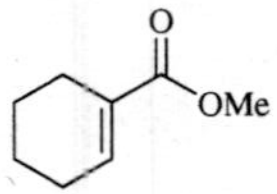	$PhSiH_3$ (1.01 eq), Ni, PPh_3 (0.02 eq)	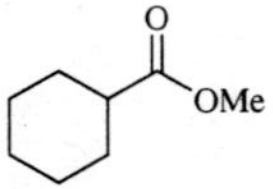 (6.4)	438
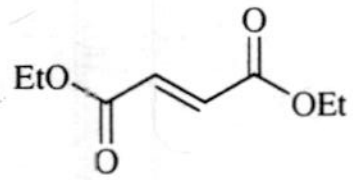	$PhMe_2SiH$ (1.2 eq), CuCl (110 mol%), PPh_3 (110 mol%), TBAF (100 mol%), DMA, 0° to rt, 2 h	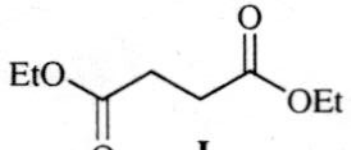 **I** (65)	444
	$PhSiH_3$ (1.3-1.5 eq), $Mo(CO)_6$ (3-5 mol%), THF, reflux, 4.5 h	**I** (100)	450
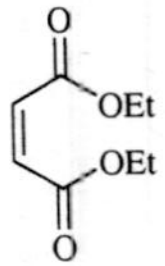	$PhSiH_3$ (1.3-1.5 eq), $Mo(CO)_6$ (3-5 mol%), THF, reflux, 10 h	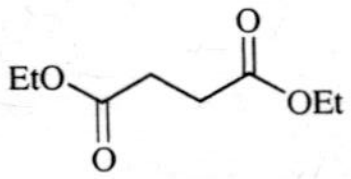 (100)	450

C_{8-13}

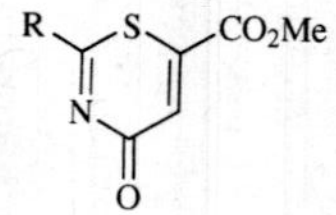

Et_3SiH (3.15 eq), TFA (19.5 eq), CH_2Cl_2, rt, 1.5 h

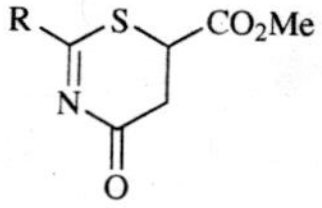

R	
EtO	(83)
Me_2N	(75)
Ph	(60)
BnS	(89)

468

C_9

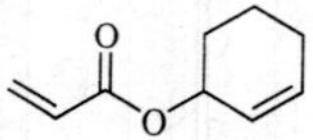

$MeCl_2SiH$, $[Rh(coc)Cl]_2$ (2.5 mol%), Me-DuPHOS (5 mol%), C_6H_6, rt, 4 h

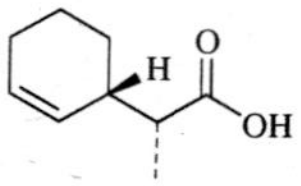

(16) dr = 3:1

474

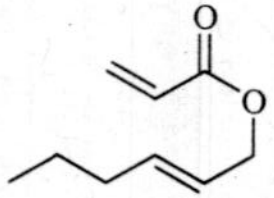

R_3SiH, $[Rh(cod)Cl]_2$ (2.5 mol%), Me-DuPHOS (5 mol %), C_6H_6, rt

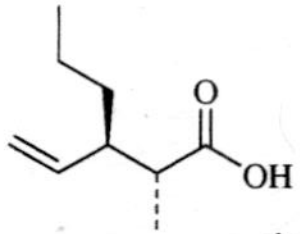

474

R_3Si	Time		dr
$MeCl_2Si$	1 h	(74)	11:1
PhMeClSi	15 h	(8)	4.3:1
Ph_2ClSi	6 h	(0)	—
$(EtO)_3Si$	2 h	(47)	21:1
$Me(EtO)_2Si$	2 h	(47)	>25:1
$Me_2(EtO)Si$	2 h	(9)	10:1
$PhMe_2Si$	1 h	(0)	—
$MeEt_2Si$	1 h	(0)	—

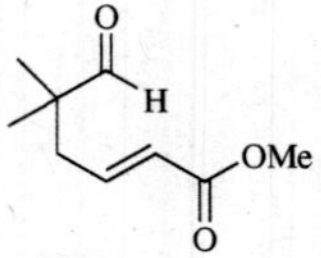

Et_3SiH (2.1 eq), $(Ph_3P)_4RhCl$ (1 mol%), MeC_6H_5, 50°, 16 h

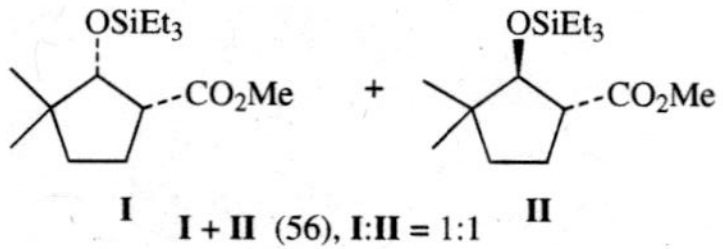

I + **II** (56), **I**:**II** = 1:1

471

Et_3SiH (2.1 eq), $(Ph_3P)_4RhH$ (1 mol%), MeC_6F_5, 50°, 16 h

I + **II** (62), **I**:**II** = 6.4:11

471

TABLE 17. ORGANOSILANE REDUCTION OF α,β–UNSATURATED ESTERS (*Continued*)

Unsaturated Ester	Conditions	Product(s) and Yield(s) (%)	Refs.
C_9 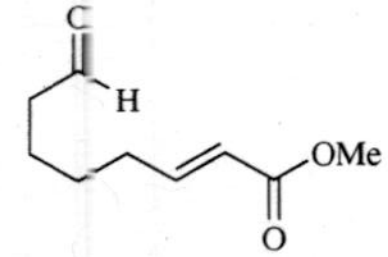	Et_3SiH (2.1 eq), $(Ph_3P)_4RhH$ (1 mol%), MeC_6H_5, 50°, 16 h	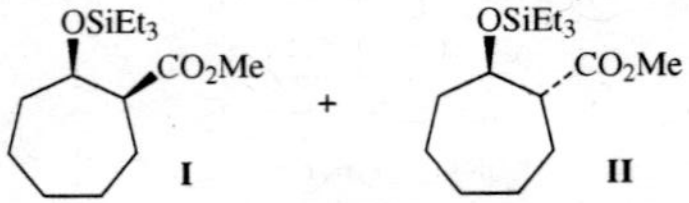 **I** + **II** (68), **I**:**II** = 2.5:1	471
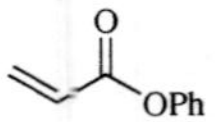	1. $EtMe_2SiH$ (1.0 eq), $[Rh(cod)Cl]_2$ (2.5 mol%), CH_3CH_2CHO (0.83 eq), (*R*)-BINAP (6.5 mol%), rt, 24 h 2. H_3O^+	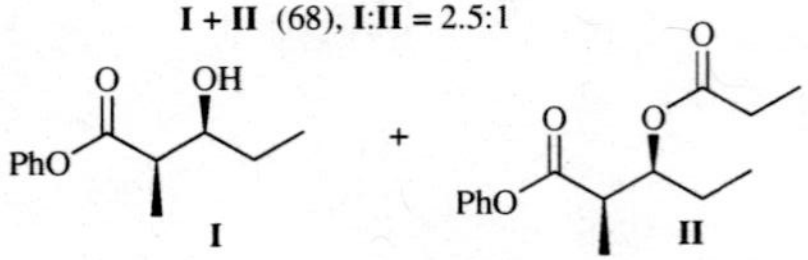 **I** (46) dr = 4.6:1, er = 94:6 + **II** (6) dr = 4.6:1, er = 94:6	470
	1. R_3SiH (0.5 eq), [Rh(cod){(*R*)-BINAP}] BF_4 (5 mol%), CH_3CH_2CHO (2.0 eq) 2. H_3O^+	**I** + **II**	470

R_3Si	I	II	er I	er II
Et_2MeSi	(23)	(42)	91:9	91:9
$PhMe_2Si$	(18)	(48)	90:10	90:10
$(EtO)Me_2Si$	(0)	(29)	—	89:11

Unsaturated Ester	Conditions	Product(s) and Yield(s) (%)	Refs.
	Et_2MeSiH (0.35 eq), [Rh(cod){(*R*)-BINAP}] BF_4 (5 mol%), RCHO (0.2 eq), rt, 12 h	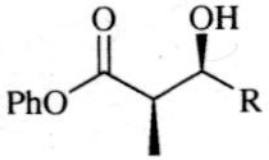	470

R	
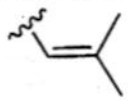	(86) syn:anti = 6:1, ee syn = 83%
	(54) syn:anti = 6:1, ee syn = 71%
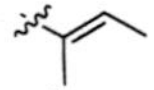	(90) syn:anti = 3:1, ee syn = 75%
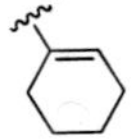	(73) syn:anti = 7:1, ee syn = 81%

Substrate	Conditions	Product(s) and Yield(s) (%)	Refs.
C_{10} 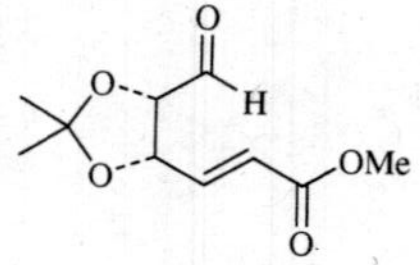	Et_3SiH (2.1 eq), $(Ph_3P)_3RhCl$ (1 mol%), MeC_6H_5, 50°, 16 h	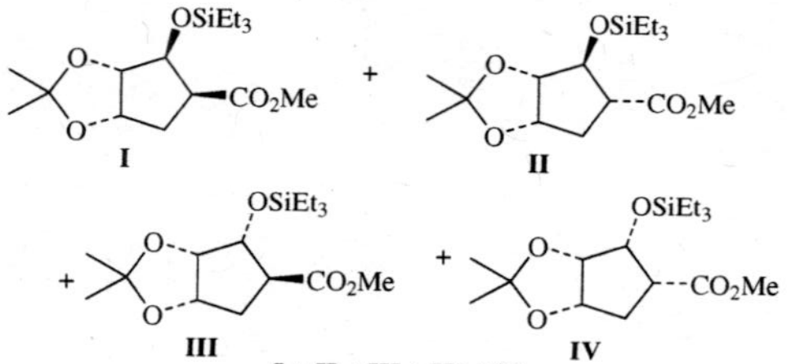 **I + II + III + IV** (65) **I:II:III:IV** = 2:5.4:2:1	471
	Et_3SiH (2.1 eq), $(Ph_3P)_4RhH$ (1 mol%), MeC_6H_5, 50°, 16 h	**I + II + III + IV** (81), **I:II:III:IV** = 5.4:4:2:1	471
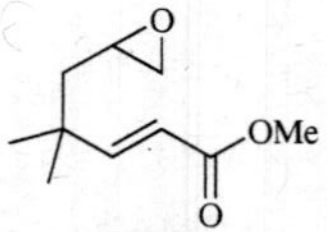	Et_3SiH (2.1 eq), $(Ph_3P)_3RhCl$ (1 mol%), MeC_6H_5, 50°, 16 h	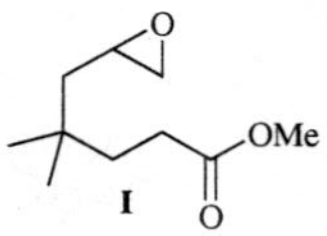 **I** (68)	471
	Et_3SiH (2.1 eq), $(Ph_3P)_4RhH$ (1 mol%), MeC_6H_5, 50°, 16 h	**I** (60)	471
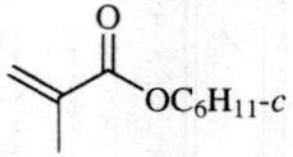	$PhSiH_3$ (1.3-1.5 eq), $Mo(CO)_6$ (3-5 mol%), THF, reflux, 35 h	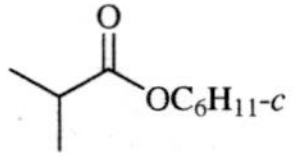 (100)	450
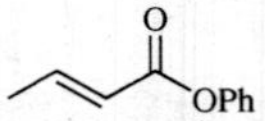	Et_2MeSiH (5 eq), $[Rh(cod)Cl]_2$ (5 mol%), (*S*)-BINAP (6.5 mol%), CH_3CH_2CHO (0.83 eq), rt, 48 h	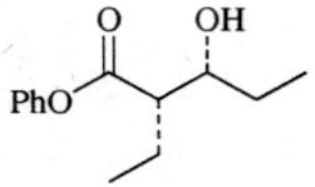 (76) syn:anti = 4.3:1, ee syn = 88%	470

TABLE 17. ORGANOSILANE REDUCTION OF α,β–UNSATURATED ESTERS (*Continued*)

Unsaturated Ester	Conditions	Product(s) and Yield(s) (%)	Refs.
C_{10}			
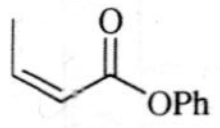	Et_2MeSiH (5 eq), $[Rh(cod)Cl]_2$ (5 mol%), (*S*)-BINAP (6.5 mol%), CH_3CH_2CHO (0.83 eq), rt, 48 h	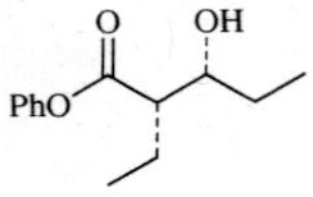 (52) syn:anti = 3.9:1, ee syn = 88%	470
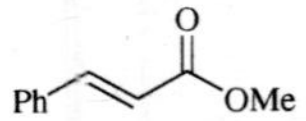	Ph_2SiH_2 (1.7 eq), $ZnCl_2$ (0.76 eq), $(Ph_3P)_4Pd$ (0.02 eq), PPh_3 (0.05 eq), rt, 48 h	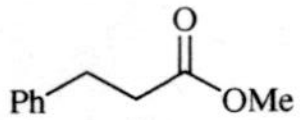 **I** (80)	436
	$PhSiH_3$ (1.5 eq), $[(Ph_3P)CuH]_6$ (5 mol%), MeC_6H_5, 20 min	**I** (95)	447
C_{11}			
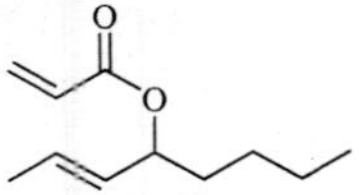	$MeCl_2SiH$, $[Rh(cod)Cl]_2$ (2.5 mol%), Me-DuPHOS (5 mol%), C_6H_6, rt, 4 h	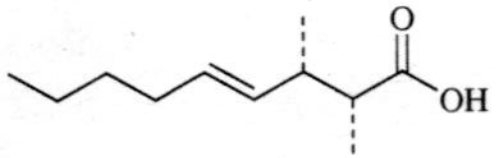 (91) dr > 25:1	474
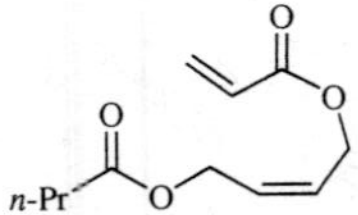	$MeCl_2SiH$, $[Rh(cod)Cl]_2$ (2.5 mol%), Me-DuPHOS (5 mol%), C_6H_6, rt, 15 h	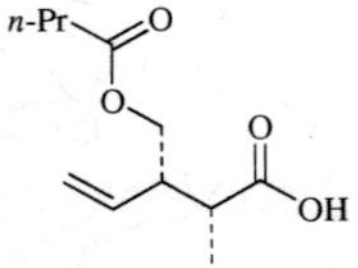 (50) dr = 23:1	474
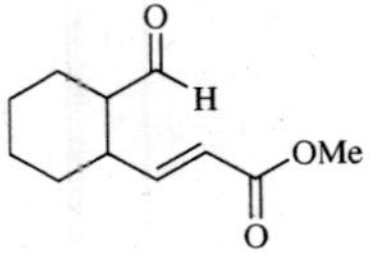	Et_3SiH (2.1 eq), $(Ph_3P)_4RhH$ (1 mol%), MeC_6H_5, 50°, 16 h	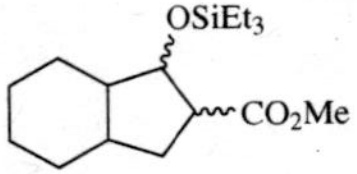 (81)	471
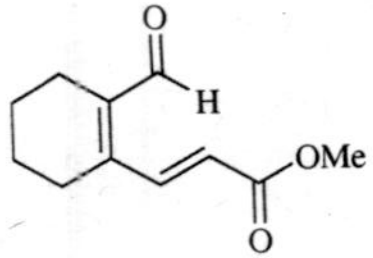	Et_3SiH (2.1 eq), $(Ph_3P)_4RhH$ (1 mol%), MeC_6H_5, 50°, 16 h	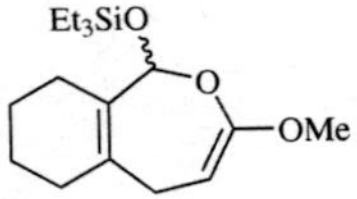 (88)	471

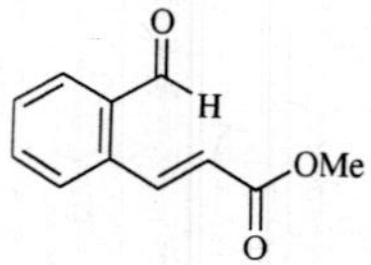	Et_3SiH (2.1 eq), $(Ph_3P)_3RhCl$ (1 mol%), MeC_6H_5, 50°, 16 h	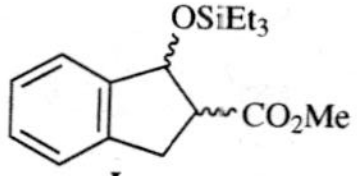 **I** (61) dr = 1.5:1	471
	Et_3SiH (2.1 eq), $(Ph_3P)_3RhCl$ (1 mol%), MeC_6H_5, 50°, 16 h	**I** (69) dr = 1:20	471
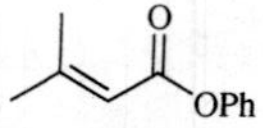	Et_2MeSiH (5 eq), $[Rh(cod)Cl]_2$ (5 mol%), (*S*)-BINAP (6.5 mol%), CH_3CH_2CHO (0.83 eq), rt, 48 h	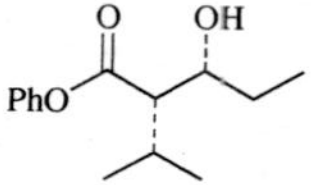 (0)	470
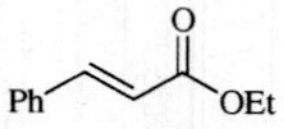	$PhMe_2SiH$ (2 eq), CuCl (x eq), DMI, rt, 5 h	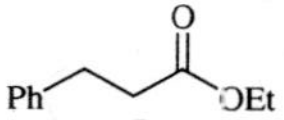 **I** x — 1 (96); 0.5 (90); 0.2 (86); 0.1 (0)	445
	$PhMe_2SiH$ (1.2 eq) CuCl (20 mol%), PPh_3 (20 mol%), TBAF (10 mol%), DMA, 0° to rt, 4.5 h	**I** (54)	444
	Cl_3SiH (2 eq), $CoCl_2$ (0.175 eq), CD_3CN, DMI (0.5 eq), 70°, 5 h	**I** (50)	451
C_{12} 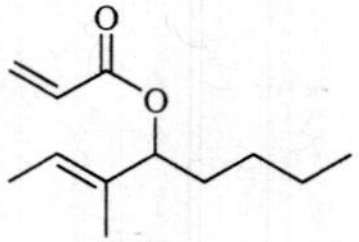	$MeCl_2SiH$, $[Rh(cod)Cl]_2$ (2.5 mol%), Me-DuPHOS (5 mol%), C_6H_6, rt, 15 h	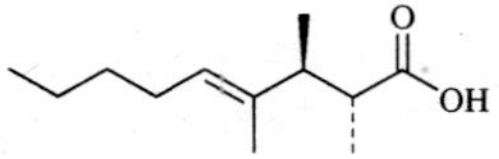 (76) dr > 25:1	474
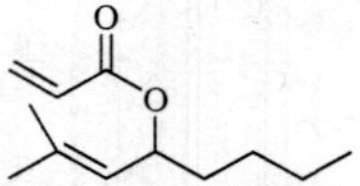	$MeCl_2SiH$, $[Rh(cod)Cl]_2$ (2.5 mol%), Me-DuPHOS (5 mol%), C_6H_6, rt, 15 h	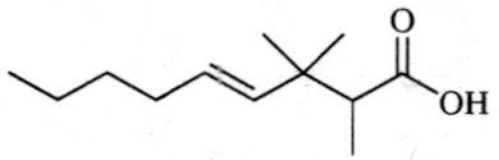 (74)	474

TABLE 17. ORGANOSILANE REDUCTION OF α,β–UNSATURATED ESTERS (*Continued*)

	Unsaturated Ester	Conditions	Product(s) and Yield(s) (%)	Refs.
C_{12}	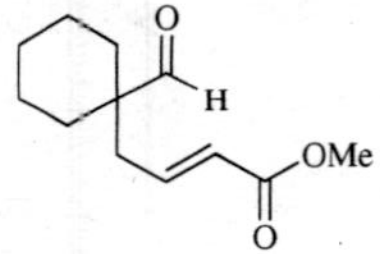	Et_3SiH (2.1 eq), $(Ph_3P)_3RhCl$ (1 mol%), MeC_6H_5, 50°, 16 h	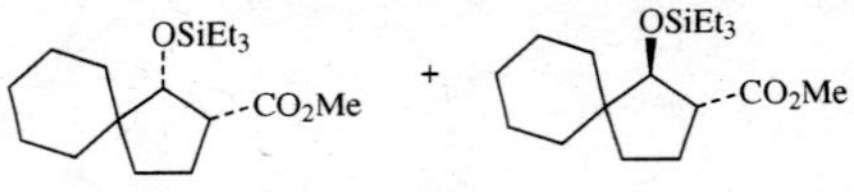 **I** **I + II** (54), **I**:**II** = 2:1 **II**	471
	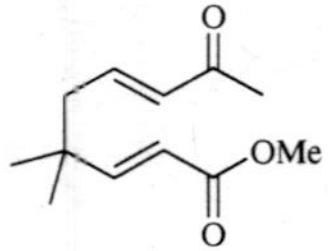	Et_3SiH (2.1 eq), $(Ph_3P)_3RhCl$ (1 mol%), MeC_6H_5, 50°, 16 h	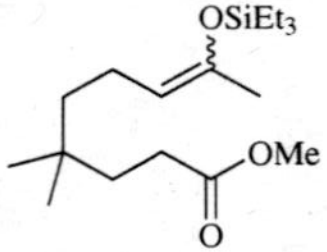 (91)	471
	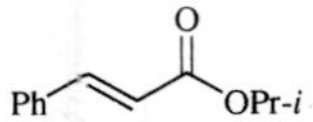	$PhSiH_3$ (1.3-1.5 eq), $Mo(CO)_6$ (3-5 mol%), THF, reflux, 24 h	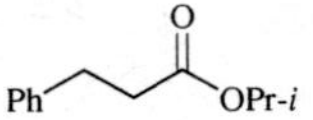 (90)	450
	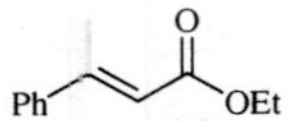	PMHS (4 eq), **209** (0.3 mol%), NaOBu-*t* (0.1 mol%), *t*-BuOH (4 eq), MeC_6H_5, rt, 1 h	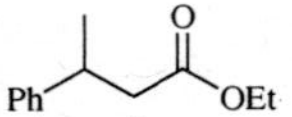 **I** (91)	454
		PMHS (4 eq), **209** (0.3 mol%), NaOBu-*t* (1.8 mol%), *t*-BuOH (4 eq), MeC_6H_5, rt, 1 h	**I** (88)	454
C_{13}	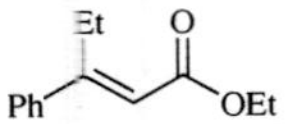	PMHS (4 eq), **209** (0.3 mol%), NaOBu-*t* (0.1 mol%), *t*-BuOH (4 eq), MeC_6H_5, rt, 1 h	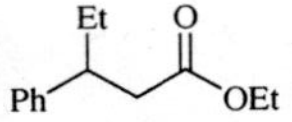 (97)	454
	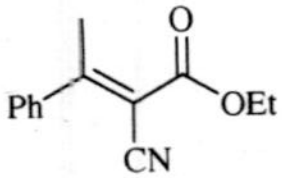	PMHS (12 eq), **209** (2 mol%), NaOBu-*t* (0.1 mol%), *t*-BuOH (12 eq), MeC_6H_5, rt, 4 h	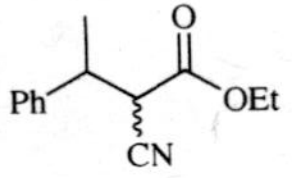 (93) dr = 1.5:1	454

C_{14}	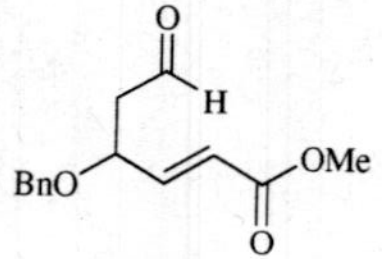	Et_3SiH (2.1 eq), $(Ph_3P)_3RhCl$ (1 mol%), MePh, 50°, 16 h	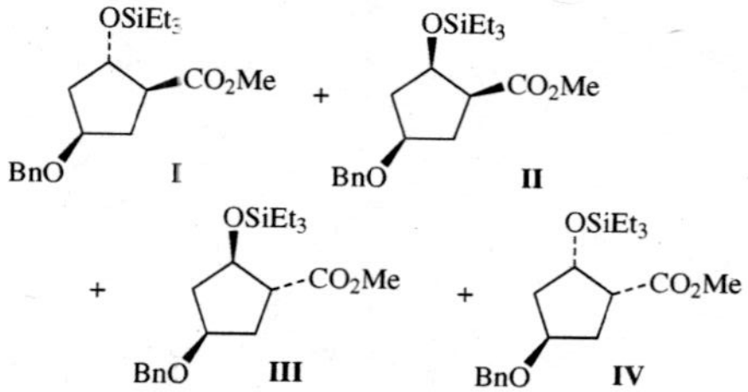 **I + II + III + IV** (81), **I:II:III:IV** = 1.7:1.5:1:1	471
		Et_3SiH (2.1 eq), $(Ph_3P)_4RhH$ (1 mol%), MePh, 50°, 16 h	**I + II + III + IV** (72), **I:II:III:IV** = 1.2:1:1.2:1.7	471
	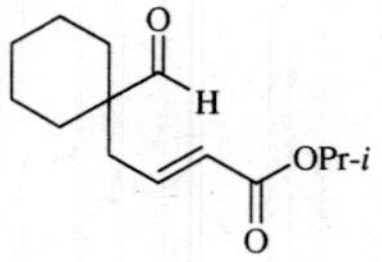	Et_3SiH (2.1 eq), $(Ph_3P)_4RhH$ (1 mol%), MePh, 50°, 16 h	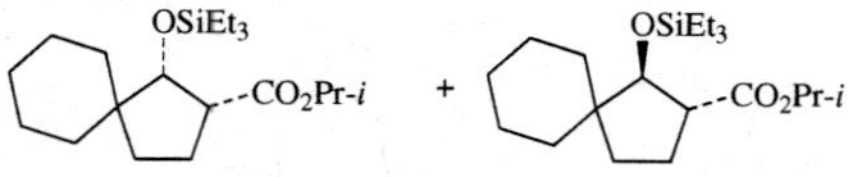 **I + II** (59), **I:II** = 1:2	471
	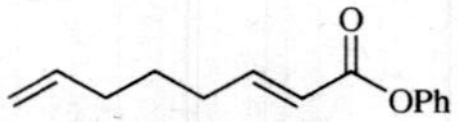	Et_2MeSiH (5 eq), $[Rh(cod)Cl]_2$ (5 mol%), (*S*)-BINAP (6.5 mol%), CH_3CH_2CHO (0.83 eq), rt, 48 h	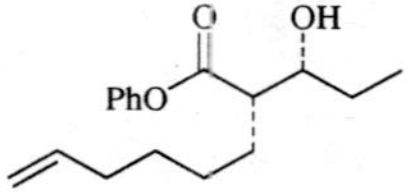 (30) syn:anti = 4.4:1, —% ee syn	470
C_{15}	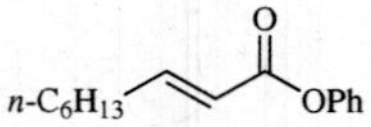	Et_2MeSiH (5 eq), $[Rh(cod)Cl]_2$ (5 mol%), (*S*)-BINAP (6.5 mol%), CH_3CH_2CHO (0.83 eq), rt, 48 h	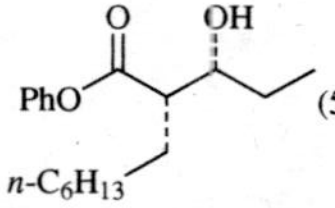 (54) syn:anti = 4.2:1, 88% ee syn	470
	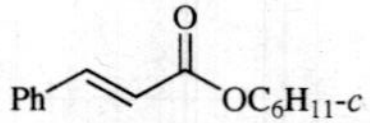	$PhSiH_3$ (1.3-1.5 eq), $Mo(CO)_6$ (3-5 mol%), THF, reflux, 30 h	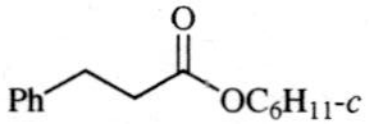 (75)	450

TABLE 17. ORGANOSILANE REDUCTION OF α,β–UNSATURATED ESTERS (*Continued*)

Unsaturated Ester	Conditions	Product(s) and Yield(s) (%)	Refs.
C_{16} 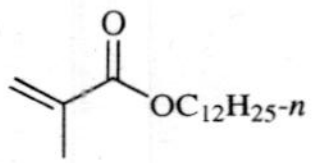	$PhMe_2SiH$ (4 eq), CuCl (2 eq), DMI, rt, 6 h	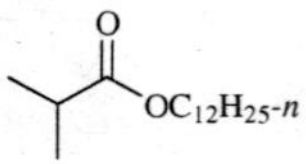 (95)	445
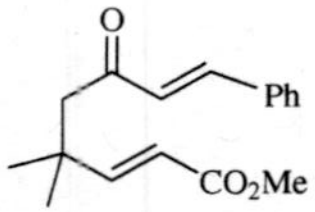	Et_3SiH (2.1 eq), $(Ph_3P)_3RhCl$ (1 mol%), MeC_6H_5, 50°, 16 h	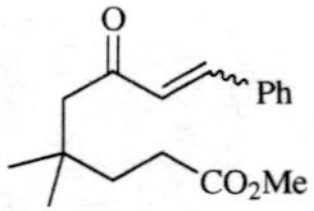 (74)	471
C_{18} 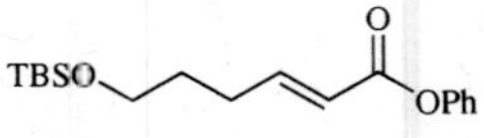	Et_2MeSiH (5 eq), $[Rh(cod)Cl]_2$ (5 mol%), (*S*)-BINAP (6.5 mol%), CH_3CH_2CHO (0.83 eq), rt, 48 h	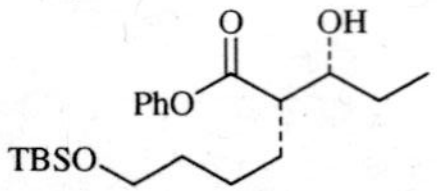 (53) syn:anti = 3.8:1, 88% ee syn	470
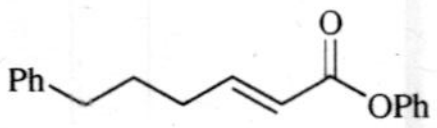	Et_2MeSiH (5 eq), $[Rh(cod)Cl]_2$ (5 mol%), (*S*)-BINAP (6.5 mol%), CH_3CH_2CHO (0.83 eq), rt, 48 h	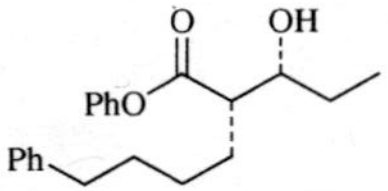 (49) syn:anti = 3.9:1, 93% ee syn	470
C_{19} 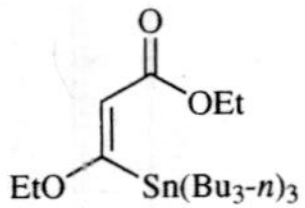	Et_3SiH (10 eq), TMSOTf (2 eq), CH_2Cl_2, –78°, 1 h	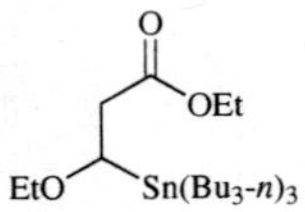 (72)	469

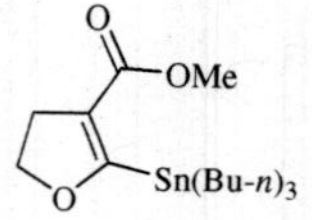

R$_3$SiH (10 eq), acid (2 eq), CH_2Cl_2, –78°, rt, 4 h

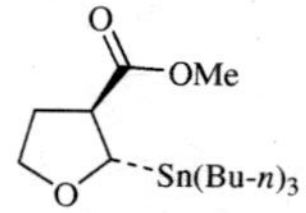

469

R_3Si	Acid	
Et_3Si	BF_3•OEt_2	(71)
Et_3Si	TFA	(96)
Et_3Si	TMSOTf	(98)
Et_3Si	HOAc	(0)
Ph_3Si	TMSOTf	(74)
Ph_3Si	TFA	(54)

Et_3SiH (10 eq), DOAc (2 eq), CH_2Cl_2, –78°, 1 h; 0°, 1 h

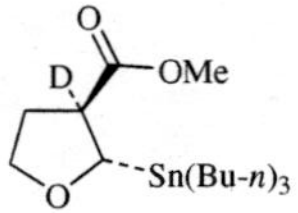

(—) 469

C_{21}

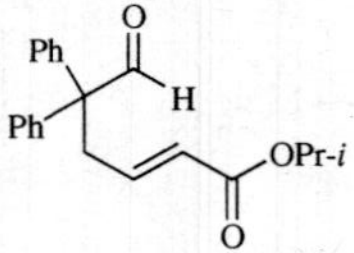

Et_3SiH (2.1 eq), $(Ph_3P)_4RhH$ (1 mol%), MePh, 50°, 16 h

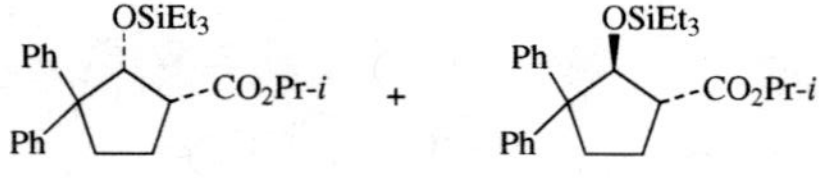

I + **II** (68), **I**:**II** = 1:2.5

471

C_{22}

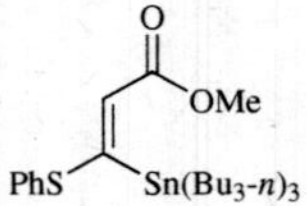

Et_3SiH (10 eq), BF_3•OEt_2 (2 eq), CH_2Cl_2, –78°, 1 h; rt, 5 d

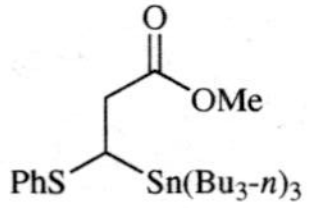

I (42) 469

Et_3SiH (10 eq), TMSOTf (2 eq), CH_2Cl_2, –78°, 1 h; 0°, 5 h; rt, 5 d — **I** (24) — 469

Et_3SiH (10 eq), TFA (2 eq), CH_2Cl_2, –78°, 1 h; 0° 1 h; rt, 4 h — **I** (24) — 469

TABLE 17. ORGANOSILANE REDUCTION OF α,β–UNSATURATED ESTERS (*Continued*)

Unsaturated Ester	Conditions	Product(s) and Yield(s) (%)	Refs.
C_{24}			
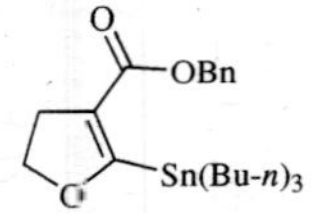	Et_3SiH (10 eq), TMSOTf (2 eq), CH_2Cl_2, –78° to rt, 1 h; 0° 1 h	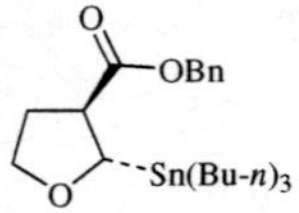 (61)	469
C_{25}			
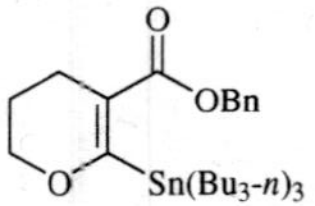	Et_3SiH (10 eq), TFA (2 eq), CH_2Cl_2, –78°, 1 h	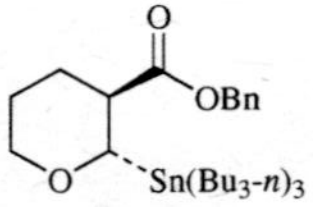 (64)	469

TABLE 18. ORGANOSILANE REDUCTION OF α,β–UNSATURATED AMIDES

Unsaturated Amide	Conditions	Product(s) and Yield(s) (%)	Refs.
C_{3-9} 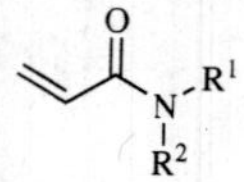	$PhSiH_3$ (1.3-1.5 eq), $Mo(CO)_6$ (3-5 mol%), THF, reflux	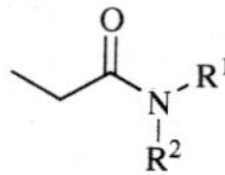	450

R^1	R^2	Time	Yield
H	H	1 h	(80)
H	*n*-Pr	0.3 h	(100)
n-Pr	*n*-Pr	0.2 h	(100)
i-Pr	*i*-Pr	0.2 h	(95)

Unsaturated Amide	Conditions	Product(s) and Yield(s) (%)	Refs.
C_{4-10} 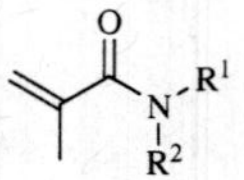	$PhSiH_3$ (1.3-1.5 eq), $Mo(CO)_6$ (3-5 mol%), THF, reflux	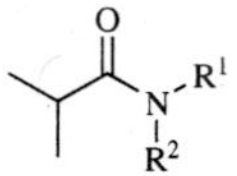	450

R^1	R^2	Time	Yield
H	H	1.7 h	(100)
H	*n*-Pr	1.3 h	(100)
n-Pr	*n*-Pr	4 h	(70)
i-Pr	*i*-Pr	15 h	(80)

Unsaturated Amide	Conditions	Product(s) and Yield(s) (%)	Refs.
C_5 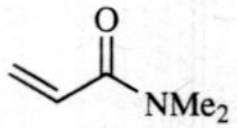	Cl_3SiH (1.3 eq), $(Ph_3P)_4Pd$, (5 mol%), RCHO, CH_2Cl_2, rt, 45 h	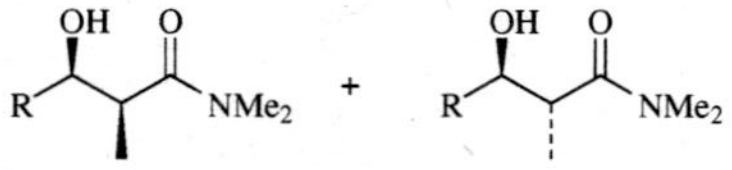 I + II	763

R	I + II	I:II
Ph	(87)	32:68
1-$C_{10}H_7$	(78)	24:76
4-MeC_6H_4	(72)	31:69
2-$C_{10}H_7$	(93)	30:70
4-$O_2NC_6H_4$	(68)	28:72
4-ClC_6H_4	(75)	36:64

Unsaturated Amide	Conditions	Product(s) and Yield(s) (%)	Refs.
$C_{5\text{-}6}$			
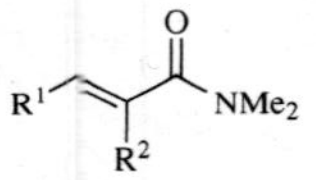	$PhSiH_3$ (2 eq), CoX_2 (0.05 mol%), R^3CHO, $ClCH_2CH_2Cl$, 20°	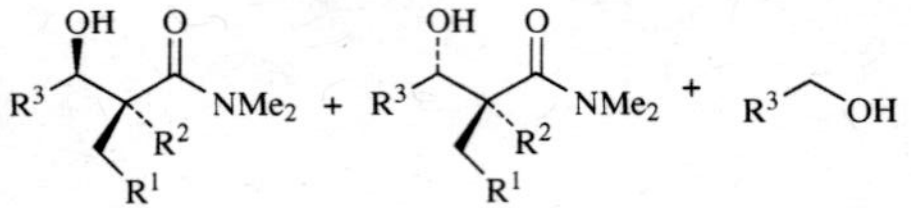 **I** + **II** + **III**	478
C_{10}			
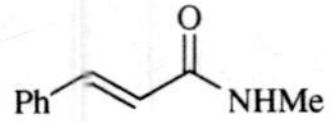	Ph_2SiH_2 (1.5 eq), $ZnCl_2$ (1.0 eq), $(Ph_3P)_4Pd$ (0.011 eq), PPh_3 (0.019 eq), rt, 10 h	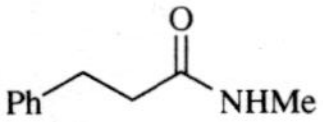 **I** (78)	436
	$PhSiH_3$ (1.3-1.5 eq), $Mo(CO)_6$ (3-5 mol%), THF, reflux, 4 h	**I** (80)	450
C_{11}			
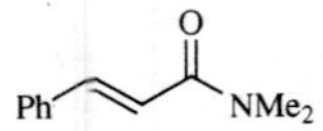	Ph_2SiH_2 (3 eq), $ZnCl_2$ (1.1 eq), $(Ph_3P)_4Pd$ (0.011 eq), PPh_3 (0.019 eq), $CHCl_3$, rt, 10 h	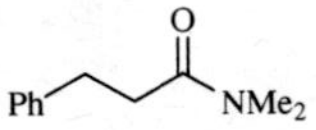 (78)	436
C_{25}			
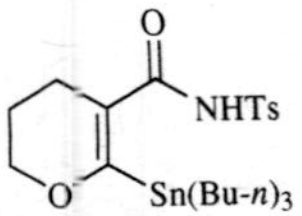	Et_3SiH (10 eq), TMSOTf (2 eq), CH_2Cl_2, –78°, 1.5 h	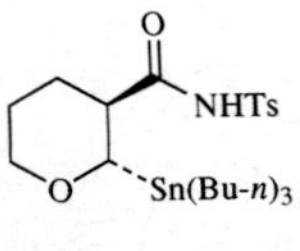 (41) + 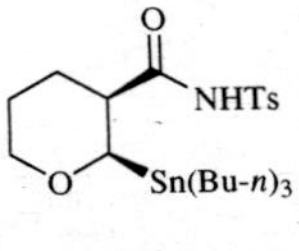 (0.5)	469

R^1	R^2	R^3	Time	I + II	I:II	III
H	H	Ph	2 h	(95)	80:20	(3)
H	H	(*E*)-PhCH=CH	4 h	(96)	72:28	(trace)
H	H	$BnCH_2$	6 h	(90)	70:30	(trace)
Me	H	Ph	3 h	(72)	70:30	(10)
Me	H	(*E*)-PhCH=CH	4 h	(68)	72:28	(12)
H	Me	Ph	5 h	(50)	—	(31)
H	Me	(*E*)-PhCH=CH	4 h	(70)	—	(14)

TABLE 19. ORGANOSILANE REDUCTION OF α,β-UNSATURATED NITRILES

Unsaturated Nitrile	Conditions	Product(s) and Yield(s) (%)	Refs.
C_{3-4}			
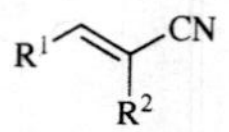	$PhSiH_3$ (2 eq), CoX_2 (0.05 mol%), PhCHO, $ClCH_2CH_2Cl$	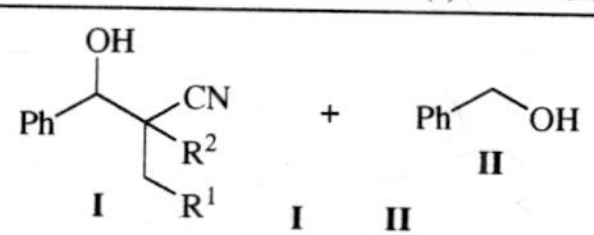	478

R^1	R^2	X	Temp	Time	I	II
H	H	acac	70°	3 h	(70)	(5)
H	H	acac	20°	24 h	(42)	(3)
H	H	dpm	20°	12 h	(93)	(2)
Me	H	acac	70°	5 h	(72)	(25)
Me	H	dpm	20°	12 h	(92)	(5)
H	Me	acac	20°	24 h	(60)	(31)

Unsaturated Nitrile	Conditions	Product(s) and Yield(s) (%)	Refs.
C_5			
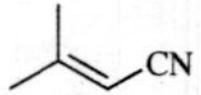	$PhSiH_3$ (1.3-1.5 eq), $Mo(CO)_6$ (3-5 mol%), THF, reflux, 12 h	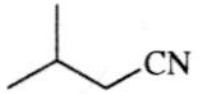 (0)	450
C_6			
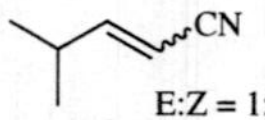 E:Z = 1:1	$PhSiH_3$ (1.3-1.5 eq), $Mo(CO)_6$ (3-5 mol%), THF, reflux, 9 h	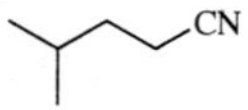 (100)	450
C_8			
	$PhSiH_3$ (1.3-1.5 eq), $Mo(CO)_6$ (3-5 mol%), THF, reflux, 4.5 h	 (100)	450
C_9			
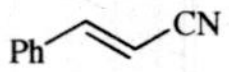	Ph_2SiH_2 (1.25 eq), $ZnCl_2$ (0.56 eq), $(Ph_3P)_4Pd$ (0.013 eq), PPh_3 (0.05 eq), rt, 72 h	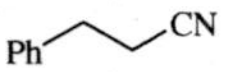 I (49)	436
	$PhSiH_3$ (1.3-1.5 eq), $Mo(CO)_6$ (3-5 mol%), THF, reflux, 12 h	**I** (90)	450
	PMHS-Pd nanocomposite, C_6H_6, rt, 5 h	**I** (90)	219
C_{16}			
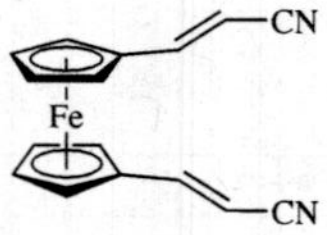	Et_3SiH (2 eq), TFA, rt, 3 h	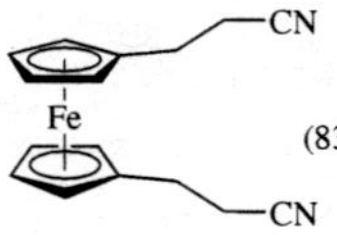 (83)	179

TABLE 20. ORGANOSILANE REDUCTION OF ACETALS, KETALS, AND HEMIKETALS

	Acetal/Ketal	Conditions	Product(s) and Yield(s) (%)	Refs.
C_5		I_2SiH_2 (2 eq), $CHCl_3$, 22°, 30 min	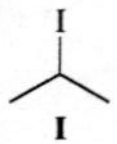 (100)	358
	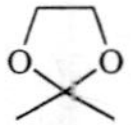	I_2SiH_2 (2 eq), $CHCl_3$, 22°, 30 min	**I** (99)	358
C_6	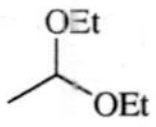	1. I_2SiH_2 (2 eq), $CHCl_3$, 22°, 30 min 2. I_2	 (100)	358
	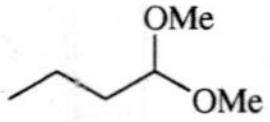	TMSH (1.1 eq), TMSOTf (0.01 eq), CH_2Cl_2, 0°, 30 min	 (100)	480
	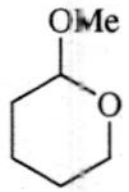	I_2SiH_2 (3 eq), $CHCl_3$, 22°, 12 h	 (100)	358
C_{6-8}	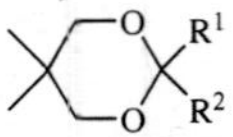 R^1,R^2 = H, H; H, Me; Me,Me	Et_3SiH, TFA, rt	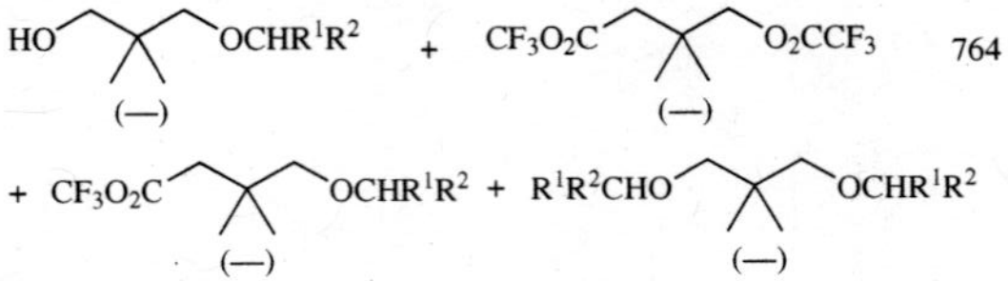	764
C_7		TMSH (1.1 eq), TMSOTf (0.01 eq), CH_2Cl_2, 0°, 30 min	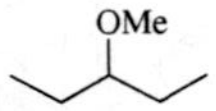 (100)[a]	480
	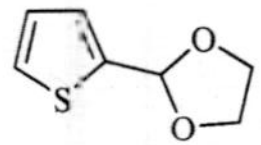	Et_3SiH, TFA, 55°, 15 h	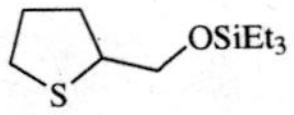 (45)	260

C$_8$

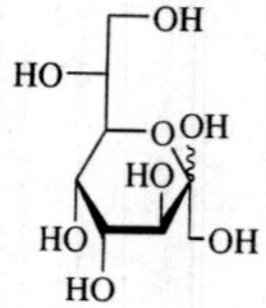	Et_3SiH, TMSOTf, MeCN	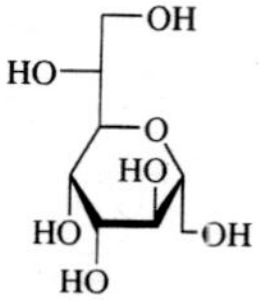 (70)	518
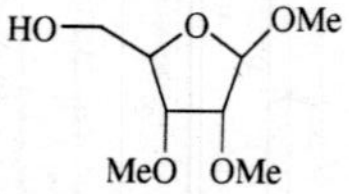	Et_3SiH (30 eq), BF_3•OEt_2 (30 eq), TFA (12 eq), 0° to rt, 16 h	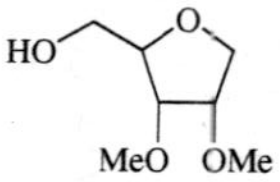 (98)	483
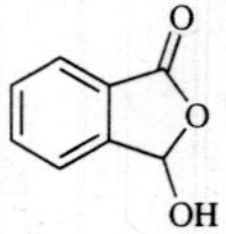	1. *i*-Bu_2AlH, hexanes, MeC_6H_5, –78° 2. Et_3SiH (1.5 eq), BF_3•OEt_2 (1.1 eq), 0°, 10 min	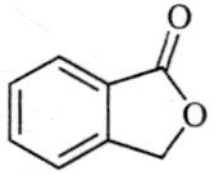 (50)	510
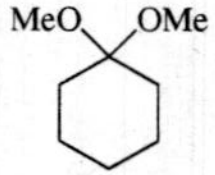	I_2SiH_2 (2 eq), $CHCl_3$, 22°, 2 h	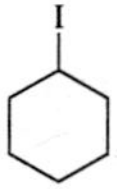 (90)	358
	Et_3SiH (1.1 eq), Nafion®-H, CH_2Cl_2, reflux, 4 h	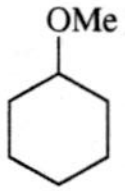 (91.7)	335
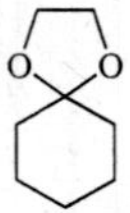	I_2SiH_2 (3 eq), $CHCl_3$, 22°, 24 h	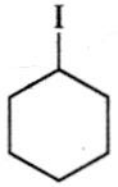 (50)	358
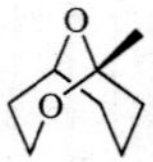	Ph_2SiH_2 (1.5 eq), $TiCl_4$ (1.2 eq), CH_2Cl_2, –78°, 15 min	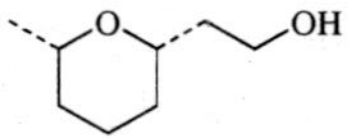 (82)	492

TABLE 20. ORGANOSILANE REDUCTION OF ACETALS, KETALS, AND HEMIKETALS (*Continued*)

Acetal/Ketal	Conditions	Product(s) and Yield(s) (%)	Refs.
C_8 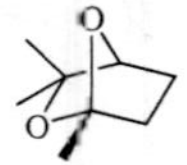	Et_3SiH (1.5 eq), $TiCl_4$ (1.2 eq), CH_2Cl_2, –78°	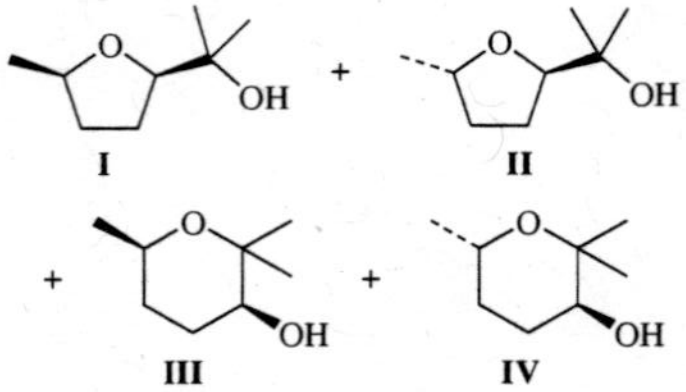 **I** + **II** + **III** + **IV** (61), **I**:**II**:**III**:**IV** = 42.6:0.8:48.7:7.9	495
	Ph_2SiH_2 (1.5 eq), $TiCl_4$ (1.2 eq), CH_2Cl_2, –78°, 15 min	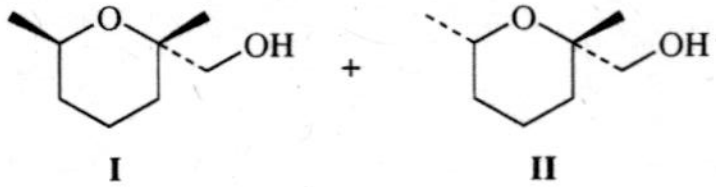 **I** + **II** (76), **I**:**II** = 7:93	495
	Et_3SiH (1.5 eq), $TiCl_4$ (1.2 eq), CH_2Cl_2, –78°	**I** + **II** (82), **I**:**II** = 1:99	495
	Et_3SiH (1.2 eq), $TiCl_4$ (1.2 eq), CH_2Cl_2, –78°, 15 min	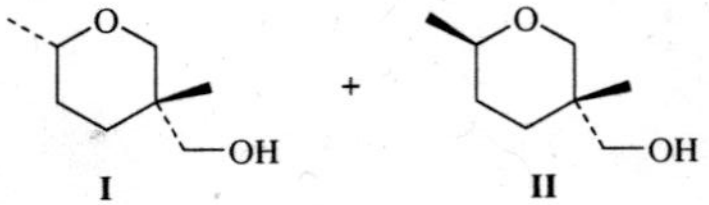 **I** + **II** (71), **I**:**II** = 63:36	495
	Et_3SiH (1.2 eq), $SnCl_4$ (1.2 eq), CH_2Cl_2, –78°	**I** + **II** (72), **I**:**II** = 76:24	495
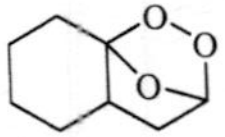	Et_3SiH, HOTf	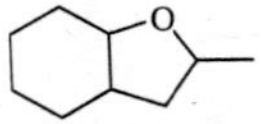 (51) single diastereomer	499

C_{8-12}

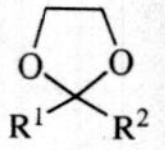

TBSH (1.4 eq), R^3OTMS (3 eq), $Sn(OTf)_2$, MeCN, −20°, 3-6 h

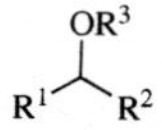

501

R^1	R^2	R^3	
—$(CH_2)_5$—		Bn	(87)
c-C_6H_{11}	H	Bn	(82)
Ph	H	Bn	(87)
Ph	H	n-C_5H_{11}	(86)
Ph	H	allyl	(80)
Ph	H	c-C_6H_{11}	(86)
4-ClC_6H_4	H	Bn	(93)
Ph	Me	Bn	(70)
3-$MeOC_6H_4$	H	Bn	(93)
3-$MeOC_6H_4$	H	allyl	(83)
4-MeC_6H_4	H	Bn	(86)
$BnCH_2$	H	Bn	(80)
$BnCH_2$	Me	Bn	(81)

TABLE 20. ORGANOSILANE REDUCTION OF ACETALS, KETALS, AND HEMIKETALS (*Continued*)

Acetal/Ketal	Conditions	Product(s) and Yield(s) (%)	Refs.
C_8			
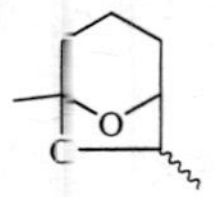 endo:exo = 40:60	Et_3SiH (10 eq), $BF_3{\bullet}OEt_2$ (10 eq), CH_2Cl_2, 0° to rt, 24 h	 **I** + **II** (85), **I**:**II** = 40:60	485
C_9			
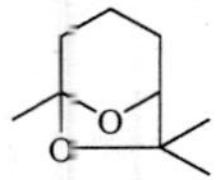	Et_3SiH (10 eq), $BF_3{\bullet}OEt_2$ (10 eq), CH_2Cl_2, 0° to rt, 24 h	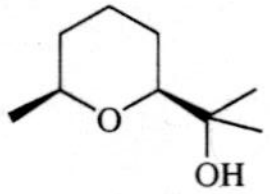 (21) + 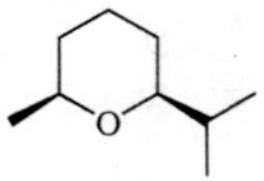 (45)	485
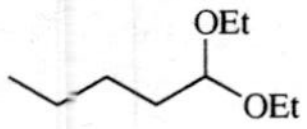	Et_3SiH (1.2 eq), TFA (30 eq), 50°, 8-10 h	 (75)	327
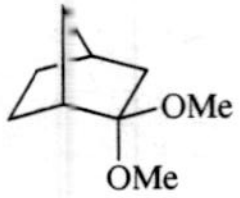	Et_3SiH (1.1 eq), Nafion®-H, CH_2Cl_2, reflux, 2 h	 (99)	335
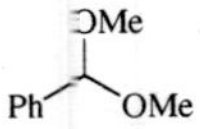	TMSH (1.1 eq), TMSOTf (0.01 eq), CH_2Cl_2, 0°, 30 min	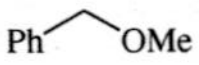 **I** (96)	480
	Et_3SiH (1.1 eq), TMSOTf (0.01 eq), CH_2Cl_2, 0°, 30 min; 29°, 16 h	**I** (90)	480
	Et_3SiH (1.1 eq), Nafion®-H, CH_2Cl_2, reflux, 2 h	**I** (96)	335

	Et_3SiH (5 eq), Sn-mont, CH_2Cl_2, rt, 10 min	**I** (93)	353
	Et_3SiH (4 eq), TCNE (0.3 eq), MeCN, reflux	**I** (46)	500
	Et_3SiH, FSO_3H, MeC(OTMS)=NHTMS	**I** (76)	487
	Et_3SiH, FSO_3H, TMSNHCONHTMS	**I** (97)	487
C_9 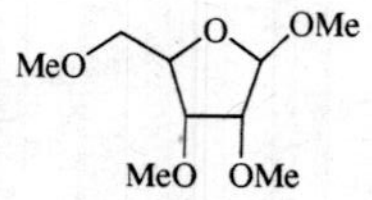	Et_3SiH (30 eq), BF_3•OEt_2 (30 eq), TFA (12 eq), 0° to rt, 16 h	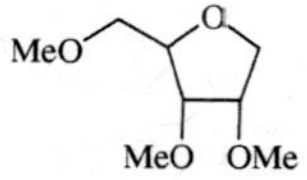 (95)	483
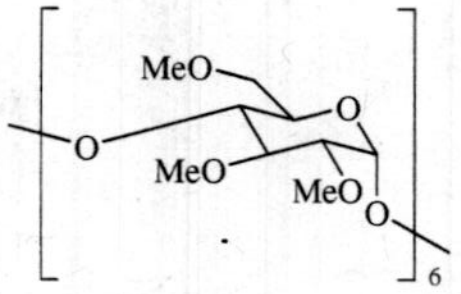	1. Et_3SiH (30 eq), BF_3•OEt_2 (30 eq), TFA (12 eq), 0° 2. Ac_2O (12 eq), rt, 24 h	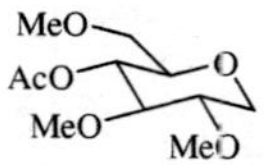 (81)	483
C_{9-10} 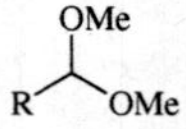	$Et_2NSiHMe_2$ (1.2 eq), $TiCl_4$ (0.2 eq), CH_2Cl_2, 0° to rt, 36 h	 R: Ph (79); Bn (31)	359
C_{9-12} 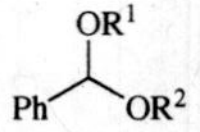	Et_3SiH (1.56 eq), $SnBr_2$ (0.056 mol%), AcBr (2.2 eq), CH_2Cl_2, rt, 3 h		506
R^1 = Me, R^2 = Me		(89)	
R^1 = Et, R^2 = Et		(91)	
—$(CH_2)_4$—		(96)	
—$(CH_2)_5$—		(82)	
R^1 = Ac, R^2 = Ac		(77)	

TABLE 20. ORGANOSILANE REDUCTION OF ACETALS, KETALS, AND HEMIKETALS (*Continued*)

Acetal/Ketal	Conditions	Product(s) and Yield(s) (%)	Refs.
C_{9-13}			
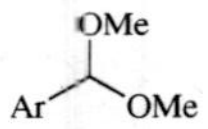	Et_3SiH (1.56 eq), SnX_2 (0.056 mol%), AcBr (2.2 eq), CH_2Cl_2		506

Ar	X	Temp	Time	Yield (%)
Ph	Cl	rt	3 h	(76)
Ph	Br	rt	3 h	(89)
3-$MeOC_6H_4$	Cl	rt	3 h	(97)
3-$MeOC_6H_4$	Br	rt	2 h	(99)
4-MeC_6H_4	Cl	rt	3 h	(89)
4-MeC_6H_4	Br	rt	3 h	(86)
4-ClC_6H_4	Cl	rt	12 h	(79)
4-ClC_6H_4	Br	rt	2 h	(87)
4-BrC_6H_4	Cl	rt	5 h	(92)
4-BrC_6H_4	Br	rt	7 h	(97)
1-$C_{10}H_7$	Cl	0	2 h	(92)
1-$C_{10}H_7$	Br	0	1 h	(96)
2-$C_{10}H_7$	Cl	0	5 h	(91)
2-$C_{10}H_7$	Br	0	3 h	(94)

Acetal/Ketal	Conditions	Product(s) and Yield(s) (%)	Refs.
C_{9-20}			
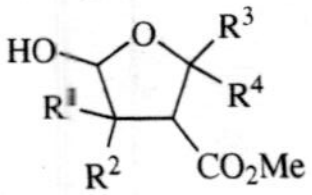	Et_3SiH, $BF_3•OEt_2$, CH_2Cl_2, –78° to rt, 4 h	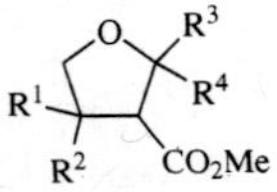	512

R^1	R^2	R^3	R^4	
Me	Me	H	Me	(61)
Me	Me	Me	Me	(43)
—$(CH_2)_5$—		H	Me	(51)
Me	Me	H	Ph	(79)
—$(CH_2)_5$—		—$(CH_2)_5$—		(48)
Me	Me	Ph	Ph	(40)

C_{9-16}

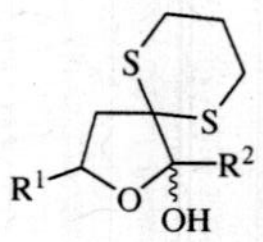

Ph$_3$SiH (1 eq), $TiCl_4$ (1.2 eq), CH_2Cl_2, −78°, 5 min

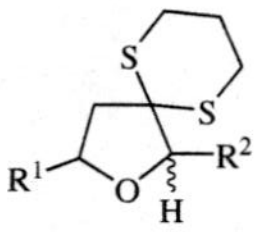

R^1	R^2	trans:cis	
Me	Me	10:1	(70)
Me	Et	7:1	(63)
Me	Ph	13:1	(83)
Ph	Me	63:1	(72)
Et	Ph	16:1	(81)
i-Pr	Ph	9:1	(82)

520

TABLE 20. ORGANOSILANE REDUCTION OF ACETALS, KETALS, AND HEMIKETALS (*Continued*)

Acetal/Ketal	Conditions	Product(s) and Yield(s) (%)	Refs.
C_{10}			
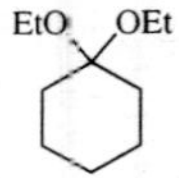	Et_3SiH (1.2 eq), TFA (30 eq), 50°, 8-10 h	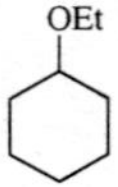 **I** (89)	327
	Et_3SiH (2 eq), $AlCl_3$ (1 eq), HCl, 20°, 1 h	**I** (93)	146
	Et_3SiH (2 eq), $AlCl_3$ (1 eq), CH_2Cl_2, rt, 1 h	**I** (93)	136
	Et_3SiH (2 eq), $AlCl_3$ (1 eq), HCl, CH_2Cl_2, rt, 1 h	**I** (50)	136
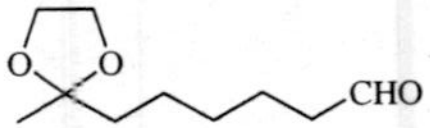	I_2SiH_2 (1.2 eq), CH_2Cl_2, rt, 4 h	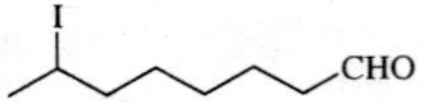 (94)	505
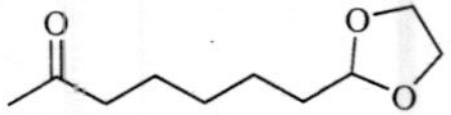	I_2SiH_2 (1.2 eq), CH_2Cl_2, rt, 4 h	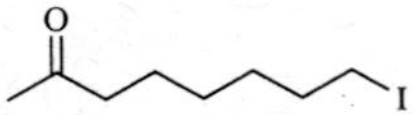 (81)	505
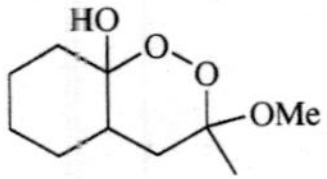	Et_3SiH, HOTf	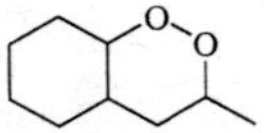 (39)[b]	499
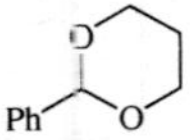	$PhSiH_3$ (3 eq), $(Ph_3P)_3RhCl$ (0.025 mol%), THF, rt, 48 h	 (91)	493
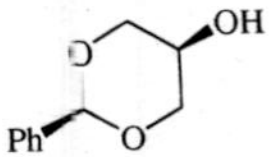	$PhSiH_3$ (4 eq), $(Ph_3P)_3RhCl$ (0.025 mol%), THF, 60°, 48 h	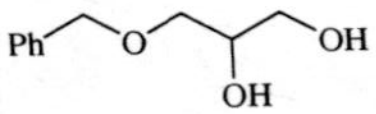 (86)	493
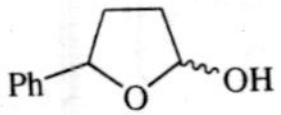	Et_3SiH, BF_3•OEt_2, –45°, 5 min	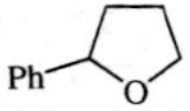 (75)	510

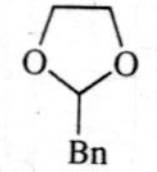	$PhSiH_3$ (3 eq), $(Ph_3P)_3RhCl$ (0.025 mol%), THF, 60°, 72 h	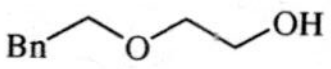 (66)	493
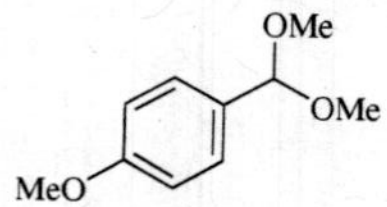	Et_3SiH (4 eq), TCNE (0.3 eq), MeCN, reflux	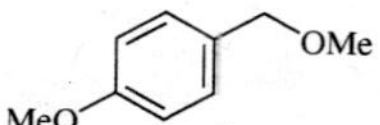 **I** (85)	500
	$PhSiH_3$ (3 eq), $(Ph_3P)_3RhCl$ (0.025 mol%), THF, 60°, 48 h	**I** (77)	493
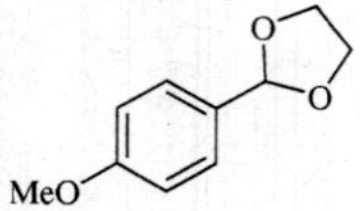	I_2SiH_2 (2 eq), $CHCl_3$, 0°, <2 min	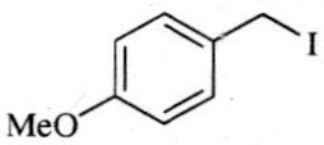 (100)	358
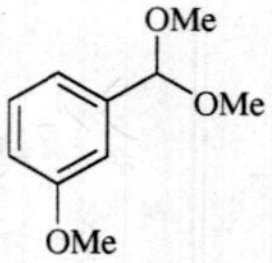	1. Et_3SiH, $SnBr_2$-AcBr, CH_2Cl_2, rt, 2.5 h 2. $(n\text{-}Bu)_3SnH$, AIBN, C_6H_6, reflux, 0.5 h	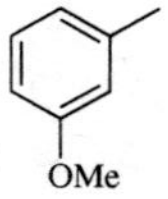 **I** (90)	479
	1. Et_3SiH, $SnBr_2$-AcBr, CH_2Cl_2, rt, 2.5 h 2. $LiAlH_4$, THF, rt, 2.5 h	**I** (92)	479
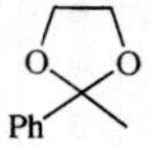	$PhSiH_3$ (3 eq), $(Ph_3P)_3RhCl$ (0.025 mol%), THF, rt, 48 h	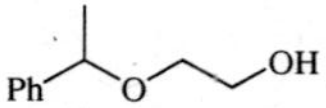 (81)	493
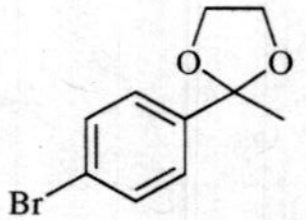	I_2SiH_2 (1 eq), $CHCl_3$, 0°, 60 min	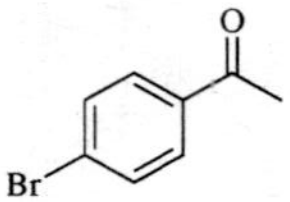 (35) + 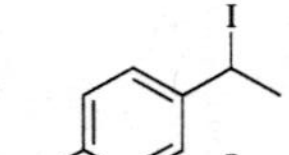 **I** (65)	358
	I_2SiH_2 (2 eq), $CHCl_3$, 22°, 2 min	**I** (100)	358

TABLE 20. ORGANOSILANE REDUCTION OF ACETALS, KETALS, AND HEMIKETALS (*Continued*)

Acetal/Ketal	Conditions	Product(s) and Yield(s) (%)	Refs.
C_{10}			
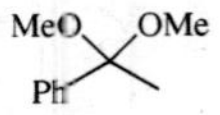	Et_3SiH (1.1 eq), Nafion®-H, CH_2Cl_2, reflux, 2 h	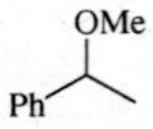 (91.2)	335
C_{10-15}			
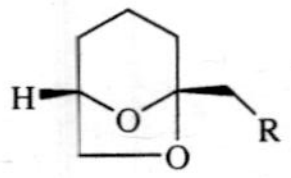 R: CO_2Et; Ph; CH_2OBn	Et_3SiH (4 eq), $TiCl_4$ (1.2 eq), CH_2Cl_2, –78°, 3 min	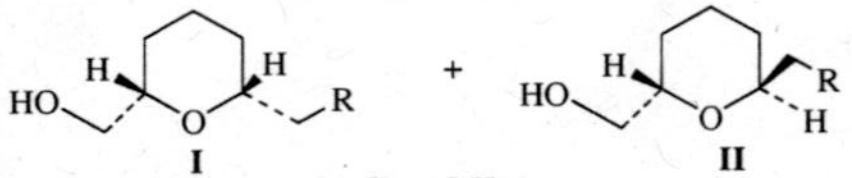 I + II / I:II (59) 99:1 (92) 99:1 (100) >99:1	489
C_{10-35}			
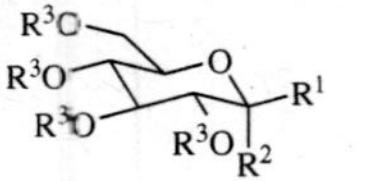 R^1 / R^2 / R^3: H, OH, Me H, OMe, Me OMe, H, Me OMe, H, Bn	Et_3SiH (30 eq), BF_3•OEt_2 (30 eq), TFA (12 eq), 0° to rt, 16 h	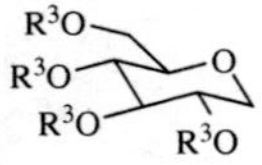 (96) (97) (91) (92)	483
C_{11}			
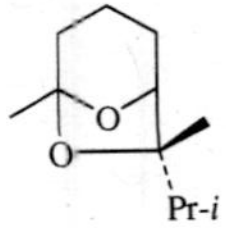	Et_3SiH (10 eq), BF_3•OEt_2 (10 eq), CH_2Cl_2, 0° to rt, 24 h	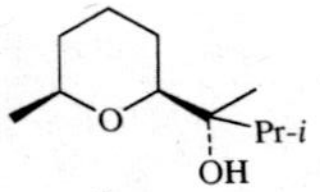 (70) + 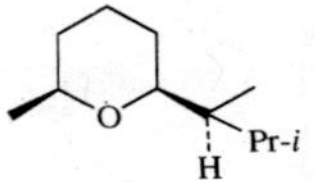 (11) + 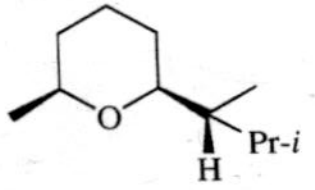 (13)	485

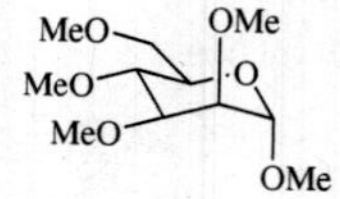	Et_3SiH (30 eq), BF_3•OEt_2 (30 eq), TFA (12 eq), 0° to rt, 16 h	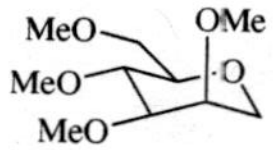	(100)	483
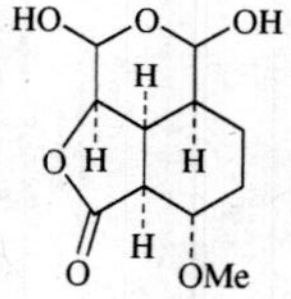	Et_3SiH, TFA, 70°, 8 h	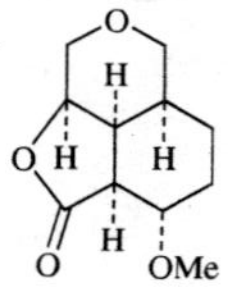	(65)	515
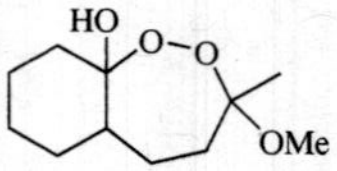	Et_3SiH, HOTf	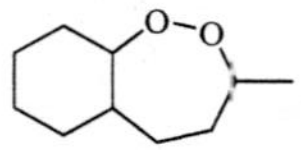	(82)	499
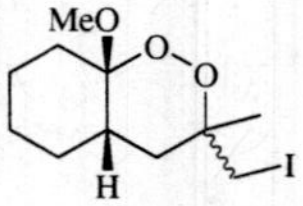	Et_3SiH, HOTf	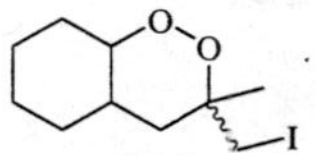	(59)	499
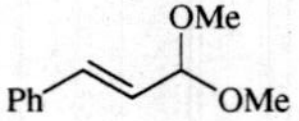	Et_3SiH (4 eq), TCNE (0.3 eq), MeCN, reflux	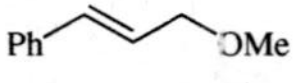	(70)	500

TABLE 20. ORGANOSILANE REDUCTION OF ACETALS, KETALS, AND HEMIKETALS (*Continued*)

Acetal/Ketal	Conditions	Product(s) and Yield(s) (%)	Refs.
C_{11}			
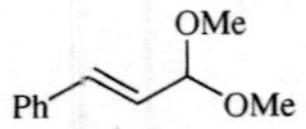	1. Et_3SiH, $SnBr_2$-AcBr, CH_2Cl_2, 0°, 5 h 2. $LiAlH_4$, THF, rt, 3.5 h	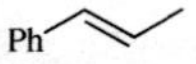 (92)	479
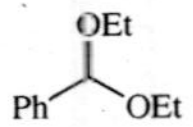	Et_3SiH (1.2 eq), TFA (30 eq), 50°, 8-10 h	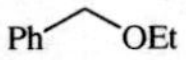 **I** (98)	327
	Et_3SiH (1.1 eq), Nafion®-H, CH_2Cl_2, reflux, 1 h	**I** (97)	335
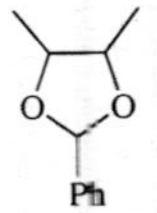	I_2SiH_2 (2 eq), $CHCl_3$, 0°, <2 min	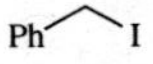 (100)	358
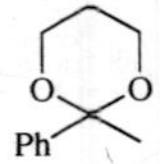	$PhSiH_3$ (3 eq), $(Ph_3P)_3RhCl$ (0.025 mol%), THF, rt, 48 h	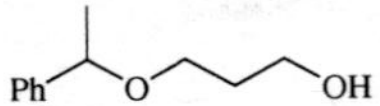 (89)	493
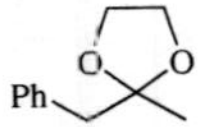	I_2SiH_2 (3 eq), $CHCl_3$, 22°, 2 h	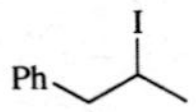 (90)	358
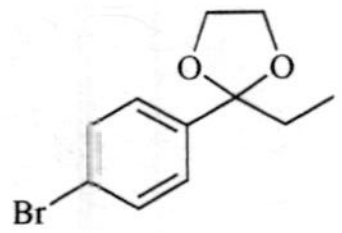	I_2SiH_2 (2 eq), $CHCl_3$, 0°, <2 min	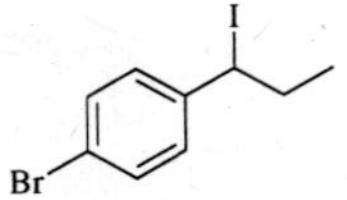 (100)	358
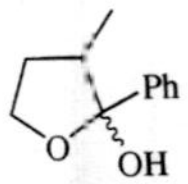	Et_3SiH (3 eq), TFA (3 eq), CH_2Cl_2, –78° to 0°	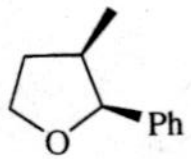 **I** + 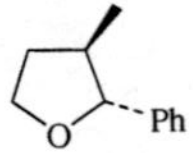 **II** **I** + **II** (74), **I**:**II** = 1:1	162, 514

C$_{11-16}$

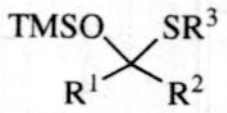

Et$_3$SiH (1.2 eq), TMSCl (50 mol%), InCl$_3$ (20 mol%), CH$_2$Cl$_2$, rt, 5 h

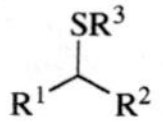

426, 502

R[1]	R[2]	R[3]	Yield (%)
—(CH$_2$)$_5$—		Et	(96)
Ph	H	Et	(83)
Ph	H	*i*-Pr	(88)
BnCH$_2$	H	Et	(68)
Ph	Me	Et	(97)
n-C$_8$H$_{17}$	H	Et	(70)
Ph	H	Ph	(77)

C$_{11-18}$

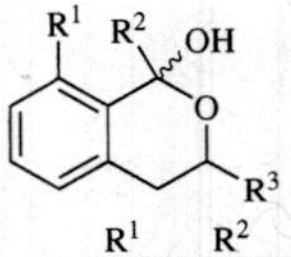

Et$_3$SiH (6 eq), TFA (6 eq), CH$_2$Cl$_2$, –78° to rt, 3 h

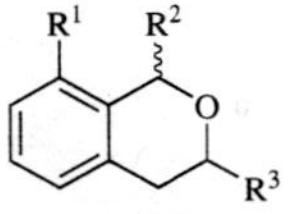

514

R[1]	R[2]	R[3]	Yield (%)	cis:trans
H	Me	Me	(88)	100:0
OH	Me	Me	(93)	100:0
H	Me	*i*-Pr	(70)	100:0
H	Ph	Me	(86)	10:1
H	Ph	*i*-Pr	(75)	100:0

C$_{12}$

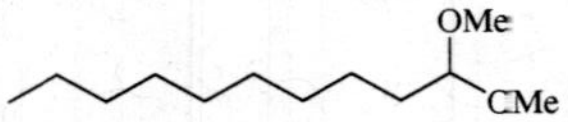

Et$_3$SiH (1.1 eq), Nafion®-H, CH$_2$Cl$_2$, reflux, 2 h

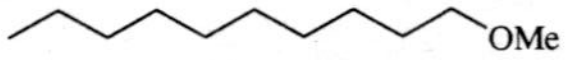

(95)

335

Acetal/Ketal	Conditions	Product(s) and Yield(s) (%)	Refs.
C_{12}			
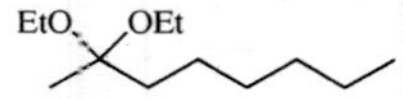	Et_3SiH (2 eq), $AlCl_3$ (1 eq), HCl, 20°, 1 h	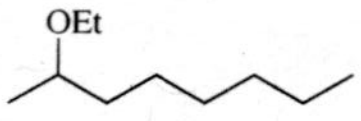 **I** (95)	146
	Et_3SiH (2 eq), $AlCl_3$ (1 eq), HCl, CH_2Cl_2, rt, 1 h	**I** (96)	136
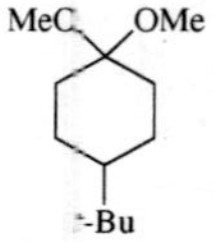	TMSH (1.1 eq), TMSOTf (3.4 mol%), CH_2Cl_2, 0°, 30 min	 (89) cis:trans = 44:56	480
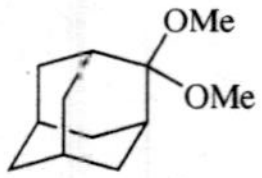	Et_3SiH (1.1 eq), Nafion®-H, CH_2Cl_2, reflux, 2 h	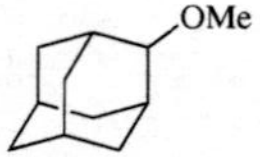 (94)	335
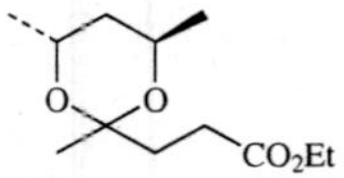	Et_3SiH (1.2 eq), $TiCl_4$ (1.0 eq), CH_2Cl_2, –78°	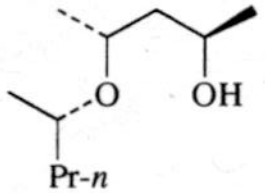 **I** + 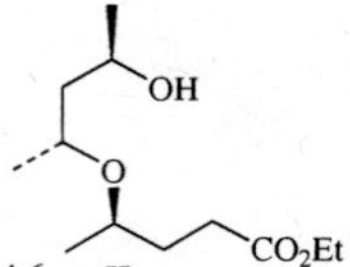 **II**; **I** + **II** (85), **I**:**II** = 94:6	491
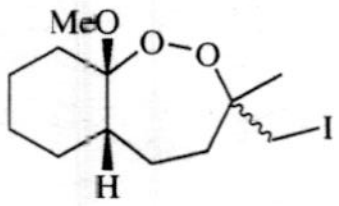	Et_3SiH, HOTf	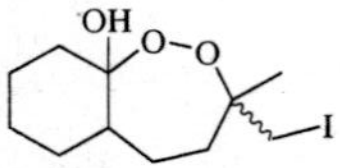 (56)	499
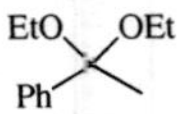	Et_3SiH (1.1 eq), Nafion®-H, CH_2Cl_2, reflux, 2 h	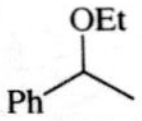 (97)	335

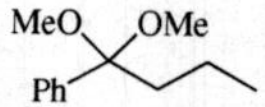	1. Et_3SiH, $SnBr_2$-AcBr, CH_2Cl_2, 0°, 2 h 2. $(n\text{-}Bu)_3SnH$, AIBN, C_6H_6, reflux, 0.5 h	 I (79)	479
	1. Et_3SiH, $SnBr_2$-AcBr, CH_2Cl_2, 0°, 2.5 h 2. $LiAlH_4$, THF, rt, 15 h	I (82)	479
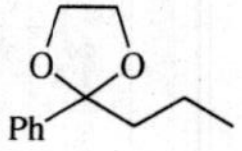	1. Et_3SiH, $SnBr_2$-AcBr, CH_2Cl_2, rt, 2 h 2. $(n\text{-}Bu)_3SnH$, AIBN, C_6H_6, reflux, 0.5 h	I (77)	479
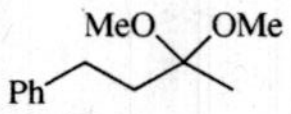	1. Et_3SiH, $SnBr_2$-AcBr, CH_2Cl_2, rt, 4 h 2. $(n\text{-}Bu)_3SnH$, AIBN, C_6H_6, reflux, 0.5 h	I (60) + 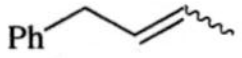 (25)	479
	Et_3SiH (1.56 eq), $SnBr_2$ (0.056 mol%), AcBr (2.2 eq), CH_2Cl_2, rt, 3 h	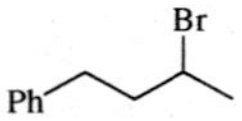 (71)	506
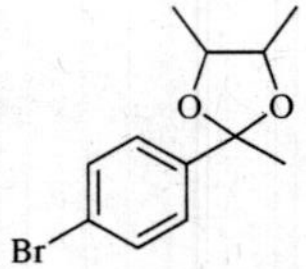	I_2SiH_2 (2 eq), $CHCl_3$, 0°, 9 min	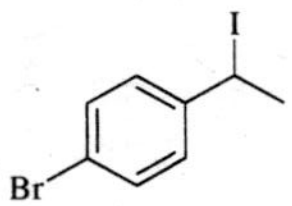 (100)	358
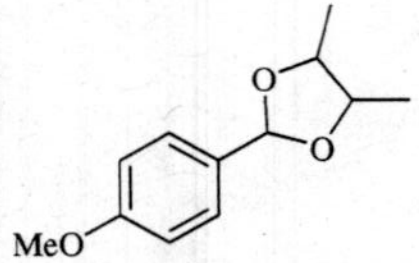	I_2SiH_2 (2 eq), $CHCl_3$, 0°, <2 min	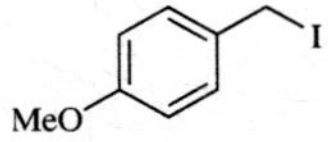 (100)	358

TABLE 20. ORGANOSILANE REDUCTION OF ACETALS, KETALS, AND HEMIKETALS (*Continued*)

Acetal/Ketal	Conditions	Product(s) and Yield(s) (%)	Refs.
C_{12}			
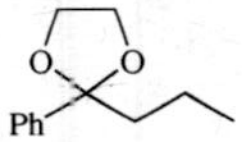	1. Et_3SiH, $SnBr_2$-AcBr, CH_2Cl_2, rt, 2 h 2. (n-Bu)$_3$SnH, AIBN, C_6H_6, reflux, 0.5 h	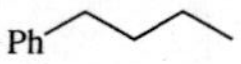 (77)	479
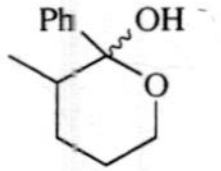	Et_3SiH (3 eq), TFA (3 eq), CH_2Cl_2, –78° to 0°	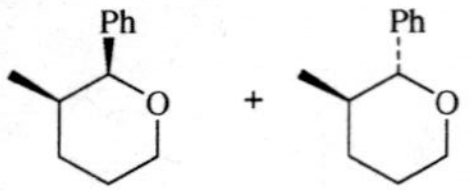 I + II **I** + **II** (75), **I**:**II** = 1:100	162, 514
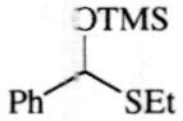	TMSH (1.1 eq), TMSI (1 mol%), TMSOTf (1.6 mol%), CH_2Cl_2, 0°, 1 h; rt, 8 h	 (100)	392
C_{13}			
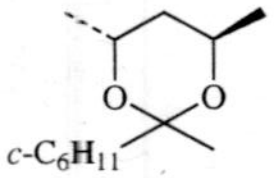	Et_3SiH (x eq), acid (y eq), CH_2Cl_2, –78°	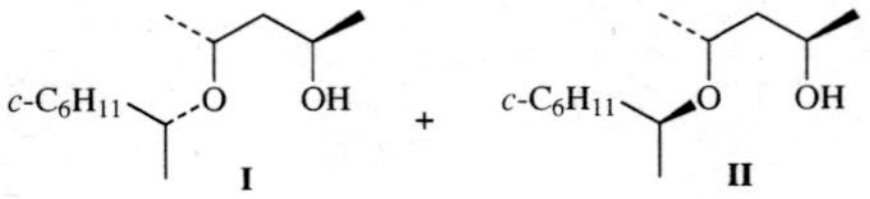 I + II	491

Acid	x	y	Time	I + II	I:II
$TiCl_4$	1.2	1.2	0.5 h	(85)	98:2
$SnCl_4$	1.0	1.0	0.5 h	(93)	65:35
$AlCl_3$	1.2	1.2	8 h	(69)	66:34
BF_3•OEt_2	1.0	1.0	5 h	(93)	93:7

Acetal/Ketal	Conditions	Product(s) and Yield(s) (%)	Refs.
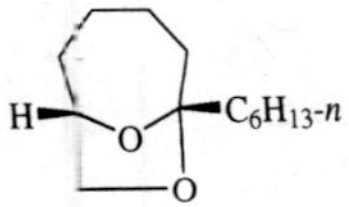	Et_3SiH (4 eq), $TiCl_4$ (1.2 eq), CH_2Cl_2, –78°, 0.5 h	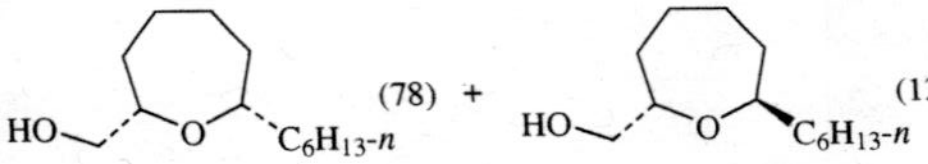 (78) + (13)	488

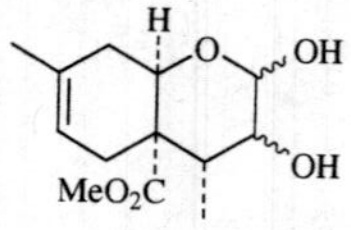	Et_3SiH, BF_3•OEt_2, –78° to –45°, 1 h	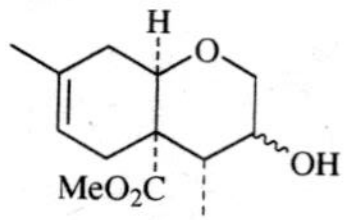	(53)	510
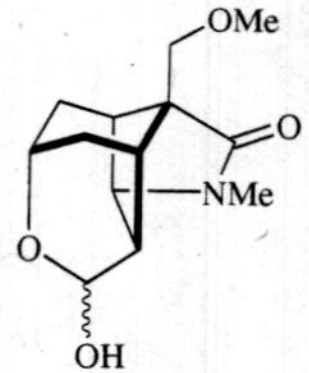	Et_3SiH, TFA, CH_2Cl_2	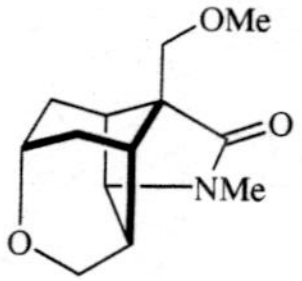	(84)	516
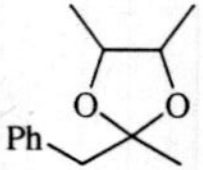	I_2SiH_2 (2 eq), $CHCl_3$, 22°, 2 h	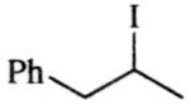	(95)	358
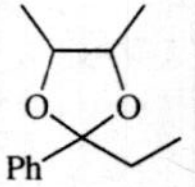	I_2SiH_2 (2 eq), $CHCl_3$, 0°, <2 min	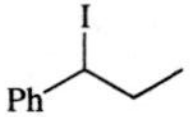	(100)	358
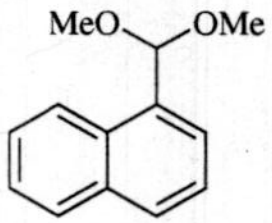	1. Et_3SiH, $SnBr_2$-AcBr, CH_2Cl_2, 0°, 1.5 h 2. (n-Bu)$_3$SnH, AIBN, C_6H_6, reflux, 0.5 h	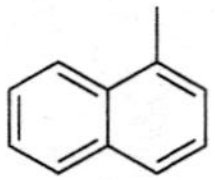 **I**	(85)	479
	1. Et_3SiH, $SnBr_2$-AcBr, CH_2Cl_2, 0°, 1 h 2. $LiAlH_4$, THF, rt, 2.5 h	**I**	(84)	479
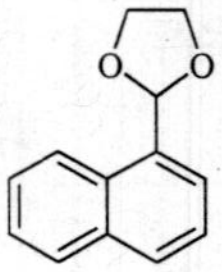	1. Et_3SiH, $SnBr_2$-AcBr, CH_2Cl_2, rt, 4 h 2. (n-Bu)$_3$SnH, AIBN, C_6H_6, reflux, 0.75 h	**I**	(72)	479

TABLE 20. ORGANOSILANE REDUCTION OF ACETALS, KETALS, AND HEMIKETALS (*Continued*)

Acetal/Ketal	Conditions	Product(s) and Yield(s) (%)	Refs.
C_{13} 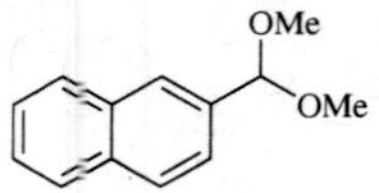	1. Et_3SiH, $SnBr_2$-AcBr, CH_2Cl_2, 0°, 3.5 h 2. $(n\text{-}Bu)_3SnH$, AIBN, C_6H_6, reflux, 0.75 h	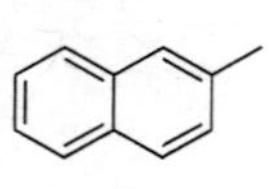 **I** (85)	479
	1. Et_3SiH, $SnBr_2$-AcBr, CH_2Cl_2, 0°, 3.5 h 2. $LiAlH_4$, THF, rt, 2.5 h	**I** (86)	479
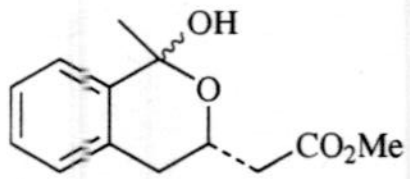	Et_3SiH (6 eq), TFA (6 eq), CH_2Cl_2, –78° to rt, 3 h	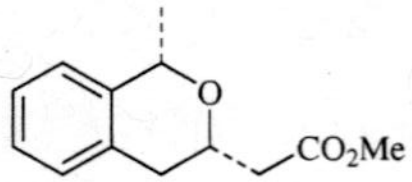 (95)	514
C_{14} 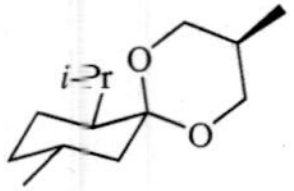	Et_3SiH, $TiCl_4$, CH_2Cl_2, –78°	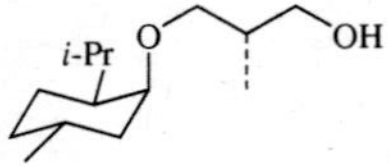 (94)	490
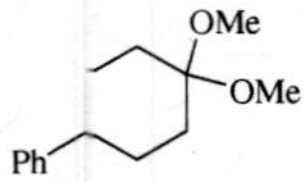	1. Et_3SiH, $SnBr_2$-AcBr, CH_2Cl_2, rt, 24 h 2. $(n\text{-}Bu)_3SnH$, AIBN, C_6H_6, reflux, 0.5 h	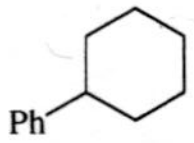 (67)	479
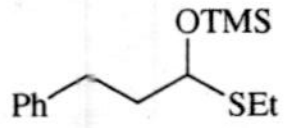	Et_3SiH (1.2 eq), TMSCl (50 mol%), $InCl_3$ (20 mol%), CH_2Cl_2, rt, 5 h	 (68)	426, 502
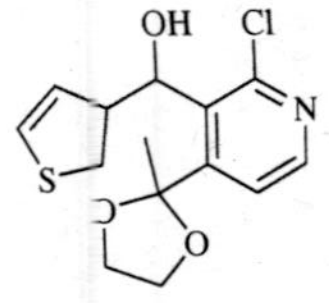	Et_3SiH (1.1 eq), TFA, rt, 18 h	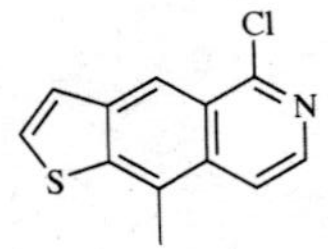 (60)	486

C_{15}

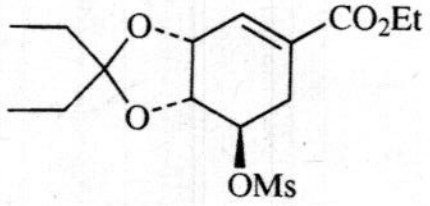

Et_3SiH (1.3 eq), $TiCl_4$ (1.1 eq.), CH_2Cl_2, –34°, 2-6 h

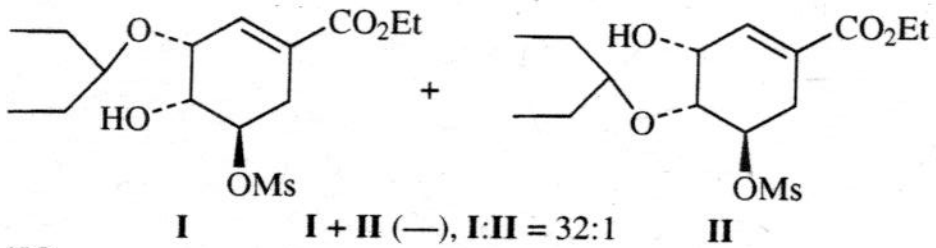

I + II (—), I:II = 32:1 — 498

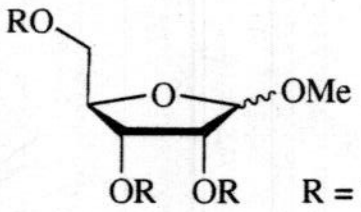

R = TMS

Et_3SiH (5 eq), TMSOTf (5 eq), MeCN, rt, 16 h

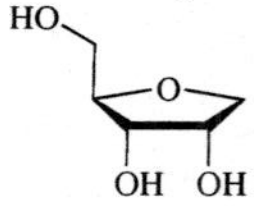

(98) — 503

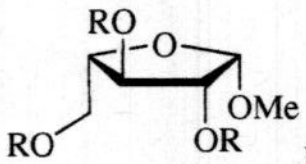

R = TMS

Et_3SiH (5 eq), TMSOTf (5 eq), MeCN, rt, 16 h

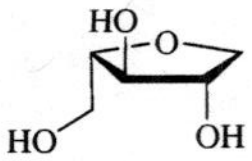

(97) — 503

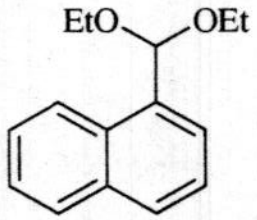

Et_3SiH, FSO_3H, TMSNHCONHTMS

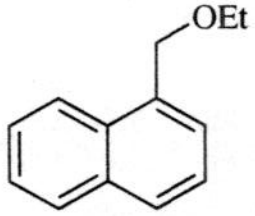

(83) — 487

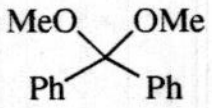

Et_3SiH (1.1 eq), Nafion®-H, CH_2Cl_2, reflux, 3 h

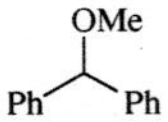

(92.3) — 335

C_{15-24}

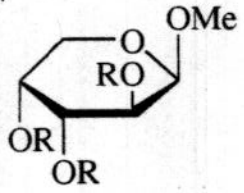

Et_3SiH (5 eq), TMSOTf (5 eq), MeCN, rt, 16 h

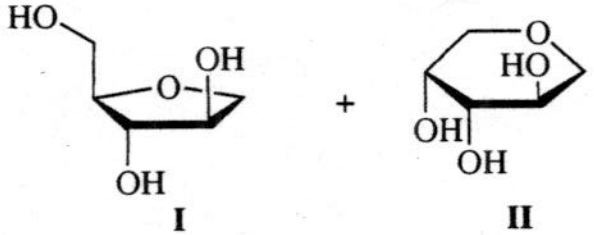

503

R	Catalyst	I	II
TMS	TMSOTf	(95)	(5)
Et_3Si	TMSOTf	(35)	(65)
TBS	TMSOTf	(25)	(75)
$ClCH_2Me_2Si$	TMSOTf	(17)	(83)
TMS	Et_3SiOTf	(77)	(23)
Et_3Si	Et_3SiOTf	(26)	(74)

Acetal/Ketal	Conditions	Product(s) and Yield(s) (%)	Refs.
C_{16}			
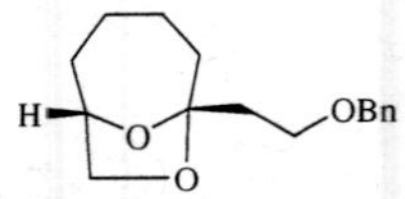	Et_3SiH (4 eq), $TiCl_4$ (1.2 eq), CH_2Cl_2, –78°, 33 h	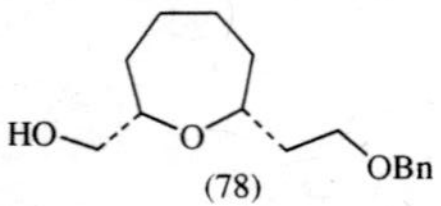 (78) + 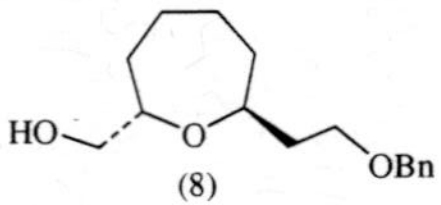 (8)	488
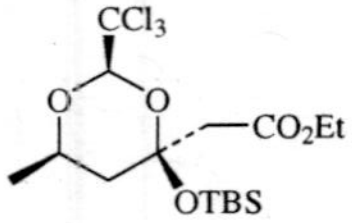	Et_3SiH (1.5 eq), Lewis acid (x eq), CH_2Cl_2, –23°	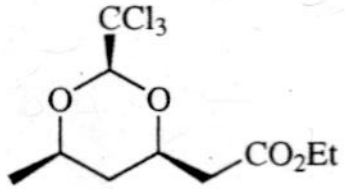	306
	Lewis acid / x	/ cis:trans	
	$TiCl_4$ / 1.1	(62) 98:2	
	$TiCl_4$ / 3.0	(91) 98:2	
	$SnCl_4$ / 3.0	(54) 97:3	
	$AlCl_3$ / 3.0	(16) 96:4	
	$SbCl_5$ / 3.0	(4) 97:3	
	$BF_3•OEt_2$ / 3.0	(1) 91:9	
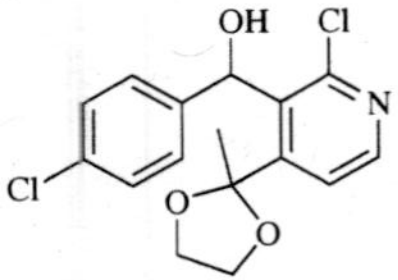	Et_3SiH (2 eq), TFA, 0°, 6 h, rt, 15 h	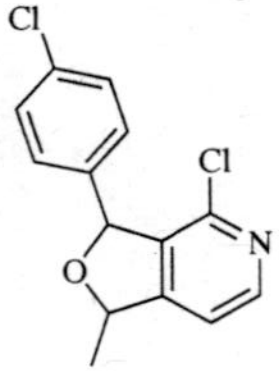 (34)	486
C_{16-25}			
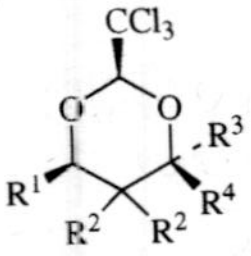	Et_3SiH (1.5 eq), $TiCl_4$ (3.0 eq), CH_2Cl_2, –23°	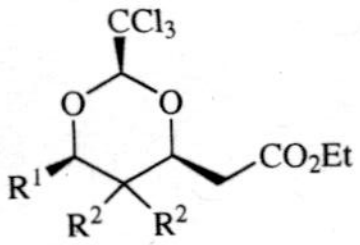	306

R^1	R^2	R^3	R^4		cis:trans
Me	H	CH_2CO_2Et	OTBS	(92)	>99:1
Me	H	OTBS	CH_2CO_2Et	(91)	98:2
Ph	H	CH_2CO_2Et	OTBS	(67)	97:3
Ph	H	OTBS	CH_2CO_2Et	(65)	98:2
Ph	Me	OTBS	CH_2CO_2Et	(97)	>99:1
n-C_7H_{15}	Me	OTBS	CH_2CO_2Et	(94)	>99:1
$BnCH_2$	Me	OTBS	CH_2CO_2Et	(98)	>99:1

C_{16-46}

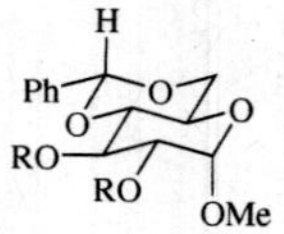

PMHS, $AlCl_3$ (1 eq), Et_2O:CH_2Cl_2 (1:1), rt, 12 h

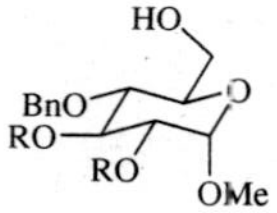

R	
Me	(78)
Ac	(72)
Bn	(80)
TBDPS	(65)

496

C_{16}

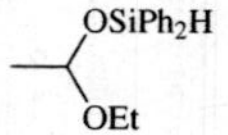

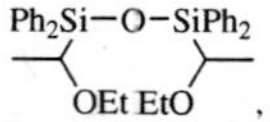

, Ph_2SiH_2 (1.2 eq), $Mn(CO)_5Ac$ (1.5-3.0 mol%), C_6D_6, rt, 4 h

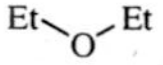

I (88) + 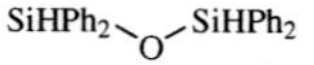 (—)

295

C_{17}

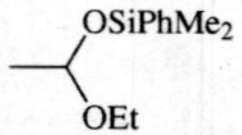

Ph_2SiH_2 (1.2 eq), $Mn(CO)_5Ac$ (3.0 mol%), C_6D_6, rt, 2 h

I (68) + $Ph_2SiHOEt$ (—) **II** + $PhMe_2SiOEt$ (—) **III**

II:**III** = 1:1

295

1. Et_3SiH, $Sn(OTf)_2$-AcBr, CH_2Cl_2, rt, 24 h
2. (n-Bu)$_3$SnH, AIBN, C_6H_6, reflux, 0.5 h

 (55) +  (9)

479

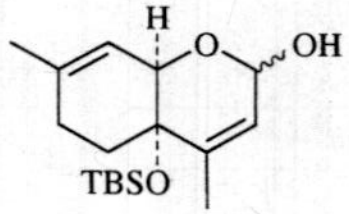

1. i-Bu_2AlH, hexanes, MeC_6H_5, –78°,
2. Et_3SiH (1.5 eq), BF_3•OEt_2 (1.1 eq), –78°, 30 min

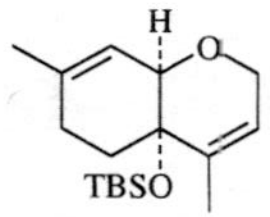

(81)

510

TABLE 20. ORGANOSILANE REDUCTION OF ACETALS, KETALS, AND HEMIKETALS (*Continued*)

Acetal/Ketal	Conditions	Product(s) and Yield(s) (%)	Refs.
C_{17}			
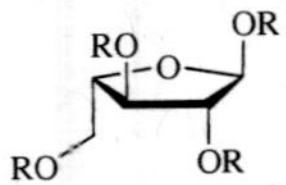 R = TMS	Et_3SiH (5 eq), TMSOTf (5 eq), MeCN, rt, 16 h	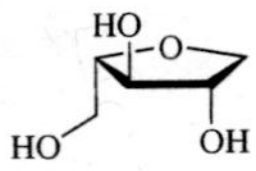 (94)	503
C_{18}			
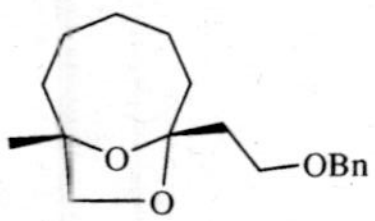	Et_3SiH (4 eq), $TiCl_4$ (1.2 eq), CH_2Cl_2, –78°, 15 h	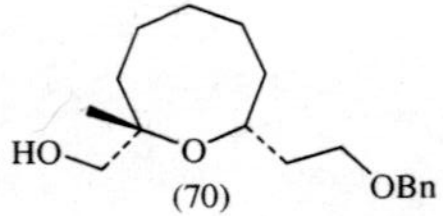 (70) + 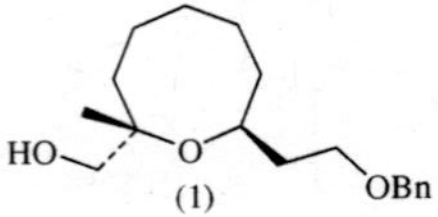 (1)	488
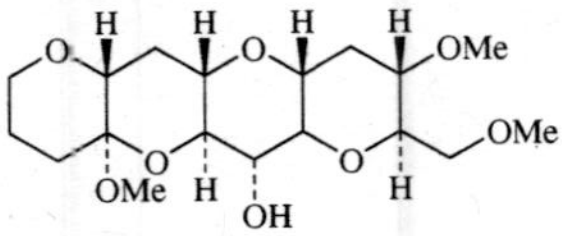	Et_3SiH, TMSOTf, CH_2Cl_2, 0°	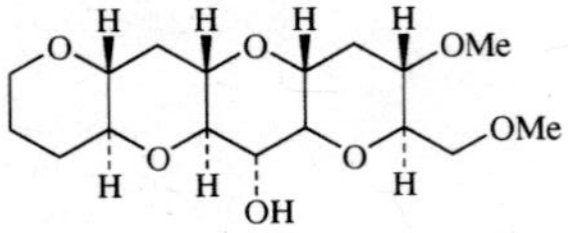 (82)	481
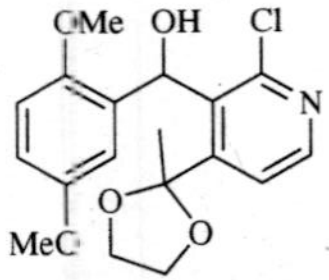	Et_3SiH (1.1 eq), TFA, rt, 18 h	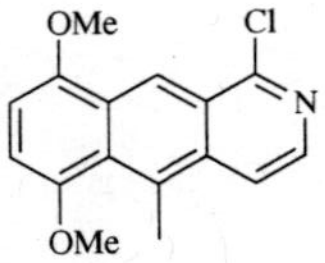 (65)	486
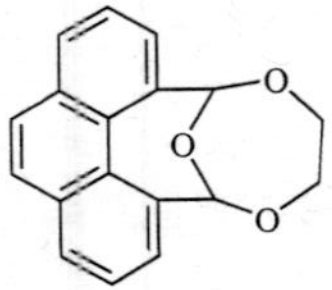	Et_3SiH (8 eq), TMSOTf (2 eq), CH_2Cl_2, –70°, 45 min	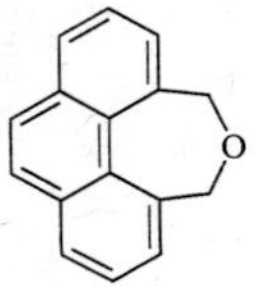 (89)	339
C_{18-20}			
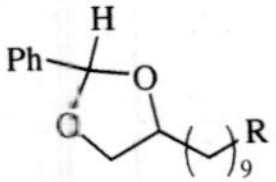	PMHS, $AlCl_3$ (1 eq), Et_2O:CH_2Cl_2 (1:1), rt, 12 h	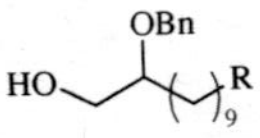 (69) R: N_3 (69); CO_2Me (70)	496

Substrate	Conditions	Product(s) and Yield(s) (%)		Refs.
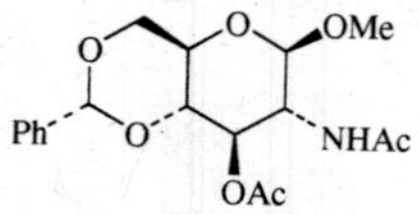	Et_3SiH (5 eq), TFA (5 eq), CH_2Cl_2, rt, 2-4 h	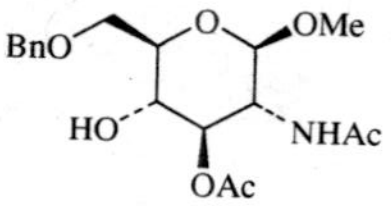	(94)	494
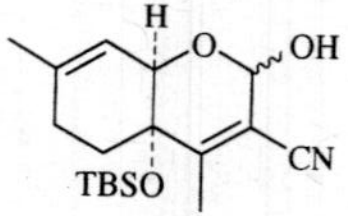	1. (i-Bu)$_2$AlH, hexanes, MeC_6H_5, –78° 2. Et_3SiH (1.5 eq), BF_3•OEt_2 (1.1 eq), –78°, 30 min	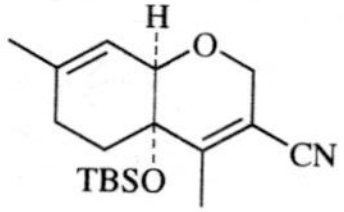	(75)	510
C_{18-19}				
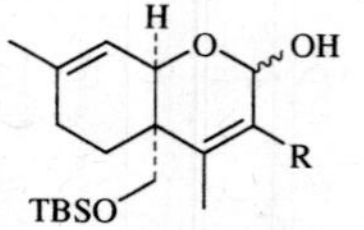	Et_3SiH, BF_3•OEt_2, –78°, 30 min	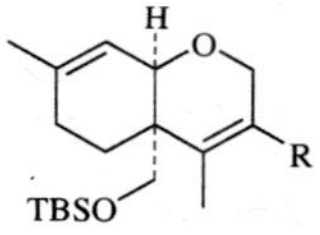	R: H (81); CN (75); SPh (60)	510
C_{18-28}				
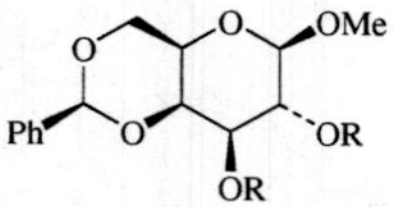	Et_3SiH (5 eq), TFA (5 eq), CH_2Cl_2, rt, 2-4 h	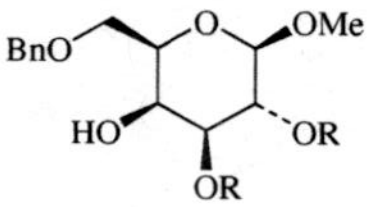	R: Ac (—); Bn (—)	494
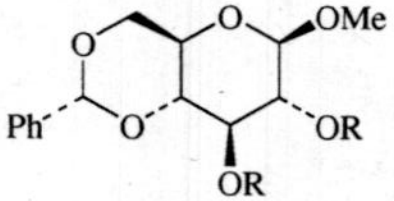	Et_3SiH (5 eq), TFA (5 eq), CH_2Cl_2, rt, 2-4 h	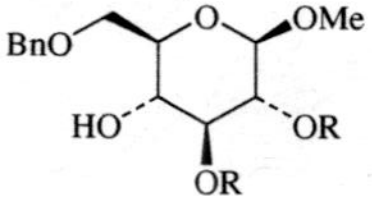	R: Ac (95); Bn (80)	494
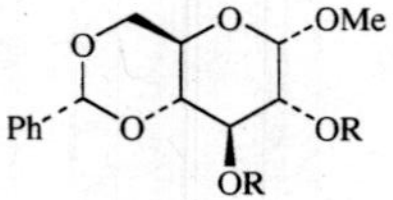	Et_3SiH (5 eq), TFA (5 eq), CH_2Cl_2, rt, 2-4 h	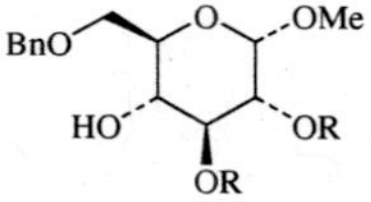	R: Ac (98); Bn (81)	494

TABLE 20. ORGANOSILANE REDUCTION OF ACETALS, KETALS, AND HEMIKETALS (*Continued*)

Acetal/Ketal	Conditions	Product(s) and Yield(s) (%)	Refs.
C_{19} 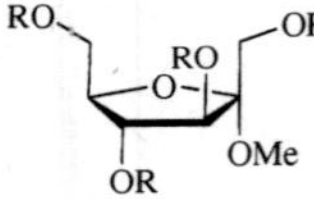 R = TMS	Et_3SiH (5 eq), TMSOTf (5 eq), MeCN, rt, 16 h	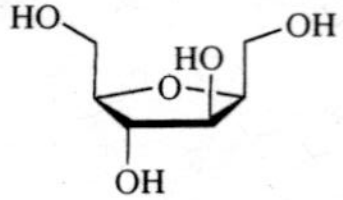 (78) + 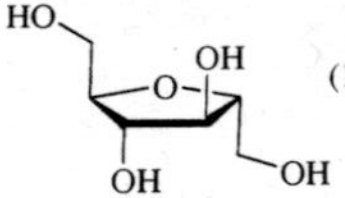 (19)	503
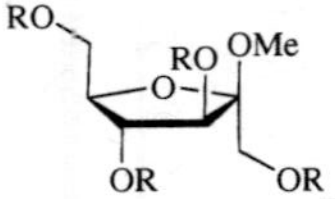 R = TMS	Et_3SiH (5 eq), TMSOTf (5 eq), MeCN, rt, 16 h	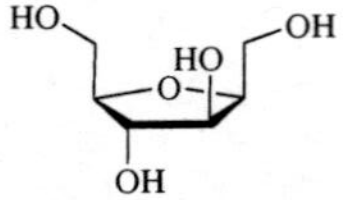 (19) + 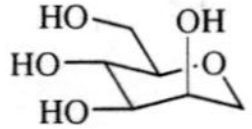 (3)	503
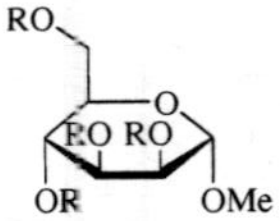 R = TMS	Et_3SiH (5 eq), TMSOTf (5 eq), MeCN, rt, 16 h	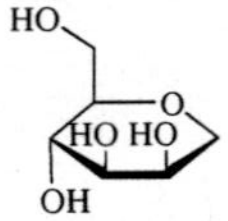 (97)	503
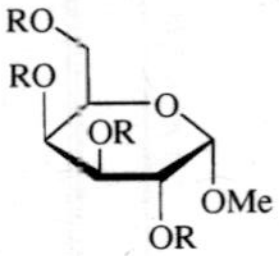 R = TMS	Et_3SiH (5 eq), TMSOTf (5 eq), MeCN, rt, 16 h	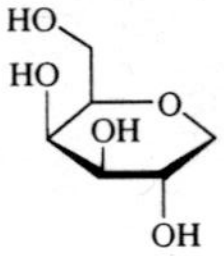 (97)	503
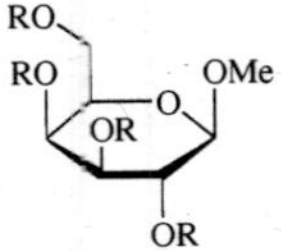 R = TMS	Et_3SiH (5 eq), TMSOTf (5 eq), MeCN, rt, 16 h	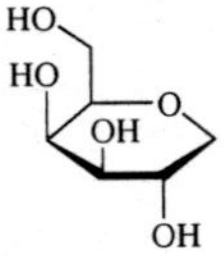 (100)	503
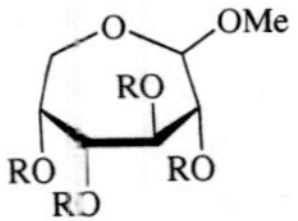 R = TMS	Et_3SiH (5 eq), TMSOTf (5 eq), MeCN, rt, 16 h	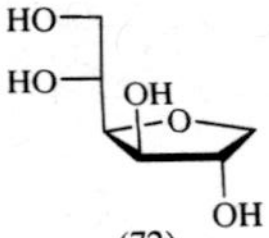 (72) + 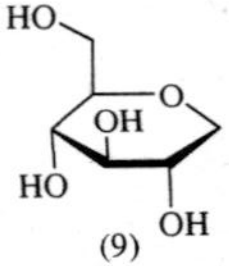 (9) + 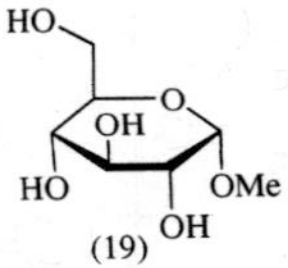 (19)	503

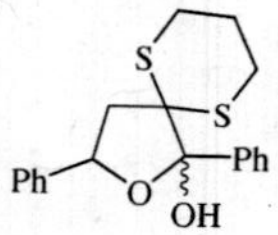

R_3SiH (1 eq), Lewis acid (1.2 eq), CH_2Cl_2, –78°, 5 min

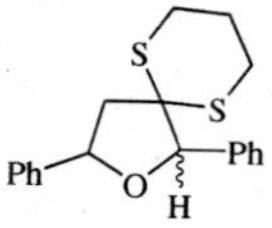

R_3Si	Lewis acid		trans:cis
Ph_3Si	$TiCl_4$	(76)	15:1
Ph_3Si	$TiCl_4$	(87)	16:1
Ph_3Si	$TiCl_4$	(88)	82:1
Ph_3Si	BF_3•OEt_2	(86)	10:1
Ph_3Si	Et_2AlCl	(82)	28:1
Ph_3Si	$EtAlCl_2$	(73)	18:1
Ph_3Si	$AlCl_3$	(76)	11:1
Ph_3Si	TMSOTf	(88)	17:1
Ph_2MeSi	$TiCl_4$	(70)	34:1
$PhMe_2Si$	$TiCl_4$	(79)	30:1
Et_3Si	$TiCl_4$	(75)	7:1

520

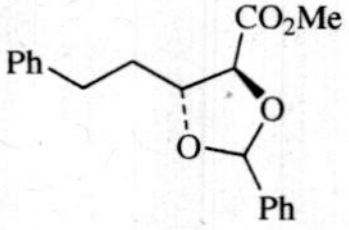

PMHS, $AlCl_3$ (1 eq), $Et_2O:CH_2Cl_2$ (1:1), rt, 12 h

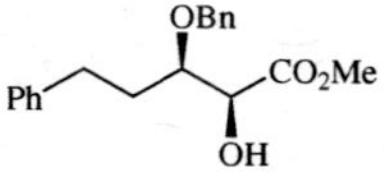

(68)

496

C_{19-21}

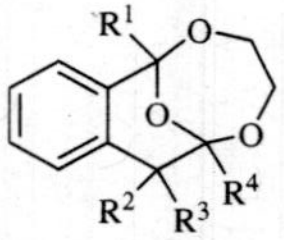

Et$_3$SiH (8 eq), TMSOTf (2 eq), CH_2Cl_2, –70°, 45 min

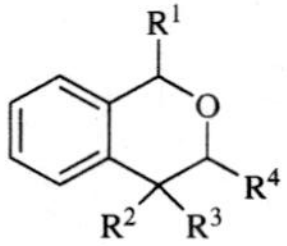

339

R^1	R^2	R^3	R^4	
Ph	Me	Me	H	(57)
H	—$(CH_2)_4$—		Ph	(42)
H	Ph	H	Ph	(53)

TABLE 20. ORGANOSILANE REDUCTION OF ACETALS, KETALS, AND HEMIKETALS (*Continued*)

Acetal/Ketal	Conditions	Product(s) and Yield(s) (%)	Refs.
C_{20} 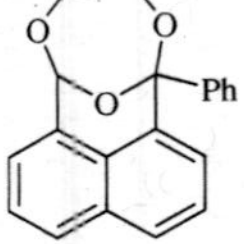	Et_3SiH (8 eq), TMSOTf (2 eq), CH_2Cl_2, –70°, 45 min	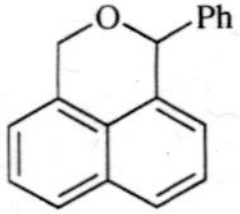 (97)	339
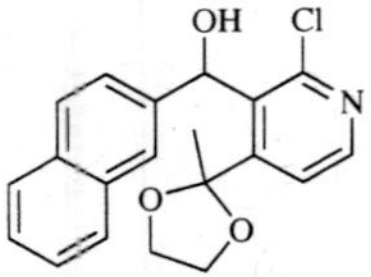	Et_3SiH (1.1 eq), TFA, rt, 18 h	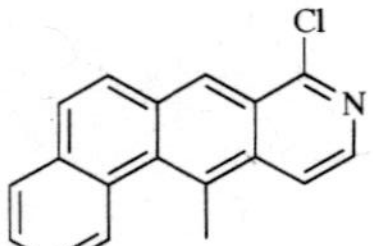 (37)	468
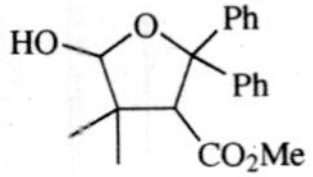	Et_3SiH (1.1 eq), BF_3•OEt_2 (1.1 eq), CH_2Cl_2, –78°, 45 min; rt, 3 h	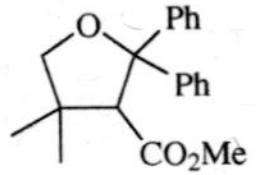 (88)	765
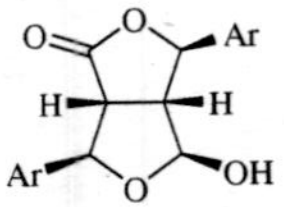 Ar = 3-MeO-4-HOC_6H_3	Et_3SiH, BF_3, Et_2O	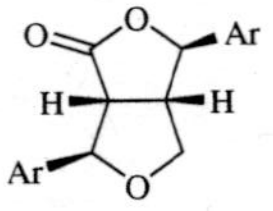 (47) + 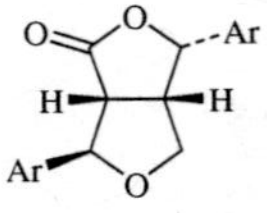 (14) + 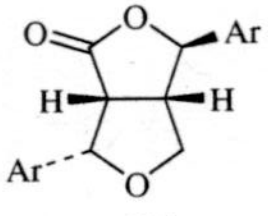 (19) + 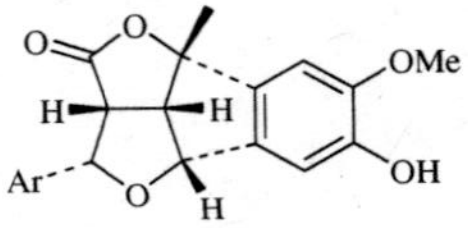 (20)	766

C_{21}

Substrate	Conditions	Product(s) and Yield(s) (%)	Refs.
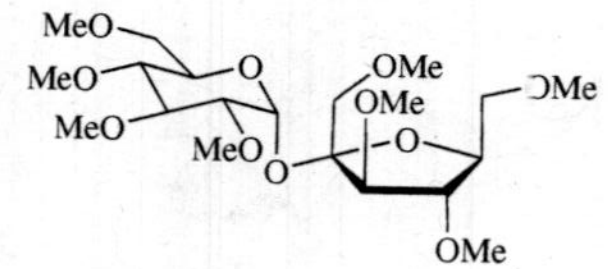	Et_3SiH (30 eq), $BF_3 \bullet OEt_2$ (30 eq), TFA (12 eq), 0° to rt, 16 h	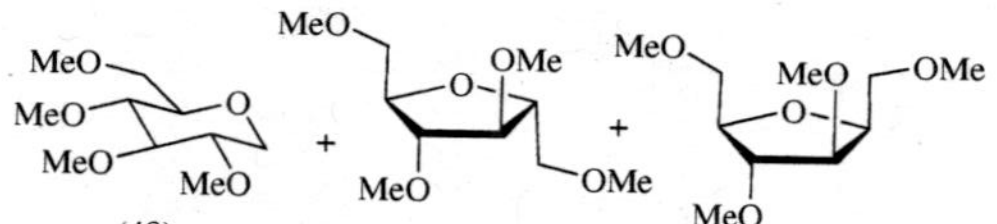 (48) + (40) + (10)	482
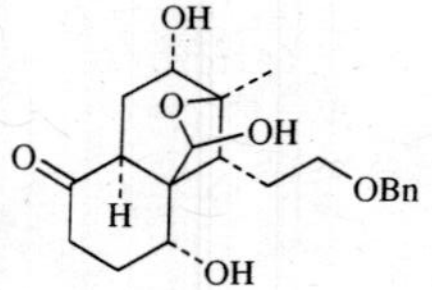	Et_3SiH, $BF_3 \bullet OEt_2$, –20°, 24 h	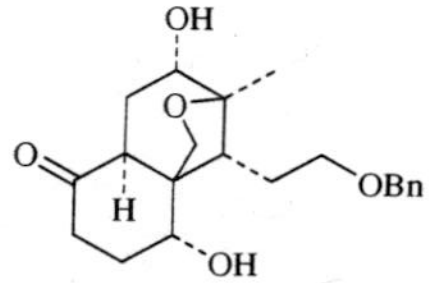 (65)	510
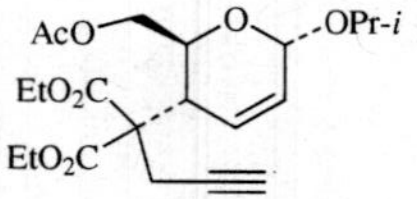	$PhMe_2SiH$, $BF_3 \bullet OEt_2$	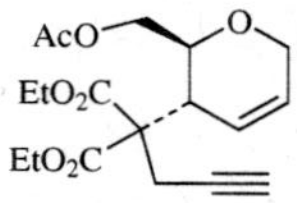 (—)	767
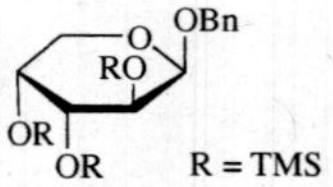 R = TMS	Et_3SiH (5 eq), TMSOTf (5 eq), MeCN, rt, 16 h	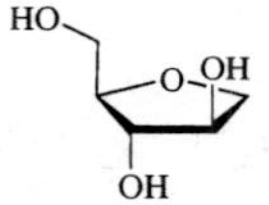 (95) + 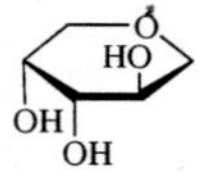 (5)	503
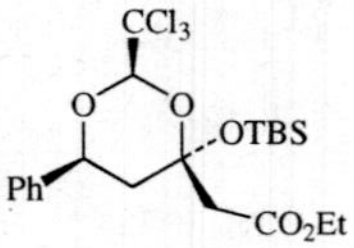	Et_3SiH (5 eq), $TiCl_4$ (3.0 eq), CH_2Cl_2, –23°	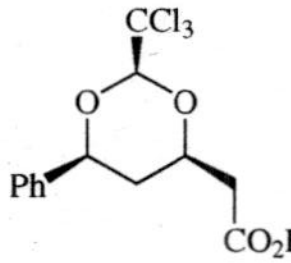 (71) + 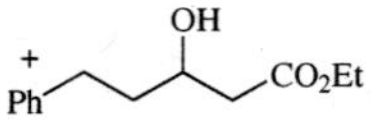 (23)	306
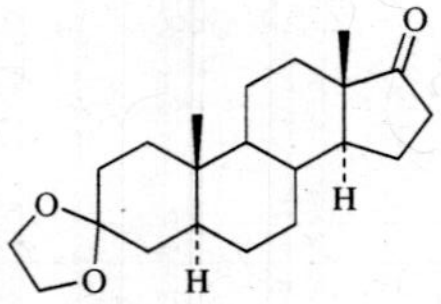	I_2SiH_2 (2 eq), CH_2Cl_2, rt, 2 h	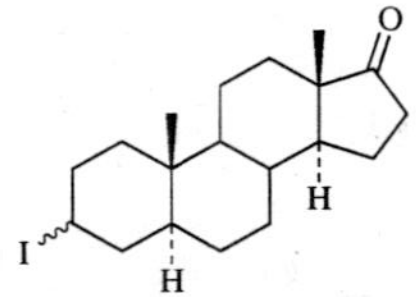 (94) α:β = 3:1	505

TABLE 20. ORGANOSILANE REDUCTION OF ACETALS, KETALS, AND HEMIKETALS (*Continued*)

Acetal/Ketal	Conditions	Product(s) and Yield(s) (%)	Refs.
C_{22}			
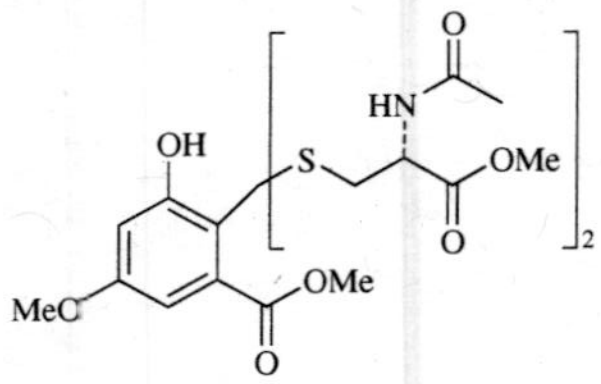	Et_3SiH (2.5 eq), TFA, 0°, 2 h	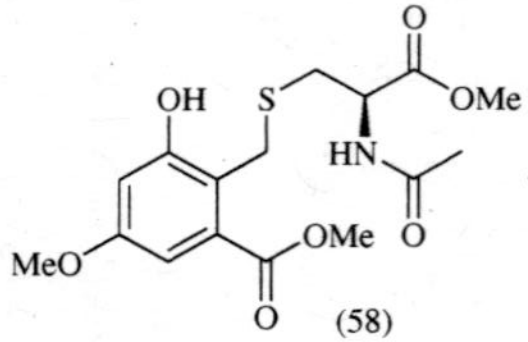 (58) + 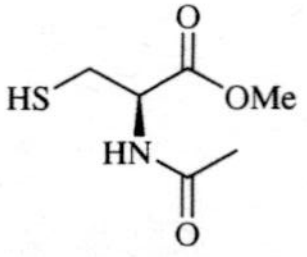 (56)	768
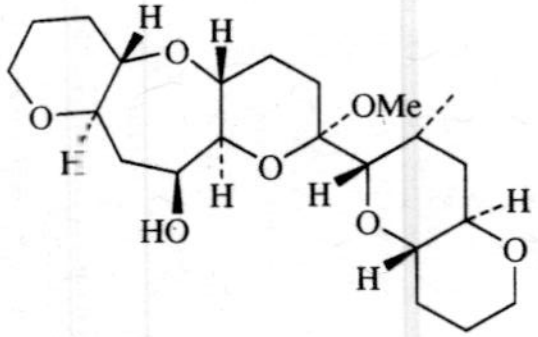	Et_3SiH, BF_3•OEt_2, CH_2Cl_2, MeCN, 0°	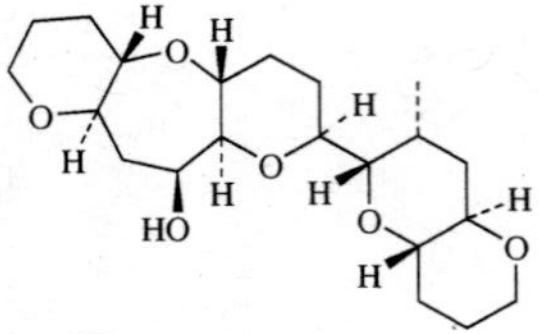 (83)	769
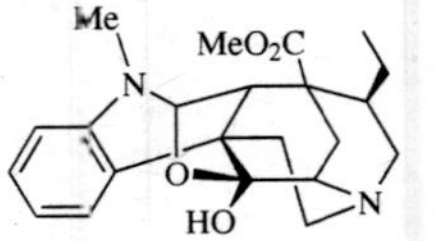	Et_3SiH, TFA	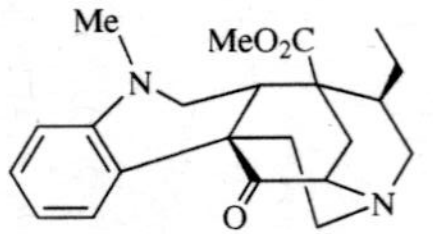 (—)	770
C_{23}			
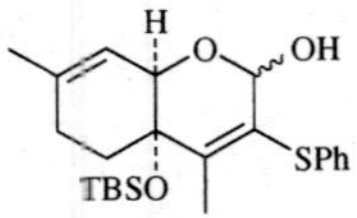	1. $(i\text{-Bu})_2AlH$, hexanes, MeC_6H_5, −78° 2. Et_3SiH (1.5 eq), BF_3•OEt_2 (1.1 eq), −78°, 30 min	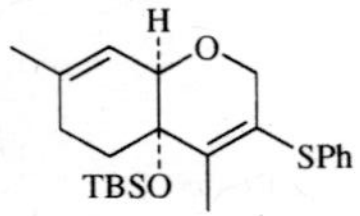 (60)	510
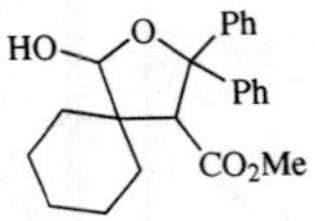	Et_3SiH (1.1 eq), BF_3•OEt_2 (1.1 eq), CH_2Cl_2, −78°, 45 min; rt, 3 h	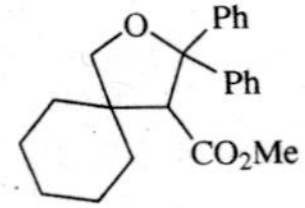 (70)	765

C_{24}				
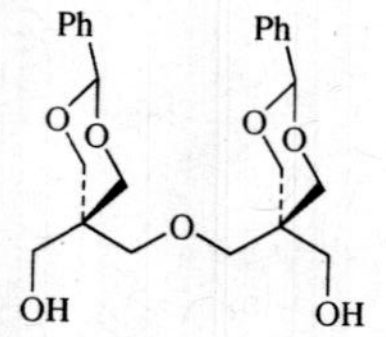	Et_3SiH (13 eq), BF_3•OEt_2 (12 eq), CH_2Cl_2	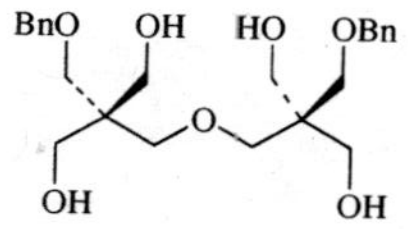	(64)	495
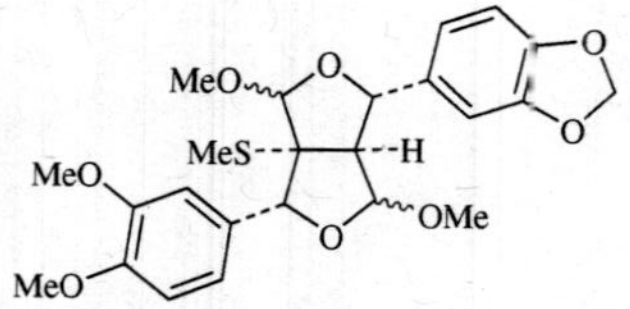	Et_3SiH (4.5 eq), BF_3•OEt_2 (1.5 eq), CH_2Cl_2, −78°; rt, 48 h	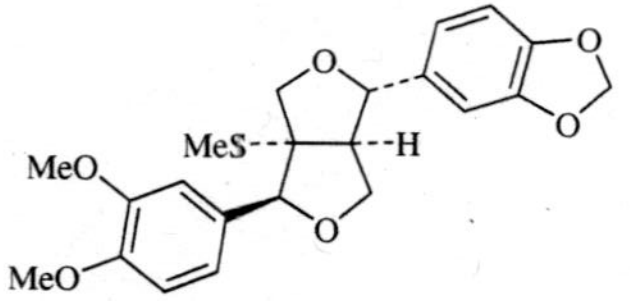	(89)	484
C_{27}				
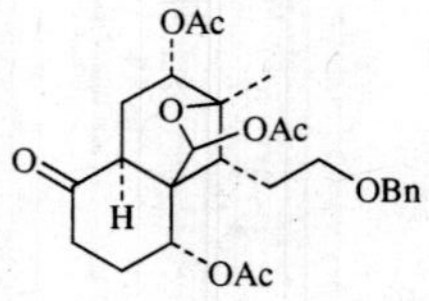	Et_3SiH, BF_3•OEt_2, rt, 3 h	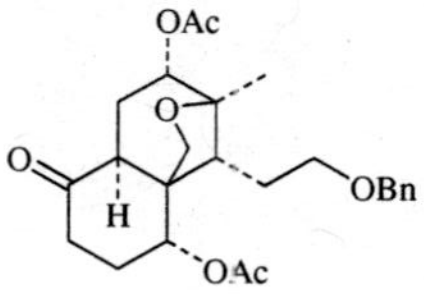	(88)	510

TABLE 20. ORGANOSILANE REDUCTION OF ACETALS, KETALS, AND HEMIKETALS (*Continued*)

Acetal/Ketal	Conditions	Product(s) and Yield(s) (%)	Refs.
C_{27} 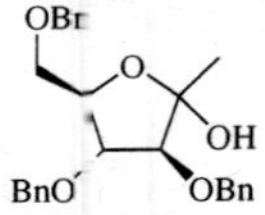	Et_3SiH, $BF_3{\bullet}OEt_2$, CH_2Cl_2, −78° 1 h; −10°, 12 h	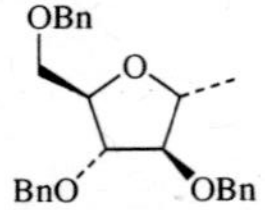 (53)	508
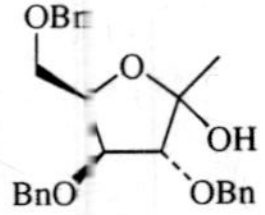	Et_3SiH, $BF_3{\bullet}OEt_2$, CH_2Cl_2, −78° 1 h; −10°, 12 h	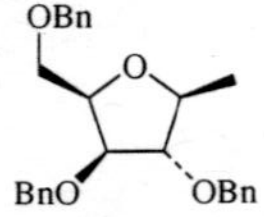 **I** + 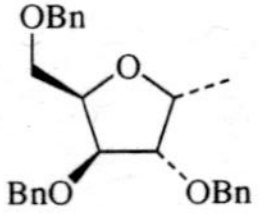 **II** **I** + **II** (76), **I**:**II** = 5:1	508
C_{28} 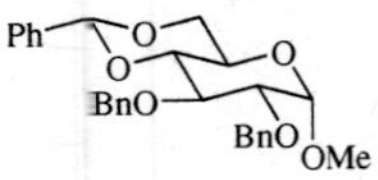	Et_3SiH (12 eq), $BF_3{\bullet}OEt_2$ (2 eq), CH_2Cl_2, rt, 4 h	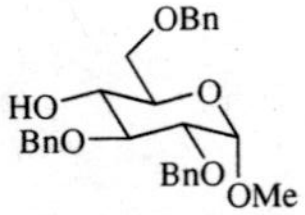 (83)	497
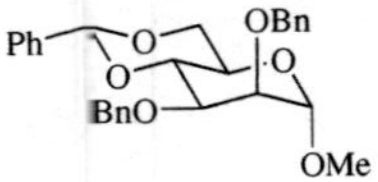	Et_3SiH (12 eq), $BF_3{\bullet}OEt_2$ (2 eq), CH_2Cl_2, rt, 4 h	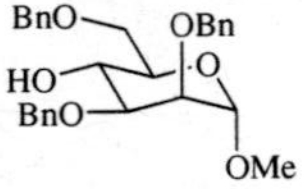 (59)	497
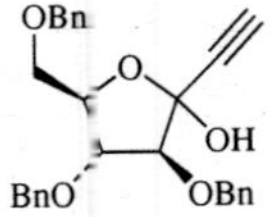	Et_3SiH, $BF_3{\bullet}OEt_2$, CH_2Cl_2, −78° 1 h; −10°, 12 h	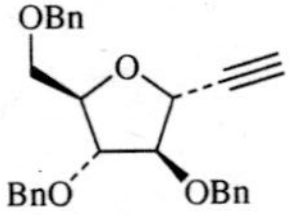 **I** + 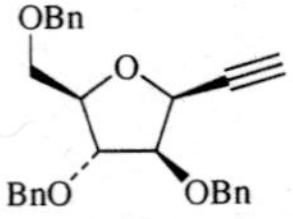 **II** **I** + **II** (81), **I**:**II** = 5:1	508
C_{29} 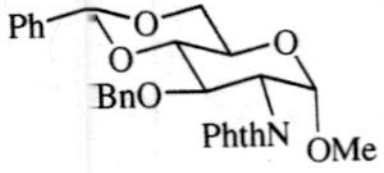	Et_3SiH (12 eq), $BF_3{\bullet}OEt_2$ (2 eq), CH_2Cl_2, rt, 4 h	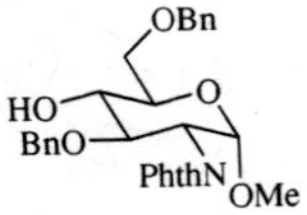 (85)	497

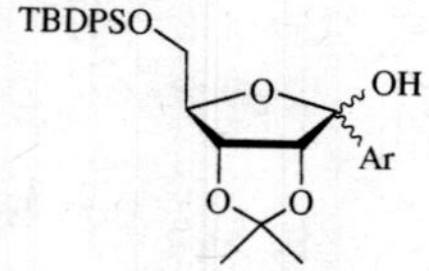

Ar = 2,6-dichloro-3-pyridinyl

Et_3SiH (x eq), catalyst

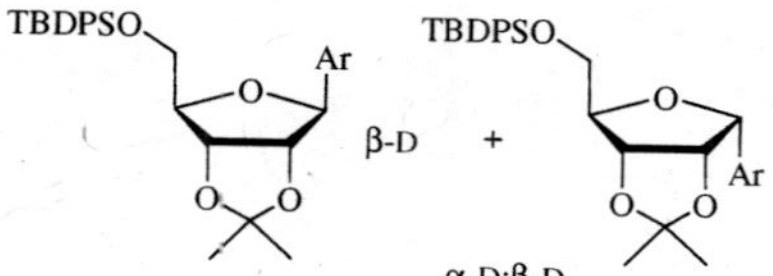

β-D + α-D — 511

Catalyst	x	Temp	Time		α-D:β-D
BF_3•OEt_2	2	–5°	1.5 h	(57)	2.6:1
$SnCl_4$	1	–5°	1.5 h	(30)	1:5
$ZnBr_2$	1	–5°	1.5 h	(—)	—
$ZnCl_2$•OEt_2	2	rt	3 h	(68)	2:1

C_{32}

Reactant	Conditions	Product(s) and Yield(s) (%)		Refs.
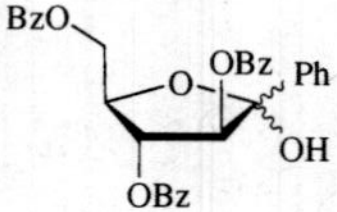	Et_3SiH, BF_3•OEt_2, MeCN, –40° to rt	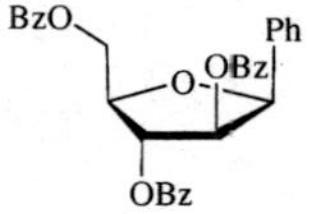	(—)	509

C_{34}

Reactant	Conditions	Product(s) and Yield(s) (%)		Refs.
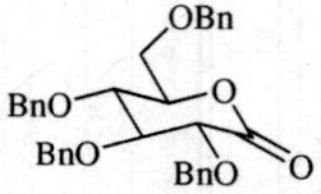	1. PhMgCl (1.5 eq), THF, –78° 2. Et_3SiH, BF_3•OEt_2, CH_2Cl_2, –78°, 1 h; rt, 3 h	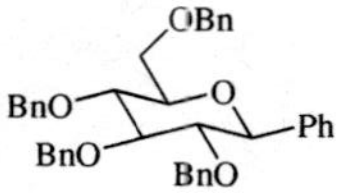	(88)	771

C_{35}

Reactant	Conditions	Product(s) and Yield(s) (%)		Refs.
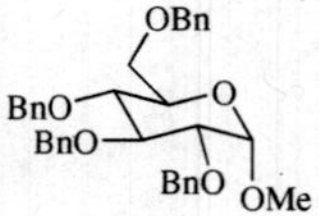	Et_3SiH (3 eq), TMSOTf (2.25 eq), CH_2Cl_2, 0° to rt, 16 h	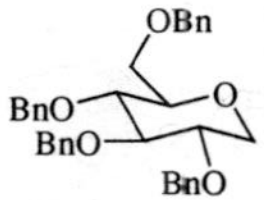	(77)	504
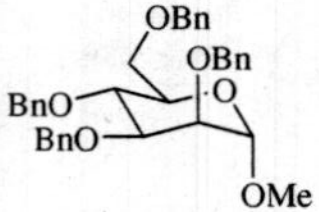	Et_3SiH (3 eq), TMSOTf (2.25 eq), CH_2Cl_2, 0° to rt, 16 h	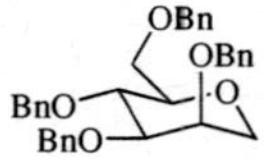	(80)	504

TABLE 20. ORGANOSILANE REDUCTION OF ACETALS, KETALS, AND HEMIKETALS (*Continued*)

Acetal/Ketal	Conditions	Product(s) and Yield(s) (%)	Refs.
C_{36} 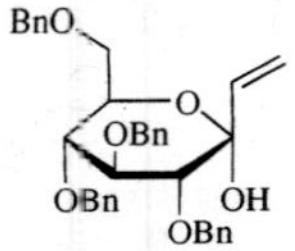	Et_3SiH, TMSOTf	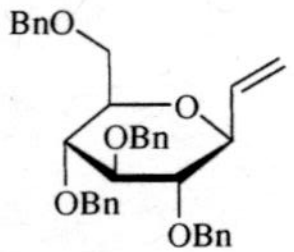 (35)	519
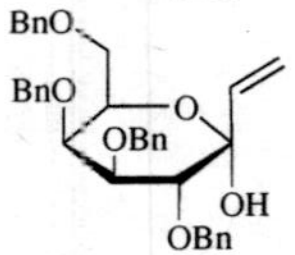	Et_3SiH, TMSOTf	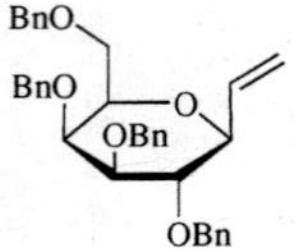 (65)	519
C_{37} 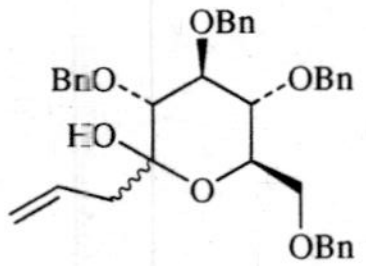	Et_3SiH, $BF_3{\bullet}OEt_2$, MeCN, 0° to rt	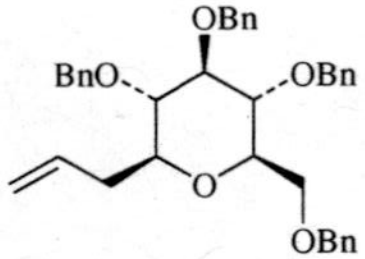 (85)	507
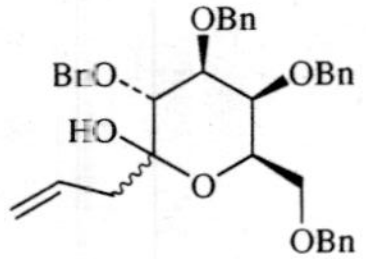	Et_3SiH, $BF_3{\bullet}OEt_2$, MeCN, 0° to rt	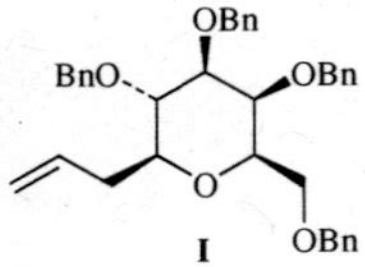 I + 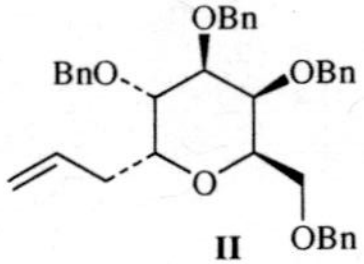 II **I** + **II** (76), **I**:**II** > 10:1	507
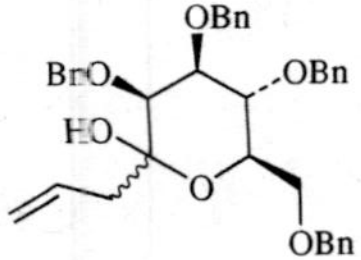	Et_3SiH, $BF_3{\bullet}OEt_2$, MeCN, 0° to rt	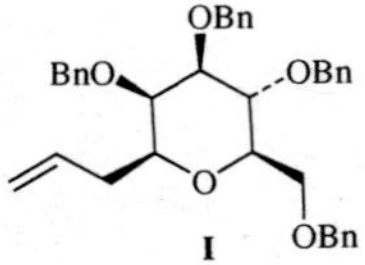 I + 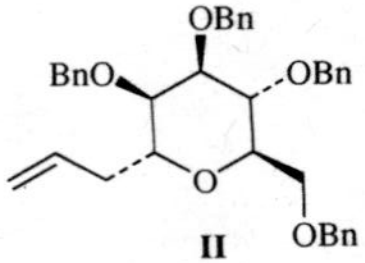 II **I** + **II** (67), **I**:**II** = ca. 1:1	507

Reactant	Conditions	Product(s) and Yield(s) (%)	Refs.
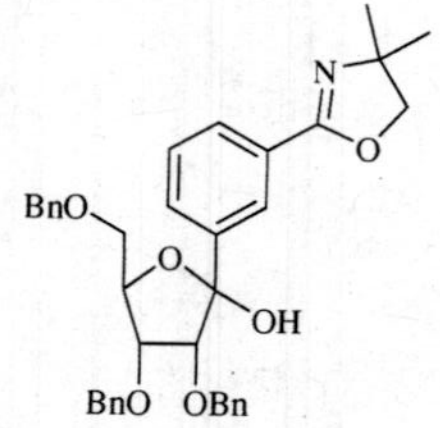	Et_3SiH (2 eq), $BF_3•OEt_2$ (2 eq), CH_2Cl_2, –78°, 1 h; 10°, 16 h	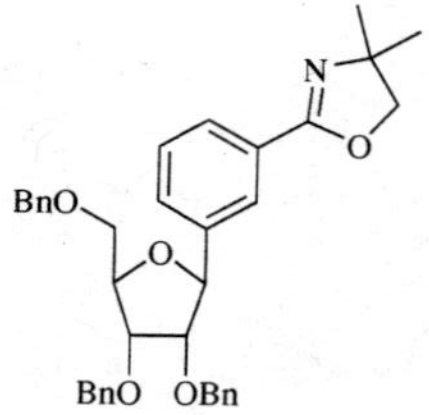 (87)	772
C_{37-39} 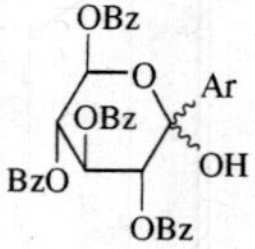	Et_3SiH, $BF_3•OEt_2$, MeCN, –40° to rt	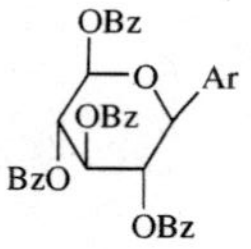 Ar: 2-C_4H_3O (77); 3-C_4H_3O (30); Ph (80)	509
C_{38} 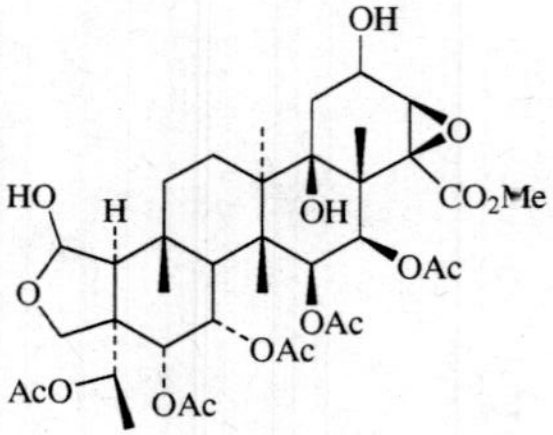	Et_3SiH, $BF_3•OEt_2$, CH_2Cl_2, rt, 1 h	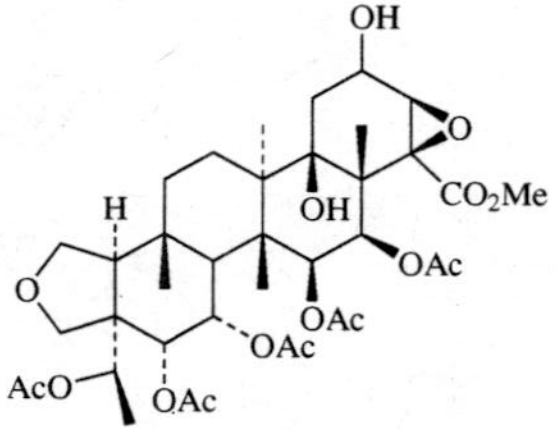 (—)	773
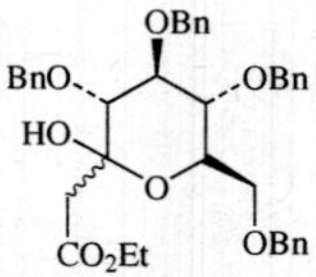	Et_3SiH, $BF_3•OEt_2$, MeCN, 0° to rt	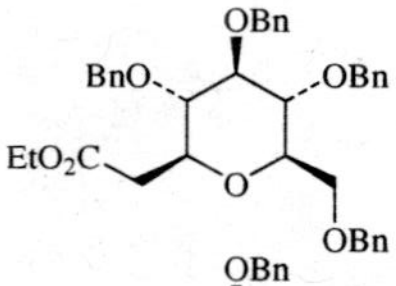 (72)	507
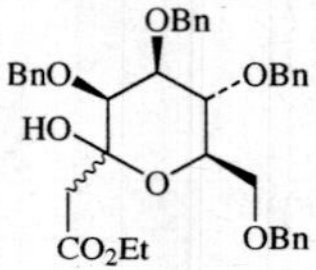	Et_3SiH, $BF_3•OEt_2$, MeCN, 0° to rt	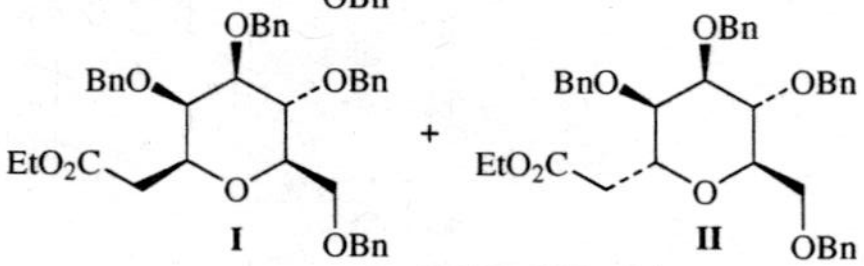 **I** + **II** (66), **I**:**II** = 3:1	507

TABLE 20. ORGANOSILANE REDUCTION OF ACETALS, KETALS, AND HEMIKETALS (*Continued*)

Acetal/Ketal	Conditions	Product(s) and Yield(s) (%)	Refs.
C_{39} 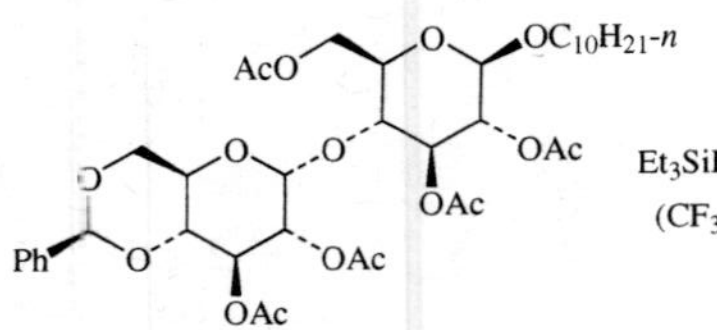	Et_3SiH (5 eq), TFA (5 eq), $(CF_3CO)_2O$ (3 eq), CH_2Cl_2, rt, 2-4 h	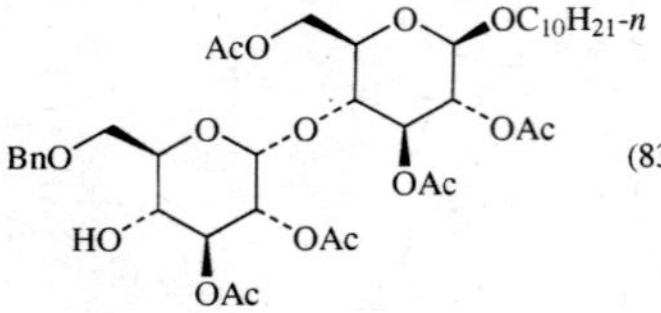 (83)	494
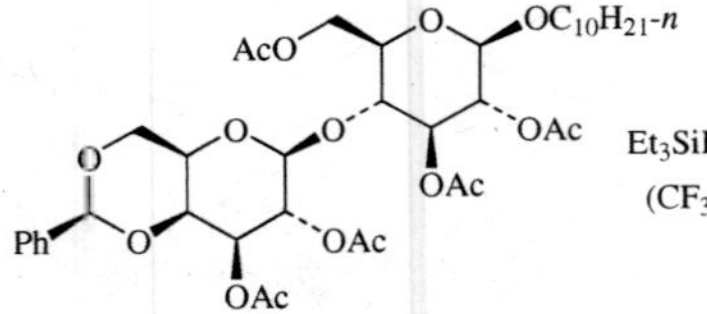	Et_3SiH (5 eq), TFA (5 eq), $(CF_3CO)_2O$ (3 eq), CH_2Cl_2, rt, 2-4 h	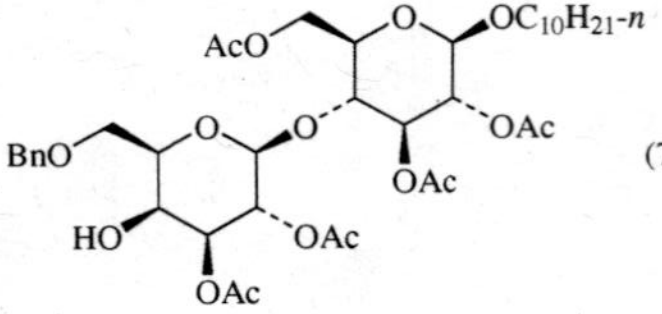 (78)	494
C_{40} 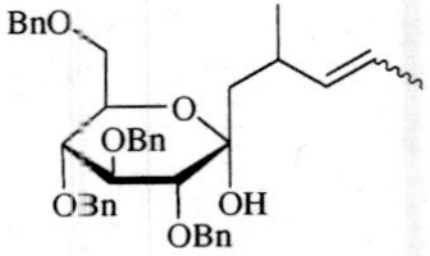	Et_3SiH, TMSOTf	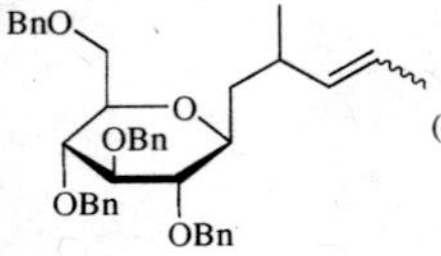 (—)	519
C_{42} 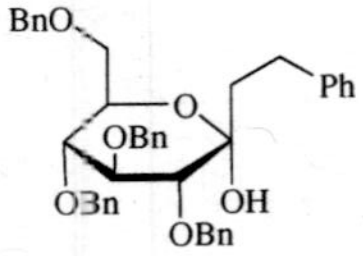	Et_3SiH, TMSOTf	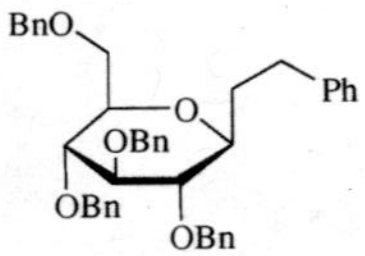 (—)	519
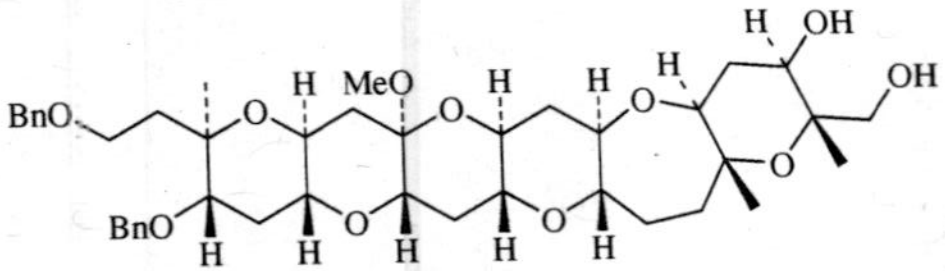	Et_3SiH, TMSOTf CH_2Cl_2, 0°	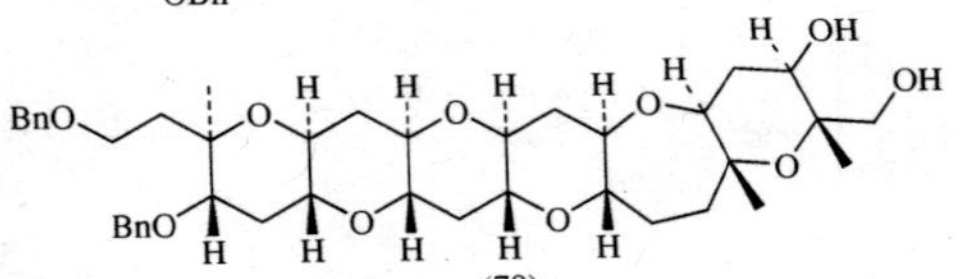 (70)	482

C_{44}

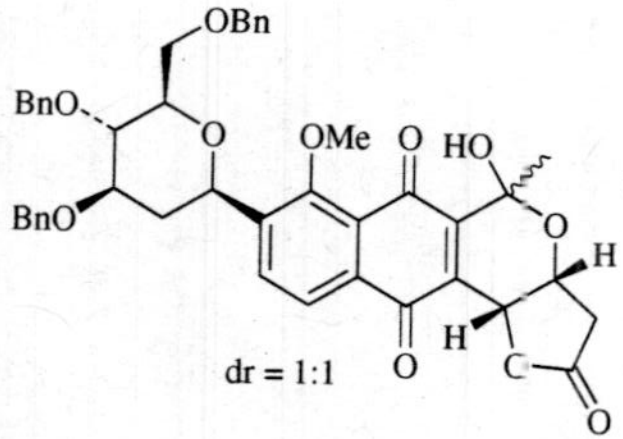

dr = 1:1

Et_3SiH, TFA, CH_2Cl_2, –10°, 72 h

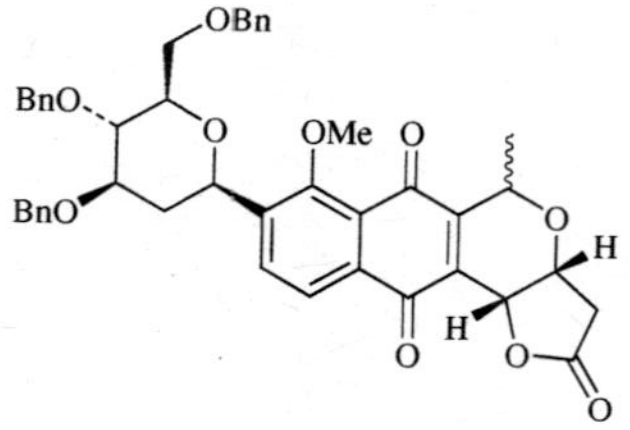

(86) dr = 1:1

513

C_{56}

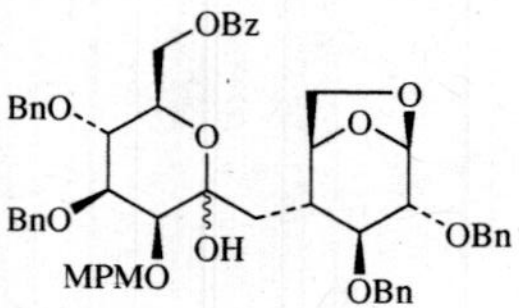

R_3SiH (10 eq), $BF_3•OEt_2$ (0.1 eq), MeCN, –20°, 20 min

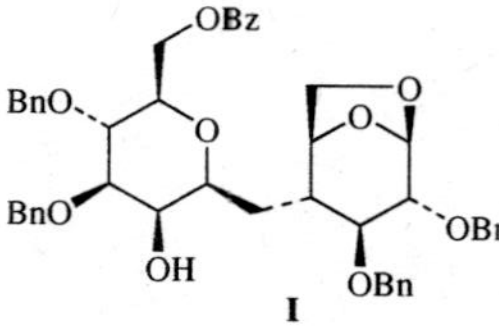

I

+

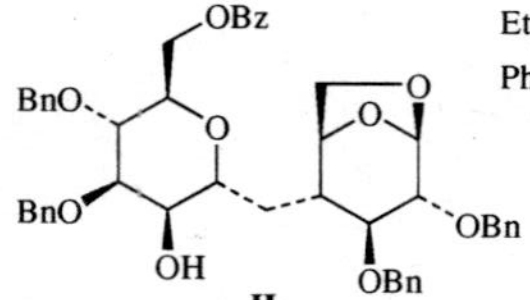

II

R_3Si	I + II	I:II
n-Pr_3Si	(82)	7:1
Et_3Si	(—)	3:1
$PhMe_2Si$	(—)	3:2

517, 774

C_{57}

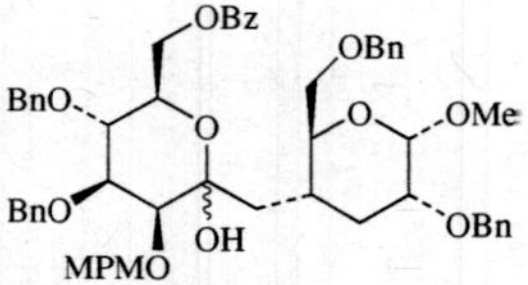

$PhMe_2SiH$ (10 eq), TMSOTf, CH_2Cl_2, –20°, 20 min

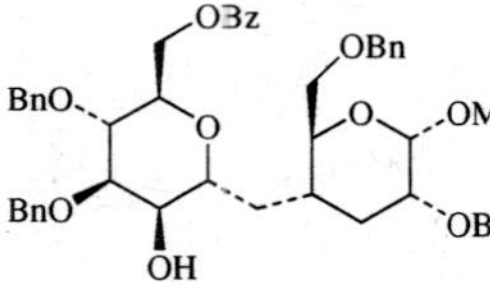

(80)

517

TABLE 20. ORGANOSILANE REDUCTION OF ACETALS, KETALS, AND HEMIKETALS (*Continued*)

Acetal/Ketal	Conditions	Product(s) and Yield(s) (%)	Refs.
C_{57} 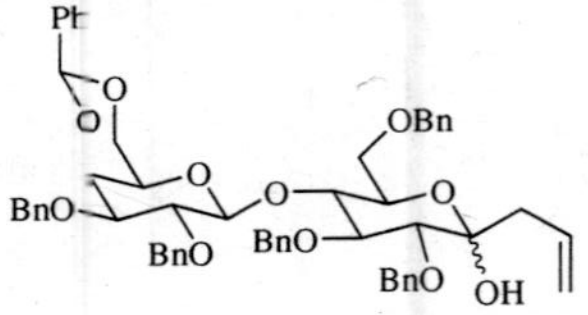	Et_3SiH (12 eq), BF_3•OEt_2 (2 eq), CH_2Cl_2, rt, 4 h	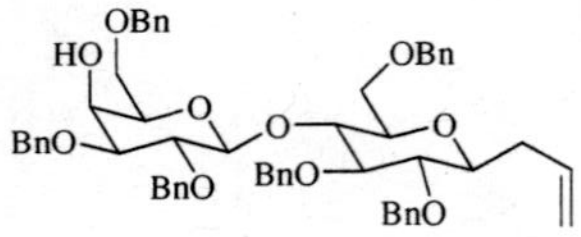 (73)	497

[a] The yield was determined by NMR spectroscopy.

[b] The product is a single isomer of undetermined configuration.

TABLE 21. ORGANOSILANE REDUCTION OF AMINALS AND HEMIAMINALS

Aminal	Conditions	Product(s) and Yield(s) (%)	Refs.
C_5 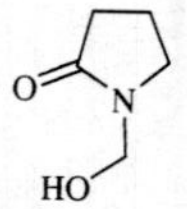	Et_3SiH (1.5 eq), TFA (10 eq), $CHCl_3$, rt, 1-4 h	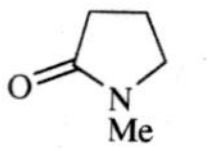 (84)	526
C_6 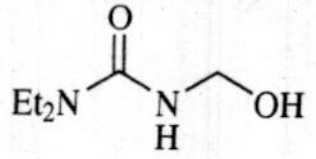	Et_3SiH (1.5 eq), TFA (10 eq), $CHCl_3$, rt, 1-4 h	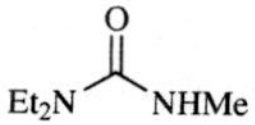 (57)	526
C_7 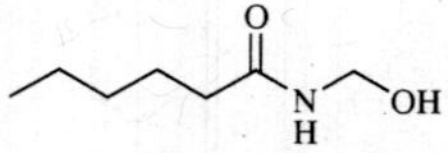	Et_3SiH (1.5 eq), TFA (10 eq), $CHCl_3$, rt, 1-4 h	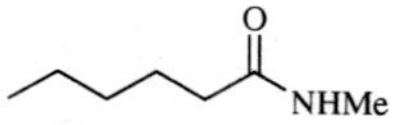 (86)	526
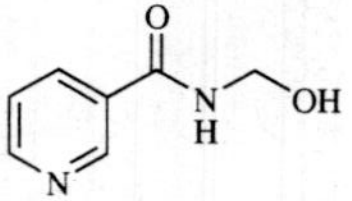	Et_3SiH (1.5 eq), TFA (10 eq), $CHCl_3$, reflux, 2 h	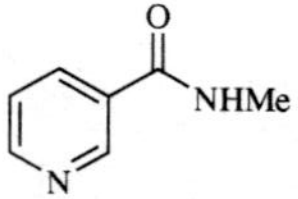 (88)	526
C_8 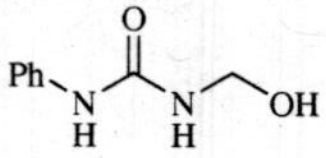	Et_3SiH (1.5 eq), TFA (10 eq), $CHCl_3$, rt, 1-4 h	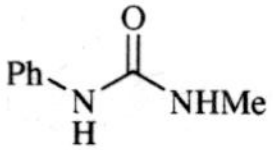 (85)	526
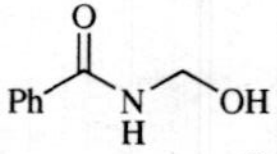	Et_3SiH (1.5 eq), TFA (10 eq), $CHCl_3$, rt, 1-4 h	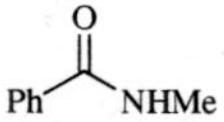 (94)	526
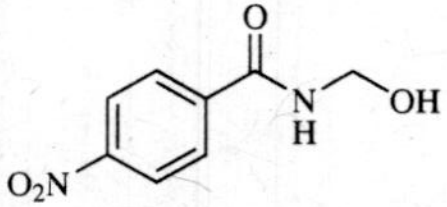	Et_3SiH (1.5 eq), TFA (10 eq), $CHCl_3$, rt, 1-4 h	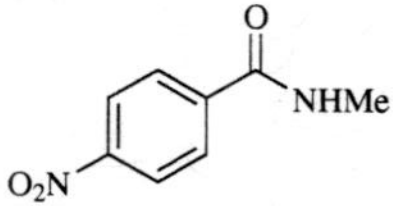 (92)	526
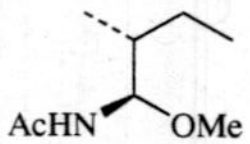	Et_3SiH (1.2 eq), $BF_3{\bullet}OEt_2$ (1.2 eq), CH_2Cl_2, 5°, 2 h	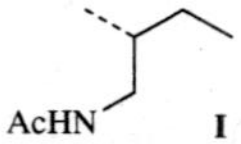 **I** (84)	521
	Et_3SiH (1.2 eq), TFA (3 eq), CH_2Cl_2, 5° to rt, 2 h	**I** (75)	521

TABLE 21. ORGANOSILANE REDUCTION OF AMINALS AND HEMIAMINALS (*Continued*)

Aminal	Conditions	Product(s) and Yield(s) (%)	Refs.
C_9 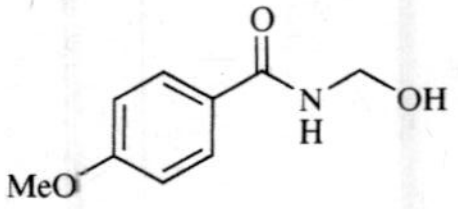	Et_3SiH (1.5 eq), TFA (10 eq), $CHCl_3$, rt, 1-4 h	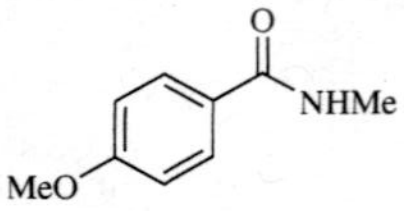 (91)	526
C_{9-15} 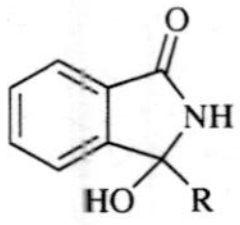	Et_3SiH (10 eq), BF_3•OEt_2 (3 eq), CH_2Cl_2, –15° to rt, 16 h	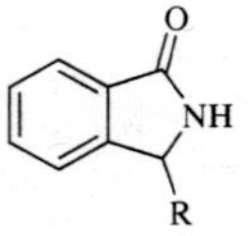	527
R			
Me		(65)	
Et		(96)	
n-Pr		(75)	
n-Bu		(97)	
i-Bu		(55)	
n-C_5H_{11}		(91)	
n-C_7H_{15}		(96)	
Ph		(90)	
Bn		(88)	
C_{10} 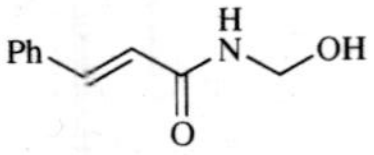	Et_3SiH (1.5 eq), TFA (10 eq), $CHCl_3$, rt, 1-4 h	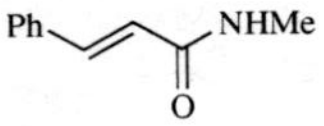 (85)	526
C_{11} 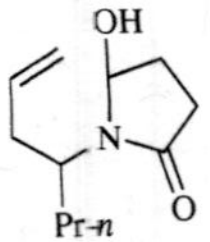	Et_3SiH, HCO_2H, CH_2Cl_2, rt, 10 h	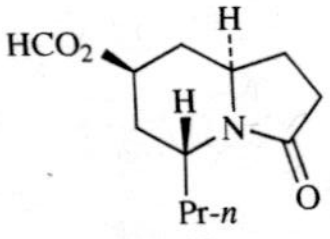 (29) + 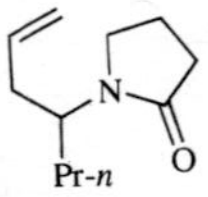 I + 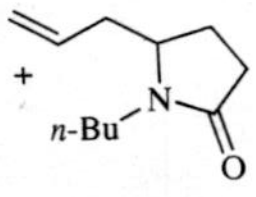 II **I** + **II** (42), **I**:**II** = 3:5	775

C_{12} 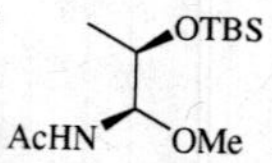	Et_3SiH (2 eq), BF_3•OEt_2 (2 eq), CH_2Cl_2, –40°, 1-2 h	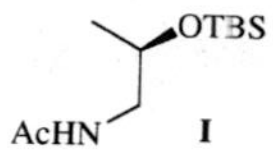 **I** (88)	521
	Et_3SiH (1.2 eq), TFA (3 eq), CH_2Cl_2, 5° to rt, 2 h	**I** (42) + 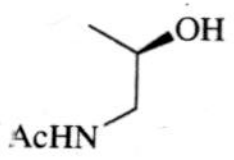 (51)	521
C_{13} 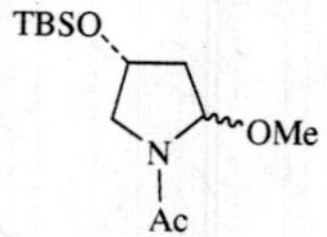	Et_3SiH (2 eq), BF_3•OEt_2 (2 eq), CH_2Cl_2, –40°, 1-2 h	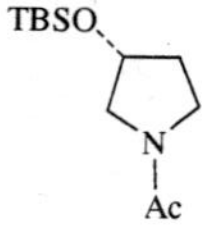 (97)	521
C_{16} 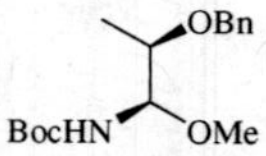	Et_3SiH (2 eq), BF_3•OEt_2 (2 eq), CH_2Cl_2, –40°, 1-2 h	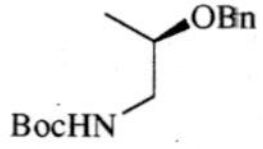 (98)	521
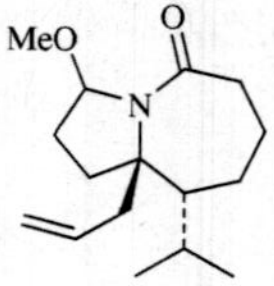	Et_3SiH, TFA, CH_2Cl_2, 15 min	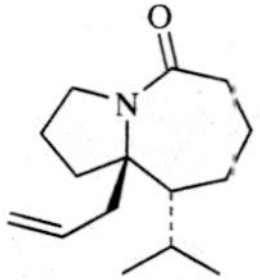 (93)	524
	1. TFA, CH_2Cl_2, 30 min 2. Et_3SiH, 0°, 30 min	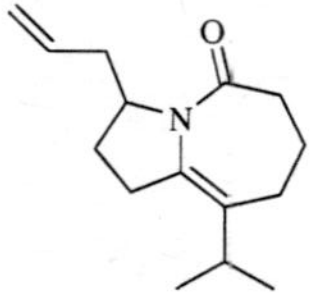 (—)	524
	1. $TiCl_4$, CH_2Cl_2, –78°, 30 min 2. Et_3SiH, 0°, 30 min	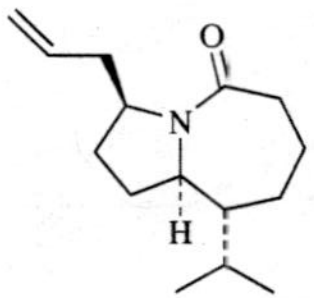 (—)	524

TABLE 21. ORGANOSILANE REDUCTION OF AMINALS AND HEMIAMINALS (*Continued*)

	Aminal	Conditions	Product(s) and Yield(s) (%)	Refs.
C_{16-17}	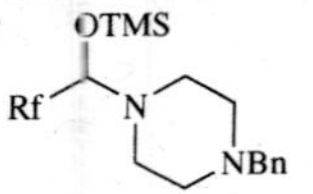	Et_3SiH (1 eq), $BF_3•OEt_2$ (1 eq), CH_2Cl_2, rt, 5 h	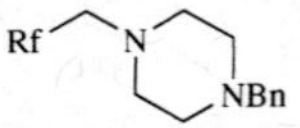 Rf CF_3 (85) $CClF_2$ (80) CF_3CF_2 (84)	522
C_{17}	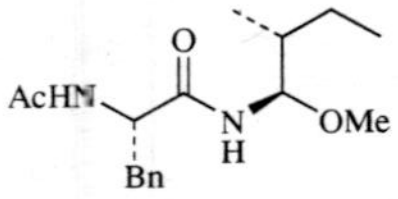	Et_3SiH (1.2 eq), $BF_3•OEt_2$ (1.2 eq), CH_2Cl_2, 5°, 2 h	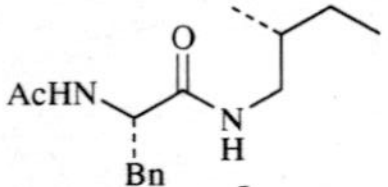 **I** (89)	521
		Et_3SiH (1.2 eq), TFA (3 eq), CH_2Cl_2, 5° to rt, 2 h	**I** (87)	521
	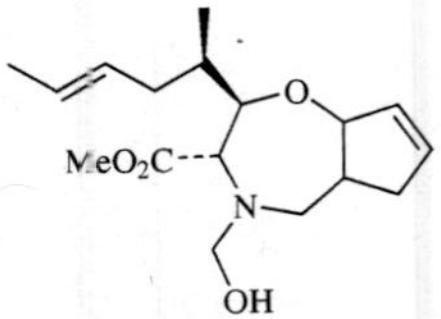	Et_3SiH (3 eq), TFA (xs), $CHCl_3$, rt, 20 h	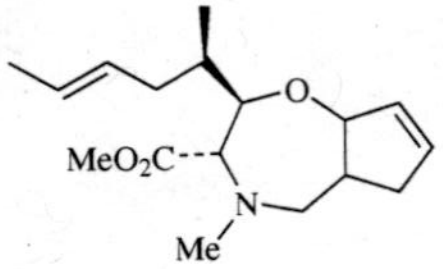 (63)	525
C_{18}	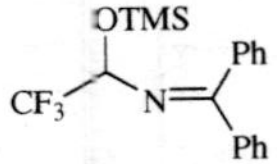	Et_3SiH (1 eq), $BF_3•OEt_2$ (1 eq), CH_2Cl_2, 50°, 24 h	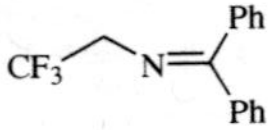 (61)	522
C_{19-28}	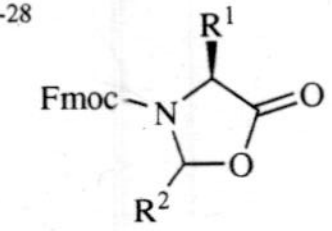	Et_3SiH (3 eq), TFA, $CHCl_3$, rt, 22 h	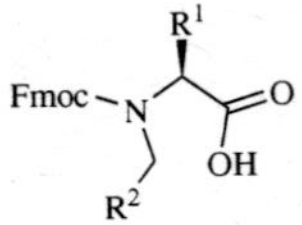 R^1 / R^2 Me / H (98) Me / c-C_6H_{11} (91) Me / Me (74) Me / Bn (95) *i*-Pr / H (100) $MeSCH_2CH_2$ / H (22) Bn / H (70) $BnOCH_2$ / H (96) $Ph(CH_2)_4$ / H (70) 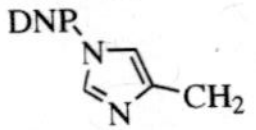 / H (67)	528

C_{20}	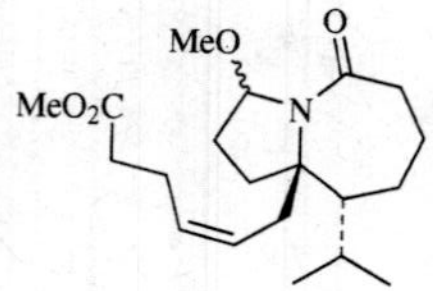	1. $TiCl_4$, CH_2Cl_2, –78° to 0° 2. Et_3SiH	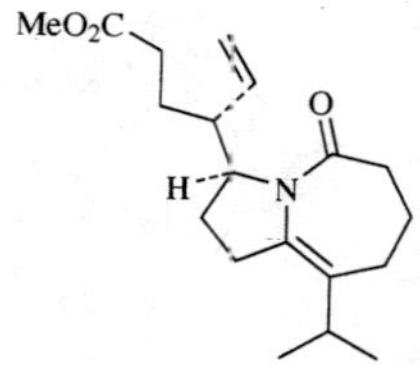 (22) + 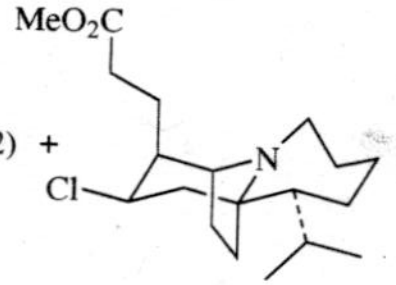 (37)	524
C_{22}	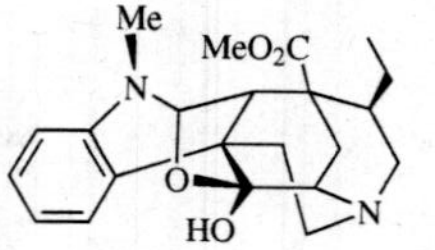	Et_3SiH, TFA	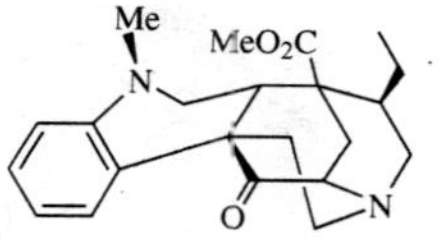 (—)	770
C_{23}	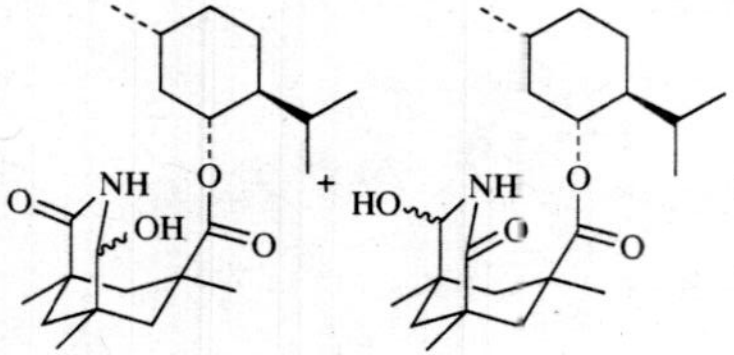	Et_3SiH, TFA	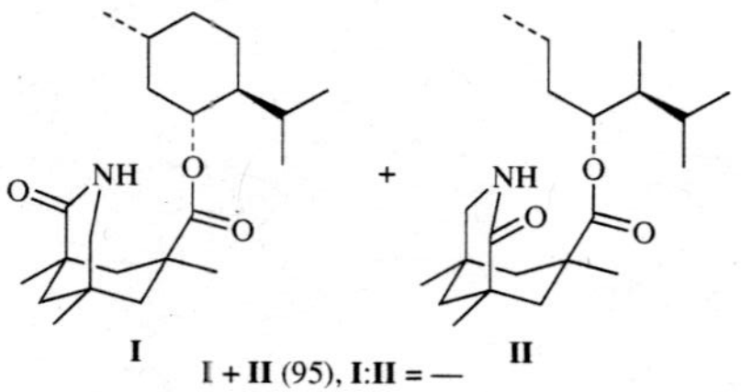 **I** + **II** (95), **I**:**II** = —	776
C_{30}	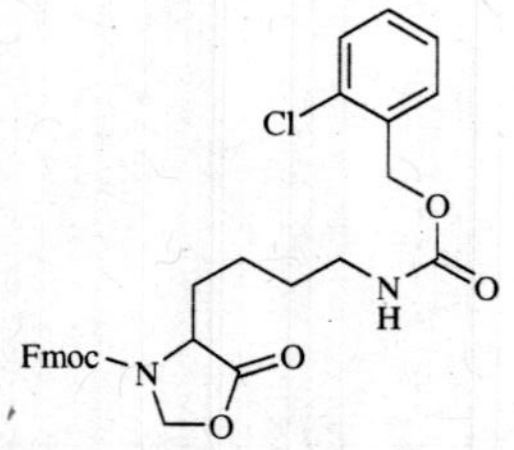	Et_3SiH (3 eq), TFA. $CHCl_3$, rt, 22 h	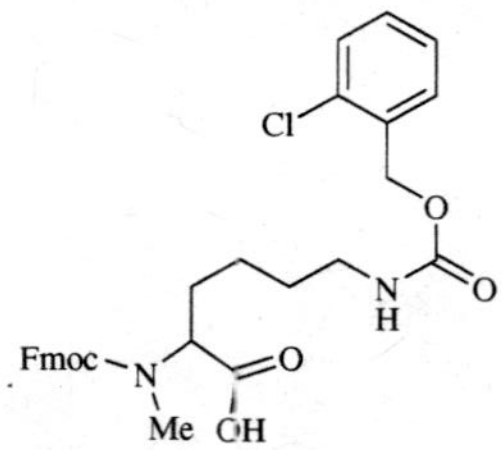 (15)	528

TABLE 22. ORGANOSILANE REDUCTION OF ENAMINES

Enamine	Conditions	Product(s) and Yield(s) (%)	Refs.
C_8 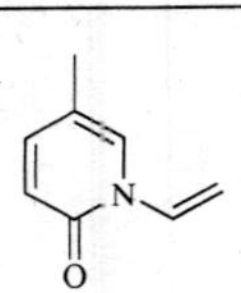	Et_3SiH (5 eq), TFA (10 eq), 65°, 25 h	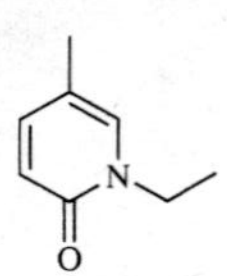 (67)	235
C_{10} 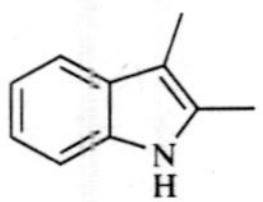	Et_3SiH (1.4 eq), TFA, 60°, 48 h	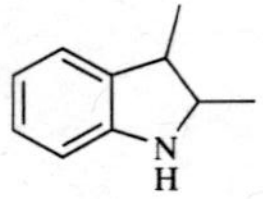 (80) cis:trans = 42:58	533
C_{11} 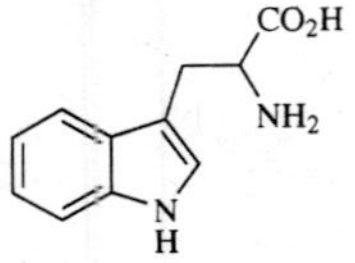	Et_3SiH (5 eq), TFA, CH_2Cl_2, rt, 2 h	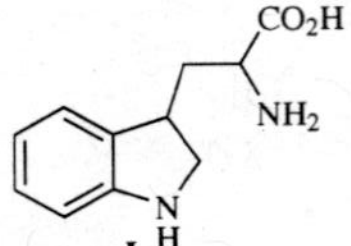 **I** (85%)[a]	532
	$(i\text{-}Pr)_3SiH$ (5 eq), TFA, CH_2Cl_2, rt, 2 h	**I** (4%)[a]	532
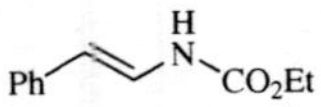	Et_3SiH (1.5 eq), TFA, −10°, 0.5 h	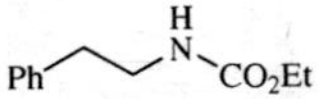 (92)	535
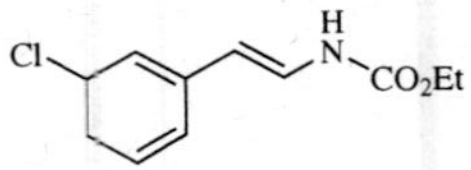	Et_3SiH (1.5 eq), TFA, −10°, 0.5 h	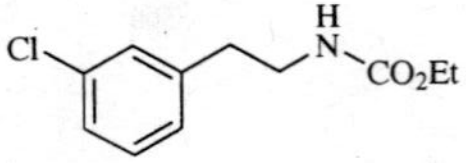 (94)	535
C_{12} 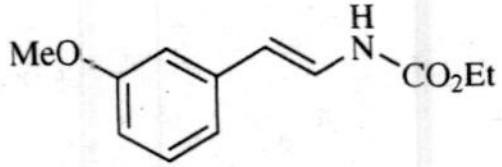	Et_3SiH (1.5 eq), TFA, −10°, 0.5 h	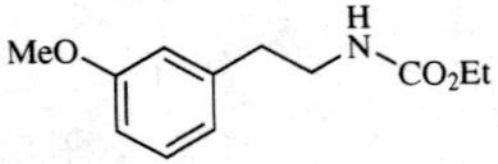 (99)	535

C_{12}

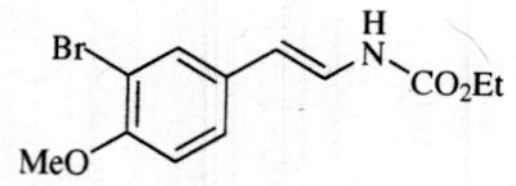

Et_3SiH (1.5 eq), TFA, –10°, 0.5 h

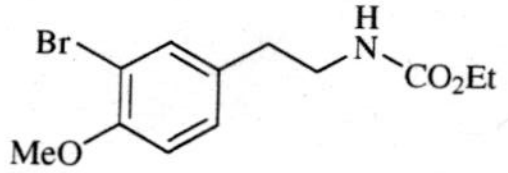

(99) 535

Et_3SiD (1.5 eq), TFA-d_1, –10°, 0.5 h

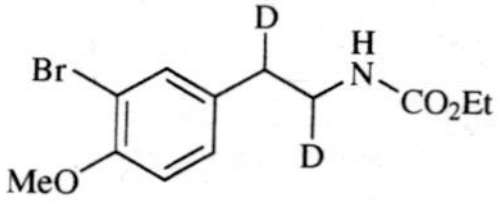

(98) 535

C_{13}

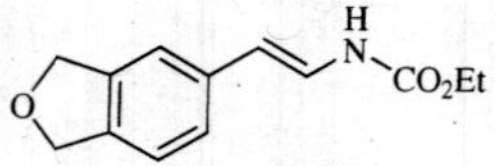

Et_3SiH (1.5 eq), TFA, –10°, 0.5 h

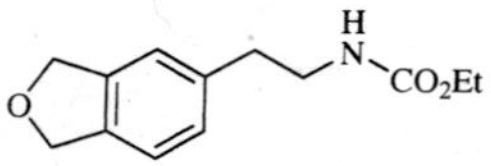

(97) 535

C_{14}

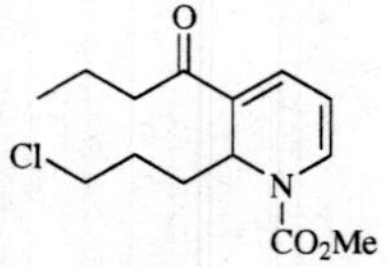

Et_3SiH (1.05 eq), TFA (1 eq), CH_2Cl_2, 0°, 6 h; rt, 6 h

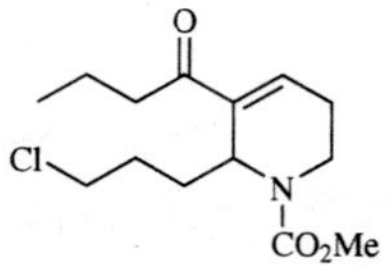

(56) 536

C_{14-21}

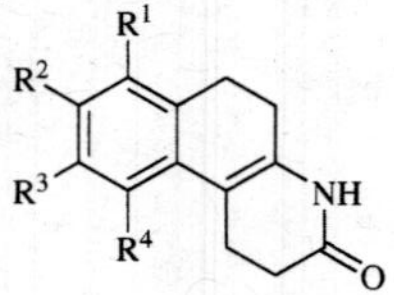

Et_3SiH (3 eq), TFA, CH_2Cl_2, rt

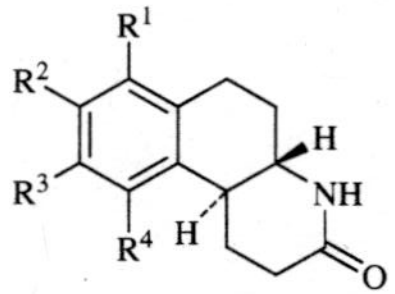

538

R^1	R^2	R^3	R^4	Time	
OMe	H	H	H	9 h	(95)
OMe	OMe	H	H	6.5 h	(80)
OMe	H	H	OMe	5 h	(80)
H	OBn	OMe	H	20 h	(33)

TABLE 22. ORGANOSILANE REDUCTION OF ENAMINES (*Continued*)

Enamine	Conditions	Product(s) and Yield(s) (%)	Refs.
C_{15} 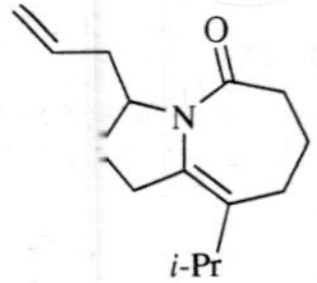	Et_3SiH, TFA, CH_2Cl_2, 15 min	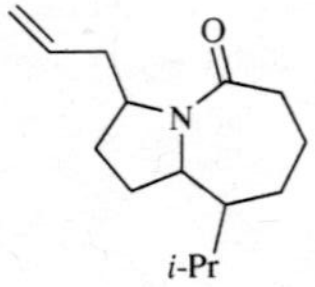 (74) dr = 1:1	524
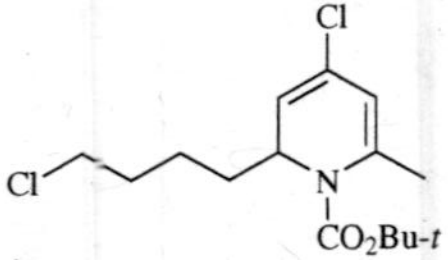	Et_3SiH, TFA, CH_2Cl_2, –42°	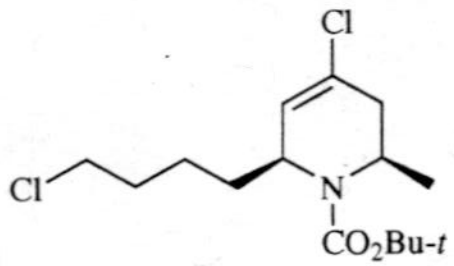 (70)	777
C_{16-17} 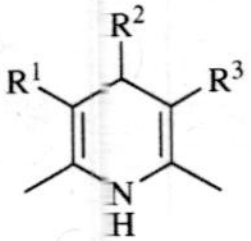	Et_3SiH (1 eq), TFA, 20°	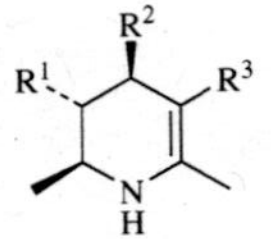	529
C_{16-19} 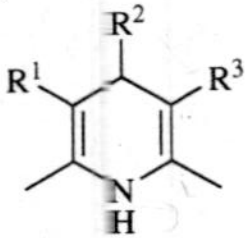	Et_3SiH (3 eq), TFA, 50°	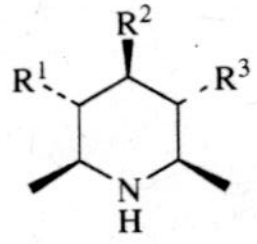	529

Products for Et_3SiH (1 eq), TFA, 20° (ref. 529):

R^1	R^2	R^3	
CO_2Me	$3\text{-}O_2NC_6H_4$	CN	(63)
CO_2Me	Ph	CO_2Me	(58)
CO_2Me	$2\text{-}O_2NC_6H_4$	CO_2Me	(66)
CO_2Me	$2\text{-}CF_3C_6H_4$	CO_2Me	(69)
Ac	$3\text{-}O_2NC_6H_4$	Ac	(64)

Products for Et_3SiH (3 eq), TFA, 50° (ref. 529):

R^1	R^2	R^3	
CO_2Me	$3\text{-}O_2NC_6H_4$	CN	(89)
CO_2Me	Ph	CO_2Me	(55)
CO_2Me	$2\text{-}O_2NC_6H_4$	CO_2Me	(94)
CO_2Me	$2\text{-}CF_3C_6H_4$	CO_2Me	(56)
Ac	$3\text{-}O_2NC_6H_4$	Ac	(60)
EtO_2C	$3\text{-}O_2NC_6H_4$	EtO_2C	(72)

C_{17} 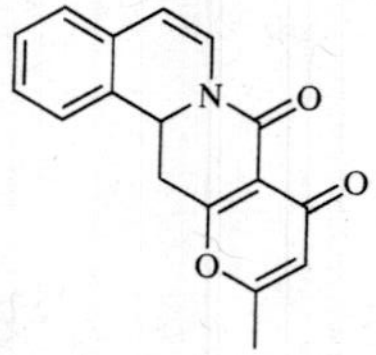	Et_3SiH (9 eq), TFA, $CHCl_3$, 20°, 24 h	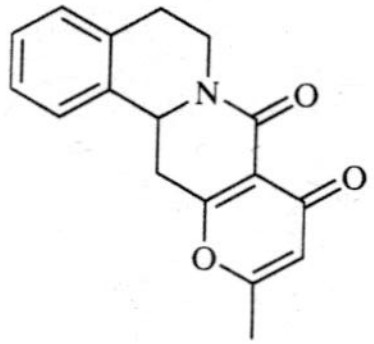 (75)	537
C_{18} 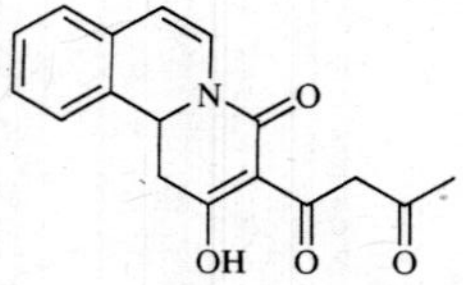	Et_3SiH, TFA, $CHCl_3$, 20°, 24 h	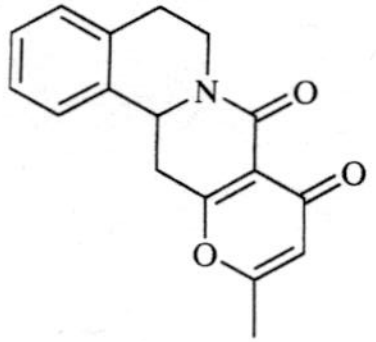 (—)	537
C_{20} 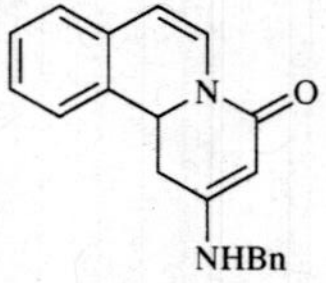	Et_3SiH, TFA, $CHCl_3$, 20°, 24 h	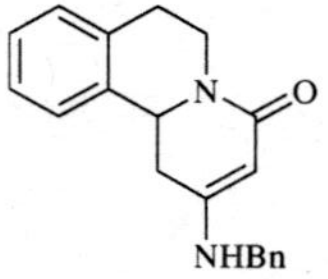 (—)	537
C_{21} 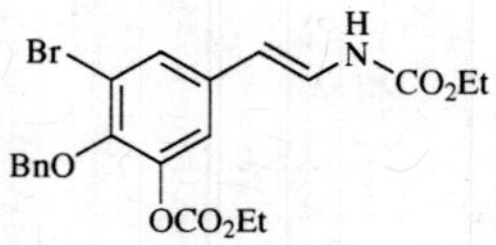	Et_3SiR^1 (1.5 eq), acid, −10°, 0.5 h	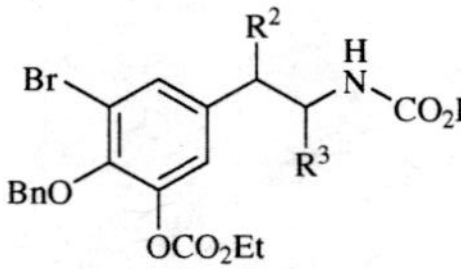	535

R^1	Acid
D	TFA
H	TFA-d_1
H	TFA

R^2	R^3	
H	D	(63)
D	H	(77)
H	H	(94)

TABLE 22. ORGANOSILANE REDUCTION OF ENAMINES (*Continued*)

Enamine		Conditions	Product(s) and Yield(s) (%)	Refs.
C_{21-23}	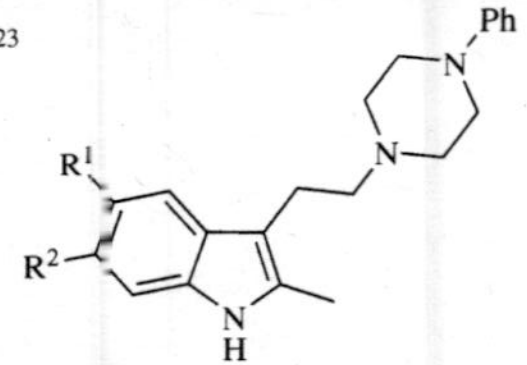	Et_3SiH (2 eq), TFA, 50°, 64 h	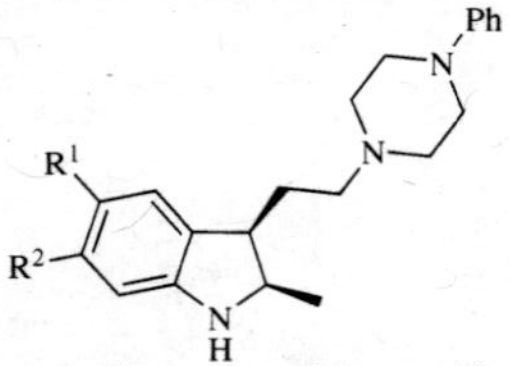 R[1] / R[2]: Br / H (25); F / H (32); Cl / Cl (14); OMe / OMe (80)	534
C_{22}	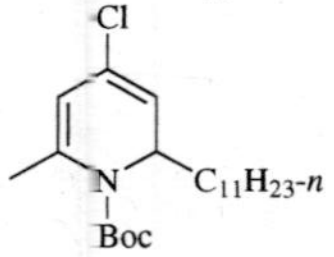	Et_3SiH (6 eq), TFA (10 eq), CH_2Cl_2, –42°, 4 h	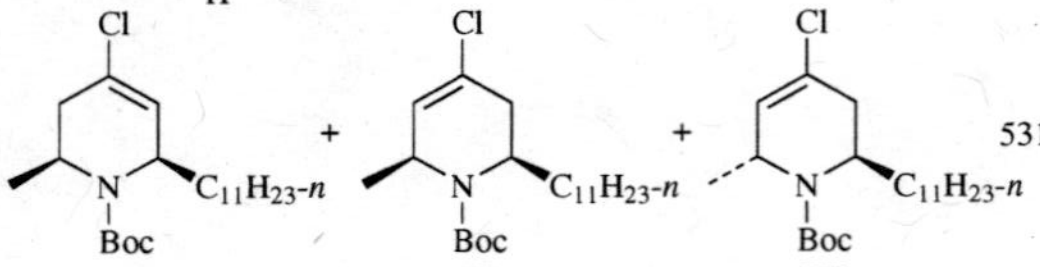 I + II + III **I + II + III** (79) **I:II:III** = 75:10:15	531
		$(i\text{-}Pr)_3SiH$ (6 eq), TFA (10 eq), CH_2Cl_2, –42°, 4 h	**I + II + III** (95), **I:II:III** = 78:9:13	531
C_{27}	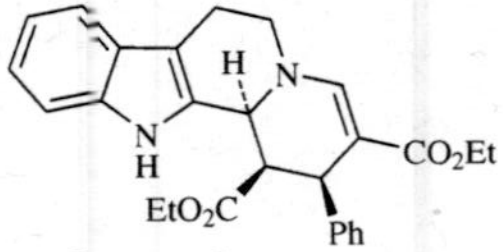	Et_3SiH (2 eq), TFA, rt, 1 h	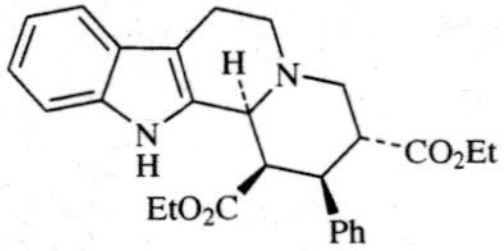 (81)	530
	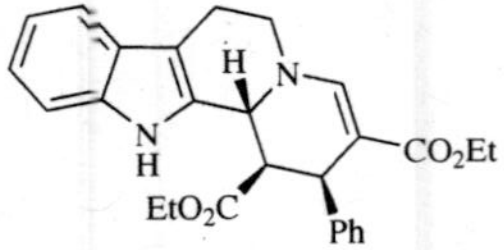	R_3SiH (x eq), TFA, 20°	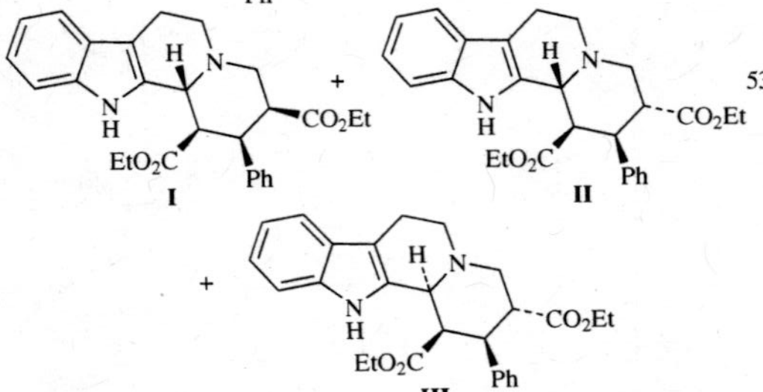 I + II + III	530

R_3Si	x	Time	I + II + III	I:II:III
Et_3Si	2	1.5 h	(—)	19:17:64
Et_3Si	2	18 h	(—)	—
Et_3Si	2	1 h	(—)	—
$PhMe_2Si$	3	18 h	(—)	46:44:10
Ph_3Si	3	18 h	(—)	—

C_{31}

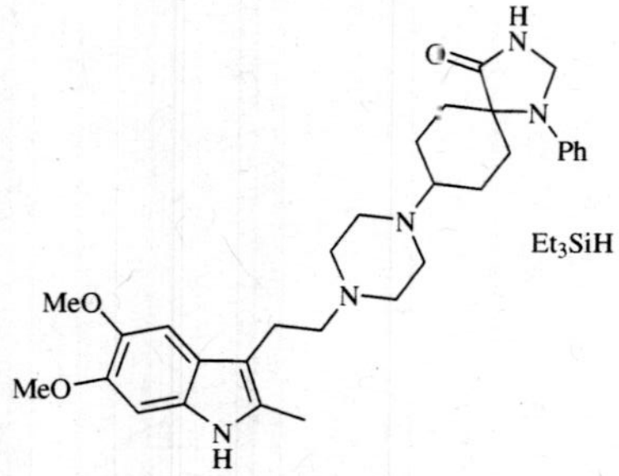

Et_3SiH (2 eq), TFA, 50°, 64 h

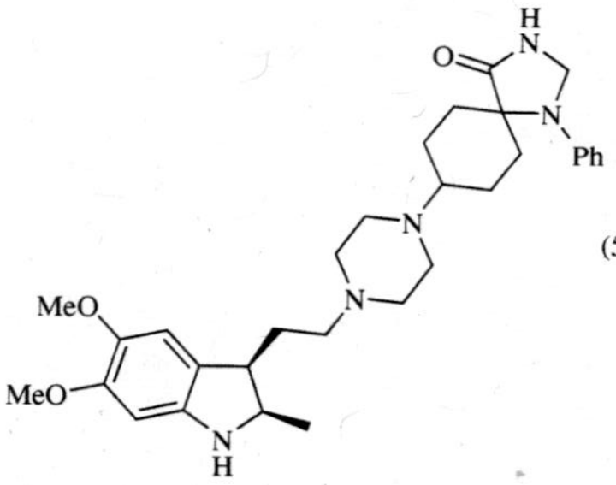

(52) 534

[a] The yield was determined by NMR spectroscopy.

TABLE 23. ORGANOSILANE REDUCTION OF IMINES

Imine	Conditions	Product(s) and Yield(s) (%)	Refs.
C_8			
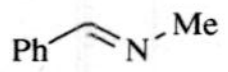	$PhSiH_3$ (1 eq), catalyst (10 mol%), THF, rt, 20 h	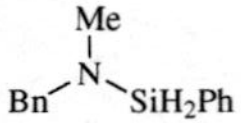 (2) + 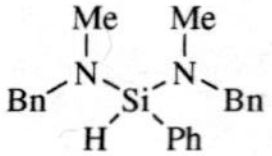 (95)	544
C_{8-13}			
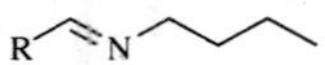	Cl_3SiH, MeCN, reflux, 4 h	 R: *n*-Pr (47); Ph (53)	542
C_{8-14}			
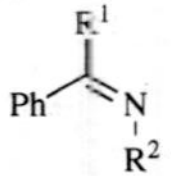	R_3SiH (1.1 eq), catalyst (0.5 mol%), C_6H_6, rt, 1 h	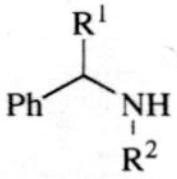	546

R^1	R^2	$R^3{}_3Si$	Catalyst	Temp	Time	
H	Me	Et_2HSi	$(Ph_3P)_3RhCl$	30°	0.5 h	(95)
H	Me	Ph_2HSi	$(Ph_3P)_3RhCl$	0°	3 h	(92)
H	Me	Et_3Si	$(Ph_3P)_3RhCl$	100°	20 h	(65)
H	Me	Et_3Si	$PdCl_2$	55°	24 h	(90)
H	*n*-Bu	Et_2HSi	$(Ph_3P)_3RhCl$	55°	1.5 h	(96)
H	*n*-Bu	PhMeHSi	$(Ph_3P)_3RhCl$	55°	0.5 h	(93)
H	Ph	Et_2HSi	$(Ph_3P)_3RhCl$	50°	2 h	(96)
H	Ph	Et_3Si	$(Ph_3P)_3RhCl$	100°	15 h	(91)
Me	Ph	Et_2HSi	$(Ph_3P)_3RhCl$	55°	72 h	(85)

Imine	Conditions	Product(s) and Yield(s) (%)	Refs.
C_{8-20}			
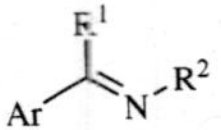	$PhMe_2SiH$ (1.05 eq), $(C_6F_5)_3B$ (5-10 mol%), MeC_6H_5	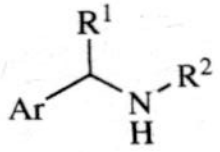	120

Ar	R^1	R^2	Temp	Time	
Ph	H	Me	70°	48 h	(0)
Ph	H	allyl	70°	26 h	(57)
Ph	H	*t*-Bu	rt	0.5 h	(80)
Ph	H	Boc	rt	0.5 h	(60)
Ph	H	SO_2Ph	rt	0.5 h	(93)
Ph	H	Ph	rt	1 h	(95)
Ph	H	Bn	70°	3 h	(91)
Ph	H	$4\text{-}MeOC_6H_4$	70°	1.5 h	(>95)
Ph	H	$2\text{-}MeOC_6H_4$	rt	2 h	(86)
$4\text{-}MeOC_6H_4$	H	Bn	70°	17.5 h	(97)
Ph	Me	Bn	rt	4 h	(95)
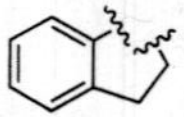		Bn	rt	23 h	(95)
Ph	Ph	Bn	rt	96 h	(96)

C_9

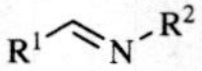

$PhSiH_3$ (1 eq), catalyst (10 mol%), THF, rt, 20 h

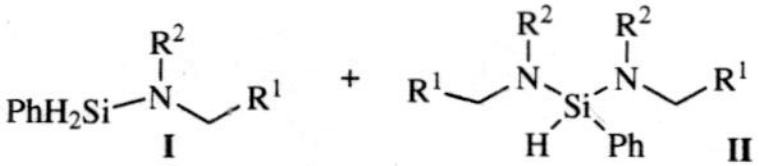

R^1	R^2	I	II
$n\text{-}C_7H_{15}$	Me	(3)	(89)
i-Pr	$n\text{-}C_5H_{11}$	(21)	(64)
n-Pr	$n\text{-}C_5H_{11}$	(28)	(43)

544

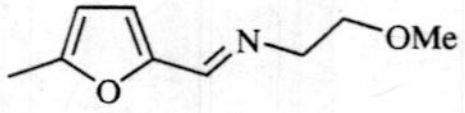

Cl_3SiH (1.2 eq), $BF_3\bullet OEt_2$, C_6H_6, reflux, 4 h

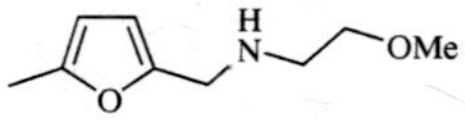 (82)

545

C_{10}

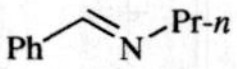

Cl_3SiH (1.2 eq), $BF_3\bullet OEt_2$, C_6H_6, reflux, 4 h

 (89)

545

TABLE 23. ORGANOSILANE REDUCTION OF IMINES (*Continued*)

Imine	Conditions	Product(s) and Yield(s) (%)	Refs.
C_{10-22}			
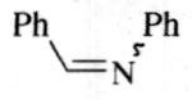	$PhMe_2SiH$ (1.1-1.2 eq), TFA, rt, 16 h	 (77)	276
C_{11}			
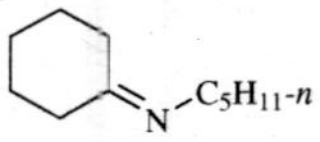	$PhSiH_3$ (1 eq), **230** (10 mol%), THF, rt, 20 h	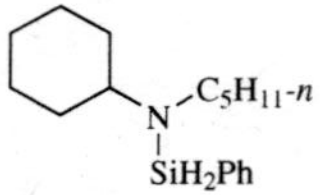 (56)	544
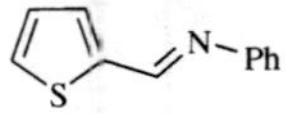	Et_3SiH (1.1-1.2 eq), HCO_2H, HOAc, KU-1, 55°, 5 h	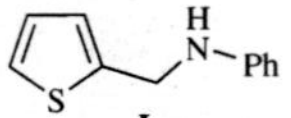 **I** (54)	208
	Et_3SiH (1 eq), TFA, (3 eq), 70°, 6 h	**I** (93)	541
	Cl_3SiH (1.2 eq), BF_3•OEt_2, C_6H_6, reflux, 4 h	**I** (37)	545
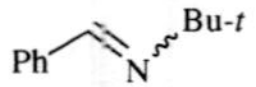	Et_3SiH (1 eq), TFA (3 eq), 70°, 30 h	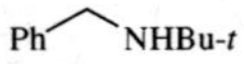 (22)	541
C_{11-14}			
	Cl_3SiH (1.5 eq), CH_2Cl_2:DMF (4:1), 0°, 4 h	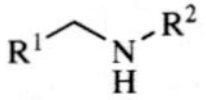	318

R^1	R^2	Yield (%)
2-C_4H_3O	Ph	(99)
2-C_4H_3S	Ph	(99)
2-C_4H_3O	Ts	(86)
Ph	Ph	(97)
Ph	4-$MeOC_6H_4$	(88)
Ph	Bn	(86)

C_{12}

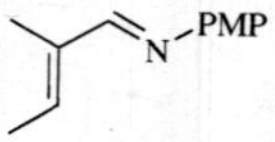	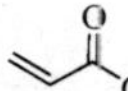 OC_6F_5 (2.5 eq), Et_2MeSiH, (2.5 eq), $[Ir(cod)Cl]_2$ (2.5 mol%), $P(OPh)_3$ (10 mol%), CH_2Cl_2, 18 h	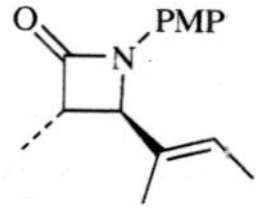 (60) trans:cis > 20:1	476
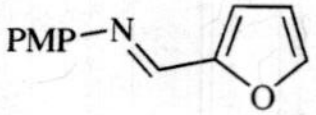	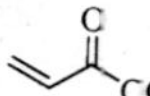 OC_6F_5 (2.5 eq), Et_2MeSiH, (2.5 eq), $[Ir(cod)Cl]_2$ (2.5 mol%), $P(OPh)_3$ (10 mol%), CH_2Cl_2, 18 h	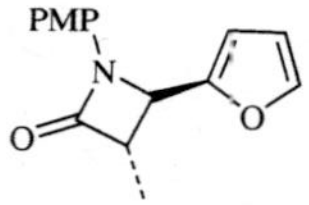 (78) trans:cis > 20:1	476
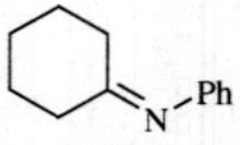	Cl_3SiH (1.2 eq), $BF_3{\bullet}OEt_2$, C_6H_6, reflux, 4 h	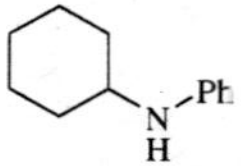 (81)	545
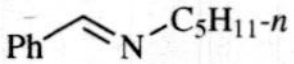	$PhSiH_3$ (1 eq), **230** (10 mol%), THF, rt, 20 h	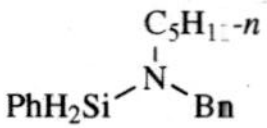 (52) + 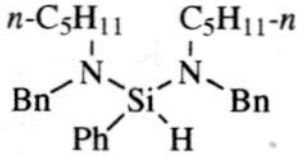 (35)	544
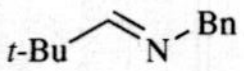	PMHS, EtOH, *n*-butyltris(EH)tin, rt	 (75)	543
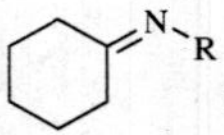 R = Ph, *c*-C_6H_{11}	Cl_3SiH, MeCN, reflux, 4 h	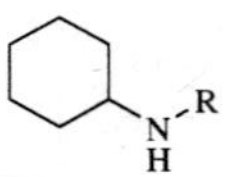 (48)	542

TABLE 23. ORGANOSILANE REDUCTION OF IMINES (*Continued*)

Imine	Conditions	Product(s) and Yield(s) (%)	Refs.
C_{12-15}			
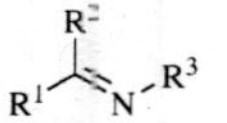	PMHS (3 eq), n-BuSn[O_2CCH(Et)Bu-n]$_3$ (0.1 eq), EtOH, rt	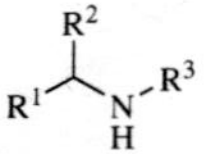	543
R^1 / R^2 / R^3	Time		
t-Bu / H / Bn	20 h	(75)	
Ph / H / Ph	7 h	(82)	
Ph / H / Bn	10 h	(76)	
n-C_6H_{13} / Me / Bn	6 h	(81)	
C_{13}			
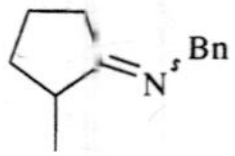	PMHS, EtOH, rt, n-butyltris(EH)tin (0.1 eq), rt, 9 h	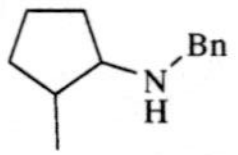 (84)	543
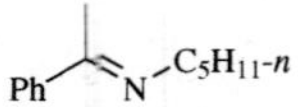	$PhSiH_3$ (1 eq), **230** (10 mol%), THF, rt, 20 h	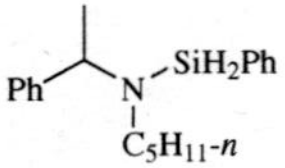 (3) + 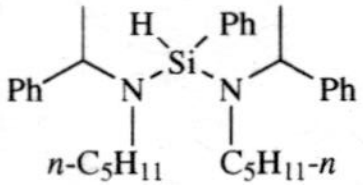 (17)	544
	Cl_3SiH, MeCN, reflux, 4 h	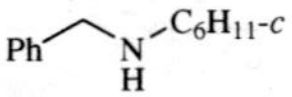 (60)	542
	Cl_3SiH (1.2 eq), C_6H_6, reflux, 4 h	 **I** (17)	545
	Cl_3SiH (1.2 eq), BF_3•OEt_2, C_6H_6, reflux, 4 h	**I** (40)	545
	PMHS, EtOH, rt, n-butyltris(EH)tin	**I** (82)	543
	Et_3SiH (1 eq), TFA (3 eq), 70°, 1.5 h	**I** (80)	541

Et_3SiH (1.1-1.2 eq), HCO_2H, HOAc, KU-1, 55°, 5 h	**I** (71)		208
PMHS (5 eq), $ZnCl_2$ (2 eq), Et_2O, rt, 10 h	**I** (67)		539
PMHS (5 eq), $ZnCl_2$ (2 eq), Et_2O, rt, 10 h	**I** (67)		539
$PhSiH_3$ (1 eq), **230** (10 mol%), THF, rt, 20 h	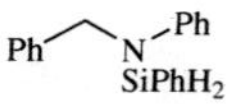 (15) + 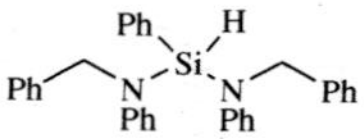 (47)		544
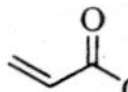 CR (2.5 eq), Et_2MeSiH, (1 eq) **230** (2.5 mol%), ligand (x mol%), CH_2Cl_2, 18 h	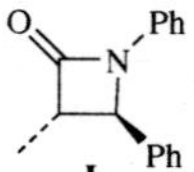 **I**	(20) dr > 20:1	476

R	Catalyst	Ligand	x	Temp	
1-$C_{10}H_7$	$[Ir(cod)Cl]_2$	$P(OPh)_3$	10	50°	(14)
Ph	$[Ir(cod)Cl]_2$	$P(OPh)_3$	10	rt	(13)
4-$O_2NC_6H_4$	$[Ir(cod)Cl]_2$	$P(OPh)_3$	10	rt	(56)
C_6F_5	$[Ir(cod)Cl]_2$	$P(OPh)_3$	10	rt	(68)
C_6F_5	$[Ir(cod)Cl]_2$	none	—	rt	(28)
C_6F_5	none	$P(OPh)_3$	10	rt	(0)
C_6F_5	$[Ir(cod)Cl]_2$	$P(OPh)_3$	5	rt	(58)

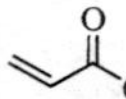 OPh (2.5 eq), Et_2MeSiH, $[Rh(cod)Cl]_2$ (2.5 mol%), dppe (5 mol%), 80°	**I** (20) dr > 20:1	476

TABLE 23. ORGANOSILANE REDUCTION OF IMINES (*Continued*)

Imine	Conditions	Product(s) and Yield(s) (%)	Refs.
C_{13}			
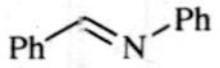	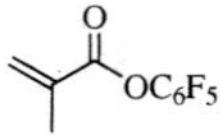 (2.5 eq), Et_2MeSiH, (2.5 eq), $[Ir(cod)Cl]_2$ (2.5 mol%), $P(OPh)_3$ (10 mol%), CH_2Cl_2, 18 h	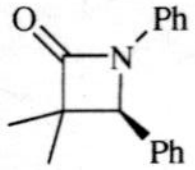 (68) trans:cis = —	476
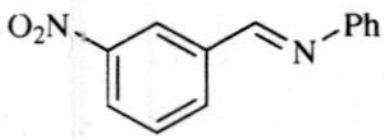	PMHS (5 eq), $ZnCl_2$ (2 eq), Et_2O, rt, 12 h	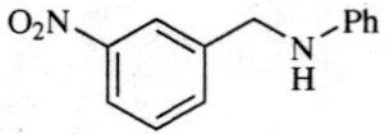 (75)	539
C_{13-14}			
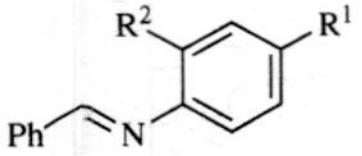	Et_3SiH (1 eq), TFA (3 eq), 60-70°, 30-60 min	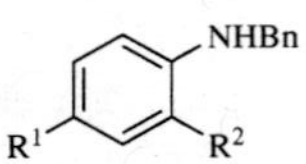	540
		R^1 / R^2: H / H (78); Me / H (86); H / Me (95)	
C_{13-15}			
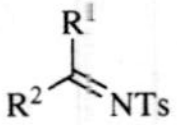	$(MeO)_3SiH$ (1.2 eq), LiOMe (4 mol%), THF, rt, 0.5 h	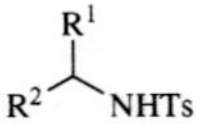	294

R^1	R^2	Time	Yield
—$(CH_2)_5$—		48 h	(62)
c-C_6H_{11}	H	67 h	(84)
Ph	H	42 h	(100)
3-$O_2NC_6H_4$	H	56 h	(92)
n-C_5H_{11}	Et	50 h	(80)
Ph	Me	59 h	(63)
4-MeC_6H_4	H	59 h	(97)
4-$MeOC_6H_4$	H	50 h	(80)

C_{13-16}			
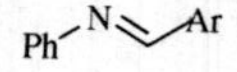	Cl_3SiH, MeCN, reflux, 4 h	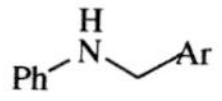	542
Ar			
Ph		(79)	
4-ClC_6H_4		(69)	
2-ClC_6H_4		(62)	
4-$MeOC_6H_4$		(78)	
2-$MeOC_6H_4$		(88)	
3-$MeOC_6H_4$		(76)	
4-(*i*-Pr)C_6H_4		(73)	
C_{14}			
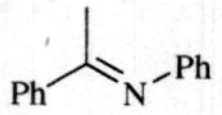	PMHS (5 eq), $ZnCl_2$ (2 eq), Et_2O, rt, 15 h	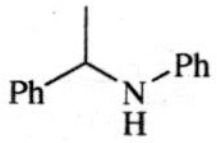 (60)	539
	PMHS, *n*-butyltris(2H)tin, EtOH, rt,	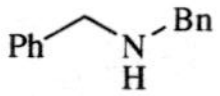 **I** (76)	543
	PMHS (5 eq), $ZnCl_2$ (2 eq), Et_2O, rt, 10 h	**I** (73)	539
	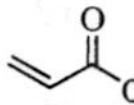 OC_6F_5 (2.5 eq), Et_2MeSiH (2.5 eq), $[Ir(cod)Cl]_2$ (2.5 mol%), $P(OPh)_3$ (10 mol%), CH_2Cl_2, 18 h	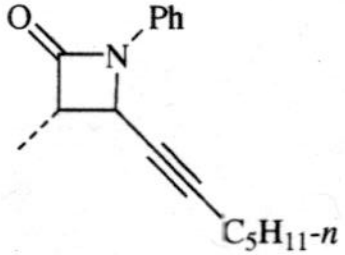 (50) trans:cis > 20:1	476
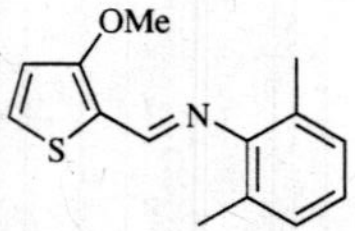	Cl_3SiH (1.2 eq), $BF_3{\bullet}OEt_2$, C_6H_6, reflux, 4 h	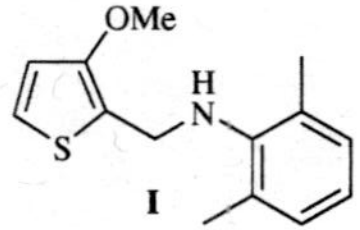 **I** (61)	545
	Cl_2SiH_2 (1.2 eq), C_6H_6, reflux, 4 h	**I** (90)	545

TABLE 23. ORGANOSILANE REDUCTION OF IMINES (*Continued*)

Imine	Conditions	Product(s) and Yield(s) (%)	Refs.
C_{15}			
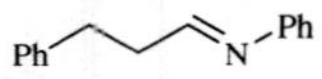	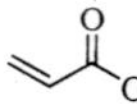 OC_6F_5 (2.5 eq), Et_2MeSiH (2.5 eq), $[Ir(cod)Cl]_2$ (2.5 mol%), $P(OPh)_3$ (10 mol%), CH_2Cl_2, 18 h	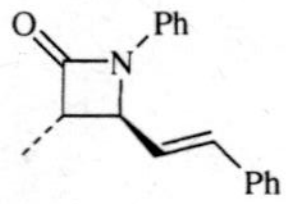 (70) trans:cis = 5.2:1	476
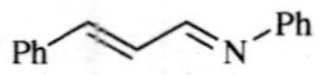	PMHS (5 eq), $ZnCl_2$ (2 eq), Et_2O, rt, 20 h	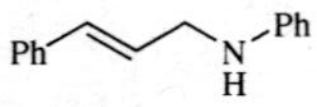 (55)	539
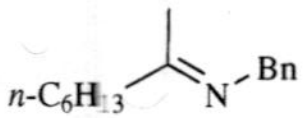	PMHS, *n*-butyltris(EH)tin, EtOH, rt	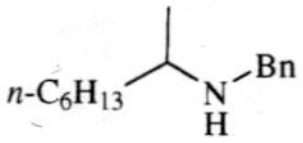 (81)	543
C_{16}			
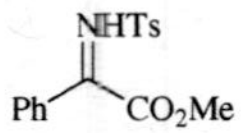	Ph_2SiH_2 (1.2 eq), catalyst (1 eq), CH_2Cl_2	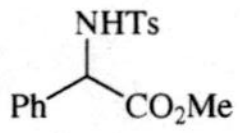	373
	Catalyst / Temp		
	$AlCl_3$ 20°	(90)	
	$TiCl_4$ 20°	(90)	
	BF_3•OEt_2 20°	(63)	
	$ZnCl_2$ 40°	(48)	
C_{17}			
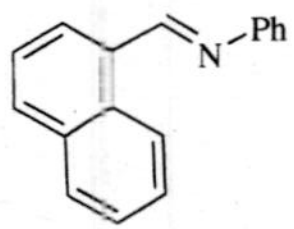	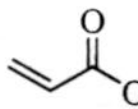 OC_6F_5 (2.5 eq), Et_2MeSiH (2.5 eq), $[Ir(cod)Cl]_2$ (2.5 mol%), $P(OPh)_3$ (10 mol%), CH_2Cl_2, 18 h	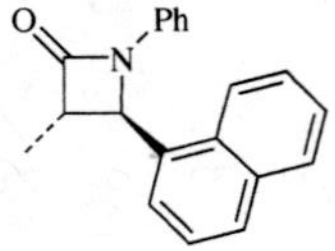 (80) trans:cis > 20:1	476

	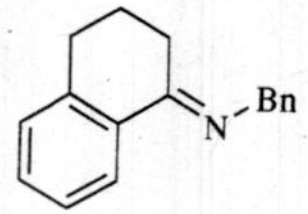	PMHS (3 eq), *n*-butyltris(EH)tin (0.1 eq), EtOH, rt, 10 h	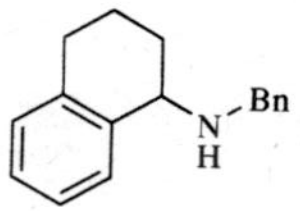 (82)	543
	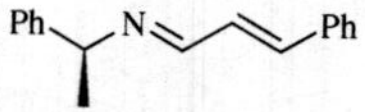	PMHS (5 eq), $ZnCl_2$ (2 eq), Et_2O, rt, 24 h	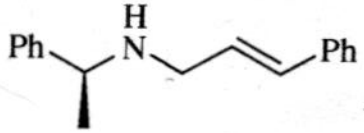 (50)	539
C_{18}	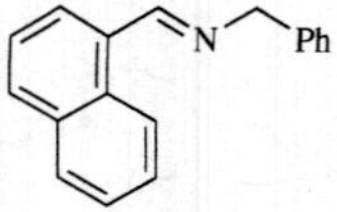	PMHS (5 eq), $ZnCl_2$ (2 eq), Et_2O, rt, 12 h	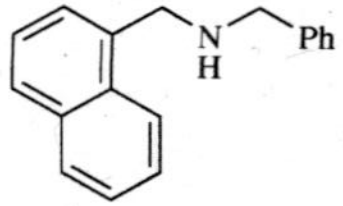 (63)	539
	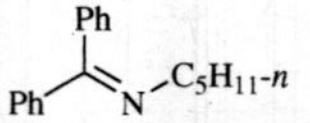	$PhSiH_3$ (1 eq), **230** (10 mol%), THF, rt, 20 h	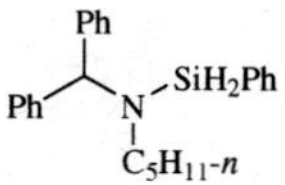 (0) + 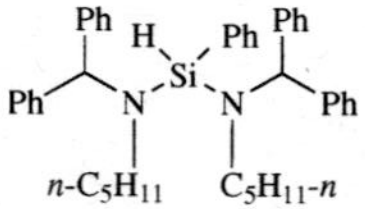 (0)	544
C_{19}	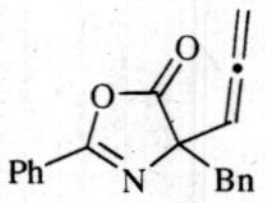	Et_3SiH, Cl_3CCO_2H	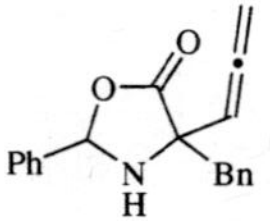 (0)	778

TABLE 23. ORGANOSILANE REDUCTION OF IMINES (*Continued*)

Imine	Conditions	Product(s) and Yield(s) (%)	Refs.
C_{23-26} 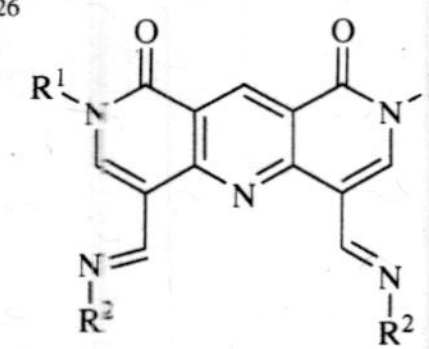	Et_2SiH_2 (2.2 eq), $(Ph_3P)_3RhCl$, CH_2Cl_2, rt, 16 h	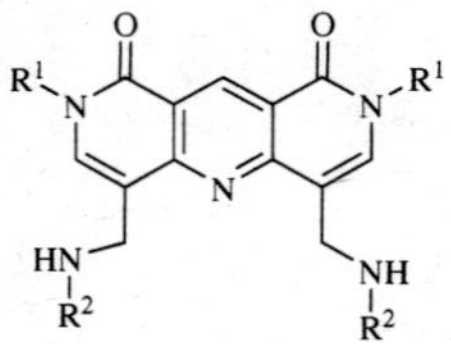	547
R^1 / R^2			
$Cl(CH_2)_4$ / Ph		(81)	
$Cl(CH_2)_4$ / 3-$O_2NC_6H_4$		(82)	
$Cl(CH_2)_4$ / PhCH(OH)CHMe		(78)	
allyl / PhCH(OH)CHMe		(69)	

TABLE 24. ORGANOSILANE REDUCTION OF HYDROXYLIMINES

Hydroxylimine	Conditions	Product(s) and Yield(s) (%)	Refs.
C_5			
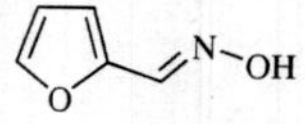	PMHS (3 eq), (t-BuCO)$_2$O (1.1 eq), 10% Pd-C, 40-50°, 3 h	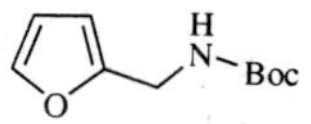 (65)	548
C_7			
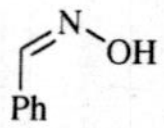	PMHS (3 eq), (t-BuCO)$_2$O (1.1 eq), 10% Pd-C, 40-50°, 3 h	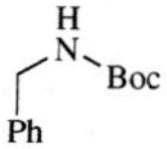 (85)	548
C_8			
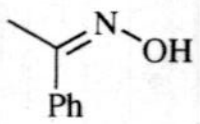	PMHS (3 eq), (t-BuCO)$_2$O (1.1 eq), 10% Pd-C, 40-50°, 4 h	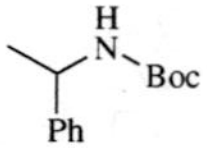 (90)	548
C_9			
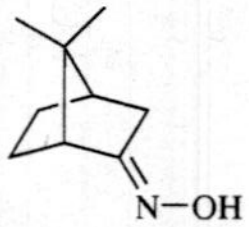	PMHS (3 eq), (t-BuCO)$_2$O (1.1 eq), 10% Pd-C, 40-50°, 5 h	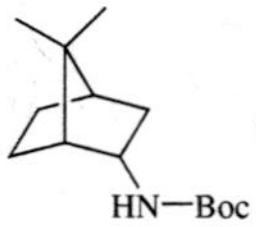 (82)	548
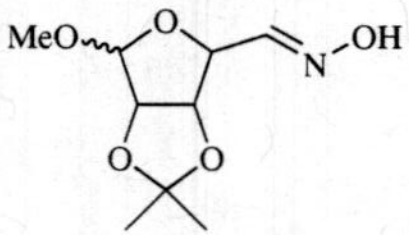	PMHS (3 eq), (t-BuCO)$_2$O (1.1 eq), 10% Pd-C, 40-50°, 6 h	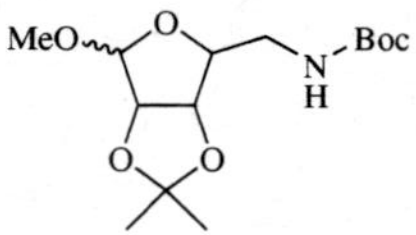 (85)	548
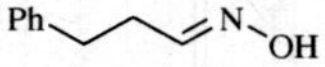	PMHS (3 eq), (t-BuCO)$_2$O (1.1 eq), 10% Pd-C, 40-50°, 4 h	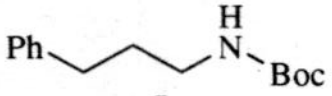 **I** (80)	548
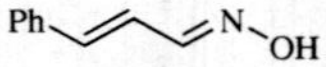	PMHS (3 eq), (t-BuCO)$_2$O (1.1 eq), 10% Pd-C, 40-50°, 3 h	**I** (75)	548

TABLE 24. ORGANOSILANE REDUCTION OF HYDROXYLIMINES (*Continued*)

Hydroxylimine	Conditions	Product(s) and Yield(s) (%)	Refs.
C_{9-15}			
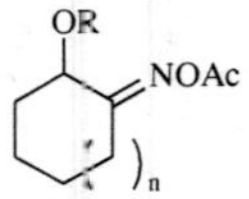	Et_3SiH (1.2 eq), TMSOTf (0.1 eq), CH_2Cl_2	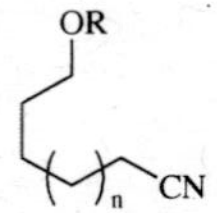	552
R n / Temp Time			
Me 1	0° 6 h	(65)	
Me 3	rt 10 h	(80)	
Bn 1	0° 2 h	(66)	
Me 7	0° 2 h	(78)	
C_{10}			
	Et_3SiH, TFA	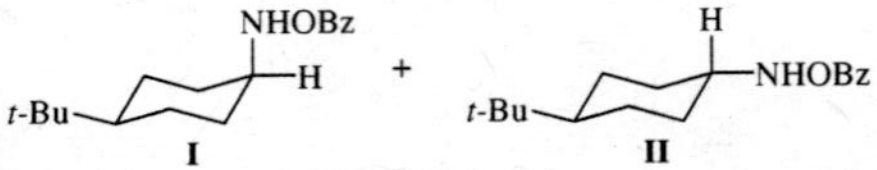 **I** + **II** (85), **I**:**II** = 5:1	550
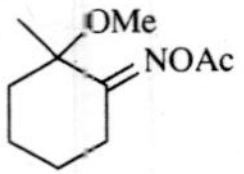	Et_3SiH (1.2 eq), TMSOTf (0.1 eq), CH_2Cl_2, 0°, 2 h	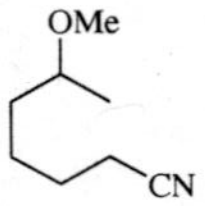 (81)	552
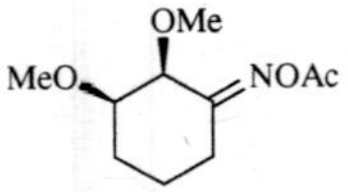	Et_3SiH (1.2 eq), TMSOTf (0.1 eq), CH_2Cl_2, rt, 9 h	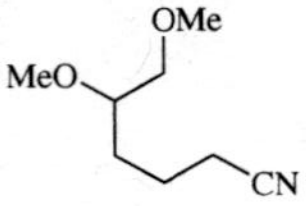 (82)	552
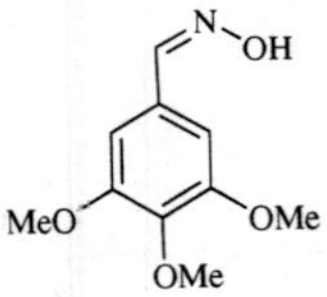	PMHS (3 eq), $(t\text{-}BuCO)_2O$ (1.1 eq), 10% Pd-C, 40-50°, 5 h	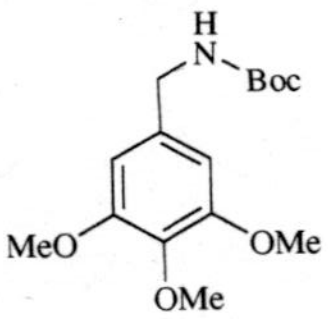 (82)	548

C_{10-16}

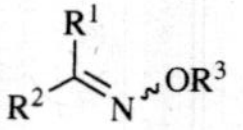

R^1	R^2	R^3
Ph	Me	Ac
—$(CH_2)_5$—		Bn
Ph	H	Bn
Ph	Me	Bz
n-C_7H_{15}	Me	Bn

$PhMe_2SiH$ (x eq), TFA or TFA-CH_2Cl_2 (1:1)

x	Temp	Time
1.2	rt	overnight
1.2	rt	24 h
1.2	rt	overnight
1.2	rt	overnight
2	50°	5 d

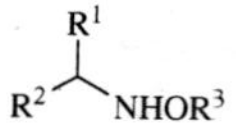

(67)
(65)
(75)
(78)
(23)

276, 551

C_{12}

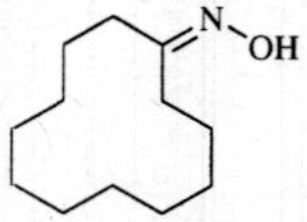

PMHS (3 eq), (t-BuCO)$_2$O (1.1 eq), 10% Pd-C, 40-50°, 7 h

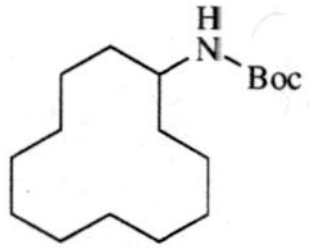

(80)

548

C_{12-13}

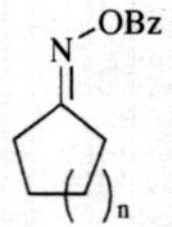

Et_3SiH, TFA

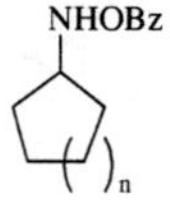

n	
1	(85)
2	(95)

550

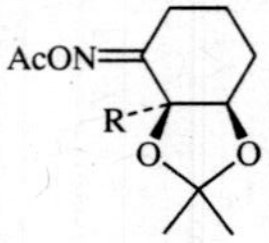

Et_3SiH (1.2 eq), TMSOTf (0.1 eq), CH_2Cl_2

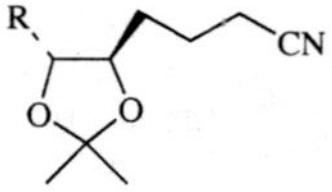

R	Temp	Time	
Me	rt	6 h	(61)
Et	0°	4 h	(81)

552

TABLE 24. ORGANOSILANE REDUCTION OF HYDROXYLIMINES (*Continued*)

Hydroxylimine	Conditions	Product(s) and Yield(s) (%)	Refs.
C_{12-23}			
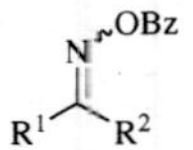	Et_3SiH, TFA, 0.5-6 h	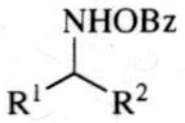	550
R^1 / R^2			
—$(CH_2)_4$—		(85)	
—$(CH_2)_5$—		(95)	
4-MeC_6H_4 / H		(54)	
Ph / Me		(0)	
$PhCH_2$ / Me		(90)	
—$(CH_2)_{11}$—		(82)	
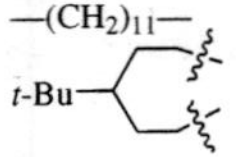		(85) cis:trans = 5:1	
C_{13}			
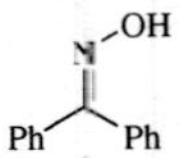	PMHS (3 eq), $(t\text{-}BuCO)_2O$ (1.1 eq), 10% Pd-C, 40-50°, 7 h	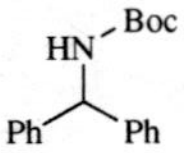 (75)	548
C_{13-20}			
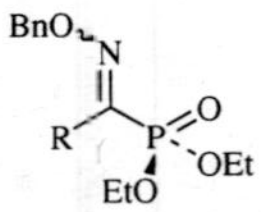	Et_3SiH (2 eq), TFA (as solvent), 40°, 12 h	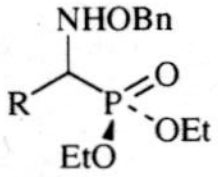	553
R			
Me		(66)	
Et		(62)	
i-Pr		(79)	
i-Bu		(73)	
Ph		(79)	
Bn		(65)	
$BnCH_2$		(54)	

C_{15} 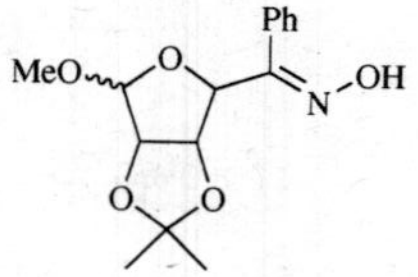	PMHS (3 eq), $(t\text{-BuCO})_2O$ (1.1 eq), 10% Pc-C, 40-50°, 6 h	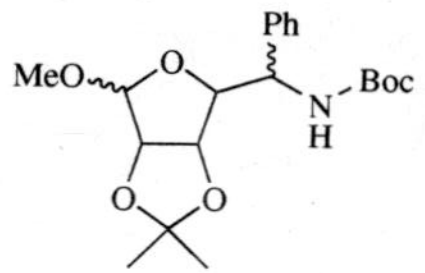 (80)	548
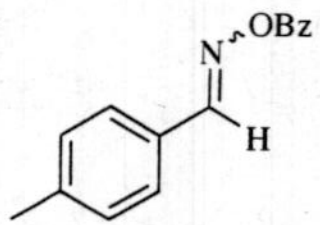	Et_3SiH (2 eq), TFA	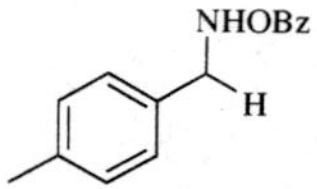 (54)	550
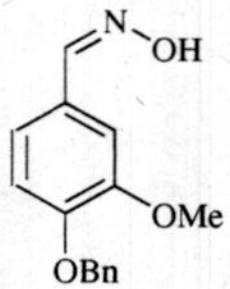	PMHS (3 eq), $(t\text{-BuCO})_2O$ (1.1 eq), 10% Pd-C, 40-50°, 3 h	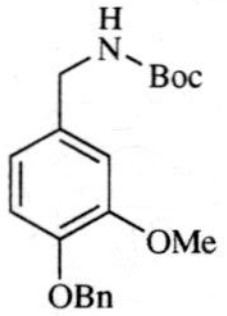 (80)	548
C_{16} 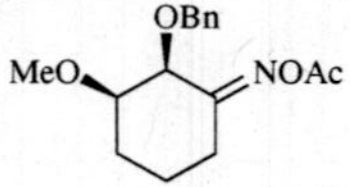	Et_3SiH (1.2 eq), TMSOTf (0.1 eq), CH_2Cl_2, rt, 3 h	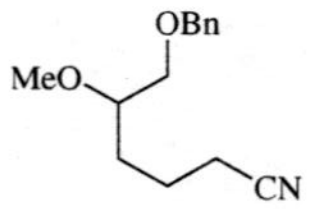 (90)	552
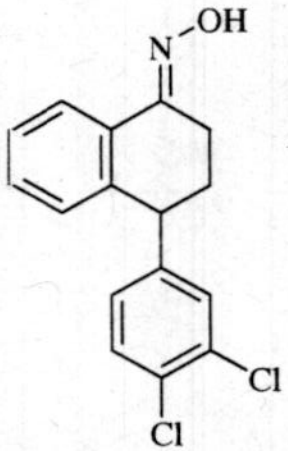	PMHS (3 eq), $(t\text{-BuCO})_2O$ (1.1 eq), 10% Pd-C, 40-50°, 6 h	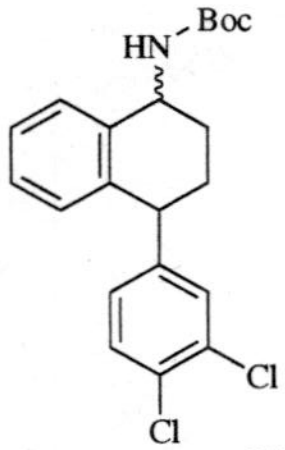 (90)	548
C_{17} 	Et_3SiH (2 eq), TFA	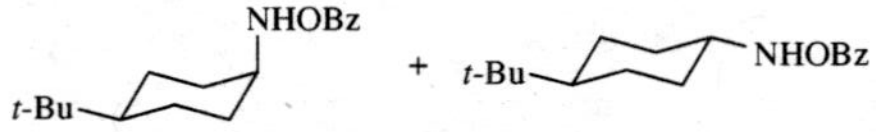 **I** **I** + **II** (85), **I**:**II** = 5:1 **II**	550

TABLE 24. ORGANOSILANE REDUCTION OF HYDROXYLIMINES (*Continued*)

Hydroxylimine	Conditions	Product(s) and Yield(s) (%)	Refs.
C_{18} 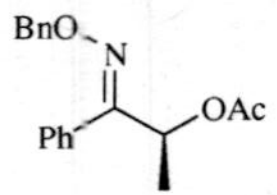	$PhMe_2SiH$ (1.2 eq), TFA, rt	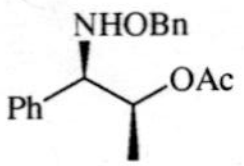 I + 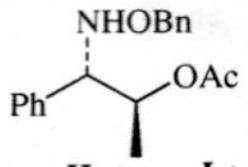 II; **I** + **II** (73), **I**:**II** = 99:1	551
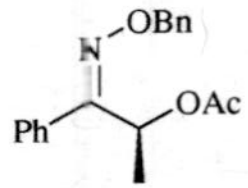	$PhMe_2SiH$ (1.2 eq), TFA, rt	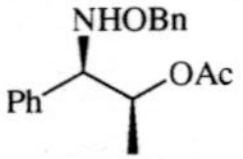 I + 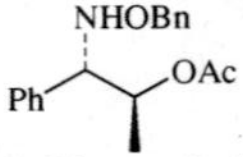 II; **I** + **II** (77), **I**:**II** = 25:75	551
C_{19} 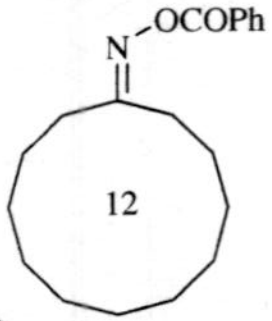	Et_3SiH (2 eq), TFA	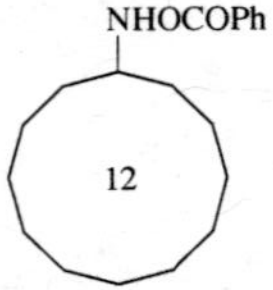 (73)	550
C_{35-42} 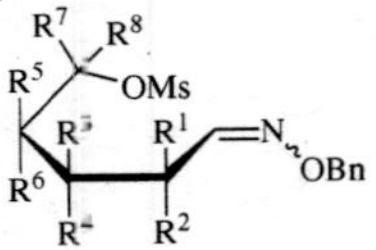	$PhMe_2SiH$ (2.3 eq), TFA	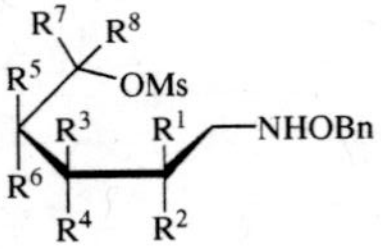	549
C_{44} 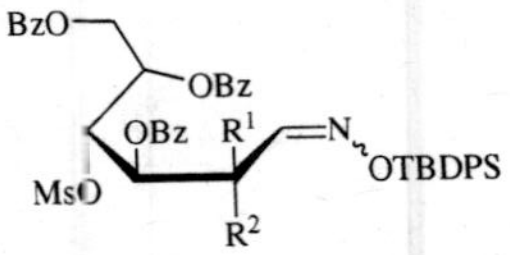	$PhMe_2SiH$ (2.3 eq), TFA, 60°, 40 min	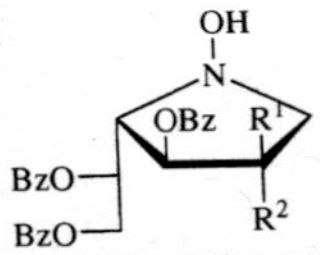 R^1 = H, R^2 = OBz (70); R^1 = OBz, R^2 = H (73)	549

C_{35-42} substituents and yields:

R^1	R^2	R^3	R^4	R^5	R^6	R^7	R^8	Yield (%)
H	OBz	H	OBz	H	OBz	H	Me	(82)
H	OBz	OBz	H	H	OBz	OBz	H	(88)
OBz	H	OBz	H	H	OBz	CH_2OBz	H	(85)
H	OBz	OBz	H	OBz	H	CH_2OBz	H	(90)

C_{44} product:

R^1	R^2	
H	OBz	(70)
OBz	H	(73)

TABLE 25. ORGANOSILANE REDUCTION OF NITROALKANES

Nitro Compound	Conditions	Product(s) and Yield(s) (%)	Refs.
C_6			
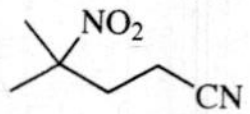	1. $PhSiH_3$ (0.5 eq), $(n\text{-}Bu)_3SnH$ (10%), ACHN (0.2 eq), MeC_6H_5, 110°, 5 h 2. ACHN (0.3 eq), MeC_6H_5, 110°, 3 h	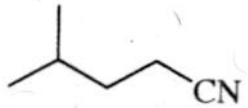 (67)	555
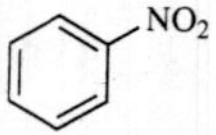	PMHS, Pd/C, EtOH, 80°	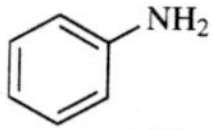 (89)	316
$C_{6\text{-}9}$			
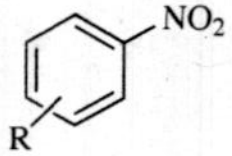	Et_3SiH (5 eq), $(Ph_3P)_3RhCl$ (0.02 eq), MeC_5H_5	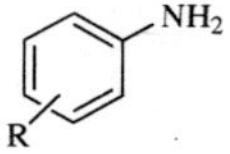	554
R	Time		
H	2 h	(86)	
4-Me	2 h	(90)	
3,4-Me_2	4 h	(82)	
2,4-Me_2	2 h	(83)	
2,6-Me_2	3 h	(83)	
3,5-Me_2	3 h	(85)	
2,4,6-Me_3	6 h	(75)	
4-MeO	2 h	(86)	
4-Cl	2 h	(71)	
4-Br	2 h	(—)	
4-CHO	3 h	(—)	
4-Ac	3 h	(—)	
3-Ac	2 h	(71)	
4-MeO_2C	2 h	·(49)	
4-CN	2 h	(0)	

TABLE 25. ORGANOSILANE REDUCTION OF NITROALKANES (*Continued*)

Nitro Compound	Conditions	Product(s) and Yield(s) (%)	Refs.
C_7			
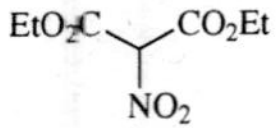	1. $PhSiH_3$ (0.5 eq), (n-Bu)$_3$SnH (10%), AIBN (0.2 eq), MeC_6H_5, 110°, 5 h 2. AIBN (0.3 eq), MeC_6H_5, 110°, 3 h	 (67)	555
C_8			
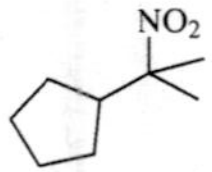	1. $PhSiH_3$ (0.5 eq), (n-Bu)$_3$SnH (10%), ACHN (0.2 eq), MeC_6H_5, 110°, 5 h 2. ACHN (0.3 eq), MeC_6H_5, 110°, 3 h	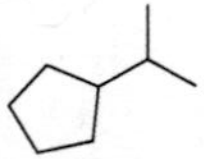 (78)	555
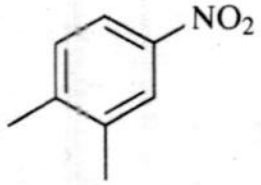	Et_3SiH (x eq), $(Ph_3P)_3RhCl$ (y eq)	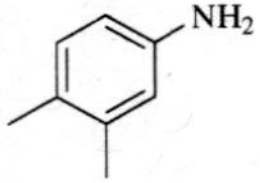	554
	x / y / Solvent / Time		
	5.0 / 0.01 / C_6H_6 / 360 h	(73)	
	5.0 / 0.01 / MeC_6H_5 / 360 h	(—)	
	8.0 / 0.01 / MeC_6H_5 / 360 h	(76)	
	8.0 / 0.01 / MeC_6H_5 / 360 h	(—)	
	5.0 / 0.02 / MeC_6H_5 / 240 h	(82)	
	5.0 / 0.05 / MeC_6H_5 / 120 h	(89)	
	5.0 / 0.02 / MeC_6H_5, EtOAc / 240 h	(78)	
C_{10}			
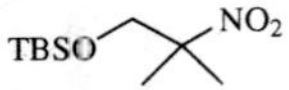	1. $PhSiH_3$ (0.5 eq), (n-Bu)$_3$SnH (10%), ACHN (0.2 eq), MeC_6H_5, 110°, 5 h 2. ACHN (0.3 eq), MeC_6H_5, 110°, 3 h	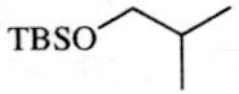 (76)	555
C_{12}			
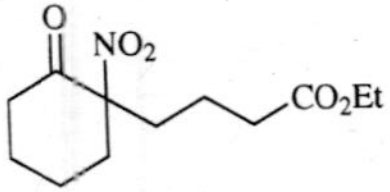	1. $PhSiH_3$ (0.5 eq), (n-Bu)$_3$SnH (10%), ACHN (0.2 eq), MeC_6H_5, 110°, 5 h 2. ACHN (0.3 eq), MeC_6H_5, 110°, 3 h	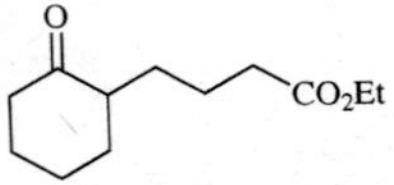 (76)	555

C_{13}				
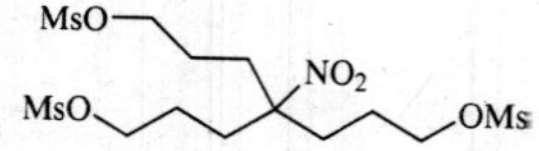	1. $PhSiH_3$ (0.5 eq), $(n\text{-}Bu)_3SnH$ (10%), ACHN (0.2 eq), MeC_6H_5, 110°, 5 h 2. ACHN (0.3 eq), MeC_6H_5, 110°, 3 h	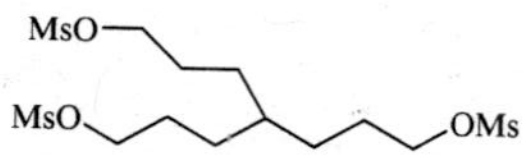	(70)	555
C_{16}				
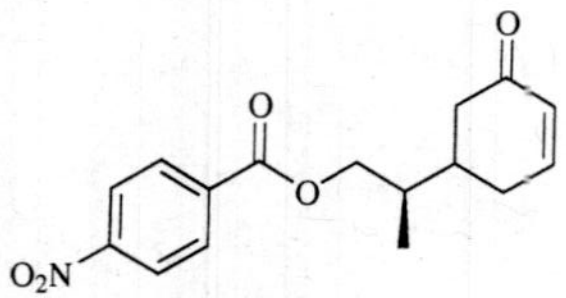	Et_3SiH (5 eq), $(Ph_3P)_3RhCl$ (2 mol%), MeC_6H_5, reflux, 4 h	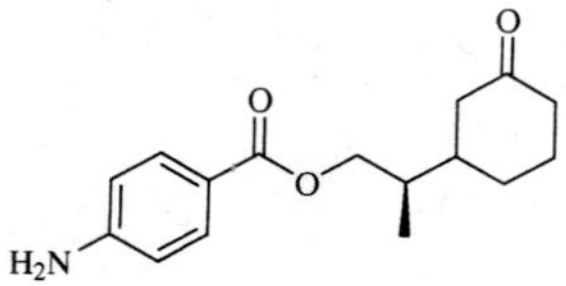	(—)	554
C_{18}				
	1. $PhSiH_3$ (0.5 eq), $(n\text{-}Bu)_3SnH$ (10%), ACHN (0.2 eq), MeC_6H_5, 110°, 5 h 2. ACHN (0.3 eq), MeC_6H_5, 110°, 3 h		(71)	555

TABLE 26. ORGANOSILANE REDUCTION OF MISCELLANEOUS NITROGEN COMPOUNDS

Nitrogen Compound	Conditions	Product(s) and Yield(s) (%)	Refs.
C_{5-10}			
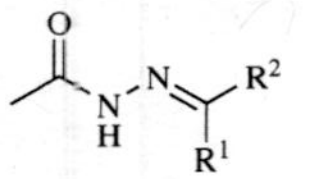	Et_3SiH (2 eq), TFA, 0°, 4 h	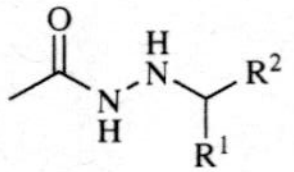	561
R^1 / R^2			
Me / Me		(86)	
H / *i*-Pr		(95)	
H / *n*-C_5H_{11}		(74)	
H / *c*-C_6H_{11}		(94)	
H / Ph		(89)	
H / Bn		(85)	
H / 3-MeC_6H_4		(84)	
Me / Ph		(56)	
C_6			
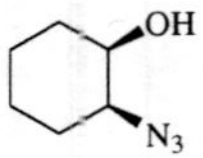	PMHS (3 eq), 10% Pd-C, $(Boc)_2O$ (1.1 eq), EtOH, rt, 5 h	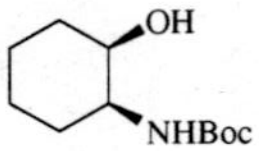 (90)	557
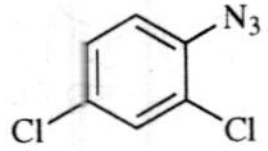	PMHS (3 eq), 10% Pd-C, $(Boc)_2O$ (1.1 eq), EtOH, rt, 3 h	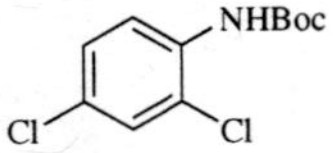 (88)	557
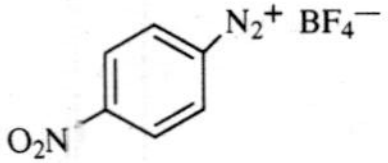	Et_3SiH, MeCN, rt, 16 h	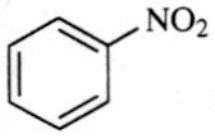 (72)	563
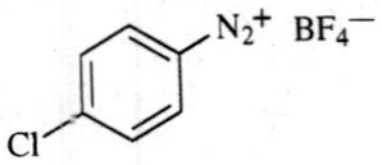	Et_3SiH, MeCN, rt, 16 h	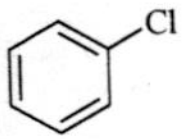 (70)	563

C_{6-13}			
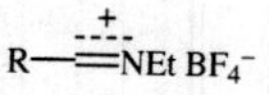 R: c-C_3H_5, i-Pr, n-Bu, t-Bu, Ph, Bn, 1-$C_{10}H_7$, 1-Ad	Et_3SiH (2 eq), CH_2Cl_2, rt, 0.5-6 h	RCHO (79) (85) (71) (61) (90) (41) (84) (83)	28, 562
C_{7-8}			
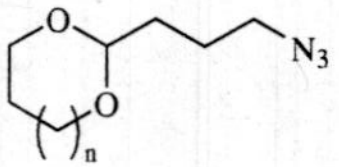	1. $NOBF_4$, SO_2 2. Et_3SiH	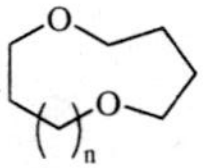 n = 1 (40) n = 2 (51)	779
C_7			
	PMHS (3 eq), 10% Pd-C, $(Boc)_2O$ (1.1 eq), EtOH, rt, 4 h	 (94)	557
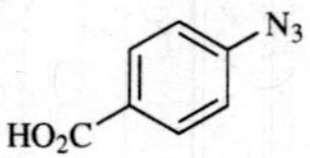	PMHS (3 eq), 10% Pd-C, $(Boc)_2O$ (1.1 eq), EtOH, rt, 3 h	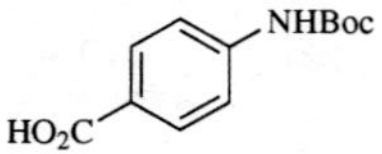 (90)	557
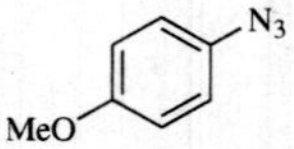	$PhSiH_3$ (1.5-2 eq), $[(n\text{-}Bu)_3Sn]_2O$ (0.025 eq), n-PrOH (2 eq), C_6H_6, AIBN (5 mol%), reflux, 90-120 min	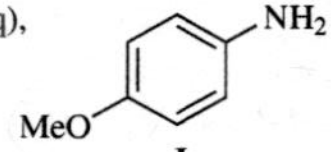 **I** (94)	556
	PMHS (1.5-2 eq), $[(n\text{-}Bu)_3Sn]_2O$ (0.025 eq), n-PrOH (2 eq), C_6H_6, AIBN (5 mol%), reflux, 90-120 min	**I** (91)	556

TABLE 26. ORGANOSILANE REDUCTION OF MISCELLANEOUS NITROGEN COMPOUNDS (*Continued*)

Nitrogen Compound	Conditions	Product(s) and Yield(s) (%)			Refs.
$C_{7\text{-}9}$ 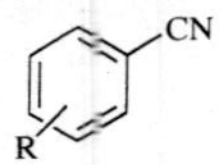	TMSH (10 eq), $Co_2(CO)_8$ (0.08-0.25 eq), CO, –20°	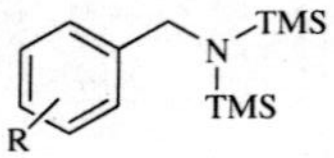	R		559
			H	(61)	
			4-Me	(91)	
			3-Me	(57)	
			2-Me	(11)	
			4-MeO	(88)	
			4-Cl	(53)	
			4-$NCCH_2$	(36)	
			4-Me_2N	(73)	
			4-MeO_2C	(46)	
$C_{7\text{-}14}$ 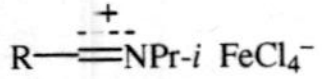	Et_3SiH (1.5 eq), CH_2Cl_2, rt, 0.5-6 h	RCHO	R		28
			c-C_3H_5	(77)	
			Ph	(87)	
			4-ClC_6H_4	(53)	
			4-$O_2NC_6H_4$	(97)	
			4-MeC_6H_4	(88)	
			4-$MeOC_6H_4$	(57)	
			4-NCC_6H_4	(19)[a]	
			4-$OHCC_6H_4$	(78)[a]	
			4-AcC_6H_4	(0)	
			PhCH=CH	(56)	
			2-$EtO_2CC_6H_4$	(42)	
			1-Ad	(91)	
C_7 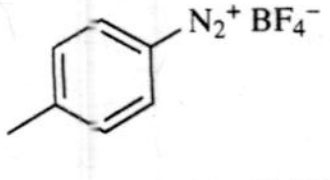	Et_3SiH, MeCN, rt, 16 h	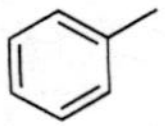 (25)			563
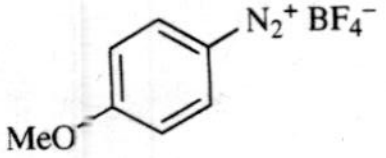	Et_3SiH, MeCN, rt, 16 h	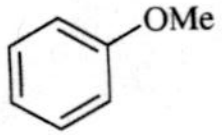 (90)			563

	Substrate	Conditions	Product(s) and Yield(s) (%)	Refs.
	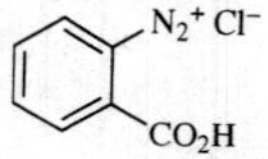	Et_3SiH, THF, reflux, 2 h	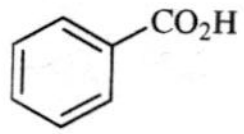 (78)	563
	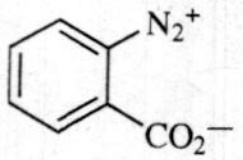	Et_3SiH, CH_2Cl_2, reflux, 2 h	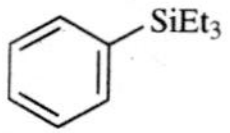 (29) + 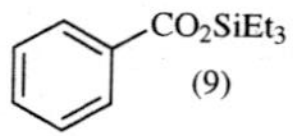 (9)	563
C_8	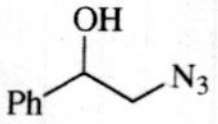	PMHS (3 eq), 10% Pd-C, $(Boc)_2O$ (1.1 eq), EtOH, rt, 4 h	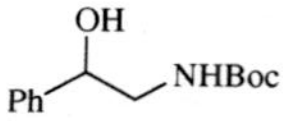 (88)	557
C_9		PMHS (3 eq), 10% Pd-C, $(Boc)_2O$ (1.1 eq), EtOH, rt, 4 h	 **I** (90)	557
		PMHS (3 eq), 10% Pd-C, $(Boc)_2O$ (1.1 eq), EtOH, rt, 4 h	**I** (89)	557
C_{9-11}	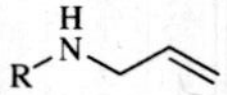	PMHS (2 eq), $(Ph_3P)_4Pd$ (0.01 eq), THF, rt, 3-5 h	RNH_2 + 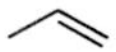 R: c-C_6H_{11} (89); Ph (90); Bn (92); 4-$EtOC_6H_4$ (88)	270
	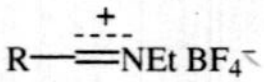	Et_3SiH (1.2 eq), CH_2Cl_2, rt, 0.5-6 h	RCHO R: 4-ClC_6H_4 (61); 4-$O_2NC_6H_4$ (21); 4-MeC_6H_4 (92); 4-$MeOC_6H_4$ (85); 4-$OHCC_6H_4$ (0); 4-NCC_6H_4 (60); 4-AcC_6H_4 (0); PhCH=CH (54)	28

TABLE 26. ORGANOSILANE REDUCTION OF MISCELLANEOUS NITROGEN COMPOUNDS (*Continued*)

Nitrogen Compound	Conditions	Product(s) and Yield(s) (%)	Refs.
C_{10} 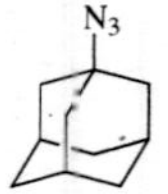	1. $PhSiH_3$ (1.5-2 eq), $[(n\text{-}Bu)_3Sn]_2O$ (0.025 eq), n-PrOH (2 eq), C_6H_6 2. AIBN (5 mol%), reflux, 90-120 min	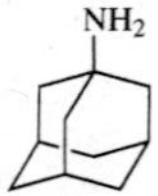 **I** (99)	556
	1. PMHS (1.5-2 eq), $[(n\text{-}Bu)_3Sn]_2O$ (0.025 eq), n-PrOH (2 eq), C_6H_6 2. AIBN (5 mol%), reflux, 90-120 min	**I** (96)	556
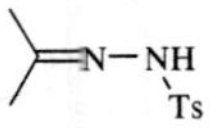	Et_3SiH (2 eq), TFA, 0°, 1 h	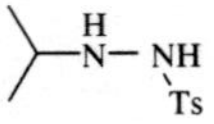 (80)	560
$C_{10\text{-}15}$ 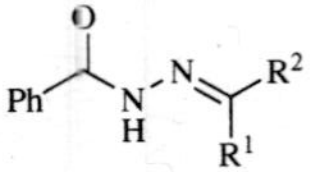	Et_3SiH (2 eq), TFA, 0°, 4 h	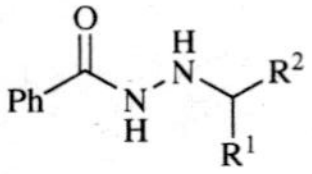	561
R^1 = Me, R^2 = Me		(86)	
R^1 = H, R^2 = i-Pr		(91)	
R^1 = H, R^2 = n-C_5H_{11}		(80)	
R^1 = H, R^2 = c-C_6H_{11}		(92)	
R^1 = H, R^2 = Ph		(82)	
R^1 = H, R^2 = Bn		(78)	
R^1 = H, R^2 = 3-MeC_6H_4		(79)	
R^1 = Me, R^2 = Ph		(46)	
C_{11} 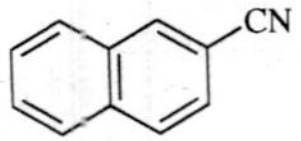	TMSH (10 eq), $Co_2(CO)_8$ (0.08 eq), CO, –20°	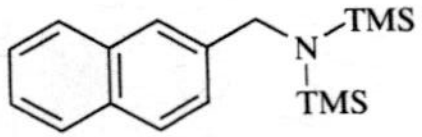 (64)	559

C_{12}	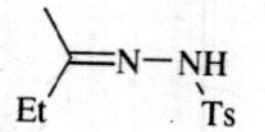	Et_3SiH (2 eq), TFA, 0°, 1 h	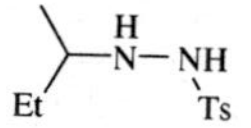 (79)	560
	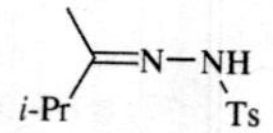	Et_3SiH (2 eq), TFA 0°, 1 h	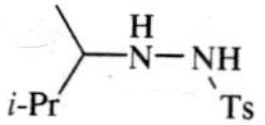 (30)	560
	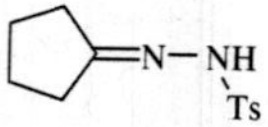	Et_3SiH (2 eq), TFA, 0°, 1 h	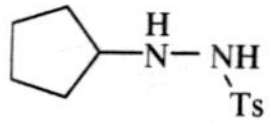 (89)	560
	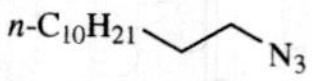	1. $PhSiH_3$ (1.5-2 eq), $[(n\text{-}Bu)_3Sn]_2O$ (0.025 eq), n-PrOH (2 eq), C_6H_6 2. AIBN (5 mol%), reflux, 90-120 min	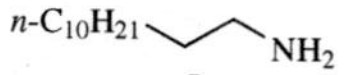 **I** (94)	556
		1. PMHS (1.5-2 eq) $[(n\text{-}Bu)_3Sn]_2O$ (0.025 eq), n-PrOH (2 eq), C_6H_6 2. AIBN (5 mol%), reflux, 90-120 min	**I** (95)	556
	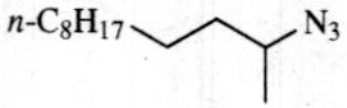	1. $PhSiH_3$ (1.5-2 eq), $[(n\text{-}Bu)_3Sn]_2O$ (0.025 eq), n-PrOH (2 eq), C_6H_6 2. AIBN (5 mol%), reflux, 90-120 min	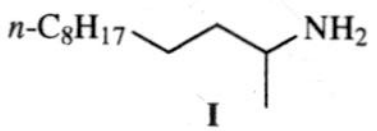 **I** (92)	556
		1. PMHS (1.5-2 eq), $[(n\text{-}Bu)_3Sn]_2O$ (0.025 eq), n-PrOH (2 eq), C_6H_6 2. AIBN (5 mol%), reflux, 90-120 min	**I** (94)	556

TABLE 26. ORGANOSILANE REDUCTION OF MISCELLANEOUS NITROGEN COMPOUNDS (*Continued*)

Nitrogen Compound	Conditions	Product(s) and Yield(s) (%)	Refs.
C_{12} 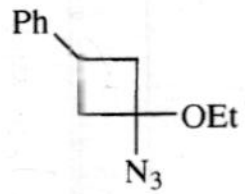	$PhSiH_3$, $(n\text{-}Bu)_3SnH$ (cat.), n-PrOH, AIBN, 80°	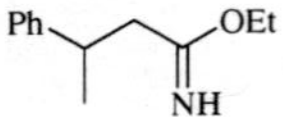 (—)	556
C_{14} 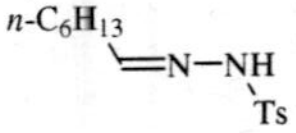	Et_3SiH (2 eq), TFA, 0°, 1 h	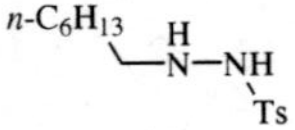 (75)	560
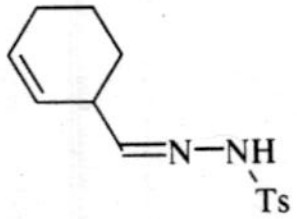	Et_3SiH (2 eq), TFA, 0°, 1 h	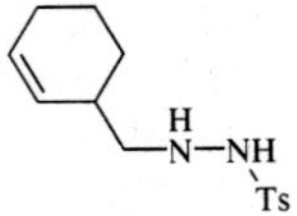 (82)	560
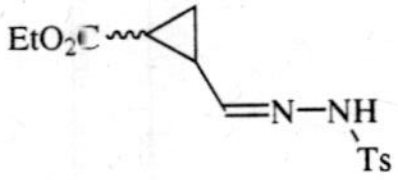	Et_3SiH (2 eq), TFA, 0°, 1 h	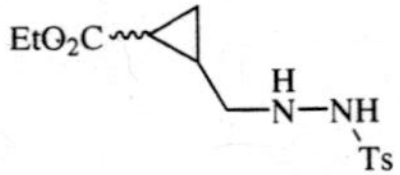 (84)	560
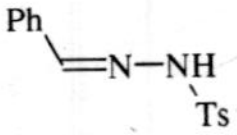	Et_3SiH (2 eq), TFA, 0°, 1 h	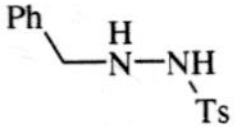 (82)	560
C_{15} 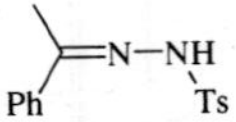	Et_3SiH (2 eq), TFA, 0°, 1 h	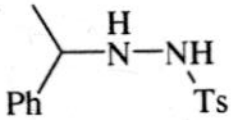 (51)	560
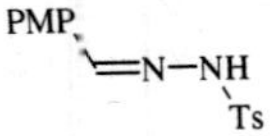	Et_3SiH (2 eq), TFA, 0°, 1 h	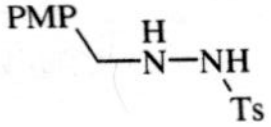 (81)	560

C_{16} 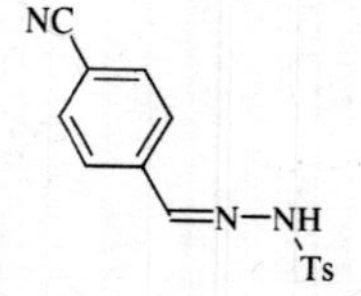	Et_3SiH (2 eq), TFA, 0°, 1 h	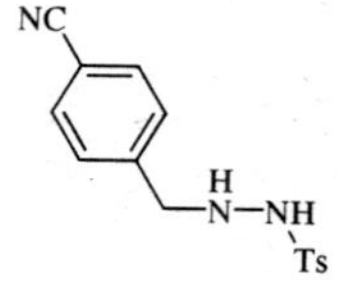 (76)	560
C_{19} 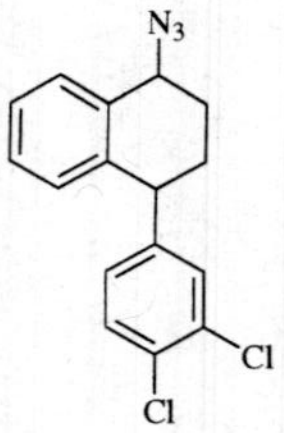	PMHS (3 eq), 10% Pd-C, $(Boc)_2O$ (1.1 eq), EtOH, rt, 5 h	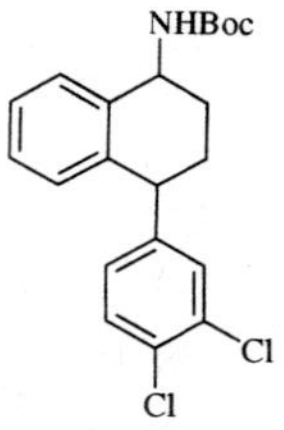 (85)	557
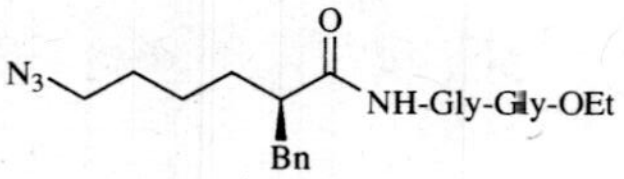	Et_3SiH, 20% $Pd(OH)_2/C$, $(Boc)_2O$	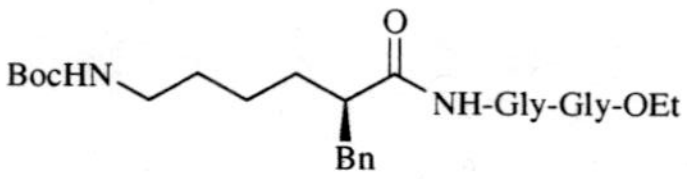 (80)	558
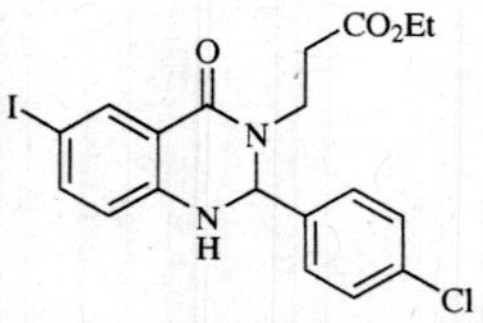	Et_3SiH, TFA, 0° to rt, 24 h	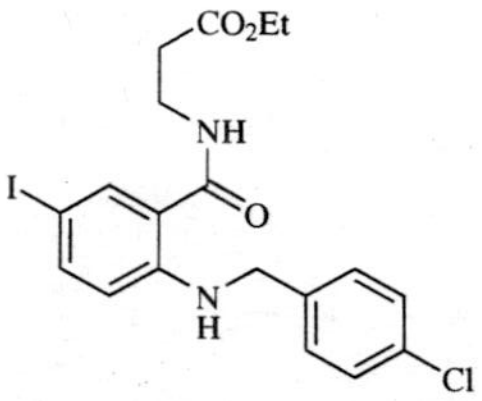 (97)	780

[a] The yield was determined by gas chromatography.

TABLE 27. ORGANOSILANE REDUCTION OF MISCELLANEOUS SULFUR COMPOUNDS

Sulfur Compound	Conditions	Product(s) and Yield(s) (%)	Refs.
C_6 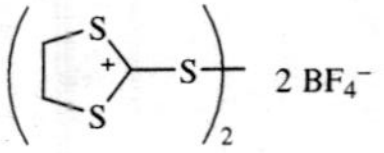	Et_3SiH (1 eq), MeCN, <5 min	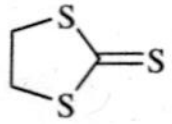 (100)	565
$C_{8\text{-}14}$ 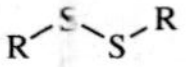	Et_3SiH (1.4 eq), TFA, 60°, 25 h	RSH R: *n*-Bu (67); Ph (45); Ph (52); $2\text{-}O_2NC_6H_4$ (0); $2,4\text{-}(O_2N)_2C_6H_3$ (0); $4\text{-}MeOC_6H_4$ (74)	564
C_{10} 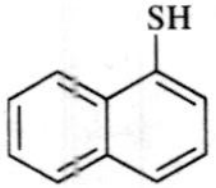	Et_3SiH (1.3 eq), $BF_3•OH_2$ (4-6 eq), CH_2Cl_2, rt, 6 h	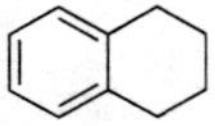 **I** (13) + 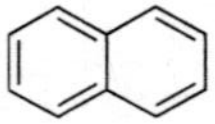 (5.5)	217
$C_{11\text{-}12}$ 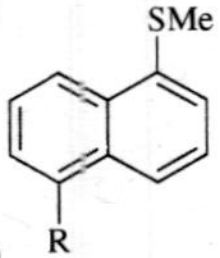	Et_3SiH (3-5 eq), $BF_3•OEt_2$, CH_2Cl_2, 0°	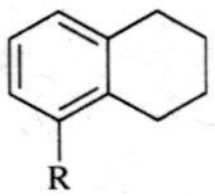 R, Time: H, 24 h (20); SMe, 92 h (70)	263
$C_{16\text{-}20}$ 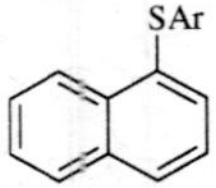	Et_3SiH (3-5 eq), $BF_3•OEt_2$, CH_2Cl_2, 0°, 20 h	**I** + ArSH Ar, Time, % Conv.: Ph, 20 h, (30); Bn, 168 h, (40); 1-Np, 20 h, (50); 2-Np, 22 h, (25)	263

TABLE 28. ORGANOSILANE REDUCTION OF SMALL RING COMPOUNDS

Small Ring Substrate	Conditions	Product(s) and Yield(s) (%)	Refs.
C_5 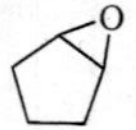	$HMe_2SiOSiMe_2H$ (1.0 eq), I_2 (cat.), CH_2Cl_2, rt, 10 min	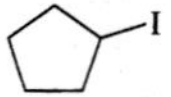 (90)	357
C_6 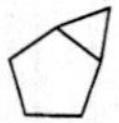	Et_3SiH (1 eq), TFA (x eq), 20°	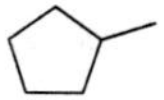 I +  II + 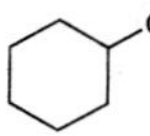 III + 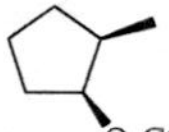 IV + 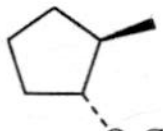 V	229
		x | I | II | III + IV + V 2 | (16) | (1-2) | (62) 4 | (28) | (3) | (70)	
	Et_3SiH (1 eq), TFA (2 eq), rt	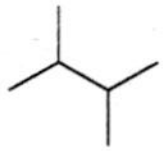 I + 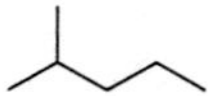 II	232
		Time | I + II 15 min | (15) 16 h | (100)	
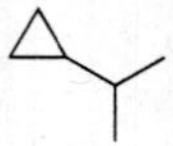	Et_3SiH (1 eq), TFA (2 eq), rt, 24 h	I + II (50)	232
	Et_3SiH (1 eq), TFA (2 eq)	II	222
		Temp | Time | II 20° | 400 h | (45) 50° | 12 h | (35) 80° | 12 h | (32)	
	Et_3SiD (1 eq), TFA (2 eq), rt, 24 d	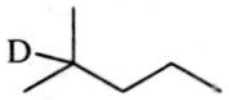	250
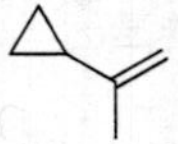	Et_3SiH (1 eq), TFA (2 eq), rt, 72 h	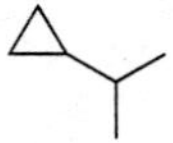 (20) + 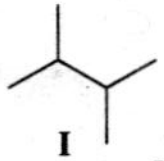 I + 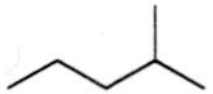 II I + II (80)	232

TABLE 28. ORGANOSILANE REDUCTION OF SMALL RING COMPOUNDS (*Continued*)

Small Ring Substrate	Conditions	Product(s) and Yield(s) (%)	Refs.
C_6 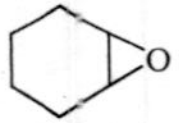	Et_3SiH, BF_3, CH_2Cl_2, several weeks	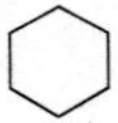 (—)	566
	$HMe_2SiOSiMe_2H$ (1.0 eq), NaI (1.3 eq), CH_2Cl_2, 5-10°, 30 min	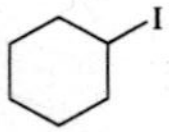 **I** (75)	357
	$HMe_2SiOSiMe_2H$ (1.0 eq), I_2 (cat.), CH_2Cl_2, rt, 10 min	**I** (100)	357
	PMHS (1 eq), NaI (1.33 eq), TMSCl (1.5 eq), MeCN, reflux, 30 min	**I** (75)	567
	PMHS (1 eq), LiBr (1.33 eq), TMSiCl (1.5 eq), MeCN, reflux, 7 h	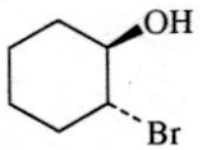 (93)	567
C_7 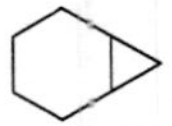	Et_3SiH (1 eq), TFA (x eq), 20°, 140 h	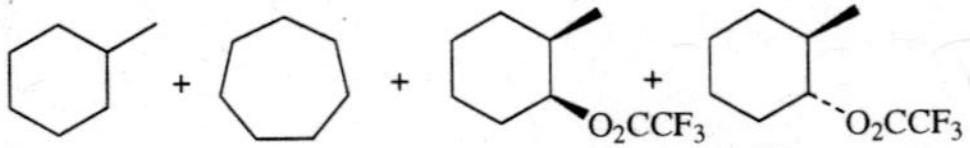 **I** + **II** + **III** + **IV** x / I / II / III + IV 2 / (36) / (1.5) / (62) 4 / (79) / (5) / (16)	229
	Et_3SiH (1 eq), TFA (2 eq), 20°	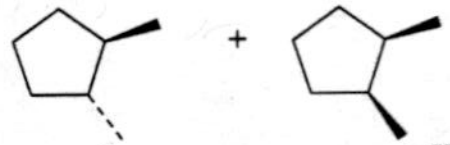 **I** + **II** Time / I + II / I:II 2 h / (26) / 11.5 24 h / (68) / 11.2 144 h / (88) / 11.2 240 h / (89) / 10.0	229
C_8 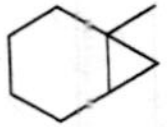	Et_3SiH (1 eq), TFA (2 eq), 20°	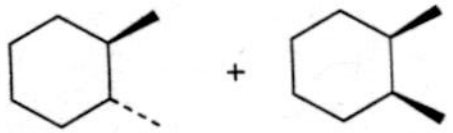 **I** + **II** Time / I + II / I:II 2 h / (13) / 1.5 24 h / (58) / 1.3 144 h / (72) / 1.5 240 h / (74) / 1.5	229

Reactant	Conditions	Product(s) and Yield(s) (%)	Refs.
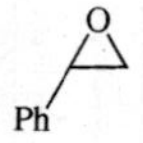	PMHS (1 eq), LiBr (1.33 eq), TMSCl (1.5 eq), HOAc, reflux, 15 min	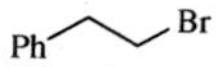 **I** (69)	567
	$HMe_2SiOSiMe_2H$ (1.0 eq), LiBr (1.3 eq), TFA, 5-10°, 15 h	**I** (69)	357
	$HMe_2SiOSiMe_2H$ (1.0 eq), LiBr (1.3 eq), TFA, 5-10°, 1 5 h	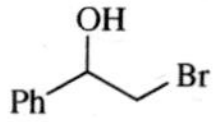 (69)	357
	PMHS (1 eq), NaI (1.33 eq), TMSCl (1.5 eq), MeCN, reflux, 45 min	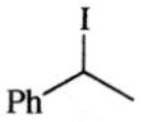 **I** (76)	567
	$HMe_2SiOSiMe_2H$ (1.0 eq), NaI (1.3 eq), CH_2Cl_2, 5-10°, 45 min	**I** (76)	357
	$HMe_2SiOSiMe_2H$ (1.0 eq), I_2 (1.6 eq), CH_2Cl_2, rt, 10 min	**I** (62)	357
C_{9-20} 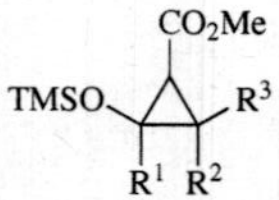	1. LDA/THF; R^4R^5CO; TBAF [a] 2. Et_3SiH (1.1 eq), BF_3•OEt_2 (1.1 eq), CH_2Cl_2, –78°, 45 min; rt, 3 h	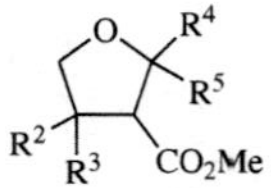	765

R^1	R^2	R^3	R^4	R^5	Yield	cis:trans
H	Me	Me	Me	H	(51)	1:3
H	Me	Me	Me	Me	(71)	—
H	—$(CH_2)_3$—		Me	H	(54)	1:1
H	—$(CH_2)_3$—		—$(CH_2)_3$—		(48)	—
—$(CH_2)_4$—		H	Me	Me	(66)	>9:1
H	Me	Me	Ph	H	(79)	2:3
Ph	H	H	Me	Me	(54)	1:1
H	Me	Me	PhMeCH	H	(71)	1:2
H	Me	Me	Ph	Ph	(47)	—

TABLE 28. ORGANOSILANE REDUCTION OF SMALL RING COMPOUNDS (*Continued*)

Small Ring Substrate	Conditions	Product(s) and Yield(s) (%)	Refs.
C_{10} 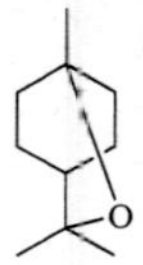	$HMe_2SiOSiMe_2H$ (1.0 eq), NaI (1.3 eq), CH_2Cl_2, 5-10°, 20 min	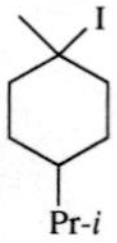 **I** +  **II** **I** + **II** (80), **I**:**II** 57:43	357, 567
	PMHS (1 eq), NaI (1.33 mol%), TMSCl (1.5 eq), MeCN, reflux, 7 h	**I** (46) + 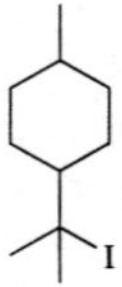 (34)	567
C_{11} 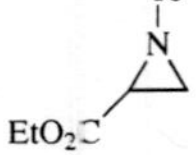	PMHS (xs), EtOH, Pd-C, rt, 6 h	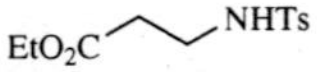 (80)	568
C_{14} 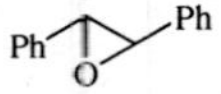	Et_3SiH (3.9 eq), BF_3, CH_2Cl_2, 20°, 7 d	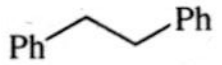 (100)	566
C_{15} 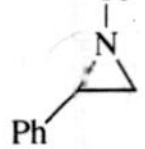	PMHS (xs), EtOH, Pd-C, rt, 6 h	 (96)	568
C_{16} 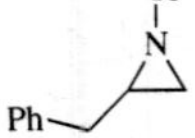	PMHS (xs), EtOH, Pd-C, rt, 6 h	 (60)	568
C_{17} 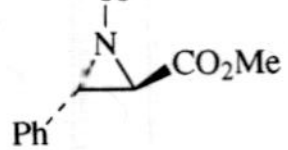	PMHS (xs), EtOH, Pd-C, rt, 6 h	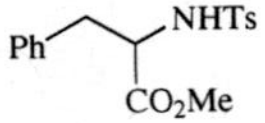 (95)	568

C_{19}	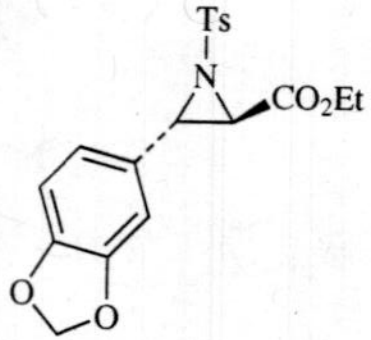	PMHS (xs), EtOH, Pd-C, rt, 6 h	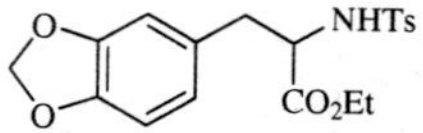 (80)	568
	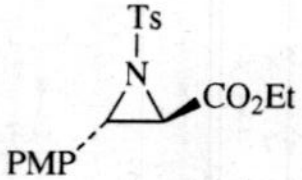	PMHS (xs), EtOH, Pd-C, rt, 6 h	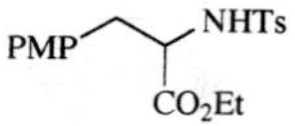 (93)	568
C_{20}	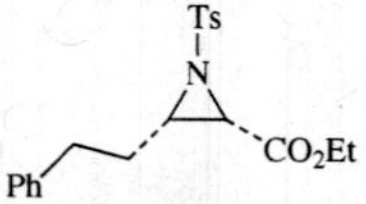	PMHS (xs), EtOH, Pd-C, rt, 6 h	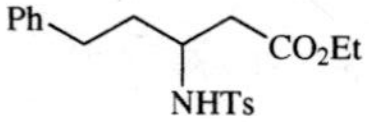 (85)	568
C_{21}	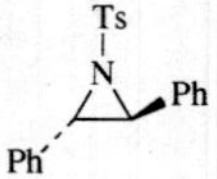	PMHS (xs), EtOH, Pd-C, rt, 6 h	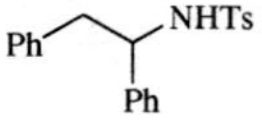 (97)	568
	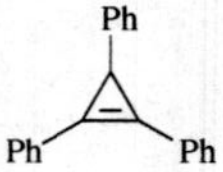	Et_3SiR (1.1 eq), TFA (2.6 eq), CH_2Cl_2, 20°, 30 h	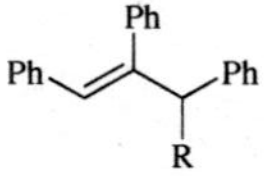 R: H (93); D (—)	781
C_{23}	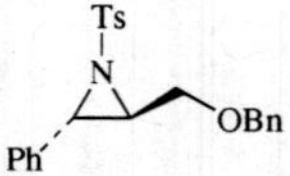	PMHS (xs), EtOH, Pd-C, rt, 6 h	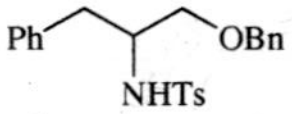 (75)	568
C_{24}	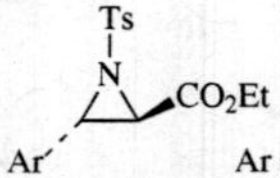 Ar = 3-$PhOC_6H_4$	PMHS (xs), EtOH, Pd-C, rt, 6 h	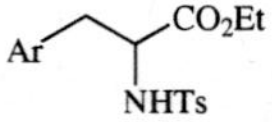 (90)	568

[a] These sequential steps generate the intermediate substituted tetrahydrofuran-2-ol, which is converted in the second sequence into the final product shown.

TABLE 29. MISCELLANEOUS ORGANOSILANE REDUCTIONS

Substrate	Conditions	Product(s) and Yield(s) (%)	Refs.
C_{18-20}			
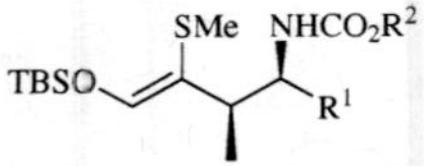	Et_3SiH, TFA, MeCN	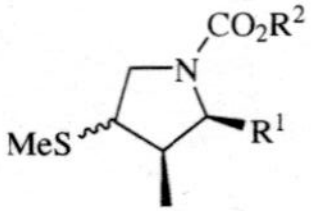	782
R^1 / R^2: *n*-Pr / Et		(95)	
i-Pr / Et		(95)	
Ph / Me		(90)	
C_{19}			
$Ph_3C^+ BF_4^-$	TMSH (0.8 eq), rt, 0.5 h	Ph_3CH (100) + Me_3SiF (100)	62
	Et_3SiH (0.8 eq), rt, 1.5 h	Ph_3CH (100) + Et_3SiF (100)	62
	Me_2SiH_2 (0.8 eq) rt, 0.5 h	Ph_3CH (100) + Me_2SiHF (100)	62
	Me_2SiH_2 (0.45 eq), rt, 8 h	Ph_3CH (100) + Me_2SiF_2 (95)	62
	Et_2SiH_2 (0.80 eq), rt, 1.5 h	Ph_3CH (100) + Et_2SiHF (100)	62
Ph_3CCl	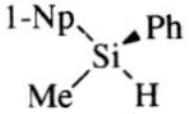 , C_6H_6, reflux, 4 h	Ph_3CH (—) + 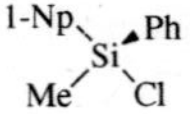 (71)	60
C_{20}			
$Ph_3C^+ {}^-Br_5\text{-}CB_9H_5$	$(i\text{-}Pr)_3SiH$, MeCN	Ph_3CH + $[(i\text{-}Pr)_3Si\text{-}NCMe][Br_5\text{-}CB_9H_5]$ (—)	46
$Ph_3C^+ {}^-Br_5\text{-}CB_9H_5$	$(i\text{-}Pr)_3SiH$, MeC_6H_5	Ph_3CH + $[(i\text{-}Pr)_3Si][Br_5\text{-}CB_9H_5]$ (—)	46

C_{43}

Substrate	Reagent	Product(s)	Ref.
$Ph_3C^+ (C_6F_5)_4B^-$	R_3SiH	Ph_3CH (—) + $R_3Si^+ B(C_6F_5)_4^-$ (—)	45

R_3Si	Solvent
TMS	C_6D_6
Et_3Si	C_6D_6
Et_3Si	MeC_6H_5
Et_3Si	MeCN
$(i\text{-}Pr)_3Si$	C_6D_6
$(TMS)_3Si$	C_6D_6

TABLE 30. ASYMMETRIC ORGANOSILANE REDUCTION OF KETONES

Ketone	Conditions	Product(s) and Yield(s) (%)	Refs.
C_4 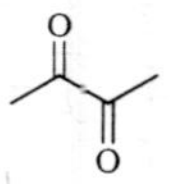	Ph_2SiH_2 (4 eq), $[Rh(cod)_2]BF_4$ (1.0 mol%), **23** (1.1 mol%), 0°, DME, 26 h	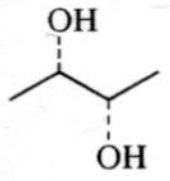 **I** + 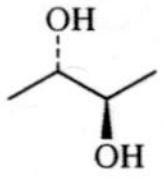 **II**; **I + II** (94), **I:II** 90:10, **I** = 96% ee	577
$C_{4\text{-}8}$ 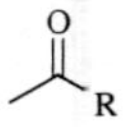	$Me(EtO)_2SiH$ (1.43 eq), **197** (0.57 mol%), MeLi (1.14 mol%), THF, rt	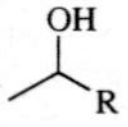 (100)	583

R	Time	% ee	Conf.
Et	1 h	7	R
i-Pr	12 h	17	S
n-Pr	12 h	6	S
t-Bu	72 h	48	R
s-Bu	12 h	50	S
Ph	12 h	47	R

Ketone	Conditions	Product(s) and Yield(s) (%)	Refs.
	[—(Me)(H)SiO—$]_4$ (1.43 eq), **197** (0.57 mol%), MeLi (1.14 mol%), THF, rt	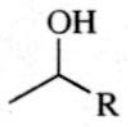 (100)	583

R	Time	% ee	Conf.
Et	1 h	7	R
i-Pr	12 h	19	S
n-Pr	12 h	5	S
t-Bu	72 h	38	R
s-Bu	12 h	45	S
Ph	12 h	41	R

Ketone	Conditions	Product(s) and Yield(s) (%)	Refs.
	$Me(EtO)_2SiH$ (3.75 eq), **197** (0.57 mol%), *n*-BuLi (1.14 mol%), THF, rt	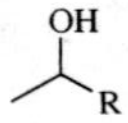	583

R	Time		% ee	Conf.
Et	1 h	(100)	2	R
i-Pr	12 h	(100)	21	S
t-Bu	140 h	(75)	42	R
s-Bu	140 h	(100)	39	S
Ph	24 h	(100)	99	R

$C_{4\text{-}10}$

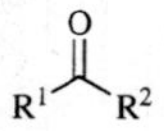

R_3SiH (3.75 eq), **197** (0.57 mol%), n-BuLi (1.14 mol%), THF, rt

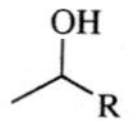

R_3SiH	Time	R		% ee	Conf.	
PMHS	1 h	Et	(100)	2	R	583
PMHS	12 h	*i*-Pr	(100)	15	S	
$(EtO)_3SiH$	12 h	*i*-Pr	(100)	17	S	
PMHS	24 h	Ph	(100)	53	R	
$MeSiH_3$	120 h	Ph	(80)	44	R	
$(EtO)_3SiH$	12 h	Ph	(100)	>99	R	

1 R_3SiH (1 eq), catalyst[a] (0.1 eq), rt
2 PhMgBr

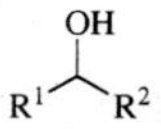

R^1	R^2	Time		% ee	Conf.	
Me	Et	18 h	(89)	10	S (+)	584
Me	*n*-Pr	18 h	(93)	19	S (+)	
i-Pr	Me	150 h	(96)	18	S (+)	
Me	*n*-Bu	18 h	(92)	5	S (+)	
Et	*n*-Pr	230 h	(97)	11	R (–)	
i-Pr	Et	230 h	(91)	65	R (–)	
Me	*s*-Bu	230 h	(99)	27	S (+)	
Me	*t*-Bu	230 h	(87)	40	R (–)	
i-Pr	*n*-Pr	680 h	(63)	60	R (–)	
Me	Et_2NCH_2	300 h	(91)	4	(–)	
Me	Ph	300 h	(90)	90	R (+)	
Et	*n*-C_5H_9	600 h	(94)	70	(+)	
i-Pr	*n*-Bu	1200 h	(75)	37	R (–)	
Et	*c*-C_6H_{11}	18 h	(92)	8	R (–)	
Ph	$MeOCH_2$	550 h	(84)	15	(–)	
Me	$BnCH_2$	25 h	(96)	11	S (+)	
i-Pr	*c*-C_6H_{11}	1270 h	(68)	(5)	R (+)	

TABLE 30. ASYMMETRIC ORGANOSILANE REDUCTION OF KETONES (*Continued*)

Ketone	Conditions	Product(s) and Yield(s) (%)	Refs.
C_{4-12} 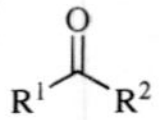	Ph_2SiH_2 (1.1 eq), $[Rh(cod)Cl]_2$ (0.5 mol%), **107** (1.2 mol%)	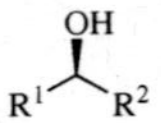 (see below)	575
C_5 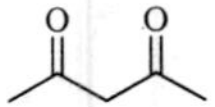	Ph_2SiH_2 (2.5 eq), $[Rh(cod)_2]BF_4$ (1.0 mol%), **23** (1.1 mol%), –30°, THF, 75 h	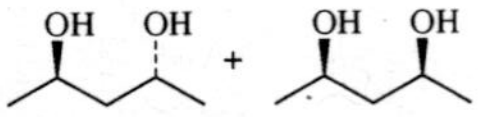 **I** + **II** (45), **I**:**II** = 42:58, **I** = 35% ee	577
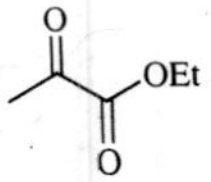	Ph_2SiH_2 (1.5 eq), $[Rh(cod)_2]BF_4$ (1.0 mol%), **23** (1.1 mol%), 0°, THF, 4 h	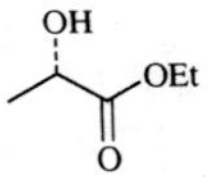 (60) 80% ee	577

R^1	R^2	Temp	Time		% ee
Et	Me	0°	24 h	(—)	56
i-Pr	Me	0°	24 h	(—)	76
t-Bu	Me	0°	96 h	(—)	95
$EtO_2C(CH_2)_2$	Me	0°	24 h	(86)	80
n-C_6H_{13}	Me	0°	40 h	(89)	42
c-C_6H_{11}	Me	0°	24 h	(82)	68
Ph	Me	0°	24 h	(89)	86
Ph	$ClCH_2$	20°	24 h	(89)	83
Ph	Et	0°	24 h	(81)	64
Ph	*i*-Pr	20°	24 h	(87)	3
Ph	*t*-Bu	20°	24 h	(69)	6
1-$C_{10}H_7$	Me	0°	24 h	(81)	89
2-$C_{10}H_7$	Me	0°	24 h	(87)	80

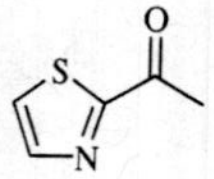

PMHS (4 eq), CuCl/NaOBu-t (1 mol%), catalyst (0.05 mol%), MeC_6H_5, 4 h

Catalyst	Temp	Time
124	−50°	4 h
121	−78°	—

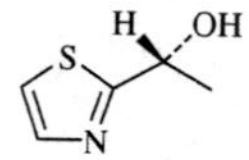

	% ee
(97)	90
(—)	90

590

C_{5-17}

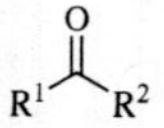

PMHS, $Sn(OTf)_2$ (10 mol%), **85** (10 mol%), MeOH, rt, 12-14 h

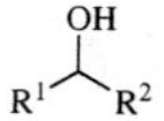

R^1	R^2	
CO_2Et	CF_3	(75)
Ph	Me	(98)
Ph	CO_2Me	(99)
Ph	CO_2Et	(99)
4-$MeOC_6H_4$	CO_2Me	(99)
4-MeC_6H_4	CO_2Me	(96)
4-$O_2NC_6H_4$	CO_2Me	(98)
1-$C_{10}H_7$	CO_2Me	(98)
2-$C_{10}H_7$	CO_2Me	(96)
Ph	CH_2-*N*-phth	(98)
4-MeC_6H_4	CH_2-*N*-phth	(88)
4-ClC_6H_4	CH_2-*N*-phth	(89)
4-$O_2NC_6H_4$	CH_2-*N*-phth	(50)
2-C_4H_3S	CO_2Et	(98)

385

C_6

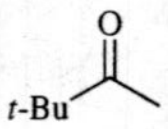

Ph(1-$C_{10}H_7$)SiH_2 (1.5 eq), [Rh(cod)]SbF_6 (1 mol%), **126** (1 mol%), THF, −20°

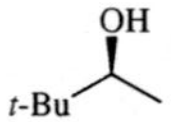

I

(75) 98% ee — 576

1. $PhSiH_3$ (0.1 eq), **197** (2 mol%), MeOH, 60°, pyrrolidine, MeOH, THF
2. PMHS, ketone, MeOH (3-7 eq), 15°, 17 h

I (30) 53% ee — 587

$PhSiH_3$ (1.1 eq), (*R*)-**195**-2*n*-BuLi (5 mol%), *n*-C_6H_{14}, 65°, 1 h

I (58) 2% ee — 783

$PhSiH_3$ (1.2 eq), CuF_2 (1%), (*S*)-BINAP (1%), MeC_6H_5, rt, 29 h

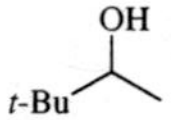

75% conversion, 20% ee[b] — 784

TABLE 30. ASYMMETRIC ORGANOSILANE REDUCTION OF KETONES (*Continued*)

Ketone	Conditions	Product(s) and Yield(s) (%)	Refs.
C_6			
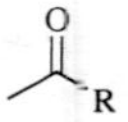	Et_2SiH_2, [(–)-(*S*)-bmpp]$_2$RhCl, C_6H_6, 5-50°, 3-48 h	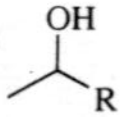 R / Opt. Yield / Conf. *n*-Bu (93) 31 S *t*-Bu (95) 39 S	785, 786
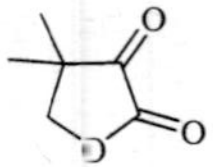	Ph_2SiH_2 (1.1 eq), $[Rh(cod)Cl]_2$ (0.5 mol%), **108** (1.2 mol%), THF, 0°, 120 h	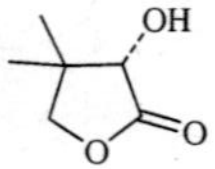 **I** (76) 84% ee	575
	Ph_2SiH_2 (1 eq), $[Rh(cod)Cl]_2$ (0.3 mol%), **44** (0.4 mol%), THF, rt, 30 min	**I** (68) 37% ee	787
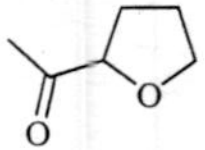	PMHS (1.1 eq), Et_2Zn (20 mol%), (*R*,*R*)-ebpe (20 mol%), vitride (3%), MeC_6H_5, rt, 24 h	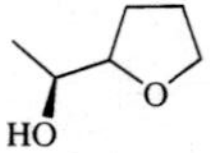 (—) 66% ee	594
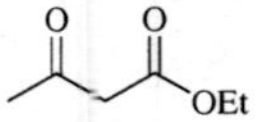	Ph_2SiH_2 (1.5 eq), $[Rh(cod)_2]BF_4$ (1.0 mol%), **23** (1.1 mol%), –30°, THF, 14 h	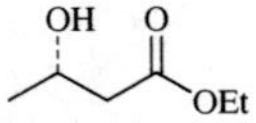 (43) 32% ee	577
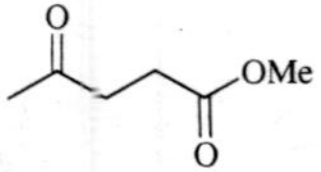	Ph_2SiH_2 (1.5 eq), $[Rh(cod)_2]BF_4$ (1.0 mol%), **23** (1.1 mol%), –30°, THF, 31 h	 (74) 88% ee	577
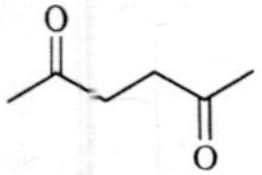	Ph_2SiH_2 (2.5 eq), $[Rh(cod)_2]BF_4$ (1.0 mol%), **23** (1.1 mol%), –30°, DME, 30 h	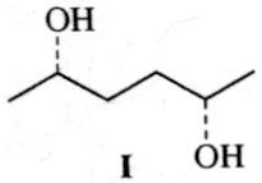 **I** + 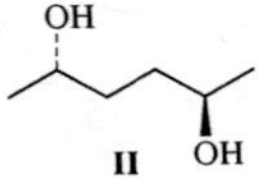 **II** **I** + **II** (97), **I**:**II** = 75:25, **I** = 97% ee	577

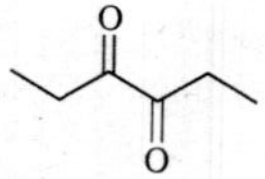

Ph_2SiH_2 (2.5 eq), [Rh(cod)$_2$]BF_4 (1.0 mol%), **23** (1.1 mol%), 0°, DME, 55 h

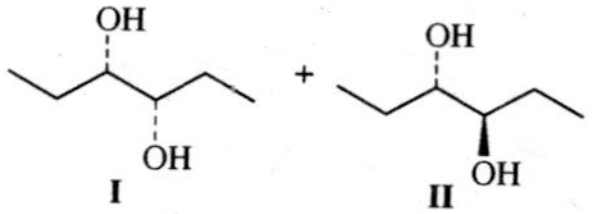

I + **II** (63), **I**:**II** 77:23, I = 70% ee

577

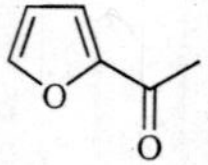

PMHS (4 eq), CuCl/NaOBu-*t* (1 mol%), **124** (0.05 mol%), MeC_6H_5, −50°, 5 h

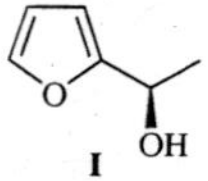

I

(85) 92% ee

590

PMHS (4 eq), CuCl/NaOBu-*t* (1 mol%), **121** (0.05 mol%), MeC_6H_5, −78°

I (—) 90% ee

590

PMHS (4 eq), CuCl/NaOBu-*t* (1 mol%), **122** (0.05mol%), MeC_6H_5, −78°

I (—) 78% ee

590

PMHS (4 eq), ligand (3 mol%), CuCl (3 mol%), NaOBu-*t* (3 mol%), THF (x%), MeC_6H_5 [(100 − x)]%

I (—)

590

Ligand	x	% ee
124	0	91
121	0	90
121	12.5	88
121	100	70
122	0	78
122	12.5	74
122	100	67

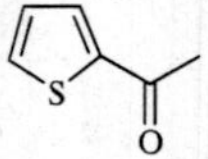

PMHS (1.2 eq), Ph_2Zn (2 mol%), **131** (2 mol%), MeC_6H_5, rt, 4 h

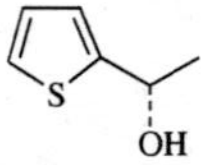

(>99) 78% ee

788

Ketone		Conditions	Product(s) and Yield(s) (%)		Refs.
C_{6-11} 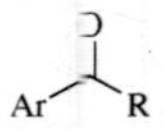		Ph_2SiH_2 (1.5 eq), $[Rh(cod)Cl]_2$ (2.5 mol%), **29** (5 mol%), THF, 0°	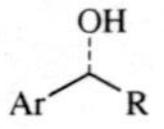		789
Ar	R	Time		% ee	
2-C_4H_3S	Me	96 h	(100)	78	
Ph	Me	24 h	(3)	85	
Ph	Et	70 h	(14)	58	
Ph	CH_2Cl	120 h	(85)	88	
Ph	CO_2Me	25 h	(31)	60	
Ph	*t*-Bu	240 h	(5)	85	
1-indanone		240 h	(5)	42	
4-$O_2NC_6H_4$	Me	72 h	(45)	76	
4-ClC_6H_4	Me	72 h	(41)	74	
C_{6-12} 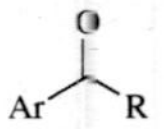		Ph_2SiH_2 (1.5 eq), $[Rh(cod)Cl]_2$, (2.5 mol%), **29** (5 mol%), MeOH, 0°	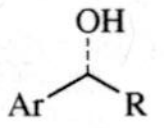		790
Ar	R	Time		% ee	
2-C_4H_3S	Me	70 h	(100)	43	
Ph	Me	70 h	(46)	48	
Ph	Et	40 h	(28)	31	
Ph	CH_2Cl	70 h	(68)	25	
Ph	CO_2Me	25 h	(95)	41	
Ph	*t*-Bu	170 h	(11)	95	
1-indanone		170 h	(15)	31	
4-$O_2NC_6H_4$	Me	50 h	(92)	25	
4-ClC_6H_4	Me	70 h	(51)	22	
4-MeC_6H_4	Me	70 h	(0)	—	
4-$MeOC_6H_4$	Me	70 h	(0)	—	
2-$C_{10}H_7$	Me	70 h	(24)	37	

C_{6-11}

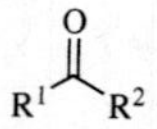

R_3SiH, [(+)-(*R*)-bmpp]$_2$RhCl, C_6H_6, 5-50°, 3-48 h

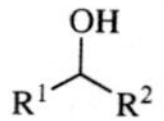

785

R^1	R^2	R_3Si		Opt. Yield	Conf.
Me	*t*-Bu	Ph_2HSi	(98)	25	R
Me	*i*-Bu	Et_2HSi	(99)	12	R
Me	*n*-C_6H_{13}	Et_2HSi	(98)	9	R
Me	*c*-C_6H_{11}	Ph_2HSi	(92)	40	R
n-Pr	*c*-C_6H_{11}	Et_2HSi	(98)	11	R
i-Pr	*c*-C_6H_{11}	Et_2HSi	(94)	4	S
n-Pr	*c*-C_6H_{11}	Ph_2HSi	(95)	10	R
t-Bu	*c*-C_6H_{11}	Ph_2HSi	(90)	43	R

C_{6-12}

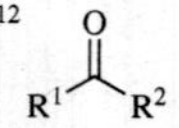

Ph_2SiH_2 (1.6 eq), **87** (x mol%), **99** (10 mol%), $AgBF_4$ (0.02 mol%), THF

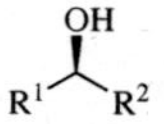

580, 581

R^1	R^2	x	Temp	Time		% ee
EtO_2CCH_2	Me	6.0	–5°	24 h	(60)	27
$EtO_2C(CH_2)_2$	Me	6.0	0°	7 h	(91)	95
$AcO(CH_2)_3$	Me	6.0	20°	24 h	(85)	68
$Me_2C{=}CHCH_2CH_2$	Me	6.0	20°	20 h	(94)	70
n-C_6H_{13}	Me	4.0	0°	2 h	(85)	63
2-ClC_6H_4	Me	4.0	0°	4 h	(74)	94
Ph	Et	4.0	5°	4 h	(73)	91
Bn	Me	6.0	0°	5 h	(95)	71
2-$MeO_2CC_6H_4$	Me	6.0	0°	14 h	(95)	96
3-$AcOC_6H_4$	Me	6.0	0°	24 h	(81)	92
Ph	*n*-Pr	4.0	0-5°	5 h	(82)	82
2-MeOBn	Me	6.0	–5°	4 h	(95)	82
$BnCH_2$	Me	6.0	0°	5 h	(92)	66
1-$C_{10}H_7$	Me	4.0	–5°	5 h	(87)	94
2-$C_{10}H_7$	Me	4.0	–5°	6 h	(93)	93

TABLE 30. ASYMMETRIC ORGANOSILANE REDUCTION OF KETONES (*Continued*)

Ketone	Conditions	Product(s) and Yield(s) (%)	Refs.
$C_{6\text{-}12}$ 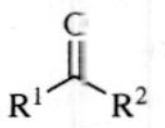	Ph_2SiH_2 (1.5 eq), $[Rh(cod)Cl]_2$ (0.25 mol%), **38** (0.5 mol%), Et_2O, rt, 15-25 h	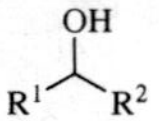	571

R^1	R^2	Yield	% ee	Conf.
t-Bu	Me	(100)	87	R
c-C_6H_{11}	Me	(45)	89	R
Ph	Me	(100)	91	R
Ph	$ClCH_2$	(19)	8	S
Ph	Et	(12)	23	S
n-C_7H_{15}	Me	(92)	60	R
2-$C_{10}H_7$	Me	(99)	88	R

Ketone	Conditions	Product(s) and Yield(s) (%)	Refs.
C_7 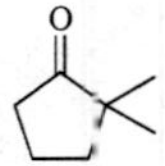	1. $PhSiH_3$ (0.1 eq), **197** (2 mol%), pyrrolidine, MeOH, THF, 60° 2. PMHS, ketone, MeOH (3-7 eq), 15°, 11 h	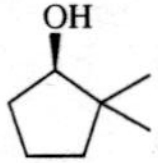 (92) 50% ee	587
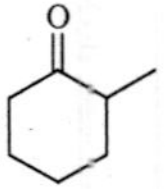	Ph_2SiH_2 (1.3 eq), $RhCl_3$ (1 mol%), **197** (5 mol%), $AgBF_4$ (2 mol%), THF, 0°, 1 d	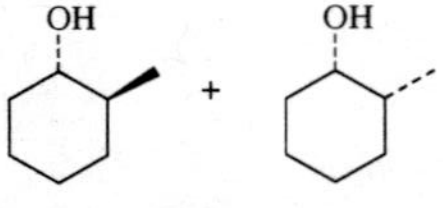 **I + II** (88), **I:II** = 41:59 **I** 1S,2S 91% ee **II** 1S,2R 89% ee	390
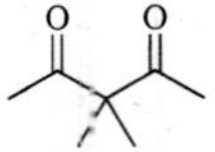	Ph_2SiH_2 (2.5 eq), $[Rh(cod)_2]BF_4$ (1.0 mol%), **23** (1.1 mol%), –30°, THF, 75 h	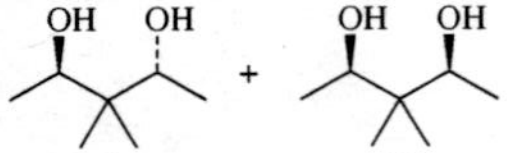 **I** **II** **I + II** (45), **I:II** = 42:58, **I** = 35% ee	577
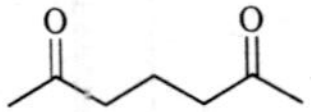	Ph_2SiH_2 (2.5 eq), $[Rh(cod)_2]BF_4$ (1.0 mol%), **23** (1.1 mol%), –30°, THF, 94 h	 **I** **I + II** (75), **I:II** = 69:31, **II** **I** = 89% ee	577

Reactant	Conditions	Product(s) and Yield(s) (%)	Refs.
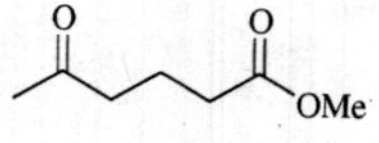	Ph_2SiH_2 (1.5 eq), $[Rh(cod)_2]BF_4$ (1.0 mol%), **23** (1.1 mol%), –30°, THF, 425 h	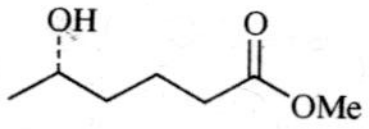 (74) 69% ee[b]	577
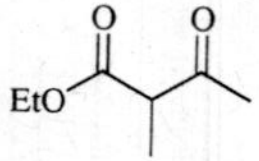	$Ph(1\text{-}Np)SiH_2$ (1.5 eq), $[Rh(cod)]SbF_6$ (1 mol%), **126** (1 mol%), THF, –20°	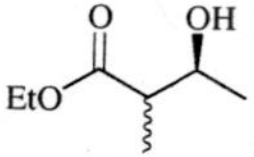 (80) 98% ee, syn:anti = 1:1	576
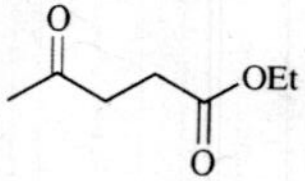	Ph_2SiH_2 (1 eq), $[Rh(cod)Cl]_2$ (0.3 mol%), **108** (0.4 mol%), THF, rt, 30 min	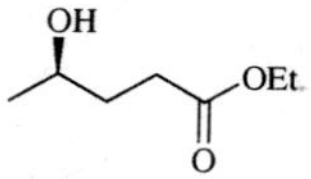 (66) 52% ee	787
	Ph_2SiH_2 (1.1 eq), $[Rh(cod)Cl]_2$ (0.5 mol%), **108** (1.2 mol%), THF, 0°, 24 h	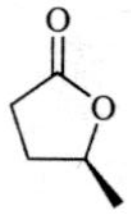 (71) 74% ee	575
	Ph_2SiH_2 (1.6 eq), **87** (4.0 mol%), complex (10 mol%), $AgBF_4$ (0.02 mol%), THF	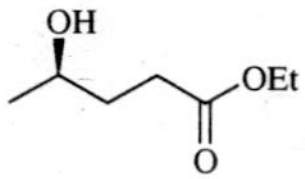	580
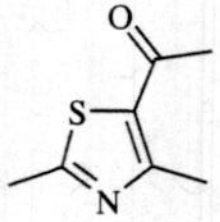	PMHS (4 eq), **124** (0.05 mol%), CuCl (1 mol%), NaOBu-*t* (1 mol%), THF/MeC_6H_5, –50°, 4 h	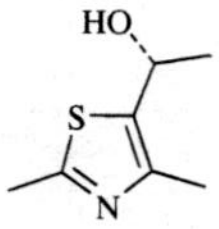 (94) 99% ee	590

Complex	Temp	Time		% ee
100	0°	7 h	(82)	94
101	10°	18 h	(79)	79
103	10°	4 h	(80)	75
104	30°	—	(0)	—

TABLE 30. ASYMMETRIC ORGANOSILANE REDUCTION OF KETONES (*Continued*)

Ketone	Conditions	Product(s) and Yield(s) (%)	Refs.
C_7			
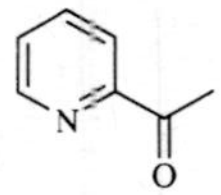	PMHS (4 eq), **124** (0.05 mol%), CuCl (1 mol%), NaOBu-*t* (1 mol%), MeC_6H_5, –50°, 2 h	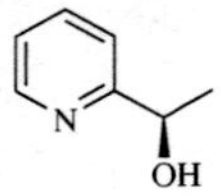 (97) 90% ee	590
	PMHS, $Sn(OTf)_2$ (10 mol%), **87** (10 mol%), MeOH, rt, 12-14 h	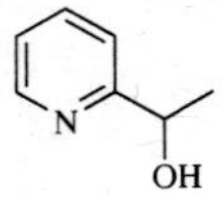 (99) 0% ee	385
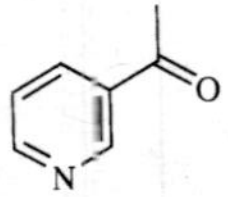	PMHS (4 eq), **124** (0.05 mol%), CuCl (3 mol%), NaOBu-*t* (1 mol%), THF/MeC_6H_5, –35°, 8 h	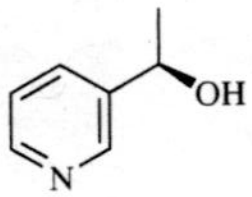 (92) 75% ee	590
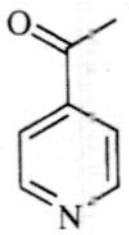	PMHS (4 eq), **124** (0.05 mol%), CuCl (3 mol%), NaOBu-*t* (1 mol%), THF/MeC_6H_5, –78°, 6.5 h	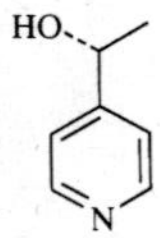 (97) 84% ee	590
C_8			
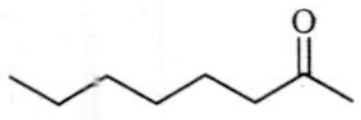	Ph_2SiH_2 (1.5 eq), $[Ir(cod)Cl]_2$ (0.25 mol%), **38** (0.5 mol%), Et_2O, 0°, 25 h	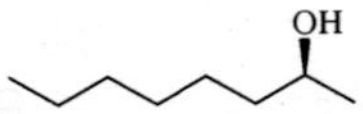 **I** (19) 19% ee	582
	Ph_2SiH_2 (1.5 eq), $[Rh(cod)_2]BF_4$ (1 mol%), **1** (1.1 mol%), THF, –50°, 48 h	**I** (82) 77% ee	578

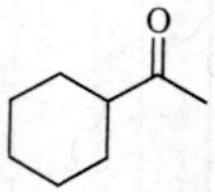	PMHS (1.1 eq), Et_2Zn (20 mol%), (*R*,*R*)-ebpe (20 mol%), vitride (3%), MeC_6H_5, rt, 24 h	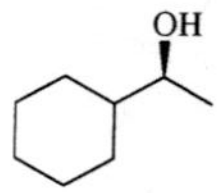 **I** (—) 20% ee	594
	1. $PhSiH_3$ (0.1 eq), **197** (2 mol%), pyrrolidine, MeOH, THF, 60° 2. PMHS, ketone, MeOH (3-7 eq), 15°, 6 h	**I** (>98) 23% ee	587
	Ph(1-Np)SiH_2 (1.5 eq), [Rh(cod)]SbF_6 (1 mol%), **126** (1 mol%), THF, –20°	**I** (90) 92% ee	576
	Ph_2SiH_2 (1 eq), $[Rh(cod)Cl]_2$, (0.3 mol%), **46** (0.4 mol%), THF, rt, 30 min	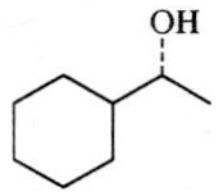 **II**	788
	Ph_2SiH_2 (1.25 eq), Rh$[(NBD)Cl]_2$ (0.5 %), ligand (0.5 %), MeC_6H_5	**I** (—)	791
	Ligand 116, Temp 0°	37% ee	
	Ligand 117, Temp 0°	33% ee	
	Ligand 118, Temp 10°	50% ee	
	Ligand 119, Temp 27°	49% ee	
	Ph_2SiH_2 (1.5 eq), $[Rh(cod)_2]BF_4$ (1 mol%), **25** (1.1 mol%), THF, rt, 10 h	**I** (62) 80% ee	606
	1. **197** (4.5 eq), *n*-BuLi (9 eq), C_6H_6 2. PMHS (5 eq) 3. Add ketone, 4 d	**II** (67) 24% ee	588

TABLE 30. ASYMMETRIC ORGANOSILANE REDUCTION OF KETONES (*Continued*)

Ketone	Conditions	Product(s) and Yield(s) (%)	Refs.
C_8			
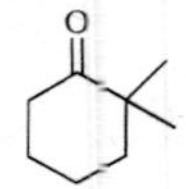	Ph_2SiH_2 (1.5 eq), $[Rh(cod)_2]BF_4$ (1.0 mol%), **25** (1.1 mol%), –40°, THF, 26 h	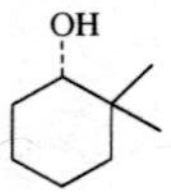 (70) 88% ee	577
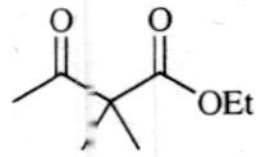	Ph_2SiH_2 (1.5 eq), $[Rh(cod)_2]BF_4$ (0.5 mol%), **23** (1.1 mol%), –30°, THF, 24 h	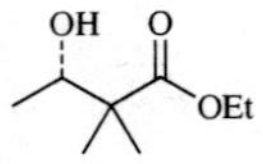 (80) 98% ee	577
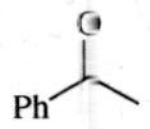	PMHS (1.2 eq), $(i\text{-}Pr)_2Zn$ (2 mol%), **131** (2 mol%), MeC_6H_5, rt, 48 h	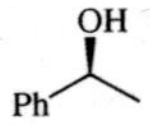 **I** (>99) 76% ee	788
	Ph(1-Np)SiH_2 (4 eq), $[Rh(cod)Cl]_2$ (1 mol%), **41**, THF, –78°	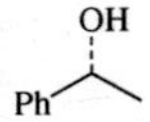 **II** (85) 86% ee	792
	Ph_2SiH_2 (1.5 eq), $[Rh(cod)Cl]_2$ (0.5 mol%), **37** (0.5 mol%), 0°	**I** (94) 60% ee	570
	Ph_2SiH_2 (4 eq), $[Rh(cod)Cl]_2$ (1 mol%), ligand	**II**	792

Ligand	Solvent	Temp		% ee
41	none	0°	(54)	73
41	none	–40°	(79)	73
41	THF	0°	(54)	73
41	THF	–78°	(86)	82
40	THF	–78°	(77)	51
40	CH_2Cl_2	–78°	(70)	48
39	CCl_4	0°	(58)	14
51	THF	–78°	(45)	0

Ph_2SiH_2 (1.5 eq), $[Rh(cod)_2]BF_4$ (1 mol%), ligand (1.1 mol%), THF, rt — **I** — 606

Ligand	Time		% ee
24	5 h	(89)	92
25	11 h	(88)	92
26	48 h	(78)	1
27	9 h	(85)	15

Ph(1-Np)SiH_2 (1 eq), $[Rh(cod)Cl]_2$ (0.3 mol%), **5** (0.4 mol%), THF, rt, 30 min — **II** (91) 76% ee — 787

Ph_2SiH_2 (1 eq), $[Rh(cod)Cl]_2$ (0.3 mol%), ligand (0.4 mol%), THF, 30 min — 787

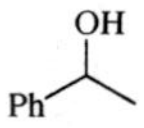

Ligand	Temp		% ee	Conf.
40	rt	(92)	44	R
41	rt	(97)	73	R
41	10°	(94)	81	R
41	0°	(95)	82	R
42	rt	(65)	40	R
47	rt	(93)	52	R
49	rt	(55)	2	R
50	rt	(56)	4	S
45	rt	(97)	69	R
44	rt	(33)	17	S
46	rt	(99)	79	R
46	–40°	(99)	86	R
43	rt	(92)	26	R
53	rt	(85)	0	—
52	rt	(78)	3	R

TABLE 30. ASYMMETRIC ORGANOSILANE REDUCTION OF KETONES (*Continued*)

Ketone	Conditions	Product(s) and Yield(s) (%)	Refs.
C_8 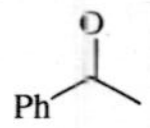	Ph_2SiH_2 (1.25 eq), **120** (0.5 %), MeC_6H_5	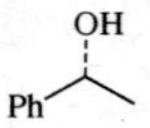 **II**	791
	Temp: −25°; −10°; 0°; 20°; 50°	% ee: 35; 25; 17; 2; 4	
	$(EtO)_3SiH$, **194** (10 mol%), Et_2O, 40°, 12 h	**II** (98) 53% ee	378
	PMHS (5 eq), CuCl, *t*-BuONa, **121** (0.0009 mol%), MeC_6H_5, or THF, or MeC_6H_5/THF, −78°	**II** (98) 92% ee	589
	Ph_2MeSiH (4 eq), CuCl (4 mol%), *t*-BuONa (4 mol%), **122** (3 mol%), MeC_6H_5, −78°	**II** (90) 95% ee	589
	$(HMe_2Si)_2O$ (5 eq), CuCl (2 mol%), *t*-BuONa (2 mol %), **122** (1 mol%), MeC_6H_5, −78°	**II** (98) 94% ee	589
	PMHS (10 eq), CuCl (3 mol %), *t*-BuONa (3 mol%), ligand (3 mol%), MeC_6H_5	**II**	589

Ligand	Time	% ee
121	5 h	94
122	1 h	90
123	6 h	95
124	<1 h	96

PMHS (1.2 eq), Et_2Zn (2 mol%),
ligand (2 mol%), MeC_6H_5, rt, 24 h

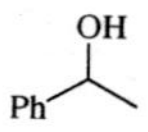

594

Ligand	% Conv.	% ee	Conf.
182	95	75	S
183	97	74	S
186	98	72	S
184	95	70	S
185	98	73	S
181	98	56	S
180	97	48	S
179	95	30	S
154	98	18	S
140	98	24	S
158	99	70	S
143	95	52	S
155	98	62	S
156	97	63	S
157	98	62	S
141	95	47	R
146	98	88	R
142	98	83	R
144	97	83	R
145	15	58	R
135	99	75	R
153	98	8	R

TABLE 30. ASYMMETRIC ORGANOSILANE REDUCTION OF KETONES (*Continued*)

Ketone	Conditions	Product(s) and Yield(s) (%)	Refs.
C_8 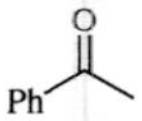	PMHS (1.1 eq), Et_2Zn (x mol%), (*R*,*R*)-ebpe (y mol%), vitride, (3%), MeC_6H_5, rt, 24 h	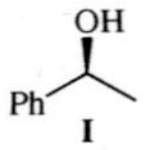 **I** (—) x / y / % ee: 20 / 20 / 76; 2 / 2 / 74	594
	PMHS (1.1 eq), catalyst (2 mol%), ligand (2 mol%), MeC_6H_5, rt, 18 h	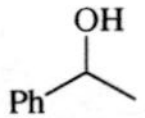	594

Catalyst	Ligand	Activator	% Conv.	% ee	Conf.
Me_2Zn	**180**	none	99	48	S
ZnH_2	**180**	none	96	48	S
PhZnH•py	**180**	none	95	49	S
Et_2Zn	**179**	none	99	75	R
Et_2Zn•ebpe	**179**	none	99	75	R
$Zn(dea)_2$	**179**	none	75	71	R
$Zn(dea)_2$•ebpe	**179**	none	70	71	R
$Zn(dea)_2$	**179**	$NaBH_4$ (2%)	100	45	R
$Zn(dea)_2$	**179**	$LiAlH_4$ (1%)	100	67	R
$Zn(dea)_2$	**179**	vitride (3%)	100	72	R
$Zn(dea)_2$	**179**	LiH (4%)	100	71	R
$Cd(dea)_2$	**179**	none	100	62	R
$Cd(dea)_2$	**146**	none	100	81	R
$Co(da)_2$	**179**	none	90	58	R
$Cu(dea)_2$	**179**	none	0	—	—
$Sn(2\text{-}EH)_2$	**179**	none	0	—	—

Ketone	Conditions	Product(s) and Yield(s) (%)	Refs.
	$(MeHSiO)_4$ (3.75 eq), **197** (0.57 mol%), *n*-BuLi (1.14 mol%), THF, rt, 24 h	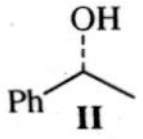 **II** (100) 99% ee	583
	PMHS (3.75 eq), **197** (0.57 mol%), *n*-BuLi (1.14 mol%), THF, rt, 24 h	**II** (100) 99% ee	583
	Ph_2SiH_2 (1.5 eq), $[Ir(cod)Cl]_2$ (0.25 mol%), **38** (0.5 mol%), Et_2O, 0°, 20 h	**I** (—) 96% ee	571

Conditions	Product	Refs.
PMHS (1.2 eq), Et_2Zn (2 mol%), ligand (2 mol%), MeC_6H_5, rt, 24 h	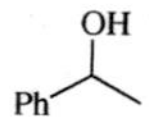	594

Ligand	% Conv.	% ee	Conf.
159	96	17	S
sparteine	60	19	R
160	98	55	S
173	67	6	R
174	100	8	S
175	60	13	R
176	9	1	S
43	88	10	S
187	100	32	S
177	10	5	R

Conditions	Product	Refs.
Ph_2SiH_2 (1.1 eq), $[Rh(cod)Cl]_2$, ligand (1 mol%), MeC_6H_5, 0°, 24 h	**II**	575

Ligand		% ee	Conf.
105	(97)	70	R
107	(86)	82	S
106	(92)	66	R
108	(89)	86	S
181	(89)	70	S

Conditions	Product	Refs.
Ph_2SiH_2 (1.5 eq), $[Rh(cod)_2]BF_4$ (1 mol%), **31** (1.1 mol%), THF, –40°, 4 h	**I** (89) 94% ee	578
Ph_2SiH_2 (1.5 eq), $[Rh(cod)_2]BF_4$ (1 mol%), **31** (1.1 mol%), THF, –30°, 24 h	**I** (96) 80% ee	578

TABLE 30. ASYMMETRIC ORGANOSILANE REDUCTION OF KETONES (*Continued*)

Ketone	Conditions	Product(s) and Yield(s) (%)	Refs.
C_8			
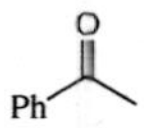	$PhSiH_3$ (1.1 eq), (*R*)-**195**-2*n*-BuLi (x mol%), MeC_6H_5, rt	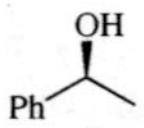 **I**	783

x	Time	% Conv.	% ee
5	24 h	100	82
4	24 h	92	68
3	24 h	100	60
2	24 h	98	64
1	24 h	97	60
0.5	3 h	21	60

Ketone	Conditions	Product(s) and Yield(s) (%)	Refs.
	$PhSiH_3$ (1.1 eq), (*R*)-**195**-2*n*-BuLi (5 mol%), rt	**I**	783

Solvent	Time	% Conv.	% ee
THF	1 h	87	6
THF	3 h	100	0
C_6H_5Me	1 h	44	59
C_6H_5Me	3 h	100	63
C_6H_6	1 h	78	36
C_6H_6	3 h	98	42
n-C_6H_{14}	1 h	67	17
n-C_6H_{14}	3 h	96	21
none	1 h	51	14
none	3 h	60	11
none	24 h	79	7

Ketone	Conditions	Product(s) and Yield(s) (%)	Refs.
	PMHS (5 eq), CuF_2, (*S*)-BINAP (4 mol%), MeC_6H_5, air, rt	**I** (93) 74% ee	784

Conditions	Product(s) and Yield(s) (%)	Refs.
FMHS (5 eq), CuF_2, (*S*)-BINAP (4 mol%), MeC_6H_5, Ar, rt	**I** (100) 72% ee	784
$Me(EtO)_2SiH$ (1.5 eq), CuF_2, (*S*)-BINAP (4 mol%), MeC_6H_5, air, rt	**I** (94) 76% ee	784
$PhSiH_3$ (1.2-1.5 eq), [Cu] (x mol%), (*S*)-BINAP, MeC_6H_5, atmosphere, rt	**I**	784

[Cu]	x	atm	Time		% ee
CuF_2, dppb	2	Ar	8 h	(90)	—
CuF_2, dppb	2	air	2 h	(99)	—
CuF_2	4	Ar	16 h	(>99)	79
CuF_2	1	air	2 h	(98)	78
CuF_2	0.5	(—)	6 h	(94)	78
$CuF(PPh_3)_3$•MeOH, dppb	4	Ar	2 h	(90)	—
$CuF(PPh_3)_3$•MeOH	4	Ar	40 h	(91)	71
$CuF(PPh_3)_3$•MeOH	2	air	1 h	(100)	74
CuI, $Ph_3SiF_2N(Bu\text{-}n)_4$	1	Ar	24 h	(40)	76
$[(Ph_3P)CuH]_6$	2	Ar	16 h	(95)	80

Conditions	Product(s) and Yield(s) (%)	Refs.
Silane (1.5 eq), **126** (1 mol%), THF, rt	**I**	576

Silane		% ee
$PhSiH_3$	(—)	0
$PhMeSiH_2$	(—)	78
Ph_2SiH_2	(—)	88
Ph(1-Np)SiH_2	(—)	95
Ph(*t*-Bu)SiH_2	(0)	—
Et_3SiH	(0)	—
PMHS	(0)	—

TABLE 30. ASYMMETRIC ORGANOSILANE REDUCTION OF KETONES (*Continued*)

Ketone	Conditions	Product(s) and Yield(s) (%)	Refs.
C_8 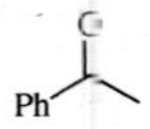	$(EtO)_3SiH$, THF, $Ti(L)_2X_2$, 2-3 d	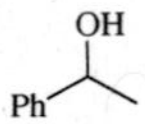 **I**	793
	L / X	(yield) / % ee / Conf.	
	51 / Cl	(60) / 50 / R	
	57 / Cl	(20) / 23 / S	
	51 / OPr-*i*	(87) / 18 / S	
	59 / OPr-*i*	(85) / 51 / S	
	62 / OPr-*i*	(13) / 65 / S	
	59 / F	(86) / 61 / S	
	60 / F	(30) / 56 / S	
	61 / F	(18) / 51 / S	
	Ph_2SiH_2 (1.5 eq), $[Rh(cod)]SbF_6$ (1 mol%), ligand (1 mol%), THF, rt	**I**	576
	Ligand	(yield) / % ee / Conf.	
	127	(36) / 62 / S	
	128	(89) / 51 / S	
	129	(50) / 20 / R	
	130	(90) / 86 / R	

PMHS (1.2 eq), Et_2Zn (2 mol%), ligand (2 mol%), MeC_6H_5, rt

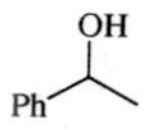

794

Ligand	Time		% ee	Conf.
131	18 h	(98)	88	S
135	18 h	(99)	75	S
139	170 h	(56)	84	R
137	48 h	(11)	37	R
147	216 h	(30)	25	R
138	288 h	(66)	91	R
151	40 h	(56)	78	R
136	44 h	(78)	67	R
149	18 h	(>99)	83	S
150	72 h	(>99)	84	R

Ph_2SiH_2 (2 eq), $(Ph_3P)RuCl_2$ (1 mol%), ligand (1 mol%), additive (1 mol%), Et_2O, 0°

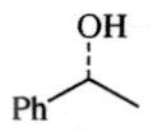

569

Ligand	Additive	Time		% ee
37	AgOTf	24 h	(56)	93
37	$Cu(OTf)_2$	24 h	(59)	95
37	—	24 h	(46)	67
35	AgOTf	5 h	(42)	90
35	$Cu(OTf)_2$	36 h	(38)	92
36	AgOTf	24 h	(52)	45
41	AgOTf	70 h	(34)	84

TABLE 30. ASYMMETRIC ORGANOSILANE REDUCTION OF KETONES (*Continued*)

Ketone	Conditions	Product(s) and Yield(s) (%)	Refs.
C_8 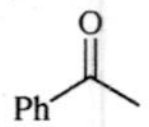	$[Rh(cod)_2]BF_4$ (5 mol%), ligand (5 mol%)	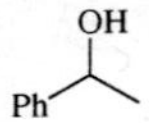	789, 795
	Ph_2SiH_2 (1.5 eq), $[Ir(cod)Cl]_2$ (2.5 mol%), ligand (5 mol%), THF	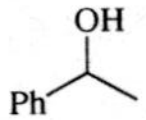	789

[Rh(cod)₂]BF₄ conditions:

Ligand	Solvent	Temp	Time		% ee	Conf.
29	none	rt	1 h	(73)	12	R
29	C_6H_6	rt	12 h	(100)	21	S
29	Et_2O	rt	2 h	(71)	8	R
29	DME	rt	3 h	(100)	0	—
29	CH_2Cl_2	rt	12 h	(100)	6	S
29	MeCN	rt	20 h	(43)	0	—
29	THF	0°	48 h	(31)	85	R
28	THF	0°	72 h	(46)	31	R
30	THF	0°	72 h	(67)	50	R

Ph_2SiH_2 / $[Ir(cod)Cl]_2$ conditions:

Ligand	Temp	Time		% ee	Conf.
28	15°	20 h	(82)	15	R
28	0°	144 h	(43)	13	R
29	40°	4 h	(100)	0	—
29	15°	1 h	(100)	23	S
29	0°	24 h	(55)	18	S
29	–20°	24 h	(40)	0	—
30	15°	20 h	(90)	0	—
30	0°	144 h	(81)	22	R

R_3SiH {RhCl(cod)}$_2$ (2.5 mol%), (–)-**37**, THF, rt

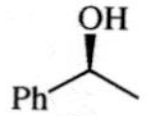

I (—)

586

R_3Si	% ee
n-$C_8H_{17}H_2Si$	3
PhH_2Si	8
Et_2HSi	1
PhMeHSi	66
Ph_2HSi	80
2-MeC_6H_4PhHSi	95
(2-$MeC_6H_4)_2$HSi	92
MesPhHSi	98
Mes_2HSi	—

Silane (1.5 eq), **126** (1 mol%), THF, rt

I

576

Silane		% ee
$PhSiH_3$	(—)	0
$PhMeSiH_2$	(—)	78
Ph_2SiH_2	(—)	88
Ph(1-Np)SiH_2	(—)	95
Ph(t-Bu)SiH_2	(0)	—
Et_3SiH	(0)	—
PMHS	(0)	—

PMHS, $Sn(OTf)_2$ (10 mol%),
ligand (10 mol%), MeOH, rt

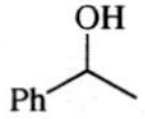

385

Ligand		% ee	Conf.
93	(95)	58	R
87	(78)	54	S
88	(50)	13	S
94	(30)	44	R
96	(50)	58	R

TABLE 30. ASYMMETRIC ORGANOSILANE REDUCTION OF KETONES (*Continued*)

Ketone	Conditions	Product(s) and Yield(s) (%)	Refs.
C_8			
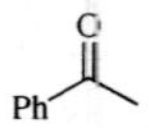	Ph_2SiH_2 (1.5 eq), $[Rh(cod)Cl]_2$ (0.5 mol%), ligand (1 or 2 mol%), THF, 0°, 20 h	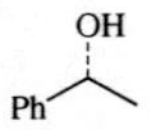 **II**	789

Ligand	Yield (%)	% ee
7	(51)	16
13	(26)	40
15	(30)	18
14	(73)	8
9	(81)	16
10	(36)	12
8	(25)	35
12	(51)	0
11	(39)	36
32	(52)	33
33	(44)	40
3	(65)	31
4	(73)	27
5	(50)	32
6	(73)	24
191	(35)	5
192	(83)	0

Ph_2SiH_2 (1.5 eq), [Rh(cod)Cl]$_2$ (2.5 mol%), ligand (5 mol%), MeOH | II | 790

Ligand	Temp	Time		% ee
29	0°	70 h	(46)	48
28	0°	40 h	(56)	10
30	0°	40 h	(72)	27

Ph_2SiH_2 (2 eq), [Rh(cod)Cl]$_2$ (1 mol%), **48** (x mol%), 0° | II | 585

Solvent	x		% ee
none	2.4	(64)	76
THF	2.4	(76)	85
Et_2O	2.4	(80)	91
MeC_6H_5	2.4	(76)	93
MeC_6H_5	4	(84)	94

TABLE 30. ASYMMETRIC ORGANOSILANE REDUCTION OF KETONES (*Continued*)

Ketone	Conditions	Product(s) and Yield(s) (%)	Refs.
C_8 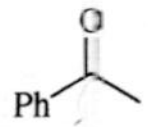	Ph_2SiH_2 (1.3 eq), catalyst (1 mol%), ligand, THF, rt, 24 h	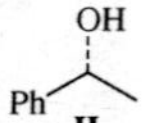 **II**	796

Catalyst	Ligand	% Conv.	% ee	Conf.
[Rh(Cl)(cod)]•L	**218**	86	4.4	R
[Rh(Cl)(cod)]•2L	**219**	90	0	—
1/2[Rh(Cl)(cod)]$_2$•2L	**218**	93	11.2	R
1/2[Rh(Cl)(cod)]$_2$•2L	**219**	94	4.4	R
1/2[Rh(Cl)(cod)]$_2$•2L	**217**	93	6.8	R
1/2[RhCl(C_2H_4)$_2$]$_2$•L	**218**	92	6.6	R
[RhCl(C_2H_4)•L]$_2$	**218**	88	5.0	R
1/2[RhCl(C_2H_4)$_2$]$_2$•2L	**218**	96	41.4	R
1/2[RhCl(C_2H_4)$_2$]$_2$•2L	**219**	92	52.3	R
1/2[RhCl(C_2H_4)$_2$]$_2$•2L	**217**	63	3.3	S
1/2[RhCl(C_2H_4)$_2$]$_2$•3L	**218**	89	51.2	R
1/2[RhCl(C_2H_4)$_2$]$_2$•3L	**219**	94	58.3	R
RhClL$_2$	**218**	90	31.0	R
[Rh(Cl)COL$_2$]	**218**	78	7.5	R
1/2[RhCl(C_2H_4)$_2$]$_2$•2L + C_2H_4	**218**	92	33.1	R
1/2[RhCl(C_2H_4)$_2$]$_2$•4L	**218**	83	27.6	R
1/2[RhCl(C_2H_4)$_2$]$_2$•5L	**218**	79	25.7	R

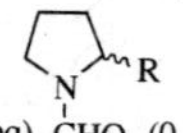

Cl_3SiH (1.5 eq), CHO (0.1 eq), CH_2Cl_2, 0° to rt, 24 h — II — 379

R	Conf.		% ee	R:S
CONHPh	S	(90)	31	65:35
CONHPh	R	(93)	27	36:64
$CONHC_{10}H_7$-1	S	(78)	43	71:29
$CONHCHPh_2$	S	(69)	20	61:39
$CONHC_6H_{14}$-*n*	S	(82)	8	54:46
CONHBu-*t*	S	(80)	11	55:45
CO_2Ph	S	(53)	21	61:39
CO_2Me	S	(39)	14	57:43
CO_2Bu-*t*	S	(88)	11	55:45
CH_2OBn	S	(29)	13	56:44

Ph_2SiH_2 (1 eq), catalyst, ligand, C_6H_6, rt, 16 h — II — 797

Catalyst	Ligand	Rh:Ligand		% ee
$[Rh(cod)Cl]_2$	TRISPHOS	1:2	(60)	0
$[Rh(cod)Cl]_2$	(*S*,*S*,*S*)-TRISPHOS	1:2	(38)	81
$[Rh(cod)Cl]_2$	(*S*,*S*,*S*)-TRISPHOS	1:3	(53)	75
(NBD)Rh(acac)	(*S*,*S*,*S*)-TRISPHOS	1:1	(67)	58
$(NBD)RhClO_4$	(*S*,*S*,*S*)-TRISPHOS	1:1	(65)	81

Table 30. Asymmetric Organosilane Reduction of Ketones (Continued)

Ketone	Conditions	Product(s) and Yield(s) (%)	Refs.
C_8 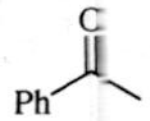	R_3SiH (1 eq), Li-(*R*)-BINOL (5 mol%)	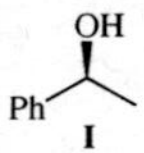 I	592

R_3Si	Solvent	Temp	Time		% ee
$(MeO)_3Si$	THF	–78°; 20°	6 h; 16 h	(88)	0
$(EtO)_3Si$	Et_2O	0°; 20°	6 h; 16 h	(33)	31
$(EtO)_3Si$	Et_2O/TMEDA 2:1	0°	24 h	(50)	64
$(EtO)_3Si$	THF/TMEDA 2:1	0°	24 h	(52)	33
$(EtO)_3Si$	Et_2O, TMEDA	0°	24h	(46)	63

Ph_2SiH_2 (1.6 eq), complex (4.0 mol%), ligand (10 mol%), $AgBF_4$ (0.02 mol%), THF — I — 580

Complex	Ligand	Time		% ee
99	**89**	3 h	(94)	22
103	**87**	2 h	(94)	31
99	**87**	2 h	(88)	62
99	**88**	3 h	(94)	84
102	**87**	7 h	(87)	87

Ph_2SiH_2 (1.6 eq), complex (4.0 mol%), ligand (10 mol%), $AgBF_4$ (0.02 mol%), THF — I — 580

Complex	Ligand	Temp	Time		% ee
100	**89**	–5°	10 h	(91)	91
101	**88**	0°	18 h	(92)	83
103	**92**	10°	2 h	(88)	54
104	**93**	20°	18 h	(82)	19
102	**91**	20°	20 h	(66)	—

Ph_2SiH_2 (1.25 eq), $Rh[(NBD)Cl]_2$ (0.5%), Ligand (0.5%), MeC_6H_5 I 791

Ligand	Temp	% ee
116	0°	52
117	10°	55
118	20°	54
119	20°	56
109	20°	55
113	20°	82

Ph_2SiH_2 (1.25 eq), $Rh[(NBD)Cl]_2$ (0.5%), **117** I 791

Solvent	$Rh[(NBD)Cl]_2$:**117**	Temp	% Conv.	% ee
MeC_6H_5	1:1	–27°	57	17
MeC_6H_5	1:1	0°	96	53
MeC_6H_5	1:1	10°	96	55
MeC_6H_5	1:1	20°	95	51
THF	1:1	–27°	39	15
THF	1:1	0°	88	52
THF	1:1	10°	79	49
C_6H_6	1:1	0°	95	46
C_6H_6	1:1	20°	86	49
MeC_6H_5	1:1	10°	86	54
MeC_6H_5	1:4	10°	97	56
MeC_6H_5	1:2	10°	96	56

TABLE 30. ASYMMETRIC ORGANOSILANE REDUCTION OF KETONES (*Continued*)

Ketone	Conditions	Product(s) and Yield(s) (%)	Refs.
C_8 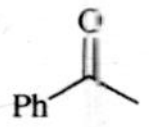	R_3SiH (1.6 eq), **87** (1.5 mol%), **99** (10 mol%), THF, –5° to 0°	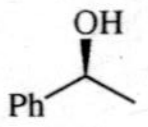 **I**	580
	R_3Si / Temp / Time	% ee	
	Ph_2HSi / –5° to 0° / —	(—) 76-78	
	PhMeHSi / 0° / 26 h	(32) 14	
	Ph(1-Np)HSi / 0° / 4 d	(62) 62	
	$(EtO)_3SiH$, **194** (10 mol%)	**I**	378
	Solvent / Temp / Time	% ee	
	Et_2O / 50° / 5 h	(>98) 55	
	MeC_6H_5 / 50° / 5 h	(78) 48	
	none / 40° / 12 h	(>98) 54	

$(MeO)_3SiH$ (1 eq), catalyst

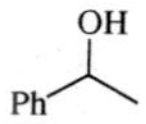

592

Catalyst	Solvent	Temp	Time		% ee	Conf.
Li-(*R*)-BINOL	Et_2O	0°	48 h	(59)	7	R
Li-(*R*)-BINOL	Et_2O/TMEDA 2:1[c]	0°	12 h	(53)	48	S
(*R*)-BINOL phosphoric acid	THF	–22°	48 h	(37)	6	R
Li-(*R*)-BINOL	Et_2O/TMEDA 2:1[c]	20°	24 h	(50)	64	S
Li-(*R*)-BINOL	Et_2O/TMEDA 30:1[c]	20°	24 h	(46)	63	S
Al-BINOL complex	Et_2O or THF	20°	24 h	(0)	—	
Li-(*R*)-2,2'-Me_2BINOL	Et_2O/TMEDA 30:1	–20°	24 h	(87)	26	R
Li-(*R*)-methoxythio-BINOL	Et_2O/TMEDA 30:1	20°	24 h	(22)	37	R
Li-(*R*)-BINOL	Et_2O/TMEDA 30:1[d]	0°	8 h	(91)	58	S
Li-(*R*)-BINOL	Et_2O/TMEDA 30:1[e]	0°	8 h	(96)	59	S
Li-(*R*)-BINOL	Et_2O/TMEDA 30:1[f]	0°	24 h	(92)	70	S
Li-(*R*)-BINOL	Et_2OTMEDA 30:1[f]	0°	24 h	(90)	70	S
Li-(*R*)-BINOL	Et_2O	0°	24 h	(47)	52	S
(*R*)-hydro-BINOL	Et_2O/TMEDA 30:1	0°	24 h	(47)	40	S
Li-(*R*)-BINOL	Et_2O/TMEDA 30:1[g]	0°	24 h	(91)	66	S

Ph_2SiH_2 (1.7 eq), catalyst (x mol%),
ligand (y mol%)

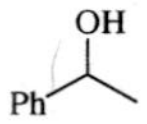

798

Catalyst	x	Ligand	Temp	Time		% ee	Conf.
CuOBu-*t*	0.3	(–)-DIOP	0-20°	15 h	(100)	19.5	R
CuO_2CPh	0.3	(–)-DIOP	0-20°	23 h	(100)	18.7	R
CuO_2CPh	0.33	(–)-DIOP	0-20°	16 h	(100)	13.4	R
CuO_2CPh	0.48	(+)-NORPHOS	20°	95 h	(51)	38.8	R
CuO_2CPh	0.5	(+)-NORPHOS	20°	120 h	(67)	16.3	S
CuO_2CPh	0.24	(–)-BPPFA	0-20°	18 h	(100)	28.9	R

Ketone	Conditions	Product(s) and Yield(s) (%)	Refs.
C_8 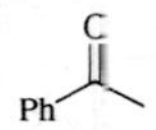	R_3SiH, ligand, THF/TMEDA, 0°, 10 min	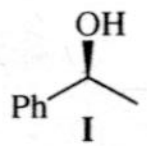 **I**	593

R_3Si	Ligand	I
$(EtO)_3Si$	**188**	(55)
$(MeO)_3Si$	none	(0)
$(EtO)_3Si$	**190**	(26)
$(EtO)_3Si$	**189**	(30)
$(MeO)_3Si$	Li-histidine	(26)
$(EtO)_3Si$	Li_2-histidine	(26)
$(EtO)_3Si$	Li-phenylalanine	(25)

Conditions	Product(s) and Yield(s) (%)	Refs.
Ph_2SiH_2 (1.7 eq), catalyst (x mol%), ligand (y mol%)	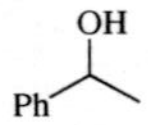	798

Catalyst	x	Ligand	y	Temp	Time		% ee	Conf.
CuOBu-*t*	0.3	(–)-DIOP	0.35	0-20°	15 h	(100)	19.5	R
CuOBu-*t*	0.55	(–)-DIOP	0.34	0-20°	20 h	(100)	18.8	R
CuOBu-*t*	0.4	(–)-DIOP	0.44	0-20°	15 h	(100)	15.3	R
CuO_2CPh	0.3	(–)-DIOP	0.30	0-20°	16 h	(100)	12.7	R
CuO_2CPh	0.16	(–)-DIOP	0.19	–10°	23 h	(100)	16.9	R
CuO_2CPh	0.04	(–)-DIOP	0.08	0-20°	40 h	(100)	16.7	R
CuO_2CPh	0.3	(–)-DIOP	0.15	0-20°	23 h	(100)	18.7	R
CuO_2CPh	0.33	(–)-DIOP	0.39	0-20°	16 h	(100)[h]	13.4	R
CuO_2CPh	0.48	(+)-NORPHOS	0.71	20°	95 h	(51)	38.8	R
CuO_2CPh	0.5	(+)-NORPHOS	0.81	20°	120 h	(67)[h]	16.3	S
CuO_2CPh	0.24	(–)-BPPFA	0.27	0-20°	18 h	(100)	28.9	R

Ph_2SiH_2 (160 mol%),
$[RuCl_2(C_6H_6)]_2$ (0.5 mol%)
ligand (x mol%), AgOTf (2 mol%)

I

Ligand	x	Solvent	Time		% ee
125	8.8	none	48 h	(94)	38
125	4.4	CCl_4	24 h	(0)	—
125	4.4	MeC_6H_5	24 h	(88)	23
125	4.4	dioxane	96 h	(45)	35
125	4.4	THF	24 h	(97)	54
(*R*)-BINAP	4.4	none	24 h	(95)	5 (R)
87	8.8	none	72 h	(16)	0

799

Ph_2SiH_2, **207** (1 mol%), rt

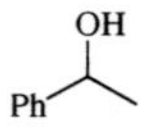

Solvent		% ee	Conf.
THF	(—)	63	R
MeCN	(8)	10	R
CCl_4	(11)	12	S
DME	(92)	12	R
MeC_6H_5	(24)	21	R

800

Ph_2SiH_2 (1.6 eq), **99** (10 mol%),
additive (x mol%), THF

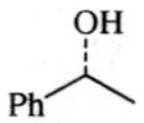

Additive	x	Temp	Time		% ee
none	—	rt	—	(0)	—
$AgBF_4$	2.0	0°	3 h	(94)	95
$AgPF_6$	2.0	−3°	5 h	(80)	87
AgOTf	1.0	−5°	27 h	(96)	89
BF_3•OEt_2	1.5	0°	14 h	(90)	82
$EtAlCl_2$	1.5	0°	18 h	(89)	67

580, 581

TABLE 30. ASYMMETRIC ORGANOSILANE REDUCTION OF KETONES (*Continued*)

Ketone	Conditions	Product(s) and Yield(s) (%)	Refs.
C_8 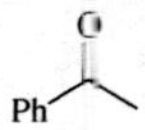	Ph_2SiH_2 (1 eq), $[Rh(cod)Cl]_2$ (0.5 mol%), **66** (2.5 mol%), 0-20°, 18 h	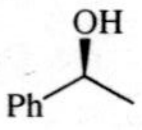	801

Solvent		% ee
none	(60)	39.8
MeC_6H_5	(66)	31.8
C_6H_6	(60)	31.0
n-C_5H_{12}	(45)	28.4
PE	(48)	24.2
Et_2O	(60)	34.0
THF	(51)	24.0
dioxane	(53)	25.2
CH_2Cl_2	(71)	23.8
$ClCH_2CH_2Cl$	(72)	22.2
$CHCl_3$	(76)	15.2
CCl_4	(93)	56.6
MeCN	(63)	16.8

Ph_2SiH_2 (1 eq), $[Rh(cod)Cl]_2$ (0.5 mol%), ligand (2.5 mol%), MeC_6H_5, 0-20°, 18 h

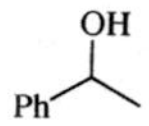

804

Ligand		% ee	Conf.
64	(42)	25.5	R
65	(66)	31.8	S
66	(66)	34.4	R
67	(40)	20.2	S
69	(56)	23.0	R
68	(77)	37.4	R
70	(44)	32.8	R
71	(63)	39.8	R
72	(77)	70.1	R
73	(63)	26.5	S
74	(69)	33.8	S
75	(58)	39.6	R
76	(57)	3.5	R
77	(19)	4.3	S
78	(16)	4.2	R
79	(66)	34.7	S
80	(69)	33.8	R
81	(68)	53.6	S
82	(51)	8.7	R
83	(69)	24.3	R
84	(48)	1.7	R

TABLE 30. ASYMMETRIC ORGANOSILANE REDUCTION OF KETONES (*Continued*)

Ketone	Conditions	Product(s) and Yield(s) (%)	Refs.
C_8 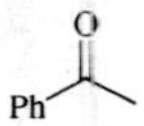	Ph_2SiH_2 (1 eq), $[Rh(cod)Cl]_2$ (0.5 mol%), ligand (2.5 mol%), CCl_4, 0-20°, 18 h	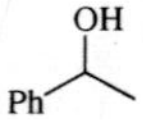	801

Ligand		% ee	Conf.
64	(80)	40.7	R
66	(93)	56.6	S
67	(75)	52.6	S
69	(90)	53.2	R
68	(85)	62.2	R
70	(67)	69.8	R
71	(89)	71.6	R
72	(90)	83.4	R
73	(70)	66.6	S
74	(85)	50.0	S
75	(95)	62.4	R
76	(10)	42.9	R
77	(27)	5.6	S
78	(63)	9.0	R
79	(90)	49.6	S
80	(90)	50.8	R
81	(89)	65.6	S
82	(17)	10.4	R
83	(49)	46.5	R
84	(22)	7.6	R

Ph_2SiH_2 (1.2 eq), $[Rh(cod)Cl]_2$ (0.5 mol%), ligand or complex (x mol%), rt

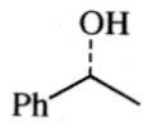

802

Ligand or complex	x	Solvent		% ee
54	5	none	(36)	7
54	5	MeC_6H_5	(61)	1
54	6	CCl_4	(—)	55
58	5	MeC_6H_5	(48)	5
58	2.5	CCl_4	(—)	15
58	5	CCl_4	(59)	84
55	5	MeC_6H_5	(68)	9
55	5	CCl_4	(21)	3
56	5	MeC_6H_5	(72)	50
56	5	CCl_4	(—)	33

TABLE 30. ASYMMETRIC ORGANOSILANE REDUCTION OF KETONES (*Continued*) [499]

Ketone	Conditions	Product(s) and Yield(s) (%)	Refs.
C_8 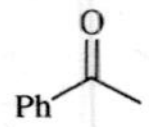	PMHS (2 eq), Et_2Zn (2 mol%), ligand (2 mol%), MeC_6H_5, rt	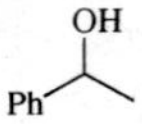	788
	Ligand / Time	(yield) / % ee / Conf.	
	131 / 18 h	(>99) / 76 / S	
	134 / 6 h	(94) / 78 / S	
	132 / 6 h	(14) / 76 / S	
	133 / 6 h	(68) / 22 / S	
	152 / 4 h	(15) / 0 / —	
	148 / 18 h	(>99) / 88 / R	
	Ph_2SiH_2 (1.5 eq), $[Rh(cod)_2]BF_4$ (1.0 mol%), ligand (1.1 mol%), –40°	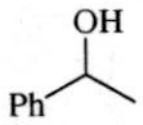	577
	Ligand / Solvent / Time	(yield) / % ee / Conf.	
	23 / THF / 24 h	(90) / 85 / S	
	24 / THF / 5 h	(89) / 92 / S	
	25 / THF / 11 h	(88) / 92 / S	
	26 / THF / 78 h	(78) / 1 / R	
	27 / THF / 9 h	(85) / 15 / S	
	25 / DME / 21 h	(96) / 90 / S	
	25 / CH_2Cl_2 / 168 h	(81) / 91 / S	
	25 / MeC_6H_5 / 4 h	(86) / 90 / S	
	25 / THF / 50 h	(95) / 89 / S	

Ph_2SiH_2 (1.5 eq), catalyst (1 mol%), **95** (x mol%), AgX (2 mol%), THF

I + II

572

Catalyst	x	AgX	Temp	Time	I	% ee	II
Rh(cot)Cl	2	none	20°	2 h	(68)	41	(32)
Rh(cot)Cl	2	AgOTf	15°	3 h	(64)	43	(33)
Rh(cot)Cl	2	$AgBF_4$	15°	3 h	(57)	70	(39)
$RhCl_3$	2	none	rt	2 h	(86)	75	(14)
$RhCl_3$	4	$AgBF_4$	5°	2 h	(98)	90	(2)

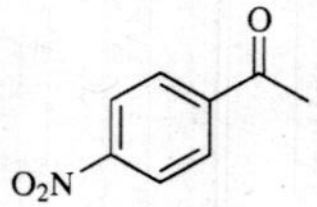

$(EtO)_3SiH$, **193** (10 mol%)

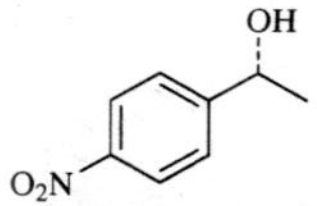

(—) 54% ee

378

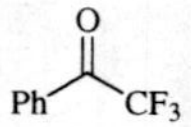

$PhSiH_3$ (1.1 eq), (*S*)-**195**-2*n*-BuLi (5 mol%), MeC_6H_5, 65°, 1 h

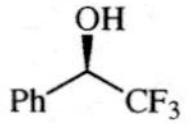

I

(9) 5% ee

783

$PhSiH_3$ (1.1 eq), (*R*)-**195**-2*n*-BuLi (5 mol%), MeC_6H_5, 65°, 1 h

I (9) 6% ee

783

PMHS (1.2 eq), Et_2Zn (2 mol%), **161** (2 mol%), MeC_6H_5, rt, [illegible] h

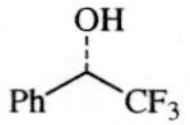

(>99) 27% ee

788

C_{8-9}

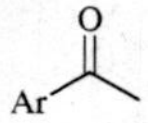

Ph_2SiH_2 (1.6 eq), $[Rh(cod)Cl]_2$ (0.25 mol%), **63** (0.5 mol%), THF, 0°

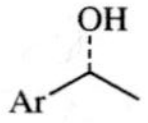

579

Ar	Time		% ee
Ph	7 h	(98)	97
4-ClC_6H_4	7 h	(97)	91
4-MeC_6H_4	3 h	(98)	93
4-$MeOC_6H_4$	4 h	(99)	21

TABLE 30. ASYMMETRIC ORGANOSILANE REDUCTION OF KETONES (*Continued*)

Ketone	Conditions	Product(s) and Yield(s) (%)	Refs.
C_{8-9} 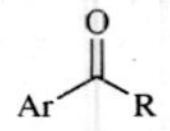	Ph_2SiH_2 (1.25 eq), $[RhCl(NDB)]_2$ (0.5 mol%), ligand (1.5 mol%), THF, 20°	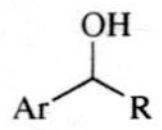	574

Ar	R	Ligand	Time		% ee	Conf.
Ph	Me	**16**	<1 h	(90)	87	S
Ph	Et	**16**	<1 h	(97)	86	S
Ph	Me	**17**	<10 min	(90)	90	S
Ph	Me	**18**	<10 min	(94)	89	S
Ph	Me	**19**	<10 min	(86)	89	S
4-ClC_6H_4	Me	**17**	<1 h	(90)	81	S
Ph	Me	**20**	<1 h	(64)	16	R
Ph	Me	**21**	2 h	(50)	39	R

Ketone	Conditions	Product(s) and Yield(s) (%)	Refs.
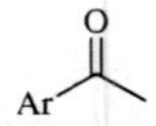	$PhSiH_3$ (1.1 eq), (*R*)-**195**/2 *n*-BuLi (5 mol%), MeC_6H_5, 65°	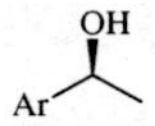	783

Ar	Time	% Conv.	% ee
Ph	1 h	100	59
Ph	24 h	100	82
4-FC_6H_4	1 h	100	66
4-ClC_6H_4	24 h	100	44
4-ClC_6H_4	1 h	96	48
4-BrC_6H_4	1 h	57	28
4-BrC_6H_4	20 h	81	20
4-$O_2NC_6H_4$	3 h	13	1
4-$CF_3C_6H_4$	3 h	99	14
4-$MeOC_6H_4$	1 h	100	63
4-MeC_6H_4	1 h	100	53

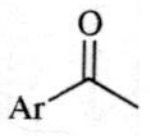

$PhSiH_3$ (1.1 eq), (*S*)-**195**-2*n*-BuLi (5 mol%), MeC_6H_5, 65°

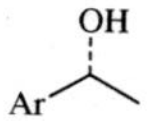

783

Ar	Time	% Conv.	% ee
Ph	24 h	100	16
4-FC_6H_4	1 h	38	12
4-ClC_6H_4	1 h	18	13
4-$O_2NC_6H_4$	1 h	9	2 (S)
2-ClC_6H_4	3 h	14	13
4-$MeOC_6H_4$	1 h	100	23
4-MeC_6H_4	1 h	100	6

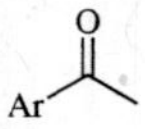

1. **197** (4.5 eq), *n*-BuLi (9 eq), C_6H_6
2. PMHS (5 eq)
3. Add ketone, time
4. TBAF, THF

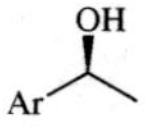

588

Ar	Time		% ee
Ph	0.9 d	(73)	97
2-BrC_6H_4	8 d	(51)	94
2-ClC_6H_4	2 d	(78)	90
2-$MeOC_6H_4$	0.8 d	(77)	91
2-MeC_6H_4	1 d	(88)	95
4-ClC_6H_4	2 d	(85)	94
4-BrC_6H_4	1 d	(75)	96
4-FC_6H_4	1 d	(89)	97
4-$CF_3C_6H_4$	8 d	(66)	65
4-MeC_6H_4	0.9 d	(84)	96
4-$MeOC_6H_4$	0.8 d	(62)	87

TABLE 30. ASYMMETRIC ORGANOSILANE REDUCTION OF KETONES (*Continued*)

Ketone	Conditions	Product(s) and Yield(s) (%)	Refs.
C_{8-9} 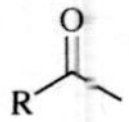	Silane (2 eq), $[Rh(C_2H_4)_2]_2$ (x mol%), (+)-DIOP (x mol%), C_6H_6, rt	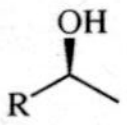	803

R	Silane	x	Time		Opt. Yield
Ph	Cl_3SiH	0.5	24 h	(0)	—
Ph	Et_3SiH	1.6	26 h	(47)	3.8%
Ph	Ph_3SiH	2	24 h	(15)	0%
Ph	$(EtO)_3SiH$	0.5	24 h	(60)	10%
Bn	Et_3SiH	1.6	20 h	(50)	1%
Bn	$(EtO)_3SiH$	1.6	48 h	(39)	5.3%
Ph	PMHS	2	24 h	(56)	3.6%
Bn	PMHS	2	48 h	(64)	0%

Ketone	Conditions	Product(s) and Yield(s) (%)	Refs.
C_{8-10} 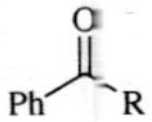	R_3SiH (2 eq), $[Rh(C_2H_4)_2]_2$ (x mol%), (+)-DIOP (x mol%), C_6H_6, rt	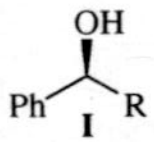 I	803

R	R_3Si	x		Opt. Yield
Me	PhMeHSi	1.6	(48)	13%
Me	Ph_2HSi	2	(100)	28%
Me	Ph(1-Np)HSi	2	(100)	58%
i-Pr	PhMeHSi	2	(87)	20%
i-Pr	Ph_2HSi	2	(100)	35%
i-Pr	Ph(1-Np)HSi	2	(70)	24%

Ketone	Conditions	Product(s) and Yield(s) (%)	Refs.
	R_3SiH, [(−)-(*S*)-bmpp]$_2$RhCl, C_6H_6, 5-50°, 3-48 h	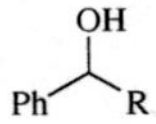	785, 786

R	R_3Si		Opt. Yield	Conf.
Me	$PhMe_2Si$	(92)	44%	R
Et	$PhMe_2Si$	(96)	50%	R
i-Pr	$PhMe_2Si$	(95)	56%	R
Et	Et_2HSi	(98)	17%	S
i-Pr	Et_2HSi	(98)	23%	S

R_3SiH (2 eq), $[Rh(C_2H_4)_2]_2$ resin (x mol%), (+)-DIOP (2 mol%), C_6H_6, rt — I — 803

R	R_3Si	x		% Opt. Yield
Me	PhMeHSi	4	(71)	12
Me	Ph_2HSi	2.8	(82)	29
Me	Ph(1-Np)HSi	2.8	(100)	58.5
Me	Ph(1-Np)HSi	5	(95)	55
i-Pr	PhMeHSi	4	(55)	6.5
i-Pr	Ph_2HSi	4	(100)	28

C_{8-12}

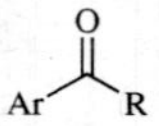

Ph_2SiH_2 (1.2 eq), C_6H_6, $[Rh(cod)Cl]_2$ (0.01 eq), ligand (0.1 eq), 0-20°, 10-15 h

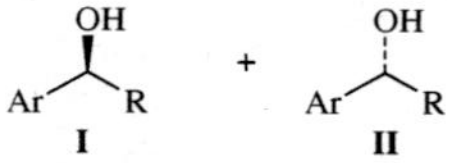

804

Ar	R	Ligand	I + II	I:II
Ph	Me	**109**	(50)	70.5:29.5
Ph	Me	**110**	(28)	55:45
Ph	Me	**111**	(59)	22.5:77.5
Ph	Me	**113**	(40)	62.5:37.5
Ph	Me	**114**	(92)	19:81
Ph	Et	**109**	(60)	55:45
Ph	Et	**110**	(74)	23.5:76.5
1-Np	Me	**109**	(91)	9:91
1-Np	Me	**110**	(76)	10.5:89.5
1-Np	Me	**113**	(46)	49.5:50.5
1-Np	Me	**114**	(99)	8:92

TABLE 30. ASYMMETRIC ORGANOSILANE REDUCTION OF KETONES (*Continued*)

Ketone	Conditions	Product(s) and Yield(s) (%)	Refs.
C_{8-12} 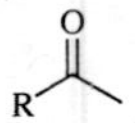	Ph_2SiH_2 (1.5 eq), $[Rh(cod)_2]BF_4$ (1.0 mol%), ligand (1.1 mol%), –40°, DME	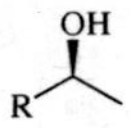	577
R	**Ligand / Time**	**(yield) / % ee**	
c-C_6H_{11}	**23** / 24 h	(76) / 81	
c-C_6H_{11}	**25** / 10 h	(62) / 80	
n-C_6H_{13}	**22** / 24 h	(75) / 70	
n-C_6H_{13}	**23** / 24 h	(88) / 65	
n-C_6H_{13}	**24** / 24 h	(78) / 60	
n-C_6H_{13}	**25** / 24 h	(81) / 55	
Bn	**25** / 24 h	(84) / 32	
$BnCH_2$	**25** / 30 h	(93) / 70	
$BnCH_2CH_2$	**25** / 20 h	(93) / 60	
1-Ad	**23** / 10 h	(78) / 96	
1-Ad	**25** / 48 h	(92) / 91	
C_{8-9} 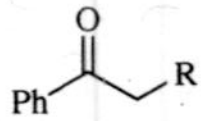	Ph_2SiH_2 (1.2 eq), catalyst (1.0 mol%), THF, 16 h	 **I** + **II**	805

R	Catalyst	Temp	I + II	I	% ee
H	**203**	rt	(91)	(89)	44
Me	**203**	rt	(82)	(79)	38
Et	**203**	rt	(81)	(80)	47
CH_2CH_2Cl	**203**	rt	(90)	(84)	48
H	**204**	rt	(98)	(95)	63
H	**204**	0°	(88)	(81)	55
Me	**204**	rt	(95)	(93)	62
Et	**204**	rt	(96)	(92)	60
CH_2CH_2Cl	**204**	rt	(95)	(89)	92
H	**205**	rt	—	—	23
Me	**205**	rt	—	—	11
Et	**205**	rt	—	—	15
CH_2CH_2Cl	**205**	rt	—	—	35
H	**206**	rt	(96)	(92)	41
Me	**206**	rt	—	—	46
Et	**206**	rt	—	—	62
CH_2CH_2Cl	**206**	rt	(88)	(77)	58

C_{8-10}

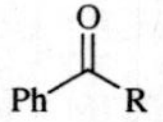

R_3SiH (x eq), $[Rh(C_2H_4)_2]_2$ (y mol%), (+)-DIOP (x mol%), C_6H_6, rt

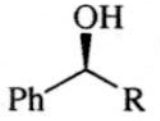

803

R	R_3Si	x	y		Opt. Yield
Me	PhMeHSi	2	1.6	(48)	13%
Me	Ph_2HSi	2	2	(100)	28%
Me	Ph(1-Np)HSi	2	2	(100)	58%
i-Pr	PhMeHSi	2	2	(87)	20%
i-Pr	Ph_2HSi	2	2	(100)	35%
i-Pr	Ph(1-Np)HSi	2	2	(70)	24%

TABLE 30. ASYMMETRIC ORGANOSILANE REDUCTION OF KETONES (*Continued*)

Ketone	Conditions	Product(s) and Yield(s) (%)	Refs.
C_{8-10} 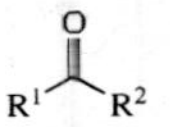	$PhSiH_3$ (1.2 eq), CuF_2 (1%), (*S*)-BINAP (1%), MeC_6H_5, rt	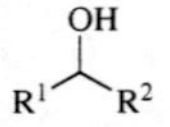	784

R^1	R^2	Time	% Conv.	% ee	Conf.
Ph	Me	2 h	98	78	S
Ph	Et	9 h	97	84	S
Ph	*n*-Pr	8 h	100	92	S
1-indanone		14 h	100	70	S
Ph	CO_2Et	3 h	100	11	R
4-ClC_6H_4	Me	7 h	100	85	S
3-ClC_6H_4	Me	24 h	97	75	S
2-ClC_6H_4	Me	20 h	96	64	S
4-$CF_3C_6H_4$	Me	3 h	100	86	S
3-$CF_3C_6H_4$	Me	3 h	95	80	S
4-FC_6H_4	Me	4 h	100	82	S
4-NCC_6H_4	Me	8 h	95	86	—
3, 5-$(CF_3)_2C_6H_3$	Me	3 h	100	85	—
$BnCH_2$	Me	8 h	100	20	R

Ketone	Conditions	Product(s) and Yield(s) (%)	Refs.
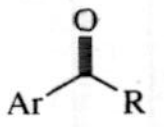	Ph_2SiH_2 (1 eq), $[Rh(cod)Cl]_2$ (0.5 mol%), **208** (2.5 mol%), 0-20°, 18 h	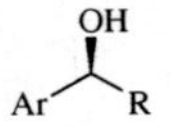	801

Ar	R	Solvent		% ee
Ph	Me	MeC_6H_5	(66)	31.8
Ph	*n*-Pr	MeC_6H_5	(45)	18.0
2-MeC_6H_4	Me	MeC_6H_5	(28)	19.6
2-$MeOC_6H_4$	Me	MeC_6H_5	(67)	31.4
Ph	Me	CCl_4	(93)	56.6
Ph	*n*-Pr	CCl_4	(26)	42.6
2-MeC_6H_4	Me	CCl_4	(59)	47.5
2-$MeOC_6H_4$	Me	CCl_4	(78)	50.4

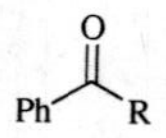

R_3SiH (1.2 eq), Li_2-2-aminoethoxide (x eq), THF

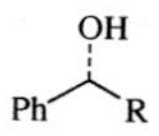

92

R	R_3Si	Li_2-2-Aminoethoxide	x	Temp	Time		% ee
Me	$(MeO)_3Si$	Li_2-alaninol	2.4	−78°	16 h	(62)	77
Me	$(MeO)_3Si$	Li_2-alaninol	0.004	rt	3 h	(89)	44
Me	$(EtO)_3Si$	Li_2-alaninol	0.004	rt	3 h	(75)	49
Me	$Me(OEt)_2Si$	Li_2-alaninol	0.004	rt	3 h	(42)	48
Me	$(MeO)_3Si$	Li_2-prolinol	0.004	rt	3 h	(89)	52
Me	$(MeO)_3Si$	Li_2-prolinol	0.004	−78°	16 h	(100)	49
Me	$(MeO)_3Si$	Li_2-valinol	0.004	rt	3 h	(99)	49
Et	$(MeO)_3Si$	Li_2-alaninol	2.4	rt	3 h	(44)	72
Et	$(MeO)_3Si$	Li_2-alaninol	0.004	rt	3 h	(88)	32
Et	$(MeO)_3Si$	Li_2-prolinol	0.004	rt	3 h	(85)	45
n-Pr	$(MeO)_3Si$	Li_2-alaninol	2.4	rt	3 h	(52)	12
i-Pr	$(MeO)_3Si$	Li_2-alaninol	2.4	rt	3 h	(55)	17
n-Bu	$(MeO)_3Si$	Li_2-alaninol	2.4	rt	16 h	(12)	84

TABLE 30. ASYMMETRIC ORGANOSILANE REDUCTION OF KETONES (*Continued*)

Ketone	Conditions	Product(s) and Yield(s) (%)	Refs.
C_{8-11} 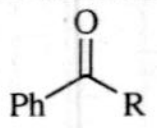	$(MeO)_3SiH$ (1.2 eq), Li_2-diolate (2.4 eq), THF, 0°, 15-20 h	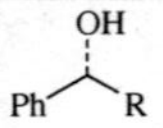	92

R	Li_2-Diolate	Yield	% ee
Me	**170**	(41)	22
Me	**171**	(66)	40
Me	**172**	(78)	44
Et	**170**	(52)	36
Et	**171**	(70)	53
Et	**172**	(66)	54
n-Pr	**170**	(67)	36
n-Pr	**171**	(67)	66
n-Pr	**172**	(71)	62
i-Pr	**170**	(48)	44
i-Pr	**171**	(63)	69
i-Pr	**172**	(55)	69
n-Bu	**170**	(51)	44
n-Bu	**171**	(59)	63
n-Bu	**172**	(64)	55

Ketone	Conditions	Product(s) and Yield(s) (%)	Refs.
C_{8-12} 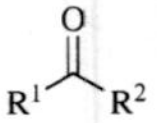	Ph_2SiH_2 (1.5 eq), $[Rh(cod)_2]BF_4$ (1.0 mol%), **25** (1.1 mol%), –40°, THF	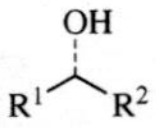	577

R^1	R^2	Time	Yield	% ee
Me	$4-MeC_6H_4$	5 h	(90)	91
Me	$4-MeOC_6H_4$	7 h	(88)	86
Me	$3-MeOC_6H_4$	6 h	(87)	85
Me	$2-MeOC_6H_4$	6 h	(79)	84
Me	$4-ClC_6H_4$	13 h	(88)	79
Me	Fc	11 h	(84)	97
Me	$1-C_{10}H_7$	9 h	(76)	82
Et	Ph	48 h	(73)	62

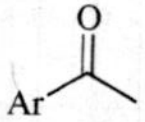

(EtO)$_3$SiH (2.5 eq), catalyst,
n-BuLi (0.010 eq to catalyst), THF, −78° to rt

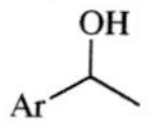

806

Ar	Catalyst	Time		% ee	Conf.
Ph	**199**	2 h	(84)	12	S
Ph	**201**	2 h	(78)	5	R
Ph	**200**	2 h	(77)	7	S
Ph	**196**	13 h	(100)	14	S
4-MeC_6H_4	**196**	58 h	(87)	0	—
4-$MeOC_6H_4$	**196**	80 h	(40)	5	S
4-BrC_6H_4	**196**	2 h	(100)	32	S
4-$CF_3C_6H_4$	**196**	0.7 h	(100)	2	S
2-$C_{10}H_7$	**199**	6 h	(84)	13	S
2-$C_{10}H_7$	**201**	6 h	(99)	6	R
2-$C_{10}H_7$	**200**	6 h	(96)	20	S
2-$C_{10}H_7$	**196**	24 h	(90)	40	S

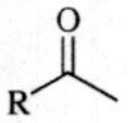

(EtO)$_3$SiH, **194** (10 mol%), Et_2O

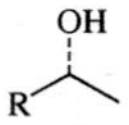

378

R	Temp	Time	
Ph	50°	5 h	(>98)
4-$MeOC_6H_4$	50°	5 h	(>98)
4-ClC_6H_4	50°	5 h	(>98)
4-$O_2NC_6H_4$	50°	5 h	(>98)
3-$O_2NC_6H_4$	50°	5 h	(>98)
n-C_6H_{13}	40°	12 h	(84)
C_6H_{11}	40°	12 h	(77)
n-BuC≡C	50°	5 h	(>98)
(E)-n-C_5H_{11}CH=CH	40°	12 h	(>98)
1-$C_{10}H_7$	40°	12 h	(50)

TABLE 30. ASYMMETRIC ORGANOSILANE REDUCTION OF KETONES (*Continued*)

Ketone	Conditions	Product(s) and Yield(s) (%)	Refs.
C_{8-12} 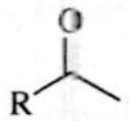	Ph_2SiH_2 (160 mol%), $[RuCl_2(C_6H_6)]_2$ (0.5 mol%), **125** (2.2 mol%), AgOTf (2 mol %), THF, rt	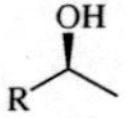	799

R	Time	Yield	% ee
Ph	24 h	(97)	54
$2\text{-}BrC_6H_4$	72 h	(87)	57
$3\text{-}BrC_6H_4$	72 h	(93)	62
$4\text{-}BrC_6H_4$	48 h	(93)	55
$1\text{-}C_{10}H_7$	30 h	(85)	48
$2\text{-}C_{10}H_7$	24 h	(98)	66

Ketone	Conditions	Product(s) and Yield(s) (%)	Refs.
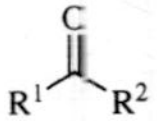	Ph_2SiH_2 (2 eq), $[Rh(cod)Cl]_2$ (1 mol%), **48** (4 mol%), MeC_6H_5, rt	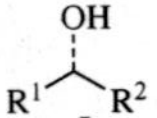 **I**	585

R^1	R^2	Yield	% ee
Ph	Me	(84)	94
Ph	Et	(91)	91
(*E*)-PhCH=CH	Me	(83)	22
$c\text{-}C_6H_{11}$	Me	(85)	87
$n\text{-}C_6H_{13}$	Me	(90)	52
$1\text{-}C_{10}H_7$	Me	(90)	92

Ketone	Conditions	Product(s) and Yield(s) (%)	Refs.
	$MesPhSiH_2$, $[Rh(cod)Cl]_2$ (2.5 mol%), (–)-**34**, THF, rt	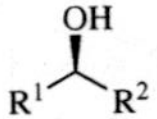	586

R^1	R^2	Yield	% ee
Ph	Me	(94)	98
$4\text{-}MeOC_6H_4$	Me	(97)	97
$4\text{-}CF_3C_6H_4$	Me	(88)	96
$2,4\text{-}Me_2C_6H_3$	Me	(97)	95
$2,4,6\text{-}Me_3C_6H_2$	Me	(99)	98
Ph	Et	(96)	98
$1\text{-}C_{10}H_7$	Me	(97)	99

Substrate	Conditions	Product(s) and Yield(s) (%), ee	Refs.
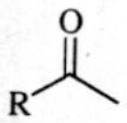	o-Tol$_2$SiH$_2$, [Rh(cod)Cl]$_2$ (1.0 mol%), **34**, THF, 0°	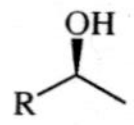	586

R	Yield	% ee
n-C_6H_{13}	(81)	72
c-C_6H_{11}	(91)	94
$BnCH_2$	(98)	82
1-Ad	(92)	96

$C_{8\text{-}13}$

Substrate	Conditions	Product(s) and Yield(s) (%), ee	Refs.
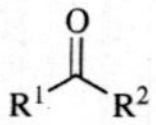	$(MeO)_3SiH$ (1 eq), Li$_2$-histidine (10 mol%), THF/TMEDA, 0°, 24 h	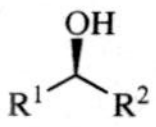	593

R^1	R^2	Yield	% ee
Ph	Me	(70)	26
Ph	Me	(85)	26
4-$CF_3C_6H_4$	Me	(86)	30
4-MeC_6H_4	Me	(80)	40
4-$MeOC_6H_4$	Me	(89)	70
Bn	Me	(90)	—
Ph	Ph	(90)	—
4-$CF_3C_6H_4$	Ph	(95)	30
4-MeC_6H_4	Ph	(82)	5
4-$MeOC_6H_4$	Ph	(78)	5

TABLE 30. ASYMMETRIC ORGANOSILANE REDUCTION OF KETONES (*Continued*)

Ketone	Conditions	Product(s) and Yield(s) (%)	Refs.
C_{8-13} 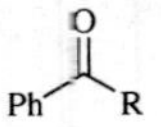	R_3SiH, [(+)-(*R*)-bmpp]$_2$RhCl, C_6H_6, 5-50°, 3-48 h	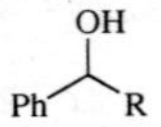	785
R	**R_3Si**	**Opt. Yield — Conf.**	
Me	Et_2HSi	(97) 1% R	
Et	Ph_2SiH_2	(98) 42% R	
i-Pr	Ph_2HSi	(96) 16% S	
i-Pr	PhMeHSi	(93) 12% S	
t-Bu	$EtMe_2Si$	(97) 56% R	
t-Bu	$PhMe_2Si$	(92) 54% S	
c-C_6H_{11}	$PhMe_2Si$	(90) 58% S	
c-C_6H_{11}	Et_2HSi	(95) 19% R	
C_{8-14} 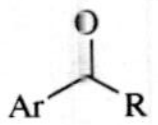	Ph(1-Np)SiH_2 (1.5 eq), **126** (1 mol%), THF, −20°	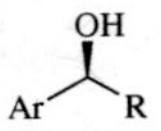	576
Ar — R		**ee (%)**	
Ph — Me		(80) 95	
2-MeC_6H_4 — Me		(98) 95	
4-MeC_6H_4 — Me		(90) 92	
2-$MeOC_6H_4$ — Me		(90) 95	
4-$MeOC_6H_4$ — Me		(56) 88	
2-ClC_6H_4 — Me		(90) 98	
4-ClC_6H_4 — Me		(95) 85	
1-$C_{10}H_7$ — Me		(99) 98	
2-$C_{10}H_7$ — Me		(99) 95	
Ph — Et		(95) 94	
Ph — *i*-Bu		(95) 94	
Ph — Bn		(75) 94	

C_{8-14}

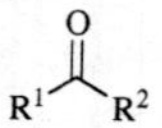

$(MeO)_3SiH$ (1 eq), Et_2O:TMEDA (30:1), Li-(*R*)-BINOL (5 mol%), 0°

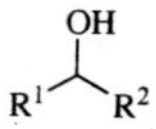

572

R^1	R^2	Time		% ee	Conf.
Ph	Me	6 h	(80)	61	S
2-BrC_6H_4	Me	24 h	(91)	66	S
$BnCH_2$	Me	24 h	(74)	46	R
Ph	$MeO_2C(CH_2)_2$	24 h	(63)	65	S
Ph	*i*-Bu	24 h	(60)	81	S
2,4,6-$Me_3C_6H_2$	Me	24 h	(57)	90	R
1-$C_{10}H_7$	Me	24 h	(67)	77	S
4-MeC_6H_4	Ph	24 h	(74)	0	—
4-$CF_3C_6H_4$	Ph	24 h	(90)	0	—

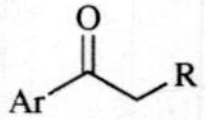

$(EtO)_3SiH$, THF, **198**, 96 h

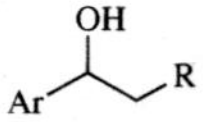

793

Ar	R		% ee	Conf.
4-BrC_6H_4	H	(61)	83	R
4-BrC_6H_4	Cl	(50)	65	R
4-EtC_6H_4	H	(60)	84	R
4-BrC_6H_4	Me	(53)	80	R
4-BrC_6H_4	OMe	(50)	78	R
4-*i*-PrC_6H_4	H	(61)	85	S
4-BrC_6H_4	Ph	(58)	75	S

TABLE 30. ASYMMETRIC ORGANOSILANE REDUCTION OF KETONES (*Continued*)

Ketone	Conditions	Product(s) and Yield(s) (%)	Refs.
C_{8-15} 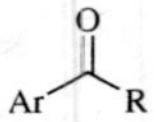	$(MeO)_3SiH$ (1.2 eq), **178**, THF, 0°, 20 h		807

Ar	R	I + II	% ee	Conf.
Ph	Me	(93)	28	S
Ph	*i*-Pr	(95)	42	S
Ph	*t*-Bu	(59)	12	S
2-$C_{10}H_7$	Me	(93)	65	S
2-$C_{10}H_7$	*i*-Pr	(>95)	82	S
2-$C_{10}H_7$	*t*-Bu	(>95)	20	S
1-$C_{10}H_7$	Me	(59)	22	S
1-$C_{10}H_7$	*i*-Pr	(50)	20	S
1-$C_{10}H_7$	*t*-Bu	(0)	—	—

Ketone	Conditions	Product(s) and Yield(s) (%)	Refs.
C_9 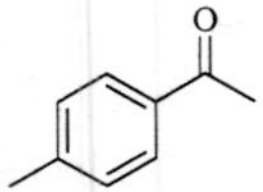	PMHS (1.1 eq), Et_2Zn (20 mol%), (*R*,*R*)-ebpe (20 mol%), vitride (3%), MeC_6H_5, rt, 24 h	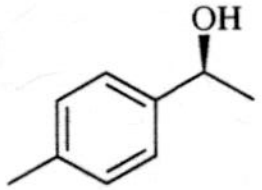 **I** (—) 71% ee	594
	PMHS (1.1 eq), Et_2Zn (2 mol%), (*R*,*R*)-ebpe (20 mol%), vitride (3%), MeC_6H_5, rt, 24 h	**I** (—) 66% ee	594
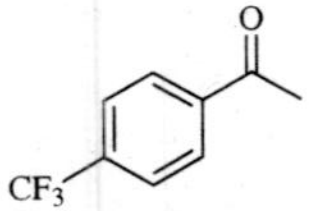	PMHS (5 eq), CuCl, *t*-BuONa, **121** (0.005 mol%), MeC_6H_5, or THF, or MeC_6H_5/THF, –78°	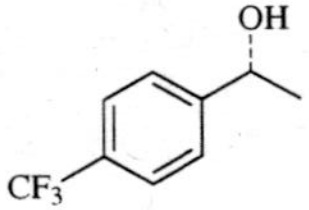 (85) 95% ee	589
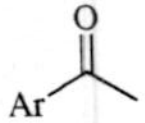	PMHS (1.1 eq), Et_2Zn (2 mol%), (*R*,*R*)-ebpe (20 mol%), vitride (3%), MeC_6H_5, rt, 24 h	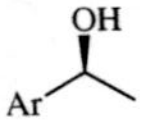 (—)	594

Ar	% ee
3-$MeOC_6H_4$	73
4-$MeOC_6H_4$	72

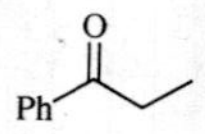	PMHS (1.1 eq), Et_2Zn (2 mol%), (*R,R*)-ebpe (20 mol%), vitride (3%), MeC_6H_5, rt, 24 h	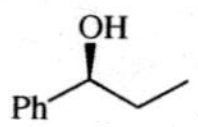 **I** (—) 81% ee	594
	PMHS, $Sn(OTf)_2$ (10 mol%), **93** (10 mol%), MeOH, rt, 12-14 h	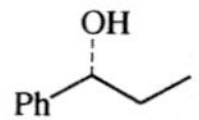 **II** (98) 48% ee	385
	1. **197** (4.5 eq), *n*-BuLi (9 eq), C_6H_6 2. PMHS (5 eq) 3. Add ketone, 1 d 4. TBAF, THF	**II** (96) 95% ee	588
	$(3\text{-}FC_6H_4)_2SiH_2$ (1.5 eq), $[Rh(cod)_2]BF_4$ (1 mol%), **1** (1.1 mol%), THF, –40°, 24 h	**I** (99) 88% ee	578
	Ph_2SiH_2 (1.25 eq), $Rh[(NBD)Cl]_2$ (0.5%), ligand (0.5%), MeC_6H_5	**I** (—)	791

Ligand	Temp	% ee
116	0°	37
117	20°	35
118	0°	45
119	20°	46
109	20°	56
113	20°	66

TABLE 30. ASYMMETRIC ORGANOSILANE REDUCTION OF KETONES (*Continued*)

Ketone	Conditions	Product(s) and Yield(s) (%)	Refs.
C_9 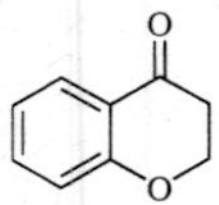	Ph_2SiH_2 (1.1 eq), $[Rh(cod)Cl]_2$ (0.5 mol%), **108** (1.2 mol%), 0°, 24 h	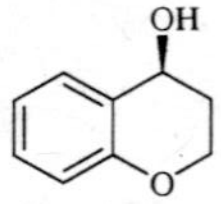 **I** (86) 87% ee	575
	Ph(1-Np)SiH_2 (1.5 eq), **126** (1 mol%), THF, –20°	**I** (90) 95% ee	576
	PMHS, $Sn(OTf)_2$ (10 mol%), **93** (10 mol%), MeOH, rt	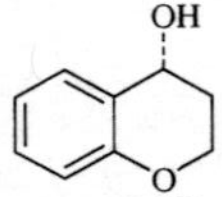 (77) 12% ee	385
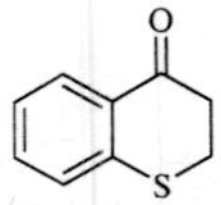	Ph(1-Np)SiH_2 (1.5 eq), **126** (1 mol%), THF, –20°	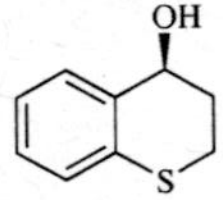 (90) 92% ee	576
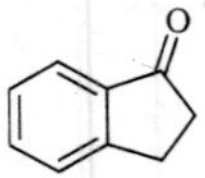	Ph(1-Np)SiH_2 (1.5 eq), **126** (1 mol%), THF, –20°	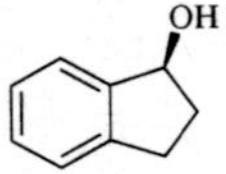 **I** (90) 85% ee	576
	Ph_2SiH_2 (1.5 eq), $[Rh(cod)_2]BF_4$ (1.0 mol%), **25** (1.1 mol%), –40°, THF, 3 h	**I** (87) 87% ee	577
	1. **197** (4.5 eq), *n*-BuLi (9 eq), C_6H_6 2. PMHS (5 eq) 3. Add ketone, 1.6 d 4. TBAF, THF	**I** (68) 92% ee	588
	Ph_2SiH_2 (1 eq), $[Rh(cod)Cl]_2$ (0.3 mol%), **41** (0.4 mol%), THF, rt, 30 min	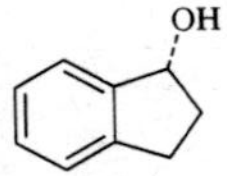 (85) 50% ee	787

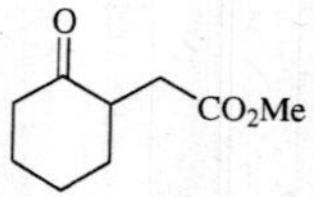

Ph$_2$SiH$_2$ (1.3 eq), RhCl$_3$ (1 mol%), **92** (5 mol%), AgBF$_4$ (2 mol%), THF, 0°, 1 d

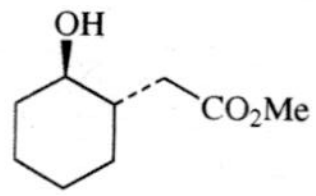

I 1S,2R 95% ee

+

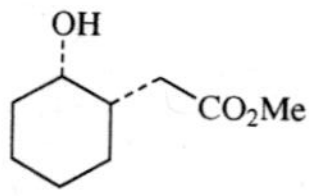

II 1S,2S 92% ee

I + II (76), **I:II** = 54:46

390

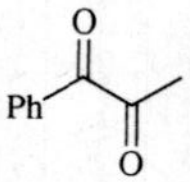

PMHS (1.2 eq), R$_2$Zn (2 mol%), **131** (2 mol%), MeC$_6$H$_5$, rt, 6 h

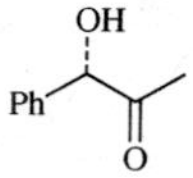

788

R	Time		% ee
Et	6 h	(>99)	28
i-Pr	6 h	(>99)	33
Ph	4 h	(31)	48

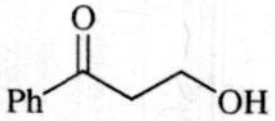

1. PhSiH$_3$ (0.1 eq), **198** (2 mol%), pyrrolidine, MeOH, THF, 60°
2. PMHS, ketone, MeOH (3-7 eq), 15°

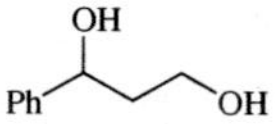

(50) 96% ee [b]

587

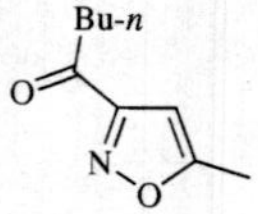

PMHS (4 eq), **124** (0.05 mol%), CuCl (1 mol%), NaOBu-*t* (1 mol%), THF/MeC$_6$H$_5$, –50°, 10 h

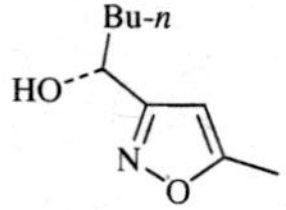

(68) 83% ee

590

C_{9-10}

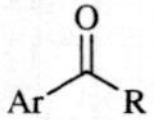

1. PhSiH$_3$ (0.1 eq), **198** (0.2 eq), MeOH, 60°
2. PMHS (6-10 eq), additive (x eq), rt

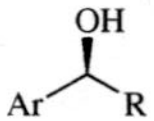

587

Ar	R	Additive	x	Time		% ee
4-MeC$_6$H$_4$	Me	MeOH	4	3 h	(97)	97
4-MeC$_6$H$_4$	Me	*i*-BuNH$_2$	4	4 h	(50)	—
Ph	*i*-Pr	none	—	39 h	(17)	90
Ph	*i*-Pr	MeOH	5	5 h	(94)	99
Ph	*i*-Pr	EtOH	5	6 h	(96)	99
Ph	*i*-Pr	*i*-PrOH	5.5	5.5 h	(65)	—
Ph	*i*-Pr	*n*-BuOH	3.8	14 h	(92)	—
Ph	*i*-Pr	*n*-BuOH	7.0	32 h	(95)	99

TABLE 30. ASYMMETRIC ORGANOSILANE REDUCTION OF KETONES (*Continued*)

Ketone	Conditions	Product(s) and Yield(s) (%)	Refs.
C_{9-13} 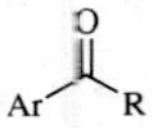	1. $PhSiH_3$ (0.1 eq), **198** (x mol%), pyrrolidine, MeOH, THF, 60° 2. PMHS, ketone, MeOH (3-7 eq), 15°	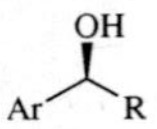	587

Ar	R	x	Time		% ee
4-MeC_6H_4	Me	1	5 h	(87)	98
Ph	Et	1	12 h	(86)	98
Ph	*i*-Pr	0.5	13 h	(86)	98
Ph	*c*-C_6H_{11}	1	8 h	(86)	99
Ph	$Me_2C{=}CH(CH_2)_2$	1	10 h	(80)	>98

Ketone	Conditions	Product(s) and Yield(s) (%)	Refs.
C_{10} 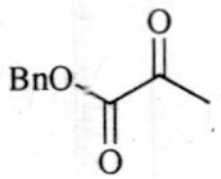	Ph(1-Np)SiH_2 (1.5 eq), **126** (1 mol%), THF, –20°	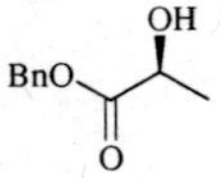 (92) 46% ee	576
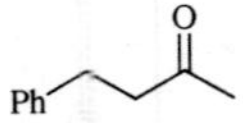	$(MeO)_3SiH$ (1 eq), Li_2-histidine (10 mol%), THF/TMEDA, 0°, 24 h	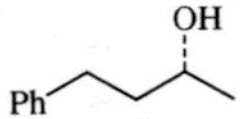 **I** (91) 28% ee	593
	1. **197** (4.5 eq), *n*-BuLi (9 eq), C_6H_6 2. PMHS (5 eq) 3. Add ketone, 0.8 d 4. TBAF, THF	**I** (88) 12% ee	588

Substrate	Conditions	Product(s) and Yield(s) (%), ee	Refs.
	Ph_2SiH_2 (1.5 eq), $[Rh(cod)_2]BF_4$ (1 mol%), **1** (1.1 mol%), THF, –50°, 48 h	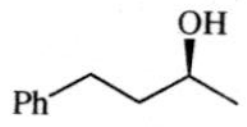 (94) 81% ee	578
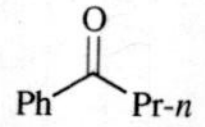	$Ph(1\text{-}Np)SiH_2$ (4 eq), $[Rh(cod)Cl]_2$ (1 mol%), **92**, THF, –78°	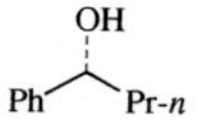 **I** (70) 59% ee	792
	PMHS (5 eq), CuCl, *t*-BuONa, **121** (0.005 mol%), MeC_6H_5, or THF, or MeC_6H_5/THF, –78°	**I** (87) 97% ee	589
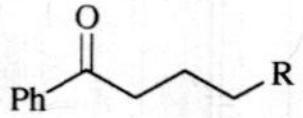	$(3\text{-}FC_6H_4)_2SiH_2$ (1.5 eq), $[Rh(cod)_2]BF_4$ (1 mol%), **1** (1.1 mol%), THF, –40°, 24 h	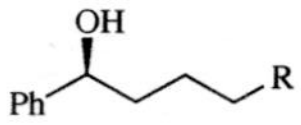 R = H (93) 89% ee; R = Cl (99) 88[b]% ee	578
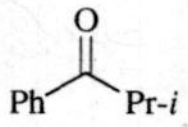	1. **197** (4.5 eq), *n*-BuLi (9 eq), C_6H_6 2. PMHS (5 eq) 3. Add ketone, 4.5 d 4. TBAF, THF	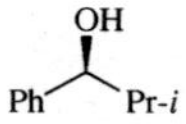 (79) 92% ee	588
	Ph_2SiH_2 (1.25 eq), $Rh[(NBD)Cl]_2$ (0.5%), ligand (0.5%), MeC_6H_5	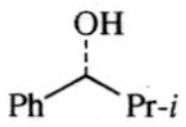 **I** (—)	791
	Ligand **116**, Temp 0°	% ee 67	
	Ligand **117**, Temp 0°	% ee 65	
	Ligand **118**, Temp 0°	% ee 66	
	Ligand **119**, Temp 20°	% ee 66	
	PMHS, $Sn(OTf)_2$ (10 mol%), **93** (10 mol%), MeOH, rt, 12-14 h	**I** (96) 44% ee	385

TABLE 30. ASYMMETRIC ORGANOSILANE REDUCTION OF KETONES (*Continued*)

Ketone	Conditions	Product(s) and Yield(s) (%)	Refs.
C_{10} 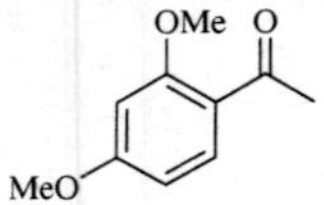	PMHS (5 eq), CuCl, *t*-BuONa, **121** (0.005 mol%), MeC_6H_5, or THF, or MeC_6H_5/THF, –78°	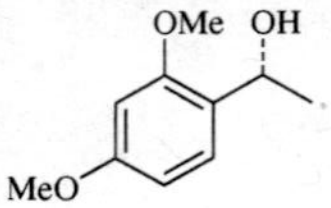 (89) 94% ee	589
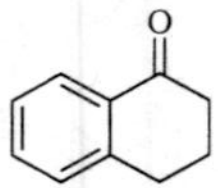	PMHS (5 eq), CuCl, *t*-BuONa, **121** (0.005 mol%), MeC_6H_5, or THF, or MeC_6H_5/THF, –78°	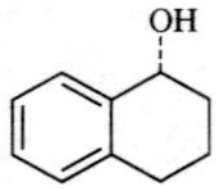 **I** (92) 99% ee	589
	$Ph(1\text{-}Np)SiH_2$ (4 eq), $[Rh(cod)Cl]_2$ (1 mol%), **92**, THF, –78°	**I** (70) 48% ee	792
	PMHS, $Sn(OTf)_2$ (10 mol%), **93** (10 mol%), MeOH, rt	**I** (80) 43% ee	385
	Ph_2SiH_2 (2 eq), $[Rh(cod)Cl]_2$ (1 mol%), (–)-**44** (4 mol%), MeC_6H_5, rt	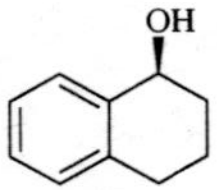 **II** (97) 92% ee	585
	$(MeO)_3SiH$ (1 eq), Et_2O/TMEDA (30:1), Li-(*R*)-BINOL (10 mol%), 0°, 24 h	**I** (39) 93% ee	592
	Ph_2SiH_2 (2 eq), $[Rh(cod)Cl]_2$ (1 mol%), **44** (x mol%), solvent, 0°	**II**	585

Solvent	x		% ee
THF	2.4	(68)	43
MeC_6H_5	2.4	(76)	59
MeC_6H_5	4	(89)	89
MeC_6H_5	10	(84)	89

	Ph_2SiH_2 (1.6 eq), **92** (4.0 mol%), $AgBF_4$ (0.02 mol%), THF, 0°, 2 h	**I** (92) 99% ee	580
	Ph_2SiH_2 (1.5 eq), $[Rh(cod)Cl]_2$ (0.25 mol%), **38** (0.5 mol%), Et_2O, rt	**II** (95) 57% ee	571
	Ph_2SiH_2 (1.1 eq), $[Rh(cod)Cl]_2$ (0.5 mol%), **106** (1.2 mol%), 0°, 42 h	**I** (70) 80% ee	575
	$Ph(1\text{-}Np)SiH_2$ (1.5 eq), **126** (1 mol%), THF, –20°	**I** (90) 91% ee	576
	Ph_2SiH_2 (1.5 eq), $[Rh(cod)_2]BF_4$ (1.0 mol%), **25** (1.1 mol%), –40°, THF, 4 h	**I** (83) 84% ee	577
	1. **197** (4.5 eq), *n*-BuLi (9 eq), C_6H_6 2. PMHS (5 eq) 3. Add ketone, 1.6 d 4. TBAF, THF	**I** (92) 91% ee	588
	$PhSiH_3$ (1.1 eq), (*S*)-**195**-2*n*-BuLi (5 mol%), MeC_6H_5, 65°, 1 h	**II** (94) 10% ee	783
	$PhSiH_3$ (1.1 eq), (*R*)-**195**-2*n*-BuLi (5 mol%), MeC_6H_5, 65°, 1 h	**I** (94) 10% ee	783
	PMHS (1.1 eq), Et_2Zn (2 mol%), (*R*,*R*)-ebpe (20 mol%), vitride (3%), MeC_6H_5, rt, 24 h	**I** (—) 64% ee	594
	$MesPhSiH_2$, $[Rh(cod)Cl]_2$ (2.5 mol%), (–)-**34**, THF, rt	**I** (95) 98% ee	586
C_{10} 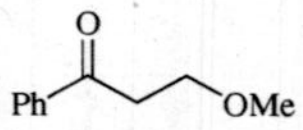	1. $PhSiH_3$ (0.1 eq), **198** (2 mol%), MeOH, 60°, pyrrolidine, MeOH, THF 2. PMHS, ketone, MeOH (3-7 eq), 15°	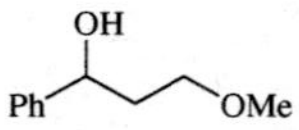 (50) 98% ee [b]	587

TABLE 30. ASYMMETRIC ORGANOSILANE REDUCTION OF KETONES (*Continued*)

Ketone	Conditions	Product(s) and Yield(s) (%)	Refs.
C_{10}			
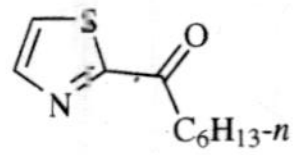	PMHS (4 eq), **124** (0.05 mol %) CuCl (1 mol%), NaOBu-*t* (1 mol%), MeC_6H_5, –50°, 4h	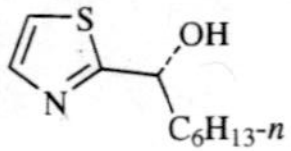 (97) 90% ee	590
C_{11} 	$PhSiH_3$ (1.1 eq), (*S*)-**195**-2*n*-BuLi (5 mol%), MeC_6H_5, 65°, 1 h	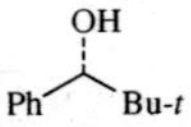 **I** (9) 5% ee	783
	$PhSiH_3$ (1.1 eq), (*R*)-**195**-2*n*-BuLi (5 mol%), MeC_6H_5, 65°, 1 h	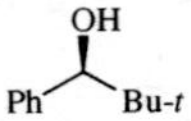 (10) 5% ee	783
	Ph_2SiH_2 (1.25 eq), Rh[(NBD)Cl]$_2$ (0.5%), ligand (0.5%), MeC_6H_5	**I** (—)	791
	Ligand / Temp	% ee	
	116 / –5°	84	
	117 / –5°	86	
	119 / 0°	86	
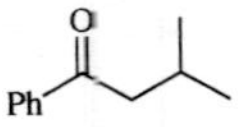	1. **197** (4.5 eq), *n*-BuLi (9 eq), C_6H_6 2. PMHS (5 eq) 3. Add ketone, 2.5 d 4. TBAF, THF	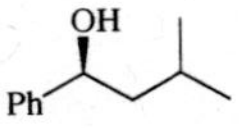 (96) 95% ee	588
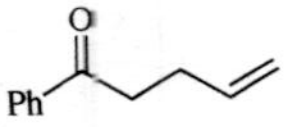	(3-$FC_6H_4)_2SiH_2$ (1.5 eq), [Rh(cod)$_2$]BF_4 (1 mol%), **1** (1.1 mol%), THF, –40°, 24 h	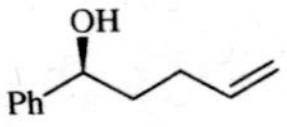 (90) 89% ee[b]	578

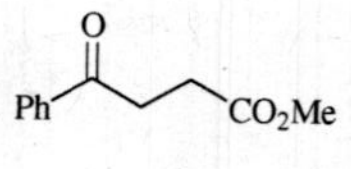

PMHS (1.2 eq), Et_2Zn (2 mol%), **131** (2 mol%), MeC_6H_5, rt, 18 h

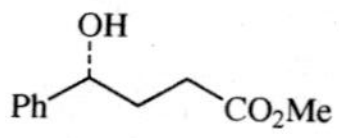

(0)

788

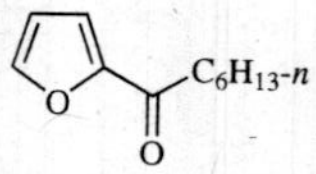

PMHS (4 eq), ligand (0.05 mol%), CuCl (x mol%), NaOBu-*t* (y mol%), MeC_6H_5

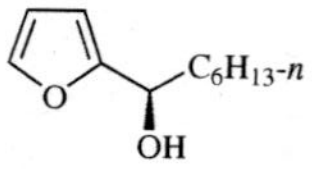

Ligand	x	y	Temp	Time		% ee
123	1	1	–50°	12 h	(98)	81
121	3	3	–78°	—	(—)	70
123	3	3	–50°	—	(—)	77

590

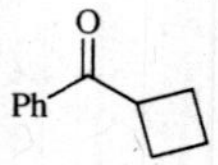

1. **197** (4.5 eq), *n*-BuLi (9 eq), C_6H_6
2. $PhSiH_3$ (1.5 eq)
3. Add ketone, 1 d
4. TBAF, THF

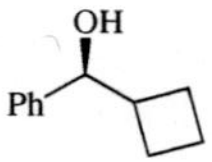

(88) 82% ee

588

TABLE 30. ASYMMETRIC ORGANOSILANE REDUCTION OF KETONES (*Continued*)

Ketone	Conditions	Product(s) and Yield(s) (%)	Refs.
C_{11} 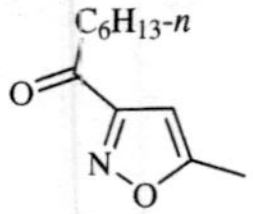	PMHS (4 eq), CuCl/NaOBu-*t* (1 mol%), **123** (0.05 mol%), MeC_6H_5, –50°, 10 h	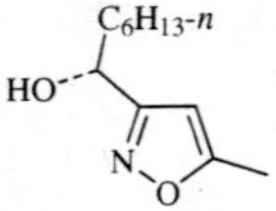 (68) 83% ee	590
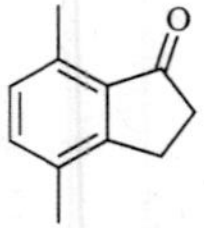	$(MeO)_3SiH$ (1 eq), Li_2-histidine (10 mol%), THF/TMEDA, 0°, 24 h	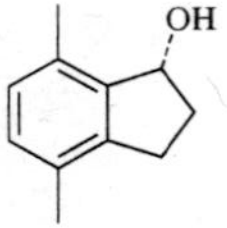 (66) 30% ee	593
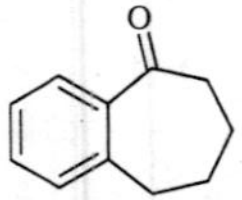	PMHS, $Sn(OTf)_2$ (10 mol%), **87** (10 mol%), MeOH, rt	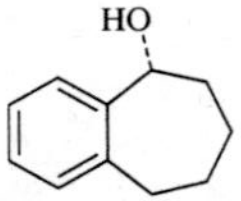 (5) 33% ee	385
	Ph(1-Np)SiH_2 (1.5 eq), [Rh(cod)]SbF_6 (1 mol%), **126** (1 mol%), THF, –20°	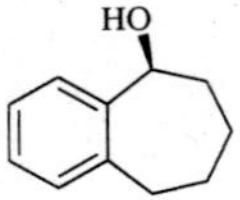 (90) 98% ee	576
	$PhSiH_3$ (1.1 eq), (*R*) or (*S*)-**195**-2*n*-BuLi (5 mol%), MeC_6H_5, 65°, 1 h	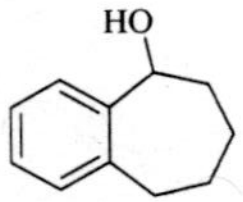 (0)	783
C_{12} 	Ph_2SiH_2 (1.5 eq), [Rh(cod)$_2$]BF_4 (1 mol%), DIOP (1.1 mol%), THF, rt, 48 h	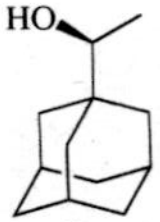 **I** (92) 91% ee	606
	Ph(1-Np)SiH_2 (1.5 eq), **126** (1 mol%), THF, –20°	**I** (98) >99% ee	576

Substrate	Conditions	Product(s) and Yield(s) (%)	Refs.
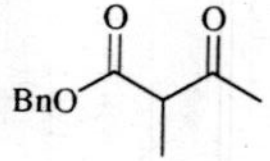	Ph(1-Np)SiH_2 (1.5 eq), [Rh(cod)]SbF_6 (1 mol%), **126** (1 mol%), THF, −20°	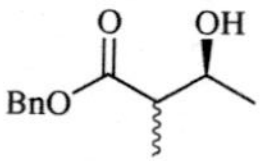 (70) 70% ee	576
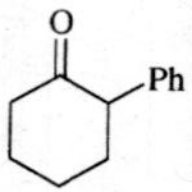	Ph_2SiH_2 (1.3 eq), $RhCl_3$ (1 mol%), **87** (5 mol%), $AgBF_4$ (2 mol%), THF, 0°, 1 d	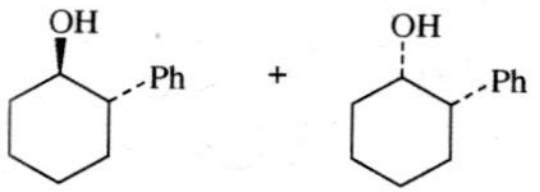 **I** 1S,2R 99% ee + **II** 1S,2S 96% ee; **I + II** (92), **I**:**II** = 51:49	390
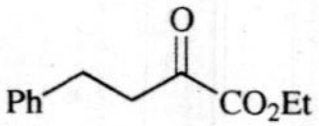	PMHS, $Sn(OTf)_2$ (10 mol%), **87** (10 mol%), MeOH, rt, 12-14 h	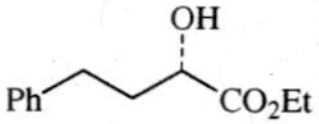 (98) 44% ee	385
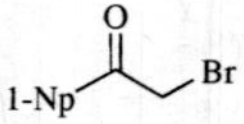	$(EtO)_3SiH$, THF, **198**, 96 h	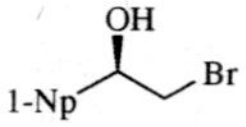 (64) 84% ee R	794
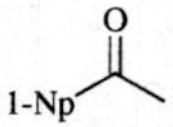	PMHS (1.1 eq), Et_2Zn (20 mol%), (*S*,*S*)-ebpe (20 mol%), vitride (3%), MeC_6H_5, rt, 24 h	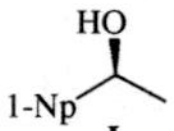 **I** (—) 75% ee	594
	$PhSiH_3$ (1.1 eq), (*S*)-**195**-2*n*-BuLi (5 mol%), MeC_6H_5, 65°, 3 h	**I** (94) 28% ee	783
	Ph_2SiH_2 (1.2 eq), C_6H_6, catalyst (0.01 eq), 0-20°, 10-15 h	**I** + 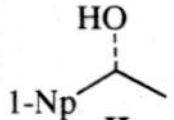 **II**	804

Catalyst	I + II	I:II
109	(82)	80.5:19.5
110	(60)	82.5:17.5
112	(56)	79:21
113	(84)	92:8
114	(95)	90:10
115	(92)	93.5:6.5

TABLE 30. ASYMMETRIC ORGANOSILANE REDUCTION OF KETONES (*Continued*)

Ketone	Conditions	Product(s) and Yield(s) (%)	Refs.
C_{12} 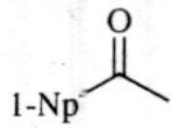	Ph_2SiH_2 (1.25 eq), $Rh[(NBD)Cl]_2$ (0.5%), ligand (0.5%), MeC_6H_5	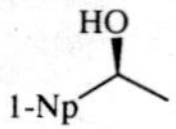 (—)	791
	Ligand: **116**, Temp: 0°	% ee: 29	
	Ligand: **119**, Temp: 20°	% ee: 42	
	Ligand: **109**, Temp: 20°	% ee: 61	
	Ligand: **113**, Temp: 20°	% ee: 84	
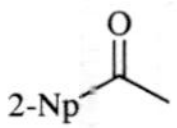	PMHS (1.1 eq), Et_2Zn (20 mol%), (*R*,*R*)-ebpe (20 mol%), vitride (3%), MeC_6H_5, rt, 24 h	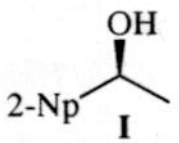 **I** (—) 80% ee	594
	1. **197** (4.5 eq), 2 *n*-BuLi, C_6H_6 2. PMHS (5 eq) 3. Add ketone, 3 d 4. TBAF, THF	**I** (84) 95% ee	588
	$PhSiH_3$ (1.1 eq), (*R*)-**195**-2*n*-BuLi (5 mol%), C_6H_5Me, 65°, 1 h	**I** (100) 62% ee	783
	$PhSiH_3$ (1.1 eq), (*S*)-**195**-2*n*-BuLi (5 mol%), C_6H_5Me, 65°, 1 h	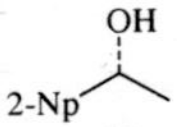 **II** (100) 25% ee	783
C_{13} 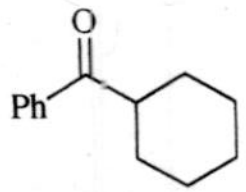	1. **197** (4.5 eq), *n*-BuLi (9 eq), C_6H_6 2. $PhSiH_3$ (1.5 eq) 3. Add ketone, 2.5 d 4. TBAF, THF	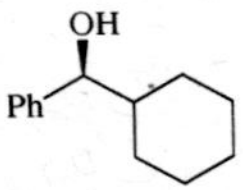 (88) 82% ee	588

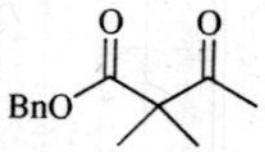

$Ph(1-Np)SiH_2$ (1.5 eq), **26** (1 mol%), THF, –20°

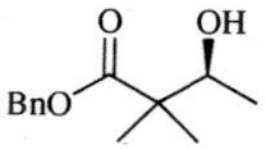

(80) >99% ee

576

C_{14-20}

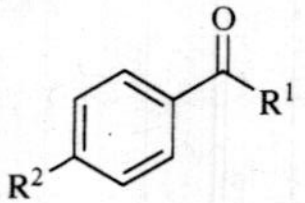

Cl_3SiH (1.5 eq), 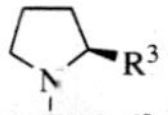 CHO (0.1 eq), CH_2Cl_2, 0° to rt, 24 h

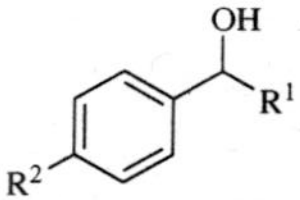

379

R^1	R^2	R^3		R:S
4-ClC_6H_4	Me	CONHPh	(87)	64:36
Ph	Et	CONHPh	(47)	61:39
$BnCH_2$	Me	CONHNp-1	(76)	54:46
Ph	*i*-Bu	CONHNp-1	(21)	68:32
Ph	Bn	CONHNp-1	(27)	75:25

C_{16}

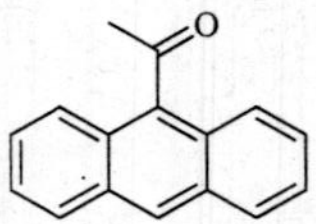

Ph_2SiH_2 (1.5 eq), $[Rh(cod)Cl]_2$ (0.25 mol%), **38** (0.5 mol%), Et_2O, rt

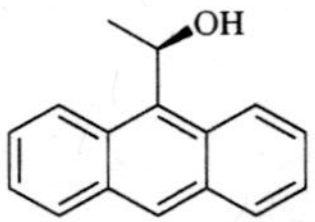

(100) 90% ee

571

TABLE 30. ASYMMETRIC ORGANOSILANE REDUCTION OF KETONES (*Continued*)

Ketone	Conditions	Product(s) and Yield(s) (%)	Refs.
C_{22}			
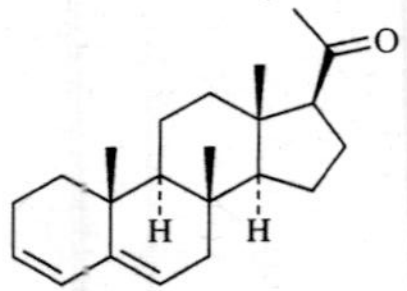	Ph_2SiH_2, catalyst, C_6H_6, 22°	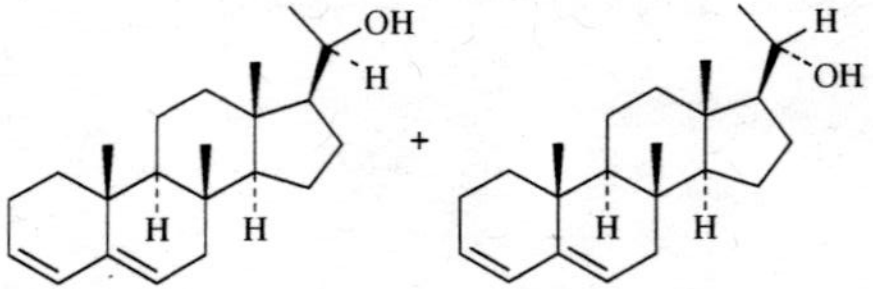 **I** (—) **II** (—)	573
	Catalyst	I:II	
	Rh-(+)-DIOP	7:3	
	Rh-(–)-DIOP	7:3	
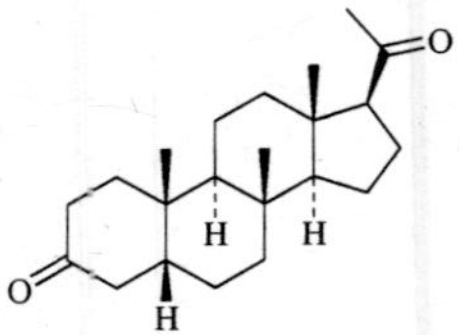	Ph_2SiH_2, catalyst, C_6H_6, 22°	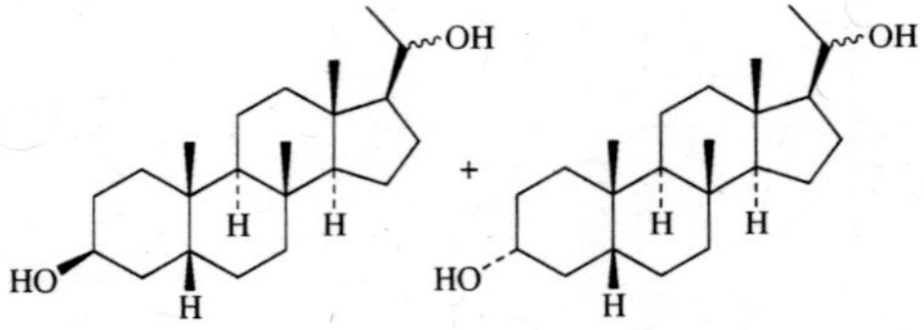 **I** (—) **II** (—)	573
	Catalyst	I:II	
	Rh-(+)-DIOP	67:33	
	Rh-(–)-DIOP	64:36	

[a] The catalyst is *S,S*-[1,2-bis(tetrahydroindenyl)ethane]titanium (IV) and derivatives.

[b] The configuration of the product was not determined.

[c] The reducing agent was $(EtO)_3SiH$.

[d] 10 mol% of the catalyst was employed.

[e] 20 mol% of the catalyst was employed.

[f] An additional 5 mol% of BINOL was added.

[g] 10 mol% of cinchonidine was added.

[h] 1-Naphthylphenylsilane was used as the reducing silane.

TABLE 31. ASYMMETRIC ORGANOSILANE REDUCTION OF α,β–UNSATURATED KETONES

Enone	Conditions	Product(s) and Yield(s) (%)	Refs.
C_6			
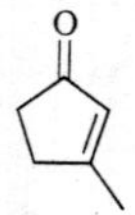	PMHS (1.05 eq), CuCl (5 mol%), t-BuONa (5 mol%) (*S*)-*p*-Tol-BINAP (5 mol%), −78°, 24 h	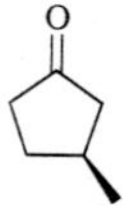 (42) 94% ee	595
$C_{6\text{-}12}$			
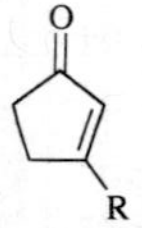	1 Ph_2SiH_2 (0.53 eq), CuCl (5%), NaOBu-*t* (5%), (*S*)-*p*-Tol-BINAP, MeC_6H_5, 0°, 2-3 h 2 BnBr (x eq), TBAT (1.2 eq), temp	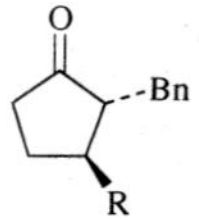	459

R	Solvent-2	x	Temp	Yield (%)
Me	CH_2Cl_2	2.0	rt	(62)
Me	THF	2.0	rt	(53)
Me	CH_2Cl_2/MeC_6H_5	2.0	rt	(62)
$BnCH_2$	CH_2Cl_2	1.2	50°	(60)
$BnCH_2$	CH_2Cl_2/MeC_6H_5	1.4	rt	(64)
$BnCH_2$	CH_2Cl_2/MeC_6H_5	2.0	rt	(67)
$BnCH_2$	CH_2Cl_2/MeC_6H_5	3.0	rt	(69)

Enone	Conditions	Product(s) and Yield(s) (%)	Refs.
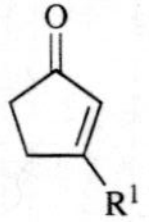	1. Ph_2SiH_2 (0.53 eq), CuCl (5%), NaOBu-*t* (5%), (*S*)-*p*-Tol-BINAP, MeC_6H_5, 0°, 2-3 h 2. R^2X, TBAT (1.2 eq), CH_2Cl_2/MeC_6H_5 (1:1), rt, 24 h	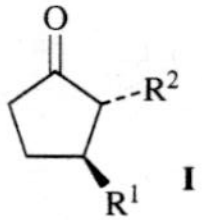 **I** + 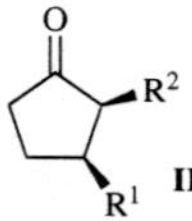 **II**	459

R^1	R^2X	I + II	I:II
Me	BnBr	(62)	92:8
Me	$BrCH_2CO_2Et$	(52)	94:6
Me	1-bromo-3-methyl-2-butene	(52)	80:20
$BnCH_2$	*n*-BuI	(42)	85:15
$BnCH_2$	BnBr	(67)	94:6
$BnCH_2$	allyl bromide	(64)	76:24
$BnCH_2$	MeI	(65)	73:27

TABLE 31. ASYMMETRIC ORGANOSILANE REDUCTION OF α,β–UNSATURATED KETONES (*Continued*)

Enone	Conditions	Product(s) and Yield(s) (%)	Refs.
C_6 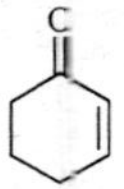	Ph_2SiH_2 (1.5 eq), $[Ir(cod)Cl]_2$ (0.25 mol%), (*R*)-**38** (0.5 mol%), Et_2O, 0°, 15 h	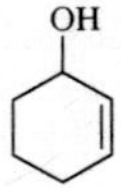 (100) 84% ee[a]	582
C_7 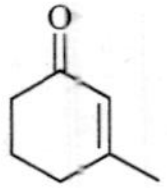	PMHS (1.05 eq), CuCl (5 mol%), *t*-BuONa (5 mol%), (*S*)-BIPHEMP (5 mol%), –78°, 24 h	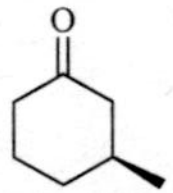 (61) 92% ee	595
	PMHS (2 eq), $(Ph_3P)CuH$ (1 mol%), **124** (0.1-0.5 mol%), MeC_6H_5, 0°, 4.5 h	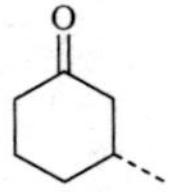 (96) 90% ee	597
C_8 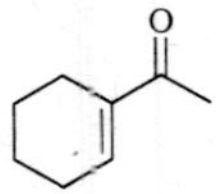	Ph_2SiH_2 (1.5 eq), $[Rh(cod)_2]BF_4$ (1.0 mol%), **25** (1.1 mol%), –40°, THF, 4 h	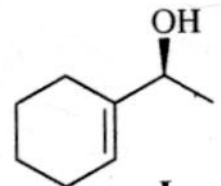 **I** (71) 95% ee	577
	Ph(1-Np)SiH_2 (1.5 eq), [Rh(cod)]SbF_6 (1 mol%), **126** (1 mol%), THF, –20°	**I** (85) 91% ee	576
	1. **197** (4.5 eq), *n*-BuLi (9 eq), C_6H_6 2. PMHS (10 eq) 3. Add ketone, 3 d	**I** (70) 85% ee	588
	Ph_2SiH_2 (1.5 eq), $[Rh(cod)_2]BF_4$ (1 mol%), DIOP (1.1 mol%), THF, rt, 10 h	**I** (71) 95% ee	606

C_9 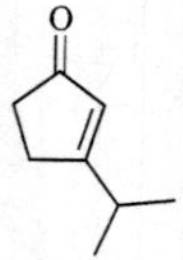	PMHS (1.05 eq), CuCl (5 mol%), *t*-BuONa (5 mol%), (*S*)-*p*-Tol-BINAP (5 mol%), 0°, 3 d	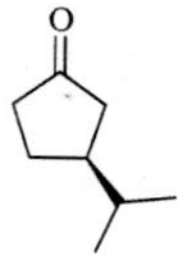 (88) 94% ee	595
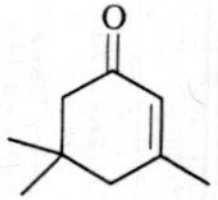	PMHS (2 eq), $(Ph_3P)CuH$ (1 mol%), **124** (0.00036 mol%), MeC_6H_5, −35°, 3 d	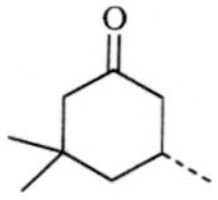 **I** (88) 98.5% ee	597
	PMHS (1 eq), $(Ph_3P)CuH$ (1 mol%), **124** (0.1-0.5 mol%), 0°, 6 h Solvent: CH_2Cl_2 / MeC_6H_5 / THF / dioxane/THF	**I** (>90) % ee: 90 / 92 / 95 / 98	597
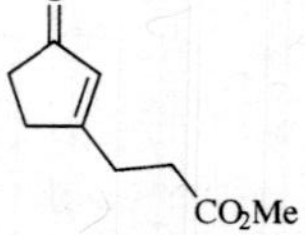	PMHS (2 eq), $(Ph_3P)CuH$ (1 mol%), **124** (0.1-0.5 mol%), MeC_6H_5, −78°, 18 h	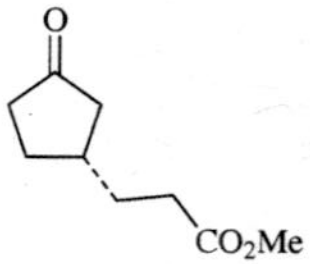 (94) 97% ee	597
	PMHS (1.05 eq), CuCl (5 mol%), *t*-BuONa (5 mol%), (*S*)-*p*-Tol-BINAP (5 mol%), 0°, 24 h	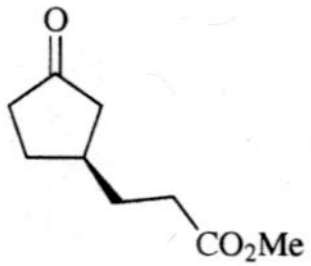 (86) 92% ee	595
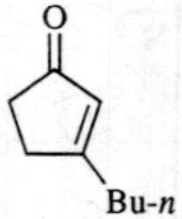	PMHS (1.05 eq), CuCl (5 mol%), *t*-BuONa (5 mol%), (*S*)-*p*-Tol-BINAP (5 mol%), 0°, 12 h	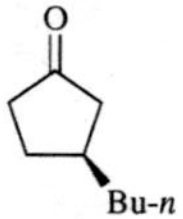 (84) 98% ee	595

Solvent	% ee
CH_2Cl_2	90
MeC_6H_5	92
THF	95
dioxane/THF	98

TABLE 31. ASYMMETRIC ORGANOSILANE REDUCTION OF α,β–UNSATURATED KETONES (*Continued*)

Enone	Conditions	Product(s) and Yield(s) (%)	Refs.
C_{10} 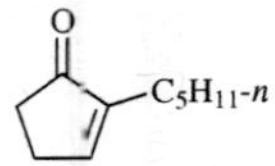	1. $PhSiH_3$ (0.1 eq), **198** (2 mol%), pyrrolidine, MeOH, THF, 60° 2. PMHS, ketone, MeOH (3-7 eq), 15°, 4 h	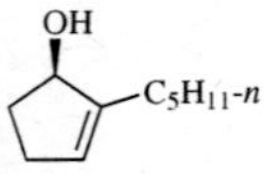 (90) 84% ee	587
	Ph_2SiH_2 (1.5 eq), $[Rh(cod)_2]BF_4$ (1.0 mol%), **25** (1.1 mol%), THF, –40°, 4 h	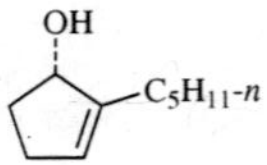 (84) 87% ee	597
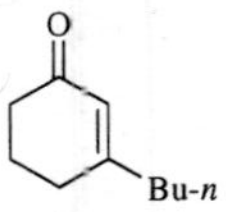	PMHS (1.05 eq), CuCl (5 mol%), *t*-BuONa (5 mol%), (*S*)-*p*-Tol-BINAP (5 mol%), –78°, 2 d	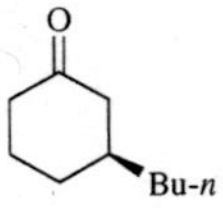 (82) 87% ee	595
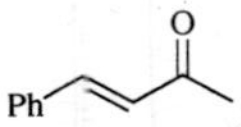	$(MeO)_3SiH$ (1 eq), Et_2O/TMEDA (30:1), Li-(*R*)-BINOL (10 mol%), 0°, 24 h	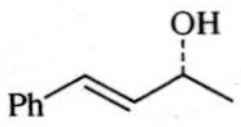 (91) 57% ee	592
	$(MeO)_3SiH$ (1 eq), Li_2-histidine (10 mol%), THF/TMEDA, 0°, 24 h	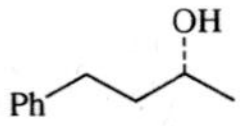 (91) 28% ee	593
	$(MeO)_3SiH$ (1 eq), Li_2-histidine (10 mol%), THF/TMEDA, 0°, 24 h	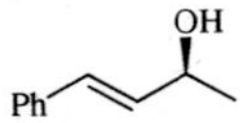 (78) 70% ee	593
C_{11} 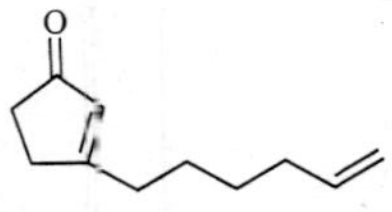	PMHS (1.05 eq), CuCl (5 mol%), *t*-BuONa (5 mol%), (*S*)-BINAP (5 mol%), 15°, 24 h	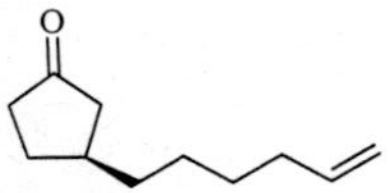 (87) 96% ee	595

C_{12}			
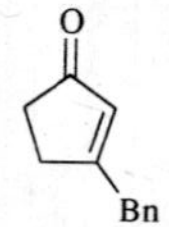	PMHS (1.05 eq), CuCl (5 mol%), *t*-BuONa (5 mol%), (*S*)-*p*-Tol-BINAP (5 mol%), 0°, 24 h	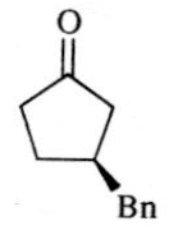 (78) 96% ee	595
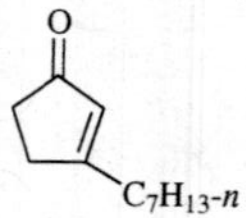	PMHS (2 eq), CuCl (1 mol%), NaOMe (0.5 mol%), **124** (0.4 mol%), MeC_6H_5, –78°	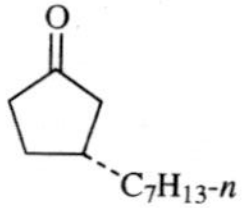 (97) 97.5% ee	597
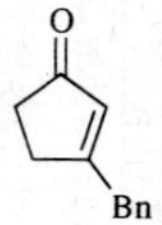	PMHS (2 eq), $(Ph_3P)CuH$ (1 mol%), **124** (0.1-0.5 mol%), MeC_6H_5, –78°, 36 h	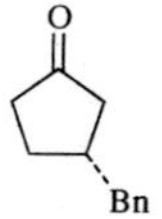 (92) 97.3% ee	597
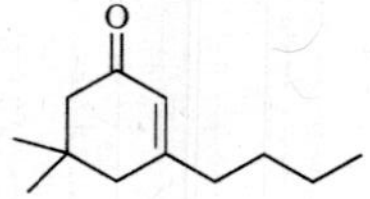	PMHS (2 eq), $(Ph_3P)CuH$ (1 mol%), **124** (0.1-0.5 mol%), MeC_6H_5, 0°, 6 h	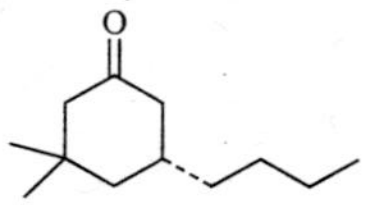 (90) 96% ee	597
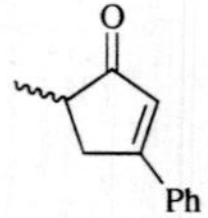	PMHS (1.1 eq), CuCl/NaOBu-*t*, (*S*)-*p*-Tol-BINAP (10 mol%), MeC_6H_5; then TBAF	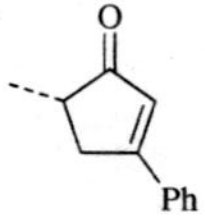 45% conv., 71.6% ee + 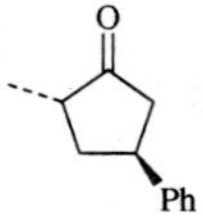 (—) + 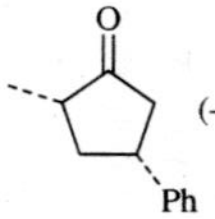 (—)	596
C_{13}			
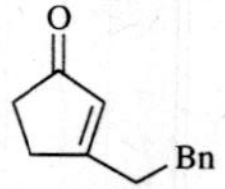	PMHS (1.05 eq) CuCl (5 mol%), *t*-BuONa (5 mol%), (*S*)-*p*-Tol-BINAP (5 mol%), 0°, 24 h	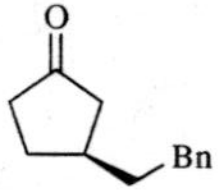 (86) 94% ee	595
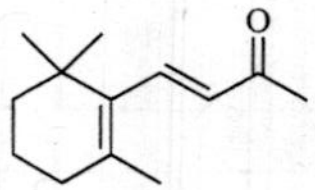	PMHS (1.1 eq), Et_2Zn (20 mol%), (*R*,*R*)-ebpe (20 mol%), vitride (3%), MeC_6H_5, rt, 24 h	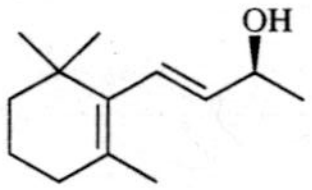 (—) 18% ee	594

TABLE 31. ASYMMETRIC ORGANOSILANE REDUCTION OF α,β–UNSATURATED KETONES (*Continued*)

Enone	Conditions	Product(s) and Yield(s) (%)	Refs.
C_{14-19}			
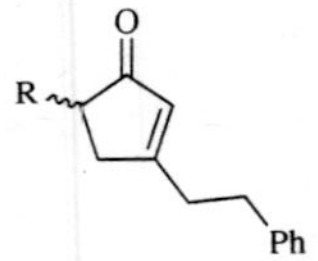	PMHS (1.1 eq), CuCl/NaOBu-*t*, (*S*)-*p*-Tol-BINAP (10 mol%), MeC_6H_5, temp; then TBAF	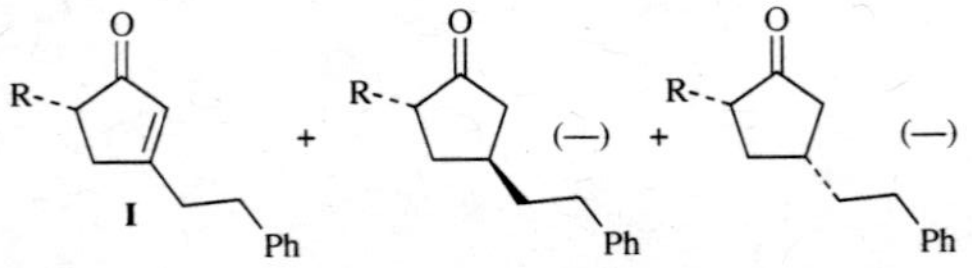 I + (—) + (—)	596

R	Temp	% Conv. I	% ee
Me	–78°	55.6	94.6
i-Pr	0°	56.5	97.4
n-Bu	–78°	53.2	90.6
t-Bu	0°	50.7	91.0
CH_2CO_2Bu-*t*	–50°	56.0	95.3

Enone	Conditions	Product(s) and Yield(s) (%)	Refs.
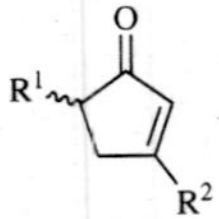	PMHS (2.2 eq), CuCl/(*S*)-*p*-Tol-BINAP (10 mol%), *t*-BuONa (1.7 eq), *t*-BuOH (5.0 eq), MeC_6H_5, temp, 26 h; then TBAF	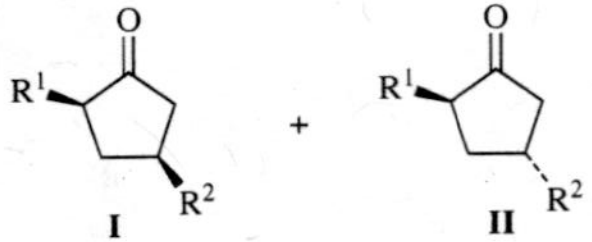 I + II	596

R^1	R^2	Temp	I + II	I:II	% ee I
Me	$BnCH_2$	–50°	(89)	91:9	91
i-Pr	$BnCH_2$	–30°	(94)	93:7	93
t-Bu	$BnCH_2$	0°	(94)	93.5:6.5	94
t-BuO_2CCH_2	$BnCH_2$	–50°	(91)	90:10	93
Bn	*i*-Pr	–50°	(95)	91.5:8.5	93
Me	Ph	–50°	(90)	96.5:3.5	91

C_{15}			
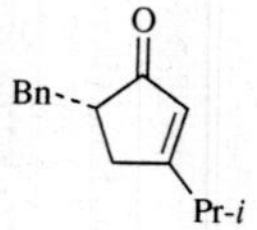	FMHS (1.1 eq), CuCl/NaOBu-*t*, (*S*)-*p*-Tol-BINAP (10 mol%), MeC_6H_5; then TBAF	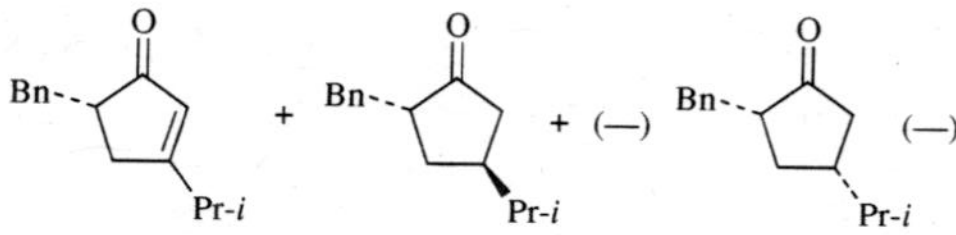 (—) 56% conv., 95.6% ee	596
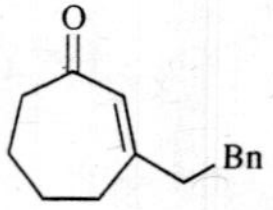	FMHS (1.05 eq), CuCl (5 mol%), *t*-BuONa (5 mol%), (*S*)-BINAP (5 mol%), –78°, 4 d	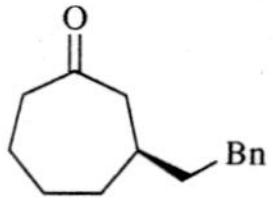 (82) 96% ee	595
C_{16}			
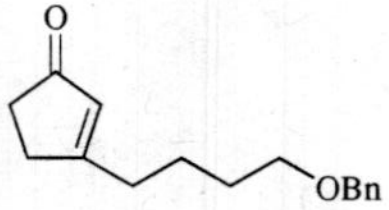	FMHS (1.05 eq), CuCl (5 mol%), *t*-BuONa (5 mol%), (*S*)-*p*-Tol-BINAP (5 mol%), 0°, 24 h	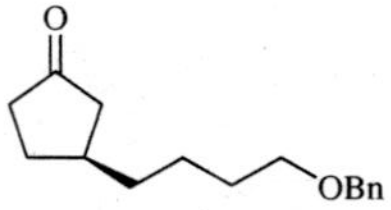 (91) 94% ee	595
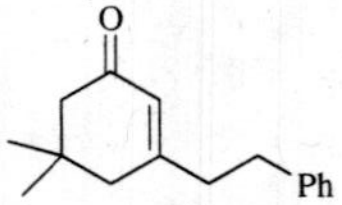	PMHS (2 eq), $(Ph_3P)CuH$ (1 mol%), **124** (0.1-0.5 mol%), MeC_6H_5, –35°, 16 h	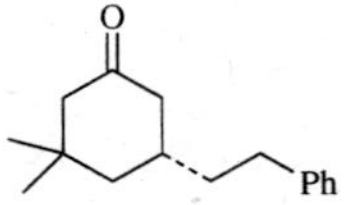 (95) 99.5% ee	597

[a] The product is the (–) enantiomer.

TABLE 32. ASYMMETRIC ORGANOSILANE REDUCTION OF α,β-UNSATURATED ESTERS

Ester	Conditions	Product(s) and Yield(s) (%)	Refs.
C_4 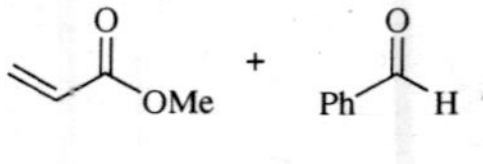	1. Et_2MeSiH, catalyst (5 mol%), ligand (7.5 mol%), CH_2Cl_2, rt, 24 h 2. H_3O^+	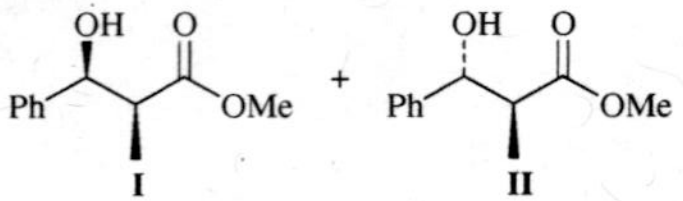	601

Catalyst	Ligand	I + II	I:II	% ee I
$[Ir(cod)Cl]_2$	**86**	(<10)	1.7:1	50
$[Ir(cod)Cl]_2$	**87**	(46)	1:1.3	0
$[Ir(cod)Cl]_2$	**88**	(32)	2.0:1	80
$[Ir(cod)Cl]_2$	**90**	(30)	1.5:1	70
$[Ir(cod)Cl]_2$	**98**	(48)	3.9:1	92
$[Ir(coe)_2Cl]_2$	**98**	(53)	4.4:1	94
$Ir(cod)BF_4$	**98**	(35)	4.0:1	86
$(CO)_2Ir(acac)$	**98**	(56)	4.0:1	82
$[Rh(cod)Cl]_2$	**98**	(9)	2.0:1	10

Ester	Conditions	Product(s) and Yield(s) (%)	Refs.
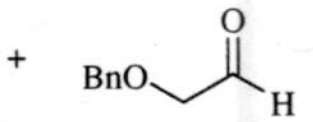	1. Et_2MeSiH, $[Ir(cod)Cl]_2$ (2.5 mol%), **98** (7.5 mol %), rt, 24 h 2. H_3O^+	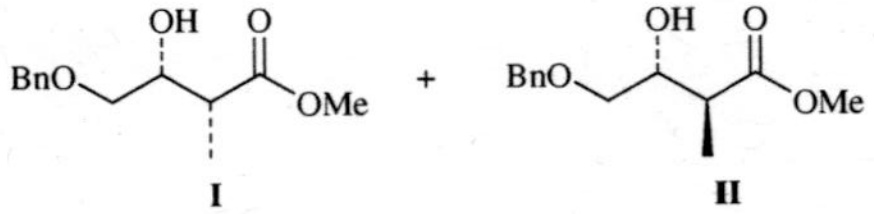 **I** + **II** (59), **I**:**II** = 9.5:1, er **I** = 98:2	601, 602
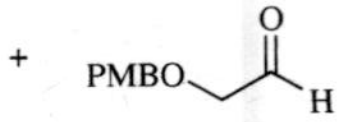	1. Et_2MeSiH, $[Ir(cod)Cl]_2$ (2.5 mol%), **98** (7.5 mol%), rt, 24 h 2. H_3O^+	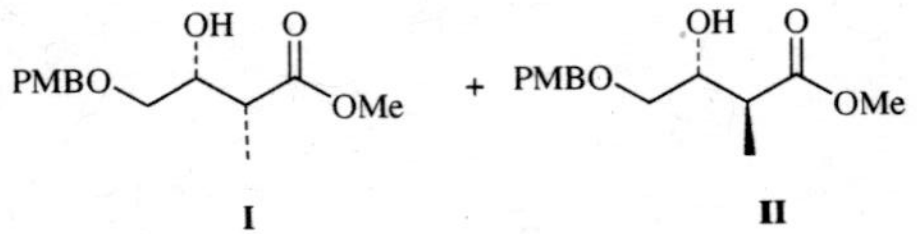 **I** + **II** (76), **I**:**II** = 6:1, er **I** > 95:5	602

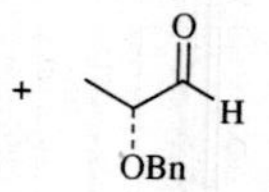	1. Et_2MeSiH, $[Ir(cod)Cl]_2$ (2.5 mol%), **98** (7.5 mol%), rt, 24 h 2. H_3O^+	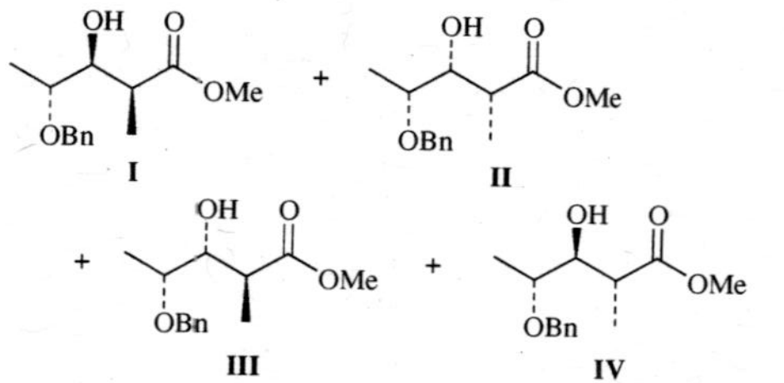 **I + II + III + IV** (50), **I**:**II** > 95:5, (**I** + **II**):**III** + **IV** > 95:5	601
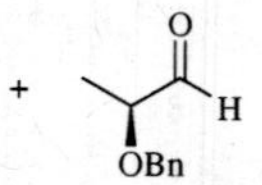	1. Et_2MeSiH, $[Ir(cod)Cl]_2$ (2.5 mol%), **98** (7.5 mol%), rt, 24 h 2. H_3O^+	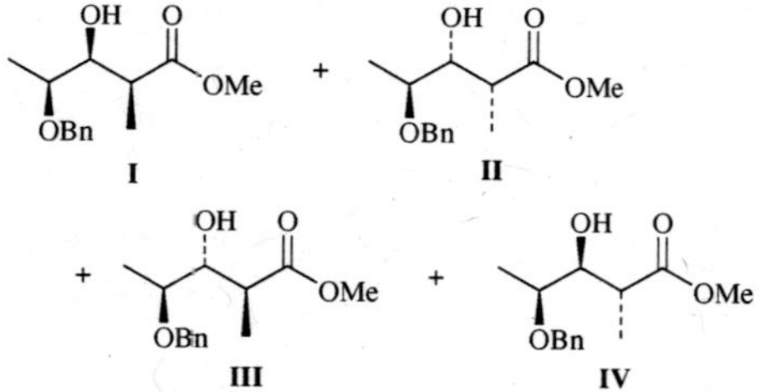 **I + II + III + IV** (< 5), **I**:**II** = —, (**I** + **II**):**III** + **IV** = —	601
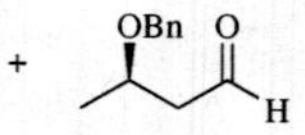	1. Et_2MeSiH, $[Ir(cod)Cl]_2$ (2.5 mol%), **98** (7.5 mol%), rt 24 h 2. H_3O^+	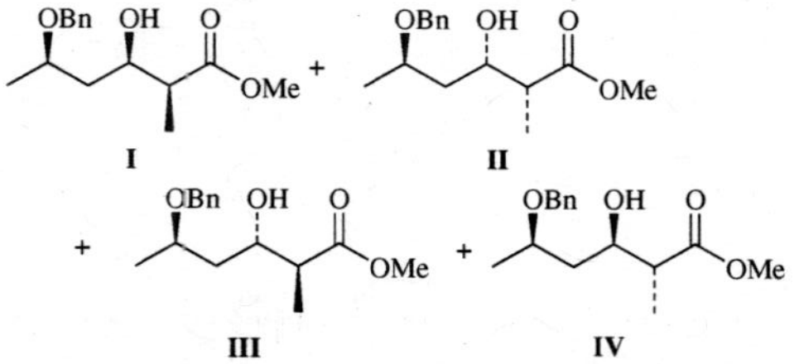 **I + II + III + IV** (65), **I**:**II** = 89:11, (**I** + **II**):**III** + **IV** = 2.7:1	601

TABLE 32. ASYMMETRIC ORGANOSILANE REDUCTION OF α,β-UNSATURATED ESTERS (*Continued*)

Ester	Conditions	Product(s) and Yield(s) (%)	Refs.
C_4 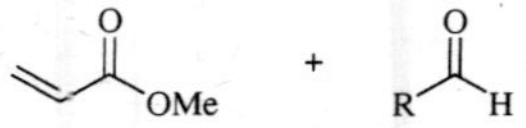	1. Et_2MeSiH, $[Ir(coe)_2Cl]_2$ (2.5 mol%), **98** (7.5 mol%), CH_2Cl_2, rt, 24 h 2. H_3O^+	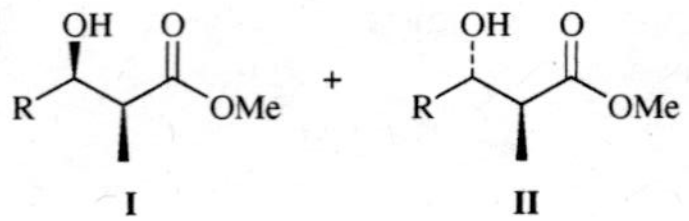	601

R	I + II	I:II	% ee I
$BnOCH_2$	(49)	9.9:1	96
$BnOCH_2$	(59)	9.5:1	96
$TBSOCH_2$	(47)	8.2:1	96
Et	(<5)	—	—
$BnOCH_2CH_2$	(65)	2.7:1	82
(*E*)-PhCH=CH	(0)	—	—

Ester	Conditions	Product(s) and Yield(s) (%)	Refs.
C_{4-9} 	1. Et_2MeSiH, $[Rh(cod)Cl]_2$ (2.5 mol%), (*R*)-BINAP (6.5 mol%), CH_2Cl_2, rt, 24 h 2. H_3O^+	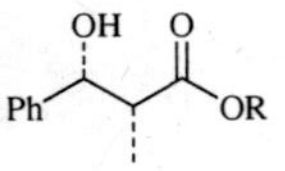 + anti diastereomer	603

R		syn:anti	% ee syn	Conf. syn	% ee anti	Conf. anti
Me	(37)	1.7:1	91	2R,3R	88	2R,3S
t-Bu	(21)	1.4:1	58	—	38	—
Ph	(72)	3.4:1	87	2R,3R	34	—

Ester	Conditions	Product(s) and Yield(s) (%)	Refs.
C_5 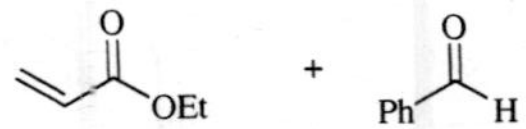	1. Et_2MeSiH, $[Ir(coe)_2Cl]_2$ (5 mol%), **98** (7.5 mol%), CH_2Cl_2, rt, 45 h 2. H_3O^+	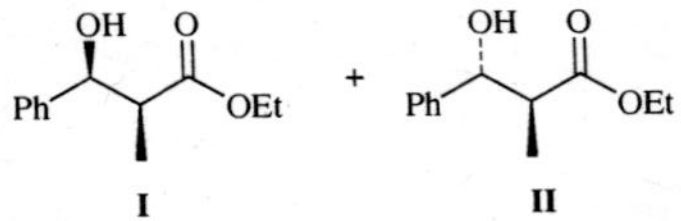 **I** + **II** (68), **I**:**II** = 6.6:1, 94% ee **I**	601

C$_{6-11}$

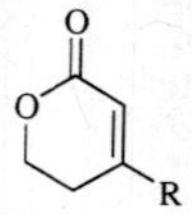

PMHS (x eq), $CuCl_2$•2 H_2O (0.1 mol%), NaOBu-*t* (20 mol%), *t*-AmOH (y eq), catalyst (5 mol%), MeC_6H_5/C_5H_{12}, 3 h

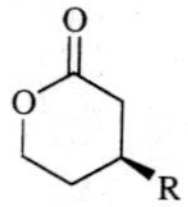

599

R	Catalyst	x	y	Temp		% ee
Me	(*S*)-*p*-Tol-BINAP	4	—	rt	(70)	82
n-Pr	(*S*)-BIPHEMP	2	2	–40°	(71)	86
c-C_6H_{11}	(*S*)-BIPHEMP	2	2	–40°	(56)	94

C$_{7-12}$

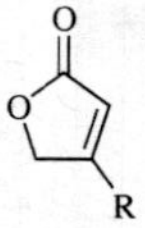

PMHS (4 eq), $CuCl_2$•2 H_2O (0.1 mol%), NaOBu-*t* (20 mol%), *t*-AmOH, (*S*)-*p*-Tol-BINAP (5 mol%), MeC_6H_5/C_5H_{12}, rt, 3 h

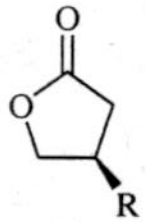

599

R		% ee
i-Pr	(87)	92
n-Bu	(89)	86
Ph	(70)	71
3,4-$(MeO)_2C_6H_3$	(85)	92

C$_9$

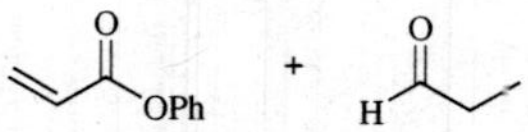

EtMe$_2$SiH, [Rh(cod)Cl]$_2$ (2.5 mol%), (*R*)-BINAP (6.5 mol%)

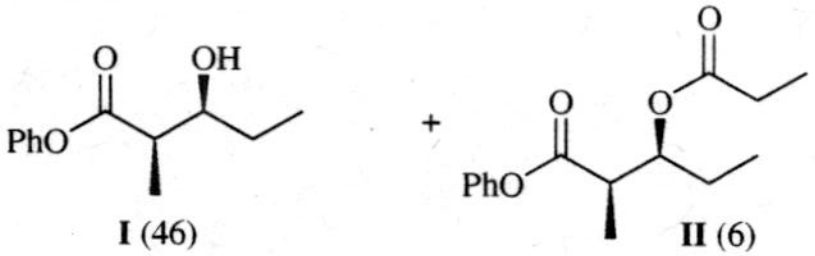

I (46)
dr = 4.6:1
er = 94:6

II (6)
dr = 4.6:1
er = 94:6

470

R$_3$SiH, {Rh(cod)[(*R*)-BINAP)]}BF_4 (2.5 mol%)

I + **II**

470

R_3Si	**I**	**II**	er **I**	er **II**
Et_2MeSi	(23)	(42)	91:9	91:9
$PhMe_2Si$	(18)	(48)	90:10	90:10
$(EtO)Me_2Si$	(0)	(29)	—	89:11

TABLE 32. ASYMMETRIC ORGANOSILANE REDUCTION OF α,β-UNSATURATED ESTERS (Continued)

Ester		Conditions	Product(s) and Yield(s) (%)	Refs.
C_9 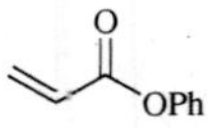	+ 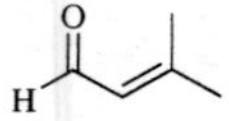	Et_2MeSiH (1.75 eq), {Rh(cod)[(R)-BINAP)]}BF_4 (5 mol%), rt, 12 h	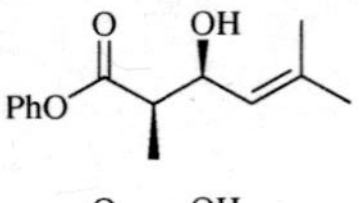 (86) syn:anti = 6:1, 83% ee syn	470
	+ 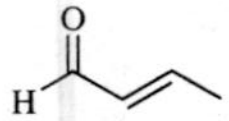	Et_2MeSiH (1.75 eq), {Rh(cod)[(R)-BINAP)]}BF_4 (5 mol%), rt, 12 h	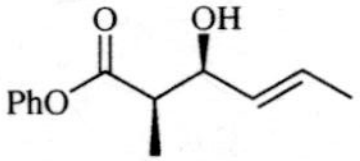 (54) syn:anti = 6:1, 71% ee syn	470
	+ 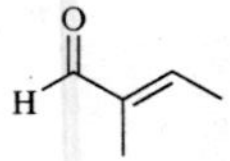	Et_2MeSiH (1.75 eq), {Rh(cod)[(R)-BINAP)]}BF_4 (5 mol%), rt, 12 h	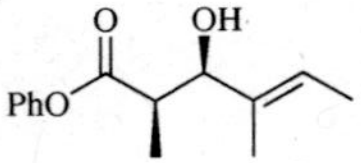 (90) syn:anti = 3:1, 75% ee syn	470
	+ 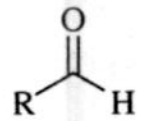	Et_2MeSiH (1.2 eq), $[Rh(cod)Cl]_2$ (2.5 mol%), (R)-BINAP (6.5 mol%), CH_2Cl_2, rt, 24 h	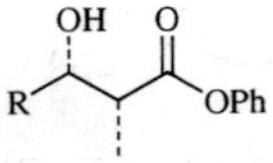 + anti diastereomer	603
	+ 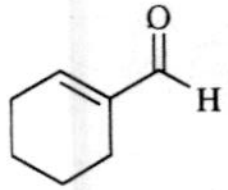	Et_2MeSiH (1.75 eq), {Rh(cod)[(R)-BINAP)]}BF_4 (5 mol%), rt, 12 h	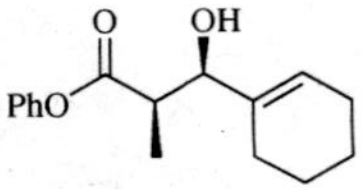 (73) syn:anti = 7:1, 81% ee syn	470
C_{10} 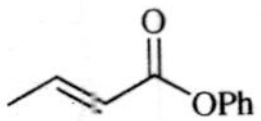	+ 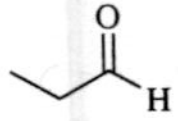	Et_2MeSiH (5 eq), $[Rh(cod)Cl]_2$ (5 mol%), (S)-BINAP (6.5 mol%), rt, 48 h	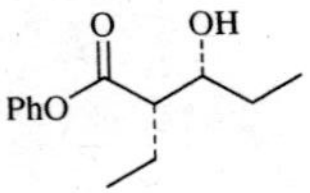 (76) syn:anti = 4.3:1, 88% ee syn	470

R	syn:anti	% ee syn	% ee anti
Ph	3.4:1	87	34
Et	5.1:1	88	7
c-C_6H_{11}	3.9:1	84	48
t-Bu	1.8:1	45	>99
1-$C_{10}H_7$	3.8:1	80	13
(E)-PhCH=CH	—	—	—

Reactant	Conditions	Product(s) and Yield(s) (%)	Refs.
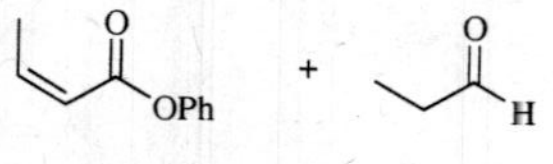	Et_2MeSiH (5 eq), $[Rh(cod)Cl]_2$ (5 mol%), (*S*)-BINAP (6.5 mol%), rt, 48 h	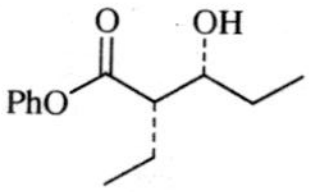 (52) syn:anti = 3.9:1, 88% ee syn	470
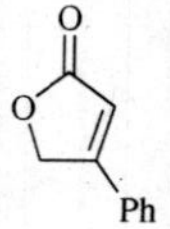	PMHS (2 eq), $(Ph_3P)CuH$ (0.5 mol%), **124** (0.1 mol%), *t*-BuOH (1.1 eq), rt, 1 h	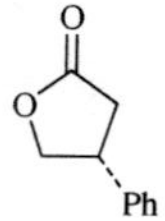 (96) 99% ee	598
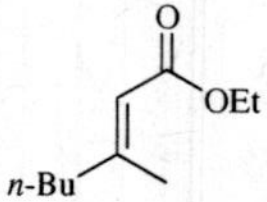	PMHS (2 eq), $(Ph_3P)CuH$ (7.3 mol%), **124** (7.3 mol%), *t*-BuOH (1.13 eq), MeC_6H_5, 0°	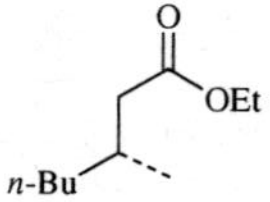 (80) 95% ee	598

C_{11}

Reactant	Conditions	Product(s) and Yield(s) (%)	Refs.
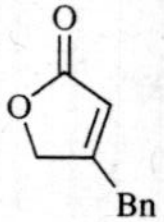	Silane (4 eq), Cu salt (5 mol%), NaOBu-*t* (5-20 mol%), additive, (*S*)-*p*-Tol-BINAP (5 mol%)	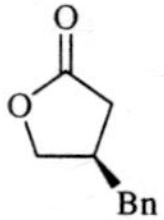	599

Silane	Cu Salt	Additive	Temp	Time		% ee
PMHS	CuCl	MeC_6H_5	rt	48 h	(50)	80
Ph_2SiH_2	CuCl	DEE/C_5H_{12}	−15°	24 h	(34)	91
PMHS	CuCl	C_6H_{12}/EtOH	rt	5 min	(89)	—
PMHS	CuCl	THF/C_5H_{12}/EtOH	−40°	4 h	(90)	93
PMHS	$CuCl_2 \cdot 2H_2O$	MeC_6H_5/C_5H_{12}/*i*-PrOH	−20°	3 h	(90)	92

TABLE 32. ASYMMETRIC ORGANOSILANE REDUCTION OF α,β-UNSATURATED ESTERS (*Continued*)

Ester	Conditions	Product(s) and Yield(s) (%)	Refs.
C_{11}			
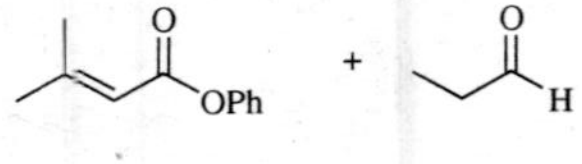	Et_2MeSiH (5 eq), $[Rh(cod)Cl]_2$ (5 mol%), (*S*)-BINAP (6.5 mol%), rt, 48 h	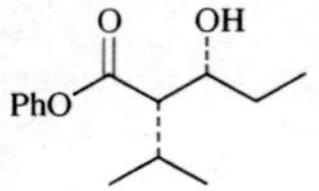 (0)	470
C_{12}			
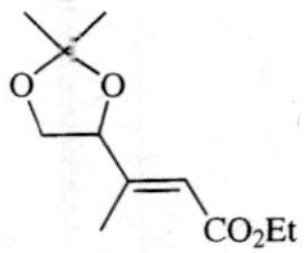	PMHS, CuH, **124** (0.2-0.4 mol%), *t*-BuOH, rt, 1 h	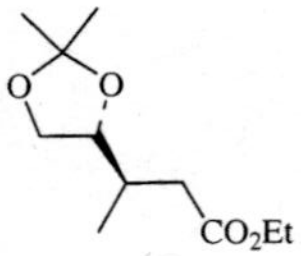 (91) 98% ee	598
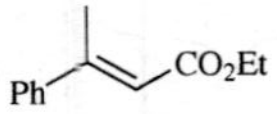	PMHS (4 eq), $CuCl_2$ (0.1 mol%), NaOBu-*t* (0.2 mol%), *t*-AmOH, (*S*)-*p*-Tol-BINAP (0.1 mol%), C_6H_{12}, rt	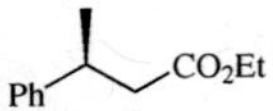 **I** (95) 86% ee	599
	PMHS (2 eq), $(Ph_3P)CuH$ (7.3 mol%), **124** (7.3 mol%), *t*-BuOH (1.13 eq), MeC_6H_5, 0°	**I** (92) 98% ee	598
	PMHS (2 eq), $(Ph_3P)CuH$ (7.3 mol%), **124** (7.3 mol%), *t*-BuOH (1.13 eq), MeC_6H_5, 0°	**I** (94) 91% ee	598
	PMHS (4 eq) CuCl (5 mol%), NaOBu-*t* (5 mol %), (*S*)-*p*-Tol-BINAP (10 mol%), MeC_6H_5, rt, 24 h	**I** (84) 90% ee	600

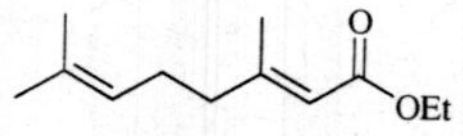	PMHS (4 eq) CuCl (5 mol%), NaOBu-*t* (5 mol%), (*S*)-*p*-Tol-BINAP (10 mol%), MeC_6H_5, rt, 25 h	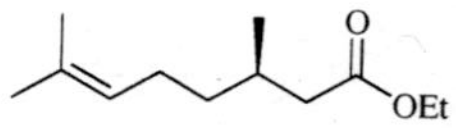 (90) 85% ee	600
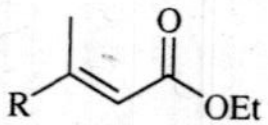	PMHS (4 eq) CuCl (5 mol%), NaOBu-*t* (5 mol%), (*S*)-*p*-Tol-BINAP (10 mol%), MeC_6H_5, rt	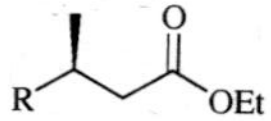 R / Time / / % ee: *n*-C_6H_{13} 24 h (94) 81; *c*-C_6H_{11} 22 h (89) 92	600
C13 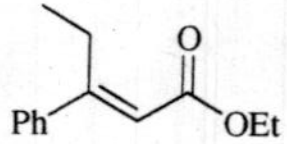	PMHS, CuH, **124** (0.2-0.4 mol%), *t*-BuOH, 0°, 1 h	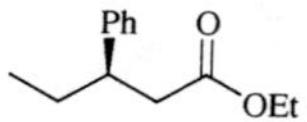 (90) 95% ee	598
	PMHS (4 eq), CuCl (5 mol%), NaOBu-*t* (5 mol%), (*S*)-*p*-Tol-BINAP (10 mol%), MeC_6H_5, rt, 25 h	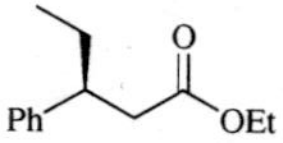 (98) 91% ee	600
C14 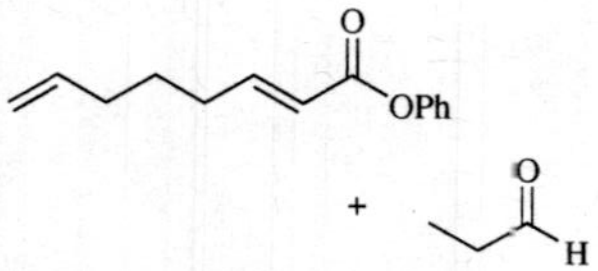	Et_2MeSiH (5 eq), $[Rh(cod)Cl]_2$ (5 mol%), (*S*)-BINAP (6.5 mol%), rt, 48 h	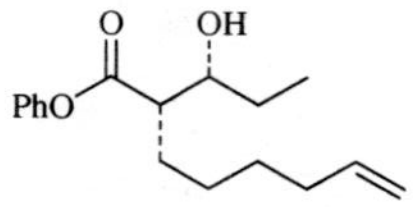 (30) syn:anti = 4.4:1, — % ee syn	470
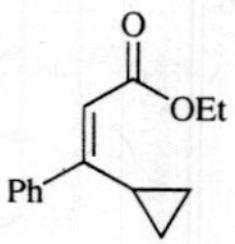	PMHS (2 eq), $(Ph_3P)CuH$ (7.3 mol%), **124** (7.3 mol%), *t*-BuOH (1.13 eq), MeC_6H_5, 0°	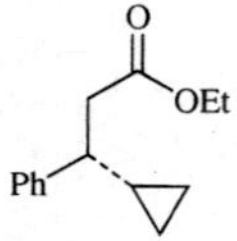 (97) 99% ee	598
	PMHS, CuH, **124** (0.2-0.4 mol%), *t*-BuOH, 0°, 1 h	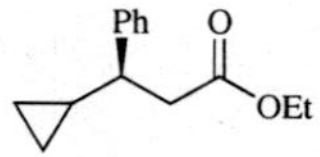 (97) 99% ee	598

TABLE 32. ASYMMETRIC ORGANOSILANE REDUCTION OF α,β–UNSATURATED ESTERS (*Continued*)

Ester	Conditions	Product(s) and Yield(s) (%)	Refs.
C_{14} 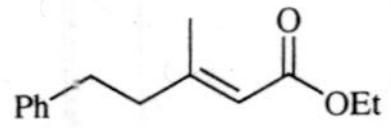	PMHS, CuH, **124** (0.2-0.4 mol%), *t*-BuOH, 0°, 1 h	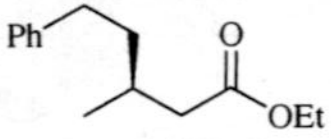 **I** (98) 99% ee	598
	PMHS (2 eq), (Ph_3P)CuH (7.3 mol%), **124** (7.3 mol%), *t*-BuOH (1.13 eq), MeC_6H_5, 0°	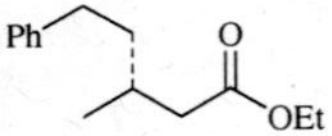 **II** (96) 90% ee	598
	PMHS (2 eq), (Ph_3P)CuH (7.3 mol%), **124** (7.3 mol%), *t*-BuOH (1.13 eq), MeC_6H_5, 0°	**II** (98) 99% ee	598
	PMHS (4 eq), CuCl (5 mol%), NaOBu-*t* (5 mol%), (*S*)-*p*-Tol-BINAP (10 mol%), MeC_6H_5, rt, 20 h	**II** (95) 84% ee	600
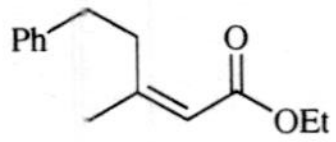	PMHS (4 eq), CuCl (5 mol%), NaOBu-*t* (5 mol%), (*S*)-*p*-Tol-BINAP (10 mol%), MeC_6H_5, rt, 18 h	**I** (96) 83% ee	600
C_{15} 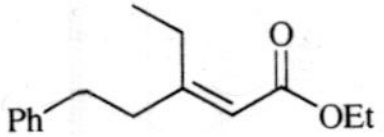	PMHS, CuH, **124** (0.2-0.4 mol%), *t*-BuOH, rt, 1 h	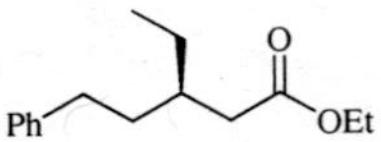 (93) 99% ee	598
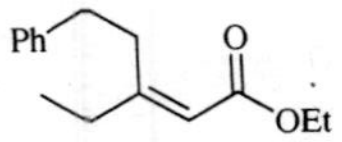	PMHS, CuH, **124** (0.2-0.4 mol%), *t*-BuOH, rt, 1 h	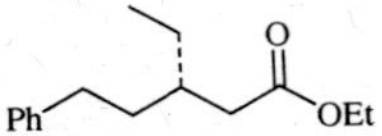 (94) 98% ee	598

C_{15-18}

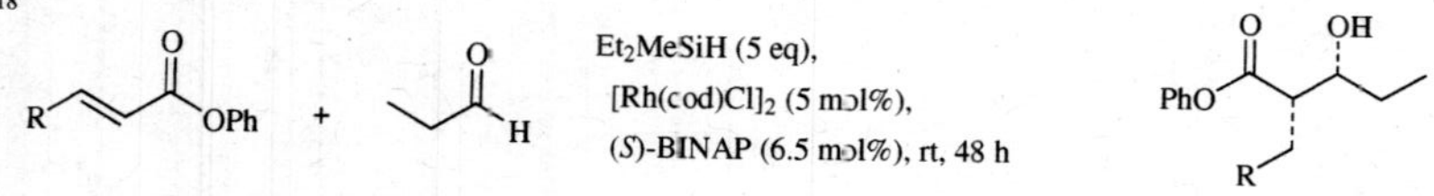

470

R
n-C_6H_{13}
TBSO$(CH_2)_3$
Ph$(CH_2)_3$

	syn:anti	% ee syn
(54)	4.2:1	88
(53)	3.8:1	88
(49)	3.9:1	93

TABLE 33. ASYMMETRIC ORGANOSILANE REDUCTION OF α,β–UNSATURATED LACTAMS

Lactam	Conditions	Product(s) and Yield(s) (%)	Refs.
C_{12}			
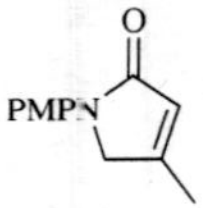	PMHS (4 eq), $CuCl_2$•2 H_2O (0.1 mol%), NaOBu-*t* (20 mol%), *t*-AmOH, (*S*)-*p*-Tol-BINAP (5 mol%), MeC_6H_5/C_5H_{12}, rt, 1 h	 (90) 92% ee	599
C_{15}			
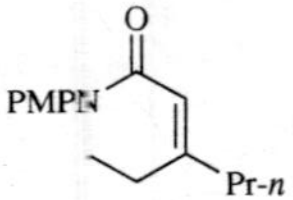	PMHS (4 eq), $CuCl_2$•2 H_2O (5 mol%), NaOBu-*t* (20 mol%), *t*-AmOH, (*S*)-*p*-Tol-BINAP (20 mol%), MeC_6H_5/C_5H_{12}, 0°, 3 h	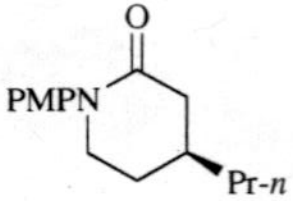 (89) 91% ee	599
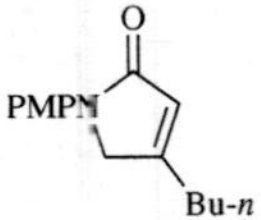	PMHS (4 eq), $CuCl_2$•2 H_2O (5 mol%), NaOBu-*t* (20 mol%), *t*-AmOH, (*S*)-*p*-Tol-BINAP (20 mol%), MeC_6H_5/C_5H_{12}, 0°, 3 h	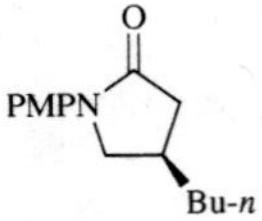 (94) 94% ee	599
C_{18}			
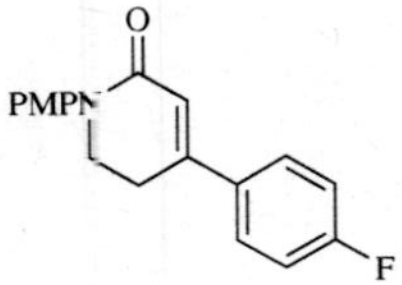	PMHS (16 eq), $CuCl_2$•2 H_2O (2.5 mol%), NaOBu-*t* (5 mol%), *t*-AmOH (16 eq), (*R*)-*p*-Tol-BINAP (0.5 mol%), C_6H_5F, air, rt, 3 h	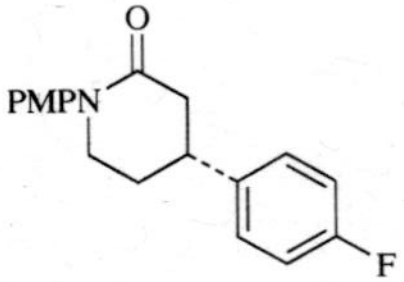 (89) 90% ee	599

TABLE 34. ASYMMETRIC ORGANOSILANE REDUCTION OF IMINES

Imine	Conditions	Product(s) and Yield(s) (%)	Refs.
C_8 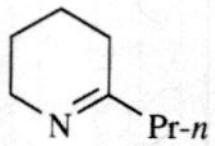	1. $PhSiH_3$ (2 eq), *R,R*-**198** (1 mol%), pyrrolidine (0.1 eq), MeOH (0.1 eq), THF, rt, 6 h 2. $(t\text{-}BuO_2C)_2O$	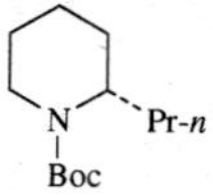 (80) 99% ee	611
	1. $PhSiH_3$ (2 eq), *S,S*-**198** (1 mol%), pyrrolidine (0.1 eq), MeOH (0.1 eq), THF, rt, 6 h 2. $(t\text{-}BuO_2C)_2O$	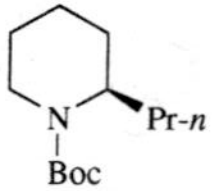 (82) 99% ee	611
C_9 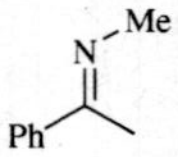	$PhSiH_3$ (4 eq), **198** (1 mol%) rt, 12 h	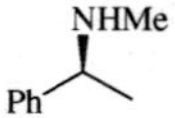 **I** (100) 97% ee	612
	PMHS (10 eq), **198** (10 mol%), rt, 48 h	**I** (50) —% ee	612
	$PhSiH_3$ (4.5 eq), **198** (1 mol%), rt, 12 h	**I** (94) 97% ee	610
	$PhSiH_3$ (4.5 eq), **198** (0.02 mol%), 35°, 12 h	**I** (95) 99% ee	610
	Ph_2SiH_2 (2 eq), **38** (10 mol%), $[(Ph_3P)RuCl_2]_2$ (10 mol%), MeC_6H_5, 0°, 90 h	**I** (51) 73% ee	605
	Ph_2SiH_2 (2 eq), **38** (10 mol%), $[Ir(cod)Cl]_2$ (10 mol%), Et_2O, 0°, 60 h	**I** (56) 89% ee	605
	Ph_2SiH_2 (2 eq), **38** (10 mol%), $[Ir(cod)Cl]_2$ (10 mol%), Et_2O, 0°, 48 h	**I** (24) 16% ee	605

TABLE 34. ASYMMETRIC ORGANOSILANE REDUCTION OF IMINES (*Continued*)

Imine	Conditions	Product(s) and Yield(s) (%)	Refs.
C_9 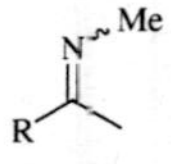	$PhSiH_3$ (4.5 eq), **198** (x mol%), 35°, 12 h	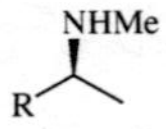 R / x / % ee: c-C_6H_{11} 0.02 (96) 92; 4-ClC_6H_4 0.1 (87) 98	610
C_{10} 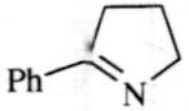	$PhSiH_3$ (4.5 eq), **198** (0.1 mol%), 35°, 12 h	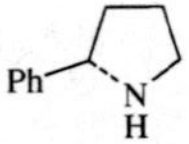 (96) 98% ee	610
	$PhSiH_3$ (2.5-3 eq), **198**, H_2 (80 psig), 65°, 7 h	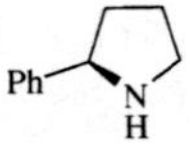 (86) 99% ee	613
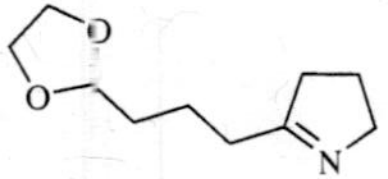	$PhSiH_3$ (2.5-3 eq), **198**, H_2 (80 psig), 65°, 16 h	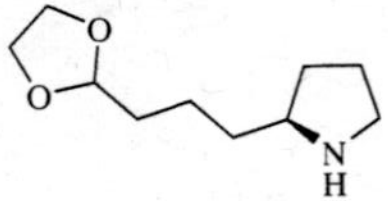 (82) 99% ee	613
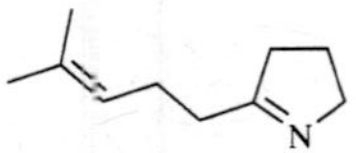	$PhSiH_3$ (2.5-3 eq), **198**, H_2 (80 psig), 50°, 23 h	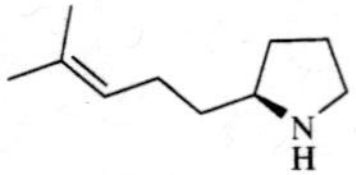 (79) 99% ee	613
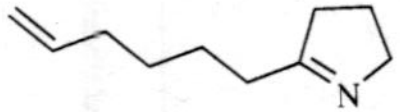	$PhSiH_3$ (2.5-3 eq), **198**, H_2 (80 psig), 45°, 23 h	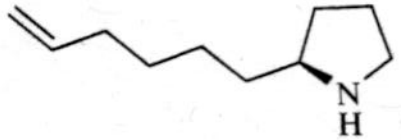 (72) 99% ee	613

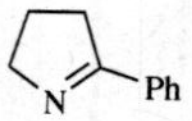

Ph_2SiH_2 (2 eq), **38** (10 mol%),
metal complex (10 mol%),
MeC_6H_5, 0°

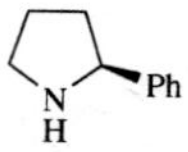

I

605

Metal Complex	Time		% ee
$(Ph_3P)_3RuCl_2$	40 h	(60)	88
$[Ir(cod)Cl]_2$	20 h	(>95)	85
$[Rh(cod)Cl]_2$	20 h	(75)	34

Ph_2SiH_2 (2 eq),
metal complex (10 mol%),
ligand (5 mol%), Et_2O, 0°

I

605

Metal Complex	Ligand	Time		% ee
$[Ir(cod)Cl]_2$	**37**	20 h	(>95)	85
$[Ir(cod)Cl]_2$	**37**	60 h	(>95)	88
$[Rh(cod)Cl]_2$	**37**	20 h	(75)	34
$[Ir(cod)Cl]_2$	**35**	50 h	(78)	88
$[Ir(cod)Cl]_2$	**36**	50 h	trace	—
$[Ir(cod)Cl]_2$	**38**	30 h	(>95)	71
$[Rh(cod)Cl]_2$	**38**	30 h	(17)	32
$[Ir(cod)Cl]_2$	**41**	40 h	(>95)	86

TABLE 34. ASYMMETRIC ORGANOSILANE REDUCTION OF IMINES (*Continued*)

Imine	Conditions	Product(s) and Yield(s) (%)	Refs.
C_{10-13} 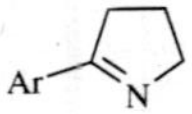	1. Ph_2SiH_2 (2 eq), $[Rh(cod)Cl]_2$ (2 mol%), (−)-DIOP (x mol%) 2. $(CF_3CO)_2O$	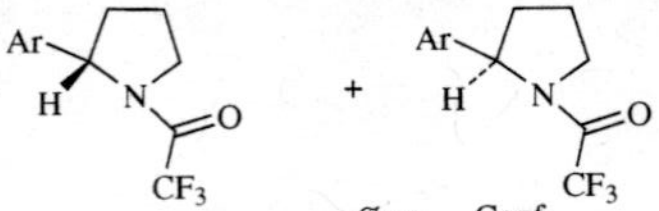	606

Ar	x	Temp	Time	Yield	% ee	Conf.
Ph	2.2	0-20°	32 h	(85)	59	R
Ph	2.6	0-20°	32 h	(83)	59	R
Ph	2.2	0°	69 h	(84)	61	R
Ph	2.2	0-20°	30 h	(84)	64	R
4-BrC_6H_4 [a]	2.0	0-20°	24 h	(82)	60	R
2-$MeOC_6H_4$	2.0	0-20°	48 h	(85)	31	R
3,4-$(MeO)_2C_6H_3$	2.24	0-20°	48 h	(87)	33	R
3,4-$(MeO)_2C_6H_3$ [a]	2.24	0-20°	20 h	(89)	33	S
3,4,5-$(MeO)_3C_6H_2$	2.0	0-20°	72 h	(81)	31	R

Imine	Conditions	Product(s) and Yield(s) (%)	Refs.
C_{11} 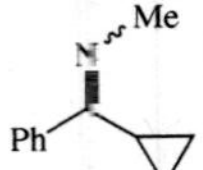	$PhSiH_3$ (4.5 eq), **198** (1 mol%), rt, 12 h	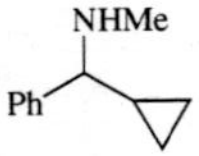 (91) 97% ee	610
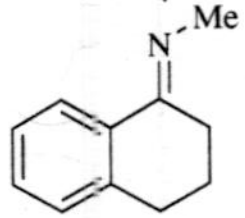	$PhSiH_3$ (4.5 eq), **198** (1 mol%), rt, 12 h	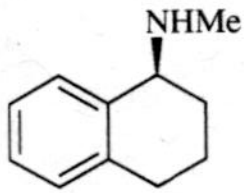 (80) 96% ee	610
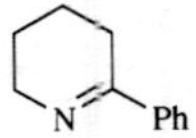	Ph_2SiH_2 (2 eq), **38** (10 mol%), $(Ph_3P)RuCl_2$ (10 mol%), MeC_6H_5, 0°, 50 h	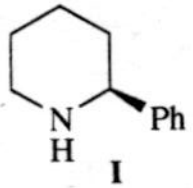 **I** (10) 25% ee	605
	Ph_2SiH_2 (2 eq), **38** (10 mol%), $[Ir(cod)Cl]_2$ (10 mol%), Et_2O, 0°, 100 h	**I** (18) 7% ee	605
	$PhSiH_3$ (2.5-3 eq), **198**, H_2 (500 psig), 65°, 24 h	**I** (78) 98% ee	613

	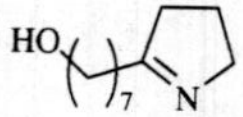	PhSiH$_3$ (1.1 eq), **198**, H$_2$ (500 psig), 65°, 8 h	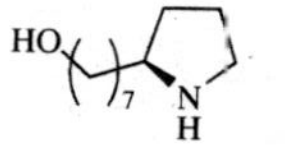 (84) 99% ee	613
	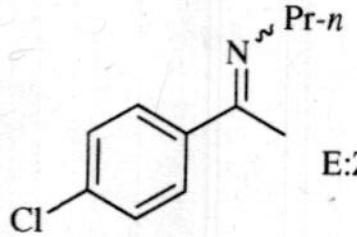 E:Z = 15:1	PMHS (9 eq), **198** (0.5 mol%), *i*-BuNH$_2$ (1.5 eq), 65°, 2.5 h	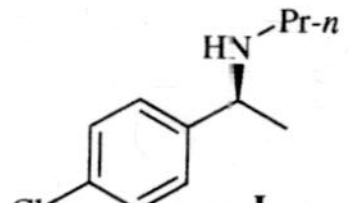 **I** (95) 98% ee	612
	E:Z = 20:1	PhSiH$_3$ (9 eq), **198** (2 mol%), *s*-BuNH$_2$ (1.5 eq), 65°, 2.5 h	**I** (73) 93% ee	612
		PhSiH$_3$ (9 eq), **198** (0.05 mol%), *i*-BuNH$_2$ (1.5 eq), 65°, 2.5 h	**I** (95) 98% ee	612
C_{12}	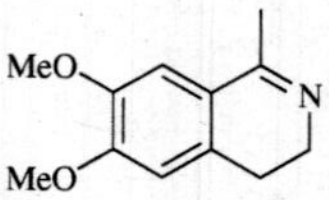	Ph$_2$SiH$_2$ (2 eq), [Rh(C$_2$H$_4$)$_2$]$_2$ (2 mol%), (+)-DIOP, C$_6$H$_6$, 1°	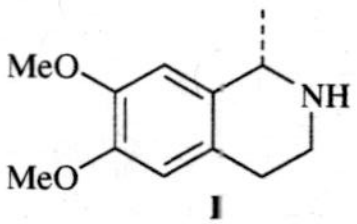 **I** (93) 5.7% ee	607
		PhSiH$_3$ (2.5-3 eq), **198**, H$_2$ (80 psig), 65°, 50 h	**I** (79) 9% ee	613
	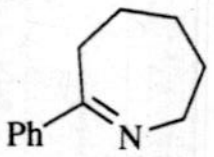	PhSiH$_3$ (2.5-3 eq), **198**, H$_2$ (80 psig), 65°, 30 h	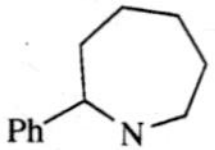 (74) 97% ee	613

TABLE 34. ASYMMETRIC ORGANOSILANE REDUCTION OF IMINES (*Continued*)

Imine	Conditions	Product(s) and Yield(s) (%)	Refs.
C_{12-18} 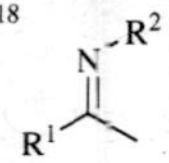	Cl_3SiH (1.5 eq), **162** (10 mol%), rt, 16 h	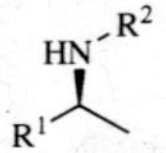	808

R^1	R^2	Solvent	Yield	% ee
Ph	*n*-Bu	$CHCl_3$	(60)	<5
Ph	Ph	CH_2Cl_2	(68)	79
Ph	Ph	$CHCl_3$	(79)	86
Ph	Ph	MeCN	(65)	30
4-$O_2NC_6H_4$	Ph	$CHCl_3$	(30)	85
c-C_6H_{11}	Ph	$CHCl_3$	(80)	37
Ph	*c*-C_6H_{11}	$CHCl_3$	(50)	<5
4-$MeOC_6H_4$	Ph	CH_2Cl_2	(62)	76
4-$MeOC_6H_4$	Ph	$CHCl_3$	(57)	80
Ph	4-$MeOC_6H_4$	$CHCl_3$	(96)	85
Ph	2-$MeOC_6H_4$	CH_2Cl_2	(36)	22
Ph	Bn	$CHCl_3$	(46)	8
4-$CF_3C_6H_4$	Ph	$CHCl_3$	(43)	87
2-$C_{10}H_7$	Ph	CH_2Cl_2	(69)	80
2-$C_{10}H_7$	Ph	$CHCl_3$	(50)	87

Imine	Conditions	Product(s) and Yield(s) (%)	Refs.
C_{13} 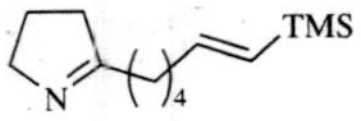	$PhSiH_3$ (2.5-3 eq), **198**, H_2 (80 psig), 50°, 27 h	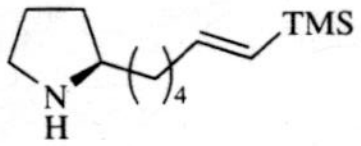 (73) 99% ee	613
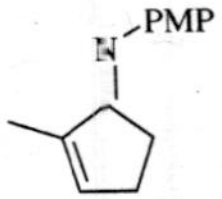	1. PMHS, **198** 2. Add imine 3. *i*-$BuNH_2$, slow addition, 60°	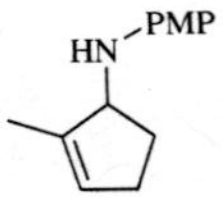 (90) 98% ee	809

	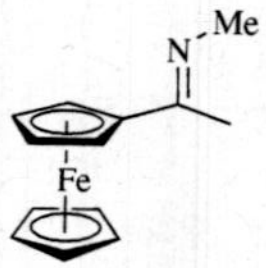	$PhSiH_3$ (4.5 eq), **198** (2.5 mol%), rt, 12 h	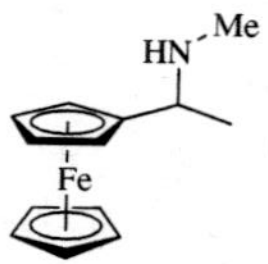 (89) 86% ee (+)	610
	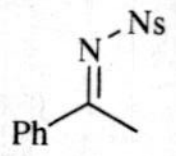	$(MeO)_3SiH$ (2 eq), LiHMDS (40 mol%), BINOL (20 mol%), THF, additive (2 eq), −20°, 48 h	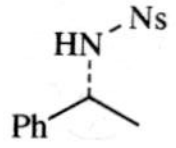 (35) 72% ee	608
C_{14}	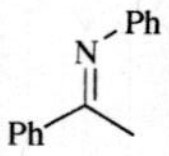	Ph_2SiH_2 (2 eq), **37** (10 mol%), $[Ir(cod)Cl]_2$ (10 mol%), Et_2O, 0°, 60 h	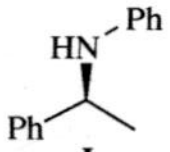 **I** (51) 73% ee	605
		1. PMHS, **198** 2. Add imine 3. *i*-$BuNH_2$, slow addition, 60°	**I** (100) 13% ee	809
	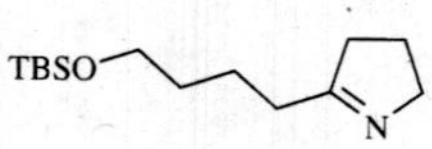	$PhSiH_3$ (1.1 eq), **198**, H_2 (80 psig), 65°, 10 h	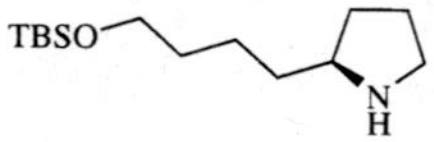 (82) 99% ee	613
C_{14-15}	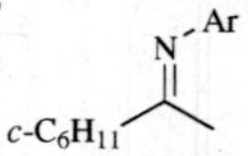	1. PMHS, **198** 2. Add imine 3. *i*-$BuNH_2$, slow addition, 60°	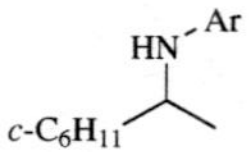	809

Ar		% ee
Ph	(63)	99
2-ClC_6H_4	(0)	—
4-MeC_6H_4	(79)	99
4-$MeOC_6H_4$	(79)	99
2-$MeOC_6H_4$	(0)	—

Imine	Conditions	Product(s) and Yield(s) (%)	Refs.
C_{14-15} 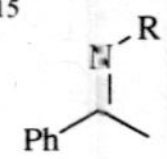	Ph_2SiH_2 (1.5 eq), $[Rh(cod)Cl]_2$ (2.5 mol%), **29** (5 mol%)	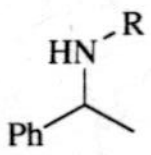	789

R	Solvent	Temp	Time	(Yield)	% ee
Ph	C_6H_6	rt	24 h	(54)	0
Ph	Et_2O	0°	140 h	(28)	53
Ph	THF	0°	140 h	(37)	18
Bn	Et_2O	0°	45 h	(52)	11
Bn	THF	0°	45 h	(47)	7

Imine	Conditions	Product(s) and Yield(s) (%)	Refs.
C_{14-16} 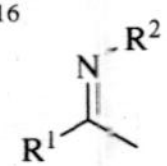	Ph_2SiH_2 (2 eq), $[Rh(C_2H_4)_2]_2$ (2 mol%), (+)-DIOP, C_6H_6	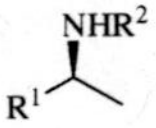	607

R^1	R^2	Temp	(Yield)	% ee
Ph	Ph	5°	(90)	47
Ph	Bn	24°	(98)	50
Ph	Bn	2°	(97)	65
Bn	Bn	24°	(40)	11.5

Imine	Conditions	Product(s) and Yield(s) (%)	Refs.
C_{14-18} 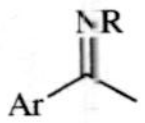	Cl_3SiH (1.5 eq), (*S*)-phenyl-*N*-formylprolinamide (0.1-0.2 eq), CH_2Cl_2, rt, 24 h	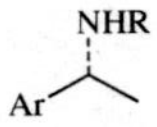	609

Ar	R	(Yield)	% ee	Conf.
Ph	Ph	(91)	55	R
4-$O_2NC_6H_4$	Ph	(>99)	49	—
4-ClC_6H_4	Ph	(95)	54	—
Ph	Bn	(97)	55	R
2-$C_{10}H_7$	Ph	(56)	49	—

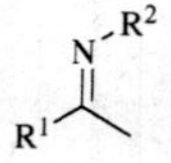

Cl$_3$SiH (1.5 eq),
catalyst (10 mol%), rt, 16 h

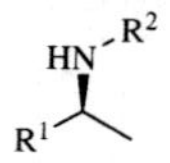

808

R^1	R^2	Catalyst	Solvent		% ee
Ph	Ph	**162**	CHCl$_3$	(49)	92
c-C$_6$H$_{11}$	Ph	**162**	CHCl$_3$	(53)	59
Ph	Ph	**163**	CH$_2$Cl$_2$	(62)	70
Ph	Ph	**163**	CHCl$_3$	(62)	85
2-C$_{10}$H$_7$	Ph	**163**	CH$_2$Cl$_2$	(53)	66
Ph	Ph	**164**	CHCl$_3$	(81)	82
4-CF$_3$C$_6$H$_4$	Ph	**164**	CHCl$_3$	(94)	77
Ph	4-MeOC$_6$H$_4$	**164**	CHCl$_3$	(82)	79
Ph	Ph	**165**	CHCl$_3$	(94)	92
4-MeOC$_6$H$_4$	Ph	**165**	CHCl$_3$	(62)	87
4-CF$_3$C$_6$H$_4$	Ph [c]	**165**	CHCl$_3$	(95)	89
Ph	4-MeOC$_6$H$_4$	**165**	CHCl$_3$	(85)	90
4-MeOC$_6$H$_4$	Ph	**165**	MeC$_6$H$_5$	(86)	85
4-CF$_3$C$_6$H$_4$	Ph	**165**	MeC$_6$H$_5$	(86)	89
Ph	4-MeOC$_6$H$_4$	**165**	MeC$_6$H$_5$	(85)	91
2-MeC$_6$H$_4$	Ph	**165**	MeC$_6$H$_5$	(90)	92
Ph	Ph	**166**	CHCl$_3$	(88)	53
4-CF$_3$C$_6$H$_4$	Ph	**166**	CHCl$_3$	(92)	69
Ph	Ph	**167**	CHCl$_3$	(35)	56
Ph	Ph	**168**	CHCl$_3$	(0)	—
Ph	Ph	**169**	CHCl$_3$	(23)	7

TABLE 34. ASYMMETRIC ORGANOSILANE REDUCTION OF IMINES (*Continued*)

Imine	Conditions	Product(s) and Yield(s) (%)	Refs.
C_{15}			
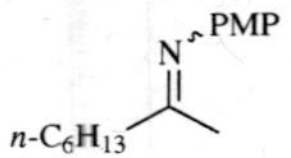	1. PMHS, **198** 2. Add imine 3. *i*-$BuNH_2$, slow addition, 60°	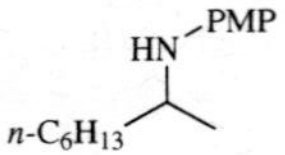 (70) 88% ee [b]	809
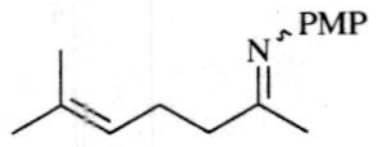	1. PMHS, **198** 2. Add imine 3. *i*-$BuNH_2$, slow addition, 60°	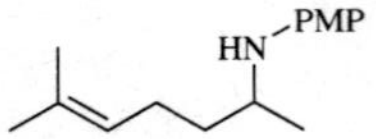 (68) 90% ee [b]	809
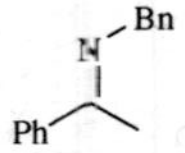	$PHSiH_3$ (10 eq), **198** (10 mol%), rt, 96 h	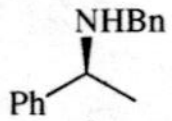 **I** (55) 47% ee	612
	PMHS (9 eq), **198** (0.5 mol%), *i*-$BuNH_2$ (1.5 eq), 65°, 2.5 h	**I** (95) 98% ee	612
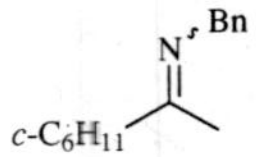	PMHS (9 eq), **198** (0.5 mol%), *i*-$BuNH_2$ (1.5 eq), 65°, 2.5 h	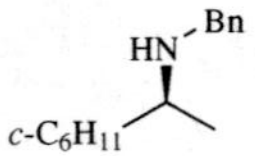 (96) 91% ee	612
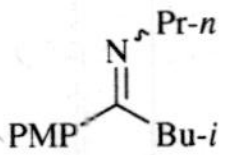	$PhSiH_3$ (9 eq), **198** (0.5 mol%), *i*-$BuNH_2$ (1.5 eq), 65°, 2.5 h	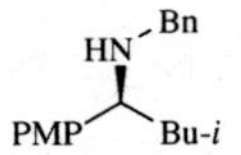 (90) 97% ee	612
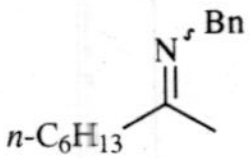	PMHS (9 eq), **198** (1 mol%), *i*-$BuNH_2$ (1.5 eq), 65°, 2.5 h	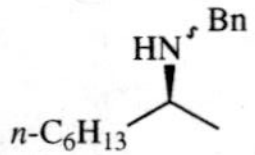 (96) 69% ee	612
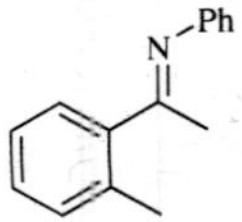	1. PMHS, (EBTHI)TiF_2 2. Add imine 3. *i*-$BuNH_2$, slow addition, 60°	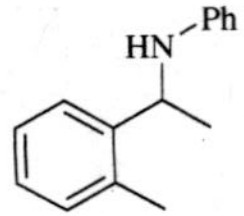 (28) 9% ee [b]	809
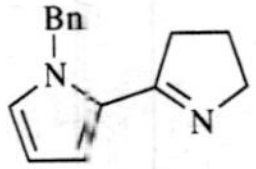	$PhSiH_3$ (2.5-3 eq), **198**, H_2 (80 psig), rt, 24 h	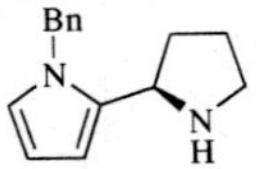 (79) 95% ee	613

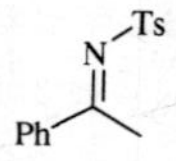

(MeO)$_3$SiH (2 eq), n-BuLi (40 mol%), catalyst (20 mol%), THF, rt, 48 h

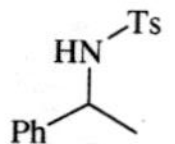

I

608

Catalyst		% ee	Conf.
(*S*)-isoleucinol	(56)	13	S
(*S*)-valinol	(28)	13	S
(*S*)-phenylalaninol	(45)	4	S
(*S*)-proline	(48)	1	S
(*R*,*R*)-2,3-butanediol	(60)	28	R
(*R*,*R*)-hydrobenzoin	(37)	32	R
TADDOL	(68)	2	R
(*R*)-BINOL	(53)	57	R
221	(37)	21	R
222	(27)	2	R
(*R*)-1-phenylethanol	(49)	0	—

(MeO)$_3$SiH (2 eq), base (40 mol%), **221** (20 mol%), THF, additive (2 eq), rt, 48 h

I

608

Base	Solvent	Additive	Temp		% ee	Conf.
n-BuLi	THF	none	rt	(53)	57	R
LiHMDS	THF	none	rt	(38)	54	R
NaHMDS	THF	none	rt	trace	—	—
KHMDS	THF	none	rt	trace	—	—
n-BuLi	THF	TMEDA	rt	(52)	51	R
n-BuLi	THF	none	0°	(43)	65	R
n-BuLi	Et_2O	none	rt	(30)	12	S
LiHMDS	Et_2O	none	rt	(65)	31	S
n-BuLi	Et_2O	TMEDA	rt	(63)	59	R
n-BuLi	$(i\text{-}Pr)_2O$	none	rt	(28)	6	S
n-BuLi	MeC_6H_5	none	rt	(4)	11	S

TABLE 34. ASYMMETRIC ORGANOSILANE REDUCTION OF IMINES (*Continued*)

Imine	Conditions	Product(s) and Yield(s) (%)	Refs.
C_{16-19}			
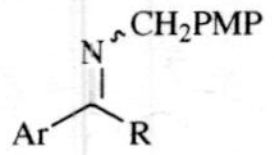	$PhSiH_3$ (9 eq), **198**, (0.5 mol%), *i*-$BuNH_2$ (1.5 eq), 65°, 2.5 h	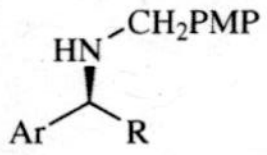 Ar / R / % ee: 4-ClC_6H_4 / Me / (92) 99; Ph / *i*-Bu / (86) 93	612
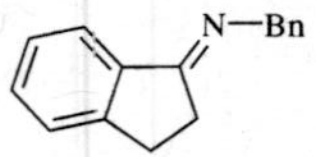	PMHS (10 eq), **198**, (1 mol%), amine, rt	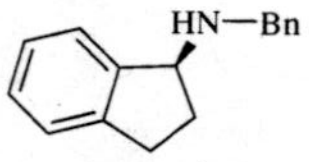	612
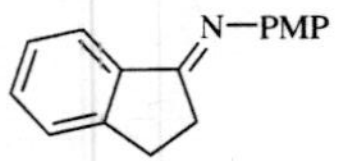	1. PMHS, **198** 2. Add imine 3. *i*-$BuNH_2$, slow addition, 60°	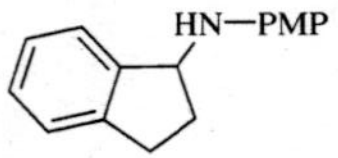 (90) 19% ee [b]	809
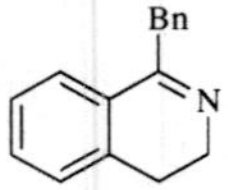	Ph_2SiH_2 (2 eq), $[Rh(C_2H_4)_2]_2$ (2 mol%), (+)-DIOP, C_6H_6, 1°	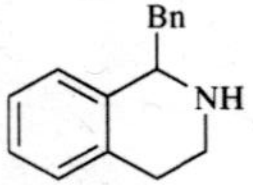 (78) 22.5% ee	607
C_{18}			
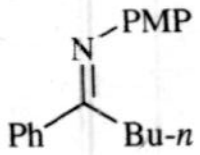	1. PMHS, **198** 2. Add imine 3. *i*-$BuNH_2$, slow addition, 60°	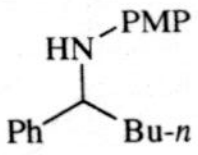 (25) 6% ee [b]	809

Amine	Time	% ee	% Conv.
none	24 h	—	5
n-$C_6H_{13}NH_2$	2 h	85	100
t-$BuNH_2$	24 h	—	39
pyrrolidine	24 h	—	4
i-$BuNH_2$	2 h	92	100
s-$BuNH_2$	2 h	78	75

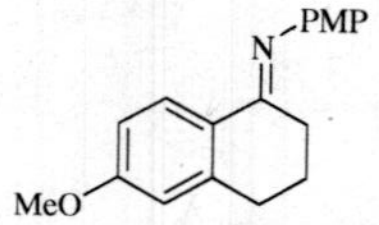	1. PMHS, **198** 2. Add imine 3. i-$BuNH_2$, slow addition, 60°	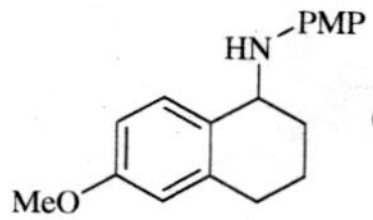 (63) 75% ee [b]	809
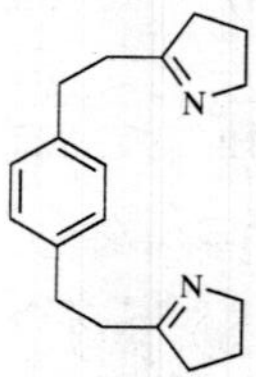	$PhSiH_3$ (4.5 eq), **198** (2 mol%), rt, 12 h	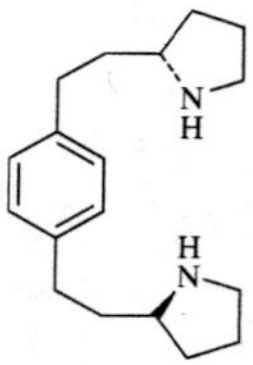 (64) 98% ee	610
C_{19}			
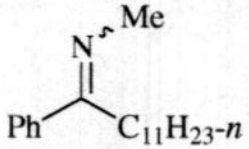	$PhSiH_3$ (4.5 eq), **198** (2.5 mol %), 50°, 12 h	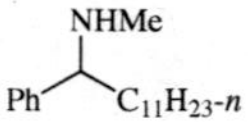 (88) 9% ee	610
C_{20}			
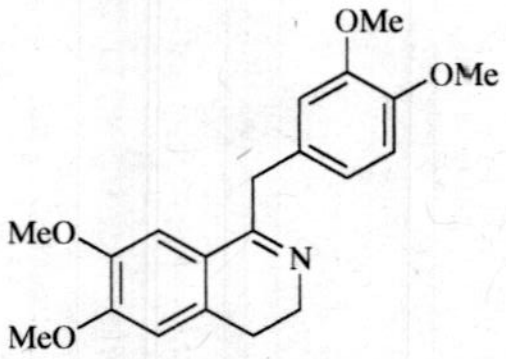	Ph_2SiH_2 (2 eq), $[Rh(C_2H_4)_2]_2$ (2 mol%), (+)-DIOP, C_6H_6, 1°	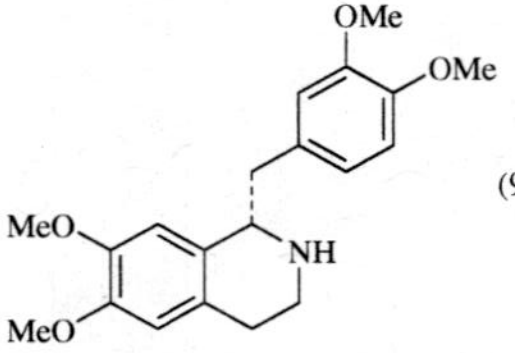 (98) 38.7% ee	607
C_{24-25}			
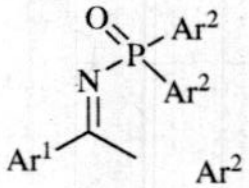 Ar^2 = 3,5-$Me_2C_6H_3$	$(HMe_2Si)_2O$ (3 eq), CuCl (1 mol%), NaOMe (1 mol%), **124** (1 mol%), t-BuOH (3 eq), MeC_6H_5, rt, 17 h	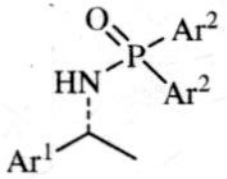	614

Ar^1		% ee
Ph	(95)	95.3
4-$MeOC_6H_4$	(87)	93.5

TABLE 34. ASYMMETRIC ORGANOSILANE REDUCTION OF IMINES (*Continued*)

Imine	Conditions	Product(s) and Yield(s) (%)	Refs.
$C_{24\text{-}26}$ 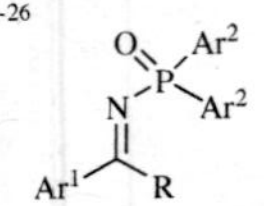 Ar^2 = 3,5-$Me_2C_6H_3$	$(HMe_2Si)_2O$ (3 eq), CuCl (6 mol%), NaOMe (6 mol%), **124** (6 mol%), *t*-BuOH (3.3 eq), MeC_6H_5, rt, 17 h	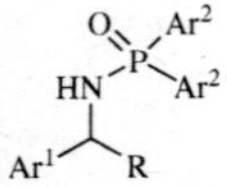	614
Ar^1 / R		/ % ee	
Ph / Me		(94) 99.3	
Ph / Et		(96) 97.5	
Ph / *i*-Pr		(93) 97.3	
4-BrC_6H_4 / Me		(95) 96.1	
4-$MeOC_6H_4$ / Me		(98) 94.2	
4-$CF_3C_6H_4$ / Me		(94) 97.3	
C_{25} 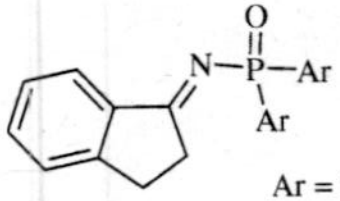 Ar = 3,5-$Me_2C_6H_3$	$(HMe_2Si)_2O$ (3 eq), CuCl (6 mol%), NaOMe (6 mol%), **124** (6 mol%), *t*-BuOH (3.3 eq), MeC_6H_5, rt, 17 h	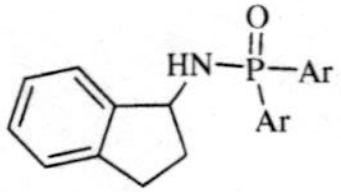 (93) 98.4% ee	614
C_{26} 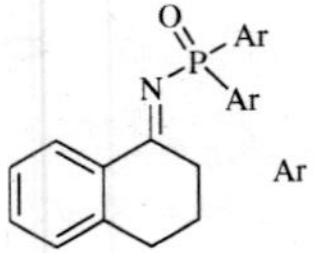 Ar = 3,5-$Me_2C_6H_3$	$(HMe_2Si)_2O$ (3 eq), CuCl (6 mol%), NaOMe (6 mol%), **124** (6 mol%), *t*-BuOH (3.3 eq), MeC_6H_5, rt, 17 h	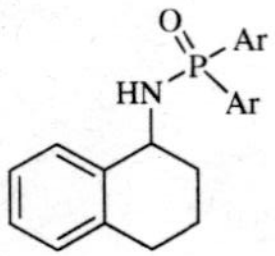 (93) 97.6% ee	614

[a] (+)-DIOP was used as the ligand.

[b] The configuration of the product was not determined.

[c] This reaction was carried out at –20°.

REFERENCES

[1] Fry, J. L.; Orfanopoulos, M.; Adlington, M. G.; Dittman, Jr., W. R.; Silverman, S. B. *J. Org. Chem.* **1978**, *43*, 374.

[2] Kursanov, D. N.; Parnes, Z. N.; Loim, N. M. *Synthesis* **1974**, 633.

[3] Nagai, Y. *Org. Prep. Proced. Int.* **1980**, *12*, 13.

[4] Kursanov, D. N.; Parnes, Z. N.; Kalinkin, M. I.; Loim, N. M. *Ionic Hydrogenation and Related Reactions*; Harwood Academic Publishers: Chur, Switzerland, 1985.

[5] Kalinkin, M. I.; Kolomnikova, G. D.; Parnes, Z. N.; Kursanov, D. N. *Russ. Chem. Rev. (Engl. Transl.)* **1979**, *48*, 332.

[6] Cheng, J. C.; Maioriello, J.; Larsen, J. W. *Energy Fuels* **1989**, *3*, 321.

[7] Nenitzescu, C. D. In *Carbonium Ions,* Olah, G. A.; von R. Schleyer, P., Eds.; Wiley-Interscience: New York, 1970; Vol. 2, pp 463–520.

[8] Singh, S.; Chhina, S.; Sharma, V. K.; Sachdev, S. S. *J. Chem. Soc., Chem. Commun.* **1982**, 453.

[9] Nelson, S. F.; Teasley, M. F. *J. Org. Chem.* **1986**, *51*, 3474.

[10] Lukevics, E. *Russ. Chem. Rev. (Engl. Transl.)* **1977**, 46, 264.

[11] Keinan, E. *Pure Appl. Chem.* **1989**, *61*, 1737.

[12] Corey, J. Y. In *Advances in Silicon Chemistry*, Larson, G. L., Ed.; JAI Press: Greenwich, CT, 1991; Vol. 1, pp 327–387.

[13] Benkeser, R. A. *Acc. Chem. Res.* **1971**, *4*, 94.

[14] Anderson, H. H. *J. Am. Chem. Soc.* **1958**, *80*, 5083.

[15] Yunnikova, L. P.; Pak, V. D.; Kozlov, N. S. *J. Gen. Chem. USSR (Engl. Transl.)* **1975**, *45*, 1368.

[16] Fry, J. L., University of Toledo, unpublished observation.

[17] Chadha, R. K.; Drake, J. E.; Neo, M. K. H. *J. Organomet. Chem.* **1984**, *277*, 47.

[18] Málek, J. *Org. React.* **1985**, *34*, 1.

[19] Málek, J. *Org. React.* **1988**, *36*, 249.

[20] Walsh, R. *Acc. Chem. Res.* **1981**, *14*, 246.

[21] Kanabus-Kaminska, J. M.; Hawari, J. A.; Griller, D.; Chatgilialoglu, C. *J. Am. Chem. Soc.* **1987**, *109*, 5267.

[22] Gronet, S.; Glaser, R.; Streitwieser, A. *J. Am. Chem. Soc.* **1989**, *111*, 3111.

[23] Allred, A. L. *J. Inorg. Nucl. Chem.* **1961**, *17*, 215.

[24] Carey, F. A.; Tremper, H. S. *Tetrahedron Lett.* **1969**, 1645.

[25] Mayr, H.; Basso, N. *Angew. Chem., Int. Ed. Engl.* **1992**, *31*, 1046.

[26] Carey, F. A.; Tremper, H. S. *J. Am. Chem. Soc.* **1968**, *90*, 2578.

[27] Raber, D. J.; Guida, W. C. *Synthesis* **1974**, 803.

[28] Fry J. L.; Ott, R. A. *J. Org. Chem.* **1981**, *46*, 602.

[29] Huszthy, P.; Lempert, K.; Simig, G. *J. Chem. Soc., Perkin Trans. 2* **1985**, 1351.

[30] Kudryavtsev, R. V.; Lyakhovetskii, Y. I.; Parnes, Z. N.; Kursanov, D. N. *J. Org. Chem. USSR (Engl. Transl.)* **1974**, *10*, 919.

[31] Fornarini, S. *J. Org. Chem.* **1988**, *53*, 1314.

[32] Goodloe, G. W.; Lampe, F. W. *J. Am. Chem. Soc.* **1979**, *101*, 5649.

[33] Goodloe, G. W.; Lampe, F. W. *J. Am. Chem. Soc.* **1979**, *101*, 6028.

[34] Corriu, R. J. P.; Henner, M. *J. Organomet. Chem.* **1974**, *74*, 1.

[35] Corriu, R. J. P.; Guérin, C.; Moreau, J. J. E. *Top. Stereochem.* **1984**, *15*, 43.

[36] Corriu, R. J. P.; Guérin, C.; Moreau, J. J. E. In *The Chemistry of Organic Silicon Compounds*, Patai, S., Rappoport, Z., Eds.; Wiley: New York, 1989; pp 305–370.

[37] Lambert, J. B.; Schulz, Jr., W. J. *J. Am. Chem. Soc.* **1983**, *105*, 1671.

[38] Lambert, J. B.; McConnell, J. A.; Schulz, Jr., W. J. *J. Am. Chem. Soc.* **1986**, *108*, 2482.

[39] Lambert, J. B.; McConnell, J. A.; Schilf, W.; Schultz, Jr., W. J. *J. Chem. Soc., Chem. Commun.* **1988**, 455.

[40] Lambert, J. B.; Schilf, W. *J. Am. Chem. Soc.* **1988**, *110*, 6364.

[41] Lambert, J. B.; Schultz, Jr., W. J.; McConnell, J. A.; Schilf, W. In *Silicon Chemistry*, Corey, E. R.; Corey, J. Y.; Gaspar, P. P., Eds.; Halsted Press: New York, 1988; pp 183–190.

[42] Lambert, J. B.; Schulz, Jr., W. J.; McConnell, J. A.; Schilf, W. *J. Am. Chem. Soc.* **1988**, *110*, 2201.

[43] Lambert, J. B.; Kania, L.; Schilf, W.; McConnell, J. A. *Organometallics* **1991**, *10*, 2578.
[44] Kira, M.; Hino, T.; Sakurai, H. *J. Am. Chem. Soc.* **1992**, *114*, 6697.
[45] Lambert, J. B.; Zhang, S. *J. Chem. Soc., Chem. Commun.* **1993**, 383.
[46] Xie, Z.; Liston, D. J.; Jelínek, T.; Mitro, V.; Bau, R.; Reed, C. A. *J. Chem. Soc., Chem. Commun.* **1993**, 384.
[47] Apeloig, Y.; Stanger, A. *J. Am. Chem. Soc.* **1987**, *109*, 272.
[48] Olah, G. A.; Laali, K.; Faroog, O. *Organometallics* **1984**, *3*, 1337.
[49] Prakash, G. K. S.; Keyaniyan, S.; Aniszfeld, R.; Heiliger, L.; Olah, G. A.; Stevens, R. C.; Choi, H. K.; Bau, R. *J. Am. Chem. Soc.* **1987**, *109*, 5123.
[50] Olah, G. A.; Heiliger, L.; Li, X. Y.; Prakash, G. K. S. *J. Am. Chem. Soc.***1990**, *112*, 5991.
[51] Eaborn, C. *J. Organomet. Chem.* **1991**, *405*, 173.
[52] Lambert, J. B.; Zhang, S.; Stern, C. L.; Huffman, J. C. *Science* **1993**, *260*, 1917.
[53] Reed, C. A. *Acc. Chem. Res.* **1998**, *31*, 325.
[54] Kim, K. C.; Reed, C. A.; Elliot, D. W.; Mueller, L. J.; Tham. F.; Lin. L.; Lambert, J. B. *Science* **2002**, *297*, 825.
[55] Gaspar, P. P. *Science* **2002**, *297*, 785.
[56] Robinson, L. R.; Burns, G. T.; Barton, T. J. *J. Am. Chem. Soc.***1985**, *107*, 3935.
[57] Bahr, S. R.; Boudjouk, P. *J. Am. Chem. Soc.* **1993**, *115*, 4514.
[58] Bertrand, G. *Science* **2004**, *305*, 783.
[59] Corey, J. Y.; West, R. *J. Am. Chem. Soc.* **1963**, *85*, 2430.
[60] Austin, J. D.; Eaborn, C. *J. Chem. Soc.* **1964**, 2279.
[61] Sommer, L. H.; Bauman, D. L. *J. Am. Chem. Soc.* **1969**, *91*, 7045.
[62] Bulkowski, J. E.; Stacy, R.; Van Dyke, C. H. *J. Organomet. Chem.* **1975**, *87*, 137.
[63] Chojnowski, J.; Mazurek, M.; Wilczek, L. *Analyst (London)* **1976**, *101*, 286.
[64] Badejo, I. T.; Karaman, R.; Fry, J. L. *J. Org. Chem.* **1989**, *54*, 4591.
[65] Carey, F. A.; Hsu, C.-L. W. *J. Organomet. Chem.* **1969**, *19*, 29.
[66] Mayr, H.; Basso, N.; Hagen, G. *J. Am. Chem. Soc.* **1992**, *114*, 3060.
[67] Chojnowski, J.; Wilczek, L.; Fortuniak, W. *J. Organomet. Chem.* **1977**, *135*, 13.
[68] Chojnowski, J.; Fortuniak, W.; Stanczyk, W. *J. Am. Chem. Soc.* **1987**, *109*, 7776.
[69] Beletskaya, N. P.; Rykov, S. V.; Vol'eva, V. B.; Buchachenko, A. L.; Kessenikh, A. V. *Bull. Acad. Sci. USSR, Div. Chem. Sci. (Engl. Transl.)* **1972**, *21*, 653.
[70] Kursanov, D. N.; Parnes, Z. N. *Russ. Chem. Rev. (Engl. Transl.)* **1969**, *38*, 812.
[71] Kursanov, D. N.; Parnes, Z. N.; Loim, N. M. *Izv. Akad. Nauk SSSR, Ser. Khim.* **1966**, 1289.
[72] Doyle, D. P.; West, C. T. *J. Org. Chem.* **1975**, *40*, 3821.
[73] West, C. T.; Donnelley, S. J.; Kooistra, D. A.; Doyle, M. P. *J. Org. Chem.* **1973**, *38*, 2675.
[74] Doyle, M. P.; West, C. T.; Donnelly, S. J.; McOsker, C. C. *J. Organomet. Chem.* **1976**, *117*, 129.
[75] Fry, J. L.; Silverman, S. B.; Orfanopoulos, M. *Org. Synth.* **1981**, *60*, 108.
[76] Deneux, M.; Akhrem, I. C.; Avetisyan, D. V.; Mysof, E. I.; Vol'pin, M. E. *Bull. Soc. Chim. Fr.* **1973**, 2638.
[77] Akhrem, I. S.; Deneux, M.; Vol'pin, M. E. *Izv. Akad. Nauk, SSSR, Ser. Khim.* **1973**, 932; *Chem. Abstr.* **1973**, *78*, 3844.
[78] Boyer, J.; Corriu, R. J. P.; Perz, R.; Reye, C. *J. Organomet. Chem.* **1978**, *148*, C1.
[79] Boyer, J.; Corriu, R. J. P.; Perz, R.; Reye, C. *J. Chem. Soc., Chem. Commun.* **1981**, 121.
[80] Boyer, J.; Corriu, R. J. P.; Perz, R.; Reye, C. *Tetrahedron* **1981**, *37*, 2165.
[81] Boyer, J.; Corriu, R. J. P.; Perz, R.; Poirier, M.; Reye, C. *Synthesis* **1981**, 558.
[82] Chuit, C.; Corriu, R. J. P.; Perz, R.; Reyé, C. *Synthesis* **1982**, 981.
[83] Corriu, R. J. P.; Perz, R.; Reye, C. *Tetrahedron* **1983**, *39*, 999.
[84] Boyer, J.; Breliere, C.; Corriu, R. J. P.; Kpoton, A.; Poirier, M.; Royo, G. *J. Organomet. Chem.* **1986**, *311*, C39.
[85] Fry, J. L.; McAdam, M. A. *Tetrahedron Lett.* **1984**, *25*, 5859.
[86] Fujita, M.; Hiyama, T. *J. Am. Chem. Soc.* **1984**, *106*, 4629.
[87] Fujita, M.; Hiyama, T. *J. Am. Chem. Soc.* **1985**, *107*, 8294.
[88] Fujita, M.; Hiyama, T. *Tetrahedron Lett.* **1987**, *28*, 2263.
[89] Yang, D.; Tanner, D. D. *J. Org. Chem.* **1986**, *51*, 2267.

[90] Kitazume, T.; Kobayashi, T.; Yamamoto, T.; Yamazaki, T. *J. Org. Chem.* **1987**, *52*, 3218.
[91] Hosomi, A.; Hayashida, H.; Kohra, S.; Tominaga, Y. *J. Chem. Soc., Chem. Commun.* **1986**, 1411.
[92] Kohra, S.; Hayashida, H.; Tominaga, Y.; Hosomi, A. *Tetrahedron Lett.* **1988**, *29*, 89.
[93] Kira, M.; Sato, K.; Sakurai, H. *J. Org. Chem.* **1987**, *52*, 948.
[94] Furin, G. G.; Vyazankina, O. A.; Gostevsky, B. A.; Vyazankin, N. S. *Tetrahedron* **1988**, *44*, 2675.
[95] Davis, L. P.; Burggraf, L. W.; Gordon, M. S.; Baldridge, K. K. *J. Am. Chem. Soc.* **1985**, *107*, 4415.
[96] Kira, M.; Sato, K.; Sakurai, H. *Chem. Lett.* **1987**, 2243.
[97] Reed, A. E.; Schleyer, P. v. R. *Chem. Phys. Lett.* **1987**, *133*, 553.
[98] Damrauer, R.; Burggraf, L. W.; Davis, L. P.; Gordon, M. S. *J. Am. Chem. Soc.* **1988**, *110*, 6601.
[99] Corriu, R.; Guérin, C.; Henner, B.; Wang, Q. *J. Organomet. Chem.* **1989**, *365*, C7.
[100] Sini, G.; Hiberty, P. C.; Shaik, S. S. *J. Chem. Soc., Chem. Commun.* **1989**, 772.
[101] Chopra, S. K.; Martin, J. C. *J. Am. Chem. Soc.* **1990**, *112*, 5342.
[102] Attar-Boshi, M. T.; Eaborn, C.; Vencl, J.; Walton, D. R. M. *J. Organomet. Chem.* **1976**, *117*, C87.
[103] Bassindale, A. R.; Stout, T. *J. Chem. Soc., Chem. Commun.* **1984**, 1387.
[104] Chuit, C.; Corriu, R. P. J.; Reyé, C. *J. Organomet. Chem.* **1988**, *358*, 57.
[105] Corriu, R. J. P.; Young, J. C. In *The Chemistry of Organic Silicon Compounds*, Patai, S., Rappoport, Z., Eds.; Wiley: New York, 1989; pp 1241–1288.
[106] Perozzi, E. F.; Martin, J. C. *J. Am. Chem. Soc.***1979**, *101*, 1591.
[107] Becker, B.; Corriu, R. J. P.; Guérin, C.; Henner, B.; Wang, Q. *J. Organomet. Chem.* **1989**, *359*, C33.
[108] Brelière, C.; Carré, F.; Corriu, R. J. P.; Royo, G. *Organometallics* **1988**, *7*, 1006.
[109] Corriu, R. J. P.; Lanneau, G. F.; Perrot, M. *Tetrahedron Lett.* **1988**, *29*, 1271.
[110] Holmes, R. R. *Chem. Rev.* **1990**, *90*, 17.
[111] Chuit, C.; Corriu, R. J. P.; Reyé, C.; Young, J. C. *Chem. Rev.* **1993**, *93*, 1371.
[112] DePuy, C. H.; Damrauer, R.; Bowie, J. H.; Sheldon, J. C. *Acc. Chem. Res.* **1987**, *20*, 127.
[113] Brelière, C.; Corriu, R. J. P.; Royo, G.; Wong Chi Man, W. W. C.; Zwecker, J. *Organometallics* **1990**, *9*, 2633.
[114] Corriu, R. J. P. In *Silicon Chemistry*, Corey, E. R., Corey, J. Y., Gaspar, P. P., Eds.; Halsted Press: New York, 1988; pp 225–234.
[115] Parks, D. J.; Piers, W. E. *J. Am. Chem. Soc.* **1996**, *118*, 9440.
[116] Parks, D. J.; Blackwell, J. M.; Piers, W. E. *J. Org. Chem.* **2000**, *65*, 3090.
[117] Roesler, R.; Har, B. J. N.; Piers, W. E. *Organometallics* **2002**, *21*, 4300.
[118] Prakash, G. K. S.; Bae, C.; Rasul, G.; Olah, G. A. *J. Org. Chem.* **2002**, *67*, 1297.
[119] Blackwell, J. M.; Morrison, D. J.; Piers, W. E. *Tetrahedron* **2002**, *58*, 8247.
[120] Blackwell, J. M.; Sonmor, E. R.; Scoccitti, T.; Piers, W. E. *Org. Lett.* **2000**, *2*, 3921.
[121] Ojima, I.; Clos, N.; Bastos, C. *Tetrahedron* **1989**, *45*, 6901.
[122] Ghosh, A. K.; Mathivanan, P.; Cappiello, J. *Tetrahedron: Asymmetry* **1998**, *9*, 1.
[123] Harrod, J. F. *Coord. Chem. Rev.* **2000**, *206–207*, 493.
[124] Reyes, C.; Prock, A.; Giering, W. P. *Organometallics* **2002**, *21*, 546.
[125] Riant, O.; Mostefaï, N.; Courmarcel, J. *Synthesis* **2004**, 2943.
[126] Adlington, M. G.; Orfanopoulos, M.; Fry, J. L. *Tetrahedron Lett.* **1976**, 2955.
[127] Carey, F. A.; Tremper, H. S. *J. Org. Chem.* **1971**, *36*, 758.
[128] Barclay, L. R. C.; Sonawane, H. R.; MacDonald, M. C. *Can. J. Chem.* **1972**, *50*, 281.
[129] Fry, J. L. U.S. Patent 4,130,574 (1978).
[130] Carey, F. A.; Tremper, H. S. *J. Am. Chem. Soc.* **1969**, *91*, 2967.
[131] Fort, Jr., R. C.; Hornish, R. E.; Liang, G. A. *J. Am. Chem. Soc.* **1970**, *92*, 7558.
[132] Ghatak, K.; Ganter, C. *Helv. Chim. Acta* **1988**, *71*, 124.
[133] Smonou, I.; Orfanopoulos, M. *Tetrahedron Lett.* **1988**, *29*, 5793.
[134] Kursanov, D. N.; Parnes, Z. N.; Tsyryapkin, V. A.; Knyazeva, Z. V.; Yunosheva, N. I. *Dokl. Akad. Nauk USSR (Engl. Transl.)* **1972**, *205*, 562.
[135] Olah, G. A.; Wang, Q.; Prakash, G. K. S. *Synlett* **1992**, 647.
[136] Kursanov, D. N.; Bolestova, G. I.; Ibatullin, U. G.; Kuramshina, E. A.; Parnes, Z. N. *J. Org. Chem. USSR (Engl. Transl.)* **1985**, *21*, 2078.

[137] Orfanopoulos, M.; Smonou, I. *Synth. Commun.* **1988**, *18*, 833.
[138] Hanaoka, M.; Yoshida, S.; Mukai, C. *Tetrahedron Lett.* **1985**, *26*, 5163.
[139] Benkeser, R. A.; Schroeder, W.; Thomas, O. H. *J. Am. Chem. Soc.* **1958**, *80*, 2283.
[140] Badejo, I. T.; Choi, H.; Hockensmith C. M.; Karaman, R.; Pinkerton, A. A.; Fry, J. L. *J. Org. Chem.* **1991**, *56*, 4688.
[141] Fry, J. L. *J. Am. Chem. Soc.* **1971**, *93*, 3558.
[142] Fry, J. L.; Adlington, M. G. *J. Am. Chem. Soc.* **1978**, *100*, 7641.
[143] Fry, J. L.; Adlington, M. G.; Dittman, Jr., W. R.; Erickson, G. W.; Orfanopoulos, M.; Silverman, S. B., University of Toledo, unpublished observation.
[144] Gevorgyan, V.; Liu, J.-X.; Rubin, M.; Benson, S.; Yamamoto, Y. *Tetrahedron Lett.* **1999**, *40*, 8919.
[145] Gevorgyan, V.; Rubin, M.; Benson, S.; Liu, J.-X.; Yamamoto, Y. *J. Org. Chem.* **2000**, *65*, 6179.
[146] Kursanov, D. N.; Parnes, Z. N.; Lyakhovetskii, Y. I.; Bolestova, G. I. *Bull. Acad. Sci. USSR, Div. Chem. Sci. (Engl. Transl.)* **1984**, *33*, 800.
[147] Olah, G. A.; Bollinger, J. M.; Cupas, C. A.; Lukas, J. *J. Am. Chem. Soc.* **1967**, *89*, 2692.
[148] Vancik, H.; Sunko, D. E. *J. Am. Chem. Soc.* **1989**, *111*, 3742.
[149] Woodworth, C. W.; Buss, V.; Schleyer, P. v. R. *J. Chem. Soc., Chem. Commun.* **1968**, 569.
[150] Fry, J. L.; Engler, E. M.; Schleyer, P. v. R. *J. Am. Chem. Soc.* **1972**, *94*, 4628.
[151] Fry, J. L.; Saba, J. A. *Tetrahedron Lett.* **1982**, *23*, 1743.
[152] Saba, J. A.; Fry, J. L. *J. Am. Chem. Soc.* **1983**, *105*, 533.
[153] Duddeck, H.; Rosenbaum, D. *J. Org. Chem.* **1991**, *56*, 1707.
[154] Choi, H.; Pinkerton, A. A.; Fry, J. L. *J. Chem. Soc., Chem. Commun.* **1987**, 225.
[155] Badejo, I. T., Ph.D. Dissertation, The University of Toledo, 1989.
[156] Badejo, I. T.; Choi, H.; Fry, J. L. *Tetrahedron Lett.* **1988**, *29*, 4787.
[157] Sundberg, R. J.; Hamilton, G. S.; Laurino, J. P. *J. Org. Chem.* **1988**, *53*, 976.
[158] Lomas, J. S.; Vaissermann, J. *J. Chem. Soc., Perkin Trans. 2* **1999**, 1639.
[159] Lomas, J. S.; Vaissermann, J. *J. Chem. Soc., Perkin Trans. 2* **1997**, 2589.
[160] Lomas, J. S.; Lacroix, J.-C.; Vaissermann, J. *J. Chem. Soc., Perkin Trans. 2* **1999**, 2001.
[161] Lomas, J. S.; Vouthier, E. *J. Chem. Soc., Perkin Trans. 2* **2000**, 417.
[162] Kraus, G. A.; Molina, M. T.; Walling, J. A. *J. Chem. Soc., Chem. Commun.* **1986**, 1568.
[163] Fallahpour, R. A.; Hansen, H.-J. *Helv. Chim. Acta* **1992**, *75*, 2210.
[164] Carey, F. A.; Tremper, H. S. *J. Org. Chem.* **1969**, *34*, 4.
[165] Bell, H. M.; Brown, H. C. *J. Am. Chem. Soc.* **1964**, *86*, 5007.
[166] Tsyryapkin, V. A.; Loim, N. M.; Parnes, Z. N.; Kursanov, D. N. *J. Org. Chem. USSR (Engl. Transl.)* **1972**, *8*, 2390.
[167] Kamikawa, T.; Kubo, I. *Synthesis* **1986**, 431.
[168] Baer, H. H.; Zamkanei, M. *J. Org. Chem.* **1988**, *53*, 4786.
[169] Tymiak, A. A.; McCormick, T. J.; Unger, S. E. *J. Org. Chem.* **1989**, *54*, 1149.
[170] Smonou, I.; Orfanopoulos, M. *Synth. Commun.* **1990**, *20*, 1387.
[171] Wiggins, J. M. *Synth. Commun.* **1988**, *18*, 741.
[172] Yasuda, M.; Onishi, Y.; Ueba, M.; Miyai, T.; Baba, A. *J. Org. Chem.* **2001**, *66*, 7741.
[173] Wustrow, D. J.; Smith, III, W. J.; Wise, L. D. *Tetrahedron Lett.* **1994**, *35*, 61.
[174] Herz, W.; Tsutsumi, T.; Prasad, J. S. *J. Org. Chem.* **1984**, *49*, 3035.
[175] Pettit, R.; Haynes, L. W. In *Carbonium Ions*, Olah, G. A., Schleyer, P. v. R., Eds.; Wiley-Interscience: New York, 1976; Vol. V, Chapter 37.
[176] Jaouen, G. In *Transition Metal Organometallics in Organic Synthesis*; Alper, H., Ed.; Academic Press: New York, 1978; Vol. II, Chapter 2.
[177] Sneedon, R. P. A. *Organochromium Compounds*, Academic Press: New York, 1975.
[178] Kazakova, L. I.; Loim, N. M.; Perevalova, E. G.; Parnes, Z. N. *J. Org. Chem. USSR (Engl. Transl.)* **1973**, *9*, 2294.
[179] Nesmeyanov, A. N.; Shul'pin, G. B.; Rybinskaya, M. I. *Bull. Acad. Sci. USSR, Div. Chem. Sci. (Engl. Transl.)* **1975**, *24*, 2714.
[180] Kazakova, L. I.; Loim, N. M.; Parnes, Z. N. *J. Org. Chem. USSR (Engl. Transl.)* **1973**, *9*, 1561.
[181] Uemura, M.; Minami, T.; Hayashi, Y. *J. Am. Chem. Soc.* **1987**, *109*, 5277.

[182] Uemura, M.; Kobayashi, T.; Hayashi, Y. *Synthesis* **1986**, 386.

[183] Seyferth, D.; Williams, G. H.; Eschbach, C. S.; Nestle, M. O.; Merola, J. S.; Hallgren, J. E. *J. Am. Chem. Soc.* **1979**, *101*, 4867.

[184] Bashilov, V. V.; Sokolov, V. I.; Reutov, O. A. *Bull. Acad. Sci. USSR, Div. Chem. Sci. (Engl. Transl.)* **1978**, *27*, 935.

[185] Whitmore, F. C.; Pietrusza, E. W.; Sommer, L. H. *J. Am. Chem. Soc.* **1947**, *69*, 2108.

[186] Doyle, M. P.; McOsker, C. C.; West, C. T. *J. Org. Chem.* **1976**, *41*, 1393.

[187] Bolestova, G. I.; Shevchenko, N. V.; Parnes, Z. N.; Vol'pin, M. E. *Zh. Org. Khim.* **1990**, *26*, 2565.

[188] Bolestova, G. I.; Shevchenko, N. V.; Parnes, Z. N.; Vol'pin, M. E. *J. Org. Chem. USSR (Engl. Transl.)* **1992**, *28*, 216.

[189] Parnes, Z. N.; Romanova, V. S.; Vol'pin, M. E. *J. Org. Chem. USSR (Engl. Trans.)* **1988**, *24*, 254.

[190] Dolgov, B. N.; Borisov, S. N.; Voronkov, M. G. *Zh. Obshch. Khim.* **1957**, *27*, 716.

[191] Newcomb, M.; Curran, D. P. *Acc. Chem. Res.* **1988**, *21*, 206.

[192] Bolestova, G. I.; Latypova, F. M.; Parnes, Z. N.; Kursanov, D. N. *Bull. Acad. Sci. USSR, Div. Chem. Sci. (Engl. Transl.)* **1982**, *31*, 1179.

[193] Parnes, Z. N.; Kalinkin, M. I.; Tsyryapkin, V. A.; Khamitova, R. F.; Kursanov, D. N. *Doklady Akad. Nauk SSSR* **1972**, *203*, 600.

[194] Hirano, K.; Fujita, K.; Yorimitsu, H.; Shinokubo, H.; Oshima, K. *Tetrahedron Lett.* **2004**, *45*, 2555.

[195] Boukherroub, R.; Chatgilialoglu, C.; Manuel, G. *Organometallics* **1996**, *15*, 1508.

[196] Keinan, E.; Greenspoon, N. *Israel J. Chem.* **1984**, *24*, 82.

[197] Perez, D.; Greenspoon, N.; Keinan, E. *J. Org. Chem.* **1987**, *52*, 5570.

[198] Hart, D. J.; Cain, P. A.; Evans, D. A. *J. Am. Chem. Soc.* **1978**, *100*, 1548.

[199] Pri-Bar, I.; Buchman, O. *J. Org. Chem.* **1986**, *51*, 734.

[200] Lipshutz, B. H.; Tomioka, T.; Sato, K. *Synlett* **2001**, 970.

[201] Kotsuki, H.; Datta, P. K.; Hayakawa, H.; Suenaga, H. *Synthesis* **1995**, 1348.

[202] Parnes, Z. N.; Zdanovich, V. I.; Kugucheva, E. E.; Basova, G. I.; Kursanov, D. N. *Doklady Akad. Nauk USSR (Engl. Transl.)* **1966**, *166*, 32.

[203] Kursanov, D. N.; Parnes, Z. N.; Bassova, G. I.; Loim, N. M.; Zdanovich, V. I. *Tetrahedron,* **1967**, *23*, 2235.

[204] Doyle, M. P.; McOsker, C. C. *J. Org. Chem.* **1978**, *43*, 693.

[205] Julia, M.; Roy, P. *Tetrahedron* **1986**, *42*, 4991.

[206] Kursanov, D. N.; Parnes, Z. N.; Tsyryapkin, V. A.; Kudryavtsev, R. V. *Dokl. Akad. Nauk SSSR* **1972**, *202*, 874.

[207] Parnes, Z. N.; Bolestova, G. I.; Intyakova, E. I.; Kursanov, D. N. *J. Org. Chem. USSR (Engl. Transl.)* **1973**, *9*, 1726.

[208] Kolomnikova, G. D.; Kalinkin, M. I.; Parnes, Z. N.; Kursanov, D. N. *Bull. Acad. Sci. USSR, Div. Chem. Sci. (Engl. Transl.)* **1981**, *30*, 921.

[209] Kolomnikova, G. D.; Kalinkin, M. I.; Makhova, I. V.; Parnes, Z. N.; Kursanov, D. N. *Bull. Acad. Sci. USSR, Div. Chem. Sci. (Engl. Transl.)* **1982**, *31*, 1677.

[210] Parnes, Z. N.; Bolestova, G. I.; Kursanov, D. N. *J. Org. Chem. USSR (Engl. Transl.)* **1977**, *13*, 434.

[211] Bolestova, G. I.; Parnes, Z. N.; Kursanov, D. N. *J. Org. Chem. USSR (Engl. Transl.)* **1979**, *15*, 1129.

[212] Kalinkin, M. I.; Parnes, Z. N.; Markosyan, S. M.; Ovsyannikova, L. I.; Kursanov, D. N. *Dokl. Akad. Nauk. USSR Chem. (Engl. Transl.)* **1980**, *254*, 439.

[213] Parnes, Z. N.; Lyakhovetsky, Yu. I.; Kalinkin, M. I.; Kursanov, D. N.; Belen'kii, L. I. *Tetrahedron* **1978**, *34*, 1703.

[214] Kursanov, D. N.; Kolomnikova, G. D.; Goloshchapova, S. A.; Kalinkin, M. I.; Parnes, Z. N. *Bull. Acad. Sci. USSR, Div. Chem. Sci. (Engl. Transl.)* **1984**, *33*, 446.

[215] Kolomnikova, G. D.; Kalinkin, M. I.; Goloshchapova, S. A.; Parnes, Z. N.; Kursanov, D. N. *Bull. Acad. Sci. USSR, Div. Chem. Sci. (Engl. Transl.)* **1984**, *33*, 997.

[216] Bullock, R. M.; Rappoli, B. J. *J. Chem. Soc., Chem. Commun.* **1989**, 1447.

[217] Larsen, J. W.; Chang, L. W. *J. Org. Chem.* **1979**, *44*, 1168.

[218] Takeda, T.; Tsuchida, T.; Fujiwara, T. *Chem. Lett.* **1984**, 1219.

[219] Chauhan, B. P. S.; Rathore, J. S.; Bandoo, T. *J. Am. Chem. Soc.* **2004**, *126*, 8493.

[220] Mizra-Aghayan, M.; Boukherroub, R.; Bolourtchian, M.; Hosseini, M. *Tetrahedron Lett.* **2003**, *44*, 4579.

[221] Parnes, Z. N.; Khotimskaya, G. A.; Lyakhovetskii, Y. I.; Petrovskii, P. V. *Bull. Acad. Sci. USSR, Div. Chem. Sci. (Engl. Transl.)* **1971**, *20*, 1463.

[222] Parnes, Z. N.; Khotimskaya, G. A.; Kudryavtsev, R. V.; Lukina, M. Y.; Kursanov, D. N. *Doklady Akad. Nauk USSR (Engl. Transl.)* **1969**, *184*, 66.

[223] Ren, L.; Crudden, C. M. *J. Org. Chem.* **2002**, *67*, 1746.

[224] Parnes, Z. N.; Beilinson, E. Yu.; Kursanov, D. N. *Zh. Org. Khim.* **1970**, *6*, 2569.

[225] McCombie, S. W.; Cox, B.; Lin, S.-I.; Ganguly, A. K.; McPhail, A. T. *Tetrahedron Lett.* **1991**, *32*, 2083.

[226] Serebryakova, T. A.; Chigir', R. N.; Zakharychev, A. V.; Ananchenko, S. N.; Torgov, I. V. *Bull. Acad. Sci. USSR, Div. Chem. Sci. (Engl. Transl.)* **1973**, *22*, 1873.

[227] Takano, S.; Moriya, M.; Ogasawara, K. *Tetrahedron Lett.* **1992**, *33*, 1909.

[228] Fedotova, O. V.; Kriven'ko, A. P.; Kharchenko, V. G. *J. Org. Chem. USSR (Engl. Transl.)* **1978**, *14*, 1655.

[229] Khotimskaya, G. A.; Kudryatsev, R. V.; Mil'vitskaya, E. M.; Platé, A. F.; Parnes, Z. N. *Izv. Akad. Nauk SSSR, Ser. Khim.* **1972**, 1989.

[230] Bolestova, G. I.; Parnes, Z. N.; Belikova, N. A.; Kursanov, D. N. *Bull. Acad. Sci. USSR, Div. Chem. Sci. (Engl. Transl.)* **1979**, *28*, 741.

[231] Kursanov, D. N.; Parnes, Z. N.; Bolestova, G. I. *Dokl. Akad. Nauk. USSR Chem. (Engl. Transl.)* **1968**, *181*, 726.

[232] Parnes, Z. N.; Khotimskaya, G. A.; Lukina, M. Y.; Kursanov, D. N. *Doklady Chem. (Engl. Transl.)* **1968**, *178*, 88.

[233] Parnes, Z. N.; Khotimskaya, G. A.; Kudryavtsev, R. V.; Kursanov, D. N. *Bull. Acad. Sci. USSR, Div. Chem. Sci. (Engl. Transl.)* **1972**, *21*, 854.

[234] Parnes, Z. N.; Bolestova, G. I.; Kursanov, D. N. *Bull. Acad. Sci. USSR, Div. Chem. Sci. (Engl. Transl.)* **1972**, *21*, 1927.

[235] Kursanov, D. N.; Kalinkin, M. I.; Gridchin, S. A.; Shatalov, G. V.; Parnes, Z. N. *Bull. Acad. Sci. USSR, Div. Chem. Sci. (Engl. Transl.)* **1979**, *28*, 746.

[236] Kolomnikova, G. D.; Prikhodchenko, D. Yu.; Gololobov, Yu. G. *Izv. Akad. Nauk SSSR, Ser. Khim. (Engl. Transl.)* **1992**, 1282.

[237] Jacob, G.; Cagniant, D. *C.R. Hebd. Sceances Acad. Sci. Ser. C* **1977**, 999.

[238] Serebryakova, T. A.; Zakharaychev, A. V.; Ananchenko, S. N.; Torgov, I. V. *Bull. Acad. Sci. USSR, Div. Chem. Sci. (Engl. Transl.)* **1972**, *21*, 1633.

[239] Serebryakova, T. A.; Zakharaychev, A. V.; Mal'gina, M. A.; Ananchenko, S. N.; Torgov, I. V. *Bull. Acad. Sci. USSR, Div. Chem. Sci. (Engl. Transl.)* **1973**, *22*, 1872.

[240] Serebryakova, T. A.; Parnes, Z. N.; Zakharychev, A. V.; Ananchenko, S. N.; Torgov, I. V. *Bull. Acad. Sci. USSR, Div. Chem. Sci. (Engl. Transl.)* **1969**, *18*, 662.

[241] Vasiyarov, G. G.; Kalinkin, M. I.; Ananchenko, S. N.; Kolomnikova, G. D.; Torgov, I. V.; Parnes, Z. N.; Kursanov, D. N. *Bull. Acad. Sci. USSR, Div. Chem. Sci. (Engl. Transl.)* **1981**, *30*, 1761.

[242] Serebryakova, T. A.; Zakharaychev, A. V.; Ananchenko, S. N.; Torgov, I. V. *Bull. Acad. Sci. USSR, Div. Chem. Sci. (Engl. Transl.)* **1974**, *23*, 713.

[243] Krasutskii, P. A.; Ambrosienko, N. V.; Rodionov, V. N.; Yurchenko, A. G.; Parnes, Z. N.; Bolestova, G. I. *J. Org. Chem. USSR (Engl. Transl.)* **1985**, *21*, 1333.

[244] Zdanovich, V. I.; Kudryavtsev, R. V.; Kursanov, D. N. *Bull. Acad. Sci. USSR, Div. Chem. Sci. (Engl. Transl.)* **1970**, *19*, 427.

[245] Trost, B. M.; Rise, F. *J. Am. Chem. Soc.* **1987**, *109*, 3161.

[246] Oh, C. H.; Jung, H. H.; Sung, H. R.; Kim, J. D. *Tetrahedron* **2001**, *57*, 1723.

[247] Oh, C. H.; Han, J. W.; Kim, J. S.; Um, S. Y.; Jung, H. H.; Jang, W. H.; Won, H. S. *Tetrahedron Lett.* **2000**, *41*, 8365.

[248] Trost, B. M.; Lee, D. C. *J. Am. Chem. Soc.* **1988**, *110*, 7255.
[249] Oh, C. H.; Park, S. J. *Tetrahedron Lett.* **2003**, *44*, 3785.
[250] Kursanov, D. N.; Khotimskaya, G. A.; Fedin, É. I.; Lukina, M. Y.; Parnes, Z. N. *Bull. Acad. Sci. USSR, Div. Chem. Sci. (Engl. Transl.)* **1969**, 679.
[251] Karakhanov, È. A.; Dem'yanova, E. A.; Bordina, L. N.; Viktorova, E. A. *Dokl. Akad. Nauk SSSR* **1974**, 584.
[252] Anisimov, A. V.; Luzikov, Yu. N.; Nikolaeva, V. M.; Viktorova, E. A. *Zh. Org. Khim.* **1979**, *15*, 172.
[253] Bolestova, G. I.; Korepanov, A. N.; Parnes, Z. N.; Kursanov, D. N. *Izv. Akad. Nauk SSSR, Ser. Khim.* **1974**, 2547.
[254] Parnes, Z. N.; Bolestova, G. I.; Dolgova, S. P.; Udre, V. E.; Voronkov, M. G.; Kursanov, D. N. *Bull. Acad. Sci. USSR, Div. Chem. Sci. (Engl. Transl.)* **1974**, *23*, 1753.
[255] Zav'yalov, S. I.; Dorofeeva, O. V. *Bull. Acad. Sci. USSR, Div. Chem. Sci. (Engl. Transl.)* **1983**, *32*, 631.
[256] Zav'yalov, S. I.; Rodionova, N. A.; Zheleznaya, L. L.; Bolestova, G. I.; Filippov, V. V.; Parnes, Z. N.; Kursanov, D. N. *Bull. Acad. Sci. USSR, Div. Chem. Sci. (Engl. Transl.)* **1975**, *24*, 1533.
[257] Kursanov, D. N.; Parnes, Z. N.; Bolestova, G. I.; Belen'kii, L. I. *Tetrahedron* **1975**, *31*, 311.
[258] Schmid, J. C.; Connan, J.; Albrecht, P. *Nature* **1987**, *329*, 54.
[259] Parnes, Z. N.; Lyakhovetskii, Yu. A.; Loim, N. M.; Belen'kii, L. I.; Petrov, P. V.; Kursanov, D. N. *Izv. Akad. Nauk SSSR, Ser. Khim.* **1976**, 2145.
[260] Roy, G. *Synth. Commun.* **1983**, *13*, 459.
[261] Parnes, Z. N.; Lyakhovetskii, Y. I.; Dolgova, S. P.; Pakhomov, A. S.; Kursanov, D. N. *Bull. Acad. Sci. USSR, Div. Chem. Sci. (Engl. Transl.)* **1977**, *26*, 2340.
[262] Eckert-Maksic, M.; Margetic, D. *Energy Fuels* **1993**, *7*, 315.
[263] Eckert-Maksic, M.; Margetic, D. *Energy Fuels* **1991**, *5*, 327.
[264] Harrod, J. F.; Shu, R.; Woo, H.-G.; Samuel, E. *Can. J. Chem.* **2001**, *79*, 1075.
[265] Hao, L.; Harrod, J. F.; Lebuis, A.-M.; Mu, Y.; Shu, R.; Samuel, E.; Woo, H.-G. *Angew. Chem., Int. Ed.* **1998**, *37*, 3126.
[266] Loim, N. M.; Baranova, V. A.; Moiseeva, L. V.; Zalukaev, L. P.; Parnes, Z. N.; Kursanov, D. N. *Bull. Acad. Sci. USSR, Div. Chem. Sci. (Engl. Transl.)* **1974**, *23*, 806.
[267] Fedorova, N. G.; Anokhina, I. K.; Zalukaev, L. P. *Deposited Doc.* **1980**, SPSTL 40 Khp-D80.
[268] Barluenga, J.; González, J. M.; Campos, P. J.; Asensio, G. *Angew. Chem., Int. Ed. Engl.* **1985**, *24*, 319.
[269] Imagawa, H.; Tsuchihashi, T.; Singh, R. K.; Yamamoto, H.; Sugihara, T.; Nishizawa, M. *Org. Lett.* **2003**, *5*, 153.
[270] Chandrasekhar, S.; Reddy, Ch. R.; Rao, R. J. *Tetrahedron* **2001**, *57*, 3435.
[271] Zdanovich, V. I.; Kudryavtsev, R. V.; Kursanov, D. N. *Proc. Nat. Acad. Sci., USSR (Engl. Transl.)* **1968**, *182*, 831.
[272] Harvey, J. A.; Ogliaruso, M. A. *J. Org. Chem.* **1976**, *41*, 3374.
[273] Keinan, E.; Greenspoon, N. *J. Org. Chem.* **1983**, *48*, 3545.
[274] Greenspoon, N.; Keinan, E. *J. Org. Chem.* **1988**, *53*, 3723.
[275] Vladuchick, W. C. U.S. Patent 4,346,219 (1982).
[276] Fujita, M.; Hiyama, T. *J. Org. Chem.* **1988**, *53*, 5415.
[277] Kalinkin, M. I.; Kolomnikova, G. D.; Parnes, Z. N.; Kursanov, D. N. *Bull. Acad. Sci. USSR, Div. Chem. Sci. (Engl. Transl.)* **1976**, *25*, 1794.
[278] Drew, M. D.; Lawrence, N. J.; Fontaine, D.; Sehkri, L. *Synlett* **1997**, 989.
[279] Breeden, S. W.; Lawrence, N. J. *Synlett* **1994**, 833.
[280] Matsubara, K.; Iura, T.; Maki, T.; Nagashima, H. J. *J. Org. Chem.* **2002**, *67*, 4985.
[281] Bajracharya, G. B.; Nogami, T.; Jin, T.; Matsuda, K.; Gevorgyan, V.; Yamamoto, Y. *Synthesis* **2004**, 308.
[282] Gevorgyan, V.; Rubin, M.; Liu, J-X.; Yamamoto, Y. *J. Org. Chem.* **2001**, *66*, 1672.
[283] Kursanov, D. N.; Parnes, Z. N.; Kolomnikova, G. D.; Kalinkin, M. I. *Bull. Acad. Sci. USSR, Div. Chem. Sci. (Engl. Transl.)* **1978**, *27*, 2147.
[284] Corriu, R. J. P.; Lanneau, G. F.; Perrot, M. *Tetrahedron Lett.* **1987**, *28*, 3941.

[285] Newman, M. S.; Sujeeth, P. K. *J. Org. Chem.* **1984**, *49*, 2841.
[286] Jenkins, J. W.; Post, H. W. *J. Org. Chem.* **1950**, *15*, 556.
[287] Kolomnikova, G. D.; Kalinkin, M. I.; Tskhurbaeva, Z. Ts.; Parnes, Z. N.; Kursanov, D. N. *Izv. Akad. Nauk SSSR, Ser. Khim.* **1979**, 1681.
[288] Corriu, R.; Guérin, C.; Henner, B.; Wang, Q. *Organometallics* **1991**, *10*, 2297.
[289] Barr, K. J.; Berk, S. C.; Buchwald, S. L. *J. Org. Chem.* **1994**, *59*, 4323.
[290] Berk, S. C.; Kreutzer, K. A.; Buchwald, S. L. *J. Am. Chem. Soc.* **1991**, *113*, 5093.
[291] Berk, S. C.; Buchwald, S. L. *J. Org. Chem.* **1992**, *57*, 3751.
[292] Reding, M. T.; Buchwald, S. L. *J. Org. Chem.* **1995**, *60*, 7884.
[293] Ohta, T.; Kamiya, M.; Kusui, K.; Michibata, T.; Nobutomo, M.; Furukawa, I. *Tetrahedron Lett.* **1999**, *40*, 6963.
[294] Hojo, M.; Murakami, C.; Fujii, A.; Hosomi, A. *Tetrahedron Lett.* **1999**, *40*, 911.
[295] Mao, Z.; Gregg, B. T.; Cutler, A. R. *J. Am. Chem. Soc.* **1995**, *117*, 10139.
[296] Tsurugi, J.; Nakao, R.; Fukumoto, T. *J. Am. Chem. Soc.* **1969**, *91*, 4587.
[297] Yato, M.; Homma, K.; Ishida, A. *Tetrahedron* **2001**, *57*, 5353.
[298] Lapkin, I. I.; Povarnitsyna, T. N. *J. Gen. Chem. USSR (Engl. Transl.)* **1968**, *38*, 620.
[299] Igarashi, M.; Mizuno, R.; Fuchikami, T. *Tetrahedron Lett.* **2001**, *42*, 2149.
[300] Fukuyama, T.; Lin, S.-C.; Li, L. *J. Am. Chem. Soc.* **1990**, *112*, 7050.
[301] Verdaguer, X.; Hansen, M. C.; Berk, S. C.; Buchwald, S. L. *J. Org. Chem.* **1997**, *62*, 8522.
[302] Verdaguer, X.; Berk, S. C.; Buchwald, S. L. *J. Am. Chem. Soc.* **1995**, *117*, 12641.
[303] Davis, A. P.; Hegarty, S. C. *J. Am. Chem. Soc.* **1992**, *114*, 2745.
[304] Newman, M. S.; Kanakarajan, K. *J. Org. Chem.* **1980**, *45*, 2301.
[305] Joshi, R. R.; Narasimhan, N. S. *Synthesis* **1987**, 943.
[306] Homma, K.; Takenoshita, H.; Mukaiyama, T. *Bull. Chem. Soc. Jpn.* **1990**, *63*, 1898.
[307] Mehta, A.; Jaouhari, R.; Benson, T. J.; Douglas, K. T. *Tetrahedron Lett.* **1992**, *3*, 5441.
[308] Winneroski, L. L.; Xu, Y. *J. Org. Chem.* **2004**, *69*, 4948.
[309] Seyferth, D.; Williams, G. H.; Hallgren, J. E. *J. Am. Chem. Soc.* **1973**, *95*, 266.
[310] Seyferth, D.; Williams, G. H.; Hung, P. L. K.; Hallgren, J. E. *J. Organomet. Chem.* **1974**, *71*, 97.
[311] Kursanov, D. N.; Parnes, Z. N.; Loim, N. M.; Bakalova, G. V. *Dokl. Akad. Nauk. USSR Chem. (Engl. Transl.)* **1968**, *179*, 328.
[312] Kalinkin, M. I.; Parnes, Z. D.; Shaapuni, D. Kh.; Kursanov, D. N. *Dokl. Akad. Nauk USSR (Engl. Transl.)* **1976**, *230*, 560.
[313] Doyle, M. P.; DeBruyn, D. J.; Donnelly, S. J.; Kooistra, D. A.; Odubela, A. A.; West, C. T.; Zonnebelt, S. M. *J. Org. Chem.* **1974**, *39*, 2740.
[314] Aizpurua, J. M.; Lecea, B.; Palomo, C. *Can. J. Chem.* **1986**, *64*, 2342.
[315] Chandrasekhar, S.; Reddy, Y. R.; Ramarao, C. *Synth. Commun.* **1997**, *27*, 2251.
[316] Lipowitz, J.; Bowman, S. A. *J. Org. Chem.* **1973**, *38*, 162.
[317] Lipshutz, B. H.; Chrisman, W.; Noson, K. *J. Organomet. Chem.* **2001**, *624*, 367.
[318] Kobayashi, S.; Yasuda, M.; Hachiya, I. *Chem. Lett.* **1996**, 407.
[319] Boyer, J.; Corriu, R. J. P.; Perz, R.; Reye, C. *J. Organomet. Chem.* **1979**, *172*, 143.
[320] Fujita, M.; Hiyama, T. *J. Org. Chem.* **1988**, *53*, 5405.
[321] Boyer, J.; Corriu, R. J. P.; Kpoton, A.; Mazhar, M.; Poirier, M.; Royo, G. *J. Organomet. Chem.* **1986**, *301*, 131.
[322] Brelière, C.; Carré, F.; Corriu, R. J. P.; Poirier, M.; Royo, G. *Organometallics* **1986**, *5*, 388.
[323] Arya, P.; Boyer, J.; Corriu, R. J. P.; Lanneau, G. F.; Perrot, M. *J. Organomet. Chem.* **1988**, *346*, C11.
[324] Izumi, Y.; Nanami, H.; Higuchi, K.; Onaka, M. *Tetrahedron Lett.* **1991**, *32*, 4741.
[325] Izumi, Y.; Onaka, M. *J. Mol. Catal.* **1992**, *74*, 35.
[326] Dubé, D.; Scholte, A. A. *Tetrahedron Lett.* **1999**, *40*, 2295.
[327] Loim, N. M.; Parnes, Z. N.; Vasil'eva, S. P.; Kursanov, D. N. *J. Org. Chem. USSR (Engl. Transl.)* **1972**, *8*, 902.
[328] Doyle, M. P.; DeBruyn, D. J.; Kooistra, D. A. *J. Am. Chem. Soc.* **1972**, *94*, 3659.
[329] Kato, J.; Iwasawa, N.; Mukaiyama, T. *Chem. Lett.* **1985**, 743.

[330] Lapkin, I. I.; Povarnitsyna, T. N.; Anvarova, G. Ya. *J. Gen. Chem. USSR (Engl. Transl.)* **1965**, *35*, 1831.
[331] Onishi, Y.; Ogawa, D.; Yasuda, M.; Baba, A. *J. Am. Chem. Soc.* **2002**, *124*, 13690.
[332] Wada, M.; Nagayama, S.; Mizutani, K.; Hiroi, R.; Miyoshi, N. *Chem. Lett.* **2002**, 248.
[333] Torii, S.; Takagishi, S.; Inokjuchi, T.; Okumoto, H. *Bull. Chem. Soc. Jpn.* **1987**, *60*, 775.
[334] Sassaman, M. B.; Kotian, K. D.; Prakash, G. K. S.; Olah, G. A. *J. Org. Chem.* **1987**, *52*, 4314.
[335] Olah, G. A.; Yamato, T.; Iyer, P.; Prakash, G. K. S. *J. Org. Chem.* **1986**, *51*, 2826.
[336] Nicolaou, K. C.; Hwang, C.-K.; Nugeil, D. A. *J. Am. Chem. Soc.* **1989**, *111*, 4136.
[337] Yokozawa, T.; Nakamura, F. *Makromol. Chem. Rapid. Commun.* **1993**, *14*, 167.
[338] Hartz, N.; Prakash, G. K. S.; Olah, G. A. *Synlett* **1992**, 569.
[339] McCullough, K. J.; Masuyama, A.; Morgan, K. M.; Nojima, M.; Okada, Y.; Satake, S.; Takeda, S. *J. Chem. Soc., Perkin Trans. 1* **1998**, 2353.
[340] Suzuki, T.; Ohashi, K.; Oriyama, T. *Synthesis* **1999**, 1561.
[341] Hatakeyama, S.; Mori, H.; Kitano, K.; Yamada, H.; Nishizawa, M. *Tetrahedron Lett.* **1994**, *35*, 4367.
[342] Evans, P. K.; Cui, J.; Gharpure, S. *Org. Lett.* **2003**, *5*, 3885.
[343] Komatsu, N.; Ishida, J.; Suzuki, H. *Tetrahedron Lett.* **1997**, *41*, 7219.
[344] Glushkova, N. E.; Kharitonov, N. P. *Bull. Acad. Sci. USSR, Div. Chem. Sci. (Engl. Transl.)* **1967**, 81.
[345] Goldberg, Yu.; Rubina, K.; Shymanska, M.; Lukevics, E. *Synth. Commun.* **1990**, *20*, 2439.
[346] Goldberg, Yu.; Âbele, E.; Shymanska, M.; Lukevics, E. *J. Organomet. Chem.* **1991**, *410*, 127.
[347] Goldberg, Yu.; Âbele, E.; Shymanska, M.; Lukevics, E. *J. Organomet. Chem.* **1989**, *372*, C9.
[348] Sato, Y.; Saito, N.; Mori, M. *J. Org. Chem.* **2002**, *67*, 9310.
[349] Yeh, M.-C. P.; Liang, J.-H.; Jiang, Y.-L.; Tsai, M.-S. *Tetrahedron***2003**, *59*, 3409.
[350] Mahaandru, G. M.; Liu, G.; Montgomery, J. *J. Am. Chem. Soc.* **2004**, *126*, 3698.
[351] Kursanov, D. N.; Anisimov, K. N.; Parnes, Z. P.; Loim, N. M.; Zlotina, I. B.; Valueva, Z. P. *Bull. Acad. Sci. USSR, Div. Chem. Sci. (Engl. Transl.)* **1972**, *21*, 683.
[352] Uemura, M.; Isobe, K.; Take, K.; Hayashi, Y. *J. Org. Chem.* **1983**, *48*, 3855.
[353] Onaka, M.; Higuchi, K.; Nanami, H.; Izumi, Y. *Bull. Chem. Soc. Jpn.* **1993**, *66*, 2638.
[354] Chandrasekhar, S.; Reddy, Ch. R.; Babu, B. N. *J. Org. Chem.* **2002**, *67*, 9080.
[355] Mahadevan, A.; Sard, H.; Gonzalez, M.; McKew, J. C. *Tetrahedron Lett.* **2003**, *44*, 4589.
[356] Aizpurua, J. M.; Palomo, C. *Tetrahedron Lett.* **1984**, *25*, 1103.
[357] Lecea, B.; Aizpurua, J. M.; Palomo, C. *Tetrahedron* **1985**, *41*, 4657.
[358] Keinan, E.; Perez, D.; Sahai, M.; Shvily, R. *J. Org. Chem.* **1990**, *55*, 2927.
[359] Miura, K.; Ootsuka, K.; Suda, S.; Nishikori, H.; Hosomi, A. *Synlett* **2001**, 1617.
[360] Chen, B.-C.; Sundeen, J. E.; Guo, P.; Bednarz, M. S.; Zhao, R. *Tetrahedron Lett.* **2001**, *42*, 1245.
[361] Apodaca, R.; Xiao, W. *Org. Lett.* **2001**, *3*, 1745.
[362] Jeanjot, P.; Bruyneel, F.; Arrault, A.; Gharbi, S.; Cavalier, J.-F.; Abels, A.; Marchand, C.; Touillaux, R.; Rees, J.-F.; Marchand-Brynaert, J. *Synthesis* **2003**, 513.
[363] Chandrasekhar, S.; Reddy, Ch. R.; Ahmed, M. *Synlett* **2000**, 1655.
[364] Same as reference 318.
[365] Olah, G. A.; Wang, Q.; Trivedi, N. J.; Prakash, G. K. S. *Synthesis* **1992**, 465.
[366] Hojo, M.; Fujii, A.; Murakami, C.; Aihara, H.; Hosomi, A. *Tetrahedron Lett.* **1995**, *36*, 571.
[367] Magnus, P.; Fielding, M. R. *Tetrahedron Lett.* **2001**, *42*, 6633.
[368] Tolf, B.-R.; Crowe, D. F.; Johansson, J. G.; Peters, R. R.; Tanabe, M. *Tetrahedron Lett.* **1984**, *25*, 4855.
[369] Solodin, I.; Goldberg, Y.; Zelcans, G.; Lukevics, E. *J. Chem. Soc., Chem. Commun.* **1990**, 1321.
[370] Török, B.; Felföldi, K.; Molnár, Á.; Bartók, M. *J. Organomet. Chem.* **1993**, *460*, 111.
[371] Palomo, C.; Cossío, F. P.; Arrieta, A.; Odriozola, J. M.; Oiarbide, M.; Ontoria, J. M. *J. Org. Chem.* **1989**, *54*, 5736.
[372] Asao, N.; Ohishi, T.; Sato, K.; Yamamoto, Y. *J. Am. Chem. Soc.* **2001**, *123*, 6931.
[373] Hojo, M.; Hojo, M.; Inoue, Y.; Tanimoto, S. *Bull. Chem. Soc. Jpn.* **1990**, *63*, 2588.
[374] Zheng, G. Z.; Chan, T. H. *Organometallics* **1995**, *14*, 70.
[375] Fujita, M.; Hiyama, T. *Org. Synth.* **1990**, *69*, 44.

[376] Nordlander, J. E.; Payne, M. J.; Njoroge, F.; Balk, M. A.; Laikos, G. D.; Vishwanath, V. M. *J. Org. Chem.* **1984**, *49*, 4107.
[377] Dioumaev, V. K.; Bullock, R. M. *Nature* **2003**, *424*, 530.
[378] Imma, H.; Mori, M.; Nakai, T. *Synlett* **1996**, 1229.
[379] Iwasaki, F.; Onomura, O.; Mishima, K.; Maki, T.; Matsumura, Y. *Tetrahedron Lett.* **1999**, *40*, 7507.
[380] Fukuzumi, S.; Fujita, M. *Chem. Lett.* **1991**, 2059.
[381] Doyle, D. P.; West, C. T. *J. Org. Chem.* **1975**, *40*, 3835.
[382] Calas, R.; Frainnet, È.; Bonastre, J. C. *C.R. Hebd. Seances Acad. Sci.* **1960**, *251*, 2987.
[383] Lapkin, I. I.; Povarnitsyna, T. N.; Kostareva, L. A. *J. Gen. Chem. USSR (Engl. Transl.)* **1968**, *38*, 1527.
[384] Doyle, M. P.; McOsker, C. C.; Ball, N.; West, C. T. *J. Org. Chem.* **1977**, *42*, 1922.
[385] Lawrence, N. J.; Bushell, S. M. *Tetrahedron Lett.* **2000**, *41*, 4507.
[386] Doyle, D. P.; West, C. T. *J. Org. Chem.* **1975**, *40*, 3829.
[387] Ojima, I.; Nihonyanagi, M.; Nagai, Y. *Chem. Lett.* **1972**, 22.
[388] Ojima, I.; Nihonyanagi, M.; Nagai, Y. *Bull. Chem. Soc. Jpn.***1972**, *45*, 3722.
[389] Niyomura, O.; Tokunaga, M.; Obora, Y.; Iwasawa, T.; Tsuji, Y. *Angew. Chem., Int. Ed.* **2003**, *42*, 1287.
[390] Nishiyama, H.; Park, S.-B.; Itoh, K. *Tetrahedron: Asymmetry* **1992**, *3*, 1029.
[391] Semmelhack, M. F.; Misra, R. N. *J. Org. Chem.* **1982**, *47*, 2469.
[392] Sassaman, M. B.; Prakash, G. K. S.; Olah, G. A. *Tetrahedron* **1988**, *44*, 3771.
[393] Loim, N. M.; Parnes, Z. N.; Brunovlenskaya, I. I.; Kursanov, D. N. *Bull. Acad. Sci. USSR, Div. Chem. Sci. (Engl. Transl.)* **1971**, *20*, 153.
[394] Rupprecht, K. M.; Boger, J.; Hoogsteen, K.; Nachbar, R. B.; Springer, J. P. *J. Org. Chem.* **1991**, *56*, 6180.
[395] Kobayashi, Y.; Ito, Y.; Terashima, S. *Bull. Chem. Soc. Jpn.* **1989**, *62*, 3041.
[396] Lakhvich, F. A.; Lis, L. G.; Rubinov, D. B.; Rubinova, I. L.; Akhrem, A. A. *J. Org. Chem. USSR (Engl. Transl.)* **1990**, *25*, 1493.
[397] Anwar, S.; Bradley, G.; Davis, A. P. *J. Chem. Soc., Perkin Trans. 1* **1991**, 1383.
[398] Anwar, S.; Davis, A. P. *J. Chem. Soc., Chem. Commun.* **1986**, 831.
[399] Anwar, S.; Davis, A. P. *Tetrahedron* **1988**, *44*, 3761.
[400] Miura, K.; Nakagawa, T.; Suda, S.; Hosomi, A. *Chem. Lett.* **2000**, 150.
[401] Nadkarni, D.; Hallissey, J.; Mojica, C. *J. Org. Chem.* **2003**, *68*, 594.
[402] Newman, M. S.; Veeraraghavan, S. *J. Org. Chem.* **1983**, *48*, 3246.
[403] Carreño, M. C.; Des Mazery, R.; Urbano, A.; Colobert, F.; Solladié, G. *J. Org. Chem.* **2003**, *68*, 7779.
[404] Evans, P. A.; Cui, J.; Gharpure, S. J.; Polosukhin, A.; Zhang, H.-R. *J. Am. Chem. Soc.* **2003**, *125*, 14702.
[405] Smith, A. B. III; Cui, H. *Org. Lett.* **2003**, *5*, 587.
[406] Carreño, M. C.; Mazery, R. D.; Urbano, A.; Colobert, F.; Solladié, G. *Org. Lett.* **2004**, *6*, 297.
[407] Fotsch, C. H.; Chamberlin, A. R. *J. Org. Chem.* **1991**, *56*, 4141.
[408] Mulholland, R. L.; Chamberlin, A. R. *J. Org. Chem.* **1988**, *53*, 1082.
[409] Tolstikov, G. A.; Lerman, B. M.; Galin, F. Z. *J. Org. Chem. USSR (Engl. Transl.)* **1977**, *13*, 208.
[410] Blazejewski, J.-C.; Le Guyader, F.; Wakselman, C. *J. Chem. Soc., Perkin Trans. 1* **1991**, 1121.
[411] Ojima, I.; Nihonyanagi, M.; Kogure, T.; Kumagai, M.; Horiuchi, S.; Nakatsugawa, K. *J. Organomet. Chem.* **1975**, *94*, 449.
[412] Kaur, H.; Zinn, F. K.; Stevens, E. D.; Nolan, S. P. *Organometallics* **2004**, *23*, 1157.
[413] Smonou, I. *Synth. Commun.* **1994**, *24*, 1999.
[414] Olah, G. A.; Yamato, T.; Hashimoto, T.; Shih, J. G.; Trivedi, N.; Singh, B. P.; Piteau, M.; Olah, J. A. *J. Am. Chem. Soc.* **1987**, *109*, 3708.
[415] Nishitani, S.; Bodor, N. *Rev. Roum. Chim.* **1991**, *36*, 635.
[416] Davis, F. A.; Clark, C.; Kumar, A.; Chen, B.-C. *J. Org. Chem.* **1994**, *59*, 1184.
[417] Smith, C. W.; Ambler, S. J.; Steggles, D. J. *Tetrahedron Lett.* **1993**, *34*, 7447.

[418] Alferova, S. I.; Kisin, A. V.; Kudryavtseva, G. A.; Zalukaev, L. P.; Parnes, Z. P. *Izv. Akad. Nauk SSSR, Ser. Khim.* **1982**, *31*, 2336.
[419] Nordlander, J. E.; Njoroge, F. G.; Payne, M. J.; Warman, D. *J. Org. Chem.* **1985**, *50*, 3481.
[420] Olah, G. A.; Arvanaghi, M.; Ohannesian, L. *Synthesis* **1986**, 770.
[421] Yato, M.; Homma, K.; Ishida, A. *Heterocycles* **1998**, *49*, 233.
[422] Popova, O. K.; Parnes, Z. N.; Kalinkin, M. I.; Markosyan, S. M.; Kopteva, N. I.; Zalukaev, L. P.; Kursanov, D. N. *Bull. Acad. Sci. USSR, Div. Chem. Sci. (Engl. Transl.)* **1981**, *30*, 1709.
[423] Lokot, I. P.; Pashkovsky, F. S.; Lakhvich, F. A. *Tetrahedron* **1999**, *55*, 4783.
[424] Kira, M.; Hino, T.; Sakurai, H. *Chem. Lett.* **1992**, 555.
[425] Seyferth, D.; Hung, P. L. K.; Hallgren, J. E. *J. Organomet. Chem.* **1972**, *44*, C55.
[426] Mukaiyama, T.; Ohno, T.; Nishimura, T.; Han, J. S.; Kobayashi, S. *Bull. Chem. Soc. Jpn.* **1991**, *64*, 2524.
[427] Yasuda, M.; Onishi, Y.; Ito, T.; Baba, A. *Tetrahedron Lett.* **2000**, *41*, 2425.
[428] Kablaoui, N. M.; Buchwald, S. L. *J. Am. Chem. Soc.* **1996**, *118*, 3182.
[429] Kablaoui, N. M.; Buchwald, S. L. *J. Am. Chem. Soc.* **1995**, *117*, 6785.
[430] Selvakumar, K.; Harrod, J. F. *Angew. Chem., Int. Ed.* **2001**, *40*, 2129.
[431] Kuwano, R.; Takahashi, M.; Ito, Y. *Tetrahedron Lett.* **1998**, *39*, 1017.
[432] Igarashi, M.; Fuchikami, T. *Tetrahedron Lett.* **2001**, *42*, 1945.
[433] Bower, S.; Kreutzer, K. A.; Buchwald, S. L. *Angew. Chem., Int. Ed. Engl.* **1996**, *35*, 1515.
[434] Parnes, Z. N.; Loim, N. M.; Baranova, V. A.; Kursanov, D. N. *J. Org. Chem. USSR (Engl. Transl.)* **1971**, *7*, 2145.
[435] Ojima, I.; Kogure, T. *Organometallics* **1982**, *1*, 1390.
[436] Keinan, E.; Greenspoon, N. *J. Am. Chem. Soc.* **1986**, *108*, 7314.
[437] Ojima, I.; Kogure, T. *Tetrahedron Lett.* **1972**, *13*, 5035.
[438] Kim, S. O.; Rhee, S.; Lee, S. H. *Bull. Kor. Chem. Soc.* **1999**, *20*, 773.
[439] Kursanov, D. N.; Loim, N. M.; Baranova, V. A.; Moiseeva, L. V.; Zalukaev, L. P.; Parnes, Z. N. *Synthesis* **1973**, 420.
[440] Hiyama, T.; Kobayashi, K.; Fujita, M. *Tetrahedron Lett.* **1984**, *25*, 4959.
[441] Akhrem, A. A.; Lakhvich, F. A.; Lis, L. G.; Sagaidak, S. U.; Garbuz, N. I.; Kurbako, V. Z. *Zh. Org. Khim.* **1981**, *17*, 1527.
[442] Dailey, Jr., O. D. *J. Org. Chem.* **1987**, *52*, 1984.
[443] Nordlander, J. E., Njoroge, F. G. *J. Org. Chem.* **1987**, *52*, 1627.
[444] Mori, A.; Fujita, A.; Kajiro, H.; Nishihara, Y.; Hiyama, T. *Tetrahedron* **1999**, *55*, 4573.
[445] Ito, H.; Ishizuka, T.; Arimoto, K.; Miura, K.; Hosomi, A. *Tetrahedron Lett.* **1997**, *38*, 8887.
[446] Mori, A.; Fujita, A.; Nishihara, Y.; Hiyama, T. *J. Chem. Soc., Chem. Commun.* **1997**, 2159.
[447] Lipshutz, B. H.; Keith, J.; Papa, P.; Vivian, R. *Tetrahedron Lett.* **1998**, *39*, 4627.
[448] Magnus, P.; Waring, M. J.; Scott, D. A. *Tetrahedron Lett.* **2000**, *41*, 9731.
[449] Yoshii, E.; Koizumi, T.; Hayashi, I.; Hiroi, Y. *Chem. Pharm. Bull.* **1977**, *25*, 1468.
[450] Keinan, E.; Perez, D. *J. Org. Chem.* **1987**, *52*, 2576.
[451] Chauhan, M.; Boudjouk, P. *Can. J. Chem.* **2000**, *78*, 1396.
[452] Ojima, I.; Kogure, T. *Tetrahedron Lett.* **1972**, *13*, 5085.
[453] Blazejewski, J. C.; Dorme, R.; Wakselman, C. *J. Chem. Soc., Perkin Trans. 1* **1987**, 1861.
[454] Jurkauskas, V.; Sadighi, J. P.; Buchwald, S. L. *Org. Lett.* **2003**, *5*, 2417.
[455] Lipshutz, B. H.; Chrisman, W.; Noson, K.; Papa, P.; Sclafani, J. A.; Vivian, R. W.; Keith, J. M. *Tetrahedron* **2000**, *56*, 2779.
[456] Gracias, V.; Zeng, Y.; Desai, P.; Aubé, J. *Org. Lett.* **2003**, *5*, 4999.
[457] Keinan, E.; Greenspoon, N. *Tetrahedron Lett.* **1985**, *26*, 1353.
[458] Yoshii, E.; Koizumi, T.; Ikeshima, H.; Ozaki, K.; Hayashi, I. *Chem. Pharm. Bull.* **1975**, *23*, 2496.
[459] Yun, J.; Buchwald, S. L. *Org. Lett.* **2001**, *3*, 1129.
[460] Wang, L.-C.; Jang, H.-Y.; Roh, Y.; Lynch, V.; Schultz, A. J.; Wang, X.; Krische, M. J. *J. Am. Chem. Soc.* **2002**, *124*, 9448.
[461] Baik, T.-G.; Luis, A. L.; Wang, L.-C.; Krische, M. J. *J. Am. Chem. Soc.* **2001**, *123*, 5112.
[462] Baik, T.-G.; Luis, A. L.; Wang, L.-C.; Krische, M. J. *J. Am. Chem. Soc.* **2001**, *123*, 6716.
[463] Giese, S.; West, F. G. *Tetrahedron* **2000**, *56*, 10221.

[464] Matsuda, I.; Takahashi, K.; Sato, S. *Tetrahedron Lett.* **1990**, *31*, 5331.
[465] Magnus, P.; Payne, A. H.; Waring, M. J.; Scott, D. A.; Lynch, V. *Tetrahedron Lett.* **2000**, *41*, 9725.
[466] Ojima, I.; Kumagai, M. *J. Organomet. Chem.* **1976**, *111*, 43.
[467] Yoshii, E.; Kobayashi, Y.; Koizumi, T.; Oribe, T. *Chem. Pharm. Bull.* **1974**, *22*, 2767.
[468] Abouelfida, A.; Roze, J.-C.; Pradere, J.-P.; Jubault, M. *Phosphorus, Sulfur, Silicon* **1990**, *54*, 123.
[469] Zhao, Y.; Quayle, P.; Kuo, E. A. *Tetrahedron Lett.* **1994**, *35*, 4179.
[470] Russell, A. E.; Fuller, N. O.; Taylor, S. J.; Aurriset, P.; Morken, J. P. *Org. Lett.* **2004**, *6*, 2309.
[471] Freiría, M.; Whitehead, A. J.; Tocher, D. A.; Motherwell, W. B. *Tetrahedron* **2004**, *60*, 2673.
[472] Emiabata-Smith, D.; McKillop, A.; Mills, C.; Motherwell, W. B.; Whitehead, A. J. *Synlett* **2001**, 1302.
[473] Revis, A.; Hilty, T. K. *Tetrahedron Lett.* **1987**, *28*, 4809.
[474] Miller, S. P.; Morken, J. P. *Org. Lett.* **2002**, *4*, 2743.
[475] Muraoka, T.; Matsuda, I.; Itoh, K. *Organometallics* **2001**, *20*, 4676.
[476] Townes, J. A.; Evans, M. A.; Queffelec, J.; Taylor, S. J.; Morken, J. P. *Org. Lett.* **2002**, *4*, 2537.
[477] Trost, B. M.; Crawley, M. L. *J. Am. Chem. Soc.* **2002**, *124*, 9328.
[478] Isayama, S.; Mukaiyama, T. *Chem. Lett.* **1989**, 2005.
[479] Oriyama, T.; Ichimura, Y.; Koga, G. *Bull. Chem. Soc. Jpn.* **1991**, *64*, 2581.
[480] Tsunoda, T.; Suzuki, M.; Noyori, R. *Tetrahedron Lett.* **1979**, *20*, 4679.
[481] Suzuki, K.; Nakata, T. *Org. Lett.* **2002**, *4*, 2739.
[482] Mori, Y.; Nogami, K.; Hayashi, H.; Noyori, R. *J. Org. Chem.* **2003**, *68*, 9050.
[483] Rolf, D.; Gray, G. R. *J. Am. Chem. Soc.* **1982**, *104*, 3539.
[484] Pelter, A.; Ward, R. S.; Collins, P.; Venkateswarlu, R.; Kay, I. T. *J. Chem. Soc., Perkin Trans. 1* **1985**, 587.
[485] Jun, J.-G. *J. Heterocycl. Chem.* **1997**, *34*, 633.
[486] Bisagni, E.; Rautureau M.; Croisy-Delcey, M.; Huel, C. *Can. J. Chem.* **1987**, *65*, 2027.
[487] El Gihani, M.; Heaney, H. *Synlett* **1993**, 583.
[488] Kotsuki, H.; Ushio, Y.; Kadota, I.; Ochi, M. *J. Org. Chem.* **1989**, *54*, 5153.
[489] Kotsuki, H.; Ushio, Y.; Kadota, I.; Ochi, M. *Chem. Lett.* **1988**, 927.
[490] Harada, T.; Hayashiya, T.; Wada, I.; Iwaake, N.; Oku, A. *J. Am. Chem. Soc.* **1987**, *109*, 527.
[491] Mori, A.; Ishihara, K.; Yamamoto, H. *Tetrahedron Lett.* **1986**, *27*, 987.
[492] Ishihara, K.; Mori, A.; Yamamoto, H. *Tetrahedron* **1990**, *46*, 4595.
[493] Ohta, T.; Michibata, T.; Yamada, K.; Omori, R.; Furukawa, I. *Chem. Commun.* **2003**, 1192.
[494] DeNinno, M. P.; Etienne, J. B.; Duplantier, K. C. *Tetrahedron Lett.* **1995**, *36*, 669.
[495] Al-Mughaid, H.; Grindley, T. B.; Robertson, K. N.; Cameron, T. S. *Can. J. Chem.* **2003**, *81*, 505.
[496] Chandrasekhar, S.; Reddy, Y. R.; Reddy, Ch. R. *Chem. Lett.* **1998**, 1273.
[497] Debenham, S. D.; Toone, E. J. *Tetrahedron: Asymmetry* **2000**, *11*, 385.
[498] Federspiel, M.; Fischer, R.; Hennig, M.; Mair, H.-J.; Oberhauser, T.; Rimmler, G.; Albiez, T.; Bruhin, J.; Estermann, H.; Gandert, C.; Göckel, V.; Götzö, S.; Hoffmann, U.; Huber, G.; Janatsch, G.; Lauper, S.; Röckel-Stäbler, O.; Trussardi, R.; Zwahlen, A. G. *Org. Proc. Res. Dev.* **1999**, *3*, 266.
[499] Tokuyasu, T.; Ito, T.; Masuyama, A.; Nojima, M. *Heterocycles* **2000**, *53*, 1293.
[500] Miura, T.; Masaki, Y. *J. Chem. Soc., Perkin Trans. 1* **1995**, 2155.
[501] Suzuki, T.; Oriyama, T. *Synth. Commun.* **1999**, *29*, 1263.
[502] Mukaiyama, T.; Ohno, T.; Nishimura, T.; Han, J. S.; Kobayashi, S. *Chem. Lett.* **1990**, 2239.
[503] Bennek, J. A.; Gray, G. R. *J. Org. Chem.* **1987**, *52*, 892.
[504] Guo, Z.-W.; Hui, Y.-Z. *Synth. Commun.* **1996**, *26*, 2067.
[505] Keinan, E.; Sahai, M.; Shvily, R. *Synthesis* **1991**, 641.
[506] Oriyama, T.; Iwanami, K.; Tsukamoto, K.; Ichimura, Y.; Koga, G. *Bull. Chem. Soc. Jpn.* **1991**, *64*, 1410.
[507] Lewis, M. D.; Cha, J. K.; Kishi, Y. *J. Am. Chem. Soc.* **1982**, *104*, 4976.
[508] Calzada, E.; Clarke, C. A.; Roussin-Bouchard, C.; Wightman, R. H. *J. Chem. Soc., Perkin Trans. 1* **1995**, 517.
[509] Czernecki, S.; Ville, G. *J. Org. Chem.* **1989**, *54*, 610.

[510] Kraus, G. A.; Frazier, K. A.; Roth, B. D.; Taschner, M. J.; Neuenschwander, K. *J. Org. Chem.* **1981**, *46*, 2417.
[511] Piccirilli, J. A.; Krauch, T.; MacPherson, L. J.; Benner, S. A. *Helv. Chim. Acta* **1991**, *74*, 397.
[512] Brückner, C.; Reissig, H.-U. *J. Chem. Soc., Chem. Commun.* **1985**, 1512.
[513] Brimble, M. A.; Brenstrum, T. J. *J. Chem. Soc., Perkin Trans. 1* **2001**, 1624.
[514] Kraus, G. A.; Molina, M. T.; Walling, J. A. *J. Org. Chem.* **1987**, *52*, 1273.
[515] Polniaszek, R.; Stevens, R. V. *J. Org. Chem.* **1986**, *51*, 3023.
[516] Choi, J.-K.; Ha, D.-C.; Hart, D. J.; Lee, C.-S.; Ramesh, S.; Wu, S. *J. Org. Chem.* **1989**, *54*, 279.
[517] Wang, Y.; Babirad, S. A.; Kishi, Y. *J. Org. Chem.* **1992**, *57*, 468.
[518] Nicotra, F.; Panza, L.; Russo, G.; Verani, A. *Tetrahedron: Asymmetry* **1993**, *4*, 1203.
[519] Xie, J.; Durrat, F.; Valéry, J.-M. *J. Org. Chem.* **2003**, *68*, 7896.
[520] Nishiyama, Y.; Tujino, T.; Yamano, T.; Hayashishita, M.; Itoh, K. *Chem. Lett.* **1997**, 165.
[521] Yamazaki, H.; Horikawa, H.; Nishitani, T.; Iwasaki, T. *Chem. Pharm. Bull.* **1990**, *38*, 2024.
[522] Billard, T.; Langlois, B. R. *J. Org. Chem.* **2002**, *67*, 997.
[523] Maier, M. E.; Evertz, K. *Tetrahedron Lett.* **1988**, *29*, 1677.
[524] Hinman, M. M.; Heathcock, C. H. *J. Org. Chem.* **2001**, *66*, 7751.
[525] Beulshausen, T.; Groth, U.; Schöllkopf, U. *Liebigs Ann. Chem.* **1992**, 523.
[526] Auerbach, J.; Zamore, M.; Weinreb, S. M. *J. Org. Chem.* **1976**, *41*, 725.
[527] Ruan, Y.-P.; Chen, M.-D.; He, M.-Z.; Zhou, X.; Huang, P.-Q. *Synth. Commun.* **2004**, *34*, 853.
[528] Freidinger, R. M.; Hinkle, J. S.; Perlow, D. S.; Arison, B. H. *J. Org. Chem.* **1983**, *48*, 77.
[529] Rostentreter, U. *Synthesis* **1985**, 210.
[530] Rosentreter, U.; Born, L.; Kurz, J. *J. Org. Chem.* **1986**, *51*, 1165.
[531] Comins, D. L.; Weglarz, M. A. *J. Org. Chem.* **1991**, *56*, 2506.
[532] Pearson, D. A.; Blanchette, M.; Baker, M. L.; Guindon, C. A. *Tetrahedron Lett.* **1989**, *30*, 2739.
[533] Parnes, Z. N.; Bolestova, G. I.; Belen'kii, L. I.; Kursanov, D. N. *Bull. Acad. Sci. USSR, Div. Chem. Sci. (Engl. Transl.)* **1973**, *22*, 1874.
[534] Lanzilotti, A. E.; Littell, R.; Fanshawe, W. J.; McKenzie. T. C.; Lovell, F. M. *J. Org. Chem.* **1979**, *44*, 4809.
[535] Masuno, M. N.; Molinski, T. F. *Tetrahedron Lett.* **2001**, *42*, 8263.
[536] Comins, D. L.; Myoung, Y. C. *J. Org. Chem.* **1990**, *55*, 292.
[537] Akhrem, A. A.; Moiseenkov, A. M.; Krivoruchko, V. A. *Bull. Acad. Sci. USSR, Div. Chem. Sci. (Engl. Transl.)* **1973**, *22*, 1745.
[538] Cannon, J. G.; Chang, Y.; Amoo, V. E.; Walker, K. A. *Synthesis* **1986**, 494.
[539] Chandrasekhar, S.; Reddy, M. V.; Chandraiah, L. *Synth. Commun.* **1999**, *29*, 3981.
[540] Loim, N. M. *Bull. Acad. Sci. USSR, Div. Chem. Sci. (Engl. Transl.)* **1968**, 1345.
[541] Kalinkin, M. I.; Markosyan, S. M.; Kolomnikova, G. D.; Parnes, Z. N.; Kursanov, D. N. *Dokl. Akad. Nauk SSSR* **1980**, *253*, 1137.
[542] Benkeser, R. A.; Snyder, D. C. *J. Organomet. Chem.* **1982**, *225*, 107.
[543] López, R. M.; Fu, G. C. *Tetrahedron* **1997**, *53*, 16349.
[544] Takaki, K.; Kamata, T.; Miura, Y.; Shishido, T.; Takehira, K. *J. Org. Chem.* **1999**, *64*, 3891.
[545] Okamoto, H.; Kato, S. *Bull. Chem. Soc. Jpn.* **1991**, *64*, 3466.
[546] Ojima, I.; Kogure, T.; Nagai, Y. *Tetrahedron Lett.* **1973**, 2475.
[547] Arany, A.; Meth-Cohn, O.; Nyerges, M. *Org. Biomol. Chem.* **2003**, *1*, 1545.
[548] Chandrasekhar, S.; Reddy, M. V.; Chandraiah, L. *Synlett* **2000**, 1351.
[549] Holzapfel, C. W.; Crous, R.; Greyling, H. F.; Verdoorn, G. H. *Heterocycles* **1999**, *51*, 2801.
[550] Sternbach, D. D.; Jamison, W. C. L. *Tetrahedron Lett.* **1981**, *22*, 3331.
[551] Fujita, M.; Oishi, H.; Hiyama, T. *Chem. Lett.* **1986**, 837.
[552] Fujioka, H.; Yamamoto, H.; Miyazaki, M.; Yamanaka, T.; Takuma, K.; Kita, Y. *Tetrahedron Lett.* **1991**, *32*, 5367.
[553] Neidlein, R.; Keller, H. *Heterocycles* **1993**, *36*, 1925.
[554] Brinkman, H. R.; Miles, W. H.; Hilborn, M. D.; Smith, M. C. *Synth. Commun.* **1996**, *26*, 973.
[555] Tormo, J.; Hays, D. S.; Fu, G. C. *J. Org. Chem.* **1998**, *63*, 5296.
[556] Hays, D. S.; Fu, G. C. *J. Org. Chem.* **1998**, *63*, 2796.
[557] Chandrasekhar, S.; Chandraiah, L.; Reddy, Ch. R.; Reddy, M. V. *Chem. Lett.* **2000**, 780.

[558] MacDonald, M.; Vander Velde, D.; Aubé, J. *J. Org. Chem.* **2001**, *66*, 2636.
[559] Murai, T.; Sakane, T.; Kato, S. *Tetrahedron Lett.* **1985**, *25*, 5145.
[560] Wu, P.-L.; Peng, S.-Y.; Magrath, J. *Synthesis* **1996**, 249.
[561] Wu, P.-L.; Peng, S.-Y.; Magrath, J. *Synthesis* **1995**, 435.
[562] Fry, J. L. *J. Chem. Soc., Chem. Commun.* **1974**, 45.
[563] Nakayama, J.; Yoshida, M.; Simamura, O. *Tetrahedron* **1970**, *26*, 4609.
[564] Kalinkin, M. I.; Parnes, Z. N.; Kursanov, D. N. *Dokl. Akad. Nauk SSSR* **1968**, *180*, 1370.
[565] Blankespoor, R. L.; Doyle, M. P.; Hedstrand, D. M.; Tamblyn, W. H.; Van Dyke, D. A. *J. Am. Chem. Soc.* **1981**, *103*, 7096.
[566] Fry, J. L.; Mraz, T. J. *Tetrahedron Lett.* **1979**, 849.
[567] Aizpurua, J. M.; Palomo, C. *Tetrahedron Lett.* **1984**, *25*, 3123.
[568] Chandrasekhar, S.; Ahmed, M. *Tetrahedron Lett.* **1999**, *40*, 9325.
[569] Nishibayashi, Y.; Takei, I.; Uemura, S.; Hidai, M. *Organometallics* **1998**, *17*, 3420.
[570] Nishibayashi, Y.; Uemura, S. *Synlett* **1995**, 79.
[571] Nishibayashi, Y.; Segawa, K.; Ohe, K.; Uemura, S. *Organometallics* **1995**, *14*, 5486.
[572] Nishiyama, H.; Yamaguchi, S.; Park, S.-B.; Itoh, K. *Tetrahedron: Asymmetry* **1993**, *4*, 143.
[573] Göndös, G.; Orr, J. C. *J. Chem. Soc., Chem. Commun.* **1982**, 1238.
[574] Hayashi, T.; Hayashi, C.; Uozumi, Y. *Tetrahedron: Asymmetry* **1995**, *6*, 2503.
[575] Heldmann, D. K.; Seebach, D. *Helv. Chim. Acta* **1999**, *82*, 1096.
[576] Evans, D. A.; Michael, F. E.; Tedrow, J. S.; Campos, K. R. *J. Am. Chem. Soc.* **2003**, *125*, 3534.
[577] Kuwano, R.; Sawamura, M.; Shirai, J.; Takahashi, M.; Ito, Y. *Bull. Chem. Soc. Jpn.* **2000**, *73*, 485.
[578] Kuwano, R.; Uemura, T.; Saito, M.; Ito, Y. *Tetrahedron Lett.* **1999**, *40*, 1327.
[579] Lee, S.; Lim, C. W.; Song, C. E.; Kim, I. O. *Tetrahedron: Asymmetry* **1997**, *8*, 4027.
[580] Nishiyama, H.; Kondo, M.; Nakamura, T.; Itoh, K. *Organometallics* **1991**, *10*, 500.
[581] Nishiyama, H.; Sakaguchi, H.; Nakamura, T.; Horihata, M.; Kondo, M.; Itoh, K. *Organometallics* **1989**, *8*, 846.
[582] Nishibayashi, Y.; Segawa, K.; Takada, H.; Ohe, K.; Uemura, S. *J. Chem. Soc., Chem. Commun.* **1996**, 847.
[583] Rahimian, K.; Harrod, J. F. *Inorg. Chim. Acta* **1998**, *270*, 330.
[584] Xin, S.; Harrod, J. F. *Can. J. Chem.* **1995**, *73*, 999.
[585] Sudo, A.; Yoshida, H.; Saigo, K. *Tetrahedron: Asymmetry* **1997**, *18*, 3205.
[586] Tao, B.; Fu, G. C. *Angew. Chem., Int. Ed.* **2002**, *41*, 3892.
[587] Yun, J.; Buchwald, S. L. *J. Am. Chem. Soc.* **1999**, *121*, 5640.
[588] Carter, M. B.; Schiøtt, B.; Gutiérrez, A.; Buchwald, S. L. *J. Am. Chem. Soc.* **1994**, *116*, 11667.
[589] Lipshutz, B. H.; Noson, K.; Chrisman, W.; Lower, A. *J. Am. Chem. Soc.* **2003**, *125*, 8779.
[590] Lipshutz, B. H.; Lower, A.; Noson, K. *Org. Lett.* **2002**, *4*, 4045.
[591] Lipschutz, B. H.; Noson, K.; Chrisman, W. *J. Am. Chem. Soc.* **2001**, *123*, 12917.
[592] Schiffers, R.; Kagan, H. B. *Synlett* **1997**, 1175.
[593] LaRonde, F. J.; Brook, M. A. *Inorg. Chim. Acta* **1999**, *296*, 208.
[594] Mimoun, H.; de Saint Laumer, J. V.; Giannini, L.; Scopelliti, R.; Floriani, C. *J. Am. Chem. Soc.* **1999**, *121*, 6158.
[595] Moritani, Y.; Appella, D. H.; Jurkauskas, V.; Buchwald, S. L. *J. Am. Chem. Soc.* **2000**, *122*, 6797.
[596] Jurkauskas, V.; Buchwald, S. L. *J. Am. Chem. Soc.* **2002**, *124*, 2892.
[597] Lipshutz, B. H.; Servesko, J. M.; Petersen, T. B.; Papa, P. P.; Lover, A. A. *Org. Lett.* **2004**, *6*, 1273.
[598] Lipschutz, B. H.; Servesko, J. M.; Taft, B. R. *J. Am. Chem. Soc.* **2004**, *126*, 8352.
[599] Hughes, G.; Kimura, M.; Buchwald, S. L. *J. Am. Chem. Soc.* **2003**, *125*, 11253.
[600] Appella, D. H.; Moritani, Y.; Shintani, R.; Ferreira, E. M.; Buchwald, S. L. *J. Am. Chem. Soc.* **1999**, *121*, 9473.
[601] Zhao, C-X.; Duffey, M. O.; Taylor, S. J.; Morken, J. P. *Org. Lett.* **2001**, *3*, 1829.
[602] Duffey, M. O.; LeTiran, A.; Morken, J. P. *J. Am. Chem. Soc.* **2003**, *125*, 1458.
[603] Taylor, S. J.; Duffey, M. O.; Morken, J. P. *J. Am. Chem. Soc.* **2000**, *122*, 4528.
[604] Evans, M. A.; Morken, J. P. *J. Am. Chem. Soc.* **2002**, *124*, 9020.

[605] Takei, I.; Nishibayashi, Y.; Arikawa, Y.; Uemura, S.; Hidai, M. *Organometallics* **1999**, *18*, 2271.

[606] Becker, R.; Brunner, H.; Mahboobi, S.; Wiegrebe, W. *Angew. Chem., Int. Ed. Engl.* **1985**, *24*, 995.

[607] Kagan, H. B.; Langlois, N.; Dang, T. P. *J. Organomet. Chem.* **1975**, *90*, 353.

[608] Nishikori, H.; Yoshihara, R.; Hosomi, A. *Synlett* **2003**, 561.

[609] Iwasaki, F.; Onomura, O.; Mishima, K.; Kanematsu, T.; Maki, T.; Matsumura, Y. *Tetrahedron Lett.* **2001**, *42*, 2525.

[610] Verdaguer, X.; Lange, U. E. W.; Reding, M. T.; Buchwald, S. L. *J. Am. Chem. Soc.* **1996**, *118*, 6784.

[611] Reding, M. T.; Buchwald, S. L. *J. Org. Chem.* **1998**, *63*, 6344.

[612] Verdaguer, X.; Lange, U. E. W.; Buchwald, S. L. *Angew. Chem., Int. Ed.* **1998**, *37*, 1103.

[613] Willoughby, C. A.; Buchwald, S. L. *J. Org. Chem.* **1993**, *58*, 7627.

[614] Lipshutz, B. H.; Shimizu, H. *Angew. Chem., Int. Ed.* **2004**, *43*, 666.

[615] Noyori, R.; Kitamura, M. In *Modern Synthetic Methods*, Scheffold, R., Ed.; Springer: Berlin, 1989; Vol. 5, p 115.

[616] Takaya, H.; Ohta, T.; Noyori, R. In *Catalytic Asymmetric Synthesis,* Ojima, I., Ed.; VCH: New York, 1993; p 1.

[617] Noyori, R. *Asymmetric Catalysis in Organic Synthesis;* Wiley: New York, 1994; Chapter 2.

[618] Ohkuma, T.; Noyori, R. In *Transition Metals for Organic Synthesis: Building Blocks and Fine Chemicals,* Beller, M., Bolm, C., Eds.; Wiley-VCH: Weinheim, 1998; Vol. 2, p 25.

[619] Brown, J. M. In *Comprehensive Asymmetric Catalysis,* Jacobsen, E. N., Pfaltz, A., Yamamoto, H., Eds.; Springer: Berlin, 1999; Vol. 1, p 121.

[620] Halterman, R. L. In *Comprehensive Asymmetric Catalysis,* Jacobsen, E. N., Pfaltz, A., Yamamoto, H., Eds.; Springer: Berlin, 1999; Vol. 1, p 183.

[621] Ohkuma, T.; Noyori, R. In *Comprehensive Asymmetric Catalysis,* Jacobsen, E. N., Pfaltz, A., Yamamoto, H., Eds.; Springer: Berlin, 1999; Vol. 1, p 199.

[622] Blaser, H.-U.; Spindler, F. In *Comprehensive Asymmetric Catalysis,* Jacobsen, E. N., Pfaltz, A., Yamamoto, H., Eds.; Springer: Berlin, 1999; Vol. 1, p 247.

[623] Ohkuma, T.; Kitamura, M.; Noyori, R. In *Catalytic Asymmetric Synthesis,* Ojima, I., Ed.; VCH: New York, 2000; p 1.

[624] Noyori, R.; Ohkuma, T. *Angew. Chem., Int. Ed.* **2001**, *40*, 40.

[625] Knowles, W. S. *Angew. Chem., Int. Ed.* **2002**, *41*, 1998.

[626] Noyori, R. *Angew. Chem., Int. Ed.* **2002**, *41*, 2008.

[627] Pfaltz, A.; Blankenstein, J.; Hilgraf, R.; Hörmann, E.; McIntyre, S.; Menges, F.; Schönleber, M.; Smidt, S. P.; Wüstenberg, B.; Zimmermann, N. *Adv. Synth. Catal.* **2003**, *345*, 33.

[628] Crépy, K. V. L.; Imamoto, T. *Adv. Synth. Catal.* **2003**, *345*, 79.

[629] Blaser, H.-U.; Malan, C.; Pugin, B.; Spindler, F.; Steiner, H.; Studer, M. *Adv. Synth. Catal.* **2003**, *345*, 103.

[630] Singh, V. K. *Synthesis* **1992**, 605.

[631] Koenig, K. E. In *Asymmetric Synthesis*; Morrison, J. D., Ed.; Academic Press: Orlando, 1985; Vol. 5, Chapter 3, p 71.

[632] Broene, R. D.; Buchwald, S. L. *J. Am. Chem. Soc.* **1993**, *115*, 12569.

[633] Waymouth, R.; Pino, P. *J. Am. Chem. Soc.* **1990**, *112*, 4911.

[634] Troutman, M. V.; Appella, D. H.; Buchwald, S. L. *J. Am. Chem. Soc.* **1999**, *121*, 4916.

[635] Lightfoot, A.; Schnider, P.; Pfaltz, A. *Angew. Chem., Int. Ed.* **1998**, *37*, 2897.

[636] Liu, D.; Tang, W.; Zhang, X. *Org. Lett.* **2004**, *6*, 513.

[637] Tang, W.; Wang, W.; Zhang, X. *Angew. Chem., Int. Ed.* **2003**, *42*, 943.

[638] Menges, F.; Pfaltz, A. *Adv. Synth. Catal.* **2002**, *344*, 40.

[639] Källstrom, K.; Hedberg, C.; Brandt, P.; Bayer, A.; Andersson, P. G. *J. Am. Chem. Soc.* **2004**, *126*, 14308.

[640] Jiang, Q.; Jiang, Y.; Xiao, D.; Cao, P.; Zhang, X. *Angew. Chem., Int. Ed.* **1998**, *37*, 1100.

[641] Wu, J.; Chen, H.; Kwok, W.; Guo, R.; Zhou, Z.; Yeung, C.; Chan, A. S. C. *J. Org. Chem.* **2002**, *67*, 7908.

[642] Pai, C.-C.; Lin, C.-W.; Lin, C.-C.; Chen, C.-C.; Chan, A. S. C.; Wong, W. T. *J. Am. Chem. Soc.* **2000**, *122*, 11513.

[643] Jiang, Y.; Jiang, Q.; Zhang, X. *J. Am. Chem. Soc.* **1998**, *120*, 3817.

[644] Ohkuma, T.; Koizumi, M.; Doucet, H.; Pham, T.; Kozawa, M.; Murata, K.; Katayama, E.; Yokozawa, T.; Ikariya, T.; Noyori, R. *J. Am. Chem. Soc.* **1998**, *120*, 13529.

[645] Sturm, T.; Weissensteiner, W.; Spindler, F. *Adv. Synth. Catal.* **2003**, *345*, 160.

[646] Zhou, Y.-G.; Tang, W.; Wang, W.-B.; Li, W.; Zhang, X. *J. Am. Chem. Soc.* **2002**, *124*, 4952.

[647] Zhang, Z.; Qian, H.; Longmire, J.; Zhang, X. *J. Org. Chem.* **2000**, *65*, 6223.

[648] Ratovelomanana-Vidal, V.; Girard, C.; Touati, R.; Tranchier, J. P.; Hassine, B. B.; Genêt, J. P. *Adv. Synth. Catal.* **2003**, *345*, 261.

[649] Kadyrov, R.; Riermeier, T. H.; Dingerdissen, U.; Tararov, V.; Börner, A. *J. Org. Chem.* **2003**, *68*, 4067.

[650] Dobbs, D. A.; Vanhessche, K. P. M.; Brazi, E.; Rautenstrauch, V.; Lenoir, J.-Y.; Genét, J.-P.; Wiles, J.; Bergens, S. H. *Angew. Chem., Int. Ed.* **2000**, *39*, 1992.

[651] Lei, A.; Wu, S.; He, M.; Zhang, X. *J. Am. Chem. Soc.* **2004**, *126*, 1626.

[652] Jiang, Q.; Xiao, D.; Zhang, Z.; Cao, P.; Zhang, X. *Angew. Chem., Int. Ed.* **1999**, *38*, 516.

[653] Wu, S.; Wang, W.; Tang, W.; Lin, M.; Zhang, X. *Org. Lett.* **2002**, *4*, 4495.

[654] Jung, H. M.; Koh, J. H.; Kim, M.-J.; Park, J. *Org. Lett.* **2000**, *2*, 2487.

[655] Cobley, C. J.; Lennon, I. C.; Praquin, C.; Zanotti-Gerosa, A. *Org. Process Res. Dev.* **2003**, *7*, 407.

[656] Uemura, T.; Zhang, X.; Matsumura, K.; Sayo, N.; Kumobayashi, H.; Ohta, T.; Nozaki, K.; Takaya, H. *J. Org. Chem.* **1996**, *61*, 5510.

[657] Hayashi, T.; Kawamura, N.; Ito, Y. *J. Am. Chem. Soc.* **1987**, *109*, 7876.

[658] Tang W.; Liu, D.; Zhang, X. *Org. Lett.* **2003**, *5*, 205.

[659] Liu, D.; Li, W.; Zhang, X. *Org. Lett.* **2002**, 4, 4471.

[660] Tang, W.; Zhang, X. *Angew. Chem., Int. Ed.* **2002**, *41*, 1612.

[661] Imamoto, T.; Watanabe, J.; Wada, Y.; Masuda, H.; Yamada, H.; Tsuruta, H.; Matsukawa, S.; Yamaguchi, K. *J. Am. Chem. Soc.* **1998**, *120*, 1635.

[662] Sawamura, M.; Kuwano, R.; Ito, Y. *J. Am. Chem. Soc.* **1995**, *117*, 9602.

[663] Li, W.; Zhang, Z.; Xiao, D.; Zhang, X. *J. Org. Chem.* **2000**, *65*, 3489.

[664] Tang, W.; Zhang, X. *Org. Lett.* **2002**, *4*, 4159.

[665] Qiao, S.; Fu, G. C. *J. Org. Chem.* **1998**, *63*, 4168.

[666] Chan, A. S. C.; Hu, W.; Pai, C.-C.; Lau, C.-P. *J. Am. Chem. Soc.* **1997**, *119*, 9570.

[667] Togni, A.; Breutel, C.; Schnyder, A.; Spindler, F.; Landert, H.; Tijani, A. *J. Am. Chem. Soc.* **1994**, *116*, 4062.

[668] Holz, J.; Quirmbach, M.; Schmidt, U.; Heller, D.; Stürmer, R.; Börner, A. *J. Org. Chem.* **1998**, *63*, 8031.

[669] Zhu, G.; Cao, P.; Jiang, Q.; Zhang, X. *J. Am. Chem. Soc.* **1997**, *119*, 1799.

[670] van den Berg, M.; Minnaard, A. J.; Schudde, E. P.; van Esche, J.; de Vries, A. H. M.; de Vries, J. G.; Feringa, B. L. *J. Am. Chem. Soc.* **2000**, *122*, 11539.

[671] Tang, W.; Wu, S.; Zhang, X. *J. Am. Chem. Soc.* **2003**, *125*, 9570.

[672] You, J.; Drexler, H.-J.; Zhang, S.; Fischer, C.; Heller, D. *Angew. Chem., Int. Ed.* **2003**, *42*, 913.

[673] Zhu, G.; Casalnuovo, A. L.; Zhang, X. *J. Org. Chem.* **1998**, *63*, 8100.

[674] Xiao, D.; Zhang, Z.; Zhang. X. *Org. Lett.* **1999**, *1*, 1679.

[675] Tang, W.; Chi, Y.; Zhang, X. *Org. Lett.* **2002**, *4*, 1695.

[676] Zhang, F.-Y.; Pai, C.-C.; Chan, A. S. C. *J. Am. Chem. Soc.* **1998**, *120*, 5808.

[677] Burk, M. J.; Wang, Y. M.; Lee, J. R. *J. Am. Chem. Soc.* **1996**, *118*, 5142.

[678] Li, W.; Zhang, X. *J. Org. Chem.* **2000**, *65*, 5871.

[679] Chan, Y. N. C.; Osborn, J. A. *J. Am. Chem. Soc.* **1990**, *112*, 9400.

[680] Spindler, F.; Pugin, B.; Blaser, H.-U. *Angew. Chem., Int. Ed. Engl.* **1990**, *29*, 558.

[681] Becalski, A. G.; Cullen, W. R.; Fryzuk, M. D.; James, B. R.; Kang, G.-J.; Rettig, S. J. *Inorg. Chem.* **1991**, *30*, 5002.

[682] Xiao, D.; Zhang, X. *Angew. Chem., Int. Ed.* **2001**, *40*, 3425.

[683] Uematsu, N.; Fujii, A.; Hashiguchi, S.; Ikariya, T.; Noyori, R. *J. Am. Chem. Soc.* **1996**, *118*, 4916.

[684] Willoughby, C. A.; Buchwald, S. L. *J. Am. Chem. Soc.* **1992**, *114*, 7562.
[685] Willoughby, C. A.; Buchwald, S. L. *J. Am. Chem. Soc.* **1994**, *116*, 8952.
[686] Willoughby, C. A.; Buchwald, S. L. *J. Am. Chem. Soc.* **1994**, *116*, 11703.
[687] Harrod, J. F.; Yun, S. S. *Organometallics* **1987**, *6*, 1381.
[688] Guillerm, G.; Frappier, F.; Tabet, J.-C.; Marquet, A. *J. Org. Chem.* **1977**, *42*, 3776.
[689] Koshutin, V. I.; Koshutina, L. L. *J. Org. Chem. USSR (Engl. Transl.)* **1976**, *12*, 1796.
[690] Anfilogova, S. N.; Andreev, V. A.; Pekhk, T. I.; Belikova, N. A. *Zh. Org. Khim.* **1989**, *25*, 1436.
[691] Takimoto, M.; Hiraga, Y.; Sato, Y.; Mori, M. *Tetrahedron Lett.* **1998**, *39*, 4543.
[692] Jones, C. D.; Audia, J. E.; Lawhorn, D. E.; McQuaid, L. A.; Neubauer, B. L.; Pike, A. J.; Pennington, P. A.; Stamm, N. B.; Toomey, R. E.; Hirsch, K. S. *J. Med. Chem.* **1993**, *36*, 421.
[693] Jogdeo, P. S.; Bhide, G. V. *Steroids* **1980**, *35*, 133.
[694] Jogdeo, P. S.; Bhide, G. V. *Steroids* **1979**, *34*, 619.
[695] Serebryakova, T. A.; Zakharaychev, A. V.; Ananchenko, S. N.; Torgov, I. V. *Bull. Acad. Sci. USSR, Div. Chem. Sci. (Engl. Transl.)* **1974**, *23*, 714.
[696] Terasawa, T.; Okada, T. *J. Chem. Soc., Perkin Trans. 1* **1979**, 990.
[697] Barraclough, P.; Jackson, W. P.; Harris, C. J. *Arch. Pharm.* (*Weinheim, Ger.*) **1991**, *324*, 473.
[698] Wilson, R. M.; Schnapp, K. A.; Patterson, W. S. *J. Am. Chem. Soc.* **1992**, *114*, 10987.
[699] Ischenko, I. V.; Shawa, A. G. *J. Org. Chem. USSR (Engl. Transl.)* **1992**, *28*, 214.
[700] Leeson, P. D.; Ellis, D.; Emmett, J. C.; Shah, V. P.; Showell, G. A.; Underwood, A. H. *J. Med. Chem.* **1988**, *31*, 37.
[701] Muratake, H.; Kumagami, H.; Natsume, M. *Tetrahedron* **1990**, *46*, 6351.
[702] Trost, B. M.; Fleitz, F. J.; Watkins, W. J. *J. Am. Chem. Soc.* **1996**, *118*, 5146.
[703] Rudakov, E. S.; Parnes, Z. N.; Osipov, A. M.; Lyakhovetskii, Y. I.; Kursanov, D. N. *Bull. Acad. Sci. USSR, Div. Chem. Sci. (Engl. Transl.)* **1976**, 1143.
[704] Zambias, R. A.; Hammond, M. L. *Synth. Commun.* **1991**, *21*, 959.
[705] Solé, D.; Cancho, Y.; Llebaria, A.; Moretó, J. M.; Delgado, A. *J. Am. Chem. Soc.* **1994**, *116*, 12133.
[706] Sole, D.; Cancho, Y.; Lleberia, A.; Moreto, J.M.; Delgado, A. *J. Org. Chem.* **1996**, *61*, 5895.
[707] Brückner, C.; Reissig, H.-U. *Angew. Chem., Int. Ed. Engl.* **1985**, *24*, 588.
[708] Corey, E. J.; Seibel, W. L.; Kappos, J. C. *Tetrahedron Lett.* **1987**, *28*, 4921.
[709] Ma, E-S. *Yak. Hoechi* **2001**, *45*, 582.
[710] Ohlsson, B.; Ullenius, C.; Jagner, S.; Grivet, C.; Wengner, E.; Kündig, E. P. *J. Organomet. Chem.* **1989**, *365*, 243.
[711] Benkeser, R. A.; Smith, W. E. *J. Am. Chem. Soc.* **1969**, *91*, 1556.
[712] Sieburth, S. McN.; Langevine, C. N.; Dardaris, D. M. *Pestic. Sci.* **1990**, *28*, 309.
[713] Waterlot, C.; Couturier, D.; De Backer, M.; Rigo, B. *Can. J. Chem.* **2000**, *78*, 1242.
[714] Battersby, A. R.; Dutton, C. J.; Fookes, C. J. R. *J. Chem. Soc., Perkin Trans. 1* **1988**, 1569.
[715] Meyers, A. I.; Fray, A. H. *Bull. Chim. Soc. Fr.* **1997**, *134*, 283.
[716] Taylor, R. J. K.; Turner, S. M.; Horwell, D. C. *J. Chem. Soc., Chem. Commun.* **1990**, 406.
[717] Semmelhack, M. F.; Bisaha, J.; Czarny, M. *J. Am. Chem. Soc.* **1979**, *101*, 768.
[718] Zhou, P.; Li, Y.; Meagher, K. L.; Mewshaw, R. G.; Harrison, B. L. *Tetrahedron Lett.* **2001**, *42*, 7333.
[719] Snider, B. B.; Duncia, J. V. *J. Am. Chem. Soc.* **1980**, *102*, 5926.
[720] Lal, K.; Zarate, E. A.; Youngs, W. J.; Salomon, R. G. *J. Org. Chem.* **1988**, *53*, 3673.
[721] Anderson, R. J.; Morris, J. C. *Tetrahedron Lett.* **2001**, *42*, 311.
[722] Sorgi, K. L.; Maryanoff, C. A.; McComsey, D. F.; Graden, D. W.; Maryanoff, B. E. *J. Am. Chem. Soc.* **1990**, *112*, 3567.
[723] Uemura, M.; Isobe, K.; Hayashi, Y. *Chem. Lett.* **1985**, 91.
[724] Semmelhack, M. F.; Herndon, J. W. *Organometallics* **1983**, *2*, 363.
[725] Uemura, M.; Take, K.; Isobe, K.; Minami, T.; Hayashi, Y. *Tetrahedron* **1985**, *41*, 5771.
[726] McCombie, S. W.; Shankar, B. B.; Ganguly, A. K. *Tetrahedron Lett.* **1987**, *28*, 4123.
[727] Dickens, P. J.; Slawin, A. M. Z.; Widdowson, D. A.; Williams, D. J. *Tetrahedron Lett.* **1988**, *29*, 103.

[728] Massiot, G.; Lavaud, C.; Vercauteren, J.; Men-Olivier, L. L.; Lévy, J.; Guilhem, J.; Pascard, C. *Helv. Chim. Acta* **1983**, *66*, 2414.
[729] Obase, H.; Nakamizo, N.; Takai, H.; Teranishi, M.; Kubo, K.; Shuto, K.; Kasuya, Y.; Shigenobu, K. *Chem. Pharm. Bull.* **1982**, *30*, 474.
[730] Vercauteren, J.; Massiot, G.; Sévenet, T.; Richard, B.; Lobjois, V.; Le Men-Olivier, L.; Lévy, J. *Phytochemistry* **1981**, *20*, 1411.
[731] Lomas, J. S.; Vaissermann, J. *J. Chem. Soc., Perkin Trans. 2* **1998**, 1777.
[732] Tong, Y.; Fobian, Y. M.; Wu, M.; Boyd, N. D.; Moeller, K. D. *J. Org. Chem.* **2000**, *65*, 2484.
[733] Furukawa, Y.; Yamagiwa, Y.; Kamikawa, T. *J. Chem. Soc., Chem. Commun.* **1986**, 1234.
[734] Lomas, J. S.; Vaissermann, J. *J. Chem. Soc., Perkin Trans. 2* **1996**, 1831.
[735] Pelter, A.; Ward, R. S.; Venkateswarlu, R.; Kamakshi, C. *Tetrahedron* **1992**, *48*, 7209.
[736] Pelter, A.; Ward, R. S.; Venkateswarlu, R.; Kamakshi, C. *Tetrahedron* **1989**, *45*, 3451.
[737] Procopiou, P. A.; Dymock, B. W.; Inglis, G. G. A.; Lester, M. G.; Roberts. A. D.; Sidebottom, P. J.; Spooner, S. J.; Srikantha, R. P.; Watson, N. S. *J. Chem. Soc., Perkin Trans. 1* **1998**, 327.
[738] Mimoun, H. *J. Org. Chem.* **1999**, *64*, 2582.
[739] Almeida, J. F.; Anaya, J.; Martin, N.; Grande, M.; Moran, J. R.; Caballero, M. C. *Tetrahedron: Asymmetry* **1992**, *3*, 1431.
[740] DeRoy, P. L.; Charette, A. B. *Org. Lett.* **2003**, *5*, 4163.
[741] Aïssa, C.; Riveiros, R.; Ragot, J.; Fürstner. A. *J. Am. Chem. Soc.* **2003**, *125*, 15512.
[742] Barton, D. H. R.; Gunatilaka, A. A. L.; Nakanishi, T.; Patin. H.; Widdowson, D. A.; Worth, B. R. *J. Chem. Soc., Perkin Trans. 1* **1976**, 821.
[743] Powell, S. Q.; Tenenbaum, J. M.; Woerpel, K. A. *J. Am. Chem. Soc.* **2002**, *124*, 12648.
[744] Corriu, R. J. P.; Moreau, J. J. *J. Chem. Soc., Chem. Commun.* **1973**, 38.
[745] Klei, S. R.; Tilley, T. D.; Bergman, R. G. *Organometallics* **2002**, *21*, 4648.
[746] Gil, J. F.; Ramón, D. J.; Yus, M. *Tetrahedron* **1993**, *49*, 4923.
[747] Bujoli, B.; Jubault, M.; Roze, J.-C.; Tallec, A. *Tetrahedron* **1987**, *43*, 2709.
[748] Nakano, T.; Nagai, Y. *Chem. Lett.* **1988**, 481.
[749] Lipshutz, B. H.; Caires, C. C.; Kuipers, P.; Chrisman, W. *Org. Lett.* **2003**, *5*, 3085.
[750] Becker, B.; Corriu, R. J. P.; Guérin, C.; Henner, B.; Wang, Q. *J. Organomet. Chem.* **1989**, *368*, C25.
[751] Smonou, I. *Tetrahedron Lett.* **1994**, *35*, 2071.
[752] Yato, M.; Homma, K.; Ishida, A. *Heterocycles* **1995**, *45*, 17.
[753] Nordlander, J. E.; Payne, M. J.; Njoroge, F. G.; Vishwanath, V. M.; Han, G. R.; Laikos, G. D.; Balk, M. A. *J. Org. Chem.* **1985**, *50*, 3619.
[754] Ookawa, A.; Soai, K. *J. Chem. Soc., Perkin Trans. 1* **1987**, 1465.
[755] Belykh, S. I.; Kudryavtseva, G. A.; Anokhina, I. K.; Parnes, Z. N.; Zalukaev, L. P.; Kursanov, D. N. *Dokl. Akad. Nauk SSSR* **1976**, *229*, 867.
[756] Alferova, S. I.; Kudryavtseva, G. A.; Zalukaev, L. P.; Parnes, Z. N. *Zh. Org. Khim.* **1982**, *18*, 1261.
[757] Revina, N. A.; Belykh, S. I.; Kudryavtseva, G. A.; Zalukaev, L. P.; Parnes, Z. N.; Kursanov, D. N. *Bull. Acad. Sci. USSR, Div. Chem. Sci. (Engl. Transl.)* **1980**, *29*, 1416.
[758] Schnaubelt, J.; Zschiesche, R.; Reissig, H.-U.; Lindner, H.; Richter, J. *Liebigs Ann. Chem.* **1993**, 61.
[759] Belykh, S. I.; Kudryavtseva, G. A.; Loim, N. M.; Zalukaev, L. P.; Parnes, Z. N.; Kursanov, D. N. *Izv. Akad. Nauk SSSR, Ser. Khim.* **1978**, 2343.
[760] Shibata, I.; Kato, H.; Ishida, T.; Yasuda, M.; Baba, A. *Angew. Chem., Int. Ed.* **2004**, *43*, 711.
[761] Dumartin, H.; Le Floc'h, Y.; Grée, R. *Tetrahedron Lett.* **1994**, *35*, 6681.
[762] Taylor, S. J.; Morken, J. P. *J. Am. Chem. Soc.* **1999**, *121*, 12202.
[763] Kiyooka, S.; Shimizu, A.; Torii, S. *Tetrahedron Lett.* **1998**, *39*, 5237.
[764] Mironov, I. V.; Kantor, E. A.; Karakhanov, R. A.; Rakhmankulov, D. L. *Zh. Org. Khim.* **1981**, *17*, 1119.
[765] Brückner, C.; Holzinger, H.; Reissig, H.-U. *J. Org. Chem.* **1987**, *53*, 2450.

[766] Quideau, S.; Ralph, J. *J. Chem. Soc., Perkin Trans. 1* **1993**, 653.
[767] Engelbrecht, G. J.; Holzapfel, C. W. *Tetrahedron Lett.* **1991**, *32*, 2161.
[768] Geiwiz, J.; Goetschi, E.; Hebeisen, P. *Synthesis* **2003**, 1699.
[769] Sasaki, M.; Ebine, M.; Takagi, H.; Takakura, H.; Shida, T.; Satake. M.; Oshima, Y.; Igarashi, T.; Yasumoto, T. *Org. Lett.* **2004**, *6*, 1501.
[770] Vercauteren, J.; Massiot, G.; Le Men-Olivier, L.; Lévy, J.; Prange, T.; Pascard, C. *Tetrahedron Lett.* **1981**, *22*, 2871.
[771] Kraus, G. A.; Molina, M. T. *J. Org. Chem.* **1988**, *53*, 752.
[772] Krohn, K.; Heins, H.; Wielckens, K. *J. Med. Chem.* **1992**, *35*, 511.
[773] Bao, J.; Baker, R. K.; Doss, G. A.; Kayser, F.; Kotliar, A.; Miao, S.; Parson, W. H.; Rupprecht, K. M. *Org. Lett.* **2002**, *4*, 1871.
[774] Babirad, S. A.; Wang, Y.; Kishi, Y. *J. Org. Chem.* **1987**, *52*, 1370.
[775] Hart, D. J.; Tsai, Y.-M. *Tetrahedron Lett.* **1981**, *22*, 1567.
[776] Bach, T.; Bergmann, H.; Grosch, B.; Harms, K.; Herdtweck, E. *Synthesis* **2001**, 1395.
[777] Comins, D. L.; Weglarz, M. A.; O'Connor, S. *Tetrahedron Lett.* **1988**, *29*, 1751.
[778] Castelhano, A. L.; Horne, S.; Taylor, G. J.; Billedeau, R.; Krantz, A. *Tetrahedron* **1988**, *44*, 5451.
[779] Alder, R. W.; Miller, A. J.; Rushbrook, D. I. *J. Chem. Soc., Chem. Commun.* **1989**, 277.
[780] Webb, II, R. R.; Barker, P. L.; Baier, M.; Reynolds, M. E.; Robarge, K. D.; Blackburn, B. K.; Tischler, M. H.; Weese, K. J. *Tetrahedron Lett.* **1994**, *35*, 2113.
[781] Kudryavtseva, G. A.; Nesmeyanova, O. A. *Bull. Acad. Sci. USSR, Div. Chem. Sci. (Engl. Transl.)* **1974**, *23*, 2271.
[782] Tohyama, Y.; Tanino, K.; Kuwajima, I. *J. Org. Chem.* **1994**, *59*, 518.
[783] Beagley, P.; Davies, P. J.; Blacker, A. J.; White, C. *Organometallics* **2002**, *21*, 5852.
[784] Sirol, S.; Courmarcel, J.; Mostefai, N.; Riant, O. *Org. Lett.* **2001**, *3*, 4111.
[785] Ojima, I.; Kogure, T.; Kumagai, M.; Horiuchi, S.; Sato, T. *J. Organomet. Chem.* **1976**, *122*, 83.
[786] Ojima, I.; Kogure, T.; Nagai, Y. *Chem. Lett.* **1973**, 541.
[787] Langer, T.; Janssen, J.; Helmchen, G. *Tetrahedron: Asymmetry* **1996**, *7*, 1599.
[788] Bette, V.; Mortreux, A.; Savoia, D.; Carpentier, J.-F. *Tetrahedron* **2004**, *60*, 2837.
[789] Nishibayashi, Y.; Segawa, K.; Singh, J. D.; Fukuzawa, S.; Ohe, K.; Uemura, S. *Organometallics* **1996**, *15*, 370.
[790] Nishibayashi, Y.; Singh, J. D.; Arikawa, Y.; Uemura, S.; Hidai, M. *J. Organomet. Chem.* **1997**, *531*, 13.
[791] Haag, D.; Runsink, J.; Scharf, H.-D. *Organometallics* **1998**, *17*, 398.
[792] Newman, L. M.; Williams, J. M. J.; McCague, R.; Potter, G. A. *Tetrahedron: Asymmetry* **1996**, *7*, 1597.
[793] Bandini, M.; Cozzi, P. G.; Negro, L.; Umani-Ronchi, A. *Chem. Commun.* **1999**, 39.
[794] Bette, V.; Mortreux, A.; Ferioli, F.; Martelli, G.; Savoia, D.; Carpentier, J.-F. *Eur. J. Org. Chem.* **2004**, 3040.
[795] Nishibayashi, Y.; Singh, J. D.; Segawa, K.; Fukuzawa, S.; Uemura, S. *J. Chem. Soc., Chem. Commun.* **1994**, 1375.
[796] Suárez, A.; Pizzano, A.; Fernández, I.; Khiar, N. *Tetrahedron: Asymmetry* **2001**, *12*, 633.
[797] Pastor, S. D.; Shum, S. P. *Tetrahedron: Asymmetry* **1998**, *9*, 543.
[798] Brunner, H.; Miehling, W. *J. Organomet. Chem.* **1984**, *275*, C17.
[799] Zhu, G.; Terry, M.; Zhang, X. *J. Organomet. Chem.* **1997**, *547*, 97.
[800] Moreau, C.; Frost, C. G.; Murrer, B. *Tetrahedron Lett.* **1999**, *40*, 5617.
[801] Brunner, H.; Obermann, U. *Chem. Ber.* **1989**, *122*, 499.
[802] Helmchen, G.; Krotz, A.; Ganz, K.-T.; Hansen, D. *Synlett* **1991**, 257.
[803] Dumont, W.; Poulin, J.-C.; Dang, T.-P.; Kagan, H. B. *J. Am. Chem. Soc.* **1973**, *95*, 8295.
[804] Sakaki, J.; Schweizer, W. B.; Seebach, D. *Helv. Chim. Acta* **1993**, *76*, 2654.
[805] Kromm, K. K.; Osburn, P. L.; Gladysz, J. A. *Organometallics* **2002**, *21*, 4275.
[806] Halterman, R. L.; Ramsey, T. M.; Chen, Z. *J. Org. Chem.* **1994**, *59*, 2642.
[807] Pini, D.; Iuliano, A.; Salvadori, P. *Tetrahedron: Asymmetry* **1992**, *3*, 693.
[808] Malkov, A. V.; Mariani, A.; MacDougall, K. N.; Kocovsky, P. *Org. Lett.* **2004**, *6*, 2253.
[809] Hansen, M. C.; Buchwald, S. L. *Org. Lett.* **2000**, *2*, 713.

Supplemental References of Recent Reviews

[810] Díez-González, S.; Nolan, S. P. *Acc. Chem. Res.* **2008**, *41*, 349.
[811] Kwong, H.-L.; Yeung, H.-L; Yeung, C.-T.; Lee, W.-S.; Lee, C.-S.; Wong, W.-L. *Coord. Chem. Rev.* **2007**, *251*, 2188.
[812] Lipshutz, B. H. *Synthesis* **2009**, 509.
[813] Rendler, S.; Oestreich, M. *Angew. Chem. Int. Ed.* **2007**, *46*, 498.
[814] Shimizu, H.; Nagasaki, I.; Saito, T. *Tetrahedron* **2005**, *61*, 5405.

Supplemental References for Table 1. Organosilane Reduction of Alkenes

[815] Gaspar, B.; Carreira, E. M. *Angew. Chem. Int. Ed.* **2007**, *46*, 4519.
[816] Kimura, M.; Horino, Y.; Mori, M.; Tamaru, Y. *Chem. Eur. J.* **2007**, *13*, 9686.
[817] Lim, C.; Evenson, G. N.; Perrault, W. R.; Pearlmen, B. A. *Tetrahedron Lett.* **2006**, *47*, 6417.
[818] Mandal, P. K.; McMurray, J. S. *J. Org. Chem.* **2007**, *72*, 6599.
[819] Mizra-Aghayan, M.; Boukherroub, R.; Bolourtchian, M. *Appl. Organomet. Chem.* **2006**, *20*, 214.
[820] Menozzi, C.; Dalko, P. I.; Cossy, J. *Synlett.* **2005**, 2449.
[821] Mizra-Aghayan, M.; Boukherroub, R.; Bolourtchian, M.; Rahimifard, M. *J. Organomet. Chem.* **2007**, *692*, 5113.
[822] Ng, S.-S.; Jamison, T. F. *J. Am. Chem. Soc.* **2005**, *127*, 7320.
[823] Saito, N.; Yamazaki, T.; Sato, Y. *Tetrahedron Lett.* **2008**, *49*, 5073.
[824] Sato, Y.; Hinata, Y.; Seki, R.; Oonishi, Y.; Saito, N. *Org. Lett.* **2007**, *9*, 5597.
[825] Yadev, J. S.; Reddy, B. V. S.; Premalatha, K.; Swamy, T. *Tetrahedron Lett.* **2005**, *46*, 2687.

Supplemental References for Table 2. Organosilane Reduction of Alkynes

[826] Barros-Francisco, R.; Garcia, J. J. *Inorg. Chem.* **2009**, *48*, 386.
[827] Chaulagain, M. R.; Sormunen, G. J.; Montgomery, J. *J. Am. Chem. Soc.* **2007**, *129*, 9568.
[828] Chrovian, C. C.; Knapp-Reed, B.; Montgomery, J. *Org. Lett.* **2008**, *10*, 811.
[829] Hayashi, N.; Shibata, I.; Baba, A. *Org. Lett.* **2005**, *7*, 3093.
[830] Herath, A.; Montgomery, J. *J. Am. Chem. Soc.* **2008**, *130*, 8132.
[818] Mandal, P. K.; McMurray, J. S. *J. Org. Chem.* **2007**, *72*, 6599.
[831] Sa-ei, K.; Montgomery, J. *Org. Lett.* **2006**, *8*, 4441.

Supplemental Reference for Table 3. Organosilane Reduction of Aromatic Hydrocarbons

[832] Voutchkova, A. M.; Gnanamgari, D.; Jakobsche, C. E.; Butler, C.; Miller, S. J.; Parr, J.; Crabtree, R. H. *J. Organomet. Chem.* **2008**, *693*, 1815.

Supplemental References for Table 4. Organosilane Reduction of Halocarbons

[833] Douvris, C.; Ozerov, O. V. *Science* **2008**, *321*, 1188.
[834] Iizuka, M.; Kondo, Y. *Eur. J. Org. Chem.* **2008**, 1161.
[835] Longshaw, A. I.; Carland, M. W.; Krenske, E. H.; Coote, M. L.; Sherburn, M. S. *Tetrahedron Lett.* **2007**, *48*, 5585.
[836] Meier, G.; Braun, T. *Angew. Chem. Int. Ed.* **2009**, *48*, 1546.
[837] Miura, K.; Tomita, M.; Yamada, Y.; Hosomi, A. *J. Org. Chem.* **2007**, *72*, 787.
[838] Narumi, T.; Tomita, K.; Inokuchi, E.; Kobayashi, K.; Oishi, S.; Ohno, H.; Fujii, N. *Org. Lett.* **2007**, *9*, 3465.
[839] Panisch, R.; Bolte, M.; Müller, T. *J. Am. Chem. Soc.* **2006**, *128*, 9676.
[840] Scott, V. J.; Çelenligil-Çetin, R.; Ozerov, O. V. *J. Am. Chem. Soc.* **2005**, *127*, 2852.
[841] Yamanoi, Y. *J. Org. Chem.* **2005**, *70*, 9607.
[842] Yang, J.; Brookhart, M. *J. Am. Chem. Soc.* **2008**, *129*, 12656.
[843] Yang, J.; Brookhart, M. *Adv. Synth. Catal.* **2009**, *351*, 175.

Supplemental References for Table 5. Organosilane Reduction of Alcohols

[844] Gu, Q.; Zheng, Y. H.; Li, Y. C. *Synthesis* **2006**, 975.
[845] Koswatta, P. B.; Sivappa, R.; Dias, H. V. R.; Lovely, C. J. *Org. Lett.* **2008**, *10*, 5055.
[846] Li, G.; Hsung, R. P.; Slafer, B. W.; Sagamanova, I. K. *Org. Lett.* **2008**, *10*, 4991.
[847] Luzzio, F. A.; Wlodarczyk, M. T.; Duveau, D. Y.; Chen, J. *Tetrahedron Lett.* **2007**, *48*, 6704.
[818] Mandal, P. K.; McMurray, J. S. *J. Org. Chem.* **2007**, *72*, 6599.
[848] Nimmagadda, R. D.; McRae, C. *Tetrahedron Lett.* **2006**, *47*, 5755.
[849] Nishibayashi, Y.; Shinoda, A.; Miyake, Y.; Matsuzawa, H.; Sato, M. *Angew. Chem. Int. Ed.* **2006**, *45*, 4835.
[850] Panda, B.; Sarkar, T. K. *Tetrahedron Lett.* **2008**, *49*, 6701.
[851] Wrona, I. E.; Gabarda, A. E.; Evano, G.; Panek, J. S. *J. Am. Chem. Soc.* **2005**, *127*, 15026.

Supplemental References for Table 6. Organosilane Reduction of Ethers

[818] Mandal, P. K.; McMurray, J. S. *J. Org. Chem.* **2007**, *72*, 6599.
[852] Qun, H.-L.; Lowe, J. T.; Panek, J. S. *J. Am. Chem. Soc.* **2007**, *129*, 38.
[853] Yang, J.; White, P. S.; Brookhart, M. *J. Am. Chem. Soc.* **2008**, *130*, 17509.

Supplemental References for Table 7. Organosilane Reduction of Allyl Esters

[854] Fernandes, A. C.; Romão, C. C. *J. Mol. Catal. A: Chem.* **2006**, *253*, 96.
[855] Sakai, N.; Hirasawa, M.; Konakahara, T. *Tetrahedron Lett.* **2005**, *46*, 6407.
[825] Yadev, J. S.; Reddy, B. V. S.; Premalatha, K.; Swamy, T. *Tetrahedron Lett.* **2005**, *46*, 2687.

Supplemental References for Table 8. Organosilane Reduction of Acids

[856] Babu, S. A.; Yasuda, M.; Baba, A. *Org. Lett.* **2007**, *9*, 405.
[857] Nimmagadda, R. D.; McRae, C. *Tetrahedron Lett.* **2006**, *47*, 3505.
[858] Ohta, T.; Kamiya, M.; Nobutomo, M.; Kusui, K.; Farukawa, I. *Bull. Chem. Soc. Jpn.* **2005**, *78*, 1856.

Supplemental Reference for Table 9. Organosilane Reduction of Acid Halides

[859] Longshaw, A. L.; Carland, M. W.; Krenske, E. H.; Coote, M. L.; Sherburn, M. S. *Tetrahedron Lett.* **2007**, *48*, 8858.

Supplemental References for Table 10. Organosilane Reduction of Esters and Lactones

[860] Depré, D.; Horváth, A.; Snissaert, W.; Van Den Bergh, L.; Dermaut, W. *Org. Proc. Res. Dev.* **2008**, *12*, 96.
[861] Lampkins, A. J.; O'Neil, E. J.; Smith, B. D. *J. Org. Chem.* **2008**, *73*, 6053.
[862] Morra, N. A.; Pagenkopf, B. L. *Synthesis* **2008**, 511.
[863] Nakanishi, J.; Tatamidani, H.; Fukumoto, Y.; Chatani, N. *Synlett* **2006**, 869.
[864] Nishimoto, Y.; Babu, A.; Yasuda, M.; Baba, A. *J. Org. Chem.* **2008**, *73*, 9465.
[865] Rainka, M. P.; Milne, J. E.; Buchwald, S. L. *Angew. Chem. Int. Ed.* **2005**, *44*, 6177.
[866] Sakai, N.; Moriya, T.; Konakahara, T. *J. Org. Chem.* **2007**, *72*, 5920.
[867] Sakai, N.; Moriya, T.; Fujii, K.; Konakahara, T. *Synthesis* **2008**, 3533.

Supplemental References for Table 11. Organosilane Reduction of Aldehydes

[868] Augustine, J. K.; Naik, Y. A.; Mandal, A. B.; Alagarsamy, P.; Akabote, V. *Synlett* **2008**, 2429.
[869] Baxter, R.D.; Montgomery, J. *J. Am. Chem. Soc.* **2008**, *130*, 9662.
[870] Chakraborty, S.; Krause, J. A.; Guan, H. *Organometallics* **2009**, *28*, 582.
[828] Chrovian, C. C.; Knapp-Reed, B.; Montgomery, J. *Org. Lett.* **2008**, *10*, 811.
[871] Chuzel, O.; Deschamps, J.; Chausteur, C.; Riant, O. *Org. Lett.* **2006**, *8*, 5943.
[872] Fernandes, A. C.; Fernandes, R.; Romão, C. C.; Royo, B. *Chem. Commun.* **2005**, 213.

873 Hashimoto, T.; I. J.; Nishiyama, H. *Tetrahedron* **2008**, *64*, 9408.
874 Ison, E. A.; Trivedi, E. R.; Corbin, R. A.; Abu-Omar, M. M. *J. Am. Chem. Soc.* **2005**, *127*, 15374.
875 Iwanami, K.; Seo, H.; Tobita, Y.; Oriyama, T. *Synthesis* **2005**, 183.
876 Iwanami, K.; Yano, K.; Oriyama, T. *Chem. Lett.* **2007**, *36*, 38.
877 Kangasmetsae, J. J.; Johnson, T. *Org. Lett.* **2005**, *7*, 5653.
878 Lantos, D.; Contel, M.; Larrea, A.; Szabó, D.; Horvath, I. T. *CSAR Comb. Sci.* **2006**, *25*, 179.
879 Lantos, D.; Contel, M.; Sanz, S.; Bodor, A.; Horvath, I. T. *J. Organomet. Chem.* **2007**, *692*, 1799.
880 Lee, D.; Yun, J. *Tetrahedron Lett.* **2005**, *46*, 2037.
881 Mirza-Aghayan, M.; Boukherroub, R.; Rahimifard, M. *J. Organomet. Chem.* **2008**, *693*, 3567.
882 Ng, S.-S.; Jamison, T. F. *Tetrahedron* **2006**, *62*, 11350.
883 Nikonov, G. I. *J. Am. Chem. Soc.* **2009**, *131*, 908.
848 Nimmagadda, R. D.; McRae, C. *Tetrahedron Lett.* **2006**, *47*, 5755.
884 Peterson, E.; Khalimon, A. Y.; Simionescu, R.; Kuzmina, L. G.; Howard, J. A. K.; Raffa, P.; Evangelisti, C.; Vitulli, G.; Salvadori, P. *Tetrahedron Lett.* **2008**, *49*, 3221.
831 Sa-ei, K.; Montgomery, J. *Org. Lett.* **2006**, *8*, 4441.
867 Sakai, N.; Moriya, T.; Fuji, K.; Konakahara, T. *Synthesis* **2008**, 3533.
885 Saito, N.; Katayama, T.; Sato, Y. *Org. Lett.* **2008**, *10*, 3829.
824 Sato, Y.; Hinata, Y.; Seki, R.; Oonishi, Y.; Saito, N. *Org. Lett.* **2007**, *9*, 5597.
886 Shaikh, N.; Junge, K.; Beller, M. *Org. Lett.* **2007**, *9*, 5429.
887 Shiomi, T.; Adachi, T.; Ito, J. Nishiyama, H. *Org. Lett.* **2009**, *11*, 1011.
888 Tondreau, A. M.; Lobkovsky, E.; Chirilk, P. J. *Org. Lett.* **2008**, *10*, 2789.
889 Tsuhako, A.; He, J.-Q.; Mihara, M.; Saino, N.; Okamoto, S. *Tetrahedron Lett.* **2007**, *48*, 9120.
890 Welle, A.; Díez-González, S.; Tinant, B.; Nolan, S. P.; Riant, O. *Org. Lett.* **2006**, *8*, 6059.
891 Wile, B.M.; McDonald, R.; Ferguson, M.; Stradiotto, M. *Organometallics* **2007**, *26*, 1069.
892 Wile, B. M.; Stradiotto, M. *Chem. Commun.* **2006**, 4104.
893 Yang, M.-S.; Xu, L.-W.; Qiu, H.-Y.; Lai, G.-Q.; Jiang, J.-X. *Tetrahedron Lett.* **2008**, *49*, 253.
894 Yun, J.; Kim, D.; Yun, H. *Chem. Commun.* **2005**, 5181.

Supplemental References for Table 12. Organosilane Reduction of Ketones

895 Asakura, K.; Chun, W. J.; Ohmiya, H.; Sawamura, M. *Organometallics* **2008**, *27*, 6495.
896 Bercot, E. A.; Kindrachuk, D. E.; Rovis, T. *Org. Lett.* **2005**, *7*, 107.
897 Bullock, R. M.; *Angew. Chem. Int. Ed.* **2007**, *46*, 7360.
898 César, V.; Bellemin-Laponnaz, S.; Wadepohl, H.; Gade, L. H. *Chem. Eur. J.* **2005**, *11*, 2862.
870 Chakraborty, S.; Krause, J. A.; Guan, H. *Organometallics* **2009**, *28*, 582.
899 Chen, L.-Z.; Peng, J.-J.; Li, J.-Y.; Bai, Y.-H.; Hu, Y.-Q.; Qiu, H.-Y.; Wu, H.; Lai, G.-G. *Lett. Org. Chem.* **2008**, *5*, 591.
900 Chianese, A. R.; Mo, A.; Datta, D. *Organometallics* **2009**, *28*, 465.
901 Comte, V.; Balan, C.; Le Gendre, P.; Moíse, C. *Chem. Commun.* **2007**, 713.
902 Diez-González, S.; Kauer, H.; Kauer Zinn, F.; Stevens, F. D.; Nolan, S. P. *J. Org. Chem.* **2005**, *70*, 4785.
903 Díez- González, S.; Kaur, H.; Zinn, F. K.; Stevens, E. D.; Nolan, S. P. *J. Org. Chem.* **2005**, *70*, 4784.
904 Diez- González, S.; Scott, N. M.; Nolan, S. P. *Organometallics* **2006**, *25*, 2355.
905 Diez-González, S.; Stevens, E. D.; Scott, N. M.; Peterson, J. L.; Nolan, S. P. *Chem. Eur. J.* **2008**, *14*, 158.
906 Domon, D.; Fujiwara, K.; Murai, A.; Kawai, H.; Suzuki, T. *Tetrahedron Lett.* **2005**, *46*, 8285.
907 Gan, L.; Brook, M. A. *Can. J. Chem.* **2006**, *84*, 1416.
908 Hamasaka, G.; Kawamorita, S.; Ochida, A.; Akiyama, R.; Hara, K.; Fukuoka, A.; Asakura, K.; Chun, W. J.; Ohmiya, H.; Sawamura, M. *Organometallics* **2008**, *27*, 6495.
875 Iwanami, K.; Seo, H.; Tobita, Y.; Oriyama, T. *Synthesis* **2005**, 183.
876 Iwanami, K.; Yano, K.; Oriyama, T. *Chem. Lett.* **2007**, *36*, 38.
874 Ison, E. A.; Trivedi, E. R.; Corbin, R. A.; Abu-Omar, M. M. *J. Am. Chem. Soc.* **2005**, *127*, 15374.
909 Kassube, J. K.; Wadepohl, H.; Gade, L. H. *Adv. Synth. Catal.* **2008**, *350*, 1155.
910 Katritzky, A. R.; Hui, T.; Rong, J.; Kazuyuki, S.; Kostyantyn, K. *J. Org. Chem.* **2007**, *72*, 407.

911 Khiebniova, T. S.; Isakova, V. G.; Lakhvich, F. A. *Russ. J. Gen. Chem.* **2008**, *78*, 1007.
912 Mitchell, T. A.; Romo, D. *J. Org. Chem.* **2007**, *72*, 9053.
913 Oishi, T.; Hasegawa, K.; Torikai, K.; Konoki, K.; Matsumori, N.; Murata, M. *Org. Lett.* **2008**, *10*, 3599.
914 Mostefai, N.; Sirol, S. Courmarcel, J.; Riant, O. *Synthesis* **2007**, *1265*.
881 Mirza-Aghayan, M.; Boukherroub, R.; Rahimifard, M. *J. Organomet. Chem.* **2008**, *693*, 3567.
848 Nimmagadda, R. D.; McRae, C. *Tetrahedron Lett.* **2006**, *47*, 5755.
864 Nishimoto, Y.; Babu, A.; Yasuda, M.; Baba, A. *J. Org. Chem.* **2008**, *73*, 9465.
915 Nishiyama, H.; Furuta, A. *Chem. Commun.* **2007**, 760.
916 Ochida, A.; Hamasaka, G.; Yamauchi, Y.; Kawamorita, S.; Oshima, N.; Hara, K.; Ohmiya, H.; Sawamura, M. *Organometallics* **2008**, *27*, 5494.
913 Oishi, T.; Hasegawa, K.; Torikai, K.; Konoki, K.; Matsumori, N.; Murata, M. *Org. Lett.* **2008**, *10*, 3599.
917 Rendler, S.; Oestreich, M. *Angew. Chem. Int. Ed.* **2008**, *47*, 5997.
918 Saito, T.; Kimishima, A.; Nakata, T. *Heterocycles* **2006**, *70*, 177.
919 Schneider, N.; Finger, M.; Haferkemper, C.; Bellemin-Laponnaz, S.; Hofmann, P.; Gade, L. H. *Angew. Chem. Int. Ed.* **2009**, *48*, 1609.
920 Šebesta, R.; Mečiarová, M.; Molnár, E.; Csizmadiová, J.; Fodran, P.; Onomura, O.; Toma, Š. *J. Organomet. Chem.* **2008**, *693*, 3131.
921 Sunada, Y.; Fujimura, Y.; Nagashima, H. *Organometallics* **2008**, *27*, 3502.
920 Šebesta, R.; Mečiarová, M.; Molnár, E.; Csizmadiová, J.; Fodran, P.; Onomura, O.; Toma, Š. *J. Organomet. Chem.* **2008**, *693*, 3131.
888 Tondreau, A. M.; Lobkovsky, E.; Chirilk, P. J. *Org. Lett.* **2008**, *10*, 2789.
889 Tsuhako, A.; He, J.-Q.; Mihara, M.; Saino, N.; Okamoto, S. *Tetrahedron Lett.* **2007**, *48*, 9120.
922 Toyooka, N.; Zhou, D.; Nemoto, H. *J. Org. Chem.* **2008**, *73*, 4575.
923 Ushio, H.; Mikami, K. *Tetrahedron Lett.* **2005**, *46*, 2903.
894 Yun, J.; Kim, D.; Yun, H. *Chem. Commun.* **2005**, 5181.

Supplemental References for Table 13. Organosilane Reduction of Amides

924 Casabona, D.; Cativiela, C. *Tetrahedron* **2006**, *62*, 10000.
925 Fernandes, A. C.; Romão, C. C. *J. Mol. Catal. A: Chem.* **2007**, *272*, 60.
926 Hanada, S.; Ishida, T.; Motoyama, Y.; Nagashima, H. *J. Org. Chem.* **2007**, *72*, 7551.
927 Hanada, S.; Motoyama, Y.; Nagashima, H. *Tetrahedron Lett.* **2006**, *47*, 6173.
928 Hanada, S.; Motoyama, Y.; Nagashima, H. *Eur. J. Org. Chem.* **2008**, 4097.
929 Li, A. R.; Johnson, M. G.; Liu, J.; Chen, X.; Du, X.; Mihalic, J. T.; Deignan, J.; Gustin, D. J.; Duquette, J.; Fu, Z.; Zhu, L.; Marcus, A. P.; Bergeron, P.; McGee, L. R.; Danao, J.; Lemon, B.; Carabeo, T.; Sullivan, T.; Ma, J.; Tang, L.; Tonn, G.; Collinss, T. L.; Medina, J. C. *Bioorg. Med. Chem. Lett.* **2008**, *18*, 688.
930 Motoyama, Y.; Aoki, M.; Takaoka, N.; Aoto, R.; Nagashima, H. *Chem. Commun.* **2009**, 1574.
931 Motoyama, Y.; Itonaga, C.; Ishida, T.; Takasaki, M.; Nagashima, H. *Org. Synth.* **2005**, *82*, 188.
932 Motoyama, Y.; Mitsui, K.; Ishida, T.; Nagashima, H. *J. Am. Chem. Soc.* **2005**, *127*, 13150.
933 Saito, M.; Matsuo, J.; Ishibashi, H. *Tetrahedron* **2007**, *63*, 4865.
934 Sakai, N.; Fuji, K.; Konakahara, T.; *Tetrahedron Lett.* **2008**, *49*, 6873.
935 Sasakuma, H.; Motoyama, Y.; Nagashima, H. *Chem. Commun.* **2007**, 4916.

Supplemental References for Table 14. Organosilane Reductive Amination of Aldehydes and Ketones

936 Kasuga, J.; Hashimoto, Y.; Miyachi, H. *Bioorg. Med. Chem. Lett.* **2006**, *16*, 771.
937 Kato, H.; Shibata, I.; Yasaka, Y.; Tsunoi, S.; Yasuda, M.; Baba, A. *Chem. Commun.* **2006**, 4189.
938 Lai, R.-Y.; Lee, C.-I.; Liu, S.-T. *Tetrahedron* **2008**, *64*, 1213.
939 Lee, O.-Y.; Law, K.-L.; Ho, C.-Y.; Yang, D. *J. Org. Chem.* **2008**, *73*, 8829.
940 Lehmann, F.; Scobie, M. *Synthesis* **2008**, 1679.
941 Malkov, A. V.; Stoncius, S.; Kocovský, P. *Angew. Chem. Int. Ed.* **2007**, *46*, 3722.

[942] Martínez, R.; Ramón, D. J.; Yus, M. *Adv. Synth. Catal.* **2008**, *350*, 1235.
[943] Menche, D.; Arikan, F.; Li, J.; Rudolph, S. *Org. Lett.* **2007**, *9*, 267.
[944] Mizuta, T.; Sakaguchi, S.; Ishii, Y. *J. Org. Chem.* **2005**, *70*, 2195.

Supplemental Reference for Table 15. Organosilane Reduction of α,β-Unsaturated Aldehydes

[945] Lipshutz, B. H.; Frieman, B. A.; Unger, J. B.; Nihan, D. M. *Can. J. Chem.* **2005**, *83*, 606.

Supplemental References for Table 16. Organosilane Reduction of α,β-Unsaturated Ketones

[946] Anada, M.; Tanaka, M.; Washio, T.; Yamawaki, M.; Abe, T.; Hashimoto, S. *Org. Lett.* **2007**, *9*, 4559.
[947] Baker, B. A.; Boskovic, Z. V.; Lipshutz, B. H. *Org. Lett.* **2008**, *10*, 289.
[948] Imao, D.; Hayama, M.; Ishikawa, K.; Ohta, T.; Ito, Y. *Chem. Lett.* **2007**, *36*, 366.
[949] Lam, H. W.; Murray, G. J.; Firth, J. D. *Org. Lett.* **2005**, *7*, 5743.
[880] Lee, D.; Yun, J. *Tetrahedron Lett.* **2005**, *46*, 2037.
[950] Li, Z.; Deng, G.; Li, Y.-C. *Synlett* **2008**, 3053.
[951] Lipshutz, B. H.; Frieman. B. A. *Angew. Chem. Int. Ed.* **2005**, *44*, 6345.
[952] Lipshutz, B. H.; Nihan, D. M.; Vinogradova, E.; Taft, B. R.; Boskovic, Z. V. *Org. Lett.* **2008**, *10*, 4279.
[953] Mu, R.; Liu, Z.; Liu, Z.; Yang, L.; Wu, L.; Liu, Z.-L. *J. Chem. Res.* **2005**, 469.
[954] Nicolaou, K. C.; Tang, W.; Dagneau, P.; Faraoni, R. *Angew. Chem. Int. Ed.* **2005**, *44*, 3874.
[955] Okamoto, K.; Hayashi, T. *Org. Lett.* **2007**, *9*, 5067.
[956] Reymond, S.; Cossy, J. *Tetrahedron* **2007**, *63*, 5918.
[957] Shan, M.; O'Doherty, G. A. *Org. Lett.* **2008**, *10*, 3381.
[958] Sugiura, M.; Sato, N.; Kotani, S.; Nakajima, M. *Chem. Commun.* **2008**, 4309.
[894] Yun, J.; Kim, D.; Yun, H. *Chem. Commun.* **2005**, 5181.
[959] Zhang, S.; Zhen, J.; Reith, M. E. A.; Dutta, A. K. *Bioorg. Med. Chem.* **2006**, *14*, 3953.
[960] Zhao, D.; Oisaki, K.; Kanai, M.; Shibasaki, M. *Tetrahedron Lett.* **2006**, *47*, 1403.

Supplemental References for Table 17. Organosilane Reduction of α,β-Unsaturated Esters

[947] Baker, B. A.; Boskovic, Z. V.; Lipshutz, B. H. *Org. Lett.* **2008**, *10*, 289.
[961] Frost, C. G.; Hartley, B. C. *Org. Lett.* **2007**, *9*, 4259.
[962] Fuller, N. O.; Morken, J. P. *Synlett* **2005**, 1459.
[963] Lam, H. W.; Joensuu, P. M. *Org. Lett.* **2005**, *7*, 4225.
[952] Lipshutz, B. H.; Nihan, D. M.; Vinogradova, E.; Taft, B. R.; Boskovic, Z. V. *Org. Lett.* **2008**, *10*, 4279.
[964] Mu, R.; Liu, Z.; Liu, Z.; Yang, L.; Wu, L.; Liu, Z.-L. *J. Chem. Res.* **2005**, 469.
[965] Nishiyama, H.; Shiomi, T.; Tsuchiya, Y.; Matsuda, I. *J. Am. Chem. Soc.* **2005**, *127*, 6972.
[966] Shiomi, T.; Nishiyama, H. *Org. Lett.* **2007**, *9*, 1651.
[967] Zheng, H.-J.; Chen, W.-B.; Wu, Z.-J.; Deng, J.-G.; Lin, W.-Q.; Yuan, W.-C.; Zhang, X.-M. *Chem. Eur. J.* **2008**, *14*, 9864.

Supplemental Reference for Table 18. Organosilane Reduction of α,β-Unsaturated Amides

[968] Lam, H. W.; Murray, G. J.; Firth, J. D. *Org. Lett.* **2005**, *7*, 5347.

Supplemental References for Table 19. Organosilane Reduction of α,β-Unsaturated Nitriles

[947] Baker, B. A.; Boskovic, Z. V.; Lipshutz, B. H. *Org. Lett.* **2008**, *10*, 289.
[969] Lee, D.; Kim, D.; Yun, J. *Angew. Chem. Int. Ed.* **2006**, *45*, 2785.
[970] Lee, D.; Yang, Y.; Yun, J. *Org. Lett.* **2007**, *9*, 2749.
[971] Kim, D.; Park, B.-M.; Yun, J. *Chem. Commun.* **2005**, 1755.
[972] Ren, Y.; Xu, X.; Sun, K.; Xu, J. *Tetrahedron: Asymmetry* **2005**, *16*, 4010.
[894] Yun, J.; Kim, D.; Yun, H. *Chem. Commun.* **2005**, 5181.

Supplemental References for Table 20. Organosilane Reduction of Acetals, Ketals, and Hemiketals

[973] Ai-Mughad, H.; Grindley, T. B. *Can. J. Chem.* **2006**, *84*, 516.
[974] Aravind, A.; Baskaran, S. *Tetrahedron Lett.* **2005**, *46*, 743.
[975] Aravind, A.; Mohanty, S. K.; Pratap, T. V. Baskaran, S. *Tetrahedron Lett.* **2005**, *46*, 2965.
[976] Brimble, M. A.; Bachu, P.; Sperry, J. *Synthesis* **2007**, 2887.
[977] Domon, D.; Fujiwara, K.; Ohtaniuchi, Y.; Takezawa, A.; Takeda, S.; Kawasaki, H.; Murai, A.; Kawai, H.; Suzuki, T. *Tetrahedron Lett.* **2005**, *46*, 8279.
[978] Gharpure, S. J.; Sathiyanarayanan, A. M.; Jonnalagadda, P. *Tetrahedron Lett.* **2008**, *49*, 2974.
[979] Gustafsson, T.; Hedenstroem, M.; Kihlberg, J. *J. Org. Chem.* **2006**, *71*, 1911.
[980] Kim, H.; Wooten, C. M.; Park, Y.; Hong, J. *Org. Lett.* **2007**, *9*, 3965.
[981] Li, K.; Vanka, K.; Thompson, W. H.; Tunge, J. A. *Org. Lett.* **2006**, *8*, 4711.
[982] Li, W.; Li, J.; DeVincentis, D.; Mansour, T. S. *Tetrahedron* **2008**, *64*, 7871.
[983] McGarvey, G. J.; LeClair, C. A.; Schmidtmann, B. A. *Org. Lett.* **2008**, *10*, 4727.
[984] Morelli, C. F.; Fornili, A.; Sironi, M.; Duri, L.; Speranza, G.; Manitto, P. *Tetrahedron Lett.* **2005**, *46*, 1837.
[985] Morita, M.; Ishiyama, S.; Koshino, H.; Nakata, T. *Org. Lett.* **2008**, *10*, 1675.
[986] Murray, T. J.; Forsyth, C. J. *Org. Lett.* **2008**, *10*, 3429.
[913] Oishi, T.; Hasegawa, K.; Torikai, K.; Konoki, K.; Matsumori, N.; Murata, M. *Org. Lett.* **2008**, *10*, 3599.
[987] Reddy, Ch. R.; Reddy, G. B.; Rao, Ch. L. *Tetrahedron Lett.* **2008**, *49*, 863.
[988] Sugawara, K.; Hashiyama, T. *Tetrahedron Lett.* **2007**, *48*, 3723.
[989] Vohra, Y.; Vasan, M.; Venot, A.; Boons, G.-J. *Org. Lett.* **2008**, *10*, 3247.
[990] Yamada, K.; Maekawa, M.; Akindele, T.; Yamamoto, Y.; Nakano, M.; Tomioka, K. *Tetrahedron* **2009**, *65*, 903.
[991] Yamada, K.; Maekawa, M.; Akindele, T.; Yamamoto, Y.; Nakano, M.; Tomioka, K. *Tetrahedron* **2009**, *65*, 903.
[992] Zhu C.-J.; Yi, H.; Chen, G.-R.; Xie, J. *Tetrahedron* **2008**, *64*, 10687.

Supplemental Reference for Table 21. Organosilane Reduction of Aminals and Hemiaminals

[993] Amat, M., Pérez, M.; Minaglia, A. T., Bosch, J. *J. Org. Chem.* **2008**, *73*, 6920.

Supplemental References for Table 22. Organosilane Reduction of Enamines

[994] Brodney, M. A.; Cole, M. L.; Freemont, J. A.; Kyl, S.; Junk, P. C.; Padwa, A.; Riches, A. G.; Ryan, J. H. *Tetrahedron Lett.* **2007**, *48*, 1939.
[995] Nolin, K. A.; Ahn, R. W.; Toste, F. D. *J. Am. Chem. Soc.* **2005**, *127*, 12462.
[996] Sato, A.; Yorimitsu, H.; Oshima, K. *Synlett* **2009**, 28.
[997] Zheng, H.-J.; Chen, W.-B.; Wu, Z.-J; Deng, J.-G.; Lin, W.-Q., Yuan, W.-C.; Zhang, X. M. *Chem. Eur. J.* **2008**, *14*, 9864.

Supplemental References for Table 23. Organosilane Reduction of Imines

[998] Conrow, R. E.; Delgado, P.; Dean, W. D.; Callen, G. R.; Plummer, S. V. *Tetrahedron Lett.* **2008**, *49*, 2348.
[999] Fernandes, A. C.; Romão, C. C. *Tetrahedron Lett.* **2005**, *46*, 8881.
[1000] Field, L. D.; Messerle, B. A.; Rumble, S. L. *Eur. J. Org. Chem.* **2005**, 2881.
[1001] Lai, R.-Y.; Surekha, K.; Hayashi, A.; Ozawa, F.; Liu, Y.-H.; Peng, S.-M.; Liu, S.-T. *Organometallics* **2007**, *26*, 1062.
[1002] Malkov, A. V.; Figlus, M.; Kocovský, P. *J. Org. Chem.* **2008**, *73*, 3985.
[1003] Malkov, A. V.; Stoncius, S.; Vranková, K.; Arndt, M.; Kocovský, P. *Chem. Eur. J.* **2008**, *14*, 8082.
[818] Mandal, P. K.; McMurray, J. S. *J. Org. Chem.* **2007**, *72*, 6599.
[1004] Onomura, O.; Kouchi, Y.; Iwasaki, F.; Matsumura, Y. *Tetrahedron Lett.* **2006**, *47*, 3751.

[1005] Park, B.-M.; Mun, S.; Yun, J. *Adv. Synth. Catal.* **2006**, *348*, 1029.

Supplemental Reference for Table 25. Organosilane Reduction of Nitroalkanes

[1006] Rahaim, Jr., R. J.; Maleczka, Jr., R. E. *Synthesis*, **2006**, 3316.

Supplemental References for Table 26. Organosilane Reduction of Miscellaneous Nitrogen Compounds

[1007] Benati, L.; Bencivenni, G.; Leardini, R.; Nanni, D.; Minozzi, M.; Spagnolo, P.; Scialpi, R.; Zanardi, G. *Org. Lett.* **2006**, *8*, 2499.
[1008] Czekelius, C.; Carreira, E. M. *Org. Proc. Res. Dev.* **2007**, *11*, 633.
[1009] Fernandes, A. C.; Romão, C. C. *Tetrahedron* **2006**, *62*, 9650.
[1010] Guo, C.; Xue, M.-X.; Zhu, M.-K.; Gong, L.-Z. *Angew. Chem. Int. Ed.* **2008**, *47*, 3414.
[1011] Itazaki, M.; Nakazawa, H. *Chem. Lett.* **2005**, *34*, 1054.
[818] Mandal, P. K.; McMurray, J. S. *J. Org. Chem.* **2007**, *72*, 6599.
[1012] Ochiai, M.; Hashimoto, H.; Tobita, H. *Angew. Chem. Int. Ed.* **2007**, *46*, 8192.
[1013] Nakazawa, H.; Kamata, K.; Itazaki, M. *Chem. Commun.* **2005**, 4004.
[1014] Rahaim, R. J.; Maleczka, Jr., R. E. *Org. Lett.* **2005**, *7*, 5087.
[1015] Sibi, M. P.; Yang, Y.-H.; Lee, S. *Org. Lett.* **2008**, *10*, 5349.
[1016] Tobisu, M.; Nakamura, R.; Kita, Y.; Chatani, N. *J. Am. Chem. Soc.* **2009**, *131*, 3174.
[1017] Yamamura, M.; Kano, N.; Kawashima, T. *Tetrahedron Lett.* **2007**, *48*, 4033.

Supplemental References for Table 27. Organosilane Reduction of Miscellaneous Sulfur Compounds

[1018] Desrosiers, J.-N.; Charette, A. B. *Angew. Chem. Int. Ed.* **2007**, *46*, 5955.
[1019] Duran, F. J.; Ghini. A. A.; Corini, H.; Burton, G. *Tetrahedron* **2006**, *62*, 4762.
[1009] Fernandes, A. C.; Romão, C. C. *Tetrahedron* **2006**, *62*, 9650.
[861] Lampkins, A. J.; O'Neil, E. J.; Smith, B. D. *J. Org. Chem.* **2008**, *73*, 6053.
[1020] Llamas, T.; Arrayas, R. G.; Carretero, J. C. *Angew. Chem. Int. Ed.* **2007**, *46*, 3329.
[1021] Paquette, L. A.; Peng, X.; Yang, J.; Kang, H.-J. *J. Org. Chem.* **2008**, *73*, 4548.

Supplemental References for Table 28. Organosilane Reduction of Small Ring Compounds

[1022] Deutsch, C.; Lipshutz, B. H.; Krause, N. *Angew. Chem. Int. Ed.* **2007**, *46*, 1650.

Supplemental References for Table 29. Miscellaneous Organosilane Reductions

[1023] Doherty, S.; Smyth, C. H.; Harrington, R. W.; Clegg, W. *Organometallics* **2008**, *27*, 4837.
[815] Gaspar, B.; Carreira, E. M. *Angew. Chem. Int. Ed.* **2007**, *46*, 4519.
[880] Lee, D.; Yun, J. *Tetrahedron Lett.* **2005**, *46*, 2037.
[1024] Matsuo, T.; Kawaguchi, H. *J. Am. Chem. Soc.* **2006**, *128*, 12362.
[1025] Sugie, A.; Somete T.; Kanie, K.; Muramatsu, A.; Mori, A. *Chem. Commun.* **2008**, 3882.
[1026] Takahashi, M.; Kamada, J.; Iwata, K.; Goto, K.; Watanabe, H.; Tamai, S. *Bull. Soc. Chem. Jpn.* **2008**, *81*, 168.

Supplemental References for Table 30. Asymmetric Organosilane Reduction of Ketones

[1027] Bette, V.; Mortreux, A.; Savoia, D.; Carpentier, J.-F. *Adv. Synth. Catal.* **2005**, *347*, 289.
[1028] Chelucci, G.; Muroni, D.; Manca, I. *J. Mol. Catal. A: Chem.* **2005**, *225*, 11.
[1029] Frölander, A.; Moberg, C. *Org. Lett.* **2007**, *9*, 1371.
[1030] Lee, C.-T.; Lipshutz, B. H. *Org Lett.* **2008**, *10*, 4187.
[1031] Imamoto, T.; Itoh, T.; Yamanoi, Y.; Narui, R.; Yoshida, K. *Tetrahedron: Asymmetry* **2006**, *17*, 560.
[1032] Issenhuth, J. T.; Dagorne, S.; Bellemin-Laponnaz, S. *Adv. Synth. Catal.* **2006**, *348*, 1991.

[1033] Kantam, M. L.; Laha, S.; Yadav, J.; Likhar, P. R.; Sreedhar, B.; Choudary, B. M. *Adv. Synth. Catal.* **2007**, *349*, 1797.
[951] Lipshutz, B. H.; Frieman. B. A. *Angew. Chem. Int. Ed.* **2005**, *44*, 6345.
[1034] Lipshutz, B. H.; Frieman, B. A.; Tomaso, Jr., A. E. *Angew. Chem. Int. Ed.* **2006**, *45*, 1259.
[1035] Lipshutz, B. H.; Lower, A.; Kucejko, R. J.; Noson, K. *Org. Lett.* **2006**, *8*, 2969.
[1036] Malkov, A. V.; Stewart Liddon, A. J. P.; Ramírez-López, P.; Bendová, L.; Haigh, D.; Kocovský, P. *Angew. Chem. Int. Ed.* **2006**, *45*, 1432.
[1037] Malkov, A. V.; Stoncius, S; MacDougall, K. N.; Mariani, A.; McGeoch, G. D.; Koèovskxcó, P. *Tetrahedron* **2006**, *62*, 264.
[1038] Matsumura, Y.; Ogura, K.; Kouchi, Y.; Iwasaki, F.; Onomura, O. *Org. Lett.* **2006**, *8*, 3789.
[1004] Onomura, O.; Kouchi, Y.; Iwasaki, F.; Matsumura, Y. *Tetrahedron Lett.* **2006**, *47*, 3751.
[1039] Rizzo, J. R.; Alt, C. A.; Zhang, T. Y. *Tetrahedron Lett.* **2008**, *49*, 6749.
[1040] Shaikh, N. S.; Enthaler, S.; Junge, K.; Beller, M. *Angew. Chem. Int. Ed.* **2008**, *47*, 2497.
[1041] Wang, Z.; Ye, X.; Wei, S.; Wu, P.; Zhang, A.; Sun, J. *Org. Lett.* **2006**, *5*, 999.
[1042] Xu, Q.; Gu, X.; Liu, S.; Dou, Q.; Shi, M. *J. Org. Chem.* **2007**, *72*, 2240.
[1043] Zhou, L.; Wang, Z.; Wei, S.; Sun, J. *Chem. Commun.* **2007**, 2977.

Supplemental References for Table 31. Asymmetric Organosilane Reduction of α,β-Unsaturated Ketones

[1044] Carreño, M. C.; Mazery, R. D.; Urbano, A.; Colobert, F.; Solladié, G. *Org. Lett.* **2005**, *7*, 2039.
[1045] Kanazawa, Y.; Tsuchiya, Y.; Kobayashi, K.; Shiomi, T.; Itoh, J.; Kikuchi, M.; Yamamoto, Y.; Nishiyama, H. *Chem. Eur. J.* **2005**, *12*, 63.
[1046] Lipshutz, B. H.; Amorelli, B.; Unger, J. B. *J. Am. Chem. Soc.* **2008**, *130*, 14378.
[951] Lipshutz, B. H.; Frieman. B. A. *Angew. Chem. Int. Ed.* **2005**, *44*, 6345.
[945] Lipshutz, B. H.; Frieman, B. A.; Unger, J. B.; Nihan, D. M. *Can. J. Chem.* **2005**, *83*, 606.
[887] Shiomi, T.; Adachi, T.; Ito, J. Nishiyama, H. *Org. Lett.* **2009**, *11*, 1011.

Supplemental References for Table 32. Asymmetric Organosilane Reduction of α,β-Unsaturated Esters

[1047] Deschamp, J.; Chuzel, O.; Hannedouche, J.; Riant, O. *Angew. Chem. Int. Ed.* **2006**, *45*, 1292.
[1048] Hashimoto, T.; Shiomi, T.; Ito, J. *Tetrahedron* **2007**, *63*, 12883.
[1049] Ito, J.; Shiomi, T.; Nishiyama, H. *Adv. Synth. Catal.* **2006**, *348*, 1235.
[1045] Kanazawa, Y.; Tsuchiya, Y.; Kobayashi, K.; Shiomi, T.; Itoh, J.; Kikuchi, M.; Yamamoto, Y.; Nishiyama, H. *Chem. Eur. J.* **2005**, *12*, 63.
[951] Lipshutz, B. H.; Frieman. B. A. *Angew. Chem. Int. Ed.* **2005**, *44*, 6345.
[1050] Lipshutz, B. H.; Frieman, B. A.; Tomaso, Jr., A. E. *Angew. Chem. Int. Ed.* **2006**, *45*, 1259.
[945] Lipshutz, B. H.; Frieman, B. A.; Unger, J. B.; Nihan, D. M. *Can. J. Chem.* **2005**, *83*, 606.
[1051] Lipshutz, B. H.; Lee, C.-T.; Servesko, J. M. *Org. Lett.* **2007**, *9*, 4713.
[1052] Lipshutz, B. H.; Tanaka, N.; Taft, B. R.; Lee, C.-T. *Org. Lett.* **2006**, *8*, 1963.
[1053] Nishiyama, H.; Ishikawa, J.; Shiomi, T. *Tetrahedron Lett.* **2007**, *48*, 7841.
[965] Nishiyama, H.; Shiomi, T.; Tsuchiya, Y.; Matsuda, I. *J. Am. Chem. Soc.* **2005**, *127*, 6972.
[1054] Shiomi, T.; Ito, J.; Yamamoto, Y.; Nishiyama, H. *Eur. J. Org. Chem.* **2006**, 5594.
[1055] Wu, J.; Ji, J.-X.; Chan, A. S. C. *Proc. Natl. Acad. Sci. USA* **2005**, *102*, 3570.

Supplemental References for Table 34. Asymmetric Organosilane Reduction of Imines

[1056] Baudequin, C.; Chaturvedi, D.; Tsogoeva, S. B. *Eur. J. Org. Chem.* **2007**, 2623.
[1057] Guizzeti, S.; Benaglia, M.; Cozzi, F.; Rossi, S.; Celentano, G. *Chirality* **2009**, *21*, 233.
[1058] Kertsnus-Banchik, E.; Kalikhman, I.; Gostevskii, B.; Deutsch, Z.; Botoshansky, M.; Kost, D. *Organometallics* **2008**, *27*, 5285.
[945] Lipshutz, B. H.; Frieman, B. A.; Unger, J. B.; Nihan, D. M. *Can. J. Chem.* **2005**, *83*, 606.
[1059] Malkov, A. V.; Figlus, M.; Soncius, S.; Kocovský, P. *J. Org. Chem.* **2007**, *72*, 1315.
[995] Nolin, K. A.; Ahn, R. W.; Toste, F. D. *J. Am. Chem. Soc.* **2005**, *127*, 12462.
[1060] Pei, D.; Wang, Z.; Wei, S.; Zhang, Y.; Sun, J. *Org. Lett.* **2006**, *8*, 5913.

[1061] Pei, D.; Zhang, Y.; Wei, S.; Wang, M.; Sun, J. *Adv. Synth. Catal.* **2008**, *350*, 619.
[1062] Wang, Z.; Cheng, M.; Wu, P.; Wei, S.; Sun, J. *Org. Lett.* **2006**, *8*, 3045.
[1063] Wang, Z.; Wei, S.; Wang, C.; Sun, J. *Tetrahedron: Asymmetry* **2007**, *18*, 705.
[1064] Wu, P.; Wang, Z.; Cheng, M.; Zhou, L.; Sun, J. *Tetrahedron* **2008**, *64*, 11304.
[1065] Zheng, H.; Deng, J.; Lin, W.; Zhang, X. *Tetrahedron Lett.* **2007**, *48*, 7934.
[1043] Zhou, L.; Wang, Z.; Wei, S.; Sun, J. *Chem. Commun.* **2007**, 2977.

INDEX

Ionic and Organometallic-Catalyzed Organosilane Reductions, by Gerald L. Larson and James L. Fry